P9-CEG-575

Publisher: Charlyce Jones Owen
Editorial Assistant: Maureen Diana
Product Marketing Manager: Tricia Murphy
Field Marketing Manager: Brittany Pogue-Mohammed
Program Manager: Seanna Breen
Managing Editor: Denise Forlow
Project Manager: Manuel Echevarria
Senior Operations Supervisor: Mary Fischer
Procurement Specialist: Diane Peirano
Art Director: Maria Lange
Cover Designer: Kristina Mose-Libon
Project Manager, Digital Studio: Pamela Weldin
Printer/Binder: RRD/Roanoke
Cover Printer: Phoenix Color
Text Font: Palatino LT Pro 9.5/13

Acknowledgements of third party content appear on pages 640–647, which constitutes an extension of this copyright page.

Library of Congress Cataloging-in-Publication Data

Hock, Roger R.
Human sexuality/Roger R. Hock.—Fourth edition.
pages cm
ISBN 978-0-13-397138-5 (alk. paper)—ISBN 0-13-397138-4 (alk. paper)
1. Sex. 2. Sex (Psychology) 3. Sex (Biology) I. Title.
HQ21.H54 2014
306.7—dc23

2014034136

10 9 8 7 6 5 4 3 2 1

Student Edition
Case Bound
ISBN-10: 0-13-397138-4
ISBN-13: 978-0-13-397138-5

Paper Bound
ISBN-10: 0-13-400356-X
ISBN-13: 978-0-13-400356-6

Books á la Carte
ISBN-10: 0-13-397165-1
ISBN-13: 978-0-13-397165-1

Human Sexuality

Fourth Edition

Roger R. Hock
Mendocino College

PEARSON

Boston Columbus Indianapolis New York San Francisco
Amsterdam Cape Town Dubai London Madrid Milan Munich Paris Montréal Toronto
Delhi Mexico City São Paulo Sydney Hong Kong Seoul Singapore Taipei Tokyo

Brief Contents

Contents

From the Author

Dear Students,

Welcome to the fourth edition of *Human Sexuality*. I know most students do not read textbook prefaces (even when you are asked to!). That's OK. I understand. I usually didn't read them either when I was a student. Obviously, however, *you* are reading this, so you are a welcome exception to the rule! I first want to take just a few minutes to summarize seven fundamental principles I followed as I planned, developed, wrote, and revised this book. I also want to highlight briefly some of the changes, updates, and revisions you will find in this edition.

My goal continues to be sharing with you the rich and complex field of human sexuality in ways that are up-to-date, engaging, authoritative, understandable, and relevant to *your life*. In addition, this preface includes descriptions and explanations of the key features throughout the textbook.

Human Sexuality is a class that offers you many opportunities to acquire knowledge and information that serves you throughout your life. With that in mind, you may want to keep this book in your personal library. I can assure you, that for one reason or another, probably sooner rather than later, you will want to look something up!

Now, read on and enjoy your journey through the fascinating world of human sexuality.

Roger R. Hock

Preface

Guiding Principles

Seven principles have influenced me every step along the way in all my teaching and writings about human sexuality. Please take a minute to review these principles and keep them in mind as you read through this text.

Choice, Awareness, Rights and Responsibilities

Personal Choice and Consent

An assumption throughout this text is that *you* are in charge of your personal sexual choices and that it is your responsibility to make decisions that are right for you, that you can feel good about, and that do not harm you or anyone else. You are in control of your sexuality and no one may ever, under any circumstances, take that control away from you. This applies to everything including others judging you, ridiculing or embarrassing you, and subjecting you to any type of sexual victimization and assault. By taking this course, you have consented to reading and learning about the topics in this book. However, if, at any time, you do not feel comfortable with the course material, you have the right to choose *not* to read or view it. If these feelings occur at any point, you should discuss them with your instructor.

Authoritative Information

The content of this text is based on the most up-to-date, scientific research available at press time. You can be assured as you read that you are receiving recent, accurate information on every topic discussed.

Real-Life Relevancy

Every effort has been made to ensure that this text is relevant and meaningful to you. Human sexuality is not hypothetical, theoretical, or abstract; it is real. In this text, you will acquire information and knowledge that you will apply and use in your life now and in the future.

Physical and Emotional Health and Wellness

In virtually all chapters, this text focuses on specific, as well as general, sexual health issues to increase your knowledge of them and enhance your level of comfort in seeking care and treatment for sexually-related health problems, if necessary.

Comfort With and Acceptance of Your Own Sexuality

One of the highlights of this book is a focus on the importance of developing your personal "sexual philosophy." Developing a clear sense of your sexual identity, needs, desires, goals, and personal sexual rights helps you navigate a happy, well-adjusted sexual life. When you are faced with sexual situations, your sexual philosophy guides you in making the choices that are right for you so that you are in charge of the sexual situations in your life instead of the other way around.

Critical Analysis of Research and Information about Sexuality

This text provides you with the skills to evaluate intelligently and critically the vast amount of sexual information you receive from the media, your friends and acquaintances, and (maybe) your parents. It should come as no surprise to you that much of that information may be flawed or just plain wrong. This book helps put to rest any misconceptions you may have about sexuality and help you become an informed consumer of sexual information.

Awareness of and Respect for Sexual Diversity

The information in this text strengthens your understanding of the complex sexual world around you and how you fit into it. Enhancing your respect for and appreciation of the full range of human diversity is a major goal of this book.

FEATURES

Choice, Awareness, Responsibility

Choice: Making Decisions That Are Right for You

Every effort has been made to ensure that this text is relevant and meaningful to you. In this text, you will acquire information and knowledge to help you make choices and decisions in your life that are right for you.

Since You Asked

1. I'm not sure about this class. I'm afraid it's going to be so embarrassing! (see page 6)
2. Do we really need a course about sex? I'm married and have a child. I'd say I probably know everything I need to know. (see page 9)
3. Is it normal to have an orgasm by masturbating but never to have one during intercourse? (see page 13)
4. I am 21 years old and had a "wet dream" two weeks ago. Is that normal at my age? (see page 13)
5. Why did my high school stop the sex ed class? Now they're saying we should just all wait until marriage to have sex. No one's really going to do that, are they? (see page 13)
6. My parents never had "the talk" with me. Were they just embarrassed, or did they think I would learn on my own? (see page 14)
7. How is it possible to study people's sexual lives when it's such a private, personal experience? (see page 17)
8. Is it really ethical to study people's actual sexual behaviors? (see page 29)

Since You Asked. Each chapter opens with a list of student questions gathered from the author's human sexuality classes. Each question is referenced to a page number in the chapter where the answer and discussion of the issue may be found.

Focus on Your Feelings

Sex is emotional. Virtually every emotion you can think of can be involved. Students studying human sexuality often experience unexpected emotional reactions. These reactions range from positive, happy, even euphoric feelings to discomfort, embarrassment, shock, anger confusion, fear, or various combinations of these and other emotions. Your personal, individual reactions will depend on your attitudes about sexual issues, your family and religious background, and your past and present sexual and relationship experiences. For example, here are two journal entries from students in the author's human sexuality classes:

> I was raised in a very strict religious household. Sex was never discussed, and even mentioning anything related to sex was not allowed. I'm learning a lot in this course, but I find it very difficult to read the material and participate in the class discussions. I'm embarrassed and feel that I am doing something wrong (guilt!). This is why I have missed so many classes.
>
> **Maya, first-year student**

> I've never told anyone this before, but from the time I was 8 until I was 12, I was sexually molested by my father. When we were in class discussing sexual abuse, it brought all this back to me. I don't know if I should get some counseling. I never thought it really bothered me, but now I'm not so sure.
>
> **Rick, sophomore**

Many (and luckily most) people have generally positive emotions and experiences about sex and sexuality. But in your human sexuality class, you should expect to experience emotions that you would not usually feel in other courses. *These emotions are completely normal.* However, if you find that you are having feelings that are bothersome, or if they interfere with your ability to enjoy and do well in this course, it would be a good idea to discuss them in private with your instructor or perhaps with a professional counselor (usually available through your college's counseling services). Remember, you should never feel *forced* to read material or attend classes that would be emotionally painful for you. Your sexuality professor should be willing and open to discuss with you ways of reducing your discomfort and maximizing your benefit from this class.

Focus on Your Feelings. The study of human sexuality sometimes evokes unexpected emotions and/or reaction to some of the more provocative topics, such as sexual violence, sexually transmitted infections, and pornography, to name just a few. This feature appears early in each chapter and is intended to alert you to these possible reactions, reassure you that they are normal, and let you know you are not the only one feeling them.

Self-Discovery. Because sexuality is a natural part of being human, your exploration will involve some self-discovery. To that end, this feature offers you many opportunities to learn about yourself as a unique sexual person. The "Self-Discoveries" may contain information to help you become better informed about your sexual body, or better understand your relationships with others, or appreciate the rich variations of human sexuality.

Self-Discovery

Guidelines for Sexual Communication

1. Know What You Want
   - Decide what you want and don't want in a relationship before becoming sexually involved.
   - Evaluate your personal sexual expectations and needs.
   - Behave in a manner that is consistent with your sexual philosophy and value system.
   - Be open with your partner by communicating your sexual philosophy to him or her.
   - Understand that two people can want different things and have differing sexual values and philosophies without one of them being "wrong."
2. Insist on Your Right to Postpone a Sexual Relationship
   - Set your own pace, one that is comfortable for you as the relationship progresses.
   - Wait until trust, comfort, and open communication exist in the relationship.
   - Be open and honest with your partner about your desires if you want to proceed slowly.
3. Be Responsible if You Decide to Engage in "Casual Sex"
   - Be honest with yourself and your partner if your commitment is limited to mutual enjoyment.
   - Consider whether or not this type of activity fits your personal sexual philosophy.
   - Be sure both partners want to be sexual.
   - Be sure neither partner feels coerced in any way by the other.
   - Respect your own and each other's feelings of self-worth.
   - Be sure that the responsibility to use precautions against sexually transmitted infections and unwanted pregnancy is mutually agreed upon.

SOURCE: Adapted from Michigan State University Counseling Center (2003).

Your Sexual Philosophy. As you proceed through these pages, this feature offers you the opportunity to develop your personal sexual philosophy by considering how the material in each chapter may relate to you and your life. It is my hope that this will help prepare and guide you as you are faced with personal choices about sexual attitudes and behaviors, or find yourself in a position to advise someone about sexual issues. The idea behind developing a clear personal sexual philosophy is that through the process of learning who you are, what you want and don't want, and planning ahead, you will be in charge of sexual situations as they arise in your life, instead of finding that the situations are in charge of you.

Your Sexual Philosophy

Studying Human Sexuality

As mentioned at the beginning of the chapter, you will have the opportunity to integrate the material throughout this text into *your own personal sexual philosophy*. You will see this feature at the close of every chapter, along with a brief explanation of how the chapter's information may fit into the sexual philosophy you are developing, and will continue to develop, throughout your life.

The basic idea behind this feature is to encourage you to do some thinking about and preparation for your life as a sexual being, now and in the future. As noted early in this chapter, your sexual philosophy is about *knowing who you are, what you want and don't want, and planning ahead*. If you take the time now to consider and explore your sexual feelings, attitudes, desires, and preferences and to form your sexual "rules for life," you will be far better equipped to be in control of your sexual life and to make healthy, informed choices about your sexual behaviors. Although no one can plan for every future event in life, having your sexual philosophy in place helps you take charge of sexual situations when they arise, rather than allowing the situations to take charge of you.

Chapter 1 contributes to your sexual philosophy in two major ways. First, it is crucial for you to gather as much accurate information as possible, now and in the future, about experiencing and understanding human sexuality, so that the choices and decisions you make about sex and relationships are based on a solid foundation of knowledge. Second, learning to analyze sexuality research *critically* will enable you to evaluate what you hear, see, and read about sex. Then you will be equipped to incorporate into your personal life the best information that will enhance your sexual enjoyment, satisfaction, health, and fulfillment. Many students have remarked that their human sexuality course was one of the most fascinating and *personally useful* courses in all of their higher education.

Sex Is More Than Intercourse. A running theme throughout the book is that sex is far more than heterosexual intercourse. It appears as a small symbol and reminder box in nearly every chapter. This idea, indicated by this symbol, is intended to remind you that a wide range of sexual activities in addition to, or instead of, intercourse can offer sexual intimacy and satisfaction without engaging in risky or unwanted behaviors. Many of you may find that this idea seems strange at first, so it appears frequently throughout the book to help you incorporate it into your thinking about sex.

Awareness: Your Health and the Sexual World Around You

Increased awareness of physical and psychological health relating to sexuality helps you make choices that are right for you. And learning to respect and appreciate the fascinating range of human diversity enhances your understanding of the complex sexual world around you and how and where you fit into it.

In Touch With Your Sexual Health

The Health Benefits of Orgasm

Recent research has revealed a number of health benefits associated with orgasm. Note that the findings cited here are based primarily on correlational research that has shown a *connection* between orgasm and these health benefits. The extent to which orgasm *causes* the benefit will require additional research.

Health Benefit	Research Findings
General health	An orgasm at least once or twice per week appears to strengthen the immune system's ability to resist flu and other viruses.
Pain relief	Some women find that an orgasm's release of hormones and muscle contractions help relieve the pain of menstrual cramps and raise pain tolerance in general.
Lower cancer rate	Men who have more than five ejaculations per week during their twenties have a significantly lower rate of prostate cancer later in life.
Mood enhancement	Orgasms increase estrogen and endorphins, which tend to improve mood and ward off depression in women.
Longer life	Men who have two or more orgasms per week live significantly longer than men who have fewer.
Greater feelings of intimacy	The hormone oxytocin, which may play a role in feelings of love and intimacy, increases five-fold at orgasm.
Less heart disease	Studies have shown that men who have at least three orgasms per week are 50% less likely to die of heart disease.
Better sleep	The neurotransmitter dopamine, released during orgasm, triggers a stress-reducing, sleep-inducing response that may last up to two hours.

SOURCES: Giles et al., 2003; Komisaruk & Whipple, 1995; Rosnick, 2002; Komisaruk, Beyer-Flores, & Whipple, 2006; Levin, 2007; Smith, Frankel, & Yarnell, 1997; Weeks & James, 1999; Whipple, 2000.

In Touch with Your Sexual Health. Increasing your awareness of sexual health issues is a guiding theme throughout this book. This frequent feature appears whenever a health issue is relevant to the topic being discussed and requires special attention. These features may involve sexual infections and diseases, problems in the functioning of the sexual body, or sexual issues that may cause psychological difficulties or emotional pain.

Sexuality and Culture

The HIV/AIDS Pandemic

HIV and AIDS have become a pandemic and one of the greatest health threats of our time (see UNAIDS, 2010). A staggering 1.6 million deaths were attributed to HIV infection in 2012, and it is estimated that over 35 million people throughout the world are currently living with HIV infection. Over 6,000 people are newly infected with HIV every day and over 4,000 die from AIDS-related illnesses *daily*. In the last decade, the number of people living with HIV infection has steadily increased, as indicated in Figure 8.5a. This appears to be a discouraging statistic, but in reality, it has somewhat of a silver lining. Although more people have become infected, more have also been living with HIV and not converting to AIDS. This is evident from Figure 8.5b, which shows a global decrease in new HIV infections over the same period. More importantly, as you

Sexuality and Culture. This feature focuses on topics from diverse cultural practices and customs. The world of human sexuality is rich in cultural, subcultural, and ethnic diversity. To increase awareness of this richness, multicultural topics are integrated throughout the text and highlighted in the "Sexuality and Culture" Features.

Responsibility: Living an Informed, Ethical, and Analytical Life as You Relate to the Complex World of Human Sexuality

Each of us is responsible for our own decisions. Making the right choices for yourself and being aware of health and diversity issues come only when you are an informed consumer of sexual information. This text provides you with the skills to evaluate intelligently and critically the vast amount of sexual information you receive almost constantly from the media, from your friends and acquaintances, and from your family; it will help put to rest any misconceptions you may have about sexuality.

Sexuality, Ethics, and the Law. Issues of ethics and the law are often an important part of discussions of sexuality, and it is our responsibility to be aware of these key topics. Examples of some issues include the ethics of informing potential intimate partners about sexually transmitted infections, the ethical consideration of sex-preselection in childbirth, Megan's Law and sex offenders, and the crime of child pornography.

Sexuality, Ethics, and the Law

Sex-Preselection Technology

The worldwide social and ethical considerations in the face of increasingly available methods for allowing prospective parents to select the sex of their children are disturbing. What, for example, might these procedures mean for male-female sex ratios in countries such as China, India, or the Middle East, where boy babies are highly prized and preferred over girls? The effect of China's one-child policy has been to increase the desire for male infants to the point that hundreds of thousands, and perhaps millions, of baby girls are secretly abandoned to orphanages or even killed every year so that the parents may try again for a boy (Banister, 2004; Lubman, 2000). Researchers estimate that over the next 20 years, these countries will see an excess of as much as 20% more young men than women. This has the potential effects of fewer men marrying, a basic cultural expectation in these cultures, leading to social exclusion of more men, increased violence, and possibly an increase in prostitution and other crimes (Hesketh, Lu, & Xing, 2011).

Recognizing the future sociological implications of infant sex selection, China and India have passed laws prohibiting abortions based on parental sex preferences (Mudur, 2002). In addition, China has begun to restrict assisted fertility clinics that may be engaged in prepregnancy sex-selection technologies and to reward families monetarily who have only one son or who have no son but have and keep two daughters (Bumgarner, 2007; Glenn, 2004; Kalb, Nadeau, & Schafer, 2004). Although the technology allowing for the *preselection* of the sex of children (prior to conception) may help reduce sex-based abortions, the consequences of such a choice to the population of countries that place a high value on male offspring could be devastating to the gender balance in those countries in future decades.

Moreover, the social and scientific ethical considerations of sex preselection are far-reaching. Even in the medical profession in the U.S., a great deal of disagreement exists (see Puri & Nachtigall, 2010). A survey of primary care physicians (PCPs) and physicians who are sex-selection technology providers (SSTPs), found that PCPs were concerned about how sex selection leads to medically unnecessary expensive and invasive medical procedures, how it fuels the fire of gender stereotyping and discrimination, and how it leads to child neglect of children who are the "wrong" sex. SSTPs believed that sex selection is a reproductive right that women should have, that it allows parents to plan for family sex balance, and helps prevent unwanted pregnancies and sex-based abortions.

This example demonstrates how the issue of sex-preselection is extremely complex and will likely remain so for the foreseeable future because it is here to stay. Individuals, couples, cultures, and countries must approach the issues in their own way and from their own legal, moral, and ethical perspectives. As with so many new and emerging technologies, scientific advances in human sexuality may be used for positive or negative ends, and the choices we make must therefore be as informed and accurate as possible.

Historical Perspectives. This feature highlights significant people or events and gives you key information about the history of the study of human sexuality.

Historical Perspectives

Anatomy in the Dark Ages

Throughout history, and even today to some extent, human sexual anatomy was seen as a shameful subject for study or discussion. Why? Because such topics might excite people and cause them to engage in "impure acts." Consequently, the study of sexual anatomy prior to the nineteenth century was wildly misguided (Stolberg, 2003). Here is a partial list of early beliefs about human sexual anatomy—all of them erroneous

Evaluating Sexual Research

"What Men and Women Really Want in Bed! Take Our Reader's Survey"

How much of your sexual knowledge have you obtained from newsstand magazines? A favorite feature, often found in magazines such as *Redbook*, *Cosmopolitan*, *New Woman*, *Playboy*, *Playgirl*, and *Esquire* (and many others), is the survey that asks readers to respond by mail or online to questions about sex. The magazines then report the findings from the survey in a subsequent issue a month or two later. Can you see a flaw in this survey methodology? Even if the items on the survey are constructed properly (and usually they are not), the responses are bound to be seriously biased. First, all respondents are readers of that particular magazine and would not, therefore, represent the general population. In fact, all magazines are intentionally targeted at a very specific audience, such as single, working women; professional men between the ages of 20 and 45; or parents. Second, only a small percentage of readers will take the time and energy to respond to the survey (see Table 1.6 on page 22 comparing volunteers to nonvolunteers in sexual research). These eager participants certainly usually are not typical of the overall population. Their responses may not even be representative of the readers of that magazine, much less you or me or most other people.

The bottom line is that surveys such as these can be fun and titillating and may, on occasion, offer some interesting information for conversation or gossip, but they should not be considered scientific and cannot be relied on for meaningful sexual knowledge.

Surveys such as those summarized in Table 1.5 are some of the larger scientific and relatively valid sexuality surveys conducted over the past 60 years or so. Most of these researchers made an effort to avoid the problems discussed in this box. You will see references to these surveys—along with other smaller survey studies that have been published in professional, scientific journals—throughout this book.

Most surveys in popular magazines rely on biased samples and lack validity.

Evaluating Sexual Research. When it comes to sexuality, stories circulating throughout the media often take on a life of their own and seem attractive and believable. However, often they simply perpetuate myths and falsehoods or create new ones. Therefore, throughout this book are features that shed some light on how you can critically analyze sexual stories and research you may come across in newspapers, magazines, online, or on TV. This, in turn, will help you become a more skilled consumer of sexual research findings by learning to evaluate sexual research before accepting what you see or read as fact.

Have You Considered?

1. Would you ever volunteer to be a participant in a study such as Masters and Johnson's research in the 1960s? Why or why not? Discuss your opinion about the ethics of such research.
2. Explain why many women do not routinely experience orgasm during heterosexual intercourse
3. Can you think of an "aphrodisiac" you have heard of that was not mentioned in the chapter? Because aphrodisiacs have never been shown scientifically to have any real effect on sexual desire, why do you think people believe in the ones you've heard of?
4. Which of the various models of sexual response discussed in the chapter do you feel is most accurate? Which seems most questionable to you? Explain your answers.
5. In your opinion, what is the most important reason for people to study and understand the process of human sexual responding? Why?
6. For many couples, the goal of lovemaking is orgasm. Explain why many sex researchers and therapists claim that this detracts from a satisfying sexual experience.

Have You Considered? At the end of each chapter are several problems relating to the chapter's content. These are designed to help you understand the material in the chapter, but, more than that, they encourage you to consider and analyze what you have learned, what the information means to you, and how you can apply it to real life.

NEW TO THE FOURTH EDITION

Although some aspects of the study of human sexuality remain fairly constant, other issues are dynamic and evolve over time. Every chapter in this text covers topics and discusses issues that are constantly changing. This new edition reflects those changes to ensure that the material is as current as possible and keeps you up-to-date on the research and historical events that compose the field of human sexuality. Here's just a sampling, chapter-by-chapter, of some of the changes in the world of human sexuality that I have added, updated, revised, or expanded for the fourth edition of this text:

A *Sampling* of Topics, Discussions, and Features, New to the Fourth Edition of *Human Sexuality*

OVERALL

- Over 450 new and updated references from the latest discoveries and research in the field.
- Revised and updated statistics throughout the book to those most recently available. These include, but are not limited to, statistics relating to: contraception, sexually transmitted infections, pregnancy and birth, sexual orientation, sexual aggression (domestic violence, rape, child sexual abuse, and sexual harassment), prostitution, and pornography
- Text-wide update to the diagnostic criteria in the American Psychiatric Association's 2013 revision of the *Diagnostic and Statistical Manual* (The DSM-5). These are especially relevant to the text's coverage of sexual problems (Chapter 7) and paraphilias (Chapter 14)
- Updates of photos, drawings, graphs, figures, and other art to maintain text freshness and currency
- Greater inclusion and integration of sexual diversity throughout the text
- Enhanced emphasis on *consent* as it relates to all sexuality discussions and activities
- Addition of topical, current-event items to enhance text currency
- Newly added learning objectives and review questions to all chapter sections

CHAPTER 1: STUDYING HUMAN SEXUALITY

- New "Sexuality and Culture" feature highlighting recent changes in sex education in China
- Enhanced emphasis on the issue of *consent* in all aspects of sexual education, research, and activities
- A new "In Touch with Your Sexual Health" feature clarifying the main categories of relevant health issues, including physical, emotional, and psychological problems
- Topical addition: New "Evaluating Sexual Research" critical-thinking feature: "How could this Happen? *Father Who Went to Hospital for Kidney Stones Discovers He Is a Woman*"

CHAPTER 2: SEXUAL ANATOMY

- Updated discussion of current controversy surrounding male circumcision
- Newly updated maps, statistics, and discussion on current global status of female genital mutilation
- Latest information concerning urinary tract infections and development of E. coli vaccine which will prevent most UTIs
- Most current guidelines on mammograms, breast self-examination, and breast health recommendations
- DSM-5 update on *premenstrual dysphoric disorder* diagnostics

CHAPTER 3: THE PHYSIOLOGY OF SEXUAL RESPONSE

- Modified conceptualizations of Masters and Johnson's model of sexual response: merging excitement and plateau phases
- Greater emphasis on alternatives to Masters and Johnson model of sexual response
- Updates on G-spot and female ejaculation discussions
- Addition of Janssen and Bancroft's "Dual Control Theory" of sexual response
- Topical addition: Reference to the new cable TV bio-pic series, "The Masters of Sex," dramatizing the careers and discoveries of William Masters and Virginia Johnson
- Revision of the "New View" approach to female sexual response

CHAPTER 4: INTIMATE RELATIONSHIPS

- Increased coverage and integration of all sexual orientations throughout chapter
- Added discussion of information technology and social networking as they relate to trust and control in relationships ("cyberspying")
- Increased emphasis on relationship abuse in gay and lesbian relationships
- New "power and control" wheel applied to non-heterosexual relationships

CHAPTER 5: CONTRACEPTION: PLANNING AND PREVENTING

- New visual conceptualization of contraception types based on overall effectiveness
- Latest research on Cowper's gland secretions, sperm cells, and the withdrawal method of contraception
- Topical addition: Legal bans and confiscation of condoms in some countries attempting to reduce sex trade activity
- Efforts to reinvent the condom to increase acceptability: Development of "origami condom" and *The Gates Foundation* condom-design competition
- Latest research suggesting reduced hormonal contraceptive effectiveness with increased body weight
- Latest, specific guidelines for missed contraceptive pills
- New reversible contraceptive IUD: *Skyla*
- Greater detail on reasons for vasectomy failure

CHAPTER 6: SEXUAL BEHAVIORS: EXPERIENCING SEXUAL PLEASURE

- New discussion of interplay of cybersex and sexual fantasy
- New research on kissing as erotic activity
- Recent data from *National Survey of Sexual Health and Behavior* (NSSHB) incorporated throughout chapter

CHAPTER 7: SEXUAL PROBLEMS AND SOLUTIONS

- Chapter significantly revised to reflect DSM-5 diagnostic guidelines including:
- Sexual disorders now divided into: Sexual desire, interest, or arousal disorders; disorders of orgasm; and sexual pain disorders
- Diagnosis of sexual disorders requires duration of problem for a minimum of six months and is causing significant psychological or emotional distress
- Disorders less focused on heterosexual intercourse
- Dyspareunia and vaginismus diagnoses eliminated and combined into *genito-pelvic pain/penetration disorder* (GPPPD)
- Disorders now classified as lifelong versus acquired and/or generalized versus situational

CHAPTER 8: SEXUALLY TRANSMITTED INFECTIONS/DISEASES

- Updated statistics on incidence and prevalence worldwide
- Greater emphasis on problem of the asymptomatic nature of most STIs
- Current cervical cancer prevention guidelines, including recommendations for HPV (genital warts) screenings
- Role of HPV in anal and oral cancers and recommendation for HPV vaccine for boys as well as girls
- Revised and greater detail of HIV pandemic globally and by region, including some new, optimistic downward trends
- Added discussion of HIV2
- New CDC recommendation for use of antiretroviral medication, Truvada, to help prevent transmission of virus

CHAPTER 9: CONCEPTION, PREGNANCY, AND BIRTH

- Enhanced discussion to clarify distinction between conception and pregnancy
- Detailed discussion of tobacco smoke as teratogen, including secondhand absorption, leading to higher rates of ectopic pregnancy, miscarriage, preterm births, and stillbirths
- New DNA-based blood test to determine sex of embryo at 7 weeks gestation (and how this relates to sex-selection debate)
- Update on high C-section rates and surrounding medical debate
- Discussion of state-level legal challenges to *Roe v. Wade* over past decade
- Expanded discussion of abortion issues

CHAPTER 10: GENDER: EXPECTATIONS, ROLES, AND BEHAVIORS

- Terminology change from "intersex" to "Disorders of Gender Development" as specified by the DSM-5
- Addition of detailed criteria for medical approval of gender reassignment surgery
- Enhanced discussion of transgender children and possible medical interventions to forestall puberty
- Topical addition: California's 2013 law requiring public schools to allow students to adopt their self-identified gender identity, participate in sports and other activities accordingly, and use the restroom and other facilities aligning with their gender identity
- Extended discussion of gender-linked aggression to include social alienation as a form of aggression in girls
- Clarification of the "overlapping curve model" of gender differences

CHAPTER 11: SEXUAL ORIENTATION

- Updates of same-sex marriage laws as recently as possible up to press time
- Recent updates of American Psychological Association's position statements on sexual orientation
- Revised discussion of legal status of homosexuality in various countries
- Topical addition: Uganda's 2014 law imposing a 14-year prison sentence for first-time offenders engaging in "homosexual acts"

- Recent research on the erroneous, yet wide acceptance, of a "gay gene" theory
- Addition of research showing anti-gay bias by students toward instructors
- Update on hate crime statistics based on sexual orientation from 2009 to present

CHAPTER 12: SEXUAL DEVELOPMENT THROUGHOUT LIFE

- Update on what constitutes troubling or problematic sexual behavior in young children
- "Final word" on comprehensive sex education versus abstinence-only programs
- Addition of new statistics on teen dating violence
- Inclusion of schools' efforts to create more open, accepting, and safe environments for teens who are gay, lesbian, transgender, or questioning their gender identity or sexual orientation
- Information on protecting teens from dangers of cybersex
- Addition of new data on teen condom use
- New trends in cohabitation

CHAPTER 13: SEXUAL AGGRESSION AND VIOLENCE: RAPE, CHILD SEXUAL ABUSE, AND SEXUAL HARASSMENT

- Greater focus on central issue of *consent* for any type of sexually related behavior
- Addition of feature on sexual violence in the military
- Discussion of new, broader FBI definition of rape that includes more forms of penetration and includes rape of men
- Cultural comparisons of child sexual abuse rates
- Addition of self-evaluation instrument for victims of predator (date-rape) drugs
- Discussion of "cyberharassment"

CHAPTER 14: PARAPHILIC DISORDERS: ATYPICAL SEXUAL BEHAVIORS

- Full chapter revision reflecting DSM-5 changes in clinical criteria of paraphilia diagnoses
- Distinction between a "paraphilia" and a "paraphilic disorder"
- Addition that non-victimizing paraphilia diagnosis must involve "personal distress"
- Explanation that for victimizing paraphilias, simple *desire* for paraphilic behavior is adequate for diagnosis
- Explanation that "sexual addiction" is not deemed a paraphilia; classified as "hypersexual disorder"
- Update on medication therapy for paraphilias, especially SSRIs

CHAPTER 15: THE SEXUAL MARKETPLACE: PROSTITUTION AND PORNOGRAPHY

- Updates on issue of legalization or decriminalization of prostitution
- Topical addition: Attempts to add male sex workers in Nevada brothels (to service women customers)
- Discussion of effectiveness (or lack thereof) of the Victims of Trafficking and Violence Protection Act (VTVPA) of 2000
- Updated statistics on crack cocaine use among sex trade workers (estimates up to 95%)
- Addition of new "Sexuality, Ethics, and the Law" feature on the use of "john school" to reduce demand for prostitution
- Discussion of the effects of pornography on intimate relationships
- Update of battle against child pornography on a global scale
- New efforts to consolidate sexually explicit sites under .xxx top-level domain
- Discussion of major Internet providers' (i.e. Google) efforts to remove child abuse sites from World Wide Web

REVEL™

Educational Technology Designed for the Way Today's Students Read, Think, and Learn

When students are engaged deeply, they learn more effectively and perform better in their courses. This simple fact inspired the creation of REVEL: an immersive learning experience designed for the way today's students read, think, and learn. Built in collaboration with educators and students nationwide, REVEL is the newest, fully digital way to deliver respected Pearson content.

REVEL enlivens course content with media interactives and assessments—integrated directly within the authors' narrative—that provide opportunities for students to read about and practice course material in tandem. This immersive educational technology boosts student engagement, which leads to better understanding of concepts and improved performance throughout the course.

Learn more about REVEL

http://www.pearsonhighered.com/revel/

TEACHING AND LEARNING

Supplementary Materials

In today's world of higher education, a good text is only one element of a comprehensive learning package. *Human Sexuality*'s supplement creators are experienced authors and professors of human sexuality who have developed high-quality supplements for instructors and students.

Supplements for Instructors

Instructor's Resource Manual (IRM)

In this abundant collection of resources for each chapter, instructors will find activities, exercises, assignments, handouts, and demos for in-class use. The material for each chapter is organized in an easy-to-use chapter "Lecture Guide" outline. This resource saves prep work and helps you make maximum use of classroom time. The IRM is available for download from the Instructor's Resource Center at http://www.pearsonhighered.com/irc

Test Item File

This test bank includes multiple-choice, true-false, short-answer, and essay questions for each chapter. All questions are noted by type—applied, conceptual, or factual—and the difficulty level is given for each. The Test Item File is available for download from the Instructor's Resource Center at http://www.pearsonhighered.com/irc

MyTest

This powerful assessment-generation program assists instructors in creating and printing quizzes and exams with ease. MyTest is available at http://pearsonhighered.com/irc, with additional information at http://www.pearsonmytest.com. Questions can be edited and authored online, providing instructors with ultimate flexibility to manage assessments efficiently, anytime, anywhere! Instructors can readily access existing questions, and edit, create, and store them using simple drag-and-drop, Word-like controls. The level of difficulty and text page numbers are provided for each question.

PowerPoint Presentation

A wide array of excellent PowerPoint slides for each chapter provides remarkable flexibility for classroom lectures and discussions. The presentations highlight all of the key topics and points from the chapters and include many graphics and photos from the text. These are available for download from the Instructor's Resource Center at http://www.pearsonhighered.com/irc

Collaboration, Expertise, Accuracy, Thank-Yous

The creation of this text and the comprehensive extras that accompany it are the result of extensive development and investment involving key content reviewers, reviewer conference participants, and student class testers. Expert reviewers in critical topic areas provided feedback on the currency and accuracy of the research. I would like to express my thanks to all those who reviewed, commented on, and helped to perfect all of the first four editions of this text:

Judi Addelston, *Valencia Community College*
Malinde Althaus, *University of Minnesota*
John Batacan, *Idaho State University*
Betsy Bergen, *Kansas State University*
Robert Boroff, *Modesto Junior College*
Saundra Boyd, *Houston Community College*
Barbara Ann Cabral, *City College San Francisco*
Sandra Caron, *University of Maine*
Glenn Carter, *Austin Peay State University*
Jane Cirillo, *Houston Community College*
Jeffrey K. Clarke, *Ball State University*
Brian Cowley, *Park University*
Elizabeth Curtis, *Long Beach City College*
Dale Doty, *Monroe Community College*
Bailey Dreschler, *Cuesta College*
Adama Dyoniziak, *California State University, Long Beach*
Steve L. Ellyson, *Youngstown State University*
Anne E. Fisher, *University of Southern Florida, Sarasota/Manatee*
Edward Fliss, *St. Louis Community College at Florissant Valley*
Debra Golden, *Grossmont College*
Debra L. Golden, *University of Hawaii*
Kathy Greaves, *Oregon State University*
Elizabeth Guillett, *University of Florida*
Terri L. Heck, *Macomb Community College*
John Hensley, *Tulsa Community College*
Suzy Horton, *Mesa Community College*
Mark Jackson, *Transylvania University*
Lyn Kemen, *Hunter College, CUNY*
Travis Langely, *Henderson State University*
Callista Lett, *Fullerton College*
Sonya Lott-Harrison, *Community College of Philadelphia*
Susan MacLaury, *Kean University*
Ticily Medley, *Tarrant County College-South*
Jose Nanin, *Kingsborough Community College, CUNY*
Shirley Ogeltree, *Southwest Texas State University*
Jennifer O'Loughlin-Brooks, *Collin County Community College*
Gina Marie Piane, *California State University, Long Beach*
Grace Pokorny, *Long Beach City College*
Brad Redburn, *Johnson County Community College*
Steve Rison, *Austin Community College*
Sonia Y. Ruiz, *California State University, San Marcos*
Edward Smith, *Georgia Southern University*
Sherman Sowby, *California State University, Fresno*
Susan Sprecher, *Illinois State University*
Nan Taylor, *Pace University*
Kristy Thacker, *University at Albany, SUNY*
Tara Lynn Torchia, *University of Maryland*
Mixon Ware, *Eastern Kentucky University*
Steve Weinert, *Cuyamaca College*
Tanya Whipple, *Missouri State University*
Midge Wilson, *DePaul University*
Paul D. Young, *Houghton College*

I want to express my deep gratitude to a dedicated and extremely talented editorial, production, and design team at Pearson Education, who combined their multitude of skills in creating this outstanding fourth edition of *Human Sexuality*. My heartfelt thanks go out to Charlyce Jones Owen, Publisher at Pearson Education, for skillfully directing this edition from the beginning; Manuel Echevarria, Project Manager at Pearson Education, for guiding the complex set of processes and people to keep the project moving forward and on schedule; Anandakrishnan Natarajan, Project Manager at Integra Software Services, for his talented and artistic design; Jonathan Fisher, Project Manager at nSight, Inc. for his expertise in ensuring the accuracy and flow of the written text; Development Editor, Tamra Orr, for her research assistance; and Rachel B. Steely for her sharp-eyed copyediting.

Finally, and saving the best for last, thanks to all my students over the past 25+ years. They have never failed to inspire and challenge me to write and teach at the highest levels I am capable of achieving. Here's to many more years in the classroom with all my future students.

Roger Hock

About the Author

Roger R. Hock, PhD, is a professor of psychology and human sexuality at Mendocino College in northern California. He received his MA in Psychology from San Diego State University and his PhD in Psychology from the University of California at San Diego. He is also the author of *Forty Studies that Changed Psychology: Explorations into the History of Psychological Research* (Pearson, 2012) and coauthor (with Meg Kennedy Dugan) of *It's My Life Now: Starting over after an Abusive Relationship or Domestic Violence* (Routledge, 2006), and (with Amy Marin) *Psychology*, a soon-to-be released digital introductory psychology textbook from Pearson Education.

Dr. Hock has been teaching psychology and human sexuality for over twenty-five years. Human Sexuality is his favorite class to teach and is consistently one of the most popular among students. He believes that Human Sexuality is an essential course for college students because the topics covered are fundamental to our lives and our identities as humans. He asserts that the human sexuality course should be a requirement for all undergraduate college and university students. Why? Because the material in this course touches everyone, in countless ways, throughout their lives. But, more importantly, students, upon entering college as adults, often lack a complete and current understanding of at least some, if not many, key issues that combine to create the complexities of humans as sexual beings. Sex education in grades K–12 tends to be incomplete, inadequate, sometimes misleading, and too often, nonexistent. Moreover, not all parents are willing or able to impart the necessary and correct information to their children. Students in higher education clearly need and deserve the knowledge they acquire in this class.

A Sexual History Time Line

The world of human sexuality has been marked by many influential historical events. To illustrate, we journey back over 150 years of significant events that have changed our views, lives, laws, and attitudes with respect to human sexuality (relevant chapter numbers are in parentheses) .

1846
U.S. patent issued for first diaphragm contraceptive device (5)

1873
Congress passes Comstock Act outlawing distribution of contraception information and devices (5)

1874
Women's Christian Temperance Union founded to oppose men's drinking of alcohol and engaging in "immoral acts" (15)

1897
Havelock Ellis publishes first of six-volume, *Studies in the Psychology of Sex*; advocating sex as pleasurable, central function in life (6)

1909
Sigmund Freud lectures at Clark University in Massachusetts; introduces U.S. to his theories of sex as primary driving force in human nature (1)

1916
Margaret Sanger arrested and jailed for opening first birth control clinic in Brooklyn signaling change in sexual norms; marks separation of sex from reproduction for women (5)

1920s
Widespread introduction of automobile offers privacy and independence for dating couples (4)

1923
John Kellogg becomes president of Battle Creek Sanitarium; promotes plain foods such as corn flakes in order to prevent sexual feelings and discourage masturbation (6)

1934
Catholic Church forms "Legion of Decency" to evaluate and rate films; Hollywood responds by reducing sexual content of movies (15)

1934
Appeals court overturns ruling of James Joyce's Ulysses as obscene, signaling liberalization of obscenity laws (15)

1936
U.S. Federal Court overturns Comstock Act's anti-contraception laws (5)

1942
Planned Parenthood Federation of America founded: advocates family planning and sexual satisfaction in marriage (5)

1948
Publication of *Kinsey Institute's Sexual Behavior and the Human Male*; sold 200,000 copies and was on *New York Times* bestseller list for 27 weeks (1, 6)

1950

1950
Existence of "zone of erogenous feeling" on wall of vagina suggested by Ernst Grafenberg (now known as the G-Spot) (2)

1953
Hugh Hefner publishes first issue of *Playboy Magazine* (15)

1953
Christine Jorgensen is among first to undergo male-to-female "sex change" operation (10)

1953
Publication of *Kinsey Institute's Sexual Behavior of the Human Female;* sells 250,000 copies (1, 6)

1960

1960
Feminine Mystique by Betty Friedan published; signals beginning of feminist movement (10)

1960 Food and Drug Administration approves first oral contraceptive: trade name: "Enovid" (5)

1966
Publication of *Human Sexual Response* by Masters and Johnson revolutionizes public understanding of sexual physiology (3, 6, 7)

1969
Huge rock concert at Woodstock marks culmination of "hippie free love" movement (6)

1969
Police clash violently with patrons of Stonewall Inn, a gay bar in New York's Greenwich Village; marks beginning of gay rights movement (11)

1970

1970
Feature film *The Boys in the Band* first wide-release movie with openly gay plot and characters (11)

1972
Rocky Horror Picture Show shakes up gender stereotypes (10)

1973
The American Psychiatric Association (APA) votes to remove homosexuality from list of psychological disorders (11)

1973
U.S. Supreme Court, in landmark decision in case of *Roe vs. Wade*; declares a woman's right to an abortion is protected by her constitutional right to privacy, effectively legalizing abortion (5, 9)

1975
U.S. Civil Service Administration lifts ban on hiring gays and lesbians (11)

1976
The Hite Report is published detailing sexual fantasies and behaviors of women (1, 6)

1978
Birth of Louise Brown, first infant conceived through in vitro fertilization, dubbed "test tube baby" by media (9)

1979
California first state to classify forced sex by husband on wife as rape (13)

1980

(Numbers in parentheses indicate text chapters for additional information)

1981
First unexplained deaths from unusual infections seen among gay men in San Francisco and New York. First dubbed "Gay-related Immune Deficiency;" street name: "gay plague" this was the beginning of the HIV/AIDS epidemic in the U.S. (8, 11)

1983
Human Immunodeficiency Virus (HIV) isolated as cause of Acquired Immune Deficiency Syndrome (AIDS); found in homosexual and heterosexual populations (8)

1990

1993
FDA approves first female condom, trade name, "Reality" (5)

1994
Publication of *Sex in America*, first large-scale survey of sexual behavior in U.S. since Kinsey Reports (1, 4, 6, 7, 11)

1997
Ellen Degeneres comes out as gay on popular T.V. show *Ellen* (11)

1998
The Clinton-Lewinsky sex scandal story breaks (13)

1998
Viagra approved by FDA for treatment of erectile disorder (7)

1999
Human papilloma virus found to be the leading cause of cervical cancer (8)

2000

2002
So called "rape drugs" become major problem on college campuses (13)

2003
The U.S. Supreme Court strikes down all Texas sodomy laws (6, 11)

2004
Massachusetts is first state to legalize gay marriages (11)

2007
Studies show "abstinence-only" sex education fails to reduce teen pregnancy or STIs (1, 12)

2007
U.S. Supreme Court upholds ban on late-term abortions regardless of health of mother; weakens *Roe vs. Wade* (9)

2007
Millions of doses of Gardasil, a new vaccine that prevents HPV (genital warts) and cervical cancer, were distributed in the U.S. Immunization recommended for girls 9–12. (8)

2008
Transgender man gives birth to healthy baby girl (10)

2008
U.S. Supreme Court declares death penalty illegal (cruel and unusual punishment) for child rape (14)

2008
New study confirms that sex remains important part of life into 70s, 80s and beyond (12)

2010

2010
The Trevor Project is formed with the goal of ending suicide among LGBTQ youth by providing life-saving and life-affirming resources: www.thetrevorproject.org (11)

2010
Gays and lesbians allowed to serve openly in the U.S. military; repeal of "Don't ask; don't tell"

2011
Teen birth rates drop to all low in California (12)

2011
New York becomes 6th (and largest) state to legalilze same-sex marriage (11)

2012
California becomes first state to ban "reparative therapy" which falsely purports to convert gay teens to heterosexuality (11)

2012
Anti-HIV drug Truvada approved as preventive for HIV infection when taken daily before and after exposure to virus (5)

2013
The revised APA disorders guide, The DSM-5, adds "Premenstrual Dysphoric Disorder" (PMDD), a severe form of PMS, to list of offcial psychiatric diagnoses

2013
Showtime TV debuts new biopic series, "Masters of Sex," about the lives and work of sex research pioneers, William Masters and Virginia Johnson (3, 6, 7)

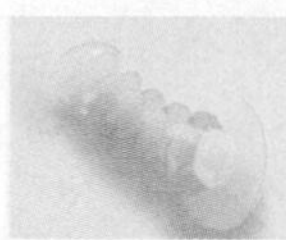

2013
New innovative contraceptive, the "Origami Condom" is announced (5)

2013
The morning-after contraceptive pill, Plan-B One Step becomes available without a prescription for girls of all ages (5)

2013
The Boy Scouts of America allow gay Scouts to join; gay Scout leaders still banned (11)

2013
California becomes first state to pass law ensuring transgender students (K-12) equal access to facilities and school activities consistent with their gender identity (10)

2014
Pope Francis officially apologizes for priest sex abuse; begins program of tougher protections and punishments (13, 14)

2014
Number of states legalizing same-sex marriage reaches 32 (plus Washington DC). (11)

Chapter 1
Studying Human Sexuality

Learning Objectives

After you read and study this chapter you will be able to:

1.1 Review the key issues people should be aware of to enjoy a healthy experience of human sexuality.

1.2 Explain the knowledge people need for a full understanding of human sexuality.

1.3 List and explain the methods used by researchers in the study of human sexuality.

1.4 Display a clear awareness of the ethical rules that researchers must follow for all research into human sexuality.

Since You Asked

1. I'm not sure about this class. I'm afraid it's going to be so embarrassing! (see page 6)
2. Do we really need a course about sex? I'm married and have a child. I'd say I probably know everything I need to know. (see page 9)
3. Is it normal to have an orgasm by masturbating but never to have one during intercourse? (see page 13)
4. I am 21 years old and had a "wet dream" two weeks ago. Is that normal at my age? (see page 13)
5. Why did my high school stop the sex ed class? Now they're saying we should just all wait until marriage to have sex. No one's really going to do that, are they? (see page 13)
6. My parents never had "the talk" with me. Were they just embarrassed, or did they think I would learn on my own? (see page 14)
7. How is it possible to study people's sexual lives when it's such a private, personal experience? (see page 17)
8. Is it really ethical to study people's actual sexual behaviors? (see page 29)

The intricacies of human sexuality play a part in everyone's life. And human sexuality, in one way or another, affects virtually everything you do. Have you given much serious thought to your knowledge, desires, or identity as a sexual being? Most people haven't. Yet few areas in your life are more important for self-reflection than sex. How can you ensure *for yourself* a physically and emotionally gratifying sexual life? Do you have a sense of what you want or don't want in terms of sexual intimacy with another person now or in the future? Under what conditions and with whom will you feel comfortable allowing that intimacy to grow? What is your vision of a healthy and fulfilling romantic relationship? Can you be sure that you will make choices and decisions that are right for you when sexual situations arise? How will you keep yourself safe from unwanted pregnancy, sexually transmitted infections (diseases), and sexual violence? How will you interact with others who are sexually different from you?

We will return to the theme of your self-knowledge, attitudes, and actions at the close of this chapter—and every other chapter in this text—and offer you the opportunity to incorporate what you have learned in the chapter into your **personal sexual philosophy**. Remember, studying human sexuality is about far more than "getting the facts." It is about knowing who you are, what you want or don't want, and planning ahead.

personal sexual philosophy
A person's unique foundation of knowledge, attitudes, and actions relating to what the person wants and who he or she is as a sexual being.

human sexuality
An area of research and study focusing on all aspects of humans as sexual beings.

Human sexuality is a complex area of study that focuses on all aspects of humans as sexual beings. This includes such topics as sexual anatomy and responses, sexual feelings and behaviors, intimate relationships, sexual identity and desires, sexual health and well-being, and the way we perceive and express our individual sexual selves. Each chapter in this text is just one piece of the rich and challenging puzzle of human sexuality.

In this chapter, we will examine the ways in which people learn about human sexuality. In the most basic sense, these learning experiences may be divided into two categories: (1) *experiencing* your own sexuality and (2) deepening your *understanding* of human sexuality issues. On the experiential side, we will explore your emotional reactions relating to sexuality, getting to know yourself as a sexual person, developing your personal set of sexual **morals** and values, making responsible choices about sexual activities, discovering the full range of sexually intimate behaviors, and enhancing your lifelong sexual fulfillment and satisfaction.

morals
A person's individual, unique attitudes about what constitutes right and wrong.

In order to attain a deeper understanding of sexuality, we will focus on its surprising complexity. We will work toward helping you:

- Reject myths and misconceptions you may hold about sexuality
- Develop an understanding and respect for sexual diversity

Table 1.1 Guiding Principles for This Text

It's no secret that most students never read the prefaces in textbooks. You know who you are! In this book's Preface, I explained seven principles for presenting human sexuality information correctly, effectively, and understandably. Because these principles have influenced me every step of the way in my teaching and writing about human sexuality and because many of you may have missed them, I will summarize them again here.

Principle	How It Applies to This Text
Personal Choice and Responsibility	The assumption throughout this text is that you are an adult and in charge of your personal sexual choices. Throughout your life, it is your responsibility to make decisions that are right for you that you can feel good about and that do not harm anyone else.
Authoritative Information	The content of this text is based on the most up-to-date, scientific, and accurate research available.
Acceptance of Your Own Sexuality	This text highlights the importance of developing your personal sexual philosophy, so that you are in charge of the sexual situations in your life instead of the other way around.
Real-Life Relevancy	Every effort has been made to ensure that this text is relevant and meaningful to you.
Physical and Emotional Health and Wellness	This text is designed to focus on specific as well as general sexual health issues and to decrease your discomfort in seeking care and treatment for sexual problems, if necessary.
Critical Analysis of Research and Information about Sexuality	This text will give you the ability to evaluate, intelligently and critically, the vast amount of sexual information you are receiving—both from the media and from your social interactions with others.
Awareness and Respect for Sexual Diversity	The information in this text will assist you in understanding the world around you and how and where you fit into it; this mosaic of diversity plays a major role in making human sexuality such a rich and fascinating study.

- Acquire a sense of what is sexually "normal" and "abnormal"
- Study what it means to stay "sexually healthy"
- Figure out how to become a critical, educated consumer of research and reporting about human sexuality
- Learn how to talk to your own children about sex if and when the time comes, and

Later in this chapter we will review the scientific methods researchers use to study human sexual behaviors, attitudes, and emotions. We will then consider the importance of ethics as it relates to sexuality research. As you read this chapter and through out this text, keep in mind the guiding principles for this book, discussed in the Preface and summarized again for you in Table 1.1.

Historical Perspectives

A Human Sexuality Time Line (in the Preface)

Many of the topics covered in this book are rooted in ancient history, some dating back millennia and others reaching back to the very beginning of humankind (or else humankind might no longer exist!). We touch on many of these events in greater detail in the Historical Perspectives section early in each chapter. Here in Chapter 1, the Historical Perspectives section offers you a glimpse of the major events of the past 150 years that have shaped sexual history in the United States and much of the Western world. Each event on the time line located at the end of the Preface to this book, notes the chapter number where you will find a more detailed discussion of that topic in its current context. Enjoy your trip through sexual time!

Experiencing Human Sexuality

1.1 Review the key issues people should be aware of to enjoy a healthy experience of human sexuality.

We experience sexuality in many personal and subjective ways. Our individual experiences regarding sexuality vary greatly from negative to positive, painful to joyful, traumatic to sublime. The factors that determine how we experience our sexuality may include any of the issues discussed next.

Enriching Self-Knowledge

First and foremost, we are born to be sexual beings. This does not imply that we all engage in any particular sexual activity or that we all have the same or even similar sexual feelings and desires. But sexuality will always be a part of what makes each of us a unique individual. From infancy through old age, we have the capacity to experience both physical and emotional sexual feelings.

It follows, then, that your perception of your own sexuality is a major part of your self-identity. To demonstrate this concept, imagine that you wake up tomorrow morning and have no idea whether you are a man or a woman. How would this make you feel? Confused? Probably. Disoriented? At least! You might think, "I no longer know who I am!" Your **gender identity**—the concept of yourself as a man or woman, masculine or feminine—is one of the most important components of your sexual identity. And you do not need to study human sexuality to know what yours is. You are already very clear about that, and it will *not* change as you read this book.

gender identity
The sex (male or female) that a person identifies himself or herself to be.

Certain components of your sexual self may not be so clear to you, however. For example, some people are confused, at least at some point in their lives, about their **sexual orientation**, whether they are more attracted to members of the same sex or the other sex romantically, emotionally, and sexually. Others may be unsure about what qualities they desire in an intimate relationship or confused about their comfort level with specific sexual activities. This book and this course will help you find answers and better understand yourself.

sexual orientation
Term specifying the sex of those to whom a person is primarily romantically, emotionally, and sexually attracted.

Sex Is Emotional

Students' personal experiences in a human sexuality course can often trigger emotional reactions, sometimes very strong ones. These feelings may include general discomfort, confusion, anxiety, embarrassment, anger, arousal, surprise, or nervousness. Because these emotions make some people uncomfortable about studying human sexuality, we will spend a moment near the beginning of each chapter, in a feature called "Focus on Your Feelings" (see p. 9), to discuss the possible emotional reactions you may experience as you read and learn.

Your Morals and Values

Part of discovering yourself sexually usually involves developing your uniquely personal set of morals and values as they relate to sexual issues. You probably already have a sense of the morals and values that were instilled in you by your parents, your religious teachings, your peers, or other factors that have influenced you throughout your early life. As you have grown into adulthood, however, you may have begun to question those beliefs and wonder if they still apply to you as an independent, mature individual. Some of you may feel the need to make modifications in your system of morals and values that are more in line with how you choose to live your life. Morality and personal values play a central role in how you experience most, if not all, of the issues discussed in this book.

It is not the job or intention of this book to encourage you to adopt anyone else's sexual morals or values (including the author's). As you study human sexuality, you will acquire or enhance the knowledge and awareness you need to develop your own sexual standards and belief systems that make sense and feel right to *you* in *your* life.

Focus on Your Feelings

Sex is emotional. Virtually every emotion you can think of can be involved. Students studying human sexuality often experience unexpected emotional reactions. These reactions range from positive, happy, even euphoric feelings to discomfort, embarrassment, shock, anger confusion, fear, or various combinations of these and other emotions. Your personal, individual reactions will depend on your attitudes about sexual issues, your family and religious background, and your past and present sexual and relationship experiences. For example, here are two journal entries from students in the author's human sexuality classes:

> I was raised in a very strict religious household. Sex was never discussed, and even mentioning anything related to sex was not allowed. I'm learning a lot in this course, but I find it very difficult to read the material and participate in the class discussions. I'm embarrassed and feel that I am doing something wrong (guilt!). This is why I have missed so many classes.
>
> **Maya, first-year student**

> I've never told anyone this before, but from the time I was 8 until I was 12, I was sexually molested by my father. When we were in class discussing sexual abuse, it brought all this back to me. I don't know if I should get some counseling. I never thought it really bothered me, but now I'm not so sure.
>
> **Rick, sophomore**

Many (and luckily most) people have generally positive emotions and experiences about sex and sexuality. But in your human sexuality class, you should expect to experience emotions that you would not usually feel in other courses. *These emotions are completely normal*. However, if you find that you are having feelings that are bothersome, or if they interfere with your ability to enjoy and do well in this course, it would be a good idea to discuss them in private with your instructor or perhaps with a professional counselor (usually available through your college's counseling services). Remember, you should never feel *forced* to read material or attend classes that would be emotionally painful for you. Your sexuality professor should be willing and open to discuss with you ways of reducing your discomfort and maximizing your benefit from this class.

Consider the value in taking the time to weigh these issues and make some conscious decisions about how you want to live your sexual life. In this way, your values and moral beliefs can help guide you through the complexities of life as a sexual being and form an important part of your sexual philosophy. At any moment, you may find yourself facing difficult decisions about your sexual behavior and interactions, without time for thought or reflection. In the absence of a personal moral compass, you may make choices that you later regret. In other words, the situation may take charge of you rather than the other way around.

Making Responsible Choices

Having a clear sense of your sexual morals and values is only one factor in making responsible sexual choices throughout your life. Choosing to be sexually active in today's world requires you to make an almost overwhelming number of crucial decisions. For example, how will you protect yourself and your partner from HIV and other sexually transmitted infections? How can you be sure to avoid an unwanted pregnancy? How can you keep yourself safe from sexual violence and coercion? What are your expectations of dating and relationships? How can you and your partner communicate your needs and desires openly and honestly to each other? How will you handle a sexual problem with your partner?

Sex Is *More* than Intercourse

Yes, you read it right: *sex* is *more* than intercourse. This theme runs throughout the text and plays a role in many of the topics in various chapters (look for the box in the margin). Why? Partly because equating sex with intercourse neglects the full range of sexual experience and pleasure that is available to us as human beings. Western cultures often tend to take *sex* as a synonym for *intercourse* (Goodson et al.,

2003). Consequently, any other pleasurable, arousing, and satisfying sexual behaviors—such as kissing, touching, massage, masturbation (solo and mutual), and oral sex—become lumped together into a single category called "foreplay," or "that which leads up to intercourse," and are not thought of by a majority of people as "having sex." But in reality many behaviors can be sexually fulfilling in themselves, and for some individuals or couples they may even be more satisfying than intercourse.

Another reason for understanding that sex is more than intercourse is that most of today's sexual problems, such as unwanted pregnancy, transmission of sexually transmitted infections, and sexual dysfunctions, stem from *insertive sexual practices*, especially vaginal, oral, and anal sexual activities. Many of these problems could be reduced if more people were increasingly comfortable with the idea of sexual intimacy *without* these activities. This does not imply becoming **celibate**—forgoing all sexual intimacy and activities—but may involve a decision to engage in only "safe" or "preferred" sexual behaviors for a while (this decision is sometimes referred to as "selective abstinence"). In addition, many situations may arise in your life that make intercourse uncomfortable, difficult, or medically inadvisable, but this does not mean that sexual intimacy must stop (Hatcher et al., 1994; Kowal, 1998a). Other intimate, sexually fulfilling activities can still be enjoyed fully.

celibate
choosing to forego all sexual activities.

Although many people have discovered the pleasures of sex without intercourse, to others this is a new and strange idea. Culturally, especially for heterosexual couples, it is not a widely accepted concept. On the contrary, many, if not most, people will argue that you haven't really had sex if you haven't had intercourse (Bogart, Cecil, Wagstaff, Pinkerton, Abramson, 2000; Hans, Gillan, & Akande, 2010; Petersen & Muehlenhard, 2007). This can be misleading and even dangerous. In one recent study, only 20% of college students believed that oral sex constituted "having sex" (Hans, Gillan, & Akande). This belief carries with it the strong possibility that these same students may believe that oral sex is also "safe sex," which it is not. Oral sexual activities have the potential to transmit nearly all sexually transmitted infections.

Since You Asked

1. I'm not sure about this class. I'm afraid it's going to be so embarrassing!

Enhancing Sexual Fulfillment

At some point in your life, nearly all of you will choose to be sexually active and to share sexual intimacy with a partner. Once that decision is made, you will desire and deserve a healthy, satisfying, and fulfilling sex life.

This text in no way intends to recommend, encourage, or promote any particular sexual behavior, feeling, or attitude. You should never feel pressured to do anything sexually that makes you feel uncomfortable. As discussed in the next section, sexuality is complex, and, consequently, a fulfilling sex life is not always easy to achieve. One route to this goal, however, involves acquiring accurate and authoritative information—such as what is provided in this text—about as many aspects of sexuality as possible. This foundation of knowledge will provide you with the tools to experience and maintain an enhanced, enriched, and exciting sexual life for yourself and your partner.

It is important to stress here that there should be no doubt in your mind that *your body belongs to you*, and that your sexual behavior is, or should be, completely in your control. This principle, of course, refers to consensual, honest, and responsible sexual behavior between adults. Behaviors such as rape, telling someone you are using contraception when you are not, or not warning a partner about having a sexually transmitted infection are contrary to this premise because these activities are nonconsensual, dishonest, and irresponsible.

Human Sexuality

1.2 Explain the knowledge people need for a full understanding of human sexuality.

Your college education in human sexuality is only partly about your *experience* of being a sexual person. Other specific topics are essential for your *understanding* of human sexuality, which we will consider next.

Sex Education or Abstinence Only?

At some point almost everyone must answer important questions about sexual issues that arise in their lives. If you make the wrong decisions owing to lack of knowledge, misinformation, or poor judgment, the consequences can be extremely serious. How do you find the answers that are right for you? First, you are off to an excellent start because you are taking this course and reading this book. Research has shown that people who take a human sexuality course tend to make better, more informed, and more thoughtful choices. For example, high school seniors who received education about HIV and AIDS were found to engage in fewer high-risk sexual activities and had a reduced risk of HIV infection (Klitsch, 1994; Underhill, Montgomery, & Operario, 2007). In another study, college students enrolled in a freshman seminar focusing on sexual health were more likely to use condoms and other forms of contraception (Turner et al., 1994). Also, students who complete a human sexuality course have been shown to be significantly *less tolerant* of rape in general, and date rape specifically, and less likely to believe common rape myths such as "most rapes are committed by strangers" and "some women ask to be raped by the way they dress" (Fischer, 1986; Flores & Hartlaub, 1998; Patton & Mannison, 1994).

Personal choice and responsibility are recurring themes throughout this book. The more accurate and complete information you have about sex, the better prepared you are to make responsible choices about your behavior. Without this knowledge, "sex" often equals "trouble." In the United States, many teens have missed out on this important information due to a complete lack of sex education in the schools or because any teaching about sex has focused on instructing students simply not to have sex at all. The **abstinence-only approach** is based on the idea that teens should resist engaging in any and all sexual activities and should wait until marriage for sexual intimacy. The abstinence methodology, therefore, assumes that teens should have no need to be educated about sexual activities, contraception, or how to prevent sexually transmitted infections. Abstinence-only programs in schools, supported and funded during the conservative political climate in Washington, DC, from 2000 to 2008, were widely seen to have failed (Boonstra, 2009). Numerous studies have demonstrated that when school districts attempted to implement these programs (rather than risk losing federal funds), the outcome was the dissemination of inaccurate information and no decrease in teen sexual activity, unintended pregnancy, or the incidence of sexually transmitted infections (STIs). In fact, in numerous school districts, unwanted pregnancy and STI rates *increased* as teens failed to receive the learning they needed in order to make responsible choices (DiCenso et al., 2002; Bruckner & Bearman, 2005; Hollander, 2007; Boonstra, 2009). Moreover, in a recent evaluation of teen pregnancy, researchers determined that the rate of pregnancy has, for years, been underestimated because the statistics took into account all teens, rather than those who were sexually active and actually at risk of pregnancy. When these statistics are adjusted for teens who are sexually active, the number of unintended teen pregnancies increased from 40 per 100,000 to 147 per 100,000 for girls age 15 to 17, and from 108 per 100,000 to 162 per 100,000 for girls age 18 to 19 (Finer, 2010). This is more than 300% and 50% higher than previously reported. Because you may not have received correct or adequate sexual

abstinence-only approach
The decision to *avoid* teaching adolescent students about sexual activity, STIs, contraception, etc., based on the theory that such education is unnecessary if students are taught to abstain from sexual behavior.

information early in your teens, it is crucial to you now. And, even though you are now an adult and in college, it's never too late!

The new, more progressive climate in Washington, DC, during the second decade of the 2000s has created changes in sex education in the United States. The abstinence-only funding expired in 2009 but was added back into the budget as part of a health care reform bill. However, the language in the abstinence-only law has been eased, allowing for more balanced and accurate information to be taught (Boonstra, 2010). The movement in sex education now is away from abstinence-only and toward more comprehensive teaching about sexuality that stresses accurate information to help teens make responsible personal decisions about sexual behavior. Table 1.2 summarizes the difference between these two approaches.

It's More Complex than You Think

One of the most important (and obvious) reasons to study sexuality is to increase your knowledge of the subject. This will not be difficult because human sexuality is a huge field. For example, in the next two chapters, you will be introduced to sexual anatomy and physiology. Right away you'll begin to see that even our sexual bodies are wonderfully complex in form and function. But in many ways the biology of sex is far simpler than the psychological and social intricacies of sexual feelings, desires, choices, interactions, and behaviors.

To acquire a general idea of how much you already know about the range of topics this book will cover, take a few minutes to complete and score the section titled: "Self-Discovery: Sexual Knowledge Self-Test." When you finish this textbook many weeks from now, you may wish to take the test again. You will have a significantly higher score—guaranteed.

Table 1.2 Comparison of Principles of Abstinence-Only and Comprehensive Sex Education in the Public School System

Abstinence-Only Approach	Comprehensive Sex Education
An abstinence-only education program during the years 2000–2008 that qualified for federal funding is a program that	*A comprehensive sex education program, encouraged following the 2008 presidential election, is a program that*
1) has as its exclusive purpose teaching the social, physiological, and health gains to be realized by abstaining from sexual activity	1) is age-appropriate and medically accurate
2) teaches abstinence from sexual activity outside marriage as the expected standard for all school-age children;	2) stresses the value of abstinence while not ignoring those young people who have had or are having sexual intercourse
3) teaches that abstinence from sexual activity is the only certain way to avoid pregnancy outside of marriage, sexually transmitted diseases, and other associated health problems	3) provides information about the health benefits and side effects of all contraceptive and barrier methods used— (a) as a means to prevent pregnancy; and (b) to reduce the risk of contracting sexually transmitted diseases infections, including HIV/AIDS
4) teaches that a mutually faithful, monogamous relationship in the context of marriage is the expected standard of human sexual activity	4) encourages family communication between parent and child about sexuality
5) teaches that sexual activity outside the context of marriage is likely to have harmful psychological and physical effect;	5) teaches young people the skills to make responsible decisions about sexuality, including how to avoid unwanted verbal, physical, and sexual advances and how to avoid making verbal, physical, and sexual advances that are not wanted by the other party
6) teaches that bearing children outside of marriage is likely to have harmful consequences for the child, the child's parents, and society	6) develops healthy relationships, including the prevention of date rape and sexual violence
7) teaches young people how to reject sexual advances and how alcohol and drug use increases vulnerability to sexual advances	7) teaches young people how alcohol and drug use can affect responsible decision making
8) teaches the importance of attaining self-sufficiency before engaging in sexual activity.	8) does not teach or promote religion.

SOURCE: Boonstra, H. (2009). Advocates call for a new approach after the era of "abstinence-only" sex education. *Guttmacher Policy Review*, 12(1). Retrieved from http://www.guttmacher.org/pubs/gpr/12/1/gpr120106.html

People Know a Lot About Sex—And Much of It Is Wrong

Each of you is unique in your level of sexual knowledge and experience. As you will discover if you answer the questions in "Self-Discovery: Sexual Knowledge Self-Test," all of you come into this course with a base of knowledge—a starting point for your

Since You Asked

2. Do we really need a course about sex? I'm married and have a child. I'd say I probably know everything I need to know.

Self-Discovery

Sexual Knowledge Self-Test

Mark each of the statements *True* or *False*. If you know someone you would like to pass the self-test along to, you may want to answer on a separate sheet of paper. Scoring instructions and interpretations are at the end of the test.

TRUE OR FALSE?

Chapter 1: Studying Human Sexuality

_____ 1. Because of ethical and personal privacy considerations, it is not possible to conduct first-hand scientific research on human sexuality.

_____ 2. The average length of a man's penis when erect is about 7.5 inches.

_____ 3. Electronic devices that measure penile and vaginal changes during sexual arousal are sometimes used by researchers to study sexual responding.

Chapter 2: Sexual Anatomy

_____ 4. Erection of the penis is caused by contracting muscles and the buildup of semen.

_____ 5. Semen is produced by the testicles.

_____ 6. When a girl is born, her ovaries contain over 400,000 immature eggs.

Chapter 3: The Physiology of Human Sexual Responding

_____ 7. The clitoris and penis both become erect during sexual stimulation.

_____ 8. Women generally say that intercourse is the most reliable and satisfying method for achieving orgasm.

_____ 9. When someone carries a sexually transmitted infection, he or she may transmit the infection even without engaging in oral sex, anal sex, or vaginal intercourse.

Chapter 4: Love, Intimacy, and Sexual Communication

_____ 10. Physical attractiveness is a relatively unimportant factor in the formation of romantic relationships.

_____ 11. In abusive or violent relationships, the victim often remains in the relationship, sometimes for years, even though the abuse continues.

_____ 12. Physical violence is virtually always present in abusive relationships.

Chapter 5: Contraception: Planning and Preventing Pregnancy

_____ 13. Vaccines now exist that can prevent most cases of cervical and anal cancers.

_____ 14. The intrauterine device (IUD) contraceptive is becoming increasingly more popular among young women.

_____ 15. A new birth control pill has been approved that is taken 365 days each year and eliminates a woman's period entirely.

Chapter 6: Sexual Behaviors: Experiencing Sexual Pleasure

_____ 16. Most sexual behaviors may be divided into one of three categories: (a) heterosexual, (b) gay, or (c) lesbian.

_____ 17. Research shows that about the same percentage of males and females masturbate.

_____ 18. Most women do not routinely experience orgasm during heterosexual intercourse.

Chapter 7: Sexual Problems and Solutions

_____ 19. Problems with erection for men and orgasm for women are rare.

_____ 20. Nearly all sexual problems are easily treated and solved.

_____ 21. Lack of sexual desire is one of the most common sexual problems couples face.

Chapter 8: Sexually Transmitted Infections

_____ 22. Bacterial sexually transmitted infections (STIs) are generally curable; viral STIs generally are not.

_____ 23. Most sexually transmitted infections (or diseases) may be spread through oral sexual activities.

_____ 24. Some strains of the human papilloma virus (HPV) that cause genital warts have also been shown to be the primary cause of cervical cancer.

Chapter 9: Conception, Pregnancy, and Birth

_____ 25. A woman can get pregnant during her period.

_____ 26. During each menstrual cycle, there are about seven to ten days during which unprotected intercourse can lead to pregnancy.

_______ 27. Miscarriage is another term for a spontaneous abortion.

Chapter 10: Gender: Expectations, Roles, and Behaviors

_______ 28. A person's degree of femininity or masculinity is usually predictive of his or her sexual orientation.

_______ 29. A small percentage of infants are born with genitals that are ambiguous—neither fully male nor female.

_______ 30. Androgyny refers to a male who is highly masculine or a female who is highly feminine.

Chapter 11: Sexual Orientation

_______ 31. The love between gay and lesbian couples is similar in emotional quality to the love between heterosexual couples.

_______ 32. Bisexuality refers to people who are actually gay, but have not yet fully acknowledged their true sexual orientation.

_______ 33. Research has shown that the sexual orientation of parents does not influence the sexual orientation of their children.

Chapter 12: Sexual Development Throughout Life

_______ 34. If a young child masturbates, it is usually a sign that he or she is being sexually abused.

_______ 35. The male sexual peak is at about age 18, while the female peaks at around age 35, causing serious difficulties in many intimate relationships.

_______ 36. Most women begin to care less about sex when they become menopausal (around their mid-40s to mid-50s).

Chapter 13: Sexual Aggression and Violence: Rape, Child Sexual Abuse, and Harassment

_______ 37. Most rapes are committed by someone the victim knows.

_______ 38. Women can resist being raped if they really want to.

_______ 39. Men who sexually abuse boys are usually gay.

Chapter 14: Paraphilias: The Extremes of Sexual Behavior

_______ 40. Sadomasochism (giving and receiving pain for sexual arousal) is clinically defined as sexual perversion.

_______ 41. If a man frequently uses women's shoes for sexual stimulation during masturbation and orgasm, this would likely be called a fetish.

_______ 42. Exhibitionism and voyeurism are noncoercive paraphilias, in that neither involves a victim.

Chapter 15: The Sexual Marketplace: Prostitution and Pornography

_______ 43. With few exceptions, most male prostitutes are gay.

_______ 44. "Sex trade worker" is a phrase often used to refer to a person who engages in prostitution.

_______ 45. Explicit sexual materials are no longer censored in the United States.

Now score your self-test using the answer key below. You might wish to compare your score with the scores from students on the first day of several past human sexuality classes, which are divided into five categories described below.

Correct Answers:

1. F, 2. F, 3. T, 4. F, 5. F, 6. T, 7. T, 8. F, 9. T, 10. F, 11. T, 12. F, 13. T, 14. F, 15. T, 16. F, 17. F, 18. T, 19. F, 20. T, 21. T, 22. T, 23. T, 24. T, 25. T, 26. T, 27. T, 28. F, 29. T, 30. F, 31. T, 32. F, 33. T, 34. F, 35. F, 36. F, 37. T, 38. F, 39. F, 40. F, 41. T, 42. F, 43. F, 44. T, 45. F

Interpreting Your Score

41–45: Sexual Wizard! However, you still have a great opportunity to learn the finer points and can probably contribute a great deal to class discussions. You should find much of the text and course material somewhat more familiar than many others in the class.

36–40: Very Sexually Knowledgeable. You should be fairly comfortable participating in class discussions and may be able to add interesting insights. However, you will be quite surprised at how many new discoveries you will make in this book and course.

30–35: Average Level of Knowledge. Most students taking this class score around this range at the beginning of the course. You will pick up many new concepts as you go through this book. You probably came into the class with some misconceptions about human sexuality, but those will be corrected soon!

25–29: Still a Lot to Learn. Like many people, you probably believe in some myths and misconceptions about sexuality that you learned from friends or parents. Perhaps you grew up in a sexually restricted household where talking and learning about sex was frowned upon or forbidden. For you, this will be an extra valuable course.

Under 25: You Need This Course! This is not a criticism. You are the person who will benefit the most from this book and this course. For whatever reason, you simply have not yet had much opportunity to acquire accurate knowledge about human sexuality. This course will give you the information you need to help you live a healthy and fulfilling sexual life.

study of human sexuality. You may also discover some important gaps or errors in your knowledge that this book can fill.

Can you think of any other course that you began with as much "advance information" (right or wrong) as this one? But consider for a moment the *sources* of your information. A great deal of your sexual knowledge probably came from parents, friends and acquaintances, books, magazines, movies, TV shows, and the Internet—what might be called your "informal sex education network." Although some of you may have had some "formal" sex education in high school, most of the sexual knowledge you possess probably came from your personal collection of casual, nonscientific sources. Unfortunately, all of these sources are subject to error, misinformation, and myths and falsehoods (yes, even parents can be wrong about sex at times).

Images from films, TV, and the Internet often provide an extreme or overly ideal view of human sexuality.

Personal experience can also be a poor teacher. Among teens and young adults, early sexual experimentation tends to be awkward, embarrassing, not necessarily pleasurable, and even frightening or painful. These experiences may lead to expectations and conclusions about sexual behavior that are fundamentally incorrect.

In this book you have a rich source of information about human sexuality that is as accurate, up to date, and scientifically based as possible. Your challenge in reading and learning the material in these pages is not only to gain new information and develop a comprehensive base of correct knowledge about human sexuality but also to *unlearn* any faulty ideas and preconceptions you may currently have about sex. Don't expect this to be easy. "Unlearning" is often more difficult than learning because many misconceptions are reinforced through your informal sex education network. It is not easy to accept that something you thought was true about sex—even something you may have used to make decisions about your behavior—may stem from faulty information from all those well-meaning but misinformed sources (Farrington, 2002; Franiuk, 2007). As you read this book, try to be open to knowledge and ideas that may be new or different from what you thought was true.

Understanding and Respect for Sexual Diversity

Sexuality is one of the most diverse of all human attributes. People's sexual "personalities" can include cultural, ethnic, and religious differences; differences in family background; differences in sexual orientation; differences in sexual attitudes, morals, and values; differences in sexual behavior, preferences, experiences, and sexual role expectations; and differences in personal comfort about sexual issues (see "Sexuality and Culture: Sex Education in China").

Humans have a tendency to fear and reject people perceived as different, especially when the differences appear strange and even extreme. Your study of human sexuality will increase your knowledge and awareness of humans' sexual richness and diversity and further develop your understanding and respect for those who may be sexually different from you. Research has demonstrated that students who take human sexuality courses increase their comfort level with various diverse groups, such as those with gay, lesbian, and bisexual orientations (Patton & Mannison, 1994; Waterman et al., 2001). Furthermore, in the author's own courses, when guest speakers who represent sexual diversity—for example, people from the gay, lesbian, and bisexual communities; guests from sexual abuse and rape crisis centers; people living with HIV or AIDS; transgender individuals—come to speak to classes about their experiences, students frequently report profound changes in their awareness and empathy for these diverse groups and often refer to these visits as among their most enlightening learning experiences of the semester.

Sexuality and Culture

Sex Education in China

China is one of many cultures that has traditionally placed taboos on sex education (including discussions about sex) in virtually any form. Nearly half of all college students in China have received no education in school about sexual behavior and health (Li et al., 2004), and sex is not generally discussed in the home. However, as China continues to modernize, an increasing number of adolescents are becoming sexually active earlier, which, in turn, has led to a dramatic rise in unplanned pregnancy and the spread of STIs, including HIV and AIDS.

A 2012 survey carried out by Beijing University's Institute of Population Research found that 60 percent of Chinese teens are comfortable with the idea of premarital sex. This is a marked change from attitudes only 10 or 20 years ago showing that the vast majority of Chinese citizens were opposed not only to premarital sexual activities, but even to the mention of them (Linfei, 2013). Moreover, the survey found that 22 percent of China's 300 million teens are sexually active. As the number of Chinese youth engaging in sexual activities increases rapidly, their level of sexual education and knowledge has not kept pace. The same survey found that of those 66 million sexually active teens, more than half had not used a condom or any other form of contraception during their first sexual experience, and 21 percent of sexually active girls reported an unwanted pregnancy with 85 percent of those resulting in abortions. The survey also found that an alarmingly low 4.5 percent of teens have accurate knowledge of human reproduction and only 14 percent understood the dangers of or precautions against HIV transmission.

Although condoms are freely sold in China today, a recent ad that depicted a condom was deemed indecent and banned. Moreover, China's ministry of communication continues to block websites from other countries that discuss STIs and other sexual health topics. Despite these ingrained attitudes, China is beginning to recognize the health dangers inherent in sexual activity (Qiaoqin, Ono-Kihara, & Cong, et al., 2009).

With one-fifth of the world's population, China's need to stem new HIV infections is crucial. Over the past several years, HIV rates have increased by 13% per year, with approximately 50 thousand new cases in 2012 (Yunting, 2013). China reports that, as of 2012, approximately 700,000 Chinese people are HIV positive. In 2009, China reported that AIDS had become the leading infectious cause of death in that country, surpassing even tuberculosis, which has been extremely common historically ("HIV and AIDS in China," 2010).

Acknowledging these many sexual and social changes, China has begun gradual efforts to educate the country's young people about sex and sexual health (Zhuhong, 2010). One of these programs involves a system of teens and young adults who are trained as peer sex educators, called the "Adolescent Reproductive Health Peer Education Society." The goal for this and other programs is to spread sexual health information not only in the major Chinese cities, but throughout China's small towns and vast rural areas.

Chinese teens in a sex education class in Beijing

SOURCE: www.chinadaily.com.cn/photo/2013-07/11/content_16760443.htm]

What Is Sexually "Normal"?

As indicated by several questions from the "Since You Asked" feature at the beginning of this chapter, one of the most common concerns people have about their sexuality, especially in young adulthood, is whether or not they are "normal." But the concept of "normal" is a slippery one. It is difficult to define exactly what we mean when we say something or someone is normal or abnormal, especially when it comes to sexuality. Furthermore, when people worry that their sexual feelings, desires, behaviors, or bodies are not normal, they are usually too embarrassed or fearful to discuss it with anyone else. Consequently, they may believe that they are alone in their particular concern. The truth is that just about anything people feel might be abnormal about themselves is probably a common concern held by many others—which in a sense makes it normal! One of the most important and reassuring benefits of studying human sexuality is that nearly everyone who feels that he or she is abnormal in some way discovers that that is not the case. For more on this issue,

Self-Discovery

Am I Sexually Normal?

Professional sex therapist Marty Klein has noted that Americans are concerned—virtually obsessed—with the normality of their sexual fantasies, preferences, responses, secrets, turnoffs, and problems; the normality of their bodies; and the frequency with which they have sex. The fear of being sexually abnormal interferes with—and can even prevent—pleasure and intimacy (1993, p. 49). As we grow and develop in our culture, the subtle (or not-so-subtle) messages most people receive during childhood and adolescence imply that sexuality is something we shouldn't discuss. This denial of our natural sexuality interferes with learning about what is considered sexually normal. In turn, this causes the development of "normality anxiety," which can negatively affect our intimate sexual relations with our partners.

Klein lists some of the most commonly asked questions about what is "normal":

1. Are my sexual fantasies normal?
2. Are my genitals the normal shape, size, and color?
3. Unlike my friends, I like/don't like watching X-rated videos. Am I normal?
4. I want sex a lot more often than my partner. Am I normal?
5. I enjoy lovemaking, but my biggest orgasms are from masturbation. Am I normal?

Klein says (and most sexuality researchers and educators would agree), "I tell them time and time again, not one of these facts or feelings is *abnormal* (in fact in a more sexually enlightened world, the question of their *normality* would not even arise)" (pp. 50–51).

SOURCE: Klein, M. (1993, May). Am I sexually normal? *New Woman* (5), 49–52.

see "Self-Discovery: Am I Sexually Normal?" By the way, the answers to "Since You Asked" questions 3, 4, and 5 are *yes*, *yes*, and *yes*. These issues are all discussed in greater detail elsewhere in this text.

Since You Asked

3. Is it normal to have an orgasm by masturbating but never to have one during intercourse?
4. I am 21 years old and had a "wet dream" two weeks ago. Is that normal at my age?
5. Why did my high school stop the sex ed class? Now they're saying we should just all wait until marriage to have sex. No one's really going to do that, are they?

Sexual Health

One of the recurring features in all the other chapters of this text, "In Touch with Your Sexual Health," is designed to highlight and draw your attention to important issues of sexual health. An awareness and understanding of these health concerns is crucial to maintaining healthy sexual functioning for yourself and your partner. **Sexual health** is a significant part of your overall health, yet it is the part that is often hidden and neglected the most due to embarrassment, denial, or lack of knowledge and information about sexual health issues or symptoms (Baber & Murray, 2001; Consedine, Krivoshekova, & Harris, 2007). Sexual health refers to a wide range of health issues relating to your sexuality, as listed in "In Touch with Your Sexual Health: Physical, Emotional, and Psychological Issues."

sexual health
A general concept referring to physical, emotional, psychological, and interpersonal well-being with regard to a person's sexuality.

Parenting

Most of you who are reading this book either have or will have children of your own. One of your important responsibilities in raising them will be to teach them about sex. The vast majority of parents acknowledge that this is an awesome responsibility, but many are hesitant to discuss sexual issues with their children because they feel they lack adequate knowledge or are unsure what to say and when to say it (Wilson, Dalberth, & Koo, 2010). Although *you* may or may not have learned about sex from your parents, most people agree that parents are the most appropriate source of this knowledge. This assumes, however, that parents are willing to talk to each other and their kids about sex and are able to impart complete and accurate information (Wilson, et al., 2010). Studying human sexuality will help prepare you for this aspect of parenthood, should you choose to become a parent, by providing you with the knowledge, resources, and degree of comfort you need to teach your children about sex.

In Touch with Your Sexual Health

Physical, Emotional, and Psychological Issues

When we talk about sexual health issues, we are referring to more than physical illnesses. Here is a list of the main categories of sexual health concerns:

- Sexually transmitted infections (e.g., HIV, genital warts, herpes, hepatitis B, gonorrhea, chlamydia, and syphilis)
- Problems with sexual functioning (e.g., premature ejaculation, erectile difficulties, problems with arousal or orgasm, loss of sexual desire)
- The health of your sexual anatomy (e.g., cancer of the breast, cervix, ovaries, testicles, or prostate; painful sex; urinary and reproductive tract infections)
- Emotional and psychological sexual health issues (e.g., abusive or controlling relationships, past sexual traumas, sexual violence, fear or guilt about sex)
- Any other symptom that looks or feels as if something is "wrong" relating to sexual body parts, behaviors, functioning, or perceptions

Research clearly supports a link between adults having accurate knowledge about sexuality and effective communication between parents and children. In a study by King, Parisi, and O'Dwyer (1993), parents who had taken a human sexuality class in college were compared to those who had not. The results indicated a striking difference between the groups. Parents who had taken the human sexuality course were found to be much more likely to discuss various aspects of sexuality with their children and to use correct terminology when doing so. Table 1.3 summarizes some of their findings. Effective communication between parents and teens is important because it leads to healthier adolescent choices. One study found that quality adolescent-parent communication is associated with lower levels of unprotected intercourse and lower incidence of teen sexually transmitted infections (Deptula et al., 2010).

Since You Asked

6. My parents have never had "the talk" with me. Were they just embarrassed, or did they think I would learn on my own?

Table 1.3 Relationship Between Taking a Human Sexuality Course and Parents' Discussions of Sexuality with Their Children

Questions Parents Were Asked	Percentage Who Took Sexuality Course Answering Yes	Percentage Without Sexuality Course Answering Yes
Have you had discussions about sexuality with your children?	45	9
Do you use correct anatomical words for genitals?	24	6
Have you discussed where babies come from? (children age 5–11)	24	5
Have you discussed inappropriate touching by others? (age 5–11)	29	22
Have you discussed menstruation? (age 12+)	30	16
Have you discussed masturbation? (age 12+)	14	2
Have you discussed intercourse and reproduction? (age 12+)	30	4
Have you discussed birth control? (age 12+)	30	6
Have you discussed homosexuality? (age 12+)	25	5
Have you discussed sexual abuse? (age 12+)	26	11
Have you discussed sexually transmitted diseases? (age 12+)	30	6

NOTES:

1. Subjects included 102 college-educated parents of children 5 years of age or older; 36% fathers, 74% mothers. All differences are statistically significant.
2. It is possible that those who choose to take a human sexuality course are already more comfortable and open about sexual communication, but the association is clear.

SOURCE: Based on data from King, Parisi, & O'Dwyer (1993).

Evaluating Sexual Research

How Could This Happen?

Father Who Went for Kidney Stones Discovers He Is a Woman

If you saw this mainstream news headline from May of 2012, you probably thought to yourself, "What the heck?" or something similar. How could someone not know his or her own gender until a trip to the hospital for a routine procedure? These types of sexuality-related stories appear in the media quite often. For people who have never studied human sexuality, reports such as this can be quite confusing. However, after taking a human sexuality course, your level of knowledge and understanding increases your ability to make sense of sexual information you come across in the media or in research findings.

What about this story? It certainly was unusual, but perhaps not as implausible as it may have sounded to you at first. Here's the explanation.

Steve Crecelius, a photographer from Denver, Colorado, went to the hospital for a test to determine if he had a kidney stone. The typical procedure to diagnose this is an ultrasound of the abdominal area. When the nurse was reading the results of the ultrasound she said,"Huh, this shows you're a female." Most people would simply laugh at the obvious mistake, but the test showed internal female anatomical structures, although his external genitals were male. And for Steve, it was a revelation. This news confirmed something Steve had always felt: that he was a woman trapped in a man's body. Steve had denied these feelings for over 40 years. At the time of this discovery, Steve had a wife and 6 children.

This news came as a relief to Steve, although he was afraid that if he told his family, he might lose them. He did not. His entire family was, and is, supportive and accepting. Steve made the decision to begin to live life as a female, and *she* became Stevie.

A person born with a combination of male and female sexual anatomy is referred to as *intersex* (see Chapter 10, "Gender" for a more complete discussion of this topic). Sometimes an intersex individual will live life as a male or female, dictated by his or her external genitals. Typically, however, that person may experience an underlying basic conflict of self-identity.

Steve to Stevie

Steve Crecelius in 2011

Stevie Crecelius with Wife Debbie Today

Stevie recalls as a child wearing her mother's clothes and makeup in secret. She said that she always felt "different," a common experience for transgender individuals. Stevie said, "When I was about 6 years old, I started having these feminine feelings . . . Wearing my mom's makeup, I thought I looked pretty" (Didymus, 2012; Jennings, 2012).

Stevie now lives life as a woman and continues to be happily married to Debbie.

The Ability to Evaluate Sexual Research

You are constantly deluged by sexual messages and information in the form of TV shows, movies, websites, rumors, gossip, stories, pictures, advertisements, catalogs, advice columns, and all the other audio or visual media you can imagine. This mass of information can be confusing and, sometimes, troubling. How do you know what to

Table 1.4 Experiencing and Understanding Human Sexuality

Experiencing Sexuality
People who take a course in human sexuality...
• Are more comfortable with the many and sometimes conflicting emotions that surround sexual issues
• Have greater knowledge and awareness of who they are as sexual individuals
• Are more confident and clear about their personal sexual morals and values
• Make more informed, responsible, and healthy choices about their sexual behavior and relationships
• Know that sex is more than intercourse and appreciate a wider range of intimate sexual behaviors
• Enjoy more satisfying sex lives overall
Understanding Sexuality
People who take a course in human sexuality...
• Understand that human sexuality is complex
• Are more likely to reject common sexual myths, falsehoods, and misconceptions
• Are more tolerant of others' sexual preferences, orientations, morals, values, customs, and differences
• Are less likely to worry about being sexually "strange" or "abnormal"
• Maintain better physical, emotional, and sexual health
• Have a greater level of skill and comfort in discussing sexual issues with their partners and their own children
• Are better able to analyze and critically evaluate sexual research and information

do with such an excess of information? (See Evaluating Sexual Research: "How Could This Happen?" for an example of media confusion.)

Your skill at sorting, analyzing, and interpreting sexual material is becoming increasingly important as more and more information becomes available in this information-packed digital age. In the next section we will focus on how human sexuality research is conducted, so that you will be able to understand better how to analyze and interpret what gets reported in the media. Moreover, this book will give you information about sexuality based on the most solid, systematic research available. As you study what is contained in these pages, you will begin to think critically about the issues and apply what you learn to assess the world around you. In short, you will become a more educated and knowledgeable consumer in the complex marketplace of sexual information.

The Value of Studying Human Sexuality

Whether considered individually or together, all of these experiential factors should make it clear that studying human sexuality may be one of the most valuable endeavors you will undertake. Many of the benefits are immediate and will affect your life now. But you will likely discover that you will put the information in this book to work for you, your relationships, and your family for years, and even generations, into the future. Table 1.4 lists many of the differences found in those who take a human sexuality course and those who do not. These differences may have been what attracted certain students to this course to begin with, or they may be the result of the wealth of information this class offers, but they are clearly related.

Methods for Studying Human Sexuality

1.3 List and explain the methods used by researchers in the study of human sexuality

Imagine for a moment that you are a researcher in a field that studies human sexuality, say, a psychologist, sociologist, biologist, physician, or nurse. You want to study some aspect of sexuality to answer a question that might further our understanding of this complex field. What sort of question might you ask? Here are some possibilities:

- On average, at what age do people first have sexual intercourse?
- How often do married couples make love?
- What percentage of college students are sexually active?
- What are the differences in sexual arousal for men and women?
- What type of therapy works best for sexual problems?
- Does an herbal aphrodisiac (so called sex stimulant) advertised in the tabloids really work?
- Does HIV awareness education in high schools reduce high-risk sexual behaviors?
- What is the most effective form of birth control?
- How common is a certain STI in a particular population?

You can probably think of many other questions (and many students have, as you will see in the "Since You Asked" section at the beginning of each chapter in this book).

Our focus here is on how you would go about *answering* these questions in ways that are accurate, meaningful, and believable to others. The study of human sexuality is scientific, and it follows the methodologies of research all scientists follow. These methods include scientific surveys, behavioral observation, correlational research, and experiments. Which of these methods might researchers use to address the questions posed in the preceding paragraph? Could they (1) ask people to tell about their personal sex lives, (2) observe people engaging in sexual behavior in a scientific setting, (3) analyze the relationships among various aspects of people's sexuality, or (4) perform actual experiments involving sexuality? The answer to all four questions is *yes!* However, we are dealing with a very intimate topic here, so you can imagine the difficulties that might be encountered when trying to carry out such studies in accurate, unbiased, and ethical ways.

In this section, we will examine scientific research methods as they apply to the study of human sexuality and consider the advantages, disadvantages, potential for error and bias, and ethical considerations associated with each.

No matter what research method a researcher chooses, information and data gathering must be planned carefully and carried out systematically. Participants to be studied should be chosen to represent, as closely as possible, the larger population of interest. Researchers must be trained to interact with participants

Since You Asked

7. How is it possible to study people's sexual lives when it's such a private, personal experience?

The 2004 film *Kinsey* (photo right) reintroduced the world to the life and work of pioneering sex survey researcher Alfred *Kinsey*, seated in the real-life photo (left). Although we must always be cautious about interpreting Hollywood's version of historical events, Kinsey provides a fascinating glimpse into one of the most influential figures in the history of sexuality research. In the late 1940s and early 1950s, when Kinsey was gathering his data about the sexual behavior of men and women in the United States, attitudes about sex were generally quite restrictive. A scene in the film shows government agents seizing a package of Kinsey's research materials, claiming they were "obscene." Not only did Kinsey pioneer the survey method of sexuality research, but he did so with conviction—some might say courage—in the highly unsupportive social and political environment of the time.

consistently and to avoid introducing their own biased attitudes into the study. Information should be gathered so that it can be analyzed using statistical methods. Participants must be made to feel comfortable, safe, and assured of confidentiality when participating in a study involving their personal sexual attitudes and behaviors. Any reliable and valid scientific study should be able to be *replicated*—repeated using the same methodology on the same or a different sample of participants—with virtually identical results. Finally, strict ethical guidelines must be followed when engaging in any research involving humans, especially when engaging in studies related to sexuality.

Surveys

survey
The scientific collection of data from a group of individuals about their beliefs, attitudes, or behaviors.

respondents
Individuals selected to respond to a researcher's request for information.

The most common form of sexual research is the **survey**. A survey is the process of collecting information from volunteer **respondents** for the purpose of explaining, describing, or comparing people's knowledge, attitudes, beliefs, and behaviors. Dr. Alfred Kinsey pioneered large-scale surveys of sexual behavior and attitudes in the 1940s. His approach and methodologies provided the foundation for most of the survey research about sexuality that has followed for over half a century. Today, the Kinsey Institute at Indiana University is one of the leading centers promoting and carrying out scientific research on human sexuality.

Conducting a survey seems simple enough, right? Just put together a list of questions, send it out to a lot of people, and ask them to fill it out and send it back. That may sound easy on the surface, but survey research is actually extremely difficult. It requires a great deal of training to carry out a valid survey and end up with results that we can trust to be consistent and accurate (see Michaels, 2013; Rea & Parker, 2005).

Types of Surveys

Four types of surveys are most often used by researchers today: self-administered written questionnaire, face-to-face interviews, interviews conducted over the telephone, and Internet surveys. Each method has inherent advantages and disadvantages (Epstein et al., 2001; Fowler, 2008; Michaels, 2013).

When dealing with sensitive issues relating to a person's sexual attitudes and behaviors, which survey method do you think would yield the most honest answers from participants? The written questionnaire offers the greatest anonymity and privacy for the respondent and may therefore produce more honest responses. However, the face-to-face interview allows the interviewer to establish a friendly, trusting rapport with the participants and to have more flexibility in asking planned and follow-up questions, which may lead to greater honesty in the answers (Fowler, 2008).

The telephone survey may fall somewhere in between these methods, allowing for both anonymity and flexibility in questioning. However, the possibility always exists that some people may not take telephone interviews seriously or may be annoyed at the intrusion into their lives, thereby reducing the completeness or accuracy of their answers. In addition, in today's world, an increasing number of people use cell phones instead of land lines, which reduces the reliability and validity of this method.

Surveys conducted over the Internet offer the opportunity for a large number of responses from a wide geographical area over a relatively short time span. However, Internet surveys also pose new and troubling concerns over research ethics and validity (e.g., Alessi & Martin, 2010; Whitehead, 2007). Some of these concerns relate to the following questions: Is it possible to inform people about the survey so that they are aware of the nature of the items before they agree to participate? Are the data truly confidential? Are all the participants of legal age (Binik, Mah, & Kiesler, 1999)? Are they all whom they claim to be relative to age, gender, attitudes, and experiences?

To what extent does the Internet provide a nonrepresentative, self-selected sample of participants; that is, how many people without computers or Internet access (and there are more such people than you may think) will be excluded (Epstein et al., 2001; Fowler, 2008)? Will Internet respondents take the survey seriously? Will they respond honestly? Can the researchers maintain their objectivity (Michaels, 2013)? The answers to these questions are not yet clear, but most researchers agree that the Internet offers far-reaching and powerful opportunities for all psychological research, including studies of human sexuality.

Several of the most important survey studies (using various methodologies) since the 1940s in the field of human sexuality are summarized in Table 1.5.

Survey Participant Sample

The entire group of people being studied in a survey is called the **target population**. Typically, this is a very large group, such as college students, teenagers, adult males, adult females, or students in human sexuality classes. Therefore, researchers usually

target population
The entire group of people to which a researcher is attempting to apply a study sample's findings.

Table 1.5 Major Surveys about Human Sexuality Since 1948

Survey	Reporting Authors	Year	Participants	Type	Samples of Findings
Sexual behavior in the Human Male ("The Kinsey Report")	Kinsey, Pomeroy, & Martin	1948	5,300 white males (United States)	Face-to-face interviews	• 56% of males between 16 and 20 years of age had experienced premarital intercourse
Sexual behavior in the Human Female	Kinsey et al.	1953	5,940 white females (United States)	Face-to-face interviews	• 30% of females between 16 and 20 years of age had experienced premarital intercourse
National Health and Social Life Survey	Michael et al.; Laumann et al.	1994	3,432 adults, age 18–59 (United States)	Face-to-face interviews	• Among married or committed respondents under 30 years of age, 55% reported that their sex life was "very exciting" • For singles under 30, 30% reported that their sex lives were "very exciting"
Durex global Sex Survey	Durex Corp.	2005	Over 50,000 men and women (worldwide)	Online, Internet survey	• Among U.S. respondents 53% reported having sex at least weekly • Of those, 48% reported that they were sexually satisfied • Among French respondents, the percentages for the same findings were 70% and 20%, respectively
National Survey of Family Growth	CDC (Centers for Disease Control and Prevention)	2002–2012 (ongoing)	Over 5,000 men and women each year, ages 15–44 (United States)	Face-to-face interviews	• In 2006–2008, 79% of females reported using contraception the first time they had sex • Among males the finding was 87%
The Toronto Teen Survey	Planned Parenthood of Toronto, ON, Canada	2009	Over 1,200 surveys from male and female teens ages 13–20	Written, anonymous survey; 29 questions	• For teens ages 13–18, 83% never accessed sexual health care from a doctor or clinic • For the same group, 27% reported ever engaging in intercourse; 25% oral sex, and 7% anal sex (4% were not sure if they had engaged in sex or not)
National Survey of Sexual Health and Behavior (NSSHB)	Center for Sexual Health Promotion, Indiana University	2010	6,000 U.S. adolescents and adults (14–94 years old)	Written, anonymous extensive survey	• Sexual interest and activity continues for many into the 8th or 9th decade of life • Over 74% of women and 80% of men, age 25–29, reported engaging in vaginal intercourse during the previous month
Youth Risk Behavior Surveillance System	Centers for Disease Control and Prevention	Biannual (2013)	15,425 students, grades 9–12 (2011)	Written, anonymous; parental permission	• 47.4% had ever engaged in sexual intercourse • 6.2% had intercourse before age 13 • 15.3% ever had intercourse with 4 or more partners (2011)

cannot practically gather data from everyone in an entire target population, so they select a smaller group, called a **sample**, from the population. To maximize the validity of a survey, this sample should *represent* as closely as possible the larger population being studied. If the sample reflects the population accurately, it is called *a representative sample*. For example, if you were interested in studying sexual attitudes of college students, you would not ask people at the local mall to fill out your surveys because they would not be representative of your target population (students are not at the mall; they're in the library studying, right?).

sample
A subset of the target population selected by researchers to represent the entire population under study.

The best way to ensure a representative sample for any survey would be to select respondents randomly from the entire population of interest. This process is called random selection or **random sampling**. To illustrate the importance of this random sampling, consider again your study on the sexual attitudes of American college students. You send a survey to 1,000 students at, say, the University of Nebraska and Texas A&M University. Do you suppose their responses would represent first-year American college students in general? How different do you think their attitudes might be from first-year students at, say, NYU or UC Berkeley? This is not to assume that students at any of these particular universities are strange or abnormal (probably not, anyway), but it points out that for your sample to be random, you would need to include participants from colleges and universities of various sizes, types, and locations throughout the country, or else you must narrow your target population.

random sampling
A method of selecting a sample of participants in such a way that each member of the population has an equal chance of being selected.

Another important point about sampling and validity is that the larger your sample, the more representative of your population it is likely to be. If your population is 10,000 people, a sample of 1,000 respondents will almost certainly represent your overall population significantly more accurately than if you have only 100 participants.

Finally, in actual research studies, these requirements for selecting a sample of participants are often not met. You can imagine how difficult, time-consuming, and expensive it would be to survey a very large random sample of a target population containing hundreds of thousands or millions of people. Therefore, many studies are published using potentially nonrepresentative samples of subjects (such as, say, first-year psychology students at a single university). Although these studies can be informative and helpful in our understanding of human nature, we must view them with a critical eye, recognizing possible weaknesses relating to the sample of subjects selected.

Self-Selection of Participants

A common problem associated with sexuality research is that some individuals will agree to participate (volunteers) and others will refuse (nonvolunteers). Do you think these two groups might be fundamentally different in some important ways? On the one hand, those who refuse to participate in a survey about sexuality might feel uncomfortable or embarrassed about revealing personal sexual issues, believe that such information is nobody else's business, consider such research to be morally wrong, or feel more guilt about sex than the volunteers. On the other hand, if a study focuses on an aspect of sexuality that most people consider positive, such as sexual attractiveness or skill as a lover, you would expect people who feel proud of possessing that quality would be more likely to volunteer for the study. If either of these situations occurs, your sample may be nonrandom and biased.

Spring break in Cancun, Mexico, was the site of one unusual research project to study penis size.

In 2001, a condom company wanted to do a study to determine average penis length. The company set up a "penis-measuring tent" outside a popular nightclub in Cancun, Mexico, during spring break. Researchers invited men to come in, look at some sexy literature, become erect, and be measured. The researchers found that the average erect penis length of the 300 volunteers was 5.877 inches. This was slightly longer than had been found in previous studies (more about this in Chapter 2, "Sexual Anatomy"). Why the difference? *Self-selection.* If you think about it, wouldn't you logically assume, considering the value placed on penis size in our culture (however misguided this emphasis may be) that men who had, or thought they had, larger penises would be more likely to volunteer to be measured that those who felt insecure about their penis size? This demonstrates how self-selection may lead to unreliable findings.

Research has uncovered a number of differences between volunteer, or self-selected, sex research participants and those who are randomly selected. These differences between volunteers and nonvolunteers, summarized in Table 1.5, may create what is referred to as a **self-selection bias** in the conclusions drawn from the data gathered in sexual research. This implies that the very nature of the people who agreed to be studied places them in a group that may not accurately represent the population as a whole. "Evaluating Sexual Research: 'What Men and Women Really Want in Bed: Take Our Reader's Survey'" focuses on how this self-selection bias renders most surveys in popular magazines seriously flawed and invalid. Of course, we cannot *force* people to participate in research about sex (or on any topic, for that matter) if they choose not to, so there is no easy solution to the self-selection bias. However, such built-in biases must be kept in mind whenever you are studying the findings of human sexuality research.

self-selection bias
The effect of allowing members of a target population under study to volunteer to participate in the study; it may compromise the randomness and validity of the research.

Evaluating Sexual Research

"What Men and Women Really Want in Bed! Take Our Reader's Survey"

How much of your sexual knowledge have you obtained from newsstand magazines? A favorite feature, often found in magazines such as *Redbook*, *Cosmopolitan*, *New Woman*, *Playboy*, *Playgirl*, and *Esquire* (and many others), is the survey that asks readers to respond by mail or online to questions about sex. The magazines then report the findings from the survey in a subsequent issue a month or two later. Can you see a flaw in this survey methodology? Even if the items on the survey are constructed properly (and usually they are not), the responses are bound to be seriously biased. First, all respondents are readers of that particular magazine and would not, therefore, represent the general population. In fact, all magazines are intentionally targeted at a very specific audience, such as single, working women; professional men between the ages of 20 and 45; or parents. Second, only a small percentage of readers will take the time and energy to respond to the survey (see Table 1.6 on page 22 comparing volunteers to nonvolunteers in sexual research). These eager participants certainly usually are not typical of the overall population. Their responses may not even be representative of the readers of that magazine, much less you or me or most other people.

The bottom line is that surveys such as these can be fun and titillating and may, on occasion, offer some interesting information for conversation or gossip, but they should not be considered scientific and cannot be relied on for meaningful sexual knowledge.

Surveys such as those summarized in Table 1.5 are some of the larger scientific and relatively valid sexuality surveys conducted over the past 60 years or so. Most of these researchers made an effort to avoid the problems discussed in this box. You will see references to these surveys—along with other smaller survey studies that have been published in professional, scientific journals—throughout this book.

Most surveys in popular magazines rely on biased samples and lack validity.

Observational Studies

A great deal can be learned about the behavior of humans simply by observing them directly. You probably engage in "direct observational research" often and can name several of your favorite "people-watching places." When you people-watch, however, you are doing it in a casual way, for fun. When social scientists engage in **observational research**, they use methods that are systematic and organized in order to obtain the most accurate and precise data possible from their observations. Many human behaviors related to sexuality can be studied using observational techniques. For example, in the 1980s, a group of researchers wanted to study flirting behavior among adults in the United States (Perper, 1985; Perper & Fox, 1981; Perper & Weiss, 1987). They went to public places such as singles bars, college pubs, and nightclubs where single people were likely to meet and flirt. Through careful observation, they identified a sequence of four steps that most people progress through when they flirt: "The Approach;" "Talk;" "Swivel and Turn;" and "Synchronization" (see Chapter 4, "Love, Intimacy, and Sexual Communication," for more details about this research).

observational research

Gathering behavioral data through direct or indirect observation using scientific techniques.

Masters and Johnson's Observational Research

Arguably the most famous and, in some ways, most important research contribution to the field of human sexuality was the observational studies conducted by William Masters and Virginia Johnson in the 1960s. Masters and Johnson first described this research in their groundbreaking book *Human Sexual Response* (1966). At their research center at Washington University's School of Medicine in St. Louis, Missouri, Masters and Johnson's research team was able to observe, literally, hundreds of volunteer participants engage in thousands of sexual behaviors both alone and as couples. The researchers' goal was to determine how the human body responds during sexual stimulation and arousal and to apply their findings to help people achieve satisfying and fulfilling sexual lives (Hock, 2012). Masters and Johnson have been at the forefront of sexuality research for decades and have published books on a wide range of topics dealing with sex and relationships. You will see their work cited throughout this book.

Masters and Johnson believed that to understand human sexuality, we must go beyond surveys that simply ask people what they do sexually (which was the focus

Table 1.6 Comparison of Volunteer and Randomly Selected Participants in Sexuality Research

Participant Characteristic	Volunteer Participants	Randomly Selected Participants
Age	Younger	Older
Education	Slightly more years	Slightly fewer years
Income level	Higher	Lower
Experienced past sexual trauma	More likely	Less likely
Masturbation	More often	Less often
Exposure to sexual erotica	More	Less
Sexual experience	More	Less
Attitudes about sex	More positive	Less positive
Attitudes about sexual research	More positive	Less positive
Sexual fear	Less	More
Sexual guilt	Less	More
Sexually permissive	More	Less
Political views	More liberal	More conservative

SOURCES: Adapted from Clement (1990); Morokoff (1986); Strassberg & Lowe (1995); Wiederman (1993); and Wolchik, Spencer, & Lisi (1983).

of nearly all prior sexuality research) and study actual physical responses to sexual stimulation. Their research objective was a therapeutic one: to help people overcome sexual problems. They expressed this goal in the following way:

> [The] fundamentals of human sexual behavior cannot be established until two questions are answered: What physical reactions develop as the human male and female respond to effective sexual stimulation? Why do men and women behave as they do when responding to sexual stimulation? If human sexual inadequacy ever is to be treated successfully, the medical and behavioral professions must provide answers to these basic questions. (1966, p. 4)

Masters and Johnson proposed that the only method by which such answers could be obtained was direct systematic observation and physiological measurements of men and women in all stages of sexual responding.

William Masters and Virginia Johnson are among the most influential researchers in the history of the study of human sexuality.

To study in detail these physiological responses during sexual activity and stimulation, various methods of measurement and observation were used. These included standard measures of physiological responses such as pulse, blood pressure, and respiration rate. In addition, specific sexual responses were to be observed and recorded. Sometimes participants were observed and measured while having intercourse in various positions, and other times they were observed and measured during masturbation either manually or with mechanical devices specially designed to allow for anatomical observation.

You can imagine that all the expectations, observations, and devices might create emotional difficulties for many of the participants. Masters and Johnson were acutely aware of these potential difficulties. To help place participants at ease with the research procedures, they ensured that

> Sexual activity was first encouraged in privacy in the research quarters and then continued with the investigative team present until the study subjects were quite at ease in their artificial surroundings. No attempt was made to record reactions... until the study subjects felt secure in their surroundings and confident of their ability to perform.... This period of training established a sense of security in the integrity of the research interest and in the absolute anonymity embodied in the program. (1966, pp. 22–23)

The results of Masters and Johnson's early work established a basic foundation and language for understanding and discussing human sexual response. This, in turn, allowed for great strides to be made in the treatment of sexual problems. Even now, four decades later, their findings relating to the sexual response cycle, sexual anatomy, and differences in sexual response between men and women form the basis for most discussions (and many controversies) about basic human sexuality.

As you read this text, you will see that the processes and theories proposed by Masters and Johnson are not the only views of human sexual response. Several related and competing theories have developed since Masters and Johnson published their early findings. These include a theory set forth by Helen Singer Kaplan, who in the 1970s proposed three stages of sexual response—desire, excitement, and orgasm—and focused far more than Masters and Johnson on the desire component of sexual arousal and satisfaction (Kaplan, 1974). More recently, researcher and psychiatrist David Reed has suggested the *Erotic Stimulus Pathway Model* of sexual response, which makes use of evocative terms such as *seduction, sensations*, and *surrender* (Greenberg, Bruess, & Haffner, 2002; Stayton, 2002).

A significant departure from Masters and Johnson's theory has become known as the "new view" model of sexual response; it strives to delineate the sexual responses and feelings of women as distinct from those of men (Tiefer, 2001). The new view model clearly states that women's sexual responses generally do not fit well into the Masters and Johnson mold and, to be understood, must take into account additional issues such as the relationship in which the sexual behavior occurs, cultural and

Evaluating Sexual Research

Evaluating a (Flawed) Study: "Lost Your Sexual Gusto? It's in Your Feet!"

Imagine that you are waiting to check out at the supermarket, and you notice on the cover of a popular magazine a headline for an article about a sex therapist from Chicago, Dr. Sylvia Jackson, who has developed a new therapeutic technique for assisting couples who are seeking help for what is called "*hypoactive* [low] sexual desire" (HSD), a chronic loss of enthusiasm and appetite for sex within a relationship (read more about this sexual problem in Chapter 7, "Sexual Problems and Solutions"). You eagerly snatch up the magazine and turn to the article. Through her own experiences and conversations with friends, the article explains, Dr. Jackson believes that if a couple who has diminished sexual desire begins to engage in a nightly pattern of mutual foot massage, focusing on a spot in the center of the arch of each foot, their sexual passion will be reawakened.

To test her new technique, Dr. Jackson suggests to a couple that has recently begun counseling with her that they try the "foot-rub path to passion." After trying the nightly foot rub for six weeks, Dr. Jackson finds that, although their sexual activity has not changed very much, the couple has noticed a steady increase in sexual desire and in satisfaction with their relationship in general. This outcome leads Dr. Jackson to publish the article you are reading, proclaiming the success of her new therapeutic technique.

Do you see any problems with Dr. Jackson's conclusions based on her case study? There are several:

1. Dr. Jackson's recording of the couple's reports may have been biased to support her theory. After all, the main objective was to increase sexual *activity*, which did not happen. Why did sexual activity not increase even though desire apparently did? Was Dr. Jackson only recording the data she wanted and expected? Was she defining "sex" to mean only sexual intercourse and not other sexual behaviors that may have increased for this couple?
2. How does she really know sexual desire increased if she was only taking the clients' word for it? She did not report any other, more objective, measures of sexual desire.
3. How could she be sure the increase in sexual desire (if it really occurred) was due to the foot massages? Maybe any type of touching would have had the same results. Or perhaps this couple would have become sexier over the six weeks with no treatment at all.
4. Maybe a foot massage worked for this couple, but this does not predict whether it will work for you, or me, or anyone else.

This case study offers some interesting facts about a *single* case, some anecdotal evidence that foot massage *may* be helpful in treating HSD in *some* people. It does not offer proof of effectiveness, and it is not scientific enough to be published in a professional journal (which supermarket magazines clearly are not). The study may have some value in that it might stimulate interest in further research. In turn, that research might demonstrate in more reliable and valid ways that foot massage really does turn people on (see "Experimental Research" later in this chapter).

Can sensual foot massage restore sexual passion?

economic factors, psychological issues, and medical factors. The Masters and Johnson model and these competing theories will be discussed in greater detail in Chapter 3, "The Physiology of Human Sexual Responding."

correlational research
A scientific research methodology that determines the extent to which two variables are systematically related to each other (how they "co-relate").

Correlational Research

Another type of research methodology often used when scientists study human sexuality, as well as many other issues, is **correlational research**. Correlational research is similar to observational research in that we are observing how two variables relate to each other (how they "co-relate"). If you know two facts about each member in a group of people, you can probably determine if those two facts are correlated, that is, if they are interrelated in a predictable and consistent way. To use a nonsexual example, height is correlated with shoe size: As one is larger, the other is larger as well. In other words, tall people tend to have larger feet than short people, and vice versa. You may find a few exceptions to this rule, but in general, this correlation holds true (except perhaps for clowns in the circus).

Three types of correlations exist among variables. A *positive correlation* indicates that we can predict that the two variables will change together in the same direction (such as shoe size and height; as one increases, the other does as well). A *negative correlation* tells us that the variables move predictably in the opposite direction. An example of a negative correlation might be amount of alcohol consumed and driving ability; as one increases, the other decreases. The third type of correlation is, simply, *no correlation* at all. That is, no consistent, predictable relationship exists between the two variables, such as the relationship between shoe size and driving ability (except perhaps among clowns in the circus).

Correlational research is very common in studies of sexuality because researchers typically cannot control people's sexual behavior as would be necessary for an experiment (discussed in the next section). Instead, what people *report* about their sexual behavior or other relevant variables that we can ethically discover about them can be studied by examining how the variables relate to each other in a predictive way; this is correlational research.

experimental method
A type of scientific research in which variables of interest are changed while all other unrelated variables are held constant to determine cause-and-effect relationships among variables.

treatment
The action performed on or by a group in an experiment.

Many, if not most, of the studies conducted in the field of human sexuality use correlational research. That is because researchers must usually take data that are already associated with their research participants and look for predictive links between them. Such relationships, when revealed, are important because scientists and clinicians are then able to predict one variable from another. An example from actual scientific studies is the correlation found between drug abuse and the spread of sexually transmitted infections (STIs). As discussed in Chapter 8, "Sexually Transmitted Infections/Diseases," one of the risk factors for contracting and transmitting STIs is the abuse of certain drugs, especially crack cocaine (Hollander, 2007; Ross & Williams, 2001). This connection is graphically illustrated in Figure 1.1. You can see that the relationship is a positive correlation; as one increases, the other climbs as well. Of course, exceptions to any correlational finding will exist (some people use crack and never contract an STI; some people contract an STI and never abuse drugs), but the graph represents the general trend of the correlational finding.

Can you conclude from this connection that researchers have demonstrated a *causal* link between the use of crack and STIs? The answer is no. All they know—and this is abundantly clear—is that the use of crack and STIs is *linked*. They are correlated. The exact factors creating this connection are not fully understood and are not revealed by the correlational finding itself. Perhaps crack cocaine reduces the brain's normal inhibitions, causing people to engage in more risky sexual behaviors when under its influence. Or maybe crack interferes in some way with the immune system, making people more susceptible to all infections, including STIs. Another explanation of the correlation, and this is probably most likely, is that some people who are addicted to crack engage in risky sexual behaviors to obtain money to feed their habit. The point is, you cannot be sure about cause and effect from a correlational finding. In other words, *correlation does not equal causation.*

Again, keep in mind that this is not to imply that correlational research is unscientific or invalid. What it means is that you must interpret correlational findings critically and not jump to an unwarranted conclusion that one variable is causing the other. In science, only one research method allows researchers to assume a cause-and-effect relationship with some degree of certainty: the experiment. As we discuss experiments in the next section, you will see that they may only be conducted when it is possible, both ethically and practically,

Figure 1.1 Positive Correlation Between Crack Cocaine Use and Sexually Transmitted Infections

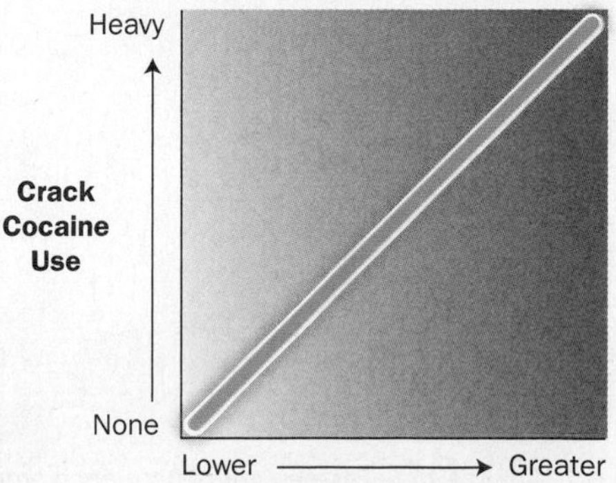

for the researcher to control or manipulate various aspects of the participants' behavior. Although ethically no one would ever be able to manipulate people's crack use or sexual practices for the sake of research, many other sexually related experiments can be, and have been, done.

experimental group
The participants in an experiment who are subjected to a variable of research interest.

control group
The participants in an experiment who receive no treatment and are allowed to behave as usual, for the purposes of comparison to an experimental group; also known as the *comparison group*.

independent variable
The variable of interest in an experiment that is allowed to change between or among groups while all other variables are held constant.

dependent variable
The result of an experiment, evaluated to determine if the independent variable actually caused a change in the experimental group of participants.

Experimental Research

Surveys, observational research methods, and correlations can tell us a great deal about *what* people do and how various characteristics or behaviors are interrelated. However, none of these methods offer very much, if any, information about *why* people do what they do. In other words, as noted earlier, they do not reveal *cause-and-effect* relationships. If we really want to understand cause and effect—what behaviors are actually triggered by certain experiences—it is necessary to do an *experiment*.

Unlike the methods discussed thus far, experiments do not rely on information about people's lives and behaviors as they already exist in the world. Instead, when researchers undertake an experiment, they stage events, set up situations, and carefully measure responses in order to determine how behavior is affected by specific conditions under the researchers' control.

In experiments, researchers employ the **experimental method**. In its most basic form, this method involves bringing together a group of participants and dividing them into two groups (subjects may be divided into more than two groups, but for simplicity's sake, we will use just two in our discussion here). One group is given some kind of **treatment**, and the other receives a different treatment or no treatment at all. Finally, the average difference (if any) between the groups on some behavior of interest is analyzed statistically. If the experiment is done carefully and correctly, the researchers can conclude that differences found in the groups' behavior were *caused* by the differences in the treatment they received. The group receiving the treatment is referred to as the **experimental group**, and the group receiving no treatment or a different treatment is the **control group** or *comparison group*. The treatment administered is the **independent variable**, and the resulting behavior is the **dependent variable**.

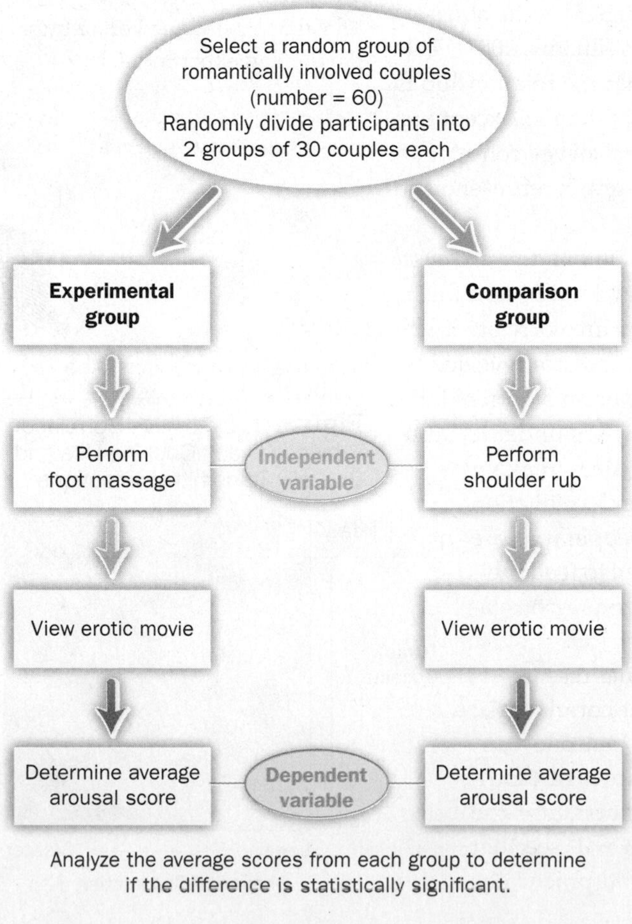

As an example of the experimental method, let's return to our hypothetical example of Dr. Sylvia Jackson and her theory of foot massage as a treatment for hypoactive sexual desire. Imagine that Dr. Jackson has hired you as her research assistant to discover if her treatment truly causes an increase in sexual appetite. She asks you to conduct an experiment, and you agree. Your experimental design would resemble the illustration in Figure 1.2. Here are the steps you might take.

First, you recruit a group of 60 or so couples that have volunteered to participate in an experiment on massage and sexual arousal. You divide them randomly (say, with a flip of a coin) into two groups of 30 couples each. You meet with each couple individually and provide training in massage techniques. To one group you teach Dr. Jackson's foot massage technique, and to the other group you teach a simple shoulder rub. All the couples are instructed that the massages should take exactly 20 minutes, 10 minutes for each partner. Each couple then enters a private room where the massage takes place. They then view an erotic movie (the same movie for all couples) while their level of sexual arousal is recorded using special devices designed to measure blood flow to the genitals (see Figure 1.3).

Figure 1.2 Experimental Design for Dr. Jackson's "Foot Rub Path to Passion"

Select a random group of romantically involved couples (number = 60) Randomly divide participants into 2 groups of 30 couples each

Experimental group | Comparison group

Perform foot massage | Independent variable | Perform shoulder rub

View erotic movie | View erotic movie

Determine average arousal score | Dependent variable | Determine average arousal score

Analyze the average scores from each group to determine if the difference is statistically significant.

When you analyze all the data using established statistical techniques, you find that the group employing Dr. Jackson's foot massage technique experienced a higher level of arousal than the shoulder rub group. This difference was *statistically significant* (this means that a large enough difference was found that we can be almost certain that it did not occur simply by chance). Because all the couples in both groups behaved approximately the same *except for the type of massage*, you may conclude that Dr. Jackson's "foot rub path to passion" probably does *cause* an increase in sexual arousal better than the shoulder rub.

Are you convinced that Dr. Jackson's method works? Are there any problems in the experimental methods you used? The answer is that there might be some problems. Although it is true that the experiment is the only way we can determine cause and effect with a reasonable degree of confidence, several potential drawbacks to all experiments must be taken into account when interpreting the results:

1. Gaining control sacrifices realism. People are very likely to behave and respond differently in an artificial setting than they would in the real world. You cannot be sure that couples using the foot massage technique in an experimental setting would actually increase their desire and arousal in the privacy of their own home under more natural conditions.
2. The act of observing changes the behavior. Simply knowing that they are part of a study and that you are analyzing their responses could alter the couples' usual behavior.
3. The subject sample is not representative. As discussed earlier in regard to surveys, the participants chosen for your experiment may not represent the general population you wish to study. To the extent that your sample of couples does not represent *all* couples, your ability to apply your findings to couples in general may be limited.

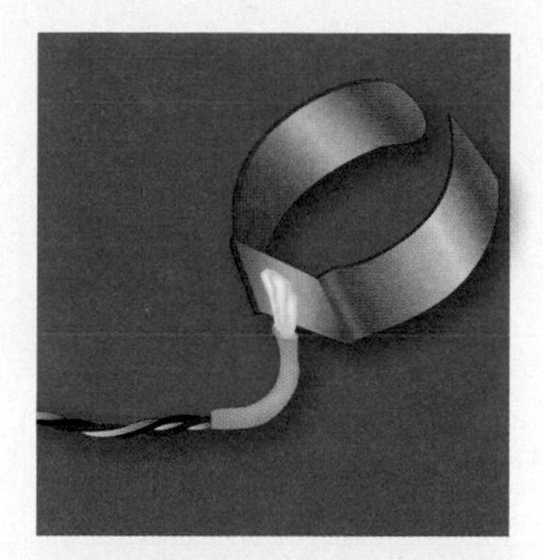

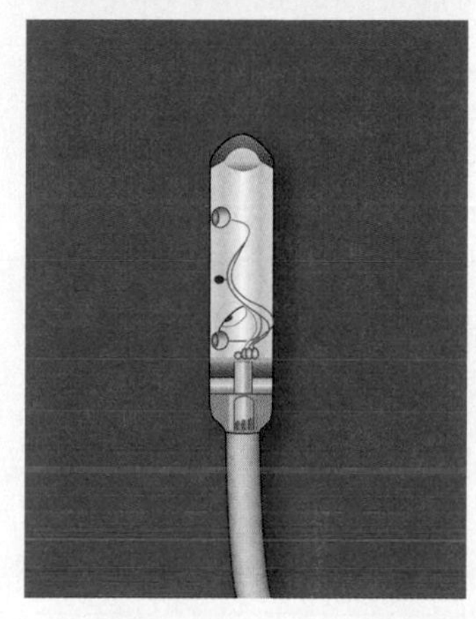

Figure 1.3 Measuring Sexual Arousal: The Penile Strain Gauge (Plethysmograph) and the Vaginal Photoplethysmograph

The devices shown here detect and measure slight changes in blood flow indicative of sexual arousal. Both devices are placed in position in private by the participants themselves, following precise instructions.

All Methods Require Reliability and Validity

When researchers in any science embark on observational, correlational, or experimental studies, their findings must meet two fundamental requirements: **reliability** and **validity**. Reliability refers to the consistency of results over time. That is, the results of a particular study should be similar if the *same* study is carried out a second, third, or numerous times with the *same* participants. If you think about it, this makes perfect sense. Imagine you surveyed 500 people about their satisfaction with their current intimate relationship (anonymously, of course). You would gather and analyze your data and draw some conclusions from their responses. Now imagine if, a week or a month later, you gave the same survey to the same participants. You would expect your results to be substantially the same. If your results changed significantly the second time, you would be very confused about what you had been measuring. Why? It's doubtful that the specifics of most people's intimate relationships would change very much over a short time. So if your survey found large differences, there must be something wrong with your survey: it was not *reliable*.

reliability
the extent to which a measurement is consistent over repeated administrations

validity
the extent to which a measurement accurately reflects the concept being measured

In addition to reliability, all research findings must be valid. This means they must be "truthful;" that is, they must measure what the researcher is trying to measure. If you set out to study couples' sexual satisfaction, your survey items must truly measure sexual satisfaction. If, by accident, you tap into a different aspect of couples' relationships, thinking you are measuring sexual satisfaction, your conclusions will be invalid. For example, a survey item such as, "When my partner and I make love, I feel a sense of deep contentment," probably relates to sexual satisfaction fairly well; however, "I feel my partner is my best friend," probably does not. Even though *relationship satisfaction* is probably connected to *sexual satisfaction*, they are different concepts and you, as a researcher, must be sure to measure the one you are studying to insure validity.

Ethics and Sexuality Research

1.4 Display a clear awareness of the ethical rules that researchers must follow for all research into human sexuality.

All research in any field involving human participants must be carried out with careful attention to ethical standards and principles. It is not difficult to see that when doing research in human sexuality, ethical guidelines relating to the safety, dignity, and anonymity of participants must be followed with extra care. Various professional groups—including the American Psychological Association (APA), the American Medical Association (AMA), and the Society for the Scientific Study of Sexuality (SSSS)—have developed guidelines for the ethical treatment of human participants, which are summarized in Table 1.7. These directives apply to all research with humans and are especially important in sexuality research. The most significant ethical guidelines are summarized below (see Elmes, Kantowitz, & Roediger, 2011).

Protection from Harm

This is the first rule for all studies employing human participants and is especially important in studies in sexuality. Researchers have the duty to protect their subjects from all physical and psychological harm. Even sexual studies that appear totally harmless on the surface may involve the potential for hidden or delayed emotional difficulty for the participants. For example, an anonymous written survey seems harmless enough, right? But what if the participant's boyfriend, girlfriend, spouse, or parent happens to see and read the answers before it is returned? Suddenly it might not seem so harmless.

Sometimes unexpected *future* harm may occur from research that appears harmless in the present. For example, imagine you have volunteered as a subject in a study involving observation of sexual activity. The study is conducted in a completely professional and ethical manner. You feel you are helping to increase our knowledge of sexuality and are completely comfortable with your decision to participate. However, suppose a few years later you become romantically involved with someone. Upon learning of your past sexual research involvement, your partner becomes uncomfortable and distant. Or perhaps worse, imagine later in your life you are in the middle of divorce proceedings and during the child custody hearings, as a legal ploy to malign your character, your spouse reveals that you participated in sex research (Masters, Johnson, & Kolodny, 1977). This is not to say that protecting research participants from harm is impossible, but it illustrates how careful researchers must be to consider all the

Table 1.7 Ethical Safeguards in Sexuality Research

All researchers conducting studies in human sexuality (and virtually any other field) are required to adhere to the principles for the ethical treatment of their human participants summarized in the table.

Principle	Summary
Protection from harm	Participants must be protected from all types of harm: physical and psychological, present and future.
Informed consent	Participants must receive an explanation of research procedures so that they may make informed choices about agreeing to participate.
Freedom to withdraw	Participants must never be coerced into participating and must be aware that they are free to withdraw from the study at any time without penalty.
Debriefing	Participants must be informed following their participation of the true purposes and goals of the study and told of any deception that may have been employed for the purposes of research validity.
Confidentiality	Participants must be guaranteed that the results of their participation will be either anonymous or kept in strict confidence.

possibilities. One common way researchers try to reduce the chance of emotional distress that might arise during or following a study is to provide participants with telephone and e-mail access to the research team so that subjects may contact them at any time with questions or concerns.

Researchers must inform potential participants of a study's procedures so that consent may be obtained.

Informed Consent

Researchers must explain to potential participants what the study is about, what procedures will be used, and what, if any, possible risks are involved. They must assure participants that the records are to be kept confidential or, if that is not possible, to explain exactly who will have access to the records and why. For sexuality research, it is also important to advise participants in advance if sensitive or potentially embarrassing topics will be part of the study (RamaRao, Friedland, & Townsend, 2007; Ringheim, 1995). All this information provided in advance to potential participants ensures that the individual is able to make an *informed* decision about whether to participate. If the person then agrees to participate, this is called **informed consent**. If a study involves a written questionnaire and then a follow-up interview, a potential participant must be told of this before the questionnaire is administered. This is because a participant might be willing to fill out a questionnaire but unwilling to be interviewed. It would not be ethical to ask individuals to complete the written survey and then spring the interview on them.

If minors are to be part of a study, informed consent must be obtained from their parents or legal guardian. If the minor is old enough to give consent, respect for the minor as a person requires that his or her consent be obtained as well (Ringheim, 1995).

informed consent
Agreeing to participate in an experiment only after having been provided with complete and accurate information about what to expect in the study.

Freedom to Withdraw

All human participants in research projects must understand that they have the freedom to remove themselves from the study *at any time*. This may seem like an unnecessary rule because it probably appears rather obvious to you that anyone in the study who becomes too uncomfortable with the procedures can simply leave. However, this is not always so straightforward. Many people feel that since they agreed to participate, it would be wrong to withdraw and risk ruining the study, so they continue even when they do not feel comfortable. Another problem with freedom to withdraw arises when participants are paid for their participation, which is a common practice. If participants are made to feel that their completion of the study is required for payment, this may produce an unethical inducement to avoid withdrawing even if they wish to do so. A common, though not universally used, solution to this problem is to pay participants at the beginning of each research session "just for showing up."

Since You Asked

8. Is it really ethical to study people's actual sexual behaviors?

Debriefing

Another important safeguard against potential harm to research participants is the ethical obligation called **debriefing**, which occurs after participants have completed their role in the study. During debriefing, the researchers explain the goals and procedures of the study to the participants and allow them the opportunity to ask questions or make comments about their experiences. If deception was employed in any way during the study, participants must be fully counseled about the form of the deception and why it was necessary. They also must be assured that they were not foolish in any way to have been deceived. The debriefing is also an opportunity for researchers to determine if any lingering negative aftereffects from the study should be addressed with the participants. This is the time when the researchers may reassure participants of the confidentiality of the data and give them phone numbers or e-mail addresses for further contact, if needed at any time in the future.

Debriefing
Explanations of the purpose and potential contributions of the findings given to participants at the end of a study.

Confidentiality

All results from research participants, especially those who agree to take part in sexuality research, must be kept in complete confidence unless participants have given permission to share their data with certain other specific individuals (such as another research team). This does not mean that *results* cannot be reported and published, but findings must be reported in such a way that individual data cannot be identified. Often no individual identifying information at all is obtained from participants, and the resulting anonymous data are combined to arrive at overall statistical findings. Today, the widespread use of computer databases, electronic storage, and transfer of information have created new challenges for maintaining the confidentiality of research data. It is crucial that researchers develop methods of guarding against any possible breach of confidentiality *before* they begin to gather data. If records of individual participant data are kept for follow-up or additional analysis, confidentiality must be guaranteed.

Your Sexual Philosophy

Studying Human Sexuality

As mentioned at the beginning of the chapter, you will have the opportunity to integrate the material throughout this text into *your own personal sexual philosophy*. You will see this feature at the close of every chapter, along with a brief explanation of how the chapter's information may fit into the sexual philosophy you are developing, and will continue to develop, throughout your life.

The basic idea behind this feature is to encourage you to do some thinking about and preparation for your life as a sexual being, now and in the future. As noted early in this chapter, your sexual philosophy is about *knowing who you are, what you want and don't want, and planning ahead*. If you take the time now to consider and explore your sexual feelings, attitudes, desires, and preferences and to form your sexual "rules for life," you will be far better equipped to be in control of your sexual life and to make healthy, informed choices about your sexual behaviors. Although no one can plan for every future event in life, having your sexual philosophy in place helps you take charge of sexual situations when they arise, rather than allowing the situations to take charge of you.

Chapter 1 contributes to your sexual philosophy in two major ways. First, it is crucial for you to gather as much accurate information as possible, now and in the future, about experiencing and understanding human sexuality, so that the choices and decisions you make about sex and relationships are based on a solid foundation of knowledge. Second, learning to analyze sexuality research *critically* will enable you to evaluate what you hear, see, and read about sex. Then you will be equipped to incorporate into your personal life the best information that will enhance your sexual enjoyment, satisfaction, health, and fulfillment. Many students have remarked that their human sexuality course was one of the most fascinating and *personally useful* courses in all of their higher education.

Have You Considered?

1. Imagine that you have been selected to give a talk about sexuality to high school juniors and seniors who probably think they already know it all. One of your goals is to convince them that they do *not* know it all and that it is important for them to learn as much as they can. What arguments would you use to convince them of the importance of being educated about human sexuality?
2. In the question-and-answer portion of your talk to the high school students, several students ask you about unusual sexual situations

and practices they have learned about on TV talk shows. What would you want to tell them about drawing conclusions about sex based on this kind of information?

3. A friend confides in you that she has been having some sexual fantasies recently that she thinks are unconventional and strange. She is feeling a lot of anxiety about these fantasies and is beginning to feel depressed about having them. She believes that these fantasies indicate that she is sexually abnormal. What would you say to try to help her?
4. Do you feel that the moral values and principles taught to you by your parents and others as you were growing up still apply to your life now as an independent adult? If so, are you comfortable living by them? If not, how have you and your values changed?
5. Do you think you will be comfortable talking about sex with your children in the future? Why or why not?
6. Suppose you want to study the progression of sexual intimacy in dating couples. You are interested in knowing how long, on average, the two people have known each other or have been dating when they first engage in various intimate behaviors (kissing, erotic touching, nudity, genital touching, oral sex, intercourse). What kind of study would you propose? How would you obtain participants? Exactly how would you go about gathering your data? What methodological problems do you think you might encounter?
7. In your study of dating and sexual behaviors in question 6, how would you ensure that your study does not violate ethical considerations?

Summary

Historical Perspectives

A Human Sexuality Time Line

- The world of human sexuality has been marked by many influential historical events. To illustrate, on the inside covers of this book, we journey back over 150 years of significant events that have changed our views of human sexuality.

Experiencing Human Sexuality

- Sexuality is a crucial part of what makes each of us unique. Learning to deal effectively with the wide range of topics that comprise the study of human sexuality is a basic foundation for living a healthy sexual life. Although we are all born to be sexual beings, knowledge of our sexuality is vital for overall understanding of ourselves and of others.
- The development of clear sexual morals and values is critical for making responsible choices. Along with self-understanding should come an exploration of your sexual morals and values—your personal "sexual philosophy." This knowledge allows for planning ahead and helps you make responsible decisions about your sexual behaviors and attitudes.
- Sex is far more than intercourse, and many people's sexual lives are enriched through an awareness of the range of activities that comprise sex. Learning about sexuality and sexual behaviors enhances sexual satisfaction throughout our lives.

Understanding Human Sexuality

- Human sexuality is far more complex than most people think. Students often enter a human sexuality class with inaccurate knowledge and various erroneous beliefs about sex. By studying and absorbing accurate information about sexuality, these misconceptions can usually be cleared up.
- An understanding of human sexuality also increases an awareness and respect for sexual diversity and allows for a clearer appreciation of what is "normal" sexually. Perhaps even more important, learning about sexuality is vital in staying sexually healthy, which includes knowing how to prevent contracting or spreading sexually transmitted infections.
- An accurate understanding of human sexuality provides the tools necessary for assessing sexual research. When it comes to sexuality, as in most areas of life, the ability to separate fact from fiction is one of our most important skills as critical thinkers. And the best way to do that is to learn as much as we can about humans as sexual beings.

Methods for Studying Human Sexuality

- Sex researchers use many methods to study human sexual behavior scientifically. Developing and administering scientific, unbiased, valid surveys requires extensive training and expertise. Scientific surveys must be reliable, providing consistent measurement over repeated administrations, and they must be valid, measuring truthfully the characteristic being studied. The respondents should represent the population being studied as closely as possible.
- Observational research may provide useful information about sexual behavior but does not indicate the cause of the behaviors.
- The work of Masters and Johnson in the 1960s was a landmark example of observational research in human sexuality. However, several diverse and competing theories of human sexual functioning have developed since then.
- Correlational research allows researchers to show relationships between variables. However, two variables may be strongly related but not necessarily causally related.
- Experimental research can demonstrate cause-and-effect connections among variables by holding constant all possible influences except those under study. Although experimental methods provide additional control of the variables under study, this control is usually at the expense of the realism of other research methods.

Ethics and Sexuality Research

- Formal ethical guidelines have been established that govern research involving human participants. Researchers must adhere to these ethical precepts, which include protecting participants from harm, obtaining participants' informed consent prior to beginning the research, ensuring participants' freedom to withdraw from the study at any time, promising the confidentiality of all findings, and offering a full debriefing for all participants following the study.

Your Sexual Philosophy

Studying Human Sexuality

- Developing your personal sexual philosophy is the key to living a fulfilling sexual life. Building this philosophy requires thinking about and preparing for your life as a sexual person, now and in the future. Ultimately, your sexual philosophy is about knowing who you are, what you want and don't want, and planning ahead. Your sexual philosophy allows you to take charge of sexual situations rather than allowing those situations to take charge of you.

Chapter 2
Sexual Anatomy

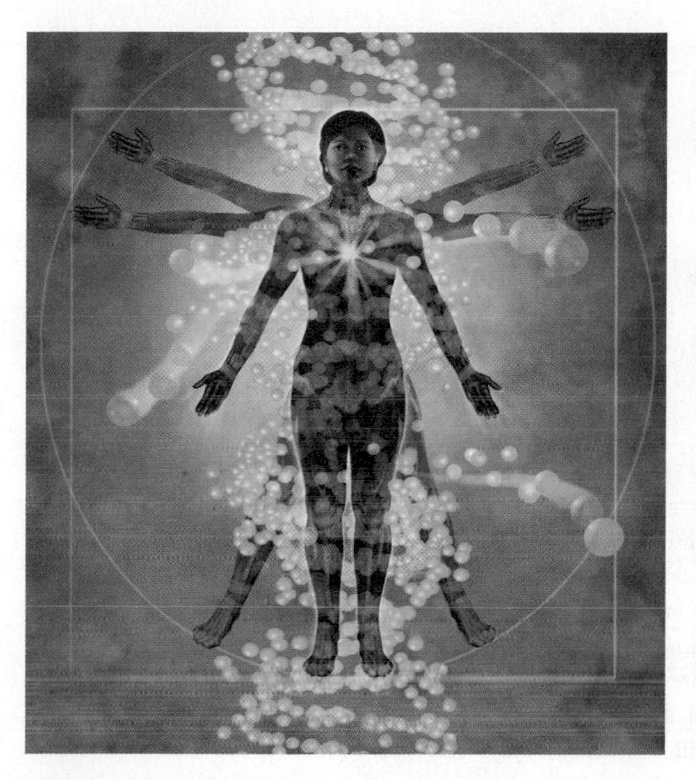

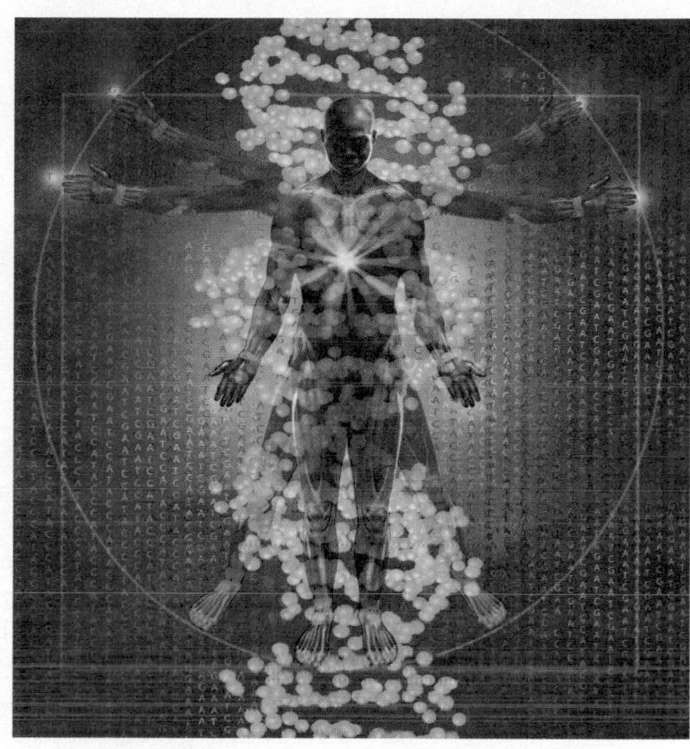

Learning Objectives

After you read and study this chapter you will be able to:

2.1 Describe the characteristics, functions, and health issues relating to male external and internal sexual anatomy.

2.2 Describe the characteristics, functions, and health issues relating to female external and internal sexual anatomy.

2.3 Explain the onset of menstruation (menarche) and review the problems associated with the menstrual cycle including PMS and PMDD.

Since You Asked

1. Is there a bone in the penis? (see page 38)
2. How do the sperm get from the testicles to the penis? (see page 45)
3. I feel that my genitals don't look right. The lips are sort of large and not pink like I see in pictures. They're more grayish. Is this normal? (see page 49)
4. I've heard that the clitoris is "the same" as the penis, but they seem totally different to me. Why do people say they are the same? (see page 50)
5. Lately I've been getting bladder infections a lot, like once or twice a month. They are awful. Why am I getting so many, and how can I prevent them? (see page 53)
6. Is there any sure way to tell if a woman is a virgin before (or after) you have intercourse with her? Is there actually something you "break" inside the vagina? (see page 53)
7. Are larger breasts more sensitive than smaller ones? (see page 56)
8. Once, a condom came off during sex and we couldn't find it. Is it possible for a condom to get "lost" up inside a woman? (see page 60)
9. If a Pap test is negative, can you be sure you are completely free of disease? (see page 62)
10. I get really bad PMS every month. Is this normal, and what can I do to stop it? (see page 68)

This chapter is going to be far more interesting than you may think. Why? For several reasons. **First**, the more you know about your sexual anatomy, the more enjoyment you will be able to derive from the many sexual activities discussed throughout this book. Of course, many sexual feelings and sensations are preprogrammed from birth, but the more you learn about your body (and your sexual responses, covered in the next chapter), the more comfortable you will feel with your personal sexuality. This, in turn, will enhance your sexual enjoyment and satisfaction, both solo and with a partner, throughout your life. Sexuality may seem to be a wondrous and mysterious gift from nature, and you may feel it should remain a mystery and not be spoiled by examining it too closely. But the truth is, with education and knowledge, sexual comfort and satisfaction increases.

Second, at the risk of sounding like a bad advertisement in the back of a nonscientific magazine, "You too can be a better lover!" The truth is, without a solid foundation in sexual anatomy and physiology, satisfying a partner or helping a partner satisfy you is difficult at best and frustrating and disappointing at worst. Even if you have a pretty good idea about what feels good and what turns you on, you need to be able to communicate that effectively and correctly to someone with whom you are sexually intimate. The more you know about sexual anatomy, the better prepared you will be to satisfy your partner and respond to his or her sexual desires. One of the most frequently cited reasons for sexual problems in relationships is poor knowledge and communication about sexual anatomy and physiology. This chapter and those that follow offer you the information you need now or in the future to be a loving and caring sexual partner.

Third, as you explore the topics throughout this book, you'll discover that many of them involve issues of sexual health and diseases. You'll be reading about various sexually transmitted infections (diseases), other bacterial infections relating to sexual anatomy and function, self-examinations of sexual parts of the body to help prevent cancers and other diseases or detect them early for more effective treatment, and many other topics relating to the health of your sexual body. Because your sexual self is the most personal and private part of you, your awareness of the signs and symptoms of

Focus on Your Feelings

Probably the most common emotion experienced by students when reading and discussing the topics in this chapter is embarrassment. We live in a paradoxical culture that is immensely preoccupied with sex, but at the same time is, generally, uncomfortable with frank, straightforward sexual discussion and education. For many—perhaps most—people, discussing and looking at drawings and photographs of sexual anatomy makes them more than a little uneasy.

Feeling embarrassed about these issues is normal and understandable. However, as you have the chance to read and discuss them, you'll likely feel more at ease. And this is one of many positive outcomes of this book and this course. Why? Because the embarrassment people feel about sex contributes to many sexual problems in life. For example, embarrassment over sexual anatomy causes couples to avoid talking about sexual difficulties they may be experiencing, which only makes the problems worse; it causes people to avoid examining their own sexual bodies, which is crucial for early detection of health problems; it causes people to avoid seeing the doctor about potential sexual medical problems that, left untreated, can become much more serious; it causes parents to fail to provide their kids with correct and much-needed information about sex, which may lead to unsafe sexual behaviors and unwanted pregnancy; it causes many people to suffer from insecurities about their bodies that are based on falsehoods rather than facts. The list goes on and on.

Sexual anatomy is simply a normal, natural part of your body and your life—one of the most important parts. The more you can move past your embarrassment or uneasiness and help others do so as well, the more we can look forward to a world in which people are sexually happy and healthy.

any potential health problems in yourself or in an intimate partner is crucial to staying well. The information in this chapter is fundamental to recognizing a possible problem, detecting abnormal changes in various sexual structures that may indicate illness or disease, and knowing when to visit a health care professional, either for routine preventive care or for treatment of a condition you may discover.

Fourth, as is obvious by now, this chapter's topics, along with those in Chapter 1, will serve as an indispensable foundation for what is to follow in this text. What you are about to explore here will provide you with the knowledge, insights, vocabulary, and understanding necessary to benefit from the material in *every* chapter throughout this book. But don't worry: This is a human sexuality text, so this chapter is *not* your typical anatomy lesson. This chapter is, first and foremost, about you and your partner (past, present, or future) as sexual beings.

After a brief look back at the history of sexual anatomy, we will explore in some detail the most important structures that constitute male and female sexual bodies. For both sexes, we will take an "outside-in" route; that is, first we will examine the external anatomy and then proceed logically inward to the various internal structures. This is not intended to be an exercise in memorization of "sexual parts" but rather a conceptual clarification of the evolution of the human body's sexual structures and their roles in reproduction and sexual responding (to be discussed more in Chapter 3, "The Physiology of Sexual Response").

Humans have not always been as enlightened as we are now about sexual anatomy (and even now, we do not know it all). In fact, in earlier times people held some pretty strange ideas, by today's standards, about the sexual body, as you will see in the next section.

Historical Perspectives

Anatomy in the Dark Ages

Throughout history, and even today to some extent, human sexual anatomy was seen as a shameful subject for study or discussion. Why? Because such topics might excite people and cause them to engage in "impure acts." Consequently, the study of sexual anatomy prior to the nineteenth century was wildly misguided (Stolberg, 2003). Here is a partial list of early beliefs about human sexual anatomy—all of them erroneous

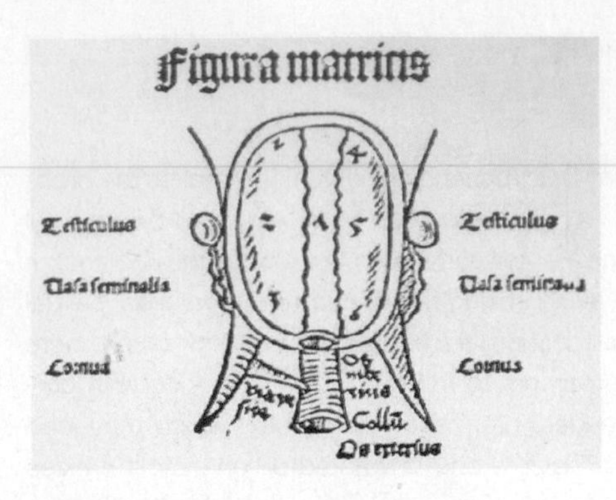

Figure 2.1 Historical View of the Uterus

An early drawing of the uterus showed seven "cells." It was thought that the location of the male sperm would determine the sex of the child: right side = male; left side = female; and middle = "hermaphrodite."

and some widely held as late as the seventeenth century (we will be discussing our *current* knowledge of all these structures next):

- Men and women have the same sexual body (the *one-sex model*), but in men the genitals have been pushed out of the body, although in the woman, they have been retained inside.
- The male body was the "norm;" the female body is merely a "variation."
- The uterus is an internal scrotum.
- The cervix is an internal penis.
- The vagina is an inwardly inverted penile foreskin.

- A sudden, violent physical movement such as jumping over a stream can turn a girl into a boy by forcing her genitals to the outside of her body.
- The "testicles" in men and women both generate sperm.
- The uterus was seen as poisonous and able to wander through the woman's body, causing illness and even suffocation—a condition referred to as *hysteria* (from the Greek word for *uterus*).
- The uterus is divided into two halves (like the scrotum) and consists of seven *cells*—three on the left, three on the right, and one in the middle (see Figure 2.1). Males are born from the right side, and females are born from the left.
- A woman has two uteri (wombs), corresponding to her two breasts.
- The penis is situated in front of the pubic bone so that it will not collapse inward during intercourse. (Leonardo da Vinci wrote, "If this bone did not exist, the penis in meeting resistance would turn backwards and would often enter more into the body of the operator than into that of the operated.")
- Sperm comes directly from the male brain.
- The fallopian tubes, discovered by Gabriel Fallopius, have no known function.
- Menstrual fluid flows directly from the uterus to the breasts, generating breast milk.
- Reproduction is similar to the manufacture of cheese: the woman's "sperm" is the milk, and the male sperm is the "clotting agent."

Even without the benefit of the wisdom you are about to obtain from this chapter, you can tell that our knowledge of human sexual anatomy has come a long way in the last few hundred years. We will now examine in detail our current and far more accurate understanding of male and female sexual anatomy.

The Male Sexual Body

2.1 Describe the characteristics, functions, and health issues relating to male external and internal sexual anatomy.

Our discussion of the male sexual body begins with an examination of the external sexual anatomy—structures on the outside of the body—primarily the penis, scrotum, testicles, and anus, along with some important related issues. Then, we move inward to the internal male structures—located inside the body cavity—including the prostate gland, seminal vesicles, urethral bulb, bladder, Cowper's glands, ejaculatory duct, and urethra. We will discuss these structures as they relate to the production of semen and its journey (usually along with sperm) through a man's sexual body, and eventually to ejaculation.

penis
The primary male anatomical sexual structure.

penile glans
The end or tip of the penis, its most sexually sensitive part.

Male External Structures

If you are looking at a naked man standing facing you, his genitals are right out there with nowhere to hide. However, if you assume the same view of a naked woman, you see virtually nothing of her genitals; they are hidden between her legs and partly, mostly, or completely concealed—but more on that later in this chapter. The external male sexual organs include the penis, the scrotum, the testicles, and the epididymis (the testicles and epididymis, although located inside the scrotum, are external relative to the man's body).

The Penis

The **penis**, the primary male sexual organ, has two jobs: to ejaculate semen and to transport urine from the inside of the body to the outside. Wide variations exist in penis appearance, as is true of all human body parts.

Part (b) of Figure 2.2 shows the basic external components of the penis. As you can see (and probably already knew), it's not a particularly complex system, consisting of the penile shaft, the foreskin (in uncircumcised men), the penile glans, the corona, the frenulum, and the urethral opening. A very sexually sensitive area of the penis, and of the male body overall, is the tip, or **penile glans**. Stimulation to the glans is primarily responsible for male orgasm and ejaculation. This is not to say that the glans is the only sexually sensitive area on a man's body; men are sexually aroused through stimulation of many body parts, but typically, the penile glans must be stimulated directly or indirectly for most men to reach orgasm. The **corona** is the raised ridge at the base of the penile glans where the tip of the penis joins the shaft. Most men report that the corona is somewhat more sexually sensitive than the rest of the tip of the penis. The **frenulum**, an area of tissue at the base of the underside of the penile glans, is typically reported to be the most sexually sensitive area of the glans.

The skin on the **penile shaft** is loose to allow for expansion during **erection**. All males are born with skin covering the penile glans, called the **foreskin**. As infants, this is a small amount of skin (because the penis is small), but in an adult male it averages approximately 3 by 5 inches, or 15 square inches. It is the largest portion of skin covering the penis. Beneath the foreskin, small glands secrete *smegma*, a substance that provides lubrication between the foreskin and the glans and assists in the separation of the foreskin from the glans in male infants.

Male **circumcision** involves the surgical removal (typically as an infant) of the foreskin covering the glans of the penis. Referring again to Figure 2.2, you can see both circumcised and uncircumcised penises. Until late in the twentieth century, most male infants born in the United States were circumcised as a matter of course. Over the past twenty years or so, however, the practice of circumcision has declined and is no longer

corona
The raised edge at the base of the penile glans.

frenulum
The band of tissue connecting the underside of the penile glans with the shaft of the penis.

penile shaft
The area of the penis between the glans and the abdomen.

erection
Rigidity of the penis or clitoris resulting from an inflow of blood during sexual arousal.

foreskin
A layer of skin covering the glans of the penis.

circumcision
Removal of the foreskin of the penis.

corpora cavernosa
Two parallel chambers that run the length of the penis and become engorged with blood during erection.

corpus spongiosum
A middle chamber running the length of the penis into the glans that engorges with blood during erection.

urethra
The tube extending from the bladder to the urethral opening, which carries urine out of the body in both women and men, as well as semen in men

Figure 2.2 The Penis

The penis varies in shape, color, and appearance, just like any part of the male body. Shown here are an uncircumcised penis (center) and two circumcised penises.

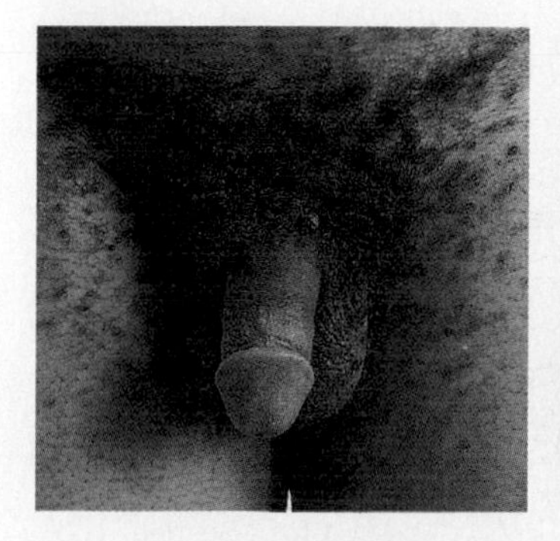

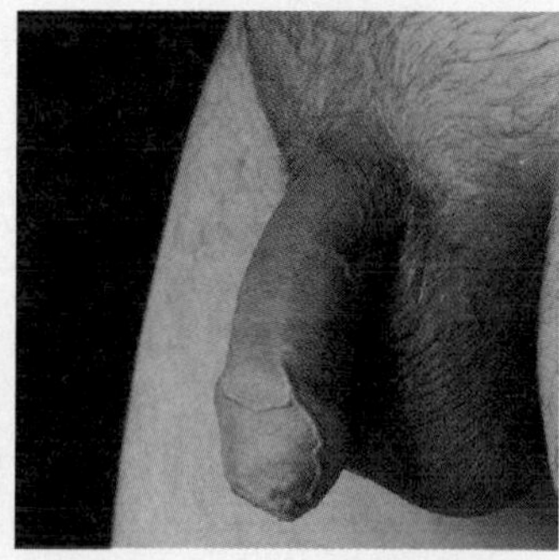

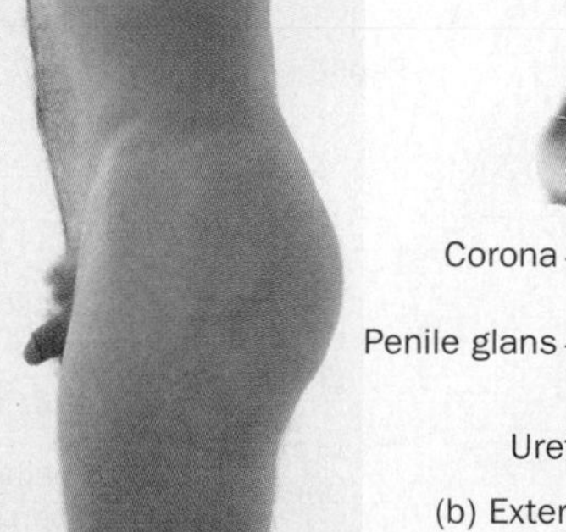

(a) Variations in Penis Appearance

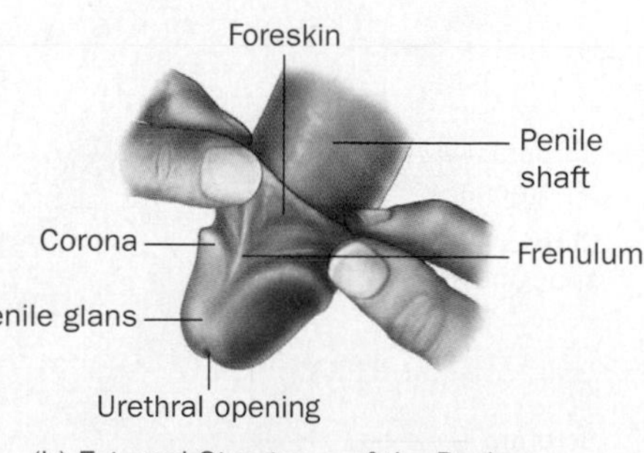

(b) External Structures of the Penis

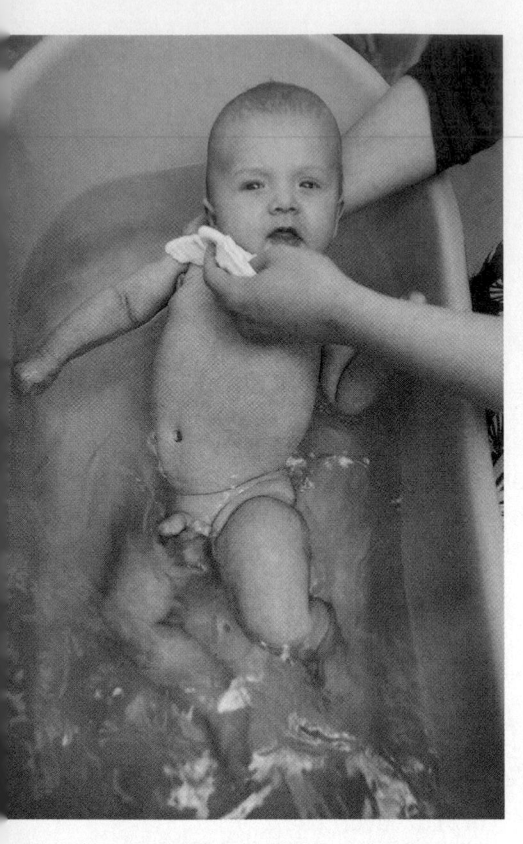

A baby's intact penis requires very little care other than gentle washing of the outside of the foreskin. No attempt should be made to retract the foreskin until it separates naturally from the penile glans months or years later.

routine. "Sexuality and Culture: Male Circumcision in the United States" summarizes the current controversy about male circumcision.

As the practice of circumcision in the United States has declined, education about caring for an intact penis has been slow to keep pace. Many parents, boys, and even some doctors are unclear about what to expect as the uncircumcised penis develops. Proper care of the intact penis can be summarized in one sentence: *Leave it alone*. The most common mistake parents make is assuming that the foreskin should be able to be retracted (pulled back) for washing the glans in early infancy, before it has separated naturally from the penile glans.

Attempting to force the retraction of the foreskin too early in a boy's development may cause tears and abrasions, which then may heal poorly and could cause the foreskin to adhere permanently to the glans. These adhesions can lead to serious problems later in life that may require medical intervention, sometimes including adult circumcision to resolve (Dalton, 2008; NOCIRC, 2010).

The age at which full retraction of the foreskin from the penile glans occurs varies greatly, from a few months to sixteen years old. Approximately 80% of foreskins will be fully retractable by age three and 99% by age sixteen, but all of these retractions are normal and should be allowed to occur naturally (Huang, 2009). Until that happens and the boy is able to retract his foreskin himself, the only hygiene necessary is gentle washing of the outside of the foreskin with soap and water (in infants, water alone is usually sufficient).

The penis consists of three spongy, cavernous tubes running along its length, as shown in Figure 2.3. The two tubes located on the top side of the penis are the **corpora cavernosa**, meaning "cavernous bodies." The third tube that runs along the underside of the penis is the **corpus spongiosum**, meaning "spongy body," and it surrounds the urethra. Despite a common slang term for the penis, it does not contain a bone. During sexual arousal, these spongy, cavernous tubes become engorged with blood, which results in the erection of the penis. When sexual excitement diminishes, blood flows back out of these structures and the erection subsides. The **urethra** is the tube that runs the length of the penis and into the body to carry semen or urine from the inside to the outside of the body.

Since You Asked

1. Is there a bone in the penis?

Penis Size

You may be surprised to learn that penises are more similar in size than they are different. Many people *think* penises vary in size more than they actually do. In fact, the

Figure 2.3 The Penis: Internal View

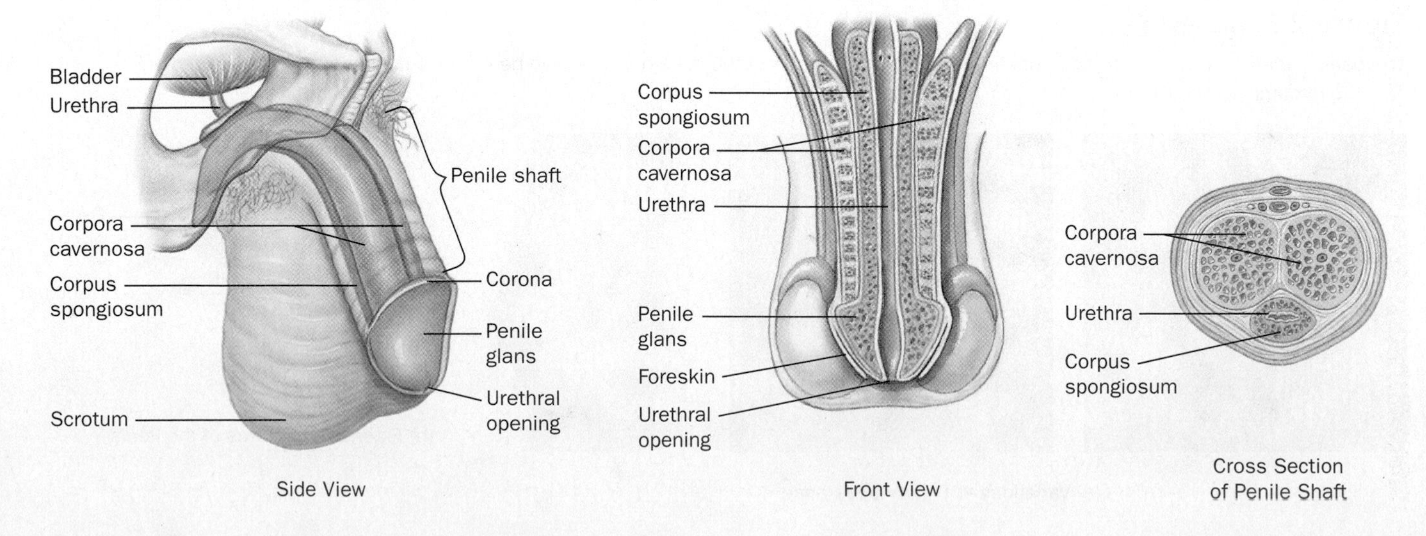

Sexuality and Culture

Male Circumcision in the United States

The Practice of Male Circumcision

Male circumcision is the surgical removal of the foreskin covering of the glans of the penis. Historians and sexuality researchers disagree as to the origins of male circumcision, but we do know that the practice has been observed in Egyptian tomb art dating back to 6000 B.C. However, many researchers believe that circumcision may have developed independently in various cultures at different times in antiquity.

Circumcisions in the United States have usually been performed on infant boys before two weeks of age and often without anesthesia. The reasoning for the lack of pain medication was based on the erroneous belief that very young infants' nervous systems have not fully developed and they do not feel much pain, and that anesthesia can cause complications with the procedure. It should be noted that the United States is one of a few countries worldwide that has routinely performed male circumcisions. Over the past twenty years or so, the practice in the United States has become quite controversial.

Historically, most circumcisions in the United States were carried out for perceived health, rather than religious, reasons. Although some variations exist among racial, ethnic, and religious groups, nearly all male babies born in the United States before 1980 were routinely circumcised. Since that time, the practice has been on the decline, especially between 2006 and 2009, when the rate plunged from 56% to 33% (of births in hospitals) nationwide, by far the lowest in United States history (Zoler, 2010). However, emerging research suggests that this rate may begin to rise once again.

The controversy currently swirling around circumcision focuses on whether a valid medical reason exists for male circumcision. Prior to 1980, the prevailing belief was that uncircumcised, or intact, males were more prone to various diseases, including urinary tract infections, penile irritations, penile cancer, and various sexually transmitted diseases. However, as these beliefs were challenged by modern scientific research, the differences between circumcised and intact males for any of these health issues were found to be nonexistent or so small that they did not support routine circumcision of infant boys (Marx & Lawton, 2008).

In 1999, the American Academy of Pediatrics (AAP), the primary governing board for medical decisions involving children in the United States, reviewed all the available literature on male circumcision and incidence of disease and concluded that there is *no medical justification for routine circumcision of newborn males*. They also added that if parents choose to have their baby circumcised, anesthesia should be used.

In 2007 and 2008, several studies conducted in Kenya and Uganda, Africa, found that men who had been circumcised had a significantly decreased incidence of HIV infection (Millet, 2007; Moses, 2009). However, questions remained in the research following these studies as to whether this research were adequate to suggest that *routine* circumcision should be recommended in the United States and other countries. Other studies have found an *increased* risk of HIV and other STI infections among circumcised men in various other countries (Moszynski, 2007; Rodriguez-Diaz, 2012). Overall, as of 2011, the controversial notion of circumcising male infants as a prevention of HIV infection in adulthood had not received wide medical endorsement in the United States (Garenne, 2008; Templeton, Millet, & Grulich, 2010). Nevertheless, in light of these African studies, the World Health Organization and the United Nations Program on HIV/AIDS began endorsing circumcision for HIV prevention, and some African countries are expanding access to and education about male circumcision (Moses, 2009). Meanwhile, in the United States, the American Academy of Pediatrics (AAP) has not changed its policies rejecting *routine* male circumcision for health purposes or their position on the use of anesthesia, but it has begun reviewing its position in light of the HIV studies in Africa. As of 2012, its revised policy statement reads:

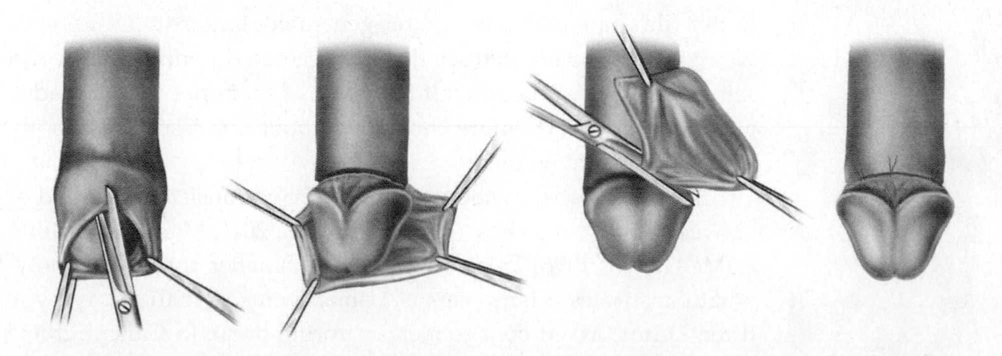

Evaluation of current evidence indicates that the health benefits of newborn male circumcision outweigh the risks and that the procedure's benefits justify access to this procedure for families who choose it. Specific benefits identified included prevention of urinary tract infections, penile cancer, and transmission of some sexually transmitted infections, including HIV. The American College of Obstetricians and Gynecologists has endorsed this statement (Policy Statement, 2012).

Confused? You're not alone. This is an ongoing, medical "work in progress." Because of conflicting research findings, strongly held attitudes, and widely diverse cultures, a final, definitive conclusion about the medical benefits of circumcision is elusive. It may be that one overarching policy that applies to everyone is unachievable and, perhaps, unwise.

The Procedure

Several procedures exist for circumcision surgery. The goal is to remove the foreskin that covers the tip, or glans, of the penis (see diagram). Methods for circumcision vary from freehand techniques, in which the foreskin is stretched away from the glans and cut away with a scalpel, to the use of various clamp devices (as seen in the figure) that reduce bleeding, limit pain, and ensure a more even surgical result.

The Results

The penile glans, protected by the foreskin of a natural, intact penis, is a moist mucous membrane one skin layer thick, much like the inside of your mouth and just as sensitive. When a penis is circumcised, the glans grows additional layers of skin as it heals to reduce the now exposed sensitive tissue. The glans of a circumcised penis eventually develops 10 to 15 layers of skin, more like the back of your hand. Therefore, the glans of a circumcised penis is less sensitive to sexual stimulation than the glans of an intact penis.

It should be noted that among some religious groups, especially in Conservative and Orthodox Judaism, circumcision is considered a necessary religious ritual for all male infants. The surgery is typically performed a week following the child's birth by a *mohel* (pronounced "moyl"), a Jewish official (a mohel is usually a man, but there are a small number of female mohels, called *mohelets*). Although this religious custom is typically not part of the larger cultural and medical debate over male circumcision, a parallel debate is occurring within some Jewish groups as well (Pollack, 2009).

large number of myths about the penis is quite remarkable, and most of them relate to penis size (see Table 2.1).

Throughout modern and ancient history, a great deal of attention has been paid to penis size. This attention has been based on erroneous beliefs that a bigger penis represents a more masculine man, and that men with bigger penises are more attractive to female partners, are better able to satisfy their sexual partners, or can even father children more readily. These myths are not based on any scientific evidence, and yet they persist. Part of the reason for these beliefs is that while growing up, most boys and young men rarely, if ever, have the opportunity to see other penises erect (see Wylie & Eardley, 2007). The exception to this is when they are able to view sexually explicit materials such as so-called pornographic materials. But if you think about it, you'll immediately see the flaw in this "education" about penis size. The men in those videos are clearly not hired because they are normal or average. They are "porn stars" precisely because they have penises that perpetuate the myths: abnormally large ones. In addition, prosthetic or even computer-generated oversized penises are sometimes used in pornographic videos. If computer-generated images can make totally realistic dinosaurs come to life or a game of Quidditch in a Harry Potter film look real, a computer-generated, larger-than-life penis should be a snap. Nevertheless, when men see these exaggerated penises on the screen and do a quick self-comparison, it is difficult for many of them not to feel inadequate. Moreover, in general, men are far more concerned about this issue (that is, the size of their own penises) than are women.

Throughout history, exaggerated penis size has been a preoccupation in many cultures, as shown in this 2,500-year-old Greek satyr sculpture.

Scientific statistics about penis size have consistently found an average length of between 5 and 6.5 inches. (Herbenick et al., 2011; Masters and Johnson, 1966; Wessells & McAninch, 1996).This is significantly smaller than commonly held beliefs about what constitutes a large versus a small penis. A 2001 study by a condom manufacturer, Durex, asked college men on spring break in Cancun, Mexico, to volunteer to have their penises measured by health professionals in exchange for a T-shirt and free condoms (Edell, 2001). In a private setting, with erotic material available for the men to use to achieve erections, each volunteer was measured by two health care professionals supervised by a physician. A total of 300 men participated. In this study, the

Table 2.1 Phallic Fallacies: Common Myths about Penis Size

Fallacy	Fact
Penis size increases with frequent sexual activity and decreases with lack of sex.	Penis size changes only temporarily with sexual activity, during erection.
Penis size can be increased with exercises, pumps, or surgery.	So far, no proven method exists for increasing penis size. Exercise of the penis (whatever that might be) does nothing for size; penis pumps simply draw blood into the penis, which makes it appear bigger temporarily, as with an erection. No surgical technique has shown consistent enough results to be approved by any professional medical association (no matter what you see online), and some can cause serious deformities.
Erect human penises have been documented from 1 inch to 18 inches in length.	A rare congenital (inborn) disorder called micropenis, usually requiring surgical intervention, may cause a penis to be an inch or less in length. However, nothing close to a penis 18 inches long has ever been documented. Overall, erect penises are between 5 and 7 inches long, and average about 5.5 inches.
A small flaccid (non-erect) penis predicts a small erect penis.	Actually, the opposite is usually true. Research has shown that smaller flaccid penises tend to grow more than larger flaccid penises upon erection. There is a significantly greater range of lengths in flaccid penises than in erect penises, which is why some analysts have called erection "the great equalizer."
The size of a man's penis can be predicted from other physical characteristics, such as physical build, race, ethnicity, or shoe size.	These have no basis in scientific fact. Contrary to popular belief, penis size is not related to a man's overall build, height, nose size, shoe size, race, or ethnicity. The only possible predictor of a man's penis size might be the size of his father's penis, due to genetic inheritance.
Large penises provide greater sexual satisfaction for a partner during intercourse.	In heterosexual intercourse, nearly all of the pleasure nerve endings in the vagina are located along the outer one-third of its length (the portion closest to the opening). Moreover, the vagina is a very elastic structure that can accommodate a penis of any size. Penises of any length are able to reach this part of the vagina. Also, for both gay and straight couples, depth of penetration depends more on sexual position than on penile length.
Most partners prefer larger penises.	Actually, most partners of men don't think or care very much about penis length. Men are much more concerned about their own penis size than their partners are. In fact, some people worry more about a partner with a very large penis than a smaller one in terms of comfort during penetration. If a preference exists, women report that penile width is more important than length.

SOURCES: Eisenman, 2001; Masters and Johnson, 1966; McGreal, 2012; Shah & Christopher, 2002; Wessells, Lue, & McAninch, 1996; Wylie & Eardley, 2007.

average erect penis length was 5.877 inches and the average girth, or circumference, was 4.972 inches (slightly less than the circumference of a D battery). Moreover, the variation in size among all 300 men was inconsequential. Length varied from longest to shortest by less than an inch (0.83 inch), and girth varied only 0.5 inches from the smallest to the largest among the 300 men. The data from the Durex study are likely to be biased due to subject self-selection, which probably accounted for the slightly longer average finding (in other words, men with smaller penises might have been less likely to volunteer to be measured). The bottom line to all these findings is that penises are much more similar than different.

Perhaps more important than all these measurements is the notion that a partner's sexual enjoyment and satisfaction depends on penis size. For heterosexual intercourse, Masters and Johnson's research found that idea to be without merit. In their careful observations, they determined that the vagina is an extremely flexible structure capable of accommodating penises of any size and that, during intercourse, a woman's sexual satisfaction is typically *not* related to the size of her partner's penis (Masters and Johnson, 1966). Furthermore, as mentioned in Table 2.1, most of the nerve endings in the vagina are concentrated in the outer third of its length, the third closest to the opening. Therefore, depth of penetration during intercourse is not a major factor in the sensations most women feel during intercourse. And, perhaps most importantly in this discussion, research has shown that penis size is significantly less important to women than it is to men ("Penis Size," 2006).

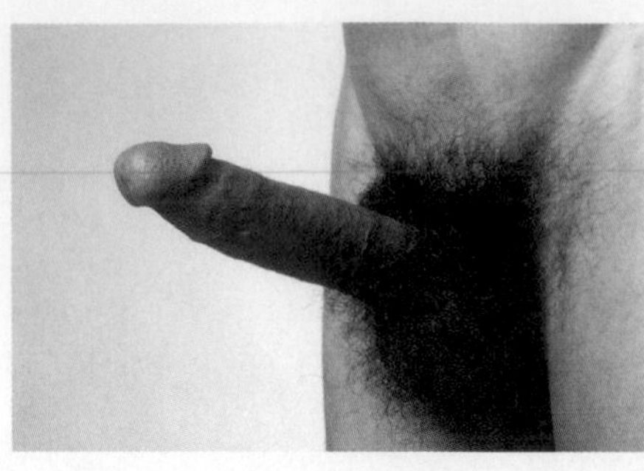

Figure 2.4 Changes in Tightness of Scrotal Skin

The scrotum tightens or relaxes in an effort to maintain a constant temperature for sperm production by the testicles. It also tightens to lift the testicles close to the body during sexual arousal, as shown here.

This is not to say that some people don't prefer, psychologically, a partner with a smaller or larger penis, just as everyone tends to be turned on by different physical characteristics in a sexual partner. However, in terms of the *anatomy* and *physiology* of sexual satisfaction, penis size appears to play no role (for more about this issue, see Chapter 3, "The Physiology of Sexual Response").

The Scrotum and Testicles

The **scrotum** is a pouch of two layers of skin that hangs below and behind the penis. Its function is to house and protect the *testicles* (also called *testes* or *male gonads*) and help provide them with optimal conditions to produce sperm cells. The scrotum is divided into two sacs, one for each testicle. Men are aware that the scrotal skin is not passive tissue that just hangs there in the same form all the time. On the contrary, the scrotum has an active role in male sexuality and reproduction. Two small muscles, one located in the walls of the scrotum and one in each of the **spermatic cords**, move the scrotum and the testicles up and down depending on specific situations, especially external temperature. The testicles require a temperature that is slightly lower than normal body temperature to maximize production of sperm cells (about 94 degrees, rather than the average body temperature of around 98.6 degrees). This is why, when a man's scrotum is hot (say, after a shower or a soak in the hot tub), the scrotal skin is very loose and the testicles are hanging down away from the body. The scrotum is cooling the testicles by moving them as far away from body heat as possible. Conversely, after a swim in cold water, the scrotal skin is very tight to hold the testicles up closely under the penis in an attempt to maintain the ideal temperature for sperm production (see Figure 2.4).

The scrotum also tightens and pulls the testicles up toward the underside of the penis during sexual arousal. As we will discuss more in the next chapter, when the testicles are drawn as far up as possible, it is a signal that orgasm is approaching.

Two **testicles** (also called testes) are housed in the scrotum, each in separate sacs, as shown in Figure 2.5. The testicles are glands referred to as **gonads**, which produce cells for reproduction (in common usage, the word *gonads* usually refers to testicles, but females have gonads too—the *ovaries*, to be discussed later). The testicles' primary functions are to manufacture sperm cells and secrete the male sex hormone, **testosterone**.

scrotum
The sac of thin skin and muscle containing the testicles in the male.

spermatic cords
Supporting each testicle and encasing the vas deferens, nerves, and muscles.

testicles
Oval structures approximately 1.0 to 1.5 inches in length made up of microscopic tubes in which sperm cells and testosterone are produced in the male.

gonads
Organs that produce cells (ova or sperm) for reproduction.

testosterone
The male sex hormone responsible for male sexual characteristics and the production of sperm cells.

Figure 2.5 Internal View of Testicles

Each testicle is suspended by the spermatic cord, which also encases the vas deferens, the tube that carries sperm into the penis for ejaculation.

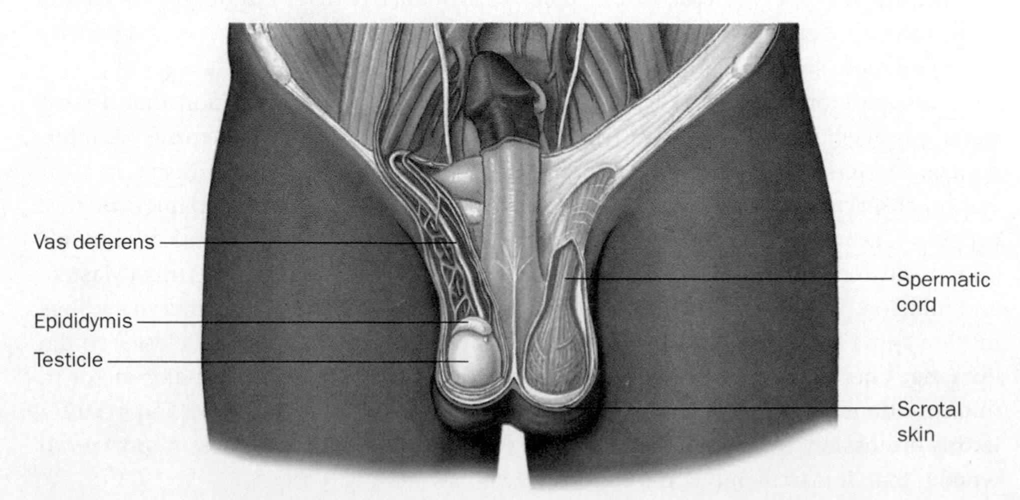

Evaluating Sexual Research

Self-Reports of Penis Size

In addition to the studies cited here, several other projects have been carried out to determine penis size. Some of these studies, rather than using trained researchers to do the measuring, asked men to measure their own penises and submit their findings (Durex, 2002b; Kinsey, Pomeroy, & Martin, 1948; Shaeer & Shaeer, 2013). Well, guess what: All those studies found larger averages, closer to 6.5 inches Not very surprising, is it? Because of the value placed (or misplaced) on penis size, men in the self-report studies fudged their answers, unconsciously measured in ways to maximize their findings, or perhaps just made something up. We should note that these studies provided directions for consistent penis measurement; had they not done so, the results might have been much more distorted.

Each testicle is composed of tightly packed, microscopic **seminiferous tubules**, or "seed-forming" tubes. To get an idea of the small size of these tubules, if they could be removed and stretched out end to end, they would be nearly 1,000 feet long. Sperm cells are continuously formed within these tubules throughout a man's life. Figure 2.6 shows (under powerful magnification) actual sperm cells forming in the seminiferous tubules.

When sperm cells are formed, they migrate in an immature state to the **epididymis**, which is a long, narrow structure attached to the back of each testicle. In the epididymis, the sperm cells mature and wait to be ejaculated. Sperm cells require approximately seventy-four days to mature before they are ready for **ejaculation**. Cells that are not ejaculated are simply reabsorbed by the man's body. Attached to each epididymis is a tube, called the **vas deferens**, through which mature sperm cells travel to the man's internal reproductive system, where they are mixed with semen and ejaculated through the penis and out the urethral opening.

The testicles are among the many sites in the human sexual anatomy that are subject to cancer. Testicular cancer, although not a common cancer overall, is the most frequently diagnosed cancer in men between the ages of 15 and 35. Approximately 9,000 new cases were identified in the United States in 2013, and approximately 380 of those men will die from the disease (American Cancer Society, 2014). In addition, the incidence of testicular cancer worldwide has increased more than 50% since the 1970s (Holmes et al., 2008), and the average age at which it is diagnosed has been steadily decreasing from 30 to under 25 years of age. The exact causes of this increase are not fully understood. However, the greater incidence of testicular cancer among men who work in certain professions—including agricultural workers, miners, firefighters, and utility workers—has led some researchers to suspect that exposure to environmental toxins may play an important role (Guillette, 2002; Meeks, Sheinfeld, & Eggener SE, 2012).

As the incidence of testicular cancer has grown, so has medicine's ability to treat it. Today, the average cure rate for testicular cancer with early diagnosis is approaching 100% (Nichols, 2013). Therefore, all men should learn to examine their testicles to help detect testicular cancer early and maximize the chances of a full cure and recovery. How to do a testicular self-exam is explained in "Self-Discovery: Testicular Self-Examination."

The Anus

The **anus** and the area around it contain nerve endings that are sensitive to stimulation and are considered by some men and women to be part of their sexual anatomy, providing pleasurable feelings when stimulated or penetrated. Some men, regardless of sexual orientation, enjoy having their anal

seminiferous tubules
Tightly wound microscopic tubes that comprise the testicles in the male, where sperm cells are generated.

epididymis
A crescent-shaped structure on each testicle where sperm cells are stored as they mature.

ejaculation
Expulsion of semen through the penis.

vas deferens
A tube extending from the testicle (epididymis) into the male's body for the transport of mature sperm cells during ejaculation.

anus
The end of the digestive tract and outlet for bodily excretions. It is also a sexually stimulating area for some people.

Figure 2.6 Development of Sperm Cells

Sperm cells are produced continuously in the hundreds of feet of seminiferous tubules that constitute the testicles.

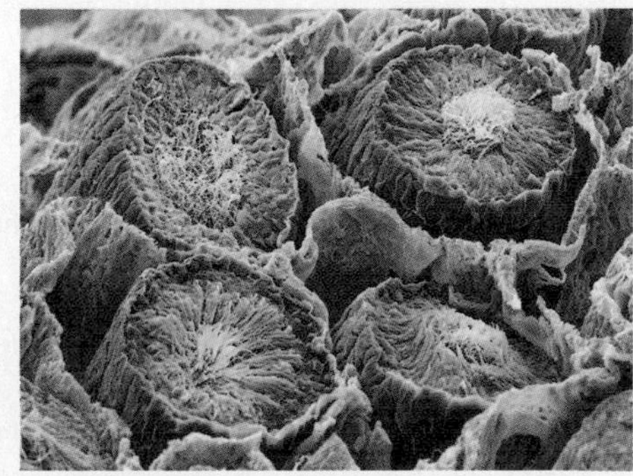

area caressed manually, stimulated orally, or penetrated during sexual activities (see Chapter 6, "The Physiology of Sexual Response," for more information about anal sexual activities). Some men find caressing or massaging of the surface of the prostate gland (to be discussed shortly) through the wall of the rectum to be sexually stimulating. These positive views of the anus are far from universal, and many people are repulsed by the idea that this area fbe considered part of sexual anatomy or behaviors.

The anal area and the walls of the rectum consist of delicate tissues that can be easily damaged during sexual activity, creating an easy route of transmission for blood-borne sexually transmitted infections (such as HIV). In addition, bacteria that exist normally and harmlessly in the anal area and rectum may cause infections if they are transferred to other parts of the anatomy, such as the urethra or digestive tract.

Male Internal Structures

Although male internal sexual, or reproductive, anatomy is fairly complex, we can simplify it here by focusing on the structures that have evolved in the human male to help ensure the reproduction of the species and those that form the foundation for the many other topics discussed throughout this book. These structures are diagrammed in Figure 2.7.

The Vas Deferens

The vas deferens is the tube connecting each testicle and epididymis with the internal reproductive structures. Upon sexual arousal and orgasm, sperm cells that have been stored in the epididymis travel up each vas deferens, where they will eventually mix with ejaculate fluids and be expelled through the urethra during ejaculation. When a man chooses to have a vasectomy so that he will no longer be fertile, each vas deferens is severed and sealed off. Obviously, if sperm cells cannot travel from the epididymis up either vas deferens, they will never be ejaculated and never be able to fertilize an ovum. The man will still ejaculate, however, because fluid in *semen* is produced by the seminal vesicles and prostate gland, to be discussed next. The biological process of sperm production and

Figure 2.7 Male Internal Sexual Anatomy

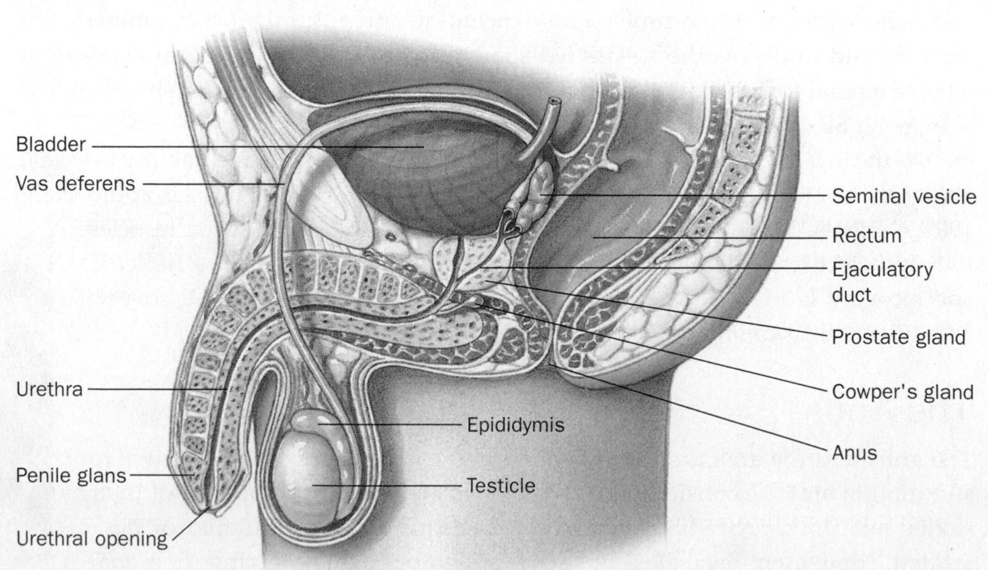

Self-Discovery

Testicular Self-Examination

Many people are surprised to learn that the most common form of cancer in males between the ages of 15 and 35 is testicular cancer. In 2014, approximately 9,000 new cases of testicular cancer were diagnosed in the United States (American Cancer Society, 2014). The three most important issues relating to this form of cancer are: (1) most young men are unaware of their risk; (2) they do not know that they should be examining their own testicles once each month for lumps, irregularities, swelling, and other changes that may be an early sign of cancer; and (3) they have no idea how to perform a testicular self-exam (TSE).

The cure rate for testicular cancer approaches 100% with early detection. Some well-known celebrities who have been diagnosed, treated, and cured of testicular cancer—comedian Tom Green and figure skater Scott Hamilton, for example—have gone public with their experiences in recent years to increase awareness of this disease and of the importance of TSE. Males should begin a monthly habit of examining their own testicles starting at about age 14 (TCRC, 2012). It is a very simple procedure.

A testicular self-exam is best performed after a warm bath or shower. Heat relaxes the scrotum, making it easier to feel anything abnormal. The Testicular Cancer Resource Center (2009) recommends following these steps every month:

1. Stand in front of a mirror. Check for any swelling on the scrotum skin.
2. Examine each testicle with both hands. Place the index and middle fingers under the testicle with the thumbs placed on top. Roll the testicle gently between the thumbs and fingers. Don't be alarmed if one testicle seems slightly larger than the other. That's normal.
3. Find the epididymis, which is the soft structure behind the testicle where sperm cells mature. If you become familiar with this structure, you won't mistake it for a suspicious lump. Cancerous lumps are usually found on the sides of the testicle but can also show up on the front or at the bottom.
4. If you find a lump, see a doctor right away. It may not be cancer, but if it is, you'll want to have it treated as soon as possible to prevent it from spreading. Only a physician can make a positive diagnosis.

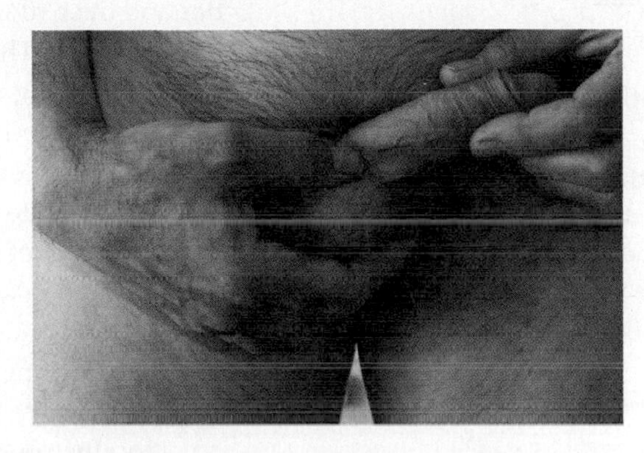

Men should perform a testicular self-exam once a month starting at age 14

function are discussed in greater detail in Chapter 9, "Conception, Pregnancy, and Birth."

Since You Asked

2. How do the sperm get from the testicles to the penis?

Semen

During sexual arousal and ejaculation, as the sperm cells are moving through each vas deferens, other anatomical structures are producing fluid, or *ejaculate*, that will mix with the sperm and carry them out of the man's body. This fluid is called **semen**. Semen is a whitish or yellowish, viscous fluid composed of water, salt, and fructose sugars designed by nature to nourish and sustain sperm cells for their journey from the vagina through the cervix, along the walls of the uterus, and into the fallopian tubes, where, if the timing is right, they may encounter and fertilize an ovum (the process of conception is covered in Chapter 9, "Conception, Pregnancy, and Birth"). Semen consists primarily of a mixture of secretions from the *seminal vesicles* and the *prostate gland*.

semen
The fluid produced primarily by the prostate gland and seminal vesicles that is ejaculated with the sperm cells by men during orgasm.

seminal vesicle
A structure that produces fluid that becomes part of the semen that is expelled during ejaculation.

ejaculatory duct
A continuation of the tube that carries semen into the urethra for ejaculation.

The Seminal Vesicles

The **seminal vesicles** are small glands that are located at the upper, internal end of each vas deferens, where the tubes feed into the **ejaculatory duct**, which directs semen into the urethra. They are connected to each vas deferens by a series of small tubules that carry fluids from the seminal vesicles to mix with the sperm cells during ejaculation. The secretions from the seminal vesicles contain high concentrations of fructose, a type of sugar that provides energy and nourishment to the sperm cells so they will be prepared to survive the arduous journey to rendezvous with the egg. Fluid from the seminal vesicles makes up approximately 60% to 70% of the volume of the semen.

The Prostate Gland

prostate gland

A gland in males surrounding the urethra that produces the largest proportion of seminal fluid (ejaculate).

prostatitis

An uncomfortable or painful inflammation of the prostate gland, usually caused by bacteria.

orgasm

The peak of sexual arousal.

urethral bulb

The prostatic section of the urethra that expands with collected semen just prior to expulsion, creating the sensation of ejaculatory inevitability.

Next, the gathering ejaculate passes through the **prostate gland**, which adds most of the remaining liquid, called *prostatic fluid*. This fluid is quite thick, giving the semen its typical viscous texture. It is also very alkaline to help counteract the normally acidic environment of the vagina, making it more hospitable to sperm and enhancing the odds of conception. You can see from Figure 2.7 that the urethra extends from the bladder and also passes through the prostate gland. When a man is sexually aroused and approaching ejaculation, the prostate contracts, shutting off the flow of urine from the bladder so that urine and semen cannot reach the urethra at the same time.

A common health problem for older men is the noncancerous enlargement of the prostate (called *benign prostatic hyperplasia*, or BPH). Fifty percent of men in their sixties and over 90% of men in their seventies will experience some prostate enlargement (Thompson & Thompson, 2010). If the prostate enlarges too much, it can squeeze the urethra, making urination slow, difficult, painful, or even impossible. This condition is usually successfully treated with medication or surgery to open up the passageway through the prostate gland.

The prostate is also a common site for cancer in older men. It is the second most common form of cancer in men (skin cancer is number one) and is more common than breast cancer in women. Most men who live to be eighty or older will develop prostate cancer, but it is also found in men in their fifties or sixties. Prostate cancer is diagnosed by a digital (finger) rectal exam (see Figure 2.8) and a blood test called the *prostate specific antigen test*.

Usually, prostate cancer grows very slowly and typically does not metastasize, or spread, beyond the gland itself. If it does spread, however, it can be deadly which is why all men over forty should be screened routinely for signs of prostate cancer. Depending on a man's age, how early prostate cancer is diagnosed, and the seriousness of the cancer, treatment options range from complete surgical removal of the prostate to chemotherapy and radiation, and sometimes no treatment at all. This last choice, called "watchful waiting," stems from the slow-growing nature of prostate cancer; for most men diagnosed in their seventies or older, prostate cancer will not be fatal. In other words, they will almost certainly die of other causes before the prostate cancer becomes life threatening. There are many effective treatments for prostate cancer; the choice of treatment depends on many factors including the patient's age and health status, the stage of the cancer, treatment side effects, and the wishes of the patient (see www.cancer.gov for more information).

Figure 2.8

All men over forty should have tests for prostate health, including the digital rectal exam (DRE), for which a physician inserts a gloved, lubricated finger into the rectum and is able to examine the surface of the prostate through the rectal wall for any signs of abnormalities.

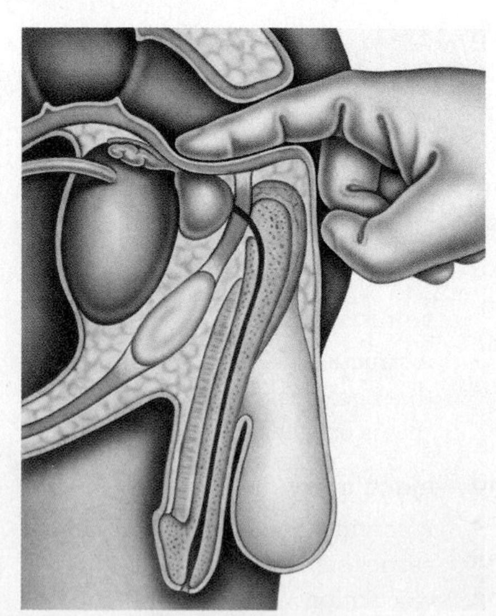

In men of any age, the prostate gland may be subject to **prostatitis**, an inflammation caused by bacterial infections. The symptoms of prostatitis include chills; fever; pain in the lower back and genital area; urinary frequency and urgency, often at night; burning or painful urination; body aches; and white blood cells and bacteria in the urine (Jennings, 2013). Prostatitis may be either acute, lasting a relatively short period of time, or chronic, coming and going over a period of months or years. Antibiotics are usually successful in the treatment and cure of prostatitis.

The Urethral Bulb

At **orgasm**, the peak of sexual arousal, the semen that has been gathering from the various structures discussed above is forced into the **urethral bulb**, a portion of the urethra surrounded by the prostate gland. This pressure causes the man to experience the feeling of impending ejaculation that is beyond his control. The muscle contractions of orgasm force the semen out through the urethra in the final stage of ejaculation. This two-stage process of ejaculation will be discussed in greater detail in Chapter 3, "The Physiology of Sexual Response."

The Cowper's Glands

As the semen passes through the urethra, the **Cowper's glands**, one on each side of the urethra, also add a small amount of fluid to the semen (see Figure 2.7). Of greater importance, however, is that the Cowper's glands often secrete fluid into the urethra and out through the penis well *before* ejaculation. This fluid is called **pre-ejaculate**. During sexual arousal, a small amount of clear, thick, slippery fluid (the actual amount varies among men) from the Cowper's glands may appear at the urethral opening at the tip of the penis. The exact function of this secretion is not fully understood, although it may help neutralize the acidity of the urethra to protect sperm cells and lubricate or flush out the urethra prior to ejaculation. However, if the man has a sexually transmitted infection such as HIV, chlamydia, or gonorrhea, the infectious microbes can be transmitted through the Cowper's gland fluid even if he does not ejaculate (Chudnovsky & Niederberger, 2007; Lampiao, 2014). Although in the past, some believed that the fluid from the Cowper's glands also contained sperm cells, more recent studies have shown that Cowper's gland secretions themselves do not contain sperm cells (Chughtai & Sawas, et al., 2005; Lampiao, 2014). However, other studies have suggested that viable sperm remaining in the man's reproductive tract from a recent ejaculation may be flushed out by the pre-ejaculate fluid during a subsequent sexual arousal. Therefore, the possibility, however small, of a pregnancy caused without ejaculation cannot be ignored (Killick et al., 2012).

Cowper's glands
Small glands near the penile urethra that produce a slippery, mucus-like substance during male sexual arousal (also referred to as the *bulbourethralglands*).

pre-ejaculate
The fluid produced by the Cowper's glands.

In "Historical Perspectives" at the beginning of this chapter, you learned how the male body was once, centuries ago, thought to be the "norm" for sexual anatomy. The sexual anatomy of women was seen as merely a variation on that of men. Today, of course, we know that this view is both naive and inaccurate. In fact, believe it or not, for a few weeks male and female sexual anatomy is the *same* (see Figure 2.9)!

The Female Sexual Body

2.2 Describe the characteristics, functions, and health issues relating to female external and internal sexual anatomy.

Male and female sexual anatomical structures are, of course, different (no one needed to tell you that!), but female sexual anatomical structures are equally, if not more, complex overall. As mentioned earlier, the external genitals of women are a bit more complicated than those of men. In addition, fewer people are as familiar with female external genitals simply because they are not as readily visible. For women to examine their own sexual anatomy typically requires some bodily contortions and the use of a mirror. Nevertheless, as for men, it is a good idea for women to become familiar with their genitals, just as they might for any other part of the body. Why? Because an intimate knowledge of your sexual body enhances your sexual satisfaction, sexual intimacy, and sexual health. The sexual anatomical organs for men and women grow from the very same cells and tissue while in the uterus and begin to differentiate as the fetus develops. Figure 2.9 illustrates how male and female anatomical structures are basically the same prior to this process of differentiation in appearance and function.

Figure 2.9 Origins of Male and Female Sexual Anatomy

If you could look carefully at a developing fetus, you would be unable to tell by looking at its genitals whether it is a boy or a girl until about the twelfth week of pregnancy. Prior to that stage of development, all fetal external genitalia resemble female structures. Normally, these genital areas only begin to develop into male external genitals when a male fetus's testicles begin to secrete androgens, or male hormones. In a female fetus, the absence of these hormones causes the genitals to continue to develop as female. This means that male and female genitalia are homologous, or originate from the very same fetal cells, but due to the presence or absence of hormones, they grow into either boy or girl sexual anatomical structures. The correspondence between the sexes is much more evident than most people realize. The following list shows which male and female genitals develop in the womb from the same underlying structures.

MALE	FEMALE
Testicles	Ovaries
Penile glans	Clitoral glans
Foreskin	Clitoral hood
Penile shaft	Clitoral shaft
Scrotum	Labia majora
Underside of penis	Labia minora

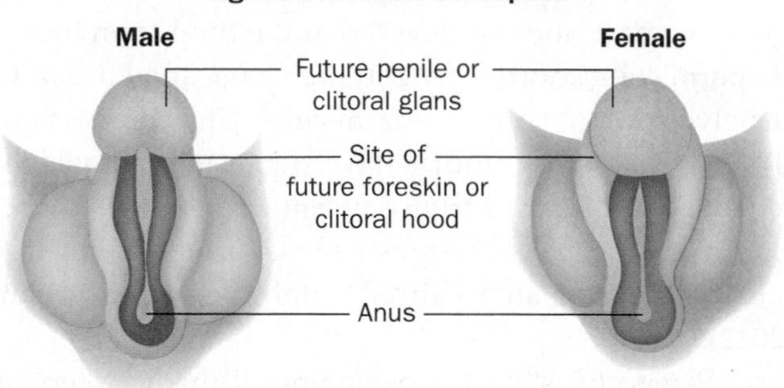

Female External Structures

vulva
The female external genitals.

The female external genitals are referred to as the **vulva** (see Figure 2.10). Components of the vulva, identified in part (a) of the figure, are the *mons veneris*, the *labia majora*, the *labia minora*, the *urethral opening*, the *clitoral glans* (or tip) of the clitoris, the *vaginal opening*, the *hymen*, the *perineum*, and the *anus*, each of which we will discuss next.

Although both men and women have breasts and nipples that may be sensitive to sexual stimulation, we are discussing them here only because in most Western cultures the breasts are more frequently associated with female sexual anatomy.

The Mons Veneris

mons veneris
A slightly raised layer of fatty tissue on the top of a woman's pubic bone, usually covered with hair on an adult.

The **mons veneris** (meaning "mount of Venus" and sometimes called the *mons pubis*) is typically the only part of female sexual anatomy easily seen when looking at a woman's body, legs together, from the front. It is simply a slightly raised layer of fatty tissue on the top of the pubic bone and is usually covered with pubic hair. Part of its evolutionary function for human reproduction is theorized to be to cushion the impact with the pubic bone during sexual intercourse (the mons pubis exists in males, too, but it is less pronounced).

The Labia Majora

labia majora
Folds of skin and fatty tissue that extend from the mons down both sides of the vulva, past the vaginal opening to the perineum.

Just below the mons veneris are the **labia majora**, meaning "major lips." These structures are folds of skin and fatty tissue that extend from the mons down both sides of

Figure 2.10 The Female External Genitals, or Vulva

The female vulva varies normally in shape, color, and appearance, just like other parts of the female body.

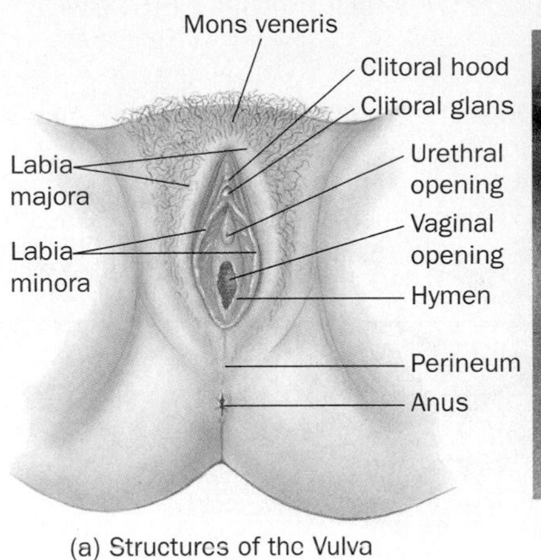

(a) Structures of the Vulva

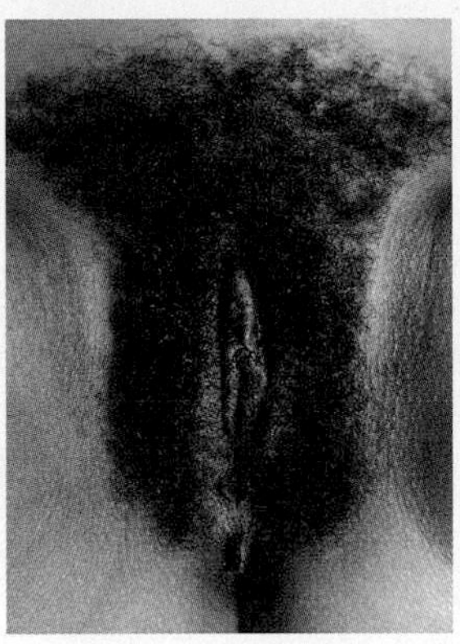

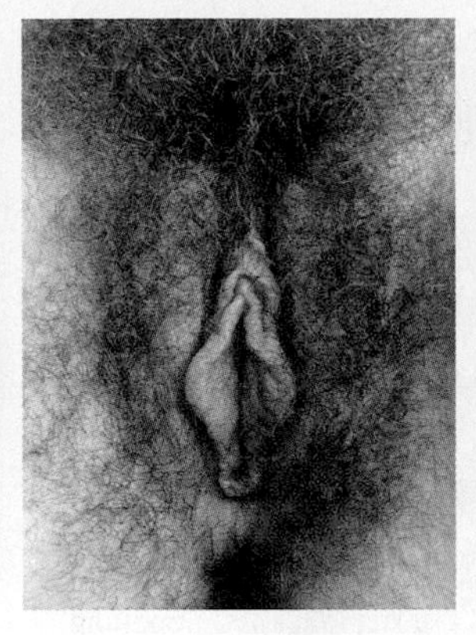

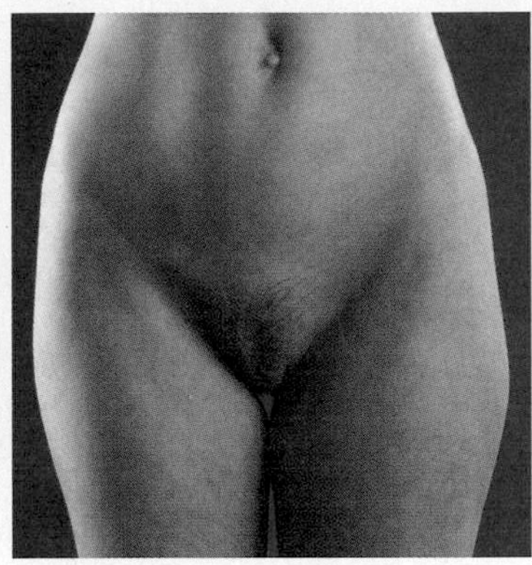

(b) Variations in the Appearance of the Vulva

the vulva, past the vaginal opening to the perineum. The labia majora usually close fully or mostly inward to protect the more sensitive and delicate genital structures underneath them. As you can see in Figure 2.10, the labia majora vary normally in size, shape, amount of hair, and skin tone.

Since You Asked

3. I feel that my genitals don't look right. The lips are sort of large and not pink like I see in pictures. They're more grayish. Is this normal?

The Labia Minora

As can be seen in part (a) and (b) of Figure 2.10, the labia majora must be parted to allow the rest of the vulva to be seen. Immediately inside the labia majora are the **labia minora**, or "minor lips." These are smooth and hairless, and again vary in size and shape from woman to woman, as you can see in Figure 2.10. Many women are insecure about this part of their anatomy because, like men concerned about penis size, women have very little opportunity for comparison. However, with rare exceptions, although women's labia minora vary in size and shape, they all fall within normal ranges. The labia minora are sexually sensitive; during sexual arousal, they become engorged with blood, become moist, and darken in color.

labia minora
The smooth, hairless, inner lips of the vulva.

The Clitoral Glans and Hood

At the top of the labia minora is the **clitoral glans**, which is the tip of the **clitoris** and the part that is visible. It is typically covered partly or completely by the **clitoral hood**. These are analogous, in terms of the fetal origin of the cells, to the penile glans and the foreskin of the penis, respectively (see figure 2.9). Stimulation of the clitoral glans, either directly or indirectly, is primarily responsible for producing orgasm in most women.

clitoral glans
The outer end or tip of the clitoris.

clitoris
An erectile sexual structure consisting of the clitoral glans and two shafts (crura) that is primarily responsible for triggering orgasm in most women.

clitoral hood
Tissue that partially or fully covers the clitoral glans.

The Clitoris

Some of you may be wondering why we are referring to this structure as the clitoral glans rather than simply as the clitoris. It is true that most people, and most anatomical references throughout history, refer to this small, visible structure as "the clitoris," but that is incorrect. Such a characterization of the clitoris would be analogous to defining the male penile glans as if it were the entire penis and disregarding the rest of it. The difference is that the rest of the clitoris is *inside* the woman's body. As shown in Figure 2.11, we now know that the clitoris is a much larger and more complex organ than has been acknowledged in the past.

Figure 2.11 The Clitoral Truth

The clitoris is a significantly larger and more complex organ than previously thought. As seen in (b), during sexual arousal the clitoris, just like the penis, engorges with blood and becomes erect.

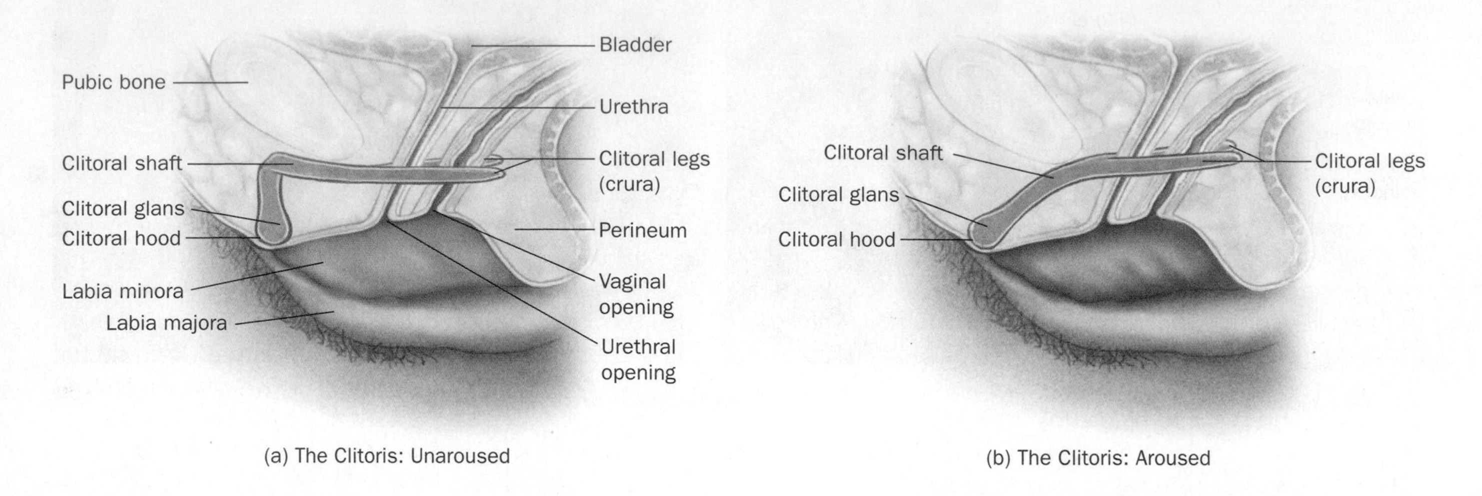

(a) The Clitoris: Unaroused

(b) The Clitoris: Aroused

A full understanding of clitoral anatomy is a relatively recent occurrence. (O'Connell & Delancey, 2005). The clitoral shaft is about 0.5 inches in diameter and divides into two legs called *crura* as it extends 3 to 4 inches into the woman's body. These shafts of the clitoris pass on either side of the urethra and vagina. The clitoris, seen in its entirety, has a similar structure to a penis, which also consists of three "sections" (the corpus spongiosum and the two corpora cavernosa), as shown in part (b) of Figure 2.3. Furthermore, we now know that the clitoris, just like the penis, engorges with blood along its entire length, straightens out, and becomes erect during sexual arousal. This typically causes the glans of the clitoris to pull up under the clitoral hood as a woman becomes increasingly sexually excited (the effect of this will be discussed in more detail in Chapter 3). Figure 2.11 offers an accurate depiction of the clitoris, both unaroused (flaccid) and aroused (erect).

Since You Asked

4. I've heard that the clitoris is "the same" as the penis, but they seem totally different to me. Why do people say they are the same?

Why have researchers discovered the true structure of the clitoris so recently, you ask? Good question! The most common explanations relate to cultural hesitancy over careful exploration of female sexuality and anatomy. One article states, "The fact that this had only been discovered in the 1990s reveals much of anatomists' prudish reluctance to accurately study female genitalia" (Williamson & Nowak, 1998). In other words, social factors trumped science (O'Connell & Delancey, 2005). In the past, the clitoris was often thought of as just a bit of extra tissue and was not even shown in some professional anatomical references. The clitoris is not technically essential to reproduction, so no one worried about it much. Instead, more emphasis was placed on female internal anatomical organs in terms of their importance in child bearing.

The discovery of the true anatomical complexity and size of the clitoris is comparable to scientists suddenly finding that the penis is several times larger than first thought—which it is not. That would be big news! However, comprehensive studies about the clitoris, which first took place in the late 1990s, have been barely a blip on the news media's radar screens. Now, at least *you* have accurate information.

Female Genital Mutilation

Earlier in this chapter, we discussed the controversy surrounding male circumcision, the surgical removal of the penile foreskin. A much more drastic practice of female genital cutting occurs in many countries throughout the world, sometimes called *female circumcision*, but more frequently, and more appropriately, referred to as

Sexuality and Culture

Female Genital Mutilation

In some countries and cultures in sub-Saharan and eastern Africa (see map), the Middle East, and Asia, ritualistic cutting of female genitals is performed on infants, girls, and women of any age, (including, in some cases, during her first pregnancy), but it is usually carried out as a girl approaches puberty (usually between the ages of nine and fifteen). It is a centuries-old practice rooted in strong cultural (rather than religious) beliefs that place a high value on female "virginity" prior to marriage, and the procedure in some cultures is a requirement for marriage. Many of these cultures also have held the belief that it is immoral for women to enjoy sexual pleasure (Abusharaf, 1998; Jaeger, Caflisch, & Hohlfeld, 2009). This procedure is referred to as "female circumcision" among some people who engage in the practice, but this is recognized as a medically incorrect and socially disturbing characterization in most of the rest of the world, where it is now referred to as *female genital mutilation*, or FGM. Researchers estimate that 2 million girls undergo some form of FGM each year throughout the world, and the percentage of girls undergoing FGM varies among the countries where the practice exists. However, in some countries, such as Somalia, Guinea, and Egypt, the rate exceeds 90% (World Health Organization, 2010). The practice of FGM is illegal in the United States, Canada, and many other Western countries, but due to emigration from countries that practice female genital mutilation, physicians in the United States, the United Kingdom, and many other countries in Western Europe are seeing an increasing number of cases in their practices. FGM is drawing increased awareness and outrage from the United States and many countries of the world, and is now seen as a violation of women's human rights.

Traditional FGM procedures vary in the degree of mutilation inflicted and are often divided into three types, although a great deal of overlap in the exact procedure exists due to the lack of surgical skills and anatomical knowledge of those performing the cutting (Adam & Bathija et al., 2010). All the procedures are typically performed without anesthesia, using nonsterile, primitive instruments such as pieces of broken glass, primitive knives, razor blades, or scissors.

African FGM instruments.

Type I involves removal of the clitoral hood and part of or the entire external clitoris (called a *clitorectomy*). In Type II, the most common form of FGM, the clitoral hood and visible portion of the clitoris are removed and part or all of the labia minora are cut away. Type III is the most extreme form but is quite common in some countries, mainly Somalia, Sudan, Mali, and Nigeria. Here the clitoris (usually) and labia minora are removed as in Type II, but in addition, the labia majora are also cut or scored. Then the vulva is effectively sealed through the stitching or scoring of the labia majora. In some cultures, the labia majora are stuck together with paste or thorns, with only a small, matchstick-sized opening left near the bottom of the vaginal opening for the passage of menstrual blood and urine. This procedure is called *infibulation.* The cutting or scoring of the outer labia causes them to fuse together during the healing process. When the woman later marries, the vulva is reopened, sometimes by the new husband, so that intercourse becomes possible. The diagrams on page 54 represent these three types of FGM.

As you can imagine, women who have been subjected to such procedures suffer great pain. In addition, medical complications stemming from the unsanitary and primitive nature of the process are extremely common. These complications include (see World Health Organization, 2010):

- Recurrent bladder and urinary tract infections
- Cysts
- Infertility
- An increased risk of childbirth complications and newborn deaths
- Need for later surgeries. For example, the FGM procedure that seals or narrows a vaginal opening (Type III above) needs to be cut open later to allow for sexual intercourse and childbirth. Sometimes it is stitched again several times, including after childbirth, hence the woman goes through repeated opening and closing procedures, further increasing both immediate and long-term risks.

Worldwide awareness of and opposition to FGM has been growing rapidly in recent decades. This is due in part to greater global intercultural awareness in general, but also because, as mentioned earlier, FGM and its consequences are being encountered increasingly in the West, including the United States, as people from cultures where it is performed immigrate in larger numbers to Western nations. Moreover, incremental progress is being made in countries that practice FGM. In the late 2000s, one African country, Eritrea, where nearly all girls have traditionally been forced to undergo this procedure, made FGM illegal, announcing, "Anyone who requests, incites or promotes female genital mutilation will be punished with a fine and imprisonment" ("Cruellest Cut," 2007).

Countries Practicing Female Genital Mutilation

Female genital mutilation is a common practice in many countries of the world, most of which are located in Africa.

SOURCE: Map, p. 26 from United Nations Children's Fund, *Female Genital Mutilation/Cutting: A statistical overview and exploration of the dynamics of change*, UNICEF, New York, 2013.

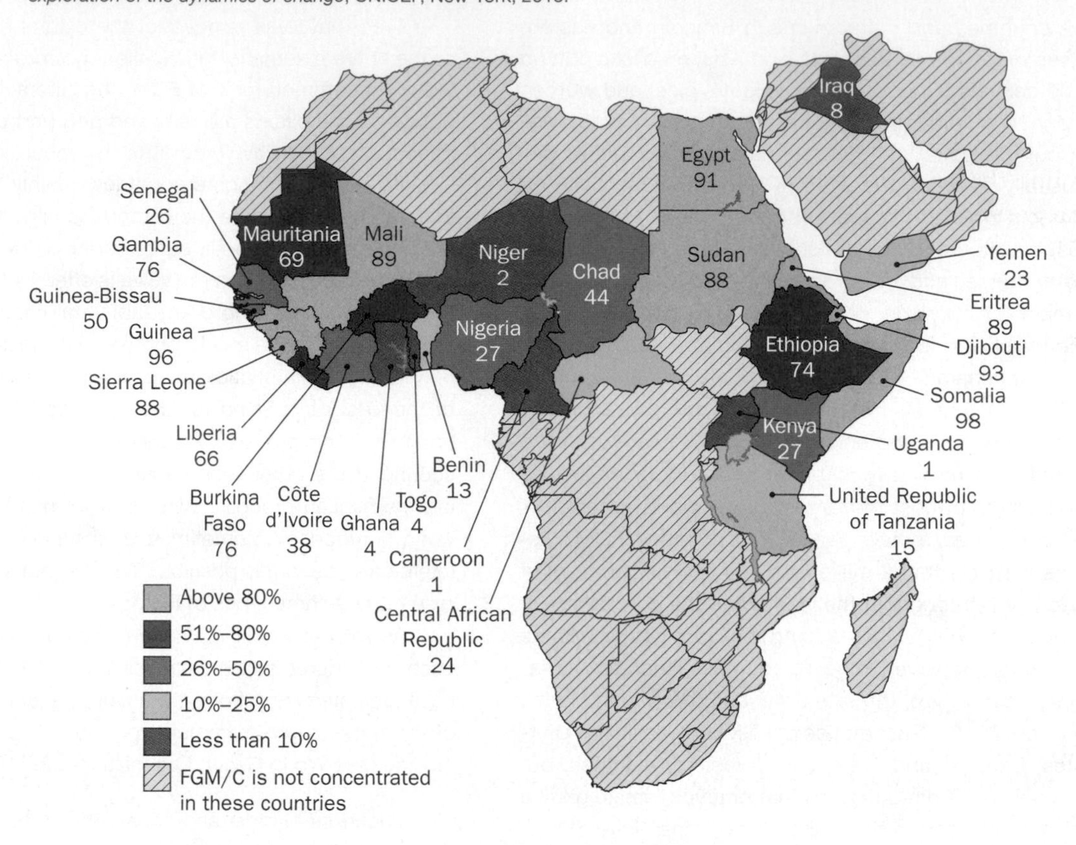

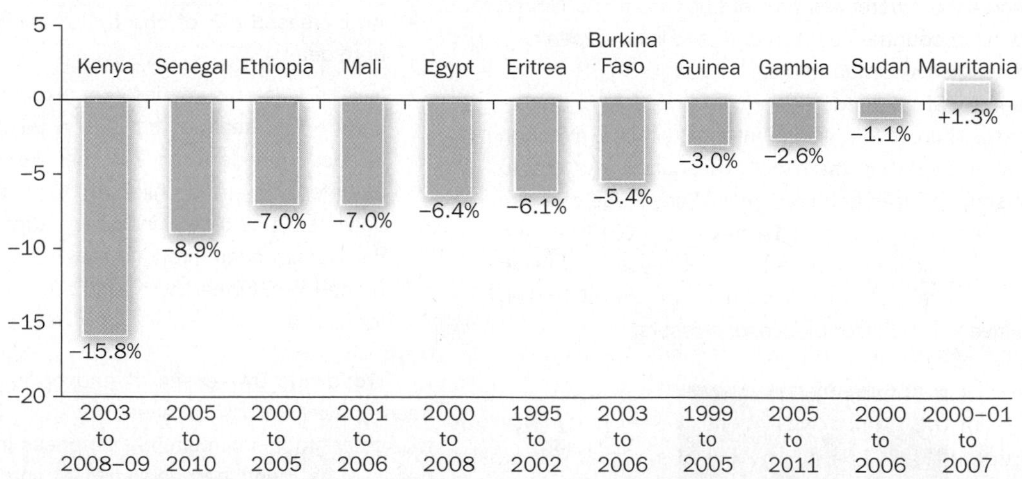

SOURCE: United Nations Children's Fund (http://www.unfpa.org/webdav/site/global/shared/documents/news/2012/FGM-C_highlights.pdf).

Although the practice is deeply embedded and remains unabated in some countries, many groups are working to educate people in those countries to reduce or eliminate FGM. This includes professional medical organizations, women's groups, and human rights groups. Most advocate organizations see that the social changes necessary to eradicate the practice of FGM will not happen quickly; however, some progress is being made. The UN Population Fund notes that during the first decade of this century, the rate of FGM has dropped in many countries, as indicated in the graphic below. A great deal of international work remains to eliminate this practice, but these small gains are cause for optimism.

female genital mutilation, or **FGM.** Found mostly in parts of Africa and the Middle East, FGM bears little resemblance to male circumcision in the procedure itself or in the purposes and philosophy underlying it. Male circumcision is performed because many parents believe (rightly or wrongly) that it promotes a healthier sexual life for the child as he grows up and will enhance his sexual pleasure. In contrast, female genital mutilation is performed with expressly the *opposite* intention of *preventing* normal sexual functioning and sexual pleasure in women, due to cultural beliefs relating to female virginity and the idea that women should not engage in sex for pleasure. The practice of FGM often leads to various health complications for these girls, including chronic pain, infections, and even death. This disturbing practice is discussed in greater detail in "Sexuality and Culture: Female Genital Mutilation" on pages 53–54.

female genital mutilation (FGM)
Removing part or most of the vulva to prevent sexual stimulation or pleasure; a cultural practice in many countries, especially in Africa.

The Urethral Opening

About halfway down the vulva, between the clitoris and the vagina, is the **urethral opening**, the outside end of the tube leading from the bladder. This is a sensitive structure and can provide pleasurable sexual sensations for some women when stimulated. The location of the urethral opening in women is important to understand for specific health reasons. When comparing male and female anatomy, one of the marked differences is the length of the *urethra*, the tube from the bladder to the outside of the body: it is considerably shorter in women than in men. The short distance from the vulva to the bladder in women means that bacteria have an easier time reaching the bladder and causing infection. This is why women are significantly more prone than men to **urinary tract infections (UTIs)**. Moreover, many women suffer from recurring UTIs, sometimes as often as once every few months or even more frequently. Urinary tract infections may occur anywhere in the urinary system, but they are usually centered in the bladder. They are caused when common bacteria travel from the outside world up the urethra and into the bladder, where they find ideal (moist, dark, warm) conditions in which to grow and proliferate. However, the good news is that UTIs are usually fairly easy to treat, and various measures exist to help prevent them or reduce their frequency. "In Touch with Your Sexual Health: Urinary Tract Infections" provides more detailed information about the causes, treatment, and prevention of UTIs.

urethral opening
An opening in the midsection of the vulva, between the clitoral glans and the vagina, that allows urine to pass from the body.

urinary tract infection (UTI)
An infection of the urethra, bladder, or other urinary structure, usually caused by bacteria.

Since You Asked

5. Lately, I've been getting bladder infections a lot, like once or twice a month. They are awful. Why am I getting so many, and how can I prevent them?

The Hymen and the Vaginal Opening

Near the lower end of the vulva is the *vaginal opening*. And at the entrance to the vagina is a structure known as the **hymen**. The hymen is a thin layer of tissue that partly covers or surrounds the vaginal opening. Although its true anatomical purpose is unclear, the hymen, in its own way, has drawn nearly as much attention as the penis in terms of sexual mythology because it is so closely tied to the notion of female virginity. Many of the most common delusions worldwide about female sexuality concern the hymen. Here is a partial list of some of the most common myths about the hymen—all of them false:

hymen
A ring of tissue surrounding, partially covering, or fully screening the vaginal opening.

Since You Asked

6. Is there any sure way to tell if a woman is a virgin before (or after) you have intercourse with her? Is there actually something you "break" inside the vagina?

1. **Myth: The condition of a woman's hymen is indicative of whether or not she is a virgin** (meaning that she has never engaged in penile-vaginal intercourse). This belief is false because the appearance of the hymen relates to many factors: for example, some girls are born without an intact hymen; nearly all hymens are at least partially separated naturally to allow for passage of normal vaginal discharge and menstrual fluid; the hymen may be separated due to strenuous athletic movements such as bicycle or horseback riding, gymnastics, or even dancing (Hegazy & Al-Rukban, 2012; Azam, 2000); the hymen may be perforated due to tampon

Figure 2.12 Normal Variations in Appearance of the Hymen

The hymen varies in appearance and degree of intactness, often regardless of a female's sexual experience.

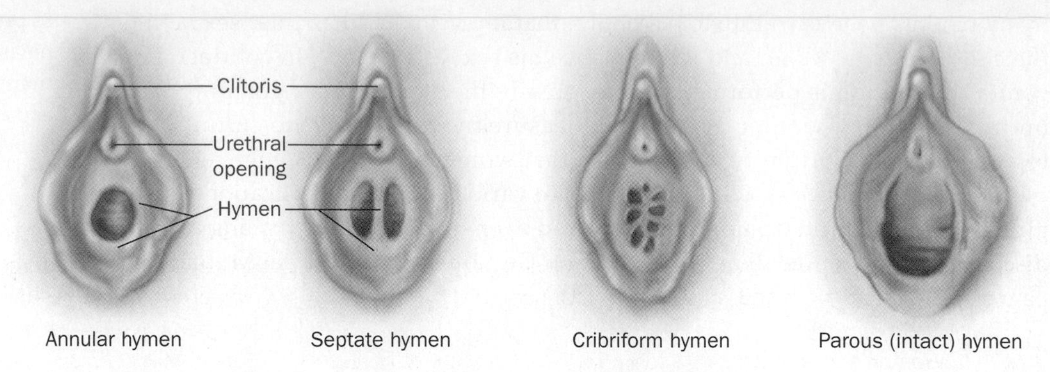

insertion; and the hymen may be separated by behaviors during masturbation. Some women's hymens are separated or perforated for no apparent reason other than normal variations among humans. Figure 2.12 shows some common, normal configurations of the hymen.

2. **Myth: Upon first sexual intercourse, the hymen will "break" and bleed**. For some women, having intercourse for the first time may indeed cause some tearing of the hymen, and this may include some minor bleeding, which typically stops quite quickly. However, for many women, first intercourse causes little or no damage to the hymen and no bleeding. In rare cases, the hymen completely covers the vaginal opening and must be opened slightly by a doctor when a girl begins to menstruate.
3. **Myth: Intercourse is very painful the first time due to the rupturing of the hymen**. As was just mentioned, the hymen often does not tear at all with intercourse, and even if it does, the trauma to the structure is minor and usually not particularly painful. In reality, the most common reason for painful intercourse, whether the first time or not, is insufficient lubrication of the vagina due to rushing the experience or lack of arousal on the part of the woman for various reasons (see Chapter 7, "Sexual Problems and Solutions," for more information about painful intercourse).
4. **Myth: If a woman has an intact hymen, she cannot become pregnant**. Even if no penetration occurs due to what is perceived as an intact hymen or for any other reason, pregnancy can still occur. If semen is ejaculated near the vaginal opening, some sperm cells may make their way past the hymen into the vagina.

Although all of these beliefs are false, the importance attached to the hymen and its role in presuming virginity persists in many cultures throughout the world. In fact, women from some Muslim cultures are seeking "certificates of virginity" and obtaining a procedure called **hymenorrhaphy**, or *hymenoplasty*, which surgically restores the hymen to an intact-appearing state, so that the women will not be deemed "unmarriageable" and will bleed on their wedding nights. If a woman does not bleed on her wedding night, the (often false) assumption will be made that she is not a virgin, and she could be subjected to societal banishment, beatings, prison, or even murder (Ahmadi, 2013; Eich, 2010; Saletan, 2009).

hymenorrhaphy

A medical procedure, common in some cultures, to reconstruct or repair the hymen to allow a woman to appear "virginal"; also known as *hymenoplasty*.

perineum

The area of skin in the female between the vulva and the anus, and in the male between the scrotum and the anus.

The Perineum

The **perineum** is the area of skin, rich in nerve endings, between the vaginal opening and the anus. Men have a perineum as well, located between the underside of

In Touch with Your Sexual Health

Urinary Tract Infections: Causes, Treatment, and Prevention

Anyone who has ever had a urinary tract infection (UTI) knows how unpleasant such infections can be. Typical symptoms include pain and stinging, often intense, during urination; the persistent feeling of a strong need to urinate even though the bladder has just been emptied; frequent urination; darker and bad-smelling urine; the sensation of pressure and pain in the lower back or abdomen; and sometimes slight fever, chills, nausea, or blood in the urine ("Urinary Tract Infections," 2008).

Women are more than 40 times more likely than men to be diagnosed with a UTI. More than 50% of women, compared to 0.5% of men, will contract a UTI at least once in their lifetime, and some women have frequent, recurring cases numerous times a year (Griebling, 2005; Walling, 1999). UTIs account for nearly 10 million doctor visits each year in the United States. The huge male-female difference in incidence is due to the fact that bacteria have an easier time reaching the bladder in women. As you can see in the diagram, the urethra in women is relatively short, measuring about 1.5 inches from the opening at the vulva to the bladder. In men, however, the urinary tract extends the full length of the penis and into the body (about 8 to 9 inches) before reaching the bladder. In addition, anatomical differences place the opening to the urethra in women in greater proximity to the anus and rectum, which contain certain bacteria (especially *Escherichia coli*) that are harmless and normal in the rectum but are the cause of nearly all (85% to 90%) cases of UTIs (Urinary Tract Infections, 2010). It is relatively easy to transfer these bacteria accidentally from the anal area to the vulva during sexual activity or when wiping the anal area following bowel movements. This by no means implies that women are somehow unhygienic or that they harbor more bacteria than men; it's just a natural anatomical difference.

Fortunately, UTIs are usually easy to cure with antibiotics that target the pathogenic bacteria. In addition to antibiotics, UTI treatments often include the recommendation to drink large amounts of water or other nonirritating fluids when symptoms appear (while avoiding alcohol, tea, coffee, or citrus juices) to flush bacteria out of the urinary system. Also, many people believe that cranberry juice has certain chemical properties that help change the acidity of the urine to fight and prevent UTIs, and some scientific evidence supports this notion (Del Mar, 2010).

Of course, it is always better to prevent illness than to treat it, especially if you are prone to recurring UTIs. With this in mind, what can women do to prevent or reduce the number of UTIs? Here is a list of suggestions from the National Institutes of Health (see Dudale 2010):

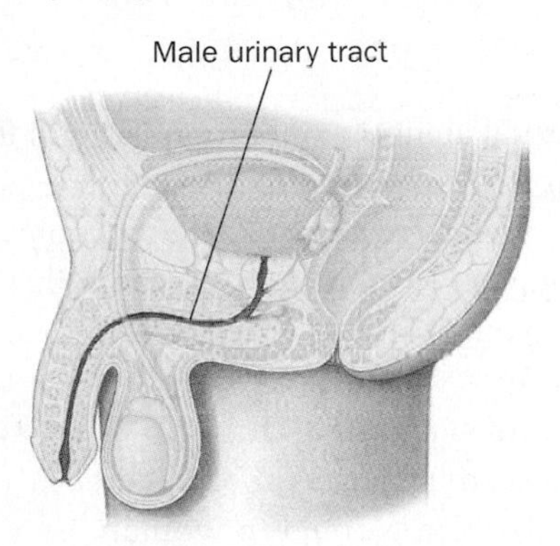

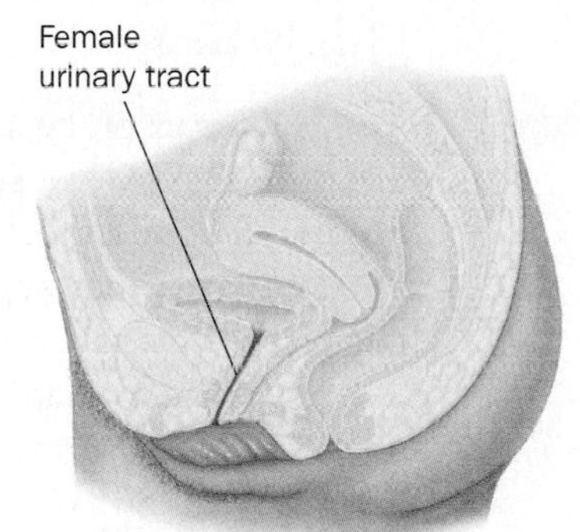

- If possible, use menstrual pads instead of tampons, which some doctors believe make infections more likely. Change the pad frequently.
- Wipe from front to back after using the bathroom, especially after a bowel movement.
- Do not douche or use feminine hygiene sprays or powders. As a general rule, do not use any product containing perfumes in the genital area.
- In general, take showers instead of baths. Avoid bath oils.
- Keep your genital area clean. Clean your genital and anal areas (genital area first) before and after sexual activity.
- Urinate before and soon after sexual activity. The urine actually serves to flush out any bacteria that may be in the urethra.
- Avoid tight-fitting pants; bacteria grow more readily in closed-off, warm areas.
- Wear cotton-cloth underwear and pantyhose (cotton breathes, keeping the vulva area less moist) and change both at least once a day.
- Drink plenty of fluids (two to four quarts each day).
- Drink cranberry juice or use cranberry tablets (but NOT if you have a personal or family history of kidney stones).
- Avoid fluids that irritate the bladder, such as alcohol and drinks with caffeine.

Finally, on a hopeful note, the possibility of a vaccine that may prevent most UTIs by targeting the E. coli bacterium, which causes 90% of all infections, is under study (Alteri & Hagan et al., 2009). If you suffer from frequent UTIs, be sure to try all the precautions listed above, but if nothing works, this vaccine may eventually be exactly "just what the doctor ordered" for you.

the scrotum and the anus. Some women and men find that manual stimulation of the perineum during sexual activity enhances feelings of arousal. The reason the structure is discussed under the heading of female external genitals is because it has been involved in childbirth. The perineum is the area that may be cut during delivery in a procedure called an **episiotomy**. This procedure was once a relatively routine part of hospital childbirths because doctors believed it allowed more room for the baby's head to move through the birth canal and reduced the chances for unplanned tearing of the perineum. Today, this procedure has become far less routine, and many health professionals now believe it is unnecessary during a routine birth (Hartmann & Viswanathan, 2005). You can read more about the newest thinking concerning episiotomies in Chapter 9, "Conception, Pregnancy, and Birth."

episiotomy
Surgical cutting of the perineum during childbirth, a procedure that was believed to allow for easier passage of the infant and less tearing of the vaginal opening. Found to be ineffective, it is rarely performed today.

The Anus

As is true for some men (discussed earlier), the anus is considered by some women to be a sexually responsive part of their anatomy. Some women find manual or oral stimulation of the anal area to be sexually arousing, and some enjoy anal penetration with a finger or penis. Statistics show that approximately one-third of women have engaged in anal intercourse at some point in their lives (Herbenick et al., 2011; Leichliter et al., 2007). It is important to stress that anal intercourse is one of the riskiest sexual activities in regard to the spread of sexually transmitted infections, especially HIV and hepatitis. Also, we should reiterate here that some bacteria normally found in the anal area and rectum may cause infections if they are transferred to other parts of the female sexual anatomy such as the vagina or urethra.

The Breasts

Breasts are part of the sexual anatomy of both males and females. However, because most cultures place a significant sexual focus on the female breast, we include the discussion here, as part of female external sexual anatomy. Breasts have three major functions: They may supply nourishment for newborn infants, they may provide sexual pleasure for both the woman and her partner, and, in most Western cultures, they play a role in a woman's perceived attractiveness and self-image. Breasts vary greatly in size, shape, coloring, and symmetry. The nipples are typically somewhat darker in pigmentation than the surrounding skin of the breast. The normal size and shape of the nipples vary greatly from woman to woman; they may protrude, lie flat, or even turn inward, which is referred to as *inverted nipples*. The darker skin encircling each nipple is called the **areola** and is actually part of the skin of the nipple. The size and color of the areolas also vary from woman to woman. Virtually all of these differences in the appearance of the breasts are normal human variations (see part (a) of Figure 2.13).

areola
The darker skin encircling each nipple; actually part of the skin of the nipple.

Internally, the female breast is a relatively simple anatomical structure consisting primarily of milk glands, milk ducts, connective tissue, and fat, as shown in part (b) of Figure 2.13. During *lactation*, the milk glands produce milk, which flows through the milk ducts to the nipple when the woman is breast-feeding.

The breasts are usually a sexually responsive part of human anatomy. Sexual stimulation of the breasts and the nipples causes most women and some men to become sexually aroused. Both the nipples and areolas contain erectile tissues that engorge with blood, causing them to become erect during sexual arousal and stimulation.

Since You Asked
7. Are larger breasts more sensitive than smaller ones?

The size and shape of a woman's breasts, nipples, and areolas are unrelated to how sexually sensitive her breasts are or to her level of sexual desire and responsiveness. Women vary greatly in the amount of pleasure they experience from breast and

Figure 2.13 The Female Breast

Female breast shape, size, and appearance vary greatly, just like other parts of the female body.

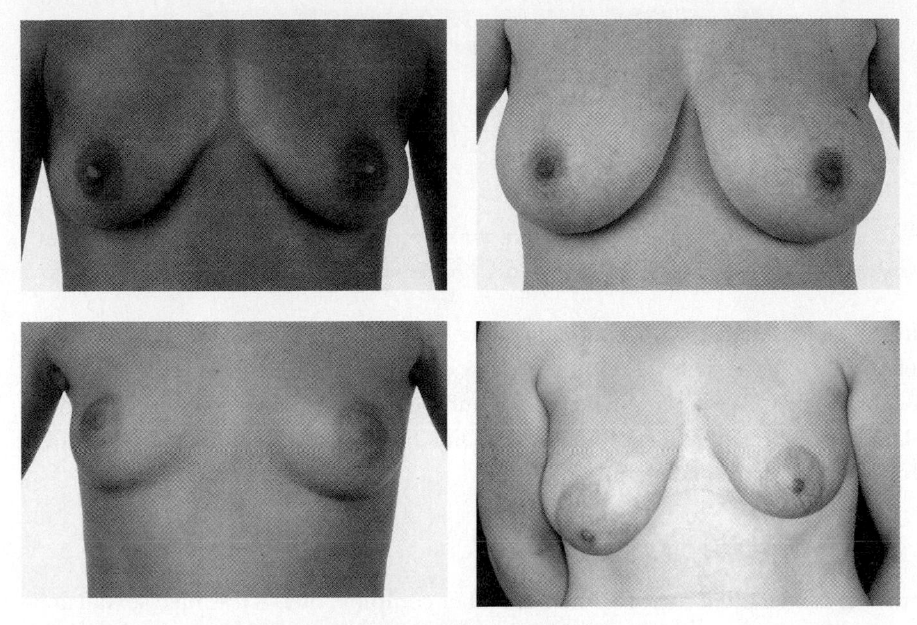
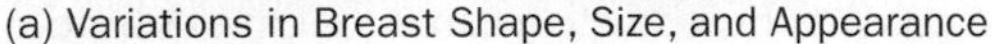

(a) Variations in Breast Shape, Size, and Appearance

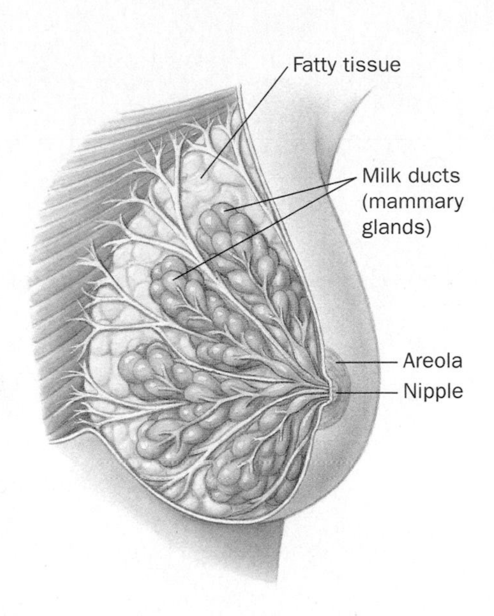

(b) Anatomy of the Female Breast

nipple stimulation regardless of their breast size or shape. Some women do not find breast and nipple stimulation particularly arousing at all, yet some women are able to experience orgasm from nipple stimulation alone. Most women fall somewhere between these extremes.

Just as men's sexuality is culturally (and mistakenly) linked to penis size, women's sexuality is often judged based on the size of their breasts. This preoccupation with breasts in most Western cultures causes many women to worry that their breasts, nipples, or areolas are too small, too large, or "incorrectly" shaped. Women's perception of their own breasts is a major determining factor in their overall body image and self-esteem. As one sign of this, consider the statistics in breast augmentation and reduction surgery. In the United States in 2012, surgeons performed approximately 317,000 breast augmentations (this was the single most common surgical cosmetic procedure) and 113,500 breast reductions were performed (American Society for Aesthetic Plastic Surgery, 2013).

The breasts are one of many human anatomical structures that are prone to cancer. In 2013, in the United States, an estimated 232,340 new cases of breast cancer were diagnosed in women, and 2,240 in men (National Cancer Institute, 2013).

Although relatively rare, breast cancer does occur in men. Peter Criss, drummer and sometimes-lead singer for the band Kiss, went public with his 2008 breast cancer diagnosis and mastectomy treatment. A diagnosis of breast cancer is a very frightening and sometimes devastating event in a person's life.

The good news is that treatments for breast cancer have become increasingly effective and five-year survival rates have steadily increased. Survival rates vary significantly relative to the stage of cancer upon diagnosis. In 2013, the survival rate for early-stage breast cancer was over 90%. Table 2.2 shows the survival percentage at various stages of breast cancer. The drop in survival for later-stage patients

Public awareness of male breast cancer was enhanced when Peter Criss of the rock group Kiss announced that he had breast cancer, which was treated with a mastectomy.

Table 2.2 Five-Year Survival Rate for Breast Cancer in Women by Cancer Stage.

Stage	Five-year Survival Rate
0	100%
I	100%
II	93%
III	72%
IV	22%

SOURCE: American Cancer Society, Breast Cancer Survival Rates by Stage, 2013.

is primarily attributable to the spread of the cancer beyond the breast (American Cancer Society, 2013a). This demonstrates how diagnosing cancer early, through breast self-awareness, regular doctor visits, and mammograms, is a crucial factor in curing it. "Self-Discovery: Breast Self-Awareness" explains the latest overall approach to breast health and breast cancer prevention.

Female Internal Structures

The internal sexual anatomy of women is rather complex, but as for male sexual anatomy, we will limit our discussion here to structures that are essential to understanding the human reproductive process and for various other topics explored throughout this book. Figure 2.14 shows the internal reproductive anatomy of the human female. We will begin our discussion with the vaginal opening and proceed inward, structure by structure, to the ovaries.

The Vagina

vagina

A flexible, muscular canal or tube, normally about 3 to 4 inches in length, that extends into the woman's body at an angle toward the small of the back, from the vulva to the cervix.

The **vagina** is a flexible, muscular canal or tube, normally about 3 to 4 inches in length when a woman is not sexually aroused. The vagina extends into the woman's body at an angle toward the small of the back, from the vulva to the cervix, as shown in part (a) of Figure 2.14. This is where the penis is inserted during penile-vaginal intercourse. When not sexually aroused, the walls of the vagina lie very close together and collapse upon one another along most of its length (the vagina is not an open "tunnel" as many people visualize it and as it is often shown, erroneously, in diagrams). During sexual arousal, the tissues lining the vagina become engorged with blood and secrete a clear, slick fluid along its entire length. This lubrication has evolved in humans to facilitate insertion of the penis for intercourse and reproduction, but it occurs during sexual excitement regardless of the particular sexual behavior a woman is engaged in, including fantasy, masturbation, or same-sex interactions. The vagina is an extremely elastic structure capable of "molding" itself around an object as small as a finger or allowing the passage of something as large as a baby's head. During childbirth, the vagina is often referred to as the *birth canal.* Most of the vagina's nerve endings that respond to sexual stimulation are located in the lower, or outer, one-third; the portion closest to the vaginal opening.

G-spot

In some women, an area of tissue on the anterior (upper) wall of the vagina that, when stimulated, may cause a woman to experience enhanced sexual arousal and more intense orgasms.

Within the vagina of some women is a structure that has become known as the Grafenberg spot or **G-spot** (named after the researcher who first discovered the structure in the 1950s). The function and even the existence of the G-spot have long been controversial. Evidence suggests that it is an area of tissue located about a third of the way in from the opening of the vagina in the anterior vaginal wall (the "upper" wall if the woman is lying on her back). Some women find that this area is sexually responsive and enhances arousal and orgasm when stimulated during sexual activities. A controversy surrounding the G-spot has arisen because not all women report having a G-spot, and researchers are still unclear as to how it functions physiologically.

Figure 2.14 Female Internal Sexual Anatomy

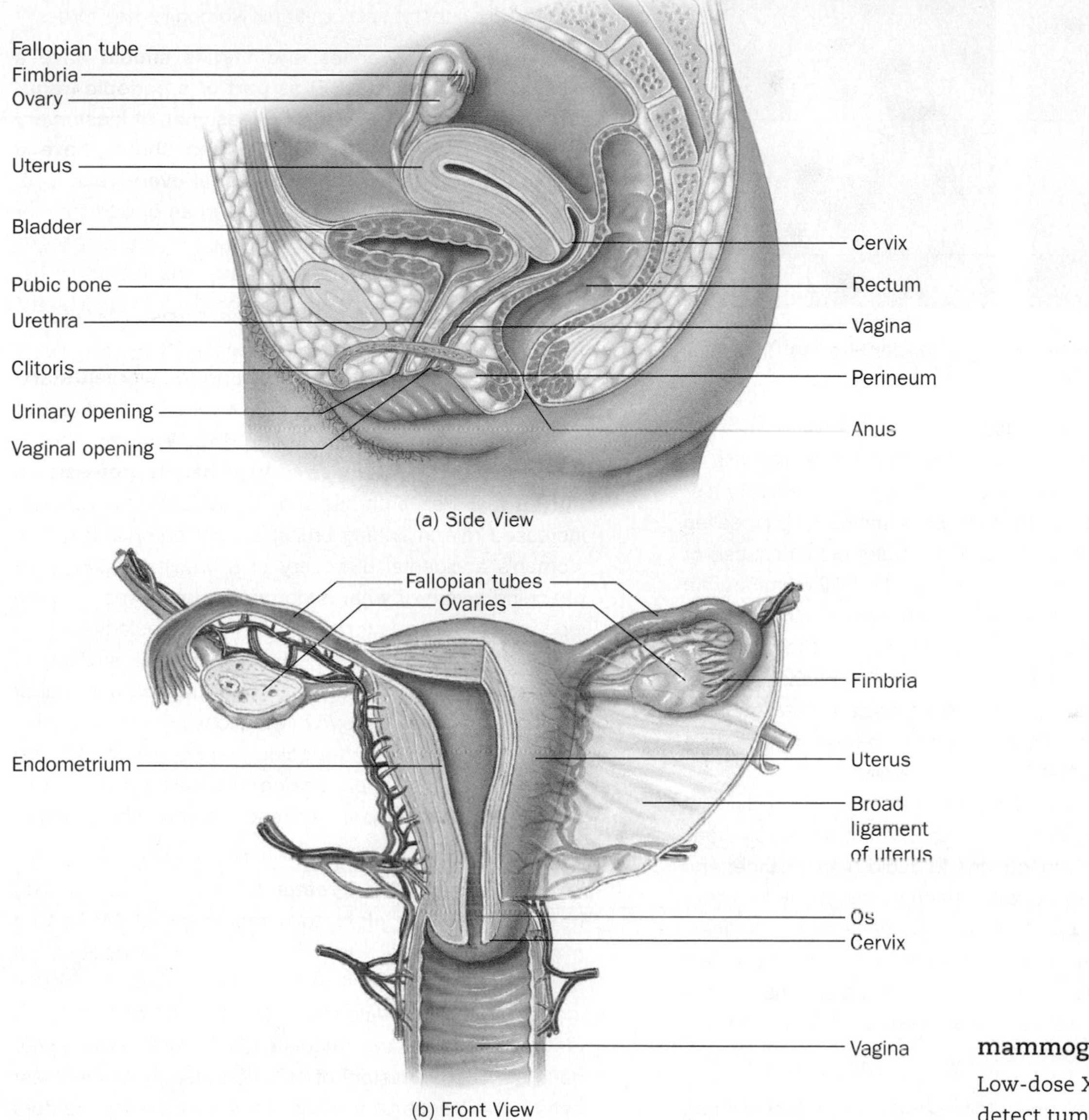

mammogram
Low-dose X-ray of the breast to detect tumors.

Self-Discovery

Breast Health Awareness

Most of you reading this have heard a great deal over the years about breast self-examination and how it is the key to early detection of breast cancer. Recently, however, the focus in medicine has shifted away from routine, monthly breast self-exams to a more overall *breast health awareness* approach. However, health awareness guidelines still include the recommendation for self-examination. The American College of Obstetricians and Gynecologists and the American Cancer Society suggest that women over the age of twenty may perform breast self-examinations as part of an overall program designed to detect breast cancer as early as possible. Breast self-exams are beneficial because a woman can become more familiar with the normal look and feel of her breasts and consequently will be more likely to notice any changes that occur, such as a lump, dimpling, or nipple discharge, which can then be reported to her physician. An excellent resource about breast health and self-examinations, including videos in English and Spanish, may be found at http://ww5.komen.org/BreastCancer/BreastSelfAwareness.html.

The newest guidelines for breast health do not place self-exams at the top of the breast cancer preventive or diagnostic list of recommendations. Medical groups agree that the **mammogram** is the gold standard in detecting breast cancer. Today, mammograms take two forms: a low-dose X-ray of the breasts, and breast MRIs. MRIs are significantly more

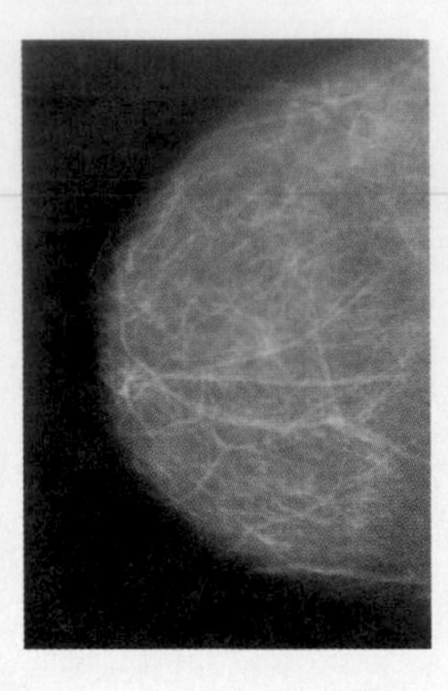
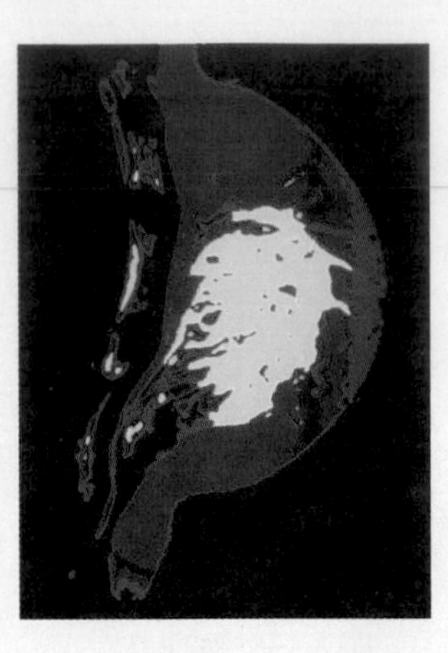

Comparison of two mammogram images: the standard X-ray method and the breast MRI.

expensive than X-ray mammograms, and sometimes they are limited to patients who are deemed to be at a high risk of developing breast cancer (especially if they have a family history of breast cancer or specific gene malformations, called BRCA1 and BRCA2 mutations, that predict breast cancer) or to confirming an unclear X-ray finding. The MRI mammogram provides significantly more detailed images, which increases the chance of detecting breast abnormalities. The downside of this greater resolution, however, is that the chance for false positive results also increases. A false positive breast cancer exam causes undue follow-up exams, biopsies, and emotional distress for the patient and the family.

If all of this seems terribly confusing to you, especially for such an important health issue, you are not alone. Changes in professional recommendations to avoid breast cancer and maintain healthy breasts have changed frequently in recent years, and each new study seems to create new guidelines. As of 2013, recommendations for breast health and testing have become increasingly specific. Here is a summary of the most recent guidelines from the American Cancer Society (American Cancer Society, 2013b):

1. **Women age forty and older should have a screening mammogram every year and should continue to do so for as long as they are in good health.** Current evidence supporting mammograms is even stronger than in the past. In particular, recent evidence has confirmed that mammograms offer substantial benefit for women in their forties.
2. **Women in their twenties and thirties should have a clinical breast exam (CBE) as part of a periodic (regular) health exam by a health professional, at least every three years. After age forty, women should have a breast exam by a health professional every year.** CBE is a complement to mammograms and an opportunity for women and a woman's doctor or nurse to discuss changes in her breasts, early detection testing, and factors in her health history that might make her more likely to have breast cancer.
3. **Breast self-exam (BSE) is an option for women starting in their twenties. Women should be told about the benefits and limitations of BSE. Women should report any breast changes to their health professional right away.** Research has shown that BSE plays a small increased role in finding breast cancer compared with a woman's accidental discovery of a breast lump or simply being aware of what is normal for her breasts. Some women feel very comfortable doing BSE regularly (usually monthly after their period) which involves a systematic step-by-step approach to examining the look and feel of their breasts (see http://www.nationalbreastcancer.org/breast-self-exam for exam guidelines). Other women are more comfortable simply looking and feeling their breasts in a less systematic approach, such as while showering or getting dressed.
4. **Women at high risk (greater than 20% lifetime risk) based on certain risk factors should get an MRI and a mammogram every year.** Women should consult with a physician to determine their level of risk. High risk factors include specific genetic mutations (BRCA1 or BRCA2 as discussed above), an immediate family relative with either gene mutation, a history of radiation therapy of the chest between the ages of ten and thirty, or other risk factors determined by medical consultation and testing.

Various studies have produced anatomical data in attempts to confirm the existence of the G-spot (see Janini & Whipple, 2010). Findings have indicated specific anatomical structures on the wall of the vagina that may define the G-spot, but these areas vary significantly from woman to woman in size, shape, and changes during sexual stimulation (e.g., Battaglia et al., 2010). Experts in the field of anatomy and female sexuality remain in disagreement about the existence and/or function of the G-spot in sexual stimulation and response (Janini & Whipple, 2010). In a surprising recent study of 1,500 women, researchers reported that the G-spot was present in all 1,500 participants and further found that it was a localized spot in 58% and a more generalized area in 42% of surveyed women (Thabet, 2013).

Regardless of solid data proving the existence of the G-spot, if a woman *experiences* a G-spot and feels that it is a sexually sensitive, it is probably not the job of science to disagree with her! However, it is also important to note that women should not be made to feel they are somehow *required* to have one. The role the G-spot plays

Since you asked

8. Once, a condom came off during sex and we couldn't find it. Is it possible for a condom to get "lost" up inside a woman?

in sexual response for some women is discussed in greater detail in Chapter 3, "The Physiology of Sexual Response."

The Cervix

At the inner, or upper, end of the vagina is the **cervix**, which is the narrow, bottom end of the uterus. A woman may feel her cervix by inserting a finger into the vagina (this is facilitated if her hips are flexed, shortening the vaginal length). The cervix connects the vagina to the uterus. The passageway through the cervix, called the **os**, is very small—about as wide as the thickness of the lead inside a pencil, on average—but it becomes slightly larger or smaller on different days during a woman's fertility cycle (and will be somewhat larger in women who have given vaginal birth). Contrary to sexual myths, an object in the vagina such as a tampon, a "lost" condom, or a vaginal contraceptive cannot accidentally travel higher up into the reproductive tract because these objects are all far too large to pass through the cervical os. Conversely, the cervix is capable of expanding greatly, in a process called *dilation*, to allow for childbirth or specific medical procedures (more on this process is found in Chapter 9, "Conception, Pregnancy, and Birth").

The cervix is a relatively common site in the female body for the formation of abnormal cells that may, if not treated, lead to cervical cancer. The medical test that is used to check the cervix for any signs of abnormal cells is called a **Pap test**, or *Pap smear.* In this simple procedure, the vagina is held open with a device called a *speculum* and a few cells are gently swabbed or brushed from the cervix. The cells are then sent to a lab to be examined microscopically for any abnormalities (see Figure 2.15). The American Cancer Society recommends that all girls and women have a Pap test three years after they begin having sexual intercourse, or at age twenty-one regardless of sexual activity. Then they should repeat the test every one to two years, depending on the type of test and a woman's risk factors (American Cancer Society, 2010b). The Pap test is fairly accurate, but it is possible for it to miss abnormal cells or to produce a false positive result. This is why it is important that the test be done on a regular basis. In general, cervical cancer, if it is found, is a very slow-growing cancer and rarely spreads beyond the cervix in the interval between tests (see Figure 2.16).

The primary cause of cervical cancer is a sexually transmitted viral infection called the **human papilloma virus (HPV)**, the virus that causes genital warts. The specific strains of HPV that are associated with cervical cancer are found in a high percentage of women regardless of whether they have ever had noticeable symptoms of genital warts. Of the approximately 12 strains of the virus that cause genital warts, four specific variants cause cervical cancer. Studies have shown the proportion of women infected with cancer-causing strains of HPV is between 11% and 17% (Dunne et al., 2013; Dunne et al., 2007). No cure yet exists for the virus, but two vaccines (*Gardasil* and *Cervarix*) are available; these provide immunity from the most dangerous strains of HPV (for more information on HPV, see Chapter 8, "Sexually Transmitted Infections/Diseases"). Because some infected people may never experience an outbreak of warts and, therefore, be unaware that they are infected, medical organizations have begun calling for routine HPV screening, along with the Pap test for effective cervical cancer prevention and detection (Saslow et al., 2012). Only a small percentage of HPV infections lead to cancer, and Pap tests showing abnormal cells usually do not indicate cancer. However, if a woman tests positive for HPV, she will then have the information she needs to be extra vigilant about her cervical health and to consider testing at more frequent intervals.

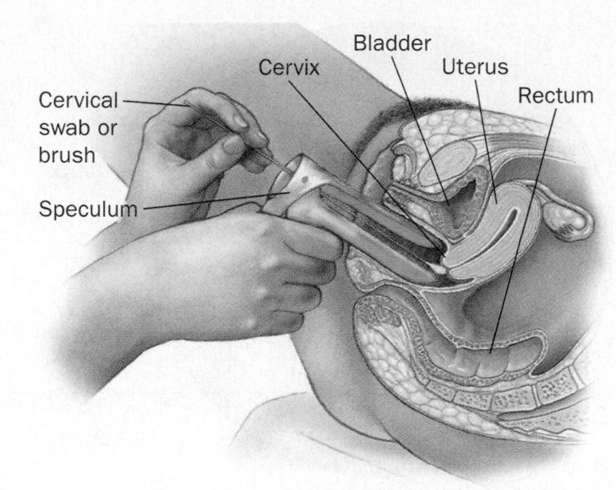

(a) Internal View of the Pap Test Procedure

Figure 2.15 The Pap Test

The Pap test detects abnormal cells in the cervix so that cervical cancer may be prevented or treated.

cervix

The lower end of the uterus that connects it to the vagina.

os

The very narrow passageway through the cervix, from the vagina to the uterus.

Pap test

A routine test in which cells from the cervix are examined microscopically to look for potentially cancerous abnormalities.

human papilloma virus (HPV)

A sexually transmitted virus that is typically characterized by warts in the genital or anal area and that may lead to some forms of cancer; also known as genital warts.

Figure 2.16

(a) Cervix showing abnormal cells.
(b) Normal cervix.

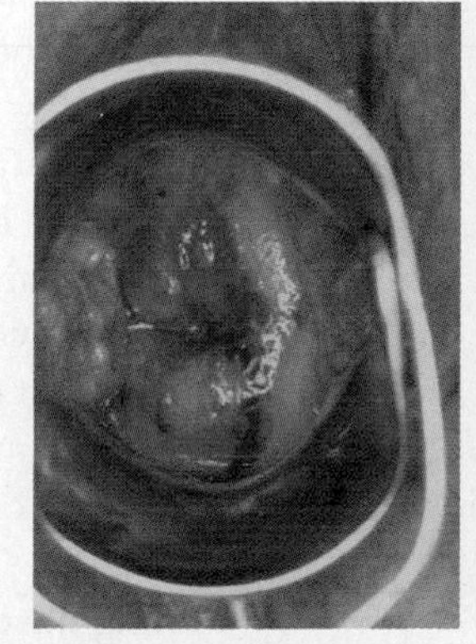

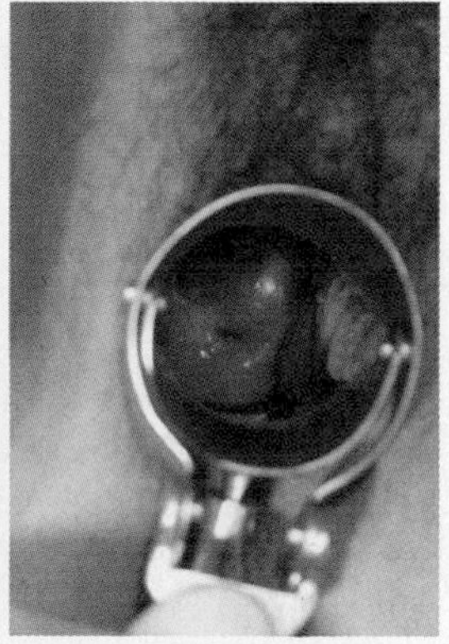

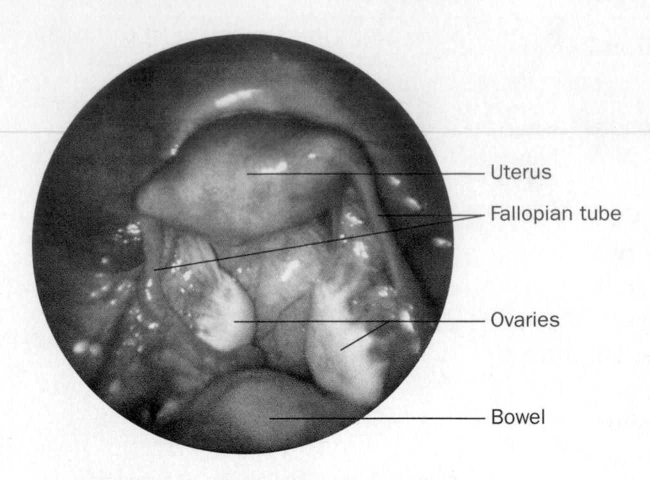

Figure 2.17 The Uterus

The Uterus

The **uterus**, or womb, is the organ in which a fertilized egg implants itself and the embryo and fetus grow (Figure 2.17). The uterus is a very flexible organ with strong muscle fibers that allow it to expand greatly during pregnancy and push the infant out during labor and delivery. In its nonpregnant state, the uterus is about the size and shape of a small pear, about 3 inches long and 1.5 inches at its widest point.

During pregnancy, the capacity of the uterus increases as much as 1,000 times its nonpregnant size. The uterus also undergoes major changes every month during a woman's menstrual cycle. Under the influence of monthly hormonal changes in a woman's body, the lining of the uterus, the **endometrium**, builds up an extra layer of blood and tissue, preparing to receive and nourish a fertilized egg. If fertilization and pregnancy do not occur, the uterus sheds this extra lining, which flows through the cervix and down the vagina as menstrual fluid (menstruation is discussed in greater detail later in this chapter and in Chapter 9, "Conception, Pregnancy, and Birth").

In some women, cells and tissue from the endometrium migrate to outside the uterus and begin to grow in the abdominal cavity, especially on the ovaries, as well as in other areas inside the body. The number of cases of this condition, called **endometriosis**, often referred to simply as "endo," has been increasing over recent decades.

The exact cause of endometriosis is not fully understood, but it is typically a painful and potentially serious health condition for women of reproductive age. The misplaced uterine cells, triggered by a woman's monthly hormone cycle, build up and bleed each month similarly to normal endometrial tissue in the uterus. The symptoms of endometriosis include severe pain before and during periods, pain during sexual intercourse, difficulty conceiving or complete infertility, general fatigue, painful urination during periods, painful bowel movements during periods, and other gastrointestinal problems such as diarrhea, constipation, or nausea (Mao & Anastasi, 2010; Jacobson et al., 2013; Tucker, 2008). Typically, endometriosis becomes worse over time if not treated. Treatment usually involves medications for pain management, hormonal treatment, or laparoscopic surgery to remove the patches of endometrial tissue within the abdominal cavity. If you or someone you know is experiencing symptoms of endometriosis, it is important to seek medical care and treatment sooner rather than later (see http://www.endometriosisassn.org for more detailed information on this condition).

Since You Asked

9. If a Pap test is negativ e, can you be sure you are completely free of disease?

uterus
A very flexible organ with strong muscle fibers where a fertilized egg implants and an embryo and fetus grow, from a few days after fertilization until birth.

endometrium
The tissue lining the uterus that thickens in anticipation of pregnancy and is sloughed off and expelled during menstruation.

endometriosis
A potentially painful and dangerous medical condition caused by endometrial cells migrating outside the uterus into the abdominal cavity.

fallopian tubes
The tubes that carry the female ovum from the ovaries to the uterus and in which fertilization occurs.

The Fallopian Tubes

At the top part of the uterus, two **fallopian tubes** (also sometimes called *ovarian tubes* or *oviducts*) extend from the uterus up to the ovaries, as shown in part (b) of Figure 2.14. Each fallopian tube is about 4 to 5 inches long and very narrow—about 2 millimeters across, or the diameter of cooked spaghetti—for most of its length. At the uterus, the passageway inside each tube is even smaller: no more than the width of a sewing needle. As the fallopian tubes approach the ovaries, they become somewhat wider and branch into many fingerlike tendrils called *fimbriae* (singular *fimbria*) that float next to and drape against the ovary. The fallopian tubes are not attached to the ovaries, but the fimbria move and create currents that draw in the **ovum**, or egg, released by the ovary and direct it down into the fallopian tube.

Fertilization of the ovum by a sperm cell occurs in the third of the fallopian tube nearest the ovary. The newly released ovum is available for fertilization in that section of the fallopian tube for about a day; if no sperm cells are present or successful in penetrating the ovum, pregnancy will not occur during that cycle.

ovum
The female reproductive cell stored in the ovaries; usually, one ovum is released approximately every 28 days between menarche and menopause. The plural is *ova*.

In approximately one out of every 50 conceptions, the fertilized egg becomes lodged in the fallopian tube, implants there, and begins to grow (Crochet et al., 2013). This is called a *tubal pregnancy* and is the most common type of **ectopic pregnancy**, meaning the growth of a fertilized egg outside the uterus (on extremely rare occasions—less than 3%—the fertilized ovum can implant on an ovary or elsewhere in the woman's abdomen). Tubal pregnancies nearly always result in the loss of the embryo and may pose a serious health risk to the mother. If the embryo grows to the point of rupturing the fallopian tube, internal bleeding occurs and, without emergency medical attention, death of the woman may result. Ectopic pregnancy is the leading (although rare) cause of maternal death (0.5%) during the first three months of pregnancy (Irish & Savage, 2008).

ectopic pregnancy
A pregnancy complication in which a fertilized ovum attaches and begins to grow outside the uterus, most commonly in the fallopian tube, which is called a *tubal pregnancy*.

The Ovaries

The **ovaries** are gonads, as are the testicles. In popular usage, the word *gonads* is used as a euphemism for testicles, but a *gonad* is technically a sex organ that produces cells for reproduction. The testicles produce sperm, and the ovaries produce eggs, or *ova* (*ovum* is the singular, meaning "one egg"). However, a major difference between male and female gonads is that the testicles manufacture sperm cells continuously throughout a man's postpuberty life, at the average rate of nearly 1,500 per second (Dell'Amore, 2010). The ovaries, on the other hand, contain their full supply of immature ova at birth—about 400,000 of them. When a girl enters puberty, hormonal changes in her body trigger the maturation of, on average, one ovum each month, a process that will continue for the next 40 years or so. If you do the math, you will see that the average woman will use 400 to 500 eggs during her fertile lifetime, but to help ensure survival of the human species, evolution has given her a huge "backup supply."

ovaries
The female organs that produce sex hormones such as estrogen and progesterone and where follicle cells are stored and mature into ova.

estrogen
The female hormone responsible for regulating ovulation, endometrial development, and the development of female sexual characteristics.

progesterone
The female hormone responsible for the release of ova and implantation of the fertilized egg in the uterine wall.

ovarian cyst
A fluid-filled sac on the surface of the ovary, formed during normal ovulation; sometimes cysts may swell and cause pain and abnormal bleeding.

The ovaries are also responsible for production of the female hormones **estrogen** and **progesterone**. These hormones are responsible for girls developing physically into mature women, inducing changes in the breasts, vagina, and uterus during puberty. They also play crucial roles in the woman's fertility cycle by signaling the woman's body to produce mature eggs, release eggs, and prepare the uterus for a possible pregnancy by building up the lining of the endometrium. The role of and functions of female hormones will be discussed in more detail in the next section and in Chapter 9, "Conception, Pregnancy, and Birth."

The ovaries, like the testicles, are subject to certain medical conditions. One of these is **ovarian cysts**—small, fluid-filled sacs, up to a half-inch across, that grow on one of the ovaries (see Figure 2.18). Usually these cysts are a normal part of ovulation, and they typically shrink and disappear in a month or two (see Chapter 9, "Conception, Pregnancy, and Birth" for more information on the reproductive function of the ovaries). Occasionally, however, a cyst will fail to resolve on its own and may persist and grow larger, or multiple cysts will develop. These may cause abdominal pain, abnormally heavy periods, or more intense menstrual cramping. Hormonal therapies often resolve the problem, but in some cases surgery (laparoscopic) may be needed. Overall, ovarian cysts do not appear to be related to ovarian cancer (Greenlee et al., 2010).

Figure 2.18 Ovarian Cyst

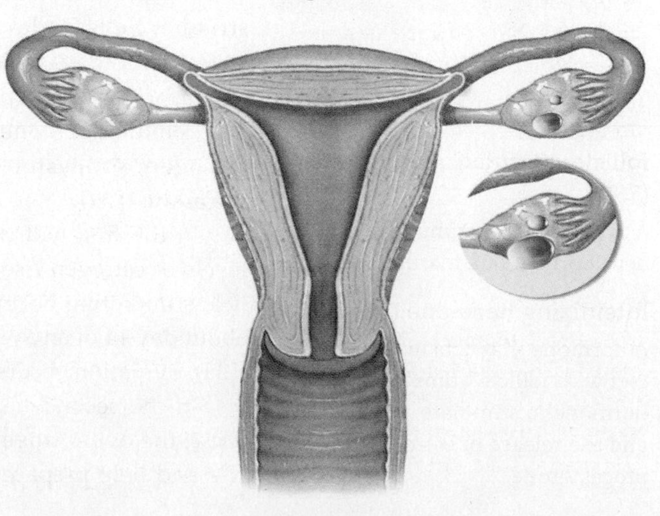

The ovaries are also a potential site in female sexual anatomy for cancer. Ovarian cancer is the fifth most common cause of cancer-related deaths in women (Stolper & Lee, 2008). In 2013, there were approximately 22,240 new cases and 14,230 deaths from the disease (American Cancer Society, 2013c). Ovarian cancer is so deadly primarily because it tends to be symptom-free until it has already spread (metastasized) beyond the ovaries, which it tends to do more easily than many other cancers. Only about 20% of ovarian cancers are diagnosed prior to becoming metastasized. Recently, however, the American Cancer Society has developed guidelines for risks and detection of ovarian cancer (see http://www.cancer.org). Risk factors include a family history of ovarian or breast cancer, a personal history of breast cancer, age (the risk increases with age), and obesity, among others. The presence of the BRCA 1 and/or BRCA 2 gene mutations, discussed earlier in relation to breast cancer, appear to be a predictor of a higher risk of ovarian cancer. The decision to be tested for this gene mutation is based on various other risk factors, and the decision to be tested must be made by each individual woman in careful consultation with her doctor (Evans et al., 2008).

Menstruation

2.3 Explain the onset of menstruation (menarche) and review the problems associated with the menstrual cycle including PMS and PMDD.

When a girl enters puberty, hormonal secretions cause her body to undergo many changes. One of these changes is called **menarche (me-NAR-ky)**, the onset of her **menstrual cycle**. In the United States, menarche occurs on average at about age twelve; however, there is a great deal of variation among girls depending on race and levels of body fat (McDowell, Brody, & Hughes, 2007). At menarche, hormones signal that a girl's body is preparing to begin **ovulation**, the releasing of a mature ovum about once each month. The first sign of menarche is a girl's first period, which is the passing of menstrual fluid from her uterus, through the cervix, and out the vagina. In the beginning, menstruation is likely to be irregular and may occur only once every two or three months. During this initial phase, ovulation may or may not occur. Then, over the next year or two, ovulation and menstruation typically becomes more regular as the young woman settles into her normal menstrual cycle.

menarche
The beginning of menstruation during puberty; a girl's first period.

menstrual cycle
The hormone-controlled reproductive cycle in the human female.

ovulation
The release of an egg, or ovum, from the ovary into the fallopian tube.

The purpose of the menstrual cycle is to create conditions in a woman's body that allow for conception and pregnancy. The menstrual cycle corresponds to a woman's *fertility cycle*, the times during each menstrual cycle when she is more or less likely to be able to conceive. The fertility cycle and menstruation are discussed in greater detail in Chapter 9, "Conception, Pregnancy, and Birth," and Chapter 12, "Sexual Development Throughout Life." However, as these processes are so closely linked to our discussion about anatomy and physiology, we will summarize some of the basics of menstruation here.

The Menstrual Cycle

The menstrual cycle starts on the first day of a woman's period. The average menstrual cycle is 28 days long; however, women's cycles can vary a great deal, from as few as 23 days to 35 days or longer. As noted earlier, regularly fluctuating levels of hormones are responsible for the changes associated with the menstrual cycle. The female hormones involved in the regulation of the menstrual cycle include estrogen, progesterone, **follicle-stimulating hormone (FSH)**, and **luteinizing hormone (LH)**.

follicle-stimulating hormone (FSH)
A hormone that stimulates the development of a mature ovum.

luteinizing hormone (LH)
A hormone that acts in concert with follicle-stimulating hormone to stimulate ovulation and the release of estrogen and progesterone.

In the first half of the menstrual cycle, called the *follicular* or *proliferative phase*, levels of estrogen rise, causing the lining of the uterus to thicken. In response to follicle-stimulating hormone (FSH), an ovum in one of the ovaries starts to mature. At about day 14 of an average 28-day cycle, in response to a surge of luteinizing hormone (LH), ovulation occurs; the egg leaves the ovary and enters the fallopian tube.

In the second half of the menstrual cycle, referred to as the *luteal* or *secretory phase*, the ovum travels through the fallopian tube to the uterus. Progesterone levels rise and help prepare the uterine lining for pregnancy (see Figure 2.19). Conception

Figure 2.19 The Uterus During the Menstrual Cycle

During the first half of the menstrual cycle (a), estrogen causes the lining of the uterus (the endometrium) to grow and thicken. In the second half of the cycle (b), if no pregnancy occurs, the lining of the uterus is shed and excreted in the form of menstrual fluid.

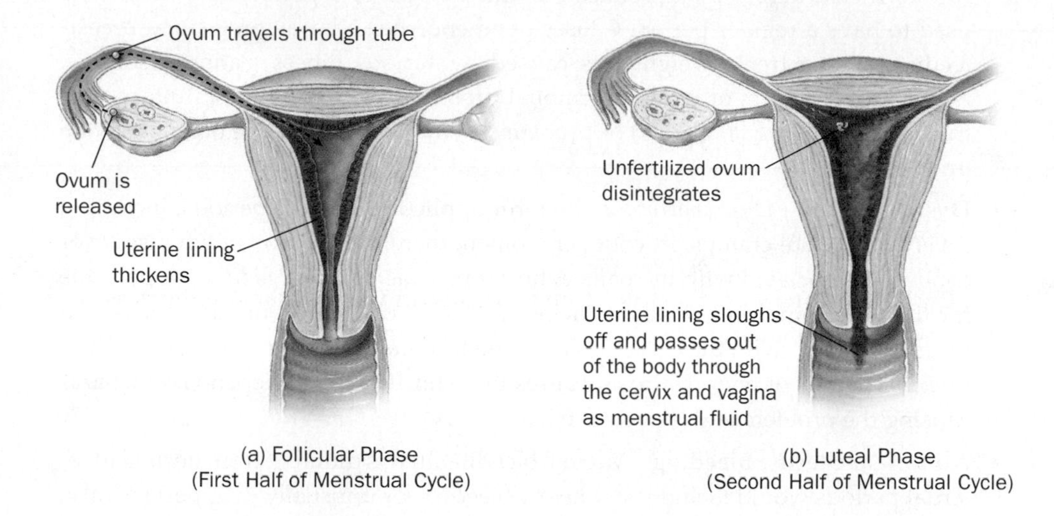

occurs if the egg in the upper region of the fallopian tube is fertilized by a sperm cell. If the fertilized ovum subsequently attaches itself to the uterine wall, pregnancy has begun. If pregnancy does not occur, the ovum moves down the fallopian tube, estrogen and progesterone levels drop, and the thickened lining of the uterus is shed, along with the ovum, during the menstrual period, and the cycle begins again (see Figure 2.20).

Menstrual Problems

During each menstrual cycle, the cells from the thickened uterine lining and extra blood are shed through the cervix and vagina. This flow is referred to as the women's *period*. A woman's period may be quite regular from month to month or may vary considerably. The amount of menstrual fluid expelled may be light, moderate, or heavy. On average, menstrual periods last from three to five days, but anywhere from two to seven days is considered normal.

As is true of all human anatomical functions, menstruation is not free from health and medical difficulties. Usually these problems are mild and little cause for concern, but some women encounter various difficulties with their periods that may require

Figure 2.20 The Phases of the Menstrual Cycle: Follicular Phase and Luteal Phase

SOURCE: Courtesy of Care Pharmaceuticals, NSW, Australia.

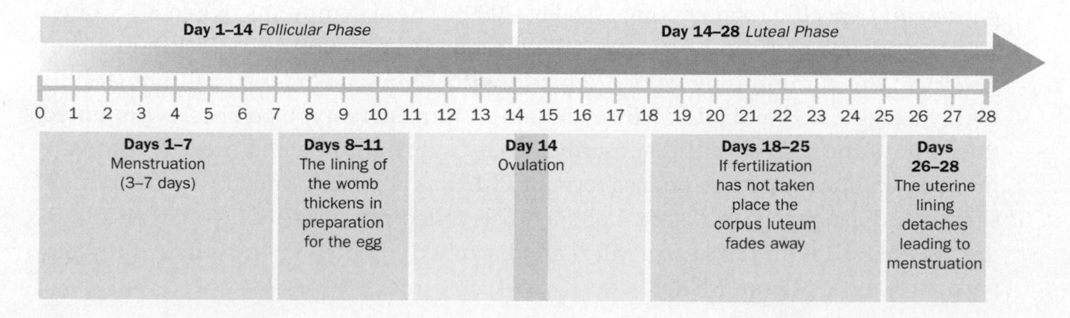

medical attention (see "In Touch with Your Sexual Health: Menstrual Problems: When to Call the Doctor"). Problems that can occur in relation to menstruation include (see U.S. Department of Health and Human Services, 2010):

- **Amenorrhea** This term describes the absence of a period in women who haven't started menstruating by age sixteen or the absence of a period in women who used to have a regular period. Causes of amenorrhea include pregnancy, breastfeeding, and extreme weight loss caused by serious illness, eating disorders, excessive exercising, or stress. Hormonal problems (involving the pituitary, thyroid, ovary, or adrenal glands) or problems with the reproductive organs may be involved.
- **Dysmenorrhea** *Dysmenorrhea* is the term applied to painful periods, including severe menstrual cramps. In younger women, there is often no known disease or condition associated with the pain. A hormone called *prostaglandin* is responsible for the symptoms. Some pain medicines available over the counter, such as ibuprofen, can help with these symptoms. Sometimes a disease or condition, such as uterine fibroids or endometriosis, causes the pain. Treatment depends on what is causing the problem and how severe it is.
- **Abnormal Uterine Bleeding** Vaginal bleeding that is different from normal menstrual periods would include very heavy bleeding or unusually long periods (also called *menorrhagia*), periods too close together, and bleeding between periods. In adolescents and women approaching menopause, hormone imbalance problems often cause menorrhagia and irregular cycles. Sometimes this is called *dysfunctional uterine bleeding* (DUB). Other causes of abnormal bleeding include uterine fibroids and polyps. Treatment for abnormal bleeding depends on the cause and may include medication or surgery.
- **Toxic Shock Syndrome** Women who use tampons to absorb the menstrual flow are advised to follow specific guidelines to avoid toxic shock syndrome, or TSS. TSS is a rare but potentially deadly bacterial infection that has been associated with tampon use. Symptoms include high fever, muscle aches, diarrhea, dizziness or fainting, sunburn-like rash, sore throat, and bloodshot eyes. Using any kind of tampon—cotton or rayon of any absorbency—puts a woman at slightly greater risk for TSS than using menstrual pads. To minimize the risk of TSS, the Food and Drug Administration (FDA) recommends that women change tampons every four to eight hours, use pads sometimes if possible, and be aware of the signs of toxic shock syndrome.

Premenstrual Syndrome

The cycling of hormones in a woman's body—in addition to forming the basic physiological foundation for fertility, conception, and pregnancy—may also affect some women emotionally and psychologically. The best known of these effects is a set of symptoms that may occur during the days leading up to the start of a woman's period, which include dysphoria (meaning a general sense of feeling unwell), mood swings, depression, irritability, tension, aggression, fatigue, headaches, breast soreness, abdominal cramping, backache, water retention (sensations of bloating), constipation, and specific food cravings (Daley, 2009). These symptoms, when associated with a woman's hormonal cycle, are called **premenstrual syndrome (PMS)**. In North America, the common image of a woman with PMS as a "menstrual monster" is highly exaggerated and false. However, this cultural stereotype persists, is promoted throughout the media, and is subscribed to by many women and men. A relatively rare but significantly more intense form of PMS has been identified by health professionals and is now an official clinical diagnosis called **premenstrual dysphoric disorder**, or **PMDD**. Both PMS and PMDD are discussed in detail in "In Touch with Your Sexual Health: PMS or PMDD?"

premenstrual syndrome (PMS)

A set of symptoms that may occur during the days just before and during the start of a woman's period, which includes irritability, depressed mood, and feelings of physical bloating or cramping.

premenstrual dysphoric disorder (PMDD)

A significantly more intense and debilitating form of PMS.

In Touch with Your Sexual Health

Menstrual Problems: When To Call The Doctor

Although most menstrual difficulties faced by girls and women, while uncomfortable, do not require medical attention, the following is a list of difficulties serious enough to seek medical attention. Consult with your health care provider if you experience any of the following (U.S. Department of Health and Human Services, 2010):

When to Call a Doctor

- You have not started menstruating by the age of fifteen
- You have not begun to menstruate within three years after the beginning of breast development
- Your period has suddenly stopped for several months
- You bleed for more days than usual or your period has become more irregular than usual
- You bleed excessively (need more than one pad or tampon every hour or two)
- You suddenly feel sick after using tampons
- You bleed between periods (more than just a few drops)
- You have severe pain before or during your period.

In Touch with Your Sexual Health

PMS or PMDD?

Almost everyone knows about premenstrual syndrome, or PMS. The old joke goes something like this: "Why does it take five women with PMS to change a light bulb? Because it just *does*! Now leave me alone!" Funny or not, the point of the joke is that a woman with PMS may be tense and irritable. But that's far from the whole story. For some women, physical and emotional changes accompanying menstruation can be quite serious and debilitating.

In 2013, the American Psychiatric Association, in its *Diagnostic and Statistical Manual,* included a new diagnosis termed *premenstrual dysphoric disorder*, or *PMDD*. This is distinguished from the more commonly known PMS in terms of the number and severity of symptoms. It is estimated that as many as 75% of women suffer some symptoms of PMS during their fertile years, but PMDD only affects between 3 and 8% of women in this age group. A diagnosis of PMDD applies only to those with the most serious and debilitating symptoms. A PMDD diagnosis requires at least five of the eleven listed criteria or symptoms below and must include at least one depression symptom (APA, 2013):

- Fatigue or low energy
- Loss of interest in daily activities and relationships
- Feelings of sadness or hopelessness, possible suicidal thoughts
- Feelings of tension or anxiety
- Feeling out of control
- Food cravings or binge eating
- Mood swings with periods of crying
- Panic attacks
- Irritability or anger that affects other people
- Physical symptoms, such as bloating, breast tenderness, headaches, joint or muscle pain
- Problems sleeping
- Trouble concentrating

How is PMDD treated? Numerous effective treatments are available for PMDD (many of which are effective for milder forms of PMS as well). The treatment chosen by a woman and her doctor will depend on many issues relating to her specific symptoms, including the seriousness of the overall symptoms, the individual patient's profile, and the treatment preferences of the doctor. The most effective treatment currently in use is one of the *selective serotonin reuptake inhibitors* (SSRIs). These drugs, with trade names such as Prozac, Zoloft, Paxil, and Celexa, were originally developed to treat depression, but some of them appear to be effective in relieving the symptoms of PMDD as well. As the ovaries produce varying levels of hormones during a woman's normal reproductive cycle, they trigger changes in the brain's balance of neurotransmitter chemicals, primarily serotonin. Numerous studies have confirmed that most of the SSRIs can significantly reduce both the psychological and the physical symptoms commonly associated with PMDD for many women (Marjoribanks et al., 2013). These drugs relieve the symptoms of PMDD for 60 to 90% of PMDD sufferers, with few negative side effects for most users (Cunningham et al., 2009). Other treatments that have been shown to be at least somewhat beneficial include increased dietary changes (especially avoiding sugar, caffeine, and alcohol; decreasing dietary fat; and increasing daily calcium intake), various relaxation and stress-reduction strategies, hormone therapy, and regular exercise (Kelderhouse & Taylor, 2013).

The primary message today about PMS and PMDD symptoms is that in virtually all cases they can be treated successfully so that women need not suffer the potentially severe effects of these conditions. If you find yourself cycling through the symptoms discussed in this section, talk to your health care professional. Effective treatments are available.

Women usually continue having regular periods until they pass through the menopause process, when ovulation and menstruation gradually cease. On average, **menopause** has been thought to occur around the age of fifty-one, but it is not a sudden change. Menopause-related physiological alterations referred to as **perimenopausal changes** occur gradually over a time frame of ten years or more. These changes may begin in a woman's early to mid-forties and extend into her mid-fifties (see Chapter 12, "Sexual Development Throughout Life," for a more detailed discussion of menopause). Some women may experience menopausal changes earlier in life due to surgery, illness, or medications that interrupt her normal hormonal cycle.

menopause
The normal, gradual change in a woman's life, typically occurring between age forty-five and fifty-five, when the ovaries produce a decreasing amount of female hormones and menstrual periods cease.

perimenopausal changes
The physical and psychological changes many women experience during the decade leading up to menopause.

Since You Asked

10. I get really bad PMS every month. Is this normal, and what can I do to stop it?

Your Sexual Philosophy

Sexual Anatomy

As discussed in Chapter 1, your personal sexual philosophy is about making choices that are right for you and planning ahead so that you are in control of your sexual life. The development of a healthy sexual philosophy depends on your willingness to become educated and comfortable about all areas of human sexuality. This comfort level begins with an accurate and reasonably complete working knowledge of the sexual anatomy of both sexes. As you will see, sexual anatomy relates to virtually every topic throughout this book, so it follows that it also relates to nearly all facets of your life as a sexual person. Many people are uneasy, embarrassed, or even ashamed to study and learn about the sexual aspects of the human body. Overcoming discomfort and being open to developing and maintaining an accurate working knowledge of sexual anatomy will assist you in living a healthy and satisfying life as a sexual person.

Most people want and expect to have a satisfying and fulfilling sexual life with their chosen partner or partners. Often one barrier to that fulfillment is a lack of knowledge about sexual anatomy. When both partners are educated about their bodies, their ability to please each other is enhanced. Knowledge of our sexual bodies allows for effective, sensitive, and meaningful sexual communication. Furthermore, this knowledge may be even more important when sexual problems arise relating to sexual responding or sexual health. Identifying, discussing, and resolving sexual difficulties or illnesses are greatly facilitated by an understanding of the physical structures that may be involved.

These are just a few of the many reasons to make an understanding of sexual anatomy a central feature of who you are as a sexual person—your sexual philosophy. That so many people resist and avoid studying sexual anatomy is rather ironic. After all, your sexual body, as a primary force in reproduction, sexual pleasure, and human intimacy, is arguably your most important anatomical system of all.

Have You Considered?

1. Imagine that you have a close friend who confides in you that he is worried and embarrassed because he thinks his penis is too small. Discuss at least three responses you could give him that might help him feel better about himself.
2. If you were the parent of a new baby boy, would you have him circumcised? Give three reasons for your answer.
3. The practice of female genital mutilation (FGM) is deeply ingrained in certain cultures throughout the world. Do you feel

that countries such as the United States that disapprove of the practice have a right to interfere in those cultures and attempt to put a stop to it? Explain your answer.

4. Many groups around the world are working to stop FGM. If they asked for your help, what three strategies might you suggest to reduce the incidence of FGM?

5. What help and advice would you give to a good friend who confides in you that she suffers from debilitating PMS every month?

6. Breast augmentation is a thriving industry in the United States. Discuss your opinion of the procedure and how you feel about women who choose to undergo breast enlargement for cosmetic reasons.

Summary

Historical Perspectives

Anatomy in the Dark Ages

- Early conceptualizations of sexual anatomy were highly inaccurate. False beliefs included the notion that the male and female bodies were the same, but that the male's genitals had simply been "pushed out;" the uterus was an internal scrotum and a girl could be turned into a boy by a sudden jarring motion, forcing the genitals to pop out; the uterus consisted of two halves (boys were born from the right half and girls from the left); and the source of sperm was the male brain.

The Male Sexual Body

- Penis size has been the central focus of male anatomy for centuries. Despite, or perhaps because of, this preoccupation among many men throughout history, most people are surprised to find that the length of the average erect penis is about 5.5 inches and varies little among the majority of men. Self-reports of penis size are larger but are probably not accurate due to inflated self-reporting or faulty measurement.
- Sperm start their journey in the testicles, which produce sperm cells and testosterone. Sperm cells mature and are stored adjacent to the testicles in the epididymis before traveling through both vas deferens into the man's body, where they mix with semen for ejaculation. Semen is produced primarily by the seminal vesicles and the prostate gland.

The Female Sexual Body

- Female external sexual anatomy, the vulva, is more complex than that of the male. The female genitals normally vary in size, shape, and color. The clitoris is a significantly larger and more complex organ than once thought; it extends several inches inside the woman's body. Women are more prone than men to urinary tract infections (UTIs) due to the shorter urethral length and its proximity to anal bacteria. Women who experience frequent UTIs should become knowledgeable about prevention techniques. A vaccine to immunize women against UTIs is under development.
- Breast augmentation is the most common cosmetic surgery procedure in the United States. Breast augmentation chosen for purely cosmetic purposes, as opposed to reconstruction of the breast following surgery or injury, reflects the value Western cultures place on that particular aspect of female sexual anatomy.
- Female internal sexual anatomy has evolved to optimize the odds of pregnancy and continuation of the human species. The passageway between the vagina and the uterus is the cervix. Certain strains of the human papilloma virus (HPV) that cause genital warts have been found to be the primary cause of cervical cancer. Many doctors and health organizations are recommending that HPV screening be done in conjunction with the yearly Pap test. Vaccines now exist that prevent contraction of the strains of HPV that are associated with cervical cancer.
- The ovaries produce estrogen and progesterone and typically release one mature ovum during each fertility cycle of, on average, 28 days from puberty until the end of menopause.

Menstruation

- A woman's menstrual cycle typically begins between the ages of ten and fifteen (the average is about twelve) and ceases during menopause in her late forties or early fifties.
- For conception to occur, the egg, or ovum, must be fertilized in the upper third of the fallopian tube. If

the fertilized egg implants successfully in the uterine wall, pregnancy begins. If neither of these events occurs, the ovum is expelled along with the uterine lining during menstruation.

Your Sexual Philosophy

Sexual Anatomy

- Accurate knowledge plays an important role in sexual health and satisfaction. A clear understanding of both male and female sexual anatomy enhances a couple's ability to communicate about sexual issues, which in turn helps promote a satisfying and fulfilling sexual life. In addition, feeling comfortable with your own sexual anatomy contributes to maintaining sexual health and wellness and seeking medical attention when it is needed.

Chapter 3

The Physiology of Sexual Response

Learning Objectives

After you read and study this chapter you will be able to:

3.1 Describe how biology and psychology interact to differentiate human sexuality from that of non-human animals.

3.2 Describe sex-research pioneers Masters and Johnson' four-stage theory of human sexual response (also called the EPOR model).

3.3 Explain the approaches proposed by Kaplan; Reed; Janssen and Bancroft, and the "The New View" to explain the process of human sexual response and how they vary from the EPOR model.

Since You Asked

1. How can we know for sure what happens during sex? You can't exactly study people while they're doing it, right? (see page 74)
2. Are aphrodisiacs for real? Which ones really work? (see page 77)
3. What is "pre-cum," and where does it come from? (see page 80)
4. I'm a woman. During intercourse, why does it take so long for me to have an orgasm? Is it unusual for women not to have orgasms at all during sex? (see page 82)
5. Do orgasms feel the same to men and women? Are men's stronger or more intense? (see page 83)
6. Are too many orgasms dangerous to your health? How many are too many? (see page 83)
7. Exactly what and where is the "G-spot"? My boyfriend and I can't seem to find it! (see page 84)
8. I've heard that some women ejaculate like men. Is this true? How is it possible? (see page 85)
9. After I have an orgasm, I can get another erection pretty fast, but I can't seem to orgasm again. Is this normal or a problem of some kind? (see page 88)
10. It seems to me that men and women are pretty different when it comes to sexual feelings. How can they be described in the same ways? (see page 93)

Your sexual anatomy is inextricably linked to sexual activities, sexual feelings, sexual interactions, sexual pleasure, and sexual intimacy. For a full understanding of the male and female sexual bodies, we must turn our attention away somewhat from the anatomical parts themselves (which are explored in Chapter 2) and refocus on the *physiology* of sex—on how the sexual structures function as a person responds physically and emotionally to sexual stimulation. In other words, *anatomy* and *physiology* are not the same, although many people use the words interchangeably. To explain the difference, let's use another, nonsexual anatomical system as an example. The *anatomy* of your digestive system includes the mouth, esophagus, stomach, intestines, anus, etc. Knowing that, however, tells us nothing about how digestion actually *works* and what role each of those structures plays in the digestive process. To understand that, we would have to study the *physiology* of digestion. And so it is with our sexual anatomy.

In this chapter, we move to a new and more personal level of appreciation of the unique complexities of human sexuality through an examination of the physiology of the body's sexual responses. We will examine what happens to the body before, during, and after sexual activities (with or without a partner). We will examine various researchers' theories and conceptualizations of sexual response. Through these theories, we will explore the many physical changes that accompany sexual desire, excitement, orgasm, and resolution (the body's process of returning to its unaroused state), and we will clarify the similarities and differences in the responses of men and women. Keep in mind that these responses are fundamentally the same whether we are discussing opposite-sex or same-sex partners and during any sexually arousing activities.

To start us off, let's go back nearly fifty years, to a time when one of the most important studies in the history of sex was being developed in a research center in Saint Louis, Missouri. Much of what we know today about sexual anatomy and response we owe to the pioneering studies of William Masters and Virginia Johnson.

Historical Perspectives

Sexual Pioneers

Prior to the 1960s, the definitive published work on the sexual behavior of humans consisted primarily of large-scale surveys, most notably those by Alfred Kinsey in the late 1940s and early 1950s. The renowned "Kinsey Reports" offered a rare glimpse into the sexual activities of humans. Kinsey and his associates (Kinsey, Pomeroy, & Martin, 1948; Kinsey et al., 1953) surveyed thousands of men and women about their sexual behavior and attitudes and reported on topics ranging from frequency of intercourse and masturbation habits to homosexual experiences. Although the research is old, data from the Kinsey Reports are still cited today as a source of statistical information about sexual behavior. Kinsey's research provided information about what people *say* they do sexually, but a conspicuous information gap remained about how the human body functions anatomically when we engage in sexual behavior (Kinsey's work is discussed more fully in Chapter 1, "Studying Human Sexuality").

In the 1960s, William H. Masters (1915–2001) and Virginia E. Johnson (1925–2013) changed all that. Their research into human sexual physiology and response (beginning with Masters' earlier research in the mid-1950s) is largely credited with revealing a great deal of our knowledge about how our sexual bodies function during sexual activity (see Hock, 2009).

As the 1960s began, the United States launched into what has become known as the *sexual revolution*, a time of sweeping social changes marked by the introduction of the contraceptive pill, a greater openness about sexuality, and the "free love" movement. This provided an ideal social environment for candid and definitive scientific exploration of our sexuality that would not have been possible previously. Until the 1960s, lingering Victorian messages that sexual behavior is something secretive, hidden, and certainly not a topic of discussion, much less study, would have precluded virtually all support, social and financial, from Masters and Johnson's research. But as men and women began to acknowledge more openly that we are sexual beings, with strong sexual feelings and desires, the social climate became ready not only to accept the explicit research of Masters and Johnson but to demand it. Statistics were no longer enough. Culturally, people seemed ready to learn about their sexual bodies and their physical responses to sexual stimulation.

Focus on Your Feelings

Topics relating to how the human body responds during sex can trigger an immense range of emotional responses, both positive and negative. These emotions are probably largely why most people tend to avoid sexual response as a popular topic of casual conversation. "Hi Lucy! Orgasms OK lately?" "Super, Pete, how about you?" Not a likely conversation starter, right? It's probably safe to say that people are generally more uncomfortable studying and discussing the *functioning* of their sexual anatomy than learning about the parts themselves as explored in Chapter 2. In some ways, this discomfort makes complete sense; after all, your sexual activities are intensely personal and are really nobody's business but your own.

Studying and thinking about your own or your partner's sexual responses may evoke pleasurable, happy, intimate, loving, or arousing feelings. There's certainly nothing wrong with this; sexual fantasy is the most common human sexual activity! On the other hand, issues relating to sexual functioning may also prompt negative emotions such as guilt, anxiety, or embarrassment. The particular emotions you experience while reading this chapter will depend on many personal factors, including your sexual experiences, your family's attitudes about sex as you were growing up, past sexual abuse or trauma, and your overall comfort level with yourself as a sexual being. No matter what emotions may arise from the material in this chapter, rest assured that they are almost certainly normal and that you are not the only one feeling them. The information you gain in these pages will help enhance the positive feelings about your sexual life and help you to find ways of overcoming the negative ones.

Our sexual anatomy plays an important role in sexual pleasure and intimacy.

It was within this social context that Masters and Johnson set about studying human sexual anatomy and physiology. Their early work culminated in the publication of their first book, *Human Sexual Response* (1966). Although, as you will read shortly, their work has created its share of controversy (various revisions or replacement theories have been suggested, and more recent research has drawn some of their findings into question), Masters and Johnson's discoveries of over forty years ago continue to form a basic foundation for helping us to understand the physiology of sexual response. In their research, they limited their participants to heterosexual couples and singles, but most of their findings apply to couples of all sexual orientations.

To study in detail the physiological responses during sexual activity and stimulation, Masters and Johnson's team of researchers developed elaborate methods of measurement and observation. These included standard measures of physiological response, such as pulse, blood pressure, and rate of respiration. In addition, the researchers observed and recorded specific sexual responses. For this, the "sexual activity of study subjects included, at various times, manual and mechanical manipulation, natural coition [intercourse] with female participants in supine (lying on the back), female superior (female on top), or knee-chest position, and, for many female subjects, artificial coition in the supine or knee-chest positions" (Masters & Johnson, 1966, p. 21). In other words, subjects were sometimes observed and measured while having heterosexual intercourse in various positions or during masturbation, either manually or with mechanical devices specially designed to allow for the clear recording of changes in external and internal sexual anatomy, such as erection, vaginal lubrication, and orgasmic contractions. In their initial studies, Masters and Johnson's team of researchers observed an estimated 10,000 complete sexual response cycles.

Masters and Johnson's contributions to our understanding and study of human sexuality is difficult to overestimate. For nearly thirty years following the publication of *Human Sexual Response*, Masters and Johnson continued to study human sexual response and apply their findings toward helping people achieve sexual fulfillment. Four years after the publication of their first book, they released *Human Sexual Inadequacy* (1970), which took their earlier research and applied it directly to helping people with sexual problems (this is the subject of Chapter 7, "Sexual Problems and Solutions"). Masters and Johnson's decades of research and writing on various topical issues in the field of sexuality may be seen in the titles of some of their subsequent books: *The Pleasure Bond* (1974), *Homosexuality in Perspective* (1979), *Masters and Johnson on Sex and Human Loving* (1986), and *Crisis: Heterosexual Behavior in the Age of AIDS* (1988).

Since You Asked

1. How can we know for sure what happens during sex? You can't exactly study people while they're doing it, right?

An interesting side note in this story is that during their early years of research collaboration, William Masters and Virginia Johnson were married and remained together for more than 20 years before their divorce in 1993. Johnson left the research clinic prior to their divorce, and Masters retired in 1995, six years before his death in 2001. Their last published work together was *Heterosexuality* (Masters, Johnson, & Kolodny, 1994).

The relationship and research of William Masters and Virginia Johnson was made into a fascinating 2013 docudrama, *Masters of Sex*, for the Showtime cable network. Although fictionalized, the facts of their research and the broad outlines of their relationship has been preserved in this TV series.

THEORIES OF SEXUAL RESPONSE

Biology, Psychology, and Human Sexual Response

3.1 Describe how biology and psychology interact to differentiate human sexuality from that of non-human animals.

For most nonhuman mammals, sexual behavior is governed primarily, or in most cases exclusively, by the biological forces of mating and reproduction. When the female of many nonhuman species is fertile, she sends out strong signals to the males of the species, indicating her readiness to mate. The males get the message clearly and respond to it with mating behavior. You know this if you have ever seen the behavior of male dogs around a female during *estrus*, when she is "in heat." At other times, however, their interest in sexual behavior is minimal or nonexistent.

Although humans are animals, too, and many aspects of our sexual responses rest on shared biological foundations, numerous differences exist between humans and other animals. Our willingness or desire to engage in mating behavior is not clearly linked to the female's fertility cycle to ensure reproduction, and in gay and lesbian sexual interactions, this consideration is virtually nonexistent. Compared to most non-human animals, our reasons for engaging in sexual behavior are far more varied and complicated because *psychological* processes play a role at least as great as, or arguably greater than, biology.

Why do humans become sexually aroused? In addition to any impulse toward reproduction that may exist, the reasons may include feeling pleasure, giving pleasure, expressing feelings of closeness and love, relieving stress, feeling valued by another person, expressing how much another person is valued, feeling more dominant or submissive, and many other reasons that you can probably cite. These are not purely biological reasons for sexual desire; they are, essentially, psychological reasons and, as such, are fundamentally *human*. As we discuss what we know about the physical side of human sexual response, we must always keep in mind how intricately connected it is to our psychological triggers and motivations (Bancroft, 2002; Trimble, 2009).

Masters and Johnson: The Excitement-Plateau-Orgasm-Resolution (EPOR) Model

3.2 Describe sex-research pioneers Masters and Johnson' four-stage theory of human sexual response (also called the EPOR model).

Although carried out fifty years ago, Masters and Johnson's observations and conceptualizations of how humans respond physically to sexual stimulation continue to form the basis of most discussions of human sexual response. Through their extensive observations of people engaged in various sexual activities, Masters and Johnson's team of researchers found certain predictable consistencies in the human body's reactions to sexual stimulation. Although we are all sexually unique in terms of the choices we make about sexual behaviors and what kinds of stimulation each of us finds most exciting, our physical patterns of sexual response are much more similar than they are different.

To facilitate explanations of how our bodies change during sexual stimulation, Masters and Johnson (1966) divided the process into four phases of sexual

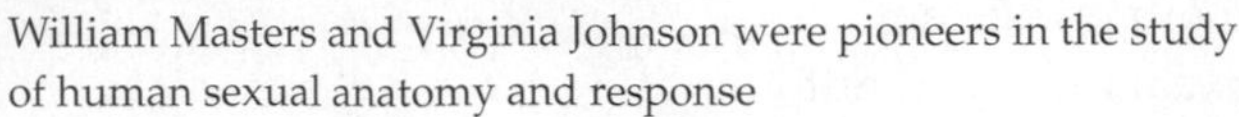
William Masters and Virginia Johnson were pioneers in the study of human sexual anatomy and response

Masters and Johnson as portrayed by Lizzy Caplan and Michael Sheen in Showtime's docudrama, "Masters of Sex"

EPOR model

Masters and Johnson's approach to explaining the process of sexual response, encompassing four arbitrarily divided phases: excitement, plateau, orgasm, and resolution.

response—*excitement, plateau, orgasm,* and *resolution*—called the **EPOR model**. They never intended for these four phases of sexual response to be conceptualized as four separate and distinct events. Masters and Johnson were careful to explain that their four-part model was designed to facilitate a meaningful and understandable description. As they readily noted, it is fundamentally impossible to identify exactly when excitement stops and plateau begins or the precise instant orgasm starts or ends. Sexual response is a much more seamless process. However, the four-phase model was easy to understand; it made sense for many people when they thought about their own and others' subjective sexual experience.

To this day, most people study sexual response against the backdrop of Masters and Johnson's four-phase model. Keep in mind that sexual responses have evolved in humans to be pleasurable, to feel good and thus (in evolutionary terms) enhance the odds for reproduction and the continuation of the species. But these physical responses do not occur only in the context of male-female intercourse, but in response to any form of sexual stimulation, whether it is a man and woman having intercourse; a same-sex couple making love; a couple sharing manual, oral, or anal sex; an individual masturbating; or even an erotic dream.

Masters and Johnson proposed that human sexual response was more understandable when conceptualized in four phases: excitement, plateau, orgasm, and resolution. Figure 3.1, adapted from their theory, shows how the female response cycle tends to be

Figure 3.1 Masters and Johnson's Four-Phase Model of the Sexual Response Cycle

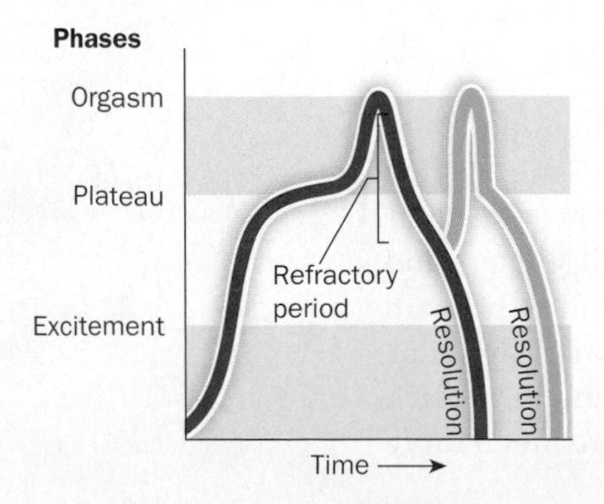

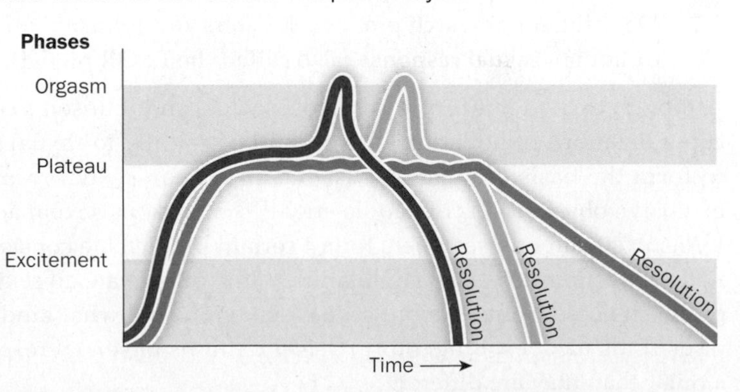

more varied than that of the male. Men generally progress fairly predictably through the four stages (red line) and require some time between orgasms, called a refractory period, before another orgasm (green line) is possible. In contrast, women may follow a similar pattern (red line) or may experience multiple orgasms without a refractory period (green line) or, as is quite common during heterosexual intercourse, may progress from excitement to plateau to resolution without experiencing an orgasm.

excitement phase
The first phase in the EPOR model, in which the first physical changes of sexual arousal occur.

plateau phase
The second phase in the EPOR model, during which sexual arousal levels off (reaches a plateau) and remains at an elevated level of excitement.

Arousal: The First Two Stages

Masters and Johnson referred to the beginning stage of the sexual response cycle as the **excitement phase**. This, they suggested, was followed by a stage they referred to as the **plateau phase**. Later, however, other researchers have argued convincingly that the plateau phase was probably an unnecessary delineation and could be more accurately described as an extension of excitement (e.g., Levin & Riley, 2007; Robinson, 1976). Today, the excitement and plateau stages are seen as accounting for a unified event that might be called "sexual arousal" (e.g., Angel, 2013)

Early arousal responses can occur from any type of pleasurable sexual stimulation, for example, kissing, erotic touching, sexual fantasy, sexually arousing visual materials, masturbation, or intercourse. Exactly what triggers sexual arousal depends on many factors, including each person's sexual history and cultural beliefs about sexuality. "Sexuality and Culture: Tantric Sexual Techniques" illustrates some of these cultural differences.

Since You Asked

2. Are aphrodisiacs for real? Which ones really work?

Throughout history, one of the most persistent and widely held beliefs about sexual response has been that certain chemicals, fragrances, foods, or other substances can produce feelings of sexual desire and arousal in humans. These substances are called **aphrodisiacs**. However, science has yet to provide any evidence that true aphrodisiacs exist; in other words, aphrodisiacs, no matter what you have heard, are a myth. See "Evaluating Sexual Research: The Legend of the Aphrodisiac", on page 79, for a more detailed discussion of this misperception.

aphrodisiac
Mythical substances that are thought to enhance sexual arousal and desire.

vasocongestion
The swelling of erectile tissues due to increased blood flow during sexual arousal.

sex flush
A darkening or reddening of the skin of the chest area that occurs in some people during sexual arousal.

During the excitement stage for both sexes, blood begins to circulate into erectile structures throughout the body, causing them to expand and enlarge, in a process called **vasocongestion**. A **sex flush** or reddening of the skin of the chest and abdomen may occur in some people, the nipples become erect, breathing becomes heavier and faster, heart rate increases, and voluntary muscles tense in a process called *myotonia*.

For men, the first and most obvious sign that sexual excitement has started is erection of the penis. The scrotal skin tightens, the testicles begin to rise up toward the underside of the penis, and the testicles themselves enlarge somewhat (see Figure 3.2). At this early stage, the man may not achieve a full erection, and his erection is easily lost if stimulation ceases or some sort of distraction occurs. Losing an erection during the excitement stage is completely normal, and usually easily regained. However, many men become concerned if they lose their erection at *any* time during sexual activity. This worry over a normal fluctuation in penile erection can become a self-fulfilling prophecy, in that the anxiety itself can prevent the return of the erection.

Figure 3.2 Physical Changes in the Male During the Excitement Phase

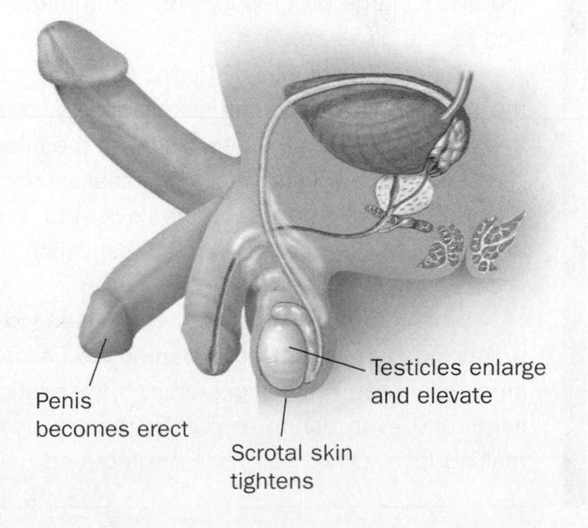

For women, the first physical change is erection of the clitoris (see Figure 3.3), but because most of the clitoris is inside the body, the most perceptible sign that sexual excitement has begun is vaginal lubrication (see Chapter 2, "Sexual Anatomy," for details on these various anatomical structures). During this phase, the shaft of the clitoris increases in size, and the labia minora swell and may separate slightly from around the opening to the vagina. Various internal changes are occurring as well. The uterus engorges with blood, enlarges, and moves slightly upward in the abdominal cavity. The vagina begins to change shape, becoming longer and widening out along the inner two-thirds of its length. Also the breasts, nipples, and areolas enlarge slightly.

Figure 3.3 Physical Changes in the Female During the Excitement Phase

Inner vagina expands
Vaginal walls lubricate
Clitoris swells

(a) Interior View

Clitoral glans and shaft increase in size
Labia minora increase in size

(b) Exterior View

Sexuality and Culture

Tantric Sexual Techniques

Tantra is a spiritual practice that originated in India thousands of years ago. Those who practice Tantrism work to achieve enlightenment or the totality of the development of their potential as humans (Lousada & Angel, 2011). Tantra involves learning to release the mind and body from their usual limited space on earth and allowing them to soar, boundlessly reaching cosmic heights of existence. More recently, Western cultures have latched on to one particular Tantric practice, commonly referred to as *Tantric sex*, which refers to a set of sexual exercises and activities that are said to transform pleasurable sex acts into ecstatic, rapturous, spiritual experiences. Some have suggested that these exercises can be useful in treating couples with sexual problems (Lorius, 2008).

Tantra relates to the Eastern religious teaching that energy flows through the body, just as blood runs through our vascular systems. In Tantric beliefs, this spiritual or psychic energy joins together seven main energy centers, called *chakras*, throughout the body, from the base of the spine to the scalp. Although no physiological studies have ever confirmed the existence of chakras, followers of Tantra and other Eastern religions believe not only that they exist but that they are also in large part responsible for human energy, healing, and sexual pleasure.

During the practice of Tantric sex, the goal is to open up the chakras and move the sexual energy, or *kundalini*, from the chakras nearest the genitals up to the heart, which is the "feeling chakra." From there, the sexual energy unites with the partner's energy channel before flowing to the crown chakra at the top of the head, creating a sensation of oneness and ecstasy.

Both partners must be committed to learning and practicing the techniques of Tantric sex. A great deal of the emphasis is on breathing exercises, meditation, yoga, massage, and even dance as part of the process leading up to making love. Other exercises are more sexual, including the woman learning to receive pleasure from stimulation of the so-called sacred spot, which appears to be what is commonly referred to in the West as the G-spot, and the man learning to have orgasms without necessarily ejaculating.

In the Tantric tradition, the joining of two people is considered a spiritual rite and leads to a unity between them that is a stronger force than the sum of their individual spiritual energy. Prolonging lovemaking serves to further increase this energy and maximize sexual pleasure for both partners. Moreover, as mentioned in "In Touch with Your Sexual Health: The Health Benefits of Orgasm" later in this chapter, Tantric philosophy suggests that lovers who have practiced these ancient techniques can learn to move sexual energy to the body's chakras, which is thought to create sensations of ecstasy throughout the body and enhance overall health.

Evaluating Sexual Research

The Legend of the Aphrodisiac

From powdered rhinoceros horn, beetles, and oysters, to alcohol, curry, and chocolate, humans throughout history have believed in the magical powers of certain foods and substances to biochemically enhance sexual arousal and desire. The problem with these beliefs is that they are false: no food, drink, drug, chemical, or any natural substance has ever been demonstrated scientifically to produce the purported effects of an aphrodisiac, no matter what you may have heard or, worse, read about on the Internet (Nordenberg, 1996; Shamloul, 2010).

In fact, some substances rumored to have such effects may be physically unhealthy or downright dangerous. The so-called Spanish Fly is made from powdered South African blister beetles, which cause major medical side effects, including mouth ulcers, urinary tract infections, and kidney and heart failure (Schmerling, 2004). As for alcohol, although its drug effect may decrease some people's sexual inhibitions, it actually serves to decrease our ability to function sexually and leads to poor sexual decision-making (e.g., unsafe sex). The worldwide belief in the sexual powers of powdered rhinoceros horn (linked simply to its "suggestive" shape) has only served to bring the rhino close to extinction. Many rhino are left to try to survive in the wild without their horns or are killed, their carcasses left intact with their horns severed (Slattery, 2003).

Some people are under the false belief that erectile dysfunction medications such as Viagra or Cialis are aphrodisiacs. This notion is also false. These medications change the way blood circulates in a man's body *during* sexual activities, which assists in his ability to achieve an erection; the drugs have no physical effect on desire.

Why, then, do so many believe the myth of the aphrodisiac? Because sexual desire and arousal in humans is very psychological, and if a food or substance becomes associated with sexual excitement, due to its appearance, texture, medical side effects, or through pure coincidence, it takes on a sexual placebo-effect life of its own. If people believe strongly enough that a so-called aphrodisiac will turn them on, it probably will, even though it has no real sexual effect.

In their early studies, Masters and Johnson observed that as sexual arousal continued, the increase in excitement responses seemed to level off. Arousal remains very high, while increasing at a slower pace. They called this the plateau phase (see Figures 3.4 and 3.5).

For both sexes, erectile tissues throughout the body are now fully engorged with blood. Respiration, heart rate, blood pressure, and muscle tension are all at high levels as orgasm approaches. A sexual flush on the skin in some people spreads and darkens.

For women, nipples maintain their erect state and the areolas continue to become larger (see Figure 3.6). The walls of the outer one-third of the vagina become engorged with blood and thicken, reducing the size of the vaginal opening (Figure 3.4). This change in the vagina may have evolved in humans due to the increase it provides in stimulation to the penis during intercourse,

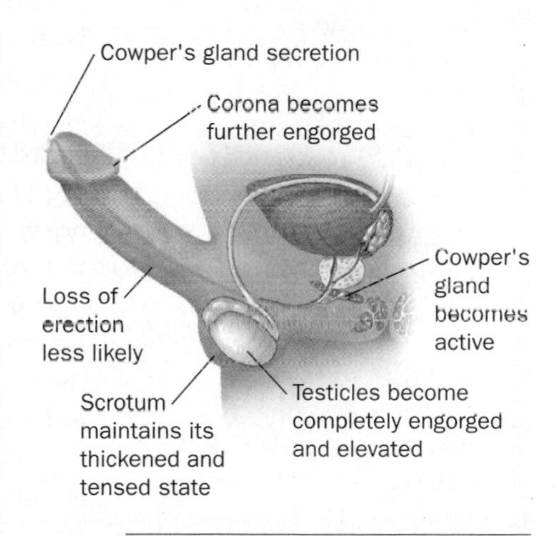

Figure 3.4 Physical Changes in the Male During the Plateau Phase

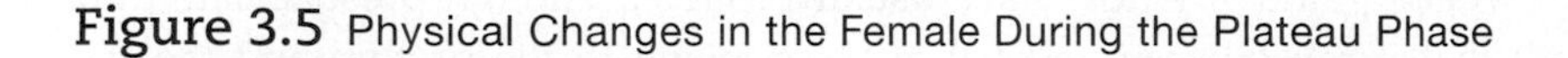

Figure 3.5 Physical Changes in the Female During the Plateau Phase

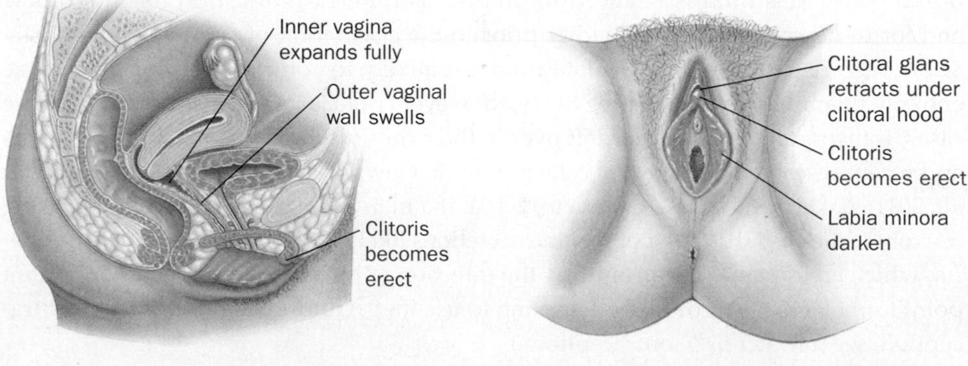

Figure 3.6 Sexual Arousal of the Female Breast

During continued arousal, the breasts, nipples, and areolas typically enlarge due to engorgement of erectile tissues.

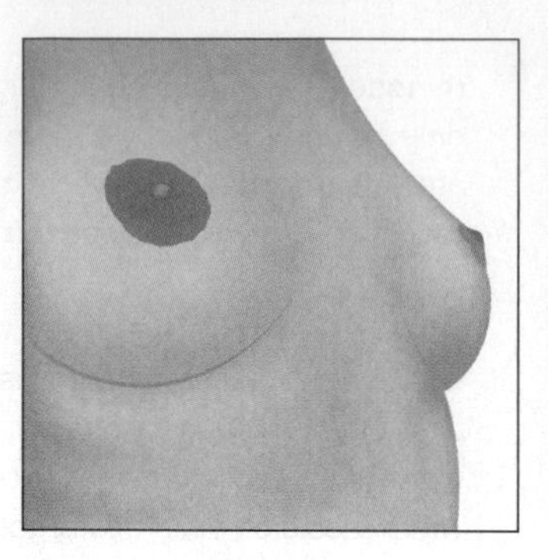
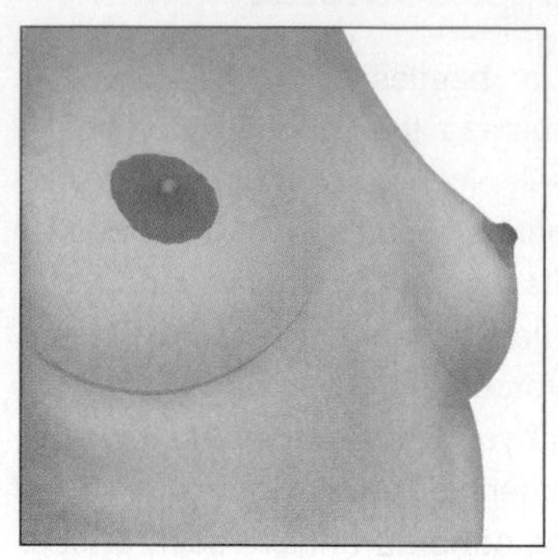
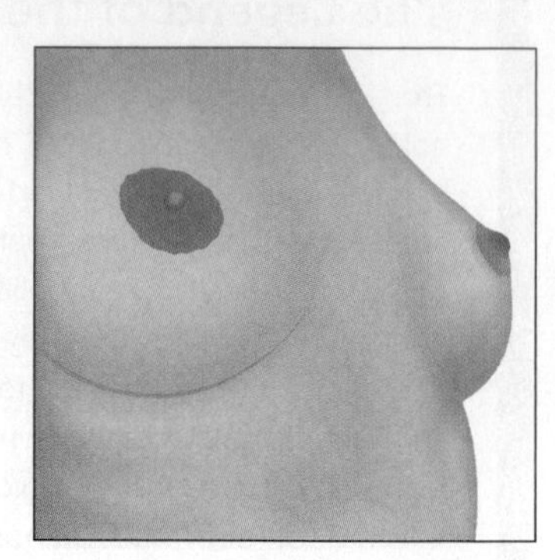

which, from an evolutionary perspective, helps ensure male orgasm and ejaculation. However, this response happens when a woman sexually aroused, whether or not intercourse is taking place. The inner two-thirds of the vagina continue to expand in a process called **tenting**. In addition to accommodating the erect penis (if intercourse occurs), many researchers suspect that tenting creates a place for semen to pool directly under the cervix helping the sperm to pass through the cervix and into the uterus on their journey to the fallopian tubes.

tenting
A widening of the inner two-thirds of the vagina during sexual arousal.

As the clitoris continues to engorge with blood, it straightens out along its length. This causes the glans of the clitoris in most women to retract closer to the body and under the clitoral hood. This is an important point, in that if you picture the changes in the clitoris at this point, you can see that the clitoral glans actually becomes *less* available for stimulation. Therefore, as explained and illustrated in "Self-Discovery: Clitoral Erection and the Myth of Female Orgasms during Heterosexual Intercourse" (on the following page), most women do not reach orgasm from heterosexual intercourse by itself and may need additional stimulation of the clitoral area either before, during, or after intercourse if they desire an orgasm. This is also why many women report that they find activities *other* than heterosexual intercourse more sexually satisfying. As the plateau stage continues, the labia minora deepen in color as they become increasingly engorged with blood, indicating that orgasm is approaching.

In men, the penis is now fully erect, the corona enlarges further, and the erection is unlikely to be lost due to anything short of a major distracting event. The scrotal skin has tightened, and the testicles are now pulled up very closely against the body at the underside of the penis, an indication that orgasm is approaching (Figure 3.5).

The Cowper's glands by now have usually secreted enough pre-ejaculate fluid ("pre-cum") that it can be seen and felt at the opening to the urethra and the end of the penis. This fluid is believed to produce additional lubrication for intercourse and/or to neutralize urethral acidity prior to the ejaculation of semen (Chudnovsky & Niederberger, 2007). Pre-ejaculate fluid is unlikely to contain sperm cells in large enough quantities to cause pregnancy; however, pre-ejaculate fluid may, in some cases, transport live sperm cells left over in the man's urethra from recent ejaculations and recent research found sperm cells present in Cowper's gland secretions (Killick et al., 2011; Zukerman, Weiss, & Orvieto, 2003). If a man is positive for infection-causing microbes, such as HIV, Cowper's gland secretions likely contain these viruses or bacteria, which are capable of transmitting the infection to his partner (this is an important point for heterosexual couples attempting to use the "withdrawal method" for contraception; see Chapter 5, "Contraception").

Since You Asked

3. What is "pre-cum," and where does it come from?

Self-Discovery

Clitoral Erection and the Myth of Female Orgasms during Heterosexual Intercourse

When Masters and Johnson wrote about the sexual response cycle in the 1960s, they placed their discussion within a framework of male–female sexual intercourse. Of course, all the phases of sexual response occur during any form of sexual stimulation, whether it's between a man and a woman, two men, two women, or just yourself, alone. However, one of the most common sexual myths that grew out of Masters and Johnson's assumption of penile-vaginal sex was the *expectation* that when a man and a woman are engaging in sexual intercourse, they both can and *should* experience an orgasm. Still today, forty years or so later, many, if not most, heterosexual couples continue to mistakenly believe that an orgasm for both is the "normal, natural, and expected" culmination of intercourse.

The problem is that, most of the time, this simply does not happen. Don't misunderstand: It nearly *always* happens for the man, but for the majority of women, heterosexual sexual intercourse, by itself, does not result in an orgasm, no matter how long that man is able to "last." In fact, only about 25% to 30% of women experience orgasms routinely through sexual intercourse exclusively, without additional stimulation of the clitoral area during intercourse (Castleman, 2009; Laumann et al., 1994). Of those women who do not have orgasms during intercourse, most are orgasmic through manual stimulation, oral sex, or masturbation, or some combination of these (Herbenick et al., 2010). What might be the cause of this male-female difference in heterosexual intercourse and orgasm?

Sex therapists and researchers have found that many physical and psychological factors may interfere with a person's sexual arousal, sexual pleasure, and orgasm. But one of the most common reasons women do not experience orgasm during intercourse relates to differences in sexual anatomy and a woman's sexual stimulation needs. As you will recall, orgasm is usually achieved through stimulation of the penile glans in men and the clitoral glans in women. The figure shows the position of the male and female sexual anatomy during sexual intercourse. If you imagine the activity of intercourse, which primarily involves the penis moving back and forth within the vagina, you can see that the glans of the penis receives a great deal of stimulation. Indeed, as the woman becomes increasingly aroused, the walls of her vagina swell, reducing the size of the vaginal opening and providing even greater friction to the penis.

Now notice the position of the clitoral glans. Many women find that they are far less likely to receive adequate, orgasmic stimulation from intercourse alone. Usually, the clitoris is stimulated when the pubic bone of the man hits it when the penis is deep in the vagina or when the motion of the penis in the vagina moves the minor labia, which in turn move the clitoral hood over the clitoris. However, as discussed and illustrated in Chapter 2 (refer to Figure 2.10), the clitoris has been moving up under the hood and closer to the woman's body, so it's even less available for such stimulation. In addition, there is far more variability in the length of the clitoral glans than of the penile glans (Wallen & Lloyd, 2008). Even if the man is able to delay his orgasm for long periods of thrusting, a large majority of women find this form of stimulation insufficient for orgasm.

This is not to imply that orgasm is or should be the goal of lovemaking. Some women are satisfied without an orgasm during intercourse or enjoy orgasms through other forms of stimulation before, during, or after heterosexual intercourse, as mentioned earlier. If a male-female couple feels somehow incomplete if orgasm isn't shared during intercourse, the easiest solution is to incorporate some additional clitoral stimulation during intercourse. Various sexual positions discussed in Chapter 7, "Sexual Problems and Solutions," allow for easier access to the clitoral area so that the man or the woman herself can provide the extra stimulation that she may want.

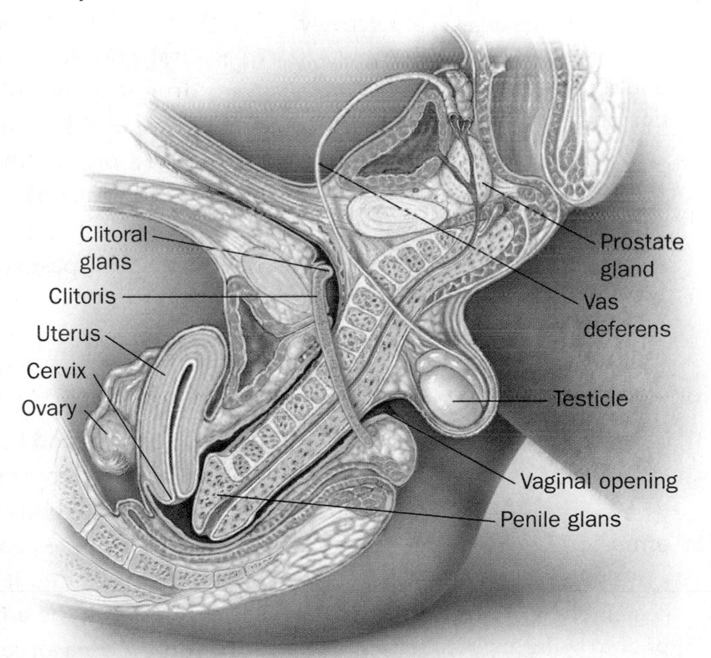

Positions of Male and Female Anatomy During Heterosexual Intercourse

orgasm
The third stage in the EPOR model, during which sexual excitement and pleasure reach a climax.

Since You Asked

4. I'm a woman. During intercourse, why does it take so long for a woman to have an orgasm? Is it unusual for women not to have orgasms at all during sex?

Orgasm

The climax of sexual arousal is the **orgasm**. Although this is the shortest of Masters and Johnson's four phases, usually lasting less than 15 seconds (Komisaruk, Whipple, Nasserzadeh, & Beyer-Flores, 2009), most people would agree that orgasm is the most intensely pleasurable experience of sexual responding. Orgasms vary in character and intensity from person to person, from one sexual experience to the next, and among various sexual acts that may lead to orgasm, including heterosexual intercourse, oral sex, manual or vibrator stimulation, anal sex, and masturbation. (These sexual activities are discussed in greater detail in Chapter 6, "Sexual Behaviors: Experiencing Sexual Pleasure.")

As noted, women tend to require a somewhat longer period of stimulation than men do to achieve orgasm with a partner, and the majority of women do not routinely experience orgasm through heterosexual intercourse exclusively. Although penile penetration may feel pleasurable for most heterosexual women, it does not generally produce orgasm, and orgasm is not necessarily always a woman's goal for intercourse (Goldstein, et al., 2006; Nicolson & Burr, 2003). Most women require additional manual or oral stimulation of the clitoral area before, during, or after intercourse. For many women, orgasm is the result of direct or indirect stimulation of the clitoris or clitoral area by a partner, manually or orally, or through self-masturbation during intercourse. For some women, direct stimulation of the highly sensitive clitoral glans may be too intense and can actually inhibit orgasm. However, Masters and Johnson found that men have one orgasm followed by a period of time, a "refractory period," which is a sexual "break" of sorts, when they are unable to have another orgasm regardless of the stimulation they receive. They found that women, however, were capable of additional orgasms with continued stimulation without a refractory period; referred to as **multiple orgasms**. This difference will be discussed more thoroughly shortly.

multiple orgasms
More than one orgasm at relatively short intervals as sexual stimulation continues without a resolution phase or refractory period in between orgasms.

Among the factors influencing the intensity and duration of orgasm are the length of arousal prior to orgasm, the length of time since the previous orgasm, alcohol or other drug use, and feelings of comfort and intimacy with a partner. Virtually everyone is *capable* of having an orgasm, although a small percentage of adults have not experienced orgasm for various reasons, and some people rarely or never have an orgasm, either alone or with a partner. The recent "National Survey of Sex Health and Behavior" found that 8.7% of men and 36.6% of women reported not experiencing an orgasm over the past year (Herbenick et al., 2010). Those who have never experienced an orgasm in their lifetime are referred to as *preorgasmic*, and although this is relatively uncommon overall, it is more common among women than men (see Chapter 7, "Sexual Problems and Solutions," for more about this issue).

For both sexes, as orgasm approaches, respiration increases dramatically, and pulse rate and blood pressure continue to rise. Any existing sexual flush may spread over more of the body. Usually, a loss of control over some voluntary muscles results in muscle contractions and spasms, especially in the hands and feet. Muscles in the pelvic area begin to contract rhythmically at the rate of once every 0.8 seconds (however, you would find it quite a challenge to try to time these as they are happening!).

emission
In males, the buildup of sperm and semen in the urethral bulb just prior to being expelled through the urethra.

For women, the anus, uterus, muscles of the pelvic floor, and walls of the outer third of the vagina all contract at intervals of 0.8 seconds during orgasm (see Figure 3.7). The number of these contractions varies from 3 to about 15 (Meston et al., 2004).

ejaculatory inevitability
In males, the sensation produced during the emission phase of ejaculation that expulsion of semen is imminent, reflexive, and cannot be stopped; often referred to as the "point of no return."

For men, orgasm also involves pelvic contractions and also usually includes ejaculation (we will discuss female ejaculation a little later). Ejaculation occurs in two stages (see Figure 3.8). The first stage, **emission**, is when semen builds up in the urethral bulb, creating the subjective sensation that ejaculation has begun and nothing can stop it. This sensation of having reached the "point of no return" is called the moment of **ejaculatory inevitability** because once the semen has collected in the urethral bulb, the rest of the ejaculatory process is reflexive and cannot be controlled voluntarily. Immediately

Figure 3.7 Physical Changes in the Female During Orgasm

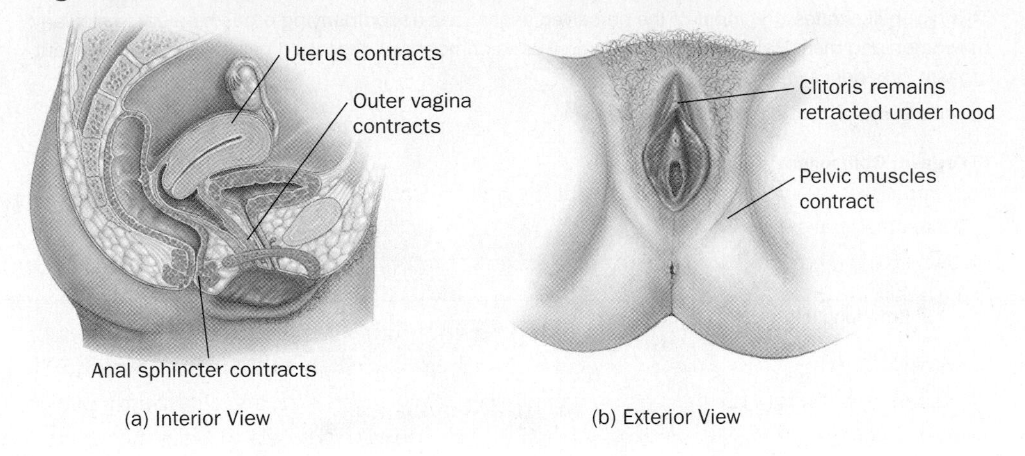

after emission, the prostate gland, urethra, and muscles at the base of the penis contract at the same intervals as with women, every 0.8 seconds, pushing the semen through the urethra and out of the penis in the second stage of ejaculation, called **expulsion**.

expulsion
In males, the contraction of pelvic muscles that force semen through the urethra and out of the body through the penis.

Explaining the physiological events that occur during male and female orgasm is not terribly difficult, but asking people to describe how orgasm *feels* is more challenging. Most people agree that the experience of orgasm is so intense and all-encompassing that they find it difficult to describe adequately in words. Also, people's conscious awareness of the experience of orgasm tends to be clouded by the intensity of the event itself, so detailed descriptions are rare. During orgasm, the person is usually very self-focused and, at the climactic peak, typically experiences some loss of awareness of his or her surroundings. Orgasm typically brings about an immense release of psychological and physical tension, leading to satisfied, warm sensations of emotional well-being, and a sense of bonding with the partner.

Since You Asked
5. Do orgasms feel the same to men and women? Are men's stronger or more intense?

You may be surprised to learn that when men and women *are* asked to describe their orgasms, their reports are quite similar. In one often-cited study, forty-eight written descriptions of orgasm, twenty-four from men and twenty-four from women, with all the sex-specific terms removed, were submitted to health care and psychological professionals, who were asked to judge whether each description was written by a man or a woman. The seventy male and female judges were unable to distinguish between men's and women's descriptions of orgasm (Vance & Wagner, 1976). In a more recent study, researchers asked men and women to rate various factors of the experience of orgasm, including satisfaction (with partner and alone), emotional intimacy, relaxation, ecstasy, and various bodily sensations (Mah & Binik, 2002). Figure 3.9 summarizes the study's results. As you can see, most rating scores, on average, were similar for the male and female participants, with women rating several of the bodily sensations only slightly higher than men. The exception was "shooting sensations," which was rated far higher by men than by women, probably due to the nature of male ejaculation.

Since You Asked
6. Are too many orgasms dangerous to your health? How many are too many?

Orgasms not only provide great physical and emotional pleasure but may actually enhance biological and psychological health as well. "In Touch with Your Sexual Health: The Health Benefits of Orgasm" on page 85, lists a number physical and psychological benefits that researchers have linked to orgasms for men and women. However, most sex educators and sex therapists stress that, although orgasm is a natural part of human sexual responding, it should not be the *goal* of sexual interactions. In fact, focusing on orgasm as the main objective of lovemaking may detract, rather than enhance, the intimate experience.

Figure 3.8 Physical Changes in the Male During Orgasm and Ejaculation

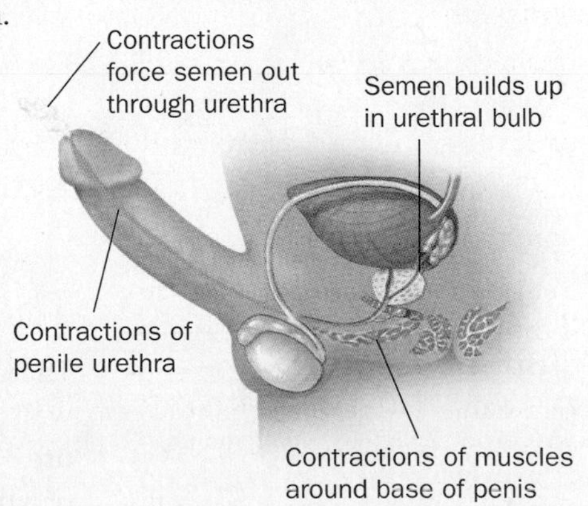

When couples value and embrace the entire lovemaking process, they often discover new and deeper levels of closeness and intimacy. An orgasm-as-the-goal approach to sexual intimacy often leads to repetitive, mechanical sexual

Figure 3.9 Average Ratings by Men and Women of Various Components of Orgasm

This graph illustrates that most of the perceived sensations accompanying orgasm are rated similarly by women and men. Ratings indicate how well each component described participants' most recent orgasm experience with a partner.

SOURCE: Based on data in Mah & Binik, 2002, p. 110.

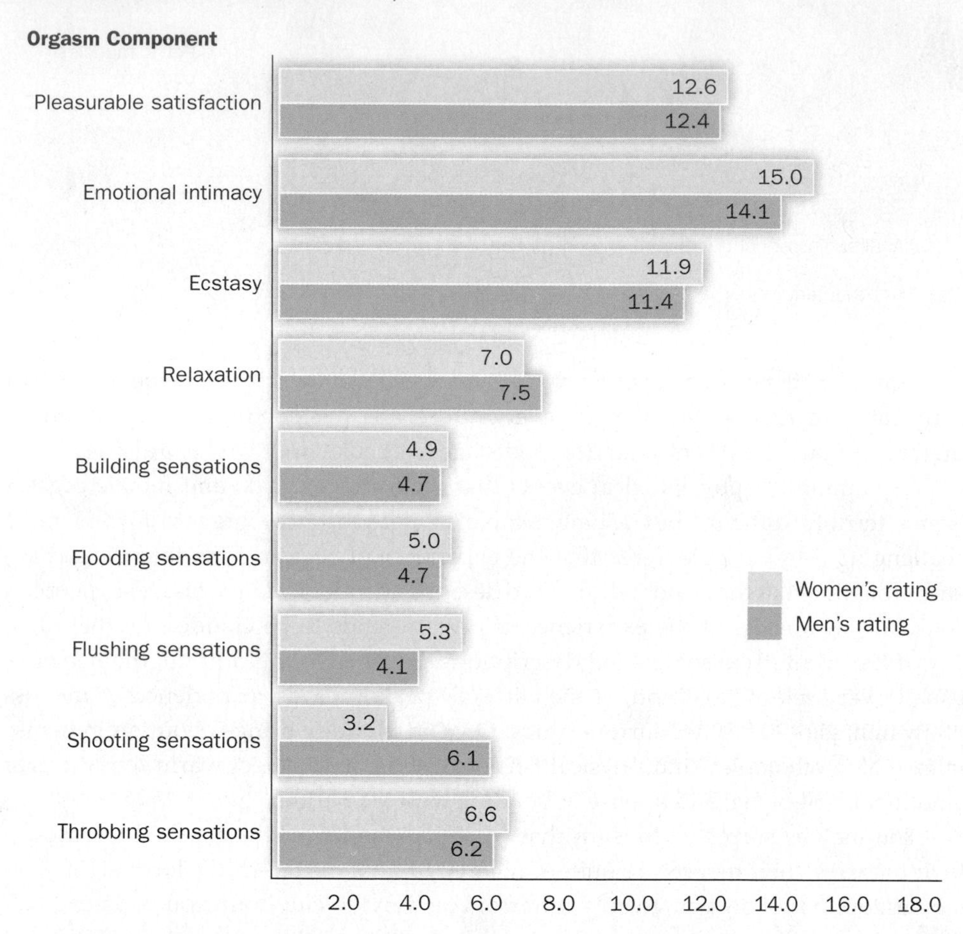

interactions that are targeted at that goal but may fail to produce the expressions of love and intimacy couples seek. Moreover, if one partner does not experience orgasm, the sexual encounter may be interpreted as a failure. Instead, sexuality educators and therapists try to assist couples in learning to experience *all* sexual activities as pleasurable in and of itself (Basson, 2001; Whipple, 1999). As one well-known sex researcher suggests:

> If the sexual experience does not lead to the achievement of the goal [orgasm], then the couple or the person who is goal-oriented does not feel good about all that has been experienced. The alternative view is *pleasure-directed*, which can be conceptualized as a circle, with each expression on the perimeter of the circle considered an end in itself. Whether the experience is kissing, oral sex, holding, etc., each is satisfying to the couple. There is no need to have this form of expression lead to anything else. If one person in a couple is goal-directed... and the other person is pleasure-directed... problems may occur if they fail to realize their goals or to communicate their goals to their partner. (Whipple, 1999)

The G-Spot Controversy

Since You Asked

7. Exactly what and where is the "G-spot"? My boyfriend and I can't seem to find it!

The G-spot is reported to be an area located on the anterior wall of the vagina (the upper wall as the woman lies on her back) about 2 inches in from the vaginal opening (see Figure 3.10). Researchers have described the G-spot as a slightly raised area about the size of a dime that increases in size during sexual stimulation (Wimpissinger et al., 2007).

Figure 3.10 The G-Spot

Although its exact nature continues to be researched and debated, some women report that stimulation of an area on the front wall of the vagina, called the "G-spot," enhances sexual arousal and pleasure.

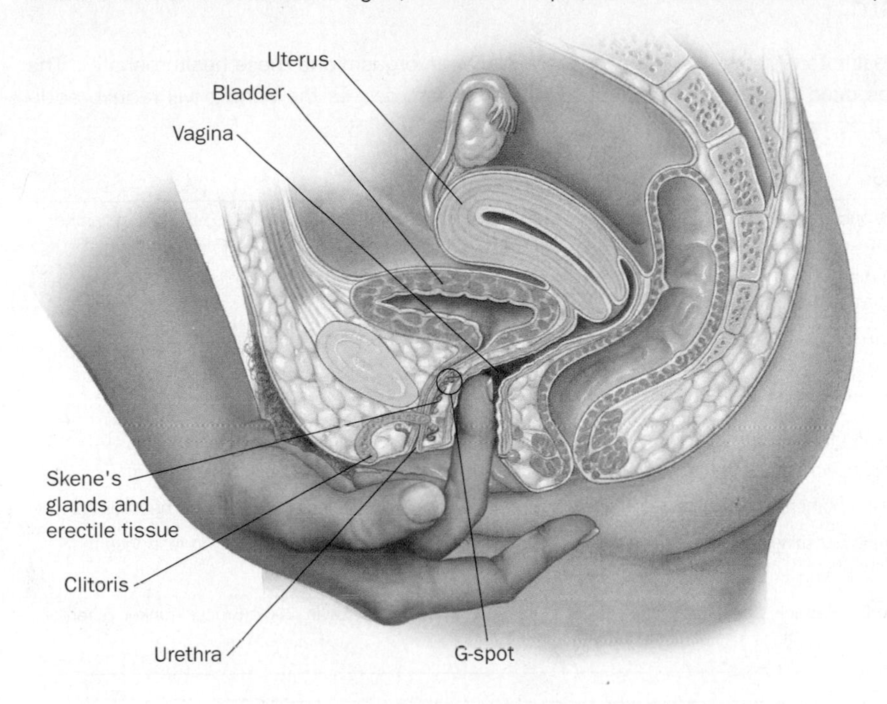

The easiest way a woman or her partner may explore the G-spot is to place one or two fingers inside the vagina and press firmly but gently against the upper wall in a massaging motion (see Figure 3.10). Some women report that when this area is stimulated during sexual activity, arousal and orgasm are enhanced and intensified.

Why is this section called the "G-spot *Controversy*?" In the 1980s, research on women's sexual anatomy claimed to have found a structure linked to the intensity of female orgasm, called the **G-spot**, or *Grafenberg spot*, named for Ernst Grafenberg, a physician who first described this structure anatomically in the 1950s (Belzer, Whipple, & Moger, 1984). Since 1981, when the findings were first published, the existence of the G-spot has been a topic of long-standing controversy among scientists and the scientific evidence for its existence has been hotly debated (Janini & Whipple, 2010). The research evidence has extended well beyond women's self-reports about the G-spot to various anatomical imaging studies, primarily using ultrasonic imaging. Some of these studies appear to lean in favor of the presence of an anatomical area than may correspond to the G-spot in some women (Foldes & Buisson, 2009; Jannini & Gravina, 2008; Ostrzenski, 2012). However, studies are equally plentiful that dispute its existence (e.g., Puppo & Gruenwald, 2012). Perhaps a useful middle ground is the one proposed by Kilchevsky in 2012:

G-spot

In some women, an area of tissue on the anterior (upper) wall of the vagina that, when stimulated, may cause a woman to experience enhanced sexual arousal and more intense orgasms.

> Objective measures have failed to provide strong and consistent evidence for the existence of an anatomical site that could be related to the famed G-spot. However, reliable reports and anecdotal testimonials of the existence of a highly sensitive area in the distal anterior vaginal wall raise the question of whether enough investigative modalities have been implemented in the search of the G-spot (Kilchevsky et al., 2012, p. 719).

In other words: we may not have found it, yet, but we should keep looking!

The Female Ejaculation Debate

Some researchers contend that another error in Masters and Johnson's observations involves something that most people take for granted: men ejaculate and women do

Since You Asked

8. I've heard that some women ejaculate like men. Is this true? How is it possible?

In Touch With Your Sexual Health

The Health Benefits of Orgasm

Recent research has revealed a number of health benefits associated with orgasm. Note that the findings cited here are based primarily on correlational research that has shown a *connection* between orgasm and these health benefits. The extent to which orgasm *causes* the benefit will require additional research.

Health Benefit	Research Findings
General health	An orgasm at least once or twice per week appears to strengthen the immune system's ability to resist flu and other viruses.
Pain relief	Some women find that an orgasm's release of hormones and muscle contractions help relieve the pain of menstrual cramps and raise pain tolerance in general.
Lower cancer rate	Men who have more than five ejaculations per week during their twenties have a significantly lower rate of prostate cancer later in life.
Mood enhancement	Orgasms increase estrogen and endorphins, which tend to improve mood and ward off depression in women.
Longer life	Men who have two or more orgasms per week live significantly longer than men who have fewer.
Greater feelings of intimacy	The hormone oxytocin, which may play a role in feelings of love and intimacy, increases five-fold at orgasm.
Less heart disease	Studies have shown that men who have at least three orgasms per week are 50% less likely to die of heart disease.
Better sleep	The neurotransmitter dopamine, released during orgasm, triggers a stress-reducing, sleep-inducing response that may last up to two hours.

SOURCES: Giles et al., 2003; Komisaruk & Whipple, 1995; Resnick, 2002; Komisaruk, Beyer-Flores, & Whipple, 2006; Levin, 2007; Smith, Frankel, & Yarnell, 1997; Weeks & James, 1999; Whipple, 2000.

not. Beginning in the 1970s, reports began appearing of women who claimed to ejaculate fluid from the urethra upon orgasm. Like the G-spot, the phenomenon of female ejaculation has been the subject of much debate but not a great deal of definitive study (Shafik et al., 2009). The research so far has tried to answer two questions: do some women truly ejaculate upon orgasm, and if so, what does the female ejaculate consist of?

If semen is produced by the seminal vesicles and prostate gland in men, as we discussed in Chapter 2, then what is it that women, who do not possess these anatomical structures, ejaculate? This is not yet fully understood, but the debate over the years has generally been between those who claim that the fluid is identical to urine and those who argue that it is not urine but something more akin to semen (see Pastor, 2013). If it is urine, that implies that the intense muscular activity during orgasm can, in some women, lead to some slight incontinence, in the same way that some people will expel a small amount of urine when laughing too hard or sneezing (referred to as *stress incontinence*). However, several other studies have found that female ejaculate actually resembles male prostatic fluid in chemical composition and that women who report ejaculating do not appear to be more prone to urinary stress incontinence (Cartwright, Elvy, & Cardozo, 2007; Moalem & Reidenberg, 2009).

Some research indicates that female ejaculate may be comprised mostly of a fluid secreted by two paraurethral glands that lie on either side of the female urethra, known as **Skene's glands**, as shown in Figure 3.10 (Chalker, 2002; Wimpissinger, Tscherney, & Stackl, 2009; Zaviacic, 2002b). These glands were once thought to be very small and nonfunctional in humans. More recent findings suggest that they may be larger, running along the length of the urethra, and are quite active in some women (Wimpissinger & Stifter, 2007; Zaviacic, 2002b). Because these glands may secrete fluid into the urethra at the moment of orgasm, a few researchers have made the case that they are a female variant of the male prostate gland (Moalem & Reidenberg, 2009; Whipple, 2002; Zaviacic, 2002a).

Skene's glands

In the female, a pair of glands on either side of the urethra that in some women may produce a fluid that is expelled during orgasm; also known as the *paraurethral glands*.

The purpose of these Skene's gland secretions are unclear. Researchers have found that the fluid is expelled from the urethral opening, not the vagina, and does not appear to add to sexual lubrication or enhance the odds of conception in any way, at least directly.

Other researchers have suggested, however, that the fluid secreted by the Skene's glands may serve a reproductive function by inhibiting the growth of bacteria that cause urinary tract infections, or UTIs (these are discussed in detail in Chapter 2, "Sexual Anatomy"). The fluid contains zinc, which is an antimicrobial element. This theory argues that as humans evolved, females who were more protected from UTIs would be more receptive to mating and would engage in sexual behavior more frequently, thereby increasing the number of pregnancies and births (Moalem & Reidenberg, 2009). In other words, early human females who ejaculated were more "fit" to secure the survival of the species.

If indeed female ejaculation occurs, not all women ejaculate, and most women who do not ejaculate do not particularly care one way or the other about it. Women who say they ejaculate report no greater overall sexual satisfaction than those who do not (Darling et al., 1990). Studies have indicated that about 40% to 50% of women report being aware of ejaculating fluid during orgasm (Cartwright, Elvy, & Cardozo, 2007), but self-reports of sexual functioning are often difficult to interpret with scientific accuracy. Lack of ejaculation is not a sign *in any way* that a woman is somehow deficient in her sexual responses, and female ejaculation should not be seen by women or their partners as something to strive for in sexual interactions. If female ejaculation is finally, clearly shown to exist in some women, this would be just one more example of how men and women are less sexually differentiated than many believe.

Resolution or Post-Arousal

Following excitement and orgasm, the body returns to its prearousal physical state. Masters and Johnson called this the **resolution phase**. Typically, this process happens fairly rapidly following orgasm but takes somewhat longer if orgasm has not occurred.

resolution phase
The fourth and last stage in the EPOR model, during which sexual structures return to their unaroused state; also referred to as *detumescence*.

For both sexes, heart rate, blood pressure, and muscle tension drop quickly. The body may be covered with perspiration. If a sexual flush was present, it now fades. Both men and women usually feel relaxed, warm, content, and sleepy. Masters and Johnson (1966) found that if a woman receives additional stimulation following orgasm, she may be capable of returning to the plateau phase and have one or more additional orgasms (multiple orgasms) without entering a resolution phase or refractory period between them.

For women, blood flows back out of erectile tissues throughout the genitals and breasts. The breasts, nipples, and areolas diminish in size and return to their unerect state. The clitoris resumes its prearousal position and shrinks slightly (see Figure 3.11).

Figure 3.11 Physical Changes in the Female During the Resolution Phase

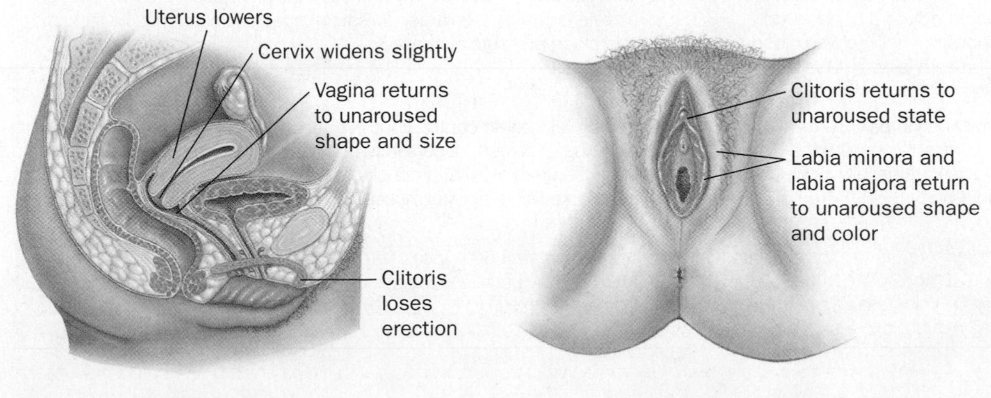

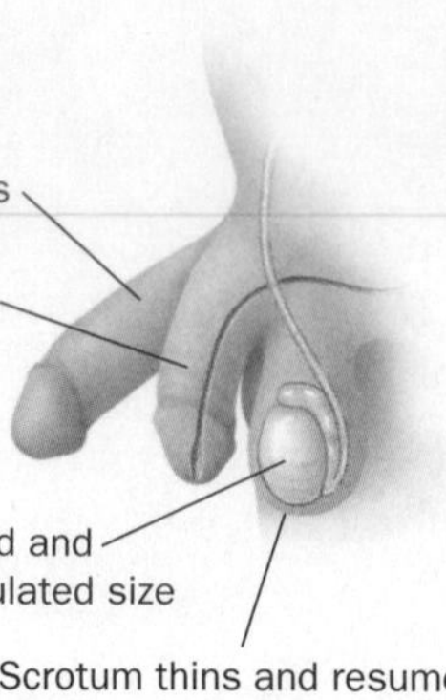

Figure 3.12 Physical Changes in the Male During the Resolution Phase

The labia minora return to their normal size and position over the vaginal opening. The walls of the vagina relax and fold in on one another. Also, the cervix widens slightly, most likely to facilitate the passage of semen and sperm into the uterus (assuming unprotected heterosexual intercourse) and narrows again about a half hour after orgasm. The uterus drops back down to just above the cervix, where the semen is pooled if ejaculation has occurred in the vagina ("Engender Health," 2003).

For men, nipples lose their erection, the penile glans lightens in color, and the penis becomes softer and smaller, returning to its unaroused, flaccid state. The scrotal skin relaxes, and the testicles drop down, away from the body (see Figure 3.12). As mentioned briefly earlier, Masters and Johnson found that, unlike women, men must "take a break" prior to becoming aroused again and reaching another orgasm. They called this break the **refractory period**. In other words, Masters and Johnson contended that men are not capable of multiple orgasms. Depending on a number of factors (including fatigue, frequency of orgasms, and age), the refractory period in men may last anywhere from a few minutes to twenty-four hours or more (DeLamater & Friedrich, 2002; Turley & Rowland, 2013). Masters and Johnson's claim of a refractory period in all men, along with some of their other findings, has been questioned, as we will discuss shortly. For a summary of the main physical changes that accompany the four phases of their model, see Table 3.1.

refractory period

A period of time following orgasm when a person is physically unable to become aroused to additional orgasms.

Questions about Masters and Johnson's EPOR Model

Although Masters and Johnson's EPOR model of sexual response has formed the foundation for discussions and explanations of human sexual response, it has not been without its challengers and competing conceptualizations. Various questions have been raised such as (a) Masters and Johnson neglect key emotional and psychological components of sexuality, especially *desire*, (b) can any valid model of sexual response emerge from purely physical reactions without considering psychological factors, and (c) is the EPOR model was far too *androcentric*—that is, it relies too heavily on a one-size-fits-all male sexual response pattern and fails to acknowledge many

Since You Asked

9. After I have an orgasm, I can get another erection pretty fast, but I can't seem to orgasm again. Is this normal or a problem of some kind?

Table 3.1 Summary of Physical Changes Accompanying Masters and Johnson's Four Phases of Sexual Response

Stage	Female Response	Male Response
Excitement*	First obvious sign: vaginal lubrication. Clitoral glans becomes erect. Nipples become erect; breasts enlarge. Vagina increases in length and inner two-thirds of vagina expands.	First obvious sign: erection of penis. Time to erection varies (with person, age, alcohol or drug use, fatigue, stress, etc.). Skin of scrotum pulls up toward body, and testicles rise. Erection may be lost if distracted but is usually regained readily.
Plateau*	Outer third of vagina swells, reducing opening by up to 50%. Inner two-thirds of vagina continues to balloon or "tent." Clitoris retracts toward body and under hood. Lubrication decreases. Minor lips engorge with blood and darken in color, indicating orgasm is near. Muscle tension and blood pressure increase.	Full erection is attained and is not lost easily if distracted. Corona enlarges further. Cowper's gland secretes pre-ejaculate fluid. Testicles elevate farther and enlarge, indicating orgasm is near. Muscle tension and blood pressure increase.
Orgasm	Begins with rhythmic contractions in pelvic area at intervals of 0.8 seconds, especially in muscles behind the lower vaginal walls. Uterus contracts rhythmically as well. Muscle tension increases throughout body.	Begins with pelvic contractions 0.8 seconds apart. Ejaculation, the expelling of semen, occurs in two phases. (1) Emission: Semen builds up in the urethral bulb, producing a sensation of ejaculatory inevitability. (2) Expulsion: Genital muscles contract, forcing semen out through the urethra.
Resolution	Clitoris, uterus, vagina, nipples, etc., return to unaroused state in less than one minute. Clitoris often remains very sensitive to touch for five to ten minutes. This process may take several hours if woman has not experienced an orgasm.	Approximately a 50% loss of erection within one minute; more gradual return to fully unaroused state. Testicles reduce in size and descend. Scrotum relaxes.

SOURCE: Adapted from Hock (2012).

*Over the past several years, most sexuality researchers have combined the excitement and plateau phases into a single excitement phase.

fundamental differences in female sexuality? Plus, many have contended that their emphasis on the four stages suggests that each stage is something couples should "strive for," when overall sexual satisfaction is far more complex than a series of goals (see Kleinplatz & Ménard, 2009). Before we discuss other, differing models that have been proposed to describe sexual responding, let's take a brief look at some of Masters and Johnson's specific findings that have been drawn into question.

Too Many—or Too Few—Stages

In reconceptualizing Masters and Johnson's EPOR model, some sexuality researchers have argued that they included too many, too few, or some incorrect stages. Beginning with Kaplan (1974), whose three-stage model will be discussed in the next section, various researchers have suggested that the plateau stage is unnecessary as an interval between excitement and orgasm and that it should be merged with the stage of excitement, building, potentially, directly to orgasm (Levin and Riley, 2007).

In addition, Kaplan and others have suggested that Masters and Johnson omitted a crucial stage in sexual responding: *sexual desire*. In other words, the question has been posed, "Would a sexual response cycle occur at all if desire is missing from the theory?" One leading researcher, Roy Levin, has proposed that two types of desire are needed to add this stage accurately: one type of desire that occurs spontaneously and then leads to sexual excitement, and another that stems from sexual stimulation, leading to excitement (Levin, 2005).

Male/Female Similarities

Many sexuality researchers over the past twenty years have argued that a single four-stage model cannot adequately describe the sexual response of both men and women. Many of these arguments have centered on the idea mentioned earlier that the EPOR model is too linear—that is, it assumes four stages that predictably follow one another in a prescribed sequence and is a pattern of response couples should "strive" for (see Kleinplatz et al., 2009). Although some researchers contend this progression may explain the male response cycle adequately, they argue that female sexual response should be interpreted as significantly more complex, involving far more than genital responding, and less linear in nature (e.g., Wylie & Mimoun, 2009).

Consequently, other models have been suggested that expand on or change the EPOR model to provide separate sexual response descriptions for men and women (Basson, 2001; Rosen & Barsky, 2006; Tiefer, 2010; Wylie & Mimoun, 2009). In the next section of this chapter, we will discuss some of these post-EPOR proposals.

Male/Female Differences

Masters and Johnson claimed that men ejaculate, but women do not; that men have a refractory period, but women do not; that women may have multiple orgasms, but men cannot. All of these generalizations, it turns out, simply do not apply to *all* men and all women.

The topic of female ejaculation has already been discussed in detail earlier in this chapter. Although the existence of female ejaculation is controversial, the evidence is mounting that some women may indeed ejaculate a small amount of non-urine fluid from the urethra upon orgasm.

Research has suggested that some women may have a refractory period that is quite similar to a man's; that is, some women require a waiting period after orgasm before they are physically capable of becoming aroused to orgasm again (Humphries & Cioe, 2009). This female refractory period may be shorter on average than men's and be less pronounced, but the categorical male–female difference found by Masters and Johnson has been drawn into question.

Helen Singer Kaplan (1929–1995) added the stage of desire to the sexual response cycle.

Although a significantly larger number of women compared to men report multiple orgasms, some men appear to be capable of multiple orgasms as well. One study found that approximately 30% of women reported multiple orgasms with a partner compared to 9% of men (Haning et al., 2008).

With these alterations and criticisms of Masters and Johnson's EPOR model in mind, we turn to alternate models of human sexual response.

Alternatives to Masters an Johnson's Four-Stage Model

3.3 Explain the approaches proposed by Kaplan; Reed; Janssen and Bancroft, and the "The New View" to explain the process of human sexual response and how they vary from the EPOR model.

Kaplan's Three-Stage Model of Sexual Response

Helen Singer Kaplan (1929–1995) followed in the footsteps of Masters and Johnson in developing many effective forms of sexual therapy and became one of the best-known sex therapists in the world. She published numerous books designed to help people enjoy sexually satisfying lives, including *The New Sex Therapy: Active Treatment of Sexual Dysfunctions* (1974), *Disorders of Sexual Desire* (1979), and *The Sexual Desire Disorders: Dysfunctional Regulation of Sexual Motivation* (1995). Her approach to sexual response mirrored that of Masters and Johnson, but, as a sex therapist, she felt they had left out one crucial factor and had divided the response process into too many rigid, sequential stages. Kaplan proposed a more fluid model consisting of *desire, excitement*, and *orgasm* that is known as **Kaplan's Three-Stage Model** (see Figure 3.13).

Kaplan's Three-Stage Model
An alternative to Masters and Johnson's EPOR model of sexual response developed by Helen Singer Kaplan that features the three stages of desire, excitement, and orgasm.

She argued that *any* sexual response is unlikely unless someone *wants* to be sexual; therefore, she conceived the first stage of sexual response as *desire*. Because of her focus on sex therapy, she also jettisoned Masters and Johnson's notions of plateau and resolution from her theory. Kaplan interpreted the plateau phase as simply a part of excitement; and resolution, although it obviously occurs, rarely poses any sexual problems and is therefore of minimal clinical interest. By contrast, difficulties with sexual desire, called *inhibited* or **hypoactive sexual desire**, are extremely common. (This disorder and Kaplan's contributions are discussed in greater detail in Chapter 7, "Sexual Problems and Solutions.")

According to Kaplan, among the many factors that may interfere with sexual desire are stress, fatigue, depression, pain, fear, prescribed medication, recreational drugs, negative past sexual experiences, power and control issues in a relationship, loss of interest in a partner, low self-image, and hormonal influences. By adding *desire* at the beginning of the stages and streamlining the sexual response cycle, Kaplan's Three-Stage Model became a popular alternative approach to Masters and Johnson's EPOR model for understanding, evaluating, and treating problems with sexual responding.

Figure 3.13 Kaplan's Three-Stage Model of Sexual Response

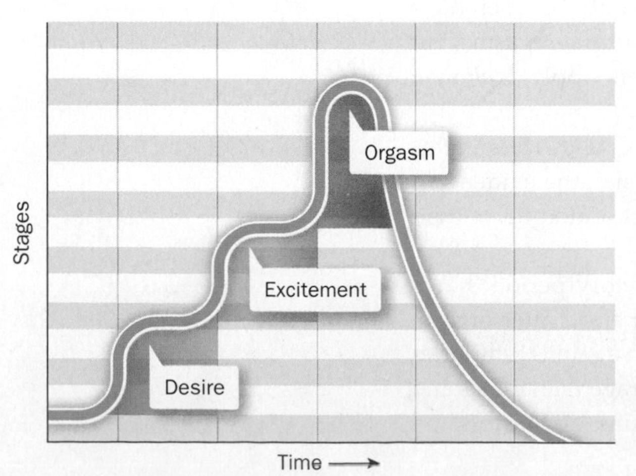

Some of Kaplan's notions about sexual response have been criticized for being overly simplified and for making the assumption that desire must *always* be present for sexual responding to occur. Some individuals and couples find that even if they are lacking in sexual desire initially, as they begin to engage in sexual activity, desire may awaken and flow out of the sexual behavior rather than necessarily preceding it. This is reflected in Levin's model mentioned earlier, which includes two different stages of desire that may occur at different times during sexual arousal (refer to Table 3.2).

Table 3.2 Comparisons of Three Theories of Sexual Response Stemming from the EPOR Model

Theorist(s)	Proposed Stages of Sexual Response			
Masters and Johnson (1966)	Excitement	Plateau	Orgasm	Resolution
Kaplan (1979)	Desire	Excitement	Orgasm	
Levin (1990)	Desire 1 (spontaneous)	Excitement	Orgasm	Resolution
Levin (1990)	Excitement	Desire 2 (Stimulation-activated)	Orgasm	Resolution

SOURCE: Adapted from Levin, 2005.

Reed's Erotic Stimulus Pathway Theory

In the 1990s, psychiatrist David Reed took the popular theories of Masters and Johnson and Kaplan and reinterpreted them from a more psychological and interpersonal perspective that he called the **Erotic Stimulus Pathway Theory** (Greenberg, Bruess, & Haffner, 2002; Stayton et al., 2007). He labeled his four stages using the more psychological terms *seduction, sensations, surrender*, and *reflection* (see Figure 3.14). His first stage, *seduction*, corresponds to Kaplan's desire stage, but in Reed's model, desire is created by the behaviors people engage in that they believe will attract another person and make themselves sexually attractive to others. These rituals may include wearing cologne and perfumes, using makeup, dressing in alluring ways, flirting, making eye contact, touching, sending love notes, buying flowers, arranging dates, engaging in self-disclosure, and signaling a desire for sex.

hypoactive sexual desire
A persistently low level of desire for sexual activity or lack of sexual fantasies; also known as *inhibited sexual desire.*

Erotic Stimulus Pathway Theory
A model of sexual response based on the psychological and cognitive stages of seduction, sensations, surrender, and reflection.

The seduction behaviors then move into the *sensation* phase, when sexual behavior and sexual arousal begin (akin to excitement and plateau in the EPOR model). Reed suggests that during this phase, our heightened senses, fantasy, and imagination combine to feed the arousal and motivate us to make it continue. Reed conceptualized the peak of sexual arousal, orgasm, as a giving over of oneself, mentally and physically—a *surrender*, as he called it—to the culmination of sexual intimacy.

Finally, Reed proposed that the after-orgasm phase is a time when both partners reflect on the experience and bring meaning to it. The importance of the *reflection* phase is that it provides an opportunity for partners to interpret the sexual

Figure 3.14 Reed's Erotic Stimulus Pathway Model

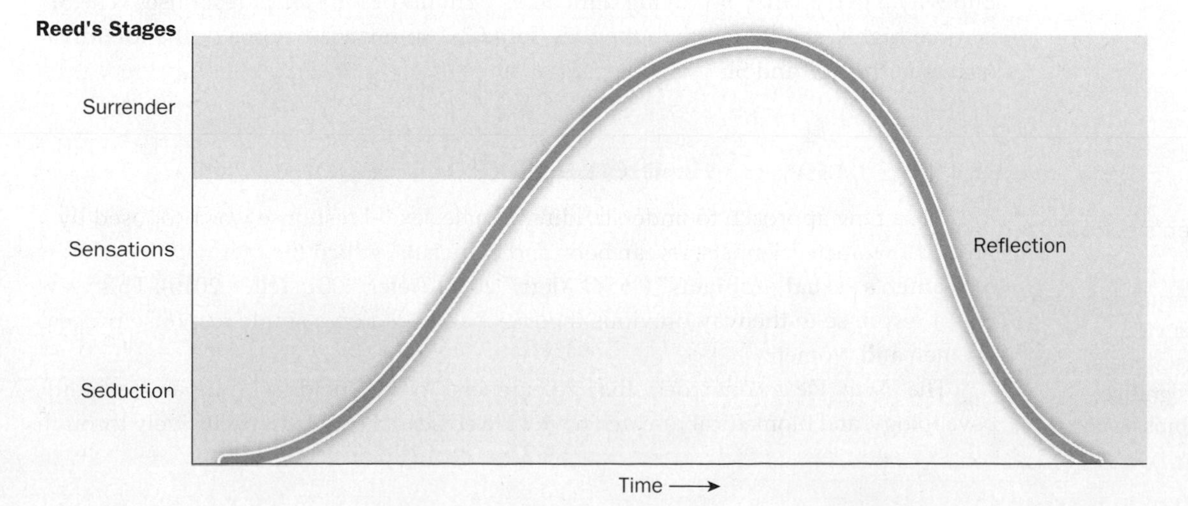

encounter in positive or negative terms, and this helps us make choices about whether or not to engage in the activity again, under the same circumstances or with the same partner.

Although Reed's Erotic Stimulus Pathway model reflected the stages of Masters and Johnson's and Kaplan's models, the focus in each phase was placed on the cognitive and psychological, rather than on, or in addition to, the physical. A great deal of current thinking about sexual response is slowly yielding to a more holistic approach to understanding the complexities of human sexual response. In this vein, some researchers are suggesting that male and female sexual responses are significantly less similar than the traditional EPOR model implies. In general, this view contends that the Masters and Johnson model applies reasonably well to male sexual response, but to understand *female* sexual response, we must consider a range of different factors.

Janssen and Bancroft's Dual Control Theory

In 2000, sexuality researchers Erika Janssen and John Bancroft at the Kinsey Institute at Indiana University proposed a model to explain sexual arousal. Although their model focuses on only one of the stages suggested in the theories already discussed (excitement), the researchers would likely suggest that all sexual response grows out of an initial process of arousal, and without arousal, other stages or phases are absent.

Janssen and Bancroft theorized that sexual responding is dependent on two opposing psychological and physiological processes: a sexual excitation system (SES) and a sexual inhibition (SIS) system. They contended that whether or not a person becomes sexually aroused is due to a combination of factors that either promote or block excitement. This theory is called the **Dual Control Model of Sexual Response** (Janssen, 2007; Bancroft, 2009).

Dual Control Model of Sexual Response
a theory that sexual arousal is controlled by a combination of excitatory and inhibitory processes

The dual control model argues that individuals vary in the relative influence of sexual excitation and sexual inhibition. And the greater our understanding of a person's SES and SIS, the greater the insight into that person's sexual responses. The theory states that both systems must form a balance for optimum sexual response. Your SES responds to environmental and mental stimuli that activate and arouse sexual feelings and reactions, such as visual sensations from looking at a sexual partner, tactile stimuli from giving or receiving sexual touch, and pleasurable sexual fantasies. Your SIS also has an important role in sexual arousal by monitoring the potential negative consequences of sexual activity such as STIs, unwanted pregnancy, or sexual coercion, and helping you to avoid them. As SES stimulation increases and SIS factors decrease, sexual response is optimized. This conceptualization may also offer insight into why a person may be having difficulties with his or her sexual responses: The SIS is more highly stimulated than the SES. Table 3.3 summarizes some of the identified factors in the SES and SIS.

A New View of Women's Sexual Response

In 2000, a new approach to understanding female sexual response was proposed by a group of women scientists, researchers, and clinicians, which they termed "**new view of women's sexual problems**" ("FSD Alert," 2008; Tiefer, 2001; Teifer, 2010). This view was a response to the way previous theories had relied on a single response process for men and women.

new view of women's sexual problems
A model of female sexual response incorporating a larger variety of factors than previous models, including physical, cognitive, social, and relationships issues.

The *New View* contended that researchers in the fields of human sexuality, psychology, and biomedical research have viewed sexual problems exclusively through

Table 3.3 Dual Control Model of Sexual Response

Sexual Excitation System (SES)	Sexual Inhibition System (SIS)
Arousability	Inhibitory Cognitions
- Sexual attraction toward partner	- General life worries
- Frequency of sexual fantasy	- Sexual response expectations
- Physical closeness to partner	- Shyness
Partner Characteristics and Behaviors	- Worry about STIs
- Witnessing partner's talents	- Worry about unintended pregnancy
- Aroused by partner showing caring	Relational Elements of the Sexual Interaction
- Confidence that partner is faithful	- Lack of trust in partner's feelings
- Being very attracted to partner	- Imbalance of sexual give and take
- Partner showing intelligence	Relationship Importance
Setting	- Lack of trust in partner's feelings
- Unusual sexual settings arousing	- Suspicion of being used for sex
- Possibility of others nearby arousing	Setting
Other	- Lack of privacy for sexual activity
- Visual stimulation	- Fear of others hearing or seeing
- Tactile stimulation	Other
- Confidence in ability to respond to partner	- Chronic stress
	- Fear of embarrassment
	- Worry about past sexual problems
	- Concern about ability to please a partner

SOURCE: Based on Bancroft & Janssen, 2009.

the lens of Masters and Johnson's EPOR model. It argued that this "one-size-fits-all" approach reduced sexual responding to mere physiological processes similar to breathing or digestion, when in reality sexuality (women's, at least) is far more complex than the EPOR model allows (see McHugh, 2006). The "new view" research team felt an urgency to differentiate between male and female sexual response and stated that "women's accounts do not fit neatly into the Masters and Johnson model; for example, women generally do not separate 'desire' from 'arousal,' [and] women care less about physical than [about] subjective arousal" (Tiefer, 2001, p. 93). They went on to discuss what they considered to be the most serious flaws in applying the traditional response models to women ("FSD Alert," 2008):

Since You Asked

10. It seems to me that men and women are pretty different when it comes to sexual feelings. How can they be described in the same ways?

1. The incorrect assumption that male and female sexuality are fundamentally the same. As you may remember from our earlier discussion in this chapter, the four-stage model assumes many similarities between men's and women's responses, so sex therapists and researchers have tended to assume that their sexual problems must be similar as well.
2. An exaggerated focus on the physiology of sexual response to the exclusion of the relationship context in which it occurs. This focus has led to assumptions that sexual response can be understood and enhanced without regard for the larger issues of the interpersonal sexual relationship.
3. The minimization of individual differences in sexual response among women. The working group proposed that women vary more than men do in their sexual responses and therefore do not fit neatly into the desire-arousal-plateau-orgasm pattern.

How do "new view" researchers propose to overcome the difficulties they see in past and current thinking about female sexuality? Here is part of their proposal:

> We propose a new and more useful classification of women's sexual problems, one that gives appropriate priority to individual distress and inhibition arising within a broader framework of cultural and relational factors.... We call for research...driven...by women's own needs and sexual realities. (Tiefer, 2001, p. 94)

More specifically, they suggest that women's sexuality, sexual response, and sexual difficulties require a radical theoretical revision that takes into account cultural, political, and economic factors (e.g., lack of sexuality education or access to contraception); a woman's partner and the relationship between them (e.g., fear of abuse, imbalance of power, overall discord); psychological factors (e.g., past sexual trauma, depression, anxiety); and medical factors (e.g., hormonal imbalances, sexually transmitted infections, medication side effects) (Graham, 2003; McHugh, 2006). Because this approach to women's sexuality is as much about women's sexual *problems* as it is about sexual response, we will revisit this idea in Chapter 7, "Sexual Problems and Solutions."

Closing Note

Although sexual responses change as we age, sexual intimacy plays an important role in our happiness, throughout our lives. Research has demonstrated that few physical reasons exist for anyone, regardless of age, to forgo satisfying sexual interactions. Changes in sexual responses and many other developmental aspects of sexuality in various life stages are discussed in Chapter 12, "Sexual Development Throughout Life."

Your Sexual Philosophy

The Physiology of Human Sexual Response

You can see how theories, approaches, and practices relating to the physiology of sexual response continue to evolve and change over time. These changes stimulate discussion, debate, and research. Although studying the various theories, both old and new, can be confusing at times, this variety is beneficial to the overall study of human sexuality. It requires researchers and clinicians to stay alert and active in their fields, and, in the final analysis, it enriches our understanding of the complexities of humans as sexual beings.

As noted at the beginning of this chapter, your knowledge of sexual anatomy from Chapter 2 is of little use without also understanding how that anatomy functions. Consequently, Chapters 2 and 3 go hand in hand to provide you with the foundation for understanding the human sexual body. You will probably find that you will be looking back at these discussions as you read the remaining chapters in this book. In fact, before moving on to our discussion of love, intimacy, and sexual communication in Chapter 4, take a look at Table 3.4, which offers some hints about how the material you've learned in Chapters 2 and 3 applies to the topics in the remaining chapters.

With this in mind, you can see how developing a clear understanding of human sexual response is a central feature in your sexual philosophy. Moreover, this understanding helps extend your knowledge of sexual anatomy to enhance sexual satisfaction and enjoyment. Your philosophical commitment to developing a personal awareness of how the human body typically reacts to sexual stimulation will allow you to become more aware of your own body's responses and acquire a greater sensitivity to your partner's responses. This knowledge, in turn, will allow you to communicate

Table 3.4 Application of Material in Chapters 2 and 3 to Other Chapters Throughout This Text

Chapter	Application of Knowledge of Sexual Anatomy and Response
Chapter 4, "Intimate Relationships"	Effective communication about love and sex often requires an awareness and understanding of your own and your partner's anatomy and how you and your partner respond to each other sexually.
Chapter 5, "Contraception: Planning and Preventing"	Choosing an effective method of birth control for yourself and, more important, using it correctly require a working knowledge of anatomy, hormones, and how the body responds during sexual arousal.
Chapter 6, "Sexual Behaviors: Experiencing Sexual Pleasure"	Developing an awareness of the sexual body will allow you to become more aware of your own body's responses and acquire a greater sensitivity to your partner's responses as well.
Chapter 7, "Sexual Problems and Solutions"	Many sexual difficulties are caused or exacerbated by a basic lack of understanding of sexual anatomy and sexual response.
Chapter 8, "Sexually Transmitted Infections/Diseases"	Knowing how STIs are transmitted and how to identify the symptoms relies on your familiarity with sexual anatomical structures.
Chapter 9, "Conception, Pregnancy, and Birth"	Virtually every topic of discussion on sexual anatomy and response relates to how conception occurs, the process of pregnancy, and the birth of a baby.
Chapter 10, "Gender: Expectations, Roles, and Behaviors"	Expectations about certain anatomical structures and about male and female sexual response play a significant role in gender roles and expectations.
Chapter 11, "Sexual Orientation"	When studying sexual orientation, it is very important to understand that a satisfying, healthy sexual life is the goal of nearly everyone, regardless of sexual orientation. A clear knowledge of sexual anatomy and response helps all of us achieve this goal.
Chapter 12, "Sexual Development Throughout Life"	Understanding and appreciating the physical side of sexual development from childhood through old age requires a clear understanding of the anatomical and sexual response changes that occur normally throughout our lives. As you will read in Chapter 12, we are sexual beings from birth to death.
Chapter 13, "Sexual Aggression and Violence: Rape, Child Sexual Abuse, and Sexual Harassment"	Understanding what constitutes sexual violence and aggression and its aftereffects often requires a clear conception of sexual anatomy and physiology.
Chapter 14, "Paraphilic Disorders: Atypical Sexual Behaviors"	When studying these unusual sexual behaviors, it often helps to know how they affect the sexual response and functioning of the people engaging in them.
Chapter 15, "The Sexual Marketplace: Prostitution and Pornography"	How erotica and pornography are defined involves issues of sexual anatomy and response, as does an understanding of the activities and dangers of prostitution.

more clearly about sexual issues with a partner or health care professional and will enhance your sexual intimacy, enjoyment, and health. That is what *your sexual philosophy* is all about—knowing who you are as a sexual person, what you want and don't want sexually, and planning ahead for a fulfilling, satisfying sexual life.

Have You Considered?

1. Would you ever volunteer to be a participant in a study such as Masters and Johnson's research in the 1960s? Why or why not? Discuss your opinion about the ethics of such research.
2. Explain why many women do not routinely experience orgasm during heterosexual intercourse.
3. Can you think of an "aphrodisiac" you have heard of that was not mentioned in the chapter? Because aphrodisiacs have never been shown scientifically to have any real effect on sexual desire, why do you think people believe in the ones you've heard of?
4. Which of the various models of sexual response discussed in the chapter do you feel is most accurate? Which seems most questionable to you? Explain your answers.
5. In your opinion, what is the most important reason for people to study and understand the process of human sexual responding? Why?
6. For many couples, the goal of lovemaking is orgasm. Explain why many sex researchers and therapists claim that this detracts from a satisfying sexual experience.

Summary

Historical Perspectives

Sexual Pioneers

- Although surveys had previously shed light on people's self-reported sexual activities, it was the pioneering studies of Masters and Johnson in the 1960s that revealed much of the physiological functioning of the human body during and after sexual activity.

Biology, Psychology, and Human Sexual Response

- One way that humans are differentiated from other animal species involves the psychological aspects of sexual feelings, functioning, and behavior. Although biological factors play important roles in the sexual functioning of all animals, human sexual desire and activity are at least as strongly influenced by psychological forces such as feeling pleasure, giving pleasure, expressing feelings of closeness and love, relieving stress, feeling valued by another person, and expressing how much another person is valued.
- Many people mistakenly equate "sex" with sexual intercourse. However, human sexual responding, from a physical perspective, follows similar patterns for all kinds of sexual behavior, including a man and a woman having intercourse, a man and a woman sharing manual or oral sex, a same-sex couple making love, an individual masturbating, or even an erotic dream.

Theories of Sexual Response

- Masters and Johnson's *Excitement-Plateau-Orgasm-Resolution (EPOR) Model* suggests that men and women typically proceed through these four predictable stages of response. Over the decades since Masters and Johnson first proposed their theory, various exceptions to their findings have been suggested, but the EPOR model continues to provide a valuable framework for discussions on human sexual responding.
- The excitement and plateau phases of the EPOR model have been combined into one continuous stage of excitement by many researchers. During these stages, the penis and clitoris become erect, the clitoral glans withdraws under the clitoral hood, and the Cowper's glands in men secrete pre-ejaculate fluid (which may contain sexually transmitted infectious microbes).
- One of the most persistent myths about sexual response is the notion of the aphrodisiac. For thousands of years, people have believed in the power of aphrodisiacs to stimulate and arouse, but no scientific evidence exists of any food or substance that truly possesses these properties.
- A common myth about sexual responding in humans is that both partners should achieve orgasm during heterosexual intercourse. However, research has demonstrated that most women do not routinely reach orgasm through intercourse exclusively, no matter how long it lasts, and require additional genital stimulation for orgasm.
- Although the existence, structure, and function of the G-spot remains embroiled in controversy, some women report that stimulation of the area called the G-spot on the wall of the vagina during sexual activity enhances and deepens the sensations of orgasm and, for some, leads to ejaculation.
- Variations in the physiology of orgasm may exist between men and women, but their subjective descriptions and perceptions of their orgasms appear to be quite similar.
- Numerous health benefits of orgasm are theorized. Various studies have suggested that orgasm, either with a partner or through masturbation, may provide many health benefits, from reduced cancer risk to pain reduction and even to an increased lifespan.
- Various researchers have suggested changes, variations, and alternatives to Masters and Johnson's EPOR model of sexual response.
- Helen Singer Kaplan's *Three-Stage Model* is designed to apply more closely to treating sexual problems than was the EPOR model. It adds the pre-excitement stage of desire and drops the stages of plateau and resolution.
- Roy Levin has proposed that there are two types of desire that may occur during sexual arousal: spontaneous desire and stimulation-produced desire.
- David Reed's *Erotic Stimulus Pathway* model of sexual response takes a more emotional and psychological approach, focusing on the experiences of *seduction*, *sensations*, *surrender*, and *reflection* during sexual responding.
- Janssen and Bancroft's Dual Control Model of Sexual Response seeks to explain sexual arousal as a combination of two human psycho-physiological processes,

a sexual excitation process and a sexual inhibition system that work in concert to create conditions for sexual responding.

- A theory known as the *New View* of women's sexual problems seeks to redefine the sexual responses of women as fundamentally distinct from those of men and argues that a single model is inadequate to explain both.
- Humans' capacity for sexual feelings, behaviors, and responses change as we age; however, our ability to live, function, and enjoy ourselves as sexual beings continues throughout our lifespan.

Your Sexual Philosophy

The Physiology of Human Sexual Response

- Familiarity with human sexual anatomy and response is crucial for understanding the complexities of human sexuality. The information contained in this chapter, along with the material in Chapter 2, "Sexual Anatomy," lays the foundation for appreciating and understanding the topics throughout this book and for applying them in meaningful ways to your life and your sexual fulfillment. Your sexual philosophy is about knowing who you are, what you want and don't want, and planning ahead for a fulfilling sexual life.

Chapter 4
Intimate Relationships

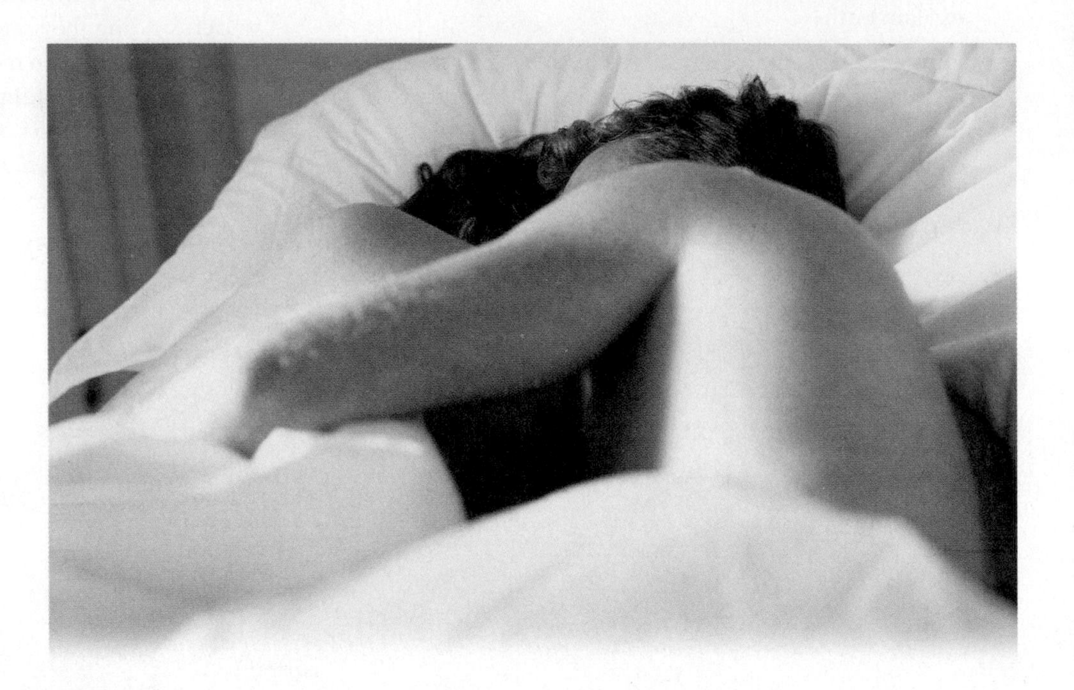

Learning Objectives

After you read and study this chapter you will be able to:

4.1 Summarize how relationship intimacy applies to couples of all sexual orientations.

4.2 Explain the factors that contribute to the early formation of a new intimate relationship.

4.3 Describe and discuss the various types of intimate relationships that may be explained by the *Triangular Theory of Love* and the concept of "love styles."

4.4 Analyze the importance of effective communication between romantic partners, patterns of effective and ineffective communication, and strategies couples can use to improve communication quality in the relationship.

4.5 Discuss the most common reasons given for why love relationships fail.

4.6 Explain the types of relationship abuse, describe the cycle of violence, and discuss the characteristics of an abusive relationship and its effects on the survivors.

Since You Asked

1. My boyfriend and I broke up almost a year ago, but I haven't been able to get involved with anyone since. In fact, I don't even like most of the guys I've met. What's going on? (see page 102)
2. My new partner (of two months) and I are very different. We disagree and argue about politics, religion, and nearly everything else. Is this going to be a problem, or is it true that opposites attract? (see page 107)
3. I feel I love my partner far more than he loves me. Is there some way to get him to love me more? (see page 109)
4. When a couple is having sex on a regular basis, does the emotional side grow over time? Does it mean the relationship is not going well if it is just for the sex? (see page 111)
5. Why do people cheat? Is it because they just need a new conquest? (see page 114)
6. How do you know when things are going wrong or if things are right between you and your partner? (see page 117)
7. My partner and I fight a lot. We yell at each other, and sometimes one of us will storm out of the house for several hours. But we love each other very much and are totally committed to each other. Is this normal? (see page 118)
8. Why is it so difficult to explain what does and does not feel good sexually? (see page 124)
9. How can I get my partner to trust me? She's jealous all the time and accuses me of cheating when I'm not doing anything. She's even been reading my texts and emails! She's convinced I don't love her, but I do—I think. (see page 128)
10. Can someone who has been in an abusive relationship ever have a normal relationship again? Is it really possible to forgive and forget? (see page 138)

For most people, love and sexual intimacy are closely linked. Research has shown that the most physically and emotionally satisfying sexual interactions occur in the context of a long-term, committed romantic relationship, especially when communication between the partners is open, honest, and intimate (Hill, 2002; MacNeil & Byers, 2009; Sprecher, 2002). You all know or have heard about people who are comfortable engaging in casual sex, without the need for any deep emotional investment, but these individuals and events appear to be the exception rather than the rule. One national survey conducted in the 1990s found that over 65% of respondents said they would not have sex with someone unless they were in love with that person (Michael et al., 1994), and that couples tend to engage in a wider range of sexual activities with a partner that they love (Kaestle, C. & Halpern, C., 2007). When young people choose to engage in casual sex, such as "hooking up" at a party or participating in sexually permissive social settings during spring break, most will readily admit that their experience(s) did not represent love and that someday they want a more meaningful emotional context in their sexual lives.

Sex, love, and intimate communication usually go hand in hand, and we would have difficulty separating them even if we wanted to, which most people do not. In this chapter, we will focus on emotional and relationship issues that typically surround sexual behaviors rather than on the behaviors themselves. This is not to say that we ignore the emotional side of sexuality in the other chapters in this book; the connection between sexual behavior and human emotion is apparent throughout this text. Even in Chapter 2, "Sexual Anatomy," the most biologically based chapter in the book, many of the topics are infused with references to emotions and attitudes. In this chapter, however, the *emotional* and *interpersonal* components of human sexuality, the good and the bad, will be our primary themes.

Focus on Your Feelings

In a way, this chapter is *all* about feelings. Many of you, as you read through its varied and complex topics relating to love, intimacy, and passion, may experience a wide range of emotional responses. Why? Because most of you either are or have been in an intimate relationship, and some of you may have experienced more than one. Therefore, at least some of the issues discussed in this chapter are likely to be familiar to you—in some cases, perhaps too familiar.

If nothing else, this chapter is likely to make you think about past, present, and future love relationships. It is difficult to read this chapter without holding your own relationships up for comparison to see how they fare. Many, if not most of you, will find validation and confirmation that your current love relationship, if you are in one, is strong and healthy. Or, you may find yourself transported back to past relationships that seemed very satisfying; nearly ideal, yet they did not last. Conversely, you may be reminded of a past love that was mismatched, unhealthy, or even abusive. The frustration and pain of that relationship may be reactivated as you read this chapter. If this happens, you should know that it is a normal reaction to the memory of an unhealthy relationship and perhaps provides an opportunity for you to work through some of those lingering negative emotions. On the other hand, some of you may feel compelled to face suspicions you have about your current relationship.

Regardless of where this chapter leads you emotionally, it should help you sort out and better understand your feelings about your romantic relationships. Keep in mind that love relationships are not easy; in fact, they are probably the most complex and mysterious of all human connections. However, through your basic awareness, sensitivity, and insight, combined with the knowledge and understanding of the issues discussed in this chapter, they can also be the most wondrous and rewarding.

We should note that a relatively small percentage of people choose to be actively involved in or seek out more than one (usually long-term) intimate relationship at the same time, with the consent and agreement of all the people involved. This relationship choice some people make is called *polyamory* (see Haritaworn et al., 2006; Wosick-Correa, 2010). However, because polyamory is the exception, not the rule, and most love relationships worldwide are between two partners, we will be focusing this chapter on romantic, intimate relationships between two people: couples, referred to as *monogamy*.

We will discuss the factors involved in early intimacy and love relationships. We will then turn to the meaning of love, exploring how intimacy and commitment between partners grows over time, how sex and intimacy intertwine, and how love relationships change as time goes on. We will consider what many people would say is the most important component of love relationships: communication. We will also explore factors involved in the decline and ending of romantic relationships and take a clear-eyed look at the tragedy and terror of abuse and violence in intimate relationships. As odd as it may sound to many of you to discuss relationship abuse and violence in a chapter on love and intimacy, it is important that we do so. Sometimes what begins as a seemingly loving, wonderful relationship turns controlling, abusive, and violent in ways that make escaping from the abusive partner extremely difficult. One effective way to prevent this from happening to you or to others you know is to learn as much as possible about these abusive relationships, which virtually never end happily. Finally, we will look at how all these issues fit into your sexual philosophy.

As we begin our journey though the complexities of love, intimacy, and sexual communication, ask yourself this question: Are love and romance timeless processes that flow naturally from human nature, or are they products of the influence of time and culture? Let's take a glimpse back to love and romance in Colonial America.

Historical Perspectives

Love and Marriage in America, c. 1750

How different are love and relationships today compared to, say, in the years approaching U.S. independence? If you said *very* different, you would be correct. For one thing, marriage agreements "in the old days" were sometimes arranged between the fathers of the potential bride and groom, not the couple themselves. A "good match" was considered one that increased the wealth of both families, through the professional prospects of the young man to earn a good living and the young woman's dowry—usually money, land, or other property given to the groom's parents in exchange for "allowing" the woman to marry their son. With all that out of the way, the courtship could begin.

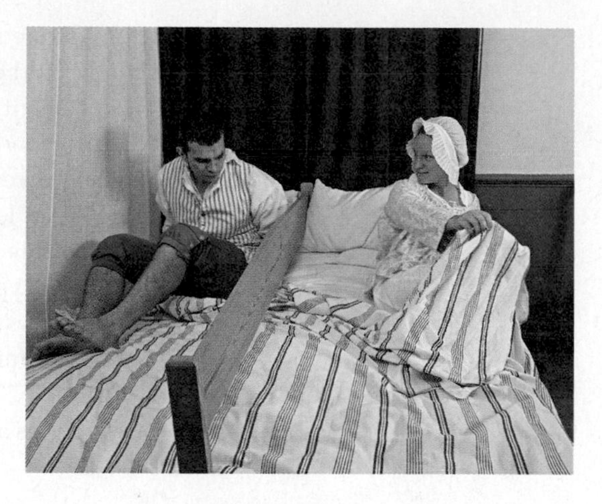

In the 1700s, the courtship process often included the practice of bundling, in which the "engaged" couple was allowed, and even encouraged by both sets of parents, to sleep in the same bed on a regular basis in one of their homes. But as you can see from this photo of a "bundle bed," the purpose of bundling was not what you might be thinking!

Once begun, "courting" tended to move quickly to marriage. This was due in part to the fact that life expectancy was significantly shorter in the 1700s, so if a couple did not marry relatively young and begin a family, they could end up without a partner for their entire lives (Dugan, 2001). Unmarried women acquired the status of unmarriageable "old maids" between the ages of twenty-two and twenty-seven; and in many of the colonies, unmarried men were forced to pay a "bachelorhood tax."

One method for hurrying the courtship process along was a practice called *bundling*, in which the engaged couple was allowed, and even encouraged, by both sets of parents to sleep in the same bed on a regular basis in one of their homes. Before you think how open-minded your ancestors were, there were major restrictions on bundling: The couple had to be fully clothed, and they were sometimes actually *sewn into* their respective bedding. To keep things "above board," a dividing board was placed vertically between them. The idea was to allow long stretches of time, sometimes all night, for couples to talk and become emotionally intimate relatively quickly. Nothing of a sexual nature was allowed. Bundling, as an integral part of courtship, was practiced in New England for nearly a century, until it was noticed (especially by the clergy) that many babies were being born much sooner than nine months after the couple married. As a result, bundling fell out of favor.

Today, in most Western cultures, long-term relationships, including marriages (with some exceptions), are motivated by love and intimacy between two people of the opposite or same sex and their desire to make a lasting commitment to each other; they are not usually based on business advantages and dowries. Moreover, the decision to make a commitment to be together ultimately rests with the couple.

Some patterns of communication and interacting may vary depending on a couple's sexual orientation, but most of the qualities of love and intimacy are fundamentally the same, regardless of sexual orientation.

Intimacy and Sexual Orientation

4.1 Summarize how relationship intimacy applies to couples of all sexual orientations.

Nearly every topic and discussion in this chapter may be applied to any couple, no matter if the partners are the same or opposite sex. Overall, the quality, depth, and commitment of the love and intimacy between loving partners do not fundamentally differ based on sexual orientation (Frost, 2011; Mackey, Diemer, & O'Brien, 2000; Strong & Cohen, 2013). Clearly, some *specific* aspects of intimate relationships covered in this chapter may apply to opposite sex couples differently than to same-sex couples. For example, flirting between men and women adheres to certain steps and stages that are specific to opposite-sex couples due to social and cultural roles and expectations. Likewise, verbal communication between men and women faces sex-specific challenges that are part and parcel of all opposite-sex interactions, both within and outside love relationships. The same barriers to effective communication are not usually present for same-sex couples. Beyond issues inherent to sex differences, our discussion of love and intimacy in this chapter does not, and needs not, distinguish among couples of any sexual orientation.

With that in mind, however, we should note that stereotypes about lesbian and gay couples have persisted, even though studies have found them to be misleading (see APA, 2008). For instance, one common belief is that same-sex relationships are dysfunctional and unhappy. However, studies have found gay and heterosexual couples to be equivalent to each other on measures of relationship satisfaction and commitment.

A second stereotype is that the relationships of lesbians and gay men are unstable. Again, however, despite continuing prejudice toward same-sex relationships, research shows that lesbian and gay relationships are at least as strong and enduring as those of opposite-sex couples.

Stereotypes such as these are fading in U.S. society, as evidenced by the rapid changes in state laws legalizing same-sex marriage and protecting gay and lesbian individuals from discriminatory practices. These issues are discussed in detail in Chapter 11, "Sexual Orientation." As U.S. culture grows more tolerant of diversity in sexual identity, the misconceptions about same-sex relationships are bound to fade as well.

Establishing Early Intimacy

4.2 Explain the factors that contribute to the formation of a new intimate relationship.

If you have experienced a romantic relationship, you can probably think back to the exact moment when it began (and it was undoubtedly *not* in a bundle bed!). Your first contact with a potential partner or your initial realization that someone might become more than "just a friend" is usually fixed in your memory, regardless of how the relationship eventually turns out. The specific circumstances of that first flash of attraction or potential romance vary greatly, and people have their own unique stories of "how it all began." These stories may the "just my type" effect; the "sooo hot" power of physical attraction; the "boy-or-girl-next-door" experience; the "we're so alike" phenomenon; or maybe, the "we started flirting and the rest is history" memory. As you read on, think about your own relationships or those of friends or relatives—you are bound to recognize the role some of these factors played during, or even before, the beginning of a romantic connection.

Since You Asked

1. My boyfriend and I broke up almost a year ago, but I haven't been able to get involved with anyone since. In fact, I don't even like most of the guys I've met. What's going on?

Your Field of Eligibles

Each of you—male, female, straight, gay—has a set of criteria that determines whether a person is a possible candidate for an intimate relationship. It's already there, in your brain. Those who meet your criteria—that is, those you perceive as having *potential* as a romantic partner—are referred to as your **field of eligibles** (FOE) (Dragon & Duck, 2005; Kerckhoff, 1962; South, 1991). To demonstrate this concept (which is not as obvious as you might think), consider for a moment how many people you meet in your life and yet how few of them become your friends. Now think about how many you would ever consider for a love relationship.

field of eligibles
All the individuals who meet a person's criteria as a potential romantic partner.

You may not be consciously aware of your own "romantic attraction criteria," but you all have them, and they are functioning all the time as you interact with others in your life. Most people typically "filter out" those outside their field of eligibles before meeting them. For example, you might be willing to consider someone older than you as a potential romantic partner. But how much older? One year? Five years? Ten years? Twenty? At some age difference, nearly everyone draws the line and eliminates people from consideration as potential romantic partners, regardless of who they are. In other words, people of a certain age or older would likely be outside your field of eligibles. That's just one example of a prerequisite that might be part of determining your field of eligibles, but many others exist. Here's another easy one—sex: Most of you filter out of your FOE about half of all humans, either males or females. You may find it interesting to examine more closely the requirements you have in your own FOE. "Self-Discovery: Who Is in *Your* Field of Eligibles?" lists some of the most common criteria, which you can rate in terms of their importance to you. You can also add criteria that you may see in yourself but may be of a more personal and less universal nature.

The fact that you have a unique field of eligibles does not imply that you are shallow, narrow-minded, or prejudiced. Instead, your FOE enables you to direct your energy into those relationships that are most likely to succeed in satisfying your needs. Keep in mind that you have an active, mentally functioning FOE even if you are not currently looking for a romantic partner. If you are in an intimate relationship, your current partner may fit your criteria so well that you feel no need whatsoever for anyone else romantically, but *psychologically*, your field-of-eligibles criteria remain intact.

A person's field of eligibles is not cast in stone. It can change, and for many people, it does change throughout their lives. Meeting people and getting to know them often leads to a relationship with someone that is, shall we say, unexpected. You have probably heard someone say, about a partner, something like, "I never expected this; when we first met, he (or she) was not my type at all!" What that person is really saying is that the partner with whom a relationship developed did not originally fall within her (or his) field of eligibles. Such an experience would change your FOE criteria, probably forever. Moreover, experiences (good and bad) with a current partner can change your criteria in terms of a future partner.

Your field of eligibles determines whom you rule out or rule in as a potential romantic partner.

Physical Attractiveness

For virtually all of you, one prerequisite in your field of eligibles is physical appearance—your perception of the attractiveness of a potential partner. Many people may want to deny that they base romantic attraction on others' physical appearance, but research has consistently shown that attractiveness plays a key role in the early formation of a potential intimate relationship. However, the criteria for what makes partner attractive varies greatly from person to person. The truth is that you probably already have your personal criteria for the level of physical attractiveness you consider necessary in a possible romantic partner; this is already part of your field of eligibles.

THE "BEAUTIFUL IS BETTER" BIAS The influence of physical attractiveness goes well beyond how drawn you may be to someone. Consider the following summary

Self-Discovery

Who Is in *Your* Field of Eligibles?

Here is a list of criteria that will help you get a sense of your personal FOE. These criteria may determine whether you include or exclude certain individuals in your field of potential romantic partners *before you get to know them or even meet them.* Feel free to add and rate other factors that are important to you.

Rate each item on its importance to you as a criterion in your own field of eligibles using the following scale:

1 = Not at all important; I would not take this characteristic into account when considering someone for a potential romantic relationship.

2 = Of minor importance; it might make a small difference; I would prefer that the person fall within my requirements, but I'm flexible.

3 = Somewhat important; I'd definitely take this into account, but it would not make or break my decision.

4 = Quite important; I doubt that I would consider someone who did not meet my requirements.

5 = Extremely important; I would never consider someone who does not meet my requirements on this factor.

Factor	Rating (1–5)
Sex: Must be either male or female?	________
Race/ethnicity: Same racial or ethnic background as yours? Certain races or ethnicities included or excluded?	________
Age: Limit on number of years older or younger than you?	________
Height: Limit on how much taller or shorter than you?	________
Religion: Must be same religion as you? Certain religions included or excluded?	________
Socioeconomic background: Must be from a family of similar income or social status?	________
Education: Must have equal or similar educational level?	________
Job or career: Must have a minimum level or specific type of job or career? Must *not* have certain jobs or careers?	________
Weight: Must be within certain limits in terms of weight (overweight or underweight)?	________
Body type: Must have (or not have) a certain body type?	________
Personality: Must have (or not have) specific personality characteristics?	________
Other: ________________	________
Other: ________________	________
Other: ________________	________
Other: ________________	________

of just some of the research findings about the effects of physical attractiveness (Forsterling, Preikschas, & Agthe, 2007; Ha et al., 2012; Regan & Medina, 2001; Wood & Brumbaugh, 2009).

- Attractive children are more popular with both classmates and teachers. Teachers give higher evaluations to the work of attractive children and have higher expectations of them (which have been shown to improve future performance).
- Attractive applicants have a better chance of getting jobs and of receiving higher salaries. (One U.S. study found that taller men earned around $600 more a year per inch of height than shorter executives.)
- In court, attractive people are found guilty less often. When found guilty, they receive more lenient sentences.
- The bias for beauty operates in almost all social situations—experiments consistently show that both heterosexual and gay individuals react more favorably to physically attractive individuals.
- The "what is beautiful is good" stereotype is an irrational but deep-seated belief that physically attractive people possess other desirable characteristics such as intelligence, competence, social skills, confidence, and even moral virtue. (In fairy tales, the good fairy or princess is always beautiful; the wicked stepmother is always ugly.)
- Overall, men tend to agree significantly more among themselves about whom they find attractive than do women, regardless of sexual orientation.
- Homosexual men and women have similar partner preferences to heterosexual men and women in terms of partner attractiveness.

The ideal of physical beauty conveyed by the mass media has changed noticeably over time.

You can see from this list of research findings that physical attractiveness is deeply embedded in cultural values and attitudes. With such a strong bias bombarding you throughout life, it is no wonder that physical attractiveness plays such a prominent role in the formation of romantic relationships.

THE EFFECTS OF MEDIA Media influences on attitudes and behavior are stronger now than in the past. TV, movies, magazines, and the Internet are constantly conveying today's standards of beauty and attractiveness in a particular culture. This repeated exposure causes the standards to become more rigid and widespread, so that even slight variations in physical appearance may fall outside a particular culture's standard of attractiveness. The more you see these images, the more normal and attainable they seem. But for most, they are not attainable. For example, the current U.S. media ideal for how a woman's body "should" look (exceedingly thin) is attainable by only about 5% of all women, and when women see current thin-idealized female bodies in the media, they become more self-conscious and depressed about their own bodies (Harper, B., & Tiggemann, M. 2008; Owen & Laurel-Seller, 2000). This media body ideal has changed drastically over time. In the early 1900s, the ideal female body was 5 feet 4 inches tall and weighed 140 pounds. In the 1970s, top fashion models weighed about 8% less than the average American woman; today the difference is 23% (see Kayhan, Baig, Mehmi, & Basra, 2010, for a review)!

Men appear to be much less self-critical of their physical appearance (Fredrick, Peplau, & Lever, 2006). As many as eight out of ten women express dissatisfaction with their bodies. However, in general, men are mostly pleased with their bodies and may actually overestimate their attractiveness and fail to see physical flaws (Fox, 1997; Garcia, Khersonsky, & Stacey, 1997). In fact, one study found no change in men's self-rated physical attractiveness following a six-week physical and strength-training program (Anderson et al., 2004). Nevertheless, recent research is showing an increase in the number of men seeking cosmetic surgery; however, it appears that men seek facial cosmetic procedures (such as Botox injections and microdermabrasion) for career reasons—to increase their professional potential and marketability in youth-focused professions (Grant, 2012; Palmquist, 2004).

THE MATCHING HYPOTHESIS Although on a cultural level, humans appear to favor a certain standard of physical attractiveness when it comes to intimate partners, beauty is clearly in the eye of the beholder. According to a psychological theory called the **matching hypothesis**, people tend to seek romantic and sexual partners who possess a similar level of physical attractiveness to their own (Feingold, 1988; Lee et al., 2008). When romantic couples are analyzed, they tend, overall, to be fairly well matched on most measures of physical attractiveness (Kalick, 1988; Pines, 2005). This may be due to the expectation that someone of similar attractiveness is an emotionally "safer" choice and is a boost to each partner's self-esteem. On the other hand, a partner of much greater physical attractiveness may be seen as more likely to reject someone who is not as attractive. How well couples are matched on physical attractiveness is also related to the success of the relationship. Matched couples tend to become closer, and their relationships tend to last longer, than couples that differ significantly in physical attractiveness (Pines, 2005).

matching hypothesis
The theory that people tend to seek romantic and sexual partners who possess a level of physical attractiveness similar to their own.

On a final note: recent research has indicated that the degree of matching among romantically linked partners depends to some degree on the nature of the couple. For example, gay male couples tend to match each other on measures of attractiveness less than lesbian couples, who, in turn, show less similarity than heterosexual couples (Swartz, 2009).

The matching hypothesis says that people tend to select romantic and sexual partners who are similar in physical attractiveness to themselves, regardless of race, ethnicity, or sexual orientation.

Proximity

Proximity, in terms of relationships, refers to how close in physical distance you are to another person over time. Various studies have shown that as you spend more time in proximity to other people, whether by living close by, working in the same office, taking many of the same classes, sitting near each other in class, and so on, you are more likely to develop positive feelings toward them. The closer you are in geographical distance, the greater the probability that you will grow to like, or even love, someone (assuming that the person does not violate too many criteria in your field of eligibles). Of course, there are exceptions to this basic theory; sometimes, the more

The proximity effect suggests that romantic relationships are more likely to occur when two people spend more time together in the same physical space.

you see someone, the *less* you like him or her. However, all things being equal, people tend to grow fonder of someone around whom they spend more time. Three reasons have been suggested for this **proximity effect**; consider how they might relate to your own situation.

First, it stands to reason that if you are sharing the same physical space with another person, you will have more opportunity to meet and get to know each other. By its very nature, this fact increases the chances of forming a romantic relationship.

Second, the more you find yourself in the same situation with another person, the more likely it is that the two of you have interests in common. Common interests pave the way for mutually interesting topics of conversation and shared activities, all of which enhance the probability that a romantic relationship may develop. To support this idea, the National Health and Social.

Life Survey of sexuality and relationships in the 1990s found that the two common locations where people meet and form romantic relationships are school and work. The survey's authors contend that these two settings are particularly conducive to the formation of relationships for two reasons, both relating to the proximity effect. First, both situations provide numerous opportunities for frequent contact with the same groups of people, which increases the odds that romantic relationships may develop. Second, both settings naturally involve people with similar backgrounds, goals, and interests, which further increase the potential for forming intimate partnerships (Laumann et al., 1994). In other words, your college or work environment, or online preferences typically contains an exceptionally large number of people who qualify for your field of eligibles.

proximity effect
The theory that the closer you are to another person in geographical distance, the greater the probability that you will grow to like or even love the person.

mere exposure effect
The psychological principle that humans appear to have a natural and usually unconscious tendency to grow fonder of a "novel stimulus" the more often they are exposed to it.

A third explanation proposed for the influence of proximity on relationships is a psychological principle called the **mere exposure effect**. Research has shown that humans appear to have a natural and usually unconscious tendency to grow fonder of a "novel stimulus" as they see it more often (Bornstein, 1989; Hicks & King, 2011; Zajonc, 1968, 2001). This is true of nearly anything you come across—a new model of car, an architecturally unusual building, a product in a TV commercial—and it's true for your perceptions of people as well. If we use the mere exposure effect to explain the impact of proximity on relationships, simply encountering a person more often, without even talking or acknowledging each other, will lead to greater familiarity, which in turn increases liking, which then enhances the probability of forming an intimate connection (Lee, 2001; Reis et al., 2011). One important exception to this theory is that if, upon first encountering a new person, your initial reaction is profoundly negative, the mere exposure effect may not occur, and instead your dislike may increase with repeated encounters.

Similarity

This concept goes beyond the matching hypothesis discussed earlier and includes many additional factors between two people that may or may not be similar. Are you more attracted to people who are similar to you or to those who are more complementary, that is, those who possess the characteristics you lack? Do "birds of a feather flock together," or do "opposites attract"? If you rely on your personal experience or common sense to answer these questions, you might be led down the wrong path. Many people believe in the saying that "opposites attract" and can point to instances in their own or others' lives when it appeared to be true. One study found that 77% of college students believed in the statement that opposites attract (Lilienfield et al., 2010). But in reality, just the opposite is true: regardless of sexual orientation, people are generally more likely to be attracted to and become romantically involved with others who possess attitudes, interests, and personality characteristics similar to their own (Decuyper, M., De Bolle, M., & Fruyt, F., 2012; Lehr & Geher, 2006; Rothblum, 2009).

Upon reflection, this makes sense. Interpersonal similarity allows two people to enjoy the same activities, to agree on most of life's major issues and decisions, to feel a mutual understanding, to reinforce and support each other's attitudes and opinions, share a joint sense of humor, and to feel that they form a united front in dealing with the world around them. All of these shared characteristics and behaviors combine to produce a stronger bond between two people than would be possible if the similarities did not exist (Barelds & Barelds-Dijkstra, 2010).

This does not imply that people are looking for a relationship with their twin or clone. Although perceived similarity appears to be important in satisfying relationships, people still seek potential partners who offer at least a few contrasts and differences to themselves (Markey & Markey, 2007). For example, someone who is a submissive person might feel more comfortable with a more dominant partner, and vice versa, even when the partners consider themselves to be quite similar in other respects.

Since You Asked

2. My new partner (of two months) and I are very different. We disagree and argue about politics, religion, and nearly everything else. Is this going to be a problem, or is it true that opposites attract?

Similarity appears to be a far more important and more powerful force than complementarity in successful, happy relationships. It is so important, in fact, that people will even make the assumption that their partners are more similar to themselves than they actually may be. A study of dating and married couples found that "people in satisfying and stable relationships assimilated their partners to themselves, perceiving similarities that were not evident in reality" (Murray et al., 2002, p. 563).

Flirting

Most of you reading this chapter have flirted with someone, have been flirted with, or both. You may have even engaged in flirtatious activities without being fully aware that it was happening. **Flirting** may be defined as subtle behaviors designed to signal sexual or romantic interest in another person. Because humans across all cultures share this tendency, flirting may be seen, to some extent, as a biologically based, genetically programmed behavior that increases the ability of the human species to "mate" and survive. However, flirting in most cases probably has very little to do with procreation. Most adults who flirt with each other (whether opposite- or same-sex) are doing so to communicate mutual sexual interest, but the flirting signals no commitment. It allows you to "check someone out" as a potential sexual partner at the earliest possible stage in a potential relationship, when rejection carries little risk of emotional pain (Fox, 2004; Rodgers & Veronsky, 1999).

flirting
Subtle behaviors designed to signal sexual or romantic interest in another person.

What is most interesting about flirting is that the behaviors involved in the flirtation process are quite similar in all humans and tend to follow fairly predictable patternsboth in-person and online (Whitty, 2004). For heterosexual individuals, these behaviors send signals not just about sex but also about the larger question of who will be the best person for mating and reproduction. You have probably not always been aware of your own or others' behaviors while flirting. After reading this section, however, you will be an informed observer of the flirting world around you. Although we will discuss flirting in heterosexual settings (where nearly all the research to date has been carried out), the behaviors and responses are often similar for gay and lesbian flirting couples. However, for same-sex flirtations, the roles of each partner are more flexible, the behaviors more varied, and the partner who takes the initiative along the flirting path is typically less well-defined (see Kiesling, 2013; Lott & Veronsky, 1999).

To study exactly how and when these behaviors occur in actual flirting situations, researchers have spent many hours in bars, lounges, restaurants, clubs, and other locales where people are likely to be flirting (Henningsen Braz & Davis, 2008; Moore, 2002; Perper, 1986; Perper & Fox, 1980). Through these observations, not only have the behaviors of flirting been identified, but a predictable pattern of flirtation has also been revealed. The five steps in a typical flirting episode have been labeled *the approach, talk, swivel and turn, touch,* and *synchronization* (see Perper & Fox, 1980). At each step, one member of the couple makes a subtle gesture designed to take the

Considerable research has explored the variety of flirting behaviors humans use to signal romantic or sexual interest in another person.

developing intimacy to the next level. It is then up to the other person to respond appropriately if the flirtation is to continue. If this back-and-forth "dance" is interrupted by either partner, the developing relationship will likely fail to materialize.

Imagine that you are observing people at a bar, lounge, or large party where people go to check out, meet, and potentially hook up with others. Here are the typical steps of flirting you are likely to see in a heterosexual setting within the space of about one hour (Martin, 2001; Perper, 1986; Perper & Fox, 1980).

1. **The Approach.** The initial contact between potential partners begins with a look, usually from the woman. The cliché "their eyes met across a crowded room" appears to contain a grain of truth. A woman will scan a room of potentially desirable men and settle on one who piques her interest the most. When he returns the eye contact, she will either glance down and back up or maintain her gaze toward him. At this point, the man may begin a slow but steady movement toward her and she may move to be more reachable.
2. **Talk.** Obviously, once side-by-side, the next step is for someone to say something. Both flirting participants would be very uncomfortable if the man were to go to the effort to reach the woman and then they just stood next to each other without speaking. But who talks first when a man and a woman are standing there next to each other? Usually, he does (Weber, Goodboy & Cayanus, 2010). He engages in small talk (too much self-disclosure at this point is a turnoff), and he typically asks a question of some sort, requiring an answer. Again, the woman must reciprocate; say something in return, or as you can imagine, the interaction may wind down rather quickly.
3. **Swivel and Turn.** Body language plays a role in most human communication, but at this stage in the flirting process, it becomes more pronounced than usual. Typically, the flirting pair will be standing or sitting side by side rather than face to face. As the couple continues to talk and become acquainted, they will begin to shift their stances, little by little, toward each other. One person will swivel and turn slightly, and within a short time, the other will reciprocate. Then they will take turns swiveling a bit more each time until, *voilà!*: they are face to face.
4. **Touch.** If all has been going well up to now, this next step is a clear escalation in the flirtation game: touching. Who do you think initiates the first touch? It's the woman. The first touch will be a subtle, seemingly "accidental" brief touch of his hand when she laughs at something witty he has said, picking a piece of lint from his shoulder, or touching his arm as she whispers in his ear. This is then reciprocated either by a subtle touch in return, a thank-you, or a smile.
5. **Synchronization.** Once again, body language becomes the focus of the final stage of flirtation. Synchronization means that the couple has established an easy, flowing unison of movement. They begin to turn their heads at the same time, pick up and put down their drinks together, shift their weight if standing, or lean forward with their chins in their hands if sitting, and they even begin to breathe in the same rhythm in a virtual dance of flirtation. Research has shown that this synchronicity appears to happen naturally and cannot be faked. When actors are asked to mimic this stage of flirting, they have great difficulty appearing realistic and natural.

After a couple completes these five steps of flirting, has a new intimate relationship been born? At this point, they cannot know. A couple reaching this point in flirting still knows very little about each other and may or may not take the relationship to the next, more intimate and private level. In heterosexual settings, the male usually makes such a suggestion. If he does not do so at this point, the relationship will usually end here.

Exceptions to this flirtation pattern often occur (especially if the flirting individuals are under the influence of alcohol or other drugs). But if you think back over situations in which you or others have been flirting, and if you keep these steps in mind when you next find yourself in a flirtatious environment, you will likely see how accurate they are.

Reciprocity of Attraction

Have you ever been in love with someone who did not share your feelings no matter what you did? Feels terrible, doesn't it? Such relationships (if they can even be called that) are unsatisfying for both people and typically do not last very long. A long-established principle of successful interpersonal relationships is that they are based on *balance* (Bagarozzi, 1990; Zimmerman et al., 2003): balance of power, balance of control, balance in money matters, balance in decision making, and so on. This type of balance does not mean that the couple must be equal in all relationship matters, but that the relationship should not be overly controlled by either partner. One type of balance, in particular, is very important at the earliest stages of a potential relationship. This balance is called **reciprocity of attraction**, which simply means that if you like or love someone, you need to feel that the other person likes or loves you back *approximately* in the same way and amount—that he or she *reciprocates* your feelings. Knowing that someone you like, likes you back is one of the key factors in the formation and continuation of new romantic relationships (Lehr & Geher, 2006; Luo & Zhang, 2009).

reciprocity of attraction
The idea that someone you like or love likes or loves you back—reciprocates your feelings—with approximately the same degree of intensity.

Since You Asked

3. I feel I love my partner far more than he loves me. Is there some way to get him to love me more?

Psychologically, this makes perfect sense. When you like or love someone, you usually want to share various aspects of yourself with him or her, share your feelings and beliefs. When you receive the openness back from a potential partner, it is much more likely that a budding relationship will deepen and grow into a romance. However, when this reciprocity is lacking, one or both people may experience negative feelings such as anger, sadness, anxiety, or confusion—or guilt, resentment, or frustration—depending on which side of the lack-of-reciprocity equation you are on.

Interestingly, reciprocity of attraction may help explain why individuals with poor self-esteem and high self-doubt have trouble forming lasting romantic relationships. The reason appears to be in part because people who doubt themselves also doubt that a partner they love could ever love them back equally. They feel unworthy of their partner's love, even when the partner really does reciprocate the love and tries to let the other know (Murray et al., 2001). In such a relationship, true reciprocity can never be achieved no matter how hard a partner tries to establish it.

What Is Love?

4.3 Describe and discuss the various types of intimate relationships that may be explained by the Triangular Theory of Love and the concept of "love styles."

Ask a hundred people to define romantic love, and you'll probably get a hundred different answers. If love is such a personal and individual phenomenon, it must be impossible to study scientifically, right? Well, not really. Researchers have attempted to develop theories of love that can encompass everyone's individual definitions into an organized set of interrelated categories or types. Two of these theories have received considerable attention and research support over many decades: Robert J. Sternberg's "Triangular Theory of Love" and John Allen Lee's "Styles of Love."

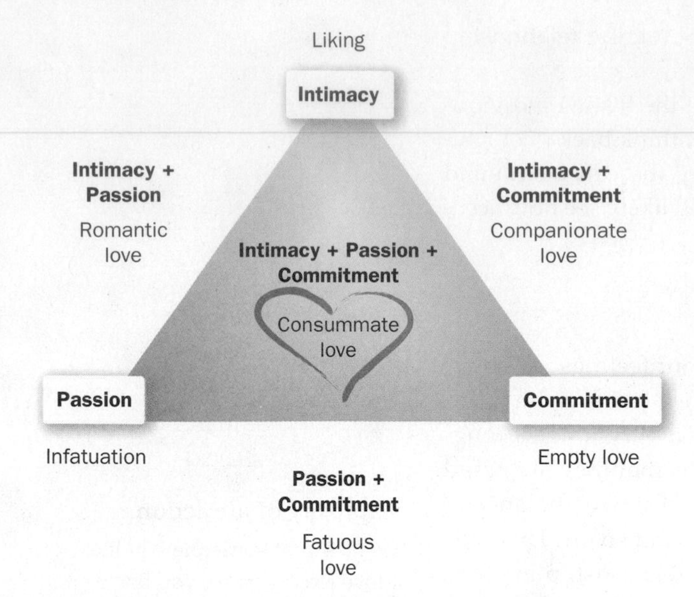

Figure 4.1 The Triangular Theory of Love

Sternberg's theory of love includes three fundamental components—intimacy, passion, and commitment—which, in various combinations, define seven types of love relationships.

SOURCE: Figure from THE TRIANGLE OF LOVE by Robert J. Sternberg. Copyright © 1998. Published by Basic Books. Reprinted by permission of Dr. Robert J. Sternberg.

The Triangular Theory of Love

No, this does not (at all) refer to the familiar "love triangle," in which three people are intimately and sexually intertwined. Rather, Robert Sternberg calls his model the **triangular theory of love** because he conceptualizes the three fundamental components of love—intimacy, passion, and commitment—positioned at the three corners of a triangle, forming various combinations that define the qualities of a relationship (Leumieux & Hale, 2002; Sternberg, 1986, 1988, 1997, 1998, 2014). A relationship may consist of any one of these components, any combination of two, or all three, and relationships that manage to maintain all three usually are longest-lasting and happiest (Acevedo & Aron, 2009).

Sternberg's *intimacy* component does not refer to sexual intimacy but rather to the emotional closeness two people feel. It includes such factors as wanting what is best for the partner, feeling the partner's happiness, holding the partner in very high regard, feeling able to count on the partner in times of need, sharing a sense of mutual understanding, giving and receiving emotional support, and being able to share private and personal thoughts and feelings with the partner.

Passion, Sternberg explains, is the physical arousal side of relationships. Passion is manifested in the increased heart rate when you are with your partner; the desire to be near your partner as much as possible; the sexual and romantic attraction you feel for your partner; the frequency of thinking about your partner; and the need to express your desire for your partner through touching, kissing, and making love.

The *commitment* component of Sternberg's model is a more rational aspect of a love relationship. It is determined by the strength of your decision to be with and stay with your partner. It is your chosen desire to be loyal and faithful and to commit to working on creating and maintaining a loving, mutually satisfying, and lasting relationship.

According to Sternberg, these three components may exist in any combination, from none of them, which is *non-love*, to all of them, which is *consummate love*. Overall, seven possible combinations can help couples see their relationship more clearly and explore what is working well or what might be causing the difficulties they have been experiencing. The combinations of the components of love are summarized in Figure 4.1; we will discuss each of these briefly. (Non-love, as noted, reflects a lack of all three components.)

triangular theory of love
Sternberg's theory that three fundamental components of love—intimacy, passion, and commitment—in various combinations define the qualities of a relationship.

infatuation
Love based on passion but lacking intimacy and commitment; usually very sexually charged but shallow and devoid of much meaning.

Intimacy Only = Liking If you imagine a relationship in which two people feel intimacy but do not experience passion or a strong sense of commitment, what sort of relationship do you see? Most people see two people who like each other quite a lot and are probably good friends. Sternberg agrees and characterizes a relationship containing intimacy only as *liking*.

Passion Only = Infatuation Now think of two people who are just bursting with passion and sexual heat for each other but who do not feel particularly intimate and are not committed to any sort of short- or long-term relationship. What would you call this type of love? Spring break, right? In a way, yes. When you experience **infatuation** with someone, you are very attracted to and focused on that person, usually in a sexual way, and you may desire to spend all your time with him or her. But the relationship does not go much beyond that. It's very sexually charged, but at the same time, it's shallow. You may know (or care) little about the other person, so you don't experience much intimacy and you aren't even thinking in terms of a commitment. It's simply a passionate connection that might be fun and sexy, but it exists only in the moment.

Commitment Only = Empty Love This corner of the triangle is a little more difficult to imagine. Can you imagine being committed to someone without feeling any intimacy or passion? This may happen when attraction is not reciprocated. Imagine that you love someone with whom you have not established any intimacy and no real passion exists between you. How would that love feel to you? That's right, it would feel empty. Such **empty love** relationships are unlikely to have much of a future, unless a person becomes too focused on the other and develops an unhealthy "fatal attraction" or "stalker" sort of obsession. However, not many would define that as love at all. Some couples with children who are experiencing empty love might stay together "for the sake of the kids," but this is typically an unsatisfactory solution for everyone involved.

empty love
Love based on commitment but lacking intimacy or real passion.

Intimacy + Passion = Romantic Love This side of the triangle connects intimacy and passion. If you get to know someone well, establish a deep level of intimacy, and then add passion to that, the result is probably going to feel very romantic. And it will probably feel romantic regardless of whether or not you have established a commitment with that person. A good example of this is the so-called shipboard romance, that short-term, intensely romantic relationship that sometimes develops between two people who meet on a cruise or at a resort or have a brief affair outside their primary relationships. It is more than mere passion because they connect on an emotional and personal level in addition to the physical attraction. But due to the circumstances of the situation or other involvements in their lives, they do not choose to commit to one another. This is **romantic love**.

Since You Asked

4. When a couple is having sex on a regular basis, does the emotional side grow over time? Does it mean the relationship is not going well if it is just for the sex?

romantic love
Love based on intimacy and passion but lacking commitment.

Passion + Commitment = Fatuous Love Now imagine, if you can, two people who are very physically attracted to each other and share a strong sexual bond between them. In addition to having "the hots" for each other, they also feel a strong commitment to making the relationship last over the long term. However, they lack intimacy. That is, they don't really like each other all that much; they don't hold each other in especially high regard (except perhaps in terms of sexual skills); and they have never achieved close, private, intimate communication with each other. Sternberg labeled this side of the triangle **fatuous love**. *Fatuous* is a fairly uncommon word that means "absurd," "foolish," or "pointless."

fatuous love
Love based on passion and commitment but lacking intimacy; a foolish or pointless love.

Commitment + Intimacy = Companionate Love **Companionate love** is a relationship characterized by two people who are truly in love and are committed to each other and who enjoy all or most of the characteristics relating to intimate love or liking. What's missing is the heat, the sexual arousal, the physical longing when apart; the *passion*. How might we describe such a couple? Without passion, it is difficult to see them as lovers, but rather they are companions, hence the term *companionate love*.

companionate love
Love based on true intimacy and commitment but lacking passion; the partners are companions more than lovers.

Intimacy + Passion + Commitment = Consummate Love Finally, what if a couple is fortunate enough to possess all three of Sternberg's basic components of love? They will have what he termed **consummate love**, meaning the most complete, most fulfilling, most ideally perfect love two people can achieve. Sternberg believes—and research has borne him out—that consummate love is not only rare but also difficult to attain, and perhaps even harder to maintain over time.

consummate love
Love that encompasses intimacy, passion, and commitment simultaneously.

Applying the Triangular Theory of Love

Sternberg's theory can be helpful in relationships in which one or both partners feel dissatisfied or sense that something is missing. Sometimes couples find it difficult to put their finger on exactly what is causing their feelings of discontent or discord with their partner. If they examine their relationship from the perspective of the triangular theory, they may make two important discoveries. First, they may feel encouraged and relieved to find that they are quite strong on one or two of the components.

This often helps validate their positive feelings for each other and motivates them to work on other areas in the relationship. Second, they will be able to identify more specifically the aspects of their relationship that may be weak, missing, or in need of work. This, then, will enable them to establish a mutual focus for enhancing and improving the bond between them.

Most people in love relationships tend to feel unhappy at times if any one of the three components—intimacy, passion, or commitment—is weak or missing. You can be in a committed relationship and feel lonely and disconnected if you do not feel that special intimacy with your partner. You can feel angry, frustrated, and betrayed if your relationship lacks commitment. And many couples find themselves dissatisfied and longing for the passion that has faded over time from their relationship, although they continue to experience a strong sense of commitment and intimacy. (This loss of sexual desire is discussed in greater detail in Chapter 7, "Sexual Problems and Solutions.")

Are you curious about *your* love triangle? "Self-Discovery: The Triangle of *Your* Love" offers you the opportunity to assess where your relationship (current or past) might fall within Sternberg's theory. Take the scale yourself or with your partner (gently, as a basis for meaningful discussion, not as an argument starter!), or you and your partner might complete the assessment separately and then discuss your results.

Self-Discovery

The Triangle of *Your* Love

Each component in Sternberg's triangular theory of love is measured by your responses on fifteen items, for a total of forty-five items. Think about a past or present relationship partner (just one!). If you have not yet been in a love relationship, think of how you might answer when you are. Respond to each item using the following key:

1 — 2 — 3 — 4 — 5 — 6 — 7 — 8 — 9
"Not at all" "Moderately" "Extremely"

The scoring key is at the end of the scale. You can take the scale on your own, with your partner, or separately before discussing your results. (Be aware that some responses may be unexpected or displeasing, so exercise caution when sharing this assessment with your partner.)

Intimacy Component

______ **1.** I am actively supportive of my partner's well-being.
______ **2.** I have a warm relationship with my partner.
______ **3.** I am able to count on my partner in times of need.
______ **4.** My partner is able to count on me in times of need.
______ **5.** I am willing to share myself and my possessions with my partner.
______ **6.** I receive considerable emotional support from my partner.
______ **7.** I give considerable emotional support to my partner.
______ **8.** I communicate well with my partner.
______ **9.** I value my partner greatly in my life.
______ **10.** I feel close to my partner.
______ **11.** I have a comfortable relationship with my partner.
______ **12.** I feel that I really understand my partner.
______ **13.** I feel that my partner really understands me.
______ **14.** I feel that I can really trust my partner.
______ **15.** I share deeply personal information about myself with my partner.

Passion Component

______ **16.** Just seeing my partner excites me.
______ **17.** I find myself thinking about my partner frequently during the day.
______ **18.** My relationship with my partner is very romantic.
______ **19.** I find my partner to be very personally attractive.
______ **20.** I idealize my partner.
______ **21.** I cannot imagine another person making me as happy as my partner does.
______ **22.** I would rather be with my partner than with anyone else.
______ **23.** There is nothing more important to me than my relationship with my partner.
______ **24.** I especially like physical contact with my partner.
______ **25.** There is something almost magical about my relationship with my partner.
______ **26.** I adore my partner.
______ **27.** I cannot imagine life without my partner.
______ **28.** My relationship with my partner is passionate.

______ **29.** When I see romantic movies and read romantic books, I think of my partner.

______ **30.** I fantasize about my partner.

Commitment Component

______ **31.** I know that I care about my partner.

______ **32.** I am committed to maintaining my relationship with my partner.

______ **33.** Because of my commitment to my partner, I would not let other people come between us.

______ **34.** I have confidence in the stability of my relationship with my partner.

______ **35.** I could not let anything get in the way of my commitment to my partner.

______ **36.** I expect my love for my partner to last for the rest of my life.

______ **37.** I will always feel a strong responsibility for my partner.

______ **38.** I view my commitment to my partner as a solid one.

______ **39.** I cannot imagine ending my relationship with my partner.

______ **40.** I am certain of my love for my partner.

______ **41.** I view my relationship with my partner as permanent.

______ **42.** I view my relationship with my partner as a good decision.

______ **43.** I feel a sense of responsibility toward my partner.

______ **44.** I plan to continue my relationship with my partner.

______ **45.** Even when my partner is hard to deal with, I remain committed to our relationship.

Scoring Key

Add your ratings for each of the three sections—intimacy, passion, and commitment—and write the totals in the blanks. Divide each score by 15 to get an average scale score or *rating.*

Intimacy score ________ ÷ 15 = ________ Intimacy rating.

Passion score ________ ÷ 15 = ________ Passion rating.

Commitment score _____ ÷ 15 = ________ Commitment rating.

Here's what your scores mean in terms of how you see your relationship:

1–3 = This component is low or lacking and could indicate a serious weakness in your relationship satisfaction or strength.

4–6 = Your relationship contains a moderate level of this component, but it could be worked on and strengthened.

7–9 = Your relationship is on a solid footing for this component.

Examining your ratings for each of the three scales will give you an idea of how *you perceive* the level of intimacy, passion, and commitment in your love relationship. If you feel your relationship is less than ideal, this knowledge may offer insights into how to make it stronger.

SOURCE: Scale, "The Triangle of Your Love" from THE TRIANGLE OF LOVE by Robert J. Sternberg. Copyright © 1998. Published by Basic Books. Reprinted by permission of Dr. Robert J. Sternberg.

One final note about the triangular theory: When college students in my human sexuality classes are asked if they feel Sternberg's theory had omitted any fundamental components of love relationships, many express that *effective communication* was at least as important as intimacy, passion, and commitment. Upon further examination and discussion, students conclude that communication was not really a separate component but an inherent part of each of the other three. Without mutually effective communication, they believed that intimacy would never be attained, passion would be less than satisfying, and neither partner would ever be sure of the other's commitment to the relationship. The students contended, therefore, that communication affects all seven types of love described by Sternberg (and even perhaps non-love). Figure 4.2 shows how they decided to revise Sternberg's model. The importance and the challenges of effective communication in love relationships will be our next topic of discussion.

Figure 4.2

A Modified Triangular Theory of Love, Incorporating Communication (based on student analysis)

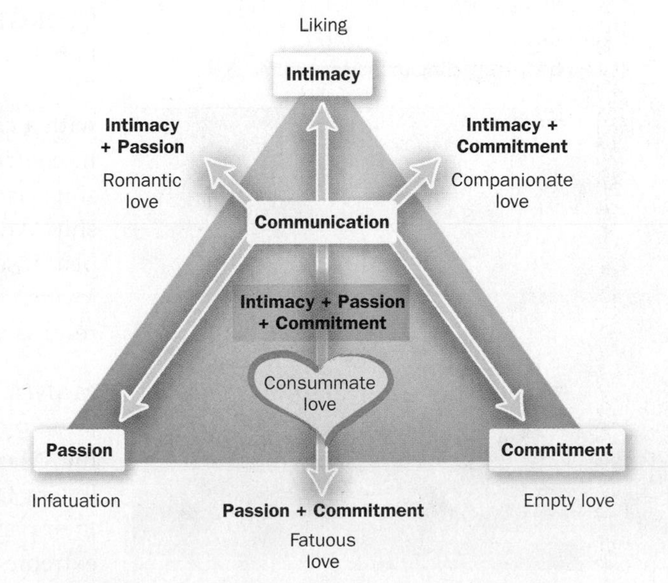

Styles of Love

Another enduring theory about the nature of love has attempted to measure what "kind" of lover you are. No, this is not referring to your skills or talents in the bedroom, but rather your approach

The Eros style of love emphasizes passion, eroticism, and sexual energy.

to and style of relating to a partner when you are in a romantic relationship. Developed by the Canadian sociologist John Allen Lee in the 1970s, this theory of love suggests that people follow various psychological motifs in relating to a love partner. Lee divided these love patterns into six major categories he called **styles of love** (Lee, 1973, 1977, 1988). Over the years, Lee's conceptualization of these styles of love has served as the basis for a great deal of research on intimate relationships for opposite- and same-sex couples (e.g., Davis et al., 2013; Goodboy & Booth-Butterfield, 2009; Kunkel & Burleson, 2003; Zamora et al., 2013). Lee applied Greek mythology and terminology to name his six identified love styles. We will briefly describe each of them here. If you would like to discover which style reflects yourself and/or your partner, pause now and fill out the assessment scale in "Self-Discovery: Styles of Love Scale" and then read the descriptions of Lee's six styles.

Since You Asked

5. Why do people cheat? Is it because they just need a new conquest?

EROS LOVE You have no doubt heard of Cupid, the god of love in Roman mythology. Cupid's counterpart in Greek mythology is Eros. In his theory of love styles, Lee conceptualized **Eros love** as erotic, passionate love. Eros lovers tend to place great emphasis on romance and physical beauty. They experience urgent sexual desires and strong physical attraction to their partners. They believe in love at first sight and have likely experienced it often. Eros lovers desire sexual intimacy earlier in a new relationship than those embracing other styles, and they value tactile (touch) sensations above the other senses. This style of love is very romantic and highly sexually charged, but typically contains a level of passion that cannot be maintained for long. Relationships based on Eros love tend to burn out quickly.

Ludus lovers enjoy playing the field and seek many sexual conquests.

LUDUS LOVE *Ludus* is Greek for "play," and **ludus love** is characterized by game playing. Ludus lovers enjoy the excitement of new relationships more than the relationships themselves—they like the "chase." They love to flirt and seduce their partners. They "play the field," moving rapidly from one relationship to another or juggling several partners at once. As you might imagine, ludus lovers are very unlikely to form a lasting commitment and tend to avoid serious relationships altogether. Often they will end a relationship just when it appears to be approaching its closest and most satisfying stage because they resist anything hinting at commitment. At times, they will even begin a new relationship before ending the current one so that they are never without the rush and excitement of the pursuit.

Mania lovers' low self-worth make them obsessively cling to their partners.

STORGE LOVE In Greek, *storge* ("STOR-gay") means "natural affection." **Storge love** is characterized by the central theme of friendship. Those who adhere to this style of romantic relationship usually begin with a close friendship and take a long time to develop feelings of love. In contrast to ludus and Eros lovers, the sexual side of storge relationships arrives late and tends to take a back seat to the emphasis on friendship. Although passion is not a central feature, storge relationships offer peace, security, and stability, all of which are greatly valued. For storge lovers, more than for any other style, if love ends, the friendship usually returns and continues over time.

MANIA LOVE *Mania* is Greek for "madness." You can probably conjure up an image in your mind of a "manic" lover—obsessed and clinging. **Mania love** is possessive, dependent, and often controlling. Mania lovers lack self-worth, feel unlovable, and are constantly fearful that their partner will leave. These relationships are characterized by turmoil, extreme and unrealistic jealousy, and sometimes true obsession. Partners

Self-Discovery

Styles of Love Scale

For each of the following statements, write T for *true* or F for *false* to reflect how the statement applies to you in your love relationships. If you are not in a love relationship, base your answers on how you acted in your last significant one. If you have never been in a love relationship, first, don't worry, the right person is out there somewhere! Second, respond to the items based on how you imagine you *will* be when you do fall in love. Try not to over think your answers. If you are not sure how to answer, respond according to how you feel *most* of the time. Answer all the items. Instructions for scoring follow the scale.

TRUE OR FALSE?

______ **1.** My partner and I were attracted to each other immediately when we first met.

______ **2.** My partner and I have great physical chemistry between us.

______ **3.** I feel that my partner and I were meant to be together.

______ **4.** I have sometimes had to prevent two of my partners from finding out about each other.

______ **5.** Sometimes I enjoy playing "love games" with several partners at once.

______ **6.** I believe it's a good idea to keep my partner a little uncertain about my commitment to him or her.

______ **7.** I find it difficult to pinpoint exactly when my partner and I fell in love.

______ **8.** The most fulfilling love relationship grows out of a close friendship.

______ **9.** It is necessary to care deeply for someone for a while before you can truly fall in love.

______ **10.** When I am in love, I am sometimes so excited about it that I can't sleep.

______ **11.** I am constantly worried that my partner may be with someone else.

______ **12.** When my partner is busy or seems distant, I feel anxious and sick all over.

______ **13.** It is best to find a partner who has similar interests to your own.

______ **14.** I try to make sure my life is in order before I choose a partner.

______ **15.** A person's goals, plans, and status in life are very important to me in choosing a partner.

______ **16.** I would rather suffer myself than allow my partner to suffer.

______ **17.** I cannot be happy unless my partner's happiness needs are met first.

______ **18.** I am usually willing to sacrifice my own needs and desires to allow my partner to achieve his or hers.

Scoring: The six styles of love discussed in this chapter are measured by each successive group of three statements. If you divide the scale into sets of three items, the six styles are as follows (refer to this chapter's discussion for a detailed explanation of the various styles):

Statements 1, 2, and 3 reflect *Eros* love.
Statements 4, 5, and 6 reflect *ludus* love.
Statements 7, 8, and 9 reflect *storge* love.
Statements 10, 11, and 12 reflect *mania* love.
Statements 13, 14, and 15 reflect *pragma* love.
Statements 16, 17, and 18 reflect *agape* love.

To determine your style, see if you have answered "True" for *all three statements* in any of these sets. If so, that is the style that best reflects your approach to love relationships. If you do not have any sets of three items to which you answered "True," look for sets where you have answered "True" to two items. These sets reflect your preferred style of love. Finally, see if you have any style for which you answered "False" to all three questions. This indicates the love style you most strongly reject; that is most unlike you.

Have you agreed with more than one style? This is possible, and it implies one of several possibilities. First, you may have shifted your style over time and behaved according to one style in a past relationship but are now behaving differently. You might want to try the scale again, focusing closely on your single most important or longest relationship. Second, some relationships can change and grow over time, and you may have changed and grown with them. Your love style earlier in a relationship may have been different from what it is now, but you responded to the scale with both in mind. Third, you may simply not be sure of your love style yet, either because you have never been in love or because you have just not thought about it very much. And fourth, some people may simply possess more than one love style that they may call on as needed in a particular relationship. Agreeing with a number of love styles on this scale might be a sign that more knowledge about yourself and your relationships is yet to come.

SOURCE: Adapted from Hendrick & Hendrick (1986).

styles of love
Lee's theory that people follow individual psychological motifs or styles in relating to a love partner.

Eros love
An erotic, passionate style of love often characterized by short-lived relationships.

ludus love
A style of love that focuses on the excitement of forming a relationship more than the relationship itself and typically moves rapidly from one relationship to another.

Individuals favoring the pragma style of love select romantic partners based on rational, practical criteria.

storge love

A love style characterized by caring and friendship.

mania love

A possessive, dependent, and often controlling style of love.

pragma love

A love style in which partners are selected in a businesslike way on the basis of rational, practical criteria.

The agape love style is a selfless, all-giving, devoted love.

of mania lovers may feel excited at first that they are so intensely loved and needed, but they soon find that they are emotionally controlled and smothered by a clinging, insecure partner. When mania lovers feel that a partner is drifting away, they may resort to such drastic measures as stalking, threats of suicide, actual suicide attempts, or physical violence to prevent the partner from leaving.

PRAGMA LOVE *Pragma* means "business" in Greek, and **pragma love** is characterized as a practical love. Pragma lovers go about selecting their partners in a businesslike way based on rational, practical criteria. You can't really say that pragma lovers fall in love; rather, they decide to love the partner who best fits their requirements. These requirements encompass some of the factors included in most people's field of eligibles discussed earlier in the chapter, but the pragma lover focuses on the most down-to-earth, *pragmatic* aspects for compatibility, such as education level, profession, social status, income, common interests, potential as a parent, and material possessions. Although on its face this may sound like an effective way to build a strong relationship, it turns out that these partnerships tend to be less mutually satisfying and often unsuccessful. You can probably see why: pragma lovers have a tendency to place too little importance on the emotional aspects of love that are so basic to bonding and forming strong attachments between people.

AGAPE LOVE *Agape* ("uh-GAH-pay") is the Greek word for "brotherly love" or "divine love." **Agape love** is a selfless love, and agape lovers offer their partners a self-sacrificing, altruistic love. This means that they strive to give to their partner whatever he or she may want or need without any expectation of receiving anything in return. The word *agape* has often been used to describe the love of God, of saints, and of martyrs. This style of love is patient and nondemanding. Sounds wonderful, right? In many ways it is, but the problem is that, although agape love may be wonderful as a way to love "all humankind" or as a way to describe the love between, say, a parent and a child, it turns out to be a rather weak form of romantic love between two adults. Why? Because agape love is only about giving, but real love involves a balance of giving and receiving by both partners.

Applying Love Styles

How can an understanding of these love styles help you to understand and improve your relationships? For one thing, discovering your and your partner's styles can allow you to communicate in new and deeper ways. Also, research has indicated that certain styles make better fits than others. Which styles do you think would work best together in a relationship? In general, people prefer to partner with others who have the same love style as their own (Hahn & Blass, 1997; Taggert, 2011). But what if two people with different love styles become romantically involved? Looking back over the descriptions of the six styles, you can see that some might lead to greater compatibility than others, some combinations just would not work well at all, and some might be downright "dangerous." For example, a storge lover and a pragma lover might get along just fine: one focused on establishing a close friendship and the other finding the perfect qualifications in the partnership. On the other hand, just imagine someone who embodies a strong mania style getting together with a pure ludus: One is maniacally in love, insecure, and clinging; the other is playing love games. What outcome might you expect from that ("homicide" might not be too far from the truth!)? Table 4.1 illustrates the quality of fit for various matches among the six love styles. Keep in mind that love relationships are extremely complex and that a couple's individual love styles tell only part of the overall story of intimacy and overall satisfaction (Hendrick, 2004; Lacy et al., 2004).

Table 4.1 Compatibility of Love Styles in Relationships

This table graphically illustrates which combinations of the love styles are more likely to result in compatible relationships, which may be successful if the relationship is strong overall, which might work despite some difficulties, and which are potentially dangerous due to the level of dissatisfaction and discord such diverse styles are likely to produce.

	Eros	Ludus	Storge	Mania	Pragma	Agape
Eros						
Ludus						
Storge						
Mania						
Pragma						
Agape						

GREEN = Good match BLUE = Possible match ORANGE = Difficult match RED = Dangerous match

Therefore, exceptions always exist, and some couples with incompatible love styles may create a successful relationship through skillful communication, clear agreements, and mutual respect.

agape love
A style of love focused on giving the partner whatever he or she may want or need without the expectation of receiving anything in return.

Communication in Love Relationships

4.4 Analyze the importance of effective communication between romantic partners, patterns of effective and ineffective communication, and strategies couples can use to improve communication quality in the relationship.

How often have you heard someone say after a relationship has ended, "We just couldn't communicate"? As mentioned earlier in our discussion of Sternberg's triangular theory, the ability of couples to communicate in clear, understandable, and meaningful ways is key to the happiness and long-term success of romantic relationships. Good communication skills allow couples to express the positive aspects of their relationship to each other, which tends to lead to more good feelings and a stronger connection. Equally important is the role communication plays in dealing with relationship problems and negative emotions when they arise.

Since You Asked

6. How do you know when things are going wrong or if things are right between you and your partner?

Look back at Figure 4.2. You can see how the students who formulated this model envisioned communication as fundamental to love relationships. This conclusion is supported by a great deal of research: The effectiveness of the pattern of communication in a relationship is one of the most accurate predictors of its future success or failure (Clements, Stanley, & Markman, 2004; Gottman & Carrere, 2000; Heyman et al., 2009). The large body of research on communication in relationships is too vast to discuss in detail here. However, examining several areas of study can provide some interesting and useful insights into effective or ineffective communication among couples and how they can learn to communicate better.

Patterns of Effective Communication

John Gottman, one of the leading researchers on why relationships fail and how couples can repair a deteriorating relationship by improving their patterns of communication has suggested that the often-perceived signs of relationship problems, such as intense verbal fights or shun-ning all conflict, are often effective strategies

A couple with a validating communication style resolves differences through quiet discussion and mutually respectful agreements.

for maintaining happy relationships. Gottman contends by airing and resolving differences loving relationships can grow closer, deeper, and long-lasting (Gottman, 1994).

Gottman's research has demonstrated that it is not how much a couple fights and argues but rather their *style* of conflict that determines the happiness and success of their relationship. Gottman claims that by examining a couple's communication and conflict styles, he can predict a relationship's outcome with around 90% accuracy (Gottman & Carrere, 2000; Gottman & Levenson, 2002). This prediction is based on the presence of healthy communication patterns versus specific warning signs that the couple has slipped into destructive modes of interaction. Interestingly, some of the effective communication patterns Gottman has identified may not sound very healthy on the surface, but they allow relationships to maintain a fundamental balance and closeness. Gottman (Gottman & Silver, 2000; Gottman & Gottman, 2008) has defined three patterns of effective communication most often seen in healthy relationships: *validating, volatile,* and *conflict-avoiding.*

Since You Asked

7. My partner and I fight a lot. We yell at each other, and sometimes one of us will storm out of the house for several hours. But we love each other very much and are totally committed to each other. Is this normal?

VALIDATING COMMUNICATION PATTERN In *validating communication*, conflicts are resolved through calm discussion and compromise. Both partners listen and seek to understand each other's problems, feelings, and points of view. Even when these couples are "fighting," they each make a point of acknowledging and respecting the validity of the other's position and emotions. Because of this mutual respect, couples with this communication style typically have fewer fights than others.

This may sound like the ideal relationship style and one that most or all successful relationships should achieve. However, research indicates that communication and conflict resolution patterns that are less validating, as discussed next, may be just as effective in keeping a relationship together.

VOLATILE COMMUNICATION PATTERN You have probably known couples (or been part of one) in which conflict arises frequently; the partners seem to fight, bicker, squabble, and even explode into shouting matches significantly more often than average. Although some of these *volatile* couples split up, others stay together over the long term and seem quite happy in the relationship, often seeming to enjoy their "combat." *Volatile* means "explosive," "flammable," and "unpredictable," and, Gottman maintains, perhaps a better term would be "passionate" couples because that is exactly what these relationships are (Gottman & Gottman, 2008). These couples don't hold back, and they readily express their feelings in strong terms. They see themselves as equals in the relationship and feel that they possess approximately equal abilities to defend their differing points of view. They solve their conflicts by fighting them out (although they avoid fighting tactics that devalue or belittle the partner) until they reach some conclusion both can accept (Gottman, 1998). On the flip side of the intense conflicts, however, is often an equally passionate and exciting love life.

The point is that "peace and quiet" are not prerequisites for relationship success. For some couples, even a seemingly stormy relationship may be strong, loving, and lasting.

Volatile couples stay together by openly and vigorously airing their feelings and fighting out their differences until they reach a mutually agreeable conclusion. They also tend to have a passionate and exciting love life.

Conflict-avoiding couples would rather downplay conflicts and avoid fights and discussions altogether.

CONFLICT-AVOIDING COMMUNICATION PATTERN What about couples that are the opposite of volatile, the ones whose main goal is to avoid conflict altogether? Sound like a bad idea? Not necessarily. Couples who are *conflict avoiders* may also find happiness and long-term success through the shared perception of a strong similarity between them and the apparent lack of importance they place on any differences or disagreements when they do occur. This might be called the "no big deal" approach in that they downplay conflicts they encounter and are not particularly concerned with reaching a mutual resolution. Although many people think this would only cause unresolved problems to fester and grow until they threaten the relationship itself, conflict-avoiding couples appear comfortable with this lack of resolution and truly seem able simply to "let it go."

Remember, all three of these communication styles allow couples to adjust to conflicts and problems in their relationships and, according to Gottman's findings, increase the likelihood for happiness and long-term success. Some research has indicated that Gottman may have overestimated the predictiveness of the various communication patterns he has examined. One study found that relationships with validating styles of communication were, overall, more successful than those with volatile styles (Holman & Jarvis, 2003). Such conflicting findings serve as an indicator of the complexities of communication in intimate relationships and are a reminder that much research remains to be done. Having said that, ineffective communication and conflict resolution are not the only reasons relationships fail; other important reasons will be discussed later in this chapter.

Communication Warning Signs

Along with the three effective communication patterns just discussed, Gottman and his research teams have also identified four communication warning signs that are common to most couples who find themselves drifting apart: criticism, contempt, defensiveness, and stonewalling. The goal in identifying these was not simply descriptive; rather, it was *prescriptive:* By understanding how relationships deteriorate, couples would be better able to intervene in the process and rekindle happiness and romance. Here is a brief description of Gottman's four warning signs, which he calls the "four horsemen of the relationship apocalypse" (Gottman, 1994; Gottman & Gottman, 2008; Gottman & Silver, 2000).

CRITICISM In successful relationships there exists a clear distinction between complaining and criticizing. **Complaining** is an expression of an unmet need, something a person desires but is not receiving. **Criticism**, however, involves an attack on the partner's actions. The unmet need begins to be blamed on the partner. Complaining, is typically healthy because it expresses each partner's needs, which allows for the need to be met. But if the needs expressed by the complaints are never met, they pile up, frustration builds, and it is then only a small step from complaints to criticisms. For example, a complaint might be "I wish we could spend more time alone, just the two of us." If the complaint is not acknowledged and dealt with in some way, over time it may become a criticism: "You don't want to be alone with me." This concept relates to the common relationship advice you have probably heard about fighting fairly: Use "I" statements, not "you" statements. If you find that many of the

complaining

Expressing an unmet need, something a person desires but is not receiving from a partner.

criticism

Verbal fault-finding, such as commenting on a character flaw in the partner.

complaints between you and your partner begin with "you," this is a sign that you may have slipped from complaining to criticizing.

contempt
Disrespect, disgust, or hate expressed when the positive feelings partners once had for each other have dissipated.

CONTEMPT If criticism is increasing in the relationship and is not averted in some way, it will eventually lead to contempt. **Contempt** implies that the true feelings the partners once had for each other have moved from liking and love to disgust, disrespect, and even hate: "You're so lazy and irresponsible; you don't even know what love is! I don't know why anyone would *want* to be with you." The difference between criticism and contempt is that criticism targets what a person does, but contempt is aimed at the person's character; comments are intentionally designed to insult and cause pain in ways only an intimate partner could know. These tactics may include specifically painful insults, name-calling, hostile humor, mockery, sneering at the partner's actions, indicating disgust for the partner, or curling the upper lip in the classic expression of contempt. Continuing this example, contempt might sound something like, "You're such a jerk—I'm glad you're not around much!" As contempt builds, the couple loses the love and attraction they once felt for each other. They no longer complement each other or interact as two people who are part of an intimate partnership; instead, their focus turns to abusiveness toward one another (Gottman, 1994; Gottman & Notarius, 2002).

DEFENSIVENESS If someone treats you with contempt or criticism, you are likely to become defensive. As natural and normal as that may be, it is bad news in a love relationship because *defensiveness* allows each partner to deny responsibility for whatever problems are causing friction. You can see that this sets up a nearly impossible obstacle for overcoming problems: If neither partner is at fault, then all that's left is blame (Gottman & Gottman, 2008). It is a good idea to be able to gauge whether your own or your partner's response to disagreements is becoming defensive, so that if this is happening in your relationship, you can be aware of it and try to reverse it. Imagine one partner saying to the other, "You're late again. You don't care about my schedule at all, do you?" Here are some signs of defensive tactics, along with an example of a defensive response to each criticism.

- Denying responsibility and blame for any problem your partner brings up: *"It's not my fault I'm late—class got out late."*
- Making excuses and claiming external forces out of your control made you behave in some way that upset your partner: *"What was I supposed to do? I had to eat lunch, didn't I?"*
- Disagreeing with your partner's attempt to attribute negative thoughts to you: *"What makes you think you know what I care about?"*
- Cross-complaining—when your partner makes a complaint or criticism, you respond with one of your own, relevant or not: *"Well, you always get upset when I want to go out with my friends."*
- Repeating yourself and your position over and over rather than trying to reach a compromise: *"I've told you before, this is about as punctual as I can get."*

stonewalling
Relying on a passive form of power and aggression by being unresponsive (erecting a metaphorical "stone wall") when disagreements and disputes erupt.

STONEWALLING Finally, as the relationship sinks deeper and deeper into hopelessness, the partners begin to give up on trying to make it work. One or both may stop engaging altogether when problems arise. This is called **stonewalling**—turning oneself into a "stone wall" when disagreements and disputes begin. In a way, stonewalling is a passive form of power and aggression. If you have ever tried discussing a personal problem with someone who simply shuts down, will not interact, and won't even offer a "hmmm" or a nod of the head, you know how powerful that silence can be. Gottman contends that stonewalling is the culmination of the three previous danger signs of criticism, contempt, and defensiveness. By stonewalling, the person is saying, "I don't care enough about you or this relationship anymore even to *try* to work things out." Stonewalling is the final event in a deteriorating relationship that leads one or both partners to say, "So, I guess it's over."

Even when a relationship reaches a seemingly unsalvageable state, however, Gottman believes that all hope is not lost and a couple can still learn how to develop the strategies they need to right the ship and closeness, intimacy, and satisfaction. Helping couples learn how to communicate their true feelings, attitudes, and desires is one of the primary paths to renewing a collapsing love relationship. With that said, we now turn our attention to strategies couples can use to improve communication patterns.

Improving Communication

One strategy that couples can use to maintain positive communication is to work on preventing any of the "four horsemen"—criticizing, contempt, defensiveness, and stonewalling—from occurring. That's easy to say, but without specific techniques to develop, preserve, and renew effective communication and positive connections between partners, this may seem like a daunting task for troubled couples. What does research have to offer in the way of specific suggestions to help couples communicate in ways that enhance intimacy?

Couples need to acknowledge that maintaining positive personal connections may be the most powerful tool they can use in building and maintaining relationship happiness and intimacy. Consequently, then, couples must understand the processes in love relationships and make an effort to incorporate that understanding into their interactions, so that their interactions bring their relationship closer rather than weakening it or tearing it apart.

UNDERSTANDING INTIMATE COMMUNICATION: FIVE "KEYS" Effective communication should be an integral, ongoing part of a successful, happy relationship, not simply something a couple makes a stab at when a big problem arises. In more than two decades of research into relationship conflict and communication processes, relationship researchers Howard Markman, Scott Stanley, and Susan Bloomberg have teamed up to explore how partners in intimate relationships either lack or lose effective interpersonal connection skills and what they can do to develop or rebuild them. These researchers have developed a plan for helping couples overcome problems and enhance their relationships. They call their program "PREP" for Prevention and Relationship Enhancement Program. They have condensed some of their findings into basic principles of communication and relating that they refer to as "the five keys" for a great relationship. These five keys have, at their core, the goal of effective connections in love relationships (adapted from Markman, Stanley, & Bloomberg, 2010, pp. 21–28).

Key 1: Decide Don't Slide. If you have experienced a relationship that has lasted for a while, you probably have noticed that when problems crop up between you and your partner, it often seemed easier to avoid them, ignore them, or push them aside instead of doing the work necessary to "fix" them; that is, you tended to let the relationship slide along expecting that things would work themselves out. Although this may seem easier, it is such a passive approach that it leaves the relationship in charge of you rather than you, as a couple, taking charge of it. Instead, making conscious, ongoing efforts to communicate about the direction you each want and expect the relationship to take and deciding how to make that happen will protect both partners from surprises and hidden individual expectations that may carry enough weight to tear a relationship apart.

Key 2: Do Your Part. A great temptation in many relationships is for each partner to see the other as the reason for problems that may arise. And, of course, one partner often does bear more responsibility than the other. However, understanding that relationships are a team effort can often help with resolutions. Each partner must take his or her share of the responsibility in making the relationship work and keeping it satisfying. The more each partner accepts this responsibility, the fewer problems are likely to arise, and, when they do, they are likely to be resolved more quickly and easily.

Many couples, as time goes on, forget to engage in behaviors that nurture a relationship, such as doing things for a partner that he or she will like and for no particular reason or expectation of receiving anything in return. Also, in most new relationships, when one partner is annoyed by something the other does or says, he or she will usually dismiss it; just let it go. This is a strategy that should be maintained throughout a relationship. If the issue is too important to be shrugged off, the annoyed partner should work to find a time to discuss it calmly and come to some agreement and solution. One way to do this is with something called "couples meetings," which we will discuss later in this chapter.

Key 3: Make It Safe to Connect. Conflict and problems in relationships are common and normal. Healthy relationships find effective strategies for working through and solving conflicts so that they don't continue to fester and form a negative undercurrent in the couple's life. However, to create an environment in which problem solving can happen requires that each partner feel safe to express his or her side of the issue. When one partner feels the other will become hostile, will respond with put-downs, or will belittle the other's concerns, this creates an emotionally unsafe environment for the positive connection between them that is necessary to work through the problem. Both partners must feel free and safe to approach the other with a problem and feel that the partner will be open to working on and solving the conflict. "When you have the skills to handle conflicts, you are able to relax, to be yourself, and to open the doors to emotional and physical intimacy" (Markman, Stanley, & Bloomberg, 2010, p. 25).

Key 4: Open the Doors to Positive Connections. The usual reasons romantic couples get together is because each perceives that the relationship offers many positive additions to his or her life: love, fun, intimacy, passion, mutual support, a special kind of friendship, etc. Although all relationships have conflicts and difficult times, it is these positive attributes that help successful, happy couples maintain in their lives together. Focusing on the good in a relationship is at least as important as knowing how to work through the bad. As a Swedish proverb says, "Shared joy is double joy; shared sorrow is sorrow halved."

Key 5: Nurture Your Commitment. This is really what Keys 1–4 are all about: Commitment. Each partner must know, beyond any doubt, that the other's commitment to the relationship is strong and must show and tell the other that this commitment is reciprocal. A solid commitment allows each partner to know that the other can be counted on to be there to help and support him or her, no matter what.

ENHANCING INTIMATE COMMUNICATION: FIVE "RULES" What sorts of rules and skills will help couples develop and preserve loving, happy relationships, or repair them when they run into serious trouble? Intimate relationship researchers Notarius and Markman (1993) suggested a strategy they called *couple meetings* that partners can incorporate into their relationships to help keep them open, on track, and communicating well. Couple meetings require an agreement between partners to meet on a regular schedule to discuss specific instances of conflict-producing behaviors.

You may think, at first glance, that a couple meeting is a silly notion because most intimate couples spend a lot of time together anyway, right? Maybe, but the

Table 4.2 Intimate Communication: Couples Meetings: Five Rules

Rule 1: Make a date. Set up a regular weekly time for half an hour that goes on each partner's calendar of very important appointments. This assures both of you that, no matter how busy your week becomes, this special time has been reserved for the two of you and your relationship.
Rule 2: Focus on the problem. The scheduled meeting must be at a time and location that allow both partners to give their full attention to the issue at hand. If this crucial time is interrupted or disturbed by TV, telephone calls, children, or other distractions, the purpose and goals of the meeting will not be met.
Rule 3: Use the "speaker-listener tool." This is an ingenious idea in which the couple simply writes the word "floor" on an index card and passes it back and forth during the meeting. This serves as a reminder that only the partner who has the floor may speak, and the other must listen attentively (but "floor time" must be kept short).
Rule 4: Do not blame or attack. The point of the meeting is to discuss relationship problems. By definition, this means that both partners share the problem and at least some of the responsibility for solving it. Each person must focus on his or her feelings and role in the problem, not the partner's. Blaming and attacking the other will do nothing except cause the meeting to fall apart.
Rule 5: Reserve the right to take a break. Nothing says that the meeting has to go straight through to a solution without a pause of any kind. Each partner needs to know that he or she can call for a break at any time. In fact, a break might be a good idea if anger, hostility, blaming, or attacking begins to arise.
Couples who follow these five rules may find that such meetings can become an effective and rewarding way to communicate and iron out problems before they have a chance to escalate and threaten the relationship itself.

SOURCE: Adapted from Notarius and Markman (1993).

point of a couple meeting is not so much about the time together but rather about *how* that time is spent. The idea here is for the couple to agree to set aside at least thirty minutes per week, and more if needed, for open and honest communication, during which it is each partner's responsibility to speak and listen as clearly and attentively as possible. Notarius and Markman (1993) offer five rules for making these couple meetings effective, which are summarized in Table 4.2.

In addition to their five rules, Notarius and Markman (1993) emphasize that one of the most fundamental attributes of happy and successful relationships is *politeness*. Interestingly, politeness is often one of the first casualties of romantic relationships. Have you ever noticed how long-term, "intimate" couples often treat each other far more disrespectfully than they would ever dream of treating a friend (or even an acquaintance)? Why do you suppose this is? It probably has something to do with the assumption that in a committed relationship, you will be loved even if you ignore rules of common courtesy. Although that may be true in some relationships, the loss of politeness often causes or exacerbates existing problems, whereas behaving politely to each other is a sign of love and respect between partners and may help prevent or diminish problems that arise.

Establishing a regular, weekly meeting time when a couple can discuss problems or concerns helps to create effective, intimate communication patterns.

Sexual Communication

Communicating effectively about sexual issues is one of the most difficult tasks in most relationships. People have a tendency to assume that the sexual side of an intimate relationship will take care of itself, naturally. This assumption may be wishful thinking because, although they may appear somewhat obsessed with sex, most Western cultures are, paradoxically, uncomfortable with open and straightforward discussion of sexual feelings and behaviors in relationships. Partners often make the assumption that they are just supposed to know, through body language or ESP or some other sort of "lover's intuition," what the other likes and wants sexually. However, these forms of pseudocommunication rarely lead to a clear understanding of *any* important

Since You Asked

8. Why is it so difficult to explain what does and does not feel good sexually?

or complex issue, least of all sex. The result is that many couples experience sexual problems that become more severe over time, in part because the partners are not willing or able to discuss them.

Research reveals that good sexual communication enhances both overall satisfaction and sexual satisfaction in intimate relationships (Butzer & Campbell, 2008; Oattes & Offman, 2007). In addition, effective sexual communication has been found to reduce unsafe sexual practices, increase sharing of information about sexually transmitted infections, and lead to more consistent use of contraception, especially condoms, in teens and young adults (Farmer & Meston, 2006; Quina et al., 2000). Many interpersonal factors are involved in effective and comfortable sexual communication.

sexual self-disclosure
Revealing private sexual thoughts and feelings to another person.

SEXUAL SELF-DISCLOSURE **Sexual self-disclosure** is an important component of intimacy with a partner and relates in significant ways to relationship success, happiness, sexual satisfaction, and sexual health (Byers, 2009; Byers & Demmons, 1999; Keller et al., 2000). Specifically, sexual self-disclosure is the process of revealing to your partner, in open and honest ways, any or all of the following:

- Sexual likes and dislikes; turn-ons and turnoffs
- Sexual needs and desires
- Sexual fears and concerns
- Questions about sexually transmitted infections
- Past positive sexual experiences
- Past negative sexual experiences or traumas
- Personal sexual values and morals
- Personal conditions for a sexual relationship

Some people believe that if you love your partner, sexual activities should automatically feel mutually wonderful and satisfying. However, the number and frequency of sexual problems reported by couples demonstrates all too conclusively that there is nothing automatic about it (read more about this in Chapter 7, "Sexual Problems and Solutions"). Communicating effectively about sex can pose a huge barrier, even for couples who communicate well about every other topic. It often feels emotionally risky to reveal yourself in that way. Consider some of the reasons people give for avoiding or feeling uncomfortable about confiding to their partner about personal sexual issues, and you will be better able to get a feel for why sexual self-disclosure is such a big stumbling block (in the next section, we will discuss ways of overcoming these barriers). Chances are good that, at some point in a relationship, you have felt some of these yourself.

LACK OF INFORMATION Of course, part of being able to discuss sexual topics is for both partners to possess a reasonable amount of knowledge and education about human sexuality. You will find that studying this book and taking a human sexuality course will help immensely with any future sexual discussions you may have. Virtually all sexual self-disclosure relates to one or more of the topics covered in this book and in your human sexuality course. Moreover, couples often possess varying degrees of sexual knowledge. When one member of a couple has a more extensive knowledge and education about sex, that person must take an understanding and gentle approach when discussing sexual issues to avoid making the partner feel ignorant and inferior.

EMBARRASSMENT Most people feel at least some embarrassment when talking about explicit sexual topics, especially when the topic relates to themselves. For some, the embarrassment is so great that they would sooner end a relationship than try to disclose personal sexual information. However, as most love relationships grow and mature, and when people are mutually supportive and willing to work at communicating sexually, this embarrassment typically begins to decrease, and two-way sexual self-disclosure gradually becomes an integral part of the relationship.

INSECURITY ABOUT USING THE RIGHT WORDS Related to these feelings of embarrassment, many people do not have a great deal of experience with sexual self-disclosure and worry about how, exactly, to say what they want to express. If the disclosure involves a sexual part of the body or a sexual act, many people are unsure what terms to use. Most of the time, a partner may know what the anatomical part or the behavior is called but may feel confused or unsure about whether to use a formal, scientific term (*penis, vulva, intercourse, cunnilingus*) or more common slang terms (I'm sure you're familiar with these). To some, the formal terms may seem too clinical or stuffy, but to others, the slang may sound shocking or degrading. Again, as you will see, the solution to this problem is to take sexual communication slowly and develop mutually agreed upon modes of discussion.

SEXUAL TABOOS Many people were brought up in families in which discussion of anything even hinting at sex was discouraged or forbidden. These individuals, as adults, often have a hard time talking about sexual issues with a romantic partner and even greater difficulty initiating such conversations. This can be especially frustrating for a couple in which one partner comes from a sexually repressed background and the other grew up in a family more comfortable with discussions of sexual topics. As trust and closeness build, often the more sexually comfortable partner can help the other open up and become more at ease with sexual self-disclosure. However, couples in which both partners are from sexually inhibited backgrounds are sometimes unable to talk about their sexual relationship at all. This is a problem that, if not addressed, may lead to dissatisfaction and even failure of the relationship.

FEAR OF JUDGMENT Even if partners are secure enough about their relationship not to fear rejection upon disclosing personal sexual information, one or both may still be concerned about negative judgments from the other. Often this fear stems from a reluctance to share sexual secrets until the relationship is well established. As discussed earlier, as relationships progress and deepen, other intimate self-disclosure has likely occurred, but sexual disclosures have been held back. The fear that a partner might be surprised, offended, shocked, or disbelieving upon learning something of a private sexual nature is understandable. However, as you will read in the next section, learning how to approach the discussion of sensitive sexual issues can relieve much of the anxiety about negative judgments.

FEAR OF REJECTION Individuals who feel they can tell their partner just about anything often stop short of exposing their most personal sexual side. Many people feel guilty and uncomfortable about their own private sexual feelings, so you can imagine how afraid they would be to tell someone else about them. Why? Because no one is ever completely sure exactly how a partner might react. Although the real impact of sexual self-disclosure on a partner is frequently overestimated, the fear of rejection is real. However, love relationships that are unable to establish clear sexual communication are unlikely to succeed anyway, so the benefits of sexual self-disclosure are usually well worth the risks.

STRATEGIES FOR IMPROVING SEXUAL COMMUNICATION With so many potential barriers to sexual self-disclosure and sexual communication in general, how can a couple ensure that they will be able to enjoy the many benefits of an open and honest dialogue about sexuality? Many of the general guidelines for intimate communication discussed earlier in this chapter—such as avoiding criticism, contempt, stonewalling, and defensiveness; practicing the six "simple truths" of love relationships; and holding couple meetings—can be applied to sexual communication as well. However, because sexual communication has its own unique set of challenges, additional specific strategies may be valuable for those conversations. "Self-Discovery: Guidelines for Sexual Communication" offers suggestions for additional strategies when the discussion turns to sex.

Losing Love: Why Relationships End

4.5 Discuss the most common reasons given for why love relationships fail.

Love relationships are among the most complex of all human connections and are rarely without conflict, stress, and periods of dissatisfaction. When a couple is unable to work through difficulties or to find other ways of resolving friction and discord when they arise, the result, all too often, is the end of the relationship. If you were to look closely at failed relationships, you would probably be able to categorize the many explanations for their breakups into specific behaviors (or lack of them) and into certain common themes.

For the purposes of our discussion here, we will consider some of the more common pitfalls couples experience that are important reasons that relationships fail.

Ten Reasons Why Relationships Fail

LACK OF SELF-KNOWLEDGE If you don't know yourself, how can you expect someone else to know you? Self-knowledge about preferred lifestyle, interests, favorite activities, morals, values, sexual attitudes and preferences, field of eligibles (discussed earlier in this chapter), and expectations about love and relationships allows you to enter into a new relationship with a conscious awareness of what you want and what is best for you. Self-knowledge also allows you to communicate who you are and what you want to potential partners, so that both of you are in the best position to decide if the relationship is right.

ACCEPTANCE OF SEXUAL MYTHS AND STEREOTYPES Here are some examples of male and female sexual myths and erroneous stereotypes that exist in many Western cultures. As you read them, you may disagree with them and perhaps feel annoyed or angry because they sound so unfair and untrue. And you would be correct. *All men are dominant; all women are passive. Men make more money than women. Women are in charge of the household. All women are multiorgasmic. All men want sex all*

Self-Discovery

Guidelines for Sexual Communication

1. Know What You Want
   - Decide what you want and don't want in a relationship before becoming sexually involved.
   - Evaluate your personal sexual expectations and needs.
   - Behave in a manner that is consistent with your sexual philosophy and value system.
   - Be open with your partner by communicating your sexual philosophy to him or her.
   - Understand that two people can want different things and have differing sexual values and philosophies without one of them being "wrong."
2. Insist on Your Right to Postpone a Sexual Relationship
   - Set your own pace, one that is comfortable for you as the relationship progresses.
   - Wait until trust, comfort, and open communication exist in the relationship.
   - Be open and honest with your partner about your desires if you want to proceed slowly.
3. Be Responsible if You Decide to Engage in "Casual Sex"
   - Be honest with yourself and your partner if your commitment is limited to mutual enjoyment.
   - Consider whether or not this type of activity fits your personal sexual philosophy.
   - Be sure both partners want to be sexual.
   - Be sure neither partner feels coerced in any way by the other.
   - Respect your own and each other's feelings of self-worth.
   - Be sure that the responsibility to use precautions against sexually transmitted infections and unwanted pregnancy is mutually agreed upon.

SOURCE: Adapted from Michigan State University Counseling Center (2003).

the time. Men can have an erection whenever and for as long as they want. The man should always initiate sex. Men like sex more than women. Men are from Jupiter; women are from Saturn—or is that Mars and Venus? True or not, many people subscribe to these stereotypes about men, women, and sex. Relationships can easily be sabotaged by these beliefs because they create false expectations. If people who have such beliefs become romantically involved, they are bound to end up disappointed and disillusioned.

When romantic partners are unable to communicate their feelings and concerns to each other, they lose any opportunity to resolve conflicts that may eventually threaten the relationship.

INEFFECTIVE COMMUNICATION We have discussed the importance of communication in love relationships throughout this chapter. When partners fail to share with each other their feelings, concerns, frustrations, needs, and desires, they have no opportunity to address them and to repair any damage that may be occurring because of them. As problems grow without resolution through effective communication, they typically lead to behaviors that may destroy the relationship. For example, if a partner is dissatisfied about the sexual side of the relationship and yet is unable to communicate this to his or her partner, what would you imagine might be the eventual outcome? One of the most common outcomes in this situation is infidelity, which eventually leads to the end of the relationship.

IMBALANCES OF DECISION-MAKING POWER Healthy, successful relationships are characterized by a balance between partners in money and finance matters, choice of friends, everyday activities, issues of family and children, and decision making in general. When a relationship is marked by an imbalance of power, it means that one partner has most or all of the power to make decisions that affect both partners (Blanc, 2001; Knudsen-Martin, 2013). You can see that such a lack of balance would render a relationship unstable. Eventually, lack of balance of power tends to lead to resentment in the lower-power partner and increasingly unhealthy levels of control by the more powerful partner.

A healthy power balance does not imply that every activity and decision must be exactly equally balanced. Partners always have different skills and preferences for dealing with various aspects of the relationship. For example, one may be good at financial details, and the other might enjoy planning social events. That partners divide up the "duties" in their relationship does not necessarily imply a lack of balance as long as such divisions are by mutual agreement and neither partner feels unwillingly excluded from the decision-making process.

A partner's insecurity and low self-esteem can lead to overdependence and possessiveness, creating a fundamental weakness in the relationship.

LOW SELF-ESTEEM, INSECURITY, AND LACK OF SELF-CONFIDENCE People with low self-esteem often feel that they are unworthy of being loved and may constantly look to their partner for validation and proof of love. This insecurity and lack of self-confidence typically leads to overdependence and possessiveness. Few relationships can bear the weight of one partner leaning so heavily on the other. The stronger partner is made to feel responsible for the weaker one's emotional well-being and begins to feel more like a parent or counselor than a lover. No matter how much the stronger partner may give to the weaker partner, it is never enough because the weaker partner's insecurity is typically not a product of external events but rather due to an internal disposition or basic personality characteristic. Moreover, the weaker partner is not in a position to give much to the relationship, so the stronger partner receives little support in return. Unless this lack of self-confidence and self-esteem are overcome, these relationships are usually doomed to eventual failure.

ISOLATION Have you ever had a good friend with whom you got together frequently who then met someone, began a romantic relationship, and seemingly just disappeared? When two people fall in love and then proceed to isolate themselves from the rest of the world, it is usually a red flag for potential relationship problems. Romantic love is a very strong force, and sometimes it can seem to block out everybody else in the couple's lives. However, no two people can meet all of each other's needs. Nearly everyone needs friends, family, colleagues, coworkers, and others who comprise a network of people who help you through life's complex challenges. People in healthy relationships understand that need in themselves and their partner and encourage those outside connections. Usually, when those in a couple isolate themselves from others, it is not a sign of a deep, complete love (although that is what they may claim), but rather arises out of fear and insecurity. Eventually, this overdependence on each other may cause the relationship to be crushed under its own weight.

FAILING TO KEEP PROMISES, LYING, OR CHEATING This one seems obvious, doesn't it? Failing to keep promises, lying, or cheating, especially relating to infidelity, are blatant betrayals of the basic trust necessary for a relationship to survive and thrive. But even more subtle forms of deception can gradually unravel the fabric of an intimate relationship. Why? Because partners in healthy relationships tend to use rational problem-solving strategies to reconcile difficulties, disputes, disagreements, and other problems as they arise. All of these effective strategies rely on both partners *agreeing* to modify certain behaviors to resolve whatever problems they may be facing. Those agreements then become the basis for renewed happiness and satisfaction in the relationship. But these strategies are only as good as each partner's promise to carry through on them and his or her trust in the other to do the same. If one or both partners fail to keep their promise, they lose the most effective tool for solving their problems. The question then becomes "Why should I bother working out our differences when my partner won't do what he or she agrees to anyway?" Unless that basic trust can be recaptured, it is probably only a matter of time until one too many unresolved problems accumulate and the relationship dissolves.

EXCESSIVE JEALOUSY Jealousy is a common reaction in romantic relationships that stems from losing all or certain aspects of a partner's exclusive love (or the fear of such a loss). Jealousy is cited as one of the most frequent causes of the breakup of romantic relationships (Knox et al., 1999; Tassy & Winstead, 2014). It appears to exist in similar forms in many cultures throughout the world (Buunk & Hupka, 1987; Edaliti & Redzuan, 2010). Perhaps most significant, jealousy is frequently the precipitating event in relationship abuse and violence, which will be discussed later in this chapter (Moore, Eisler, & Franchina, 2000; Puente & Cohen, 2003). Jealousy may be defined in many ways, but for our purposes here, it is useful to divide it into two main types—normal and pathological—and the distinction between them is important in understanding the effect jealousy can have on a love relationship (e.g., Harris, 2003). **Normal jealousy** is based on a real threat to the relationship, as when one partner discovers that the other is attracted to, is in love with, or has been unfaithful with someone else. Normal jealousy is a reaction to factors that already indicate that a relationship is probably in some sort of trouble, although the intensity of the reaction may play a role in whether the couple is able to work through the episode and stay together. **Pathological jealousy** is a reaction that emanates from within the jealous person when no threat or infidelity actually exists. Often called the "green-eyed monster," pathological jealousy is the form that is more likely to destroy an intimate relationship on its own.

normal jealousy
Jealousy based on a real threat to the relationship, as when one partner discovers that the other has been sexually unfaithful.

pathological jealousy
Jealousy felt within one partner despite the fact that no threat to the relationship actually exists.

Since You Asked

9. How can I get my partner to trust me? She's jealous all the time and accuses me of cheating when I'm not doing anything. She's even reading my texts and emails! She's convinced I don't love her, but I do—I think.

Pathological jealousy typically stems from low self-esteem and overdependence in one or both partners. People with a low opinion of themselves who look to a partner for validation, strength, and legitimacy are, understandably, terrified of losing that partner. In their eyes, without the partner, they are unworthy members of society. In these cases, it makes no difference how devoted and faithful a partner might be; pathologically jealous partners are incessantly suspicious. They constantly accuse the other of cheating. They check up on him or her by following, "spying," or through the many and varied electronic means available today such as: social networking sites, phone call logs, *69 call-backs, and text and email records (Muise, 2009). In fact, research has suggested that jealousy-based monitoring of a romantic partner is more common online than offline (see Sexuality and Culture: Jealousy and the Culture of Social Media). Regardless of how pathological jealousy manifests itself, it is a no-win situation for the partners, who are forced to live in an atmosphere of suspicion, distrust, and invaded privacy when they have done nothing to cause or deserve it.

Ironically, pathological jealousy is often a self-fulfilling prophecy. The beleaguered, persecuted partner usually gives up trying to prove his or her fidelity and innocence because nothing ever seems to convince the jealous partner. Some individuals may eventually decide to stray from the relationship, saying to themselves, "My partner is convinced I'm guilty anyway, so why not?" More commonly, the relationship cannot withstand the pressure of the constant suspicion, and the couple breaks up. The breakup then confirms the jealous partner's fears and strengthens the tendency for similar pathological jealousy patterns in subsequent relationships. In the worst-case scenario, pathological jealousy may lead the jealous partner to controlling, abusive, or violent actions (these are discussed next).

Normal jealousy, like most other relationship problems, can sometimes be overcome, but doing so requires a strong desire and commitment to work on the jealousy issue specifically. This may require individual and joint counseling to help the partners reestablish the trust and commitment that has been lost and gain the confidence that the cause of the jealousy will not happen again. When dealing with pathological jealousy, the solutions can be much more difficult in that the underlying problems are generally not relational, but instead deeply embedded in the nature of the jealous partner.

CONTROLLING BEHAVIORS A student in one of my classes once remarked with a laugh, "My boyfriend got really mad at me last night because he said that there were twenty miles on the odometer of my car that I hadn't accounted for. Is that really weird?" No one else in the class laughed. Instead, most were shocked that anyone would keep such close tabs on someone they supposedly love, and they seemed uncomfortable that this student was laughing about it. This is an example of a controlling behavior in a potentially dangerous relationship.

When one partner seeks to control the actions of the other, the relationship becomes weak at best and dangerous at worst. The need to control is inextricably linked to many of the other factors we have been discussing in this section, including belief in gender myths, lack of communication, unequal power, low self-esteem, isolation, overdependence, and jealousy. A relationship can remain healthy and happy only when both partners know that they have freely chosen to be in the relationship and are free to be themselves. In a controlling relationship, one partner attempts to take that freedom away from the other and control virtually everything the other does. This controlling behavior may take many forms, including dictating where, when, and with whom the partner may go out; requiring the partner to report every detail of every minute whenever they are apart; demanding that the partner never deviate from an expected schedule and never arrive home late; threatening the partner with property damage or physical injury for any "disobedience"; frequently checking up on the partner at work, where the partner lives, or waiting outside the partner's classes; taking charge of all the couple's money;

Sexuality and Culture

Cyberspying—Jealousy and the Culture of Social Media

The data below represent some of the findings from a study of college students conducted in 2011. The research examined to what extent students monitored their romantic partners' activities both off- and online. Some monitoring may be innocent: simply a way of staying in touch with a partner, but the way the questions are worded makes it clear that most of this monitoring is of a suspicious nature. Furthermore, if you compare the likelihood of monitoring a partner's behavior offline versus online, you will see that online monitoring is far more common.

Percentages of respondents who monitor their partners offline

	1 = Never	2	3	4	5	6	7 = All the time
How often do you look through your partner's drawers, handbag, or pockets?	52.4	26.2	4.9	3.9	9.7	2.9	—
How often do you secretly read the SMS messages on your partner's mobile phone?	42.7	27.2	7.8	4.9	15.5	1	1
How often do you secretly read your partner's e-mails?	64.1	14.6	3.9	6.8	7.8	1.9	1.0

Percentages of respondents who monitor their partners on the Social Networking Sites (SNS)

	1 = Very unlikely	2	3	4	5	6	7 = Very likely
Check your partner's profile on a regular basis.	24.7	17.3	6.2	14.8	21.0	14.8	1.2
Look at your partner's profile page if you are suspicious of his or her activities.	27.2	18.5	13.6	8.6	25.9	2.5	3.7
Monitor your partner's activities on the SNS.	27.2	14.8	9.9	19.8	22.2	6.2	—
Add your partner's friends as friends to keep tabs on your partner.	50.6	18.5	8.6	9.9	8.6	1.2	2.5

SOURCE: Utz, & Beukeboom, (2011).

using intimidation to prevent the partner from leaving (the relationship or simply to go out); and name-calling, put-downs, or sudden explosions of anger.

Controlling partners are weak individuals with low self-worth who believe that the only chance they have of keeping their partners is to prevent them from having any opportunity to meet anyone else who might be a relationship possibility. They are convinced that if their partner does meet someone else, he or she will choose that person over the controlling partner.

As with pathological jealousy, discussed earlier, relationships in which one partner is excessively controlling often fall victim to the self-fulfilling prophecy. No one wants to be controlled by another person. Eventually, the partners of controlling individuals will want out and will want their freedom. The problem is that as a controlling partner becomes more and more controlling, the other partner becomes less and less happy, which, in turn, makes the controlling partner even more insecure, leading to stronger and more dangerous controlling tactics until the relationship begins to resemble that of a prisoner and a prison guard.

The most serious outcome of a controlling relationship occurs when the line is crossed from control to abuse or violence. In fact, as you will see, many abusive behaviors fall under the heading of control. Not all controlling relationships become violent, but as you can imagine, if a controlling partner begins to perceive that nonviolent control tactics are becoming ineffective, turning to violence is the next obvious step to maintain control.

ABUSE AND VIOLENCE Obviously, abuse and violence are sure ways to destroy a relationship. Abuse between romantic partners is often referred to as *relationship abuse, domestic violence,* or *intimate partner violence.* Unfortunately, all too often this type of violence does not end a relationship fast enough, and victims may become trapped in abusive and violent relationship cycles over long periods of time: years, or even decades. The violence in intimate relationships is for one purpose: to exert ultimate and total control over the victim. The abuse or violence and the threat of more in the future typically create such fear in the victim that attempting to escape becomes impossible. Sadly, some violent relationships end in the death of the victim. Both men and women may be victims of relationship violence, but violence by men against women is significantly more common. Statistics show that between 2002 and 2011, over 800,000 women and nearly 175,000 men were victims of violence at the hands of an intimate partner (Catalano, 2013). Approximately one-third of all women who are murdered in the United States are killed by their male intimate partners (Catalano, Smith, Snyder, & Rand, 2009). Intimate partner violence (IPV) may occur in any relationship, whether heterosexual, gay, or lesbian, but is most commonly male abuse perpetrated on female victims.

Many people believe that as soon as a relationship becomes violent, the victim should simply leave. However, leaving an abusive relationship is typically far from simple. Violence in intimate relationships is a very serious and complex matter from which escape is far more difficult than it sounds. We will discuss this issue along with many others relating to IPV next.

Abusive and Violent Relationships

4.6 Explain the types of relationship abuse, describe the cycle of violence, and discuss the characteristics of an abusive relationship and its effects on the survivors.

Intimate partner abuse and violence comprise what might be called the *dark side* of love and intimacy, and unfortunately, they are far more common that most people believe. That is why these topics are being discussed here, in the relationships chapter, in greater detail than in Chapter 13, "Sexual Aggression and Violence," which is where we discuss rape, child sexual abuse, and sexual harassment. Everyone needs to be aware of these dangerous and destructive aspects of relationships in order to increase their own chances of recognizing and avoiding them and to be able to help those they care about who may not perceive the risks in which they find themselves.

People who have never experienced an abusive or violent relationship find it nearly impossible to imagine how a connection that seems to begin with love and happiness can transform into horror, pain, and despair. As noted earlier, violence between partners not only destroys their relationship, but is also devastating to the victim's emotional wellbeing. This form of abuse and violence most commonly entails a male abuser and a female victim. However, it is very important to be aware that intimate partner abuse and violence may involve a female abuser and male victim and also exists in same-sex relationships as illustrated in Figure 4.3.

If you or someone you know is caught in the trap of a violent relationship, help is available. Anyone in North America may call the National Domestic Violence Hotline (NDVH) at (800) 799-SAFE (799-7233) or go to their website at www.thehotline.org. The hotline is staffed 24/7/365 by trained counselors who can provide crisis assistance and information about local shelters, legal assistance, health care resources, and counseling. The NDVH may be contacted via e-mail at ndvh@ndvh.org (with access to translators for 140 languages).

Relationship abuse may include physical violence, intimidation, or restraint, or it may involve verbal or emotional attacks such as threatening, ridiculing, or humiliating. The abuser's goal is to gain complete control over the partner and the relationship.

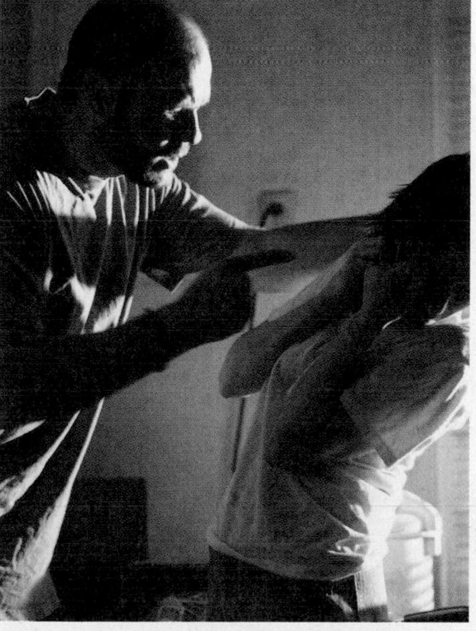

Types of Relationship Abuse

No all-encompassing definition of relationship abuse exists. Each intimate relationship is unique, and so each abusive relationship is unique as well. However, if you suspect that you may be in an abusive relationship, you will notice

Figure 4.3 The Power and Control Wheels of Abusive Relationships

When one person in a relationship repeatedly scares, hurts, puts down, or injures the other person, it is abuse. These Power and Control Wheels illustrate examples of these forms of abuse in heterosexual and same-sex relationships. Remember, abuse is much more than grabbing, slapping, or hitting.

SOURCE: "Power & Control in Dating Relationships" and "Lesbian/Gay Power and Control Wheel" as published by Domestic Abuse Intervention Programs. Reprinted by permission of DAIP, 202 E. Superior St., Duluth, MN 55802.

Heterosexual Power Wheel

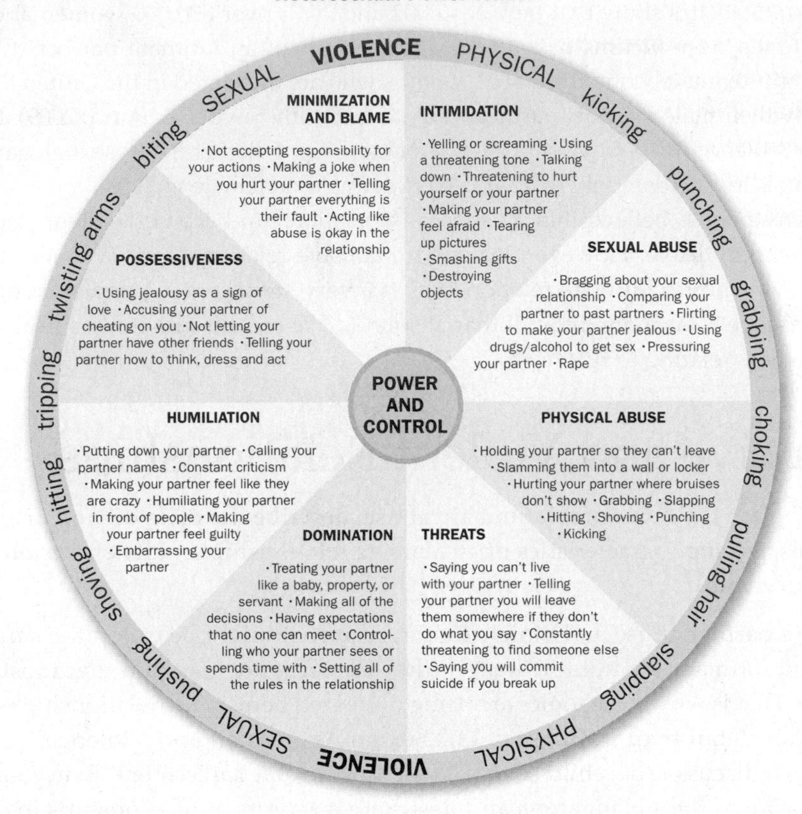

Same Sex Power Wheel

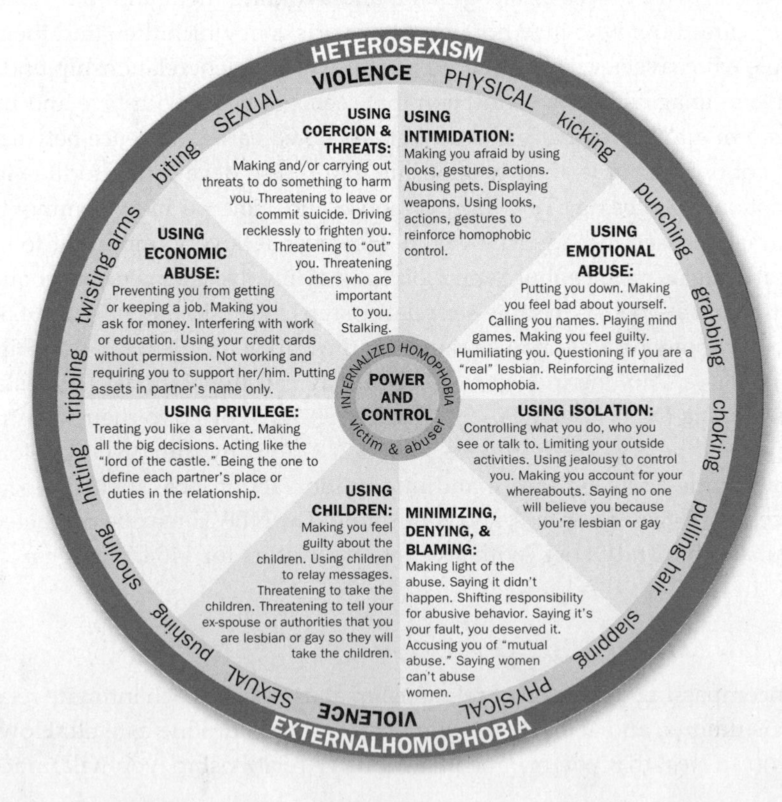

certain signs that demonstrate clearly that your relationship is unhealthy and abusive. Two of these signs are common to nearly all cases of relationship abuse (Dugan & Hock, 2006).

First, abusive relationships are all about the use of power and control that is achieved through a wide range of tactics. The abuser's goal is to take complete control of the partner and of the relationship. The controlling tactics may be subtle and not easily recognized. It may seem to the victim that the partner's gradual, increasing control over his or her time, friends, and daily activities is a sign of love and caring. As time goes by, however, the abuser is able to render the partner totally powerless and acquire complete control of the relationship. Figure 4.3 illustrates the many behaviors that fall under the umbrella of relationship abuse and violence.

Second, relationship abuse usually involves a *pattern* of abusive events. A single abusive or violent act may or may not constitute abuse. However, abusive relationships are generally defined by a pattern of destructive behaviors that are cyclical and escalate over time.

Although nearly all abusive relationships share these two characteristics, the specific behaviors used by abusers to achieve these goals vary greatly and are not limited to physical violence. We will examine the three most common abuse strategies: physical, verbal, and emotional (Dugan & Hock, 2006). It is important to remember that at the beginning of most abusive relationships, the potential victim may not notice the negative behaviors. Some of the behaviors may even seem loving and caring. The jealousy may seem extreme, but it feels endearing and protective. The constant attention and togetherness may seem like a sign of the intensity of the love. However, paradoxically, these may also be signs of the violence to come.

PHYSICAL ABUSE It seems hard to believe, but some victims have experienced intimate physical violence that they didn't perceive as abuse. It is quite common when a woman describes a past relationship and is asked if it was abusive to reply with a statement along the lines of, "Well, he never punched me." But not all physical abuse entails punching, broken bones, bleeding, or stitches. Most relationship violence causes damage that others cannot see, so they don't even know it has occurred if the victim does not choose to talk about it or does not label it as violence.

Violent partners may hit hard enough to leave bruises, but only on parts of the victim's body where others would not see them. Or the violence might entail grabbing and twisting an arm sufficiently to cause real pain without leaving a noticeable mark. If an abuser's partner is thrown against a wall, pushed to the floor, or strangled without leaving marks or bruises, no one else may ever know simply by looking. Moreover, abusers may not even recognize their behavior is particularly deviant. One study found that male perpetrators of intimate partner violence estimated the number of men who engage in similar behavior at three to four times the actual percentage (Neighbors et al., 2010).

The truth is that many of these behaviors are considered physical abuse, including: pushing, hitting, grabbing, choking, slapping, punching, biting, cutting, hitting with objects, pinching, physically restraining (holding partner down, pinning partner against a wall, etc.), rape or other sexual abuse, or physical intimidation (blocking partner's exit, waving a fist or a weapon, etc.). Although these behaviors may be easy for most victims to identify, some other forms of abuse, such as verbal or emotional abuse, may not be so easy to recognize.

VERBAL ABUSE Abusers may attack with words instead of—or in addition to—physical violence. Attacks of verbal abuse may be directed at the victim or may take the form of belittling or humiliating comments or untrue accusations said to others about the victim.

Verbal abuse is extremely painful, and its effects are long lasting. As with physical violence, verbal abuse can take many different forms, but its goal, once again, is to gain control over the victim, to crush the victim's self-image, and to cause fear and powerlessness. It often is designed to create feelings of worthlessness in the partner and transfer blame onto the victim for the abuser's attacks.

Behaviors that can be characterized as verbal abuse include yelling; threats; intimidation; ridiculing; name-calling; criticizing; accusing; insulting; humiliating; swearing; blaming; belittling; mocking; sarcasm; put-downs; and trivializing the victim's ideas, opinions, or wishes. All of these are abusive behaviors and have no place in a loving partnership.

EMOTIONAL ABUSE Physical and verbal abuse take a terrible psychological toll. However, another category of abuse focuses on the victim's feelings. This is referred to as *emotional abuse.* Emotional abuse is so insidious and psychologically devastating that it can take the longest time to identify and heal. Bruises, cuts, broken bones, or hurt feelings almost always mend faster than the wounds of emotional abuse. Abusers use many tactics to take control of and manipulate the partner's feelings and emotions. They attempt to make the victim feel unworthy of being loved, unattractive, sexually unskilled, and at fault for causing the abuse that is occurring. They intimidate and threaten their victims with bodily harm, damage to personal property, and injury to loved ones and pets.

Abusers also tend to minimize the extent of their abuse in numerous ways, trying to make the victim feel discounted, as though the seriousness of the abuse was not really so bad. Often this includes an implied or overt threat that the abuse could get much worse.

Actions that may constitute emotional abuse include entitlement ("I expect you to do what I say"); withholding information ("Why should I tell you what I'm thinking or feeling?"); withholding sex ("Why would I want to make love with someone like you?"); emotionally misrepresenting the victim's feelings ("You're not hurt—stop complaining"); risk taking (doing drugs, engaging in high-risk behaviors such as driving recklessly or not seeking medical care when needed); withholding help from the partner; excessive jealousy; threats of suicide; threatening to hurt or kill the partner, friends, relatives, or pets; and taking charge of all the decisions in the relationship.

How can you know if you are in an abusive relationship? The answer may not be as obvious as it seems. Many levels and types of violence exist, and some are not as immediately recognizable as others. "In Touch with Your Sexual Health: Signs of Relationship Abuse" will help you assess whether you are in an abusive relationship or whether you are engaging in behaviors commonly seen in abusers.

The Cycle of Violence and Abuse

Relationship abuse and violence typically increase gradually over time, often causing the victim to be unsure of exactly when the abuse began. This is in part because no violent relationship is violent all the time. Even if a relationship is violent on a frequent and ongoing basis, the couple still experiences many moments, hours, or even days or weeks when no overt violence occurs. This pattern can make it difficult for a victim to recognize the abusiveness of the relationship until the violence becomes extreme (Dugan & Hock, 2006). Nevertheless, the abuse usually follows a fairly predictable pattern, often called the *cycle of abuse.*

cycle of abuse

The repetitive pattern of stages that define most abusive and violent relationships, cycling through the honeymoon stage, the tension-building phase, and the explosion of violence, followed by a return to the honeymoon stage and the beginning of a new cycle.

The **cycle of abuse** describes how a violent relationship typically develops unless something is done to intervene. For most relationships, even potentially violent ones, the early stages are happy, exciting, and even idyllic. Everything seems wonderful. Partners are loyal, devoted, and caring and seem to have eyes only for each other. It feels like a dream come true. This is the *honeymoon phase* of an intimate relationship.

In all relationships (healthy and unhealthy), after a while, something happens that creates tension between the partners. It may be a simple difference of opinion, a disagreement over a purchase, an activity, or just an argument about a controversial subject. Whatever the issue is, even if is relatively trivial, it temporarily disrupts the harmony of the new relationship. This is completely normal. In *healthy* relationships, the partners typically resolve the problem by talking about it and working it out between themselves.

In contrast, in abusive relationships, this rational, problem-solving approach fails, and the tension persists and builds. Soon one partner realizes that the only way to regain the harmony is to give in to the other's position. Although giving in may cause this partner to feel resentful and discounted, he or she usually decides that if giving in relieves the tension, it is worth it so that the loving, harmonious atmosphere of the honeymoon phase can return. However, at some point, another event occurs that once again creates tension between the partners. This is the *tension-building* phase of the cycle. This time, instead of one partner surrendering, an abusive or violent explosion occurs. Typically, the abusive partner yells, ridicules, threatens, insults, or engages in other acts of bullying and intimidation. Now the victim is afraid and does whatever is necessary to calm things down. Then, again, for days, weeks, or even months, relations return to a happy, loving, honeymoon-like period.

As time passes, this cycle continues, but typically the explosive events become more extreme and dangerous, and the buildup of tension nearly always ends in abuse. What began as insults and threats may now become physical violence. But still, after the violence, the honeymoon begins again. The abuser is sorry, promises it will never happen again, asks forgiveness, is repentant, perhaps buys gifts for the victim. The victim, in turn, tries to make things better and does everything humanly possible to make the relationship happy and tension free. As the relationship reenters the honeymoon phase, emotions are soothed, but by now the victim lives in fear of another explosion and goes to great lengths (characterized as "walking on eggshells") to avoid any behavior that might trigger more violence. Compared to the other phases of the cycle, this time between outbursts truly feels like another honeymoon, and the victim hopes it will last. But it does not, and sooner or later the abuse and violence return. Figure 4.4 illustrates this cycle of violence and abuse.

Figure 4.4 The Cycle of Violence and Abuse in Intimate Relationships

Violent and abusive relationships tend to progress in a predictable cycle of repeating and increasing volatility.

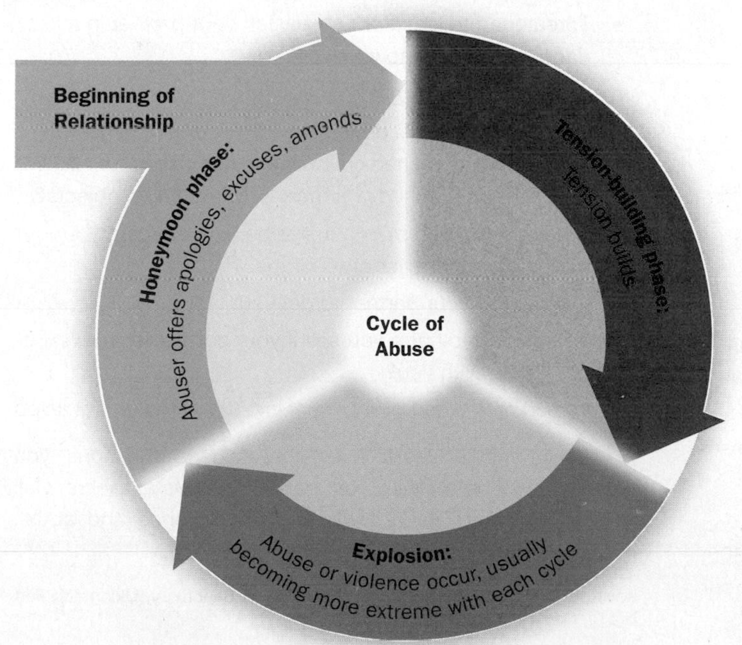

Leaving a Violent Relationship

Relationship abuse is far more complex a problem than most people realize. Strange as it may seem to many people, those caught up in a violent or abusive relationship may not even realize what is happening until the abuse has escalated to the point of real physical danger. Even then, a victim may find escape from the violent setting nearly impossible because of overdependence on the abuser, fear of being left alone and without financial means, terror over worse or even deadly violence, or concern for children who may be part of the relationship.

Unfortunately, relationship violence tends not to diminish or go away over time; on the contrary, it is more likely to become worse. Usually, the only way to stop relationship violence is for the victim to leave. Ironically, however, a victim of an abusive relationship is at the greatest risk of violence when trying to leave the abuser. The vast majority of serious assaults and murders in violent relationships

In Touch with Your Sexual Health

Signs of Relationship Abuse

Signs That You May Have an Abusive Partner

If you think you may be the victim of a violent relationship, ask yourself the following questions:

Are you...

- Frightened at times by your partner's behavior?
- Afraid to disagree with your partner?
- Often apologizing to others for your partner's behavior toward you?
- Verbally degraded by your partner?
- Unable to see your friends or family due to your partner's jealousy or control over you?
- Afraid to leave your partner because of threats to hurt you or commit suicide if you do?

Do you...

- Sometimes feel as if you have to make up excuses and justify your behavior to avoid your partner's anger?
- Avoid attending family or social functions because you are afraid of how your partner will behave?

Have you been...

- Hit, shoved, thrown down, choked, grabbed, physically restrained, threatened, intimidated, humiliated, put down, ridiculed, or attacked by your partner with thrown objects?
- Forced by your partner to engage in sexual acts against your will?

If you answer yes to even *one* of these questions, you may be in an abusive relationship. Help is available: call (800) 799-SAFE (799-7233) at a time and place unknown to your abuser.

Signs That You May Be an Abusive Partner

If you think you are being abusive to your partner, ask yourself the following questions:

Do you...

- Frequently check on your partner's whereabouts, friends, or activities?
- Criticize or insult your partner?
- Believe that you are permitted to hit, shove, or slap your partner for actions you do not like?
- Believe that your inappropriate or violent behaviors are caused by your partner's actions?

Have you...

- Threatened or broken things to frighten your partner?
- Threatened to leave or actually left your partner in a dangerous or unknown place?
- Driven too fast or recklessly to frighten your partner?
- Hit, shoved, thrown down, choked, grabbed, physically restrained, threatened, intimidated, humiliated, put down, ridiculed, or attacked your partner with thrown objects?
- Blamed your violent or abusive behavior toward your partner on alcohol or other drugs?
- Intimidated your partner to get your way?
- Threatened to harm yourself if your partner leaves or breaks up with you?
- Forced or coerced your partner into unwanted sexual acts?

If you answer yes to even *one* of these questions, you may very well be abusing your partner. Help is available: Call (800) 799-SAFE (799-7233) before the violence and abuse become more serious and more dangerous.

SOURCE: Adapted from publications by Project Sanctuary, Ukiah and Fort Bragg, California.

occur while the victim is attempting to leave the relationship or has already left the abuser (Gottleib, 2008; Oths & Robertson, 2007). If you think about it, this makes sense. Remember, the goal of the abuse is *control* over the other person, and typically this need for control stems from the abuser's pathological fear that the victim will leave the relationship. When that fear becomes a reality, the only recourse an abuser knows is to become more violent in a desperate attempt to regain control over the departing partner.

What should a victim do to be reasonably safe after leaving an abusive relationship? No one can make any guarantees of safety, but several guidelines can help (Dugan & Hock, 2009):

1. Assess the abuser's danger level. Analyze past violent behavior, threats made about the consequences for attempting to leave, possession or past use of weapons, and so on.
2. Have an escape plan. Develop a route of escape if the abuser attempts access at home, at work, at relatives' homes, or anywhere else a violent approach might be attempted. Know how to get away and where to go. This helps a victim stay one step ahead of an abuser out for control and revenge.
3. Create a safety network of trusted people. Keep their phone numbers handy at all times. Work out how they can provide a place to hide or other forms of shelter from the abuser.
4. Obtain a court-ordered restraining order that legally requires the abuser to stay away from you. Even this is no guarantee of safety, but many abusers will not violate a restraining order because they can be arrested and jailed if they do.

These measures may sound drastic, but severe violence and even murder are very real possibilities in the aftermath of violent relationships.

The Temptation to Go Back

Anyone who has not experienced an abusive relationship probably has difficulty understanding how a survivor would ever consider, even for a moment, returning to the abusive partner. When survivors consider returning and try to discuss it with friends or family, they may encounter shock, astonishment, disbelief, anger, and recriminations. Some survivors become very upset with themselves at the thought of returning to their abuser. They may wonder if they are crazy to contemplate such a seemingly irrational act.

But returning can be very tempting. The vast majority of women who are victims of relationship violence return to their abusing partner at least once, abuse survivors return an average of five times, and nearly a fifth of victims have returned ten times before they finally leave permanently (Anderson, 2003; Roberts, Wolfer, & Mele, 2008). Survivors' reasons for returning range from lack of financial resources to continuing feelings of love for the abuser; from loneliness to fear of being harmed even worse or killed by the abuser (Griffing et al., 2005). Reconstructing a life alone after any relationship can be extremely difficult, and the same is true of abusive relationships. Some survivors even say they were happier in the relationship than they are without it. Numerous new sources of stress may tend to eclipse the memory of the abuse. For many survivors, the horror of the abuse does not completely erase the positive memories of the relationship (recall that the violence and abuse is cyclical, with good times as well as bad). However, regardless of the stresses of life alone and these selective positive memories, survivors must guard against returning to a violent and potentially deadly relationship. When a victim returns to an abuser, the violence does not stop (no matter how much the abuser has insisted that it will). Instead, the abuse and violence resumes and nearly always becomes worse with each successive abusive episode (Anderson, 2003; Dugan & Hock, 2006).

Loving Again

Can someone who has survived a terribly abusive relationship ever find the strength to risk an intimate relationship again? The answer is yes, but often the road back to love requires a great deal of careful thought and self-assessment. One common fear survivors face is becoming involved with another abusive or violent person. When victims look back on a violent relationship, they typically report that they never saw it coming; they fell in love with a person who showed no signs of an abusive, violent nature that would later be revealed. However, some signs that a person *may* become a violent partner have been identified and are listed in Table 4.3. Knowing these signs is no guarantee that an abusive partner can be identified in advance, but they provide a backdrop against which to judge a potential partner early on in a relationship.

Since You Asked

10. Can someone who has been in an abusive relationship ever have a normal relationship again? Is it really possible to forgive and forget?

When survivors of violent relationships begin to consider a new relationship, they must apply what they have learned about themselves and about relationships. A reliable sign that survivors are ready to love and be loved in a happy, healthy relationship is that they are able to articulate what they want from that relationship and what they

Table 4.3 Warning Signs of a Potential Abuser

The following characteristics have been identified as more common in individuals who become abusive or violent partners. No one can know for sure if a relationship may turn violent, but an awareness of these features may help avoid a future tragedy.

Characteristic of Potential Abuser	Description
Has a history of battering	May admit to hitting previous partners but will blame the partner for provoking the attacks; partner may learn about past abuse from relatives or ex-partners; battering behavior is independent of situation or partner.
Uses threats of violence	Makes threats of physical violence meant to control the partner; will attempt to excuse behavior by claiming "everybody talks like that."
Breaks or hits objects when angry	May break partner's favorite possessions or hit walls or furniture near partner to terrorize; action implies that "this could be you."
Uses force during an argument	May involve grabbing, pushing, physically restraining, blocking exit from situation, and so on.
Displays excessive jealousy	Is intensely jealous without reason or provocation, even in the very early stages of the relationship; will often claim jealousy is sign of deep caring.
Engages in controlling behavior	Attempts to take over all decision making; checks up on where partner is at all times; becomes angry at any lateness; takes charge of the money.
Becomes romantically involved very quickly	"Falls in love" very soon after meeting; pressures partner for commitment; desires to get married or live together after very short time (less than six months); comes on very strong with proclamations of undying and ultimate love.
Has unrealistic expectations	Expects relationship to fulfill both partners' every need and desire; assumes that partner automatically knows and will attend to all the abuser's physical and psychological needs.
Isolates partner from social contacts	Attempts to separate partners from all personal and social connections (work, friends, relatives) and establish sole and complete control.
Blames others for own problems	Any problem (job loss, car accident, etc.) is partner's or someone else's fault; never takes responsibility for any mistake or bad decision.
Blames others for own feelings	Blames partner for own anger and violence; believes partner is responsible for all negative actions or feelings.
Is hypersensitive to criticism	Is easily insulted and claims hurt feelings when partner is the one who is hurt and angry; interprets any small problem as a personal attack; sees world as basically unfair.
Displays cruelty to animals or children	May punish animals brutally or be insensitive to their pain; expects children to be capable of behaviors far beyond their abilities and is angered when they fail; may tease young children until they cry.
Uses "playful" force in sex	May desire to tie or hold partner down; demands sex regardless of partner's mood, desire, or willingness.
Engages in verbally abusive behaviors	Makes comments intended to be cruel and hurtful; verbally degrades partner; diminishes partner's accomplishments.
Subscribes to rigid, traditional sex roles	Male batterer expects woman to serve him; may require that she stay at home; expects obedience in every way; sees women as inferior to men.
Displays "Jekyll and Hyde" mood swings	Has instantaneous mood swings; is pleasant one minute, but the next minute is explosive or "crazy."

SOURCE: Adapted from The Project for Victims of Family Violence, Inc. Reprinted by permission of The Project for Victims of Family Violence, Inc.

Self-Discovery

Your Criteria for a Healthy, Nonabusive Relationship

The ease with which a survivor of an abusive or violent relationship is able to complete this exercise offers an important clue about his or her readiness for a new relationship. Those who have survived such an unhealthy relationship should be clear about what they require and what is unacceptable in a partner *before* they begin a relationship with someone new.

A. List at least five qualities you will absolutely require in a new partner. These are nonnegotiable.

Qualities I will *require:*

1.
2.
3.
4.
5.

B. List at least five characteristics or behaviors you will avoid in a new partner. These you will not accept under any circumstances.

Characteristics I will *not accept:*

1.
2.
3.
4.
5.

SOURCE: Adapted from Dugan and Hock (2006, p. 242).

Survivors of abusive relationships can be more confident of finding a healthy relationship in the future if they are clear about their personal criteria for an intimate partner.

will never again tolerate from a partner. "Self-Discovery: Your Criteria for a Healthy, Nonabusive Relationship" provides a framework for survivors to test their level of clarity about potential future partners.

Your Sexual Philosophy

Love, Intimacy, and Sexual Communication

Falling in love is one of the most emotionally charged events in life. Incorporating the issues discussed in this chapter into your working sexual philosophy *frees* you to enjoy love and intimacy to the fullest because you know that you have the tools to avoid being swept away to a "destination" that is wrong for you. How might the information in this chapter fit into your developing sexual philosophy?

First, the more you know about your requirements for an intimate relationship, the better your ability to judge whether or not someone you meet and feel attracted to is likely to be right for you. This awareness does not *guarantee* that every potential relationship is going to work out and make you happy, but the more you develop a real sense of your needs and desires for a potential intimate partner, the better your chances of avoiding becoming involved with the wrong person, realizing too late that the person is totally wrong for you, and then enduring the pain of breaking up.

Second, effective communication can be difficult in love relationships, especially when the topics are about sensitive or sexual issues. Many, if not most, people enter into intimate relationships without understanding the basic strategies and barriers for intimate communication. This places them at a disadvantage from the start. If you have made an effort to familiarize yourself with at least a few proven techniques for successful communication, you will be better prepared and one step ahead when those first difficulties with your partner arise (and they always do).

Third, most of you will, unfortunately, experience at least one breakup of an intimate relationship, and many of you probably already have. The intense emotions that usually surround the ending of a relationship often leave people confused, bewildered, and asking, "What went wrong?" "How did something that seemed so great at the beginning end up causing so much pain?" Adding an understanding of why relationships end to your sexual philosophy will better prepare you to prevent a breakup of a good relationship; perhaps more important, you will "see the writing on the wall" sooner and reject potentially abusive relationships.

Your sexual philosophy will need to be crystal clear that you will not tolerate abuse or violence in your relationships. Most people who find themselves in a violent situation with a partner say they were completely shocked that their relationship transformed into such a nightmare. Knowing what these relationships are like and how they develop does not ensure completely that it can't happen to you, but it does allow you to be alert to warning signs and, ideally, escape at the earliest hint of violence or abuse before you become trapped by violence, danger, and fear.

Have You Considered?

1. Think of at least three people who are *not* in your field of eligibles. Briefly describe each one, and explain what characteristics cause you to exclude the person from any possibility of a romantic involvement.
2. Now think of a romantically involved couple, either people you know or celebrities, who violate the matching hypothesis. Explain why you think they are together and why their relationship is successful despite significant differences in level of attractiveness.
3. Imagine that you are in a romantic relationship that is lacking one of Sternberg's three basic components. If you had to accept less than a consummate relationship, which component would be the *least difficult* to live without? Which would be the *hardest* to live without? Explain your answers.
4. As discussed in this chapter, John Gottman believes that complaints can be positive in a relationship because they express an unmet need felt by one of the partners. Consider the following complaint in an intimate relationship: "I feel as if I'm doing all the yard work, and I wish we could share those responsibilities." Rewrite how that complaint would sound as a relationship begins to exhibit each of Gottman's four communication warning signs—(a) criticism, (b) contempt, (c) defensiveness, and (d) stonewalling.
5. Think about an unsuccessful relationship you have had or observed in others. Discuss which of the ten relationship destroyers summarized in the chapter most likely caused those relationships to fail. Explain your answer.
6. Explain what you think are the three most common reasons survivors of abusive relationships return to their batterers, often many times, before they finally leave for good. Suggest possible interventions that might help survivors of an abusive relationship stay out of it once they are out.
7. Summarize abusive, controlling tactics that may be seen in same-sex relationships that are different from those in opposite-sex relationships.

Summary

Historical Perspectives

Love and Marriage in America, c. 1750

- In the 1700s, most marriages were undertaken for financial gain and social standing. Marriages were often arranged by the bride and groom's fathers to enhance the family's position in the community. Engagements, or courtships, tended to be relatively brief. The process of betrothed couples "getting acquainted" was speeded up through the practice of "bundling," or sleeping in the same bed fully clothed, sometimes with a board between the young couple to prevent them from "losing control."

Establishing Early Intimacy (Today)

- Today, courtship is more under the control of the couple themselves in most cultures. Many factors influence romance and intimacy and whether a couple makes a lifelong commitment to each other. These factors include a person's field of eligibles, physical attractiveness, proximity, similarity, flirting, and reciprocity of attraction.

What is Love?

- Sternberg's triangular theory of love identifies the three primary components of love as intimacy, passion, and commitment. Relationships with all three components are called consummate love; when one or more of these components is lacking, the relationship may be characterized as empty love, mere infatuation, romantic yet short-lived love, friendship, companionate love, or fatuous love.
- Individuals may be characterized by one or more of six "love styles" in intimate relationships: Eros (intensely romantic love), ludus (game-playing love), storge (friendship love), mania (obsessive love), pragma (practical love), and agape (selfless love). Couples with differing love styles are less likely to sustain a relationship over time.

Communication in Love Relationships

- Effective communication is fundamental to successful relationships. Researchers have identified three distinct communication styles in intimate relationships: validating, volatile, and conflict-avoiding. Poor communication patterns may signal that a relationship is weak and unstable. Warning signs of poor communication include criticism, contempt, defensiveness, and stonewalling. However, couples may improve their ability to communicate by learning new, more effective communication styles, such as holding regular "couples meetings," avoiding blame, and improving their understanding of effective communication in intimate relationships.
- Communicating about sex is one of the most difficult challenges for couples because of embarrassment over sexual topics, lack of accurate information, cultural sexual taboos, and fear of being judged or rejected by the partner. Communicating about sex is a learning process that takes time as partners' comfort levels increase and their emotional intimacy deepens.

Losing Love: Why Relationships End

- Specific behaviors have been identified that contribute to the end of relationships. The most common factors include one or more of the following: lack of self-knowledge, acceptance of false relationship myths, poor communication, power imbalances between the partners, low self-esteem of one or both partners, social isolation of the couple from others, lying and cheating, excessive jealousy, controlling behaviors, and violence.

Abusive and Violent Relationships

- Abuse and violence are the dark side of love relationships. Relationship abuse includes physical violence, verbal abuse, and emotional abuse.
- Abusive patterns typically repeat in a cycle: the honeymoon phase, the tension-building phase, and the explosion of abuse or violence, followed by a return to the honeymoon phase. The pattern then repeats, with the abuse and violence typically escalating.
- The fear of this escalating violence, concern over children who may be part of the relationship, lack of money, loss of emotional support, and even the possibility of being murdered by the abusive partner, are a few of the reasons abuse victims become trapped in these relationships and why, among those who do leave, most return to their abusers one or more times. Learning the signs of potential abusers may help prevent initial involvement or allow for earlier escape.

Your Sexual Philosophy

Love, Intimacy, and Sexual Communication

- Incorporating the issues discussed in this chapter into your working sexual philosophy frees you to enjoy love and intimacy to the fullest because you will have the knowledge to avoid being swept into an intimate relationship that may be wrong for you. You will also have the tools to make a good relationship better or to help heal a relationship that may be having problems. Perhaps most important, you will be less likely to become trapped in unhealthy relationships that are abusive or violent.

Chapter 5
Contraception
Planning and Preventing

Learning Objectives

After you read and study this chapter you will be able to:

5.1 Explain the considerations a couple or individual must include in the selection of the most effective method of contraception.

5.2 Discuss the contraceptive methods that help prevent *both* pregnancy and STIs, how they accomplish these tasks and the advantages and disadvantages of each.

5.3 List and explain the function, effectiveness, advantages, and disadvantages of the various types of contraceptive methods in each category discussed (withdrawal, hormonal, barrier, fertility awareness, intrauterine devices, and surgical methods).

5.4 Explain how abortion is *not* contraception and how the two are linked.

Since You Asked

1. Which method of birth control is best? (see page 146)
2. Are the safest condoms the ones with spermicide in the lubrication in case the condom breaks? (see page 155)
3. I don't like using condoms because they diminish sensation too much. Is there another safe method of contraception and STI prevention? (see page 156)
4. Does the female condom work as well as regular male condoms? (see page 159)
5. When my girlfriend and I are having intercourse, I always pull out before I ejaculate. That means there's no chance of pregnancy, right? (see page 160)
6. My girlfriend was on the pill, but she still got pregnant. How did that happen? (see page 163)
7. Is there really a "morning-after pill," and does it cancel out the need for my boyfriend and me to use condoms? (see page 169)
8. When is it safe to have sex without contraception, before or after a woman has her period? (see page 173)
9. I've heard that the IUD is a painful and dangerous method of contraception. Why would anyone use it? (see page 176)
10. After a man has a vasectomy, does it feel different to have an orgasm without ejaculating sperm? (see page 179)

Throughout this book, we discuss how "sex" includes a far greater number of activities than male-female intercourse. To many of you, this may be a new idea, and yet it is an important one for helping couples to discover the full range of pleasurable sexual behaviors that can enhance intimacy *and* allow for safer sexual choices. In reality, however, most heterosexual couples want to engage in sexual intercourse, and most of the time that they do, they want to prevent pregnancy. Most sexual activities (to be discussed in the next chapter) cannot lead to pregnancy, yet about half of all pregnancies in the United States are *unplanned*. This does not mean that they are all necessarily *unwanted*, but it does mean that a discussion of contraception is appropriate *before* we begin our discussion of sexual activities.

If you are having heterosexual intercourse, or planning to, and you do not want that activity to result in pregnancy or in a sexually transmitted infection (STI), you must use contraception (and the condom is the only method that offers protects from both pregnancy and STIs). Although most college students are aware of this fact, many do not adhere to it. One study of community college students found that 49% did not use a condom during their most recent vaginal intercourse, and 26% did not use a condom during their last anal intercourse (Trieu, Bratton, & Hopp-Marshak, 2011). In a national study, among young adults 18–24 years of age, 57% of males and 63% of females did not use a condom during their most recent heterosexual intercourse (Fortenberry et al., 2010). Clearly, many of these heterosexual couples used other forms of contraception. However, too many rely on luck or chance to avoid pregnancy, a problematic behavior because over a year's time, a couple using no contraception has an 85% chance of pregnancy.

Even if you are not currently sexually active, and even if you do not plan to be in the foreseeable future, an accurate working knowledge of contraception is still important. Chances are that the day will come when you *will* want to be sexually intimate with someone. If that intimacy involves intercourse, you should begin thinking about contraception in advance (like, now), rather than trying to make such decisions "in the heat of the moment." The point here is that even if you don't need contraception immediately, careful consideration now will help you make the best decisions about contraception later, or will provide accurate information to others who may seek your advice. Acquiring the information you need to make the right choices about contraception may be one of the most important decisions in your sexual life.

contraception
The process of preventing sperm cells from fertilizing an ovum.

conception
The moment a single sperm cell penetrates the wall of an ovum and the sperm's and ovum's DNA fuse together.

pregnancy
The period of growth of the embryo and fetus in the uterus.

Simply put, **contraception** refers to preventing conception (*contra* means "against"). This, of course, also prevents pregnancy, but conception and pregnancy are different. **Conception** occurs at the moment the ovum (egg) and a sperm cell join in the fallopian tube; **pregnancy** is when this fertilized ovum implants on the wall of the uterus (this is discussed in detail in Chapter 9, "Conception, Pregnancy, and Birth"). Contraception is about preventing conception, but it's also about allowing individuals and couples to plan, to decide when, or if, to conceive and bear children. It is about *choice*. If you choose to have sexual intercourse with a member of the opposite sex and you choose not to use contraception, these choices will almost certainly, eventually, lead to pregnancy. For many couples, this may be exactly what they want. But for many others, especially teens, young adults, and those of traditional college age, pregnancy is something they want to avoid, at least for now. An unwanted pregnancy is a difficult and often traumatic experience for everyone involved. With very few exceptions, the *only* way for a couple who is having heterosexual intercourse to prevent conception (and pregnancy) is to make contraception a routine part of their sexual behavior.

You are probably thinking about another important reason for using contraception other than avoiding unintended pregnancy: reducing the risk of STIs. As explored in detail in Chapter 8, "Sexually Transmitted Infections/Diseases," an alarming number of potentially serious and even deadly infections can be contracted through unprotected sexual behavior. Some of these are painful and uncomfortable, some can have serious lifelong consequences if left untreated, others have no cure, and still others can kill you. The good news is that some contraceptive methods, while helping to prevent pregnancy, also help protect you from these infections.

In this chapter, we will consider what an individual or couple must think about when choosing a personally comfortable and effective method of contraception. We will then examine in some detail the methods of contraception currently available, their degree of effectiveness in preventing pregnancy and STIs, and the advantages and disadvantages of each. We will also preview some future contraceptive options

Focus on Your Feelings

Contraception may not be one of the most emotionally charged topics in this book, but you might be surprised at how often strong feelings can enter into the discussion of contraception methods, the choice of a method, and issues about using (or not using) contraception.

You may one day find yourself (if you haven't already done so) in an intimate relationship in which you and your partner must agree not only about the importance of using contraception, but, perhaps more importantly, about which type of contraceptive method to choose. These decisions can be an enjoyable part of a new, intimate romance, but sometimes may be frustrating when you both desire an enjoyable sexual relationship, but you cannot agree on a contraception method that protects both of you from pregnancy or STIs. If this happens, couples *must* work out these differences, no matter how emotional they become, if they are to make smart and healthy decisions about their sexual activities together. Although for some, this discussion may be uncomfortable or embarrassing, it's too important to ignore. The more educated you both are about the methods that are available and how effective and safe they are, the more likely you will be able to come to a mutually happy decision that is acceptable to both partners.

Emotions can also run high if contraception fails—that is, when a woman becomes pregnant or when an STI is transmitted. Incorrect or inconsistent use of the contraceptive method is usually the culprit, but even with perfect use, no method is 100% failure-proof. If contraception has failed for anyone you know of (or for you!), you know the range of emotions that are possible; frustration or anger, anxiety, or even panic about the eventual outcome of the situation.

You can see, then, that even though contraception seems on the surface to be a relatively straightforward, rational topic, it has its emotional side, which is important to keep in mind as you read this chapter.

that are currently in various phases of research and testing. We will conclude, as always, by looking at how you can incorporate the information in this chapter into your sexual philosophy. First, however, let's take a brief look at the fascinating history of how contraception first became available in the United States. It is a shorter, more recent history than you might imagine.

Historical Perspectives

The Politics of Contraception

The early history of contraception in the United States is a story of radical politics. That's right, *politics*. Prior to 1930, federal laws prohibited nearly everything related to contraception, including providing information about it (D'Emilio & Freedman, 1988; Jütte, 2008). These laws were contained in a bill referred to as the Comstock Act, named after its primary supporter, Anthony Comstock (1844–1915), an "anti-obscenity," right-wing radical who considered all contraceptive information and devices obscene and indecent. His act made it illegal to sell or distribute contraceptive information or devices in any form. In addition, the act made it a crime for *any mention* of sexually transmitted diseases to appear in print. If that were not enough, doctors and nurses were prohibited from discussing these issues with their patients. Anyone found to be in violation of the provisions of the Comstock Act was subject to arrest and imprisonment (some unfortunates were arrested by Comstock himself).

Consequently, the only form of contraception available to most people in those years was the so-called rhythm method, a relatively unreliable method that will be discussed later in this chapter. At the time, even this method was largely misunderstood, and doctors often mistakenly advised couples who wanted to avoid pregnancy to engage in sexual activity at times during a woman's fertility cycle when they were *most* likely to become pregnant, not least!

Perhaps the most amazing fact about the history of contraception in the United States is that, essentially, one individual was single-handedly responsible for overcoming the Comstock Act and making contraception widely available. That person was Margaret Sanger (1883–1966), who is credited with coining the phrase "birth control."

Sanger was a nurse and mother of three children. In her thirties, she moved to New York City and became involved in socialist politics and the fledgling women's rights movement. Sanger believed that women's freedom to make choices about their personal and sexual lives depended on their ability to make choices about their reproductive lives. In 1912, she began to write and publish articles on female sexuality and on a woman's right to control her own reproductive body and to express herself sexually without the fear of pregnancy (D'Emilio & Freedman, 1988; Jütte, 2008). She was proposing the separation of sex from reproduction. At the time, this was a very radical idea.

While in Europe to avoid arrest in the United States on obscenity charges (her writings about contraception were considered obscene and illegal), Sanger visited a birth control clinic in the Netherlands, where women were being fitted by trained nurses with early versions of what we now know as the diaphragm. Convinced that this was the best and most effective form of contraception for women, Sanger returned to the United States and opened the first birth control clinic, located in Brooklyn, New York. After less than two weeks in operation, her clinic was raided by the police. Sanger was arrested on obscenity charges and spent 30 days in jail. Because of the publicity surrounding these events, Sanger became a celebrated and widely recognized leader of the contraceptive movement in the United States and in many countries throughout the world. By 1916, some 100,000 copies of her booklet, "Family Limitation," had been distributed, and birth control clinics had begun to open (Katz, 2002). This marked the

Anthony Comstock (1844–1915) created laws making all forms of contraception and contraceptive information illegal.

Margaret Sanger (1883–1966), crusader for the legalization and distribution of contraceptives in the United States.

start of a new sexual age in the United States and of the modern era of contraception and family planning. Sanger's later work to help fund contraceptive research resulted in the development of the first hormonal contraceptive pill in the late 1950s. Sanger is considered one of the original founders of the organization known today as Planned Parenthood.

As you will see in this chapter, in the decades since Sanger's early activism, the availability and variety of contraception methods has steadily grown. However, controversy and politics continue to complicate the contraception picture. Consider, for example, the debate over making condoms available in public high schools or even middle schools and whether doing so would promote an increase in sexual activity among teens (which research has shown it does not) or reduce disease and unintended pregnancy among sexually active teens (which it does) (Brenner et al., 2007; Kinnaman, 2007). The goal in this chapter is not to attempt to resolve these controversies over contraception, but rather to provide you with the information you need to make your own informed choices about this aspect of your sexuality.

Choosing a Method of Contraception

5.1 Explain the considerations a couple or individual must include in the selection of the most effective method of contraception.

Once you have made the decision to use contraception, you must determine what method is (or will be) best for you. To do this, most of you will need to do some "homework." First, you must educate yourself about the many available methods of contraception. For each method, you need to understand its level of effectiveness, how to use it correctly, how well it is likely to fit with your sexual lifestyle, how comfortable you will be using it, and whether the method is one you can commit to using not only correctly but also consistently, *every time* you have intercourse.

Since You Asked

1. Which method of birth control is best?

An important part of this education will be to learn about the many common myths and erroneous beliefs about contraception so that you can avoid assuming you are safe when you are not. Research has shown that 53% of unplanned pregnancies occur in couples that are using contraceptives! Most of these contraceptive failures are due either to incorrect use of the contraceptive method or to reliance on false information about pregnancy and contraception (Gabelnick, Schwartz, & Darroch, 2008; Henshaw, 1998).

An Individual or Shared Decision?

Is choosing a method of contraception something that each individual should decide or is it a shared decision between both partners? The answer is either, or both. Men and women certainly have the option of making an individual choice about their method of contraception. However, if you are in an ongoing, intimate relationship for which contraception either is or will be necessary, it is probably best to discuss the options with your partner and decide on a method of contraception that will be effective and comfortable for *both* of you. This will help ensure that you will use it consistently and correctly. If you cannot discuss and agree about conception with a partner, deciding to have sexual intercourse with that person is probably a mistake. Moreover, research has demonstrated that couples who are able to communicate effectively about contraception are more likely to use contraception consistently and are therefore at lower risk of unwanted pregnancy and contracting STIs (Beckman et al., 2006; Ryan, Franzetta, Manlove, & Holcombe, 2007).

On the other hand, most college students who are sexually active are not in stable relationships, and some may have more than one sexual partner at a given point in time or several partners sequentially. Individuals may have little opportunity to make a joint decision about contraception with a prospective partner before sexual activity begins. In these situations, each person needs to take personal responsibility for his or her own contraception method and for using it correctly and consistently.

Lifestyle Considerations

Not all contraception methods are equally appropriate for all sexual lifestyles and life stages. To make a rational and informed decision about contraception, you need to do some self-analysis about personal sexual life issues and activities, both now and in the future. For example, ask yourself the following questions:

1. Am I currently engaging in sexual activities that involve any risk of pregnancy or transmission of sexually transmitted infections?
2. Am I confident that I am in a one-on-one, exclusive relationship in which both of us are free of diseases that may be sexually transmitted?
3. How often do I engage in potentially risky sexual activities?
4. Do I have (or am I likely to have) more than one concurrent sexual partner?
5. How seriously would an accidental pregnancy disrupt my life?
6. Will the method I choose conflict with my religious beliefs?
7. Do I need to be in charge of contraception, or can I trust my partner to take care of it?
8. Do I feel comfortable overall with using this method of contraception?

How you answer these questions will help you determine exactly what features you need in the method of contraception you choose. For example, if you answered yes to question 2, you may feel comfortable and safe selecting a method that is effective in preventing pregnancy, even though it may not provide protection from STIs. On the other hand, if your answer to question 4 is yes, prevention of pregnancy *and* STIs needs to be part of your decision process.

To get a better idea of whether your contraceptive method of choice is working (or will work) for you, see "Self-Discovery: Contraceptive Method Comfort and Confidence Scale." (on the following page).

Unreliable Methods

Flawed information and beliefs about contraception often lead to poor decisions about contraception and safer sex. Many people still believe in contraceptive strategies that historically were thought to be effective but that, in reality, are not only ineffective but are based on fundamentally false assumptions. Table 5.1 summarizes some of these myths about contraception.

Table 5.1 Contraceptive Myths

Contraceptive Myth	The Contraceptive Facts
Douching	The stream of water or other solution used to wash out the vagina can actually push the remaining sperm through the cervix even faster. Not only is douching not recommended for contraception, but health professionals recommend that women do not douche at all as a routine hygiene practice because it may force bacteria higher into a woman's reproductive system and *increase* the risk of infections, including bacterial vaginosis and STIs.
Cola douche	Cola or other beverages might be lethal to some sperm, but they are very unhealthy for the internal environment of the vagina and are not effective as a contraceptive.
Standing up during intercourse	Sperm are very skilled at swimming in any direction, up or down; with or against gravity.
Plastic wrap in place of male condoms	Plastic wrap might keep a penis fresh if it were in the fridge, but it will almost certainly slip off, leak, or break during intercourse and will not reliably protect anyone from pregnancy or STIs.
Toothpaste as spermicide	Toothpaste does not kill sperm or infectious microbes and works very poorly as a sexual lubricant as well.
Pregnancy is not possible the first time	This is just wrong. A fertile girl or woman can become pregnant the first or any other time she has unprotected intercourse.

Self-Discovery

Contraceptive Method Comfort and Confidence Scale

To help you determine whether a contraceptive method suits your lifestyle and is a realistic choice for you, answer the following questions (yes or no) for each method you are considering.

_______ **1.** Have I had problems using this method before?

_______ **2.** Have I ever become pregnant while using this method?

_______ **3.** Am I afraid of using this method?

_______ **4.** Would I really rather not use this method?

_______ **5.** Will I have trouble remembering to use this method?

_______ **6.** Will I have trouble using this method correctly?

_______ **7.** Do I still have unanswered questions about this method?

_______ **8.** Does this method make menstrual periods longer or more painful?

_______ **9.** Does this method cost more than I can afford?

_______ **10.** Could this method cause me to have serious complications?

_______ **11.** Am I opposed to this method because of my religious or moral beliefs?

_______ **12.** Is my partner opposed to this method?

_______ **13.** Am I using this method without my partner's knowledge?

_______ **14.** Will using this method embarrass my partner?

_______ **15.** Will using this method embarrass me?

_______ **16.** Will I enjoy intercourse less because of this method?

_______ **17.** If this method interrupts lovemaking, will I avoid using it?

_______ **18.** Has a nurse or doctor ever told me *not* to use this method?

_______ **19.** Is there anything about my personality that could lead me to use this method incorrectly?

_______ **20.** Am I at any risk of being exposed to HIV (the AIDS virus) or other sexually transmitted infections if I use this method?

Scoring: How many total yes answers did you have?

A FEW (0–3): Most people will have a few *yes* answers. *Yes* answers mean that potential problems may arise with that method. All methods have their strong points and their drawbacks. If you only have a few, chances are good that the method you've picked will work for you.

MORE THAN A FEW (4 OR MORE): If you have more than a few *yes* answers, you may want to talk with a physician, counselor, partner, or friend to help you decide whether to use this method or how to use it so that it will really be effective for you.

In general, the more *yes* answers you have, the less likely you are to use this method consistently and correctly each and every time you have sex. You might want to consider using another method.

SOURCE: From CONTRACEPTIVE TECHNOLOGY 18th edition, R. Hatcher, J. Trussel, F. Stewart, W. Cates, G. Stewart, F. Guest & D. Kowal (Eds.).

Reliable Contraception Methods

As noted earlier, your choice of a successful method depends on your sexual lifestyle and your comfort with and confidence in the method itself. If you trust that you and your partner are in a mutually exclusive, monogamous relationship, then you probably do not have to be concerned about STIs, unless one partner is already infected. Therefore, you may feel free to choose from *all* the methods discussed in this chapter for the prevention of pregnancy. However, most sexually active college students are not in monogamous relationships that are STI-free and long-term. If that is, or may be, true for you, then you must consider *both* contraception and prevention of disease in your choice of effective contraception. We will turn now to methods that help prevent both unwanted pregnancy and STIs (which are discussed in detail in Chapter 8, "Sexually Transmitted Infections /Diseases").

Choosing a method of contraception that is right for you requires knowledge of the different types available, including abstinence, hormonal methods, barrier methods, fertility awareness approaches, intrauterine devices, and surgical methods (sterilization). Table 5.2 offers an overall summary of these methods' effectiveness in preventing pregnancy and Table 5.3 provides more detailed information on each.

Table 5.2 Contraceptive Effectiveness Summary Table

Method Group	Most Effective	Average in-Use Failure Rate*	How to Maximize Method Effectiveness
– Surgical: Tubal Procedure, Vasectomy – IUD (Intrauterine Device) – Hormonal: Implant		0.3%	• No Further Action Needed • Vasectomy: Use Back-Up Method for First 3 Months
– Hormonal: Oral (Pills), Injection, Ring, Patch		8.0%	• Injection: Update Injection Promptly • Pill: Take Pill Daily at Approximately the Same Time • Patch; Ring: Keep In Place; Change On Time
– Condoms (Male and Female)		19.5%	• Use for *every* Act of Intercourse • Use Early in Sexual Interaction • Use Correctly • Do not Reuse
– Vaginal Barriers (Diaphragm, Cap, Sponge, Spermicides)		20.6%	• Do not use Without Spermicide • Insert Correctly for *every* Act of Intercourse • Do not Insert too Long Before Intercourse • Keep in Place Six Hours after Intercourse • Consider Combining with Condom
– Withdrawal		22.0%	• Male Focus on Ejaculation Sensations • Do not Ejaculate near Vaginal Opening
– Fertility Awareness		24.0%	• Use Method with Precision—do not Guess • Use Barrier Backup or Abstain on Fertile Days
– Abstinence		45.0%	• Exercise Commitment and Willpower • Have Backup Method Available
– Chance		85.0%	• Cannot be Maximized
	Least Effective		

* This Table does not Address Sti Prevention

SOURCE: Based on Hatcher, et al. (2011). Figure 3-1, p. 52

Coming up next, we explore each of these methods in detail. For our purposes, the methods have been divided into two categories relating to the methods' most important functions: those that help to prevent pregnancy *and* STIs, and those that help to prevent pregnancy, but offer no protection against STIs.

Methods for Preventing Pregnancy *and* STIs (STDs)

5.2 Discuss the contraceptive methods that help prevent *both* pregnancy and STIs, how they accomplish these tasks, and the advantages and disadvantages of each.

Many individuals want contraception that accomplishes two important jobs: preventing pregnancy *and* protecting against sexually transmitted infections. Contraception methods that help do both of these include complete or "selective abstinence" and two barrier methods—male and female condoms.

Selective Abstinence

The only way to prevent pregnancy and STIs 100% of the time is to avoid all sexual behaviors that can result in pregnancy or the spread of infections. This strategy is

Table 5.3 Available Contraceptives by Type

Contraceptive Method	Typical Failure Rate (%)	Perfect Use Failure Rate (%)	Approximate Cost[1]	Advantages and Possible Positive Side Effects	Disadvantages and Possible Negative Side Effects
No method (chance)	85	85	Average cost of full-term pregnancy delivery in the United States: $7,000–$12,000 (this does not include prenatal medical costs)	Best chance of pregnancy	No protection against pregnancy or sexually transmitted infections, near certainty of "accidental" pregnancy or spread of STIs
METHODS THAT REDUCE RISK OF PREGNANCY *AND* STIs					
Selective abstinence (no intercourse)	45 (estimate)	0	Free but requires commitment and self-control by both partners	No cost, no risk of pregnancy, protects against STI spread with committed and consistent use	Requires self-discipline, partner agreement, strong commitment; risk of unprotected intercourse if commitment weakens
Male condoms	18	2	$0.20–$1.00 each[1]	Accessible, portable, inexpensive; usable with many other methods; help protect against STIs and pelvic inflammatory disease (PID)	Rarely, latex may cause irritation (avoided with use of polyurethane condoms); possible reduced male sensitivity for some; may create interruption of sexual "flow" to put condom on
Female condoms (*Reality; FC2*)	21	5	$1.00–$3.00 each	Does not require male involvement in decision, may be inserted up to eight hours before intercourse; less interruption in sexual activity	Slightly higher failure rate than male condom; some irritation and allergic reactions reported (rare); additional lubrication may be needed for comfort (provided with condom)
METHODS THAT REDUCE RISK OF PREGNANCY BUT *NOT* STIs					
Withdrawal: "Coitus Interruptus"	22	4	Free, but requires accurate bodily self-knowledge	May be used when no other method is available	Requires male to be acutely aware of impending orgasm and withdraw prior to any ejaculation; can fail if ejaculation occurs near vaginal opening
Hormonal Methods					
Oral contraception (birth control pills)	9[2]	0.3	$25–$60 per month of protection (Rx)[1]	Very effective when taken consistently and correctly; reversible; safe for most healthy women; may decrease menstrual cramps and pain and regulate periods; may reduce premenstrual symptoms; may treat acne; may decrease risk of ovarian and endometrial cancer, ectopic (tubal) pregnancy, and PID; decreased PMS symptoms	Must remember to take pills daily; some forms may be associated with breakthrough bleeding or spotting between periods; possible weight gain, increased breast size, nausea, headaches, depression; most minor side effects decrease after first few months of use; very rare complications of cardiovascular disease
90-day pills (*Seasonale* and *Seasonique*)	2[2]	0.1	$40–$75 per month of protection (Rx)[1]	Four periods per year instead of 12; fewer menstruation-related problems, including migraine headaches, cramping, mood swings, and bloating; may reduce uterine fibroids and endometriosis. [Seasonique: lower hormone dose and shorter periods]	Greater chance of intermenstrual bleeding and spotting; breast tenderness; low but slightly increased risk of blood clots, heart attack, and stroke, especially in women over 35 who smoke
365-day pills (*Lybrel*)	2[2]	1.0	$40–$60 per month of protection (Rx)[1]	Continuous use oral contraceptive; most women will experience no menstrual periods; advantages similar to 90-day pills, but expanded to 12 months	Same as those for 90-day pills
Hormonal implant (*Implanon*)	0.05	0.05	$400–$800 for insertion; $11–$22 per month of protection over three years (Rx)[1]	Long-term, effective contraception; reversible; light or no periods and decreased menstrual cramps and pain; decreased risk of developing endometrial cancer, ovarian cancer, and PID	Higher initial costs; headaches; menstrual irregularities (breakthrough bleeding or cessation of periods); breast tenderness; rare cases of prolonged periods
Injectable (*Depo-Provera*)	6	0.2	Approx. $90 per 90-day injection; $30 per month of protection (Rx)[1]	Very high contraceptive efficacy; each injection provides three months of contraception; light or no periods; decreased menstrual cramps and pain; decreased risk of developing endometrial cancer, ovarian cancer, and PID	Return of fertility may take up to a year; discomfort of injections for some; return visits every three months; menstrual irregularities (breakthrough bleeding or cessation of periods); weight gain; breast tenderness; possible loss of bone mass (recovers when discontinued)

Table 5.3 Continued

Contraceptive Method	Typical Failure Rate (%)	Perfect Use Failure Rate (%)	Approximate Cost[1]	Advantages and Possible Positive Side Effects	Disadvantages and Possible Negative Side Effects
Contraceptive patch (*Ortho Evra*)	9[2]	0.3	Approx. $60–$80 per month of protection (three-pack) (Rx)[1]	Continuous, consistent dose of hormones; one patch per week eliminates need to remember daily pill; no interruption of sexual activity; helps regulate periods; fewer abdominal side effects than pill (hormones do not pass through digestive tract)	Breast tenderness and enlargement; headache; nausea; spotting or break-through bleeding; abdominal cramps and bloating; vaginal discharge; may expose women to level of hormones 60% higher than most oral contraceptives; serious risks include blood clots, stroke, or heart attack
Contraceptive ring (*NuvaRing*)	9	0.3	Approx. $70 per ring (one per month) (Rx)[1]	Convenient once-a-month insertion; no medical fitting required; no interruption in spontaneous sexual activity; lowest dose of hormones due to release directly into mucous membranes	Headaches; vaginal irritation and discharge; may slip out during sexual intercourse (rare); weight gain; headache; nausea; serious risks include blood clots, stroke, or heart attack
Emergency contraception	25–40[2, 3]	20[3]	Approx. $20–$50 per kit (usually OTC)[1] over for *Plan B* and *Next Choice*; required for *ella*[1]	Provides protection from conception after unprotected intercourse or failure of primary method; highly effective in reducing risk of pregnancy; provides backup method for emergency situations; should not be used for routine contraception; does *not* cause an abortion	Should not be used if woman is already pregnant; nausea; vomiting; lower abdominal pain; fatigue; headache; dizziness; breast tenderness; and menstrual timing changes in month of use (side effects may last several days)
Barrier Methods					
Diaphragm with spermicide	12	6	$50–$200 for exam; $25–$75 for diaphragm. Lasts up to two years; $3–$11 per month of protection (Rx)[1]	Does not require male involvement in decision, need not interrupt intercourse (may be inserted up to six hours in advance); may protect against cervical abnormalities and cancer	Possible skin irritation from spermicide or latex; may cause increased risk of female urinary tract infections; additional spermicide must be used for repeated intercourse
Cervical cap with spermicide (rarely used in U.S.)			$50–$200 for exam; $60–$75 for cap. Lasts two years; $5–$11 per month of protection (Rx)[1]	Does not require male involvement in decision; need not interrupt intercourse; protects for 48 hours	Low effectiveness, increased risk of urinary tract infections; should not be used during menstruation or left in place longer than 48 hours to avoid possible toxic shock syndrome; possible skin irritation from spermicide or latex; low effectiveness for women who have previously given birth
Women who have never given birth	16	9			
Women who have previously given birth	32	26			
Contraceptive sponge (*Today*)			$2–$4 each[1]	Controlled by woman; can be inserted ahead of time; provides 24 hours of protection	Possible temporary skin irritation due to spermicide; low effectiveness for women who have previously given birth
Women who have never given birth	12	9			
Women who have previously given birth	24	20			
Vaginal spermicides (not recommended for use without additional barrier device)	28	18	$0.50–$1.50 per use	Simple to use; available over the counter; provides increased effectiveness when used with other methods; provides extra lubrication for intercourse	Low effectiveness rate; recommended for use with other barrier method; possible temporary skin irritation or vaginal infection
Fertility Awareness Methods					
Overall	25	0.4–5	Either no cost or purchase of thermometer, $10–$50; ovulation predictor kit, $10–$40	Inexpensive; may be effective when used perfectly; no serious side effects; increases knowledge of reproductive physiology; enhances self-reliance	Requires careful record keeping; unforgiving of imperfect use; irregular menstrual cycles make use more difficult

(continued)

Table 5.3 Continued

Contraceptive Method	Typical Failure Rate (%)	Perfect Use Failure Rate (%)	Approximate Cost[1]	Advantages and Possible Positive Side Effects	Disadvantages and Possible Negative Side Effects
Standard Days Method	12	5	Cycle beads, $10–$20		
Two-Day Method	14	4			
Symptothermal Method	20	0.4			
Intrauterine Devices (IUDs)					
Copper T (*Paragard T380A*—up to 12 years)	0.8	0.6	$200–$700 for insertion; ($1.50–$5.00 per month of protection when used for full effectiveness term (Rx)[1]	Highly effective; safe; long-acting; hassle-free; no interruption of intimacy; no hormone level change	May pose slight increase in risk of PID; may cause menstrual problems; more risk for women who have never had children; requires trained medical professional for insertion and withdrawal
Levonorgestrel (*Mirena*—up to 5 years)	0.2	0.2	$200–$700 for insertion; $3–$12 per month over full term (Rx)[1]		
Levonorgestrel (*Skyla*—up to 3 years)			$200–$700 for insertion; $5–$19 per month over full term (Rx)[1]		
Surgical Methods					
Female sterilization (fallopian tube procedures)	0.5	0.5	$1,000–$5,500 (no further expense)	Highly effective; permanent; cost-effective over time; reversal may be possible in some cases	Surgical procedure; expensive at time of surgery; permanent; reversibility difficult and very expensive
Male sterilization (vasectomy)	0.15	0.10	$250–$1,000 (no further expense)	Highly effective; inexpensive in the long run; very safe; considered permanent, but successful reversals are common; far simpler procedure than female sterilization	Permanent; minor surgical procedure; moderately expensive at time of procedure; reversal expensive

[1] The cost may be significantly lower (or free) at some public health clinics and campus health services, or paid fully or in part by medical insurance. The Affordable Care Act, which took effect in 2013, mandates coverage for contraceptives.

[2] Effectiveness may be slightly reduced in overweight or obese women.

[3] Effectiveness depends on the interval between unprotected intercourse and use of emergency contraceptive pills; the shorter the interval (0–5 day maximum), the greater the effectiveness.

SOURCES: Chan, Harting, & Russell (2009); Corinna, 2010; Hatcher, Trussell, Nelson, Cates, & Kowal (2011); Pallone & Burgess (2009).

selective abstinence
Choosing to engage in or avoid certain sexual behaviors on the basis of their risks of STIs or pregnancy.

celibacy
Choosing to forego all sexual activity.

outercourse
A form of abstinence in which a couple chooses to engage only in sexual behaviors that are unlikely to result in pregnancy or infection and to avoid all others, such as vaginal or anal intercourse and oral sex.

called **selective abstinence**. But what does abstinence really mean? You are probably thinking it means "no sex." But throughout this book we stress the idea that "sex" is more than intercourse.

Abstinence does *not* necessarily mean avoiding *all* sexual activity—that's known as **celibacy**. In our current context, abstinence is choosing not to engage in *specific* sexual behaviors that involve any risk of pregnancy or STI transmission. Abstinence is different from celibacy. "Selective" abstinence (sometimes called **outercourse**) means engaging only in sexual behaviors that are "safe" and avoiding those that are risky, such as vaginal or anal intercourse and oral sex (see Chapter 6, "Sexual Behaviors: Experiencing Sexual Pleasure," for a discussion of these activities). Many couples freely chose to abstain from all risky sexual activities, but still want to express the physical intimacy they may feel for one another. People who equate sexu al behavior with intercourse hold a rather limited view of the range and variety of sexual interactions that can provide intense feelings of intimacy and sexual satisfaction without the risk of pregnancy or STIs (Hatcher et al., 2008).

If a couple wants to be abstinent, and sexually intimate, what can they do? They can hold hands, kiss, touch each other in intimate ways, and explore each other's bodies (Dailard, 2003; Hatcher et al., 1994; Hatcher et al., 2008). Touching genitals, masturbating, mutual masturbation, and orgasm are all included in these activities, as long as the couple ensures that semen or vaginal fluids do not touch a partner's mucous membranes or breaks in each other's skin, and that semen is not ejaculated near the vaginal opening.

Many couples who choose selective abstinence as their method of birth control find that their sexual interactions are especially satisfying because they are free of the worry of pregnancy or STIs. However, everyone knows how easy it is for judgment to be clouded in moments of great passion and sexual excitement. Therefore, deciding on abstinence as your method of contraception takes a strong commitment combined with clear communication between partners. "In Touch with Your Sexual Health: Abstinence—Can You Make It Work for You?" summarizes the steps that must be taken to ensure effective and successful reliance on this method.

Male Condoms

The **male condom** is a thin sheath of latex (rubber) or polyurethane (plastic) that is placed over an erect penis prior to and during intercourse. Condoms are also sometimes referred to as "safes," "rubbers," or prophylactics.

male condom
A thin sheath of latex (rubber), polyurethane (plastic), or animal tissue that is placed over an erect penis prior to and during intercourse.

Condoms are a barrier method of contraception—the condom forms a barrier, preventing sperm from entering the woman's reproductive tract. Several additional forms of barrier contraceptives are available; these will be discussed later in the chapter because they do not offer protection from STIs.

Condom use has steadily increased over the past twenty-five years (Mosher & Jones, 2010). Most people use condoms primarily for contraception, although some report using them strictly for the prevention of STIs, either because they are relying on another highly effective method of contraception or because they are in a gay male relationship (Pilkinton, Kern, & Indest, 1994). Used correctly, condoms are an

In Touch with Your Sexual Health

Abstinence—Can You Make It Work for You?

When individuals or couples choose selective abstinence as their method of choice of contraception, the effectiveness of this choice depends on their ability to commit to the plan and maintain the necessary amount of discipline and self-restraint required for success. Here are some ways to ensure that abstinence is a successful means of contraception.

1. Learn everything you can about sexual anatomy and response, and what sexual behaviors are risky for pregnancy or STIs.
2. Decide about your specific sexual choices at a time when you feel clear headed, confident, secure, and not under the influence of alcohol or other drugs that can cloud judgment and decision making.
3. If you have a regular current or potential sexual partner, make decisions together at a time when you feel close to each other and comfortable talking, but *not* while you are being sexual. For example, talk as you take a walk holding hands or across the dinner table after a romantic dinner. Be sure you have plenty of time, so you won't need to rush the conversation.
4. Decide in advance what sexual activities you will engage in and those that you will not. Discuss these clearly and specifically with your partner in advance; don't wait until the last minute or when you are already engaging in intimate sexual behavior.
5. Avoid high-pressure sexual situations and stay sober in sexual settings.
6. Do not waver in your decisions. If you have decided a certain behavior is off limits, do not change your mind in the heat of passion. This will undermine your commitment on all of your decisions.
7. Be as informed as possible about safe sex and contraception, so you can be prepared if you change your mind about abstinence. Have condoms or another barrier method available, just in case.
8. Obviously, if you are using abstinence, you must use self-control to refrain from intercourse, especially if you do not have a ready means of contraception available.
9. Some people or couples, regardless of their intentions to practice abstinence, do "give in" and engage in intercourse, often without contraception (because they were planning on being abstinent). If this happens, be informed about *emergency contraception pills*, which are quite effective in preventing pregnancy if taken within 72 hours of unprotected intercourse.

SOURCE: Adapted from Kowall, 2008.

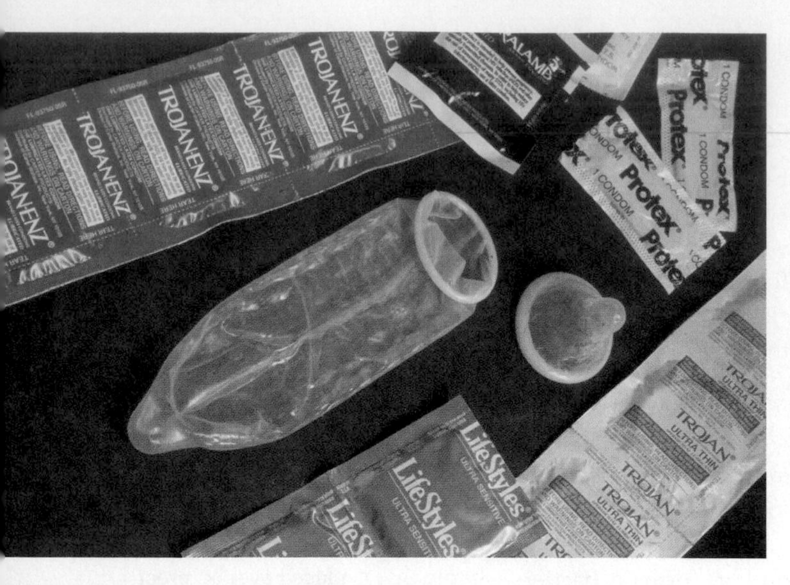

Today's condom market offers an amazingly wide selection to satisfy all preferences.

extremely effective method of contraception *and* of reducing the risk of all sexually transmitted infections.

There are many varieties of condom: with or without lubrication, with or without spermicide (chemicals that kill sperm cells), with plain or reservoir tips, ultrathin, extra-strength, assorted colors, ribbed, textured, "snugger" or larger sizes, with carrying case included, and yes, even flavored and glow-in-the-dark models. Entire stores, both physical and online, are devoted exclusively to selling condoms and condom-related products. Some examples are www.condomania.com, Condom World in Boston (www.condomworldboston.com), or Condom Sense in Dallas (www.condomsenseusa.com). More than a hundred different brands and styles of condom are available in the United States. And research is continuing into new condom designs (see "Evaluating Sexual Research: Reinventing the Condom?")

This level of availability of this popular form of contraception and STI protection is far from universal. Until recently, China, for example, banned advertising of contraception and even discussion of related topics was taboo. However, as China has acknowledged and attempted to deal with its growing problem of sexually transmitted infections (including HIV), its repressive attitudes about contraception have undergone nothing short of a revolution. As we first noted in Chapter 1, China's efforts to increase safer sex have even included the installation of condom vending machines in various locations throughout the country, including 9,000 new machines dispensing *free* condoms in Shanghai (IPPF, 2010), not to mention that China is one of the world's leading exporters of condom vending machines.

In other countries, anticondom laws are much stricter. One study found that law enforcement officials actively confiscate and/or destroy condoms, especially among sex workers (prostitutes), those who need them most for preventing the spread of HIV (Kardas-Nelson, 2012). The study found that 80% of sex workers surveyed in Russia reported that police had confiscated condoms; in Namibia, 50% of sex workers reported that condoms had been destroyed by police; and in Zimbabwe, sex workers have been arrested, not on prostitution charges, but on charges of condom possession. Even in the Western world, police are naively moving to ban condoms from saunas in Edinburgh, Scotland (Withnall, 2014). This is especially troubling in that using condoms as a pretext in antiprostitution enforcement will have no effect on the sex trade (sex workers simply keep working without condoms), but only serve to make it far more dangerous and deadly due to HIV transmission.

An "assembly line" in a condom factory.

HOW MALE CONDOMS WORK The way condoms work is not very mysterious. For contraception, they act as a barrier to prevent semen from entering the vagina. They also prevent the most contact of skin with mucous membranes and the exchange of genital fluids (semen, vaginal secretions) that can transmit various sexually transmitted infections (see Chapter 8, "Sexually Transmitted Infections /Diseases," for details).

For the greatest protection against pregnancy and STIs, it is recommended that you follow these five basic rules:

1. Use latex or polyurethane condoms only. Natural membrane ("lambskin") condoms have small pores that may allow disease-causing viruses to pass through.
2. Use lubricated condoms (but *not* with spermicide). Inadequate lubrication during intercourse may cause a condom to break or tear. Using lubricated condoms greatly reduces this possibility.

Evaluating Sexual Research

Reinventing the Condom?

The Bill and Melinda Gates Foundation, which supports many important health initiatives around the world, decided that we need a "better" condom, one that would help encourage their use and increase prevention of unwanted pregnancies and STIs (Doucleff, 2013). To this end, as part of their Grand Challenges in Global Health, in 2013, they announced worldwide challenges for research proposals into the development of "a more fun and pleasurable condom." The prize? $100,000. This contest was designed to address two issues with the current male condom: (1) many people find that they interfere with sexual pleasure in some way, and (2) there are shortages of condoms in some parts of the world.

The foundation was looking for a "new generation of condoms" that involved:

- Application of safe new materials that may preserve or enhance sensation
- Development and testing of new condom shapes/designs that may provide an improved user experience
- Application of knowledge from other fields (e.g., neurobiology, vascular biology) to new strategies for improving condom desirability
- Ideas that address loss of erection while donning condoms, including decreased application time or increased ease and clarity of application direction
- New ideas for female condoms that are easier to apply and/or increase pleasure

As of the writing of this chapter, the challenge is continuing into 2015, but a significant number of proposals had been chosen for funding. Some examples of the creativity this challenge stimulated and funded are (Grand Challenges Explorations Grants, 2013):

- Benjamin Strutt and a team from Cambridge Design Partnership in the United Kingdom will design a male condom made from a composite anisotropic material that will provide universal fit and is designed to gently tighten during intercourse, enhancing sensation and reliability.
- Willem van Rensburg of Kimbranox, Ltd., in South Africa, will test a condom applicator—the Rapidcom—that is designed for easy, technique-free application of male condoms.
- Jimmy Mays of the University of Tennessee in the U.S. will develop a prototype male condom made from superelastomers (a highly elastic polymer). This will enable the manufacture of thinner and softer condoms that will enhance user experience.
- Mark McGlothlin of Apex Medical Technologies, Inc., in the U.S., will produce a male condom with enhanced strength and sensitivity using collagen fibrils. Collagen fibrils would provide a hydrated micro-rough skin-like surface texture that facilitates heat transfer to produce a more natural sensation.

Origami Condoms based in Los Angeles is one of the companies working on the Global Health Grand Challenge to reinvent the condom.

Condoms lubricated with a spermicide are no more effective and no longer recommended. This is because research has found that the spermicide chemical (usually nonoxynol-9) may cause irritation to the mucous membranes of the vagina or anus, which actually *increases* the possibility of HIV transmission (Moniz & Beigi, 2012). A new, less-irritating spermicide called *C31G* is currently under investigation but, as of this writing, is not available (Kling, 2010). Therefore, *nonspermicidal lubricated* condoms are the best all-around choice.

Since You Asked

2. Are the safest condoms the ones with spermicide in the lubrication in case the condom breaks?

3. Use condoms with a reservoir tip (or nipple). This feature allows space at the end of the condom to catch the semen upon ejaculation so there is less chance of the condom breaking or leaking.
4. Store condoms in a drawer, medicine cabinet, cupboard, etc. Avoid hot or crushed storage such as a glove compartment or wallet.
5. Use a new condom for every act of intercourse—they are *not* reusable.

Incorporating condom use into lovemaking can enhance the couple's intimate experience.

EFFECTIVENESS OF MALE CONDOMS As you can see in Table 5.3, the effectiveness of condoms depends a great deal on proper use. When condoms are used consistently and correctly for vaginal intercourse, they are very effective in preventing pregnancy. The 3% failure rate indicated in the table means that if 100 couples have intercourse an average of twice each week for a year and use *condoms correctly each and every time*, only three of the couples will become pregnant. This 3% failure rate is based on the couples, not condoms themselves. Think about it. If you do the math, this means for that group of 100 couples, 10,400 condoms were used in a year and only three pregnancies occurred. That's an incredibly low per-condom failure rate of 0.028% (Warner & Hatcher, 1998).

However, looking further at Table 5.2, you can see that the pregnancy rate for condoms in the way couples *typically* use them is nine times greater than for perfect use (2% versus 18%). These failures are not generally due to product failure but to incorrect use. Why is typical use so poor? Because using condoms correctly every time takes some learning and, well, practice!

PROPER USE OF MALE CONDOMS What is the *correct* way to use condoms? The procedures may sound somewhat complicated, but it's simple once a person becomes accustomed to them. The more a couple uses condoms, the easier and more natural feeling they will become.

Many people argue that condoms interrupt or interfere with the flow and romance of making love. However, for many others, placing the condom on the man's penis (either by the man himself or by his partner) can be viewed as an enjoyable part of the sexual experience that can actually enhance rather than reduce intimacy and erotic feelings.

Using condoms correctly is the most important determinant of their effectiveness. Remember, condoms themselves rarely fail, but failure due to the incorrect use of condoms is all too common. "In Touch with Your Sexual Health: Using Male Condoms Correctly" summarizes the recommended steps for using condoms effectively. (Instructions for using female condoms will be discussed shortly.)

CONVINCING YOUR PARTNER TO USE CONDOMS If you choose condoms as your preferred method of contraception, it does not automatically mean your partner will agree. Generally, men have less favorable attitudes toward condom use than women do (Conley & Collins, 2005; Franzini & Sideman, 1994). Many men are convinced that condoms will significantly reduce sensitivity during intercourse, interfere with erection, create an uncomfortable interruption in the lovemaking process, or make them seem "less macho." These and other objections, however, are not exclusive to males. In fact, when undergraduate students were surveyed, 30% of men and 41% of women reported that a partner had attempted to dissuade them from using a condom on at least one occasion (Oncale & King, 2001; Reece, 2010).

Since You Asked

3. I don't like using condoms because they diminish sensation too much. Is there another safe method of contraception and STI prevention?

Convincing a partner to use a condom remains a common problem, especially among college students. Often when a couple is in an ongoing relationship, concerns about using condoms can be overcome by experimenting with different brands and styles of condoms and incorporating their use into erotic touching and foreplay. If a partner steadfastly refuses to use a condom, various strategies have been reported that may be effective in overcoming the resistance and ensuring safer penetrative sexual activities, including explanation of risks of unprotected intercourse, refusing unprotected intercourse, negative emotional appeals ("I'm upset and angry about your refusal to use a condom"), positive emotional appeals ("I'll like you way more if you agree to use condoms"), and even deception about it being her most fertile "time of the month" (De Bro, Campbell, & Peplau, 1994). Students were also asked to rate the

In Touch with Your Sexual Health

Using Male Condoms Correctly

The number of steps below may make condom use seem pretty complicated, but, with a bit of practice, the whole process becomes very easy and natural.

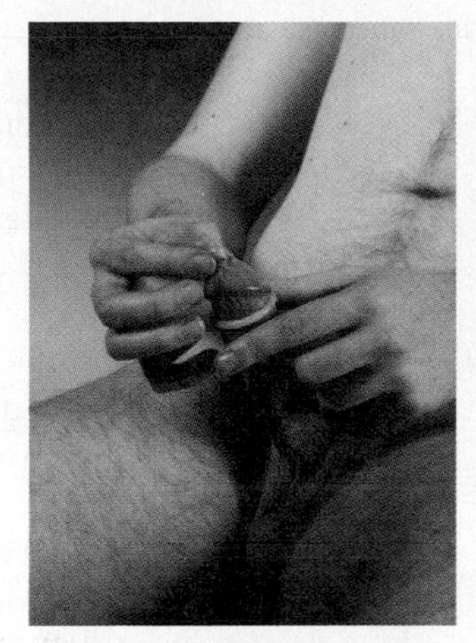

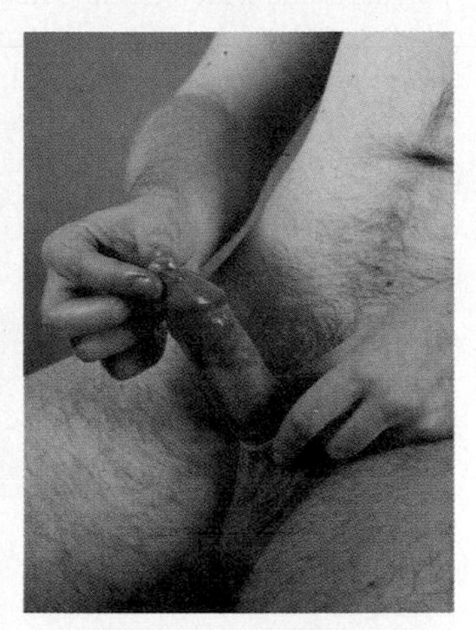

1. Open the condom package carefully. Sharp fingernails or teeth can tear or puncture the condom.
2. Don't try to test the condom by filling it with air or water, as this will weaken the latex. All condoms are pretested at the factory.
3. The condom should be placed on the erect penis before any contact between the penis and the partner's genital area.
4. To be sure which way the condom unrolls, place it over a finger and unroll it slightly. If the condom is placed upside down on the penis and then turned over, some pre-ejaculate fluid, and possibly sperm cells, may be left on the outside of the condom after it is unrolled.
5. Pinch the end of the condom as you unroll it to leave a small amount of empty space at the tip. This may help prevent breaking and may allow for more movement of the glans of the penis within the condom for increased sensitivity.
6. Unroll the condom all the way to the base of the penis.
7. Be sure there is adequate lubrication, either produced naturally from sexual arousal, from the lubricated condom, or from another source, prior to penetration.
8. If you are adding lubrication, do not use oil-based products such as hand creams, lotions, Vaseline, baby oil, cooking oil, massage oils, suntan lotions, or edible oils, as these will weaken the latex (they do not harm polyurethane condoms). Always use water-based lubricants such as K-Y Jelly, spermicidal creams, or other lubricants made especially for sexual activity (available at most pharmacies).
9. If it feels as if the condom has broken or slipped off during intercourse, stop and check. If it has, replace it with a new one.
10. After ejaculation, hold the rim of the condom all the way around the base of the penis and withdraw the penis *before* loss of erection. This will help ensure that no semen comes in contact with the partner's genital area internally or externally. Accidentally leaving the condom inside your partner or spilling semen into or near the opening of the vagina greatly increases the risk of pregnancy and STIs. Sperm do *not* die immediately when exposed to the air, as some people believe.
11. Holding the penis away from your partner, remove the condom and check it for signs of damage. Then wrap the used condom in tissue and throw it in the trash.
12. If you desire to have intercourse again, you must use a new condom. Condoms should never be reused.
13. If the condom breaks or slips off without your knowledge, consider using emergency contraception pills (discussed later in this chapter) to help prevent unwanted pregnancy.

All these steps might make condoms seem to be a difficult method of contraception, but they're not. After a short time of using condoms as your chosen method of contraception, the process becomes easy, automatic, and even enjoyable.

SOURCE: Adapted from Warner & Steiner (2008). Male condoms. In *Contraceptive Technology 18th Revised Edition*, R. Hatcher, J. Trussel, F. Stewart, W. Cates, G. Stewart, F. Guest, & D. Kowal (Eds.), pp. 297–316. Reprinted by permission of the publisher, Ardent Media, Inc.

effectiveness of the various strategies they used. All of the strategies were found to be moderately effective, but negative emotional appeals received the lowest ratings by both sexes. Women rated refusing sexual intercourse, providing risk information, and deception as most effective (in that order); men indicated deception as most effective, followed by refusing sexual intercourse.

Among most adults, none of these strategies and manipulations should be necessary. Most agree that open, honest, and respectful communication between partners is the best path to effective contraceptive use.

The Female Condom

female condom
A tube or pouch of thin polyurethane with a flexible ring at each end. One end is sealed and the other is open. The condom is inserted into the vagina to protect against pregnancy and the transmission of STIs.

The **female condom** (inserted vaginally) is available over the counter virtually anywhere male condoms are sold. Just as the male condom covers the penis to prevent genital contact or exchange of sexual fluids, the idea behind the female condom (the newest version is the female condom 2, brand name *FC2*) is to create a lining on the inside of the vagina to accomplish basically the same results. The female condom consists of a tube (or pouch) made of thin polyurethane, or a new thinner and more flexible material called *nitrile*, with a flexible ring at each end. One end is sealed and the other is open. The flexible ring at the closed end is squeezed in half, inserted into the vagina, and moved up against the cervix, as shown in Figure 5.1. The internal ring holds the condom in place. The ring at the open end is positioned on the outside of the body, around the opening to the vagina. This type of condom may also be used for STI protection during anal intercourse using the same insertion technique.

The female condom consists of a tube of thin polyurethane, similar to the polyurethane male condom, but with a flexible ring at each end, as shown in the photo (a). The three-step procedure (b) illustrates how to insert a female condom. Care must be taken when removing the condom to prevent semen from spilling out.

The condom is lubricated (with a nonspermicidal lubricant) and is sold with extra lubricant included, in case it is needed for easier insertion or more comfortable movement. During intercourse, the female condom stays inside the vagina and the penis moves within it. After ejaculation, the condom must be removed by twisting the outside ring to trap the semen inside and gently pulling the condom out of the vagina. This should be done before the woman stands up, to be sure none of the semen spills

(a)

Figure 5.1 The Female Condom
The female condom consists of a tube of thin polyurethane, similar to the polyurethane male condom, but with a flexible ring at each end, as shown in the photo (a) The three-step procedure (b) illustrates how to insert a female condom. Care must be taken when removing the condom to prevent semen from spilling out.

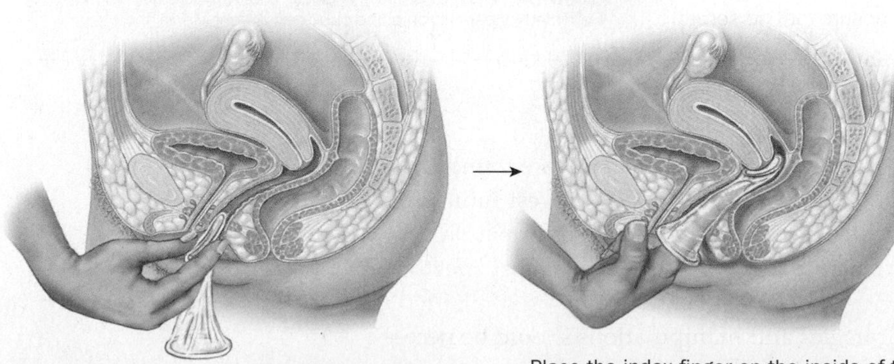

Gently insert the inner ring into the vagina. Feel the inner ring go up and move into place.

Place the index finger on the inside of the condom, and push the inner ring up as far as it will go. Be sure the sheath is not twisted. The outer ring should remain on the outside of the vagina.

Be sure the penis is inserted into condom, not next to it.

(b)

out of the condom into or around the vaginal area. As with male condoms, each female condom should be used only once.

The female condom has the advantage of allowing the woman to have more control of contraception and preventing sexually transmitted infections. It also allows for greater spontaneity than the male condom because the female condom may be inserted up to eight hours before intercourse. For greater contraceptive effectiveness, the female condom may be combined with hormonal contraceptives, but it should *not* be used at the same time as a male condom (this can create too much friction and cause the condoms to break). The female condom costs somewhat more than male condoms—approximately $1 to $2 each (depending on where they are purchased and the quantity)—but are often provided free of charge at public health clinics and university health centers.

Since You Asked

4. Does the female condom work as well as regular male condoms?

An additional advantage of the female condom is that, when used correctly, it may provide slightly *greater* protection from some STIs than a male condom. This increased protection is because the polyurethane material does not degrade with the use of oil-based lubricants, and the shape of the external ring of the condom protects a wider area of skin between the base of the penis and the vulva or anus during intercourse, helping to prevent skin-to-skin transmission (Hudson, 2003; Nicolette, 1996). As mentioned, male and female condoms are not designed to be used simultaneously, as the effectiveness of each will be reduced. However, either the male or female condom may be combined with other contraceptive methods (such as hormonal or other barrier methods—see below) for increased protection against pregnancy and STIs.

The main disadvantage of the female condom is that typical-use contraception failure rate is slightly higher than for male condoms. Errors in use include removing the condom incorrectly, allowing semen to spill out in or near the vagina, or mistakenly inserting the penis next to, rather than inside, the condom. However, if used correctly and consistently, the female condom is nearly as effective as the male condom (refer to Table 5.3).

Methods for Preventing Pregnancy (But *Not* STIs)

5.3 List and explain the function, effectiveness, advantages, and disadvantages of the various types of contraceptive methods in each category discussed (withdrawal, hormonal, barrier, fertility awareness, intrauterine devices, and surgical methods).

All of the remaining methods of contraception discussed in this chapter are designed only to help prevent pregnancy. Although a few of them may offer minimal protection against some STIs, they should not be used for that purpose. However, all of the methods discussed in this section may be combined with the condom for protection from STIs and added insurance against pregnancy.

Withdrawal

The **withdrawal method**, in which the penis is withdrawn from the vagina prior to ejaculation (also called *coitus interruptus* or "pulling out"), has been a widely practiced method of birth control for centuries. Many of you may be surprised to see it discussed here because, in general, it has been rejected for decades as being very unreliable and risky for both pregnancy and the transmission of STIs. Although these risks are real, if a couple is able to use the withdrawal method consistently and correctly, it *can* be an effective way of preventing pregnancy (but *not* STIs) (see Table 5.2).

withdrawal method
Removing the penis from the vagina just prior to ejaculation—a usually unreliable method of contraception; also called coitus interruptus and "pulling out."

The fluid produced by the Cowper's glands, well before ejaculation (as discussed in Chapter 2, "Sexual Anatomy"), is likely to contain HIV and other STI microbes if the man is infected. Consequently, this method will not prevent him from infecting a partner, even if he successfully withdraws before orgasm and ejaculation. As for contraception, contrary to previous conventional wisdom, pre-ejaculate fluid does not typically appear to contain sperm cells, but this does not mean that semen with sperm cannot enter the vagina when a couple decides on the withdrawal method. The main problem for contraceptionwith using the withdrawal method is that there are too many "ifs." *If* the man withdraws his penis from the woman's vagina before he ejaculates, and *if* he ejaculates well away from the vaginal opening, and *if* no semen enters her reproductive tract, and *if* no infectious microbes are present in the pre-ejaculate, then the method may be fairly effective. All of these ifs, however, combine to make the withdrawal method, in actual use, quite unreliable, with an estimated 22% failure rate (see Kowall, 2008b).

Since You Asked

5. When my girlfriend and I are having intercourse, I always pull out before I ejaculate. That means there's no chance of pregnancy, right?

Hormonal Contraceptives

Female hormonal methods of contraception prevent pregnancy by altering a woman's ovulation cycle or reproductive tract. Hormonal contraception's primary method of preventing conception is preventing ovulation. Because hormones signal a woman's body to release an ovum, altering hormone levels blocks that signal and no egg is released by the ovaries. No ovum, no conception. Hormonal contraceptives also appear to thicken the cervical mucus, which creates a barrier against sperm. The various hormonal contraceptives on the market differ primarily in the specific hormonal formulation and the method used to deliver the hormone to the body.

As this book goes to press, five types of hormonal contraceptive delivery systems are on the U.S. market: oral (pills), implants (inserted under the skin), an injection, a skin patch, and a vaginal ring. All hormonal methods of contraception have both advantages and disadvantages—positive and negative side effects—so women and couples should strive to become as educated as possible about all the methods before selecting one.

All hormonal contraceptive methods are for women. A male hormonal method (a pill or shot or patch, etc.), appears to be a very distant possibility, if a possibility at all. The primary reason for this is due to a basic reproduction difference between men and women. Women release one ovum per month; men ejaculate hundreds of millions of sperm with each orgasm and sperm are produced at the rate of about 1,000 per minute. No effective hormonal method exists that can reduce sperm production to an infertile level without also altering other male sex characteristics.

One last note should be mentioned about hormonal contraception in general. Research is showing that some hormonal contraceptives may decrease slightly in effectiveness for women whose weight is considered to be seriously overweight or obese (Murthy, 2010; Robinson & Burke, 2013; Stacey, 2012). In particular, evidence suggests that the pill, the patch, and emergency contraceptive pills have increased failure rates in obese women compared with normal-weight women. The risk of failure appears to be greatest in those women who weigh over 198 pounds or have a body mass index (BMI) greater than 27.4, with the greatest increased risk among those with a BMI greater than 35.0. If you want to determine your BMI, which is a combination of your height and weight, an easy calculator may be found at www.bmi-calculator.net.

oral contraceptives

Tablets containing female hormones that are ingested every day. They constitute the most popular reversible contraception method used by women in the United States; also known as birth control pills.

ORAL CONTRACEPTIVES In general, **oral contraceptives**, also referred to as "birth control pills," (referring to this method as simply "birth control" is inaccurate) are among the most popular reversible contraception methods (Hatcher et al., 2008). The first oral contraceptive appeared on the U.S. market over fifty years ago. When used correctly, they provide a convenient and extremely effective method of preventing pregnancy. But they offer no protection against STIs. However, we should note that all hormonal contraceptives may be combined with the male or female condom to add protection from sexually transmitted infections and the effectiveness of the prevention of pregnancy.

Numerous brands and formulations of oral contraceptives offer a great many options for women who choose this form of hormonal contraception.

How Oral Contraceptives Work Oral contraceptives work in the same manner as all hormonal methods. Combination oral contraceptives contain both estrogen and progestin. The contraceptive effect on a woman taking these hormones is that her body's normal hormonal cycle is altered, and ovulation (the release of an egg) is typically prevented. In addition to preventing ovulation, oral contraceptives cause a thickening of the mucus secreted by the cervix (Hatcher & Nelson, 2004). This thickening of the mucus sets up a barrier that is very difficult for sperm to penetrate.

Different brands of **combination pills** have varying doses of hormones. In general, lower-dose pills are recommended because they are effective in preventing pregnancy and are less likely to produce negative side effects sometimes associated with oral contraceptives, such as spotting between periods, headaches, and breast tenderness. Although serious negative side effects are relatively rare, a woman's decision about combination oral contraceptives should be an informed one made in close consultation with her health care provider.

combination pill
An oral contraceptive containing a combination of estrogen and progestin.

Birth control pills are sold in various types of packaging that number and label each pill so that it is easier to keep track of the pills taken and avoid missing doses. Most oral contraceptives are taken for 21 consecutive days and then stopped for seven days (called a "hormone-free interval," or HFI), during which time the woman will usually have a menstrual period (for an exception to this rule, see the discussion on "extendedextended dose pills" next). We should note that the bleeding that occurs when a woman in using hormonal contraception is not the same as a regular period. The bleeding is usually lighter and of a shorter duration because the change in hormones has also caused less uterine lining thickening (see Chapter 9, "Conception, Pregnancy, and Birth" for a complete discussion of the fertility cycle). Oral contraceptives are packaged for all 28 days of the cycle, but the last seven pills in the pack contain no hormones and are supplied as "placeholders" during the HFI so that the habit of taking a pill every day will not be interrupted.

The progestin-only "**minipill**" has approximately the same low failure rate as the combination (progestin and estrogen) pill (Hatcher & Trussell, 2008; De Melo, 2010). The main advantage of the progestin-only oral contraceptive pill is that it avoids some of the potential negative side effects associated with estrogen in the combination pill. However, many of the health benefits associated with the estrogen in the combined pill are also not present in the progestin-only pill. Also, the "minipill" must be taken on a stricter time regimen each day. The pros and cons of oral contraceptives will be discussed shortly.

minipill
An oral contraceptive containing progestin only.

Extended-Dosage Contraceptive Pills In the past ten years, new ways of taking oral contraceptives were approved by the Food and Drug Administration and introduced to the U.S. market under the brand names *Seasonale, Seasonique*, and *Lybrel*. Instead of the traditional 21-day hormone regimen, these pills provide a daily dose of hormones for longer time frames, thereby reducing the number of menstrual periods a woman experiences. Seasonale and Seasonique provide pills for 84 consecutive days, during which a woman is protected against pregnancy and does not menstruate. With Seasonale, after the 84 pills, the woman takes the usual placebo pills for the seven-day HFI, during which she will typically have a period. The hormonal formulation in Seasonale is basically the same as for standard 28-day oral contraceptives, but by taking the hormone pills continuously for 84 days, a woman will have only four menstrual periods per year instead of the average of 13 (Mangan, 2003; Sulak, 2007). Seasonique is also a "90-day" pill, but differs from Seasonale in that, instead of the typical seven-day break in hormones, it provides the standard hormonal dose for 84 days, but then continues with a reduced dose of estrogen for the usual seven hormone-free days, reducing the length of a woman's period to four days, four times a year. Seasonique is a slightly lower dose of estrogen, thereby reducing potential negative side effects.

Expanding the concept of extended-dosing hormonal contraception to maximum is Lybrel, a 365-day oral contraceptive approved by the FDA in 2007. Lybrel has a lower dose of hormones compared to most other contraceptive pills and is taken on a continuous basis. A woman takes one Lybrel pill every day and virtually eliminates her periods. The safety and effectiveness of Lybrel approximates that of other oral contraceptives. Nearly all women using Lybrel no longer experienced monthly bleeding, although some experienced minor spotting or breakthrough bleeding, but this number decreases significantly during the first year of use (from 47% of women during the first three months to 20% at the end of the year). **Amenorrhea** (the cessation or suppression of a woman's period) rates increased during the first year from 27% to 59% (Casey & Pruthi, 2008). Current research has found no evidence of health dangers from menstrual suppression (see Casey & Pruthi, 2008; Hicks, 2010).

amenorrhea
Cessation of a woman's period.

Some studies have shown beneficial effects from Lybrel, including decreased hormonally related headaches, reduced feelings of menstrual bloating, and less menstrual pain compared to standard 28-day cyclic pills (Hicks, 2010).

All of these newer, extended-dose oral contraceptives offer several potential advantages for some women. Doctors and many women have been aware for decades that menstrual periods could be delayed or reduced in frequency if a woman skipped the placebo pills in the pack of oral contraceptives and went to a new pack immediately upon completing the 21 days of hormonal pills. This strategy has been recommended by doctors for women with menstruation-related health problems and has been used by women themselves to delay or skip a period that was due to come at an inconvenient or inopportune time (vacation, sports competition, honeymoon, etc.).

For many women, reducing the number of periods has various advantages beyond the convenience of less frequent menstruation. Women who experience especially painful periods, heavy bleeding, cycle-related migraine headaches, endometriosis, or other debilitating menstruation-related symptoms may benefit from reducing the annual number of periods. Moreover, medicall research has found no increase in medical side effects for extended-cycle oral contraceptives compared to the standard 21-day pills (Edelman et al., 2010; Portman et al., 2014). In modern society, women have, on average, 450 periods during their lifetime. This is nearly three times as many as humans during the early hunter-gatherer era, when women spent far more time either pregnant or nursing their young. This has led some researchers to conclude that nature did not intend for women to have as many lifetime periods as they do today (e.g., Kalb, 2003).

In studies of extended oral contraceptives, the most serious side effect was slight spotting (also called "breakthrough bleeding") between periods. In one major study, 7.5% of women discontinued use of Seasonale due to breakthrough bleeding, compared to only 1.8% using standard 28-day birth control pills (Kalb, 2003). These extended-dose oral contraceptives carry approximately the same health risks as other contraceptive pills.

Effectiveness of Oral Contraceptives Oral contraceptives offer one of the most effective methods of preventing pregnancy. Nevertheless, a very small percentage of women have become pregnant while taking oral contraceptives, either because of the very unlikely event of ovulating in spite of the hormone shifts or, more often, due to incorrect use (such as forgetting pills). However, when used correctly, as explained in the next section, only the hormonal implant and sterilization have lower failure rates than oral contraceptives. With conscientious and careful use of the pill (meaning never accidentally forgetting to take a pill and being sure to take it at approximately the same time each day), the effectiveness rate is 0.3%, meaning that only three woman in 1,000 should expect to experience an accidental pregnancy.

Even with less-than-perfect use, such as the occasional missed pill, all oral contraceptives have high effectiveness rates (Hatcher & Trussell et al., 2011). The pill's failure rate for the average user is about 9% (mostly due to inconsistent dosing). As with nearly all forms of contraception, the most important factor in effectiveness is correct and consistent use.

Using Oral Contraceptives Correctly The most common incorrect use of oral contraceptives is forgetting to take them every day. Just how serious a problem is it to miss a pill or two? Potentially very serious, if the goal is to prevent conception. Over the years since birth control pills first appeared on the market, the dose of hormones contained in them has been steadily reduced. This decrease has made the pill much safer in terms of negative side effects, but it has also increased the importance of regular doses. Women using oral contraceptives must take one pill every day of the 21-day course (in the 28-day pack, as noted earlier, the last seven pills do not contain hormones and are there as "placeholders" to maintain the habit of taking one pill each day). Furthermore, women should try to take a pill at approximately the same time each day. This will maintain constant hormone levels in the bloodstream. Even the most conscientious of women may forget to take a pill. Because of this, and because the pill does not protect against STIs, most sexuality educators and health care professionals strongly recommend that women who choose oral contraception keep an STI-fighting backup method of contraception (such as condoms) readily available. If a woman misses a pill, she need not panic, but she should be aware of the steps to take to maintain maximum contraception protection. "In Touch with Your Sexual Health: Missed Your Pill? Here's What You Should Do" summarizes the steps a woman should take if she misses one or more pills during a cycle.

The start of protection depends on when in her cycle a woman begins taking oral contraception. The time after beginning birth control pills until she is safe from conception is typically about a week. This assumes, however, that she begins taking combination pills during the first seven days after her menstrual period begins or, for progestin-only pills (the minipill), during the first five days. Although she is unlikely to be fertile during these days, a backup method of contraception, such as condoms, is recommended during the first week after beginning the pills. In addition, she should consider emergency contraception pills (discussed later in this chapter) if she has had unprotected intercourse recently. However, if a woman begins taking birth control pills at any other time during her cycle, she should use a backup contraceptive during the first month of pills. If a woman is at all concerned about exposure to sexually transmitted infections, she should continue to insist on condom use, even though she is protected from pregnancy by the birth control pills.

Since You Asked

6. My girlfriend was on the pill, but she still got pregnant. How did that happen?

In Touch with Your Sexual Health

Missed Your Pill? Here's What You Should Do

Oral contraceptives are most effective when taken regularly, at about the same time each day. If you miss a pill, the effect on fertility depends on many factors, such as when in your cycle the pill was missed, how many pills during a cycle were missed, and the timing of intercourse relative to the missed pill(s). Usually, pregnancy can be avoided even if pills have been missed through use of a backup barrier method (i.e., condoms) or by using emergency contraception (EC) *especially if intercourse has occurred in the week-prior to the forgotten pill(s).* The correct procedure is a bit complicated, but the following chart offers recommendations about exactly what actions should be taken in the event of missed pills to maintain effective contraception (for all combination pills).

Timing and Number of Pills Missed	What to do...	Seven-Day Backup or EC Needed?
1 pill; remembered within 12 hours	– Take a pill as soon as you remember. – Take the next pill at the usual time. – Continue with daily pills.	No
1 pill remembered after 12 hours	– Take missed pill as soon as you remember. – Take the next pill at the usual time. – Continue with daily pills. – Use condoms or abstain for 7 days.	7-day backup needed; no EC necessary
2+ pills forgotten with 7 or more hormonal pills left in packet	– Take current day's pill and last forgotten pill as soon as you remember (two pills at once). – Take remaining pills at the usual time. – Use backup for 7 days. – Take EC if intercourse occurred within previous week.	7-day backup needed; EC necessary if had Intercourse within previous week
2+ pills with fewer than 7 pills left in packet (option 1)	– Take the rest of the hormonal pills in the packet. – Skip the placebo pills and begin next packet of pills without interruption. – Use condoms or abstain for 7 days.	7-day backup needed; no EC necessary
2+ pills with fewer than 7 pills left in packet (option 2)	– Take remaining pills on schedule. – Use condoms or abstain until 7 hormonal pills have been taken daily. – Take EC if intercourse occurred within previous week.	7-pill backup needed; EC necessary if had Intercourse within previous week

Progestin-Only Pills

You could become pregnant if you take your progestin-only pill more than three hours past your regular time. If you do, you must use a condom, diaphragm, sponge, or emergency contraception. Emergency contraception is an effective backup method if you had vaginal intercourse before you realized you missed pills.

- Take a pill as soon as you remember.
- Take the next pill at the usual time.
- Continue to take the rest of the pack on schedule.
- Use a backup (i.e., condoms) for 48 hours after taking the late pill.

Many women have spotting or light bleeding when they miss a birth control pill—even if they make it up later. Women also sometimes feel a little sick to their stomachs if they take two pills to make up for a missed pill. If you do feel a bit sick after taking two pills in a day, don't worry: the nausea won't last long.

SOURCES: www.plannedparenthood.org (2010); Hatcher et al. (2011).

Risks and Benefits of Oral Contraceptives When the first oral contraceptive pills came onto the U.S. market in the early 1960s, they were associated with some rare but potentially serious health dangers, such as blood clots and increased risk of stroke or heart attack. However, over the years, with changes in dose and type of hormones used in oral contraceptives, these dangers have been minimized, and today oral contraceptives are more likely to protect a woman from some serious diseases. One point is clear: For nearly all women, the health risks associated with pregnancy are, overall, greater than those associated with oral contraceptives. It is impossible to address fully all of the advantages, disadvantages, health benefits, and dangers of birth control pills in this rather brief discussion. See "In Touch with Your Sexual Health: Pros, Cons, and Cautions of Using Oral Contraceptives" for a summary of considerations when choosing oral contraceptives as your contraceptive method.

THE HORMONAL IMPLANT Two hormonal implant contraceptives are currently available to women in the U.S.: *Implanon* and *Nexplanon.* These consist of a small bioneutral polymer tube containing a progestin hormone that is implanted under the skin of a woman's upper arm. The implant slowly releases a hormone similar to that found in many oral contraceptives. The implant is easily inserted in about one minute, is continuously effective for three years, and may be removed at any time to allow fertility to return.

The contraceptive action of a contraceptive implant involves a gradual measured release of hormones into a woman's body. The hormones prevent pregnancy (as do all hormonal methods) by inhibiting ovulation and thickening the cervical mucus.

In Touch with Your Sexual Health

Pros, Cons, and Cautions of Using Oral Contraceptives

Here is a list of the major advantages and disadvantages of birth control pills. Also included is a list of warning signs of potentially serious side effects that may require immediate medical attention. Keep in mind that most of these pros and cons are also associated with other types of hormonal contraceptives discussed in this chapter.

Pros

- Very effective method of contraception
- No interruption of lovemaking
- Generally very safe
- Often makes menstrual cycle more regular and reduces or relieves symptoms of PMS and menstrual cramps
- May reduce length of period and amount of bleeding
- May protect against pelvic inflammatory disease (PID), which can lead to infertility
- Appears to reduce risk of certain cancers, especially ovarian and endometrial cancers
- Does not interfere with future fertility
- May help clear up acne
- May reduce unwanted hair growth
- May reduce incidence and number of ovarian cysts

Cons

- Does not protect against any sexually transmitted infections
- Easy to miss doses
- Relatively expensive (depending on where purchased)
- Possible spotting (slight bleeding between periods)
- Possible nausea (infrequent and usually only during first cycle of use)
- Possible headaches and breast tenderness (usually solved by changing pill dose)
- Depression (rare)
- Possibility, in some women, of reduced sexual desire and response
- Cardiovascular disease (heart attack, stroke, blood clots; very rare and virtually always associated with women over thirty-five who smoke)

Cautions

- Oral contraceptives provide no protection against sexually transmitted infections. Many women choose to combine condoms with the pill for maximum prevention of both pregnancy and STIs.
- You should never begin taking oral contraceptives without a complete understanding of all the risks listed here and the correct use of the pills.
- You should be aware that the hormones in birth control pills may interact with other prescription medications (e.g., some antiseizure medications and certain antibiotics) and herbal remedies (e.g., Saint John's wort), causing a decrease in contraceptive effectiveness. Be sure to keep your health care provider informed of all medications and remedies you are taking when being prescribed any hormonal method of contraception.

Danger Symptoms

- When you are on the pill, you should be alert to all the side effects listed here. You should also be vigilant about warning signs of serious problems that may be associated with the pill, especially if you smoke cigarettes. If you experience any of the following symptoms while taking hormonal contraception, you should contact your doctor immediately. To help you remember, these danger signs are summarized by the acronym ACHES:

A—Abdominal pain (severe)

C—Chest pain (severe) with a cough or shortness of breath

H—Headache (severe), dizziness, weakness, numbness

E—Eye problems, vision loss, blurring, slurred speech

S—Severe leg pain

SOURCE: Adapted from Hatcher et al., 2008, pp. 254–255.

Using a local anesthetic, a physician inserts the implant under the skin, usually on the underside of the upper arm (see Figure 5.2). The contraceptive effect is immediate (within twenty-four hours). Upon removal of the implant, fertility typically returns quickly.

Effectiveness of the Implant One of the greatest advantages of implant contraceptives is that once inserted, it is virtually impossible to use the method incorrectly. The contraceptive hormone is released over time so that protection is provided without any action at all on the woman's part. Therefore, as you can see in Table 5.3, no difference between typical and theoretical effectiveness exists. Perhaps more important is the extremely low failure rate of implants. Less than one woman in 2,000 became pregnant

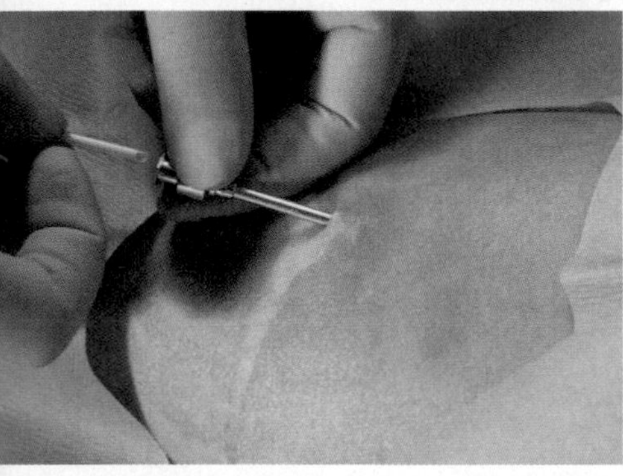

Figure 5.2 The Hormonal Implant

The hormonal implant (Implanon) is a small tube that is implanted under the skin of a woman's upper arm. Hormones are slowly released into the woman's body over several years, thus preventing ovulation.

Depo-Provera

a hormonal contraceptive in the form of an injection that provides 90 days of protection from conception.

in the first year using the implant, a remarkable 0.05% failure rate (Hatcher & Trussel et al., 2011). Moreover, among nearly 1,000 women participating in one large study of Implanon, *not a single pregnancy* was reported over a two-year period (Edwards & Moore, 1999; O'Connell, 2001).

Advantages and Disadvantages of the Implant The advantages of the hormonal implant are its ease of use and near-perfect effectiveness. Still, the hormonal implant, as is true of all hormonal contraceptives, is not without negative side effects. A small percentage of women report negative experiences with Implanon, principally involving changes in menstrual bleeding patterns, such as cessation of menstrual periods, spotting, or more frequent, unpredictable periods. Other less common (and often temporary) side effects include headaches, weight gain, acne, and vaginal infections (Belden, 2006a). Studies of women using Implanon in Europe have found relatively high user-acceptance rates, with 84% of women continuing to use the implant method after two years or longer (Hatcher & Trussell et al., 2011).

THE INJECTABLE HORMONAL CONTRACEPTIVE Another method of administering hormonal contraceptives is through an injection (given by a physician or nurse) in the arm or hip. The injectable form of contraception offers convenience, in that a woman does not need to remember to take a pill every day but is continually protected from pregnancy at a high level of effectiveness for three months. As of 2014, only one injectable contraceptive, ***Depo-Provera***, was available in the United States.

Often referred to simply as "Depo," Depo-Provera has been in use throughout the world since the 1970s. It was approved for use in the United States in 1992. Each injection imparts three months of highly effective contraception. Like all hormonal forms of contraception, Depo-Provera does not provide protection from sexually transmitted infections.

Depo-Provera consists of the hormone DPMA (medroxyprogesterone acetate), which is injected into the hip or arm. DPMA (a form of progestin) prevents pregnancy in the same manner as implants: by inhibiting ovulation; thickening the cervical mucus to prevent passage of sperm; and thinning the endometrial lining so that, should fertilization occur, implantation (pregnancy) is unlikely. A single dose of DPMA provides contraceptive protection for 90 days, at which time another injection is required for continued effectiveness. This is a reversible method of contraception, although, unlike other hormonal methods, the return of fertility can take up to a year after discontinuing the method (Barron, 2013; Hatcher, 1998).

Effectiveness of the Injectable Contraceptive Depo-Provera is another extremely effective hormonal contraceptive with a failure rate of only 0.2%. This rate assumes, of course, that a woman returns every three months for another injection. A buffer exists so that effectiveness is maintained for at least two weeks past the three-month recommended injection schedule (Hatcher & Trussell et al., 2011). The contraceptive action of the method takes effect almost immediately following the first injection.

Side Effects of the Injection Method Negative side effects associated with the use of Depo-Provera are similar to all hormonal methods and include: menstrual changes such as spotting and breakthrough bleeding (common), headaches (17%), nervousness (11%), decreased sex drive (5%), breast tenderness (3%), and depression (2%) (Hatcher, 1998). From those relatively low percentages, you might assume that the majority of women using Depo are free of side effects. However, one poll asked women on Depo if they had ever experienced side effects; only 20% said no, and only about 44% of women choosing Depo continued to use the method after one year (CDC, 2008; Hatcher & Trussell, 2011).

Another side effect common to Depo-Provera is amenorrhea, the cessation of a woman's period, as discussed above in relation to the use of extended oral hormonal contraceptives. The health risks of amenorrhea appear to be negligible, and some women feel that having fewer or no periods may provide various benefits, such as the elimination of monthly symptoms of menstrual pain and bloating, and the emotional swings often associated with PMS (Casey & Pruthi, 2008; Hicks, 2010).

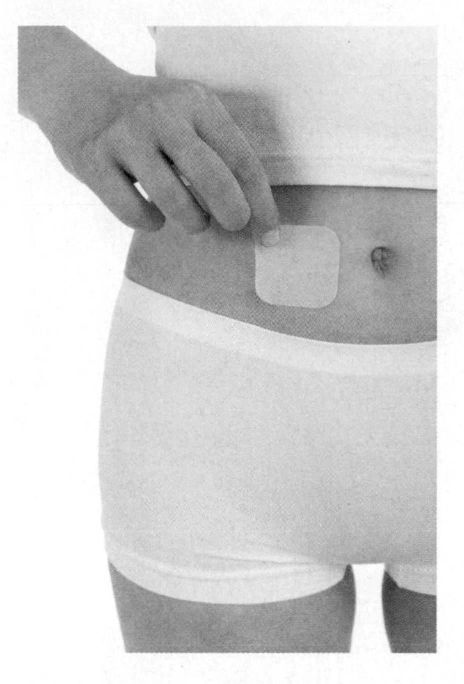

The contraceptive patch functions physiologically in basically the same way as other hormonal contraceptives, but delivers hormones through the skin, preventing pregnancy.

Courtesy of John Nebraska.

THE CONTRACEPTIVE PATCH No doubt you have heard about the nicotine patch to help people stop smoking. It delivers, through the skin, a continuous dose of nicotine that helps the smoker avoid withdrawal symptoms while trying to give up cigarettes. In a similar delivery process, the weekly hormonal **contraceptive patch** provides a precise dose of two hormones into a woman's body, through the skin, preventing ovulation using basically the same principle as all hormonal methods. The contraceptive patch was introduced to the U.S. market in early 2002 under the name *Ortho Evra*. This hormonal contraceptive is a 1-inch-square patch that a woman wears like an adhesive bandage on her abdomen, buttocks, upper body (front or back, but not on the breasts), or upper arm. Each patch is worn for one week, for three consecutive weeks, and then the fourth week is patch free (the hormone-free interval, during which she will have a light period). The adhesive in the patch allows it to stay on the skin throughout various activities, including swimming, bathing, and exercising.

The patch's mechanism of action follows the same pattern as 28-day oral contraceptives: three weeks of hormones and one week without, during which a menstrual period occurs. The primary difference here is the hormone delivery system. Side effects of the patch are similar to those discussed earlier for the combination pill: breakthrough bleeding between periods, weight gain or loss, breast tenderness, nausea, headaches, decrease or increase in sexual desire, and temporary symptoms of depression.

It is important to note that recent studies found that the patch may expose women to higher levels of hormones than most oral contraceptives. This may slightly increase the risk of dangerous side effects, including blood clots in the legs and lungs and problems related to clotting, such as strokes and heart attacks. One study found the risk of blood-clotting side effects to be higher than that of the pill, although the risk remains very small (Carusi, 2008). The normal risk for this health problem among women using *no* hormonal contraception is about 1 in 10,000 women per year. This was found to increase to 1.5 per 10,000 women for the pill and an estimated 3 to 5 per 10,000 for the patch. As medical health risks go, these numbers are among the smallest. As for all hormonal contraceptives, risks increase with age (as a woman reaches age thirty-five and above) and with cigarette smoking. Although no actual increase in the incidence of these side effects has been seen among women younger than forty using the patch (Jick et al., 2010), the manufacturer was required to revise the product's label, advising women of the possibility of the increased hormonal exposure and potential side effects (Mechcatie, 2006). Most recent research indicates that the patch is well within safety parameters of all hormonal contraceptives (see Casey & Pruthi, 2008).

contraceptive patch
A stick-on patch that delivers a precise dose of two hormones into a woman's body through the skin, preventing ovulation.

THE CONTRACEPTIVE RING Another hormonal delivery system is the **contraceptive ring**. Called *NuvaRing*, it was released in the United States in 2002 and has become very popular among hormonal contraceptive choices. No, it's not a ring worn on your finger (that would be a new twist on the engagement ring!). NuvaRing is a colorless, flexible, transparent silicone ring about 2 inches in diameter, similar to the ring inside the female condom. When the ring is inserted into the vagina, up near the cervix, it releases a continuous low dose of estrogen-like and progestin-like hormones into the bloodstream through the vaginal walls for three consecutive weeks. These are basically the same hormones found in other combination hormonal contraceptives and the ring's mechanism of action in preventing conception is the same as all hormonal methods. Are you beginning to see a pattern here? Researchers

contraceptive ring
A colorless, flexible, transparent silicone ring about 2 inches in diameter that is inserted into the vagina and releases a continuous, low dose of estrogen- and progestin-like hormones into the bloodstream.

have found an effective, reasonably safe, and even beneficial combination of hormones to prevent pregnancy, and now various pharmaceutical companies are working to find the best method for delivering them to a woman's body.

Here's how the vaginal ring works. The woman inserts the ring into her vagina (in much the same way as inserting a female condom) and leaves it there for three weeks (see Figure 5.3). At the end of the three weeks, she removes it for a week, when she will typically have a light period; then she inserts a new one for the next three weeks (Dunaway, 2008). Because it is not a barrier device, such as a female condom, exact fit and position are not necessary for effectiveness. NuvaRing's effectiveness is excellent, with a failure rate of 0.3% when used according to the package instructions (Hatcher & Trussell et al., 2011). Most couples are unaware of the ring during sexual activities, but it may be removed during intercourse as long as it is replaced within three hours. If the ring is removed for more than three hours, an additional form of barrier contraception should be used until the ring has been in the vagina for seven days.

The contraceptive ring offers the same pluses and minuses as other combined hormonal contraceptives—pill, implant, or patch. Research has found that the ring may be effective in reducing spotting and breakthrough bleeding, which are common side effects of other hormone contraceptive methods; few women report spotting or breakthrough bleeding with NuvaRing (Smith & Coffee, 2008; Stevenson, Crownover, & Mackler, 2009). The contraceptive ring is as effective as other combination hormonal methods but may require smaller amounts of hormones due to its location in the vagina, which targets the hormones exactly where they are needed for greatest effectiveness. One study reported that the ring releases up to 250 times fewer hormones into a woman's body compared to the contraceptive patch (Chan, Harting, & Rosen, 2009).

On the down side, in addition to the overall negative side effects of combined hormonal contraceptives, a small number of women have experienced vaginal irritation

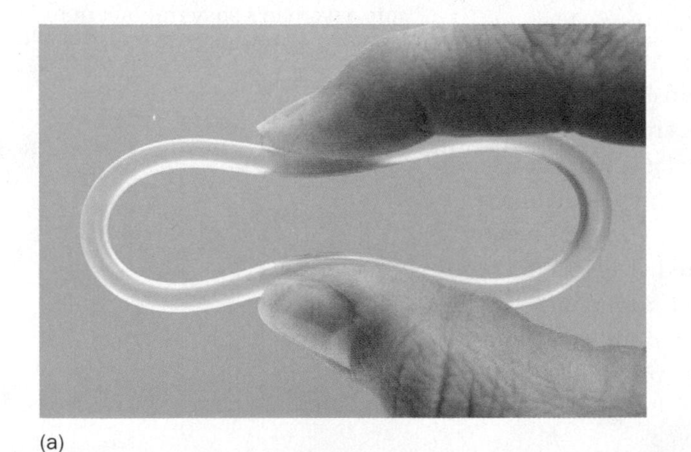

(a)

Figure 5.3 The Vaginal Ring

The vaginal ring is a colorless, flexible silicone ring (a) about 2 inches in diameter, which prevents ovulation by secreting hormones through the vaginal walls continuously for three weeks. The three-step procedure (b) illustrates how to insert a vaginal ring.

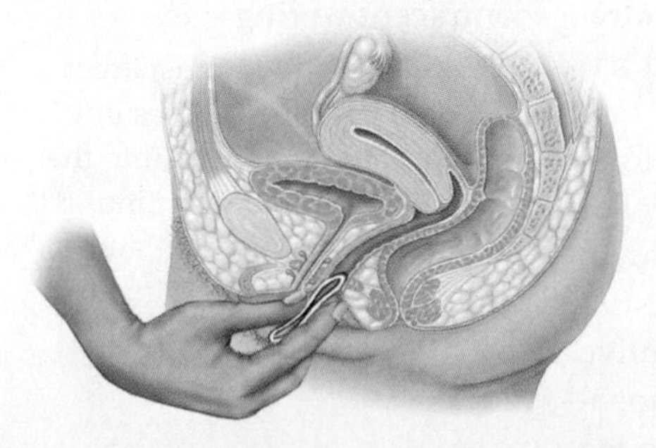

Squeeze the sides of the flexible ring together.

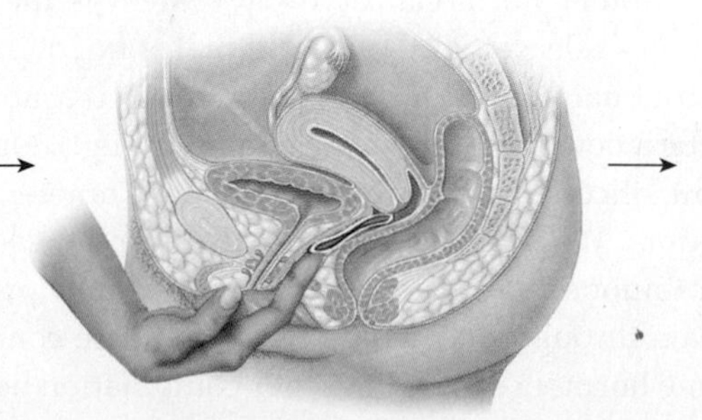

Insert the ring into the vagina.

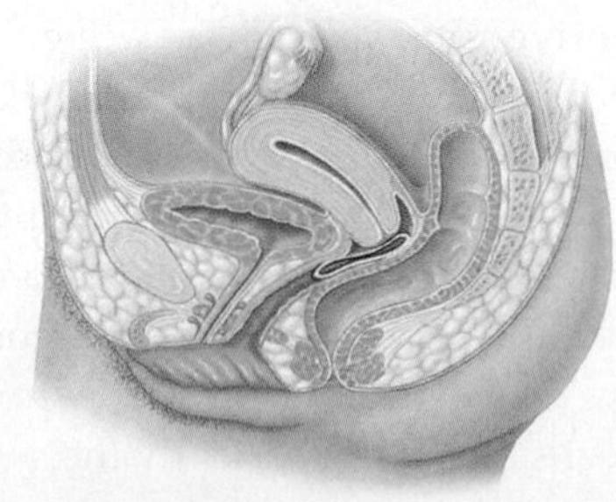

Position the ring deep in the vagina adjacent to the cervix.

(b)

and inflammation (vaginitis) related to the ring. For ongoing updates on NuvaRing, visit the manufacturer's website (www.nuvaring.com), where the company offers access to the newest research on the ring, a voucher for a free NuvaRing (a doctor's prescription is also needed), and even an app to remind you when to remove and insert a ring each month.

As this edition of this book is going to press, two developments about the contraceptive ring should be noted. First, in 2013, a class-action lawsuit was brought against the manufacturers of NuvaRing (Organon, USA, and Merck) by 1,500 women who claimed to have been harmed by due to serious blood clots stemming from the ring's dose of hormones. The suit claims that the ring imparted far greater doses of hormones, despite the fact that scientific research disputes this assertion. Lawsuits against hormonal contraceptives are relatively common due to acknowledged, yet rare side effects. Without commenting on the merit of the suit, it should be noted that no medical treatment is free of potential negative side effects and when the treatment is used by millions of women, some will experience these health problems. The issue of whether these risks were known by the manufacturers and properly publicized to patients was left to the courts to decide. After a year-long battle, Merck, who has denied any wrong-doing, has agreed to pay 100 million dollars to settle the claims.

The other recent development relating to NuvaRing involves an attempt by a competitor to produce a generic (and therefore less expensive) contraceptive ring. An Irish company began to manufacture and market what amounted to a knockoff prior to the expiration of Merck's patent in 2018. Merck is suing to prevent what it sees as a copyright infringement; a decision should come in 2016 (Pearson, 2013).

EMERGENCY HORMONAL CONTRACEPTION (EC) **Emergency hormonal contraception (EC)** uses various combinations of drugs to interrupt a woman's normal hormonal patterns and prevent pregnancy *after* unprotected intercourse has already occurred. These have often been referred to as the "morning after pill," but it is not necessary to wait until the next morning to take it. This is *not* a form of abortion; as with other hormonal methods, EC helps *prevent* conception (Sanfilipo & Downing, 2008). It is approximately 75% effective in preventing pregnancy following unprotected intercourse.

emergency contraceptions (EC)
Hormonal contraceptive that helps prevent pregnancy after an unprotected act of intercourse; also known as the "morning-after" pill.

Since You Asked

7. Is there really a "morning-after pill," and does it cancel out the need for my boyfriend and me to use condoms?

The emergency contraception pills currently on the market by brand name are: Plan B *One Step, Next Choice, and MyWay*. These consist of one pill that should be taken as soon as possible after unprotected intercourse, but within five days maximum. These products contain hormones similar to the oral contraceptives discussed earlier in this chapter, but birth control pills should *never* be substituted for these. Emergency contraception pills are available at pharmacies without a prescription for $30–$50 and are provided at little or no cost by most public health clinics and university health services.

Why would a woman or couple need emergency contraception? Some possible reasons might include these:

- A couple had unprotected intercourse, for any reason.
- A condom broke or slipped off.
- The man forgot to withdraw before ejaculation.
- A diaphragm may have slipped out of place or was used with inadequate spermicide.
- The woman has missed too many birth control pills.
- The woman was raped.
- The woman believes she may have been a victim of a date rape drug.
- The woman was exposed to a toxin that might harm a newly conceived fetus.

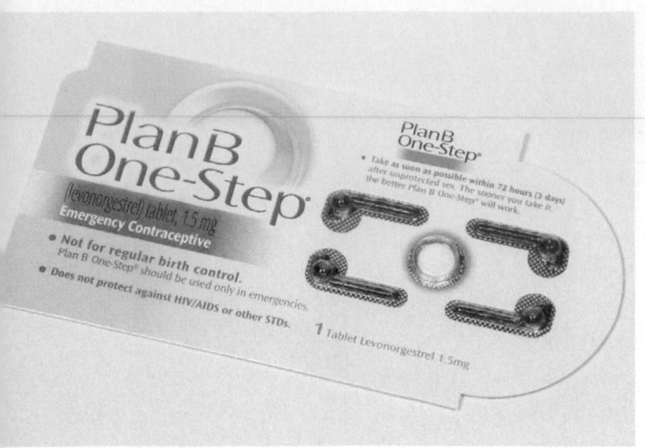

Emergency contraception pills offer protection from conception up to five days after unprotected intercourse. These are *not* "abortion pills"; they help prevent *conception*.

When taken correctly, EC reduces the chance of pregnancy by up to 75%, depending on how soon and when during the fertility cycle the pills are taken. The dose of hormones in emergency contraceptive pills causes some common and unpleasant side effects, such as nausea, vomiting, headache, and breast tenderness. Women should be thoroughly informed of the possible side effects and should discuss them with their health care provider or pharmacist before taking EC.

With correct use, EC has the potential to decrease the number of abortions, a goal shared by people on all sides of the abortion debate. Conversely, many have voiced concern that EC will facilitate irresponsible and unsafe sexual activity among some who will rely on them in place of condoms or other forms of contraception that are used *before* intercourse.

To date, neither of these effects of EC has been clearly demonstrated scientifically. Nevertheless, public health officials and sexuality educators are stressing as clearly as possible that EC is *not as effective* as other forms of contraception and should never be used in place of effective, before-the-fact birth control. As the name implies, they are intended for *emergencies only*—that is, when *"Plan A" failed!*

Barrier Methods

barrier method

Any contraceptive method that protects against pregnancy by preventing live sperm from entering the woman's reproductive tract

Early in this chapter, we discussed two common **barrier methods** of contraception, male and female condoms, because of their ability to help prevent both pregnancy and the spread of STIs. Other vaginal barrier contraceptives are discussed here because they are helpful in preventing conception, but they are not considered useful in preventing STI transmission. In fact, some evidence suggests that some spermicidal formulations used with many of these barriers may actually cause minor irritations in the vaginal tissues of some women, making transmission of some STIs *more* likely, especially among high-risk populations (Dixon, 2007; Schneider, 2008).

The barrier methods discussed in this section are the diaphragm, the cervical cap, and the contraceptive sponge. All barrier methods help protect against pregnancy by preventing live sperm from entering the woman's reproductive tract to fertilize an *ovum. These barrier methods do not offer protection from STIs.* The effectiveness rates of the various vaginal barrier methods of contraception are similar when used perfectly, but all have a higher failure rate than most of the other methods discussed so far. In general, if a woman or a couple places a particularly high value on preventing conception, these methods should not be the first choice or should be used *in combination* with male or female condoms. Moreover, these methods have far worse effectiveness rates for women who have given vaginal birth (refer to Table 5.3) and are not recommended for these women.

THE DIAPHRAGM AND CERVICAL CAP Variations of current diaphragms and cervical caps were among the earliest methods of contraception used by women in this country. As discussed early in this chapter, the diaphragm was introduced to America in the early 1900s by Margaret Sanger. She discovered the device in Europe, and her husband opened the first U.S. diaphragm manufacturing company (D'Emilio & Freedman, 1988). Both the latex diaphragm and the cervical cap must be obtained by prescription because proper fitting by a health care professional is required. The use life of these barrier devices ranges from six months to two years with proper care and storage. Neither method is considered to be effective without the addition of spermicide prior to insertion.

diaphragm

A flexible ring of latex or silicone inserted into the vagina that impedes conception by preventing sperm from getting past the cervix.

The **diaphragm** is a thin latex membrane attached to a flexible ring, forming a shallow cup. After applying a spermicidal cream or jelly along the circumference and in the center of the diaphragm, the ring is flexed and the diaphragm is inserted into the vagina so that it nests around the cervix (see Figure 5.4). It is held in place by the

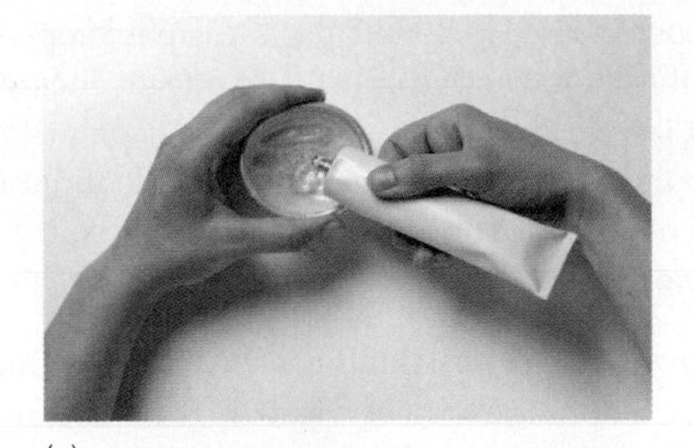

Figure 5.4 The Diaphragm

The diaphragm is a thin latex membrane attached to a flexible ring, forming a shallow cup, as shown in (a) It can be inserted up to two hours prior to intercourse and must be left in place for at least six hours (but not more than 24 hours) after. For effective contraception, the diaphragm must always be used with spermicide. The three-step procedure (b) illustrates how to insert a diaphragm.

(a)

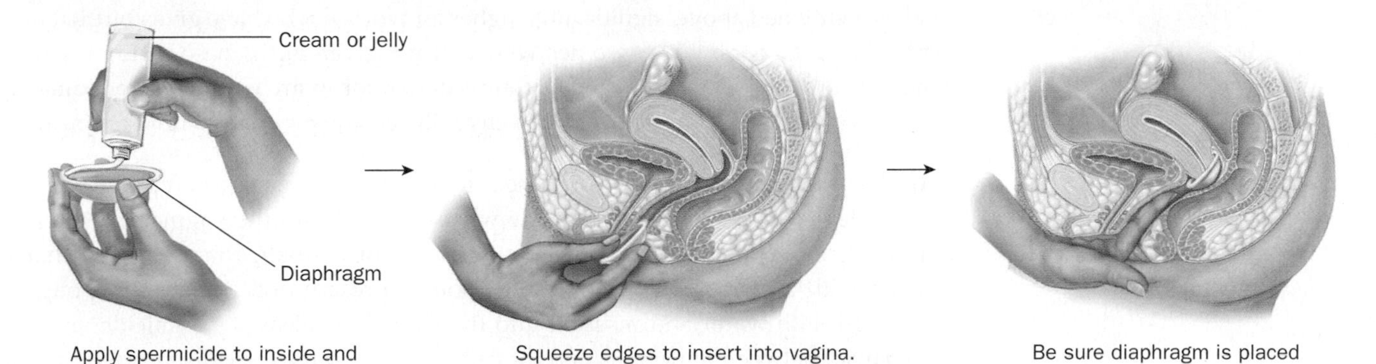

Apply spermicide to inside and around edge of diaphragm.

Squeeze edges to insert into vagina.

Be sure diaphragm is placed snugly against cervix.

(b)

action of the snug fit of the flexible ring behind the pubic bone. Following ejaculation, sperm are blocked from reaching the cervix by the diaphragm and destroyed by the action of the spermicide.

The diaphragm must *always* be used with spermicide and may be inserted up to six hours before intercourse (ACOG, 2007; ARHP, 2010). Following intercourse, the diaphragm must be left in place for at least six hours, but not longer than 24 hours. If intercourse is to be repeated within six hours, additional spermicide must be added using a plunger-type applicator (often supplied with the spermicide) *without removing the diaphragm*.

The **cervical cap** works similarly to the diaphragm except that it is designed to fit more snugly over the cervix itself (see Figure 5.5). The cervical cap currently available

cervical cap
A device similar in function to the diaphragm that fits more snugly over the cervix.

Figure 5.5 The Cervical Cap

The cervical cap (FemCap), shown in (a), is used with spermicidal cream and inserted much like the diaphragm. It is designed, however, to fit more snugly over the cervix itself (b).

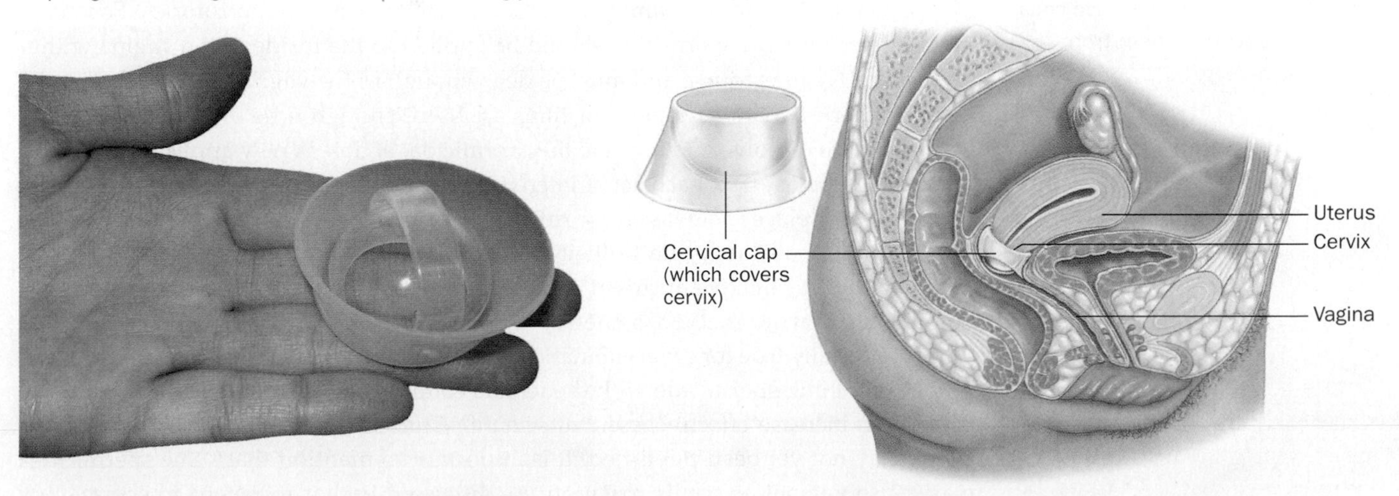

(a) (b)

in the United States is the *FemCap*. Once spermicide is applied and the cap is properly placed in the vagina and over the cervix, the contraceptive effect (and effectiveness) is basically the same as for the diaphragm. Like the diaphragm, the cervical cap must be left in place at least six hours after intercourse. However, the cap provides continuous protection for up to 48 hours without the necessity of additional spermicide, so it can be placed in the vagina up to two days before intercourse.

Effectiveness of the Diaphragm and Cervical Cap Looking back at Table 5.3 once again, you can see that the failure rate of the diaphragm and cervical cap, both in actual use and in theoretically perfect use, is higher than many of the other available methods (and as mentioned above, significantly higher for women who have given birth). However, when used correctly by women who have not given birth, these barrier methods can provide effective and convenient contraception for many women. Approximately 6% to 9% of women who use them correctly will become pregnant within one year.

THE CONTRACEPTIVE SPONGE The contraceptive sponge is a soft porous contraceptive device that releases spermicide when inserted into the vagina. The sponge, marketed under the brand name *Today*, consists of a small, round sponge that is infused with a spermicide. It is sold without a prescription. When the sponge is moistened with water and inserted into the vagina, it releases spermicide and sets up a barrier between the vagina and the cervix. One sponge provides protection from pregnancy for 24 hours regardless of frequency of intercourse. The sponge is a convenient vaginal barrier method that is about as effective as the diaphragm or cervical cap (see Table 5.3) and requires no prescription. As with other barrier methods that use spermicide, the sponge must be left in place for at least six hours after intercourse, but it is effective for twenty-four hours and repeated intercourse. The sponge has the same decreased effectiveness as the diaphragm and sponge for women who have given birth.

The main disadvantages of the sponge include the cost (on average about $2 to $3 each) and the fact that, on its own, it is not as effective in preventing pregnancy as many other methods. However, on the plus side, the sponge does provide a method of contraception that may be used on an "as-needed" basis and is under the woman's control. Moreover, the sponge can be combined with other methods, such as male or female condoms, for increased effectiveness and protection from sexually transmitted infections.

spermicide

Any substance containing a chemical (most commonly nonoxinol-9) that kills sperm cells, thereby preventing them from fertilizing an egg.

SPERMICIDES Vaginal **spermicides** contain a chemical (most commonly nonoxinol-9) that kills sperm cells, thereby preventing live sperm cells from entering the woman's reproductive tract. Spermicides are available without a prescription in many forms, including foams, creams, films, foaming tablets, and suppositories. For maximum effectiveness, a spermicide should be applied to the inside of the vagina within an hour before intercourse and must be deep enough in the vagina to reach the cervix. If suppositories, foaming tablets, or films are used, enough time must be allowed for the product to dissolve and activate the spermicidal action. A new application of spermicide must be used for each act of intercourse. Following intercourse, the spermicide must be left in place (not washed or rinsed out in any way) for at least six hours.

Generally speaking, due to their rather large failure rate, spermicides by themselves are *not* an effective means of contraception. However, by combining spermicide with another barrier method, somewhat higher rates of effectiveness can be achieved. This is especially true for the combination of spermicide and a correctly used male condom. Combining spermicide with the female condom (on the outside of the condom) is likely to increase effectiveness, but scientific studies demonstrating this combined effect have not yet been published. It is important to mention that some spermicides may cause vaginal or penile irritation. As discussed earlier in regard to spermicidal condoms, this may increase the chances of transmission of some STIs.

Fertility Awareness Methods

Fertility awareness methods of contraception are based on the knowledge that a couple's ability to become pregnant varies during the woman's menstrual cycle. A woman is fertile for only about one day each month after an egg is released from one of her ovaries and enters the fallopian tube (see Chapter 9, "Conception, Pregnancy, and Birth," for a more detailed discussion of this process). Sperm cells ejaculated into the vagina may survive in the woman's reproductive tract for an average of three to five days. Combining the woman's fertility period and the maximum life span of sperm, you can estimate that an average couple's "window of fertility" in a given month is about six days. Theoretically, then, if a couple could know *exactly* when she was going to ovulate and agree to abstain from intercourse or use a barrier method of contraception for around five days prior to and one day after ovulation, they would be at very low risk of pregnancy and would not need to use any other form of contraception during the other days of the month. When intercourse is avoided entirely during this fertile time, this methods is sometimes referred to *natural family planning*. Natural family planning is the only contraception method sanctioned by the Catholic Church. Historically, older, less reliable forms of this technique, have been referred to as the "rhythm method."

fertility awareness
A method of contraception based on ovulation prediction and the viability of sperm; intercourse is timed to avoid fertile days, or a barrier method is used during those days.

This is the *theoretical* idea behind fertility awareness methods of contraception (see Figure 5.6 for an illustration of this method for an "average" hypothetical couple). The problem lies in determining when a woman will ovulate. Women vary a great deal in terms of timing of ovulation. This is true among women and also an individual woman may be ovulating on an irregular basis from month to month. Therefore, attempting to "guess" at ovulation is pretty much the same as using no contraception at all.

Since You Asked

8. When is it safe to have sex without contraception, before or after a woman has her period?

Many couples around the world have used fertility awareness as their method of choice with good success; however, it is a challenge to use it correctly. Knowing when a woman will ovulate is a complex process requiring education, organization, and cooperation between partners, and willingness to become familiar with the woman's hormonal cycle. Moreover, fertility awareness methods provide no protection from STIs. In general, most people should not attempt to use this method of contraception without serious research or guidance from a health care professional.

Several methods of fertility awareness or ovulation prediction are currently in common usage, including the Standard Days method, the TwoDay method, and the basal body temperature method.

THE STANDARD DAYS METHOD The **Standard Days method** of ovulation prediction (a new variation of what was called the "rhythm" or "calendar" method in the past) can be a reliable form of contraception if used consistently and correctly. This method is based on the fact that most (80%) women's fertility cycles are between 26 and 32 days long (the number of days between her periods). This means that 80% of women ovulate between days 11 and 18 of their fertility cycle; that is, day 11 through day 18, counting from the first day of her period (Arévalo et al., 2004; Hatcher & Trussell et al., 2011; Sinai et al., 2006).

Standard Days method
A fertility awareness technique for tracking fertile and infertile days during a woman's menstrual cycle.

Figure 5.6 The *Average* Window of Fertility

This is an example of the average window of fertility in a woman's monthly cycle, calculated on the assumption that ovulation occurs on day 14 and that sperm may live as long as seven days. Note that the exact fertile days vary among women and, for many women, from cycle to cycle.

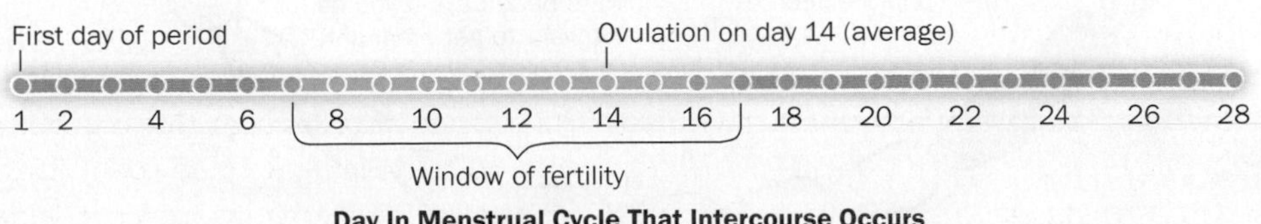

To be safest from pregnancy, a couple using this method should assume that their "window of fertility" is from five days before ovulation to a day after. This takes into account the life of the sperm cells (5 days) and a day when an ovum is most likely to be available for fertilization. For 80% of couples, then, this fertility window will occur between day 8 and day 19 of her cycle. A couple must either abstain or use a barrier method of contraception during this window.

Careful track of these days is paramount for reasonable success. Some women in the U.S. and many other countries have learned to use a set of colored beads with a ring (called "CycleBeads") that serves as a marker to help a woman keep track of her fertile and less fertile days (see Figure 5.7). A woman moves the ring over one bead each day starting from the first day of her period. The colors of the beads indicate when she is more or less likely to be fertile. One red bead indicates the first day of her period and is a starting point. The white beads indicate the days of her cycle when pregnancy is most likely. A set of beads may be purchased for about $10.00–$20.00 or can be homemade using exacting instructions available on the Internet.

This method of contraception is only recommended for women who have fairly regular menstrual cycles, typically *between 26 and 32 days every month*. If a woman is less regular, the ability to predict ovulation is significantly decreased. A couple must avoid unprotected intercourse on days when pregnancy is likely. This method, like all fertility awareness methods, does not protect against transmitted infections.

TwoDay method

A fertility awareness technique that relies on careful observation of secretions from the cervix to predict ovulation.

Figure 5.7 CycleBeads

Some women choosing the Standard Days method of contraception find that the use of CycleBeads facilitates keeping track of their fertile days during their cycle.

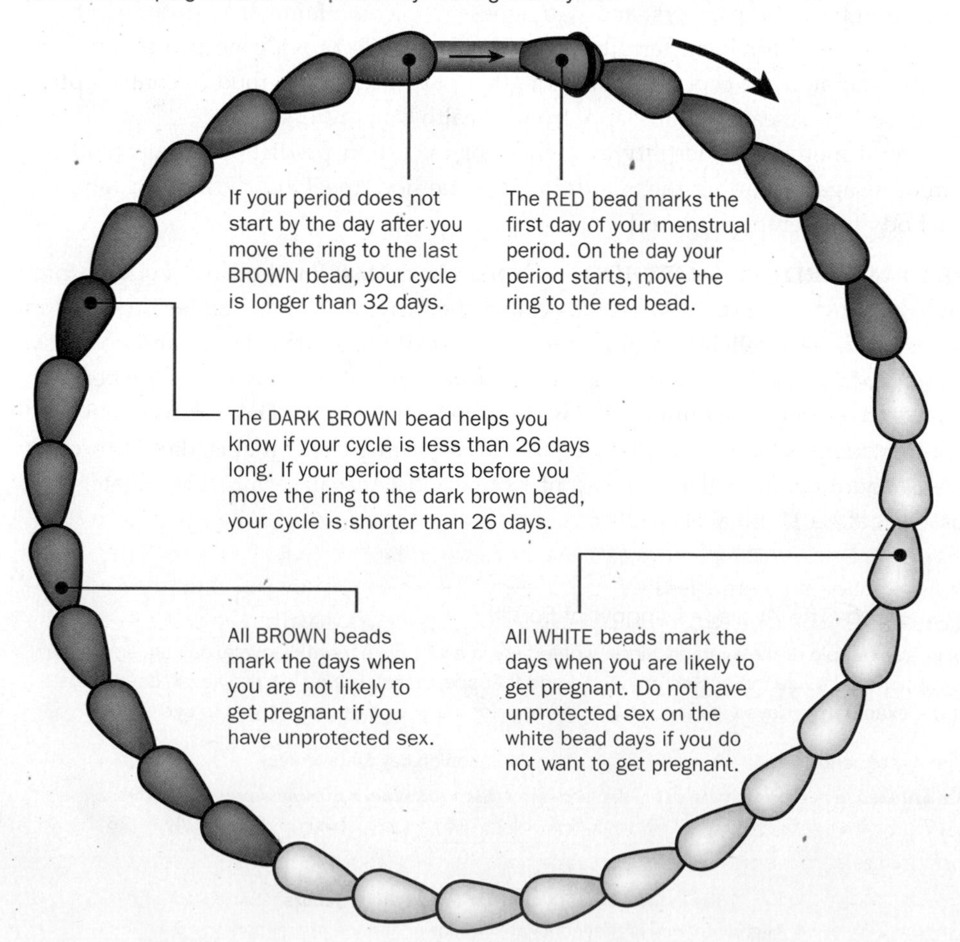

THE TWODAY METHOD The **TwoDay method** of fertility awareness allows a woman to tell when she is ovulating by carefully observing changes in the mucus that is secreted from the cervix into the vagina. Mucus secreted from the cervix begins to liquefy and flow around the time she is ovulating. This cervical mucus transformation enhances the ability of sperm cells to swim through the cervix at the precise time of ovulation. These cervical changes are working to maximize the possibility of conception, but they can also serve as a signal of ovulation to help couples avoid pregnancy if they wish.

For most of the month, the mucus is minimal and has a sticky texture, indicating that ovulation is not occurring at that time. As ovulation approaches, the amount of mucus increases, and it becomes cloudy and then clear and slippery. A woman can usually observe changes in her cervical mucus by wiping a tissue over the vaginal opening from front to back or by inserting a finger into her vagina. If the mucus appears thick and "sticky," ovulation is probably not occurring. If it is white and creamy in texture, ovulation is approaching. If, however, the mucus has a slippery, stretchy texture (similar to egg white), this indicates that she either is or will be ovulating very soon.

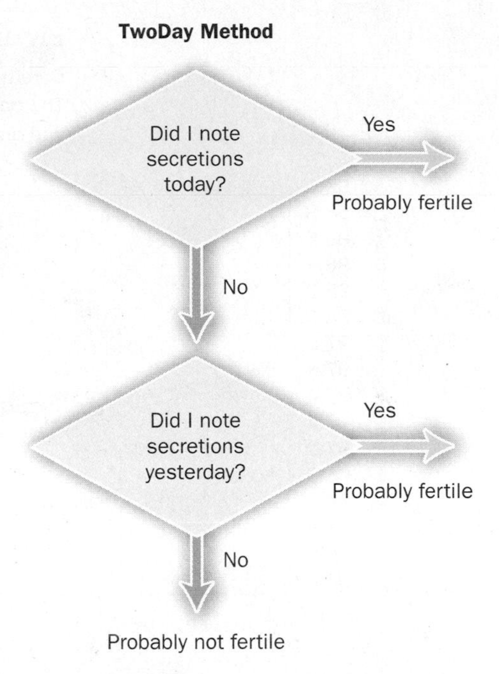

Figure 5.8 The TwoDay Fertility Awareness Method of Contraception

SOURCE: Based on Institute for Reproductive Health (2008).

When she notices the presence of clear, slick mucus for two days (either visually, using a finger to check, or seeing in the toilet paper) the woman should abstain from unprotected intercourse because she is likely to be ovulating or will soon. The method suggests that the woman ask herself two questions: "Did I notice the mucus yesterday?" and "Is the mucus present today?" If the answer to *either* question is "yes," she should consider herself fertile and avoid unprotected intercourse (see Figure 5.8). If the answer to both questions is "no," she is probably not fertile and is unlikely to become pregnant from unprotected intercourse (Jennings & Landy, 2006; Hatcher & Trussell et al., 2011; Sinai et al., 2006). While a woman is becoming familiar with this method, she should not be using any hormonal contraceptive because this can change her usual patterns of mucus secretion. If she is having intercourse while she is learning to use the mucus method, a woman should use condoms so that the mucus is not confused with semen or spermicides.

symptothermal method
A fertility awareness method based on monitoring a woman's cervical secretions and internal body temperature upon awakening in the morning.

THE SYMPTOTHERMAL METHOD In addition to changes in cervical mucus associated with ovulation, a woman's body temperature also changes in relation to ovulation. Basal body temperature refers to the internal temperature immediately upon awakening in the morning. Special, highly accurate thermometers are available that measure exact body temperature more precisely than a standard fever thermometer. If a woman takes her temperature every morning, she will usually notice a slight drop the day before she ovulates. Then a noticeable rise in temperature of between 0.4 and 0.8 degrees (F) will occur during or immediately after ovulation and will remain elevated until her next period. This temperature increase, in combination with the changes she observes in her cervical mucus, will indicate when her fertile period is beginning and ending (see Figure 5.9 on the following page for an example of a BBT chart). The combination of these two measurements is called the **symptothermal method** of fertility awareness. For preventing pregnancy, this method is best used to determine when ovulation has ended, at which time unprotected intercourse will most likely be free of the risk of pregnancy.

The mucus secreted by the cervix changes to a clear, slippery, stretchy fluid just prior to and during ovulation. Often this mucus is noticeable at the vaginal opening when a woman is ovulating.

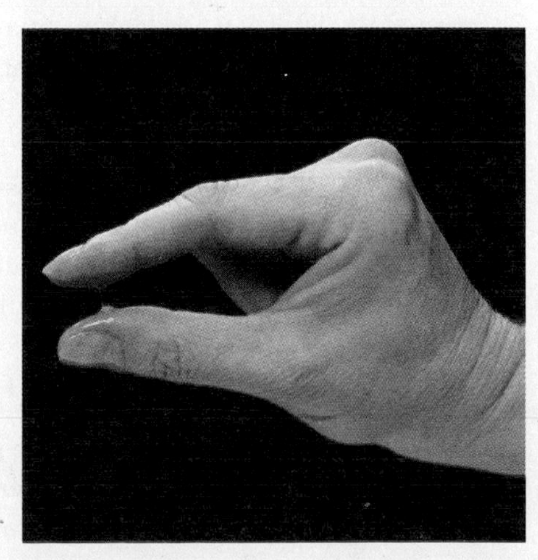

OVULATION PREDICTOR KITS Many companies market various types of ovulation predictor kits that use either hormonal testing (based on urine or saliva hormonal content) or temperature recording to help a woman pinpoint exactly when she will ovulate each month. These kits can make fertility awareness methods of contraception more effective. The kits are readily available without a prescription

Figure 5.9 Charting Your Basal Body Temperature

Careful monitoring of basal body temperature allows a woman to determine when ovulation occurs. The process involves taking her temperature first thing every morning using a special thermometer and graphing the changes on a chart. A sharp rise in BBT indicates that ovulation is imminent.

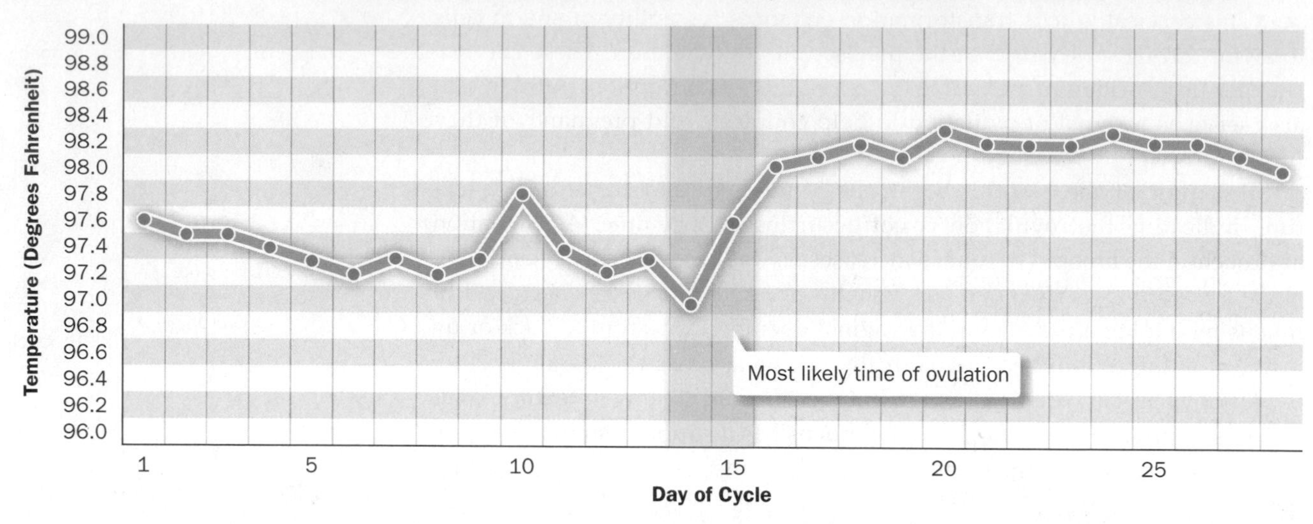

at pharmacies and online and include simple urine tests (much like a home pregnancy test), saliva-based testing kits, and computerized electronic monitoring devices. They cost between $5 and over $100 dollars depending on how high-tech they are.

EFFECTIVENESS OF FERTILITY AWARENESS METHODS The effectiveness of fertility awareness methods varies considerably, depending on the method used and the study consulted. You can see from Table 5.3 that the overall success rate is rather poor, ranging from 12% to 20% failure rate with typical usage. This is primarily because perfect use is difficult, requiring careful self-observation, record keeping, and the willingness and self-control to abstain from intercourse or use a barrier method during fertile days. However, correct and consistent use of the various fertility awareness methods discussed will lead to far greater effectiveness and can provide a reliable and natural method of contraception for couples not concerned about STIs.

ADVANTAGES AND DISADVANTAGES OF FERTILITY AWARENESS Fertility awareness offers many advantages for preventing pregnancy. Probably most important to people who choose this method is that it involves no hormones, chemicals, or devices, so there are no negative side effects to worry about. Another advantage is that these methods are very inexpensive. Many couples actually enjoy the imposed break from intercourse because it allows them to explore other forms of sexual expression and provides a honeymoon effect each month following the fertile period. In addition, an awareness of a couple's fertility is helpful for couples who wish to become pregnant to increase their odds by increasing the frequency of intercourse prior to fertile days of the woman's cycle.

Several important disadvantages should be reiterated as well. Fertility awareness provides no protection from sexually transmitted infections. These methods also require a high degree of planning, self-discipline, and cooperation by both partners. Finally, the most important disadvantage for most couples is the high failure rate of fertility awareness methods if they are used incorrectly.

Since you asked

9. I've heard that the IUD is a painful and dangerous method of contraception. Why would anyone use it?

intrauterine device (IUD)

A small plastic device in the shape of a T that is inserted by a doctor into the uterus through the cervix, via the vagina. It then may remain in place for three to ten years, during which time pregnancy is effectively prevented.

Intrauterine Devices

The **intrauterine device (IUD)** is a small plastic device in the shape of a T that a doctor inserts into the uterus through the cervix, via the vagina. It then remains in the uterus for an extended period of time, up to ten years.

While the IUD is in place, pregnancy is very effectively prevented, with a failure rate of less than 1% (see Figure 5.10).

Most researchers and medical professionals worldwide consider today's IUDs to be a safe, effective, and hassle-free method of contraception with very few side effects (e.g., Bhathena & Guillebaud, 2008). IUDs are a popular contraception device throughout the world and are becoming increasingly popular in the United States for women of all childbearing years.

HOW THE IUD WORKS The intrauterine device (IUD) is inserted into the uterus by a health care professional and may remain in place up to ten years, depending on the type of IUD. Three IUDs are currently available in the United States: *ParaGard, Mirena,* and *Skyla.*

The ParaGard IUD contains copper that repels sperm, preventing them from reaching the ovum. Mirena and Skyla contain a small amount of progesterone-like hormone (levonorgestrel) that is slowly released into the uterus and contributes to pregnancy prevention by thickening the cervical mucus, reducing passage of sperm cells from the vagina. The ParaGard IUD protects against pregnancy for up to ten years, Mirena up to five years, and Skyla up to three. They may be removed at any time and fertility quickly returns.

ADVANTAGES AND DISADVANTAGES OF THE IUD The main advantages of the IUD are that it is a hassle-free, effective method of contraception that does not require the user to remember to take a daily pill or have a method of contraception readily available prior to intercourse. Once inserted, the IUD is a relatively inexpensive method when calculated on a monthly basis.

Negative side effects of the IUD are rare and include spotting and heavy periods. A small percentage of women experience spontaneous expulsion of the IUD during the first year of use, requiring reinsertion. In very rare cases (one-tenth of 1%), perforation of the uterine wall occurs during insertion. This typically heals quickly and is not a medical emergency (Kaunitz, Grimes, & Stier, 2007). But remember, the IUD offers no protection from sexually transmitted infections and is usually recommended for women in monogamous relationships with a lower risk of STIs.

Figure 5.10 The IUD

The intrauterine device (IUD) is inserted into the uterus by a health care professional and may remain in place up to ten years, depending on the type of IUD.

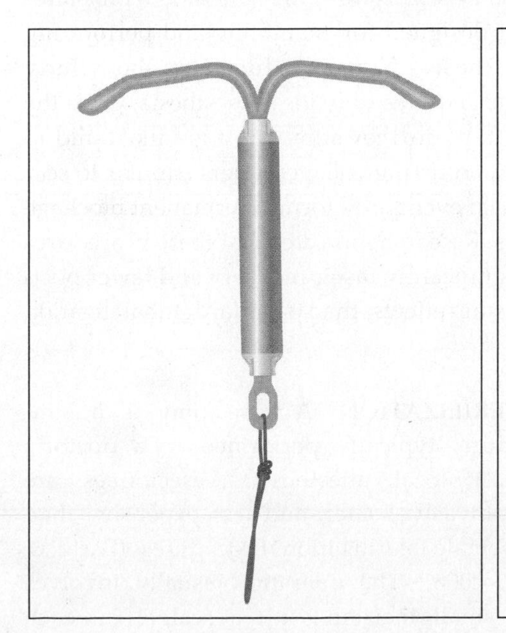

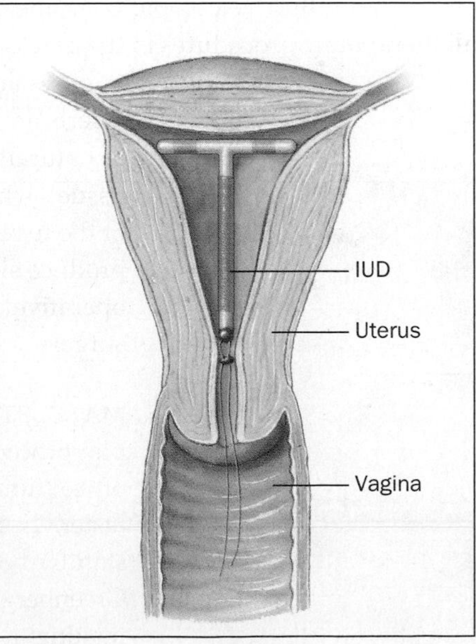

Surgical Contraceptive Methods

sterilization

Any surgical alteration that prevents the emission of sperm or eggs; also referred to as voluntary surgical contraception.

tubal ligation

A permanent method of contraception involving tying, cutting, clipping, or otherwise blocking the fallopian tubes to prevent passage of an ovum.

vasectomy

Cutting and tying off or sealing each vas deferens so that sperm produced by the testicles can no longer mix with semen in the ejaculate.

Sterilization, also referred to as voluntary surgical contraception (VSC), is one of the most widely chosen methods of contraception in the United States and throughout the world, accounting for 39% of contraception choice in the U.S. (Bedaiwy, Barakat, & Falcone, 2011). Of these, approximately 30% of these procedures are fallopian tube procedures and 70% are vasectomies. In the United States, around 700,000 tubal procedures and 500,000 vasectomies are performed annually (Bartz & Greenberg, 2008). Female sterilization prevents the egg from passing through the fallopian tubes by cutting, tying, or otherwise blocking the tubes (most commonly, this is a procedure called a **tubal ligation** or *tubectomy*). Male sterilization, or **vasectomy**, usually involves cutting and tying off or cauterizing (sealing with heat) the vas deferens so that sperm produced by the testicles can no longer mix with semen in the ejaculate. VSC is considered a permanent method of contraception, although reversal is becoming increasingly common and successful, especially for vasectomies (Bartz & Greenberg, 2008). Those who choose VSC are usually older (in their thirties or forties), typically have already had at least one child and do not desire more, and are in a stable, long-term relationship. A small percentage of people later regret their choice of sterilization, so this is a decision that must be made very thoughtfully and carefully (Bartz & Greenberg, 2008).

laparoscopy

A surgical procedure in which a tube with a tiny camera and light is inserted through a small incision in the abdomen.

FEMALE STERILIZATION Surgical sterilization is typically done by means of **laparoscopy** or minilaparascopy (the main difference is the size and length of the surgical device). Both of these procedures require two very small incisions in the abdomen, through which a viewing scope and a surgical instrument are inserted (see Figure 5.11). The fallopian tubes are then cut, tied, or clamped off using various medical techniques. Depending on the surgeon and the procedure, either local or light general anesthesia is used. This procedure usually takes less than thirty minutes and is nearly always performed on an outpatient basis. The woman can usually resume her normal activities within a few days. Voluntary surgical sterilization for women involves the cutting, tying, or clamping of the fallopian tubes.

Two new methods of nonsurgical sterilization for women, *Essure PBC* (permanent birth control) and *Adiana PCS* (permanent contraceptive system) have been introduced over the past decade. Essure is a tiny spring-like device that is inserted into each fallopian tube at the uterine end. Adiana is a small, rice-size silicone insert similarly inserted via the cervix and uterus using a standard hysteroscope in a doctor's office (see Figure 5.12). No incision is necessary for either method. A hysteroscope is a very thin, telescopic instrument designed for examining and performing small therapeutic procedures in the uterus. The insertion procedure takes about thirty minutes and requires only local anesthesia. Once the inserts are in place, they stimulate a gradual buildup of naturally occurring collagen fibers (similar to scar tissue), which eventually form a permanent blockage of the tubes. Research indicates that these procedures produce significantly faster recovery and fewer postoperative side effects than standard tubal ligation surgery.

Figure 5.11 Tubal Sterilization

Voluntary surgical sterilization for women involves the cutting, tying, or blocking of the fallopian tubes.

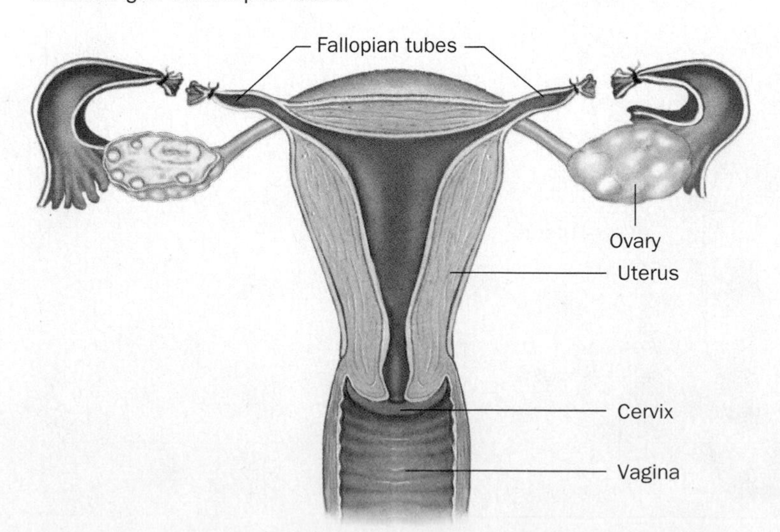

MALE STERILIZATION A vasectomy is a simple procedure, typically performed in a doctor's office under local anesthesia. Vasectomies are faster, less invasive, safer, and less expensive than standard female sterilization procedures (Bartz & Greenberg, 2008). The operation usually involves making a small incision with a scalpel on each

Figure 5.12 The Essure and Adiana Tubal Insert methods

Essure and Adiana are relatively new methods of female sterilization that use small inserts to block the fallopian tubes. Essure is a coil device and Adiana is a small silicone insert about the size of a grain of rice. These devices are inserted via the cervix into the uterine end of the fallopian tubes. This is a quick, outpatient procedure using local anesthetic. In approximately three months, the body creates tissue in and around the inserts, permanently blocking off the fallopian tubes.

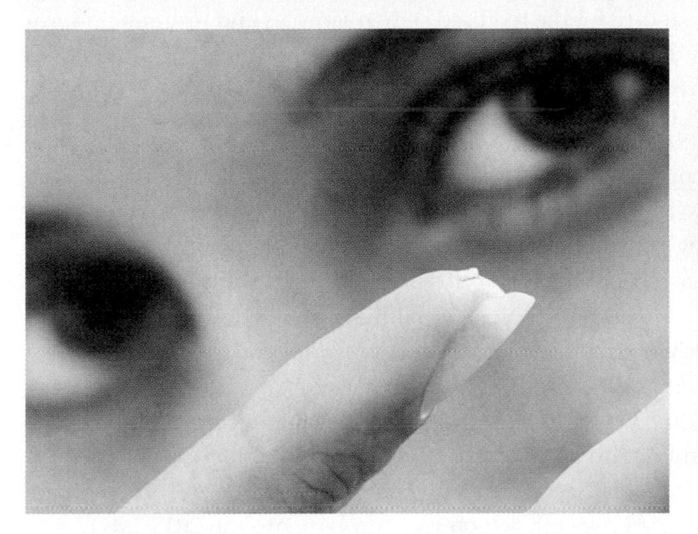

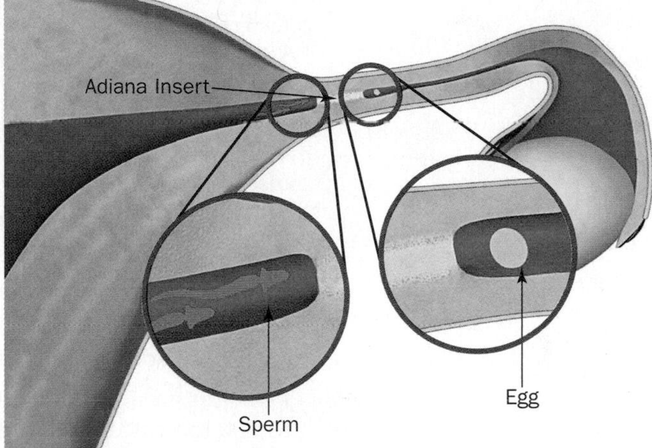

side of the scrotum to remove a small section of the vas deferens and close the ends, either by tying or cauterizing (see Figure 5.13). This makes it impossible for sperm to travel into the man's body from the testicles, rendering the male infertile. However, this does not mean that his sexual functioning or ejaculation is changed. As discussed in Chapter 2, "Sexual Anatomy," semen that is ejaculated upon orgasm is composed primarily of fluids from the prostate gland and seminal vesicles. Following a successful vasectomy procedure, the man still has orgasms and ejaculates as before, but the semen will no longer contain sperm cells.

Since You Asked

10. After a man has a vasectomy, does it feel different to have an orgasm without ejaculating sperm?

For many men, using the word *scalpel* in the same sentence as *scrotum* strikes fear deep into their hearts (and elsewhere). Consequently, some men for whom a vasectomy might be an option will often opt to avoid it at all costs. In response to these real, if somewhat irrational, perceptions, a "no-scalpel" technique of vasectomy is available (see Zini, 2010). For this procedure, also under local anesthesia, only a small puncture is made in the scrotum to allow access to the vas deferens, which is then snipped and closed off as in the scalpel method. No sutures are required, the method is faster, and the effectiveness is equal to the standard incision method. Vasectomy complications are rare for either method but seem to be lower for the no-scalpel technique (Zini, 2010).

Figure 5.13 Vasectomy Procedure

A vasectomy severs each vas deferens tube so that sperm will no longer be able to pass through and be ejaculated with semen.

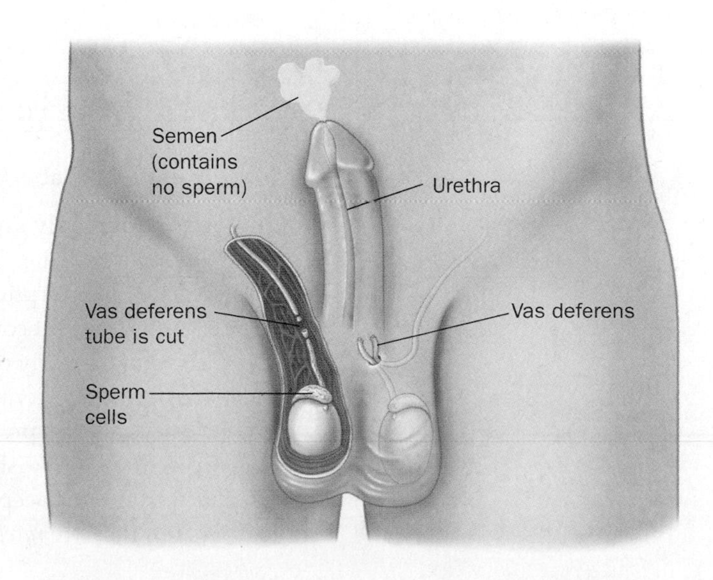

EFFECTIVENESS OF VOLUNTARY SURGICAL CONTRACEPTION Obviously, if you surgically prevent an ovum from ever encountering a sperm, you would have a 100% effective method of contraception. However, a very small failure rate (0.1% to 0.4%) is associated with this method. Why? Several reasons are possible. In women, the main reasons for this rare occurrence are: the unknown occurrence of pregnancy prior to the surgery, an incorrect surgical technique, or the spontaneous reconnection of a severed fallopian tube.

In men, the main reason for failure is usually impatience. That is, a man is not considered infertile post-vasectomy until he has submitted two semen samples after two to four months, and at had least twenty (non-vaginal) ejaculations following surgery. During this time, he and his partner must use a backup method of contraception before resuming unprotected intercourse because sperm cells may be left over in the male reproductive tract following surgery (Sharlip et al., 2012; Zini, 2010). In rare instances, women have become pregnant before the male's reproductive structures were fully purged of sperm cells. These instances are usually the result of the man's failure to return for the required semen samples and resuming unprotected intercourse too early. Excluding these errors in procedure or patient compliance, VSC is as near to a 100% effective method as exists today, other than celibacy.

ADVANTAGES AND DISADVANTAGES OF STERILIZATION Interestingly, the main advantage of VSC is probably also its main disadvantage: it is permanent (see the discussion in the next section regarding reversal). This means that a couple no longer needs to be concerned about using contraception. However, some people regret their decision and later wish that they were fertile. The reasons for this regret include: a change in desire for more children, divorce and remarriage, or loss of children through death. Regret is more common among individuals who choose VSC at earlier ages (Jamieson, 2007).

Other advantages of VSC include a lack of long-term side effects and the low cost per month of protection, after the initial expense of the permanent procedure (about $1,500 to $2,000 for a tubal ligation and $350 to $1,000 for a vasectomy). Both procedures are often covered by medical insurance and are automatically covered by plans available under the Affordable Care Act, which was implemented in 2013.

The Reversibility of VSC

In some cases, reversal of a vasectomy or tubal sterilization is possible. For women pregnancy rates after tubal ligation reversal range between 40% and 85% (Tubal Ligation Reversal, 2012), depending on the woman's age, the type of blockage used in the sterilization, remaining scar tissue, and the skill and experience of the surgeon. The Essure and Adiana procedures discussed above are not reversible.

For men, pregnancy rates following vasectomy reversal surgery range from 30% to 90% depending on the type of original surgery, the female partner's age, the skill and experience of the surgeon, and the length of time since the original vasectomy—the longer the time, the lower the success rates (Hatcher et al., 2011).

The most important point to consider is that reversal is a more difficult and more extensive surgery than the original procedure, and success is not guaranteed. Reversal is also far more expensive. Reversing any of the methods discussed here in the U.S. costs between $4,000 and $15.000. Those considering voluntary surgical contraception should not make such a decision unless they are prepared for it to be permanent.

Abortion is *Not* Contraception

5.4 Explain how abortion is not contraception and the two are linked.

You are undoubtedly aware that abortion is one of the most controversial topics in the United States today. The complexities of the abortion issue are discussed in detail in Chapter 9, "Conception, Pregnancy, and Birth." However, abortion must be mentioned here because it comes about most often as the result of a couple's failure to use contraception correctly or making the mistake of not using contraception at all.

Abortion is not contraception! It occurs *after* conception. It is the termination of a pregnancy, not the prevention of fertilization of an ovum by a sperm. If a woman or couple does not wish to have a child, their best course of action is, of course, using effective contraception. However, everyone knows that unwanted pregnancies happen, but they are *rarely* due to a failure of a properly-used method of contraceptive.

They are nearly always the result of incorrect, inconsistent, or failure to use contraception. Consider these statistics (see Guttmacher Institute, 2013):

- Fifty-four percent of women who have abortions had used a contraceptive method (usually the condom or the pill) during the month they became pregnant. But, among those women, 76% of pill users and 49% of condom users report having used their method *inconsistently*.
- Forty-six percent of women who have abortions had not used a contraceptive method during the month they became pregnant.
- Of these women, 33% had perceived themselves to be at low risk for pregnancy, 32% had had concerns about contraceptive methods, 26% had had unexpected sex.
- Eight percent of women who have abortions have *never* used a method of birth control.

Abortion is never an easy decision and is usually a very emotional and psychologically painful experience. It is safe to say that virtually everyone on all sides of the abortion debate is opposed to the use of abortion as a birth control method. One of the best reasons for using contraceptives responsibly whenever deciding to have intercourse is never to have to face the emotionally painful and potentially life-changing decision of whether to have an abortion.

One relevant fact is clear: As contraception use rises, abortion rates decline. Researchers have demonstrated in numerous countries around the world that as modern forms of contraception increase in acceptance and prevalence, the number of abortions declines at virtually the same rate, over the same time periods (e.g., Marston & Cleland, 2003).

Your Sexual Philosophy

Contraception: Planning and Preventing

Over the past twenty years, contraception research has exploded with innovative, simpler, and more effective methods. Never before have individuals and couples wishing to prevent pregnancy or engage in effective family planning had so many options. If issues of contraception concern you, someone you care about, or someone who may rely on your counsel in such matters, consider it your personal goal to stay up to date and informed about existing, new, and upcoming methods of controlling fertility. Make use of the many online resources you can find at this book's website (www.prenhall.com/hock). This is a field of knowledge in which making informed, safe choices is crucial to everyone's sexual lives.

Deciding on a method for contraception that is best and most comfortable for you is one of the fundamental components of your sexual philosophy. As mentioned at the outset of this chapter, if your choices about sexual behavior ever include heterosexual intercourse, that choice, by definition, includes the possibility of conception and pregnancy. If you carefully consider your contraception options now, based on all the available information from this chapter and other reliable sources, you will be able to make educated decisions about having a child or preventing an unwanted pregnancy. Regardless of your pregnancy plans, consider the importance of protecting yourself from sexually transmitted infections. This is equally important for men *and* women. If your decisions about contraception are based on a solid foundation of knowledge that is a part of your personal sexual philosophy, you can avoid the uncomfortable and risky position of trying to make the right choices about contraception and STI protection when you are unprepared or when your judgment may be clouded by the sexually charged events of an intimate sexual moment. Remember, your sexual philosophy is about knowing who you are, what you want or don't want, and planning ahead so that you can be in control of your sexuality rather than allowing it to control you.

Have You Considered?

1. What would you do if contraception were illegal, as it was less than a hundred years ago in this country and still is in others? Would you defy the law and obtain contraceptives on the black market? What other actions might you take?
2. Would you ever consider voluntary surgical sterilization as your contraception method of choice? Why or why not?
3. Imagine that you are the parent of a nineteen-year-old daughter who is home from college for a holiday weekend. You've always had a close relationship with her, so she comes to you explaining that she has met the man of her dreams and would like your advice about contraception. What would you want to be sure to tell her? What if your college student were a son? Would your advice to him be different? If so, in what ways? If not, why not?
4. Suppose that heterosexual partners who have been in an intimate, loving relationship for nearly a year decide to go camping together. They're sitting by the fire under the stars feeling very romantic and sexually excited. Suddenly they realize they forgot to put the condoms in the backpack! Discuss how they might best deal with this situation.
5. Imagine a situation in which you truly want to make love, but it is extremely important to you to use condoms. However, your partner doesn't want to use them. Discuss at least three strategies you might employ to resolve this disagreement.

Summary

Historical Perspectives

The Politics of Contraception

- Margaret Sanger led the contraception revolution in the early years of the twentieth century. It was largely through her efforts discussing contraception with health care providers that contraception for women became legal in the United States. Sanger believed that women could never achieve equal rights with men unless they were able to control their own reproductive bodies.

Choosing a Method of Contraception

- When choosing a method of contraception, both sexual lifestyle and the various pros and cons of the available methods must be considered. This important decision may be made by an individual or is often a shared decision by a couple who are becoming involved in an intimate relationship.

Methods for Preventing Pregnancy and STIs

- Only two methods, selective abstinence and condoms (both male and female), are effective in helping prevent both pregnancy and STIs.

Methods for Preventing Pregnancy (but Not STIs)

- Hormonal methods for preventing pregnancy include various types of oral contraceptives (birth control pills), the implant, the contraceptive patch, the injectable method, and the vaginal ring. Differences among these methods involve primarily how the hormones are delivered to the woman's body. All hormonal contraceptives are highly effective in preventing pregnancy; however, they are not without some negative side effects. Having a clear understanding of the pros, cons, and cautions of hormonal contraceptives is important in deciding whether or not to use them.
- Vaginal barrier methods of contraception include the diaphragm, cervical cap, and the *Today* sponge. All are effective forms of contraception, but they have somewhat higher failure rates than some other methods.
- Fertility awareness methods of contraception may be effective for some couples. However, predicting ovulation accurately may be difficult for some women and requires very careful planning. Techniques for predicting ovulation include the Standard Days

method, the TwoDay method, and the symptothermal method. Ovulation prediction kits are now available that aid significantly in the timing of intercourse relative to ovulation so as to reduce (or enhance) the possibility of pregnancy.

- The intrauterine device (IUD) is considered a safe, effective, hassle-free contraceptive choice. Three types are available in the United States: *ParaGard, Mirena,* and *Skyla*.
- Voluntary surgical contraception (VSC), commonly referred to as sterilization, is a popular, permanent form of contraception. Methods include tubal ligation or blocking for women and vasectomy for men. Vasectomy is usually considered safer and less expensive than female tubal procedures and results in faster recovery time.
- Emergency contraception pills, now available without a prescription, allow a woman who has had unprotected intercourse within the past three to five days to reduce her chances of conception significantly.

Abortion Is *Not* Contraception

- People on both sides of the abortion debate generally agree that relying on abortion as a method of birth control is seriously misguided.
- Greater availability and proper use of contraception significantly reduce the need for and number of abortions.

Your Sexual philosophy

Contraception: Planning and Preventing Pregnancy

- Your decisions about contraception should be based on a foundation of accurate and complete information that is a part of your personal sexual philosophy. With this knowledge, you will be prepared to make choices about contraception and safe sex practices with forethought and planning.

Chapter 6

Sexual Behaviors

Experiencing Sexual Pleasure

Learning Objectives

After you read and study this chapter you will be able to:

6.1 Explain the importance of individual differences in people's comfort level with sexual issues and topics, how these can be measured, and the possible sources of the variations.

6.2 List and discuss the mainstream sexual behaviors covered and how they relate to the risk of sexually transmitted infections.

6.3 List the positive and negative effects of sexual fantasy; discuss male and female fantasy similarities and differences; explain how fantasy is separate from actual behavior; and how fantasy may be used to enhance intimate relationships.

6.4 Discuss historical and present attitudes about masturbation, how this behavior affects couples in intimate relationships, and the possible health benefits of masturbation.

6.5 Summarize the role of erotic touch (kissing; breast and nipple touching; and genital touching) as an integral part of couples' sexual activities.

6.6 Summarize the sexual activities of fellatio and cunnilingus, people's feelings about them, and the risks of spreading sexually transmitted infections through these behaviors.

6.7 Review anal sexual activities, attitudes about these behaviors, and the potential health concerns related to them.

6.8 Summarize the most common positions used for coitus (heterosexual intercourse), the advantages of each, how the positions relate to orgasm, and the frequency of coitus by age and by country.

Since You Asked

1. I think about sex a lot. Does that make me weird? (see page 191)
2. Some of my sexual fantasies involve things I would never really do, so why do I feel guilty just thinking about them? (see page 193)
3. When someone is feeling very sexually frustrated, how can a person satisfy their sexual needs? (see page 198)
4. Does masturbation detract from one's sexual drive or pleasure? (see page 199)
5. I believe that foreplay is the best part of sex. Sometimes I like it more than intercourse. Is that strange? (see page 201)
6. Do women get more pleasure from intercourse or oral sex? How and why? (see page 206)
7. My boyfriend says it's OK if we have oral sex because it doesn't spread sexual diseases. Is that true? (see page 207)
8. Is there anything wrong with anal sex? (see page 208)
9. What is the best position for intercourse? (see page 209)
10. Is it abnormal to have intercourse twice or even three times a day with a partner whom you have been seeing for a long time? (see page 213)

Sexual intimacy with the right person at the right time is one of the most physically and emotionally satisfying experiences in life. Approaching sexual activities responsibly—with accurate knowledge, mutual respect, and care—can enhance a person's self-esteem and life happiness in general (Branchflower & Oswald, 2004; Stephenson & Meston, 2013). Conversely, engaging in sexual acts irresponsibly—for example, at too young an age, when a person does not feel ready, or when care and respect between the partners are lacking—can lead to regret, guilt, and decreased self-esteem (Starks & Morrison, 1996; Wight et al., 2000).

Who decides when and with whom you are going to be sexual? *You* should, of course. You should be in charge of your body and what you do sexually. Unfortunately, some individuals have had sexually related experiences in the past, such as child sexual abuse or rape, that have taken that control over their body away from them through no fault of their own. This does not mean that they do not or cannot reclaim that control and take steps to prevent such painful events from ever happening again to them or to the people they care about.

In this chapter, we will explore what might be referred to as *mainstream* sexual behaviors: activities that are engaged in and enjoyed by a majority or large percentage of couples and individuals. These include sexual fantasy, masturbation, touching and kissing, oral-genital sexual activities, sexual intercourse (coitus), and anal sexual activities. All of these activities can be, and are, engaged in by couples of all sexual orientations, whether straight, gay, lesbian, or bisexual (with the exception of *coitus*, which is, by definition, a heterosexual behavior). Although more research has been conducted pertaining to heterosexual than gay or lesbian sexual activities, most of our discussions in this chapter assume readers of any sexual orientation. Furthermore, issues specifically pertaining to nonheterosexual orientations are the focus of Chapter 11, "Sexual Orientation." The goal of this chapter is to inform you about the variety of sexual activities so that you can make educated and responsible choices that are comfortable for you, whatever your sexual orientation, about your behavior now and in the future.

Focus on Your Feelings

The idea of sexual intimacy may produce a wide range of emotional responses. Usually, for most people these emotions are positive and pleasurable, such as feelings of excitement, happiness, desire, love, and contentment. Sometimes, however, a person may experience negative emotions associated with sexual behavior, such as embarrassment, anxiety, fear, or even aversion. These negative emotions can interfere with enjoyment of intimacy and create barriers to fulfilling sexual experiences. However, when individuals and couples who are experiencing unpleasant feelings about sex come to understand the source of their emotions and can communicate about them, they discover ways of overcoming them. *Emotional* responses to sex stem from experiences, either personal or what someone has seen and leaned from others throughout life. If an individual has experienced negative or traumatic sexual experiences in the past, current or future sexual behavior may become associated with those earlier events. Even though his or her *rational* self may know *this* current situation, *this* partner, *this* sexual experience now may be far different from the negative events in your past, the *emotional* self may involuntarily react negatively. Psychologically speaking, present sexual cues tend to reactivate your emotional past.

Although no one can change the past, people can take steps to reduce the present emotional effects of a past negative sexual event. Often a caring, trusting relationship with a partner who is willing to be patient and understanding can allow a person to move beyond negative sexual associations. Approaching sexual intimacy slowly, without demands or expectations, can help a partner feel safe, in control, and less anxious. Sometimes, however, past events exert such a strong influence that the person may require more focused help through counseling with a trained professional therapist. Therapy for these kinds of emotional-sexual problems is extremely effective and nearly always successful in helping individuals overcome their sexual fears and find the intimacy they desire. If you feel that negative emotions are preventing you from enjoying a gratifying sexual life, counseling is usually available through your college or university health or counseling services, public and private counseling agencies, and therapists in private practice.

It's no accident that this chapter appears at this point in the book. Earlier chapters have discussed sexual anatomy, sexual responding, emotional intimacy in relationships, and contraception. These discussions have given you a foundation of knowledge for making your best personal choices about the behaviors you choose for yourself.

One last point for you to keep in mind as you begin this chapter: People's knowledge and experience of sexual activities varies greatly, owing to our wide diversity of family backgrounds, cultural heritages, religions, past relationships, childhood events, sexual orientations, personal interests, and previous sexual study or education. Although social norms in college often assume that everyone is sexually knowledgeable, this assumption is not true. Significant differences in sexual knowledge exist, and respecting those differences in each other is extremely important. As mentioned in the preface of this book, whether you feel you have learned a little or a lot about sex, what you know will grow and change as you read and think about the issues in this chapter and throughout this book.

Although sexual activities themselves have not changed much throughout history, cultural attitudes toward them have undergone and are undergoing substantial social changes. Many sexual activities that were considered immoral, illegal, unhealthful, and even evil less than a century ago are now commonly practiced, openly discussed, and even recommended by health professionals and therapists. Probably no better example exists of these social and cultural changes than masturbation. Let's take a brief look back at how the act of "self-pleasuring" was regarded by some a mere one hundred years ago.

Historical Perspectives

Masturbation Will Be the *Death* of You...

Few areas in Western cultural beliefs have changed as radically or as quickly as our attitudes about sexual behavior. Take one example, the M-word, *masturbation*. For many people, this is an embarrassing and highly personal topic. But in contrast to most people's attitudes today, consider the following excerpts from *Sexology*, a book by William H. Walling, M.D., published in 1904. The first quote is taken from his chapter titled "Masturbation, Male."

> Perhaps the most constant and invariable, as well as earliest signs of the masturbator are the downcast, averted glance, and the disposition of solitude. Prominent characteristics are loss of memory and intelligence, morose disposition, indifference to legitimate pleasures and sports, and stupid stolidity. The masturbator gradually loses his moral faculties; he becomes listless, incapable of all intellectual exertion; he is taken by surprise if required simply to reply to a child's question. (pp. 38–39)

Dr. William H. Walling

You can see from this quote how extreme some professionals' views were in the early 1900s concerning the practice of masturbation among boys and men. However, their condemnation of the activity reached an even more negative tone when such behavior was suspected in women. Here's what Walling had to say in his chapter titled "Masturbation, Female":

> Alas that such a term is possible! O, that it were as infrequent as it is monstrous, and that no stern necessity compelled us to make the startling disclosures which this chapter must contain! We beseech, in advance, that every young creature, if she yet be pure and innocent, will at least pass over this chapter, that she may not know the depths of degradation into which it is possible to fall.
>
> The symptoms which enable you to recognize or suspect this crime [of masturbation] are the following: A general condition of weakness and loss of flesh; the absence of freshness and beauty; of color from the complexion; of the vermillion from the lips; and whiteness from the teeth; which are replaced by a pale, lean, puffy, flabby, livid physiognomy;...bluish circles around the eyes, dry cough, panting on the least exertion. (pp. 42, 46)

For both male and female masturbation, Walling warned that, if left unchecked, masturbation would lead to sickness, physical wasting away, and eventually death.

We will discuss today's prevailing views about masturbation in more detail in this chapter. The point of these quotes is to demonstrate the extent to which times have changed. Slightly more than a century after the publication of Walling's *Sexology*, masturbation has become a widely accepted sexual activity. As you will read, masturbation is today one of the key components of sex therapy and sexual satisfaction, and it has even been included in the plot lines of numerous television dramas and sitcoms. This does not imply that everyone masturbates or that everyone should. Masturbation, like all other sexual activities, is an individual choice. And for people to make those choices, they need to be as aware as possible of what sexual activities and sexual situations make them feel comfortable. Each person has his or her ideal "sexual comfort zone."

Your Sexual Comfort Zone

6.1 Explain the importance of individual differences in people's comfort level with sexual issues and topics, how these can be measured, and the possible sources of the variations.

The extent to which you experience positive or negative reactions or emotions to sexual issues, discussions, and behaviors determines what may be called your *sexual comfort zone*. People with wide sexual comfort zones are able to respond to various sexual topics and behaviors with relative ease and positive feelings and openness. Those with narrower comfort zones are more likely to approach sexual topics with hesitation, negative emotions, anxiety, and even fear.

Erotophobia–Erotophilia: A Measure of Sexual Comfort

How would you describe *your* sexual comfort zone? In the 1980s, a group of researchers developed the Sexual Opinion Survey (SOS) to measure and study people's reactions to sexual issues and behaviors (Fisher, 1988; Fisher et al., 1988). This scale places people on a scale or continuum the researchers called **erotophobia–erotophilia**. *Erotophobic* individuals have narrow sexual comfort zones, respond negatively and uncomfortably to sexual topics, and may tend to avoid sexual information and activities. *Erotophilic* individuals feel more comfortable with sexual issues, seek out sexual information, enjoy sexual behavior, and respond with positive emotions to sexual cues.

erotophobia
An attitude toward sexuality in which individuals are generally uncomfortable with sexual topics, respond negatively and uncomfortably to sexual issues, and tend to avoid sexual information and activities.

erotophilia
An attitude toward sexuality in which individuals are comfortable with sexual issues, seek out sexual information, enjoy sexual behavior, and respond with positive reactions to sexual topics.

The scale asks respondents to rate various questions about sexual attitudes on a 7-point strongly agree to strongly disagree scale. Their answers to the 21 statements on the scale result in an erotophobia–erotophilia score. Examples of items on the scale are, "If people thought I was interested in oral sex, I would be embarrassed," "Swimming in the nude with a member of the opposite sex would be an exciting experience," and "Seeing an erotic (sexually explicit) movie would be sexually arousing to me." You can see how people with varying comfort zones regarding sex and sexuality would likely rate these statements and the others on the scale quite differently.

Why Your SOS Score Is Important

Scores on the SOS have been linked to numerous sexual attitudes and behaviors (e.g., Macapagal & Janssen, 2011; Rye et al., 2012). Whether you score high (more erotophilic) or low (more erotophobic) or somewhere in the middle, you should understand what your score might predict about your sexual attitudes and choices. Table 6.1 on page 189 summarizes some of the differences among people who score higher and lower on the SOS.

The measurement of erotophobia–erotophilia is not intended to place value judgments on people according to their score. It simply points out our differences about sexual issues. Nevertheless, someone who is highly erotophobic may want to be alert to some potentially negative consequences. As you can see in Table 6.1, erotophobic individuals are more uncomfortable about sexual issues and topics in general, so they are less likely to take precautions against sexually transmitted diseases, use birth control, or seek medical attention for sexual health problems. In addition, they tend to avoid opportunities to become more educated and thus less uneasy about sexuality. For these reasons, a person's score on the SOS can serve as a gauge of how emotionally and psychologically ready the person is to be involved in sexually intimate relationships and how a course and text such as this one can be valuable in a person's life.

Sources of Your SOS Score

Your sexual feelings and attitudes develop throughout your life and are reflected, at least in part, in your SOS score. Humans are probably not born preprogrammed with any of these specific attitudes. Rather, you acquire them from your experiences and interactions with others (parents, mainly) as you move through childhood and adolescence into adulthood. To demonstrate this, researchers asked subjects who had taken the SOS to fill out detailed questionnaires about their childhood and adolescent sexual influences.

Some of the factors found to influence participants' development of erotophobia include strict parental attitudes about sex play in childhood, religious training that inhibits free sexual expression, conservative family sexual attitudes, and guilt about

Table 6.1 Erotophobia–Erotophilia: Effects on Attitudes and Behaviors

People who score high or low on the Sexual Opinion Survey (SOS) tend to differ in many ways relative to their sexual attitudes and behaviors. This table summarizes some of the most common differences overall.

Issue	Erotophobic Attitudes and Behaviors	Erotophilic Attitudes and Behaviors
Traditional sex-role behaviors	Adhering	Flexible
Men (in general)	More	Less
Women (in general)	Less	More
Enjoyment of sexual activities	Less	More
Attitude toward masturbation	Negative	Positive
Sexual self-concept and attractiveness	Hesitant	Confident
Attitudes toward nonheterosexual orientations	Negative, intolerant	Accepting, tolerant
Development of sexual problems	More likely	Less likely
Sexual health care (doctor visits, self-exams, etc.)	Less conscientious	More conscientious
Use of measures to prevent sexually transmitted infections	Less likely	More likely
Correct and consistent use of birth control	Less likely	More likely
Attitude toward learning about sexual topics	Reluctant	Eager
Attitude toward sexual discussions between parents and children	Uncomfortable, avoidant	Open, comfortable
Attitude toward sexual discussions with partner	Uncomfortable, avoidant	Open, comfortable
Willingness to initiate new sexual behaviors with partner	More likely	Less likely
Attitude toward engaging in sexual fantasy	Avoidance, guilt	Comfortable
Attitude toward breast-feeding in public	Uncomfortable	Comfortable
Viewing sexually explicit Internet sites	Less likely	More likely

SOURCES: Barak et al. (1999); Byers, Purdon, & Clark (1998); Fisher et al. (1988); Forbes et al. (2003); Garcia (1999); Geer & Robertson (2005); Humphreys & Newby (2007).

masturbation in adolescence. Factors that lead to more erotophilic attitudes in adulthood include having parents who are good sexual educators and who are not embarrassed to discuss sex, more frequent masturbation in childhood and adolescence, and a factual understanding of sexual topics in adolescence (see Fisher et al., 1988; Macapagal & Janssen, 2011; Rye, Meany, & Fisher, 2011).

If you think about these influences and compare them with your own memories of childhood and adolescence, you may gain some important insights into the sources of your sexual attitudes and feelings and how they relate to your current sexual philosophy. With that in mind, we turn to our discussion of sexual behaviors themselves.

The Sexual Inhibition and Sexual Excitation Scales (SIS/SES)

In the early 2000s, a group of researchers developed a new scale to measure sexual attitudes in a different way. The *Sexual Inhibition Scale (SIS)* and the *Sexual Excitation Scale (SES)* propose to go beyond measuring what people report about their sexual attitudes, values, and desires, and seeks to measure individual differences in sexual *response* patterns, thereby helping to explain the source of their SOS scores (Janssen et al., 2002). The researchers proposed that human sexual activities and responses are mediated by excitatory and inhibitory messages that have evolved in us through evolutionary survival mechanisms, such as avoiding threats and maximizing reproduction.

The SIS/SES ask respondents to rate, on an *agree-disagree scale* (1 = strongly agree, 2 = agree, 3 = disagree, and 4 = strongly disagree), 73 statements describing sexual situations that involve some sort of threat (an inhibitory statement) and sexual situations that are free from any threat (an excitatory statement—no threat). An example of an excitatory statement from the scale is "When I am taking a shower or a bath, I easily become sexually aroused"; an example of an inhibitory statement, containing a threat is, "If I think I can be heard by others while having sex, I am unlikely to stay sexually aroused" (Janssen et al., 2002, p. 126).

Overall, men's and women's scores on the SIS/SES have been found to be similar, but women generally scored higher on the *Sexual Inhibition Scale* and lower on the *Sexual Excitation Scale* than did men (Carpenter et al., 2008). Furthermore, although the SIS/SES mirrors in many ways how people score on the SOS, the SIS/SES helps us to understand the underlying *causes* of people's levels of erotophobia or erotophilia.

Solitary and Shared Sexual Behaviors

6.2 List and discuss the mainstream sexual behaviors covered and how they relate to the risk of sexually transmitted infections.

As mentioned earlier, our discussion here focuses on "mainstream" sexual behaviors, relatively common behaviors that are engaged in by a majority or a relatively large minority of people and couples. Moreover, these behaviors would not be considered particularly strange or deviant in most Western cultures (nonmainstream behaviors are discussed in Chapter 14, "Paraphilic Disorders"). Table 6.2 offers a glimpse of sexual activities in the United States. One sexual activity is conspicuously missing from the list in Table 6.3 on page 1941. It is one that many people fail to realize is the most common sexual activity of all and the one that we will discuss first: *sexual fantasy*.

We will follow our discussion of sexual fantasy with a discussion of masturbation. Fantasy and masturbation, which typically go hand-in-hand, are, for the most part,

Table 6.2 Sexual Experiences of Americans by Selected Age Ranges*

	Males					Females				
Age range ⇨	16–17	18–19	20–24	25–29	30–39*	16–17	18–19	20–24	25–29	30–39*
Activity experienced at least once during past year ⇩										
Solo masturbation	75%	81%	83%	84%	80%	45%	60%	64%	72%	63%
Masturbation with partner	16%	42%	44%	49%	45%	19%	36%	36%	48%	43%
Received oral sex from female partner	31%	54%	63%	77%	78%	5%	4%	9%	3%	5%
Received oral sex from male partner	3%	6%	6%	5%	6%	24%	58%	70%	72%	59%
Gave oral sex to female partner	18%	51%	55%	74%	69%	7%	2%	9%	3%	4%
Gave oral sex to male partner	2%	4%	7%	5%	5%	22%	59%	74%	76%	59%
Vaginal intercourse	30%	53%	63%	86%	85%	30%	62%	80%	87%	74%
Insertive anal intercourse	6%	6%	11%	27%	24%	–	–	–	–	–
Receptive anal intercourse	1%	4%	5%	4%	3%	5%	18%	23%	21%	22%

*Statistics from *The National Survey of Sexual Health and Behavior,* Indiana University (2010). This study included over 5,000 respondents, ages 14 to 94. All age groups were found to be sexually active (see Chapter 12, "Sexual Development Throughout Life," for more information about sexuality and aging). The age ranges included in this table were selected from the total male and female groups who were part of the study. Also refer to the NSSHB website for additional statistics and additional age ranges, at www.nationalsexstudy.indiana.edu.

SOURCE: Data adapted from: D. Herbenick et al. (2010a). *National Survey of Sexual Health and Behavior*, Indiana University, 261–263.

personal and solitary activities. When most people think of "sex," however, they think of interpersonal or *shared* behaviors; these will be the focus of our discussion in the balance of this chapter.

Most shared sexual activities carry some level of risk for contracting and transmitting sexually transmitted infections (STIs). See Table 6.3 for a summary of STI risks associated with different sexual behaviors (and see Chapter 8, "Sexually Transmitted Infections/Diseases," for more in-depth coverage of STIs). The importance of this point cannot be overemphasized: sexual intimacy can be one of life's most profound and enjoyable experiences if you are comfortable, confident, informed, and safe in the sexual choices you make. Doing everything possible to avoid the risks of STIs (and unwanted pregnancy) must be part of everyone's sexual decisions.

Sexual Fantasy

6.3 List the positive and negative effects of sexual fantasy; discuss male and female fantasy similarities and differences; explain how fantasy is separate from actual behavior; and how fantasy may be used to enhance intimate relationships.

It has been said that the most important sexual organ in the human body is the brain. And fantasizing about sex the most common of all sexual activities. It is so prevalent among college students that no one would be very surprised if you are doing it right now! And the difference in sexual fantasy frequency appears not to be very different for men and women. One older study found that, on average, college men reported having sexual fantasies approximately seven times a day and college women almost five times a day (Jones & Barlow, 1990). However, other studies have that, overall, the percentage of people who report sexual fantasies is over 95% and the prevalence appears to be about the same for men and women (Carpenter et al., 2008; Fisher, Moore, & Pittenger, 2012).

Sexual fantasies may emerge virtually anytime, anywhere. They may begin spontaneously or through a conscious decision to fantasize about something sexual. They may or may not be accompanied by physiological sexual arousal, and emotionally

Since You Asked

1. I think about sex a lot. Does that make me?

Table 6.3 Sexually Transmitted Infection Risk for Various Sexual Activities

Activity	Risk of STI Transmission When One Partner Is Infected
Sexual fantasy	Safe.
Masturbation (solo)	Safe.
Erotic touching	Safe.
Genital touching to orgasm (mutual masturbation)	Safe, unless semen or vaginal secretions touch broken skin or mucous membranes. Possible, though unlikely, to transmit HPV (genital warts) or herpes simplex virus.
Cunnilingus (unprotected)*	*Giving partner:* risk of transmitting or contracting genital herpes in mouth area; genital warts in mouth and throat; gonorrhea; syphilis; hepatitis B; chlamydia; small risk of HIV. *Receiving partner:* risk of transmitting or contracting herpes from mouth to genitals; genital warts; gonorrhea; syphilis; hepatitis B; chlamydia; small risk of HIV transmission; may cause candidiasis infection (yeast from mouth to vagina) and possible nonspecific urinary tract infections.
Fellatio (unprotected)*	*Giving partner:* risk of transmitting or contracting genital herpes in mouth area; genital warts in mouth and throat; gonorrhea; syphilis; hepatitis B; chlamydia; very small risk of HIV. *Receiving partner:* risk of transmitting or contracting herpes from mouth to genitals; genital warts; gonorrhea; syphilis; hepatitis B; chlamydia; small risk of HIV transmission.
Coitus (unprotected)*	Risk of contracting and transmitting *all known STIs*. Risk generally greater for female than male.
Anal stimulation	Risk of vaginal, urinary, and reproductive tract infections if normal rectal bacteria are spread from anal area to vagina, vulva, or urethra (men and women). Risk of HIV, hepatitis B, HPV (from genital warts in anal area), chlamydia, and other bacterial risks from oral-anal contact (anilingus).
Anal intercourse (unprotected)*	Highest risk of transmission of HIV. Risk greatest to receiving partner due to semen contacting damaged rectal tissue. May also transmit genital warts, syphilis, chlamydia, and hepatitis B. Risk with condom higher than vaginal intercourse with condom.

*The use of protection such as a male or female condom significantly reduces, but does not entirely eliminate, the risk of transmission of these STIs.

they may be either positive or negative (Byers, Purdon, & Clark, 1998; Fisher, Moore, & Pittenger, 2012). Most people report using sexual fantasy as part of masturbation (Brody, 2003), and many people fantasize while engaging in intercourse or other sexual activities with a partner (Davidson & Hoffman, 1986; Lunde et al., 1991; Stockwell & Moran, 2013). Although some people may think these fantasies indicate a problem with the relationship (as indicated by statements such as, "If you need to fantasize when we're together, you must not find me attractive"), research has shown that this is usually not the case. In fact, the majority of people (over 70% of men and women) report having sexual fantasies during sexual activities with a partner (Reinisch, 1990). Moreover, those who fantasize more tend to be better sexually adjusted overall, have fewer sexual problems, and report the greatest overall satisfaction in their sexual relationships (Gagnon & Simon, 2011; Zurbriggin & Yost, 2004).

Today, virtually all health professionals and sexuality educators consider sexual fantasy in all its contexts to be a normal and healthy part of a person's sexual life (Cato & Leitenberg, 1990; Janus & Janus, 1993; Leitenberg & Henning, 1995; Zurbriggin & Yost, 2004). As shown in Table 6.4, sexual fantasy has been shown to enhance sexual enjoyment and satisfaction in many different ways, whether with a partner or by oneself.

When Sexual Fantasies May Cause Problems

Sexual fantasies are usually harmless, enjoyable activities; however, under some circumstances, they can be a source of discomfort. Sexual fantasies may cause a problem in various specific ways.

Some people feel that their fantasies are somehow wrong, sinful, or immoral, and this can create a great deal of guilt and worry (Renaud & Byers, 2001). One study found that 25% of male and female college students reported feeling guilty about having sexual fantasies during intercourse (Cato & Leitenberg, 1990; Kahr, 2008). Other research has revealed that 84% of college students report experiencing intrusive or unpleasant sexual fantasies (Byers et al., 1998). This is not surprising, considering that many religions typically teach that a sinful *thought* is morally equivalent to the act itself. To demonstrate this fact, 95% of conservative Christian college students reported substantial guilt related to their sexual fantasies (Cato & Leitenberg, 1990; see also Ahrold et al., 2011). Interestingly, the same study found that those who felt the greatest guilt about

Table 6.4 Positive Functions of Sexual Fantasy

Function	Positive Effects
Provides a safe sexual outlet	By definition, a fantasy itself does not involve any physical contact, so there is no danger of pregnancy or STIs. By itself or combined with masturbation, fantasy can provide a completely safe sexual experience.
Enhances arousal	Probably the most common reason people fantasize is to increase and enhance their feelings of sexual excitement. Fantasy in a nonsexual setting has the potential to enhance future sexual encounters, and fantasy during masturbation or lovemaking can strengthen physical and psychological responses.
Relieves boredom	Sex therapists often recommend fantasy to spice up a couple's sex life that has become routine and predictable.
Treats sexual problems	Fantasy (usually combined with masturbation) is a cornerstone of sex therapy. It has been shown to increase sexual desire, arousability, orgasm, and sexual satisfaction in general.
Relieves sexual anxiety, guilt, and doubt	Nearly everyone at one time or another feels anxiety, guilt, or self-doubt about sexual issues. Fantasy offers an opportunity to think about sexual values, to replay sexual scenes, and to learn about one's sexual identity.

SOURCES: Alfonso, Allison, & Dunn (1992); Byers, Purdon, & Clark (1998); Carpenter et al. (2008); Cato & Leitenberg (1990); Davidson & Hoffman (1986); Leitenberg & Henning (1995); Masters, Johnson, & Kolodny (1995); Reinisch (1990); Renaud & Byers (1999); Striar & Bartlik (1999); Wilson (2010); Zurbriggin & Yost (2004).

their sexual fantasies also reported higher levels of sexual dissatisfaction and sexual problems in their relationships.

For a small percentage of people, fantasies can become so dominant in their thinking that they interfere with the activities of daily life. Others may find that they are unable to function sexually or respond to their partner without a specific sexual fantasy (Janus & Janus, 1993; Reinisch, 1990). An example of this might be the person who must fantasize about a previous lover in order to become aroused with the current partner. Although these are relatively rare occurrences, they can be very disturbing to the person experiencing them. Fortunately, counseling is usually successful in helping these individuals redirect their sexual thoughts in directions that are more acceptable and comfortable for them (Reinisch, 1990).

Sometimes when a person decides to share a sexual fantasy with a partner (such as thinking about a previous lover or a mutual acquaintance), the partner may as a result feel embarrassed, uncomfortable, angry, or threatened. This may create tension and has a negative impact on the relationship (Heubeck, 2007; Masters, Johnson, & Kolodny, 1995). Therefore, if sharing of fantasies is to occur in a relationship, couples should do so with great care and sensitivity to each other's feelings and sexual attitudes. The flip side of this is that under some circumstances, sharing fantasies can enhance intimacy and trust between two people and open up new areas for a couple to explore sexually. This positive aspect of fantasy sharing will be discussed shortly.

Most people have fantasies of "forbidden" or even illegal sexual acts but would never consider acting on them in real life (Byers et al., 1998). However, evidence exists that some criminal sex offenders (such as child molesters, rapists, and exhibitionists) fantasize about these illegal behaviors and activities more often than nonoffenders (Bartles & Gannon, 2013; Kleinplatz, 2007; Renaud & Byers, 2005), and this is especially true of repeat, predatory sex offenders (Deu & Edelmann, 1997; Woodworth et al., 2013). We cannot know for sure if the fantasy led to the criminal behavior or if the behavioral tendency created the fantasies, but it is important to be aware that the association exists in this population of sexual offenders.

Since you asked

2. Some of my sexual fantasies involve things I would never really do, so why do I feel guilty just thinking about them?

Sex Differences in the Experience and Content of Sexual Fantasy

If you were to ask around, most people would probably say that men and women differ a lot in how much they think about sex and in the content of their sexual fantasies. But is this common view supported by the available research? Several studies on sexual fantasy have focused on male–female differences, and the results have been mixed. As mentioned earlier, research has not found significant overall differences between the *number* of men and women who engage in sexual fantasy during intercourse or experience sexual daydreams.

Many sexual fantasy topics appear to be common for women and men. However, some differences in fantasy content have been revealed. When students at a large university in the western United States were asked whether they had ever experienced certain sexual fantasies, their reports revealed a more precise picture of these differences (Hsu et al., 1994). Table 6.5 lists some of the *specific* fantasies that were equally common for men and women and some that were reported more often by one sex or the other (usually more often by men).

Bear in mind that the differences in Table 6.5 are based on *averages* and that many people who have sexual fantasies contradict these statistics. Furthermore, although we see these differences in sexual *fantasy*, we do not necessarily find similar differences in *real-life* sexual experiences. This brings us to the important discussion of the relationship between fantasy and reality.

Table 6.5 Comparing the Content of Men's and Women's Sexual Fantasies

Sexual Fantasy Content	Men Experiencing (%)	Women Experiencing (%)
Fantasies with No Significant Sex Differences		
Touching or kissing sensuously	98	97
Naked caressing	93	92
Sex in unusual locations	89	82
Intercourse in unusual positions	87	82
Masturbating your partner	83	75
Performing sexual acts in a mirror	61	61
Having your partner watch you masturbate	56	53
Being tied up during sex	41	43
Fantasies With Statistically Significant Sex Differences		
Watching your partner undress	98	81
Oral sex	96	84
Sex with a virgin	85	32
Two or more lovers at the same time	76	45
Sex with a mysterious stranger	70	47
Getting married	52	71
Being rescued by a future lover	26	46
Homosexual fantasies if heterosexual or heterosexual fantasies if homosexual	19	33

SOURCE: Adapted from Hsu et al. (1994).

Fantasy Versus Reality

Sometimes sexual fantasies are about a sexual situation or activity that the fantasizer has experienced or would like to experience in real life. Some couples will share their fantasies with each other as a way of expanding their sexual repertoire and enhancing their relationship. However, many people report having sexual fantasies that they would never truly want to experience in real life (Byers et al., 1998; Dutton, 2009; Renaud & Byers, 1999). These fantasies may involve famous people; unusual, "forbidden," or "taboo" sexual activities; or situations that, in reality, might be frightening, dangerous, or illegal. The vast majority of those who experience these fantasies have no desire to act them out in real life (Masters et al., 1995).

One important example of a fantasy that is *not* a real wish is the forced-sex or "rape fantasy." Although some women (and some men) do have these fantasies, they have *no* wish whatsoever to be raped in reality—*no one ever does*! A *fantasy* of being seduced into a sexual act bears no resemblance to the terror, violation, and violence of real sexual aggression. In a fantasy, the *woman* is completely in charge; she experiences no pain; she suffers no real violence; and she has no reason to fear being injured or killed, as accompanies many rapes in real life (Shulman & Horne, 2006; Strassberg & Lockerd, 1998).

Often in fantasy rapes, the woman actually sees *herself* as the powerful one, so irresistible that the man cannot control himself, and she allows his *illusion* of his dominance to give *her* pleasure (Bivona, Critelli, & Clark, 2012; Strassberg & Lockerd, 1998). Obviously, in a real-life rape, none of this is true. For the victim, real-life rape is entirely about violence, dominance, and humiliation (we will discuss this topic in greater detail in Chapter 13, "Sexual Aggression and Violence: Rape, Child Sexual Abuse, and Sexual Harassment").

Clearly, it would be a serious mistake to assume that simply because people have certain sexual fantasies, they want to act them out in real life. On the other hand, are there ways in which some types of sexual fantasies may be shared or be acted out to enhance a couple's sexual life? The answer for most couples is "maybe."

Men and women use online sexual sites alone and together to enhance sexual arousal and enjoyment.

Using Sexual Fantasy to Enhance Your Relationship

One common problem in long-term relationships is sexual boredom and loss of sexual desire. A technique frequently suggested by relationship counselors and sex therapists (and often discovered by couples themselves) is for the couple to share and experiment with each other's sexual fantasies (Newberry et al., 2012; Striar & Bartlik, 1999). However, many couples are hesitant to share their fantasies out of fear of embarrassment or rejection. In addition, many people have not taken the time to think about exactly which of their sexual fantasies might be exciting to actually try out with their partner. "Self-Discovery: What Turns *You* On?" (on following page) offers you the opportunity to analyze your own fantasies and, if you are in a close, trusting relationship, to discover whether some of them might add some variety and spice to your sexual life with your partner.

Sexual Fantasy Online

Today, sexual fantasy has taken on significant new meaning due to the massive number of sexually explicit sites and images online. The majority of people of all ages have accessed these images, to varying degrees and for many different reasons.

The availability of these sites plays directly into the notion of sexual fantasy because virtually anything a person can imagine can be found on the Internet. You might argue that if someone is viewing sexual images, then they aren't really fantasizing, but, if you think about it, the image probably serves as a stimulus for greater fantasy rather than replacing the mind's ability and desire to fantasize. The possibility for online sexual content to cause relationship problems is well known, and is a separate discussion from its interplay with fantasy. The main point is that the availability of online sexual images and sexual fantasy go hand-in-hand.

Research has found that among college students, men and women access sexual sites online for several reasons (Shaughnessy, Byers, & Walsh, 2011). One reason is for solitary sexual arousal, often during masturbation. This use of the internet is more common among men than women. Another reason is not arousal-related per se, but involves learning about sex and seeking information. Men and women appear to go online with equal frequency for this purpose. Finally, the third reason: using the Internet to enhance the couple's arousal also appears to occur with equal frequency for men and women.

Masturbation

6.4 Discuss historical and present attitudes about masturbation, how this behavior affects couples in intimate relationships, and the possible health benefits of masturbation.

Masturbation usually refers to sexual activities performed on oneself, typically focusing on stimulation of the genitals to orgasm. Another term for this is *autoeroticism* ("auto" meaning "self"). Most people masturbate, but it is one of the most secretive, embarrassing, and taboo of all sexual topics (Halpern et al., 2000; Laqueur, 2004). Why? Because most Western cultures are still feeling the effects of hundreds of years of extremely negative attitudes about masturbation (see "Historical Perspectives" at the beginning of this chapter). We will address this history in a moment, but first let's look at a few statistics.

masturbation
Any sexual activity performed on oneself by oneself, typically focusing on manipulation of the genitals to orgasm.

Self-Discovery

What Turns *You* On?

Below is a list of various intimate activities that are the focus of some people's sexual fantasies. To get an idea of your own fantasies, rate each activity, and if you so choose, share your answers with your partner, now or in the future. If you are in a trusting, caring relationship, you may want your partner to complete this exercise as well. This exercise requires a great deal of mutual trust and respect, so approach it carefully, and be sure to read *all* the instructions!

Use the following ratings key to respond to each item on the scale. Be honest with yourself; no one ever has to see your answers but you.

Key

1 = I think I would like this a lot.

2 = I think I would probably like this.

3 = I'm not sure, but I might be willing to try it.

4 = I'm sure I would not like this at all.

NOTE: This is an intensely personal exercise, so you might want to use a separate sheet of paper for your answers.

Intimate Activity	Your Rating			
	1	2	3	4
1. New or unusual positions for intercourse				
2. For me to initiate lovemaking more often				
3. For my partner to initiate lovemaking more often				
4. A "quickie" once in a while				
5. More touching before intercourse				
6. Lovemaking to last longer				
7. To watch my partner undress				
8. For my partner to watch me undress				
9. More oral stimulation of me				
10. More oral stimulation of my partner				
11. To watch my partner masturbate				
12. For my partner to watch me masturbate				
13. To take showers or baths together				
14. To be touched to orgasm				
15. To touch my partner to orgasm				
16. Some anal activities				
17. To experiment with sex toys together				
18. To explore erotic materials together (videos, etc.)				
19. To make love in new and different locations				
20. To make love in front of a mirror				
21. More sex talk during lovemaking				
22. To wear sexy clothes for lovemaking				
23. To have my partner wear sexy clothes for lovemaking				
24. To make love outdoors				
25. To share sexual fantasies				
Add a few of your own, if you wish:				
26.				
27.				
28.				

How to evaluate your ratings: There is no overall score on this exercise. The point is to consider how you rate each item individually. You will begin to discover for yourself what fantasies turn you on, which ones you are willing to share with your partner, and which ones you might want to try.

Sharing your answers with your partner: If you are currently in an intimate relationship and both you and your partner have filled out this scale, you can decide together whether you want to share your answers. If you do, remember to be open and nonjudgmental of each other's answers. Some of your answers will match and some probably won't. But you will learn something about each other that may add to your enjoyment of your relationship and shared sexual intimacy.

Ever since Kinsey's 1940s and 1950s survey data on men and women, researchers have been trying to determine just how many people masturbate. Percentages vary a great deal from study to study, depending on who the subjects were and how they were surveyed. Kinsey's early work found that 92% of men and 58% of women reported that they had masturbated to orgasm (Kinsey, Pomeroy, & Martin, 1948; Kinsey et al., 1953). Another survey conducted at a Southern California urban university found rates of 71% for women and 83% for men (Hsu et al., 1994). The National Health and Social Life Survey, published in the mid-1990s, found that about 60% of men and 40% of women between the ages of 18 and 60 reported masturbating during the previous year (Das, 2007). The most recent data available, from the National Survey of Sexual Health and Behavior (Herbenick, et al., 2010b; Reese et al., 2010), reported that 81% of men and 61% of women between the ages of 16 and 40 reported masturbating (see Table 6.2). How would you account for the variations in all these findings? Think about this for a moment and then read "Evaluating Sexual Research: Self-Reporting of Personal Information" (on following page).

Changing Attitudes about Masturbation

Few, if any, human sexual behaviors have a history of such terrible "press" as has masturbation, as demonstrated by the quotes from William Walling at the beginning of this chapter. But times have changed, right? When discussing masturbation, the answer to that question is, "Yes and no." Few people believe today, as many did a century ago, that masturbation will lead to hairy palms, genital cancer, blindness, insanity, hysteria, sterility, impotence, asthma, reduction or increase in penis size, loss of future enjoyment of sex with a partner, or instant death from too much sexual excitement. And fewer people still would buy the notion suggested by J. H. Kellogg in 1888 that masturbation can be prevented by eating plain, whole-grain foods such as cornflakes (Michael et al., 1994).

Nevertheless, negative views and taboos concerning masturbation can still be found in far more recent events. It was not until 1972 that the American Medical Association declared masturbation to be a normal sexual behavior. Furthermore, even though most people consider masturbation to be a natural part of life (Dodson, 2002; Janus & Janus, 1993), a significant percentage of teens and adults feel guilty about masturbating (Kaestle & Allen, 2011; Michael et al., 1994). In 2003, a *Doonesbury* cartoon strip that made reference to the medical news that masturbation may have a preventive effect on prostate cancer was pulled from 300 newspapers in the United States because it was considered offensive and in poor taste (Evans, 2003).

Perhaps the most serious example of continuing negative attitudes about masturbation in modern history occurred in 1994 when then U.S. Surgeon General Joycelyn Elders was addressing a United Nations delegation on the spread of sexually transmitted diseases. Following her talk, a member of the audience asked if she would support teaching masturbation as a way of discouraging children from trying riskier forms of sexual behavior. Elders replied, "With regard to masturbation, I think that is something that is part of human sexuality and part of something that should perhaps be taught" (Popkin, 1994). The storm of controversy that her comment set off was so intense that President Clinton, who had appointed her

Centuries ago, many people believed that masturbation would lead to various forms of mental illness, including a condition known as "masturbatory hysteria."

Evaluating Sexual Research

Self-Reporting of Personal Information

Imagine that you've been asked to be part of a sex survey and a researcher comes to your house to interview you and ask you a lot of questions. One line of questioning goes something like this: "Have you ever masturbated to orgasm? When was the last time you masturbated? How often do you masturbate?" How likely is it that you would be completely honest in your answers? What if the same questions were asked in an anonymous, written questionnaire? Might your answers be different? Many people are embarrassed talking about (or even thinking about) sexual topics, and many are especially uncomfortable discussing masturbation (Catania, 1999; Halpern et al., 2000). Their discomfort, guilt, or shame about this or other behaviors may cause them to withhold information or to be less than truthful in their answers, creating bias in the survey's results.

This bias may help explain some of the differences seen in the statistics about masturbation or other behaviors reported in the various studies cited in this chapter. Perhaps students in the Midwest really do masturbate less than students in California or New York, and maybe people in general are masturbating less today than in past decades. But some of these differences across studies could also be due to higher levels of guilt or embarrassment among Midwestern or New England students, leading to a greater hesitancy to report that they have masturbated or to underreport how often they do so. In the case of a well-known 1990s survey, *The Social Organization of Sexuality: Sexual Practices in the United States* (Laumann et al., 1994), researchers actually went to people's homes for interviews, and in some cases, it was reported that spouses and children of the respondents were also present at the time. You can see how this situation might lead to less than truthful responses given to the interviewers to sensitive questions about behaviors such as masturbation (Catania, 1999; Reiss, 1995). This is not to say that the researchers' findings in this survey or in any survey are wrong. The point is that the statistics reported in this survey and in all surveys of sexual behavior do not tell us how many people masturbate or engage in any other sexual activity but only how many are *willing to report* that they do.

15 months earlier, had to dismiss her from her post. Apparently, the notion of teaching young people about masturbation was still too radical to receive "official" support in American society. Since that time, no U.S. surgeon general has discussed the topic of masturbation. Elders went on to write her autobiography and is coauthor of a book about the value of a comprehensive sex education, *Health and Healing for African Americans*, which includes information about masturbation (Levine & Elders, 2002).

In spite of all the past negative attitudes and news about masturbation, an increasing amount of evidence suggests that the United States and many other countries are beginning to recognize its many beneficial and healthful aspects, and acceptance of the practice is growing. In 1999, even the Catholic Church, while continuing to declare masturbation wrong, no longer deemed it a mortal sin (Sked, 1999). These cultural changes can also be seen in various popular-press books (the most notable: *Fifty Shades of Grey*), magazine articles, and TV shows such as "The Big Bang Theory" and one particularly well-known episode of "Seinfeld." Furthermore, beyond these changing popular beliefs and attitudes, sexuality researchers, therapists, and educators, recognizing that no evidence exists that masturbation is harmful in any way for most people, have begun to promote the physical and psychological health *benefits* of masturbation. These are detailed in Table 6.6.

Masturbation Within an Intimate Relationship

A common belief about masturbation is that it serves as a substitute for sexual activities with a partner. This belief then leads to other assumptions—that masturbation within an intimate relationship must be a sign of dissatisfaction with one's partner, that people masturbate only when they don't have a sexual partner, that people who masturbate are sexually deprived, or that masturbation is simply an outlet for people who do not have any other form of sexual release. The truth is that *all* of these assumptions are *wrong*. Instead of *substituting* for sex with a partner, masturbation is typically part of a healthy sexual life regardless of an individual's relationship status and, for many, is a comfortable way to feel sexually satisfied.

Since You Asked

3. When someone is feeling very sexually frustrated, how can a person satisfy their sexual needs?

Table 6.6 The *Benefits* of Masturbation

Benefit	Function
Sexual self-discovery	Masturbation provides a "laboratory" for exploring and experimenting with one's own sexual feelings and sensations. This learning process is important for achieving sexual satisfaction with a partner and in life in general.
Release of sexual tension or frustration	Masturbation may be used to alleviate feelings of sexual frustration resulting from romantic and sexual activities with a partner that do not lead to orgasm or general sexual tensions that may build up over time.
Enhancement of sexual interactions with a partner	Masturbation appears to be related to a person's sexual satisfaction with a partner now and in the future. Research has shown that people in relationships who masturbate are more sexually satisfied with their partner and their relationship.
Resolution of sexual problems, including inhibited sexual desire, premature ejaculation, difficulty with orgasm, erectile problems, general arousal difficulties, and delayed orgasm (these are discussed in Chapter 7, "Sexual Problems and Solutions")	Many common sexual problems are treated with the help of masturbation. The therapeutic benefits of masturbation exercises may be part of a treatment program prescribed by a professional counselor or a self-help strategy individuals or couples may discover on their own.
Orgasm	Masturbation has been reported, especially by women, as providing the most reliable and most intense orgasms.
Relief from stress and depression	Many people who masturbate find it helpful in relieving stress, tension, sleeplessness, and depressive symptoms that are not necessarily related to sex activities per se.
Relief from menstrual pain	For some women, masturbating before or during their period reduces or even eliminates menstrual cramps.
Compensation for a disparity in a couple's levels of sexual desire	A common problem couples face is that one wants more frequent sexual contact than the other. While couples may solve this problem using various strategies, masturbation can often help reduce the guilt, demand, and frustration often felt by couples in this situation.
Safe sex	Solo masturbation poses zero risk of STIs or pregnancy. You cannot give yourself an STI or make yourself pregnant. Even when masturbation activities are shared by a couple, it is one of the lowest-risk sexual behaviors.

SOURCES: Davidson & Moore (1994); Kay (1992); Kelly, Strassberg, & Kircher (1990); Levin (2007); Mahoney (2006); Masters & Johnson (1974); "The Politics of Masturbation" (1994); Tiefer (1998).

Research clearly demonstrates that many people do not stop masturbating when they enter a sexual relationship. In fact, most people continue to masturbate after entering a sexual relationship, living together, or marrying (Herbenick et al., 2010a). Moreover, those who are having the most frequent sexual intercourse with their partner are the most likely to masturbate (Michael et al., 1994). People who masturbate are more likely than those who don't to be both physically and emotionally satisfied with their overall sex lives (Davidson & Moore, 1994; Hook, 2013). Taken together, the evidence is quite strong that masturbation is far more than an activity people simply do in place of other sexual activities or a sign of a sexual problem; rather, it is an integral part of the diverse repertoire of normal human sexual activities. "In Touch with Your Sexual Health: Techniques of Masturbation," explores this activity in greater detail.

Since You Asked

4. Does masturbation detract from one's sexual drive or pleasure?

In Touch with Your Sexual Health

Techniques of Masturbation

One important point should be made before discussing techniques of masturbation: It is perfectly OK *not* to masturbate. Although masturbation is discussed more openly in society and its benefits are more widely known, no one should interpret this as evidence that something is wrong or abnormal with choosing not to masturbate. As for all sexual behaviors, this is each person's individual choice. If you are not comfortable with the idea of masturbation and you have chosen to refrain from that activity for any reason (or for no reason at all), that is a completely legitimate decision, and you should never feel pressured to do anything sexual that feels wrong or uncomfortable to you. If you already masturbate regularly, you probably know what feels best and how you prefer to do it, so the techniques discussed here may be of less value to you than to others who have never masturbated or are in the early stages of experimentation. When people masturbate, their bodies typically progress through the same pattern of sexual responses as would happen if they were with a partner.

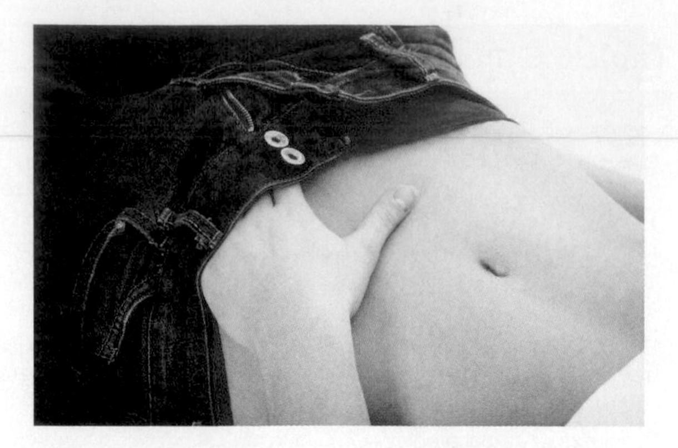

Masturbation has many benefits for men and women, including greater sexual self-awareness, release of sexual tension, and enhanced sexual interactions with a partner.

For a better understanding of the bodily changes that usually accompany masturbation, you may want to review Chapter 3, "The Physiology of Sexual Response."

Following are the aspects often reported as part of many people's pleasurable masturbation experiences.

- **Privacy.** First, be sure you have a comfortable place to be by yourself where you are sure you will not be interrupted or intruded upon.
- **Time.** Allow plenty of time so that you do not need to feel rushed or pressured in any way.
- **Sexual Self-Awareness.** To allow yourself pleasurable masturbation experiences, become familiar with your own sexual anatomy. Take a few minutes and examine your own body. You may find that a small hand mirror is helpful for seeing everything. As you do this exam, touch parts of your sexual anatomy gently and allow yourself to become familiar with them. If you find this exercise difficult and embarrassing, as many of you will, just proceed slowly and do as much self-examination as feels comfortable for now. As time passes and you continue this activity, you will become increasingly at ease with your body.
- **Arousal.** As you examine and touch yourself, you may begin to become sexually aroused. The physical changes accompanying masturbation parallel those during sex with a partner. A woman may notice her clitoris becoming enlarged or an increase in vaginal lubrication. A man's penis becomes erect, and he may be aware of changes in the skin of his scrotum and the position of his testicles.
- **Techniques of Stimulation.** Of course, each individual has unique preferences, but some similarities may be found in typical styles of masturbation among men and women relating to position, touch, and orgasm.
- **Position.** Most men and women masturbate while lying face up. However, variations may include sitting in a chair or on the edge of a bed, kneeling, standing, or in the shower or bath (see Figure 6.1). Some people masturbate lying face down and rubbing against the bed or an object.

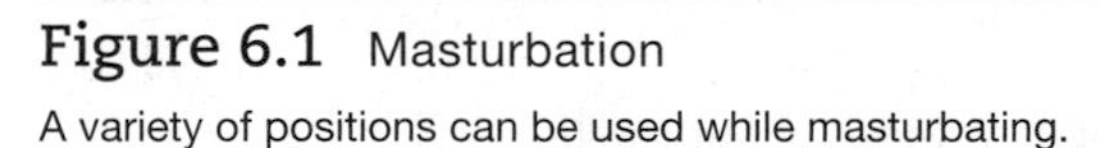

Figure 6.1 Masturbation

A variety of positions can be used while masturbating.

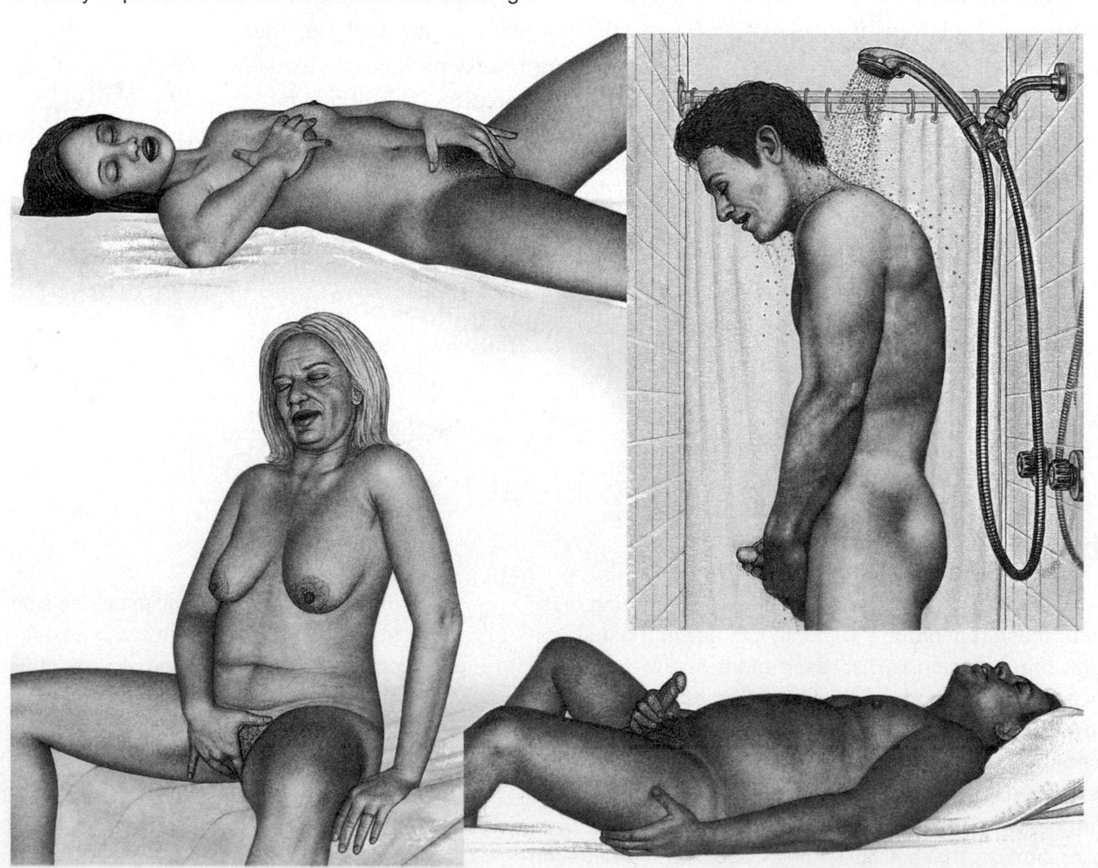

- **Touch (Men).** In general, men grasp the penis with the whole hand or between the thumb and one or more fingers. They usually stroke the shaft and glans of the penis in an up-and-down motion. Men who are uncircumcised will often use the foreskin as a "sleeve" for movement over the glans.

The pressure and speed of stroking varies among men, but usually these will both increase as the man becomes increasingly aroused and closer to orgasm. A man should be aware that as he becomes aroused, the Cowper's gland will secrete a clear, slick fluid that may appear at the urethral opening before he ejaculates. This pre-ejaculate is completely normal. Some men will stimulate their testicles, anus, or nipples while masturbating, though this is the exception rather than the rule.

- **Touch (Women).** Techniques women use for masturbation are often significantly more varied. Most women touch, stroke, rub, massage, or otherwise stimulate the outer and inner labia and the area on or near the clitoral glans. Some women enjoy direct stimulation of the glans; others find this uncomfortable due to the extreme sensitivity of the clitoral glans. Contrary to popular belief, very few women ever insert anything into the vagina while masturbating. Also, only a small percentage of women stimulate their breasts and nipples during masturbation. As a woman becomes aroused, she will notice an increase in vaginal secretions. The woman may use this fluid to lubricate the rubbing or massaging of the vulva area.
- **Orgasm.** Some people may not experience orgasm the first few times they try masturbating. Eventually, however, nearly everyone who masturbates on a regular basis has an orgasm every time, and most people feel that orgasm is the goal, if not the main point, of masturbating (for most men, this will invariably include ejaculation of semen). For someone who has never experienced an orgasm, the strength of the physical and psychological sensations during masturbation as orgasm approaches can sometimes feel too intense and even a little frightening. Remember, masturbation is about self-pleasuring. If at any time you feel uncomfortable, just stop. As you experiment over time, the sensations will become increasingly predictable and enjoyable.
- **Artificial Lubrication.** Many men and women will often add some lubrication while masturbating to allow for easier movement of the hand against the genitals, which often intensifies sensations. Water-based lubricants specifically designed for sexual use are probably best for this purpose because they have excellent lubricating properties and will usually not cause irritation to sensitive body areas. However, many people have found that hand creams, lotions, or body oils work just as well.
- **Vibrators.** Many people enjoy the sensations produced by incorporating a vibrator into their masturbation experiences. In addition, some women who have had difficulty reaching orgasm report that using a vibrator (usually near the clitoral area, not inserted into the vagina) has allowed them to have orgasms more easily, which in turn helps them to experience orgasms more readily without the vibrator (Lankveld, 2009). Electric and battery-operated vibrators have become increasingly popular and more readily available in recent years. They can be purchased at various retail outlets, through mail order catalogs, on the Internet, and at "sex-toy sales parties" in homes or in college residences. They come in a variety of shapes and sizes, some of which are shown in Figure 6.2. In addition, many research articles and entire books have been written describing vibrators and their sexual advantages (Lankveld, 2009: Venning & Cavanah, 2003; Winks & Semans, 2002).

Figure 6.2 Sexual Vibrators

Many people find that sexual vibrators enhance their masturbation (or partnered) experiences.

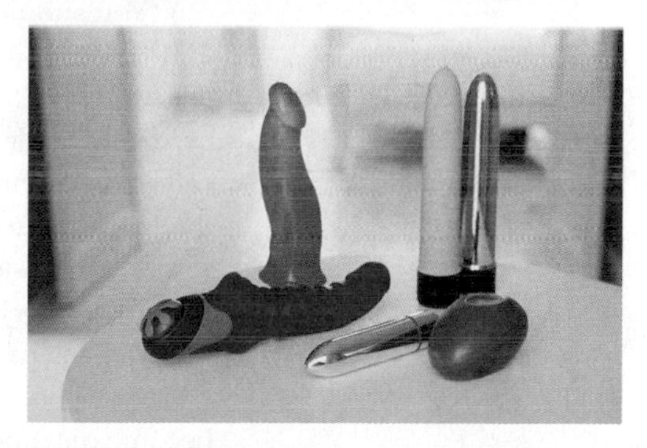

Erotic Touch

6.5 Summarize the role of erotic touch (kissing; breast and nipple touching; and genital touching) as an integral part of couples' sexual activities

Erotic touch refers to intimate or sexual touching between partners, usually with the hands, for the purpose of sexual arousal and sharing sexual or sensual pleasure. Specifically, the activities discussed in this section include kissing, sensual massage, caressing of the breasts and nipples, and genital touching. Keep in mind, however, that any type of touch in any location on the body may be sexy and arousing to couples, depending on their personal preferences.

Many people view erotic touching as "foreplay," a routine step on the path to a "sexual goal" (typically, intercourse for heterosexual couples or anal or oral sex for same-sex couples), as expressed in this statement: "We kiss for a while, then we touch each

erotic touch

Intimate or sexual touching between partners, usually with the hands, for the purpose of sexual arousal and sharing sexual or sensual pleasure.

Since You Asked

5. I believe that foreplay is the best part of sex. Sometimes I like it more than intercourse. Is that strange?

other for a while, and then after we touch each other's genitals for a while, it's time for 'real sex.'" However, touch is far more than that. People need to touch and be touched. Research has clearly demonstrated that human infants need close physical contact with others in order to thrive or even to survive. Throughout life, touching continues to be an important means of communicating closeness, affection, and sexual intimacy. The famous sexuality researchers Masters and Johnson, in their book *The Pleasure Bond* (1974), are very clear on this point. They believed that couples committed to a long-term relationship should value touch in itself as an important part of intimacy and not "simply" one step in the process that leads up to intercourse or any other sexual behavior. Erotic touch, in their conceptualization, promotes a special "skin-to-skin" bond that brings enhanced closeness to an intimate relationship, and couples should not expect that the touching should necessarily flow into other sexual activities (Masters & Johnson, 1974).

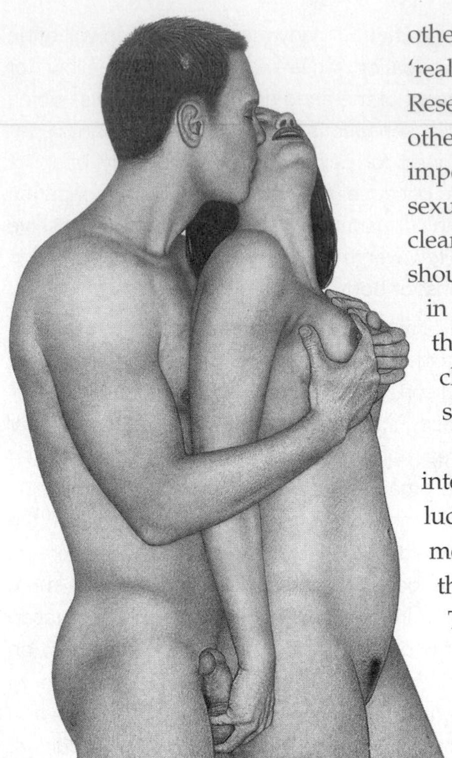

Figure 6.3 Erotic Touch

Erotic touch is a pleasurable, intimate behavior by itself and need not necessarily lead to other sexual activities.

This does not mean that touch should not lead to other sexual activities, such as intercourse, and for many couples, it often does. But when touching is seen as *only* a prelude to other sexual activities, it often loses much of its pleasurable value and becomes no more than a way of signaling a desire or demand for something else. Many couples report that early in their relationship, touching was central to their physical intimacy together. Touch was extremely erotic, almost electrically charged. Then, as intercourse became a more common part of the sexual relationship, touching began to take a backseat, so to speak, and became a much more perfunctory or superficial part of their lovemaking. This shift in focus from whole-body sensuality to purely genital-focused behaviors sometimes leads to sexual problems discussed in the next chapter. Interestingly, one of the exercises recommended for helping couples who are experiencing sexual problems is to return to a focus on touch, rather than on intercourse.

How to Touch Your Partner

How should you touch your partner sexually? The answer to that question is obvious. You should touch your partner in the way your partner likes to be touched! (See Figure 6.3.) But how do you know exactly what your partner likes? Sometimes you can tell by changes in movements, facial expressions, breathing, and vocalizations of pleasure. However, unless you are a mind reader (and you're not), these nonverbal clues can often be misleading. Consider the following common scenario of a couple's sexual interaction. He enjoys a firm, rapid style of touching, similar to a massage, but she prefers a lighter, slower, stroking touch. In an attempt to communicate to each other what they like, he will touch her more and more firmly and she will touch him increasingly lightly. Each has the feeling that the other just isn't getting it. So they try harder to demonstrate what they like. He touches harder still, she lightens up even more, until both become disappointed and frustrated (this is called "touch mirroring" and is discussed further in Chapter 7). What's the solution? The best way to discover your partner's preferences for being touched (or for any intimate activity) is to *ask*.

One way to communicate what you like is a novel idea: *talk*. Tell your partner in words when he or she does something that feels especially good. Saying, "That feels really good," or "What you were just doing is exactly how I like it," helps your partner learn the behaviors that most satisfy you. Another strategy is to bring up the issue at a time when you are feeling close and comfortable but not being sexual. Be positive and talk about your feelings. Say, "I really like it when you touch me slowly and lightly," *not* "You really don't touch me the way I like." Criticism will get you nowhere! Hopefully, then, the next time you are involved in erotic touching with your partner, your discussion will be put into practice. Finally, another way to let your partner know what you like is to demonstrate. An exercise devised by Masters and Johnson (1970) and often suggested by sex therapists is for you to guide your partner's hand over your body to illustrate how and where you most enjoy being touched. Many people

Mutual sensual massage often enhances sexual feelings between partners.

are embarrassed by this at first, but slowly it will become more comfortable and will actually open up new lines of sexual communication (see Chapter 4, "Intimate Relationships," and Chapter 7, "Sexual Problems and Solutions," for more discussion of these issues).

Figure 6.4 Kissing
Kissing is one of the most common intimate activities in most Western cultures.

Kissing

In most Western cultures, kissing is a very common and popular form of erotic touching (see the feature "Sexuality and Culture: Cultural Variations in Kissing Behavior"). Many forms of kissing are popular, from a friendly or familial peck on the cheek to more arousing and romantic kissing known as French, deep, or soul kissing (see Figure 6.4 on page 203). Kinsey's work in the 1940s and 1950s found that nearly all couples (87%) in the United States include deep kissing as part of lovemaking (Kinsey et al., 1948).

More recent research has shed some interesting insights into romantic kissing. Many couples are aware that the desire, frequency, duration, and preferred type of kissing are not the same for each partner, and this is true in same- and opposite-sex couples. And, indeed, there are many differences in individual tastes (yes, it's a pun) in kissing. One study looked at kissing behavior following lovemaking (Hughes & Kruger, 2011). Apparently men tend to enjoy and initiate kissing before lovemaking, whereas women desire and initiate more kissing after intercourse. In addition, intimate talk and kissing were rated by both sexes as more important before intercourse with a long-term partner, whereas cuddling and professing one's love was rated more important after sex.

Another provocative study examined the perceived purpose or function of intimate kissing (Wlodarski & Dunbar, 2013). Results indicated that women, far more than men, use kissing early in a relationship to help them judge the potential of a male partner. They claimed that an initial kiss influenced their level of attraction toward a potential mate (no pressure, though!). Also, kissing was important in long-term relationships (especially by women) and a greater level of kissing was linked to overall satisfaction with the relationship for both sexes. Probably, the most surprising finding in this study was that kissing was not seen as a major pathway to sexual arousal.

Kissing is, obviously, one of the most popular sexual activities among college students. One study of college students that did include kissing found that 88% reported deep kissing within the past month and over 93% said they fantasized about it, making deep kissing the most common fantasy (Hsu et al., 1994). Moreover, couples that kiss more report greater relationship and partner satisfaction (Gulledge et al., 2003). Kissing is a very safe behavior in terms of STI transmission; the herpes virus (usually oral herpes—Type 1) may possibly be transmitted if one person has active sores on the lips or around the mouth. Without an obvious outbreak, transmission is still possible (called *asymptomatic transmission*), but the chance is greatly reduced (Johns Hopkins, 2014).

Touching the Breasts and Nipples

Manual or oral stimulation of the breasts and nipples can be very sexually arousing for both men and women (see Figure 6.5 on following page). Women, in general, are more sexually aroused by having their breasts and nipples touched than are men, but the difference is not as great as most people might believe. Over 80% of women and 50% of men report that breast stimulation during lovemaking increases sexual arousal (Levin & Meston, 2006). A small percentage of women have reported that they are capable of having an orgasm through breast and nipple stimulation alone (Levin, 2006; Masters & Johnson, 1966). A commonly believed myth is that larger female breasts are more sexually sensitive to touch, but some research indicates just the opposite: that smaller breasts may be slightly more sensitive to erotic touch than are larger breasts (Levin, 2006).

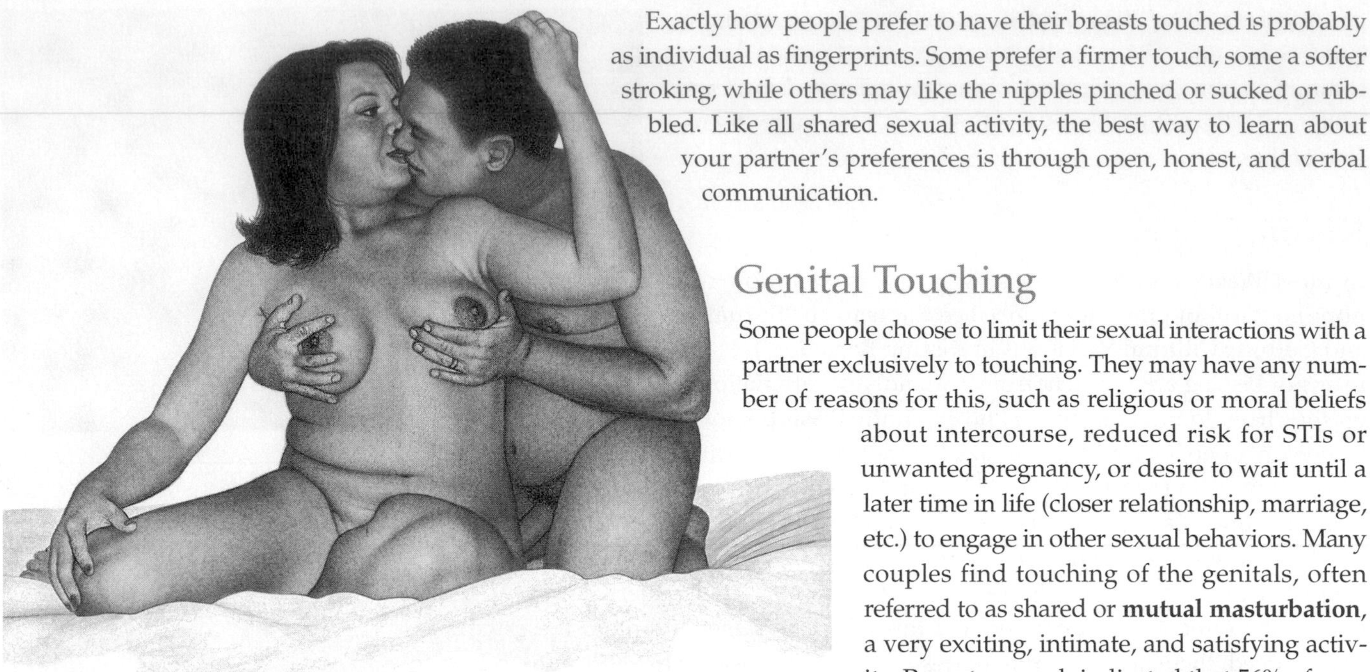

Exactly how people prefer to have their breasts touched is probably as individual as fingerprints. Some prefer a firmer touch, some a softer stroking, while others may like the nipples pinched or sucked or nibbled. Like all shared sexual activity, the best way to learn about your partner's preferences is through open, honest, and verbal communication.

Figure 6.5 The Breasts and Sexual Behavior

Most women and many men find breast and nipple stimulation very sexually exciting.

Genital Touching

Some people choose to limit their sexual interactions with a partner exclusively to touching. They may have any number of reasons for this, such as religious or moral beliefs about intercourse, reduced risk for STIs or unwanted pregnancy, or desire to wait until a later time in life (closer relationship, marriage, etc.) to engage in other sexual behaviors. Many couples find touching of the genitals, often referred to as shared or **mutual masturbation**, a very exciting, intimate, and satisfying activity. Recent research indicated that 56% of men and 32% of women between 18 and 49 reported experiencing mutual genital touching to orgasm (Herbenick et al., 2010a).

mutual masturbation

Partners' touching of each other's genitals, often to orgasm and enjoyed as a sexually intimate and satisfying activity.

Again, as with all sexual behavior with a partner, everyone has specific preferences for genital touching. As mentioned earlier, you should avoid assuming that your partner likes the same style of genital touching as you because you'll probably be wrong. As usual, the best way for a couple to learn about each other's preferences is to *talk* about them and, if the relationship is open and comfortable, to show and tell. That said, here are some general guidelines for genital touching.

Both men and women usually prefer a gentle, slower touch at first, with increasingly firmer pressure and faster stroking as excitement builds. Most men like to have the erect penis encircled by their partner's hand and stroked up and down. Usually, there will be some contact with the glans, but some men find the tip of the penis too sensitive for focused stimulation. Some men enjoy gentle touching and massaging of the scrotum, testicles, and anal area as well. Most women enjoy having the outer and inner labia gently massaged or stroked and the area of the clitoris stimulated. Some women like more focused touching of the clitoris, while others prefer more general, overall massaging of the vulva. The glans of the clitoris is extremely sensitive for some women, making direct stimulation uncomfortable or even painful. Many women also enjoy some stimulation of the outer part of the vagina and anus (the vulva and vagina should not be touched by the same part of the hand that has stimulated the anus because bacteria that exist normally in the rectum may be transferred from the anal area, creating a risk of vaginal or urinary tract infections). Most women do not prefer a finger or fingers inserted deeply into the vagina during mutual masturbation, and some even find it uncomfortable or distracting ("Female Masturbation," 2001; Hite, 1976; Masters & Johnson, 1979).

Figure 6.6 Touch and Orgasm

A couple can experience intense pleasure, including orgasm, from mutual erotic touching.

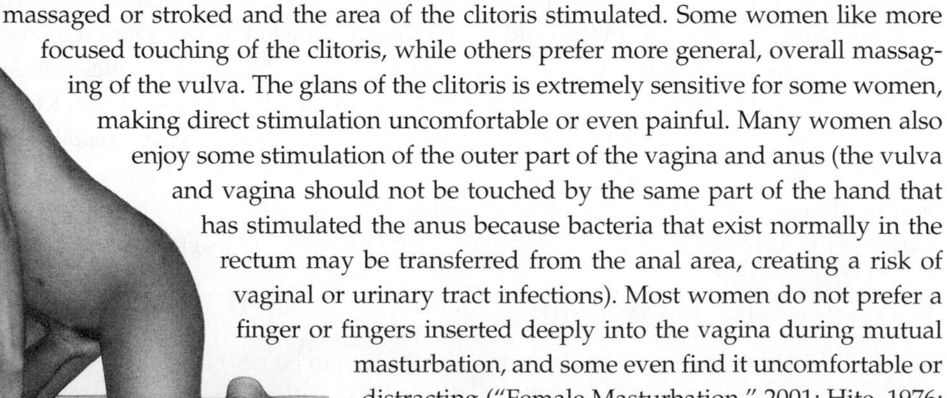

Often, but not always, genital touching leads to orgasm. Many couples find this to be a very pleasurable and satisfying form of sexual intimacy and release. It allows those who are in relationships but do not feel ready for other activities such as oral sex or intercourse to enjoy the intense pleasure and closeness that sexual intimacy can provide (see Figure 6.6).

Finally, genital touching, even to orgasm, is relatively free from the risks of sexually transmitted infections. To be confident of this, however, couples must make sure that semen and vaginal fluids come into contact only with skin that is healthy and unbroken and not with the genitals.

Oral Sex

6.6 Summarize the sexual activities of fellatio and cunnilingus, people's feelings about them, and the risks of spreading sexually transmitted infections through these behaviors.

Oral sex, like masturbation, is something most people have experienced but few are comfortable talking about openly. As you can see in Table 6.2, a large percentage of people between the ages of 16 and 40 have engaged in oral sex at least once during the past year. Other research has found that nearly 80% of people have engaged in oral sex and almost 90% feel that it is a normal and acceptable sexual activity (Billy et al., 1993; Janus & Janus, 1993; Leichliter et al., 2007; Laumann et al., 1994). In research with college students, the percentages are even higher, with over 90% of men and women reporting experience with oral sex (Hsu et al., 1994; von Sadovszky, Keller, & McKinney, 2002). In another study of over 1,600 young adults (18 to 24 years old), 78% reported receiving oral sex and 57% reported performing oral sex (Ompad et al., 2006).

Oral sex involves two distinct behaviors, depending on the recipient of the activity. **Cunnilingus** is oral sex performed on a female, and **fellatio** is oral sex performed on a male. Each of these behaviors also involves two components: giving oral sex and receiving it. Approximately the same percentage of college-age men and women report giving and receiving oral sex with opposite-sex partners (Herbenick et al., 2011).

cunnilingus
Oral sex performed on a female.

fellatio
Oral sex performed on a male.

Cunnilingus

The word *cunnilingus* is derived from the Latin *cunnus*, meaning "vulva," and *lingere*, meaning "lick." The behavior itself, however, often involves many other activities besides licking. Techniques of cunnilingus are extremely varied, and each woman is different in her preferences for receiving oral sex. In general, the vulva is stimulated by the mouth, lips, and tongue of the woman's partner (see Figure ww). This may include kissing, licking, sucking, and sometimes light biting of the outer and inner labia, the clitoral hood, and the shaft and glans of the clitoris. Some women also enjoy the tongue inserted into the outer part of the vagina. The contact of the mouth with the vulva may range from very light, slow stroking with the lips and tongue to faster, firmer movements of the tongue and strong sucking of the labia and clitoris. Many women enjoy

Figure 6.7 Cunnilingus

Cunnilingus is the stimulation of the vulva with the mouth, lips, and tongue.

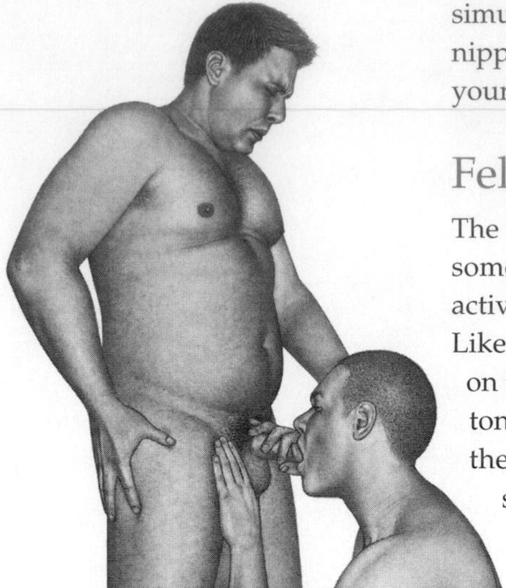

simultaneous manual stimulation of the outer vagina, G-spot area, anus, or breasts and nipples during oral sex (if necessary, refer to Chapter 2, "Sexual Anatomy," to refresh your memory on these structures for both cunnilingus and fellatio).

Fellatio

The word *fellatio* originates from the Latin *fellare*, meaning "to suck." This is actually something of a misnomer because sucking is usually not a major component of this activity, and vigorous sucking may even become uncomfortable or painful to the man. Like women, men vary a great deal in exactly how they like oral sex to be performed on them. In fellatio, a man's genitals are stimulated by his partner's mouth, lips, and tongue (see Figure 6.8). This may include kissing and licking of the penis, including the shaft, glans, and frenulum. Often the entire penis is taken into the mouth and stimulated by moving the penis in and out of the mouth, caressing it with the tongue, and perhaps sucking lightly. Some men also enjoy oral stimulation of the scrotum and testicles, including gentle sucking of the testicles through the scrotum. Some men enjoy simultaneous stimulation of other parts of the body during oral sex, including the testicles, anus, or breasts and nipples.

Figure 6.8 Fellatio

Fellatio involves stimulation of the male penis and surrounding genital area with the mouth, lips, and tongue.

Since You Asked

6. Do women get more pleasure from intercourse or oral sex? How and why?

Why People Like Oral Sex

Most people who engage in oral sex, whether giving or receiving, find it to be an extremely intimate, pleasurable, and exciting activity. A clear majority of men *and* women rate giving and receiving oral sex as somewhat or very appealing (Gates & Sonenstein, 2000; Laumann et al., 1994). Moreover, oral sex has become far more common over the past 20 years. Research indicates that over 40% of young heterosexual adults view oral sex to be as intimate as intercourse and 9% see it as more intimate (Malacad & Hess, 2010). In addition, the majority of those who engage in oral sex view it in positive terms and enjoy it.

People report a wide variety of reasons that they enjoy oral sex. Some couples feel that giving and receiving oral sex creates more intense feelings of intimacy than any other sexual behavior, even intercourse. Others enjoy the feeling of the mutual enjoyment and acceptance of their partner's entire body that oral sex represents. Most people who engage in oral sex find it a highly physically arousing behavior, in which stimulation can be focused exactly where and how it feels the best. Some find that orgasms through oral sex feel different or better than masturbation, touching, or even intercourse. Many women and some men report that the best or only way they have orgasms with a partner is through oral sex. Shere Hite, in her reports on female and male sexuality, quotes women as saying:

> "A tongue offers gentleness and precision and wetness and is the perfect organ for contact. And besides, it produces sensational orgasms!"
>
> "It really puts me in orbit, and I always have an orgasm!" (Hite, 1976, p. 234)

Hite also offers the following quotes by men about oral sex:

> "This is by far my favorite sexual activity. I always orgasm and it is more intense for me than an intercourse orgasm."
>
> "Fellatio is the best. Besides the physical sensations, I feel the woman really loves and appreciates my body." (Hite, 1981, p. 539)

Another reason that many couples enjoy oral sex is that it can serve as a substitute for intercourse. Some couples who have made a decision to avoid intercourse due to various reasons including personal moral beliefs, the choice to be safe from unwanted pregnancy, or a desire to wait until a later point in time for intercourse, will engage in oral sex as a way of exploring sexual intimacy without intercourse. However, contrary to the belief held by many people, especially adolescents, oral sex is *not* safe sex and can easily transmit most sexually transmitted infections (see "In Touch With Your Sexual Health: Oral Sex and STIs").

Why Some People Object to Oral Sex

In contrast to people who enjoy oral sex, others are uncomfortable with oral sex and some may be averse to it. If you do not feel comfortable about giving or receiving oral sex, that is your choice, and you do not need a justification for refusing. No one should ever be coerced into any sexual behavior that is not freely chosen. In one national study from the 1990s, about 10% to 30% of people surveyed considered oral sex unusual, kinky, or unappealing (Michael et al., 1994). At times, oral sex can become a problem in a relationship if one person desires to give or receive it and the other does not.

Since You Asked

7. My boyfriend says it's OK if we have oral sex because it doesn't spread sexual diseases. Is that true?

Many reasons have been suggested as to why someone may feel uncomfortable with oral sex and wish to avoid it. One of the most obvious relates to an individual's morals and attitudes about sex. Some believe that oral sex is simply wrong and conflicts with their personal code of acceptable behavior. Other reasons concern the mechanics of the behavior itself. Some partners are often concerned that the penis will be pushed too deep into the mouth and throat or the inserting partner will ejaculate in the mouth, causing a gag reflex or other unpleasant reactions. Although swallowing semen is not harmful or dangerous in itself (assuming no STIs), many people are uncomfortable or repulsed by the idea. Some men and women may worry that the smell or taste of the female genitals will be unpleasant or offensive in some way. Western society has encouraged this view with the numerous "feminine hygiene" products on the market that send the message that a woman's *natural* genital odor is bad and should be covered up. In reality, female genitals that are washed with normal regularity have a natural fragrance that most people find attractive and enjoyable. A strong or unpleasant odor from a woman's vagina may be a sign of either poor hygiene or, possibly, an infection that may require treatment.

The bottom line is that if two people desire oral sex and one or both have some reservations about it, their concerns can usually be overcome through sensitivity, respect, and open communication.

In Touch with Your Sexual Health

Oral Sex and STIs (STDs)

Recent research has seen a shift in the way young adults and teens view oral sex. Most consider vaginal and anal intercourse as "having sex," but only 20% consider oral sex as "having sex" (e.g., Hans, Gillen, & Akande, 2010). This is a troubling trend, in that oral sex carries a clear risk of transmission of nearly all sexually transmitted infections, including human papilloma virus (genital warts, which may lead to oral or throat cancer), herpes, hepatitis B, gonorrhea, syphilis, chlamydia, and HIV (see Chapter 8, "Sexually Transmitted Infections/Diseases," for more information on all STIs). In addition, research has found that oral sex may be a major cause of recurring vaginal yeast infections because the mouths of one-third to one-half of all adults normally contain yeast bacteria that can overwhelm the normal balance of yeast and other organisms in the vagina (Geiger & Foxman, 1996; "Oral Sex, Masturbating," 2004).

The risk of transmission of STIs during oral sex is reduced by using condoms during fellatio (nonlubricated condoms are usually preferred for this) and dental dams (thin sheets of latex placed over the vulva) for cunnilingus. Also, a condom may be cut lengthwise, opened up, and placed over the vulva for STI protection during cunnilingus. While the risk for transmission of HIV through oral-genital contact appears to be relatively low, unprotected oral sex should never be thought of as safer than vaginal or anal intercourse in terms of the risk for STIs overall.

Anal Stimulation

anal intercourse
A sexual position in which the penis is inserted through the partner's anus into the rectum.

anilingus
stimulation of the anus.

Since You Asked

8. Is there anything wrong with anal sex?

Figure 6.9 Anal Intercourse
Some couples include anal intercourse as part of their intimate sexual activities together.

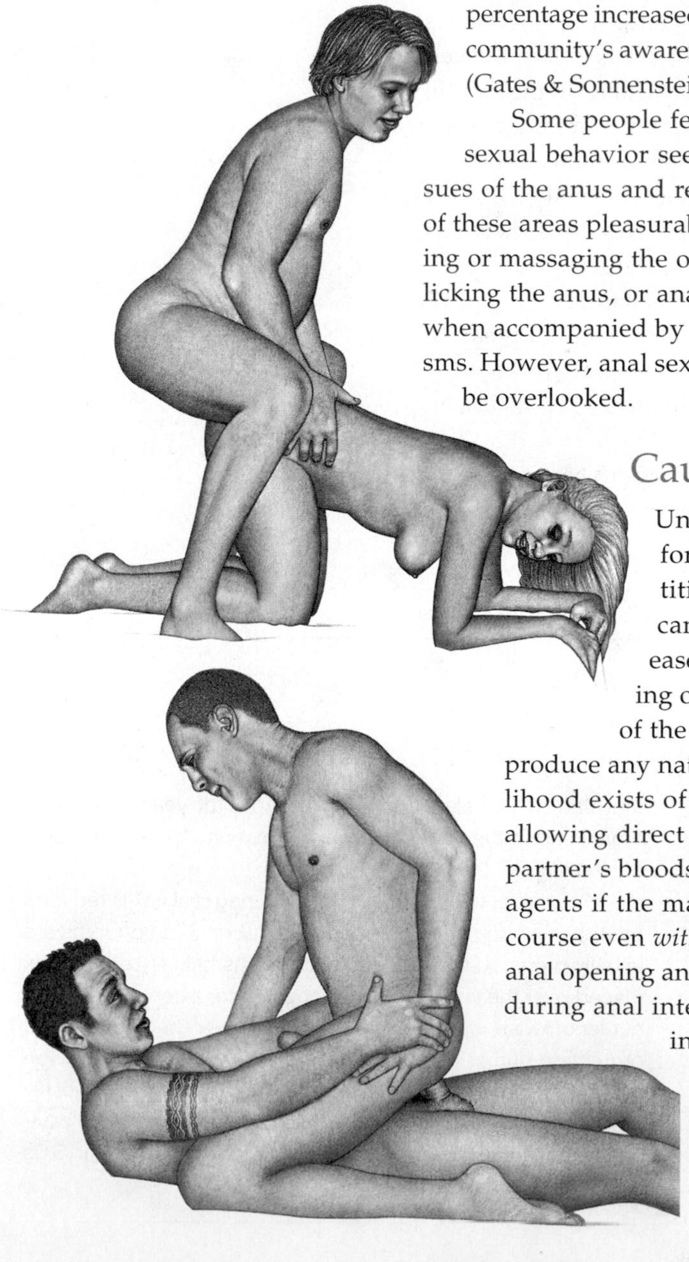

6.7 Review anal sexual activities, attitudes about these behaviors, and the potential health concerns related to them.

Anal stimulation, which includes **anal intercourse**, manual stimulation of the anal area, and oral-anal stimulation **(anilingus)**, is the least common of the sexual activities discussed in this chapter. A widespread belief among heterosexuals is that anal sex is practiced only by gay men and that most gay men engage in anal intercourse. However, as you can see in Table 6.2, the incidence of anal sexual behavior among heterosexual couples is significantly higher than most people would expect, ranging from 1% to 26% during the previous year, depending on the age group. A large study from 2011 estimates that 36% of women and 44% of men have experienced anal intercourse (Chandra et al., 2011). Among college students, approximately 20% report experiencing anal intercourse, and over 50% have engaged in manual anal stimulation. (Baldwin & Baldwin, 2000; Flannery & Ellingson, 2003)

To debunk further the myths about anal sex, not all gay men engage in anal intercourse with their partners. Studies have shown that between 10% and 40% of gay men do not include anal intercourse in their lovemaking activities, and that percentage increased dramatically during the 1980s and early 1990s due to the gay community's awareness of the high risk of HIV transmission from anal intercourse (Gates & Sonnenstein, 2000; Goldbaum, Yu, & Wood, 1996).

Some people feel that the anus is dirty or unappealing, and to include it in sexual behavior seems to them unpleasant or simply wrong. However, the tissues of the anus and rectum are very sensitive, and many people find stimulation of these areas pleasurable (see Figure 6.9). Anal erotic activities may include touching or massaging the outer anal area, insertion of a finger into the anus, kissing or licking the anus, or anal intercourse. For some people, anal stimulation, especially when accompanied by stimulation of the penis or clitoris, may lead to intense orga sms. However, anal sexual activities involve some unique health risks that must not be overlooked.

Cautions Regarding Anal Intercourse

Unprotected anal intercourse is the *highest-risk sexual behavior* for the transmission of virtually all STIs, including HIV, hepatitis B, and the human papilloma virus (which may cause anal cancer—see Chapter 8, "Sexually Transmitted Infections/Diseases") if the inserting partner is infected. This is because the lining of the rectum is thin and fragile, and lacks the elastic qualities of the vagina. Also unlike the vagina, the anus and rectum do not produce any natural lubrication upon sexual arousal. Therefore, a high likelihood exists of damage to rectal walls (small tears, abrasions, or fissures), allowing direct contact of the inserting partner's semen with the receiving partner's bloodstream. Semen is a blood product and contains infectious STI agents if the man is infected. Many sex educators suggest that anal intercourse even *with* a condom is risky if you consider that the tightness of the anal opening and the lack of natural lubrication may cause condoms to break during anal intercourse. If a couple who are free of STIs wishes to engage in anal intercourse, generous amounts of artificial lubricant should be used and penetration should be done very slowly and carefully. Some evidence suggests that frequent anal intercourse over time may lead to large and painful anal fissures, hemorrhoids, and potential weakening of the anal sphincter muscles.

Finally, if the normally harmless bacteria in the anus and rectum somehow transfer to the urethra (in men or women), the vagina, or the vulva, they can cause vaginal, reproductive, and urinary tract infections. Therefore, it is extremely important that a couple having engaged in anal stimulation or anal intercourse carefully avoid any vaginal stimulation or intercourse until the hand, mouth, or penis has been washed or the condom has been changed.

Coitus

6.8 Summarize the most common positions used for coitus (heterosexual intercourse), the advantages of each, how the positions relate to orgasm, and the frequency of coitus by age and by country.

All of the sexual activities we have discussed so far—sexual fantasy, masturbation, erotic touching, oral sex, and anal sexual activities—are engaged in and enjoyed by people in relationships with opposite- or same-sex partners. The one exception is **coitus**, which refers to penis-vagina intercourse and is, therefore, by definition a *heterosexual* behavior. In most cultures around the world, coitus, commonly referred to as *sexual intercourse* (a more general term), is the shared sexual behavior that is considered most appealing, most acceptable, and most common among heterosexuals. One large survey in the 1990s found that 95% of heterosexual men and women of all ages rated vaginal intercourse as either very or somewhat appealing (Michael et al., 1994). The 2011 NSSHB national study of sexuality found that among men, 63% had engaged in vaginal intercourse by age 19, 70% by 24, and, 89% by 29. Among women, 64% had experienced intercourse by age 19, 86% by 24, and 91% by 29. (Herbenick et al., 2010a)

Coitus

Penis–vagina intercourse.

When discussing coitus, three issues appear to be of greatest interest to researchers and people in general. The first involves the various positions for intercourse and the pros and cons of each. The second relates to duration and frequency of intercourse (probably so that people can compare averages with their own duration and frequency). And the third focuses on intercourse and female orgasm. Let's examine each of these in turn.

Since You Asked

9. What is the best position for intercourse?

Positions for Coitus

Keeping in mind that coitus is, by definition, an opposite-sex behavior, how many positions for coitus are there? The easy answer is, as many as anyone can imagine: man on top, woman on top, side by side, facing each other, facing away from each other, sitting, standing, hanging from the chandelier, and numerous variations on each of these themes. Which position is best? This depends completely on the couple engaging in the activity. Each person and each couple are different in their preferences for coital positions. It would be impossible here to even try to discuss or picture all the possible variations. Instead, we can observe some sketches of the basic positions and offer a brief description of some of the advantages and disadvantages of each. As we discuss some of these, remember that coitus in *any* position without a condom is a high-risk behavior for the transmission of STIs if either partner is infected.

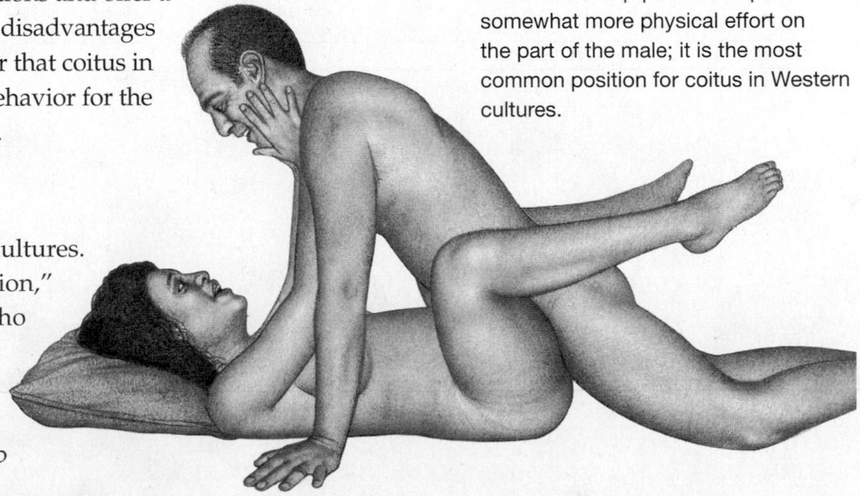

Figure 6.10 Man-on-Top Coital Position

The man-on-top position requires somewhat more physical effort on the part of the male; it is the most common position for coitus in Western cultures.

Man on Top, Face to Face

This is the most common position in Western cultures. It is often referred to as the "missionary position," so named after the Christian missionaries who introduced the position to the people of the Polynesian Islands (see Figure 6.10). Prior to the missionaries' arrival, these island cultures traditionally practiced woman-on-top coital positions.

Pros

- Ease of insertion of penis into vagina
- Either partner can guide penis into vagina
- Good for kissing and eye contact
- Woman can touch and stroke man's body, buttocks, scrotum
- Wide variation of leg positions for woman
- Man feels dominant (often an expectation in Western cultures)

Cons

- Difficult for man to caress woman's body
- Difficult for clitoral stimulation by man
- Highly stimulating for man, less control over ejaculation
- May be tiring for man to support weight on arms and legs
- May be too much weight on woman
- Woman may feel too passive and submissive

Woman on Top, Face to Face

female-superior position
A position for heterosexual intercourse in which the woman is sitting on or crouching over the male.

The woman-on-top, or **female-superior position**, is one of the most common positions for coitus worldwide (see Figure 6.11). People use many variations of the female-superior position: usually the woman is kneeling or squatting over the man, but she may also lie down on him, face-to-face. By the way, the expression "female-superior" does not imply anything about power or sex roles; it simply means "woman above."

Pros

- Woman has more control over angle of penetration, movement, and angle and speed of thrusting
- Usually considered best position for woman's orgasm
- Partners may caress each other's body

Figure 6.11 Woman-on-Top Coital Position

The female-superior position allows the woman more control over angle and depth of penetration and the man greater control over his orgasm.

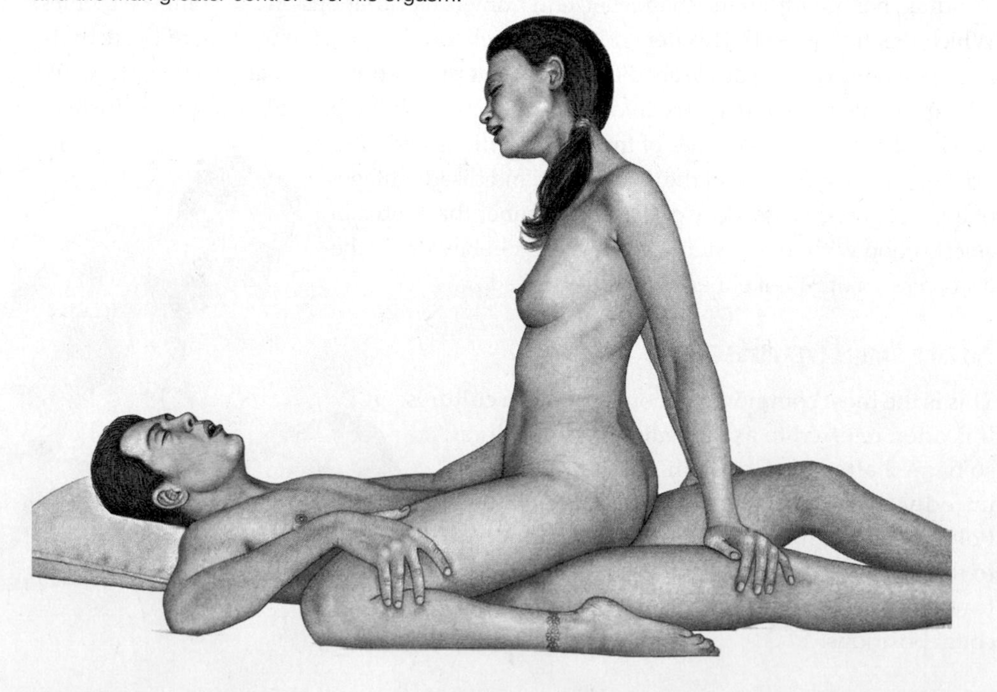

Figure 6.12 Side-by-Side Coital Position

The side-by-side position requires less physical effort for both partners and allows more freedom of movement for mutual caressing during intercourse.

- Either partner may stimulate clitoris during intercourse
- Man can stimulate woman's breasts manually or orally
- Less physically stimulating for male; easier to control ejaculation
- Woman feels more active and "in-charge" both physically and psychologically (which may be desirable for some women and men)

Cons

- Some men (and some women) feel man is too passive in this position
- Less stimulation for man if he has difficulty reaching orgasm

rear-entry position
A position for heterosexual intercourse in which the penis is inserted into the vagina while the man is behind the woman.

Side by Side, Face to Face

In this position, both partners lie on their side, facing each other with their legs intertwined (see Figure 6.12). Insertion of the penis into the vagina can sometimes be difficult in this position, so some couples will begin intercourse in another position and roll into a side-by-side posture.

Pros

- Neither partner is supporting own or other's weight
- Not physically demanding
- Good for extended coitus
- Shallower penetration
- Caressing is easy with free (top) arm
- Easy to control movements; good for controlling ejaculation
- Less hot and sweaty in warm temperatures

Cons

- Difficult to insert penis into vagina
- Shallower penetration
- Difficult clitoral stimulation
- Chance of penis slipping out of vagina during thrusting
- More difficult position for exuberant thrusting
- Arms underneath may fall asleep

REAR ENTRY In the **rear-entry position**, the penis is inserted into the vagina from behind the woman (see Figure 6.13). Regardless of similarities in appearance, this coital position should not be confused with anal intercourse. Several variations on the

Figure 6.13 Rear-Entry Coital Position

Some people enjoy the sensations provided by the rear-entry position; it may also be more comfortable if the woman is pregnant.

position may be used, including lying side by side, front to rear ("like spoons"); sitting with the woman in the man's lap, facing away; or the woman kneeling and the man behind her (often called "doggy style").

Pros

- Deeper penetration of penis in vagina possible
- Man may reach around woman for breast or clitoral stimulation
- Woman may reach back to stimulate man's scrotum
- Some men and women report different and enjoyable feelings of stimulation
- Good position during pregnancy

Cons

- Penetration may be uncomfortably deep
- Kissing and eye contact difficult
- Lack of face-to-face contact may feel too impersonal
- For some, association with animals may be objectionable
- For some, association with anal intercourse may feel uncomfortable

Frequency and Duration of Coitus

In the 1940s and 1950s, Alfred Kinsey's surveys (see Chapter 1) revealed that the average number of times married couples reported having intercourse (coitus) was 2.8 times per week (for those in their twenties). Suddenly, nearly everybody in the country had a "yardstick" with which to compare their own sex lives. And because

2.8 was only an *average*, many people saw themselves as "abnormal": "Oh, no! We're sex fiends, Honey! We do it twice a day!" or "Something must be wrong with our marriage, dear. We only do it once a week." Since then, several studies, including *The National Health and Social Life Survey* (Laumann et al., 1994; Michael et al., 1994), an annual survey conducted by the Durex condom company (2003), and the 2011 *National Survey of Sexual Health and Behavior* provided some newer averages to ponder, such as intercourse frequency at various ages and the average frequency in various countries (see Tables 6.7 and 6.8).

Such statistics are interesting to know, but they should not be taken too seriously by most couples. The *ideal* frequency of sexual intercourse (or any sexual activity) is as often as feels good and comfortable to each individual and couple. One couple might have intercourse twice a week, another twice a month, and still another twice a year; as long as they are happy with their sexual life together, that is their "normal" frequency. Furthermore, a couple's frequency of sexual encounters is unlikely to be very stable over time. Depending on stress, fatigue, illness, presence or absence of kids, vacations, and so on, some days, weeks, months, or years are likely to be more sexually active than others, and sometimes this variation can be large. Moreover, the frequency (although not the pleasure) of most sexual activities decreases as we age, a topic we will discuss in greater detail in Chapter 12, "Sexual Development throughout Life."

What about duration of coitus? How long should intercourse last? This is another question that each couple must answer for themselves. For some couples, a fast, superficial lovemaking session (commonly known as a "quickie") is perfectly fine with both of them, while another couple might feel cheated if sex is less than an hours-long event. Most couples who are comfortable and happy with their sexual life find that duration of lovemaking varies from one time to the next, depending on many of the same issues that influence frequency.

Of course, lovemaking usually includes behaviors other than coitus itself, and the entirety of the experience is important to most people. Laumann and colleagues (1994)

Since You Asked

10. Is it abnormal to have intercourse twice or even three times a day with a partner whom you have been seeing for a long time?

Table 6.7 Frequency of Coitus in the United States

Age	Not at All in the Past Year (%)	A Few to Twelve Times in the Past Year (%)	A Few to Four Times a Month (%)	Two or Three Times a Week (%)	Four or More Times a Week (%)
Men					
18–24	40	12	23	18	7
25–29	14	12	40	28	5
30–39	15	16	39	25	6
40–49	27	16	37	17	5
50–59	42	18	26	12	2
60–69	46	17	25	11	<1
70+	57	19	20	4	1
Women					
18–24	25	16	24	25	10
25–29	14	14	40	27	6
30–39	26	14	39	16	5
40–49	30	18	31	18	4
50–59	49	15	23	11	1
60–69	58	13	23	5	1
70+	69	10	9	1	2

SOURCES: Adapted from: M. Reece (2010), Table 5, p. 299; D. Herbenick et al. (2010a), Table 5, p. 285.

Table 6.8 Average Intercourse Frequency per Year in Selected Countries (online survey data)

Country	Frequency	Country	Frequency
Over 130		Slovakia	106
Greece	138	Austria	105
Croatia	134	Spain	105
120–129		Germany	104
Serbia & Montenegro	128	Switzerland	104
Bulgaria	127	Israel	100
Czech Republic	120	**90–99**	
France	120	Denmark	98
110–119		Norway	98
United Kingdom	118	Ireland	97
Netherlands	115	Thailand	97
Poland	115	China	96
New Zealand	114	Sweden	92
United States	113	**80–89**	
Chile	112	Taiwan	88
Turkey	111	Vietnam	87
100–109		Malaysia	83
Iceland	109	**70–79**	
South Africa	109	Hong Kong	78
Australia	108	Indonesia	77
Canada	108	India	75
Portugal	108	Singapore	73
Belgium	106	**Overall Average = 103**	
Italy	106		

SOURCE: Adapted from the Durex Global Sex Survey (2005).

found that about 70% of men and women reported that their last lovemaking (referred to in the survey as a "sexual event," meaning the entire lovemaking session, not just coitus) lasted from 15 minutes to an hour, and 20% of men and 15% of women said it lasted an hour or more. Moreover, approximately 41% of people report that their lovemaking sessions involved three or more distinct sexual activities (Herbenick et al., 2010a). Only 11% of men and 15% of women said that their most recent sexual event was less than 15 minutes. In terms of intercourse itself, researchers have found that sex therapists' estimation of the normal length of time of actual penile-vagina thrusting varies from approximately 3 to 13 minutes (Corty & Guardiani, 2008).

Coitus and Orgasm

As was discussed in Chapter 3, virtually all men will experience orgasm through the act of coitus itself. However, most women do not. This is due primarily to the type of stimulation that is produced by the thrusting of the penis in the vagina. The glans of the penis is surrounded by the vaginal walls and usually receives more than adequate stimulation for orgasm. In fact, many men report that intercourse provides *too much* stimulation, causing them to reach orgasm and ejaculate sooner than they would like (see Chapter 7, "Sexual Problems and Solutions," for a discussion of this).

However, for women, this thrusting by itself may not provide much stimulation to the clitoris, especially as the clitoris retracts during increasing arousal (see Chapter 3, "The Physiology of Sexual Response"). For most women, this makes orgasm difficult, if not impossible, to achieve from intercourse when it does not include other types of stimulation before, during, or after intercourse. Although the duration of intercourse is related to reaching orgasm for some women, at least half of all women (some research reports the number is as high as 75%) do not experience orgasm from coitus by itself *regardless of the duration of thrusting* (Laumann et al., 1994; Richters et al., 2006; Slob, Van Berkel, Van der Werfften, & Bosch, 2000). On the other hand, when people report whether or not they experienced an orgasm during their sexual encounters over the past year, 91% of men, and 64% of women reported that they did (Herbenick et al., 2010a).

We should keep in mind, however, that very few sexual interactions consist of intercourse without any other sexual behaviors, and this is reflected in the most recent research. In fact, nearly all partnered sexual acts involve various combinations of sexual activities that are bound to increase either partner's experience or orgasm. In fact, when the sexual encounter included three different sexual acts (e.g., oral sex, partnered masturbation, and penile-vaginal intercourse), orgasm rates increased to 92% for men and 72% for women, and this percentage increased as more sexual behaviors were added into the sexual interaction until, among those (few) couples that engaged in six sexual acts, orgasm was experienced by 100% of both men and women (Herbenick et al., 2010a, p. 358).

Consequently, although most women do not routinely reach orgasm solely through coitus, many men and women believe that a woman *should* achieve orgasm with penile thrusting alone, and this perception can lead to problems in the relationship. Once a couple understands the facts about female orgasm and intercourse, the "problem" is easily solved. For women who do not receive the stimulation during lovemaking that they may need and want for orgasm, additional stimulation through a wider variety of sexual activities can be included in a sexual encounter so that both partners may achieve an orgasm, if that is their goal. Consequently, the reality of male and female sexual responsiveness lies not in intercourse, but in the full range of sexual expression.

Your Sexual Philosophy

Sexual Behaviors: Experiencing Sexual Pleasure

The information in this chapter has offered you the knowledge you need for an enhanced appreciation of the range of behaviors available for experiencing and sharing sexual pleasure, expressing your sexual desires, and conveying feelings of closeness, intimacy, and love between two people. A clear understanding of these activities is crucial when people are faced with sexual choices and decisions, as they are throughout their lives. Without a foundation for making sound choices that are right for you, these usually pleasurable activities can sometimes become uncomfortable, frightening, and even dangerous.

This might be a good time to pause again and consider how all this information might fit into your sexual philosophy. Think for a moment about the sexual choices you have made up to this point in your life. Do you feel comfortable with your sexual decisions? Have you made mistakes? Do you have regrets? Do you feel guilty or ashamed? Would you change anything if you could? Next, consider the sexual choices you are making now and may make in the future.

The decisions you make about sex (or anything else, for that matter) throughout your life should be based on applying the sum of your experience, knowledge, and past and present feelings. Exploring who you are sexually and what path you want your sexual life to take allows you to make clear choices that feel right to you when the time to act on those decisions arrives. To clarify your personal feelings about the behaviors discussed in this chapter, ask yourself the following questions:

1. Is this behavior something I want to do and would feel comfortable and positive about doing?
2. What sort of relationship with a partner do I need in order to feel comfortable sharing this activity (casual, monogamous, marriage)?

3. What if my partner wants to do something that makes me uncomfortable? Will I agree anyway? If so, why? If not, how will I handle the situation if my partner continues to pressure me?
4. If the behavior involves the risk of STIs or unwanted pregnancy, how will I be sure I am protected?
5. Will I be able to discuss my feelings about this activity with my partner?

This kind of self-exploration provides you with opportunities to place the many aspects of human sexual behavior into the context of your personal life and gain valuable insights into yourself as a sexual being now and in the future. Remember, your sexual philosophy is about discovering who you are, knowing what you want and don't want, and planning ahead. The information in this chapter, perhaps more than any other in this book, plays a crucial role in the development of your personal sexual philosophy, so that you can take charge of your sexual life rather than having it take charge of you.

Have You Considered?

1. Assume for a moment you are in a sexually active relationship with someone you love deeply and with whom you are hoping to spend the rest of your life. One evening, your partner expresses a desire to engage in a sexual activity that you are not interested in and that you find somewhat offensive (you pick the activity). How would you react? Explain what you think would be the best way to handle this situation in your relationship.
2. Suppose that a heterosexual couple is feeling a strong mutual desire to be sexually intimate, but they've agreed that intercourse (coitus) should be saved for marriage. Explain how they might balance their desire for sexual intimacy with their moral convictions regarding sex and marriage.
3. Imagine for a moment that you have a fourteen-year-old daughter and you are discussing sexual issues with her. What would you want her to know and understand about the behaviors discussed in this chapter? Explain your answer.
4. Now imagine for a moment that you have a fourteen-year-old son and you are discussing sexual issues with him. What would you want him to know and understand about the behaviors discussed in this chapter? Explain your answer. Were your answers different for a daughter and a son? Why or why not?
5. Consider a situation in which sexual partners are discussing experimenting with various sexual positions. Explain why some couples might find the various positions appealing while others might reject some of them.
6. Discuss the influences or experiences from childhood or the teen years that you feel often shape people's adult feelings and attitudes toward sexual activities.

Summary

Historical Perspectives

Masturbation Will Be the *Death* of You…

- Historically, masturbation was regarded as an unhealthy, immoral activity that could lead to serious health consequences and even death. Today, most health professionals and sexuality educators view masturbation as a normal, healthy sexual activity; it is sometimes incorporated into therapy for certain sexual problems.

Your Sexual Comfort Zone

- The Sexual Opinion Survey measures a human characteristic called erotophobia–erotophilia, which helps to predict people's individual sexual comfort zones. It measures a person's level of ease in learning and talking about sexual issues.

- The Sexual Opinion Survey may also help predict how likely people are to use contraceptives and to engage in risky sexual behaviors.
- The *Sexual Inhibition Scale* and the *Sexual Excitation Scale* extend the SOS to help explain the underlying, adaptive sources for people's individual differences in sexual responding.

Solitary and Shared Sexual Behaviors

- The sexual activities covered in this chapter are engaged in by a majority or large minority of people. These behaviors would not be considered strange or deviant by most people in our culture.

Sexual Fantasy

- Sexual fantasy is by far the most common sexual "behavior." Estimates are that 95% of people have engaged in sexual fantasy. Sexual fantasies may occur as daydreams, while masturbating, or during sex with a partner. Although sexual fantasy has many useful functions, it can create guilt in some people, and a small percentage of people may become obsessive.
- Some sexual fantasies, such as sex in unusual locations and positions, are equally common among men and women; others, such as having sex with two partners at once or with a mysterious stranger, show significant differences between the sexes. For some couples, sharing sexual fantasies may enhance their intimate relationship.

Masturbation

- Masturbation, contrary to old beliefs, is a common behavior. It does not cause hairy palms, genital cancer, sterility, blindness, or a decrease in enjoyment of sex with a partner.
- Masturbation has been found to have many health benefits. For example, research has found that people who masturbate while in an intimate relationship have more active and more satisfying sex lives with their partner.

Erotic Touch

- Erotic touching is an enjoyable sexual behavior in itself and need not always lead to intercourse.
- Individual preferences for where and how to be touched vary greatly. In general, the breasts and nipples are sexually sensitive areas for women *and* men. Genital touching can be a safe source of great pleasure, intimacy, and sexual satisfaction and does not necessarily lead to other less-safe sexual activities.

Oral Sex

- While most people find oral sex to be an appealing, intimate behavior, some find the activities involved to be morally wrong or physically offensive.
- Oral contact can transmit most sexually transmitted infections.

Anal Stimulation

- Anal sex among heterosexual couples is more frequent than is commonly believed. Among gay male couples it is less frequent than is commonly believed.
- Unprotected anal intercourse has the highest risk of transmission of HIV and other sexually transmitted infections of all sexual behaviors.

Coitus

- Coitus, defined as penile-vaginal intercourse, is the only strictly heterosexual sexual behavior. All other sexual activities may be engaged in by straight, gay, lesbian, or bisexual individuals. The man-on-top coital position is the most common in Western cultures, but the woman-on-top position is also very common worldwide. Various coital positions have specific advantages and disadvantages for each partner.
- The movements of coitus typically do not, in themselves, provide adequate stimulation for most women to reach orgasm routinely through intercourse without additional stimulation, usually through the addition of more varied sexual activities.

Your Sexual Philosophy Sexual Behaviors

Experiencing Sexual Pleasure

- Understanding the full range of sexual activities available to humans in intimate relationships enhances one's ability to experience and share sexual pleasure; express sexual desires; and convey feelings of closeness, intimacy, and love with another individual.

Chapter 7

Sexual Problems and Solutions

Learning Objectives

After you read and study this chapter you will be able to:

7.1 Explain how sexual disorders are defined, diagnosed, and evaluated.

7.2 List and explain the possible sources and causes of sexual problems.

7.3 Analyze and explain general approaches to solving individual and couple-based sexual disorders.

7.4 Discuss the description, causes and suggested solutions for each specific sexual disorder.

Since You Asked

1. What is "bad" sex? How do you know if you are doing it right? (see page 221)
2. My girlfriend is from India. We've been together for nearly a year, but she doesn't seem interested in sex at all. Could this be due to growing up in a different culture, or is it something in our relationship? (see page 230)
3. How can a couple who has been together for a long time (married three years) get their sexual feelings back? We have a good relationship, but we hardly ever make love anymore. (see page 230)
4. I'm a male and I'm curious if it is odd that I hardly ever masturbate. Is masturbation harmful for you? (see page 233)
5. I can't tell if my partner is satisfied or not when we make love. What are the signs that a person is sexually satisfied? (see page 234)
6. Is impotence always psychological? Does Viagra work for everyone? (see page 238)
7. I love my boyfriend and want to have sex with him, but I just seem to have no interest—I feel guilty, but he is trying to be understanding. What's wrong with me? (see page 243)
8. Are there ways for a guy to control orgasm and last longer? (see page 249)
9. What causes a woman to be unable to experience orgasm during sexual intercourse even though she can have an orgasm during oral sex? Is it a physical thing or mental? Maybe both? (see page 250)
10. My girlfriend says sex hurts (physically). Why is it painful, and what can we do to relieve the pain? (see page 253)

Most people assume that sex is just *natural*; that our bodies are somehow "preprogrammed" for sexual activity. From this assumption, many people take for granted that human sexual behaviors and responses should simply occur *naturally*—that the mind and body should do what they are "programmed by nature" to do whenever they are supposed to do it. After all, that is how sex works for *nonhuman* animals, right? The female animal sends out strong mating signals, and the male animals sense these signals and eagerly seek out the female for mating. This is true—true, that is, for *nonhuman* animals. For humans, it's a different story. For us, sex is *not* a purely natural, "preprogrammed" event.

Our special human abilities to think, feel, and reason add layer upon layer to the human sexual experience until it hardly resembles that of nonhuman animals at all. In other words, humans have an intellectual and emotional investment in sex—not just physical. On the one hand, this infuses people's sexual interactions with immense joy and passion; on the other, it can create psychological insecurities, doubts, or other negative perceptions that may lead to sexual problems. As you will see in this chapter, such problems, and the emotions surrounding them, are more common than most people think.

You may have noticed that this chapter focuses on the idea of "sexual problems" rather than "sexual disorders" or "sexual dysfunctions." The reason for this word choice is that nearly all sexual problems are not serious, are readily treatable and, with very few exceptions, no one should have to endure chronic, long-term sexual difficulties. Words such as *disorder* and *dysfunction* imply that these conditions are more serious, that they are illnesses of some sort, and difficult to solve. Sometimes the words themselves can discourage individuals from seeking the help they need to solve whatever sexual problem they are facing. Nevertheless, you might see words such as *disorder* or *dysfunction* used occasionally in this chapter because they are often incorporated into formal psychological or medical diagnoses for specific sexual problems. However, they are *just words*, and you should not be put off by them. As the title of this chapter suggests, sexual problems are no different from most other difficulties in life, and although they may be emotionally or physically painful when they happen, they all have solutions.

Sexual problems seldom confine themselves to the bedroom. They have a tendency to intertwine with other aspects of a relationship and with each partner's dissatisfactions with life in general. Determining which came first, the sexual problem that led to the relationship issues or other nonsexual conflicts that affected the couple's sexual connection, is often difficult. Regardless, the bottom line is that sexual and relationship problems nearly always go together. This deterioration of the relationship overall, then, often causes a couple to feel an increased sense of concern and urgency to solve the sexual problem, which frequently, but not always, helps them overcome other difficulties, too (see Chapter 4, "Intimate Relationships," for more about this connection).

This chapter will examine many facets of sexual problems and their solutions, starting with how common sexual difficulties are and how to know whether you have a sexual problem. We will then move on to how sexual problems can be evaluated and classified in ways that help you understand them as clearly as possible. Once a person or couple has a clear understanding of a particular sexual problem, he or she (or they) can better pinpoint the possible causes that led up to the condition. Then, with strategies for a clear understanding of the problem and its possible causes, we will be in the best position to discuss treatments and solutions to specific sexual problems and methods to enhance sexual enjoyment and satisfaction in general.

Before we begin our journey through sexual problems and solutions, let's place them into a historical frame by looking back to a time when sexual problems were, for the most part, invisible. The same basic problems have always existed and people suffered from them as they do now, but in the past, cultural attitudes dictated that people should not talk about or acknowledge them to their partners or even to themselves, and certainly not to anyone else—even to a doctor or a therapist. Furthermore, people saw little point in discussing such private matters because nobody really knew very much about diagnosing or treating them. If this sounds like ancient history to you, it isn't; we only have to go back about 50 years.

Focus on Your Feelings

For some people, the topics covered in this chapter evoke intense emotions. Sexual feelings and responses are closely linked to our *expectations* about sex. Most people *expect* sex to work correctly and feel good. They *expect* their bodies to do what they are supposed to do during sexual activities. They *expect* that they and their partners will be sexually satisfied and content. But when expectations are not met, people have a tendency to react with stronger-than-usual emotions: disappointment, frustration, anger, sadness, withdrawal, and so on. When your sexual expectations are not met, when your body or your partner's body does not "behave" the way it "should," you are likely to become emotional about it in one way or another.

What emotions are you likely to feel? Almost anything *except* happiness and contentment. You might be disappointed in yourself or worried that you have disappointed your partner. You might feel embarrassed because, well, sexual problems are embarrassing! You might feel afraid because you are unsure what is wrong with you and how long it's likely to last. You might be frustrated because you really want to enjoy sexual intimacy with your partner, and this problem is getting in the way of that. You might feel guilt over the frustration you perceive in your partner. You might experience anger—at yourself, at your body, at your partner, or at the world in general for the unfairness of it all.

Becoming educated about sexual problems and their solutions *before* they occur (or before the next one occurs) is the key to minimizing the intensity and duration of your potential negative feelings. When they do occur, with your emotions in check and with the knowledge gained in this chapter, you will be able to think rationally about your problem and take the necessary steps to find a satisfying solution. Moreover, worrying less about these problems goes a long way toward helping you resolve them.

Historical Perspectives

Before Masters and Johnson

A careful examination of the history of sexual dysfunction prior to the 1960s reveals little published literature and a quagmire of myths and faulty information. Prior to Masters and Johnson's pioneering research into the physiology of sexual response (discussed in greater detail in Chapter 3, "The Physiology of Sexual Response"), most sexual difficulties were attributed—depending on how far into the past we want to go—to anything from demonic possession to the Freudian notion of unresolved, unconscious childhood conflicts stemming from poor parenting practices (Wiederman, 1998). Based on these old models of sexual problems, most attempts at helping people overcome them generally met with poor success, due in large measure to the difficulty of understanding and treating sexual problems without the benefit of a road map of human sexual anatomy, physiology, and response.

The work of William Masters and Virginia Johnson in the early 1960s revolutionized our understanding of human sexual anatomy and sexual responding. They revealed, for example, how erection of the penis and clitoris prepare the body for sexual interactions and what types of stimulation produce (or fail to produce) sexual excitement and orgasm. But the goals of their careful laboratory observations of human sexual response, far beyond simply *describing* sexual structures and their functions during sexual activity, were to help couples and therapists apply that knowledge to solving sexual problems and to enhancing people's sexual lives (Masters & Johnson, 1970). Although some flaws and errors in Masters and Johnson's early work have been pointed out over the decades since they first published their findings, they clearly accomplished the goal of bringing sexual problems and solutions into the open, offering people solutions and hope.

The insights from the sex laboratories of Masters and Johnson awakened a new era of treatment methods for sexual dysfunctions (Annon, 1974; Kaplan, 1974). As you will see in this chapter, today's successful treatments of sexual difficulties rely heavily on an *interaction* of psychological and physical factors. Prior to the work of Masters and Johnson, the overall ineffectiveness of the approaches to treating sexual difficulties led to a widespread belief that such problems and conditions must be endured in silence. Today, thanks to these researchers and the many advances in psychological and biomedical sciences that have followed their work, most sexual problems are relatively easy to treat and cure.

What is a Sexual Problem?

7.1 Explain how sexual disorders are defined, diagnosed, and evaluated.

Probably the best response to the question, "Do I have a sexual problem?" is another question: "Do *you* think you do?" Although this may seem like a frivolous response, it is not, because most sexual problems are self-diagnosed. This means that sexual problems are, to a large degree, a matter of an individuals' or couples' *perception* that a problem exists. For example, imagine a couple who have been in a happy, intimate relationship for several years and make love once every couple of months. Would you assume that they have a sexual problem? Or consider a man who makes love with his partner several times a week but is able to achieve an erection only once or twice a month. Does he have a sexual problem? How about the couple in which one partner always experiences an orgasm during their lovemaking but the other never does? Is this a sexual problem?

Your first thought might be, "Yes!" But when you stop to think about it, although you may have found yourself surprised by these couples' situations, you stop short of

Since You Asked

1. What is "bad" sex? How do you know if you are doing it right?

assuming that they were experiencing sexual problems. You probably thought something along the lines of, "It depends. If they feel happy and satisfied with their sexual lives, then these are probably not sexual problems *for them*." Sex is so personal, how can anyone make a judgment for *them*? But if infrequent lovemaking, problems with erections, lack of orgasm, or any other sexual difficulties are causing frustration, emotional pain, or relationship discord, then they would probably be diagnosed as sexual problems.

Remember, however, that not everyone knows for sure if he or she is experiencing a sexual problem. It is quite common for someone to feel sexually unfulfilled but not understand why or lack an awareness of what constitutes a satisfying sexual life. For this reason, becoming knowledgeable about sexual activities and about problems that may arise in one's sexual life is important for living a healthy and satisfying sexual life.

In addition to this process of self-diagnosis, physicians and sexual health professionals use a more formal system to diagnose sexual problems. The purpose for such a system is that it creates consistency in the identification of specific sexual difficulties and allows for more effective treatment. Health professionals turn to a guidebook published by the American Psychiatric Association that lists the criteria for diagnosing most health problems (sexual or not) that have a psychological component, and includes detailed descriptions of recognized sexual disorders. We will use this system as a framework to discuss the problems in this chapter.

Clinical Diagnosis of Sexual Problems—The DSM-5

The formal diagnostic criteria for sexual problems are described in the *The Diagnostic and Statistical Manual of Mental Disorders, Fifth Edition,* published by the American Psychiatric Association. The manual is referred to in the mental health world as *The DSM-5* (DSM-5, 2013). The DSM-5 divides sexual disorders into three general categories: sexual desire, interest, or arousal disorders; disorders of orgasm; and sexual pain disorders (see Table 7.1). In general, a diagnosis on any of these requires that the problem last for at least six months and cause significant psychological or emotional distress. Each of the disorders will be discussed in detail later in this chapter.

Table 7.1 Categories of Sexual Problems

Category	Problem	Main Indication
		Male
Sexual Desire, Interest, or Arousal	Male Hypoactive Sexual Desire Disorder	A lack of sexual fantasies and desires
	Erectile disorder	Difficulty achieving or maintaining an erection for sexual activity
Orgasm	Premature Ejaculation	Reaching orgasm and ejaculation immediately upon (within one minute of) penetration
	Delayed Ejaculation	Unwanted delay in reaching orgasm or absence of orgasm regardless of stimulation
		Female
Sexual Desire, Interest, or Arousal	Female Sexual Interest/Arousal Disorder	Lack of interest in sexual activity, absent sexual fantasies, little or no pleasure from sexual activity, or muted sensations during sexual stimulation
Orgasm	Female Orgasmic Disorder	Delayed, infrequent, or absent orgasms during sexual activity or minimal orgasmic sensations
Sexual Pain	Genito-Pelvic Pain/Penetration Disorder (GPPPD)	Pain before or during vaginal penetration and fear or anxiety about painful intercourse

SOURCE: American Psychiatric Association. (2013). Diagnostic and statistical manual of mental disorders (5th ed.). Arlington, VA: American Psychiatric Publishing.

How Common Are Sexual Problems?

Sexual problems are more common than people think. Why? Because most people typically do not discuss their sexual lives, and especially their sexual difficulties, with each other. When people are experiencing a sexual problem, they often assume that they are the only ones suffering from that particular difficulty. They assume that "everybody else's sex life must be just fine, but ours is in trouble!" They keep the problem to themselves, acting as if everything is normal and suffering in silence. In the meantime, numerous others are experiencing the same sexual problems and are thinking the very same thing, hiding their problems, and also pretending everything is OK! If people could be more open about sexual difficulties, they would find they have plenty of company, and they would feel less isolated and anxious. Fortunately, the Internet now offers individuals and couples a safe and reasonably private mode of communication and information about sexual problems.

Noted sexual researchers Masters and Johnson (1970), estimated that half of all married couples experience a "diagnosable" sexual problem at some point in their marriage. Other studies have found that rates for various sexual problems range from 10% to 52% for men and 24% to 63% for women (Heiman, 2002; Rosen et al., 1993; Spector & Carey, 1990). A U.S. national survey conducted in the 1990s found that over the previous 12 months, the incidence of individuals who reported at least one sexual problem lasting at least one month ranged from 3% to 33%, depending on the particular problem and the person's sex (Laumann et al., 1994). Table 7.2 summarizes these findings. Looking at the *prevalence* of sexual problems, that is, the number of people reporting a problem at one specific point in time, research has shown the range to be between 0% and 10%, depending on the study methodology and the specific sexual problem.

The main point here is simply to let you know that sexual problems are *very* common. This is an important piece of information because when people are suffering from sexual problems and understand that many others experience similar difficulties, they are more likely to seek help; find solutions; and live happier, more satisfied lives (Rosen & Laumann, 2003).

Table 7.2 Prevalence of Sexual Problems*

Percentage of Individuals Reporting Each Problem—All Age Groups (Ranges Reflect Varying Findings in Studies Using Different Forms of Questions, Descriptions of Problems, and Measurement Methods)

Problem	Men (%)	Women (%)
Low sexual desire	9.5–17.6	16.0–33.9†
Erectile problems	11.2–20.6	N.A.
Premature ejaculation	15.7–24.7	N.A.
Difficulty becoming aroused (indicated by vaginal lubrication)	N.A.	7.0–25.2
Inability to reach orgasm	8.1–14.5	8.0–14.0
Pain during sex	1.0–9.70	8.1–14.0
Performance anxiety††	17.0–20.0	11.5–16.0
No pleasure in sex††	8.1	21.2

N.A. = not applicable.

*Additional detailed statistics are discussed for each specific problem later in the chapter.

†In some studies, this included women who reported being too tired for sexual activity.

††May be the cause or the result of another sexual problem.

SOURCES: Balon & Seagraves (2009); Hayes (2008); Laumann et al. (1994); Mischianu & Pemberton (2007); Montorsi (2005); Saigal et al. (2006); Townsend (2006); Reissing, Laliberte, & Davis (2005).

Evaluating a Sexual Problem: Duration and Context

To understand and treat a specific sexual problem effectively, it can be helpful to consider three distinct dimensions of the problem: its duration and its context (the settings in which the problem occurs). This seemingly simple approach can be a great help in explaining the nature of any specific sexual difficulty. We'll discuss each dimension separately and then see how they combine to "paint a picture" of a sexual problem (see Table 7.3).

DURATION: LIFELONG VERSUS ACQUIRED This first dimension helps clarify how long a person has been experiencing the problem. A *lifelong* sexual problem is just what it sounds like: one that has always existed in the person's sexual life; an acquired problem is one that is occurring now but was not always present in the person's past sexual experiences. For example, if a man is having difficulties achieving and maintaining an erection, it is important to determine whether his problem is lifelong or acquired. If it is lifelong, this implies that he has *always* had difficulty with erections; if it is acquired, this would indicate that his erections have been fine in the past—say, in a previous relationship or at a younger age—but are not so fine now. Likewise, if a woman is unable to experience an orgasm and this problem is lifelong, it means she has never experienced one. However, if in the past she was orgasmic and now is not, her problem would be diagnosed as acquired.

CONTEXT: GENERALIZED VERSUS SITUATIONAL A *generalized* problem is one that occurs for an individual or couple in virtually *all* settings. On the other hand, a *situational* problem is experienced in specific settings but is absent in other contexts. For example, if a man has a generalized erectile problem, he is unable to achieve an erection in all situations, including, say, during masturbation, spontaneously during sleep, and with a partner. If, however, his erection problem is situational, we can assume he is capable of having erections in some settings, such as masturbation, but not others, such as while making love with his partner. Similarly, if a woman is orgasmic when she masturbates or during oral sex but not during intercourse with her partner, this would constitute a situational problem. Alternately, if she is unable to reach orgasm regardless of the setting or activity, her problem would be classified as generalized.

Sources of Sexual Problems

7.2 List and explain the possible sources and causes of sexual problems.

Now that you have some idea of how sexual problems may be evaluated and understood better, the next step in resolving them may lie in determining where or how the problem may have originated. Although many treatments for sexual problems are often effective *regardless* of the cause, finding the *best* treatment for a specific difficulty will usually require some clarification of the source of the problem. The source of most sexual problems usually falls into one or more of the following four categories: medical or biological causes, individual psychological causes, relationship issues, and differing cultural expectations. If the reasons for a sexual problem are not immediately evident (as is often the case), a skilled therapist can usually help individuals or couples uncover the true causes through careful questioning and discussion. Here is a brief summary of each of these potential causes of sexual problems.

Medical and Biological Causes

Sexual problems stemming from biological or physiological sources typically occur when the physical body is *incapable* of responding appropriately, regardless of the partner, the setting, or the sexual activities that may be occurring. These causes can be neurological, such as spinal cord or other nerve damage; hormonal, involving

Table 7.3 Classification of Specific Sexual Problems

Example of Problem	Duration		Context		Suspected Causes
	Lifelong	Acquired	Generalized	Situational	
A thirty-year-old woman has never experienced an orgasm with or without a partner in any setting	✔		✔		Physical, organic, or hormonal disorder; psychological issues from past sexual abuse and trauma; repressive family environment
A forty-year-old man who has had erections in the past is now completely unable to achieve an erection in any setting		✔	✔		Circulatory (vascular) illness; complications from prostate surgery; physical injury; nicotine, drug, or alcohol abuse; severe psychological trauma; depression
A forty-five-year-old woman finds that she rarely achieves orgasm regardless of the sexual setting. This has not been a problem for her in the past		✔	✔		Relationship difficulties; hormonal imbalances; drug or alcohol abuse; depression
A twenty-five-year-old man is never able to achieve or maintain an erection with his current partner but has no such difficulty when he masturbates		✔		✔	Relationship problems such as anger, resentment, or guilt; sexual orientation issues; performance anxiety; fear of pregnancy or STI transmission; drug or alcohol abuse in settings with partner
A twenty-five-year-old woman finds that she often fails to reach orgasm during lovemaking with her partner. This has not been a problem for her with past partners, and she is fully capable of orgasm through masturbation		✔		✔	Relationship problems such as anger, resentment, guilt, or abuse; sexual orientation issues; pregnancy or STI concerns; lack of adequate stimulation during sexual encounter; fatigue, stress, or anxiety in settings with partner; unrealistic expectations of intercourse and female orgasm

imbalances in the hormones that control and activate sexual responding; or vascular, meaning that the circulatory system is malfunctioning so that blood is not circulating as it should for sexual arousal. Such circulatory deficiencies may be due to disorders such as heart disease, diabetes, or atherosclerosis (hardening of the arteries) arising from cigarette smoking or poor diet. Physiological causes may also arise from physical injuries or trauma to the genitals or to the brain and nervous system.

Included in this category of sources of sexual problems is the use and abuse of alcohol and other recreational drugs. Use of alcohol and other drugs often accompanies sexual interactions because many people believe that drugs enhance the sexual experience. Such enhancement may be true for some people some of the time, usually in very small doses. In general, however, most recreational drugs *interfere* with normal, desirable sexual functioning. Probably the most common "antisex drug" is alcohol.

Calling alcohol an antisex drug may surprise you because many people believe that drinking alcohol makes people feel *more* sexual, not less. Indirectly, this may be true in some ways. Small to moderate amounts of alcohol might cause some people to become less sexually inhibited and to engage in various sexual behaviors more readily than they would if sober. However, beyond this depressant effect, alcohol also causes blood vessels throughout the body to dilate, which can reduce blood flow to the genitals and inhibit such responses as penile erection, clitoral engorgement, and vaginal lubrication (Arackal & Benegal, 2007; George et al., 2008).

Although alcohol may reduce inhibitions against sexual behavior, it may also interfere with sexual responding.

Alcohol is not the only culprit in drug-related causes of sexual problems. Most so-called recreational drugs, including amphetamines, heroin, cocaine, and marijuana, produce negative sexual side effects, especially with heavy or long-term abuse. The nicotine in heavy cigarette use has been found to interfere with erections in men (and possibly arousal in women) by causing plaque to build up in the arteries and reducing blood flow to the genitals. In fact, sometimes a male patient's complaint of erection difficulties is a physician's first hint of possible heart disease (Sadovsky, 2005).

SSRIs
Selective serotonin reuptake inhibitors; drugs administered to treat depression that may cause various sexual side effects, especially inhibited or delayed arousal or orgasm.

Probably the most common sexual side effects from prescription medications today relate to the widespread use of drugs that have revolutionized the treatment of depression over the past 20 years: the **SSRIs** (which stands for *selective serotonin reuptake inhibitors*). These drugs, known by the various trade names including *Prozac, Zoloft, Paxil, Luvox, Lexipro,* and *Celexa,* are highly effective in treating depression, usually without most of the negative side effects associated with earlier antidepressants. Unfortunately, however, for some patients taking these medications, one common side effect involves sexual problems, usually relating to inhibition of orgasm. A more complete discussion of these drugs and their effect on sexual functioning appears later in this chapter. Table 7.4 summarizes some of these drugs and their effects on sexual response.

Individual Psychological Causes

Although sexual problems may stem from various or multiple causes, it is safe to say that most are either *psychogenic* (having primarily a psychological cause) or involve one or more psychological components. Although the cause of a sexual difficulty may be psychological, the effect is usually physically manifested in the form of no erection, no arousal, no orgasm, reaching orgasm too rapidly or too slowly, and so on. This is not surprising in light of the many recent discoveries in psychology and biomedical research demonstrating a link between mind and body.

We know that strong emotions such as stress, fear, guilt, anxiety, and depression trigger responses in your nervous and endocrine systems that are incompatible with sexual arousal. If you stop to think about it, there's a very good reason for this. When your brain senses that you are feeling threatened in some way, it activates some systems such as heart rate, blood pressure, and adrenalin secretion to prepare you to meet the challenge of the perceived danger. This is called the "fight or flight response." At the same time, however, your brain shuts down networks in your body that are unnecessary for fighting or fleeing, such as digestion, salivation, and, yes, sexual responding. These built-in responses have helped humans to survive as a species through millions of years of evolution; we probably wouldn't be here without them. After all, if you are being chased across the plains by a wooly mammoth, wasting energy on sexual responding will do you no good at all!

However, for most of you in today's world, the threats you face over life's issues, such as relationships, money, jobs, family problems, school, grades, deadlines, and computer crashes, are not the sort of dangers that require you to fight or run away (physically, that is). Rather, the difficulties and challenges in your life cause you to worry, experience anxiety, become overly stressed, feel fearful, feel guilty, or become depressed. Nevertheless, when your responses to those threats become strong enough, your brain

Table 7.4 Effects of Selected Drugs on Sexual Functioning

Substance	Desire	Arousal	Orgasm
Tobacco		Inhibits—Especially in Males	Inhibits Due to Lack of Arousal
Alcohol	Inhibits—in High Doses—Especially in Males	Blocks in High Doses	Blocks in High Doses
Methamphetamine		Inhibits	Inhibits
Cocaine	Inhibits	Inhibits—Especially in Men—May Cause Priapism*	Inhibits
Marijuana**	Decreases		Inhibits
SSRI Antidepressants	Reduces	Inhibits—Especially in Males	Delays
MDMA (Ecstasy)	Decreases	Decreases—Especially in Males	Delays

* Priapism: Painfully long lasting erection.

** Most evidence is self-report; scientific evidence weak.

SOURCES: McKay, 2005; Finger, Lund & Slagle, 1997.

Table 7.5 Common Individual Psychological Causes of Sexual Problems

Fear of...	Anxiety About...	Guilt Over...	Stress Due to...	Other Causes
Pregnancy	Relationship issues	Lack of feelings for partner	Job or work	Poor self-image
Infertility	Body image	Unfaithfulness	Money	Low self-esteem
Abuse	Money	Masturbation	Nonsexual relationship problems	Feelings of powerlessness
Partner violence	Ability to respond	Sexual fantasies	Family problems	Feeling trapped
Transmitting or contracting STIs	Ability to please partner	Past behaviors	Illness	Past or current abuse
Pain	Lack of sexual experience or knowledge	Betrayal of partner	Loss or grief	Posttraumatic Stress Disorder (from combat or other traumatic experiences)
Being walked in on during sex	Infertility	Lying	Children and child rearing	Loss or grief
	Pregnancy	Cheating	Other family responsibilities	Serious illness (self or loved ones)
	Past sexual trauma	Other deception		Infertility
	Childhood sexual abuse	Repressive family environment in childhood		Physical or sexual abuse by partner

senses the danger you perceive and orders your nervous system to react much as it would to that wooly mammoth. Your body becomes ready to fight or run away (whether you need to or not), and it also becomes very unready to respond sexually. Table 7.5 lists some of the most common psychological causes of sexual problems. We will touch on many of these causes again in later discussions of the various specific sexual problems.

One of the growing recent issues relating to sexual problems concerns post-traumatic stress disorder in veterans returning from combat in Vietnam, Iraq, and Afghanistan. Clearly, this problem has existed for veterans of all wars, but our understanding of the sexual effects of PTSD are better understood today than in the past. Research has shown that both male and female combat veterans diagnosed with PTSD have significantly higher rates of marital and sexual problems. As the military's and medicine's awareness of the link between PTSD and sexual problems increases, greater steps are being introduced to address these issues along with those that many others veterans of war face (Price & Stevens, 2009; Schnurr et al., 2009).

Relationship Issues

Because nearly all sexual problems occur in the context of a relationship between two people, it is logical to suspect that the causes may lie in issues within the relationship.

Sometimes, problems will arise that neither partner has experienced before, or for some people there may be a pattern of similar difficulties in past relationships. Whatever the case, determining if specific relationship factors are contributing to the couple's sexual problem and then defining and resolving the issues may be crucial in solving their sexual troubles. Although individual couples experience a wide variety of specific relationship problems, difficulties in six specific areas may be the most likely ones to affect the sexual side of a relationship: trust, communication, anger or resentment, sexual expectations, respect, and love.

One of the most common causes of sexual problems today is the stress and anxiety of our fast-paced, multitasking lives.

LOSS OF TRUST Successful intimate relationships rely on trust, and the same is true of healthy, satisfying sexual interactions (Besharat, 2003; Kaplan, 1974). Most couples need to trust that their partner is faithful, is honest in expressing feelings, will not inflict emotional or physical pain, and is not withholding important information that might negatively affect the relationship. Sexual problems such as desire disorders, arousal difficulties, erectile problems, and orgasm difficulties may sometimes be linked to the lack or loss of trust in a relationship. When this basic trust is lost, it stands to reason that the overall relationship will suffer, and the sexual side especially may begin to deteriorate (Smyth, 2002).

POOR COMMUNICATION How often have you heard someone who is experiencing relationship problems say something like, "We just don't communicate!"? Effective communication is a cornerstone of a good relationship, and good sexual communication is usually the foundation of a good sexual relationship (Montesi et al., 2013; Weeks & Gambescia, 2000). Unfortunately, many couples—even those who enjoy excellent communication about most topics—find communicating about sex difficult.

As discussed at the beginning of this chapter, people tend to expect that good sexual interactions simply happen "naturally"; that we should be able to "sense" our partner's needs, wishes, and desires; and that something must be amiss if we have to *talk* about it. In reality, just the opposite is true. Sure, sex just "clicks" for some couples, but for most, the truth is that *words enhance sex*. The ability to express sexual feelings, sexual desires, or insecurities about sex is crucial in establishing and maintaining a sexually satisfying relationship. Often simply setting aside time or increasing the amount of time a couple allows for talking to each other about intimate issues and sharing feelings can go a long way toward increasing physical and emotional intimacy. Many couples complain of a gradual loss of intimacy over time; making the romantic side of the relationship a greater priority can help rekindle feelings of pleasure and intimacy.

ANGER AND RESENTMENT Strong negative emotional reactions, especially anger and resentment, work directly against sexual responding. If you are feeling angry or resentful toward your partner, regardless of the reasons, responding sexually is going to be difficult, at best (Bélanger, Laughrea, & Lafontaine, 2001; Kaplan, 1974; Wincze & Carey, 2012). In anger, two barriers to sexual intimacy are functioning simultaneously. One barrier is the specific physiological responses of the autonomic nervous system, which are incompatible with sexual arousal; the other is the psychological distance and loss of desire for intimacy with the partner that accompanies the anger.

One of the most difficult tasks in sex therapy is helping angry and resentful couples find strategies for resolving their conflicts and moving past the hostility (see the discussion on defining and resolving anger later in this chapter). If they refuse to do so, the therapy they are seeking for their sexual difficulties is likely to fail (Weeks & Gambescia, 2000).

CONFLICTING SEXUAL EXPECTATIONS Everyone grows up with expectations about sex, and no two people's expectations are the same. When people enter into a sexually intimate relationship, they are sure to have some differing expectations about sex that will need to be negotiated and reconciled. For example, one of the most common differing expectations about sex is how often they "should" make love. If one partner desires and expects that having sex is a daily event, but the other feels that once a week is ideal, they are likely to experience some very difficult conflicts if they fail to work out a compromise. Or if one partner expects to be the initiator of sex but the other often enjoys making the first move, these differing expectations, if not reconciled, can lead to a serious imbalance in the relationship and potential sexual problems.

When both partners are willing and able to communicate and discuss differences in expectations, chances are good they can work them out. However, sometimes partners have expectations of each other or about the relationship that they keep hidden, each assuming that the other should "just know what I want" and behave accordingly. This has been called the *myth of the lover as mind reader*. These hidden expectations are often extremely problematic and can lead to difficult conflicts, sexual, and otherwise (Weeks & Gambescia, 2000).

LACK OF RESPECT Lack of respect in an intimate relationship will invariably undermine sexual feelings, desire, and responses. Two distinct yet related types of respect are fundamental to successful and satisfying relationships: *self-respect* and *mutual respect*. A lack of self-respect, and the low self-esteem that invariably accompanies it, usually cause one or both partners to feel unworthy of experiencing sexual pleasure and undeserving of sexual pleasure that may be offered by the other. A person lacking self-respect may actually lose respect for a partner who attempts to initiate sexual intimacy (Leiblum & Rosen, 2000). Mutual respect is equally important, in that each partner needs to feel that his or her wishes, ideas, attitudes, desires, abilities, and unique characteristics as a person are honored and valued by the other. When this mutual respect is absent, the foundation for successful and satisfying sexual interactions crumbles away, and along with it goes other keys to intimacy such as trust, communication, and love.

LOSS OF LOVE Clearly, sexual interactions sometimes occur outside a context of a love relationship. However, in an intimate love relationship, sexual satisfaction and functioning often falter when those feelings of love between the partners decline. Of all the factors in intimate relationships that can influence sexual functioning, love is probably the most difficult to define. It is at least as important as the others we've discussed, possibly the most important of all (Sprecher, 2002). Perhaps love is a blend of the various aspects of relationships we have been discussing. Or maybe it doesn't really matter how we define love here. It probably suffices to say that love is an emotional, psychological, and physical condition about which most people are able to say, "I know it when I'm in it." What does matter here—most people would agree—is that sex combined with love is very different from sex without love. People generally make very clear distinctions between "making love" and "having sex" (Faulkner, 2003). Even sex therapists are careful about helping couples not focus too much on the sexual component or lovemaking may be reduced to "sexmaking" (Weeks & Gambescia, 2000).

Interactions of the Six Factors

Trust, communication, anger and resentment, sexual expectations, respect, and *love* are so closely intertwined that it is virtually impossible for a couple to be experiencing a problem with one factor without also feeling the effects on at least some of the others. For example, if a couple lacks effective communication skills, they will likely have difficulty reconciling conflicting sexual expectations. Their conflicting expectations could in turn lead to feelings of anger and resentment, a loss of respect for each other, and the eventual loss of feelings of love. You can start from any one of the relationship problems we have been discussing and logically trace a chain of escalating barriers to sexual intimacy to all or most of the other five. Figure 7.1 visually depicts this interdependent circle of relationship issues that may lead to sexual problems.

Figure 7.1 The Circle of Relationship Issues and Sexual Problems

Any single relationship factor on the circle may lead to, or be caused by, any other factor, which may in turn cause another relationship problem, and so on, until sexual intimacy is lost.

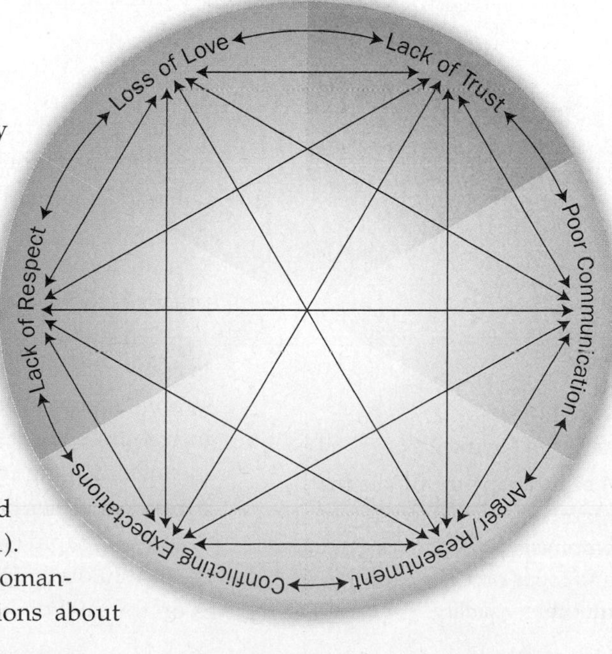

Cultural Expectations

If we are to have any hope of accurately studying and correctly interpreting any human behavior, including human sexual behavior, it must be considered within its cultural context (Bancroft, 2002; Triandis, 1989; Triandis et al., 1988). Moreover, few complex human interactions are influenced by culture more than sexuality.

As we become more global in our human interactions through freer economic trade, the Internet, and the ease of travel, intimate relationships are forming across cultural and national boundaries. Consequently, the sexual problems sometimes experienced by partners from diverse ethnic, social, and religious backgrounds have come into sharper focus (see "Sexuality and Culture: Latino and Caucasian Relationship Attitudes" on page 231).

When people from different cultures become involved in a romantic relationship, their deeply ingrained attitudes and expectations about sex and romance may clash, causing relationship or sexual problems. For example, cultures differ greatly in attitudes and expectations about nudity, modesty, acceptable sexual behaviors, sexual roles and responsibilities of men and women, privacy, parts of the body that are considered sexual and nonsexual, religious teachings about sex, and even expectations about sexual responding (Rye & Meany, 2007). Moreover, cultural sexual attitudes and teachings lead to wide variations of levels of guilt about sex, which in turn affects many aspects of sexual responding (Woo, Brotto, & Gorzalka, 2011). This is not to say that cross-cultural unions are doomed. On the contrary, the recognition of cultural sexual diversity has led today's sex therapists and researchers to pay significantly more attention to cultural differences and conflicts in their study and treatment of sexual problems (Ribner, 2003; Riela et al., 2010; Sunger, 1999). Beyond the professionals, however, this cultural awareness is growing throughout the general population. As our globe continues to shrink, cultural barriers to intimate relationships are increasingly likely to fade with time and increased interpersonal contacts across cultural boundaries.

Since You Asked

2. My girlfriend is from India. We've been together for nearly a year, but she doesn't seem interested in sex at all. Could this be due to growing up in a different culture, or is it something in our relationship?

In the next section, we begin discussing some general principles for helping individuals and couples resolve sexual problems. You may be wondering why a discussion of treatments appears *before* considering the specific problems themselves. The answer to that will become clear as you read on.

General Guidelines for Solving Sexual Problems

7.3 Analyze and explain general approaches to solving individual and couple-based sexual disorders.

To discuss *solutions* before discussing the various problems may seem backward. However, because most sexual problems share certain characteristics, you will find it helpful to learn a bit about several general approaches that have been shown to be successful in helping individuals and couples overcome their difficulties. We will refer back to these general treatment principles later as we review the various sexual problems and the *specific* solutions for each. The solutions or treatments for most sexual problems typically incorporate one or more of these basic strategies, combined with some specific methods designed for each individual difficulty. The most prominent of these general principles are *sensate focus*, therapeutic masturbation, and enhanced communication.

Sensate Focus

In their groundbreaking book, *Human Sexual Inadequacy*, Masters and Johnson (1970) described a comprehensive sex therapy program for couples dealing with various sexual problems. Mutual "pleasuring exercises" were a central feature of their program. They referred to this approach as **sensate focus**. Sensate focus, in various interpretations, forms, and applications, continues to play an important role in sex therapy practices today (Albaugh & Kellogg-Spadt, 2002; Lankveld, 2009; Pereira et al., 2013).

Since You Asked

3. How can a couple who has been together for a long time (married three years) get their sexual feelings back? We have a good relationship, but we hardly ever make love anymore.

Sensate focus is a paradoxical treatment for sexual problems because it assumes, in essence, that the sex will become better if a couple *stops doing it*! The idea behind this strange-sounding notion relates to Western cultures' insistence on equating sex with intercourse. For most couples, especially among heterosexual partners, "having sex" means engaging in sexual intercourse. However, most sexual problems occur in the context of either having or leading up to having intercourse. Erectile difficulties, lack of arousal, rapid ejaculation, lack of orgasm, inhibited orgasm, and painful sex usually occur while a couple is having or trying to have intercourse. In other words, the focus is on the goal of intercourse. Sensate focus exercises require that the partners redirect

sensate focus
A sex therapy technique that requires a couple to redirect emphasis away from intercourse and focus on their capacity for mutual *sensuality*.

Sexuality and Culture

Latino and Caucasian Relationship Attitudes

Attitudes Toward Romantic Behaviors: United States and Puerto Rico

As one example of how different expectations in romantic relationships have the potential to cause misunderstandings and discord, the research illustrated in the figure below sheds some light on what may appear to be minor variations, but, if you think about it, could produce serious consequences.

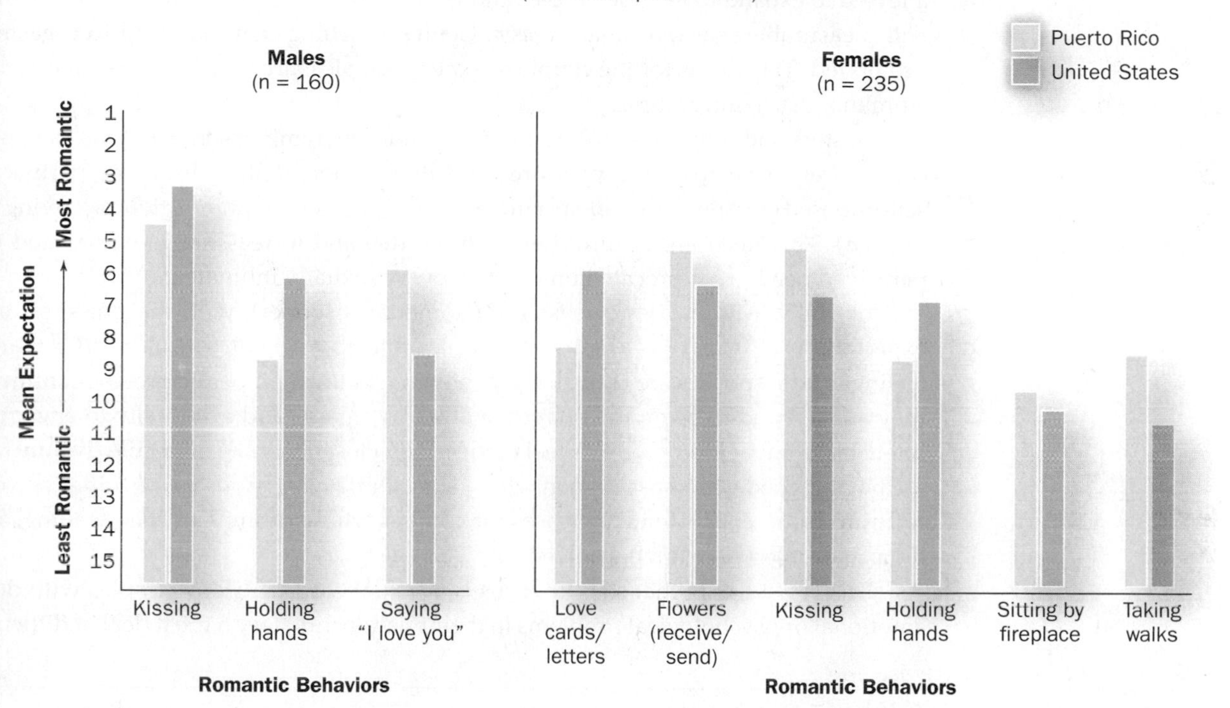

SOURCE: Data from table, "Significant Differences for Expectations for Romantic Behavior: University Students in Puerto Rico and the United States," from J. Quiles (2003). Romantic Behaviors of University Students: A Cross-Cultural and Gender Analysis in Puerto Rico and the United States. *College Student Journal, 37*, pp. 354–366. Copyright © 2003. Reprinted by permission of the publisher.

their emphasis away from intercourse and focus on their capacity for *sensuality* (see Albaugh & Kellogg-Spadt, 2002; Joanning & Keoughan, 2005; Pereira et al., 2013).

One way you may be able to relate to this idea is this: Most couples can remember times in their lives when they were being sexual with a partner, perhaps early in the relationship, but there was an agreement, spoken or unspoken, that their sexual behavior would not lead to intercourse for any number of reasons. Usually, people recall those sexual encounters as intensely arousing; the possibility that the body might not respond "correctly" was the farthest thing from their minds because their bodies were responding powerfully and effortlessly. Often, however, when a relationship progresses to having intercourse, much of that sensual, intensely sexy behavior disappears and becomes simply "foreplay," behaviors leading up to intercourse. If later on, either person begins to experience a sexual problem, the couple begins to feel anxious about sex, worried that the problem, whatever it is, will happen again when they have intercourse. This anxiety may in turn lead to other sexual problems or even the avoidance of sex altogether because any sexual activity triggers the awareness that intercourse is approaching. Therefore, the first rule of sensate focus is "no intercourse."

Here's how sensate focus typically works. Assuming any underlying physical causes have been ruled out or corrected, a couple who is experiencing a sexual problem is instructed to find quiet, private time together one or more times each week. The length of time may vary, depending on the individual couple's needs. During these sessions, they are to remove their clothing and spend the time taking turns touching and caressing each other, focusing on the pleasure they feel in touching and being touched. However, they must not touch each other's breasts or genital areas. "Self-Discovery: An Example of the Steps in a Heterosexual Couple's Sensate Focus Exercise" on the following page, provides one example of the sensate focus process, step-by-step in the style of one of the most famous sex therapists of the twentieth century, Helen Singer Kaplan.

As the therapeutic process proceeds, usually over several weeks, the caressing is allowed to expand to include nipples and genitals, but the goal continues to be sensual and pleasurable sensations, *not orgasm*. Genital touching that may lead to orgasm is prohibited. The idea is for the couple to experience pleasure for pleasure's sake, not as a preliminary to intercourse.

Masters and Johnson (1970) found that while engaging in sensate focus exercises, couples felt freed from the pressures and expectations (either from the partner or self-imposed) to "perform" in specific ways (i.e., achieving an erection or having an orgasm). This freedom, in turn, removed the demand to reassure a partner and the perceived need to reciprocate immediately every sexually intimate gesture.

Typically, within a few weeks (and sometimes sooner), with the pressures and expectations of intercourse removed, couples find that sexual, sensual, and romantic feelings and responses are reawakening in them. As the therapeutic process continues, they are allowed to begin to touch each other to orgasm and eventually to engage in intercourse once again. Couples feel emotionally closer, men with erectile dysfunction are having good erections, women who were experiencing arousal difficulties become fully aroused, sexual desire increases for couples whose desire had been waning, and orgasm problems are often resolved.

This is not to say that sensate focus is a sexual cure-all. Many couples with deep emotional or psychological problems in their relationship have a great deal of difficulty

Figure 7.2 Sensate Focus Exercises

Sensate focus exercises are designed to help a couple reconnect on a sensual, rather than sexual, level, without expectations or demands. After a while, the couple may begin to include touching of genitals and breasts, but without orgasm.

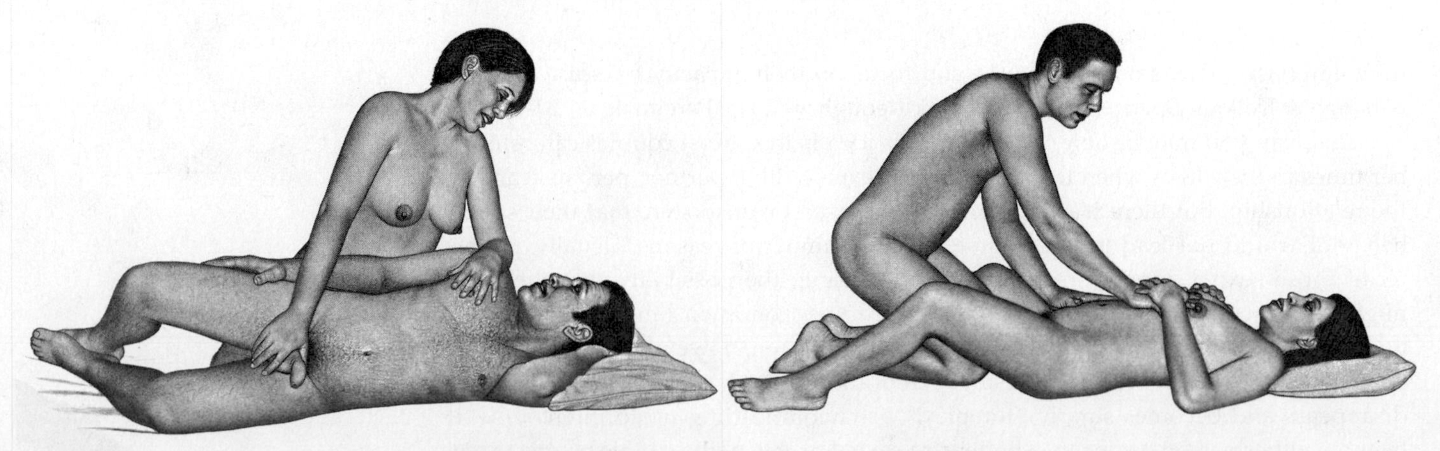

Self-Discovery

An Example of the Steps in a Heterosexual Couple's Sensate Focus Exercise (Kaplan, 1974)

1. Find about an hour when you can be alone together without distractions.
2. Take a relaxing shower, either together or separately.
3. While you're both still naked, the woman should lie on her front on a comfortable surface (your bed or a soft pad or cushion).
4. The man should now begin to caress her body from head to toes, very slowly and gently, focusing on how it feels to touch her skin. He may use his hands, lips, or both.
5. During this time, the woman should focus her attention on the sensations of his touches and try to stay in the moment without being distracted. She should not worry whether he is becoming tired or enjoying himself—she should only focus on her sensations.
6. Verbal communication should be kept to a minimum, but it may be necessary. If something feels either particularly good or too uncomfortable, she should let him know verbally (and gently).
7. She should try to identify which areas of her body feel especially pleasant and are the most responsive to his touch.
8. After ten or fifteen minutes, the woman should turn over onto her back so the man can caress the front of her body in the same way. The breasts and genital areas should be avoided.
9. Both partners should focus only on how it feels to touch and be touched.
10. After ten to fifteen minutes, it's the man's turn to be caressed by the woman, following the same process.

agreeing to and beginning the exercises. Other couples may find the intimacy and arousal of the exercises too threatening to continue. Still others, as they find themselves becoming aroused during the sessions, will depart from the instructions and have intercourse. As Kaplan (1974) notes, it is not unusual for a therapist to hear the following comment: "He started to touch me and we got so excited that we thought we shouldn't waste it, so we had intercourse" (p. 213). This demonstrates the power of sensate focus to enhance sexual feelings, but "breaking the rules" can lead to less successful long-term outcomes. Sensate focus approaches continue to be a mainstay of sexual therapy. You will see later in this chapter how it is often applied to specific sexual problems.

Masturbation as Treatment

As noted in Chapter 6, "Sexual Behaviors: Experiencing Sexual Pleasure," most medical and other clinical professionals today, along with most people in Western cultures, consider masturbation a normal and healthy sexual activity. Indeed, far from its long-past reputation as leading to a loss of sexual performance and enjoyment, masturbation today is a component in the *treatment* of various sexual problems. Why? Several reasons have been suggested: (1) masturbation allows individuals to become more aware of their own bodily sensations during sexual arousal and orgasm, which in turn (2) helps partners explain to each other what kinds of stimulation feel best; (3) it serves as a sexual release while couples are working on solving their sexual problems; and (4) it helps individuals work on specific sexual difficulties without a partner or issues they are not comfortable exploring with a partner.

This does not imply that a person or couple can simply "masturbate their way out of a sexual problem." The use of masturbation in sex therapy is usually referred to as **directed masturbation** (this does *not* mean that the therapist offers suggestions *while* you do it!). The idea is for the therapist to advise the client on how masturbation activities might effectively be used to help overcome a sexual problem (A. Kaplan, 2004; H. Kaplan, 1974).

Directed masturbation has been demonstrated to be an effective therapeutic component in the treatment of most common sexual problems, such as premature ejaculation,

Since You Asked

4. I'm a male and I'm curious if it is odd that I hardly ever masturbate. Is masturbation harmful or good for you?

directed masturbation
A sex therapy strategy in which the therapist advises the client on how to use masturbation activities to help overcome a sexual problem.

erectile dysfunction, arousal disorder, and especially inhibited or lack of orgasm (Frank et al., 2008; Lankveld, 2009). Specific applications of masturbation in treating these problems will be discussed shortly in the context of each specific problem. For now, here is the key to understanding why masturbation is such a powerful tool in treating sexual disorders: People who are experiencing a sexual problem with their partner rarely encounter the same problem when masturbating. For example, among the 510 couples that participated in Masters and Johnson's early studies, only 11 women and no men were unable to achieve orgasm through masturbation. Therefore, we can assume this particular problem is usually related to some aspect of the sexual interactions between the partners. On the other hand, when a sexual problem exists with a partner *and* during masturbation as well, this is often a sign that the problem is generalized and is more likely to stem from either physical or deep-seated psychological causes.

The use of masturbation in therapy is not always appropriate for all people, even if their sexual problem might be helped by it. Not everyone feels comfortable with masturbation no matter what the reasons. Furthermore, in some relationships, one or both partners' attitudes about masturbation may create a situation in which the therapy serves to compound their sexual and relationship problems rather than solve them. To avoid this, therapists must be aware of when and with whom recommendations of masturbation as therapy are appropriate.

Communication

The importance of effective communication in intimate relationships is discussed in detail in Chapter 4, "Love, Intimacy, and Sexual Communication," and was also touched upon in relation to the causes of sexual problems earlier in this chapter. Communication issues between intimate partners comes up so often it seems almost a cliché, but this only demonstrates communication's significant role in the success or failure of relationships. A common misconception is that if a couple has to *talk* about their sexual interactions, they must be doing something wrong. Nothing could be further from the truth. If you don't communicate your sexual desires, likes, dislikes, preferences, and feelings to your partner, how is he or she supposed to know them? Mind reading? No scientific evidence exists that people are telepathic about sex (or anything else, for that matter).

Since You Asked

5. I can't tell if my partner is satisfied or not when we make love. What are the signs that a person is sexually satisfied?

What about nonverbal communication? Certainly nonverbal forms of communicating, such as moans and groans and body movements, play a part in sexual interactions, but these signals are often vague and easily misread. What you might hear as a moan of pleasure from your partner might actually be a sign of encouragement to do something more intensely because it feels so good, or a request to stop whatever it is you are doing because it has become overly stimulating or even painful! Sometimes, you simply cannot be sure without using *words*. "Self-Discovery: Touch Mirroring: A Potential Problem with Nonverbal Communication," demonstrates how relying on nonverbal communication can backfire. It may sound familiar to you.

Virtually all sex therapy strategies involve an element of enhancing communication between partners (Carey, 1998; Kaplan, 1974; Lankveld, 2009; Masters & Johnson, 1970; Striar & Bartlik, 1999). Poor communication is an important factor in the development, escalation, and persistence of many sexual problems. If you think about it, how can anyone be a good sexual partner if he or she cannot communicate sexual likes and dislikes; turn-ons and turn-offs; and preferences for how, when, and where sexual interactions occur? These individual sexual specifics usually need to be communicated using words, because nonverbal communication, as shown in the "Self-Discovery" on page 235, can often lead a couple in exactly the wrong direction.

In the next section, you will see how some basic treatment principles—sensate focus, therapeutic masturbation, and communication enhancement—combine in the treatment of specific sexual problems.

Self-Discovery

Touch Mirroring: A Potential Problem with Nonverbal Communication

Here is an example of how a lack of verbal communication about the seemingly simple act of sexual touching can lead to frustration and dissatisfaction; it is called *touch mirroring*. A couple is beginning to make love. She prefers to be touched in soft, slow strokes, but he likes to be touched in a firm and more rapid rubbing motion. As they explore each other's bodies, she attempts to signal her touch preferences to him by demonstrating, touching him as she wants to be touched, softly and slowly, so that he can "mirror" the behavior back to her.

But this gentle touching is not what feels best to him, so he begins to touch her more firmly and rapidly, hoping that she will mirror his behavior in how she is touching him. She then perceives that her message is not getting through, so she touches him even more lightly and slowly, to which he responds with even firmer and faster rubbing. This continues to escalate until the pleasure evaporates in a cloud of frustration and neither feels much like touching at all anymore. What's the solution to this problem? Communication . . . using *words*.

Specific Problems and Solutions: the DSM-5 Revisited

7.4 Discuss the description, causes, and suggested solutions for each specific sexual disorder.

As mentioned earlier in this chapter, sexual problems are usually divided into three categories: problems with desire, interest, and/or arousal; problems with orgasm; and problems involving sexual pain. We will discuss each category in turn and focus on the most common problems for men and women in each category. For each specific sexual problem, you will find a description of the problem, an overview of possible causes of the difficulty, and a brief discussion of treatment methods that have been shown to be most effective.

Because this is a textbook and not a self-help book, we won't discuss the treatments in great depth or detail. We will, however, touch on a suggestion or two for each problem, in addition to referring to guidelines for solving sexual problems discussed earlier. If you are concerned about a sexual problem in your life, many resources are available from qualified sex therapists, in self-help books, and on the Internet. Going online for any sort of health problem is risky and it's important to seek out trustworthy, quality sites. One such site relating to sexual problems is www.sextherapyonline.org.

Problems with Sexual Desire, Interest, or Arousal

As outlined in Table 7.1, three specific sexual problems fall into the category of sexual desire, interest and/or arousal; for men, the problems are: *male hypoactive sexual desire disorder* and *erectile disorder*; for women, the problem is called *female sexual interest/arousal disorder*. These disorders are parallel for men and women because they describe, in a fundamental way, the same issues: either an overall loss of interest or desire to be sexual with a partner or, in the presence of interest or desire, an inability or difficulty responding to sexual stimulation with a partner.

MALE HYPOACTIVE SEXUAL DESIRE DISORDER When a man experiences very low or absent desire for sex with a partner, this is referred to as **male hypoactive sexual desire disorder (MHSDD)**. According to the DSM-5, this difficulty is diagnosed when it occurs on at least 75% of sexual occasions and has been a problem for at least six months.

male hypoactive sexual desire disorder (MHSDD)
A persistently low level or lack of sexual fantasies or desire for sexual activity

Description *Hypo-* means "under," "low," or "not enough" (as opposed to *hyper-*, which indicates "over," "high," or "too much"). Male hypoactive sexual desire is formally defined as "deficient or absent sexual fantasies and desires, as judged by a clinician" (DSM-5, 2013). In other words, when a man simply does not think about sex very much, he rarely, if ever, feels desirous of sexual activities, and if he experiences distress over this problem, he may be diagnosed with MHSD.

Typically, a man with hypoactive sexual desire disorder is sexually functional—that is, he is able to respond, become aroused, and reach orgasm—but he feels little or no desire for sex. In some cases, this lack of desire exists for sexual interactions with a partner only, but often desire fades for all forms of sexual activity, including masturbation. Low sexual desire has historically been assumed to afflict women more than men; however, this difference may be due to perceptions based on cultural expectations and stereotypes and not on actual numbers. Many therapists have reported approximately equal numbers of men and women who experience desire difficulties (Leiblum & Rosen, 2000). The American Psychiatric Association has recognized this fact in that, as of 2013, a similar disorder exists for men and for women, as will be discussed shortly.

Causes In attempting to determine the possible cause of male hypoactive sexual desire disorder, the criteria of duration and context (refer to Table 7.3) are very important. In cases involving low desire that is lifelong (has always existed) and generalized (occurs in all settings) underlying biological causes such as hormonal imbalances, and nervous system issues must be considered (Feldhaus-Dahir, 2009). However, for desire difficulties that are acquired (current, but not in the past) and situational (desire is present only in some settings), nonphysical factors such as psychological, relationship, or cultural issues are more likely to be in play. And virtually *any* of the factors listed in Table 7.5, alone or in combination, may play a role in a man's loss of sexual desire. Among the most commonly suggested nonphysical causes are anxiety over relationship and other sexual issues; depression; fear (of pregnancy, STIs, performance); and past sexual victimization (Atlantis & Sullivan, 2012; Kaplan, 1979; Leiblum & Rosen, 2000)

Solutions Due to the complexity of desire disorders in general in terms of diagnosis and causal factors, the most important piece of the treatment puzzle may be uncovering the reasons for the low desire. Examples of such issues might be: relationship discord; fear of unwanted pregnancy or STIs; past sexual trauma, sexual orientation confusion, loss of trust in partner; or a combination of these. In addition, the therapist must determine that both partners independently perceive that a problem exists and that they both truly wish to work toward a solution.

A therapist may suggest various therapeutic interventions. All three general treatment principles discussed earlier offer potential benefits for MHSDD. *Sensate focus*, in reducing anxiety about intercourse and fears about performance, may reawaken sensuality and, consequently, sexual desire. Masturbation exercises combined with desire-enhancing fantasy may assist in transferring fantasies to an increasing desire for a partner. Of course, enhanced communication between partners is crucial for keeping desire alive or rekindling it. Often, when a couple is dealing with a desire problem, they begin to avoid *all* forms of affection because touching, kissing, and even holding hands triggers negative feelings that have become associated with the low desire. And when affection and sexual desire fade, the communication that comes from sexual intimacy deteriorates as well (Leiblum & Rosen, 2000).

cognitive-behavioral therapy
A therapeutic approach designed to gradually eliminate specific thoughts and associated behaviors that may be contributing to sexual problems.

Some researchers and therapists have suggested varied yet similar versions of **cognitive-behavioral therapy** in treating low sexual desire (Immanuel & Phill, 2011; Stinson, 2009; Trudel et al., 2001). As the name implies, cognitive-behavioral therapy for sexual desire problems is designed to (1) assist an individual or a couple in identifying and exploring irrational, faulty, and self-defeating beliefs and attitudes (*cognitions*) that underlie sexual difficulties; (2) develop strategies to discard those ineffective cognitions and replace them with new, more accurate, and more constructive thought processes; and (3) gradually eliminate undesirable *behaviors* that were based on the old ways of thinking and replace them with new and effective actions stemming from the new belief systems. These approaches to overcoming low-desire problems have been shown to result in relatively high rates of success for both increasing sexual desire and enhancing overall relationship satisfaction. This approach will be discussed again in addressing female arousal/interest disorder below.

ERECTILE DISORDER A common sexual myth is that the penis should be ready for sexual intercourse at any moment; it should become hard and erect whenever a man (or his partner!) wants it to; and it should stay that way until he (or his partner) is finished. This expectation has been linked historically to a man's overall strength and masculinity as a person. Male erectile disorder used to be called "impotence," which implied that the potency or vitality of a man somehow vanishes if his penis doesn't become erect well enough, fast enough, and for a long enough period of time. Over the past decade, sex therapists, physicians, and educators have phased out "impotence" as a label for this sexual problem, even though it is still quite common to hear it in nonprofessional settings.

Difficulties with erections, called **erectile disorder (ED)**, are extremely common and occur at some point in most men's lives. It's estimated that between 5% and 20% of men have some degree of ED (Hatzimouratidisa, 2010). Over half of all men responding to surveys say they have experienced problems with erections during intercourse on at least some occasions (Feldman et al., 1994), and one study estimates that by age 40, fully 90% of men have experienced erectile difficulties at least once (McCarthy, 1992). Erectile problems increase with a man's age. Approximately 10% of men in the 20–40 age group report frequent erectile difficulties, increasing to just over 30% for men 40–60, although this problem is easily treated at any age (Saigal et al., 2006).

erectile disorder (ED)
Recurring or persistent difficulty in achieving or maintaining an erection.

Description The clinical definition of erectile disorder is: "The experience of at least 1 of the following 3 symptoms: (a) marked difficulty in obtaining an erection during sexual activity; (b) marked difficulty in maintaining an erection until the completion of sexual activity; (c) a marked decrease in erectile rigidity" (DSM-5, 2013). In addition, the problem must occur on at least 75% of sexual occasions over a period of six months or more, and must be causing emotional or psychological distress.

What does this clinical definition really mean? As discussed early in this chapter, the mere fact that a man is having trouble with his erections does not automatically imply that he has a sexual problem; he and his partner may be happy and satisfied engaging in other sexual activities together when an erection problem is occurring. In real life, erectile problems usually mean that the man is having difficulty achieving an erection and maintaining it for the penetration, thrusting, and orgasm (his and maybe hers) involved in sexual intercourse. In fact, vaginal and anal intercourse are really the only sexual activities for which an erect penis is necessary. Some couples do *not* find the occasional or even the frequent lack of penile erection to be an important problem in their sexual life together and have developed ways of "working around it" without feeling deprived of sexual satisfaction or intimacy. Remember, intercourse is not the only path to sexual fulfillment.

However, erectile problems are typically quite upsetting to most men and most couples. A complex set of psychological and emotional reactions typically accompanies erectile disorder—embarrassment, rejection, shame, frustration, anger, fear, and anxiety, just to name a few. Such reactions may lead to withdrawal from all intimacy in the relationship in order to avoid the possibility of further failure. Even when erectile problems have a physical basis, they usually are intertwined with psychological, relationship, or intimacy problems.

CAUSES The current consensus among sex researchers and therapists is that erectile problems usually stem from some combination of physical, psychological, and social factors, and rarely is only one of those three causes at work (Chillot, 2002; Weeks & Gambescia, 2000; Wincze & Carey, 2012). If you look back to the section on "sources of sexual problems" earlier in this chapter, you can see how a biological factor such as blocked coronary arteries (preventing adequate blood flow to the penis) or other circulatory problems (often caused by cigarette smoking); a psychological factor such as anxiety or depression; and social factors, including relationship problems or cultural issues, could combine to produce recurring erectile problems.

Owing in part to societal expectations placed on the penis's assumed *boundless* ability for erections, many men (and often their partners) are concerned over *any* loss of erection during lovemaking. The irony of this concern is that it is quite normal for penile erection to vary during a sexual encounter (as discussed in Chapter 3: "The Physiology of Human Sexual Resonse"). Nevertheless, if a man's erection subsides, a common reaction is, "Ohmygod! What is wrong with me?!" Consequently, one or more instances of erectile difficulty, regardless of the initial reason, frequently lead to a new cause: *performance anxiety*, the fear of not being able to perform "as expected." The development of performance anxiety was first described in a therapeutic setting by Masters and Johnson (1970) and has been seen in numerous studies over the years since their early work (Vares et al., 2003; Wincze & Carey, 2012). Masters and Johnson (1970) summarized the problem as one of the "impotent" man's "obsession" with his erection as each episode of sexual interaction approaches. He becomes worried, stressed, and anxious that he will not be able to function like a "normal" man, and will be unable to perform intercourse, thereby disappointing and perhaps embarrassing his partner and himself.

This is often the cause, not the result, of erectile problems. When the human body's stress mechanisms are activated, the nervous system *inhibits* sexual arousal, including erection. Therefore, it is the very anxiety felt from past erectile difficulties that often fuel subsequent dysfunction. Men who become entangled in this cycle do not approach lovemaking with thoughts of pleasure, but with anxiety so strong that they sometime break out into a cold sweat. This process often turns into a vicious cycle: perceived sexual failure leads to fear and anxiety, which leads to more failure, which increases the fear and anxiety, which produces more failures, and so on.

spectatoring

Mentally observing and judging oneself during sexual activities with a partner; may cause sexual problems.

Performance anxiety may also play a role in producing another common factor in ED and other sexual problems referred to as **spectatoring** (Van Lankveld, Geigen, & Sykora, 2008; Wincze & Carey, 2012). Spectatoring is not about watching your partner during sex or your partner watching you. It is about you "watching" *yourself*, mentally. When a person feels sexually insecure, a common and perhaps natural reaction is to become very self-aware during sexual interactions. If a man becomes anxious about his sexual performance and fearful about achieving or maintaining an erection, he begins to second-guess everything he does during sex. Instead of allowing himself to be spontaneous and enjoy the feelings and responses of sexual interaction, he's watching and judging himself: "Am I becoming turned on fast enough?" "How's my erection doing?" "Oh, am I losing it?" Spectatoring serves to increase his anxiety, create an emotional and psychological distance from the pleasure of lovemaking, and turn his erectile fears into self-fulfilling prophecies.

SOLUTIONS I know what you're thinking—*Viagra! Levitra! Cialis!* True, the treatment of ED has undergone a major revolution in recent years, in the form of a little blue pill called Viagra (sildenafil citrate), approved by the FDA in 1998 as the first pill for the treatment of erectile problems. Within a few years after Viagra, second-generation erectile medications—including Cialis (tadalifil) and Levitra (vardenifil)—began appearing on the market. They produce similar effects but are different mainly in terms of dosing; that is, they stay active in the man's body longer, allowing him to

Since You Asked

6. Is impotence always psychological? Does Viagra work for everyone?

"Now, that's product placement!"

SOURCE: Robert Mankoff. Copyright © *The New Yorker Collection*, 1999, from cartoonbank.com.

be more spontaneous as to when sexual activity may occur (and a newer version of Cialis is designed for daily use). These drugs are helping millions of men overcome problems with erection, but it would be a mistake to assume that they are a "cure" for every case of erectile disorder. Some men find that the drugs have little to no effect. These medications enhance the body's ability to *respond* to sexual stimulation with increased blood flow to the penis. These drugs are not "erection pills"; they do not *induce* an erection or increase desire in and of themselves. If a man were to take one of these medications before a final exam in a class or mowing the lawn, he would *not* be taking the exam or mowing with an erection.

Many men with ED who have been prescribed Viagra, Cialis, or Levitra, especially those with nonphysical causes of erection problems, report using it once or twice and then returning to satisfying sexual activity and perfectly fine erections without it. Simply knowing that the drug is handy, sort of an "erection good luck charm," is enough to alleviate their fears and return them to satisfactory functioning. At this moment, several additional erectile enhancement medications are in various phases of research and testing. These range from quick-acting pills to nasal sprays, from topical creams that are applied to the penis to gene therapy. Some of these may be on the market within several years (Jennings, 2013). If you or someone you know is wondering whether he may have ED and might need a medication such as Viagra, the "Self-Discovery: Self-Screening for Erectile Dysfunction," which appears on the following page, can provide a rough idea of the seriousness of the problem.

Any man experiencing problems with erections should see a physician for a physical exam to rule out potential physical causes. Such an exam goes beyond simply determining if the "sexual plumbing" is functioning properly. Loss of erections may be a sign of a more pervasive health problem such as arterial blockages, heart disease, or diabetes. Often, the man himself knows whether his difficulty is mainly physical or psychological. How does he know? Because he has erections in some settings, such as when he masturbates, when he awakens in the morning, or when he is with different partners.

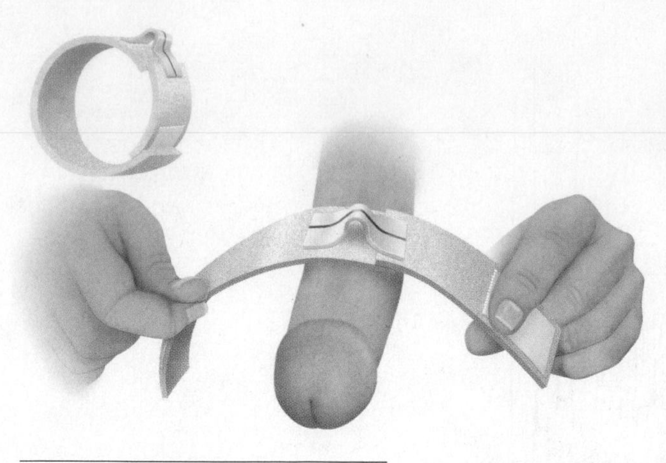

Figure 7.3 Home Screening for Erectile Disorder

Devices such as these allow men to determine if they are having normal erections during sleep.

nocturnal penile tumescence
Erection of the penis while a man is asleep.

A common test for determining the underlying cause of ED is to monitor **nocturnal penile tumescence (NPT)**, that is, erections during nighttime sleep. Normally, men experience three to six erections per night as a normal part of rapid-eye-movement or dreaming sleep (REM). These erections are a normal, routine physiological response during REM sleep and do not imply that the man is always dreaming about sex. However, if a man's penis is becoming erect during REM sleep, blood flow to the penis is probably functioning normally, and his problems with erections likely stem from nonphysical sources. One problem with this self-study is: How can he know if he is having erections at night if he's asleep? If he has a bed partner (or very close friend), he could ask this person to watch his penis during the night, but for most people, that's probably asking a lot! Today, inexpensive devices called penile tumescence monitors are available that are worn on the penis during the night and indicate in the morning whether the man has had an erection during sleep (see Figure 7.3).

Many, if not most, erectile problems, especially if they are primarily psychological and not physical, can be treated effectively without drugs. ED is a perfect example of a sexual problem that may be helped by sensate focus exercises such as those discussed earlier in this chapter. When the pressure to "perform"—to achieve

Self-Discovery

Self-Screening for Erectile Dysfunction

This quiz can help you figure out if you may have erectile dysfunction (ED). It is designed for men who are, or have been, sexually active, including vaginal or anal intercourse. If you think you may be having a problem with ED, be sure to take your results to your doctor. Only your doctor can decide if you have ED and if medication is right for you.

Pick the answer for each question that fits you best. Then add up the point values for your answer choices as indicated at the end of the scale.

1. How do you rate your confidence that you could get and keep an erection?
   - **a.** Nonexistent
   - **b.** Very low
   - **c.** Low
   - **d.** Moderate
   - **e.** High
   - **f.** Very high
2. When you have had erections with sexual stimulation, how often were your erections hard enough for penetration (entering your partner)?
   - **a.** Avoided sexual intercourse
   - **b.** Almost never or never
   - **c.** A few times (much less than half the time)
   - **d.** Sometimes (about half the time)
   - **e.** Most times (much more than half the time)
   - **f.** Almost always or always
3. During sexual intercourse, *how often* were you able to maintain your erection after you had penetrated (entered) your partner?
   - **a.** Avoided sexual intercourse
   - **b.** Almost never or never
   - **c.** A few times (much less than half the time)
   - **d.** Sometimes (about half the time)
   - **e.** Most times (much more than half the time)
   - **f.** Almost always or always
4. During sexual intercourse, *how difficult* was it to maintain your erection to *completion* of intercourse?
   - **a.** Avoided sexual intercourse
   - **b.** Extremely difficult
   - **c.** Very difficult
   - **d.** Difficult
   - **e.** Slightly difficult
   - **f.** Not difficult
5. When you attempted sexual intercourse, *how often* was it satisfactory *for you*?
   - **a.** Avoided sexual intercourse
   - **b.** Almost never or never
   - **c.** A few times (much less than half the time)
   - **d.** Sometimes (about half the time)
   - **e.** Most times (much more than half the time)
   - **f.** Almost always or always

6. When you attempted sexual intercourse, *how often* was it satisfactory *for your partner*?
   - **a.** Avoided sexual intercourse
   - **b.** Almost never or never
   - **c.** A few times (much less than half the time)
   - **d.** Sometimes (about half the time)
   - **e.** Most times (much more than half the time)
   - **f.** Almost always or always

Scoring

For each letter you circled, add up your points according to the following scoring key:

a = 0, b = 1, c = 2, d = 3, e = 4, f = 5 Total = ______

Interpreting your score

Along with discussions with his doctor, this quiz can help a man determine if he may have erectile dysfunction (ED). According to the pharmaceutical company that makes and markets Viagra, a score below 24 *may* indicate some degree of ED. However, this is only an estimate, and if any man's sexual response is troubling him, he should discuss it with his doctor, regardless of his score on this scale.

SOURCE: Adapted from Pfizer (2005).

and maintain an erection for intercourse—is removed, many men find that erections cease to be a problem. When the agreement has been made between partners to engage in "nondemand pleasuring," such as kissing, touching, and caressing, for the sake of pleasure itself, and *not as a prelude to intercourse*, many men who have been dealing with erectile disorder find themselves very aroused and very erect throughout the entire session.

Of course, other factors that may be contributing to the erection difficulties should be resolved so that the overall quality of the relationship that may have deteriorated either prior to or during the period of the specific erectile disorder may be improved. As mentioned earlier, these issues may be psychological, such as depression, fear of pregnancy or STIs, or the effects of past sexual traumas. They may be social problems relating to the current relationship, such as anger, resentment, guilt, loss of trust, and so on. Table 7.6 (on the following page) provides some important information relating to male arousal and erections that can help couples in understanding and overcoming ED.

The important point here is that psychological and medical science have advanced to the point where no man need resign himself to life without erections. Virtually every case of erectile difficulty can now be successfully treated.

Female Sexual Interest/Arousal Disorder

In the past, women suffering from **female sexual interest/arousal disorder (FSIAD)** were often referred to as "frigid," but this negative, judgmental term has disappeared from the professional literature, as has "impotence" when discussing ED. These sexual problems faced by women involve one or more issues including low interest and desire in sexual activities in general, or difficulty becoming sexually aroused when engaging in desired sexual interaction. This disorder parallels (but is not identical to) male sexual desire disorder and erectile disorder.

female sexual interest/arousal disorder
A woman's frequent or persistent inability to attain or maintain sexual arousal.

DESCRIPTION The formal, clinical definition of female interest/sexual arousal disorder (or FSIAD) involves a collection of several symptoms as follows: A lack of, or significantly reduced, sexual interest/arousal with *at least* 3 of the following: (a) absent or reduced interest in sexual activity; (b) absent or reduced sexual/erotic thoughts or fantasies; (c) no or reduced initiation of sexual activity and unreceptiveness to a partner's attempts to initiate; (d) absent or reduced sexual excitement/pleasure during sexual activity in almost all or all (at least 75%) sexual encounters; (e) absent or reduced sexual interest/arousal in response to any internal or external sexual/erotic stimuli (written, verbal, visual); and (f) absent or reduced genital or nongenital sensations during sexual activity in almost all or all sexual encounters (DSM-5, 2013).

Table 7.6 Understanding Male Arousal and Erections

Issue	Interpretation
It's common.	By age 40, fully 90% of all men experience at least one instance of erectile difficulty. This, then, is normal and does not indicate the presence of an erectile disorder.
Don't overreact.	The key to overcoming an erection problem is to avoid overreacting and labeling yourself "impotent" or some sort of sexual failure.
Men are *people*.	Reject the myth of the "male machine" who should be ready, willing, and able to have and use an erection at any moment, with any partner, and in any situation. A man and his penis are human, not performance machines.
Erections come and go.	Remember, during sexual activity of, say, 45 minutes, it is completely normal for an erection to come and go several times. You cannot *will* it to return, and trying too hard will backfire. Simply relax and enjoy the moment, and chances are your erection will return.
Sex is far more than intercourse.	It is not necessary to have an erect penis to be a good lover and to satisfy a partner. Even if you are having difficulty achieving and maintaining an erection during lovemaking, pleasure and satisfaction may be realized through a multitude of sexual behaviors such as manual or oral stimulation or with the use of "sex toys."
Stop thinking about an erection.	Instead of focusing on your erection, try to involve yourself physically and mentally in giving and receiving sexual pleasure. Remember, lovemaking is a participatory, shared activity, not a "spectator sport." Through participation and involvement, an erection will often follow in the natural course of events.
Communicate, communicate, communicate.	Try to become comfortable communicating your desires and needs to your partner. Tell your partner if you feel pressured, if things are moving too fast, or if you need more or different stimulation. Your partner will probably appreciate the feedback, and overall intimacy will be the result.
Try it by yourself.	Masturbation alone is a good method for overcoming ED. You can practice having an erection, losing it, and gaining it again through masturbation exercises. Through these exercises, you can become more familiar with your body's responses and more confident during lovemaking.
Erection is separate from ejaculation.	Although they usually occur together, it is not necessary to have an erection in order to ejaculate. They are separate physiological functions.
Not all erections are for sex.	Avoid assuming that a morning erection means you should immediately have sex. An erection upon awakening is a result of routine physiological arousal during the sleep cycle, and a "use it or lose it" attitude often leads to loss of the erection.
Think of sex as pleasure, not sex as performance.	Rather than focusing stimulation on a totally flaccid penis, engage in "nondemanding" sensate focus exercises instead. Become involved in the sensual feelings, and the sexual responses will often follow. The heart of good sex is pleasure, not performance.
Not all sexual encounters are perfect.	"Feelings about a sexual experience are best measured by your sense of pleasure and satisfaction rather than whether you got an erection. Accept that some sexual experiences will be great for both of you, some will be better for one than the other, some will be mediocre, and there will be others that are poor or downright failures. Do not put your self-esteem on the line for each sexual encounter" (McCarthy, 1992, p. 31).

SOURCE: Table, "Understanding Male Arousal and Erections," from "Erectile Dysfunction and Inhibited Sexual Desire: Cognitive-Behavioral Strategies," by B. McCarthy, in *Journal of Sex Education and Therapy, 18,* (1992), pp. 22–34. Copyright © 1992. Reprinted by permission of American Association of Sex Educators, Counselors & Therapists.

Sexual arousal in women does not have the easily observable signs of ED. Engorgement of the genitals in women is not as easily "measured," as is an erection, and a lack of natural genital lubrication may be difficult to determine, especially if the couple is using sexual lubricants in their lovemaking. Furthermore, engorgement and lubrication from arousal are not *technically* "required" for sexual intercourse, but an erection is. Historically, sexual problems revolved narrowly around whether or not a heterosexual couple was able simply to complete the act of intercourse; therefore, arousal problems in women received relatively little attention by researchers and therapists in past decades. In contrast, erection problems have been a primary research and treatment focus (Bartlik & Goldberg, 2000). This discrepancy, however, has changed, and greater research and therapeutic attention are appropriately being placed on female arousal and pleasure issues.

Female sexual interest/arousal disorder may be considerably more common than most people believe. One large survey found that nearly a fifth of women under the age of sixty reported difficulties with lubrication, and the percentages are considerably higher for women in their fifties or older (Laumann, Paik, & Rosen, 1999; Sutton et al., 2012). Although for men, ED may be a distinct and separate problem, women typically experience arousal problems in combination with other sexual disorders, such as orgasm difficulties or sexual pain issues (to be discussed shortly). It stands to reason that past difficulties reaching orgasm or painful sexual experiences might contribute to a woman's lack of sexual interest or arousal with a partner. Moreover, many leading researchers and sexuality educators are contending that female sexual problems, such as interest and arousal difficulties, are fundamentally different, physically and psychologically, from those of men and may require a separate diagnostic scheme if they are to be understood fully (Bancroft et al., 2003). Consequently, you may notice as you read this that some of the issues discussed relative to the causes and treatments for female problems do not necessarily correspond to those discussed for men.

Since You Asked

7. I love my boyfriend and want to have sex with him, but I just seem to have no interest—I feel guilty, but he is trying to be understanding. What's wrong with me?

CAUSES FSIAD may stem from biological, psychological, relationship, or cultural causes or any combination of these. One biological issue relates to imbalances in hormonal levels (Frank, Mistretta, & Will, 2008). A woman's hormonal levels are cyclical and may vary considerably during her fertility cycle, with pregnancy, during breastfeeding, with certain medications, and with age. Hormone levels are associated with sexual responses in women. One in particular that has been receiving increased attention is testosterone. You're probably thinking, "Testosterone? That's a male hormone, isn't it?" Yes it is, but women have some too, although only about one-tenth as much as men. In women, testosterone is produced by the ovaries along with estrogen, and it appears to be an important factor in female sexuality. Abnormally low levels of testosterone in women seems to be linked to low levels of desire, arousal, and even low sexual sensitivity in the nipples, vagina, and clitoris (Bartlik & Goldberg, 2000; Polan, 2007).

Yet, many women experiencing FSIAD may have no physiological problems at all, and their difficulties in becoming sexually aroused may stem from psychological, emotional, or relationship issues. Many psychological difficulties involved in various sexual disorders may play a role in FSIAD (refer back to Table 7.5). The most common of these difficulties include depression, anxiety, fear, and guilt. Loss of sexual desire is one of the hallmarks of clinical depression for men and women. Depressed people typically withdraw from human contact altogether and from sexual intimacy especially. Other psychological obstacles to arousal include anxiety and fear from past sexual abuse or rape, concern about one's ability to reach orgasm with a partner, fear of pregnancy or STIs, or other sources of personal or sexual distress. "In Touch with Your Sexual Health: Uncovering Possible Causes of Female Sexual Interest/Arousal Disorder" reveals how one woman's loss of sexual interest or desire may be linked to a very rational, yet not realized, fear of becoming pregnant.

In Touch with Your Sexual Health

Uncovering Possible Causes of Female Sexual Interest/Arousal Disorder

This dialogue between a sex therapist and her client demonstrates how fear and guilt may be contributing to (or causing) the client's sexual difficulties, but she has not come to this realization on her own. Through dialogue, the therapist uncovers this probable barrier to her sexual desire issue and helps her to come up with a possible solution.

CLIENT: My biggest concern is that I just don't seem to want to have sex with my husband.

THERAPIST: Are you and your husband having sex at all?

CLIENT: Well, yes, but not very often.

THERAPIST: When was the last time?

CLIENT: Maybe a month ago. I feel bad. And guilty. I love him and I know he wants to... but I'm just not interested.

THERAPIST: Did you and your husband have sex more often in the past?

CLIENT: Oh, yeah! At the beginning, we made love every day, at least. And even after two or three years, we still had sex a few times a week.

THERAPIST: So do you have a sense of when that changed?

CLIENT: Around the time of our second baby, about a year ago. After the first one, we got back to making love as soon as we could. But not this time. I'm pretty exhausted all the time with both kids.

THERAPIST: So, you don't want more children?

CLIENT: No! We're clear about that. In fact, the second one wasn't planned. We love her dearly, but she was definitely a surprise. I'm at my limit with two.

THERAPIST: So now, what form of birth control are you using?

CLIENT: (*pausing, sheepish*) We, uh, don't.

THERAPIST: You're not using any birth control at all?

CLIENT: We try to time it right, you know, during the month.

THERAPIST: Have you ever used birth control in your marriage?

CLIENT: Actually...just a little at the beginning, when we first met—he used condoms, but nothing since we've been married.

THERAPIST: Have you thought about what will probably happen now if you have sex with your husband on a regular basis, like before?

CLIENT: Well, I guess I'll probably get pregnant. Seems we're pretty fertile!

THERAPIST: And it sounds to me from what you said that you don't want to have another child....

CLIENT: That's true. I just can't handle another child, and I don't think he wants anymore either!

THERAPIST: Let's think about your situation. Do you think your concerns about avoiding pregnancy could have anything to do with your lack of sexual desire?

CLIENT: So, you're saying that I have no sexual desire because I'm afraid of getting pregnant?

THERAPIST: Well, let's think about it. No one is ever eager to do something if she knows that it will lead to undesirable or negative consequences. It makes perfect sense, don't you think?

CLIENT: [long pause, looking down] So, I guess I better get on the pill... ?

Women who have grown up in a sexually restrictive culture or very religious family setting often find it difficult to enjoy what they were taught was wrong, bad, and sinful, and instead, they may feel guilty. To be able to relax and enjoy the pleasure of lovemaking is the key to becoming aroused for women and for men. Just as a man cannot "will, wish, or demand an erection," neither can a woman command herself to become aroused.

Most sexuality educators agree that the quality of a relationship is key to the enjoyment of sexual interactions. Relationship difficulties are a common reason that a woman may have difficulty becoming aroused with her partner. Suppressed anger, fear of rejection, imbalances of power, poor sexual communication, lack of trust, infidelity, specific deep conflicts between the partners, and conflicting expectations may interfere with a woman's ability to respond sexually with her partner. Beyond these relationship problems, however, it is important to note that when a woman ceases to become aroused with a partner who is abusive or who employs coercive or violent tactics to obtain sex from her, she does *not* have a sexual disorder. The issue of relationship abuse is discussed in detail in Chapter 4, "Intimate Relationships," and Chapter 13, "Sexual Aggression and Violence: Rape, Child Sexual Abuse, and Harassment."

SOLUTIONS Therapy for FSIAD typically includes a combination of the general treatment guidelines discussed throughout this chapter: enhanced communication, specific sensate focus exercises (see Table 7.7), and directed masturbation. These

Table 7.7 Five Stages of Sensate Focus Exercises for Women Experiencing Female Sexual Arousal Disorder (FSIAD)

Helen Singer Kaplan, one of the pioneers in applying Masters and Johnson's theory of *sensate focus* to solving sexual problems, suggested a multi-stage sensate focus process for addressing FSIAD specifically. These stages should take place over several weeks, not in one or two intimate sessions. The goal of this form of treatment for FSIAD is to alleviate anxiety that a woman may be feeling about impending intercourse or other sexual activities. By removing that anxiety through this specific form of sensate focus exercises, she begins to relax and her capacity to become sexually excited returns (Kaplan, 1974).

Exercise States	
Stage 1:	As with most of these types of treatments for sexual problems, the couple should first ban intercourse or other activities that stimulate the genitals or breasts and nipples.
Stage 2:	The couple should engage in gentle touching and caressing of each other's bodies, avoiding the genitals and breasts.
Stage 3:	The couple may begin to engage in genital and breast touching without orgasm. In fact, if the woman feels an orgasm approaching, she should signal that the activity stop, until she becomes less aroused, before continuing. She should do this each time she feels she may be approaching orgasm.
Stage 4:	The couple may begin to have orgasms through methods other than intercourse if sexual frustration or desire becomes too intense.
Stage 5:	A gradual return to intercourse (for heterosexual couples) or other sexual activities (for same-sex couples) may now begin. This should happen very slowly, perhaps over several love-making sessions. It should seem almost as if the return to these activities feels "teasing," similar to the beginnings of sexual relations early in many relationships. The woman with the arousal difficulties should be the initiator and in charge of these activities.

treatment strategies are all designed to help a woman understand her body's responses to sexual stimulation, acquire a sense of mastery over her ability to become aroused, and enhance intimacy between her and her partner.

If a lack of normal levels of testosterone in women contributes to FSIAD, then shouldn't medically prescribed small doses of testosterone help return women to normal levels of desire? Current research indicates that the answer for some women to that question may be yes, especially for those who are near *menopause*, the period of time in a woman's life when she gradually ceases to produce estrogen and her menstrual cycle ends (see Chapter 12, "Sexual Development Throughout Life," for more information about menopause). Although negative side effects must be carefully monitored and avoided, testosterone replacement therapy in pill, sublingual (under the tongue), patch, or cream form has been shown to increase sexual desire and genital responsiveness in women with FSIAD (Chu & Lobo, 2004; Frank et al., 2008; K. Johnson, 2004; Poels et al., 2013).

Research continues on the effectiveness of Viagra and similar drugs for women. A few studies have found increases in sexual fantasy, desire, and function in women using Viagra (K. Johnson, 2002; Salerian et al., 2000). One larger study, however, found that Viagra significantly helps women experiencing sexual side effects from serotonin-targeted (SSRI) antidepressant medications (Nurnberg et al., 2008). Why this medication would be effective for this particular group of women and not those who are not using these antidepressants is as yet unclear. Efforts continue among major drug companies to develop medications targeted specifically for female sexual problems. However, an increasing number of researchers argue that sexual functioning in women is fundamentally distinct from that of men and that the development of new medicines to solve female sexual problems may be a misguided approach (McHugh, 2006) (see "A 'New View' of Female Sexual Problems" later in this chapter).

Desire: A Shared Couples' Problem

Although difficulties with desire and arousal have been distinguished and described separately in the DSM-5 for men and women, it is important to point out that, as with most sexual problems, this is typically a *couples' problem*. That is, a loss of desire or decrease in arousal are issues that *both* partners are experiencing; it is affecting both and may be impacting, in negative ways, the relationship overall.

In the 1990s, Barry McCarthy, a professor of psychology at the American University/Washington Psychological Center and certified sex therapist, developed a cognitive-behavioral approach for helping couples understand what might be occurring when desire declines and how they might address it. To help you grasp a sense of how cognitive-behavioral techniques help with sexual desire problems, Table 7.8 on the following page, summarizes his program (McCarthy, 1995). McCarthy reminded us that hypoactive sexual desire is usually a problem for the couple as a unit:

> Conceptualization, assessment, and treatment of [low desire] needs to focus on this as a couple issue and broadly examine cognitions, knowledge, scripts, family of origin, sexual socialization, systems issues, comfort, skills, and flexible variable sexual scenarios and techniques. [Low desire] is a changeable pattern as the couple identifies and confronts inhibitions and builds bridges to desire. (p. 140)

Problems with Orgasm

Sexual problems involving orgasm are somewhat different in character for men and women. For men, the most common problem is *premature* (or rapid) *ejaculation*—when orgasm happens too quickly and before the man wants it to. However, a small

Table 7.8 A Cognitive-Behavioral Approach to Hypoactive Sexual Desire

Barry W. McCarthy at the American University/Washington Psychological Center developed a cognitive-behavioral approach for treating couples dealing with low sexual desire. He introduced his method as follows:

> The optimal sexual process involves positive anticipation [of sex], an emotionally stable relationship, non-demand pleasuring [such as sensate focus], multiple stimulation, sharing orgasm, emotional satisfaction, and a regular rhythm of sexual activity. The dysfunctional pattern is negative anticipation, blaming and alienation, performance anxiety, goal-oriented stimulation, intercourse focus, failure, and sexual avoidance. (1995, p. 133)

McCarthy described a five-component model of treatment that focused on sexual anticipation, owning your own sexuality, deserving sexual pleasure, arousal and orgasm problems, and valuing emotional *and* sexual intimacy. Here is a brief description of each component.

Component	Description
Sexual anticipation	It's tough to desire sex if you are not looking forward to it. When one or both members of a couple feel pressured to desire sex, the result is usually the opposite: loss of desire and even avoidance of sex. Couples can build greater positive anticipation of sexual interactions in two ways. First, a couple should set specific "sexual dates" that they can plan for and look forward to. These should be when both partners are at their best in terms of mood, energy, and time. Second, couples should avoid falling into predictive or boring lovemaking routines by communicating their sexual desires and planning special sexual turn-ons.
Owning your own sexuality	You are the one who is in charge of your own sexuality. It's your responsibility, not your partner's, to be sexually satisfied; to develop a respectful, trusting, and intimate relationship; and to insist on your sexual rights, such as not fearing unwanted pregnancy, being safe from STIs, and rejecting abusive behaviors. You must accept your sexual history, including any negative sexual experiences you may have suffered in your past, and work to resolve any lingering guilt. You must be willing to communicate to your partner your sexual feelings, desires, and needs, as well as activities that turn you on.
Deserving sexual pleasure	Not all people feel that they deserve to experience sexual pleasure. This may be due to such issues as family upbringing, past negative sexual experiences, low self-esteem, or negative body image. If you don't feel you deserve sexual pleasure, how are you going to desire sexual interactions? The answer is, you're not. All people deserve sexual pleasure because they are human beings, and that pleasure should not depend on the presence or absence of any particular personal characteristic or experience.
Arousal and orgasm problems	Problems with arousal or orgasm go hand in hand with HSD. "For males, dysfunction problems almost always precede desire problems. The male apologizes or tries to cover up his erection, early ejaculation, or [delayed ejaculation] problem. He feels he is failing sexually, avoids [sex], and this inhibits his desire .... For females, the mistaken assumption was if the woman could only learn to have orgasms, or better orgasms, or more orgasms, then the desire problem would take care of itself—a seductive, but false assumption" (pp. 138–139).
Valuing emotional *and* sexual intimacy	Emotional intimacy in a relationship is not the same as sexual intimacy. When emotional intimacy is *combined* with sexual intimacy, a couple usually does not have a problem. Emotional intimacy problems tend to be more of a male problem, contends McCarthy, and men must alter their cognitions about emotional involvement if desire problems are to be avoided or overcome. "Both the male and his partner need to value intimate, interactive sex.... The major psychological aphrodisiac is an involved, aroused partner" (p. 140).

SOURCE: Table, "Cognitive-Behavioral Approach to Hypoactive Sexual Desire," by B. McCarthy, in *Journal of Sex Education & Therapy, 21,* (1995), pp. 132–141.

percentage of men experience what might be considered the opposite problem of *delayed ejaculation,* when it is difficult to achieve an orgasm regardless of the type of amount of stimulation. When a woman is experiencing a problem with orgasm, it is virtually never a problem of overly fast orgasms, but usually means she is not having orgasms often enough or "quickly" enough, or her orgasms are unsatisfying. This is known as *female orgasmic disorder.*

delayed ejaculation
Unwanted delay in reaching orgasm or absence of orgasm regardless of amount of stimulation

DELAYED EJACULATION A small percentage of men (an estimated 8% or less) experience **delayed ejaculation** (Corona et al., 2010). Delayed ejaculation is characterized by an unwanted delay in ejaculation or the absence of ejaculation during sexual activity with a partner (DSM-5, 2013). Men with this problem find that they are unable to reach orgasm after long periods of stimulation, which in most cases would be adequate in terms of focus, intensity, and duration for producing an orgasm and ejaculation. To be diagnosed, this must have been occurring on at least 75% of sexual occasions for a minimum of six consecutive months. Although this problem may sound advantageous to some men (and to some women), the real effect is extreme frustration, usually for both partners, which may lead to other relationship problems and avoidance of lovemaking. This difficulty is usually acquired and situational, in that the problem typically has not been present in the past or with other partners, and is not a problem during masturbation. Men who experience delayed ejaculation typically have no difficulty obtaining or maintaining an erection, and their erections typically last throughout the extended period of time they are attempting to reach orgasm.

CAUSES The causes of delayed ejaculation are unclear but may include many of the psychological and relationship issues discussed earlier in this chapter, such as anger, fear of pregnancy, past traumatic events, or sexual guilt. Probably the most common cause of this complaint, and the reason cases of delayed ejaculation have increased in incidence over the past 20 years, involves the sexual side effects of commonly prescribed SSRI antidepressants, such a Prozac, Paxil, Luvox, and Celex.

Solutions Treatment for delayed ejaculation usually consists of identifying and resolving such psychological and/or relationship issues, or adjusting the antidepressant medication or dose (Ferentz, 2007; Hall-Flavin, 2011). Although sexual side effects due to these antidepressants are common, they occur in a relatively small percentage of people. Often, after taking the medication for a month or two, the sexual side effects will improve. Moreover, different users of these medications react in various ways to different SSRI formulations. In nearly every case, a specific antidepressant in a specific dose can be effective without causing serious sexual side effects. Sometimes, this requires some experimentation and trial-and-error in consultation between the patient and his doctor. Moreover, some men find that taking a break for a day or two from the medication (sort of a medication-free "sexual holiday") will allow the body to return to normal functioning, but this should be done only under a physician's supervision.

PREMATURE EJACULATION By contrast, when a man is reaching orgasm *too* quickly, it is called **premature ejaculation (PE)**. Reaching orgasm and ejaculating "too soon" with a partner is the most common sexual complaint of men (especially younger men) and may be the most frequently encountered sexual problem (other than low desire) in sex therapy. Estimates are that 20% to 35% of men are affected by this sexual problem (Balon & Seagraves, 2009; Byers & Grenier, 2003).

premature ejaculation (PE)
A man's tendency to have an orgasm suddenly with little penile stimulation, typically within one minute of penetration; also referred to as *rapid or early ejaculation*.

Description Premature ejaculation (PE), also often referred to as *rapid* or *early ejaculation*, has always been difficult to define. Masters and Johnson (1970) diagnosed PE for heterosexual relationships when the man reaches orgasm before his partner more than 50% of the time during intercourse. But that is way off the mark because half or more of all women do not have orgasms routinely during intercourse *regardless* of how long intercourse takes. Some researchers have attempted to define PE as the man's *subjective* experience of having an orgasm sooner than he wishes. This definition is also problematic because many men define their own ability to control orgasm based on the myth of "lasting long enough" for their partner to have one.

Today, the formal definition of PE is quite specific: it is a pattern (over six months on 75% of sexual occasions) of ejaculation during sexual activity with a partner within approximately one minute following vaginal penetration and before the man wishes it (DSM-5, 2013).

Like most sexual problems discussed in this chapter, PE revolves around sexual activities with a partner, usually vaginal or anal sex. Rarely is a man concerned with reaching orgasm too fast during masturbation, oral sex, or other activities with a partner. PE is a common problem in part because, as has been discussed throughout this book, intercourse has been made the singular goal of lovemaking. Obviously, if a man ejaculates before intercourse can even get underway, both he and his partner may see this as a serious problem in their sexual lives. For heterosexual couples, the man's inability to delay his orgasm may be interpreted (often erroneously) as the cause of whatever orgasm problems she may be experiencing. Consequently, he feels inadequate as a lover, ashamed of his perceived lack of control, and filled with anxiety and even dread as each new sexual encounter approaches. These reactions may in turn lead to erectile difficulties, loss of sexual desire, and potentially avoidance of sex with a partner altogether (Wincze & Carey, 2012). Fortunately, PE is one of the easiest sexual problems to treat, as you will see shortly.

Causes Many causes for rapid ejaculation have been hypothesized, but none has been clearly demonstrated by scientific research. With this lack of solid evidence for any specific cause in mind, Table 7.9 offers a brief list of some of the *suggested* origins of PE.

What's the bottom line on the cause of premature ejaculation? No one really knows for sure. Perhaps a more important question, then, is: how can a man acquire better ejaculatory control? Fortunately, treatment strategies and outcomes for PE are much more definitive than its origins.

Solutions Men who consistently ejaculate very quickly typically strive to delay their orgasm during intercourse using strategies that, unfortunately, are minimally effective (Althof, 2006; Grenier & Byers, 1997). The most common of these strategies involves attempts to distract themselves from the sexual act and their physical sensations of pending orgasm, usually by focusing on antiarousing thoughts such as visualizing an ugly teacher, thinking about STDs, counting backward from one hundred, or fantasizing about dead animals on the side of the road or parents walking in, to name just a few. Another common behavioral strategy involves the use of local anesthetics to numb the penis (these penis-numbing products are sold at most pharmacies, but usually contain benzocaine or other chemicals that will numb the body no matter where applied). These are usually unproductive tactics for two reasons. One, few men would want to employ these strategies on every sexual occasion, and two, they detract from the pleasure of the sexual encounter. Think about it: who would want to be making love with a

Table 7.9 Suggested Causes of Premature Ejaculation

Research has proposed the following characteristics of men who experience premature ejaculation. This is not to suggest that a man with the disorder will possess all of these; any one factor could be a cause in itself.

Men who experience long-term rapid ejaculation:

1. May be more sensitive physically to sexual stimulation, which causes them to become aroused more quickly (some research support, some refutation; more research needed).
2. May have faster-acting reflexes in the pelvic and genital muscles that control orgasm and ejaculation, so that ejaculation may occur prior to full arousal (some research support; further study needed).
3. May be less able to perceive the moment in their arousal sequence that signals the onset of orgasm and ejaculation ("the point of no return") (some clinical support, some refutation; findings unclear).
4. May be angry with or dislike women in general or their partners specifically and unconsciously wish to deprive them of their pleasure (little to no support).
5. May tend to have longer periods of abstinence between sexual encounters, leading to faster arousal and ejaculation (some research support; findings unclear).
6. May be conditioned over time to ejaculate quickly due to rushed masturbation and sexual encounters during their developmental years when the threat of getting caught placed a premium on finishing quickly (some anecdotal and clinical support; little scientific evidence).
7. May be normal. This theory contends that as humans evolved, ejaculating quickly had survival value in that males who were able to mate quickly could impregnate more females and were in a condition of increased vulnerability (that is, easy to attack while mating) for shorter periods of time (generally refuted).
8. May have higher levels of testosterone (convincingly refuted).

SOURCES: Grenier & Byers (1995); Hong (1984); Masters & Johnson (1970); O'Donohue, Letourneau, & Geer (1993); Polonsky (2000); Waldinger (2007); Wincze, & Carey (2012).

numb penis while thinking of road kill? Not to mention that if the anesthetic agent is not used *inside* a condom, the man's partner will become numb, too.

What are effective solutions for PE? Once again, if you remove *intercourse* from the sexual equation, most men do not have a rapid ejaculation problem. *Sensate focus* exercises (discussed earlier) in which intercourse is banned for a while often help the man with PE feel more in control and allow for greater intimacy between partners by extending each overall lovemaking session. It is important for couples to remember that pleasure, closeness, intimacy, and sexual satisfaction can be achieved through various sexual behaviors and not through intercourse alone.

Since You Asked

8. Are there ways for a guy to control orgasm and last longer?

Beyond sensate focus, other treatments for PE approach the problem from exactly the opposite direction than most men attempt. The solution is found not in *distraction* from sexual sensations, but by *focusing* on them. Most treatments employ one of many variations on a theme of interrupting sexual stimulation as the man approached orgasm, called the **start-stop method**. This involves stimulation of the man's penis just up to the beginning of orgasm, followed by stopping the stimulation and backing off from that moment. This is repeated, over and over, to allow the man to focus on his physical sensations just prior to an orgasm and to develop conscious control of his orgasm. This starting and stopping can be accomplished through manual or oral stimulation or with intercourse with his partner-on-top to allow for easier starting and stopping (Figure 7.4, a and b). Furthermore, it is not absolutely necessary for a man to do this with a partner. He can begin to learn about his responses and acquire some conscious control over his orgasms by engaging in "start-stop" techniques while masturbating by himself (Polonsky, 2000).

start-stop method
A technique used in the treatment of premature ejaculation involving intermittent increases and decreases in arousal

After several sessions using the squeeze or start-stop technique, the next step is for the couple to begin to engage in slow, nondemanding intercourse with the receptive partner on top. When the man feels he is nearing orgasm, he lets his partner know this verbally, and the movement stops or the penis is withdrawn until the sensation of imminent orgasm passes, usually in less than 10 seconds. The couple then practices this process until the man gains increased conscious control over his body's progression toward ejaculation. These techniques allow most men (70% to 80%) to develop greater control over orgasm and ejaculation within a relatively short time (Rowland & Motofei, 2007). The same techniques have been shown to be successful for heterosexual and same-sex male couples.

Figure 7.4 Treatment for Premature Ejaculation

The start-stop Masters and Johnson's (a) "squeeze technique" and newer versions (b) called "start-stop techniques" help some men overcome problems with rapid ejaculation.

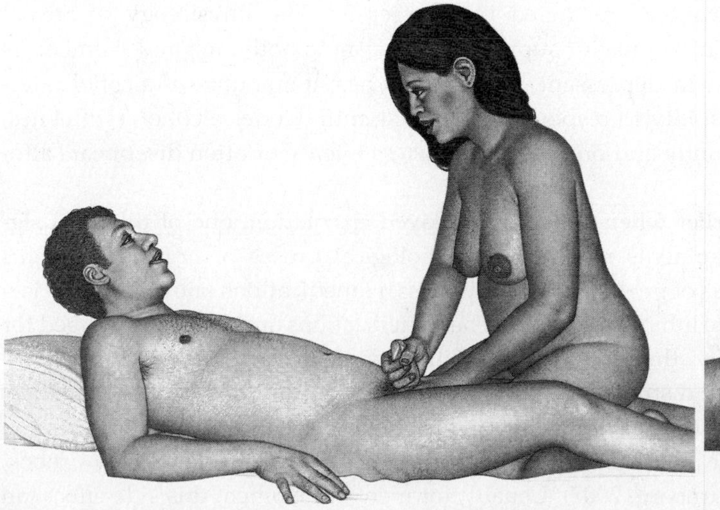

These treatment strategies for PE are effective because the man is becoming conscious of his *moment of ejaculatory inevitability*, as discussed in Chapter 3: "The Physiology of Sexual Response." This is the moment in his response cycle when the physical mechanisms of ejaculation kick in, after which the ejaculation is reflexive and out of his conscious control. To acquire *conscious* control over ejaculation, he must learn to identify, with some precision, the sensations immediately *preceding* that moment. That is why focusing *on* his arousal, rather than distracting himself *from* it, is the key to overcoming premature ejaculation. This procedure echoes the cognitive-behavioral therapy (CBT) technique discussed earlier, in that a man learns to focus mentally (cognitively) on his bodily sensations, is aware of how his body responds, and becomes more comfortable with his sexual ability to control the behavior of ejaculation (Abdo, 2013).

FEMALE ORGASMIC DISORDER Women whose orgasms are delayed or absent on most sexual occasions (75%) over at least the past six months, are experiencing **female orgasmic disorder (FOD)**. As for all sexual problems, FOD must be causing personal or relationship distress to be diagnosed as a clinical problem. Difficulties with orgasm are among the most common of women's sexual complaints. The diagnostic criteria are that a woman experiences a marked delay in, marked infrequency of, or absence of orgasm, *or*, she feels a markedly reduced intensity of orgasmic sensations (DSM-5, 2013).

female orgasmic disorder
A sexual problem in which a woman rarely or never reaches orgasm or orgasms are delayed; also known as *inhibited female orgasm or anorgasmia.*

Description The circumstances surrounding a woman's orgasm difficulties are especially important. A woman who has never experienced an orgasm under any conditions has a lifelong problem that is sometimes referred to as "preorgasmic," implying that all women have the capacity for orgasm, even if they have never had one.

Lifelong absence of orgasm is relatively rare and accounts for only 5% to 15% of all cases of FOD (Comer, 1998; Hite, 1976; Simons & Carey, 2001). Most instances of female orgasmic disorder are acquired and situational. Many women are capable of achieving orgasms through masturbation, oral sex, manual stimulation by a partner, or with the use of a vibrator, but rarely or never do so during penis–vagina intercourse. Other women reporting orgasm difficulties may be orgasmic through masturbation, but never or rarely with a partner, regardless of the type of stimulation.

Causes The most common cause of FOD is simply lack of adequate or desired sexual stimulation. Often, a woman (or a couple) believes she is experiencing an orgasmic disorder, when her sexual responses are perfectly normal. "In Touch with Your Sexual Health: Is Lack of Orgasm During Intercourse *Normal* for Women?" discusses this issue in detail.

Medical or biological causes of FOD may include hormonal imbalances, as discussed previously, use or abuse of alcohol or other recreational drugs, or side effects of prescription medications. As noted in Chapter 3, "The Physiology of Sexual Response," alcohol is notorious for suppressing orgasm in both men and women. As a central nervous system depressant, moderate to heavy amounts of alcohol interfere with the body's ability to respond to sexual stimuli. Under alcohol's influence, arousal slows significantly and orgasm becomes less intense or often disappears altogether (McKay, 2005).

As alluded to earlier when discussing delayed ejaculation, one of the most significant recent developments relating to physiological causes of orgasm problems involves the side effects of prescription antidepressant medications known as the selective serotonin reuptake inhibitors (SSRIs). These medications are widely prescribed for depression and various other psychological difficulties, such as obsessive-compulsive disorder, premenstrual distress, and even smoking cessation. Most of these medications have been shown to produce sexual side effects in some users, the most common of which is delayed or inhibited orgasm in both men and women (Clayton, Warnock, & Kornstein, 2004; Nurnberg, 2007). Usually, for men and women, this side effect can

Since You Asked

9. What causes a woman to be unable to experience orgasm during sexual intercourse even though she can have an orgasm during oral sex? Is it a physical thing or mental? Maybe both?

be reversed by changing the specific medication, altering the dose, or combining other medications with the SSRI (Keltner, McAfee, & Taylor, 2002).

As for all sexual problems, various additional causes for FOD must be considered. For example, research has found that age, marital status, and education level appear to be linked to orgasm. Women who are younger, are not married, and have less education report greater problems with orgasm (Laumann et al., 1999). Past sexual traumas including sexual harassment, childhood sexual abuse, sexual assault, and rape have also been shown to relate to orgasm difficulties (McHugh, 2006). And of course, as has been mentioned for all sexual problems discussed here, psychological issues, such as depression, anxiety, and fear; relationship factors, such as infidelity, anger, and resentment; and cultural factors, including religion, values, and role expectations may all play a role in sexual responding and orgasms (refer to Table 7.5).

Solutions Treatment strategies for female orgasmic disorder generally incorporate a combination of the treatment guidelines discussed earlier in this chapter, especially masturbation and *sensate focus* exercises. Historically, in Western cultures, women have been given significantly less "permission" than men to think and act in sexual ways. The effects of this double standard are still felt today. Consequently, many women are insufficiently familiar with their own bodies and with their individual sensations when responding sexually. Sexual self-exploration through masturbation can be a powerful tool for a woman to discover that she is capable of achieving orgasms and where and how she needs to be touched to achieve them. Sometimes the incorporation of a vibrator into her masturbatory repertoire can assist a woman who has never or rarely experienced orgasm by providing maximal stimulation, leading to more predictable orgasms and thereby enhancing self-awareness (Brassil & Keller, 2002; Laan, Rellini, & Barnes, 2013).

Once a woman is reasonably comfortable with her own body and her ability to have orgasms, nondemanding sexual exercises, such as sensate focus, can begin to include her partner. In some cases, she may masturbate in her partner's presence as a way of demonstrating, educating, and involving both as a team. As part of their sensate focus

In Touch with Your Sexual Health

Is Lack of Orgasm During Intercourse *Normal* for Women?

As discussed in Chapter 3, for most women stimulation of the clitoris is necessary for orgasm. During intercourse, the thrusting of the penis in the vagina may not provide much contact with the clitoris. Consequently, it is quite common for women not to experience orgasm during intercourse without additional clitoral stimulation (Balon & Seagraves, 2009; Kaplan, 1974). One study found that women reported experiencing an orgasm during sexual activity with a male partner between 11% and 34% of the time over the previous month. Sixteen separate studies found that an average of only 25% of women were regularly orgasmic from the stimulation provided by intercourse exclusively (see Lloyd, 2006). In another, more recent study, over 90% of men reported experiencing orgasm during their most recent sexual encounter, compared to 64% of women (these sexual encounters may have included various behaviors other than intercourse, which were likely to provide stimulation of the clitoral area in addition to intercourse) (Herbenick et al., 2010).

However, the vast majority of women who do not achieve orgasm from intercourse are orgasmic from manual or oral stimulation and from masturbation. The problem here is the continued widespread misunderstanding of this basic physiological concept, and many women (and men) incorrectly believe that both *should* experience orgasm during intercourse. An all-too-common expectation is that if the man can "last long enough," that is, can hold off his own orgasm and continue thrusting, the woman *should* reach orgasm. When this does not happen, she may feel distressed that she has failed to please *him* (Lloyd, 2006; Nicholson & Burr, 2003). In reality, lack of orgasm in women during intercourse is *normal.* Most women do not routinely experience an orgasm through intercourse, regardless of how long their male partner "lasts." However, over 90% of all women are orgasmic through activities other than, or in addition to, intercourse, such as masturbation or manual or oral stimulation. Today, women who do not have orgasms during intercourse are usually *not* considered to have a sexual problem based on that symptom alone (Lloyd, 2006).

Figure 7.5 Back-to-Front Sensate Focus Position

The back-to-front position allows the woman to guide the man's touching movements—to help him understand what feels best to her.

program, Masters and Johnson developed a specific position to allow the woman to guide her partner in ways of touching and pleasuring that provide the sensations and stimulation that is most effective for her (see Figure 7.5). Not only does this exercise help inform her partner about what feels best to her, but it also tends to enhance closeness and communication in general (Masters & Johnson, 1970; see also Giraldi et al., 2013).

For heterosexual couples, these exercises are usually carried out during a period of time when intercourse is banned, as discussed earlier, so the pressure of performing according to old expectations is removed. When the woman is comfortable with her ability to have orgasms through these exercises, intercourse may be resumed. If the goal for the couple is for both to reach orgasm during intercourse, this is now usually accomplished through the addition of manual stimulation by either partner to the clitoral area during penetration (see Figure 7.6). These methods, and various

Figure 7.6 Clitoral Stimulation During Intercourse

Many women need and desire simultaneous clitoral stimulation during intercourse in order to achieve orgasm.

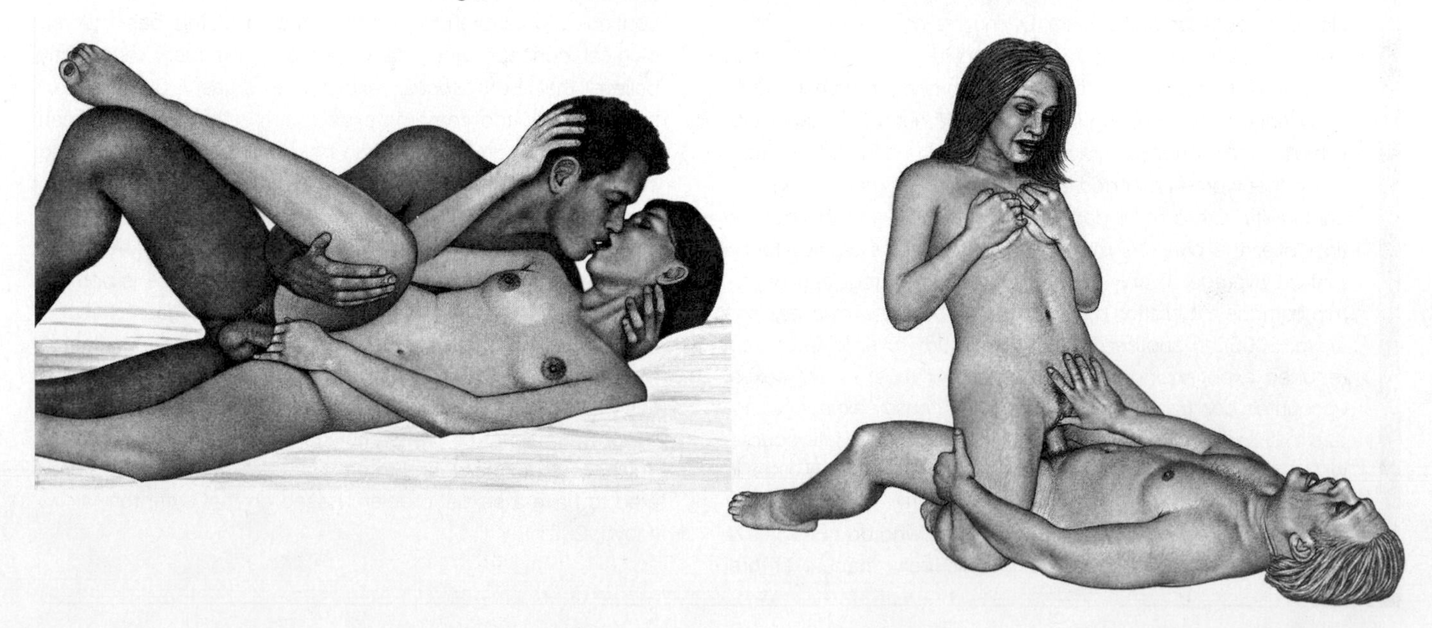

adaptations of them, appear to be quite successful in helping most women with female orgasmic disorder—as much as 90%—become orgasmic alone, with a partner, and sometimes during intercourse (Albaugh & Kellogg-Spadt, 2002; Segraves & Althof, 1998).

Since You Asked

10. My girlfriend says sex hurts (physically). Why is it painful, and what can we do to relieve the pain?

PAINFUL SEX If sex hurts, something is wrong. That may sound overly simple, but it is very difficult to think of an exception (well, except for the consensual, voluntary inflicting or receiving of pain for sexual pleasure, called *sadomasochism*, which will be discussed in Chapter 14, "Paraphilic Disorders, Atypical Sexual Behaviors").

The human body has evolved to ensure that the act of heterosexual intercourse works smoothly and feels good (or, at least, does not feel *bad*!). After all, it is through intercourse that we survive as a species. If we had evolved so that sex was painful, the human race would probably be extinct by now. That is why pain associated with intercourse is not normal. Painful intercourse is a problem experienced primarily by women, and the diagnostic guidelines for sexual pain focus on women (DSM-5, 2013; Laumann et al., 1994; Simons & Carey, 2001). This sexual problem is called, **Genito-Pelvic Pain/Penetration Disorder (GPPPD)**. This disorder is clinically defined as intercourse-related pain or discomfort that has occurred on at least 75% of sexual occasions over at least six months. Specifically the pain or discomfort involves one or more of the following characteristics: (a) difficulty with vaginal penetration; (b) marked genital pain during intercourse or attempted penetration; (c) marked fear or anxiety about painful intercourse, and/or (d) marked [involuntary] tensing or tightening of the pelvic muscles interfering with attempted vaginal penetration (DSM-5, 2013).

Genito-Pelvic Pain/Penetration Disorder (GPPPD)
intercourse-related pain that interferes physically or emotionally with sexual activity and/or enjoyment.

DEFINITION Genito-Pelvic Pain/Penetration Disorder, abbreviated, GPPPD, is a complex disorder that is difficult to define in simplistic terms. For clarity, let's look at each of the four diagnostic criteria separately (DSM-5, 2013).

If a woman has difficulty with vaginal penetration (symptom a), this means that she is unable to tolerate penetration of the vagina during sexual activity. For some women, this may extend beyond sexual situations to any attempted vaginal penetration, including gynecological exams and the use of tampons.

Pain during penetration and/or intercourse (symptom b) implies a more general symptom of intercourse-related pain. The pain may be experienced at the vaginal opening, in the vagina, or deeper in the abdomen or pelvis. The pain may occur at the start of intercourse or may last throughout intercourse and may linger after the completion of intercourse. Women's descriptions of this pain include "burning," "cutting," "shooting," and "throbbing" (DSM-5, 2013, p. 438).

Of course, it is not difficult to understand that if intercourse causes pain, a woman would likely develop anxiety and even fear of engaging in the act (symptom c). Anxiety about, and fear of, pain is a perfectly normal reaction. This anxiety is one cause of female sexual interest/arousal disorder and/or the avoidance of any situation in which vaginal penetration might occur.

Finally, in response to any of the first three symptoms, a woman's body begins to (or perhaps "learns" to) react muscularly to the possibility of vaginal penetration. The pelvic muscles may flex and tighten involuntarily, like a fear reflex, as the moment of penetration approaches (symptom d). This reflex typically has the effect of closing the vaginal opening, making penetration difficult if not impossible.

We should note that a common complaint of women (and couples) who experience painful intercourse is the incorrect *perception* of a disproportion in vaginal and penile size—that is, the sensation that his penis is too large and/or that her vaginal opening is too narrow, even when lubrication seems adequate for intercourse. However, genital size discrepancies are highly unlikely as a cause of sexual difficulties. As explained in Chapter 2, "Sexual Anatomy," erect penis size does not vary greatly among men, and the vagina is a very flexible structure, capable of accommodating something as small as a tampon or as large as a baby's head during childbirth. Because no penis approaches the size of a baby's head, and assuming adequate lubrication, women who perceive an anatomical size discrepancy during intercourse may be experiencing some degree of pelvic muscle tension without realizing it, because this is not a well-known condition. Even therapists and physicians may not think of pelvic muscle contractions to explain these complaints from women.

Furthermore, beyond causing a difficult sexual problem, GPPPD may interfere with a woman's sexual and reproductive health. Routine exams such as the Pap test for cervical cancer, screening for certain STIs, and other medical procedures require medical access to the vagina. If a woman's body will not tolerate these exams, the potential health consequences can be serious (Heim, 2001; Reissing et al., 2004).

Causes Painful intercourse is one of the most common sexual problems in women, with up to 40% of women reporting that such pain occurs at least occasionally and approximately 15% experiencing the problem over a period of several months (Butcher, 2003; Laumann et al., 1994; Simons & Carey, 2001).

The most common cause for vaginal pain during sexual intercourse is not muscle contractions, but rather lack of adequate vaginal lubrication prior to and during intercourse. As you know from Chapter 3, "The Physiology of Sexual Response," sexual arousal causes the vaginal walls to secrete a fluid that lubricates and protects the vagina during intercourse. If penetration is attempted before a woman is fully aroused, as may happen if the sexual encounter is rushed or if she is not feeling sexually excited for any reason, lubrication will be insufficient and the thrusting of the penis can irritate the vagina, resulting in burning or shooting kinds of pain. Usually, this cause of sexual pain can be overcome through increased arousal prior to attempted penetration (to increase natural lubrication) or the use of a sexual lubricant. However, it is important to note that lack of lubrication may be a sign of a woman's fear or anxiety about the sexual setting, emotional issues stemming from negative feelings about her partner, or past sexual trauma that is interfering with her sexual arousal and vaginal lubrication.

Other sources of vaginal pain that are related to sexual activity may include irritation due to vaginal infections such as yeast infections, bacterial vaginosis, trichomoniasis, or sexually transmitted infections, including the herpes simplex virus or the human papilloma virus (HPV) (refer to Chapter 8, "Sexually Transmitted Infections/Diseases," for a more detailed discussion of STIs). If a woman has any of these infections, the friction of intercourse is likely to aggravate the symptoms, causing increased irritation and pain. In addition, some women have a sensitivity or allergy to the latex or spermicides used in condoms and other forms of contraceptives. This allergy may cause an inflammation in the vagina that can be further irritated by intercourse.

The deeper, pelvic or abdominal pain some woman with GPPPD experience is often the result of deep thrusting during intercourse, when the partners' bodies make vigorous pelvic contact. The abdominal jostling may impact the ovaries. The ovaries are very sensitive to trauma, much like the testicles are in men, and when impacted during sexual activity, even indirectly, this deep ache may result.

The issue of pelvic muscle contractions demonstrates the powerful link between mind and body. Researchers and sex therapists are in general agreement that this sexual problem is psychological and probably based on a deeply conditioned fear that penetration will be painful or traumatic in some way (Crowley, Goldmeier, & Hiller, 2009; Reissing et al., 2004). The origins of that fear can vary widely and may include childhood sexual abuse; painful past intercourse experiences; rape and sexual assault trauma; belief in myths about painful and bloody first intercourse; or deep fears about unwanted pregnancy or STIs. Most physicians and sex therapists who treat this problem agree that it is rare to see a case that lacks at least one serious past sexual trauma.

SEXUAL PAIN IN MEN Sexual pain in men is uncommon and less well researched and understood (Luzzi, 2003). When it does occur, it is often due to a localized infection either on the penis or in the reproductive tract, including the scrotum, testicles, prostate, or anus (Bunker, 2010). Also, pain upon intercourse may be caused by inflammation of the skin of the penis due to vigorous masturbation without adequate lubrication, a sensitivity to contraceptive chemicals or latex condoms, or infections such as herpes or HPV.

Solutions In treating sexual pain disorders, once again the duration and context must be determined. GPPPD is generally treated using a combination of psychological and physical strategies (Frank et al., 2008). Usually, treatment includes sensate focus exercises to enhance intimacy between the partners and help the woman feel more relaxed and familiar with her sexual arousal without worrying about approaching penetration.

For pelvic muscle contractions, the woman practices exercises to learn voluntary control of the muscles that surround the vagina. a gradual process of nonthreatening vaginal insertion that allows her to become less sensitive to penetration. Traditionally, this step-by-step therapy has been accomplished using a set of *dilators*: smooth, round-ended cylinders in graduated sizes from very small (the width of a pencil) to the approximate size of an erect penis. In conjunction with self-exploration, relaxation, and masturbation exercises, the woman inserts these dilators sequentially into her vagina, beginning with the smallest, to become accustomed to penetration. Over time, assuming she is comfortable with her partner, her ability to accept vaginal penetration comfortably will gradually increase. Today, the use of dilators is decreasing, and women are often instructed simply to use the tip of a little finger at first, and slowly increasing finger insertion up to two or three fingers at once (see Figure 7.7). When she feels ready, her partner may join the process if desired and continue to proceed slowly until eventually penile penetration becomes possible. Using these techniques, success rates approaching 100% have been reported (Jeng et al., 2006). A newly approved treatment for women who have had poor success with traditional methods is the use of Botox injections into the vaginal muscles. The Botox relaxes the muscles and thus allows the woman to use the step-up dilators without pain. Research has shown a 75% success rate (defined as pain-free intercourse) within two to six weeks of this treatment ("Botox Vaginismus Treatment," 2011).

Treatment of painful sex in men must address the underlying physical causes to reduce inflammation. This may include enhanced proper hygiene; and, where indicated, treatment of infections through antibiotics or other medications.

The most important point to keep in mind is that sex should *not* be painful. If you experience any of the symptoms described here, you should consult your doctor for appropriate consultation, diagnosis, and treatment. Solutions are virtually always successful. And, in the meantime, assuming

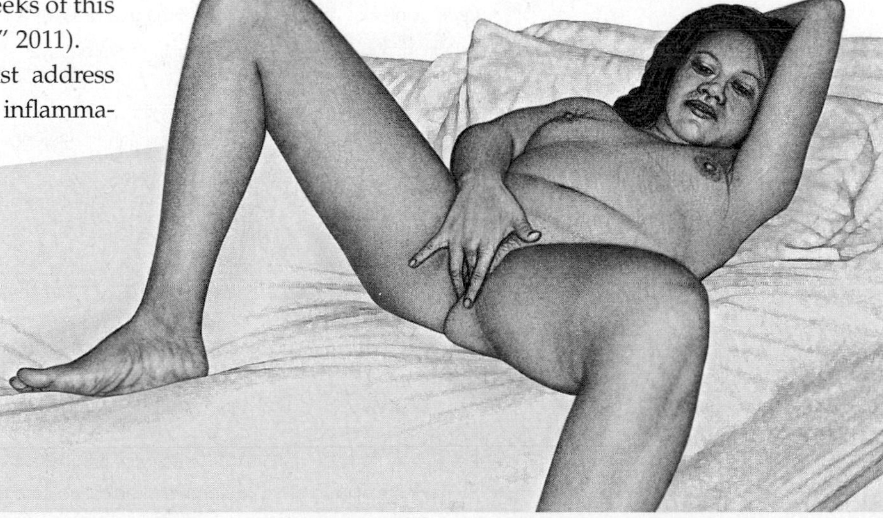

Figure 7.7 Treatment for Pelvic Muscle Contractions

Treatment exercises for vaginismus typically involve gradual, smaller-to-larger vaginal penetration using special dilators or finger insertion in a relaxed, fear-free setting, sometimes in conjunction with therapy to deal with the source of the fear.

both partners are comfortable and desirous of sex, many nonintercourse activities are very intimate and satisfying.

A "New View" of Female Sexual Problems

In Chapter 3, "The Physiology of Sexual Response," you read about an addition to the literature about women's sexual problems called "A New View of Women's Sexual Responding." This approach has been suggested as an alternative to classifying and explaining both male and female sexual problems using a single model that focuses on the physical functioning of sexual behavior. The "New View" model is a response to what the researchers saw as a trend to "medicalize" sexual problems and attempt to find "cures" in the form of drugs, symbolized by the huge success of male erectile dysfunction medications such as *Viagra, Cialis,* and *Levitra.* Proponents of the "New View" contend that, although a medical model of sexual problems may be effective for men, women's sexual problems are fundamentally different, involving complexities that cannot be treated with a simple pill or patch. Instead, they suggest that women's sexuality, sexual response, and sexual difficulties require a radical theoretical revision that takes into account cultural, political, and economic factors (i.e., lack of sexuality education or access to contraception); a woman's partner and the relationship between them (i.e., fear of abuse, imbalance of power, overall discord); psychological factors (i.e., past sexual trauma, depression, anxiety); and medical factors (i.e., hormonal imbalances, sexually transmitted infections, medication side effects) (Graham, 2003; McHugh, 2006). The New View model is described in greater detail in Table 7.10.

Table 7.10 Summary of "A New View of Women's Sexual Problems"

The following classification system was proposed by researchers who have suggested a "New View of Women's Sexual Problems" as an alternative to the traditional model of assigning problems to such categories as desire, excitement, and orgasm. The group's model defines sexual difficulties for women as "discontent or dissatisfaction with any emotional, physical, or relational aspect of sexual experience that may arise in one or more of the following interrelated aspects of women's sexual lives."

Classification Category	Definitions
Sexual problems due to sociocultural, political, or economic factors	• Ignorance and anxiety due to inadequate sex education, lack of access to health services, or other social constraints. • Sexual avoidance or distress due to perceived inability to meet cultural norms regarding correct or ideal sexuality. • Inhibitions due to conflict between the sexual norms of one's subculture or culture of origin and those of the dominant culture. • Lack of interest, fatigue, or lack of time due to family and work obligations.
Sexual problems relating to partner and relationship	• Inhibition, avoidance, or distress arising from betrayal, dislike, or fear of partner, partner's abuse or couple's unequal power, or arising from partner's negative patterns of communication. • Discrepancies in desire for sexual activity or in preferences for various sexual activities. • Ignorance or inhibition about communicating preferences or initiating, pacing, or shaping sexual activities.
	• Loss of sexual interest and reciprocity as a result of conflicts over commonplace issues such as money, schedules, or relatives, or resulting from traumatic experiences such as infertility or the death of a child. • Inhibitions in arousal or spontaneity due to partner's health status or sexual problems.
Sexual problems due to psychological factors	• Sexual aversion, mistrust, or inhibition of sexual pleasure due to past experiences of physical, sexual, or emotional abuse; general personality problems with attachment, rejection, cooperation, or entitlement; depression or anxiety. • Sexual inhibition due to fear of sexual acts or of their possible consequences, such as pain during intercourse, pregnancy, sexually transmitted disease, loss of partner, or loss of reputation.
Sexual problems due to medical factors	• Local or systemic medical conditions affecting neurological, neurovascular, circulatory, endocrine, or other systems of the body. • Pregnancy, sexually transmitted infections, or other sex-related conditions. • Side effects of drugs, medications, or medical treatments.

SOURCES: Kaschak & Tiefer (2002); McHugh (2006); New View Campaign (2008); Wood, Koch, & Mansfield (2006); "FSD Alert" (2008); Teifer (2010).

Your Sexual Philosophy

Sexual Problems and Solutions

By, now you know that sexual problems are quite common and usually are not particularly serious or difficult to overcome. The biggest obstacles to solving sexual problems are ignorance, embarrassment, and acceptance of the many myths and misconceptions that usually accompany sexual difficulties and poor communication. After reading this chapter, you now have the tools to overcome those hurdles and are better prepared to deal rationally and effectively with sexual difficulties when and if they arise.

Most people never give sexual problems even a passing thought until one happens to them or their partner. This is a bad idea. Although you may never experience a diagnosable sexual disorder, most people do at some point in their lives. An awareness and understanding of sexual problems and their solutions is a vital component of your sexual philosophy. Why? Here are just a few good reasons.

First, knowledge is prevention. If you realize that sexual problems are both common and easily treated, your anxiety level at the first potential sign of one will be much lower than if you lacked this knowledge. That reduction in anxiety may help you head off the problem before it worsens or leads to relationship complications.

Second, knowledge is treatment. After reading this chapter, you now know many strategies to help you or your partner deal with sexual problems as soon as they appear. You will be less likely to feel embarrassment, confusion, frustration, or various other emotions that often lead to denial of a problem, discomfort with intimacy, loss of desire, and avoidance of lovemaking altogether. You will be better able to communicate about the problem with your partner and help your partner through it, too. You will likely feel more comfortable discussing the problem with your doctor, and you will be more open to the possibility of seeking the help of a counselor or sex therapist. In other words, should you ever need to confront sexual difficulties, you will be better equipped to handle the situation in rational, mature, and effective ways.

Finally, knowledge is support. With this base of knowledge added to your sexual philosophy, you may be able to help others who might seek your advice and reassurance at a time when they are experiencing sexual difficulties. Who knows? The person in need of your assistance might be a close friend, a brother or sister, a son or daughter, or even a parent (I know that talking to a parent about these issues is difficult to imagine, but it could happen!). Even if you never need all your new information about sexual problems, someone else you know and care about just might.

Have You Considered?

1. Imagine that you begin to experience one of the sexual problems discussed in this chapter (your choice). Explain how you think your would react. Assuming that the problem does not have a physical cause, what strategies would you use to understand and resolve the problem?
2. Now imagine that your *partner* (who has not read this book) begins to experience one of the sexual problems discussed in this chapter (again, your choice). Explain how you think you would react. Assuming no physical cause, how would you help your partner understand and resolve the problem?
3. Suppose that a man is having trouble with premature ejaculation. Describe how his problem might manifest itself if it was (a) lifelong, and generalized, total; (b) acquired and generalized; (c) acquired and situational; (d) lifelong and situational.

4. Explain *why* a man with erection problems and a woman with arousal difficulties are both able to become very excited (he has firm, long-lasting erections, and she easily becomes aroused and lubricated) during sensate focus exercises but not during lovemaking.
5. Do you think a couple could have a satisfying sexual relationship without ever having sexual intercourse? Explain your answer.
6. Which sexual problem do you think would be most difficult to solve? Explain your answer.

Summary

Historical Perspectives

Before Masters and Johnson

- Masters and Johnson's research during the 1960s led the way into modern sex therapy. Their research was the first to examine and explore the typical physical responses of the human body to sexual stimulation. Their goal was to achieve greater understanding of these responses in order to improve couples' and therapists' ability to solve sexual problems.

What is a Sexual Problem?

- Most sexual problems are self-diagnosed. This means that generally, if individuals or couples perceive that they have a sexual problem, they probably do. However, feeling that a sexual problem exists is very different from knowing what it is and how to treat it successfully. Fortunately, most sexual problems are quite common and usually easily treated.
- Sexual problems may be classified into two dimensions for more accurate diagnosis: duration of the problem (lifelong versus acquired), the context in which the problem occurs (generalized versus situational). Classifying the problem into these dimensions assists in discovering the origins of the problem and how best to treat it.

Sources of Sexual Problems

- The origins and causes of sexual problems vary widely. They may include physical, psychological, relationship-based, or cultural issues, or a combination of these factors.

General Guidelines for Solving Sexual Problems

- Sexual problems may be reduced by temporarily banning the sexual behavior during which the problems occur. Using the *sensate focus* technique, the sexual activity, such as intercourse or oral sex, that is at the center of a particular sexual problem may be regarded as "off limits" while the couple focus on exploring other sensual activities to assist them in overcoming their anxiety.
- Masturbation "exercises" are sometimes employed by therapists as an effective strategy in helping individuals overcome sexual problems.
- Effective communication is the most basic key to successful relationships and satisfying sex. Successful sexual communication must rely primarily on words. Nonverbal communication may often be misinterpreted and has been found to be an unreliable and even harmful strategy for attaining sexual compatibility.

Specific Problems and Solutions

- Problems with sexual desire include *Male Hypoactive Sexual Desire*, *Erectile disorder*, and *Female Sexual Interest/Arousal Disorder*.
- Treatments for arousal disorder include sensate focus exercises; directed masturbation; and sometimes (mainly for men) medications such as *Viagra*, *Cialis*, or *Levitra*.
- Difficulties with orgasm for men are *premature ejaculation* and its converse, *delayed ejaculation*. Premature ejaculation is the most common complaint among younger men and is usually effectively treated with variations of the technique called the "start-stop method." This technique allows a man to become in touch with the physical sensations of approaching ejaculation and control them. Mental distraction strategies, attempted by many men (thinking about sports or car accidents during intercourse) often fail to provide dependable help with the problem.
- Problems with attaining orgasm for women is referred to *female orgasmic disorder*. The most common reason for this problem is inadequate stimulation to the clitoral area from sexual intercourse alone.

- SSRI antidepressant medications (such as *Prozac, Effexor*, and *Paxil*) have been found to inhibit and delay, or even prevent, arousal and orgasm in some men and women.
- Sex should never be painful. Painful sexual experiences are far more common for women than men. Sexual pain disorders are referred to as genito-pelvic pain/penetration disorder. These may include difficulty with vaginal penetration, pain during sexual intercourse, fear and anxiety about painful sexual intercourse, and involuntary pelvic muscle contractions making intercourse difficult or impossible. Sexual pain difficulties can almost always be treated successfully.
- In general, all sexual problems are common and usually easy to treat and cure. No one should suffer unnecessarily or for long periods of time with sexual problems.

Your Sexual Philosophy

Sexual Problems and Solutions

- Knowledge and understanding of the characteristics, causes, and treatments are the most important keys to solving sexual problems if and when they arise in a person's or a couple's life. Therefore, the information in this chapter plays an important role in the development of your comprehensive sexual philosophy.

Chapter 8
Sexually Transmitted Infections/Diseases

Learning Objectives

After you read and study this chapter you will be able to:

8.1 Explain the extent of the STI pandemic and what groups are at greatest risk.

8.2 List and discuss the factors that contribute to increased risk for contracting, transmitting, and the spread of STIs.

8.3 Summarize the symptoms, mode of transmission, treatment, and complications for each of the STIs presented.

8.4 Review the similarities among viral STIs and summarize the specific symptoms, mode of transmission, diagnosis, and treatment methods for each.

8.5 Summarize the differences between bacterial and viral STIs and the similarities among the various bacterial STIs.

8.6 Explain what parasitic STIs are and the symptoms, causes, and treatments for each.

8.7 Discuss the various specific strategies for preventing or reducing the risk of contracting or spreading STIs.

Since You Asked

1. How would I know if I had a sexually transmitted disease? (see page 264)
2. Other than AIDS, what is the worst sexually transmitted infection? (see page 272)
3. If I had sores on my genitals from an STI, but the sores went away, does that mean I'm cured? (see page 273)
4. My boyfriend told me he had an outbreak of herpes two years ago, but no signs since then. Could I still catch it from him? (see page 275)
5. Is it true that genital warts can cause cancer? (see page 276)
6. I have an uncle who says he had hepatitis B, but he is now cured. I thought viruses could not be cured, so how is this possible? (see page 280)
7. I heard that AIDS has been nearly eliminated in the United States. Is this true? (see page 283)
8. Can people get STIs from having oral sex? (see page 287)
9. There is a rumor that people with AIDS are swimming in the community pool in my neighborhood. Should I be concerned about letting my kids swim there? (see page 289)
10. Are the STIs caused by viruses and bacteria basically the same? (see page 295)
11. What are crabs, and how can I tell if someone has them? (see page 304)
12. Once I start having sex, how can I be sure to never catch an STI? (see page 305)

Throughout this book, you have read about many issues related to your sexual health. In this chapter, the *primary* focus is on sexual health—in particular, on a group of viral, bacterial, and other infections that are directly caused and spread by sexual behaviors. Most people know these as *sexually transmitted diseases (STDs)*, but the terminology among health professionals and sexuality educators today is **sexually transmitted infections (STIs)**. The name change is meant to emphasize an important message—that these are *infectious* conditions (unlike many diseases, such as heart disease or diabetes, for example) that are spread from one person to another through certain behaviors, and also that these infections are treatable or curable.

Paradoxically, the same activities that can bring us so much pleasure and allow us to express our most intimate feelings toward another person can also spread infections that may be uncomfortable or embarrassing at best and deadly at worst. By the time you finish reading about all of the many sexually transmitted infections, you may feel as though you never want to touch another person again, at least not sexually (see "Focus on Your Feelings"). Of course, that's far from the goal of this chapter. No one has to give up being sexual or sharing sexual pleasure to avoid sexually transmitted infections. However, your ability to protect yourself from contracting an STI or to keep from infecting others if you happen to become infected depends on becoming as educated as possible about STIs. *That* is the purpose of this chapter. Even though you may be troubled by the sheer number of these infections and the disturbing symptoms that may accompany them, ignoring them won't make them go away. Knowing all you can about them can help you avoid them and reduce their spread overall.

sexually transmitted infections (STIs)
A group of viral, bacterial, and other infections that are spread primarily by sexual behaviors; also called sexually transmitted diseases (STDs).

The worldwide STI **pandemic** is not showing many encouraging signs of slowing. Consequently, everyone, whether sexually active currently or not, must possess accurate and complete information to make informed, healthy sexual choices. In this chapter, we will review each of the most common STIs: the infection's symptoms and diagnosis, how it is spread, how to know if you have it, and how it can be treated or cured. We will also focus on prevention, which is always more desirable than treatment

pandemic
A sudden outbreak of an illness that spreads relatively quickly over entire continents or throughout the world.

Focus on Your Feelings

STIs tend to stir up a variety of negative emotions. One, of course, is fear of contracting an STI. This fear is real—*if* you don't know how to prevent an STI. As you will see in this chapter, everyone has the ability to avoid STIs by understanding how they are spread and how contracting them can be avoided.

It is not unusual for people to react with disgust to any disease that produces symptoms such as ulcerations of the skin or bodily discharges. With STIs, the disgust is often extended beyond the disease to the infected person, which increases the stigma that already surrounds these infections and in turn contributes to people's hesitation to be tested or seek treatment. STIs are a fact of life (albeit a preventable one) and should be seen in the same light as any other curable or treatable illness.

Anxiety over STIs typically peaks *after* people have engaged in risky sexual behavior. The distress over the possibility that they may have contracted an STI can be overwhelming. If you find yourself in this situation, the best way to alleviate your anxiety is to be tested. The relief of receiving a negative result (which is the most likely outcome) may motivate you to avoid unsafe behaviors in the future. If, however, you do discover that you have contracted an STI, you may feel angry at the person who transmitted the infection to you. Although this hostility is understandable, remember that most STIs have no symptoms, and the person who spread the infection may have had no idea he or she was infected.

For most STIs, the physical pain is minor and subsides fairly quickly once treatment begins. However, the emotional pain of dealing with the stigma attached to having an STI and, in the case of viral infections, the prospect of living with an STI for the rest of one's life, may be devastating. This pain may be lessened through education about exactly what to expect from the disease, obtaining the best treatments and care available, and joining support groups with others who are in the same situation. Sometimes a person who has been diagnosed may experience severe depressive symptoms or even thoughts of suicide. Anyone experiencing such extreme emotional reactions should seek professional help immediately. All STIs can be treated and managed effectively to allow the infected person to live a happy and fulfilling life in spite of the infection.

for any health problem. Before we begin our discussion of STIs today, let's take a look back at how one common sexual infection, syphilis, was at the center of a shameful, racist event in the medical history of sex.

Historical Perspectives

The Tuskegee Syphilis Study

One of the darkest episodes in the history of medical research was the Tuskegee Syphilis Study carried out over 70 years ago; the details of which only came to light publicly in the late 1990s. In the early 1930s, the U.S. Public Health Service set about studying the short- and long-term effects of the sexually transmitted infection caused by the syphilis bacterium (which we will discuss in detail later in this chapter). The methodology employed by the researchers at that time was to *deny treatment* to 400 low-income African American men who tested positive for the disease. In fact, the researchers decided not to tell the men that they even had syphilis and conspired to prevent them from obtaining treatment elsewhere. Finally, when it was discovered that penicillin could cure the illness, they did not inform the patients of this because, the researchers decided, to treat them would destroy the potential findings of their long-term study of untreated syphilis.

The Tuskegee Syphilis Study was one of the most unethical studies in the history of modern medicine. In 1997, President Bill Clinton offered the victims and their families an official apology and compensation for the injustice done to them.

As you will see later in this chapter, without treatment, syphilis is a particularly insidious infection that slowly, over years and decades, invades various systems of the body, causing tumors of the skin, bones, or liver; aneurisms of the heart valves and arteries; and major disorders of the central nervous system leading to symptoms of severe mental illness (referred to historically as "syphilitic insanity"). Most of the men in this group faced shortened lives, full of illness and pain for themselves and their families. In addition, some of these men undoubtedly passed the disease on to their sexual partners.

Today, everyone agrees that this was a horribly inhumane abuse of science and research, clearly reflective of the state of bigotry and discrimination at that time.

It violated all of the ethical research guidelines discussed in Chapter 1 and it has been determined that any benefit gained from the study was minimal compared to the harm inflicted on the participants (see http://www.cdc.gov/tuskegee/timeline.htm for most detailed information). Although evidence of the study began to come to light in the 1970s, it was not until 1997 that President Bill Clinton made an official apology to the victims and their families and made funds available to compensate the families of the men for their suffering. The last surviving widow receiving compensation died in 2009. However, no amount of money or regret could change the legacy of mistrust the Tuskegee study created between the African American community and the medical establishment in the United States (CDC, "The Tuskegee Timeline," 2013; Fairchild & Bayer, 1999).

The STI Pandemic

8.1 Explain the extent of the STI pandemic and what groups are at greatest risk.

Sexually transmitted infections are among the most common causes of illness and disease worldwide. This **pandemic** is so vital to the world's health, that the 2008 Nobel Prize in medicine was awarded for research relating to STIs, specifically the discovery of HIV (the virus that causes AIDS) and the discovery of the human papilloma virus (HPV) that causes genital warts and leads to cervical and other cancers (Altman, 2008). These infections will be discussed in detail later in this chapter.

pandemic
A sudden outbreak of a disease (an epidemic) of major proportions that affects a large region, a continent, or spreads worldwide

On any given day, more people are suffering from STI-related symptoms, discomfort, or illness than the total number of people with a common cold. Every year, according to the Centers for Disease Control and Prevention (CDC), approximately 20 million new cases are diagnosed in the United States alone, and over half of these are among those between 19 and 24 years of age (CDC, "STD Trends," 2013). These account for over $16 billion in cost to the health care system in the United States. Before discussing the various infections in detail, let's briefly summarize some statistics for the more common STIs, so you can get a feel for the sheer size of this pandemic. In 2014, the CDC reported the following number of cases of three STIs in 2013 (the most recent tracking year as of this writing): chlamydia: 1,422,976; gonorrhea: 334,826; and syphilis: 15,667 (CDC, "Reported STDs," 2014).

Approximately half of all people will contract an STI at some point in life, and an estimated 20 million people are infected with an STI each year (CDC, "STD Trends," 2013). People between the ages of 15 and 24 are at greatest risk for becoming infected with an STI. The most common STIs in this age group are *gonorrhea, chlamydia*, and the *human papillomavirus* (HPV), which is the cause of genital warts. However, this age group is vulnerable to all the STIs discussed in this chapter. Among women, the estimated rate of chlamydia in the United States is three times higher than for men (CDC, "STD Trends," 2013). This sex difference is true for most STIs, for reasons we will discuss later in this chapter. Figure 8.1 illustrates the infection rate comparisons for two common STIs.

Figure 8.1 Chlamydia and Gonorrhea Infection Incidence by Age Group

SOURCE: CDC, "Reported STDs" (2014).

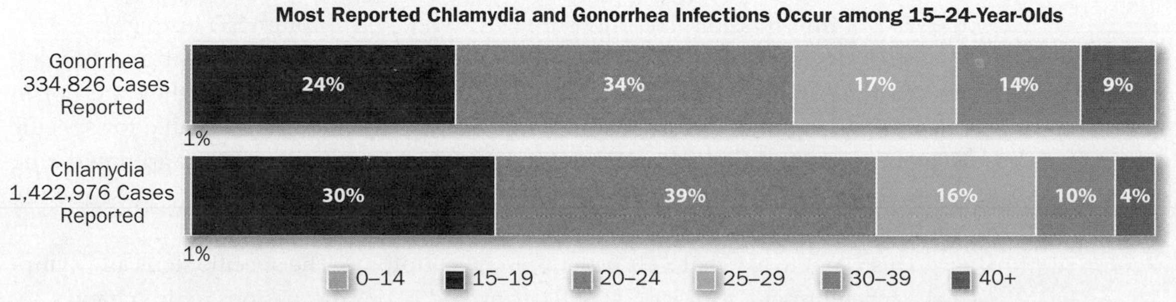

Yes, I realize that we are free agents, but I have to take on the additional risk of pregnancy and am more susceptible to certain sexually transmitted diseases, so I think you should pay for the movie.

Rates of STIs such as herpes and HPV also appear to affect young people disproportionately. Although representing only 25% of the sexually active population, those 15 to 24 years of age acquire nearly half of all new STDs (CDC, "STD Trends," 2013). Accurate estimates are difficult to make because, as you will see shortly, many STIs, such as herpes, genital warts, and chlamydia, may not present any obvious symptoms. The CDC estimates that there are approximately 20,000,000 current cases of genital HPV and over 6 million new cases are reported each year. Moreover, HPV is so common that 50% of all sexually active people will be infected during their sexual lifetime (CDC, "STI Surveillance," 2010).

HIV and AIDS are major pandemics. In the United States alone, approximately 1.2 million people are living with HIV and AIDS, and the number of new cases per year is estimated at over 56,000 (CDC, "HIV in the United States," 2013). The cumulative number of U.S. deaths from AIDS-related illnesses as of 2010 was over 610,000, and more than 15,000 people died from AIDS in 2010. Worldwide, the numbers are even more staggering: Over 33 million men, women, and children are infected with HIV; 2.6 million new HIV infections occurred in 2009, 50% of them among women. Over 30 million people have died from AIDS-related illnesses since the pandemic began officially in 1981 (Avert.com, 2008).

Risk Factors for STIs

8.2 List and discuss the factors that contribute to increased risk for contracting, transmitting, and the spread of STIs

If you ask people why the transmission of STIs is so common, they will probably say something such as "unsafe sex, of course!" and they would be right. However, that is only a small part of the story. For a full understanding of STI risks, we must examine the medical, educational, psychological, and emotional factors behind people's decisions to engage in risky sexual activities. Later in the chapter, we will discuss how you can avoid contracting STIs altogether.

No Warning Signs or Symptoms

A common misconception about STIs is the assumption that if you become infected, the signs will be obvious: a rash, a discharge, a sore, an odor, urination pain, and so on. But, some STIs produce no clear symptoms at all; they are **asymptomatic**. So, when a person is infected, he or she may be completely unaware of it. This is more likely among women in whom STI signs and symptoms are more likely to go unnoticed than for men. But the lack of symptoms does *not* imply lack of infection. Infected people who are asymptomatic may unknowingly transmit the infection to their sexual partners. In addition, these infected individuals are unlikely to seek testing or treatment because they show no signs and are unaware of any infection. Without treatment, virtually all STIs become increasingly serious over time and lead to additional health problems. If you have an STI that is asymptomatic or have risky sexual contact with someone with an asymptomatic STI, you have no way of being fully aware of the risks of transmission. The only sure way to know is through STI testing.

asymptomatic
Having no noticeable symptoms despite the presence of an infectious agent.

Since You Asked

1. How would I know if I had a sexually transmitted disease?

Most doctors and sexuality educators recommend periodic, routine screening for various STIs for people who are sexually active, especially if they are not in faithful, monogamous relationships. We will discuss the tests available for specific STIs throughout this chapter. Knowing the risks of contracting STIs and preventing them altogether is a far better health strategy than diagnosing and subsequently treating them.

This is not to say that STIs never present symptoms. The specific signs and symptoms, when present, for each STI will be covered in our discussions in this chapter.

Lack of Accurate Information

Currently, in the United States, the average age of first intercourse is approximately 17 years old for both boys and girls ("American Teens," 2014). In the early 2000s, researchers reported a significant increase in early teen oral sex, based on a widespread misconception among young people that oral sexual activities do not pose significant risks of STI transmission, and that teen girls can engage in oral sex and still claim that they are *technically* "virgins" (O'Brien, 2003; Prinstein, Meade, & Cohen, 2003). Oral sex carries approximately the same risk of STI transmission as vaginal intercourse. Subsequent research found that about 50% of teens had engaged in heterosexual oral sex, about the same proportion who had engaged in sexual intercourse, and 20% of self-proclaimed "virgins" reported having oral sex (Lindberg, Jones, & Santelli, 2008).

These numbers, combined with inconsistent, poor, or nonexistent sex education programs in schools—for example, *abstinence-only* programs provide no information about sexual behaviors and their consequences—create an environment that not only fails to prevent, but actually *increases* the spread of STIs in teens, especially when teens fail to understand the ease with which STIs are spread through nonintercourse sexual behaviors, including oral and anal sex. In Chapter 12, "Sexual Development Throughout Life," you can find a more detailed discussion of these teen sexual issues.

Lack of accurate STI information is not limited to teens. Other studies have found that significant numbers of university students do not believe that oral sex is "having sex" and therefore may underestimate the STI risks that accompany the behavior (see Knox, Zusman, & McNeely, 2008). Individuals and couples in all age groups may place themselves at risk of STIs owing to various misconceptions, ignorance, or gaps in their knowledge about how these infections are spread (Hyde & Forsythe, 2006; Shelby, 2003). When people, for whatever reason, are poorly educated about sexuality and STIs, they are likely to make risky decisions about sexual behavior based on false or distorted beliefs. Table 8.1 (on the following page) summarizes some of the common and persistent myths about STIs. You can see that faulty beliefs, such as those listed in the table, may lead individuals and couples to engage in risky behaviors without even being aware of the dangers.

"Unhealthy" Sexual Emotions and Attitudes

In addition to faulty beliefs and misinformation, negative emotions surrounding sexual behavior, such as guilt, shame, fear of stigma, and embarrassment, may create a risky environment for the spread of STIs. Often these unpleasant emotions about sex stem from life experiences such as traumatic past events, restrictive parental attitudes toward sex, cultural expectations, or religious teachings about sex. These feelings are often completely normal, but when people who hold these emotions become sexually active, the psychological distress they experience may decrease their ability to make safe choices and take appropriate actions that will protect them from contracting or transmitting STIs.

For example, people who feel shame, guilt, embarrassment, or fear of being stigmatized over sexual activities or contracting an STI are more likely to avoid discussing sexual risks with a partner, resist STI testing, delay or avoid seeking medical care for suspected STI infection, and shy away from sexuality information and STI education (East et al., 2010; Hampton, 2008). In addition, those who hold negative attitudes and emotions about sexual issues often resist availing themselves of authoritative education and information about risks and prevention. This resistance, in turn, typically leads to a reliance on often inaccurate sources of information about STIs, such as common myths, peers, or the popular media (movies, TV, Internet, etc.).

Table 8.1 Common Myths about STIs

The following statements about STIs are widely believed; however, they are all false.

Myth	Truth
Only people who have sex with many partners get STIs.	A single sexual encounter can transmit an STI; someone's only partner may have other partners.
Condoms don't prevent STIs.	Condoms, used correctly, *greatly* reduce the chances of STI transmission.
A lack of any symptoms of an STI probably means no infection.	Many STIs "hide" in the body and do not show any obvious or visible symptoms.
I don't know anyone who has an STI; it must not be a problem in my town.	Unfortunately, STIs are everywhere, and often an infected person is unaware of the infection or avoids telling anyone.
You can get an STI only from vaginal or anal sex.	Vaginal and anal intercourse are common routes of transmission, but other activities such as oral sex, intravenous drug use, and do-it-yourself tattooing may transmit STIs as well.
I'm on the pill, so I'm pretty well protected from STIs.	Hormonal forms of contraception offer no protection from STIs (refer to Chapter 5, "Contraception: Planning and Preventing").
People cannot get an STI the first time they have sex.	Yes, they can.
STIs can only be transmitted if the man ejaculates inside his partner.	STIs may be transmitted through oral, vaginal, or anal sex without ejaculation of semen; vaginal secretions and pre-ejaculate fluid from the Cowper's glands may contain STI microbes (see Chapter 2, "Sexual Anatomy").
Getting an STI is no big deal; the infections are easy to treat, and then you're immune from them in the future.	Some STIs, if untreated, may infect other areas of the reproductive tract and cause problems with fertility later in life; other STIs have no cure, and some may lead to death.
A person who is clean, well-dressed, and neatly groomed is at low risk of STIs.	Bacteria, viruses, and parasites do not discriminate based on a person's appearance.
The risk of STIs is very low except in certain geographical or low-income areas.	STIs are common in all socioeconomic groups.
Male circumcision reduces the spread of STIs.	Some studies in African countries have suggested this link (for HIV), but findings remain controversial and are not conclusive enough to support routine male circumcision. (See Chapter 2, "Sexual Anatomy," for more about this issue.)

Poor Sexual Communication

As discussed in detail in Chapter 4, "Intimate Relationships," talking about sexual issues in meaningful and effective ways is one of the most difficult areas of communication for many intimate couples, even in well-established, long-term relationships. You can imagine, then, how uncomfortable many people in new or more casual relationships would feel talking openly about sex, negotiating safer sex agreements, insisting on condom use, or avoiding certain high-risk behaviors. Nevertheless, such communication is crucial in preventing STIs. Research has consistently demonstrated that communication between partners is one of the most important factors predicting safer sex practices, including condom use (Palmer, 2004; Ryan et al., 2007). Interestingly, *your* parents' communication styles and openness about discussing sexuality with *you* may have molded your ability to engage in safe sex discussions with potential or current sexual partners. A study by Troth and Peterson (2000) found the following links between parents' communication styles and their children's ability to talk about safe sex and use of condoms in their early sexual relationships:

1. Children of parents who openly discussed sexual matters are more willing to discuss safe sex with their dating partners.
2. Children whose parents were less involved in their sexual education are less willing to discuss safer sex with their partners.
3. Parents who use avoidance as their primary strategy for resolving conflicts have children who are less willing to discuss condom use and negotiate safer sex practices with their partners.

4. In general, most parents did not provide meaningful sex education or communicate effectively with their children about safe sex. Children in the study who held more positive attitudes about discussing safer sex were more likely to have talked to their partners about HIV/AIDS and to have used condoms upon becoming sexually active.

As an indication of your personal attitudes toward safer sex practices and condom use, the "Self-Discovery: STI Risk Scale" provides an opportunity to measure your own sexual risk, based on your specific attitudes about sexual activities and STIs.

Self-Discovery

STI Risk Scale

This brief scale will help you assess your attitudes and knowledge about the spread of STIs. To the extent that our beliefs influence our actual behaviors, you can use your score to get an idea of your STI risk level. However, for that judgment to be accurate, *your sexual behaviors in your life must coincide with your answers to this scale*. It won't come as news to you that many people fail to "practice what they preach," and with STIs, that can be dangerous. Be honest in your answers; only you will know the outcome of the scale.

Rate each statement using the following scale: 5 = strongly agree; 4 = agree; 3 = not sure; 2 = disagree; 1 = strongly disagree.

_____ **1.** Sexual behavior choice plays a major role in your risk of contracting STIs.

_____ **2.** Obtaining and using methods of STI prevention are really pretty simple.

_____ **3.** Being tested for STIs is a good idea for most young, sexually active people.

_____ **4.** Getting to know your sexual partner's history before becoming sexually intimate is a good STI prevention strategy.

_____ **5.** Insisting on testing for a potential sexual partner makes a lot of sense.

_____ **6.** If you found out you had an STI, you would get treated as fully and as quickly as possible.

_____ **7.** You believe condoms are a good way to reduce a person's risk of contracting or transmitting an STI.

_____ **8.** You are comfortable talking to your sexual partner about sex and STIs.

_____ **9.** If you had physical symptoms that you thought might be an STI, you would see the doctor immediately.

_____ **10.** You would be likely to use "selective abstinence" (avoiding specific, risky sexual behaviors) to avoid STIs.

_____ **11.** If you were having sex with more than one partner, you would get tested for STIs more often.

_____ **12.** You are comfortable examining your sexual anatomy for signs of an STI.

_____ **13.** If you were at all unsure of a potential sexual partner's STI status, you would not have sex with that person.

_____ **14.** You feel that everyone should become educated about the risks and prevention of STIs.

_____ **15.** You believe that anyone may possibly be infected with an STI.

_____ **16.** You know that even when no symptoms are present, a person may still be infected.

_____ **17.** You would tell your sexual partner if you are or were diagnosed with an STI.

_____ **18.** You are careful to avoid excess alcohol or other drugs when in a potential sexual situation.

_____ **19.** You are reasonably comfortable discussing STIs with your doctor or other health professional.

_____ **20.** If your partner will not agree to your requirements for safe sex, you will not have sex.

Add up your score, which will range between 20 and 100. *The higher your score, the safer you are from contracting an STI.*

Here is what your score means:

Scores below 40 indicate that you may be at a very high risk for contracting one or more STIs.

Scores between 41 and 60 indicate a moderate to high risk.

Scores between 61 and 80 indicate a moderate to low risk.

Scores over 80 indicate that you are at a fairly low risk for STIs.

Regardless of your score on this scale, everyone must be educated about STIs to minimize risks of transmission.

Substance Abuse

Another behavior inextricably linked to the spread of STIs is substance abuse. The nature of this link involves the use of drugs, especially alcohol, that cause the user to take sexual risks that he or she would be unlikely to take when sober, such as unsafe sexual activities, especially lack of condom use (Dunn, Bartee, & Perko, 2003; Dye & Upchurch, 2006). One in five teens reports engaging in unprotected intercourse after drinking alcohol or using other drugs (Calvert & Bucholtz, 2008; Davis et al., 2007). A study of alcohol consumption and STIs between 1983 and 1998 in all 50 states found a clear association between the two (Chesson, Harrison, & Stall, 2003). For every 1% rise in alcohol consumption, the researchers found about a 0.5% increase in gonorrhea incidence rates and about a 2.5% increase in the rate of syphilis infection.

Why is alcohol associated with these effects on sexual behavior and the spread of STIs? A hypothesis frequently proposed to explain this connection is known as **alcohol myopia theory** (Davis et al., 2007; Griffin et al., 2010). This theory suggests that under the influence of alcohol, people are more likely to focus on immediate, "feel-good" behaviors (such as sexual arousal) and discount the more distant, long-term consequences of the activity (such as the possibility of contracting an STI). In other words, when it comes to making good decisions about sexual activity, alcohol causes your brain to be myopic or nearsighted, able to see up close but not at a distance (Davis et al., 2009). Beyond this decreased ability to make clearheaded decisions, recreational drug use also facilitates the spread of STIs by creating enhanced routes of transmission. You are probably aware that blood-borne STIs, such as HIV and hepatitis, are often transmitted through the sharing of needles by intravenous drug users. However, drug use is commonly associated with a greater number of sexual partners. And these partners are more likely to be drug users themselves and therefore more likely to be infected with one or more STIs (Ramisetty-Mikler et al., 2004). The bottom line is that STIs, drug use, and sexual behavior combine to maximize risk. Together, they create a complex environment conducive to many negative sexual experiences and outcomes, including poor sexual decision making, unsafe sex, and the rapid spread of STIs.

alcohol myopia theory

The belief that under the influence of alcohol, people are more likely to focus on immediate, "feel-good" behaviors (such as sexual arousal) and ignore future negative consequences.

High-Risk Sexual Behaviors

Nearly all sexual behaviors with a partner (phone sex, masturbation, and mutual touching are exceptions) carry some risk of transmitting STIs, and some activities are far riskier than others. As you read through the discussion of specific STIs in this chapter, you will see in greater detail how certain sexual behaviors pose risks for contracting or transmitting that particular infection. However, we can discuss some general guidelines here.

Overall, the more a behavior involves bodily fluids and/or sexual penetration, the more risk it carries. However, abstaining from vaginal or anal intercourse does not ensure that you are safe from infection. Skin-to-skin and oral sexual contact may transmit STIs regardless of any obvious exchange of bodily fluids. Table 8.2 summarizes the specific STI risks of various mainstream sexual behaviors.

Your ultimate defense against sexually transmitted infections is education, knowledge, awareness of unsafe behaviors, and *planning ahead* to ensure your sexual health.

This issue is discussed further near the end of this chapter in "Your Sexual Philosophy: Sexually Transmitted Infections."

One explanation for why alcohol use leads to poor sexual decision making is called the *alcohol myopia theory*.

Table 8.2 Sexual Behaviors: Risk Levels for Specific Infections

Behavior	Infection	Level of Risk
Fantasy, solo masturbation, phone or Internet sex	None	None
Kissing	Herpes	Very low (no sores present); high (sores present or recently healed). Deep, intense kissing with blood in saliva may pose risk of HIV or hepatitis transmission
Mutual masturbation	HIV, syphilis, hepatitis	Very low, unless sores, cuts, or abrasions are present on hands that make contact with semen or vaginal secretions
Fellatio, with condom	Herpes, genital warts	Low to moderate, depending on presence of sores or warts. Virus may transmit skin-to-skin around condom
Fellatio, without condom	Chlamydia, herpes, genital warts, gonorrhea, hepatitis, syphilis, HIV	Low to high, depending on infection and presence of lesions. Risk is *not* reduced by avoiding ejaculation in mouth
Cunnilingus	Chlamydia, herpes, genital warts, gonorrhea, hepatitis, syphilis, HIV	Low to high, depending on infection and presence of lesions. Risk may be higher during menstruation. Risk may be reduced significantly by use of dental dam
Vaginal intercourse with condom	Herpes, genital warts, pubic lice	Moderate to high, depending on presence of sores or warts and extent of lice infection
Vaginal intercourse without condom	All	Moderate to high
Oral-anal contact	All	Moderate to high, depending on presence of infection and routes of transmission through skin
Anal intercourse (insertive or receptive) with condom	All	Moderate to high
Anal intercourse (insertive or receptive) without condom	All	High

Specific Sexually Transmitted Infections

8.3 Summarize the symptoms, mode of transmission, treatment, and complications for each of the STIs

Now we are ready to begin our discussion of the STIs themselves. The various infections may be divided into categories based on the type of pathogen (germ) that causes the infection: viral, bacterial, and parasitic. For each category, we will discuss the characteristics and symptoms of specific infections, the typical modes of transmission, the tests available for diagnosing the infections, and the cures or treatments that are available now and those that may become available in the near future. These are summarized in Table 8.3.

Viral STIs

8.4 Review the similarities among viral STIs and summarize the specific symptoms, mode of transmission, diagnosis, and treatment methods for each.

All STIs are serious infections to varying degrees, and certainly no one *wants* any kind of STI! Nevertheless, some STIs are more serious than others. Among various criteria we might use to judge the seriousness of a particular infection, perhaps the most important is our ability to cure someone who is diagnosed with it. Using this as a bright-line test, many researchers would argue that the most serious STIs are those caused by viruses. Why? Because medical science has not yet discovered how to cure viruses.

Table 8.3 Sexually Transmitted Infections

Infection	Typical Symptoms (If Present)*	Mode of Transmission	Treatment	Vaccine Available?	Potential Complications
VIRAL STIs					
Herpes simplex virus (HSV), type 1 and type 2	Often asymptomatic. Clusters of small, painful blisters in genital and anal areas or around mouth. General flu-like symptoms: mild fever, fatigue, and tenderness in lymph nodes, perhaps urinary pain.	Oral, vaginal, or anal sexual activities; kissing.	No cure. Treated with antiherpetic drugs (acyclovir, famciclovir, valacyclovir).	No vaccine available as of 2011. Research is underway.	Sores create route of transmission for other STIs. May be passed from pregnant mother to child during birth.
Hepatitis B (HBV)	Usually no symptoms unless virus becomes active. Jaundice, a deep yellowing of the skin and eyes. Loss of appetite, fatigue, abdominal pain, nausea, vomiting, darkening of the urine, rash, and joint pain.	Contact with blood, serum, semen, vaginal fluids, and in rare instances, saliva. Routes of transmission are sharing needles, sexual contact, and sometimes tattooing and body piercing.	No cure. Acute infection: bed rest, increased fluid intake, good nutrition, and avoiding alcohol (which may exacerbate liver inflammation). Chronic infection: Alpha interferon boosts the body's immune system; lamivudine, antiviral medication, slows liver infection.	Yes. Series of three injections over six months. Prevents infection for life.	In small percentage of chronic cases: serious liver disease, including cirrhosis and liver cancer, which may lead to liver failure and either a liver transplant or death.
Human papilloma virus (HPV): genital warts	Often asymptomatic. Visible warts on the penis, opening to the vagina, cervix, labia minora, or anal area. Warts may also appear inside the vagina or anus and may not be noticed at all. Warts are flat or lumpy; they may be single or in groups and vary in size. Sometimes the warts may be too small or flat to be seen or felt.	Skin-to-skin contact; vaginal, oral, or anal sexual activities. Transmission possible with or without visible warts.	No cure. Removal of warts through cryo-surgery (freezing), laser therapy, surgery, and topical medications applied by the patient. Removing warts does not cure infection.	Vaccine, trade name *Gardasil,* released in 2007. Series of three injections. The vaccine imparts lifelong immunity from the four cancer-causing strains of HPV.	Major cause of cervical, anal, and throat cancers.
HIV/AIDS	HIV: Initially no symptoms or mild fever, headache, fatigue, and rash. As virus progresses: loss of energy; unexplained weight loss; frequent fevers and night sweats; frequent yeast infections; enlarged lymph glands; persistent skin rashes; short-term memory loss; mouth, genital, or anal sores; and blurred vision. AIDS diagnosed when CD4+ T-cell count dips to below 200 or when the person becomes infected with one or more opportunistic infections.	Transmission requires that semen, vaginal fluids, or blood from an infected person enter the bloodstream of an uninfected person. The most common routes of HIV are through sexual contact, primarily vaginal and anal intercourse and, less commonly, oral sex.	No cure. More than twenty specific medications for fighting HIV are currently on the market, including protease inhibitors, nucleoside-nucleotide reverse transcriptase inhibitors, nonnucleoside reverse transcriptase inhibitors, and fusion inhibitors. Also available are immune system enhancers and specific treatments for AIDS-related infections.	No. Research is continuing. Medication is now available that help to prevent transmission of virus (trade name: *Truvada).*	With proper treatment, lifespan of HIV-infected individuals is rapidly increasing (approaching twenty-five years in the United States). Survival following diagnosis with AIDS is approaching ten years.

(continued)

Table 8.3 Continued

Infection	Typical Symptoms (If Present)*	Mode of Transmission	Treatment	Vaccine Available?	Potential Complications
BACTERIAL STIs					
Chlamydia	Thick, cloudy discharge from the vagina or penis occurring one to three weeks after exposure; less common: pelvic pain, irregular periods, increased pain during menstrual periods, discomfort during urination, or irritation of the vaginal and/or anal area; often asymptomatic, especially in women.	Oral, vaginal, and anal sex. Extremely contagious bacterium; may be transmitted during a single sexual encounter with an infected partner.	Oral antibiotics will cure most cases. All sexual partners must be treated concurrently to prevent reinfection.	No. Research is ongoing; recent studies promising	Untreated may lead to infections of reproductive system and pelvic inflammatory disease in women and epididymitis in men, a major cause of infertility.
Gonorrhea	Men: painful, burning sensation during urination or bowel movement, and/or a cloudy discharge from the penis or anus. Women: often no noticeable symptoms. Cloudy vaginal discharge, irritation during urination, or vaginal bleeding between periods. Rare symptoms: lower abdominal pain or pain during intercourse.	Vaginal, anal, or oral sexual contact. Transmission to newborns can occur at delivery and cause eye infections.	Oral antibiotics will cure most cases. Some strains are becoming antibiotic-resistant. All sexual partners must be treated concurrently to prevent reinfection.	No. Research is ongoing. New treatment method may prevent future infections (see Feltman, 2013).	Untreated may lead to infections of reproductive system and pelvic inflammatory disease and possible infertility in women and epididymitis or prostatitis in men.
nongonococcal urethritis (nGU)	Similar to gonorrhea infection: discharge from the urethra; burning and itching at the end of the urethra or upon urination.	Vaginal, anal, or oral sexual contact.	Oral antibiotics will cure most cases. All sexual partners must be treated concurrently to prevent reinfection.	No.	Untreated may lead to infections of reproductive system and pelvic inflammatory disease and possible infertility in women and epididymitis or prostatitis in men.
Syphilis	Primary stage: characterized by a sore called a chancre that appears at the point of infection. Usually heals completely within a few weeks. Secondary stage: occurs within weeks or months; symptoms include low-grade fever, sore throat, fatigue, headache, alopecia (hair loss), and skin rashes, typically on the hands or feet. Late stage:	Sexual contact, including vaginal, anal, and oral sexual practices.	Antibiotics, primarily penicillin, typically cure the infection. All sexual partners must be treated concurrently to prevent reinfection.	No. Research is on-going.	Untreated, bacterium spreads to all bodily systems, causing new symp toms and eventual destruction of organs and brain tissue and death.
Mycoplasma genitalium	Extensive spread of the bacterium leading to organ damage, neurological damage, paralysis, mental illness, and death. Often asymptomatic. Men: urethral discharge, burning on urination, arthritic-like pain and swelling in various joints. Women: vaginal itching, pain or burning while urinating, or painful lymph nodes located in the groin area.	Vaginal or anal intercourse; possibly oral sex.	Specifically targeted antibiotic; most effective to date is moxfloxacin.	No	Associated with pelvic inflammatory disease, cervical and uterine inflammation, infertility, preterm labor in women; may be linked to infections of the epididymis and prostate gland in men.

(continued)

Table 8.3 Continued

Infection	Typical Symptoms (If Present)*	Mode of Transmission	Treatment	Vaccine Available?	Potential Complications
Pelvic inflammatory disease (PID)	Not an STI but a condition resulting from several STIs. Lower abdominal sensitivity and pain, pain during inter-course, irregular periods, cervical discharge and tenderness, fever, nausea, and vomiting.	Direct result of various STIs transmitted through sexual contact.	Various oral or injected antibiotics.	No. Vaccines that will prevent STIs such as chlamydia and gonorrhea will have the effect of preventing millions of cases of PID.	PID can affect the uterus, ovaries, fallopian tubes, or other related structures. Untreated, PID causes scarring and can lead to in-fertility, tubal pregnancy, chronic pelvic pain, and other serious consequences.
PARASITIC STIs					
Trichomoniasis	Men: usually asymptomatic. Possibly: irritation of the urethra, slight discharge, pain after urination or ejaculation. Women: yellow or greenish vaginal discharge accompanied by an unpleasant odor, genital irritation, and pain upon urination.	Sexual contact, including vaginal, anal, and oral sexual practices.	Oral antibiotics will cure most cases. All sexual partners must be treated concurrently to prevent reinfection.	No.	Without proper treatment, may cause PID among women and nGU among men.
Pubic lice	Genital irritation and extreme itching, often said to be the most intense itching imaginable, caused by lice biting the skin to feed on blood.	Sexual contact, sharing bedding, sharing clothing.	Prescription shampoos, creams, and ointments containing chemicals that are toxic to the lice. Washing of all bedding and clothes in hot water.	No.	No serious complications. Difficult to cure completely and easily spread.

*Many STIs are highly likely to be asymptomatic (having no noticeable symptoms).

Viruses, one-hundredth the size of bacteria, invade the body's cells and essentially transform those cells into virus factories that produce still more viruses that invade still more cells. Any agent that will destroy the virus will likely have to destroy the cell as well, leading to highly undesirable side effects, including the death of the host (the person). This is why we do not have a cure for the common cold (which is caused by various strains of the rhinovirus), the flu (caused by various influenza viruses), or viral STIs. Consequently, if you contract a viral STI, no pill, injection, or other treatment can cure it. Because a person cannot be cured of a viral STI, these infections present unique challenges in learning to live with the virus. "In Touch with Your Sexual Health: Living with an Incurable STI" discusses important considerations for helping those who have contracted a viral STI to live a full and satisfying life.

Since You Asked

2. Other than AIDS, what is the worst sexually transmitted infection?

In Touch with Your Sexual Health

Living with an Incurable STI

Although STIs caused by viruses are treatable, they are not curable at this time. Consequently, people who contract a viral STI are often strongly affected psychologically and emotionally with the thought that they now must endure this infection for a lifetime. However, all viral STIs are treatable, and infected individuals must become educated about the best strategies for living a healthy life with the virus. Here, we present information for living with herpes simplex, hepatitis B, and the human papilloma virus. Our discussion of the unique challenges of living with HIV and AIDS is included in the body of the chapter.

Emotional distress is a common reaction when a person discovers that he or she has been infected with an incurable STI. However, stressing out does little good and may, in fact, exacerbate outbreaks of certain infections. Stress, anxiety, attitudes about being infected, and lack of social and information support have all been linked to outbreaks of various STIs (Chida & Mao, 2009; Goldmeier, Garvey,

& Barton, 2008). People living with STIs have been shown to have lower self-esteem, experience greater daily challenges, and demonstrate greater psychological problems (Boonstra, 2006).

Disclosing an infection to potential sexual partners is another issue in the lives of people with STIs. Studies have found that some STIs can spread even in the absence of sores or other symptoms (Mertz, 2008). Despite such evidence, many people continue to believe that they are contagious only during obvious symptoms or an active outbreak of lesions. A key question in the minds of most people is when and to whom they should disclose their STI status. This is a question with no easy, one-size-fits-all answer. Some argue that in casual sexual encounters, when condoms are used consistently, disclosure may not be necessary. Others contend, however, that disclosure should be automatic with all potential sexual partners, regardless of the nature of the relationship. Ultimately, this becomes an ethical and philosophical decision each person living with these viruses must make. The reality, however, is that research has shown a substantial percentage of people have either lied or failed to disclose their STI infection status to a prospective sexual partner (Marelich et al., 2008).

People who discover they are infected often find counseling to be helpful in developing effective coping strategies for living with the infection. Feelings such as depression, fear of being "found out," anxiety about being seen as "damaged goods," social rejection, and anger can be debilitating, even for those who have lived with the virus for many years (O'Connor, 2007). Owing to the chronic nature of these viruses, patients may need one-on-one therapy during the first few months to the first year after the initial diagnosis. Social support, particularly support groups made up of people living with viral STIs, can be found in most parts of the country, and there is a national STI hotline, set up by the CDC, to help people find the resources they may need: 800-232-4636. Such support groups can be very helpful for people who need reassurance that they are not alone and who may need the help of others in the same situation. Support groups can also help with issues of disclosure, including when to disclose to partners. Last but not least, support groups provide a safe environment for socializing with understanding others and the possibility for new intimate relationships.

A diagnosis of an incurable STI is never easy. However, health care, information, resources, and support all work together to create and sustain a quality of life that can be rich and satisfying despite the infection.

As you will see, however, just because a viral STI is not curable does not mean that viral STIs are not treatable or preventable; they are. And the symptoms of these infections may be controlled and limited to varying degrees. The viral STIs that we will focus on here are herpes, genital warts (HPV), hepatitis, and HIV, which leads to AIDS.

Herpes Simplex Virus (HSV)

Researchers have identified at least eight strains of the human herpes virus. One of these strains causes chickenpox, another causes shingles, and a third causes mononucleosis. Two viruses in the herpes family are STIs: *herpes simplex virus type 1* (HSV-1) and *type 2* (HSV 2) (Bernstein et al., 2013). HSV-1, the more common of the two, is typically responsible for cases of oral herpes: cold sores or fever blisters around or in the mouth (although, not all mouth sores are caused by herpes). HSV-2 is the primary cause of **genital herpes**, which is characterized by painful sores and blisters (much like cold sores) that occur in the genital or anal area. HSV-1 and HSV-2 are extremely similar and often interchangeable; that is, either may produce infection in either the mouth or genital areas, or both (Tronstein et al., 2011). In fact, studies have shown that among college students, more cases of genital herpes are caused by HSV-1 than HSV-2 (Bernstein et al., 2013; Horowitz et al., 2010).

genital herpes

An STI caused by the herpes simplex virus and characterized by painful sores and blisters, usually in the genital or anal area.

Herpes infections are extremely common. Approximately one out of six people (over 16%) between the ages of sixteen and forty-nine is currently infected with genital herpes in the United States alone ("Genital Herpes," 2010). HSV-2 prevalence is nearly double in women (20.9%) what it is in men (11.5%), and three times higher in African Americans (39.2%) than in whites (12.3%) (CDC, "Study," 2010). Heterosexual women are at greater risk of all STIs due to greater exposure to vaginal mucous membranes (in both duration and area) from an infected male partner. African Americans tend to have higher rates because the population has higher existing rates, which leads to higher transmission rates.

An estimated 90% of those currently infected with HSV-2 are asymptomatic, meaning they have never noticed any signs of the virus and are unaware that they are infected. Because HSV may be transmitted without active lesions, this lack of awareness of infection contributes significantly to its spread (Tronstein et al., 2011).

Since You Asked

3. If I had sores on my genitals from an STI, but the sores went away, does that mean I'm cured?

Figure 8.2 Herpes Simplex Virus

Shown here are male and female examples of active genital herpes lesions.

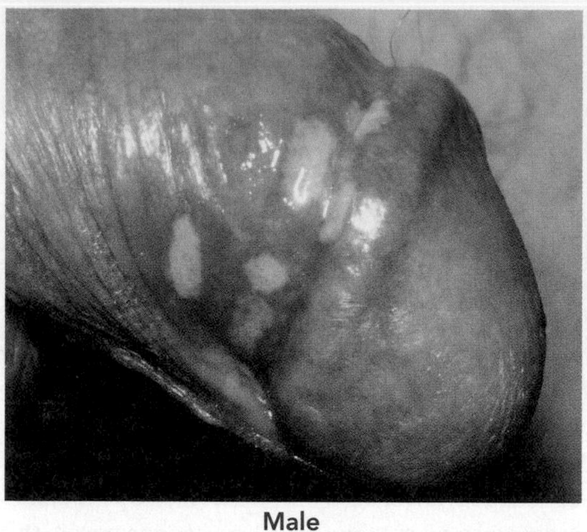

Male

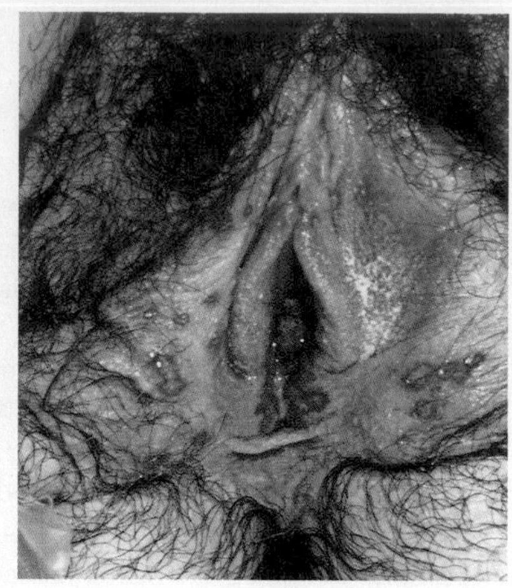

Female

Approximately one in six women and one in five men in the United States are infected with genital herpes (CDC, "Genital Herpes," 2010).

SYMPTOMS When symptoms of HSV are present, they occur two to ten days after exposure to the virus. Some people may experience general flu-like symptoms, including a mild fever, fatigue, tenderness in lymph nodes in the groin or neck, and perhaps urinary pain or discomfort. The primary symptom of genital herpes is the appearance of clusters, or *crops*, of small, painful blisters in the genital area—labia, vaginal opening, vagina, anal area, penis, or scrotum. Less commonly, the sores may appear in other parts of the body, such as the thighs or buttocks, and tend to be less painful than the genital outbreaks. Other symptoms sometimes include itching, pain during intercourse, vaginal or urethral discharge, and sensitivity in the abdominal area. Typically, the first episode of a genital herpes outbreak is the most severe, in terms of both symptoms and duration of the outbreak (see Figure 8.2).

Within a few days, the blisters rupture and ooze fluid. This fluid contains of millions of herpes virus particles, and the risk of transmission is highest at this point. Over the next week or two, the blisters scab over and heal without leaving a scar. If the lesions are inside the vagina, anus, or mouth, healing may take longer. We should mention that these sores are sometimes confused with other STIs such as syphilis (to be discussed later). Therefore, diagnosis by a qualified health care professional is important for proper treatment and management of the infection.

After the blisters heal, HSV typically enters a latency period when the person is free of outbreaks. However, this does not mean that the virus is gone. Latency simply implies that the infection is suppressed, usually temporarily. Recurrent outbreaks of herpes are common, but the severity of the symptoms tends to become less pronounced over time and is often limited to blisters, without the accompanying bodily discomforts. Many people with herpes learn to recognize the early warning signs of an impending attack. These **prodromal symptoms** include itching, burning, pain, and sensitivity, often in the same areas where the initial outbreak occurred.

prodromal symptoms
Warning signs, such as itching, burning, or pain, that an outbreak of an infection such as herpes may be impending.

Most people infected with HSV-2 will experience recurring outbreaks ranging from once a year or less to twice a month, with an average of five new outbreaks in the first year (CDC, "Genital Herpes," 2010c). The number of recurring episodes tends to decrease over time and may be significantly limited or even eliminated with medications (to be discussed shortly). The exact cause of these recurrences remains somewhat unclear. Some of the factors to do with herpes outbreaks that have received research support to varying degrees include hormonal changes associated with a woman's menstrual cycle; emotional stress; physical stress; vigorous sexual intercourse; exposure to ultraviolet light, especially sunlight; another illness (especially with fever); steroidal medications; or poor nutrition. One of the most serious health concerns about HSV is that those infected have been shown to have a twofold risk of contracting HIV most likely because outbreaks of herpes sores provide a convenient route of transmission for HIV (Xu et al., 2010).

TRANSMISSION Herpes is nearly always transmitted through oral, vaginal, or anal sexual activities. Herpes is readily transmitted via oral sex, which is largely responsible for the high number of cross-infections of oral and genital herpes (Horowitz et al., 2010). People who become sexually active at younger ages, who have had numerous sexual partners, or who have a history of drug abuse or heavy use of alcohol when engaging in sexual activities are at increased risk of becoming infected with HSV, probably due to their overall sexual behavior patterns (Cowam et al., 2002; Dempsey et al., 2008).

Some anecdotal evidence has suggested that the herpes simplex virus may be transmissible in nonsexual ways, through toilet seats, shared towels, contaminated surfaces, hot tubs, and so on. However, the herpes virus can survive outside the body for only about ten seconds. In hot, moist environments, survival time may be somewhat increased, but the likelihood of nonsexual transmission of HSV is virtually nonexistent, and no case of transmission from towels or toilet seats has ever been documented. Therefore, medical professionals assume that *all* cases of genital herpes have been transmitted the "old-fashioned way," through sexual contact.

In the past, most people believed that if no herpes sores were visible, the virus was not contagious. However, more recent research has found that many people who are infected with HSV may release virus particles, referred to as **viral shedding**, when they have no symptoms of the infection. This is called **asymptomatic shedding** and is far more common than most people believe. Studies have found that among people diagnosed with genital herpes, and among many who have not been diagnosed, the virus is shed on days when no clinical signs (blisters or sores) of infection are present ("Genital Herpes," 2010). Moreover, as mentioned earlier, most people (75% to 90%) who are infected with genital herpes are unaware of their infection. Consequently, these individuals, who have never noticed any symptoms, are also unaware of their risk of transmission and may be less likely to take precautions to reduce the possibility of transmission.

viral shedding
The release of virus particles that can potentially spread the infection to others.

asymptomatic shedding
Release of infectious virus particles when no symptoms of infection are present.

DIAGNOSIS AND TREATMENT Herpes, when there are symptoms, is typically diagnosed visually by the person who has the infection or by a health care provider. Diagnostic tests may be done directly on the fluid from the sores if the infection is active or a blood test can reveal herpes antibodies.

Since You Asked

4. My boyfriend told me he had an outbreak of herpes two years ago, but no signs since then. Could I still catch it from him?

As was pointed out earlier, herpes is not curable, but it is treatable with medications. Several antiviral drugs, known as **antiherpetics**, are currently available and are quite effective in suppressing the action of the herpes virus, as well as preventing outbreaks. The most common medication is *acyclovir* (trade name Zovirax; also available in generic form). Others include *famcyclovir* (trade name Famvir) and *valacyclovir* (trade name Valtrex). All of these drugs are taken orally and appear to be equally effective. The newer drugs, famcyclovir and valacyclovir, differ from acyclovir in the convenience of the dosing schedule (fewer doses each day). These medications may be taken at the first preinfection (*prodromal*) sensations or at the earliest sign of an outbreak of herpes. They are generally taken for seven to ten days during the first outbreak or for five days during subsequent episodes. All of these antiherpetic medications, when taken on a continuous daily basis, are effective in suppressing outbreaks of herpes altogether. In general, suppressive therapy is recommended for individuals who suffer six or more outbreaks per year, but it is often prescribed for people who experience less frequent recurrences (Mechcatie, 2006; CDC, "Genital Herpes," 2010).

antiherpetics
Medications developed to treat (reduce or prevent but not cure) outbreaks of the herpes virus.

When episodes of blisters do occur, these medications can reduce the severity and duration of the symptoms. Also, various nondrug treatments, including warm baths, loose-fitting clothing, and avoidance of activities that might further irritate the affected area, such as some sports or sexual activities, may be helpful in managing the pain and discomfort that often accompany an outbreak.

Research has also found that suppressive, or continuous, dosing of antiherpetic medications appears to significantly reduce the rate of *transmission* of HSV from an infected to an uninfected partner (CDC, "Genital Herpes," 2010; Elliot, 2004). Finally, research has been underway to develop a vaccine to prevent HSV. Unfortunately, one of the first large-scale clinical trials to test the effectiveness of a promising vaccine produced disappointing results, showing no significant reduction in the transmission of herpes type 1 or type 2 infection (Cohen, 2010; NIAID, 2010). Vaccine research and FDA approval of vaccines typically move very slowly, and the research to produce a widely available and effective HSV vaccine continues.

Human Papilloma Virus (Genital Warts)

human papilloma virus (HPV)
A sexually transmitted virus that is typically characterized by warts in the genital or anal area and that may lead to some forms of cancer; also known as *genital warts*.

The **human papilloma virus (HPV)**, which causes genital warts, is the most common STI in the United States today (CDC, "Genital HPV," 2014). The CDC estimates that 79 million people in the United States are currently infected with genital HPV and 14 million new infections occur each year. Nearly all men and women will contract at least one strain of HPV during their lifetime (CDC, "Genital HPV Infections," 2014). What makes this epidemic even more worrisome than simply the vast numbers is that *most* people infected with HPV have never noticed an active outbreak of warts and are unaware they are infected with the virus.

HPV AND CANCER Over 140 types of HPV, the virus that causes all types of warts, are present in humans, and approximately 69% of all healthy adults is infected with at least one them. About 40 of them are transmitted through sexual contact and appear primarily in the genital or anal area. But before you freak out, nearly all of the HPV strains are relatively harmless and do not cause any major health problems (Ma et al., 2014). Only four strains are associated with potentially serious diseases as we will discuss next.

Many people who become infected with genital strains of HPV are unaware of the infection because they never show any symptoms. However, the virus lives in the body's mucous membranes and can be transmitted to a sexual partner without the presence of any visible warts. If warts do appear in the genital area, the outbreak may be weeks or even months after exposure (CDC, "Genital HPV," 2014). The infected partner may not have been aware that he or she carried the virus. Although most of the strains of the virus are benign, a few types have been shown to increase the risk of other potentially serious diseases, and two strains are the primary cause of potentially deadly diseases such as cervical cancer, or other cancers of the reproductive tract, mouth, throat, and anus ("HPV and Cancer," 2010; Chesson et al., 2008).

Two strains of HPV (types 16 and 18) in particular have been implicated in causing these cancers. People infected with other strains of genital HPV may still develop genital warts but are unlikely to be at increased risk for cancer. Based on this clear connection between HPV and cancerous cellular abnormalities, the American Cancer Society has issued new guidelines for the detection and prevention of cervical cancer with which HPV is most closely connected. These are explained in the "In Touch with Your Sexual Health: HPV, Cervical Cancer, and the Pap Test: Advances in Cervical Cancer Screening."

Since You Asked

5. Is it true that genital warts can cause cancer?

HPV infection may be discovered when a person notices the presence of visible warts or clusters of warts in the genital area. The warts may be on the penis, opening to the vagina, cervix, labia minora, or anal or mouth area. However, warts may also be *inside* the vagina or anus, or may not be present at all. In such cases, the person may be

Figure 8.3 Genital Warts

Shown here are typical, observable cases (when symptoms are present) of male and female genital warts.

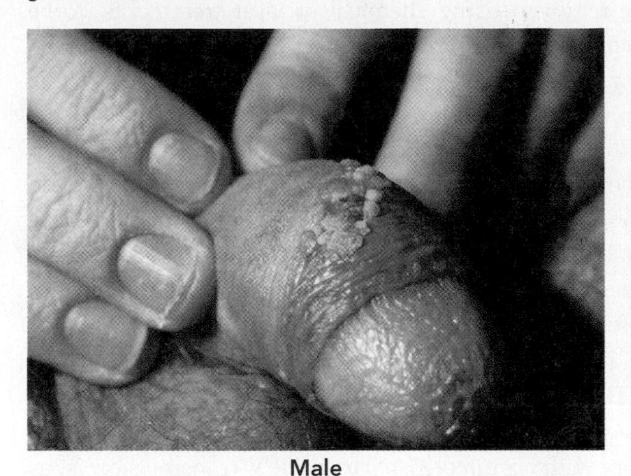

Male

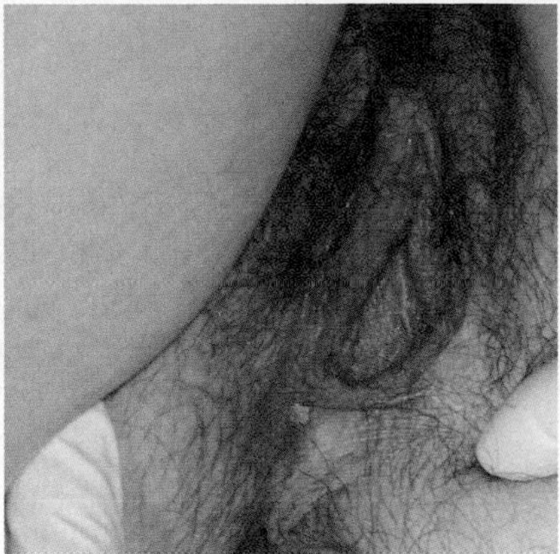

Female

unaware of the infection. The majority of women and some men who are infected with HPV have never seen or been aware of a genital wart.

If warts appear, this usually occurs about four to twelve weeks after exposure. The warts are typically painless, but in some sensitive areas such as the urethra, they may be uncomfortable or cause pain. The warts can have a flat, lumpy, or cauliflower appearance; they may occur singly or in groups; and they may vary in size. Sometimes the warts may be so small or flat that they cannot be seen or felt (see Figure 8.3).

In Touch with Your Sexual Health

HPV, Cancer, and the Pap Test: Cervical Cancer Screening Guidelines

Over the past fifty years, use of the Pap test to screen for cervical cancer has reduced the incidence of the disease by 70% in the United States. Yet despite this enormous success, each year thousands of American women develop cervical cancer and some will die unnecessarily. Although the Pap test is an important tool for cervical cancer screening and detection, the current understanding of the HPV's role in this form of cancer has prompted the addition HPV screening to cancer prevention and diagnosis.

In 2012, the American Cancer Society (ACS) published new guidelines for cervical cancer screening. These measures are echoed by the current recommendations from the American College of Obstetricians and Gynecologists (ACOG). Both sets of guidelines recognize new cervical cancer screening and detection techniques and contain important information concerning the prevention and early detection of cervical cancer. These guidelines were updated in 2009 and include the following:

- **Cervical cancer screening (the Pap test) should begin at age 21** regardless of a woman's sexual history.
- **Between ages 21 and 29**, Pap tests should be done every three years.
- **Starting at age 30**, women should have a Pap test and an HPV screening every five years.
- **Women 65 years of age and older** who have had adequate negative prior screenings may choose to stop screening for cervical cancer.
- **Annual screening Pap tests are recommended for women 21 years and over whose immune systems may be weakened** by HIV or organ transplant medications, who have a history of abnormal cervical cells found in previous Pap tests, or who were exposed to DES in utero, an anti-miscarriage medication prescribed between 1938 and 1971 with previously undiscovered negative side effects for the female later in her life (see CDC, "About DES," 2012).
- It is recommended that these guidelines be followed regardless of whether the woman has been vaccinated against HPV.

In women aged 21–29, when a Pap test finds abnormal cervical cells, a follow-up HPV screen is recommended. The HPV screening test detects the strains of HPV that are most commonly related to cervical cancer. Not all women infected with these strains of HPV will develop cancer, and the risk is small. The key to preventing cervical cancer and its progression is regular Pap tests and HPV screens using the current guidelines summarized above.

SOURCES: Broder (2012); National Cancer Institute (2012); Saslow et al. (2012).

TRANSMISSION Although HPV in humans has existed for decades and perhaps even centuries, only recently has it been elevated to "high-threat" status. Three reasons account for this new danger. First, as mentioned earlier, HPV cannot be cured. When the warts are visible, they can be removed using chemical or laser treatments (to be discussed next), but the virus remains in the body for at least a year until, in most cases, the immune system is able to fight and eradicate it (*if no reinfection has occurred*). While the virus is in the body, it can express itself with a new outbreak at any time. The second and more important issue is the discovery that certain types of HPV are the major cause of various cancers, as listed earlier. The third factor generating increased concern is that HPV may be transmitted *despite the use of condoms*, although condom use greatly decreases the risk of transmission. However, in some cases, the virus can shed from the infected person to his or her partner during sexual acts, through skin-to-skin contact in genital areas that are not well protected by a condom (Long, Laack, & Gostout, 2007).

Nevertheless, HPV is *most* likely to be transmitted when warts are visible (Sanfilipo, Cox, & Wright, 2007). The virus usually transmits from genitals to genitals, but may also be spread through oral and anal sexual activities. It can be spread to the mouth and throat through oral sex with an infected partner. HPV transmission is, as for all STIs, also linked to the number of sexual partners a person has; the more partners, the higher the risk.

DIAGNOSIS AND TREATMENT Genital warts, when present, are typically diagnosed visually by a health care professional. However, given that HPV infection is frequently asymptomatic, the presence of the virus is often not suspected by the infected person *or* by a health care provider. Tests are now available to screen women for HPV and to reveal the exact strain of the virus if a person is infected and to determine if the strain is one of the high-risk types associated with cervical cancer. No HPV test is currently available for men, although this is being intensely researched and should be available soon (CDC, "HPV and Men," 2012). Additionally, a saliva-based test is now available to dentists to help diagnose oral cancer in their patients as early as possible (see Nagler, 2009).

Current medical guidelines recommend an HPV test for women who received Pap test results indicating abnormal cervical cells (referred to as ASCUS, for *atypical squamous changes of unknown source*). Approximately 5% of all Pap smear tests find such abnormalities, and only a small percentage of these women are eventually diagnosed with cervical cancer upon follow-up testing. The PAP test itself does *not* detect HPV or any other STIs. A Pap test method called the *ThinPrep* collects enough cervical cells to allow for an HPV test to be done immediately if abnormal cells are found. If strains of HPV associated with cancer are discovered, the woman can be monitored more closely and more often for signs of cervical cancer and receive early, effective treatment (Etingen, 2007). However, such testing is not routine, so women need to be aware of HPV testing and request the appropriate tests from their doctors.

One more crucial point must be made about HPV and cancer: In addition to HPV's link to cervical cancer, evidence exists that the virus may also play a role in causing *anal cancers, mouth and throat cancers, penile cancers*, and *vaginal cancers* (CDC, "HPV and Cancer," 2010). In the United States, there are four times as many new cases of these cancers diagnosed each year, and twice as many deaths from them compared to cervical cancer. Most HPV infections will be cleared by the body within one to two years and not cause any serious additional health problems. But HPV infection that is long term and persistent increases the risk of one or more cancers. Consequently, tests for these cancers are recommended for anyone, female or male (when available), infected with immune-resistant, cancer-related strains of HPV.

Many visible outbreaks of genital warts resolve on their own (but the virus typically remains in the body). In cases where this does not happen, various methods are available for removal of the warts, including cryosurgery (freezing with liquid nitrogen), laser surgery, and several topical medications that are designed to be applied by the patient.

In pregnant women with genital warts, surgical removal is recommended to prevent transmission to the infant during delivery. Perhaps most important, HPV may be spread without any obvious signs of infection. However, for most girls and young women with fully functioning immune responses, HPV is cleared from the body naturally in 6–24 months (assuming no reinfection during that time). When the infection is persistent and does not clear, the risk of cancers increases ("Cervical Cytology Screening," 2009).

Although no medical cure exists for HPV, and prospects for one in the foreseeable future are not optimistic, biomedical researchers have accomplished something as beneficial, or perhaps more so: *prevention*. Two effective genital HPV *vaccines*, trade named *Gardasil* and *Ceravix*, immunizes females against the four most common and dangerous strains of genital HPV, providing they have not already been infected. By preventing infection with HPV, these vaccines also prevent HPV-related cancers as well including cancers of the cervix, anus, throat, vulva, vagina, and penis (DeNoon, 2010; National Cancer Institute, 2012). These vaccines are major events in the history of sexual medicine in that, not only do they prevent the most common strains of genital HPV, but, by doing so, they prevent the HPV strains that are most responsible for causing the cancers discussed above. Therefore, these are the first vaccines to *prevent cancer in humans*.

An important update about Gardasil is that research has now determined that the vaccine is effective in boys and men as well as girls and women. This is an important finding because males have a role in the spread of HPV and are at risk of the HPV-related oral, penile, and anal cancers discussed earlier.

The development of this vaccine is good news on the sexual health front, and you might assume that girls, and now boys, would be lining up to receive the vaccine. However, since its release, Gardasil has not been an easy "sell" to many consumers (you may have seen the national advertising campaign that uses the catch phrase "one less," meaning one less potential case of cervical cancer). As of late 2008, two years after its release, the CDC estimated that approximately 25% of teenage girls had received at least the first dose of the Gardasil vaccine. Data for boys is not yet available.

The controversy surrounding the vaccine is centered on the public health recommendation that it be given to children between the ages of nine and twelve. Why so young? Because immunization must occur before all children have any chance of becoming infected with genital HPV (the vaccine will *not* cure HPV once infection has occurred). We all know that *some* very young children are and will be, unfortunately, exposed to STIs through sexual abuse or early sexual activity. This is a relatively small percentage, yet for a vaccine to be effective in fighting an epidemic, it must be administered from the perspective of prevention in the *total* population. Therefore, although most children will not be exposed to genital HPV, from a public health perspective, the optimal age range for immunization is nine to twelve years. And that's the problem. Many parents and social advocacy groups are reluctant (or adamantly opposed) to immunizing their young children, who, in the parents' view, are nowhere near becoming sexually active and are being taught "abstinence until marriage," and, therefore, they (the parents) believe they have no chance of exposure to HPV. On the other hand, if the United States and other countries are not able to vaccinate a large enough percentage of children and young adults, the vaccine will do little to stop the epidemic of genital HPV. "In Touch with Your Sexual Health: The Genital HPV Vaccine: The Controversy" (on the following page) summarizes this debate.

HEPATITIS B VIRUS (HBV) *Hepatitis*, meaning inflammation of the liver, is a disease marked by an impairment of liver functioning due to various causes such as alcohol abuse; some recreational and prescription drugs; and specific bacteria. In addition, one of the most common causes of hepatitis is a group of viruses. Although at least seven strains of hepatitis are found in humans, you are probably most familiar with those designated hepatitis A, B, and C. Currently, the only viral strain of hepatitis categorized as a *sexually* transmitted infection is the **hepatitis B virus (HBV)**, which accounts for about a third of all cases of hepatitis in the United States.

hepatitis B virus (HBV)
A virus that may be sexually transmitted and may lead to inflammation and impaired functioning of the liver.

In Touch with Your Sexual Health

The HPV Vaccine for Girls and Boys: The Controversy

Although the relatively new HPV vaccines prevent the most common types of genital warts (HPV) infection, two of which are responsible for causing cervical, oral, anal, penile, and vaginal cancers, the release of the vaccines has stirred up a great deal of controversy, especially because many medical and health organizations are proposing that these vaccines become a mandatory part of the current immunization program for children. The debate centers on the fact that the immunization is recommended for children between the ages of nine and twelve. This early age for vaccination is important in order to immunize all children *before* they contract genital HPV. But a vaccine for a *sexually* transmitted infection at such a young age is objectionable to many parents. Here is a brief summary of the arguments on both sides of the HPV vaccine controversy.

Anti-Vaccine Arguments	Pro-Vaccine Arguments
The vaccines will give children "permission" to have unprotected sexual intercourse.	These are among the most important vaccines in history in that they are, essentially, anticancer vaccines. Research demonstrates that the vaccines will not alter sexual behavior in children in any direction; STIs are, rightly or wrongly, an uncommon reason that children avoid sexual activity.
They send "mixed messages" to children. Children hear the message to wait to become sexually active and yet they are receiving a vaccine against an STI.	The vaccines are virtually 100% effective in preventing the four most common and dangerous strains of genital HPV.
Mandatory vaccination infringes on parental rights in deciding health matters for their children.	Medically, school-aged children are the appropriate group for vaccination against diseases. All fifty states allow parents to opt out of childhood vaccinations. In addition, the vaccine will motivate parents to be more active in the children's sexual health.
They are not the same as other vaccines. Because genital HPV is only transmitted through sexual activity, it does not pose the same risk as, say, measles, chicken pox, and polio, which are highly contagious airborne viruses.	As for many vaccines, the risk of infection is not large at the time of the vaccine, but the protection is aimed at later, potential exposure. If all children were immunized from genital HPV, over 1,300 deaths from cancer would be avoided per year.
These will "medicalize" the problems of risky behaviors and impede the ability to instill positive morals and values in children.	They will significantly reduce health costs nationally by preventing both HPV and the associated cervical, oral, penile, throat, and vaginal cancers. Currently, the cost of genital HPV and its associated illnesses is estimated at $5 billion per year.
These vaccines would be discriminatory in that access to them will be more difficult for certain low-income, difficult-to-reach, and uninsured girls and women.	Making the HPV vaccine mandatory for schoolchildren is the best way to ensure it reaches the largest percentage of the population. Current school health programs will defray the cost, as is currently the case for other vaccines.
These may undermine current efforts at abstinence-only sex education programs.	Research indicates that parental attitudes and communication are the most powerful influence on children's sexual behavior and that a genital HPV vaccine will not alter this. In addition, abstinence-only programs have been unsuccessful in preventing risky teen sexual behavior.
The vaccines have not yet been adequately tested for safety, side effects, and duration of effectiveness.	Both Gardasil and Cervarix have been tested in tens of thousands of people globally. No serious side effects have been shown to be caused by the vaccines.

Since You Asked

6. I have an uncle who says he had hepatitis B, but he is now cured. I thought viruses could not be cured, so how is this possible?

We should note that although *hepatitis* C (HCV) may be transmitted through sexual activity, that is not the primary route of transmission and designation of HCV as an STI is not currently suggested. More likely routes of infection for HCV include sharing IV drug needles, accidental needle sticks in medical settings, mother to fetus transmission, and a history of multiple tattoos, especially when received under less-than-sanitary conditions or given by friends or others who do not have proper training (CDC, "Hepatitis C," 2014; Senecal & Morelli, 2007).

Researchers estimate that around 5% of the U.S. population is infected with HBV. Over 18,000 new cases of hepatitis are estimated to have occurred in 2011 (CDC, "Viral Hepatitis," 2014).

SYMPTOMS People infected with hepatitis B typically show no symptoms until the virus becomes active and begins to affect the liver. When this occurs, symptoms usually appear between one and six months after exposure. The most noticeable symptom is

jaundice, a yellowing of the skin and eyes (see Figure 8.4). Other symptoms include loss of appetite, fatigue, abdominal pain, nausea, vomiting, darkening of the urine, rash, and joint pain. In approximately 90% of cases, the illness runs its course over a month or two, after which the person recovers fully, is no longer contagious, and may not even test positive for antibodies to the virus.

The remaining 10% of adult patients become lifelong *chronic carriers* of the virus (for infants and young children, infection leads to chronic carrier status in 50% to 90% of cases). Approximately 10% to 15% of chronic carriers of hepatitis B will develop serious liver disease, including **cirrhosis of the liver** and liver cancer, which may lead to liver failure and, without a liver transplant, death.

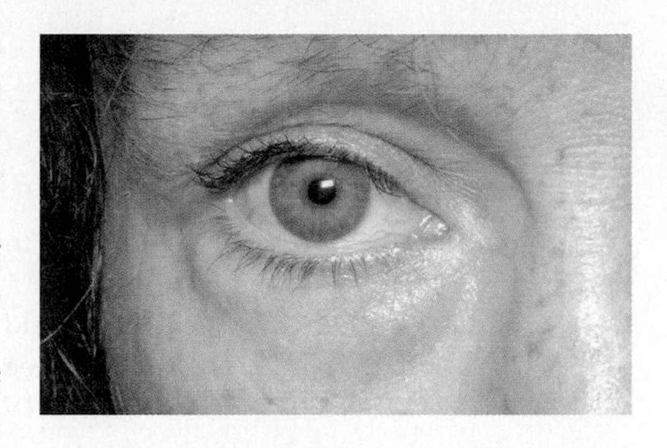

Figure 8.4 Jaundice Caused by Hepatitis

One of the most obvious signs of an active hepatitis infection is a yellowing of the skin and eyes.

TRANSMISSION HBV is spread by direct contact with contaminated blood products: semen, vaginal fluids and, in rare instances, saliva. These bodily fluids of an infected person must enter the bloodstream of another person for the virus to be transmitted. The most common routes of transmission are sharing needles, insertive sexual contact and, occasionally, tattooing and body piercing. HBV is not spread by casual contact or by respiratory droplets (sneezing or coughing). HBV may also be transmitted from an infected woman to her infant during pregnancy, birth, or breast-feeding.

Any sexual activity that allows the blood or blood product of an infected partner to enter the bloodstream of an uninfected partner may transmit HBV. The riskiest behaviors are unprotected vaginal and oral sex, intercourse without a condom, anal intercourse with or without a condom, and oral-anal activities (refer to Table 8.2).

jaundice
A symptom of hepatitis characterized by a yellowing of the skin and eyes.

cirrhosis of the liver
A potentially serious liver disease that may lead to liver cancer.

DIAGNOSIS AND TREATMENT Jaundice, in combination with reports of other symptoms, is the most common indicator of HBV infection. The diagnosis is usually then confirmed by a blood test that detects antibodies to the hepatitis virus. After a person is initially exposed to the virus, antibodies will become detectable in one to four months. During this **incubation period**, although the virus may not be detectable, the person may transmit the virus.

incubation period
The time between infection and the appearance of physical symptoms of illness.

Once a person has been diagnosed with acute HBV, treatment options are limited. As mentioned earlier, the body's immune system usually resolves an acute infection on its own. For those with symptoms of acute infection, most doctors recommend bed rest, increased fluid intake, good nutrition, and avoidance of alcohol (alcohol may exacerbate liver inflammation). The symptoms of acute hepatitis usually last four to twelve weeks and the body usually eliminates the virus entirely within six months (Lin & Kirchner, 2004).

In cases of chronic infection, various antiviral medications may be prescribed when and if a person shows signs of liver infection ("Approved Treatments," 2010). These medications, which do not *cure* hepatitis B, are used, often in combination, to slow the rate at which the hepatitis B virus multiplies, which helps prevent inflammation that can lead to liver damage.

Of course, *preventing* any illness is preferable to treating it. To that end, an HBV vaccine that will prevent infection with near certainty has been available since the early 1980s. The vaccine is recommended for *everyone* by most medical organizations in the United States, and it is now part of the routine vaccination schedule recommended for infants. Moreover, the HBV vaccine is required routinely for public school admission in most states. The vaccine is given in a series of three injections over a six-month period and imparts lifetime immunity to the virus. As a testament to the medical power of vaccines, during the period between 1991, when the vaccine was made routine for young children, and 2008, the annual number of new cases of hepatitis B in the United States dropped 82% to 3300 and continued to fall to 2900 in 2011 (CDC, "Viral Hepatitis," 2014; CDC, "Hepatitis B," 2009). There are now two vaccines available, neither of which contains live HBV viruses, so they *cannot* "accidentally" cause hepatitis. Studies have shown the vaccine to be one of the safest on the market, with no serious negative side effects ("Hepatitis B Vaccine," 2009).

HIV and AIDS

One of the most profound and frightening public health crises of the last century is the HIV/AIDS pandemic. It is safe to say that HIV/AIDS has forever changed how the world views sexuality and sexual behavior. The staggering number of people who have died (and continue to die) as a result of AIDS, the vast number who are infected with HIV (the virus that causes AIDS), and the effect of HIV on human sexuality necessitate a somewhat deeper and wider discussion than we have provided for the other STIs in this chapter. Therefore, in this section, we will focus on the background and origins of HIV/AIDS, the *prevalence* and *incidence* of the infection, symptoms and progression of HIV/AIDS, the proven routes of transmission, myths and misconceptions, diagnosis and testing for HIV, current treatment options, and the challenges of living with HIV. We will focus our attention on HIV and AIDS in the United States, but the devastating toll AIDS is taking around the world is discussed in "Sexuality and Culture: The HIV/AIDS Pandemic."

human immunodeficiency virus (HIV)
The virus that causes AIDS.

acquired immune deficiency syndrome (AIDS)
A gradual failure of the immune system, leading to serious, opportunistic infections, which, in turn, may lead to death.

opportunistic infections
Diseases that establish themselves in the human body only when the immune system is weakened and incapable of fighting them off.

BACKGROUND AND ORIGINS OF HIV AND AIDS Most likely, everyone reading this chapter is familiar, to some extent, with the **HIV (human immunodeficiency virus)** and **AIDS (acquired immune deficiency syndrome)**, which is caused by the HIV. What may surprise many people is that, although HIV and AIDS were not identified until the early 1980s when the infection first appeared in the United States, they have been present in humans since 1900, and perhaps earlier. Recent discoveries suggest that the virus probably crossed from chimpanzees and other nonhuman primates infected with a simian form of HIV (called SIV) to humans between 1894 and 1924, and began to spread as cities developed in an area of Africa now known as Kinshasa, Congo (Worobey et al., 2008).

Although SIV is harmless to the nonhuman primate population, in humans it converts to HIV, which leads to the weakening and destruction of the immune system and, eventually, to AIDS. The cause of this animal-to-human crossover is thought to relate to human practices in some parts of Africa of hunting and butchering primates, leading to repeated exchanges of blood products from simians to humans and the transfer of the virus, which then evolved as HIV (Heeney, Dalgleish, & Weiss, 2006; Lemey et al., 2003; "Where HIV Began," 2006).

HIV and AIDS orphans outside an orphanage in Changhai Mai, Thailand. HIV infections continue to increase worldwide.

Awareness of HIV and AIDS in the United States dates back only to the early 1980s. On June 5, 1981, the CDC released their first report that a few cases of a rare form of pneumonia (*pneumocystis carinii pneumonia*, or PCP) had been diagnosed among gay men in New York, Los Angeles, and San Francisco. Soon other mysterious diseases, including a rare form of skin cancer (*Kaposi's sarcoma*, or KS) and infections normally thwarted entirely by the human immune system, were being diagnosed among more and more men in the gay community. Because these diseases only establish themselves in the human body through the opportunity of a weakened immune system, they are referred to as **opportunistic infections**. At that time, doctors and medical researchers were bewildered by the array of symptoms and diseases exhibited by the patients who had first contracted the infection that would become known as AIDS; they were helpless to stop the disease's progression. These symptoms included unexplained weight loss, rare forms of pneumonia, rare cancers, and various infections caused by microbes not normally found in healthy human populations. Within one year, over 1,500 cases of this new illness had been diagnosed, and over 600 of the infected men had died.

In the first few years immediately following the discovery of AIDS in the United States in 1981, the illnesses and deaths were occurring primarily within the gay male community. On the street and among the population in general, the disease was initially referred to as the "gay plague," a misnomer adding fuel to the already hot fire of prejudice and discrimination that existed against nonheterosexual individuals. Even to the professional health community, AIDS did indeed appear, at first, to be limited to gay men. Consequently, the first official, medical designation for AIDS was GRID, for *gay-related immune deficiency*. Within less than a year, however, AIDS was discovered in populations beyond gay males, including women, intravenous drug users, and large groups of people in other countries, especially Haiti and Africa. Consequently, in 1982 the CDC officially changed the name of the disease to *acquired immune deficiency syndrome*, or AIDS. Even then, no one knew that the cause of AIDS was a virus.

However, in late 1983 and early 1984, American and French researchers discovered that the causal organism for AIDS was a virus, or more specifically, a **retrovirus**. Retroviruses are especially insidious because they survive by using their RNA to invade and destroy the DNA of normal body cells and then copy the virus's DNA into the host cells' chromosomes. This process replicates over and over, manufacturing more viruses, which in turn migrate into additional cells. Only a handful of human retroviruses have been discovered, including those that cause certain cancers and two that cause HIV.

retrovirus

A type of virus, such as HIV, that survives and multiplies by invading and destroying the DNA of normal body cells and then replicating its own DNA into the host cell's chromosomes.

PREVALENCE AND INCIDENCE OF HIV/AIDS As of the end of 2011, the cumulative number of diagnosed AIDS cases in the United States (the **prevalence**) was estimated at 1,159,752. Approximately 15% of these individuals (174,000) are unaware that they are infected. Over 636,000 have died from the effects of the disease (CDC, "HIV in the United States," 2013). In 2011 (the most recent statistics available), the number of new HIV diagnoses (the **incidence**) was approximately 50,000. This number has remained steady each year over the past decade.

prevalence

The total cumulative number of cases of a disease in a given population.

incidence

The number of new cases of a disease in a given population over a specific time period.

Although HIV infection crosses all ethnic, age, and economic boundaries, it is diagnosed disproportionately in certain populations. Nearly 50% of new HIV infections are among African Americans. In comparison, white Americans account for 32% of new HIV infections. Table 8.4 details the number of new HIV infections by ethnicity and sex. The rate of AIDS is nearly four times higher among men than women; however, the number of new HIV infections is growing alarmingly fast among women, particularly African American and Hispanic women. Moreover, women are twice as likely as men to contract HIV from heterosexual contact (CDC, "Diagnoses of HIV Infection," 2013).

Since You Asked

7. I heard that AIDS has been nearly eliminated in the United States. Is this true?

Table 8.4 Estimated Number of New HIV/AIDS Diagnoses by Race/Ethnicity and Sex: United States (2011; most recent data)

Race/Ethnicity	New HIV/AIDS Cases
White	13,846
Black/African American	23,168
Hispanic/Latino	10,159
Native Hawaiian/Other Pacific Islander	78
American Indian/Alaska Native	212
Multiple Races	827
Sex	**New Hiv/Aids Cases**
Females	10,257
Males	38,824

SOURCE: CDC, "Diagnoses of HIV Infection" (2013).

Sexuality and Culture

The HIV/AIDS Pandemic

HIV and AIDS have become a pandemic and one of the greatest health threats of our time (see UNAIDS, 2010). A staggering 1.6 million deaths were attributed to HIV infection in 2012, and it is estimated that over 35 million people throughout the world are currently living with HIV infection. Over 6,000 people are newly infected with HIV every day and over 4,000 die from AIDS-related illnesses *daily*. In the last decade, the number of people living with HIV infection has steadily increased, as indicated in Figure 8.5a. This appears to be a discouraging statistic, but in reality, it has somewhat of a silver lining. Although more people have become infected, more have also been living with HIV and not converting to AIDS. This is evident from Figure 8.5b, which shows a global decrease in new HIV infections over the same period. More importantly, as you

Figure 8.5 Estimated numbers of people living with HIV, new HIV infections, and AIDS deaths, 2001–2012, globally.

SOURCE: UNAIDS, Global Report (2013). www.UNAIDS.org.

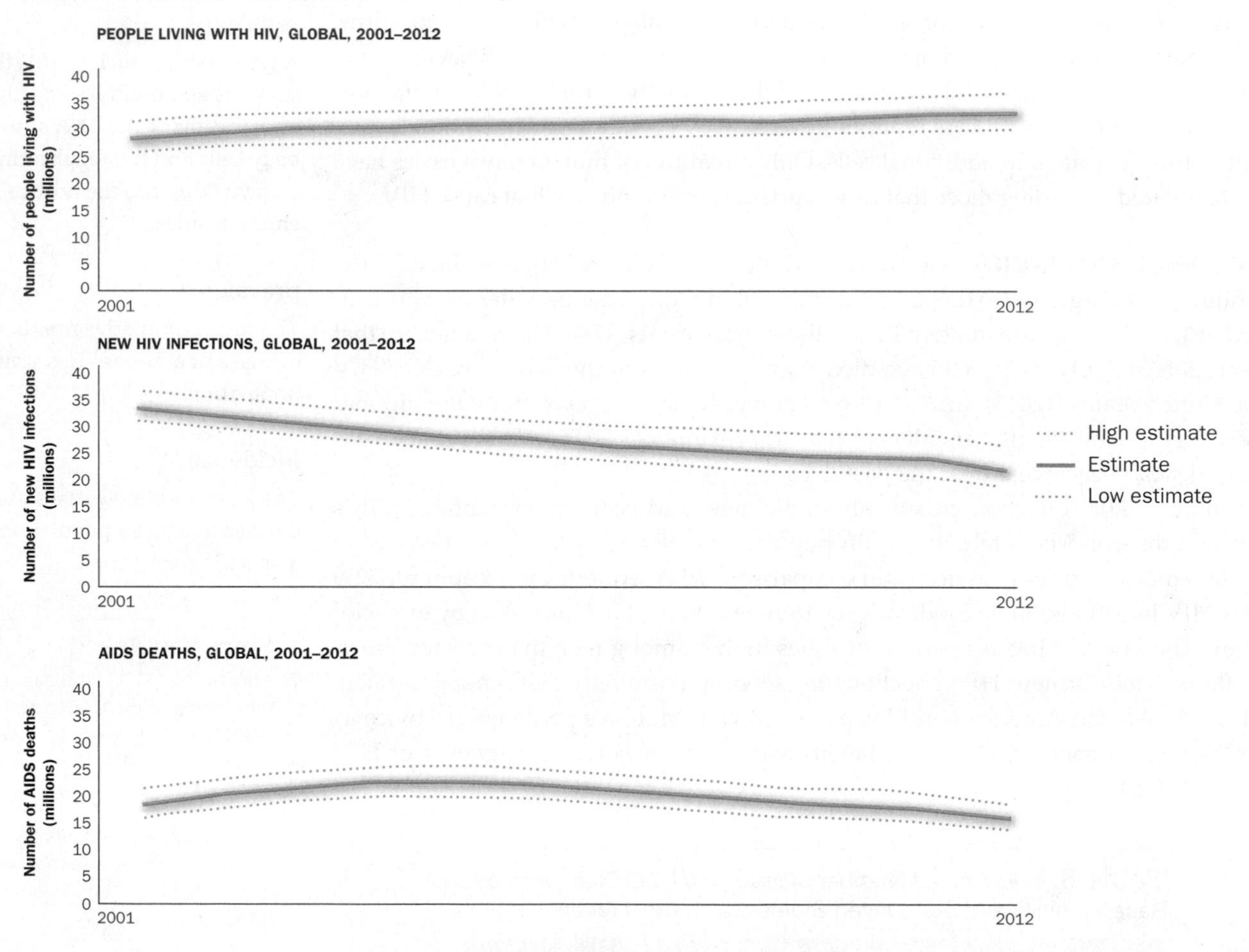

Public health officials have reported disturbing increases in HIV infections in the United States, especially among men who have sex with men, African American men, and Hispanic men. Researchers have attributed these increases to two primary shifts in attitudes among the public about prevention. One of these shifts is the growing *erroneous* belief that the new, more effective drugs now available to treat HIV infection can actually cure the disease or render it harmless and noninfectious (which they cannot). Combined with this growing confidence that HIV is no longer a serious health threat—a "death sentence"—is a phenomenon that has been termed **safe-sex fatigue**, the feeling of being fed up with always practicing safer sex behaviors. These changing attitudes are causing some people, especially within the gay male and college can see in Figure 8.5c, the number of deaths from AIDS, after a continued upswing during the mid-2000s, showed a clear drop from the mid-2000s to 2012 (UNAIDS, Global Report, 2013). Although these numbers indicate progress in the global fight against HIV and AIDS, the battle against HIV and AIDS is far from over as the epidemic improves and worsens at different rates from country to country.

The United States and European countries were among the first to report AIDS cases. However, less economically developed countries, especially those in sub-Saharan Africa, are the most severely affected and were among the first to report epidemic levels of HIV infection and AIDS-related deaths. AIDS is the leading cause of death in sub-Saharan Africa, where an estimated 25 million people are HIV-positive; most victims are between 15 and 49 years of age (UNAIDS/WHO, 2009). Sub-Saharan African countries accounted for 70% of new HIV infections and for a stunning 75% of AIDS deaths in 2012.

As the accompanying maps show, rates of HIV infection and AIDS are major problems in Asia, Europe, and Latin America as well. World health organizations are focusing intensively on the extreme danger of an even larger and more widespread increase of HIV through the huge populations of India, China, and Southeast Asia, where nearly one-quarter of the world's population resides.

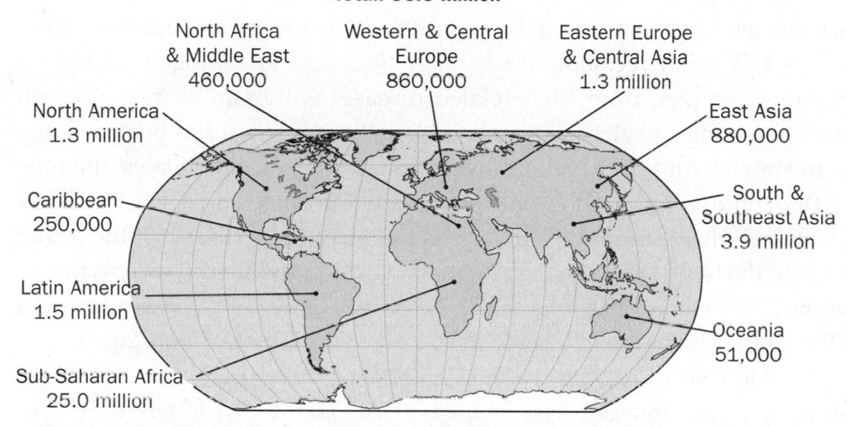

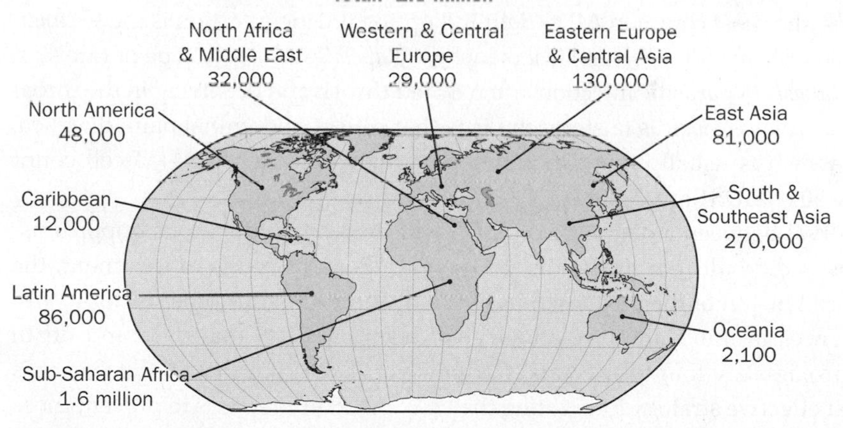

These maps show the number of people estimated to be living with HIV/AIDS worldwide as of the end of 2012, and the estimated number of new HIV infections around the world recorded in 2012.

communities, to abandon safer sex practices and return to riskier sexual behaviors (see van Griensven, et al., 2009; Rowniak, 2009). Furthermore, many young adults in the United States today have not experienced firsthand the staggering number of deaths from AIDS that occurred in the earlier years of the disease and seem to fear the virus far less. Paradoxically, advances in the treatment and management of HIV and AIDS seem to be undermining public health efforts to promote safer sex behaviors.

safe-sex fatigue
A loss of tolerance for practicing and a decrease in safer sex behaviors.

SYMPTOMS Most people who become infected with HIV at first experience no symptoms at all. If symptoms do occur, they may resemble those of the common cold or flu, including fever, headache, fatigue, and rash—all the usual signs that the immune system is responding to a foreign microorganism. These signs of infection typically occur within a few days to several weeks after exposure and usually resolve within one to three weeks; however, these symptoms are too general to serve as reliable indicators of HIV infection. The only accurate diagnosis of HIV is an HIV antibodies test (to be discussed in the next section).

Once the initial symptoms of infection, if any, subside, the infected person usually feels normal and in typically good health. However, he or she now carries the virus, and the immune system is busy manufacturing antibodies in an attempt to fight the infection. More importantly, the person is now capable of transmitting the virus to others, primarily through the exchange of blood or blood products during sexual contact or injectable drug use.

Over time, HIV microbes continue to multiply in the body, attacking and killing immune system cells, especially CD4+ T cells, the body's first line of defense against most infections. As the immune system weakens, new and more severe symptoms begin to appear. These include loss of energy; unexplained weight loss; frequent fevers and night sweats; frequent yeast infections; enlarged lymph glands; persistent skin rashes or flaky skin; short-term memory loss; mouth, genital, or anal sores; and blurred vision.

Earlier in the HIV epidemic, individuals diagnosed with HIV progressed to full-blown AIDS status and died from AIDS-related diseases within an average of eight to twelve years following initial HIV diagnosis. This time interval has been steadily increasing as treatments for HIV have improved. Today, in the United States, the time between an HIV diagnosis and AIDS is, on average, eight to ten years (AIDS.org, 2014). The median survival (that is, from HIV diagnosis and conversion to AIDS) for young people has risen to thirty-five years, assuming the infected individual has access to current medications (Llibre et al., 2009; Lohse, 2007). In fact, as of 2008, those diagnosed with HIV in the United States are *unlikely* to die because of AIDS (DeNoon, 2008).

Generally, a diagnosis of AIDS is made in an HIV-positive person when two criteria are met: the person's immune system CD4+ T-cell count dips to below 200 per cubic millimeter of blood, which is less than half the normal lower limit, and when the person becomes infected with one or more of over 20 opportunistic infections. In the United States, the most common AIDS-defining opportunistic infection is the *pneumocystis carinii pneumonia* (PCP) virus. Others include *Kaposi's sarcoma* (a type of rare skin cancer), *candidiasis* (a parasitic infection, often called thrush and occurring in the throat or lungs), and *cytomegalovirus* (a stomach virus that causes abdominal pain, diarrhea, and fever), which is usually found in AIDS patients when their CD4+ T-cell count drops below 50 ("AIDS Signs," 2011).

The interval between a diagnosis of AIDS and death due to various opportunistic infections is difficult to estimate due to the variations of access to treatment, the age and overall health of the diagnosed individual, and numerous other factors. With those differences in mind, the median survival from an AIDS diagnosis and death today is approximately four to six years (Gadpayle et al., 2012). What all this means is that the most effective strategy for fighting the HIV/AIDS epidemic is to prevent those who are HIV positive from converting to AIDS (see "Treatment" below).

TRANSMISSION The human immunodeficiency virus in transmissible amounts is found in the semen, vaginal fluids, and blood of infected persons. Transmission of HIV requires that one of those virus-containing bodily fluids from an infected person enters the *bloodstream* of an uninfected person. The most common sexual routes of HIV transmission are through unprotected vaginal and anal intercourse. The act of vaginal intercourse itself does not *automatically* ensure the transmission of HIV if one partner is infected. The probability of transmission of HIV from a single instance of unprotected vaginal intercourse between monogamous couples, where one is infected, is approximately 1 in 500 from an infected male to a female partner and 1 in 1,000 from an infected female to a male partner (Cairns, 2011; Garnett et al., 2008). However, the probability increases greatly as the number of viral cells in the blood, called *viral load*, increases, especially in the early or *acute stage* of HIV.

The acute stage of HIV infection (also called *primary HIV*) is the first few months following a person's initial infection, when the virus is multiplying fastest and the individual is most likely to be unaware of his or her HIV-positive status (thereby enhancing the overall chances of transmission). During the first five to six months of infection, sexual transmission is eight to ten times more likely compared to later stages of infection (Pilcher, 2003; Smith et al., 2013). The odds of infection also increase dramatically with the presence of genital lesions, such as herpes sores, or genital warts, on either partner, because the sores or warts may provide easier access to the bloodstream of either or both partners. Regardless of these statistics, the consequences of contracting HIV are far too great to "play the odds" because transmission *can* occur during *any single* instance of vaginal intercourse when one partner is infected, including the first time.

As mentioned earlier, overall, women are at greater risk of contracting HIV through heterosexual intercourse than men are during any given sexual encounter. This difference is primarily due to biological and anatomical factors, including the larger surface area of mucous membrane exposed during heterosexual intercourse in women than in men, the greater volume of bodily fluids transferred from men to their partners during vaginal or anal intercourse, the higher total number of viruses in male sexual fluids, and the small abrasions or other lesions that may occur in vaginal (or rectal) tissue during sexual penetration. In addition, research has demonstrated that the viral load required to transmit HIV is considerably lower for male-to-female transmission than for female-to-male transmission. Findings have indicated that women who transmit HIV to men have a four times greater viral load (number of HIV microbes in the blood) than women who are HIV positive and do not transmit the virus. In contrast, men who transmit HIV to women have only a 1.5 times greater viral load than HIV-positive men who do not transmit the virus (Gomez, 2004; Lingappa et al., 2011).

Anal intercourse presents a significantly different picture relative to HIV transmission risk. The overall risk of transmitting HIV through *unprotected anal* intercourse is eighteen times greater than unprotected vaginal intercourse (Baggaley et al., 2010). Various factors account for the high risk of transmission through anal sex. The mucous membrane lining the rectum is only one cell layer thick and does not become lubricated during sexual activity as the vagina does. This means that rectal tissues can be damaged easily. Slight tears or abrasions in the rectal walls allow an easy route of transmission from the semen of an HIV-positive insertive partner to the receptive partner. The reverse may also be true: If the receptive partner is HIV-positive, the virus may enter through damaged mucous membranes at the opening to the urethra or through a sore or abrasion on the penis. Moreover, condoms offer less protection during anal intercourse than during vaginal intercourse because the lack of natural lubrication and the tightness of the anal opening may cause condoms to tear or break.

Since You Asked

8. Can people get STIs from having oral sex?

Evidence suggests that the risk of transmission of HIV during oral sex is lower than for vaginal or anal activities but it is a real risk (CDC, "HIV Transmission," 2010; Flannigan, 2008). This is an extremely important finding because many people, regardless of sexual orientation, hold the *mistaken* belief that oral sex is safe sex, a misconception that leads to further spread of HIV (*as well as most other STIs*). The relative rates of HIV transmission risk of oral, vaginal, and anal sex are shown in Figure 8.6 (on the following page).

HIV may also be spread through various nonsexual routes, including the sharing of needles by intravenous drug use or accidental needle sticks among health care workers. When an HIV-positive person injects a drug, some of his or her blood

Sharing needles used to inject drugs remains a major route for HIV infection.

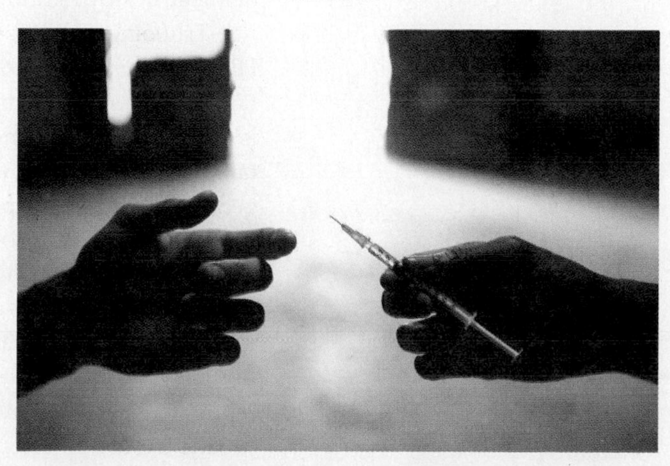

Figure 8.6 Relative Risk of Various Sexual Behaviors in the Transmission of HIV

SOURCE: CDC, "Risk Chart." Adapted from www.cdc.gov/hiv/topics/treatment/PIC/pdf/chart.pdf.

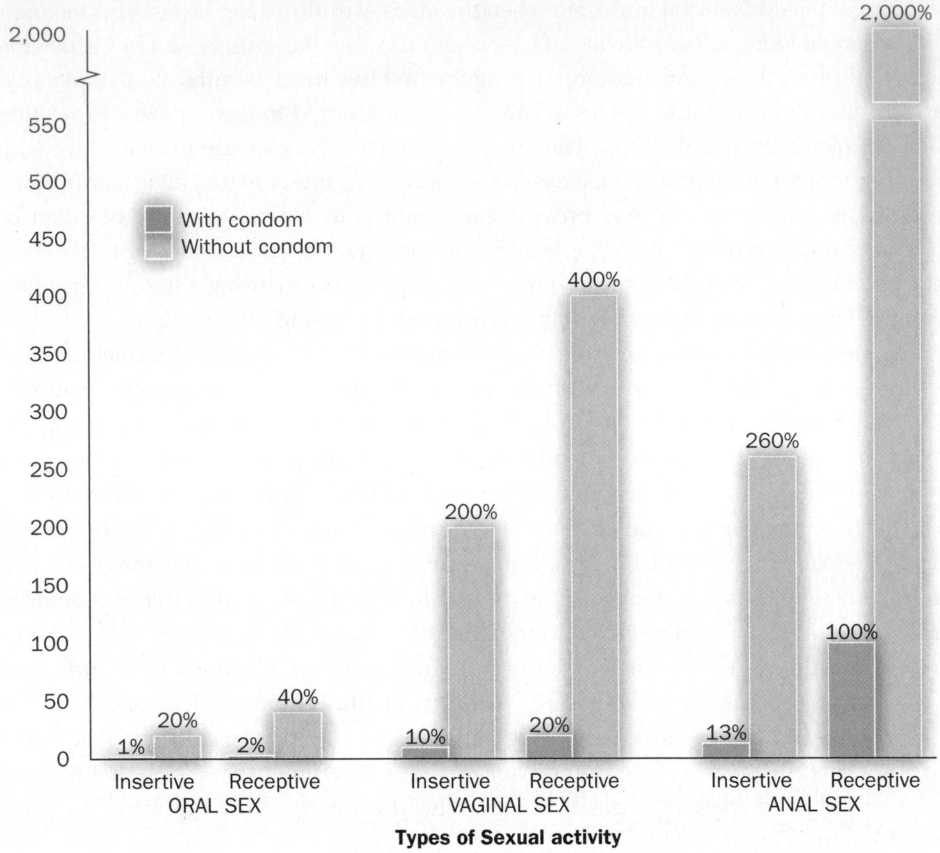

is likely to be drawn into the syringe. If another person then uses the same needle, a highly infectious route of transmission is established. Health care professionals have contracted the virus when caring for an HIV patient by accidentally sticking themselves through gloves or other protective clothing with a needle that has been used to give an injection or draw blood. This mode of transmission has decreased over time as greater precautions have been taken and better protective materials have been developed.

Early in the HIV outbreak (prior to 1985), one of the most common routes of transmission was through transfusions of blood donated by people who were unaware of their positive HIV status. Today, in most industrialized countries, the blood supply is routinely screened for HIV and numerous other diseases and is very safe. The current risk of an HIV-infected donated blood sample is less than one in 2,000,000 (less than being struck by lightning) ("What Are the Risks?" 2011).

Tattooing or body piercing is theoretically a risky activity in terms of potential for HIV transmission. If nonsterile or unsanitary utensils are used for piercing or tattooing, HIV may be spread among customers, especially if the time interval between procedures is short. Most tattoo and piercing businesses are required by law to use universal sanitary precautions to prevent infections of all kinds, and these precautions, if followed, are generally effective in preventing HIV transmission (CDC, "HIV Transmission," 2010). More worrisome are do-it-yourself piercing and tattooing because groups or individuals may share whatever sharp tools they may be using, creating a route of HIV transmission if one of them is infected with the virus, just as with the risk of sharing drug needles.

HIV may be transmitted from a mother to her infant during pregnancy, during the delivery process through exposure to HIV-infected blood, or through breast-feeding. In 2005, 92% of all cases of pediatric AIDS in children under age 13 were attributed to mother-to-child transmission of HIV (Mofenson et al., 2006). This form of HIV transmission continues to be a major public health problem in poor and developing countries. Nearly 1,000 infants infected with HIV are born each day, primarily in Sub-Saharan Africa (CDC, "Eliminating HIV," 2012). This means that a total of almost 370,000 new cases of childhood occur each year in the world. However, in stark contrast to this terrible statistic, the transmission from mother to fetus can be virtually eliminated with proper diagnosis and medical treatment. In the United States, mother-to-child transmission has been reduced to less than 1%. And in African countries where these diagnostic and prevention strategies have been instituted, such as Botswana and South Africa, the transmission rate has declined to 3.5% (CDC, "Eliminating HIV," 2012).

On a reassuring note: many of the routes of transmission for HIV that some people worry about are nothing but myths and misconceptions. For example, no evidence exists that HIV can be transmitted through insect bites (such as mosquitoes or flies); sharing dishes, food, or even toothbrushes; donating blood; sharing hot tubs or swimming pools; contact with pets or other animals; or nonsexual contact with HIV-positive people. Contact with saliva, tears, or sweat does not transmit HIV. Using the same toilet, restroom, towels, or washcloths as an HIV-positive person will not transmit the virus. HIV is not airborne, so coughing and sneezing are not risk factors. And probably most importantly, caring for a person with AIDS is not a risk factor for transmission, especially if no needles are involved.

Since You Asked

9. There is a rumor that people with AIDS are swimming in the community pool in my neighborhood. Should I be concerned about letting my kids swim there?

HIV-2 Our discussion of HIV would be incomplete without including a second type of infection known as **HIV-2**. This strain of HIV is separated out from the overall discussion because it is a far more limited type of HIV both in numbers and geography. This is not to say HIV-2 is unimportant; between 1 and 2 million infections have been recorded. This HIV strain, however, is confined to West African countries, including Gambia, Mali, Mauritania, Nigeria, Sierra Leone, Angola, and Mozambique (this is a partial list; see http://www.hivworkshop.com/hiv-2.htm). A small number of cases have been observed in other countries, mostly those with close social ties to West Africa (Campbell-Yesufu & Gandhi, 2011).

Other important differences between HIV 1 and HIV-2 include symptoms, progression of the disease, and its history (Gottlieb, 2013). Compared to HIV-1, HIV-2 has a longer time frame between infection and any obvious symptoms; fewer numbers of the virus in the bloodstream, and causes a weaker attack on the immune system, Consequently, HIV-2 leads to lower death rates due to AIDS and does not transmit as easily from person to person. Perhaps the most profound difference is that HIV-2 appears to be declining, both in incidence and prevalence over the past ten years, seemingly of its own accord. Due to all of these factors, our discussion of HIV here is limited to the global pandemic of HIV-1.

HIV TESTING Simple, accurate, and noninvasive tests for HIV are widely available throughout most of the developed world. The vast majority of these tests detect antibodies that the body's immune system has manufactured in response to HIV infection, not the virus itself (which is very difficult, though not impossible, to detect). After a person is exposed to HIV, the immune system goes to work (as it does for any invasive microbe) attempting to fight off the newly introduced virus by manufacturing antibodies designed to attack and kill the viral invader. However, because HIV is so skilled at hiding in the body's cells, the human immune system is unsuccessful in destroying it. The antibodies remain in the blood indefinitely and are detectable with an HIV antibodies test.

The number of antibodies needed to be detected by HIV tests may not build up to detectable levels in the blood for weeks or even months after exposure (on average, this takes about three to four weeks). This is called the "window period." Therefore, if a person obtains a negative HIV antibodies result within the first three months after the time of possible infection, another test should be taken after three months following exposure to guard against a false negative outcome. After three months nearly everyone (97%) will have developed measurable levels of antibodies (CDC, "HIV

Testing Basics," 2010). If a person is still uneasy about that 3% false negatives, virtually everyone can receive truly accurate results after six months from exposure. A person planning on being tested for HIV should refrain from *all* risky activities during the window period prior to the test. Although chances are quite high that the result will be negative, when an HIV antibody test produces a positive result, it is always followed up by at least one additional confirmatory test using an alternate method, to confirm the result.

The most common method for HIV testing uses a small blood sample or mouth fluid (more concentrated than saliva) that is subjected to an antibody screening test (EIA, or enzyme immunoassay) analysis. Results from a blood EIA can detect HIV antibodies sooner than the test using oral fluid. Negative results are usually returned to a person in a day or two. In some cases (such as the aftermath of a rape or an accidental needle-stick to an HIV health care provider), the much more expensive HIV RNA test may be used, which can detect the presence of the virus itself in as little as ten days (CDC, "HIV Testing," 2010). HIV antibodies are present in oral fluids and urine, but the virus itself is not present in sufficient quantities to be transmissible from these bodily fluids.

Another type of HIV testing is the "rapid test" (CDC, "HIV Basics," 2014). Using blood or oral fluids, these tests are able to detect antibodies to HIV in about 20 minutes with 99.6% accuracy. In addition to the rapid results, this test does not require refrigeration or specialized equipment, so it may be used in situations outside the standard medical office or clinic setting, such as schools, bars, social service agencies, counseling centers, detention centers, and jails. Tests that provide nearly immediate results are important in the fight against the spread of HIV and AIDS because thousands of people who are tested for HIV each year using conventional tests never follow up or check their results. Expanded use of new rapid-results testing technologies that provide while-you-wait findings greatly increases the effectiveness of screening, counseling, and prevention. Because some rapid tests have been shown to produce a slight increase in false positives, any rapid test showing the presence of HIV is followed up with another type of test for confirmation (Wright et al., 2008).

The vast majority of HIV tests are negative. If a test returns positive results, followup testing is done to confirm the results because there is always the rare possibility of a false positive (in less than 6 people out of 1,000). The followup testing usually used a different technique that the initial testing for increased accuracy, such as the HIV RNA test. The rare possibility of a false negative also exists (up to 9 in 1,000 individuals). If a person has an HIV test during the antibody window period (about a month from possible exposure), and the result is negative, it is cause for optimism, but the person should return again after the window for a confirming followup (Nettleman, 2014).

The HIV testing process often includes counseling. Helping individuals who are concerned about their HIV risk is important in the fight against HIV because it reduces future risky sexual behaviors (regardless of the test results), determines the most common risk behaviors for various populations, provides timely and effective treatments for those with positive test results, helps patients cope with the emotional and psychological stresses of a positive test, and reduces the probability of the spread of the virus. Many venues offer HIV (and other STI) testing. If you are in the United States, the CDC in Atlanta now offers an easy way to locate an HIV testing site near you. Simply go to http://hivtest.cdc.gov and search for a testing service according to your zip code. In nearly all locales in the United States, you will find a testing center (if not many) within a maximum of a 30-mile radius.

confidential testing

Test recipients' names are kept on file in the lab's records, but with the assurance of full confidentiality.

anonymous testing

Tests administered without collecting any personal information about clients, who are identified only by an assigned code number.

Making the decision to be tested for HIV (or any STI) is an important part of taking responsibility for one's own health and health care. Test sites usually offer either **confidential testing** (in which the patient's name and other identifying information are obtained for statistical purposes but kept confidential) or **anonymous testing** (in which the patient is assigned a number and provides no identifying information at all). Anyone who has *any* concern about HIV status should be tested—the sooner, the better. The CDC recommends testing if you answer "yes" to any of the following questions (CDC, "HIV Testing," 2010):

- Have you injected drugs, steroids, or shared equipment (such as needles, syringes, works) with others?
- Have you had unprotected vaginal, anal, or oral sex with men who have sex with men, multiple partners, or anonymous partners?
- Have you exchanged sex for drugs or money?
- Have you been diagnosed with or treated for hepatitis, tuberculosis (TB), or any sexually transmitted disease (STD)?
- Have you had unprotected sex with someone who could answer yes to any of the above questions?

A negative HIV test result, which is *very likely* for most people, provides a deep sense of relief and allows for a fresh start in avoiding risks and *staying* negative. A positive result, though psychologically and emotionally difficult, allows for early medical intervention and emotional support to help the person develop and sustain a long, healthy, and satisfying quality of life while living with HIV, as well as take steps to avoid transmitting the virus to others (more on this issue shortly).

TREATMENT As of 2014, despite substantial research efforts, no cure or vaccine for HIV or AIDS had been developed, and the prospects for either are not optimistic. However, *treatments* for HIV, focusing on preventing the progression of the infection to AIDS, have improved greatly since the virus was discovered (see Mayer & Venkatesh, 2010). Treatment consists of a three-pronged approach: attacking the virus itself, strengthening the immune system, and preventing and controlling opportunistic infections and diseases. The treatment for each HIV-positive individual depends on the person's specific health status and involves working with doctors to tailor a program to the patient's needs. The typical treatment for HIV involves combinations of various medications in conjunction with healthy lifestyle changes.

The FDA has approved more than thirty medications for fighting HIV (FDA, 2013). The most effective treatments usually involve a combination of medications. As a group, these are referred to as antiretroviral agents, and the application of them for those diagnosed with HIV is called **antiretroviral therapy (ART)**. Various combinations of these medications, depending on the infected person's immune strength (determined by an immune system T-cell count) and presence of other infections, optimize management of HIV for most people who are HIV positive. The drugs help to reduce the body's viral load (number of HIV cells in the person's system), and thereby keep the infected person healthy and able to maintain a satisfying quality of life; in most cases, this therapy extends survival time significantly and reduces, but does not eliminate, the risk of transmission to sexual partners ("Panel on Antiretroviral Guidelines," 2010).

antiretroviral therapy (ART)
A combination of several medications prescribed for people who are HIV-positive to delay the onset of AIDS.

Often these medications are prescribed in specific combinations (usually three or more) to maximize their effectiveness in reducing an infected person's HIV viral load (this is sometimes referred to as an antiretroviral "drug cocktail"). Two single-pill, one-dose-per-day ART medications have been approved by the FDA, *Atripla* and *Stribild*). These combine several previously approved ART medications and provide similar effectiveness to a multi-pill regimen, but are less expensive, easier to self-administer, and enhance adherence to treatment (Laurence, 2007). They also have the potential advantage of increasing the distribution and correct use of effective HIV therapy to a larger cross section of infected individuals worldwide. Figure 8.7 shows the dramatic improvement in survival rates from the moment in history when ART was introduced.

Figure 8.7 The Impact of ART on HIV Survival Rates

SOURCE: NIH (2012). DrugFacts: HIV/AIDS and Drug Abuse: Intertwined Epidemics. *National Institutes of Health, National Institute on Drug Abuse*. Retrieved from www.drugabuse.gov/publications/drugfacts/hivaids-drug-abuse-intertwined-epidemics.

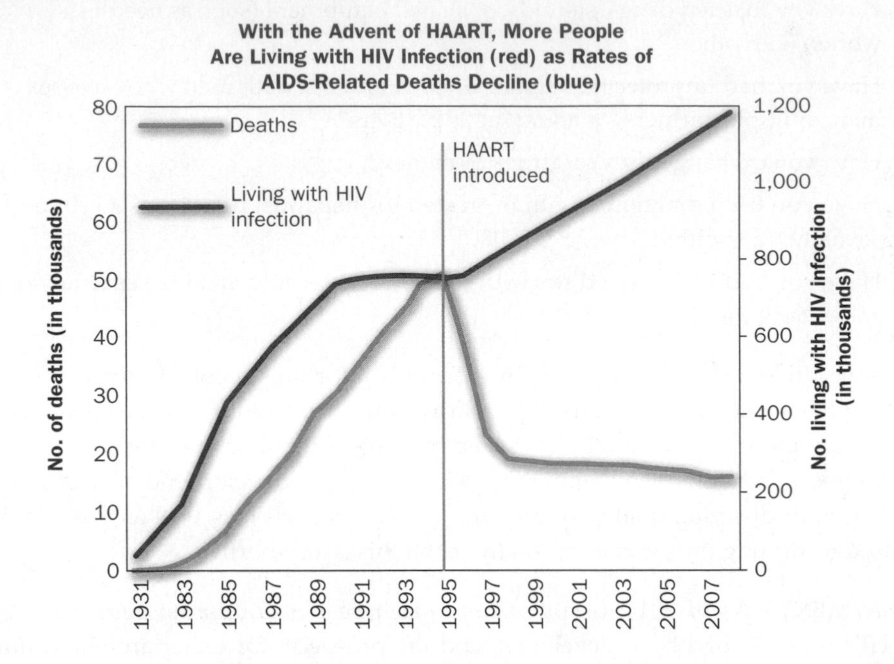

Although ART appears to be quite effective in slowing and even reversing the progression of HIV, the medications are not without negative side effects, some of which may be serious. Research has indicated that long-term use of ART therapy for HIV may increase a person's chances of developing coronary heart disease. Researchers have reported a significant increase in heart attack risk among HIV-positive individuals, with the risk increasing with the length of time on the ART medications (Bocarra et al., 2013; Friis-Moller et al., 2003). However, researchers point out that even this increased CHD risk rate represented a small absolute risk and must be weighed against the significant survival and quality-of-life benefits of ART.

Today, the larger variety of medications and advancements in testing patients for variations in the likelihood of developing resistance is helping to maintain a high degree of effectiveness of this form of combination drug therapy (Kutzer et al., 2008). The many available antiretroviral agents and the various ART combinations, as effective as they are in controlling the progression of HIV and postponing a diagnosis of AIDS, are not a *cure* for HIV or AIDS. Even when an infected person's viral load is reduced to undetectable levels, he or she is not cured and is still considered able to transmit the virus to others (Flannigan, 2008).

As noted earlier, HIV weakens a person's immune system, allowing various opportunistic infections to invade the body. Part of a comprehensive treatment program for people with HIV must include specific treatments for these infections if and when they arise. The increasing use of ART treatment modalities has changed the profile of the most common opportunistic infections observed among HIV-positive individuals. Some of these infections are virtually never seen in people with normally functioning immune systems; others exist in non-HIV populations but tend to be far less common or less debilitating. Some of the more commonly observed opportunistic infections among the HIV-positive population worldwide are tuberculosis, pneumocystis pneumonia (a fungus- or parasite-based lung infection affecting only humans with compromised immune systems), mycobacterium avium complex (MAC, a bacterial lung infection), septicemia (blood poisoning), and cryptococcal meningitis (a yeast-like fungus that infects the lining of the brain and the spinal cord). Also frequently diagnosed are other strains of fungal diseases such as candidiasis, cryptococcosis, and penicilliosis, and viral infections caused by herpes simplex and herpes zoster virus.

Certain cancers are also associated with HIV infection, including skin cancers such as Kaposi's sarcoma (typically found only in HIV-positive individuals), squamous cell carcinoma (a more common form of skin cancer), and lymphoma (cancer of the lymphatic system). Hepatitis B and C (viral liver infections) are diagnosed more frequently in HIV-positive individuals than in the general population (Avert, 2008).

The incidence of opportunistic infections associated with HIV and AIDS has decreased dramatically in recent years owing in part to the success of antiretroviral treatments. These infections are more likely in HIV-positive individuals who are newly diagnosed, who are not receiving or not adhering to medical HIV care, or who develop resistance to antiretroviral therapies (James & Szabo, 2006). Medical treatment for opportunistic infections requires a correct diagnosis of the specific infectious agent to determine the appropriate medical regimen, as well as careful monitoring

Evaluating Sexual Research

Babies Born with HIV Cured?

In 2013 and 2014, medical researchers announced that two babies born with HIV to HIV-positive mothers appeared to be cured of the virus after aggressive medical treatments. Scientists suggest that this very early intervention may eliminate the virus before it has a chance to establish itself permanently in the baby's system. This, as you can imagine, was huge medical news. No one, prior to these babies, had ever been cured of this terrible virus. How did this happen and what does it mean for the possibility of a more far-reaching cure of HIV? This is one of those attention- and media-grabbing medical stories that create excitement both within and outside of the medical community. And the nonmedical population also tends to overreact and leap to the conclusion that a cure for HIV has finally been found. If true, few medical advances of the last hundred years would be as important. And maybe it is indeed the start of a breakthrough; we should be optimistic. Here's a summary of the stories:

- In 2011, a baby was born to a mother who had not known she was HIV positive until just prior to the birth.
- The baby was given high doses of several ART medications thirty hours after birth, even before being subsequently being diagnosed with HIV.
- A little over a year later, the mother, for unknown reasons, took the baby off the ART therapy, something that is *not* medically approved with HIV babies.
- After testing, doctors determined that the girl, three years later and without treatment, was "functionally cured" of HIV (this means that the virus cannot be detected in the blood and continued treatment is unnecessary).
- In 2013, a girl was born to a mother with AIDS and received similar ART treatment only four hours after birth.
- Within two weeks, no sign of the virus was detectable in the baby's system and the virus was still completely cleared of the virus nine months later—doctors said it was still too soon (as of early 2014) to say for sure if the baby has been cured, but they are optimistic.

What do you think? Based on these stories, it was easy to assume that these babies had been cured. If so, perhaps a medical breakthrough may have been found for curing HIV, right? Well, not to dash all of our hopes, but that's still very unclear. Any *single* case that demonstrates a new medical treatment (for anything) proves *nothing* scientifically. It could have been an error in diagnosis or testing after treatment or some sort of medical fluke. When two similar cases occur, that's a bit more convincing that something real may have happened, but it's still far from conclusive. With over 200,000 new cases of HIV in children each year in the world, you can see how insignificant two possible "cures" become. But two is often enough to move to the next step. Unfortunately, in July of 2014, the girl born in 2011 relapsed and HIV was once again present in her body. Nevertheless, biomedical researchers remain optimistic that very early, aggressive treatment for newborns infected with HIV may significantly slow, if not necessarily cure their infections.

Researchers will now design and carry out rigorous scientific studies to demonstrate for clearly whether this approach works or not. Soon, a "clinical trial" of this treatment will begin involving sixty newborns in various countries. The results will not be known quickly because it is medically unethical to remove babies from HIV treatment due to the likelihood that they will relapse and possibly convert to AIDS. Therefore, in this study, HIV babies will be left on the ART treatment for two years and tested regularly. If all tests of infection are negative, the treatment will be discontinued and the babies monitored closely for any return of the virus. It's a tricky ethical study, but done properly, with all safeguards, it can be done. And if it can be demonstrated that the HIV positive can indeed be cured, we must follow through; the lives of hundreds of thousands may be at stake.

SOURCES: Marchione (2014); Wilson & Young (2014).

of the patient during treatment, as the progression of most infections in HIV-positive patients will differ from that of noninfected individuals.

LIVING WITH HIV For about the first ten years of the HIV epidemic in the United States, most people thought of HIV as a death sentence, and with good reason. Until the availability of the more effective medications discussed in the previous section, people diagnosed with HIV commonly died of AIDS-related infections within a year or two (and sadly, this is still the case in many less developed countries where HIV and AIDS are rampant). Today, as you can see in Figure 8.7, anti-retroviral treatments are allowing people with HIV to live far longer and maintain a near-normal quality of life. This is evidenced by the fact that the number of HIV infections in the United States is remaining approximately stable, but deaths from AIDS are dropping. A diagnosis of HIV is no longer considered a death sentence in the United States. This is a major medical advance; however, it has produced a dangerous trend. Some individuals are interpreting these medical advances as "cures" for HIV and AIDS and therefore have begun to return to unsafe sexual practices. Increases in various STIs, other than HIV, especially in some large U.S. cities, are troubling to public health officials. It is as important as ever for the public to understand that, although treatments for HIV are effective in reducing or eliminating symptoms, they are not cures, and HIV-positive individuals, regardless of how healthy they feel, are infectious and may spread the virus to others.

Psychologists and other health professionals have made extensive efforts to change negative attitudes about HIV and AIDS. Many people now understand that HIV is an infection that one can live with but, like all serious health issues, it is something that must be treated and carefully monitored. This reframing of HIV has begun to reduce the prejudice and discrimination against infected individuals and to focus attention on prevention, treatment, and a cure. Improvements in treatment and survival have had yet another very important effect: they have given people living with HIV and their loved ones *hope*. The mind–body connection can be a powerful healer in itself, and hope is a key element in that survival process.

HIV, LIFESTYLE, AND HEALTH Obviously, adopting safer sexual behaviors is particularly important for people living with HIV in order to prevent the spread of the virus to others. Unfortunately, studies indicate that as many as 64% of people who know they are HIV positive continue to engage in unsafe sexual activities and fail to inform their partners of their HIV status (Reilly et al., 2010). In addition to the danger this poses to their sexual partners, unsafe sex is also a serious health threat to HIV-positive individuals themselves. At one time, people believed that HIV-positive couples (called HIV-seroconcordant couples) had little reason to worry about unprotected sexual activities; after all, what could they possibly transmit that was worse than HIV?

People who have been diagnosed with HIV continue to be at risk for all other STIs, too, and for many, the risk and seriousness of infection is much greater than for those whose immune systems are already at risk by HIV. Many seroconcordant couples appear to be aware of these risks. A study conducted with HIV-concordant couples found that the primary reasons these couples used condoms was to reduce their risk of other viral STIs such as herpes, HPV, and hepatitis (Williams, 2001). Thus, the importance of safer sex does not end with a couple's HIV-positive test results; instead, safer sex practices become even *more* essential.

THE FUTURE OF HIV/AIDS Over three decades have passed since the first cases of AIDS were identified in the United States. In that time, enormous effort and resources have been poured into HIV education, treatment, and prevention. However, far more resources are needed to fight this disease. For every new case of HIV that is diagnosed and treated, many more have yet to be identified. Clearly, defeating HIV is one of the single greatest challenges the world has ever faced. Worldwide, factors such as poverty, drug use, poor health care, violence, famine, and hopelessness all contribute to a lack of adequate HIV education, which in turn perpetuates an epidemic that is far from being controlled.

Fortunately, we can find some cause for optimism as newer and more-effective HIV preventatives are being developed and being made available more widely, driving the sickness and death rate from HIV/AIDS downward. Research for an HIV vaccine and new medications continues on many paths (NIAID, "HIV Vaccine Research," 2010). A recently developed medication that has been shown previously to help prevent the spread of HIV among gay men has recently been found to reduce heterosexual transmission as well. Researchers still do not know the eventual impact of this drug on infection rates. The medication (trade name, *Truvada*) is new, expensive (without insurance, $900 per 30-day supply in the United States), and the demand may quickly outpace the supply. In addition, a person must take the pill every day indefinitely if he or she is at any risk of contracting HIV, and researchers are not sure how widely people will follow this regimen or whether it might encourage riskier sexual practices (see Stobbe & Cheng, 2011). Research over the next few years will help determine to what extent this medication and others under development are effective in preventing the spread of HIV.

Effective treatments and preventions for HIV and AIDS are significantly more available in Western and first world countries than in developing nations, especially Africa, where the disease is having the most devastating impact (see "Sexuality and Culture: The HIV/AIDS Pandemic" earlier in this chapter). This discrepancy in resources and treatment must be addressed globally if we are to rid the world of this deadly virus.

Bacterial STIs

8.5 Summarize the differences between bacterial and viral STIs and the similarities among the various bacterial STIs.

Although the various viral infections we just finished discussing are quite common, *most* sexually transmitted infections in the United States are caused by bacteria, not viruses. The main bacterial STIs are chlamydia, gonorrhea, and syphilis. Chlamydia and gonorrhea are very common STIs and they are diagnosed in the largest numbers by far in young adults between the ages of fifteen and twenty-four (refer back to Figure 8.1 on page 263). These bacteria are harbored in bodily fluids (saliva, mucous membranes, vaginal secretions, semen, blood, etc.) and may be transmitted during oral, vaginal, or anal sex, when contact with these fluids occurs. They may be transmitted between heterosexual or same-sex couples.

Since You Asked

10. Are the STIs caused by viruses and bacteria basically the same?

Medical science is far more advanced in its ability to cure bacterial infections than those caused by viruses. Think about it: When you have a common cold, which is caused by a virus, no cure exists—you must let the illness run its course. However, if your upper respiratory symptoms are caused by a bacterial infection, a week or two of treatment with an *antibiotic* usually cures you. So it is with bacterial STIs—for the most part, they are relatively easy to treat and cure, provided that they are diagnosed early, have not had a chance to spread to other areas within the body, and have not become resistant to antibiotics. However, even in the most difficult cases, requiring repeated treatment and stronger antibiotics, a cure is usually assured. The most important factor in successful treatment of bacterial STIs is obtaining medical attention at the first signs of infection.

For each bacterial STI, we will follow the basic format used for viral infections: description, signs and symptoms, routes of transmission, and available tests. However, because treatments for all bacterial STIs are similar, we will examine bacterial STI treatment issues in a single section toward the end of our discussion.[1]

[1] A bacterial STI called "Chancroid" (a genital ulcerative infection) has been omitted from this text edition because it has virtually been eradicated. In 2009, The CDC reported a total of twenty-eight cases in the United States, down from a high of approximately 5,000 in the late 1980s (see CDC, "Other STDs," 2010).

Chlamydia

chlamydia
A sexually transmitted bacterium, often causing a thick, cloudy discharge from the vagina or penis; may be asymptomatic, especially in women.

Chlamydia is an infection caused by the *Chlamydia trachomatis* bacterium, which survives in the cells of bodily fluids such as vaginal secretions and semen. The bacterium is transmitted primarily through sexual contact between an infected person and an uninfected partner. Chlamydia is one of the most commonly diagnosed and reported STIs in the United States, where an estimated 1.2 million new infections were reported in 2009; a nearly 3% increase from 2008 (CDC, "Chlamydia," 2014). Anyone can be infected with chlamydia, but the infection rate among women is three times that of men. One reason for these troubling high numbers is that chlamydia is one of several STIs that often produce no noticeable symptoms (especially in women), so many people are unaware they are infected. However, the bacterium is highly contagious whether or not it produces symptoms.

SYMPTOMS Seventy-five percent of women and 50% of men infected with chlamydia experience no noticeable symptoms and may have no idea that they have been exposed to an STI. This is a major reason this infection is so common and has spread so easily. Often, a diagnosis is made during a routine medical examination or while obtaining care for an unrelated medical issue.

If signs are present, the most common symptom of chlamydia infection is a thick, cloudy discharge from the vagina or penis occurring one to three weeks after exposure to the bacterium. This discharge is typically less noticeable in women than in men. Most women experience some normal fluid discharge from the vagina associated with hormonal changes of the menstrual cycle. However, the fluid associated with chlamydia is typically greater in quantity, cloudier in appearance, and has an unpleasant odor. Other symptoms women may experience include pelvic pain, irregular periods, increased pain during menstrual periods, discomfort during urination, or irritation of the vaginal or anal area (see Figure 8.8).

pelvic inflammatory disease (PID)
A painful condition in women marked by inflammation of the uterus, fallopian tubes, and ovaries; typically caused by one or more untreated STIs.

Men who become infected may experience a discharge from the penis and a burning sensation when urinating. Men may also notice irritation or inflammation at the opening to the urethra, and in the morning, the urethral opening may be red and sealed together with dried secretions.

If left untreated, any symptoms that may be present in either sex usually diminish and disappear within about a month. At that point, however, the infection itself is not cured, remains highly transmissible, and may lead to more serious health problems.

If left untreated, chlamydia is associated with reduced fertility in both men and women. As the infection grows and moves up the reproductive tract, women may develop **pelvic inflammatory disease (PID)**, a painful condition marked by inflammation of the uterus, fallopian tubes, and ovaries. PID has the potential to

Figure 8.8 Chlamydia Infection

Discharge commonly observed in males with chlamydia. Most females who contract the chlamydia bacterium show no symptoms at all. The chlamydia bacterium is shown on the right.

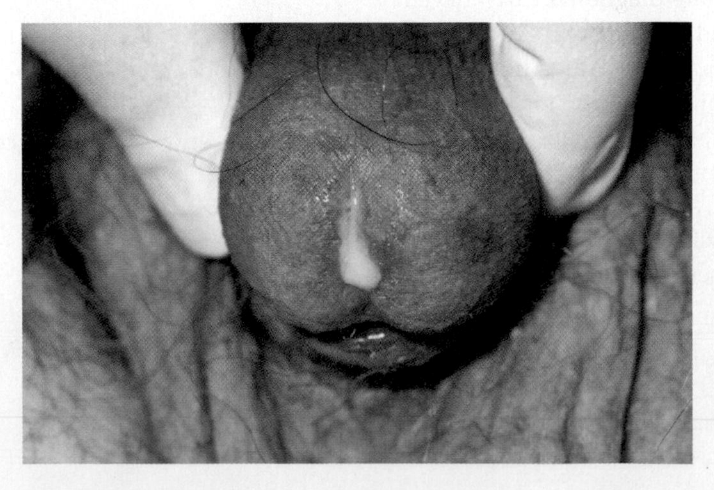

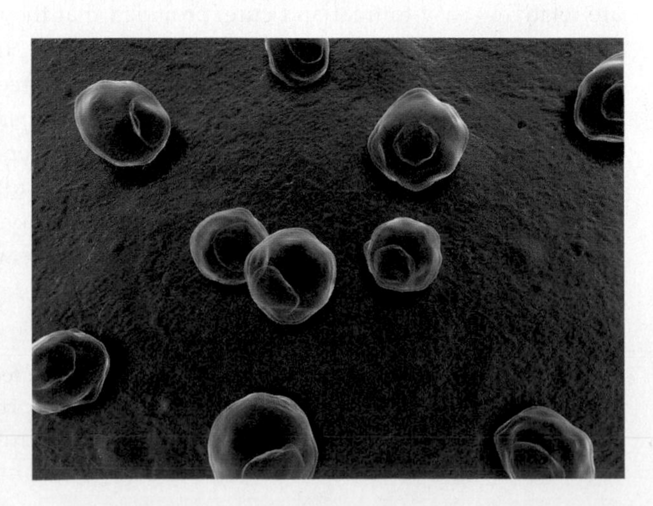

cause infertility, chronic pelvic pain, and ectopic (tubal) pregnancy (see Chapter 9, "Conception, Pregnancy, and Birth"). We will discuss PID in greater detail later in this chapter.

Among men, untreated chlamydia infection may lead to **urethritis**, an uncomfortable condition that is marked by the inflammation of the urethra, or **epididymitis**, a painful swelling and inflammation of the epididymis, the structure at the back of each testicle that stores maturing sperm (see Chapter 2, "Sexual Anatomy"). Chlamydial infection may play a role in altering the formation or blocking the transport of sperm cells, leading to infertility in men (Noruziyan et al., 2013; "Previous *Chlamydia*," 2004).

urethritis
A painful inflammation of the urethra; often caused by one or more untreated STIs.

epididymitis
A painful swelling and inflammation of the epididymis, the structure at the back of each testicle that stores maturing sperm; often caused by one or more untreated STIs.

TRANSMISSION Chlamydia, like other STIs, is transmitted through sexual contact, primarily oral, vaginal, and anal sex. The bacterium is extremely contagious and can easily be transmitted during a single sexual encounter. Vaginal and anal intercourse are the most common routes of transmission. Chlamydia infection of the throat may also result from oral sex, although this type of transmission occurs less frequently than transmission through vaginal or anal intercourse. The infection can also be spread to the eyes if a person touches an infected area on themselves or a partner and then touches the eyes. Chlamydia may also be passed from a pregnant mother to her infant with potentially serious health consequences for the infant.

DIAGNOSIS Several types of tests may be used to detect *C. trachomatis*. The best and most widely used testing is performed by a health care professional at a doctor's office or clinic. A urine sample is examined microscopically for the presence of the *C. trachomatis* bacterium (bacteria are ten to twenty times larger than viruses and are far easier to see under a microscope). This noninvasive test has contributed to the increase in diagnosis of chlamydia in men and women (CDC, "Chlamydia," 2010).

Home tests for chlamydia (as well as herpes and gonorrhea) are now available from numerous sites on the Internet (e.g., http://home-bio-test.com or www.privatediagnostics.com). The "do-it-yourself" portion for some of these self-tests consists only of the collection of the specimen (a urine sample or vaginal swab for women; a urine sample and/or penile swab for men), which must then be mailed to a lab for analysis and results. Other home tests indicate findings at home without sending the sample to the lab.

Given the high prevalence of asymptomatic chlamydia infections, yearly screening regardless of the presence of symptoms is recommended for sexually active girls and women under the age of twenty-five, all women who have new or multiple sexual partners, and all pregnant women (due to the complications the bacterium may cause to both infant and mother) (CDC, "Chlamydia," 2011). Finally, chlamydia infection is found to coexist in many people diagnosed with gonorrhea. Because of this link, many tests for chlamydia include testing for gonorrhea using a single cellular or urine sample.

GONORRHEA **Gonorrhea** is a sexually transmitted infection caused by the bacterium *Neisseria gonorrhoeae*. The vernacular names for gonorrhea include *clap, drip,* and *burn.* Gonorrhea has been infecting humans for centuries and millennia; it is even mentioned in the Bible (Leviticus 15:1–15).

gonorrhea
A sexually transmitted bacterium typically producing pain upon urination and a thick cloudy discharge from the penis or vagina; often asymptomatic, especially in women.

Gonorrhea is the second most common reportable STI in the United States (CDC, "Gonorrhea," 2010). During the 1980s and 1990s, owing in part to an increase in public awareness campaigns to prevent all STIs, gonorrhea rates declined at a rate of 10% per year. According to the CDC, rates stabilized around 1996 and then began to increase again. In 2009, there were 301,000 new cases of gonorrhea reported. However, incidence climbed again and reached 334,826 cases in 2012. As can be seen in Figure 8.1 (on page 263), gonorrhea rates are highest among young people, particularly among those between the ages of 15 and 24. At one time, rates were substantially higher among men than women, and they still are, but the gender gap is closing due primarily to increased screening programs for women. Between 2009

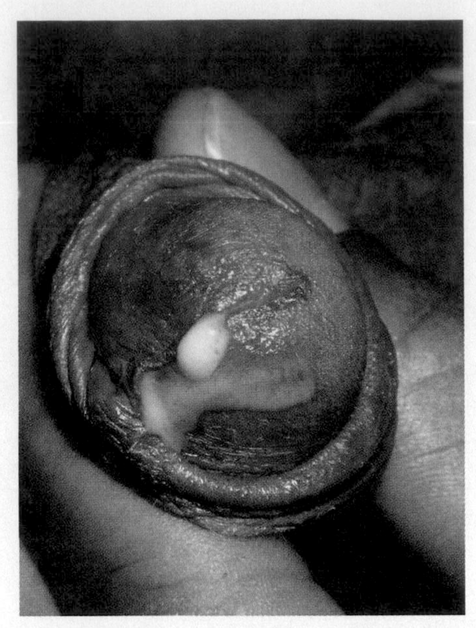

Figure 8.9 Gonorrhea Symptoms in the Male

The most common symptoms of gonorrhea in the male are a cloudy discharge from the penis and painful urination. In women, as with chlamydia, the infection usually has no obvious symptoms, but when symptoms do appear, a similar discharge from the vagina may be observed, accompanied by painful urination.

and 2012, cases of gonorrhea decreased by 11% among both sexes and all age groups (CDC, "Gonorrhea," 2014).

SYMPTOMS As with chlamydia, symptoms tend to be more obvious among men than among women. Most infected women do not experience obvious symptoms at all, yet they are contagious.

In men, symptoms are likely to occur within two to five days of exposure to gonorrhea, although the incubation period can range from one to fourteen days. The infection in men may occur in the urethra, anus, rectum, mouth, or throat (see Figure 8.9). Symptoms typically include a painful, burning sensation during urination or bowel movement and a cloudy discharge from the penis or anus. In the case of oral infection, the main symptom is a severe sore throat. Untreated, gonorrhea may spread to other regions of the reproductive tract and can cause epididymitis (discussed in conjunction with chlamydia), acute or chronic prostatitis (inflammation of the prostate gland), and potential scarring of the urethra (Bridges, 2010; Ludwig, 2008).

Among women, the vagina and cervix are the primary sites of infection for the gonorrhea bacterium and, as for chlamydia, most women have no observable symptoms. The asymptomatic nature of the infection in women contributes to delayed diagnosis and treatment, increased chance of transmission to others, and more complicated and more serious medical conditions, such as pelvic inflammatory disease, fallopian tube damage leading to infertility, and an increased risk of contracting HIV (CDC, "Gonorrhea," 2010).

When symptoms are present (remember, most of the time they are not), women may notice a cloudy vaginal discharge, irritation during urination, or vaginal bleeding between periods. Other symptoms, such as lower abdominal pain or pain during intercourse, may be indicative of a more complicated infection. Unlike men, who have a shorter and more defined incubation period, symptoms in women may not appear until months after initial exposure to the microorganism. The longer period between exposure and detection in women contributes to the increased complications discussed above.

Sometimes a woman may confuse gonorrhea with non-STIs that may produce a vaginal discharge. The most common of these is *candidiasis*, better known as a yeast infection, which is not an STI but rather an inflammation of the vagina caused by a yeast fungus that may occur in women regardless of sexual activity.

Pharyngeal gonorrhea (infection of the throat) can occur as a result of oral sex. Fellatio (oral sex performed on a man) is a more likely means of transmission than cunnilingus (oral sex performed on a woman), which explains the greater prevalence of pharyngeal infection among individuals who have receptive oral sex with men.

A relatively rare systemic complication of gonorrhea that may afflict both men and women is *disseminated gonococcal infection* (DGI). The most common clinical manifestations of DGI are skin rashes and joint pain, usually in only one joint in the hands or wrists. This form of gonorrhea often presents with minimal genital symptoms as described for more common gonorrhea conditions (CDC, "Gonococcal Infections," 2010).

TRANSMISSION Gonorrhea thrives in the moist mucous membrane of the genitals, throat, and anus, and is easily transmitted through vaginal, anal, or oral sexual contact. Infection with gonorrhea during pregnancy increases a woman's chances of miscarriage or preterm birth (see Chapter 9, "Conception, Pregnancy, and Birth"). Transmission from mother to fetus during the birthing process poses serious risks to the newborn, including miscarriage, blindness, infections of the joints, or incurable blood infection (Koumans et al., 2012; Nihira, 2010). These risks may be minimized with appropriate treatment of the infant soon after birth.

DIAGNOSIS The American College of Obstetricians and Gynecologists recommends that all females who fall into the following categories be tested for gonorrhea (CDC, "Sexually Transmitted Diseases," 2010):

- women under age twenty-five
- all sexually active teen girls
- women over the age of twenty-five with risk factors such as previous gonorrhea infection, other STDs, or new or multiple sex partners
- women with a history of inconsistent condom use

Gonorrhea infection is diagnosed by examining fluid from the urethra, vagina, cervix, rectum, or throat. A cotton swab is used to collect a sample of the fluid from the infected areas. The specimen is then either cultured or stained to reveal the presence of the gonorrhea bacterium. The results are usually known in one to two weeks and are 90% to 95% accurate. Another test now in widespread use is a urine-based screening test ("Gonorrhea Diagnosis," 2011). As with chlamydia, urine-based tests are easier to perform in most health care offices, patients appreciate their less invasive nature, and results may be known in as little as twenty minutes. Urine-based tests tend to be more expensive than traditional forms of testing, but they may lead to wider testing, better treatment, and reduced spread of infection.

Nongonococcal Urethritis (NGU)

As the name implies, **nongonococcal urethritis (NGU)** is an inflammation of the urethra *not* caused by the gonorrhea bacterium. NGU is considered to be an STI, but it is of a less specific nature. It is caused by various other bacteria, the most common of which are chlamydia and **Mycoplasma genitalium** (often called *M. genitalium*), a relatively new, emerging sexuallyy transmitted bacterium to be discussed next. Both of these bacteria are more common than gonorrhea (see Manhart et al., 2007). The transmission, symptoms, and treatments of these STIs are similar to those of gonorrhea for men and women. NGU is typically characterized by a discharge from the urethra or vagina as well as burning and itching at the end of the urethra or upon urination. It is more common in men than in women. NGU is usually transmitted through insertive sexual activity, primarily vaginal and anal intercourse. NGU can be diagnosed with a microscopic examination of cells from the urethra or with a urine test and is treated and cured with antibiotic regimens, usually a course of azithromycin or doxycycline.

nongonococcal urethritis (NGU)
A sexually transmitted bacterial infection of the urethra characterized by urethral inflammation and discharge, but not caused by the gonorrhea bacterium.

Mycoplasma genitalium (M. genitalium)
A sexually transmitted bacterium that is responsible, along with chlamydia, for the largest percentage of nongonococcal urethritis.

Mycoplasma Genitalium

As if there were not enough already, another microbe, well known for its role in various nonsexual infections, has recently been classified as a sexually transmitted pathogen (see CDC, "Urethritis," 2011). *Mycoplasma genitalium*, commonly referred to as "*M. genitalium*," is an extremely small bacterium-like organism that appears to be involved in 15% to 25% of diagnosed cases of NGU. It may also play a role in pelvic inflammatory disease and bacterial vaginosis in women. M. genitalium infection has been on the rise and is now more common than diagnoses of gonorrhea (but less than chlamydia). The prevalence of M. genitalium in young adults has been found to be 2% in women and 7% in men (with higher rates in high-risk populations) (Anagrius, Lore, & Jensen, 2005; McGowan & Anderson-Smits, 2011).

Many people infected with M. genitalium have no symptoms; however, a severe infection may cause urethritis or cervicitis (inflammation of the urethra or cervix), which, in turn, will present similar symptoms to gonorrhea and chlamydia. In women, these may include vaginal itching, pain or burning while urinating, or painful intercourse. In men, if symptoms are present, they include urethral discharge, burning on urination, and arthritic-like pain and swelling in various joints. In most cases, M. genitalium can be cured with a single dose of 1 gram of Azithromycin (CDC, "Urethritis," 2011).

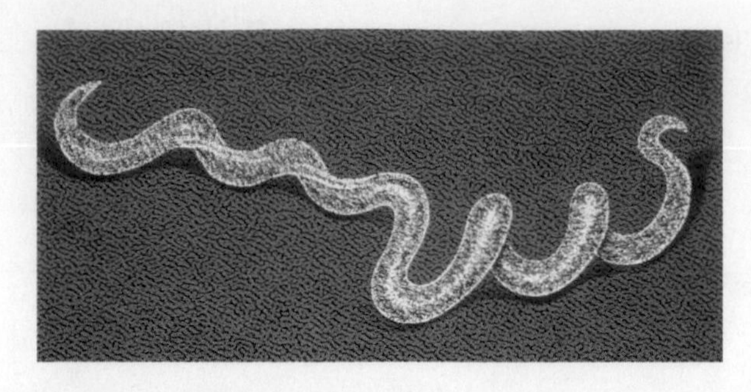

Figure 8.10 Syphilis Bacterium

The syphilis bacterium, magnified here over 1,000 times, is called a spirochete due to its spiral or corkscrew shape.

syphilis

A sexually transmitted bacterium characterized by a sore, or chancre, at the point of infection; untreated, it may progress to more serious stages and even death.

Syphilis

Syphilis is one of the earliest recorded and probably the most infamous of the STIs. The list of famous people who purportedly died of untreated syphilis includes King Henry VIII, Napoleon, John Keats, Paul Gauguin, and Al Capone. Today, almost no one dies of syphilis, but many cases occur that are typically cured relatively early. Approximately 31,700 cases of all stages of syphilis were reported in 2013 (Patton et al., 2014). Syphilis is caused by the bacterium *Treponema pallidum* or *T. pallidum*. First identified in 1905, the bacterium is also sometimes referred to as a *spirochete* because of its spiral or corkscrew shape (see Figure 8.10).

Prior to the discovery and widespread use of penicillin during the 1940s, more than 50,000 people became infected with syphilis each year, and thousands of those died from the infection. When penicillin was found to be an effective treatment, rates dropped drastically. In the early 1990s, a resurgence of syphilis occurred in the United States, especially in the South and large urban areas. At that time, the CDC increased attention on syphilis prevention in the affected areas, including more efficient partner notification and better testing and treatment. In response to these stepped-up efforts, syphilis rates began to drop again. From 1990 to 2000, the rate of syphilis decreased nearly 90%. By the late 1990s, syphilis had decreased to a historic low in the United States of less than 8,000 cases, and the CDC saw an opportunity to eradicate the infection entirely from the United States. In September 1999, a national plan was initiated to eliminate syphilis in the United States. Initially, the plan appeared to be working. As of late 2000, syphilis rates were decreasing, but then by mid-2001, a new increase was once again observed, which has continued to the present (Patton et al., 2014).

The increases have occurred primarily among men, especially among men who have sex with men. This increase reflects trends discussed earlier in this chapter about a return to unsafe sexual practices among the gay male population. This increase is disturbing for many reasons, not the least of which is that infection with syphilis and the unprotected sexual behaviors that cause it also increase the incidence of more serious, incurable STIs, including HIV and hepatitis.

chancre

A sore that typically appears at the site of infection with syphilis.

SYMPTOMS Symptoms of syphilis occur in four distinct stages. The *primary stage* is characterized by a sore called a **chancre** (pronounced "shank-er") that appears at the point of infection within ten days to three months (with an average of three weeks) after exposure to the virus (see Figure 8.11). Chancres have hard edges with a watery center

Figure 8.11 Syphilis Chancre

The first sign of syphilis infection for males and females is a chancre, or sore, at the point of initial infection.

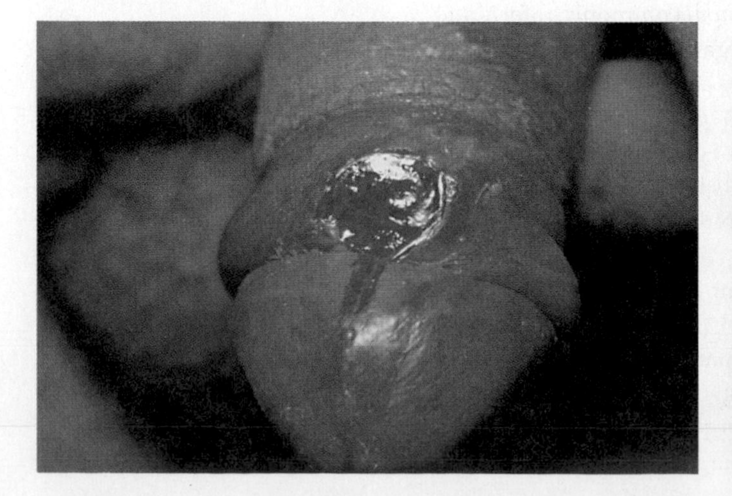

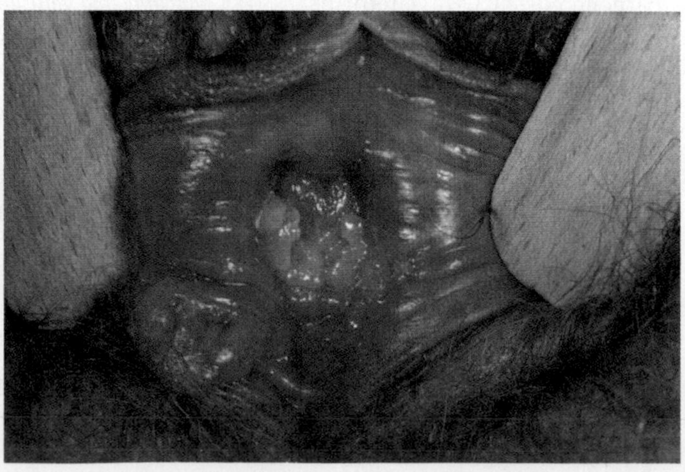

and vary in size. The center of the chancre is teeming with syphilis bacteria, making transmission highly probable, and greatly increasing the risk of HIV transmission (CDC, "Syphilis Fact Sheet," 2014). Often the chancre may go unnoticed because it is painless and may be hidden inside the mouth, vagina, or anus. After a few weeks without treatment, the chancre heals and disappears on its own. However, without treatment, the bacterium stays in the body, continuing to multiply, and is highly transmissible to others through sexual contact.

The *secondary stage* of syphilis occurs within a few weeks to up to six months of the initial infection. Secondary symptoms include low-grade fever, sore throat, fatigue, headache, alopecia (hair loss), and skin rashes, typically on the hands or feet (see Figure 8.12). Secondary indications are more apparent and more uncomfortable than primary symptoms, but they are likely to be misdiagnosed as various other health problems. Untreated, secondary symptoms also resolve on their own, but the person is still infected and can transmit the organism to others.

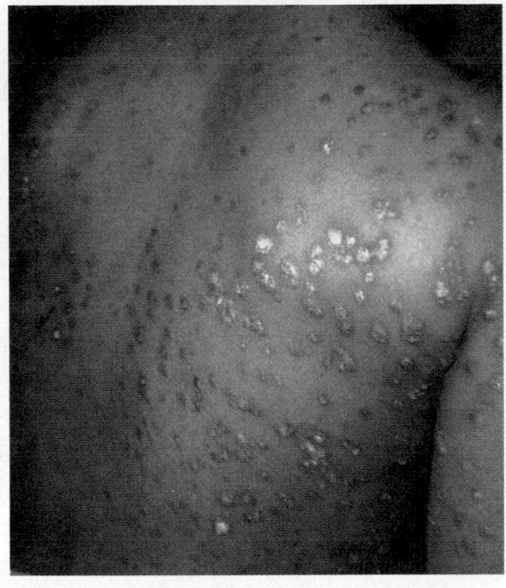

Figure 8.12 Secondary Syphilis Rash

If untreated, syphilis progresses to more serious symptoms, including a pronounced skin rash that may appear anywhere on the body.

Following the disappearance of secondary-stage symptoms, the infected person enters the *latent stage* of syphilis. This stage has no symptoms, but blood testing would indicate the presence of *T. pallidum.* This stage can last for years or even decades. Ten to thirty years after initial infection, the disease enters the *tertiary stage* or *late stage* of syphilis. In this stage, extensive damage has occurred, often leading to organ failure and neurological injury resulting in paralysis, mental illness, and death. The long-term neurological damage caused by the syphilis bacterium was once known as "syphilitic insanity" but is now called *neurosyphilis*. Today, with antibiotic treatment for syphilis, the infection rarely moves beyond the primary or in rare cases the secondary stage before being treated and cured.

TRANSMISSION As with all bacterial STIs, transmission occurs through sexual contact, especially vaginal, anal, and oral sexual practices. Syphilis may be transmitted between heterosexual, gay, or lesbian couples. As noted earlier, the sore that typically appears soon after infection is located at the point of the bacterium's entry into the body, so the sore may be found on or in the genitals, the anus, or the mouth.

DIAGNOSIS During the primary and secondary stages, diagnosis can be made with visual inspection, if symptoms are present. Fluid from the chancre can be sampled and examined microscopically for bacteria. In the absence of symptoms, a blood sample can be taken and tested for the presence of *T. pallidum* antibodies. Although the microscope test indicates infection with the organism, a laboratory blood test can reveal infection *and* how much of the organism is present in the blood. This finding may be important because even after people infected with syphilis are cured, they will always have small amounts of antibodies in their blood. If the amount of antibodies is very low, a past, inactive infection is indicated. Home test kits are available for syphilis.

Pelvic Inflammatory Disease (PID)

Aside from HIV, the most common and most serious *complication* of sexually transmitted infections among women is *pelvic inflammatory disease* (PID), an infection of the upper genital tract. PID can affect the uterus, ovaries, fallopian tubes, or other related structures. If left untreated, PID may cause scarring within the reproductive tract and can lead to infertility, tubal pregnancy, chronic pelvic pain, and other serious consequences (CDC, "PID," 2014).

Each year in the United States, approximately 750,000 women are diagnosed with PID and about 10% of those will experience subsequent infertility due to the infection. In addition, a significant proportion of tubal pregnancies (ectopic pregnancies) are thought to be related to PID (see Chapter 9, "Conception, Pregnancy, and Birth," for more information about this condition).

PID results when bacteria from the vagina move through the cervix and migrate up into the deeper regions of the woman's reproductive tract. Most cases of PID are caused by STIs: either chlamydia or gonorrhea bacteria. These bacteria then settle into the uterus, fallopian tubes, and ovaries and continue to multiply, causing inflammation and scarring and sometimes leading to fallopian tube blockage and hence infertility. Younger women and girls are the most susceptible to PID because the cervix is still maturing, making infection from these STIs more likely (CDC, "PID," 2014).

PID may have very subtle symptoms, causing a woman to postpone treatment, which allows the bacteria to grow and spread. When symptoms are present, they include lower abdominal sensitivity and pain, pain during intercourse, irregular periods, cervical and vaginal discharge and tenderness, fever, nausea, and vomiting. Often, these symptoms are not felt until the infection has progressed to a dangerous stage.

Recommended treatment typically consists of one or more of several antibiotics. Usually the infection will resolve with this treatment. Follow-up is necessary within three days of treatment and then again four to six weeks after treatment is complete to be sure the infection has been cured. It is also recommended that sexual partners of the diagnosed woman be treated for infections even if they have no symptoms to help ensure against reinfection.

If the infection has progressed significantly, sometimes hospitalization may be indicated if the woman (CDC, "PID," 2010):

- is severely ill (e.g., nausea, vomiting, and high fever);
- is pregnant;
- does not respond to or cannot take oral medication and needs intravenous antibiotics;
- has an abscess in the fallopian tube or ovary (tubo-ovarian abscess); or
- needs to be monitored to ensure that her symptoms are not due to another condition that would require emergency surgery (e.g., appendicitis).

Treatment for Bacterial STIs

As mentioned earlier, virtually all bacterial STIs are cured with a single course of antibiotic medication. The most commonly used antibiotics for uncomplicated cases of bacterial STIs are *azithromycin* and *doxycycline*, both of which can be taken orally. Other oral antibiotic treatments include *cefixime, ceftriaxone, ciprofloxacin, erythromycin, levofloxacin, ofloxacin*, and *penicillin*. With the exception of penicillin, which continues to be used primarily to treat syphilis, almost all these medications have been effective in curing gonorrhea and chlamydia, whereas the most effective drug for treating *M. genitalium* infection is *Azithromycin* (see the next section about antibiotic-resistant gonorrhea).

Prescribing considerations include the severity of the infection, the age of the patient, coinfection with other diseases, and pregnancy. All of these antibiotics may be taken orally, over a course of one day to two weeks. In most cases, the correct antibiotic will cure the infection, but little may be done to reverse any physical damage to the infected area or scarring due to ulcerations. It is important to stress that after a bacterial STI has been successfully cured, reinfection is no less likely than was the original infection. Many people infected with one of these bacteria become reinfected multiple times if they do not adopt safer sex practices.

Follow-up examination is very important when treating bacterial STIs to ensure that the infection has been cured and that no reinfection has occurred. Reinfection occurs in about 25% of cases and may be even higher among teens and young adults (Rietmeijer et al, 2002; Slater & Robinson, 2014). Women are at particular risk of becoming reinfected with an STI that was previously treated. One study examined the rates of repeat infections among women who tested positive and were treated for chlamydia. Among the women who had a repeat infection, the average time between the first infection and a repeat infection was a little over seven months

(Burstein et al., 2001). The percentage of women found to be reinfected with chlamydia or gonorrhea at 9 months from diagnosis was 14% and 12%, respectively (Hosenfeld et al., 2009). The results of these and other studies indicate that a considerable number of patients' sexual *partners* are not being properly treated or treated at all for STIs. Therefore, treatment of all sexual partners of an infected person is of paramount importance, to prevent both reinfection and further spread of the bacteria. Thus, taking an antibiotic for a bacterial STI is only part of a complete treatment. Effective treatment also requires notification and treatment of the person's sexual partner or partners, a conscious decision to avoid anonymous or unprotected sexual encounters, and the careful use of safer sex practices.

Antibiotic-resistant strains of some bacterial STIs, especially gonorrhea, have been diagnosed with increasing frequency in recent years. For example, in the early 2000s, 14% of all cases of gonorrhea in Hawaii and 0.4% on the West Coast of the United States were found to be resistant to the usual ciprofloxacin antibiotic treatment. Moreover, these numbers and locations of resistant gonorrhea are increasing: in 2008, approximately 24% of new gonorrhea cases in the United States were found to be resistant to treatment with the usual classes of antibiotics, including penicillin, tetracycline, ciprofloxacin, or a combination of those antibiotics (CDC, "Antibiotic Resistant Gonorrhea," 2010). In 2007, the CDC changed their treatment guidelines for gonorrhea and recommended that all cases be treated with one of the cephalosporin antibiotics (e.g., Ceftin, Vantin, or Maxipime) due to the concern that other traditional antibiotic treatments are no longer 100% effective. If a person is treated with an ineffective antibiotic, symptoms, if any, may decrease and he or she may assume the medication cured the infection, when in reality, that person is still actively infectious and may unknowingly continue to spread the infection.

antibiotic-resistant strain
A strain of bacteria that has mutated and is no longer treatable with standard antibiotic therapy.

Parasitic STIs

8.6 Explain what parasitic STIs are and the symptoms, causes, and treatments for each.

A *parasite* is an organism that attaches itself to a host and uses the host's biological resources to survive. A common example of a parasite in the plant world is mistletoe, which spreads through various species of trees (you may have observed large patches of it high in trees, especially in fall or early winter). In animals, an all-too-familiar parasite is the common flea. And in humans, you are probably familiar with the leech. Some parasites found in food can cause serious illnesses in humans, such as Giardia, which causes illness of the intestinal tract. None of these, of course, is an STI, but in the same fashion, some sexually transmitted infections are caused by parasites. The two most common of these are trichomoniasis and pubic lice.

trichomoniasis
A common sexually transmitted protozoan parasite causing symptoms in women, including genital irritation, painful urination, and a foul-smelling vaginal discharge; infected men are typically asymptomatic, yet contagious.

Trichomoniasis

Trichomoniasis, also known as "trich," is caused by a protozoan parasitic organism called *Trichomonas vaginalis* (see Figure 8.13). *T. vaginalis* is the most common nonbacterial, nonviral STI, and it accounts for one-third of all STI diagnoses in the United States each year, or an estimated 7.4 million new cases each year (CDC, "Trichomoniasis," 2012). Estimates are that approximately 170 million people worldwide are infected with the *T. vaginalis* parasite and that one in four will be infected at some point in their lifetime (Tucker, 2003). Most men who contract trichomoniasis have no symptoms, but they are contagious and can pass the infection on to their sexual partners. If men do experience symptoms, they are usually irritation of the

Figure 8.13 Trichomoniasis Parasite

Trichomoniasis is caused by this microscopic protozoan parasite, *Trichomonas vaginalis*.

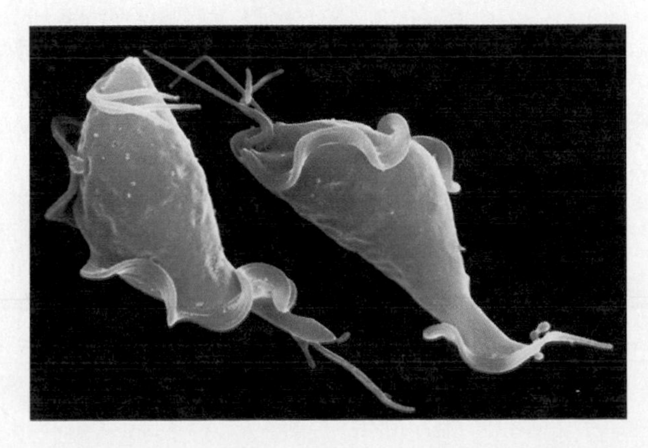

urethra, a slight discharge, and pain after urination or ejaculation. Among women, common symptoms include a yellow or greenish vaginal discharge accompanied by an unpleasant odor, genital irritation, and pain upon urination. Some women may have very mild symptoms or no symptoms at all. Transmission of *T. vaginalis* is almost exclusively through sexual contact. Without proper treatment, trichomoniasis may cause PID among women and NGU among men.

Trichomoniasis may be successfully cured with a single dose of *metronidazole* (brand name Flagyl), an antibacterial, antibiotic-like drug taken orally. Because there is a significant risk of reinfection, all sexual partners of an infected person should be treated for the infection. Metronidazole is not recommended for pregnant or nursing women. A follow-up visit is not typically necessary; however, if symptoms persist, a higher dose of the same medication is usually prescribed.

Pubic Lice

pubic lice

Small, bug-like parasites, usually sexually transmitted, that infest the genital area, causing extreme itching; often referred to as "crabs" because of their resemblance to a sea crab.

This is an STI that no one wants to hear or think about. **Pubic lice** are tiny, bug-like parasites often referred to as "crabs" because they look somewhat like a tiny version of the crustaceans. Pubic lice are *not* the same as head lice, which commonly occur among children of school age and are *not* sexually transmitted. Pubic lice are not found in head hair, but as the name implies, they reside in the pubic hair (although very rarely may be found in chest, beard, or armpit hair). They are spread from person-to-person through intimate sexual contact.

Since You Asked

11. What are crabs, and how can I tell if someone has them?

Estimates of how many people are affected by pubic lice are difficult because treatments are readily available without a prescription. However, sales of those treatments have indicated more than 3 million cases a year may occur in the United States (CDC, "Parasites," 2010).

Pubic lice feed off the blood of their host in the moist, warm areas of the genitals, anus, and abdomen (see Figure 8.14). They invade the pubic area and live among the pubic hairs. They can survive up to thirty days on a host, but only twenty-four hours "on their own." Adult pubic lice are about the size of a sesame seed and can be seen with the unaided eye. Their only goal in life appears to be to procreate because once they reach adulthood, they mate until they die. The female louse lays about four eggs a day (called "nits"), attaching them with a sticky fluid to the base of the pubic hairs. These eggs hatch in five to ten days. Even away from the host, pubic lice can survive on clothing, sheets, or towels and continue to multiply.

The most obvious symptom of lice infestation is irritation and extreme itching, said to be the most intense itching imaginable. The itching is a result of the lice biting the skin to feed on the host's blood and attaching themselves to the shaft of the pubic hairs. When an infested person scratches, it irritates the skin and creates further opportunities for the lice to feed and spread.

Figure 8.14 Pubic Lice

Pubic lice infest the genital area and lay eggs at the base of the pubic hairs. This is a highly magnified photograph of a pubic louse attached to pubic hairs.

Pubic lice are diagnosed by visual inspection. A skilled examiner will be able to see the adult lice and maybe even the eggs. The lice may also look like scabs from scratching, but when they are removed and placed under a microscope, the "scabs" may actually scamper away!

Pubic lice and their eggs are difficult to remove completely. The typical treatment is with prescription and nonprescription shampoos, creams, and ointments (with brand names such as Elimite or Kwell) containing chemicals that are toxic to the lice. Following treatment, the pubic hair should be combed with a special fine-mesh comb to remove any remaining eggs, and the area examined for any remaining nits (this is where the term "nit-picking" comes from). After treatment, shaving of the pubic hair may be helpful, but this will not clear lice by itself. Typically, one treatment is insufficient to rid the body of all lice, and a follow-up treatment a week later is recommended to be sure any newly hatched lice are killed. In addition, all clothing, towels, bedding, and other materials that may have come in contact with the lice should be washed in hot water or sprayed with special medicated spray and sealed in plastic bags for two weeks. Anyone who has had recent intimate contact with the infected person should follow the same procedure before any sexual contact is resumed (CDC, "Parasites, Treatment," 2013).

Preventing STIs

8.7 Discuss the various specific strategies for preventing or reducing the risk of contracting or spreading STIs.

Throughout this chapter, we have talked about the various behaviors that transmit STIs. With the exception of coercive situations, such as rape, your STI risks are determined by your individual *choices* of behavior. To sum up prevention strategies, we must return to the five major risk categories discussed at the beginning of this chapter and examine how you can reduce or eliminate each of them. As you may recall, the five risk categories are lack of accurate information, negative emotions and faulty beliefs, poor sexual communication, substance abuse, and risky sexual activities.

Since You Asked

12. Once I start having sex, how can I be sure to never catch an STI?

Seeing or Feeling STI Symptoms Is Not Enough

Remember, for most of the STIs we have discussed in this chapter, you can be infected and be totally unaware of it. This is because the infectious agent (a virus or bacterium) may be in your body but has not produced any noticeable symptoms. And it's possible to have an infection indefinitely with no outward sign to make you aware that you have it or to alert you that you could infect someone else. You often have to judge your likelihood of infection based on your sexual practices and behaviors. You now know the various sexual activities that put you at risk. If you have engaged in any of them, there is a chance you may be infected. And the riskier and more frequent your sexual behavior, the greater your likelihood of infection.

As discussed earlier in this chapter, if you are sexually active, the only way you can know for sure than you are free of STIs is to be tested. Odds are good that you do not have an STI, but knowing for sure can be a huge relief and allow you to "start fresh" practicing safer sex and keeping yourself infection free. And if you do test positive for an STI, you can begin treatment immediately to cure it or to prevent it from becoming more serious. Plus, you will be better able to avoid passing the STI on to anyone else.

Obtaining Accurate Information about STIs

The research evidence is clear: people who are effectively *educated* and informed about the causes, symptoms, routes of transmission, and treatment options for all STIs are *far* more likely to make healthy sexual decisions and avoid infection (Ross & Williams, 2002).

As Table 8.2 (on page 269) makes clear, most sexual behaviors that involve physical contact with a partner carry at least some risk of STI transmission. Therefore, each person must make an educated and thoughtful assessment of the amount of risk he or she is willing to take when choosing to be sexually active. This is referred to as a person's **acceptable level of risk**. In this sense, sexual behavior is no different from other health-related behaviors that involve choices made on the basis of what we perceive to be an acceptable level of risk. For example, some people smoke cigarettes, eat a high-fat diet, or fail to wear a seat belt, even though they know these behaviors are clear risks to their health. For them, however, these are acceptable risks. Sexual behavior should be based on a similar principle. Like it or not, sex is a *health behavior*, and each

acceptable level of risk
The level of risk one is willing to accept when making behavioral choices about one's health and wel-lbeing.

person's level of acceptable risk should determine what sexual activities the person chooses to engage in relative to STIs (this concept also relates to other sexual decisions such as birth control and pregnancy).

Here is the most important point about your acceptable level of risk: No one can wisely determine his or her acceptable risk without knowing what all the risks are! Now that you have read this chapter, you are better equipped to make decisions about your sexual behaviors based on your level of acceptable risk, but it doesn't end here. You have a responsibility to *stay* informed and educated about sexual risks throughout your life.

A thorough understanding of the symptoms, modes of transmission, diagnosis, and treatment of all STIs is crucial for individuals and couples to determine their acceptable risk levels and make effective choices about their sexual behavior. The key here is that the knowledge be as accurate as possible and not based on myths, such as those in Table 8.1 (on page 266). Some individuals and couples are able to obtain solid factual information about STIs on their own, through reading, researching, and talking to knowledgeable professionals such as counselors, sexuality educators, and physicians. A perhaps more far-reaching source for this information is high-quality, effective sex education in school and college classes.

The solution does not lie in merely *offering* sex education to young people, however. Research has shown that even among teens who receive STI education in school or from parents, their *working knowledge* of sexual risks remains poor (Clark, Jackson, & Allen-Taylor, 2002). And students in school systems that teach "abstinence only" programs receive next to no information on STIs or pregnancy prevention. When teens were asked about their sex education history, 97% said they had received specific education about STIs. Most of the participants knew that HIV is a serious STI, but only 2% could name all eight major STIs (HIV, gonorrhea, syphilis, herpes, chlamydia, hepatitis B, trichomoniasis, and genital warts), 9% could name the four curable bacterial infections, and 3% could name the four viral incurable infections. The study suggested that STI education efforts may be spending a disproportionate amount of time and resources on HIV education at the expense of other STIs that are much more common among young people.

Overcoming Unhealthy Emotions about Sex

As people become educated and knowledgeable about human sexuality in general and about STIs specifically, the negative STI emotions discussed earlier in this chapter (shame, guilt, fear of stigma, embarrassment) all tend to decrease. The reduction in these emotions opens the door for better communication between partners, greater comfort with their own and each other's bodies, greater awareness of the symptoms of STIs, greater willingness to seek medical attention for possible infection, less discomfort with STI testing (see "Self-Discovery: STI Testing"), and an enhanced ability to make intelligent choices about living a healthy sexual life. We all have the responsibility to take charge of our sexual lives and assert our needs, desires, and rights as sexual people. The negative emotions discussed here undermine that goal. Those who are unable, for whatever reason, to face and overcome negative sexual feelings may end up—literally—risking their lives.

Communicating Effectively about Sex

Many people fail to realize that effective sexual communication is an important *safe sex behavior*. The ability to communicate openly and honestly is key to avoiding contracting or transmitting all of the STIs we have discussed in this chapter. The topics of communication that relate directly to STI prevention include issues such as number of previous sexual partners, past or current STI infections, and

Self-Discovery

STI Testing

No one likes to think that he or she is or will be at risk for an STI. The reality is, however, that most sexually active people will, at some time in their lives, come in contact with a sexual partner who is or was infected with an STI. For this reason, it is important to practice safer sex, such as consistent condom use, as well as to consider STI testing part of your preventive health care.

Currently, STI testing is usually not a standard preventive medical practice. In fact, most physicians do not routinely test for STIs during annual or biannual health visits. If you want to be tested for STIs, you usually have to ask, which in itself can be an uncomfortable experience. However, the stigma of being tested for STIs is disappearing, and doing so is now often seen as a smart move.

When STIs are diagnosed and treated early, their transmission and potential negative side effects can usually be reduced or prevented. You may feel uncomfortable requesting STI testing, but it's your health, and unless you take charge, you may not get this important health care service. You can be your own best health advocate, so consider the following the next time you're scheduling a health exam.

______ 1. **Am I comfortable talking about my sexual health with my health care provider?** The experience of STI testing can be easier when your health care provider is someone with whom you feel comfortable discussing your sexual history and activities. This is particularly important if one is diagnosed with an STI such as herpes or HIV that requires long-term management.

______ 2. **Have I requested STI testing?** Because STI testing is usually not a standard health care practice, you must request it. Avoid assuming you are not infected with an STI if your health care provider does not identify an infection or recommend STI testing.

______ 3. **Should I consider being tested for STIs routinely during my annual health exams?** Because STIs can have few or no observable symptoms, if you are sexually active, routine testing during annual health exams may identify an unknown infection. Annual testing is not considered an STI *prevention* method but may allow an STI to be identified and treated in its earliest stages.

______ 4. **Has my partner been tested? Can we both be tested at the same time?** It is often not enough for one person in the relationship to be tested for STIs. The results of one partner do not necessarily apply to the other partner. Being tested with your partner and being sexually faithful are signs of mutual caring and among the safest sexual practices of all.

agreements to use condoms or to avoid specific risky sexual behaviors. Disclosing infection with an incurable STI to a partner can be the most difficult communication challenge of all. "In Touch with Your Sexual Health: Disclosing an STI to a Sexual Partner" discusses the importance of disclosure and offers some tips for feeling less uncomfortable.

In Touch with Your Sexual Health

Disclosing an STI to a Sexual Partner

Discovering you are infected with an STI, especially one that has no cure, can be devastating. It can shake your trust in others and in yourself. However, in addition to obtaining as much information as possible on the infection and seeking out the best treatments available, you also have to prepare yourself for disclosing your status to current or potential sexual partners. Informing a sexual partner of your STI status is an ethical and potentially a legal responsibility.

If you have a steady partner at the time of your diagnosis, you should not assume that he or she intentionally infected you with the STI. As you know from your reading of this chapter, many people who are infected are unaware of it because they have never had symptoms or they mistakenly assume they have been cured. Moreover, you may have been infected *prior* to your current relationship. If so, you may have transmitted it to your partner rather than the other way around. From a personal health perspective, the issue of who infected whom is secondary to the immediate need for screening and treatment for your sexual partner or partners.

So what is the best way to break the news to your partner? This dilemma is so individual that it's difficult to make one overall recommendation. However, here are a few suggestions that may help.

Be Honest with Yourself

It's easy to get treated for *an STI, especially* one that is curable, and forget that it ever happened or to have genital warts removed or wait until herpes blisters heal and hope (mistakenly) that you then have no chance of spreading the infection. However, once you accept the fact that you presently have or once had an STI, breaking the news to your partner will be easier.

Say It, Don't "Show It"

Often people are too scared or embarrassed to tell their partner about the infection, so they attempt to make the partner figure it out. Some indirect strategies people use to disclose STI status include leaving STI pamphlets around, testing the partner's reaction by telling stories about "other people," or suggesting that they both be tested together (that way someone else can tell the partner). However, using a direct, honest approach, talking or perhaps writing a personal note (and then following up with a discussion; an e-mail would probably be too impersonal!) is likely to be a more effective way of initiating a discussion about these health issues, ensuring effective treatment for both partners and preserving the closeness and trust of the relationship.

Avoid the Blame Game

As mentioned earlier, wasting energy on determining who gave the STI to whom can be destructive and may not be in the best interest of your relationship. When people are blamed for something that they may or may not be responsible for, they become defensive. Most people would never intentionally set out to infect another with an STI. When transmission does occur, it is often because the infected person is unaware of having been infected or did not fully understand the risk of transmitting the infection to others.

Telling a partner that you have an STI is never easy. The goal of sharing positive (or negative) STI test results leads to a responsible course of action that will include treatment and prevention strategies for the future. By being honest with yourself, talking directly about your infection status, and encouraging your partner to get immediate testing and treatment, you and your partner can move past the diagnosis and focus on developing the positive aspects of your relationship.

Discussing these issues does not come easily to most people, and many avoid it out of embarrassment, fear of rejection, or assumptions about their risks of contracting or transmitting STIs. One study of college students found that, although 99% felt knowledgeable about safe sex practices, only 77% considered disclosure of past or current STI infections a safe sex practice, only 58% regarded disclosure of the number of previous sex partners as a safe sex practice, and 37% did not feel that disclosure of the number of partners should be a requirement prior to becoming sexual with a new partner (Lucchetti, 1999). Moreover, the same study found that over 30% of students had *actively avoided* such discussion with at least one sexual partner, and over 20% admitted to misrepresenting their prior sexual history to their new sexual partners. On the other hand, people who do disclose their STI status tend to have more positive feelings about themselves and about their sexual self-concept (Newton & McCabe, 2008).

Your ability to communicate and make agreements with a current or potential sexual partner about the risks of STIs relies in part on successfully navigating the two skills discussed previously: becoming knowledgeable about STIs and overcoming negative emotions about them. Both of these skills allow you to obtain clarity in your own mind, which in turn smooths the way for communicating about these issues with a potential partner. Here are some general tips that often help people add communication to their arsenal of STI prevention strategies:

- Become as educated and knowledgeable as you can about the symptoms, causes, and treatments of all STIs.
- Do not allow embarrassment, guilt, shame, or fear stop you from protecting your health and your life. If you need to, practice with a friend, make notes, rehearse.

- Communicate in person, face-to-face. These are not the sort of issues to be discussed on the phone, e-mailed, or texted.
- Do *not* wait until you are sexually intimate to start communicating.
- Think about what you want to say, and how you will say it, ahead of time.
- Be clear about your limits, and decide what you are comfortable doing sexually.
- Keep your sense of humor, and avoid becoming overly serious, but don't avoid saying what you need to say.
- Be flexible about when you will talk. Find a private, quiet, nonsexual setting when you have plenty of time so you won't feel rushed.
- Communicate with your partner honestly, so that you can both make clear, informed choices.
- Give your partner time to think about what you have said, and listen carefully to his or her responses.
- Decide *in advance* what you both feel comfortable doing sexually, based on all the information you have communicated, and stick to your agreements.
- Never allow yourself to be seduced, cajoled, persuaded, or coerced into agreeing to behaviors that you feel are too risky or make you uncomfortable.
- If you and your partner cannot agree to respect each other's level of acceptable sexual risk, you may need to reevaluate your relationship.

Avoiding Mixing Sex and Drugs (Especially Alcohol)

The best-laid plans for sexual communication, however, often collapse when people's ability to communicate and make rational decisions is compromised by alcohol or other drug use. Research has left little doubt that, in terms of STIs, drugs and sex do not mix. Drugs, especially alcohol and marijuana, often accompany sexual interactions. In college and university settings, they nearly *always* occur together. However, as noted early in this chapter, this is a dangerous and potentially deadly mix. No matter how educated and informed a person may be, regardless of that person's level of comfort with sexual issues, and notwithstanding that person's ability when sober to communicate about sex with a partner, alcohol or other drug use may lead to poor sexual choices and the spread of STIs. Interestingly, efforts in the United States to stop people from drinking and driving ("Friends don't let friends drive drunk") have found significantly greater support than similar campaigns to reduce the dangers of mixing alcohol and sex. Yet the connection is equally strong. Perhaps one day billboards and TV ads will read, "Friends don't let friends have sex drunk." Of course, a "designated sex partner" program is unrealistic, but perhaps people will get the message nevertheless.

Refraining from Risky Sexual Behaviors

You already know that the absolute best method for avoiding STIs is abstinence. However, the reaction of most college students to the suggestion of abstinence is along the lines of "Yeah, right." Indeed, as you can see in Chapter 6, "Sexual Behaviors," the vast majority of college students choose to be sexually active. "Abstinence only" education is not sex education, and has been proven to be *ineffective* in reducing risky and unsafe sexual practices (e.g., Springer-Hall

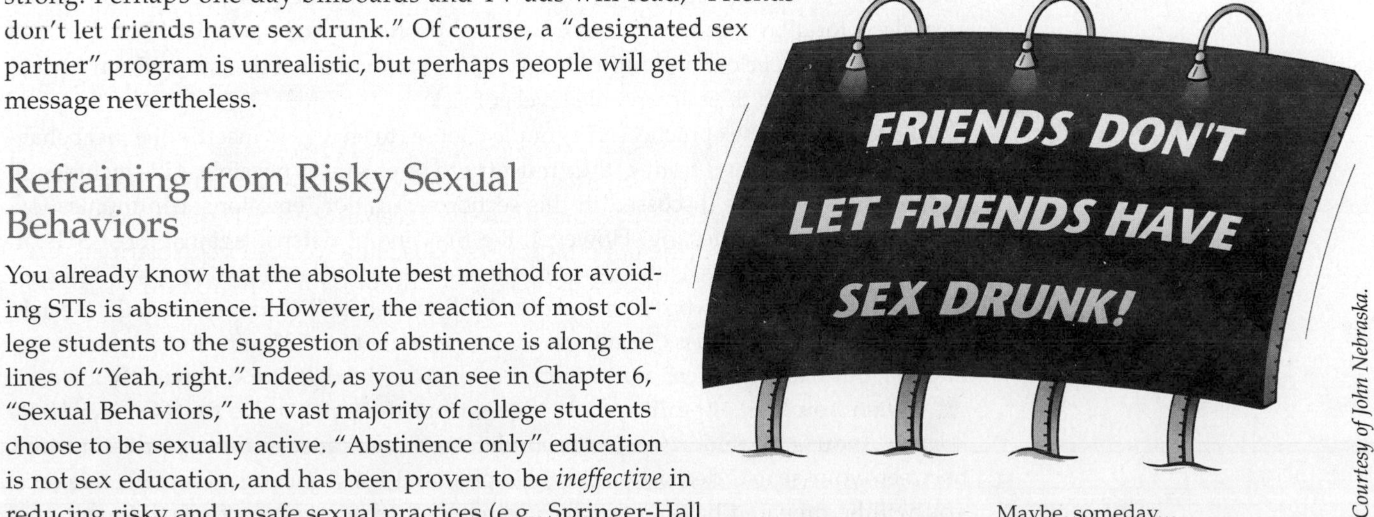

Courtesy of John Nebraska.

Maybe, someday...

& Hall, 2011). However, abstinence must be part of any STI prevention discussion, if for no other reason that it is, obviously, the *most* effective means of avoiding STIs.

As we discussed in Chapter 5, "Contraception: Planning and Preventing," not everyone interprets abstinence in the same way. To some people, it implies no sex of any sort, a concept closer to **celibacy**. To others, abstinence is the process of picking and choosing which sexual behaviors a person is willing to engage in, either alone or with a partner, in order to enjoy sexual feelings and activities while avoiding exposure to infections or pregnancy. From your reading of this chapter and looking at Table 8.2, you may be thinking, "That's great, but what safe sexual behaviors are left?" Well, there are many; you just have to be smart and, sometimes, *creative*! Here are five suggestions from various sexuality educators on how to have sexually intimate relationships while minimizing your risks for STIs.

celibacy
Choosing to engage in no sexual activities whatsoever.

First, limit your *number* of sexual partners. The more sexual partners a person has, the greater the chances of contracting an STI, no matter how careful the person may be in pursuing safer sex practices.

Second, get to know a potential sexual partner before initiating a sexual relationship. If more couples would do this, many STI risks could be avoided. This point relates to our earlier discussion about sexual communication. As difficult as clear, honest sexual communication may be, it becomes increasingly easier as two people become emotionally closer. Making psychological intimacy a higher priority than sexual intimacy is a very effective safer sex strategy.

Third, ensure that you and your potential sexual partner are free of STIs. This means agreeing to be tested *prior* to engaging in any risky sexual behavior together. For many people, that means waiting six months from their last sexual encounter before being tested so that the presence of all STIs, including HIV, is reasonably sure to be detected. Six months is a long time for many couples to postpone their sexual involvement, especially at the beginning of a new relationship when excitement is running high. However, the sexual side of the relationship does not have to be put completely on hold, as will be explained next.

Fourth, practice risk-free sexual behaviors. This does not refer to condom use (that's safer, but not totally risk-free – see the next paragraph). It is discussed in the contraception chapter as **selective abstinence**. If a couple are clear about their intentions to be safe and wait until they can both be tested, they may still engage in intimate, low- or no-risk *noninsertive* sexual activities such as kissing, hugging, massaging, body exploration through touch, rubbing genitals together clothed, mutual genital touching, and mutual masturbation. These activities can be extremely exciting and allow the couple to get to know each other's bodies and responses. If all goes well, then, when their tests come back negative (which is likely), they will be (very!) ready for further sexual exploration together. If one or both test positive for one or more STIs, they will then probably feel close and intimate enough to make rational decisions about further sexual activity and their acceptable levels of risk.

selective abstinence
Choosing to engage in or avoid certain sexual behaviors on the basis of their risks of STIs or pregnancy.

Fifth, use safer sex practices. If a couple chooses to engage in insertive sexual behavior without waiting and testing, they must try to be as safe as possible. This includes all of the issues we have discussed in this section: education, emotions, communication, substance abuse, and testing. However, the first line of defense against most STIs, in most real-world settings, is the use of condoms for vaginal, anal, and oral sexual activities. You will notice the word *safer*, not *safe*. As discussed in "Evaluating Sexual Research: The Condom Effectiveness Controversy," condoms are not 100% effective in preventing any STI, but with consistent and correct condom use, the risks are significantly lower.

When you look at your sexual choices from the perspective that sex is a *health behavior*, you will find it easier to understand the cause-and-effect relationships between your sexual decisions and their potential consequences. Through this lens, you will be prepared better to protect yourself and your partner from STIs and help reduce the spread overall.

Evaluating Sexual Research

The Condom Effectiveness Controversy

Male condoms have been available for over a century, with the first condom dating back to the late 1800s. However, the rise of HIV in the early 1980s changed the way we perceived the utility of condoms, and their use became the single most effective method of preventing STI and HIV transmission among sexually active people.

Throughout the HIV epidemic, some individuals and groups have opposed the safer sex message (preferring the teaching of abstinence only) and pointed to limitations of condom use in preventing HIV, such as breakage, slippage, and punctures. However, such limitations were and continue to be due primarily to *human error*, not faulty condoms. And the minuscule number of condom failures represents a *much* smaller risk than the failure to use condoms at all.

Today, many groups that support abstinence-only prevention programs, including various religious groups, have spread erroneous information that condoms do not prevent STIs and that promoting condoms for safer sex *encourages* unsafe sex practices and gives people a false sense of security. However, that is untrue. The effectiveness of male and female condoms in reducing the risk of STI transmission is indisputable, based on all the scientific evidence. Claims that HIV and other viruses may pass through the latex in the condom are absolutely false. Some groups have even gone so far as to claim that condoms are the *cause* of people dying from HIV. This is simply a vacuous claim.

Although no one has ever asserted that condoms are 100% effective at preventing HIV or other STIs, they are *far* more effective than unprotected sexual activities. Therefore, unless a strategy can be found to stop all unsafe sexual activity worldwide, referred to as "universal abstinence" (a goal that is clearly impossible), the importance of condom use must continue to be stressed as a crucial part of STI prevention (CDC, "Condoms and STDs," 2013).

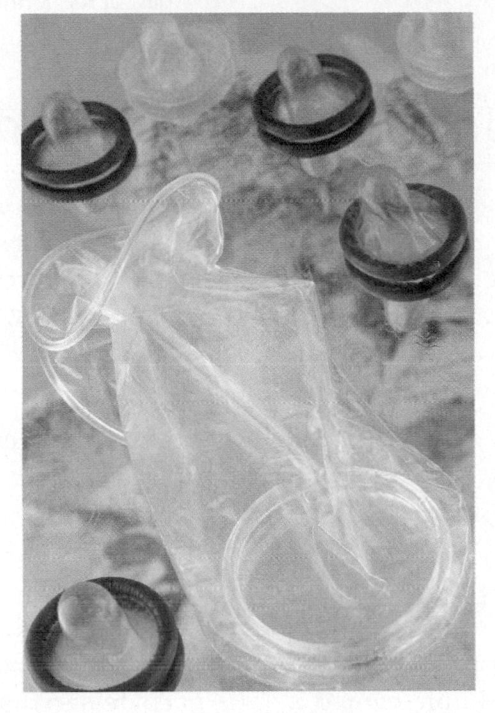

Your Sexual Philosophy

Sexually Transmitted Infections/Diseases

From our earliest stirrings of sexual feelings in puberty, we often make the misguided and potentially dangerous assumption that sex will just "take care of itself." We plan for college, we plan a major, we plan a career, we may even plan ahead for marriage, but too often we fail to plan for life as a sexual being. Then, when we are faced with sexual decisions, sexual situations, sexual pressures, or the heat of sexual moments, instead of being in control of those situations, we allow them to take control of us. The result: poor sexual choices, including an increased likelihood of exposure to STIs.

The information in this chapter is essential to an effective sexual philosophy. Combining accurate information and education with carefully considered attitudes and goals will help you know who you are, know what you want and don't want, and *plan ahead*. Although it is true that sometimes the best-laid plans go awry, a strong personal sexual philosophy serves as the best foundation for a healthy, disease-free sexual life.

The goal of this chapter was not to frighten you away from sexual intimacy altogether. On the contrary, the objective has been to provide you with a solid base of knowledge about sexually transmitted infections: what they are, their symptoms (or lack thereof), how they are transmitted, how they are diagnosed, and available treatment options. When you incorporate the information in this chapter into your sexual philosophy, you will help to *enhance* sex and intimacy in your life by making decisions and choices that provide a satisfying and healthy sexual life. With this knowledge, you will be able to be part of the solution to stopping the spread of STIs, rather than another STI statistic.

Have You Considered?

1. Imagine that you are the public health director of a major city. You have been hired to stem the STI epidemic in the city. Explain three strategies you feel would be most effective in reaching your goal (and keeping your job) and how you might go about implementing them.
2. Do you think sex education that includes accurate information on contraception and STIs should be part of the curriculum in junior high and high schools? Why or why not? If a school system decided to add comprehensive sex education to its standard curriculum for high schools, how would such a program be implemented in light of some parents' objections to such teaching?
3. Suppose for a moment that you are a parent of a fourteen-year-old daughter. She has a boyfriend, and although she hasn't said anything specific to you, you suspect she may be on the verge of becoming sexually active. What would be the three most important pieces of advice and information you would want to be sure to communicate to her about STIs? Discuss how you might approach the subject with her.
4. Take a look back at the feature "In Touch with Your Sexual Health: Genital HPV Vaccine: The Controversy." Which side of the debate do you lean toward? What would you say are the three best points among those listed to support your beliefs?
5. Discuss why a diagnosis of HIV is no longer considered a death sentence in the United States.
6. Imagine that you meet someone you really like and you begin to date. As the relationship progresses, you become closer and more intimate, but the relationship still has not progressed beyond the kissing and touching stage. Your feelings for each other are deepening, and you know you are falling in love. Then one night, this person sits you down for a serious talk and discloses to you, very gently and lovingly, that he or she is HIV-positive. What do you think your reaction would be? Would you continue seeing the person? Why or why not?
7. Discuss three reasons sexually active people should be tested for STIs regardless of any obvious symptoms.
8. For most college and university students, alcohol and sexual activity are usually connected. This is often a dangerous mix. Explain why alcohol increases the risk of STI transmission and how colleges and universities might reduce alcohol use in potential sexual situations.

Summary

Historical Perspectives

The Tuskegee Syphilis Study

- In the 1930s, in Tuskegee, Mississippi, doctors hid the diagnosis and denied treatment to a group of black men with syphilis in order to study the long-term effects of the disease. The study was publicly revealed in the 1970s, and President Clinton in 1997 issued a national apology to the victims and their families.
- The racist aspects of the Tuskegee Syphilis Study still have repercussions today, as many minorities continue to mistrust the U.S. medical system.

The Sti Pandemic

- Approximately half of the population of the United States will contract an STI at some point in life, and an estimated 19 million people are newly infected with an STI each year. Sexually active people between the ages of 15 and 24 are at greatest risk of contracting STIs.
- Gonorrhea, chlamydia, and the human papillomavirus (HPV), which causes genital warts, are the most common agents causing STIs among teens and young adults.

Risk Factors for STIs

- STIs are among the most common illnesses throughout the world. Many STIs may have no obvious symptoms and may go undiagnosed and untreated, allowing them to spread more widely.
- Lack of education, negative emotions about sex, poor communication, substance abuse, and risky sexual activities all contribute to the current worldwide STI epidemic.

Specific Sexually Transmitted Infections

- STIs may be categorized according to the type of microbe causing the infection.

Viral STIs

- STIs that are caused by viruses currently have no cure. (This is true of virtually all viruses.)
- Millions of people in the United States are infected with viral STIs. The most common of these are herpes, hepatitis B, and the human papilloma virus.
- A vaccine (trade name *Gardasil*) is now available that will prevent genital HPV and the potential for related cancers of the cervix, anus, and throat.
- HIV, the virus that causes AIDS, has caused a pandemic, with an estimated 35 million individuals infected worldwide. In the United States, at least 48,000 new cases are reported each year.
- No cure for HIV or AIDS exists. A recent increase in new cases of HIV and other STIs may reflect the mistaken belief that new HIV treatments and medications actually cure AIDS, which they do not.
- A simple oral saliva test is now available for HIV, and researchers are working around the clock to find an effective HIV vaccine, which still appears many years away from becoming a reality.

Bacterial STIs

- Although most bacterial STIs are readily cured with antibiotics, when they are asymptomatic and left untreated, they can lead to more serious conditions of the reproductive tract, including infertility.
- Virtually all bacterial STIs are spread primarily through unprotected oral, vaginal, or anal sexual activities. The most common bacterial STIs include chlamydia, gonorrhea, and syphilis.

Parasitic STIs: Sexually Transmitted Bugs?

- Parasitic STIs are caused by organisms that feed off the human body. The most common of these are trichomoniasis and pubic lice. The main symptoms of sexually transmitted parasites are vaginal irritation and/or intense itching. These infections are relatively easy to treat and cure.

Preventing STIs

- Prevention is always preferable to curing STIs. This is true of any infection or disease.
- Strategies for preventing STIs include becoming educated about their causes, symptoms, and treatments; overcoming negative sexual feelings that often interfere with treatment; maintaining effective communication between partners to help avoid transmission; resisting the use of mind-altering substances that cloud judgment and may lead to unsafe sexual practices; and engaging in safer sexual activities.

Your Sexual Philosophy

Sexually Transmitted Infections/Diseases

- Incorporating an accurate and working knowledge of STIs into your sexual philosophy is the best way for you to stay safe and help others to avoid contracting or transmitting one of these infections. In terms of sexual health, this may be the most important area for accurate information in a person's sexual philosophy of life.

Chapter 9

Conception, Pregnancy, and Birth

Learning Objectives

After you read and study this chapter you will be able to:

9.1 Review the issues relating to a person's or couple's decision whether or not to have children.

9.2 Summarize the process of conception from ovulation and sperm production to fertilization of the ovum and the development of the zygote.

9.3 Describe the development of the embryo and fetus from implantation in the uterine wall through the three trimesters of pregnancy.

9.4 List and discuss the medical problems that may occur during pregnancy.

9.5 Summarize the issue of abortion including relevant statistics, the various methods of pregnancy termination (surgical and medical), and the socio-political controversy that surrounds the practice of abortion in the U.S.

9.6 Discuss the stages of labor, the birthing process, and the choices and decisions that accompany child birth including, the selection of a doctor or midwife, birthing settings, pain management decisions, and the current C-section controversy.

9.7 Review the issues that may be present following child birth, including postpartum depression, the importance of breast feeding, and resuming sexual activities after delivery.

9.8 Explain impaired fertility (or infertility) and summarize the scope of the problem, causes of low or absent fertility, how fertility is tested, and the possible solutions to impaired fertility.

Since You Asked

1. My older sister (she's twenty-eight) really wants to get pregnant, but her husband is saying he's not ready to be a father. Is there any way she can convince him to agree? (see page 318)
2. I feel that I really do not want to have children ever! Does that make me totally weird? (see page 318)
3. After a woman's period, how soon is the next egg released, and when is she most fertile? (see page 322)
4. Last week, I took two pregnancy tests that were both negative. But I still haven't started my period. Could I be pregnant? (see page 325)
5. A few months ago, I started my period very late and had a lot of blood and very bad cramps (worse than usual). My friend said it was probably a miscarriage. Is that possible? (see page 331)
6. My aunt's baby was born after only twenty-five weeks of her being pregnant. The baby is probably going to be OK, but will the baby have any problems later on? (see page 332)
7. How does the abortion pill work? Is it really just a pill you take and then you have an abortion? (see page 340)
8. My sister wants to have a midwife deliver her baby. Is this dangerous compared to having a doctor handle the birth? (see page 347)
9. Why do some women get so depressed after giving birth? That seems so opposite of how a new mother should be feeling, if she really wanted the baby. (see page 351)
10. What is infertility and what causes it in males and females? (see page 355)

Most college students are more concerned about *preventing* pregnancy than about the reality of a conception, pregnancy, and birth. And it's probably safe to say that most young adults probably see the processes of conception, pregnancy, and birth as far simpler than they are. The reality is that reproduction is one of the most intricate and elegant of all human biological functions. It is also one of the most important reasons we are sexual beings—without it, the human species would have ceased to exist a long time ago. Biomedical science has progressed to the point that we now understand most of the human reproductive processes; yet still today, we do not understand all of them completely. As you will see throughout this chapter, conception, pregnancy, and birth remain, in very real ways, mysterious and miraculous.

It is very likely that you will, one day, have one or more children, if you don't already. Sometimes pregnancy and the birth of a baby happen when the parents plan for it to happen. You may be surprised to learn, however, that in the United States nearly half (49%) of all pregnancies are unplanned (Trussell & Wynn, 2008). This is not to say that all unplanned pregnancies are *unwanted*. On the contrary, many unplanned pregnancies are greeted with acceptance, pleasure, and even joy, at least once the initial surprise wears off. However, the fact that over two million unplanned pregnancies occur in the United States each year makes a strong case that the general public has a poor understanding of the realities of exactly how, when, and why conception occurs. This lack of understanding, perhaps combined with poor decision making about sexual behavior, is probably linked to an incomplete education regarding conception, pregnancy, and birth. This very important and very basic information has traditionally been referred to as "the birds and the bees"—clearly indicating a desire to sidestep the straightforward details of *human* reproduction.

Although you may know more about human reproduction than the average person who has received the "birds and bees" talk, let's put an end to the cycle of incomplete information right here. After you read this chapter, you'll know a lot more, and

Focus on Your Feelings

Becoming pregnant, being pregnant, and giving birth are times of great emotion. For those who *want* to get pregnant, who want to be parents, the emotions experienced upon learning the news are typically joy and excitement, perhaps mixed with some nervousness about what's to come. As the pregnancy progresses, many women acquire a kind of glow that accompanies the happiness of being pregnant, and they may also grow weary of the inescapable discomforts of pregnancy. Pregnancy is also the time when most women and couples are preparing emotionally (and economically!) for parenthood. They may feel some concern that the fetus is developing properly and hope that no complications are arising.

Perhaps one the most difficult emotional events a person or couple can experience is the loss of an unborn child. When a couple must endure the early, spontaneous loss of their unborn baby, it is rarely "just a miscarriage." To most couples, it is the profound loss of a future child whom they had already grown to love and cherish; the emotions they feel are similar to those felt upon losing any loved one through death. If you focus on what your feelings might be if you ever have to deal with a miscarriage (although chances are good that you won't), you will be much better prepared to offer the kind of understanding and support close friends or relatives may need should they ever have to face this difficult event.

And talk about feelings! Few people are ambivalent about the topic of abortion. Many fiercely stake out and defend ideological territories about the morality of choosing to terminate a pregnancy. Whatever your personal attitudes about abortion, you probably feel strongly about them. When or how this contentious issue will be resolved is anybody's guess, but if we can work together to reduce the *number* of unwanted pregnancies that will be one step in a direction on which everyone can agree.

For most couples, the decision to have a child, and the realizing of that decision when a birth happens, is one of the greatest joys of human existence. Sometimes, however, when a woman or couple who wants children has difficulty conceiving, they experience intense feelings of disappointment and frustration as time passes without a pregnancy. They often feel that for some reason they are being denied one of their most cherished and anticipated life goals, and many couples feel a deep emotional loss of a child who has not yet been conceived. Luckily, these couples can nearly always find a solution to their infertility issues that is right for them.

one day you may be in a position to pass your knowledge on to others, perhaps to your own children.

For perspective, we'll begin with a look back at some early and rather disturbing childbirth practices. Then, returning to the present, we will cover the sometimes complex issues people face in life as they make decisions relating to whether and when to have a child. Whether or not such a conscious decision is made, pregnancies happen. And they all begin with the fascinating series of events leading to conception: the meeting and joining of the male sperm with the female egg, or ovum. If the fertilized ovum survives, the next step in understanding reproduction is taking a close look at the many changes in the growing embryo and fetus, as well as in the mother, during approximately nine months of pregnancy. It is also important to be aware of some of the problems and decisions that sometimes accompany pregnancy, including miscarriage (also called *spontaneous abortion*); voluntary, or induced, abortion; and possible complications in pregnancy and birth. Next, we will turn to the final stage in pregnancy: childbirth, and the issues accompanying the birthing process, both during and after. Finally, we will examine infertility issues some couples face and solutions available today to assist them in becoming pregnant or, if they so choose, to adopt. Many of the issues discussed in this chapter (such as abortion or infertility) may trigger some strong emotional responses in you. This is normal and to be expected.

Historical Perspectives

The Pain of Giving Birth

Everyone knows that physical pain is usually a part of childbirth. Today, medical science in most developed nations has advanced to the point that a woman has the option of giving birth with minimal or tolerable levels of pain while remaining awake and fully aware of the profound moment when her baby enters the outside world.

These modern pain-reducing techniques, however, are relatively new to the history of human birthing practices. In fact, the use of medical pain control for childbirth did not come into common practice until the mid- to late nineteenth century. The earliest methods required the mother to be deeply sedated under *general anesthesia*.

In the early 1800s, the pain-reducing qualities of ether, a mildly sedating and pain-reducing gas, had been discovered, and it was beginning to be used during childbirth. Although ether was often effective for pain, large amounts were required for full pain relief, and many medical professionals feared that such high levels of the drug might lengthen the birth process and potentially harm the fetus.

In the mid-1800s, an obstetrician and professor of **midwifery** at Edinburgh University, Dr. James Young Simpson, began experimenting (on himself!) with a new anesthetic he had discovered. As you can see from the sketch, his early experiments were quite "successful." In 1847, he invited two of his medical colleagues to dinner. After dinner, he asked them to try his new anesthetic. A couple of hours later, the story goes, Dr. Simpson awoke underneath his dining room table and observed his fellow physicians still asleep and snoring peacefully next to him (Fissell, 1999b). This was exactly the medical result he had hoped for (although the dinner party itself may have fallen short of expectations), and within a month, he had used his new discovery, *chloroform*, on more than fifty women during the birth process (*Gazeteer*, 1995). The method of administering the drug was primitive: A handkerchief with a few drops of the liquid was held over the woman's nose and mouth (today, we tend to see this in movies when someone is kidnapped). Sedation was quick and deep and did not have many of the negative side effects associated with ether.

The earliest form of childbirth pain medication was deep sedation using a handkerchief containing a few drops of chloroform, a chemical discovered in the mid-1800s by Sir James Simpson, who, after experimenting on the drug himself, was discovered unconscious in his study by his butler.

midwifery

The practice of trained midwives assisting women through normal pregnancy and childbirth.

A few years after Simpson's discovery, acceptance of the use of chloroform for childbirth grew rapidly when Queen Victoria demanded and received chloroform during the birth of her eighth child in 1853. For the next 50 years, chloroform became widely accepted for the pain of childbirth and for other painful medical procedures. The drug, however popular at the time, was not entirely free of potentially serious side effects. The handkerchief method of delivery was imprecise, and overdose was not uncommon. In addition, around 1900 the drug was found to cause liver damage in some patients, and its use decreased drastically. Chloroform is still available today, but it is used primarily as a solvent in paint thinner and not at all in modern medicine.

About the time chloroform was losing favor in the early 1900s, a new method of childbirth pain control, called *twilight sleep*, was being developed. Pioneered by Dr. Bertha Van Hoosen, a gynecology professor at the University of Illinois Medical School, twilight sleep used a combination of two injectable drugs: morphine and scopolamine (Fissell, 1999a). These two drugs combined have the advantage of pain relief (morphine) and amnesia (scopolamine), so the woman awoke without any memories of labor or delivery (which most people back then considered a positive side effect). Furthermore, if the morphine administered was kept to a minimum, the mother was capable of hearing, speaking, and participating emotionally in the birthing process, although, owing to the effects of the scopolamine, she would have no memory of it. Even though the mother's memories of childbirth, if any, were vague, many women at the time saw twilight sleep as the most natural method of childbirth available. Because in twilight sleep the pelvic and uterine muscles are not paralyzed, as they are with ether and chloroform, the woman was able to assist in the birthing process and help push the baby out. Overall, twilight sleep made the entire birth process easier on the mother and the newborn, and women began to demand this method of childbirth as part of the growing women's rights and suffrage movement of the early 1900s.

Over the past several decades, childbirth has undergone nothing short of a total revolution as the emphasis has turned to natural or other methods of childbirth that reduce pain while allowing the woman to be awake for, actively participate in, and have a full recollection of the delivery of her baby. Current childbirth methods and pain-reduction techniques will be discussed in detail later in this chapter.

Deciding Whether or Not to Have a Child

9.1 Review the issues relating to a person's or couple's decision whether or not to have children.

Most people perceive that having children (sooner or later) is an important and necessary part of their lives. It is important to realize as well that becoming a parent is not a decision to be taken lightly. Having and raising children carries with it challenges and responsibilities that are, arguably, the greatest that we humans face in a lifetime. However, most parents will tell you without hesitation that, along with those challenges and responsibilities, parenting is also their greatest source of joy, love, pride, and meaning. Couples who desire children but have difficulty conceiving often experience turmoil and emotional pain, and some couples are unable to sustain their marriage after the discovery that they are infertile. A huge biotech industry known as "assisted reproductive technology" (ART) has developed to assist infertile couples to conceive (to be discussed later in this chapter). In addition, many couples turn to adoption (also discussed later) to create a family.

Since You Asked

1. My older sister (she's twenty-eight) really wants to get pregnant, but her husband is saying he's not ready to be a father. Is there any way she can convince him to agree?

How can you know if and when you are ready to have a child? The truth is, most people probably can't know, at least not for sure. Indeed, as mentioned at the beginning of this chapter, many individuals or couples are not even planning on becoming parents, at least not at that particular point in their lives, when they discover that a baby is on the way. Nevertheless, adults can ask themselves some important questions as they consider their future as parents. Some of these questions are discussed in "Self-Discovery: Are You Destined to Be a Parent?" You may find the questions and your answers to them quite revealing, whether you already have children or parenthood still seems far in the future.

Choosing *Not* to Have a Child

Is it OK to decide not to have children? Of course it is. Not everyone feels destined to be a parent. In fact, a small but increasing percentage of Americans are *choosing* not to have children. These individuals and couples are often fertile and biologically capable of conceiving, but they feel comfortable and content to live lives that do not include parenting. They often prefer to be called "childless by choice" or "child-free," rather than "childless," which tends to carry negative connotations.

Since You Asked

2. I feel that I really do not want to have children ever! Does that make me totally weird?

Among women of childbearing age in the United States, approximately 6–7% define themselves as "voluntarily childless" (CDC, "Fertility," 2012; Hollander, 2007). That's approximately four million women. Such a decision is often seen as a violation of expectations in a child-oriented society, and these women and couples may be viewed negatively by some (Gillespie, 2003; Mollen, 2006). Many working child-free couples (sometimes referred to as *DINKs*, for "Dual Income, No Kids") face subtle and sometimes not-so-subtle forms of discrimination, disapproval, and even punishment from various groups in their lives (Mollen, 2006; Roy, Schumm, & Britt, 2014). Some people view a decision to be child-free as a great loss, assuming that a life without kids is incomplete or lacking in some way. People who choose not to have children, however, view that decision very differently. Child-free couples cite many advantages of this important life decision. Consider for yourself some of the pros and cons of a child-free life in Table 9.1. (on page 320)

Self-Discovery

Are You Destined to Be a Parent?

If you are wondering if and/or when you might be ready to become a parent, here is a list of questions and considerations that may be helpful. Even if you already have children or if you know for sure you do not want them now (or ever), a little self-analysis never hurts. If you are in a relationship that you see as potentially long-term, you might want to go through these items with your partner.

Your Expectations about Yourself as a Parent

- **Do you currently spend time with children and teens? Do you enjoy it?** Whether you answer yes or no doesn't predict how you'll feel about your own children, but giving some thought to the issue can highlight some of your assumptions and attitudes about life with children.
- **What are your thoughts on the responsibilities and commitment of parenthood?** This question is just a way to help you reflect on the demands of parenting and whether you're comfortable with them. Parenthood is permanent; you can't just raise your kids during the "fun" years.
- **How do you cope with stress? Is your stressed-out self something you would want your child to witness?** Research shows that your level of stress can affect your children and your ability to parent effectively. If you feel you don't have a good handle on managing your stress, now is a great time to start learning some new coping mechanisms, regardless of your future as a parent.
- **What are your hopes and fears about parenthood?** Being a parent isn't all shared hugs and fits of giggling. You will have tough times and disappointments, and your children will not always meet your expectations. Some aspects of parenting can be frightening—it's a big responsibility. But voicing your fears and examining them now can help. Our own parents are typically the strongest models we have for raising children. Some of their lessons are positive and others negative. Examine your life with your parents, and think about what you can learn from their successes and shortcomings. Think about what you'd like to emulate from your own childhood and what you'd like to change or avoid altogether. If you're having trouble deciding whether you want children, it *may* have something to do with unresolved issues from your own childhood.

Your Values and Expectations about Life

These questions will help you pinpoint the personal attitudes and values you'll bring to the role of parent. They will also help identify differences that may exist between you and your current or future partner.

- **What values and morals would you like to pass on to your children? What attitudes would you want to be sure your child avoids?** This question helps you verbalize what you think is important to bring to the role of parent.
- **What are your priorities for your children? For example, do you want them all to have a college education? What are your expectations about their social lives, relationships, marriage, and careers?** We all come to parenthood with a set of expectations, often unspoken. This question helps you clarify your hopes and dreams for your children.
- **What are your thoughts about disciplining children? How strict or permissive do you think you would be as a parent? Why?** This is an area where many new parents are unprepared and partners often disagree. Thinking about these issues now will help you in planning strategies for setting limits, establishing acceptable and unacceptable behaviors, and developing consistent consequences for your child's actions.

Your Lifestyle and How It Will Change

Answering these questions will give you insight into the practical realities of your situation, which you should consider before taking on parenthood.

- **Talk to people you know who've decided not to have children; talk to others who've decided to have children. How does what they tell you make you feel?** This is not to suggest that you should base your decision on what others say, but hearing friends and relatives talk about their own parenthood choices can raise new issues for you to consider.
- **What does your support system look like?** You don't need a *whole* village to raise your child, but a few people to lean on can really help. Child rearing is difficult to do on your own. Do you have a partner or family and friends nearby to whom you can look for assistance? A circle of support isn't a prerequisite for parenthood, but it's a wonderful addition.
- **What do you do when you have free time? What will you do when you don't have any?** This is one of the practical realities of parenthood. Your needs, goals, and activities will become secondary to those of your child most of the time. He or she will become your number one priority almost all the time—are you ready for that?
- **How do you think your life will change?** If and when you have kids, it will. Big time. Most parents say it's for the better, but the effect on your time and energy can be enormous. Take a moment now to think seriously about the new life you are considering.

SOURCE: Adapted from "Evaluate Your Parenting Readiness" (2006). www.babycenter.com/0_evaluate-your-parenting-readiness_7311.bc.

Table 9.1 Advantages and Disadvantages of Choosing a Child-Free Life

Advantages	Disadvantages
Greater discretionary income	Potential feelings of loneliness and missing an important part of life
More free time	Possible lifelong conflict or regret over decision to be child-free
Greater education opportunities	Lack of child's unconditional love
Greater freedom to have second job	Lack of feeling needed by child
More time for hobbies and passions	Negative judgment by others (selfish, immature, unhappy)
No interference with career path	Disappointment of potential grandparents
More satisfying marriage*	Discrimination at work (being expected to do more than others due to "no kid" status)
Sense of not contributing to world overpopulation	Less ability to enjoy more flexible hours and possible shorter workweeks
Not subjecting children to dangers and perils of modern life	Lack of support when ill or elderly
Fewer worries in general	Difficulty finding other adults and couples for socializing
Calm and uncluttered home life	Lack of child tax credits
More time and energy for spouse	Tendency to become more rigid, less flexible
Better sex life	Risk of becoming overly independent and isolated from society at large
Opportunities for travel	More difficult search for self-knowledge without the "education" of having children
More time for charitable volunteer work	Greater chance of divorce*
Energy for spiritual and stress-relieving activities	
More time to explore personal goals and values	

*Child-free couples who stay together consistently report great marital satisfaction compared to traditional married couples, but divorce rates are higher among child-free couples, probably owing to the greater financial independence of both partners and the lack of perceived obligation to stay together "for the sake of the children."

SOURCES: "Childless by Choice" (2001); Connidis & McMullin (1999); Hollander (2007); Kopper & Smith (2001); La Mastro (2001); Mcquillan et al. (2008); Mollen (2006); Morell (2000).

The Influence of the Child's Sex

If you were deciding whether or when to have children, do you think you might be influenced by being able to *choose* the sex of the child? Just one girl? A boy first and then a girl? How about two girls and then a boy? What if you could take your pick? *Just drink the blue medicine to make a boy baby and the pink medicine for a girl.* Of course, this is just a fantasy. Prospective parents can't actually choose the sex of their children; they just have to take what they get. Or do they? Although a method of prenatal sex selection as simple as a pink or blue beverage does not exist (yet!), researchers have developed various laboratory and genetic techniques that allow parents to choose, with a high degree of success, the sex of their child *prior to pregnancy*. Also, as of 2010, a new, non-invasive blood test exists that can determine an embryo's sex as early as seven weeks after conception (Rattue, 2011).

The theory behind prenatal sex preselection is simple. The egg released from a woman's ovary contains only female (X) sex chromosomes. Each of the hundreds of millions of sperm in the man's ejaculate, however, may carry either X or Y chromosomes. An egg that is fertilized by an X-bearing sperm cell will be a girl (XX = female) and one fertilized by a Y-bearing sperm cell will be a boy (XY = male). Consequently, if there was a way to allow only X or only Y sperm cells to reach the ovum in the fallopian tube, the sex of the resulting child would be a sure thing.

It turns out that there is! Because of inherent DNA differences, X-bearing sperm cells are slightly larger than Y-bearing cells. Based primarily on this difference, techniques have been developed to separate X- and Y-bearing sperm cells in the lab. Once the sperm cells have been sorted, samples rich in the X or Y cells that will determine the desired sex of the future child are used to artificially inseminate the woman, or are incorporated in other assisted reproductive techniques (discussed later in this chapter).

Methods of prenatal selection are not perfect, but they have over 90% success rates in producing girl babies and over 80% for boys (Karabinus, 2009). In another sex-selection method, called preimplantation genetic diagnosis, couples use in vitro fertilization to create embryos in the lab. Then the embryos' DNA is analyzed for any medical abnormalities and for sex. The embryos may then be sorted into males and females, and only the ones of the desired sex are implanted into the uterus (Bumgarner, 2007; Wells & Fragouli, 2013).

As you probably guessed, sex preselection technology is quite controversial and provides much material for debate (Dickens, 2002; Hollingsworth, 2005; Puri & Nachtigal, 2010). In purely medical terms, however, some very serious and even deadly inherited diseases are linked to the X and Y chromosomes, and these risks may be reduced through sex-selection technologies. Beyond medical considerations, many parents feel that the procedures provide them with the opportunity to achieve the desired gender makeup of their family: to have all of one sex, to create a balance between the sexes of their children, or whatever combination they prefer. But what are the ethical and social considerations of this technology? These are discussed in "Sexuality, Ethics, and the Law: Sex-Preselection Technology."

Sexuality, Ethics, and the Law

Sex-Preselection Technology

The worldwide social and ethical considerations in the face of increasingly available methods for allowing prospective parents to select the sex of their children are disturbing. What, for example, might these procedures mean for male-female sex ratios in countries such as China, India, or the Middle East, where boy babies are highly prized and preferred over girls? The effect of China's one-child policy has been to increase the desire for male infants to the point that hundreds of thousands, and perhaps millions, of baby girls are secretly abandoned to orphanages or even killed every year so that the parents may try again for a boy (Banister, 2004; Lubman, 2000). Researchers estimate that over the next 20 years, these countries will see an excess of as much as 20% more young men than women. This has the potential effects of fewer men marrying, a basic cultural expectation in these cultures, leading to social exclusion of more men, increased violence, and possibly an increase in prostitution and other crimes (Hesketh, Lu, & Xing, 2011).

Recognizing the future sociological implications of infant sex selection, China and India have passed laws prohibiting abortions based on parental sex preferences (Mudur, 2002). In addition, China has begun to restrict assisted fertility clinics that may be engaged in prepregnancy sex-selection technologies and to reward families monetarily who have only one son or who have no son but have and keep two daughters (Bumgarner, 2007; Glenn, 2004; Kalb, Nadeau, & Schafer, 2004). Although the technology allowing for the *preselection* of the sex of children (prior to conception) may help reduce sex-based abortions, the consequences of such a choice to the population of countries that place a high value on male offspring could be devastating to the gender balance in those countries in future decades.

Moreover, the social and scientific ethical considerations of sex preselection are far reaching. Even in the medical profession in the U.S., a great deal of disagreement exists (see Puri & Nachtigall, 2010). A survey of primary care physicians (PCPs) and physicians who are sex-selection technology providers (SSTPs), found that PCPs were concerned about how sex selection leads to medically unnecessary expensive and invasive medical procedures, how it fuels the fire of gender stereotyping and discrimination, and how it leads to child neglect of children who are the "wrong" sex. SSTPs believed that sex selection is a reproductive right that women should have, that it allows parents to plan for family sex balance, and helps prevent unwanted pregnancies and sex-based abortions.

This example demonstrates how the issue of sex-preselection is extremely complex and will likely remain so for the foreseeable future because it is here to stay. Individuals, couples, cultures, and countries must approach the issues in their own way and from their own legal, moral, and ethical perspectives. As with so many new and emerging technologies, scientific advances in human sexuality may be used for positive or negative ends, and the choices we make must therefore be as informed and accurate as possible.

Conception

9.2 Summarize the process of conception from ovulation and sperm production to fertilization of the ovum and the development of the zygote.

Conception requires a single sperm cell to penetrate and fertilize an ovum (egg). Sounds pretty basic, doesn't it? And in many ways it is: after all, it may be the single most important event in the survival of the human species. However, as you will see, the journey of that sperm and that ovum prior to reaching the moment of conception is amazing and complex. Our story begins with the woman, before she is born.

Ovulation: The Ovum's Journey

When a female fetus is born, her ovaries already contain approximately one to two million immature eggs (called **oocytes**). As the girl develops through childhood, the number of oocytes in her ovaries decreases until puberty, when the first mature ovum is released, at which point her ovaries contain approximately 400,000 oocytes. This turns out to be hundreds of thousands more than a woman will actually release in her lifetime, but evolution is taking no chances with reproduction: Most women today will have approximately 35 to 40 fertile years between the onset of puberty and menopause (when the release of ova and fertility ceases). Releasing one ovum each month (as is typical) amounts to a lifetime total of 400 to 500 ova released from a woman's ovaries.

oocyte
An immature reproductive egg, or ovum.

menarche
The beginning of menstruation during puberty; a girl's first period.

follicular phase
The early period during a woman's monthly fertility cycle when the pituitary gland secretes *follicle-stimulating hormone* (FSN) to enhance ovum development.

During puberty, a set of preprogrammed hormonal changes occur in a girl's body, triggering **menarche** (pronounced MEN-ark-y), her first menstrual period, and, soon after, the beginning of ovulation, the release of one ovum each month. (Puberty and menarche are discussed in greater detail in Chapter 12, "Sexual Development Throughout Life.")

Since You Asked

3. After a woman's period, how soon is the next egg released, and when is she most fertile?

Following the onset of a woman's menstrual period each month, a combination of hormones released over the next two weeks promotes the growth of *ovarian follicles*, small sacs on the surface of the ovaries, each of which contains a developing ovum. (A more thorough explanation of menstruation and the menstrual cycle may be found in Chapter 2, "Sexual Anatomy," and Chapter 5, "Contraception: Planning and Preventing Pregnancy.") As this **follicular phase** of her fertility cycle begins, the pituitary gland at the base of the brain begins to secrete *follicle-stimulating hormone* (FSH), which further enhances follicle growth and ovum development. The many ovum-containing follicles that have been growing now release their own hormone, *estradiol*, a powerful form of estrogen (see Table 9.2 for a summary of hormone production during a woman's fertility cycle). At this point, the follicles are in a kind of "biological competition" because only one follicle will eventually rupture and release the ovum into the fallopian tube for possible fertilization (more than one ovum may be released in relatively rare cases, resulting, if fertilized, in fraternal twins or multiple births).

Figure 9.1 A Mature Ovarian Follicle

As an ovum matures, its follicle on the wall of the ovary may grow to nearly three-quarters of an inch in diameter.

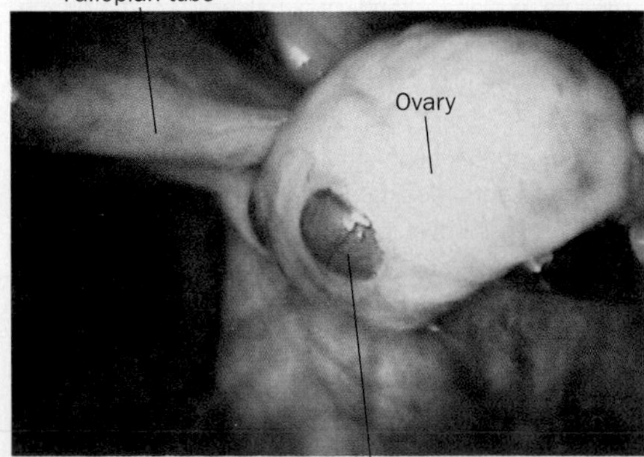

By about a week after the first day of a woman's period, the follicle producing the most estrogen receives an extra dose of FSH, causing it to become the "dominant follicle," while the others deteriorate. This dominant follicle then continues to grow, produces increasing amounts of estrogen, forms a pouch that looks somewhat like a water-filled blister on the ovarian wall, and will grow to nearly 0.75 inch in diameter just prior to rupturing and releasing the ovum (see Figure 9.1).

The continuing rise in estrogen levels signals a further increase in secretion of FSH, along with the release of *luteinizing hormone* (LH). These two hormones now combine to cause the wall of the follicle at the edge of the ovary to break down, and about a day later, the follicle ruptures, releasing the ovum into the abdominal cavity. This is the precise moment of ovulation. A human ovum is the largest cell in the

Table 9.2 Hormonal Changes in Women During a Typical Fertility (Menstrual) Cycle

Menstrual Phases	Days in Phase	Hormonal Actions
Follicular (proliferative) phase	Days 1–6: Beginning of menstruation to end of blood flow Days 7–13: Endometrium (inner lining of the uterus) thickens to prepare for egg implantation	Estrogen and progesterone start out at their lowest levels. Follicle-stimulating hormone (FSH) levels rise. Ovaries start producing estrogen and levels rise, while progesterone remains low.
Ovulation	Day 14	Level of luteinizing hormone (LH) surges. Largest follicle on ovary bursts and releases egg into fallopian tube.
Luteal (secretory) phase, also known as the premenstrual phase	Days 15–28	Ruptured follicle develops into corpus luteum, which produces progesterone. Progesterone and estrogen stimulate a blanket of blood vessels to prepare for egg implantation.
If fertilization occurs:	Days 15–28	Human chorionic gonadotropin (hCG) hormone begins to increase in the woman's body. Fertilized egg attaches to blanket of blood vessels and becomes placenta.
If fertilization does not occur:	Days 15–28	Corpus luteum continues to produce estrogen and progesterone. Corpus luteum deteriorates. Estrogen and progesterone levels drop. Blood vessel lining sloughs off, and menstruation begins.

human body: it measures 0.004 inches across (about the size of a sharp pencil point) and is visible to the naked eye (see Figure 9.2).

As explained in Chapter 2, "Sexual Anatomy," the fallopian tubes are not physically connected to the ovaries; rather, the strand-like structures (the *fimbriae*) at the ovarian end of the tubes rest adjacent to and against the ovaries. The back-and-forth motion of these fimbriae set up a current in the abdominal fluid that draws the newly released ovum into the fallopian tube (see Figure 9.2). Once the ovum has entered the fallopian tube, the woman is "officially" fertile. For a natural pregnancy to occur, sperm cells and ovum must meet here, and one sperm must penetrate the ovum's outer membrane during its brief journey along the upper region of the fallopian tube (more on this shortly).

After the ovum is released, estrogen production decreases and progesterone and LH secretions begin to rise. Over the next 14 days of a woman's cycle, called the **luteal phase**, the growth of new follicles in the ovaries is suppressed, and the lining of the uterus (the *endometrium*) thickens with tissue and blood in preparation to receive and nourish a fertilized ovum.

luteal phase
The later period of a woman's monthly fertility cycle, when the lining of the uterus thickens in preparation for receiving a fertilized ovum if conception has occurred.

If no fertilization occurs, changes in hormone secretions cause a breakdown in the lining of the endometrium, which then drains, typically over several days, through the cervix and out the vagina (see Figure 9.3). This draining of the uterine lining and fluid is what is commonly known as menstruation: a woman's period.

Figure 9.2 Ovulation

Ovulation occurs when the ovarian follicle containing a mature ovum ruptures and releases the ovum toward the opening of the fallopian tube.

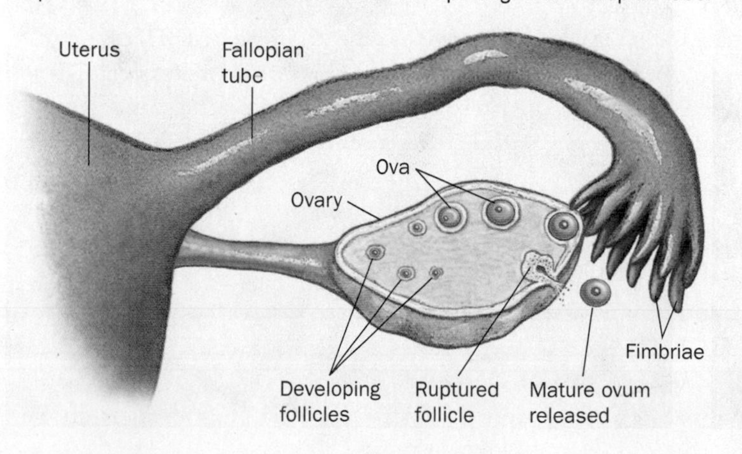

Figure 9.3 Menstruation

If fertilization of the ovum does not occur, the lining of the uterus is shed as menstrual fluid during a woman's period.

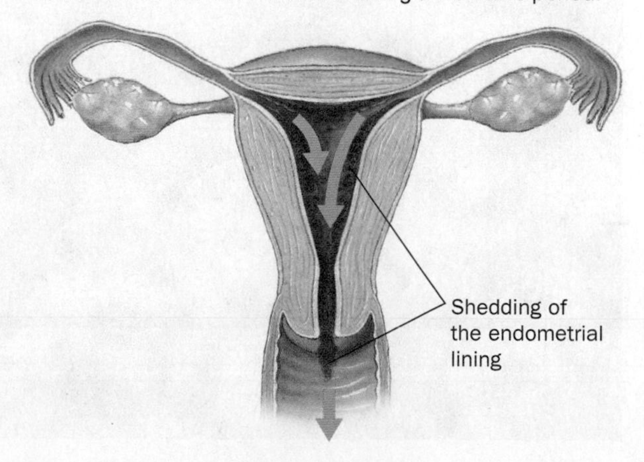

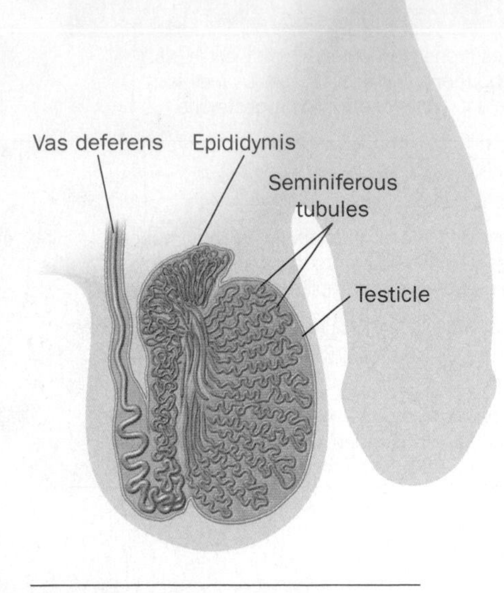

Figure 9.4 Tubules in the Testicle

Each testicle contains hundreds of feet of tightly coiled seminiferous ("seed-bearing") tubules.

seminiferous tubules

Tightly wound microscopic tubes that comprise the testicles in the male, where sperm cells are generated.

The Sperm's Journey

Unlike girls, boys are not born with a lifetime supply of reproductive cells. Upon reaching puberty, boys normally begin to produce sperm cells in astonishingly large numbers and continue to produce them throughout their lives. Sperm cells are produced in the testicles. As covered in Chaper 2, "Sexual Anatomy," the testicles are made of tightly packed **seminiferous tubules**, literally "seed-bearing tubes" (see Figure 9.4). When these tubules are stimulated by male secretions of follicle-stimulating hormone (FSH, which is produced by the male system just as by the female) and testosterone, millions of primitive cells that line the tubules begin to develop into sperm cells. These cells then migrate from the testicles to the epididymis, the structure attached to the back side of each testicle (see Figure 9.5). Once in the epididymis, the sperm cells mature and develop their ability to swim.

The length of time needed for the sperm cells to develop from formation in the tubules to migration to the epididymis and, finally, to ejaculation is approximately 68 to 72 days. In other words, the sperm cells that are beginning to develop in a man's testicles today will not be ready to be ejaculated and, potentially, fertilize an ovum for ten weeks.

Sperm cells are minuscule, among the smallest cells in the human body. And compared with the ovum, they are appear quite scrawny. What sperm lack in size, however, they make up for in sheer numbers. In normal adult men, the testicles produce about 50,000 sperm cells per *minute*; the number of sperm cells in a single ejaculate of a fertile man ranges from approximately 30 million to 200 million per milliliter (cubic centimeter) of semen, with an average of about 60 to 70 million (see Figure 9.6). The normal volume of semen ejaculated ranges from 1.5 to 5.0 milliliters (a teaspoon holds about 5 milliliters). This means that a single ejaculation typically contains somewhere between 45 million and 800 million sperm cells. The average is between 200 and 400 million. The sperm cells themselves make up only about 5% of the total volume of the ejaculate (Zieve & Miller, 2011).

Why so many sperm? This is a very good question, especially when you consider that an ovum is fertilized by only a single sperm. Probably the most important reason is that, although sperm are numerous, they are not particularly hardy. Some may be abnormally formed, some may become trapped among other body cells, some may be

Figure 9.5 Sperm Cells in the Epididymis

Sperm are transported from the testicle to the epididymis, where they continue to mature at varying rates in preparation for ejaculation and potential fertilization of an ovum.

Figure 9.6 Sperm Cells in the Ejaculate

The typical male ejaculate contains an average of around 400 million sperm cells.

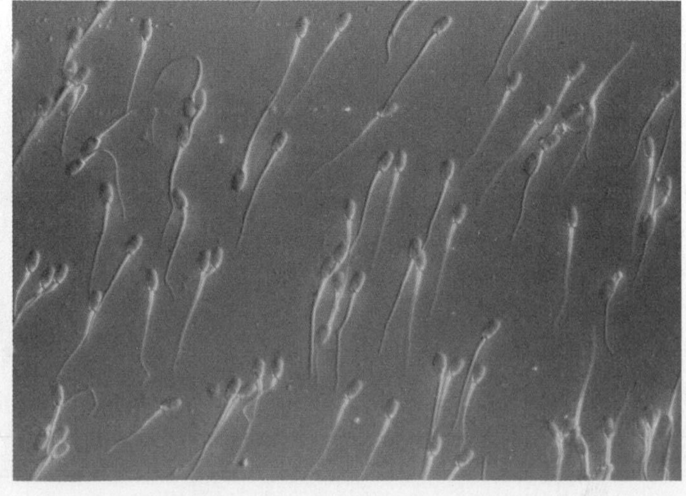

attacked by antibodies in the woman's body, many will die in the normal acidic environment of the vagina, and many just don't have the necessary stamina to make the 3- to 4-inch trip (a long way for a microscopic sperm) from the vagina, through the cervix, into the uterus, and on into the fallopian tubes. Therefore, to help ensure the survival of the human species, we have evolved in such a way that the male produces more than enough reproductive cells.

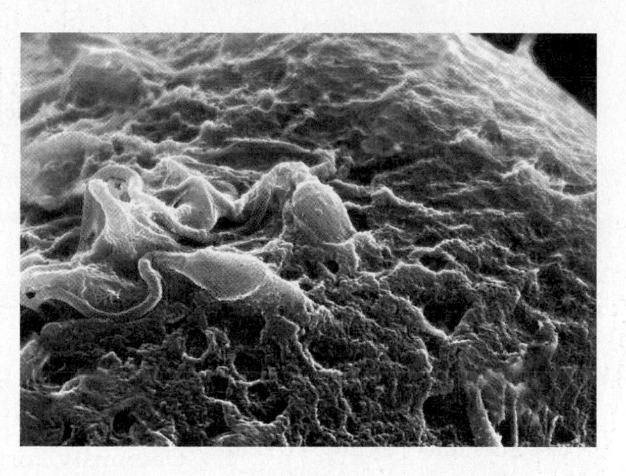

Figure 9.7 Sperm Cells on the Surface of the Ovum

Only one sperm cell will be allowed to penetrate the ovum's membrane.

The Ovum–Sperm "Rendezvous"

Assume that the ovum is in the fallopian tube and the sperm cells have recently been ejaculated into the vagina. The ovum is gently whisked along the tube by *cilia*, tiny hair-like structures that create a current in the *intrafallopian fluid*. The ovum can be fertilized during the first 12 to 18 hours after entry into the tube (after which it is no longer viable), so the best chance for conception occurs when sperm cells are already high up in the fallopian tube when the ovum arrives. On average, sperm cells survive in the female reproductive tract for about two to five days, so the average *couple* is fertile for approximately six to seven days each month. In other words, if ejaculation occurs in the vagina from four days before to approximately one day after the moment of ovulation, live sperm cells may have an opportunity to encounter the ovum in the fallopian tube. This does not, in itself, guarantee conception. In fact, the odds of conception are only about one in five from a single act of intercourse and only one in three when a couple is *trying* to become pregnant and the timing of intercourse is right.

Moreover, only about a hundred to a thousand of all those hundreds of millions of sperm manage to make it all the way to the upper portion of the fallopian tubes where fertilization occurs. On their way there, the sperm cells have undergone physical and chemical changes that make them capable of penetrating the outer membranes of the ovum by releasing a mix of chemicals that literally dissolves the ovum's layers of membranes (Silber, 2007). These survivor sperm then surround the ovum in a sort of competition to be the first to penetrate and fuse their chromosomes with those of the ovum (see Figure 9.7). Immediately after the first sperm successfully penetrates the ovum's outer membrane, biochemical changes triggered in the ovum create a barrier that effectively blocks any other sperm cells from entering because only one sperm cell's DNA can enter and fuse with the DNA contained in the ovum.

zygote

A fertilized ovum (or egg) moving down the fallopian tube.

Since You Asked

4. Last week, I took two pregnancy tests that were both negative. But I still haven't started my period. Could I be pregnant?

Upon successful fertilization, the DNA of the man's sperm and the woman's ovum join together, and a new organism, combining the genetic material of both, is created. Within hours, this organism, now called a **zygote** (see Figure 9.8), divides into two cells, then four, and eight, and so on, as it continues its week-long journey along the remainder of the fallopian tube to the uterus (see Nihira, 2009). Immediately upon fertilization, a hormone called *human chorionic gonadotropin* (hCG) begins to increase in the woman's body. Within about two weeks after conception, this hormone has reached high enough levels that conception can usually be determined using a standard home pregnancy test that measures the level of the hormone in the woman's urine. Because the level of hCG continues to increase day by day, the test becomes more reliable each day following the initial two weeks.

Figure 9.8 Zygote Moving Through the Fallopian Tube

The fertilized egg, now known as a *zygote*, begins to divide as it travels along the fallopian tube.

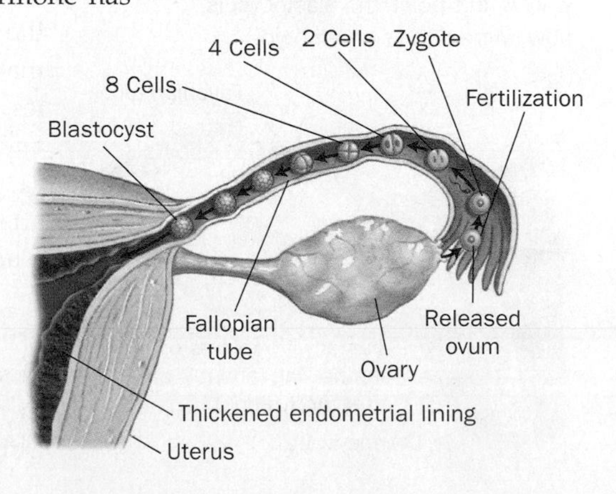

However, the woman is not pregnant, yet. Although *conception* may have occurred, *pregnancy* is not established until about a week after conception when the zygote divides into many cells (called a *blastocyst*) and implants into the lining of the uteruine wall (this will be discussed in the next section). Therefore, what we call a *pregnancy* test is often, in reality, a *conception* test. If a woman has a positive test result, she should assume that she has *conceived* and may be, or become, pregnant until proven otherwise through a doctor visit or additional tests. She should not use alcohol or other drugs, be sure to take a prenatal vitamin containing folic acid (an important B-vitamin for normal fetal development), and stop smoking. If a woman is planning on conceiving or is not taking steps to prevent an unwanted pregnancy, she should engage in a healthy lifestyle and avoid activities that are dangerous to a fetus (such as ingesting alcohol and all other drugs) *before* conception occurs.

blastocyst
The developing zygote, with cells surrounding a fluid-filled core, upon entering the uterus and before implanting in the uterine wall.

embryo
A blastocyst that has implanted in the uterine wall.

pregnancy
The period of growth of the embryo and fetus in the uterus.

placenta
An organ that develops on the uterine wall during pregnancy and joins the developing embryo to the mother's biological systems, transferring nourishment, oxygen, and waste products between the fetus and the mother.

umbilical cord
A structure approximately 22 inches in length, consisting of one large vein and two arteries that transport nutrients, oxygen, and fetal waste products back and forth between the fetus and the placenta.

Pregnancy

9.3 Describe the development of the embryo and fetus from implantation in the uterine wall through the three trimesters of pregnancy.

Obviously, for all or nearly all of the 400 to 500 fertility cycles in an average woman's fertile lifetime, no pregnancy will happen. As mentioned earlier, if the ovum is not fertilized by a sperm cell, progesterone and estrogen levels diminish during the second half of a woman's cycle, blood flow to the uterus decreases, and the lining of the uterus is shed during the woman's period (this is what comprises her menstrual fluid).

If fertilization has occurred, however, the lining of the uterine wall, the *endometrium,* signaled by hormones, continues to thicken in preparation for the arrival of the zygote. About a week after fertilization, the zygote enters the uterus. It has already divided many times and consists of approximately 50–100 cells surrounding a fluid-filled core. It is now called a **blastocyst**. On the sixth or seventh day after fertilization, assuming that the endometrium lining has formed properly, the blastocyst may implant into the wall of the uterus. Once implantation occurs, the blastocyst is referred to as an **embryo**. The blastocyst's successful attachment to the uterine lining is the beginning of **pregnancy** (see Figure 9.9). Many people confuse conception and pregnancy, although they are distinct, separate events. As discussed previously, conception occurs when a sperm cell penetrates an ovum and the genetic material from each parent merges. Then, if all proceeds normally and successfully, pregnancy begins when the fertilized ovum implants into the wall of the uterus. This process is shown in Figure 9.10.

As pregnancy is established, a temporary organ called the **placenta** begins to develop on the uterine wall adjacent to the embryo. The placenta contains a rich supply of blood vessels and is, in essence, the life-support system that biologically unites the developing embryo to the mother. As it grows along with the embryo, the placenta serves as a "transfer station" for nourishment and oxygen from the mother's body to the fetus and for waste products from the fetus back to the mother's bloodstream for disposal. During childbirth, the placenta, often referred to as the *afterbirth,* is typically expelled from the uterus following the newborn and (usually) examined for any abnormalities that could signal problems for the infant.

Figure 9.9 Blastocyst Implanting on Uterine Wall

Pregnancy begins when the developing zygote has left the fallopian tube, grown to become a blastocyst, and embedded itself in the uterine wall. At this point, the blastocyst is now referred to as an embryo.

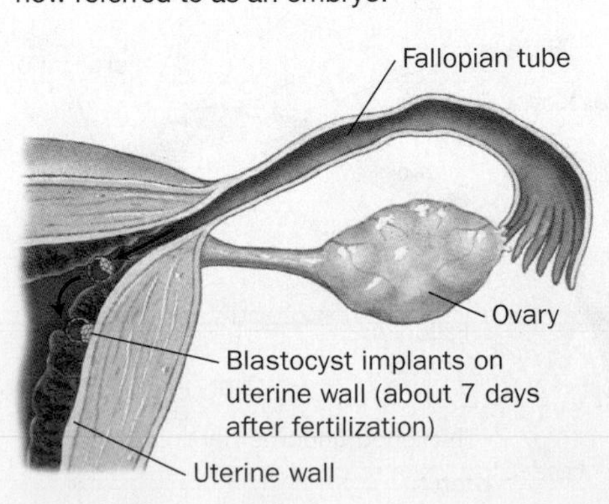

Throughout pregnancy, the placenta is attached to the abdominal wall of the fetus by the **umbilical cord**, which reaches an average length of approximately 22 inches during pregnancy and consists of one large vein and two arteries. Its function is to transport nutrients, oxygen, and fetal waste products back and forth between the fetus and the placenta. When a baby is born, the umbilical cord is tied off near the baby's belly and severed. The small portion still attached to the infant shrinks and detaches within about two weeks, leaving a scar we all know as the navel.

Assuming a normal, healthy pregnancy, the embryo will now begin to grow in predictable stages over the next nine months (Bainbridge, 2003; "Fetal Development," 2005; Nihira, 2009). The photographs and brief explanations in this section highlight the remarkable process of the transformation from a tiny cluster of cells to the birth of a new life. For the sake of discussion,

Figure 9.10

From conception to Pregnancy. Conception and pregnancy are separate events occurring about a week apart.

CONCEPTION (FERTILIZATION)

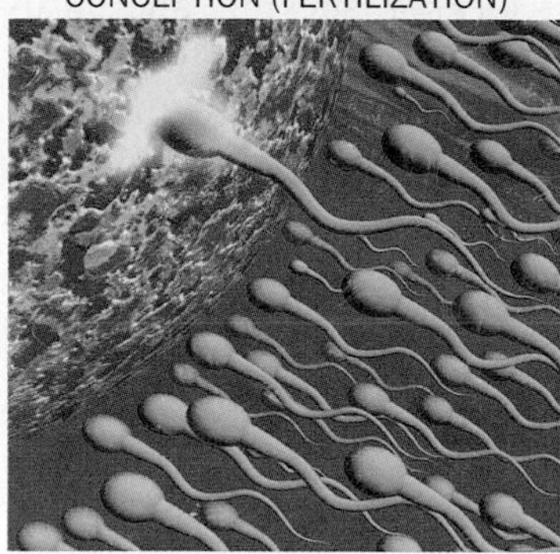

A single sperm cell penetrates the wall of the ovum and DNA from the ovum and sperm fuse—this is *conception.*

FERTILIZED OVUM DIVIDING

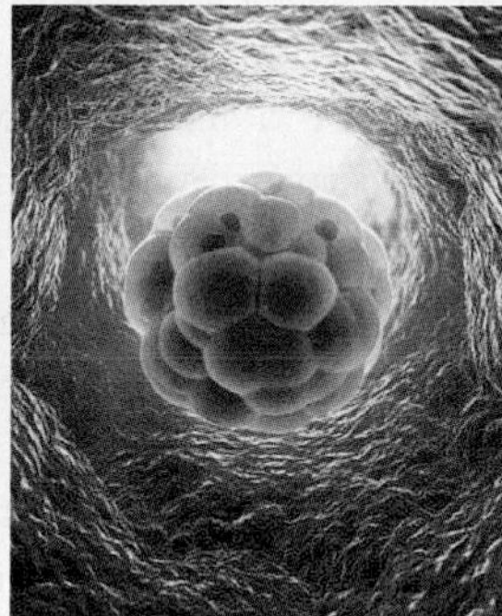

The fertilized ovum (the *zygote*) travels through the fallopian tube toward the uterus and continues to divide for seven to ten days.

PREGNANCY BEGINS

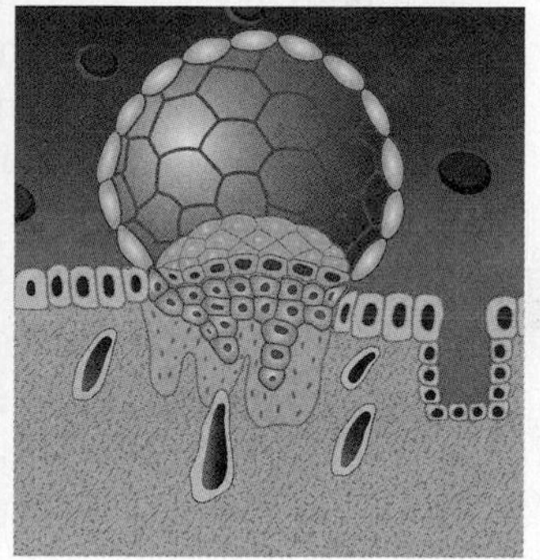

The fertilized ovum which has divided into 50–100 cells (now called a *blastocyst*) enters the uterus and implants into the thickened uterine wall; this is the start of *pregnancy.*

the 40 weeks of a full-term pregnancy are often divided into phases of about three months each, called **trimesters**.

trimester
One of three periods of about three months each that make up the phases of a full-term pregnancy.

First Trimester

The first three months of pregnancy begin with the **embryonic period**: the initial eight weeks following implantation of the zygote in the wall of the uterus. At the end of this eight-week period, the embryo becomes known as a **fetus**. This first trimester is a critical time for the developing embryo. During these months, the woman is at highest risk of *miscarriage* (also referred to as *spontaneous abortion*) owing to various genetic, structural, hormonal, or environmental causes. Approximately 15% of *confirmed* pregnancies (women who know they are pregnant) end in miscarriage ("Research on Miscarriage," 2014). You will find a more complete discussion of miscarriages a bit later in this chapter. Barring miscarriage or early voluntary termination of pregnancy, here are some of the milestones reached during this first trimester.

embryonic period
The initial eight weeks of pregnancy following fertilization.

fetus
An embryo after eight weeks of pregnancy.

- Month 1: The embryo resembles a tadpole or tiny shrimp; the buds of arms and legs begin to appear; the heart begins to beat; the embryo is about the size of the eraser on a pencil and weighs less than 1 ounce.
- Month 2: The embryo grows to about 1 inch long; distinct fingers and toes appear; the heart muscle has divided into chambers; facial features are visible; the placenta is functioning, bringing nutrients to the embryo and carrying wastes away; the embryo takes on a more human appearance.
- Month 3: The embryo, now called a fetus, grows to about 3 inches in length; fingers and toes have nails; the fetus begins to move in the uterus, but movements cannot usually be felt by the mother; all vital organs and muscles are formed and functioning by the tenth week. The fetus now weighs about 1.2 ounces (the equivalent of a dozen U.S. pennies).

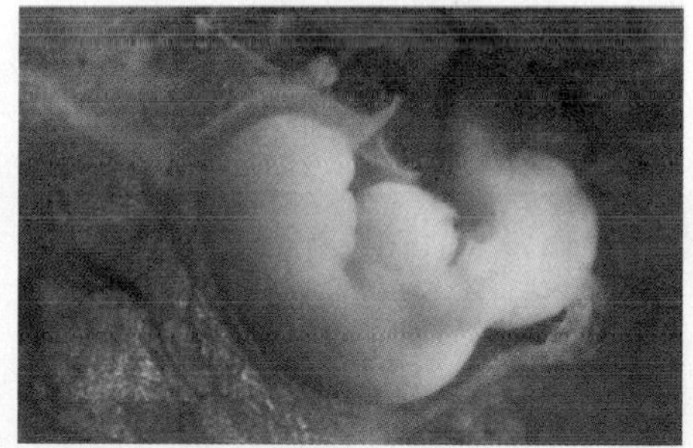

Embryo at one month.

Fetus at three months.

Second Trimester

Usually, by the fourth month the pregnancy is firmly established and the risk of miscarriage decreases significantly. It is early in this second trimester that

quickening
The first movement of the fetus that is felt by the mother.

the mother will begin to feel the fetus move, which is called **quickening**. During these months, the growth of the fetus accelerates greatly. Here are some of the developmental changes during this phase.

- Month 4: Fetal movement increases and includes kicking, sucking, and swallowing; about 20 tooth buds develop; fingers and toes are more clearly defined; the sex of the fetus may now be determined by observing the growth of the genitals with ultrasound. The fetus grows to 6 or 7 inches long and now weighs around 6 or 7 ounces.

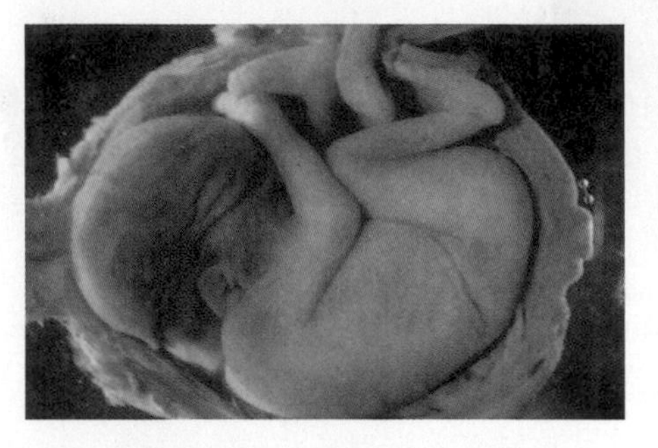

Fetus at five months.

- Month 5: Fetal activity increases further and may even include rolls, flips, and somersaults; the fetus enters a predictable waking–sleeping cycle; the mother will feel frequent unmistakable movements of the fetus; fetal eyelashes and eyelids appear; growth accelerates to 8 to 10 inches long; and weight increases to nearly a pound by the end of the fifth month.
- Month 6: Fetal skin is red and wrinkled and covered with a protective coating called the *vernix*; the eyes open and close; the fetus's lungs are "breathing" the amniotic fluid in and out; the fetus is capable of hearing sounds from the outside world; rapid growth continues as the fetus reaches 11 to 14 inches in length and 1.5 to 2 pounds. If born in this month, the baby, though very premature, will have a good chance of survival with specialized intensive care (see the discussion of premature births later in the chapter).

Fetus at seven months.

Third Trimester

The last three months of pregnancy are marked by a dazzling increase in the rate of growth of the fetus. During the third trimester, the fetus will typically add up to a foot in length and at least 7 more pounds. Also, changes will accelerate that prepare the baby for birth, including internal organ maturation, layers of fat, muscle tone, and rotation into a head-down position in the womb. Here are the major events as birth approaches.

- Month 7: The fetus's taste buds have developed and it may be observed sucking its thumb; muscle tone develops through kicking and stretching; clear responses to external sounds may be observed; layers of fat are forming beneath the skin; internal organs are maturing quickly. If born prematurely at this time, the baby has a high chance of survival and a normal life with proper medical care. The fetus reaches 14 to 16 inches long and about 3 pounds by the end of the seventh month.
- Month 8: The fetus rotates to head-down position; the brain grows quickly; overall growth is now at a rate of a half pound per week; the fetus's kicks and elbow thrusts now become visible from the outside surface of the mother's abdomen; the fetus's internal organs, except for the lungs, are nearly fully developed; the skull bones are flexible and not yet connected to allow them to compress for passage through the birth canal; size is approximately 18 inches long and weight is 4 to 6 pounds at the end of this month.
- Month 9: The fetal lungs develop fully and are ready to breathe air; the skin typically becomes pink and smooth; the fetus now turns and settles head down, low in the mother's uterus, in preparation for birth; activity decreases, producing fewer noticeable movements; on average, the fetus reaches 18 to 22 inches in length and 6 to 9 pounds as the onset of labor approaches (labor and birth will be discussed shortly).

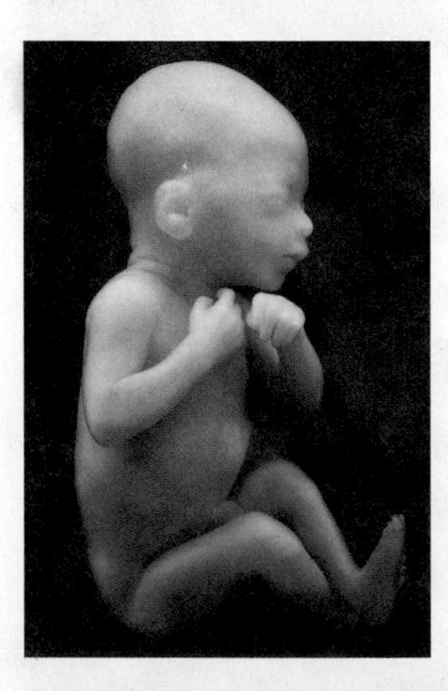

Nearly full-term fetus (eight to nine months).

Potential Problems in Pregnancy

9.4 List and discuss the medical problems that may occur during pregnancy.

The vast majority of pregnancies proceed virtually trouble-free from conception to birth, except for some common discomforts for the expectant mother, such as nausea and vomiting, backache, swollen feet and ankles, and hemorrhoids. Nonetheless, these "normal" discomforts can make for a difficult eight to nine months for many women, and the burdens on her body and overall wellbeing should not be ignored. Table 9.3 summarizes some of these effects on the expectant mother, along with some causes and possible remedies. As you can see from the table, these typical problems that may occur during pregnancy usually do not threaten the eventual wellbeing or survival of either mother or fetus. Much less frequently, serious and

Table 9.3 Common Discomforts of Pregnancy and Some Remedies

Many women proceed through pregnancy with very little distress and few problems. However, as the body changes to accommodate a new human life growing within it, certain discomforts are bound to occur. Here is a list of the more common discomforts associated with pregnancy for the woman, along with some possible causes and suggested remedies.

Symptom	Cause	Remedies	Situations to Avoid
Nausea and vomiting	Common discomforts of early pregnancy, affecting between 50% and 80% of all pregnant women. Although often referred to as *morning sickness*, nausea and vomiting can occur at any time and may persist throughout the day.	Increase amount of sleep. Avoid triggering odors. Exercise more. Eat smaller amounts of food more often. Take more vitamin B_6 or ginger (which combats nausea).	Some women may need medical support to avoid severe dehydration. They should contact their doctor if they experience persistent vomiting and are unable to keep any fluids down for more than 24 hours.
Leg cramps	Caused by poor circulation, not enough calcium, pressure on nerves.	Increase calcium in the diet. Try a calcium supplement like Tums. Stretch the calf muscles by flexing the foot to relieve cramps. Apply a warm cloth or heating pad to leg muscles. Put feet up.	Avoid too much phosphorus in foods (phosphorus is linked to the depletion of calcium).
Breast tenderness	Hormonal changes are preparing milk ducts behind the nipples. Soreness will probably continue through first trimester.	Increasing bra size by one or more sizes and wearing support bras.	Avoid conditions that will cause excessive breast tissue disturbance.
Backache	Caused by strain of increased uterine weight on back muscles and ligaments, made worse by poor posture.	Stand straight. Wear flat shoes with arches. Sleep on a firm mattress. Try pelvic-rocking exercises while on hands and knees. A maternity belt or sling might help as well.	Avoid high heels; avoid becoming overly fatigued or overexertion using the back muscles.
Shortness of breath	Pressure of the growing uterus on the diaphragm; anemia can exacerbate.	Stand up straight. Sleep with extra pillows to prop upper body up.	Quit smoking. Do not exercise to the point of fatigue.
Varicose leg veins	Progesterone, one of the pregnancy hormones, causes the veins to dilate or relax and not return blood to the heart as efficiently.	Elevate legs frequently. Walk daily. Put on support hose immediately upon awakening.	Avoid standing for long periods, sitting with crossed legs, and knee socks with tight tops.
Sleeplessness	Usually in the last months, caused by difficulty getting comfortable, frequent trips to the bathroom, worries, and the baby's movements.	Take a warm bath. Drink warm milk or chamomile tea at bedtime. Increase B vitamins by eating whole grains. Ask someone for a relaxing back rub. Lavender aroma may help with sleep.	Do not take sleeping pills or tranquilizers; avoid caffeinated drinks such as coffee, tea, and colas.
Frequent urination	Due to the fetus (even when still quite small) exerting pressure on the bladder.	Do not avoid fluids; they are important. Caffeine products can increase frequency of urination. Pregnant women should go to the bathroom when they feel the need and not hold it in.	Avoid situations in which bathroom facilities will be unavailable for long periods of time.

SOURCES: Adapted from Chang (2010); Obstetrics-Gynecology-Infertility Group (2008).

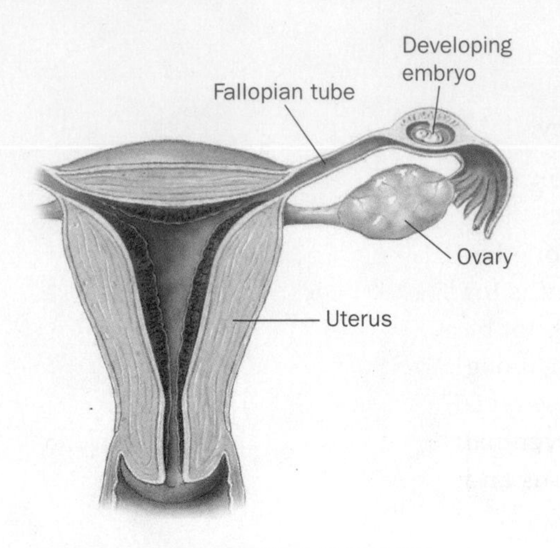

Figure 9.11 Ectopic (Tubal) Pregnancy

An ectopic pregnancy occurs when the zygote implants outside the uterus, usually in the fallopian tube.

ectopic pregnancy

A pregnancy complication in which a fertilized ovum attaches and begins to grow outside the uterus, most commonly in the fallopian tube; also called a *tubal pregnancy*.

even life-threatening difficulties may arise during pregnancy. Here, we will examine some of the more serious problems of pregnancy and then look at some of the prenatal tests that can be used to screen for genetic or physical defects.

Ectopic Pregnancy

An **ectopic pregnancy** (*ectopic* means "out of place") occurs when a zygote implants somewhere in the woman's body other than the uterus. Nearly all ectopic pregnancies occur in one of the fallopian tubes, where fertilization has just taken place, and are often referred to as *tubal pregnancies* ("Ectopic Pregnancy," 2009; Vorvic et al., 2010). Invariably, ectopic pregnancies are short-lived and a health risk for the woman because, as you might imagine, the fallopian tube, although ideal for conception, is far too small for the developing embryo (see Figure 9.11).

Tubal pregnancies carry a high risk of rupture of the fallopian tube, causing internal bleeding, hemorrhaging, and even death of the pregnant woman if not diagnosed and treated quickly. Ectopic pregnancies occur in approximately 1% to 2% of all pregnancies and accounts for 0.1% of pregnancy-related deaths in the United States (Vorvic et al., 2010).

Among the most common causes of ectopic pregnancy is infection with one or more of various sexually transmitted infections, especially chlamydia and gonorrhea. As discussed in Chapter 8, "Sexually Transmitted Infections/Diseases," these bacteria, if not properly treated, may lead to pelvic inflammatory disease (PID). In some women, PID may damage a woman's fallopian tubes, limiting passage of a fertilized ovum and increasing the risk of ectopic pregnancy (CDC, "PID," 2014; Bakken et al., 2007).

An ectopic pregnancy often produces all the signs and symptoms of a normal pregnancy, including absence of the menstrual period, nausea, and positive results on a pregnancy test. Six to eight weeks after the missed period, however, as the zygote grows in the fallopian tube, symptoms begin to appear, most commonly vaginal bleeding and abdominal pain ("Ectopic Pregnancy," 2009). Usually, a physician will be able to diagnose a tubal pregnancy using a combination of physical examination, hormonal blood tests, and ultrasound imaging techniques.

A common treatment for ectopic pregnancies is a single injection of a medication called *methotrexate*, which inhibits cell growth in the zygote and allows it to be expelled via the fallopian tube, uterus, and vagina. This treatment typically resolves approximately 90% of ectopic pregnancies. If the zygote is too large or the medication is ineffective, laparoscopic surgery will be performed to remove the cell mass and part or all of the fallopian tube. In the past, ectopic pregnancies usually caused the loss of the fallopian tube and sometimes reduced future fertility. Today, the use of methotrexate and tube-preserving surgery techniques has increased the rate of a normal future pregnancy and delivery among women who have had one ectopic pregnancy to 60% (Silipian & Wood, 2011). Moreover, as medical methods of preventing and treating ectopic pregnancies, along with assisted fertility techniques (to be discussed later) continue to develop, the number and most negative effects of ectopic pregnancies are likely to decrease significantly.

Miscarriage

miscarriage

The loss (without any purposeful intervention) of an embryo or fetus during the first 20 weeks of pregnancy; also called *spontaneous abortion*.

Miscarriage, or *spontaneous abortion*, is the loss (without any purposeful intervention) of an embryo or fetus during the first 20 weeks of pregnancy ("Miscarriage," 2011). As mentioned earlier in this chapter, 10% to 20% of all *known* pregnancies end in miscarriage. The actual number may be much higher (up to 75%) due to what are sometimes called *silent miscarriages*, which occur soon after uterine implantation and often before the woman is aware that she is pregnant. In these cases, the spontaneous abortion typically manifests itself as a somewhat late and often unusually heavy menstrual period (American Pregnancy Association, 2014). Typically, the symptoms of a miscarriage will include one or more of the following symptoms (Nordqvist, 2013a):

- Cramping and pain in the abdomen
- Fluid discharge from the vagina
- Tissue discharge from the vagina
- Feeling faint or light-headed

Since You Asked

5. A few months ago, I started my period very late and had a lot of blood and very bad cramps (worse than usual). My friend said it was probably a miscarriage. Is that possible?

Most commonly, miscarriage is the result of one or more serious fetal or placental abnormalities. In these cases, the spontaneously aborted embryo would probably not have survived beyond the first trimester, no matter what. Other factors that have been linked to miscarriage include the mother's age (the chances of miscarriage increase as parental age increases), abnormalities of the uterus, hormonal imbalances, and environmental toxins such as recreational drug use by the mother (especially alcohol and cocaine), cigarette smoking, radiation, or heavy caffeine intake (Griebel et al., 2005; Nordqvist, 2013a). On the other hand, some risk factors that many people *believe* are related to miscarriage, such as stress and sexual activity during pregnancy, have not been supported by scientific research.

Those who have never experienced the loss of a pregnancy often find it difficult to understand the deep sadness experienced by the prospective parents. Indeed, most miscarriages happen very early in pregnancy, well before the embryo would have any chance of survival outside the womb. Moreover, the vast majority of miscarriages happen for important biological reasons that would probably have rendered the fetus nonviable under any circumstances. All the same, these seemingly rational "justifications" for the miscarriage provide little comfort for the parents, who must deal with the loss of a future child whom they had already incorporated into their psychological and emotional lives. When a miscarriage occurs, the parents may have already selected names for the baby, told friends and relatives the happy news of the pregnancy, and even begun decorating the baby's room.

Often doctors, nurses, and friends of the parents try, with all the best intentions, to offer condolences and support (Murphy & Merrill, 2009). But sometimes friends and even medical professionals will try to offer comfort with statements such as "Don't worry; you'll be able to become pregnant again. You'll still be able to have a child," or "But you were hardly even pregnant at all!" These attempts at comfort frequently occur too soon after the loss, sometimes even while the woman is still in the hospital receiving treatment for the miscarriage. She and her partner at that point may not have had even a moment for the grieving process, a process that is very real, very painful, and very personal. To consider, even for a moment, the idea of "replacing" the baby they have just lost usually provides no comfort at all.

Preterm Birth

preterm birth
Birth of an infant less than 37 weeks after conception.

Preterm birth, also known as *premature birth*, is the birth of a fetus well before the 9-month, 280-day pregnancy period is complete. More specifically, preterm birth is defined as any delivery of a normally formed infant that occurs fewer than 37 weeks (259 days) after conception, which is three weeks fewer than a full-term pregnancy ("Preterm Labor," 2010). It is the leading cause of fetal death in the United States and occurs in approximately 11.5% of births each year, which is a significant decrease from a high of 12.5% in 2008 and the lowest rate in 15 years (Weber, 2013). However, the March of Dimes has set a goal of only 9.6%, which has been reached in only six states: Alaska, California, Maine, New Hampshire, Oregon, and Vermont.

The rate of premature births has been increasing in the United States, but survival rates of preterm babies is on the rise as well.

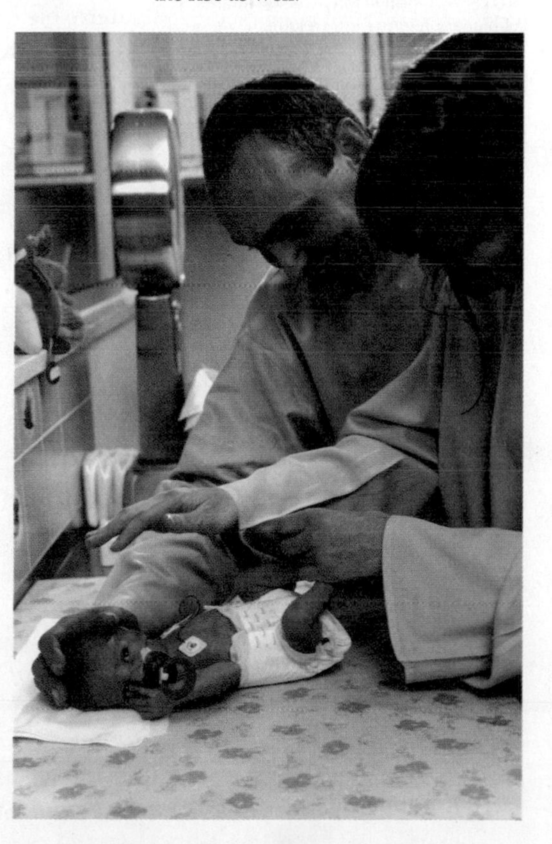

Giving birth to an early-term infant is a difficult and stressful challenge for any parent. Typically, premature infants require special and often highly intensive care in a hospital setting before they are strong and healthy enough to be taken home. Premature births are often associated with a wide range of developmental difficulties for the child and consequently present extra challenges for the parents (LaHood & Bryant, 2007). Preterm infants often exhibit early deficits in cognitive, attention, perceptual, and motor skills, some of which may continue into later childhood. Moreover, preterm infants are more prone to learning disabilities, diabetes, hearing impairment, coronary heart disease, and cerebral palsy ("Preterm Births," 2010; Weber, 2013). They tend to be less alert and slower to acquire early visual and auditory skills compared to full-term babies. These cognitive and perceptual difficulties may interfere with interactions between mother and infant, which may in turn disrupt the normal mother–child bonding process. This lack of social interaction and bonding between the preterm infant and the mother may further intensify the developmental delays for the child (Feldman, Rosanthal, & Eidelman, 2014).

Since You Asked

6. My aunt's baby was born after only twenty-five weeks of her being pregnant. The baby is probably going to be OK, but will the child have any problems later on?

Recent research findings from a study of over one million births found that as preterm infants grew up, their risk of death in childhood and adolescence was higher than for full-term babies. In addition, being born prematurely appears to predict a somewhat lower rate of reproduction for men and women and a greater chance of preterm delivery for women who were born prematurely (Swamy, Ostbye, & Skjaerven, 2008).

Significant medical advances now allow many more preterm infants not only to survive but also to live normal, healthy lives. Today, even infants born as early as 23 weeks have a small (12% to 20%) but very real chance of survival. That's 17 weeks, or more than *four months*, before full term! Figure 9.12 illustrates the rates of survival of premature infants by weeks of gestational (developmental) time in the womb. You can see how every week in the womb after 21 weeks greatly increases survival odds.

Figure 9.12 Predicted Survival Rates for Preterm Infants

NOTE: Full-term is forty weeks. Infants of higher preterm birth weights have, on average, higher survival rates.

SOURCES: Based on data from Draper et al. (2003); Ross (2007); Tyson et al. (2008).

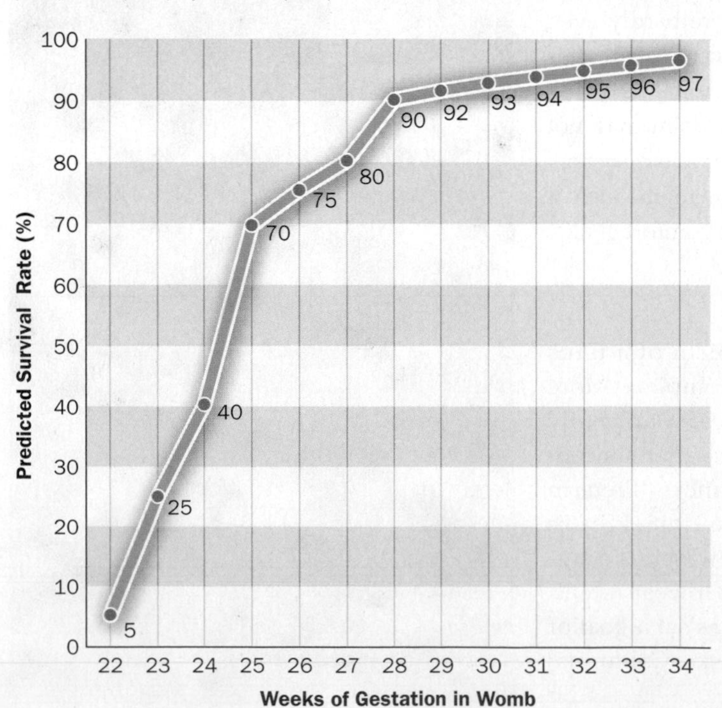

Predicting if or when a pregnancy is likely to end in a preterm delivery has been generally unsuccessful (Colombo, 2002; Lockwood, 2002). Research has attempted to define the causes of preterm birth. Some of the factors found to be associated with preterm birth are drug use, cigarette smoking, or various infectious agents, poor nutrition, and environmental toxins such as exposure to traffic-related air pollution (Paddock, 2009). However, approximately half of all preterm births occur for unknown reasons.

A newly developed blood test appears to be able to predict with nearly 80% accuracy whether a pregnant woman will give birth early. The test measures two proteins in her blood, combined with a cervical length measurement. Still in the research stages, if this test is approved, it will allow for targeted treatments for at-risk women to prevent preterm birth (Glynn, 2012).

One often-mentioned "cause" of premature births relates to intercourse during pregnancy. Contrary to popular belief, however, sexual activity during pregnancy does not increase the risk of preterm labor (or any other pregnancy problems). Research has demonstrated that sexual activity, including intercourse, throughout the duration of a normal pregnancy poses no risk to the developing fetus or to the mother. Sexual activity does not normally increase the chances of premature birth or birth defects, does not affect birth weight or birth length, and won't cause uterine membranes to rupture (Aziken et al., 2007; Schaffir, 2006). In fact, sexual intercourse and orgasm for the woman later in pregnancy appear to be associated with *lower* rates of preterm deliveries (Brown, 2001). An exception to these findings, however, is if the woman has or contracts a sexually transmitted infection such as gonorrhea, chlamydia, or trichomoniasis during pregnancy (see Chapter 8, "Sexually Transmitted Infections/Diseases"). These infections have been associated with increased rates of preterm labor and birth and may be passed to the fetus (Jancin, 2000; Majeroni & Ukkadam, 2007).

Of course, it's quite possible that the woman, the man, or the couple will not *feel* particularly sexually desirous in the later stages of pregnancy. Sexual desire varies a great deal among pregnant women and their partners. Some may feel *more* sexual and more sexually eager during pregnancy, but others may feel that sex is the last thing on their mind. Clearly some of the normal female side effects of pregnancy, such as nausea, fatigue, and body changes, may interfere with sexual feelings for some pregnant couples. Moreover, sometimes it is the woman's partner who is hesitant to engage in sexual activities for fear of somehow harming her or the baby, even though those risks are virtually nonexistent.

Heterosexual couples who choose to have intercourse during pregnancy may need to experiment with various positions to accommodate the changes in the woman's body as the fetus and the woman's abdomen grow (see Chapter 6, "Sexual Behaviors," for more about various intercourse positions). As you see so often throughout this book, sexual intimacy and satisfaction need not always involve sexual intercourse. Many pregnant women and couples find that they are able to fulfill their sexual needs through other intimate and satisfying activities, such as oral sex and mutual masturbation, if they choose to avoid vaginal penetration during pregnancy but still desire sexual closeness and sexual release.

Fetal Abnormalities (Birth Defects)

One of the worst fears prospective parents face is that their baby will be born with a condition commonly referred to as a **birth defect**. Fortunately, the odds of any these fetal problems occurring are extremely low, as you can see in Table 9.4 (on following page); all fetal abnormalities occur in *less than* 1% of births. A birth defect is a physical abnormality or a dysfunction of metabolism (body chemistry) that is present at birth and results in physical or mental disability. Some rare birth defects may be serious enough to cause the death of the infant. Over 4,500 specific birth defects have been identified, and they are the second leading cause of infant mortality, after preterm birth and occur in 1 out of every 33 births in the U.S. (CDC, "Birth Defects," 2014; "March of Dimes Update," 2003). Testing of the fetus while in the uterus for various abnormalities is becoming ever safer and more accurate, which is even further decreasing the chances of an infant being born with a serious problem. Early embryonic testing for genetic abnormalities will be discussed in greater detail later in this chapter.

birth defect

A physical abnormality or metabolic dysfunction that is present at birth and may result in physical or mental deficits.

Teratogens

The developing fetus needs a safe, natural, toxin-free environment in the uterus to have the best chance of developing into a normal, healthy baby. To accomplish this internal environment, the pregnant woman must be careful to avoid putting or allowing any substance into her body that may cross through the placenta into the fetus and cause complications in its physical development. Any outside agent, whether it is a drug (including many prescription medications), a microbe (such as measles), a chemical (such as antifreeze), radiation, or certain metals (such as lead or mercury), that has the potential to cause a fetal abnormality is called a **teratogen**. Teratogens can be ingested by a pregnant woman by any means, including eating, smoking, injecting, or coming into contact with the teratogenic substance. The most commonly used substances that have been clearly defined as teratogens are alcohol, nicotine (cigarette smoking), and

teratogen

Any agent that has the potential to cause a fetal abnormality.

Table 9.4 Incidence of Birth Defects (United States)

Fetal Abnormality	Percentage of Births (All < 1%)
Structural, Chromosomal, and Metabolic Problems	
Heart and circulation defects	.80%
Muscle and skeleton defects	.77%
Cerebral palsy	.25%
Genital and urinary tract defects	.74%
Eye defects	.02%
Chromosomal syndromes	.66%
Club foot	.10%
Down syndrome (retardation; physical abnormalities)	.13%
Respiratory tract defects	.11%
Cleft lip or palate	.12%
Spina bifida (defect in spinal vertebrae)	.04%
Metabolic disorders	.03%
Anencephaly (lack of brain matter)	.02%
PKU (absence of a digestive enzyme)	.004%
Congenital (Present at Birth) Problems	
Congenital syphilis	.05%
Congenital HIV infection	.04%
Congenital rubella ("German measles") syndrome	.0001%
Other Problems	
Rh disease (mismatched mother-fetus Rh factor)	.071%
Fetal alcohol syndrome	.10%

SOURCES: Adapted from CDC, "Birth Defects" (2011); March of Dimes Foundation, Perinatal Statistics (2008); March of Dimes, Birth Defects (2010).

all recreational drugs. A complete list of specific teratogens is far too long to include here (more than 2,000 known or suspected teratogens have been identified).

For legal, prescription medications, the United States Food and Drug Administration (FDA) has developed a coding system for over-the-counter and prescription drugs to rate their safety during pregnancy. This system uses the code letters A (safest), B, C, D, and X (most harmful) to indicate the danger of a drug to a fetus. Many physicians rely on this lettering system when prescribing medications to their pregnant patients. However, various teratogens are more or less toxic depending on the dose, the timing of exposure during the nine months of pregnancy, the characteristics of the pregnant woman, and various other factors. In addition, biomedical research, although acknowledging the teratogenic effects of many substances, has not identified the exact biochemical mechanisms of the adverse effects (Holmes, 2010). Consequently, many researchers have suggested that the FDA's ABCDX system is overly simplistic and confusing and should be replaced with a more scientifically researched and accurate teratogen identification system.

Probably the most common, and therefore the most dangerous, teratogens worldwide are tobacco and alcohol. Cigarette smoking during pregnancy has been shown in hundreds of studies to predict all of the problems described above: ectopic pregnancies, miscarriage, preterm births, and fetal abnormalities (Rai & Regan, 2006; Rosenthal, Melvin, & Barker, 2006). And the dangers of cigarettes are not limited to active smoking during pregnancy. A study of over 26,000 women who had *ever* smoked since their teens were 44% more likely to deliver a stillborn baby, 43% more likely to experience an ectopic pregnancy, and 16% more likely to have a miscarriage compared to women who had never smoked (Whiteman, 2014). Even more surprising was the finding that these risks also significantly increased in women who had never smoked but had been exposed to secondhand smoke (even as children); the longer the exposure, the greater the risks. In spite of these facts, and in spite of women's near-universal awareness of the dangers, nearly 11% of women smoke cigarettes during pregnancy (CDC, "Tobacco Use," 2014).

When a woman drinks alcohol while pregnant, her child may be born with any number of serious, often permanent disorders referred to as **fetal alcohol spectrum disorders**. These disorders involve a wide spectrum of problems including (CDC, "FASDs," 2014):

fetal alcohol spectrum disorders
A variety of disorders than may occur in a person whose mother drank alcohol during pregnancy.

- Abnormal facial features
- Small head size
- Shorter-than-average height
- Low body weight
- Poor coordination
- Hyperactive behavior
- Difficulty paying attention
- Poor memory
- Difficulty in school (especially with math)
- Learning disabilities
- Speech and language delays
- Intellectual disability or low IQ
- Poor reasoning and judgment skills
- Sleep and sucking problems as a baby
- Vision or hearing problems
- Problems with the heart, kidneys, or bones

The fundamental lesson to be gained from leaning about teratogens is that a woman who is pregnant or planning to become pregnant should avoid ingesting or coming into contact with any substance other than a standard diet of healthy foods and beverages, unless she has discussed a need for a specific medicine with her doctor.

Embryonic and Fetal Testing

Prenatal tests for birth defects and fetal abnormalities are used widely in the United States, and medical science's ability to screen for genetic or physical defects of the embryo or fetus is growing rapidly (Chachkin, 2007; Williams, 2007). Moreover, many of these tests can now be carried out far more simply (e.g., with a simple blood test) earlier in pregnancy than was previously possible, often within the first 8 to 12 weeks, and are less invasive than more traditional tests (Williams, 2007). Blood tests performed on the expectant mother measure levels of proteins and hormones that can predict an increased risk of some disorders, such as *spina bifida* (a neural tube defect in which the spinal column fails to close completely during fetal development) and *Down syndrome* (a chromosomal condition characterized by some degree of intellectual disability, often combined with specific physical and medical problems such as small stature, lowered resistance to infection, vision and hearing difficulties, and, in some, heart defects that are usually correctable). Such tests are becoming increasingly accurate during very early stages of pregnancy (Bianchi, 2014).

Figure 9.13 Three-Dimensional Sonogram of a Fetus in the Uterus

Ultrasonic visualization of the fetus inside the uterus lets doctors and parents literally see and monitor the progression of pregnancy and can alert them to the presence of any birth defects.

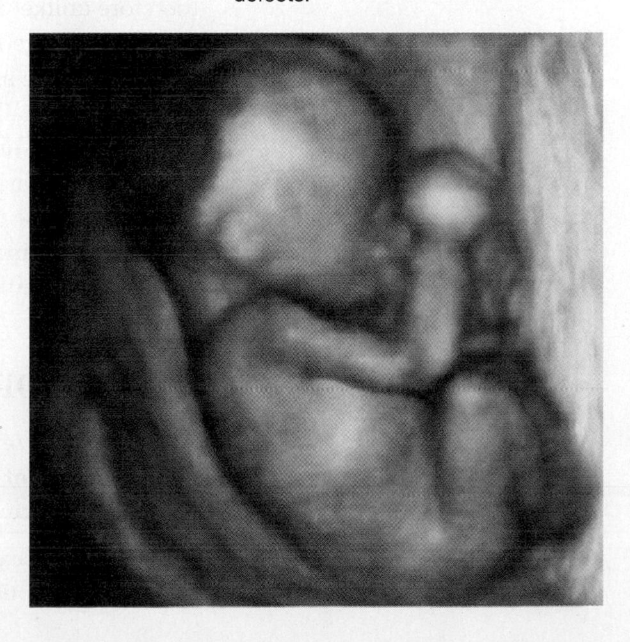

Today, electronic scans such as ultrasound, three-dimensional ultrasound, and magnetic resonance imaging (MRI) allow physicians to see the fetus in the uterus directly (as shown in Figure 9.13);

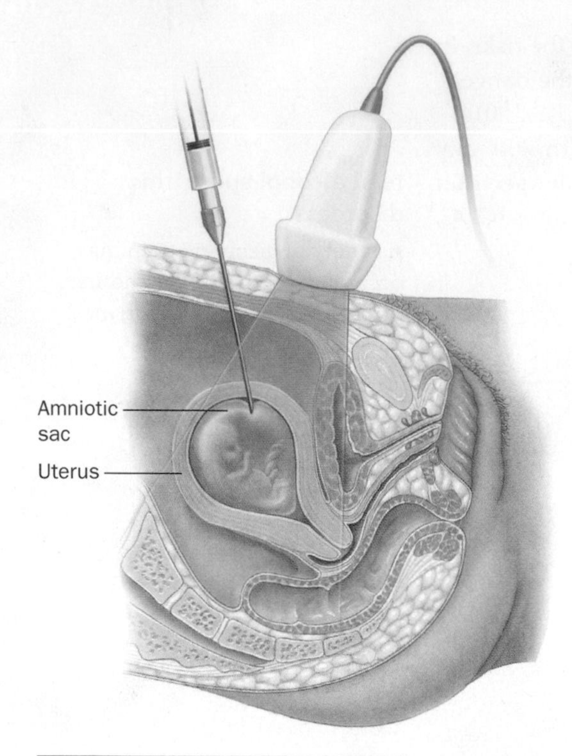

Figure 9.14 Amniocentesis Procedure

During amniocentesis, fluid from the sac surrounding the fetus is extracted and tested for genetic characteristics and is capable of detecting many potential abnormalities.

not only does this ability and assist greatly in diagnosing fetal abnormalities earlier and with greater accuracy. These safe, noninvasive exams may be performed as early as the 12th week of pregnancy and can assist greatly in diagnosing or ruling out various fetal abnormalities with far fewer errors than blood tests.

Ultrasonic visualization of the fetus inside the uterus lets doctors and parents literally see and monitor the progression of pregnancy and can alert them to the presence of any birth defects.

Amniocentesis and *chorionic villus sampling* (CVS) are tests that require the extraction of a small amount of fluid from the sac surrounding the fetus (in amniocentesis) or a sample of placental tissue (in CVS). The cells in these substances contain the full genetic code for the fetus and can be used to diagnose many genetic abnormalities, such as Down syndrome, cystic fibrosis, hydrocephalus (fluid pressure in the skull), sickle-cell anemia, and hemophilia.

Amniocentesis requires that a long needle guided by ultrasound be inserted into the uterus through the abdomen (see Figure 9.14), and CVS uses a small tube inserted through the cervix or a thin needle passed through the abdominal skin. Most women report only minor discomfort from these procedures. It was long believed that the disruption to the uterine balance during the test carried a small increased risk of miscarriage. However, newer research has shown that this risk has been reduced virtually to zero due primarily to more advanced procedural techniques and the ability to use ultrasonic guidance in performing the tests (Kuehn, 2006). Therefore, these prenatal tests are no longer reserved only for women at higher risk of pregnancy complications (such as those aged 35 or older or those with a history of pregnancy problems). Although the risk of all birth defects, and especially Down syndrome, increases with maternal age, the American College of Obstetricians and Gynecologists now recommends that *all* pregnant women be offered the option of these tests, especially if an analysis of blood factors indicates an increased risk of fetal abnormalities ("Down Syndrome Screening," 2007). In addition, if other risk factors are present, such as a family history of abnormalities or exposure to environmental toxins, these genetic tests are more likely to be prescribed.

New research is also focusing on the father's role in contributing to fetal abnormalities and birth defects (see Saey, 2008). In the past it was assumed that because men create new sperm cells throughout their lives the cells are "forever young" and therefore unlikely to cause any problems. However, evidence is mounting that increasing paternal age may contribute to birth defects. Recent research evidence points to the father's age as a potential factor in birth defects, miscarriage, and certain mental illnesses, especially schizophrenia (see Teich, 2007; Wyrobed et al., 2006).

Prenatal testing allows parents to better prepare and plan for an infant that may be born with an abnormality or to make an informed decision to terminate the pregnancy, if they so choose. The most frequent outcome of prenatal testing, however, is the relief and peace of mind that come with knowing that the development of their baby will be normal and healthy.

Abortion

9.5 Summarize the issue of abortion including relevant statistics, the various methods of pregnancy termination (surgical and medical), and the sociopolitical controversy that surrounds the practice of abortion in the U.S.

Many human sexuality textbooks include the discussion of abortion in the chapter about contraception. However, virtually everyone agrees, regardless of their personal attitudes about abortion, that abortion should never be used as birth control. Contraception, by definition, is about *preventing* pregnancy, whereas, also by definition, abortion occurs *after* a pregnancy has been established. Therefore, we discuss abortion in this chapter because it is one of the many issues that may arise for some women or couples during pregnancy.

As you learned earlier in the chapter, termination of a pregnancy from natural causes prior to the 20th week of pregnancy is referred to as a *spontaneous abortion* or *miscarriage*; between 20 and 23 weeks, it is usually called "fetal death"; and between 23 and 37 weeks, it is referred to as a *preterm birth*. However, *spontaneous abortions* are clearly distinguished from our everyday understanding of the meaning of **abortion**, which is used by most people to mean the *voluntary* termination of a pregnancy, or *induced abortion*. Although induced abortions are often unrelated to any physical health problems of the embryo or the mother, it is an important and common event that must be included in our examination of conception and pregnancy.

abortion
Termination of a pregnancy; in common usage, assumed to be the result of an a voluntary medical procedure as opposed to a miscarriage.

Approximately half of all pregnancies in the United Stated are unplanned, and about 40% of theses unplanned pregnancies are voluntarily terminated. Approximately 22% of all pregnancies in the United States (not including miscarriages) are terminated through voluntary abortion. Slightly over one million legal abortions occur in the United States each year, and about 2% of women between the ages of 14 and 44 have an abortion every year ("Facts on Induced Abortion," 2011; "Incidence of Abortion," 2008). Abortion rates in the United States have been falling since 1981, when nearly 30 abortions were performed per 1,000 women, to 2008, when the number was under 28 per 1,000 ("Facts on Induced Abortion," 2011).

The Guttmacher Institute, which carries out continuous research on all facets of human sexuality, has provided the following statistics on the characteristics of girls and women who choose to have an abortion (Facts on Induced Abortion, 2011):

- Eighteen percent of U.S. women obtaining abortions are teenagers; those aged 15–17 obtain 6% of all abortions, teens aged 18–19 obtain 11%, and teens under age 15 obtain 0.4%.
- Women in their twenties account for more than half of all abortions; women aged 20–24 obtain 33% of all abortions, and women aged 25–29 obtain 24%.
- Thirty percent of abortions occur with non-Hispanic black women, 36% with non-Hispanic white women, 25% with Hispanic women, and 9% with women of other races.
- Thirty-seven percent of women obtaining abortions identify themselves as Protestant and 28% as Catholic.
- Women who have never married and are not cohabiting account for 45% of all abortions.
- About 61% of abortions are obtained by women who have one or more children.
- Forty-two percent of women obtaining abortions have incomes below 100% of the federal poverty level ($10,830 for a single woman with no children).
- Twenty-seven percent of women obtaining abortions have incomes between 100% and 199% of the federal poverty level.
- The reasons women give for having an abortion underscore their understanding of the responsibilities of parenthood and family life. Three-fourths of women cite concern for or responsibility to other individuals; three-fourths say they cannot afford a child; three-fourths say that having a baby would interfere with work, school, or the ability to care for dependents; and half say they do not want to be a single parent or are having problems with their husband or partner.

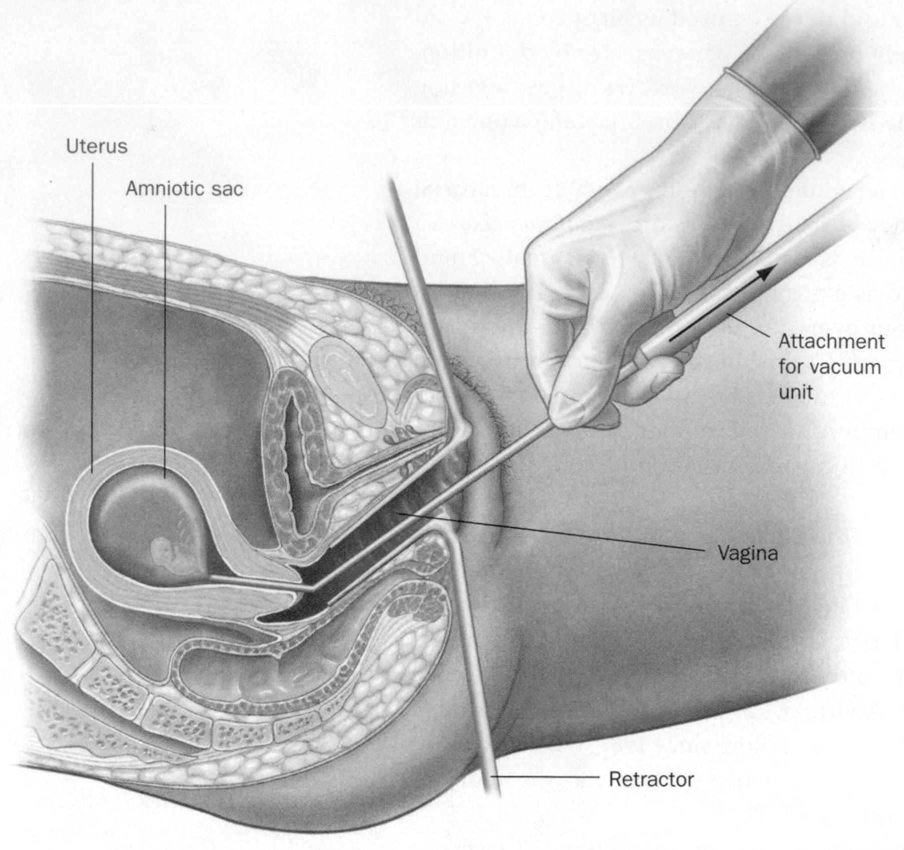

Figure 9.15 Vacuum Aspiration Procedure

In vacuum aspiration, the most common abortion technique for early pregnancies, a small tube is inserted through the cervix. Suction is then applied to the tube, creating a slight vacuum effect that draws out the contents of the uterus, including the uterine lining and the embryo.

The debate over induced abortion is, and has been for decades, one of the most polarizing controversies in the United States and elsewhere in the world. The extremes of the two sides of this controversy in the United States, the "**pro-choice**" and the "**pro-life**" positions, are summarized in "Sexuality, Ethics, and the Law: The Abortion Controversy." Many individuals subscribe to positions on abortion that have some elements of both the pro-life and the pro-choice extremes (e.g., a pro-life individual supporting abortion only in pregnancies that result from rape or incest, or a pro-choice supporter opposing abortion for sex-selection purposes). Beyond this impassioned debate that continues to rage in the United States and many other countries as well, various issues relating to abortion are important to discuss briefly in a human sexuality textbook. These topics include how abortions are performed; the possible psychological and emotional effects on the woman, and often on the father as well; how and when the choice is made to end a pregnancy; and suggested strategies to reduce the number of abortions, a goal on which people on both sides of the controversy can agree.

Abortion Procedures

The number of options available to a woman for choosing to terminate her pregnancy decreases as the time since conception increases. In general, a decision to abort in the first eight weeks of pregnancy offers the woman the greatest choice of methods and carries the lowest risk of complications. However, termination of a pregnancy may also occur after eight weeks and at any time during the first two trimesters (the first 24 weeks of pregnancy) with relatively little risk of serious complications for the woman. Nearly 90% of all surgical abortion in the United States occur within the first trimester of pregnancy, and only 0.3% of women who have abortions in the United States experience medical complication that require hospitalization ("Facts on Induced Abortion," 2011; Fox & Hayes, 2007). In general terms, the two types of procedures for terminating a pregnancy are surgical abortions and medical abortions.

vacuum aspiration
A method of abortion in which a small tube is inserted through the cervix to extract the contents of the uterus, including the endometrium lining and embedded embryo.

pro-choice
The belief that a woman has the moral and legal right to choose freely to abort her pregnancy.

pro-life
The belief that voluntary abortion is akin to murder and that it should be illegal.

SURGICAL ABORTIONS Until the late 1990s, virtually all abortions performed in the United States used surgical procedures, and nearly all of those used a technique called **vacuum aspiration** (see Figure 9.15). An older method of abortion called "dilation and curettage" or a "D&C" is rarely used today because it presents a greater chance of complications. The vacuum aspiration technique incorporates a syringe device that is operated by hand (called *manual vacuum aspiration*) or a small hand-held electrical medical vacuum device (do *not* ever try this with other types of vacuums!). Both vacuum methods have been shown to be equally safe and effective (Berer, 2007; Hussein, 2007).

Vacuum aspiration, as the term implies, incorporates a small tube that is inserted through the cervix to which suction is applied, creating a slight vacuum that draws out the contents of the uterus, including the endometrium lining and the embedded embryo. The procedure takes 5 to 15 minutes and, where abortion is legal, is typically

Sexuality, Ethics, and the Law

The Abortion Controversy

In the United States and elsewhere, intense debate and even violence surround the issue of abortion. Some people are ambivalent, unsure, or stake out some middle ground about abortion. But most fall fairly clearly into one of two camps, commonly referred to as pro-choice (those who believe that a woman has the moral and legal right to choose to abort her pregnancy) and pro-life (those who are morally opposed to abortion, believe it is akin to murder, and hold that it should be made illegal). Here is a summary of the central ideological points of these two sides of the abortion issue along with some relevant facts about abortion rates from the World Health Organization.

The Pro-Life Position

- Pro-life ideology is generally based on the belief that human life begins at conception and that the embryo or fetus, from conception to full gestational term, has the same inalienable right to survival as any other human.
- The pro-life position tends to encompass more than an antiabortion position; many of those who believe in basic pro-life principles also tend to oppose assisted suicide and stem cell research but, paradoxically, tend to favor capital punishment.
- Although those adhering to a pro-life position may do so for many reasons, most pro-life individuals have a Christian religious orientation (they are sometimes referred to as "fundamentalist" or "evangelical" Christians); many claim that the basic assumptions of the pro-life movement are based on Christian ideology.
- Some people within the pro-life movement allow for little, if any, gray area in the debate. They hold that any and all abortions are wrong, immoral, and tantamount to murder. This often includes the abortion of pregnancies caused by rape or incest and pregnancies that may have medical complications that threaten the life of the mother. Others, however, who consider themselves strongly pro-life are willing to make exceptions in some of these cases.
- Pro-life supporters tend to support reducing the number of abortions by making the practice illegal for physicians and/or women.

The Pro-Choice Position

- Pro-choice ideology is centered on the belief that a woman has the right to control her own body and reproductive life.
- Pro-choice individuals are not pro-abortion, per se; that is, they do not want pregnant women to have abortions. Rather, they believe in the right of every woman to make an educated decision about her reproductive rights and responsibilities and to do so in consultation with her partner, family, or doctor if she so chooses.
- Some of the gray areas within the pro-choice movement relate to the father's degree of responsibility for the decision to abort a pregnancy and his moral and legal duty to support a child born of a pregnancy that he would have chosen to abort. Pro-choice advocates tend to be more liberal politically in general and may also support assisted suicide legislation and stem cell research. They also tend to oppose the death penalty.
- Another controversial aspect of the pro-choice movement is the topic of late-term abortions (which the pro-life movement calls "partial-birth abortions"). Some disagreement exists within the pro-choice movement regarding the ethics of these late-term abortions, unless the health or life of the mother is at stake.
- Pro-life supporters seek to reduce the number of abortions through education about sexuality and greater availability and use of contraception.

Relevant Facts about Abortion Rates

- An estimated 13% of maternal deaths worldwide are due to unsafe, illegal abortions, which is 350 times more than in countries where abortion is legal.
- Approximately 8.5 million women in developing nations, where abortion is commonly restricted and illegal, experience serious medical complications from unsafe and unsanitary abortions. Of those, 3 million do not have access to the post-abortion care they require
- The number of abortions does not appear to be linked to laws against it. In 2008, the abortion rate in Western Europe, where abortion is generally allowed, was 12 abortions per 1,000 childbearing-age women. In that same year, in Latin America, where abortions are widely banned, the rate was approximately 30 per 1,000 women.
- South Africa, where abortion laws are the most liberal in all of Africa, has the world's lowest rate with 11 abortions per 1,000 women.
- In Eastern Europe, the abortion rate is 44 per 1,000 women, likely due to lack of modern contraceptives such as the pill and the IUD.

NOTE: Research assistance: La Sara W. Firefox.

performed in a doctor's office or medical clinic under local anesthetic. Usually, over-the-counter pain relievers (such as ibuprofen) are adequate for postsurgical pain and cramping. Normal side effects of the procedure include abdominal cramping and usually some bleeding. If pregnancy has progressed beyond the first trimester, the usual termination procedure is **dilation and evacuation**, commonly called **D&E**. D&E is a more invasive and extensive procedure than MVA, largely because by the second trimester (which is 13 to 26 weeks, although only about 1% are performed after 23 weeks of gestation), the pregnancy is more firmly established, and the developing fetus in the uterus is larger. The procedure therefore typically requires greater dilation of the cervix than a first-trimester abortion. Prior to the procedure, the woman is usually given sedatives and a local or general anesthetic. Then the cervix is dilated so that a vacuum tube may be inserted to remove the fetus and most of the remaining contents of the uterus. Next, a curved surgical instrument called a *curette* is inserted to scrape the lining of the uterus to free any additional tissue. Finally, suction may be applied to be sure the uterus has been fully emptied.

dilation and evacuation (D&E)
A method of abortion commonly used when a pregnancy has progressed beyond the first trimester, involving scraping of the uterine walls and suctioning out of the contents.

A D&E is typically performed as "day surgery" in a hospital setting. A woman should expect to experience cramping for several hours to several days and bleeding for about two weeks following the procedure. Postsurgical pain is usually relieved with over-the-counter medication such as acetaminophen or ibuprofen. Rare but serious complications of this procedure may include damage to the uterine wall, severe bleeding (hemorrhaging), and infection.

Surgical abortions, since becoming legal in the United States in 1973, are considered quite safe, with risks to the health and life of the mother significantly lower than the risks of pregnancy and giving birth. However, in countries where abortion is illegal, abortion is often one of the leading causes of maternal deaths (Haddad & Nour, 2009). This was also true in the United States before the legalization of abortion. In 1972, the year before abortion was legalized in the United States, the number of deaths attributed to illegal and legal (terminations for health reasons) abortions reported to the CDC was 39 and 24, respectively (Strauss et al., 2004). Eight years later, in 1980, the number of illegal abortion deaths had dropped to 1 and the number of deaths from legal abortions fell to 9. Jumping ahead nearly 30 years to 2010, these numbers have held steady with no deaths due to illegal abortions and eight resulted from legal abortions (Pazol et al., 2013).

MEDICAL ABORTIONS For the past 12 years or so, the United States has seen a clear trend away from surgical abortions and toward medical methods for the termination of early pregnancies. A **medical abortion** relies on specifically targeted drugs to terminate a pregnancy, rather than surgical procedures. The medications used in abortions are either *mifepristone* (brand name Mifeprex; formerly referred to as RU-486) or *methotrexate,* both of which are combined with *misoprostol* (a synthetic form of the hormone *prostaglandin*).

medical abortion
A method of abortion using drugs rather than surgery to terminate a pregnancy.

Since You Asked
7. How does the abortion pill work? Is it really just a pill you take and then you have an abortion?

Medical abortions are most appropriate for pregnancies of less than nine weeks. When a woman chooses to have a medical abortion, the doctor administers either mifepristone or methotrexate to the pregnant woman during an office visit. Mifepristone is administered either orally or as an injection. Misoprostol is administered concurrently either orally or as a vaginal insert ("Medical Abortion," 2010). This combination of medications causes the uterus to contract, the embryo to detach from the uterine wall, and the cervix to soften somewhat to allow the contents of the uterus to be expelled. After two or three days, the woman returns to the doctor for a checkup to be sure the abortion was complete. In rare cases (5–10%), an abortion will not have occurred or will be incomplete. In those cases, an additional dose of misoprostol or a surgical abortion may be necessary.

The most common negative side effects of medical abortions are cramping and bleeding. Usually, more vaginal bleeding will occur following a medical abortion than

Table 9.5 Pros and Cons of Medical and Surgical Abortion Methods

Medical Abortion	Surgical Abortion
Pros	**Pros**
High success rate (greater than 95%)	High success rate (99%)
Allows woman more control over abortion process	Available early and later in pregnancy
Seems more natural to some women, similar to a heavy period	Usually requires only one visit to doctor, hospital, or clinic
May be less emotionally painful than surgery for some women	Does not necessarily require follow-up doctor visits (although usually recommended)
Avoids invasive procedure and anesthesia	Requires less involvement by woman; may choose general anesthesia
Affords more privacy; much of process occurs at home	More certainty and less doubt about abortion outcome
Little to no risk of physical injury	Typically shorter duration of cramping following procedure
	Procedure typically completed in a single day (usually less than fifteen minutes)
Cons	**Cons**
Available only during first trimester of pregnancy	Allows for less sense of control by woman over the abortion process
Requires at least two or possibly more doctor visits	Involves an invasive procedure, sedation, or general anesthesia
Requires follow-up to ensure abortion is complete	Less natural-feeling, similar to any surgical procedure
Requires woman's active involvement and participation in abortion process (may be seen as a plus by some women)	Sometimes greater sense of grief and loss follows the procedure
Can take days to weeks to complete the abortion process	Slight risk of injury or infection during or following the procedure
Less certainty that abortion is occurring properly and fully	Possible distress caused by the noise of the electrical vacuum aspirator (manual aspirators are quiet)
Typically more cramping during and following procedure	May be inappropriate for women with allergies to anesthesia or serious anemia
Usually more bleeding following procedure	
Possible sight of blood clots and fetal tissue being expelled	
Not appropriate for women with certain medical conditions such as anemia or liver disease	

SOURCES: Berer (2005); Kapp et al. (2013); Healthwise (2004); Hollander (2000).

a surgical one, as the uterus contracts and passes the embryo and other uterine contents through the cervix. Typically, blood clots will be passed during the abortive process, but at the time these abortions are carried out, the embryo is usually still too small to be seen. Most women will be unaware exactly when it has been expelled. A smaller percentage of women report additional side effects, including nausea, diarrhea, heavy bleeding, and fever ("Medical Abortion," 2010).

SURGICAL OR MEDICAL ABORTION: PROS AND CONS If a woman has chosen to have an abortion, she will need to decide which method she prefers. This decision should be made in consultation with her partner (if appropriate) and with her health care provider. Her choice may be based on many factors, such as the time since conception, possible side effects of the procedure, and her own emotional and psychological preferences. Table 9.5 summarizes the pros and cons of surgical versus medical abortions. It should be noted that, contrary to common belief, abortion is *not* related to a higher incidence of any type of cancer in women (Tobin, 2008).

The Psychological and Emotional Experience of Abortion

Some people mistakenly believe that a woman's decision to terminate her pregnancy is a relatively easy, uncaring, or detached decision. Yet many women who choose to have an abortion are conflicted about the decision and may experience stressful indecision, along with complex, painful, and confusing emotional reactions about the unwanted pregnancy and their decision to terminate it. Some may experience negative emotions such as loss, guilt, and depression (e.g., Fergusson et al., 2005). However, other research has shown that women do not necessarily experience negative emotions, and some may feel a sense of relief at solving a stressful episode and returning to the normal routine of life.

Of course, many women who have undergone an abortion may feel a mix of any or all of these emotions (Astbury-Ward, 2008). A 2006 study conducted by the American Psychological Association reviewed all available research since 1989 of the psychological impact of abortion on women and concluded that there was no scientific evidence to support the idea that the termination of an unwanted pregnancy leads to mental health problems in women (American Psychological Association, 2008). Following up on this research, a study of adolescents who had abortions found a similar lack of depression or loss of self-esteem during the five years following the decision to have an abortion (Warren, Harvey, & Henderson, 2010).

Most mental health professionals today agree that when a woman is considering her options for dealing with an unwanted pregnancy, doctors, clinicians, and pregnancy counselors should attempt to provide her with nonjudgmental and empathetic support. Doctors and nurses should offer referrals to a patient for psychological or religious counseling services if they suspect she is having difficulty with the emotional effects of her decision to abort her pregnancy. She should be given as much information about *all* her pregnancy options (keeping the baby, adoption, abortion) as she personally needs and wants so that she can make as informed and educated a decision as possible. She should be encouraged to discuss her feelings about the pregnancy, the potential of becoming a parent or bringing a new child into the family, her ability and desire to provide for a child, and her attitudes about abortion and adoption. The medical or counseling setting should be open to discussing issues relating to the father of the baby and, if the woman wishes, he should be allowed to participate in the decision process and the termination of the pregnancy, if that is what the woman ultimately decides to do. And after an abortion has occurred, attention must be placed on *nonjudgmental* postabortion support and counseling for those who may experience negative reactions following the decision to terminate pregnancy.

Reducing the Number of Abortions

Perhaps one area of common ground exists between the pro-choice and pro-life factions of the abortion debate, and that is the desire to reduce the number of abortions performed each year. The route to this goal, however, often leads to further disagreements. Pro-life advocates endeavor to limit the *availability* of abortions by making them illegal and at times engaging in tactics that frighten, intimidate, or even physically harm individuals who seek or perform abortions.

Since abortions were legalized in 1973 as a result of the well-known U.S. Supreme Court decision in *Roe v. Wade*, numerous legal attempts have been undertaken on the state legal level to overturn or limit the Court's decision. Many of these state provisions have been overturned by higher courts who have deferred to the central provisions of the *Roe v. Wade* decision. However, the frequent attempts by state legislatures to circumvent legalized abortions are seen by some as a slow erosion of a woman's legal right to choose an abortion. Obviously the "pro-life" side of the debate hails these efforts as working to save the life of the embryo and fetus, but "pro-choice" advocates worry about a building political movement to overturn *Roe v. Wade* and return the U.S. to a country where abortions are illegal and criminal. Here is a partial list of state legislation over the past several years that has weakened the right of a woman to choose an abortion as determined by the Supreme Court ("State Policies," 2014):

- Thirty-nine states require an abortion to be performed by a licensed physician. Twenty states require an abortion to be performed in a hospital, and eighteen states require the involvement of a second physician's opinion before the procedure is allowed.

- Thirty-two states and the District of Columbia prohibit the use of state funds for abortions except in those cases when federal funds are available, where the woman's life is in danger, or the pregnancy is the result of rape or incest.
- Eight states restrict coverage of abortion in private insurance plans, most often limiting coverage only to when the woman's life would be endangered if the pregnancy were carried to term.
- Forty-six states allow individual health care providers to refuse to participate in an abortion; forty-three states allow institutions to refuse to perform abortions.
- Seventeen states mandate that women be given counseling before an abortion that includes information on at least one of the following unproven notions: the link between abortion and breast cancer (five states), the ability of a fetus to feel pain (twelve states), or long-term mental health consequences for the woman (eight states).
- Twenty-six states require a woman seeking an abortion to wait a specified period of time, usually 24 hours, between when she receives counseling and when the procedure is performed. Nine of these states have laws that effectively require the woman to make two separate trips to the clinic to obtain the procedure.
- Thirty-nine states require some type of parental involvement in a minor's decision to have an abortion. Twenty-one states require one or both parents to consent to the procedure, while thirteen require that one or both parents be notified and five states require both parental consent and notification.

Many physicians, sexuality counselors, educators, and clinicians find these laws distressing because, based on past experience in the United States and still today in countries where abortion is a crime, making abortion illegal does not *prevent* abortion, but rather drives it underground or to other countries, where the procedures are often far more dangerous and potentially fatal for the mother.

An alternate route to reducing abortions is to decrease the number of unwanted pregnancies. To do that, efforts must be increased to educate young people about making effective and safe choices about their sexual behaviors and to develop and encourage the effectiveness and consistent use of contraceptives (see Chapter 5, "Contraception: Planning and Preventing"). Research has demonstrated a direct and unmistakable link between increased use of contraception and decreases in abortion rates (Marston & Cleland, 2003; Peipert, 2013).

As mentioned earlier, The United States is making progress in reducing the number of abortions. The number of abortions in the United States has been dropping steadily from a high of approximately 1.6 million in 1990 to about 1.2 million in 2008, a decline of 7.5%. This decline is due in part to increased awareness and use of contraception (which may be related in part to a greater awareness of STIs, leading to changes in sexual behavior and more consistent condom use).

Birth

9.6 Discuss the stages of labor, the birthing process, and the choices and decisions that accompany child birth including, the selection of a doctor or midwife, birthing settings, pain management decisions, and the current C-section controversy.

Exactly what mechanism triggers the beginning of labor and the birth process in humans is not fully understood. We do know that under normal pregnancy conditions, changes occur in the uterus, cervix, and especially in the placenta at about 40 weeks of gestation, typically just before the onset of labor, signaling the mother's brain to increase production of various chemicals and hormones, including estrogen, steroids, prostaglandin, and a chemical called *corticotropin-releasing hormone* (Bainbridge, 2003; Smith, Smith, & Bisits, 2010). Corticotropin-releasing hormone is produced naturally in the human body and is related to our physical responses to stress. It is also intimately linked to the beginning of childbirth and signaling the uterus to begin contractions. Amazingly, the woman's body "knows" the right time to produce these physical changes in almost all births—not too early and not too late—to provide for the best chance of survival for the infant.

Labor

The various chemical changes send an unmistakable biological signal to the baby and the mother that it is time for them to meet "face to face." The first sign of labor is usually the pain of a contraction indicating that the uterus is beginning the process of expelling the fetus. Women may also experience some blood from the vagina, called "the show," as the cervical "plug" of mucus and blood is expelled. Some women may also experience a heavy flow of clear liquid (referred to as her "water breaking") as the amniotic sac containing the fetus ruptures and the fluid flows from the uterus, through the cervix, and out the vagina (Romito et al., 2010a). A woman's water breaking does not necessarily imply that labor has begun, but it usually signifies that the birthing process will begin soon.

stage one labor
The first stage of the birth process, involving the beginning of contractions of the uterus.

STAGE ONE LABOR At the onset of labor, called **stage one labor**, contractions begin as the cells of the uterus, now the largest muscle in the mother's body, contract in sequence along its length, pulling and straightening the uterus and pushing the fetus down toward the cervix (see Figure 9.16a). At this point, the cervical canal is beginning to *dilate* (become larger in diameter) and *efface* (become thinner). During the beginning of stage one labor, the cervix is about 1 to 2 inches dilated, and as stage one labor continues, it will continue to dilate to approximately 10 centimeters, or about 4 inches (Romito et al., 2010b). The pains women experience during labor are primarily due to the muscles forcing open the cervix, which has been tightly closed until now to hold the fetus in the uterus.

Figure 9.16 The Stages of Childbirth
Crowning of the baby's head (top) and delivery of the baby's head (bottom).

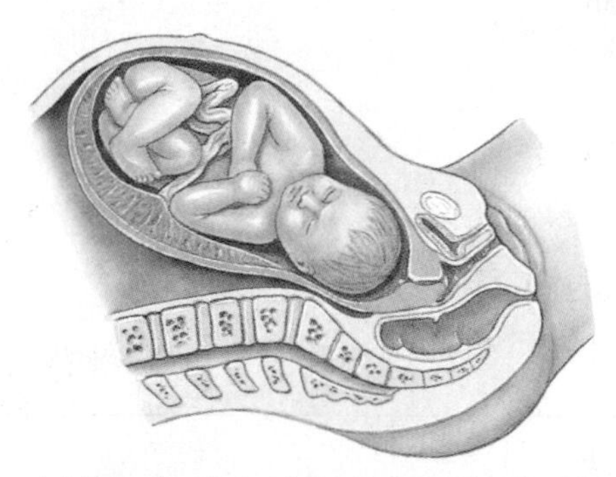
(a) Fully developed fetus before labor begins

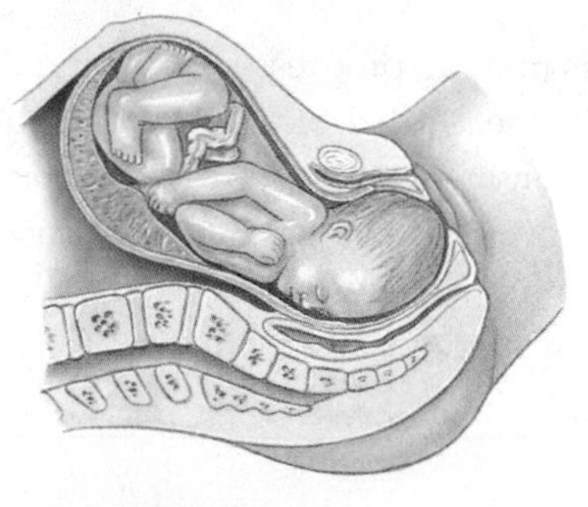
(b) Stage one labor

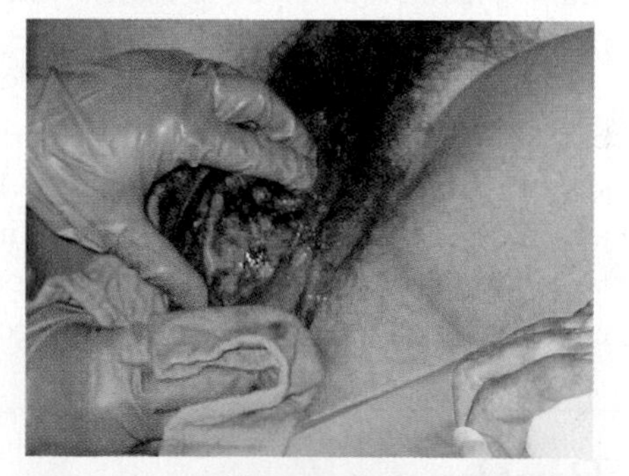

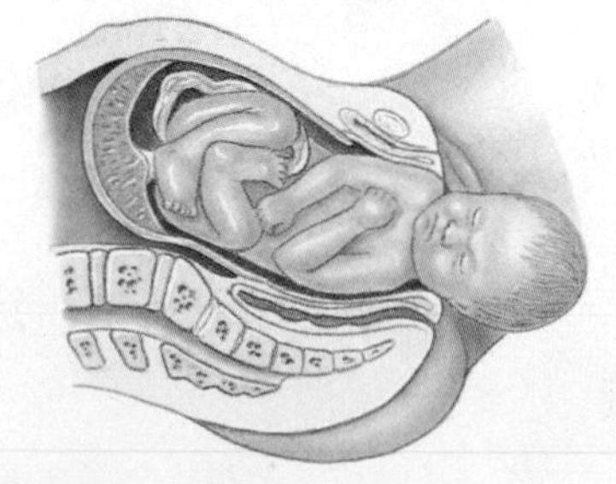
(c) Stage two labor

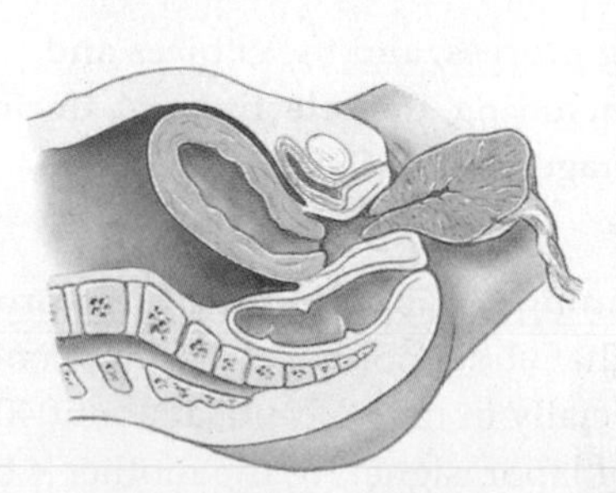
(d) Stage three labor; delivery of placenta

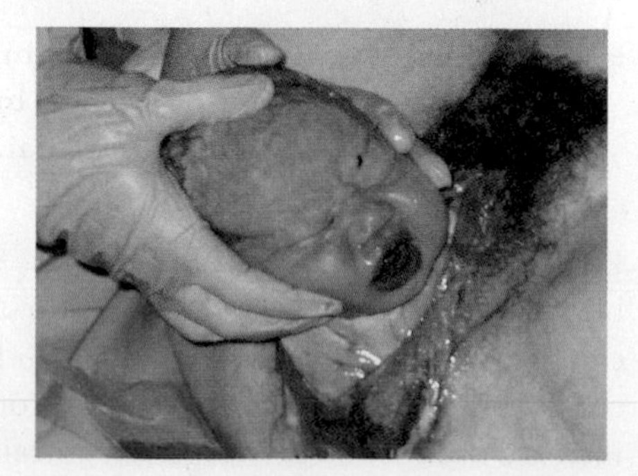

As stage one labor continues, it enters the **transition phase**, when her contractions begin to come more rapidly—typically 2 to 3 minutes apart—and the fetus begins its movement through the cervix and down the birth canal. The top of the baby's head will "crown," that is, become visible in the vagina (see Figure 9.16b). This phase is typically the most difficult physically and emotionally for the mother and may last 3 or more hours for a first-time vaginal birth (Romito et al., 2010b). Typically, the baby will turn in the uterus so its face is turned toward the mother's spine, slightly to the left or right of the spinal column itself. This is the most common and best fetal position for moving through the birth canal, although not every infant follows the rules.

About 3% to 4% of infants present themselves for birth in various other positions, such as bottom first or leg first (called a **breech birth**, shown in Figure 9.17). Although some breech babies are delivered vaginally without major complications, some research indicates that this position presents a greater chance for negative health outcomes for the infant (Gallagher, 2006). For the health of the baby and the mother, the majority of breech births are delivered by C-section (Marshall & Gilbert, 2010) (breech birth will be discussed more later in this chapter).

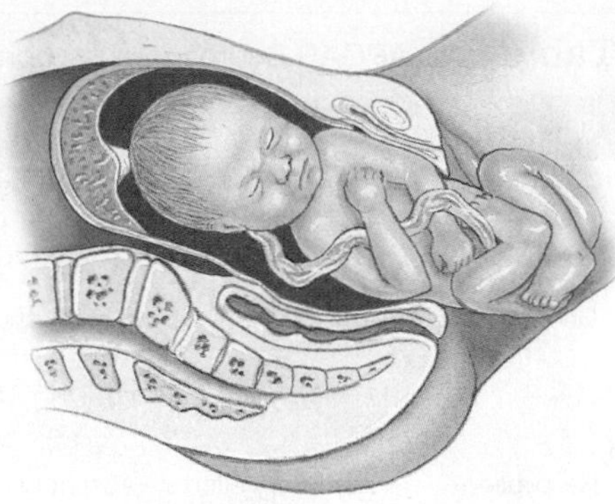

Figure 9.17 A Breech Birth

In a breech birth, the infant presents itself for birth in positions other than head-first, such as legs-first, as shown here.

The contractions continue, becoming increasingly frequent. Overall, first stage labor lasts, on average, about 12 to 14 hours for first-time mothers (although most women will tell you it *seems* considerably longer!); typically, the duration of labor decreases with subsequent births, down to eight or even fewer hours. Keep in mind, however, these are averages, and labor can in some cases proceed as quickly as a few hours or take longer than 20 hours, but these are exceptions rather than the rule. Usually, during the entire delivery process, the infant's heart rate is electronically monitored for any signs of distress. This monitoring allows the doctor or midwife and the mother to know immediately if the infant is experiencing any major difficulties during the strenuous birth procedure; most often, it reassures everyone that all is well.

transition phase
The end of stage one labor as the fetus begins moving through the cervix and down the birth canal.

breech birth
Delivery of a fetus emerging with buttocks or legs first rather than head first.

stage two labor
The stage of the birth process in which contractions occur closer together than in stage one, involve the muscles of the abdomen as well as the uterus, and continue until the infant has been expelled from the mother's body.

STAGE TWO LABOR **Stage two labor** leads to the birth of the baby. Contractions begin to involve the muscles of the abdomen as well as the uterus itself. The mother has somewhat more control over these contractions, and many women report that this sense of control seems to reduce the amount of pain they perceive (Bainbridge, 2003; Holvey, 2014). At this point, the woman will feel a great deal of abdominal pressure, as her body is working to deliver her baby.

Typically, the woman now feels the need to become an active participant in the delivery process: to bear down and to push the baby out (although the doctor or midwife will usually signal her when the best times occur for pushing). With each contraction and each push, the infant rocks back and forth and inches slowly along until the head is ready to emerge from the vagina. The passage of the head is typically the most painful moment of the birth process, and once that is completed, the rest of the baby usually emerges with somewhat less difficulty. In the past, doctors at this stage would routinely perform an *episiotomy*, the surgical cutting of the perineum (the area between the vagina and the anus), presumed to expand the passageway for delivery. During the 1990s, however, this practice decreased dramatically, and today, most medical professionals believe that routine episiotomy offers no advantages and may even make various postpartum physical problems worse. Many researchers and most national health organizations have called for an end to the routine use of episiotomies (Hartmann et al., 2005; Romito et al., 2010c).

As stage two continues, the infant's head usually exits first (see Figure 9.16c), followed by one shoulder and then the other, and at that point, the baby emerges rather quickly—so fast, in fact, that the job of the doctor or midwife is often referred to as "catching" the baby. If you have (or have had) the chance to see an actual birth, you may see exactly why this term is used.

Table 9.6 APGAR Scores for Newborns

Quality	Sign	Score = 0	Score = 1	Score = 2	Total for Item
Appearance	Color	Blue all over or pale	Acrocyanosis (blue and red patches on skin, sweating)	Pink all over	
Pulse	Heart rate	Absent	Below 100	Above 100	
Grimace	Reflex/irritability	No response	Grimace or weak cry	Good cry	
Activity	Muscle tone	Flaccid	Some flexing of extremities	Active flexing or movements of extremities	
Respiration	Respiratory effort	Absent	Weak, irregular, or gasping	Good, crying	

Score ranges: 7–10: Infant displays healthy, normal newborn responses.
4–6: Infant will need close watching and repeated APGAR evaluations.
0–3: Infant is in serious distress and his or her survival is in danger.

The APGAR score should be calculated at one minute and five minutes after delivery, finding the total score (0–10) each time by adding up the points in the table. Scores should be calculated every five minutes thereafter as long as the total score is less than 7.

Following delivery, the umbilical cord is cut, and the newborn infant is immediately examined by the doctor, nurse, or midwife for any signs of difficulty. One widely used immediate assessment of a **neonate** (a newborn) is the **APGAR score**, developed by Dr. Virginia Apgar in 1953 and presented in Table 9.6 (see Rubarth, 2012). This test measures the crucial signs of infant health with respect to skin color, heartbeat, reflexes, movement, and breathing (colleagues of Dr. Apgar assigned the acronym of her name to the test as a way of teaching it to medical and nursing students, the letters APGAR standing for appearance, pulse, grimace, activity, and respiration, respectively). As you can see in the table, the maximum score is 10 (which is fairly rare because most newborns tend to be somewhat deficient in at least one area). However, most newborns will score between 7 and 9. This is a very basic test and, except for scores of under 4, does not predict problems in the future health or abilities of the child. Furthermore, with today's advances in neonatal care, even babies with relatively low APGAR scores, such as between 4 and 6, typically develop without major difficulties (Haidari et al., 2010; O'Reilly, 2011).

neonate
A newborn infant.

APGAR score
A test that analyzes infant health at birth on the basis of skin color, pulse, reflexes, movement, and breathing.

STAGE THREE LABOR In **stage three labor**, within a few minutes following the completed birth of the infant, the placenta (which has joined the fetus and mother throughout pregnancy) is expelled from the uterus with the umbilical cord attached (see Figure 9.16d). The doctor or midwife should carefully inspect the cord and both sides of the placenta to be sure that they are intact and that the placental tissue appears normal and shows no signs that might indicate problems for the mother or the baby. A placental examination typically takes about a minute, and if any serious abnormalities are discovered (which is rare), the placenta is sent on for further pathological examination (Hargitai, Marton, & Cox, 2004; Kaplan, 2007; Roberts & Oliva, 2006).

stage three labor
The final stage of the birth process, when the placenta is expelled from the uterus with the umbilical cord attached.

Choices and Decisions in Childbirth

In Western cultures in past decades, giving birth was a passive activity, with expectant mothers heavily sedated (as described early in this chapter) and fathers, pockets stuffed with cigars, anxiously pacing in the hospital waiting room or even at work, waiting for a call from the hospital. In stark contrast, today the process of labor and delivery has become an active, shared event for most couples, especially in industrialized countries. Today, women and couples are not only deeply involved in their child's birth, but they are taking over a great deal of the control and decision making relating to labor and delivery.

Pregnant women and couples now often expect and demand detailed information, education, comfort, and various amenities to enhance the experience of birth itself. Parents, rather than health professionals, may be the ones to make decisions about medical procedures, such as when to shift course from a vaginal birth to a cesarean section, what type and how much pain medication the mother desires, whether an episiotomy should be performed, and whether their boy baby should be circumcised. In addition to the presence and participation of fathers in the delivery room, some parents are inviting even more people to attend their baby's birth, including family members such as grandparents, siblings, aunts, and uncles of the new infant, and close friends. Let's take a brief look at some of the issues surrounding these various decisions.

DOCTOR OR MIDWIFE? Women and couples may choose among several types of trained specialists to assist them in the actual process of delivery. Typically the choices are from among a physician, usually an **OB/GYN** (obstetrician-gynecologist) or family practitioner; a **midwife**, usually a woman who has been trained in most aspects of pregnancy, labor, and delivery, but is not a physician or registered nurse; or a **nurse-midwife**, who is a registered nurse who also has completed an accredited midwifery program and has been certified by the state to become a certified nurse midwife, or CNM.

Both physicians and midwives may legally attend births in hospitals in most states, but in general, it is midwives who deliver babies in birthing centers and for home births, and the popularity of midwife-assisted births appears to be on the rise. For some deliveries, a midwife and a physician may be present simultaneously. Research has found that births managed by physicians and by midwives do not differ in terms of delivery problems and complications, but patients sometimes report greater satisfaction with the personal involvement and care they receive from midwives (Janssen et al., 2009; Johnson & Daviss, 2005). These comparisons assume a normal, relatively uncomplicated delivery. Physician intervention is usually needed when complications arise that require medical training. Studies have found that between 30% and 40% of births present complications that may require medical intervention by an obstetrician ("Midwives," 2011). On the other hand, women who give birth with midwives are significantly less likely to have common medical interventions such as an episiotomy, cesarean section birth, or artificially induced labor (Johnson & Daviss, 2005).

OB/GYN
Short for obstetrician-gynecologist, a physician specializing in pregnancy and childbirth.

midwife
A person (usually a woman) who has been trained in most aspects of pregnancy, labor, and delivery and who is not a physician or registered nurse.

nurse-midwife
A registered nurse who has completed an accredited midwifery program and has been certified by the state to deliver babies.

Since You Asked

8. My sister wants to have a midwife deliver her baby. Is this dangerous compared to having a doctor handle the birth?

BIRTHING SETTINGS In North America, delivery venues include hospitals, birthing centers, and homes. Most major hospitals have maternity services, and the vast majority of births occur in hospital settings, where many women and couples feel more comfortable knowing that a full range of medical care is immediately available should any problems arise during delivery. A hospital setting is especially recommended for women who have a history of medical problems during childbirth or who have medical issues with their current pregnancy (such as knowledge that a cesarean section may be necessary, or that there may be multiple births, diabetes-related issues, or a breech delivery).

Over the past twenty-five years, women and couples have created a demand for more personalized, parent-controlled childbirth, and in response to this demand, many hospitals have renovated their childbirth departments to make them feel more comfortable and less institution-like. In addition, freestanding **birthing centers** have grown in number and popularity throughout the United States. Birthing centers are facilities with basic, hospital-like medical care equipment but are not formally part of a hospital. They offer a home-like setting and a natural, family-centered approach to the birth process. These centers tend to promote midwife-tended births (typically with on-call physicians readily available) and encourage less use of pain medications during labor and delivery (epidural and certain other medical analgesics are ordinarily not available). Birthing centers offer more choices of food and drink (no alcohol, of course), more activities are permitted and even encouraged during labor (walking, stretching, etc.), and various comfort measures such as massage, relaxation exercises, and even warm tub soaks are often available. Of a total of over four million births per year in

birthing center
A hospital-like facility with basic medical care equipment, focusing on a natural, family-centered approach to the birth process in a home-like setting.

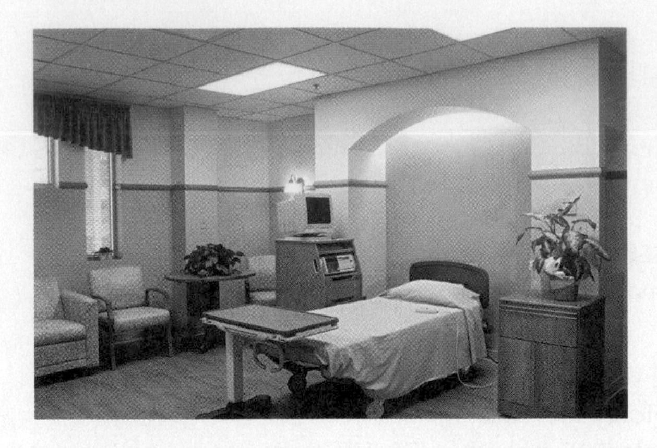

Many hospitals try to create a more comfortable, "home-like" birthing environment.

the United States, only about 1% were delivered outside a hospital. Of those, approximately 28% were delivered in free-standing birthing centers (Martin et al., 2009).

Another option that some prefer is **planned home birth**. Fewer than 1% of births occur in home settings (MacDorman et al., 2014). This is not usually a "do-it-yourself" birthing process but rather involves the use of a home birth service that provides necessary equipment and personnel in the parents' home (Robertson, 2007). A home birth resembles a delivery in a birthing center in that the materials necessary for most normal births are present, including fetal monitors, oxygen, medications to slow bleeding, and sutures to repair tears or an episiotomy (if one is performed). A well-planned home birth requires backup plans, including prearranged transportation to a hospital and an on-call physician in case serious complications arise that may require medical attention beyond that available to the midwife in the home. Home births are recommended only for women who are deemed low risk for delivery problems. Complications are more common in home births, including a higher percentage of stillbirths, seizures, and neurological disorders (Ellis, 2013). The advantages of home birth is that it places the woman or couple in familiar surroundings, where they feel more comfortable and in control of the birthing process, and under normal circumstances, its total costs are lower.

planned home birth
Delivery of an infant in a private home setting, usually with necessary equipment and personnel provided by a professional service.

PAIN MEDICATION DURING LABOR AND DELIVERY In the now distant past, women were placed under general anesthesia to ease the pain and trauma of childbirth. This, of course, also removed the mother from the psychological and emotional joy of seeing her baby come into the world, as well as her help in pushing the baby out. It also subjected the fetus, along with the mother, to the potentially dangerous sedating effects of the drugs.

In the early 1950s, the French obstetrician Ferdinand Lamaze, on a trip to Russia, discovered a drug-free method of childbirth that used psychological techniques to allow women to limit the pain of childbirth and deliver babies naturally. When he brought the techniques back to Europe and the United States, they began to catch on. Here was a way for women to give birth not only fully conscious but also, if they chose, with no drugs at all, using special breathing exercises, visualization, massage, focused attention, and relaxation training, along with the help of a "coach" (husband or partner, friend, relative) to ease pain and increase the mother's comfort during labor and delivery (Lothian, 2011).

Today, variations on the *Lamaze method* are used by millions of women worldwide. Most women desire to be conscious, alert, and aware during the birth of their children. They want to be an active participant in giving birth, and they want to protect their baby from the negative effects of medications. Using techniques related to Lamaze's work, many women are able today to make the journey through labor and birth with less or no pain medication (see Lamaze.org for more detailed information).

Many other women, however, will tell you that "talk is cheap," and that the *idea* of childbirth without medication is much more attractive than actually enduring it. Lamaze never truly took a stand *against* pain medication; he simply felt that women should understand the available options for pain management and make an informed decision. One of the many decisions a woman or couple must make concerns the availability and type of pain medication, if any, that they will use during labor and delivery.

It is not at all unusual for a woman to plan for or desire a medication-free, or "natural," childbirth, but at some point in the emotionally intense, physically demanding, and painful birthing process, she may wish to change that plan. Statistically, less than 10% of all women in the United States use no medication during labor and delivery ("Women Are Receiving," 2006). Many childbirth experts suggest that the expectant mother be given the *option* to receive pain medication during birth even if she fully intends to attempt an "all-natural" birth.

Once the decision has been made to use pain medication or have it available during the birth process, the next consideration is the type of medication preferred. The most common pain medications in use today are *sedatives*, such as *antianxiety medications; local anesthesia*, numbing agents used in specific areas of the birth canal; *regional anesthesia*, the numbing of the entire birth canal and pelvic area; and the *epidural*, currently the most popular method in the United States.

An epidural involves the insertion of a small tube, called a *canula*, near the base of the woman's spine into the fluid that surrounds the spinal cord. This is typically done prior to the beginning of hard labor. With the canula in place, pain-relieving anesthetic can be administered easily as needed during labor and delivery. The administering of the anesthesia may be controlled by the mother herself, as she experiences various levels of pain (within medically safe parameters). Research has found that when the mother is given control over the amount and frequency of the epidural pain medication, she will, on average, use less than when it is administered by the delivery room medical staff (Kraft, 2011).

These targeted approaches to pain during childbirth usually provide enough relief to allow for labor and delivery with tolerable levels of pain and anxiety. This, in turn, leaves her able to participate in the birthing experience, while staying awake, alert, and active enough to help push the baby out when the time comes to do so.

BIRTH BY CESAREAN SECTION As most of you are aware, a **Cesarean section** or *C-section birth* (from the Latin verb *caedere*, "to cut") involves removing the fetus from the woman's uterus surgically, through an incision in her abdomen (see Figure 9.18). Historically, this procedure has been performed when complications arise that might make a vaginal delivery too dangerous for the health of the mother or baby, such as the baby's size; maternal high blood pressure; active maternal sexually transmitted infections; or the position of the baby during labor, such as a breech position, as discussed earlier ("C-Section," 2010). The World Health Organization suggests that a C-section rate of 5% to 15% of births is optimal for the health of the mother and infant. A higher rate indicates overuse of C-sections (with the related risks discussed below) and less than 5% points to the lack of medical care available when a C-section is medically necessary. As you can see in Table 9.7 (following page), many countries far exceed the maximum recommended C-section rate, although many other countries are dangerously below 5%.

cesarean section
Removal of a fetus from the mother's uterus surgically, through an incision in her abdomen; also called a *C-section birth*.

In the United States and many other industrialized countries, the number of C-section births has been steadily rising over the past ten years or so, but the number of natural childbirth complications has not. In 2009, the U.S. cesarean birth rate reached its highest point in history, accounting for 33% of all births, which represented a 60% increase since 1996 (Hamilton et al., 2010). Then, in 2009, the C-section rated dropped and leveled off to 31% through the end of 2011, still significantly higher than the 21% in 1996 (Nordqvist, 2013b). Why so many C-sections? Various possible reasons are discussed in the birthing literature: (1) More mothers are choosing to have C-sections; (2) doctors are more willing to recommend them due to fears of malpractice litigation if any problems arise from a potentially complicated vaginal birth; (3) the average age at which women give birth has risen significantly, increasing the possibility for complications that may lead to cesarean births; and (4) C-section births are approximately twice as profitable for the hospital as uncomplicated vaginal births. Research has shown that women are significantly more likely to deliver by C-section at private compared to nonprofit hospitals (see Johnson, 2010). As of 2013,

Figure 9.18 C-Section Birth
A cesarean section is performed when medical complications prevent a vaginal delivery or, in some cases, it is the method of delivery elected by the mother.

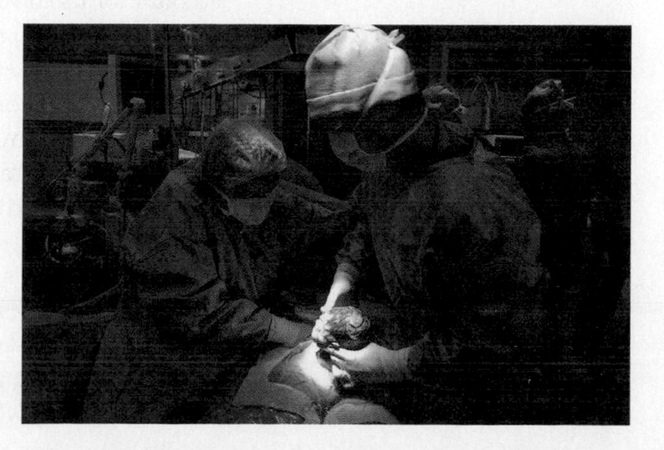

Table 9.7 Global Rates of Cesarean Section Births: Selected Countries (2010)

Country	Births by C-Section (%)
Ten Countries with Dagerously Low Rates	
Madagascar	1.0
Niger	1.0
Guinea	1.7
Cambodia	1.8
Cameroon	2.0
Uganda	2.1
Zambia	2.1
Rwanda	2.9
Haiti	3.0
Tanzania	3.2
Ten Countries with Excessively High Rates	
Brazil	41.3
Republic of Korea	37.7
Mauritius	37.0
Mexico	36.1
United States	31.0
Australia	30.8
Chile	30.7
Germany	28.8
Cuba	28.5
Hungary	28.0

SOURCE: Adapted from World Health Organization, World health statistics 2010, Part II, TABLE 4, pp. 92–96 (partial).

the American College of Obstetricians and Gynecologists (ACOG) recommended that, "In the absence of maternal or fetal indications for cesarean delivery, a plan for vaginal delivery is safe and appropriate and should be recommended. Cesarean delivery on maternal request should not be performed before a gestational age of 39 weeks. Cesarean delivery on maternal request should not be motivated by the unavailability of effective pain management" (ACOG, 2013).

Pregnant women are choosing to have C-section births (called *elective C-sections*) for various reasons, including avoidance of labor and childbirth pain, the convenience of giving birth on a schedule set by the mother rather than "by the baby," or the ability to ensure the father's presence at the birth. In some countries, elective C-sections are even more common than in the United States. In Brazil, for example, where C-sections account for up to 90% of all births in private hospitals, a commonly heard joke is that the only way a woman has a vaginal delivery is if her obstetrician gets stuck in traffic (Stalburg, 2008).

The medical community in the United States and other countries are debating this issue because many doctors consider an elective C-section unnecessary surgery, and avoiding unnecessary surgery is a basic tenet of the medical code to "do no harm." Other physicians disagree and believe a woman has the right to choose how to have her baby.

Overall, the risks of a C-section delivery to the mother are greater than those of a vaginal delivery because a C-section is major abdominal surgery, requiring stronger anesthesia and carrying a higher risk of blood loss, infection, and damage to the reproductive tract. Moreover, if a C-section occurs too early in pregnancy (prior to 37 weeks gestation, which could be due entirely to a slight miscalculation in the fetus's due date), these are preterm births and are subject to all the complications discussed earlier in this chapter.

Most importantly, C-section births have been associated with significantly higher infant and mother morbidity and mortality rates (see Nordqvist, 2013b). Mothers who have C-sections are at greater risk of infections, bladder and bowel injuries, and longer hospital stays. Infants delivered via C-section are more likely to develop allergies and asthma in childhood; are at greater risk of obesity in adulthood; and more likely to lack certain beneficial bacteria necessary for digestion, immune functioning, and healthy bowel functioning. Nevertheless, the worldwide trend toward more pregnant women electing C-section deliveries is growing and is likely to intensify as a major controversy in the expanding multicultural complexities of childbirth (see "Sexuality and Culture: Cesarean Section Births Around the World").

THE CHALLENGES OF CHILDBIRTH CHOICES Childbirth seemed so much simpler in the past; people just "had babies." Now prospective parents face a multitude of choices and decisions. This is not necessarily a negative development, but how do people cope with all they must learn and decide for a safe and personally satisfying birth process? It takes a great deal of education, awareness, and knowledge to cope with all the choices parents-to-be now are given. Luckily, many thorough and informative resources are available from doctors, nurses, midwives, hospitals, and birthing centers, not to mention the hundreds of excellent websites, including, to name just two: http://womenshealth.gov/pregnancy and www.childbirth.org.

No two pregnancies are exactly alike, and perhaps the most important resource for women and couples about their very personal experience of childbirth is the comfort and trust they are able to establish with their doctor or midwife. Those are the professionals who are in the best position to provide support and consultation in the complex process of bringing a new life into the world. If a woman or couple do not feel a trusting connection with their birth facilitator, they should not hesitate to obtain a referral to someone who can better meet their needs.

Postpartum Issues

9.7 Review the issues that may be present following child birth, including postpartum depression, the importance of breast feeding, and resuming sexual activities after delivery.

The delivery of a baby is, of course, only the first step in a lifetime of challenges presented to parents. Effective parenting is an important discussion; one that is well beyond the scope of this chapter and this book. Nevertheless, we should touch on a few issues that many parents face immediately **postpartum**, that is, following the birth of a child. These include the very real and all-too-common problem of postpartum depression; the resumption of sexual activities following birth; and issues relating to the return of ovulation, fertility, and menstruation.

postpartum
Literally "following birth"; typically refers to the months or first year following the birth of a child.

Postpartum Depression

After giving birth, some new mothers experience what are commonly referred to as the "baby blues," a normal, relatively mild depressed mood, usually stemming from the exhaustion from labor, hormonal shifts, the fatigue of nights with little and poor sleep, and the stress of a major lifestyle upheaval with the addition of a new family member. Although women often worry about these feelings of sadness, crying spells, and extra anxiety at a time when they feel they should be happy, the baby blues are short-lived and usually pass within a week or two at the most. And this condition does not appear to be limited to mothers. Some research has indicated that the symptoms of PPD may affect new fathers as well (Harmon, 2010).

Since You Asked

9. Why do some women get so depressed after giving birth? That seems so opposite of how a new mother should be feeling, if she really wanted the baby.

For as many as one in five women, postpartum depression (PPD) is a psychological disorder that, in addition to all the typical signs of depressive illness, also may involve a lack of interest in the new baby, overly intense worry about the baby's welfare, and fears that she might harm the infant in some way. If not treated, this may cause the relationship between the mother and infant to suffer, and the wellbeing of both may deteriorate.

postpartum depression (PPD)
A psychological depressive disorder that begins within four weeks after childbirth.

postpartum psychosis
A severe postpartum psychological disorder that may include delusions, hallucinations, and extreme mental disorganization.

Over 14% of new moms, however, suffer from a much more serious problem called **postpartum depression (PPD)**, many of whom are not properly diagnosed or treated (Fitzgerald, 2013). Postpartum depression is a clearly defined psychological disorder that begins within four weeks after childbirth and shares many of the symptoms of other forms of depression, such as deep sadness, emotional apathy, withdrawal from family and friends, loss of interest in favorite activities, fatigue, changes in eating habits, and feelings of failure or inadequacy. PPD may also typically involve a troubling lack of interest in the new baby, overly intense worry about the baby's welfare, fears that she might harm the infant in some way, and often extreme anxiety or panic attacks (Merrill & Zieve, 2010). Even more concerning are the findings that a fifth of all mothers with PPD consider harming themselves, and their babies are at greater risk of attachment disorders and inhibited cognitive development during infancy. Although postpartum depression is a well-recognized issue among women, evidence is increasing that new fathers may suffer from similar symptoms nearly as commonly. Postpartum depression among fathers appears to negatively affect the mother's mood and the healthy development of the infant (Aldhous, 2008). "Self-Discovery: A Postpartum Depression Checklist" offers a useful guide for evaluating if you or someone you know might be suffering from PPD.

Lack of recognition and diagnosis of PPD can, in some of the most serious cases, lead to negative consequences, including neglect of the infant that may be life-threatening. A rare, severe form of PPD called **postpartum psychosis** occurs in one to two out of every 1000 new mothers and may be responsible for temporary mental illness that may cause such extreme postpartum events as infant abuse, or even infanticide (the murder of an infant Berga, 2007; Sit, Rothschild, & Wisner, 2006; Stoltz, 2013). Postpartum psychosis, in addition to many of the symptoms of PPD, may be characterized by delusions, hallucinations, and extreme mental disorganization, leading to extreme and often violent behavior.

Treatments for PPD are similar to established treatments for other forms of depressive illness and include psychotherapy and antidepressant medication. Any medication a nursing mother takes has the potential to enter the breast milk and possibly the infant's system. Therefore, many physicians and new mothers avoid antidepressants for PPD in nursing mothers. This has been one reason why some women with PPD have gone untreated or have been forced to stop nursing sooner than they wish so that they could start (or resume) medication. This creates a dilemma of weighing the risks of PPD to the mother and the infant against the risks of the medications or not breast-feeding to the infant (see *Breastfeeding* below). Studies are ongoing, but

Self-Discovery

A Postpartum Depression Checklist

The following list of statements relates to postpartum depression. Keep in mind that this disorder typically lasts for at least ten days after childbirth and may continue for up to a year. If you suspect that you or someone you know may be suffering from postpartum depression, take this completed form to the doctor for further evaluation and treatment. *The more statements checked, the more likely a diagnosis of postpartum depression.*

_______ I don't enjoy holding and cuddling my baby.

_______ I have had depression or bipolar disorder before.

_______ I have frequent crying spells for little or no reason.

_______ I have difficulty sleeping, even when the baby is not keeping me awake.

_______ I feel extremely anxious, even when caring for my baby.

_______ I have thoughts of injuring myself or the baby.

_______ I have thoughts of injuring others.

_______ I sometimes feel as if I just don't want to live anymore.

_______ I dread or avoid nursing, feeding, or changing my baby.

_______ I feel as if I'm a poor mother, even though others say I'm doing fine.

______ I have tremendous feelings of guilt (about anything).

______ I feel worthless.

______ I'm having difficulty eating (or I'm seriously overeating).

______ I have headaches that don't resolve with over-the-counter painkillers.

______ I have difficulty concentrating or remembering even the simplest things.

______ I have a history of PMS.

______ I have a family history of depression or bipolar disorder.

______ I hate or resent the baby's father.

______ I feel that if that baby doesn't stop crying, I'm going to explode!

______ I don't have a supportive relationship with the baby's father.

______ I don't have family or friends to support me emotionally.

______ I don't have support to help out with baby care or other daily tasks.

______ I had a traumatic pregnancy or delivery; it was one of the worst experiences of my life.

______ I just can't seem to get out of bed.

______ I've experienced PPD in the past.

SOURCE: Based on D. Moore (2002), http://www.drdonnica.com/display.asp?article=154.

a trend in the findings is that for at least some of the newer SSRI antidepressants—especially paroxetine (Prozac) and sertraline (Zoloft)—the adverse effects on infants are minimal, and the benefits of treating a mother's PPD outweigh any potential risk to the normal development of the baby (Berle & Spigset, 2011; Meltzer-Brody, Payne, & Rubinow, 2008).

Breast-Feeding

One of the most healthful activities a new mother can do for her child is to breast-feed. Breast-feeding provides a rich bonding experience between mother and child. In addition, the benefits of breast feeding for mother and infant are many. "In Touch with Your Sexual Health: The Many Advantages of Breast Feeding" summarizes these.

In Touch with Your Sexual Health

The Many Advantages of Breastfeeding

The California Department of the Federal Women, Infants, and Children Supplemental Nutrition Program (WIC) offers this comprehensive list of the many and varied benefits of breastfeeding.

Human milk is uniquely suited for human infants

- Human milk is easy to digest and contains more than 200 components that babies need in the early months of life.
- Factors in breast milk protect infants from a wide variety of illnesses.
- Children who have been breastfed have less risk of becoming overweight or obese, even as adults.
- Research has shown that children who have been breastfed have higher IQs.

Breastfeeding saves lives

- Lack of breastfeeding is a risk factor for sudden infant death syndrome (SIDS).
- Human milk protects premature infants from life-threatening gastrointestinal disease.
- Breastfed children have lower risk of dying before their first birthday.

Breastfed infants are healthier

- Formula fed infants have twice the risk of having ear infections in the first year than infants who are exclusively breastfed for at least four months.
- Breastfeeding reduces the incidence and lessens the severity of a large number of infections, including pneumonia and meningitis in infants.
- Breastfeeding protects infants against a variety of illnesses, such as diarrhea and infant botulism.
- Breastfed babies have less chance of allergies, asthma, and eczema.
- Evidence suggests that exclusive breastfeeding for at least two months protects susceptible children from Type I diabetes.
- Breastfeeding may reduce the risk for subsequent inflammatory bowel disease, multiple sclerosis, rheumatoid arthritis, and childhood cancers.

Breastfeeding helps mothers recover from childbirth

- Breastfeeding helps the uterus to shrink to its prepregnancy state and reduces the amount of blood lost after delivery.
- Mothers who breastfeed for at least three months may lose more weight than mothers who do not breastfeed.
- Breastfeeding mothers usually resume their menstrual cycles 20 to 30 weeks later than mothers who do not breastfeed.

Breastfeeding keeps women healthier throughout their lives

- Breastfeeding reduces the risk of breast and ovarian cancer.
- Breastfeeding may reduce the risk of osteoporosis

Breastfeeding is economical

- The cost of infant formula has increased 150 percent since the 1980s.
- Breastfeeding reduces health care costs.

Breastfeeding is environmentally sound

- Unlike infant formula, breastfeeding requires no fossil fuels for its manufacture or preparation.
- Breastfeeding reduces pollutants created as byproducts during the manufacture of plastics for bottles and metal for cans to contain infant formula.
- Breastfeeding reduces the burden on our landfills, as there are no cans to throw away.

Breastfeeding is the single most important health behavior for an infant and mother.

On a related note, one common fertility myth is that a woman cannot become pregnant while she is breastfeeding (lactating). Although the chances of pregnancy during lactation may be reduced, it is far from a reliable method of contraception. The world is full of biological siblings who are only 10 or 11 months apart in age! The best advice for most heterosexual couples is to resume some form of contraception rather than relying on lactation when they start having intercourse following childbirth, just to be on the safe side (assuming that they do not wish to become pregnant again soon after the birth).

Many nonintercourse sexual activities offer ways for couples to express love and intimacy following childbirth.

Barrier methods of birth control, used correctly, provide the safest contraception for new parents (see Chapter 5, "Contraception: Planning and Preventing"). In general, the use of hormonal contraceptives should be avoided by women who are nursing for at least the first six weeks after birth because combined hormonal contraceptives may interfere with the production of breast milk and small amounts of the hormones may be passed along to the infant during breastfeeding (Lesnewski & Prine, 2006; Teal, 2011).

Sexual Activities after Childbirth

As mentioned earlier in this chapter, for most couples, sexual activity is usually considered safe during pregnancy nearly up to the onset of labor, if the couple desires and feels comfortable with it. But what about sexual intimacy *after* childbirth? Most women who have recently given birth will probably not be eager to engage in some sexual activities because they are, at the very least, exhausted and sore. The exact length of time before a woman feels ready for sexual intimacy depends on many factors, including the length and difficulty of labor, whether she had an episiotomy, and whether the birth was vaginal or via C-section. For heterosexual couples, most doctors and midwives recommend waiting at least six weeks after birth to have intercourse in order to allow adequate time for the woman to heal and for the cervix to close fully, so that bacteria that may enter the vagina during sexual activity cannot travel up into the uterus On average, couples typically wait a little longer than that—about seven weeks—before resuming lovemaking involving sexual intercourse, and they wait even longer if complications experienced during delivery make healing slower (see Hipp et al., 2012).

This can be a very trying time for some couples, especially if intercourse was suspended for a while prior to the birth as well. Sometimes, a new mother's partner may feel excluded from all the attention the mother is lavishing on the baby. At the same time, couples need to wait an appropriate period of time to resume intercourse, and they clearly do not want to rush into something that might hurt or harm the woman. So what is a couple to do?

As mentioned throughout this book, if new parents understand and accept the full range of sexual expression, many behaviors can allow them to feel close, intimate, and sexually satisfied without physical discomfort. What kinds of behaviors? With the exception of intercourse itself, most or all of the behaviors discussed in Chapter 6, "Sexual Behaviors: Experiencing Sexual Pleasure"—including kissing, touching, masturbation (individual, shared, and for each other), and oral sex—are common sexual behaviors couples use to feel close and sexually intimate during the postpartum period (Delvin & Webber, 2005). To reduce the risk of injury or infection, the couple should avoid inserting anything into the vagina, including a penis, fingers, "sex toys," or the tongue until the end of the postpartum recovery period. None of these should come into contact with any stitches or other areas that are still healing (McKain, 2002). Other than those cautions, most sexual behaviors are safe and usually very enjoyable for couples during the postpartum "waiting period" or, for that matter, anytime. Some couples even report that their sex lives become more varied and experimental during this time and create lasting, positive changes.

Impaired Fertility

9.8 Explain impaired fertility (or infertility) and summarize the scope of the problem, causes of low or absent fertility, how fertility is tested, and the possible solutions to impaired fertility.

As this chapter draws to a close, we arrive at the discussion that is at the opposite end of nearly everything about conception we have been discussing so far: the *lack* of conception, pregnancy, and birth. As mentioned at the very beginning of this chapter, most of you who are sexually active have probably not thought very much about trying to get pregnant; you've been more concerned with *avoiding* or *preventing* pregnancy. You are undoubtedly aware of the emotional turmoil of unwanted pregnancy. You might be surprised, therefore, to learn just how many couples struggle with the emotional and psychological pain of *not* becoming pregnant when they desperately want to.

Since You Asked

10. What is infertillity and what causes it in males and females?

Scope of the Infertility Problem

Infertility, sometimes called "impaired fertility," is typically defined as the lack of conception by a heterosexual couple after 12 consecutive months of trying to become pregnant; that is, having regular, unprotected intercourse (CDC, "Infertility," 2009). In the United States, the number of couples who meet this definition of infertility is estimated to be approximately 10%, which works out to approximately five million couples. These statistics do not imply that all of these couples will *never* become pregnant with continued trying; on the contrary, 80% of couples conceive within one year, but that rises to 90% over two years of trying (Jenkins, 2006).

infertility

A failure to conceive for 12 consecutive months despite persistent attempts.

Figure 9.19 Probability of Pregnancy for a Couple with Normal Fertility
(when woman is under thirty years of age)

SOURCE: Adapted from Petrozza (2006).

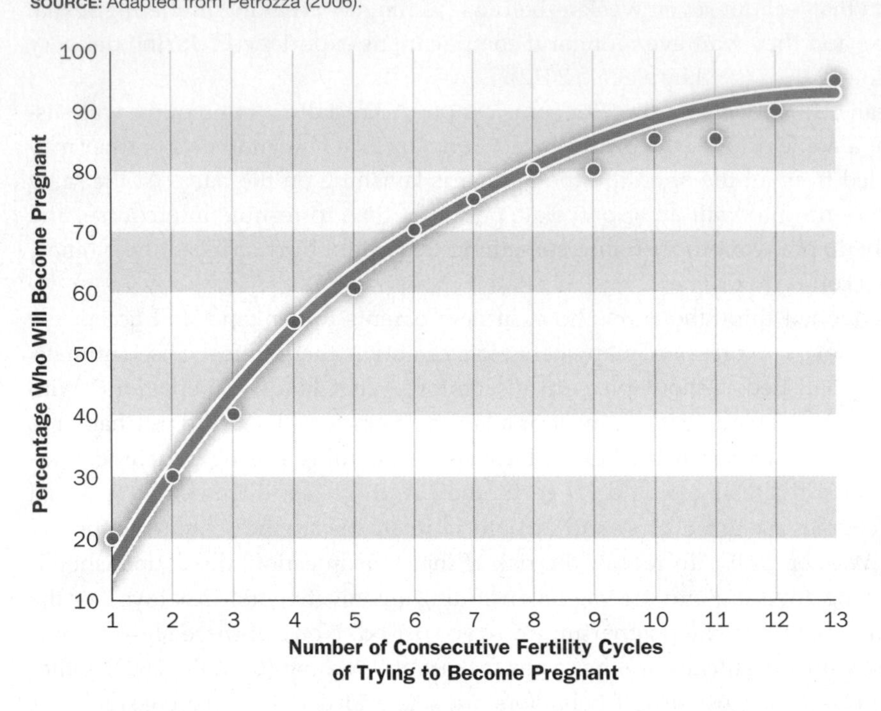

Even when both partners are fertile and making love regularly, the odds of a pregnancy occurring in any particular month are only about 20% to 30% (although it would be a mistake to rely on these odds for contraception!). Figure 9.19 shows the cumulative percentage rates for a normal pregnancy occurring over 13 consecutive fertility cycles, assuming a 20%-per-cycle rate. As you can see, pregnancy isn't a near certainty for fertile couples until after about a year of trying. Nevertheless, when partners desire and expect to get pregnant and stop using birth control, they may begin to feel very frustrated if pregnancy does not occur over the next few months. If a couple has lower-than-normal fertility, the odds of pregnancy for each cycle will be lower, and their overall pregnancy rate will rise more slowly over time. In other words, it will take them longer than average to become pregnant. Typically, most fertility experts suggest that couples try to conceive for at least a year before they consider seeking medical fertility intervention.

Infertility is usually very frustrating and emotionally difficult for a couple. They want to become pregnant, have likely been hoping to start a family for some time, and have assumed that when the time came, becoming pregnant would not be a problem. When they experience difficulty conceiving, the passing months and even years of trying can be one of the most challenging periods in a couple's life, for both the potential mother and father (Glynn, 2013).

Causes of Impaired Fertility

One of the first questions couples ask themselves and their doctor when they seek help for difficulty conceiving is some version of "Whose *fault* is it?" Usually, this is not intended to assign blame; they are simply asking whether the problem lies with the man or the woman. If we consider all cases of impaired fertility, research has shown that the main causes of infertility break down as follows: 25% due to a male sperm factor (e.g., low number, poor quality, or low motility), 20% due to a female ovarian problem (e.g., no ovulation or poor quality ova), 14% due to a female tubal problem (e.g., fallopian tube blockage), and 41% due to other problems, or there is no identifiable problem in either partner (unexplained infertility) (Jose-Miller, Boyden, & Frey, 2007). The most common causes of impaired fertility will be discussed in more detail next.

OVUM AND SPERM PROBLEMS If you think about the basics of conception as discussed earlier in this chapter, you will probably be able to figure out the most common causes on your own. For a pregnancy to occur naturally, sperm cells must encounter and be able to penetrate the ovum in the fallopian tube. If the woman is not ovulating or if her fallopian tubes are blocked in some way, conception cannot occur. Typically, the lack or irregularity of ovulation is mirrored by a similar lack or irregularity of a woman's menstrual period. Because the menstrual cycle is controlled by various hormones (discussed earlier in this chapter), any condition that interferes with normal hormonal balance, such as stress, extreme dieting, overexercising, certain medications, or the abuse of recreational drugs, may inhibit ovulation. Damaged or blocked fallopian tubes are most commonly caused by one of two conditions: Untreated sexually transmitted infections, such as chlamydia or gonorrhea, may lead to pelvic inflammatory disease, which may in turn damage and block the fallopian tubes (see the discussion of these issues in Chapter 8, "Sexually Transmitted Infections/Diseases"). The other most common cause of impaired fallopian tube functioning is *endometriosis* (discussed in Chapter 2, "Sexual Anatomy"), when uterine tissue cells are growing outside the uterus, potentially affecting the passage of sperm through the fallopian tubes ("Infertility," 2010).

On the male side, the problem of infertility is related to abnormalities in sperm cells. If too few sperm cells are ejaculated, if the sperm cells cannot swim well enough to move up into the fallopian tube, if the vas deferens does not allow passage of sperm cells, or if the sperm lack the chemicals required to penetrate the ovum, fertility will be impaired. Many factors may contribute to poor sperm quality, including illnesses such as mumps in adulthood, sexually transmitted infections (most commonly gonorrhea and chlamydia), testicular infections or abnormalities, the abuse of tobacco or drugs (alcohol, cocaine, marijuana), and exposure to environmental toxins ("Male Infertility," 2010).

Sometimes infertility is not related to conception but rather to pregnancy. If an ovum is successfully fertilized and the zygote travels through the fallopian tube normally but for some reason fails to implant in the wall of the uterus, no pregnancy will be established. The zygote will be expelled, and the woman will begin her next period without even being aware that conception occurred during her previous cycle.

AGE Increasing age is associated with decreasing fertility, especially in women. As women enter their thirties and early forties, it may take longer to conceive, but this does not necessarily imply that they cannot become pregnant. Approximately 45% of women over 40 who wish to become pregnant eventually do. However, after age 45, the odds become very slim (Glazer, 2007). Of 4.2 million births in the United States during 2009, only 2% were to mothers 45 or older, and the chances of miscarriage, Down syndrome, or other pregnancy complications increased dramatically (Hamilton et al., 2010; Seidel, 2011). The reasons for the drop in fertility during a woman's forties are not clear, but most researchers agree that it is probably related to a decline in the viability of the aging eggs, which have been present in the woman's body since before birth (Klotter, 2002; Liu & Case, 2011).

A man's age, on the other hand, does not appear to affect fertility as significantly as does the woman's. Overall, most men's sperm retain their ability to fertilize an ovum throughout the male lifespan, although the number, quality, and motility (swimming ability) may decrease after age 35 (Teich, 2007). And one other issue should not be overlooked: for most men, the frequency of intercourse decreases with age, which naturally affects a couple's overall chances of pregnancy. This brings us to the next issue relating to infertility.

FREQUENCY OF INTERCOURSE Obviously, if a couple wants to become pregnant, they need to have intercourse often enough for sperm to be present in the fallopian tube around the time of ovulation. Therefore, you will not be surprised to learn that the percentage of couples who conceive increases right along with frequency of coitus. In the past, a common belief was that for optimum fertility, a man should abstain from ejaculation and "save up" his semen for a few days to increase the number of sperm cells in his ejaculate. This does not appear to be the case, however. Research indicates that intercourse once every day or two on a regular basis is probably best for optimizing the probability of conception. Moreover, sperm quality may even decrease with long periods of abstinence ("Study Refutes," 2003). However, keep in mind that a woman is fertile only for about one day each month and sperm may survive in the woman's reproductive tract for up to five days. Therefore *when* during her fertility cycle a couple has intercourse affects fertility rates as well as we will discuss next.

TIMING OF INTERCOURSE The timing of intercourse in relation to ovulation is key to conception. It follows, then, that one reason a couple might be having difficulty conceiving is that they may not be engaging in intercourse near the time when the woman is ovulating. This problem may stem from the couple's lack of understanding of the fertility cycle. If a couple believes that the best time for conception is immediately after, immediately before, or during the woman's period (and many people do have such misunderstandings), they may be focusing their lovemaking energy on precisely the times of the month when pregnancy is *least* likely to result. Remember, as discussed earlier in this chapter, each woman is unique relative to her fertility cycle, but *on average*, frequent intercourse near the middle of her cycle, when most women ovulate, will enhance the odds of conception, especially if she has fairly regular periods. Some couples use the techniques of "fertility awareness" discussed in Chapter 5, "Contraception," not to prevent conception, but to enhance their changes of conceiving.

FERTILITY TESTING Many couples are not terribly eager to seek medical attention for fertility problems. Some people feel that these issues are intensely personal and find them difficult to discuss with anyone, even a doctor. Moreover, they are likely to feel some trepidation about the nature of any tests that might be prescribed and fearful of perhaps discovering that they may be permanently infertile (which is rarely the case). However, the problem of impaired fertility is common enough that couples having difficulty conceiving should not feel alone and should know that many dedicated professionals are available to help them.

Typically, a family doctor or a physician specializing in fertility issues will take a detailed medical history from each partner and discuss in detail what the two of them have done so far in their efforts to conceive a child. At this point, the couple is often reassured, given accurate and helpful information about their chances of becoming pregnant without any intervention, informed about the frequency and timing of intercourse, and advised to continue "doing what comes naturally" a while longer. In other cases, depending on the length of infertility, the age of the couple, and the individual couple's needs and desires, some testing may be initiated to try to determine a cause of their fertility problem.

The main test for the male is a *semen analysis*, in which a sample of the man's semen (usually gathered through masturbation into a sterile cup) is analyzed microscopically to check for adequate numbers, movement, and formation of the man's sperm cells (a do-it-yourself home semen analysis kit now exists). For the woman, tests may include a *hysterosalpingogram*, also (thankfully) referred to as the HSG, an X-ray that allows the doctor to view the interior of the uterus and fallopian tubes to check for malformations or blockages. For this test, a special dye is infused into the uterus under slight pressure so that it flows through the fallopian tubes, allowing any tubal problems to be seen on an X-ray monitor. The HSG is somewhat invasive and can cause some temporary cramping for the woman during the test itself.

Sometimes, doctors need to have a direct, internal look at the ovaries, fallopian tubes, and uterus. This is accomplished with a **laparoscopy**, in which a tube with a tiny camera and light is inserted through a very small incision in the abdomen (see Figure 9.20). Laparoscopies are usually performed under general or spinal anesthesia. By moving this camera inside the abdomen, the doctor is able to visually examine the woman's reproductive anatomy to check for any abnormalities of the ovaries or uterus, cysts, endometriosis, or damage to the fallopian tubes.

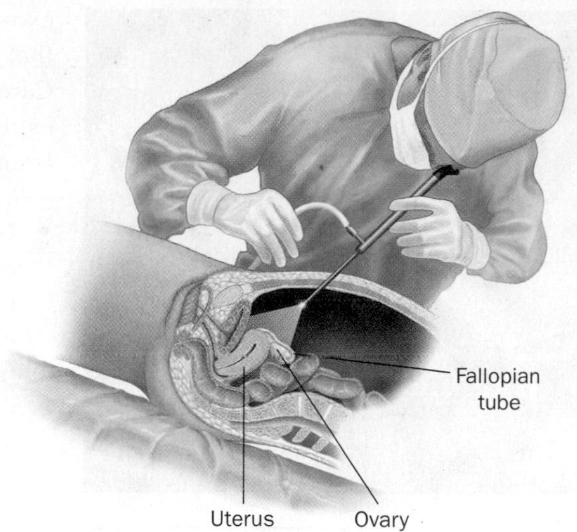

Figure 9.20 Laparoscopy

A laparoscopy allows a physician, through a tiny incision, to check visually a woman's ovaries, fallopian tubes, and surrounding structures for any problems that may contribute to infertility.

laparoscopy

A surgical procedure in which a tube with a tiny camera and light is inserted through a small incision in the abdomen.

assisted reproductive technology (ART)

Various treatments to help infertile women or couples to become pregnant and have a child.

Solutions to Impaired Fertility

Individuals and couples who want to have children but who have difficulty becoming pregnant for any of the reasons discussed in the previous section often face a difficult emotional road. They make love with the goal and the hope of pregnancy, but each time the woman's period begins, they must deal with a new wave of disappointment and start all over again. Couples in this situation often report that they are living their lives in two-week fragments: two weeks or so until ovulation is pending when they "must" engage in frequent intercourse and then about two weeks of waiting and hoping that the next scheduled menstrual period does not arrive. For some couples on this path, lovemaking becomes a "duty" that must be performed with a dictated frequency and at predetermined days of the month to optimize the odds of conception. They feel that the spontaneity, joy, and fun of their love life have given way to a tightly scheduled sexual obligation.

Fortunately, numerous paths to parenthood are available today to infertile couples. These include various treatments to enhance natural fertility, assisted reproductive techniques that allow pregnancies to occur using laboratory fertilization techniques, and adoption.

INFERTILITY TREATMENTS Various treatments have been developed that enhance the chances of conception via sexual intercourse. Several fertility drugs using *clomiphene citrate* (brand names Clomid and Serophene) have been developed to stimulate ovulation in women, and the drug may also assist in sperm production and quality in men (Murdock, 2004). Surgical procedures may be effective in some cases to unblock fallopian tubes, repair testicular or related disorders, or remove cysts or scar tissue from the uterus, ovaries, or fallopian tubes.

Intrauterine insemination (IUI), often referred to as "artificial insemination," is a rather simple and low-cost process of inserting a small tube through the woman's cervix and injecting sperm cells directly into the uterus (see Figure 9.21). Usually, the sperm for this procedure is from the woman's male partner; however, the cells may be from a sperm donor if her partner does not produce viable sperm, if she is not in a relationship but wants to conceive, or if she is in a same-sex relationship in which children are desired.

Figure 9.21 Intrauterine Insemination

Intrauterine insemination, or IUI, introduces sperm cells directly into the uterus, thereby placing them closer to the fallopian tubes and enhancing the odds of conception.

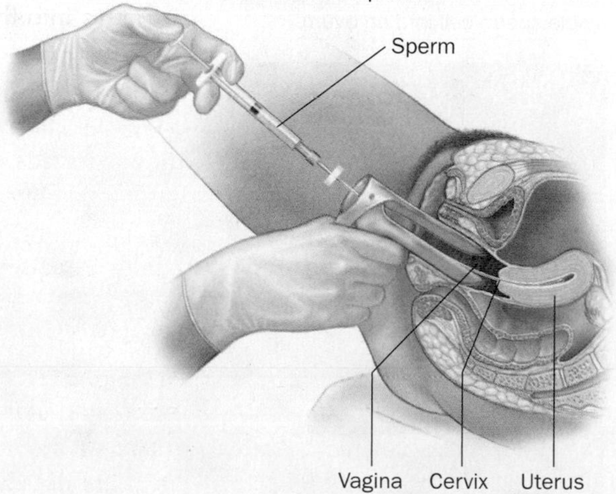

Another group of treatments for infertility are collectively referred to as **assisted reproductive technology (ART)**. These technologies represent a "brave new world" of reproduction and are constantly being researched, modified, and improved.

Four methods of ART are in current use. The average cost of these methods in 2014 was $12,400 per cycle (i.e., per attempt). Although a successful pregnancy may occur on the first IVF try, the average number of cycles is three, meaning the average cost

Figure 9.22 The World's First "Test Tube Baby"

The first in vitro fertilization baby, Louise Brown, born in 1978, is shown here with her husband, Wesley Mullinder, and son, Cameron, in 2013.

exceeds $30,000 to achieve a pregnancy (ASRM, 2014). Except in a few states that mandate it, most health insurance will not cover ART and the Affordable Care Act of 2013 does not mandate this coverage. The four methods are briefly explained in the following paragraphs (see CDC, "ART," 2011; Eisenberg & Brumbaugh, 2009).

- **In Vitro Fertilization (IVF).** This technique is probably the best known as the one said to produce "test tube babies." Its first successful use in 1978 led to a normal pregnancy and birth (see Figure 9.22). It does not involve growing a baby in a test tube, of course, but rather a process in which ova (eggs) are extracted from the woman's ovaries during a laparoscopy, fertilized with sperm in a shallow dish in the lab (*in vitro* means "in glass"), and then placed into her uterus through her cervix approximately three days later. Typically, three to five zygotes are transferred to help ensure at least one pregnancy, and thus this procedure increases the chances of multiple births. IVF is typically used when a woman's fallopian tubes are blocked or missing. Here, either the eggs or sperm cells, or both, may be from the couple themselves or from individuals outside the relationship (egg donors or sperm donors). Donor eggs or sperm are sometimes used for this and other ART procedures when the woman is unable to produce viable eggs, if the man cannot provide functional sperm, or when a gay or lesbian couple desire to have children (the issue of gay and lesbian couples having children is discussed in greater detail in Chapter 11, "Sexual Orientation").
- **Gamete Intrafallopian Transfer (GIFT).** This procedure is nearly the same as IVF, except that fertilization occurs naturally in the fallopian tube. As in IVF, eggs are retrieved from the ovaries and mixed with the father's sperm in the lab. The sperm and eggs are then transferred to the fallopian tubes for conception using laparoscopic surgery. This method can be used only for women who have at least one healthy fallopian tube.
- **Zygote Intrafallopian Transfer (ZIFT).** ZIFT combines IVF and GIFT. The process for harvesting and fertilizing the ova is exactly the same as in IVF, but instead of transferring the zygotes to the uterus, they are placed into the fallopian tube using the laparoscope.
- **Intracytoplasmic Sperm Injection (ICSI).** This ART method was developed to assist a couple when the man's sperm production is very low, the cells have poor shape or motility, or some other problem exists with his sperm or semen that makes the other ART methods ineffective. In ICSI, a single sperm cell from the man is injected using a microscopic needle directly into each harvested egg from the woman (see Figure 9.23). The resulting zygote is then transferred after a few days into the mother's uterus. All this method requires is a minimum of one viable sperm cell, which nearly all men with functional testicles can provide, even if they are unable to ejaculate.

Figure 9.23 Intracytoplasmic Sperm Injection

Intracytoplasmic sperm injection involves fertilization by injecting a single sperm cell into an ovum.

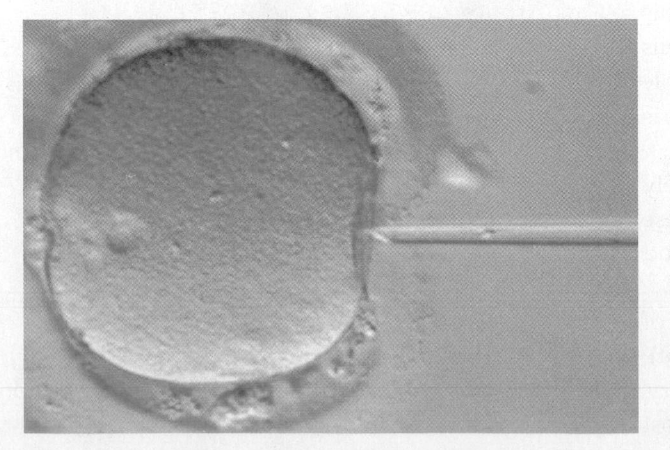

How effective is ART? Successful pregnancies from various ART methods have been steadily climbing over the past quarter century and have doubled over the past decade, and the Centers for Disease Control and Prevention estimates that 1% of all babies born in the United States are the result of ART (CDC, "Success Rates," 2010). As you can see in Figure 9.24, the birth rate for all ART methods is approaching 50% for women under age 35. Success rates drop significantly as the age of the woman increases, until they are quite low for women in their forties (CDC, "Assisted," 2013).

Figure 9.24 Percentages of ART Cycles Using Fresh Nondonor Eggs or Embryos That Resulted in Pregnancies, Live Births, and Singleton Live Births, by Age of Woman, 2011.

SOURCE: CDC. Assisted Reproductive Technology Surveillance, United States, 2006. FIGURE 2, p. 15.

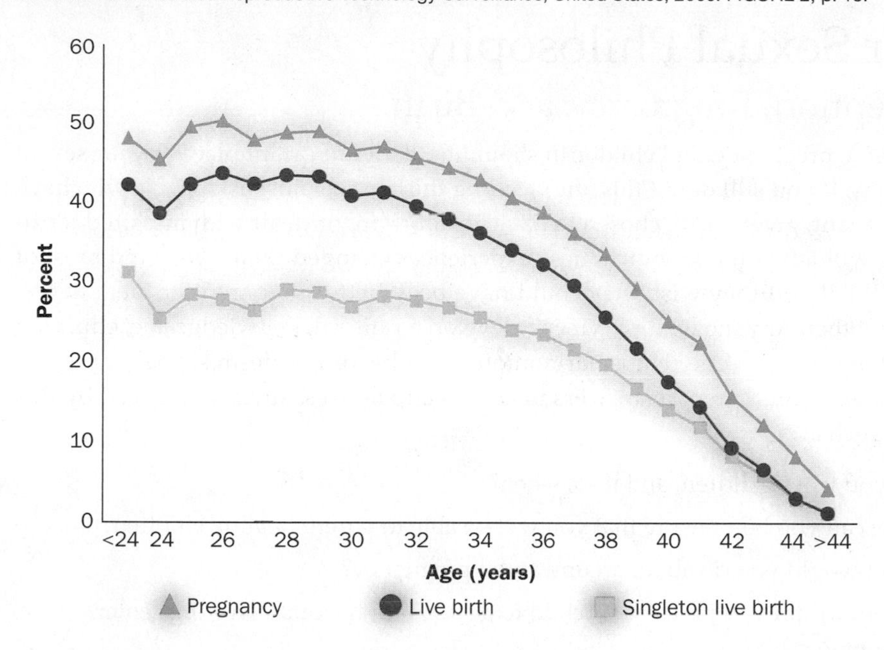

ADOPTION One very common and popular route to parenthood is adoption. Many potential parents—those who are able to conceive and have conceived children, those dealing with fertility difficulties, singles, and gay and lesbian couples—choose to start or expand their families through adoption. In the United States and many other countries, one of the first decisions a couple must make once they consider adopting a child is whether they prefer to adopt a child from their home country or from a foreign country. It is important to consider the numerous pros and cons for either decision so that prospective adoptive parents make the choice that's right for them. Many excellent websites exist that advise and guide couples (and singles as well) through the complex adoption process.

As with fertility treatments, adoption can be expensive. Prospective adoptive parents must consider attorney fees, agency fees, and potential medical fees if the birth mother has not yet given birth, social worker home-study fees, document fees, fees charged by orphanages and agencies in other countries, and travel expenses in the case of foreign adoptions.

In general, total costs for an adoption from start to finish are between $10,000 and $20,000, with most domestic adoptions at the lower end of the range and foreign adoptions at the higher end. Because there are so many orphaned children, Congress has passed tax laws that provide financial help to people who choose to adopt children. In 2014, these benefits amounted to approximately $13,000 in tax credits for most couples.

Although adoption may seem expensive, if you look back at our discussion of the costs of fertility treatments, which come with no guarantee of success, you will see why many couples struggling with conception begin to look at adoption in a new light. This is not to say adoption is for all infertile couples. To many, the need to have a baby who carries their own genes is so deep and strong that adoption does not seem an acceptable option. For many other couples, however, it is the ideal solution, benefiting both the parents and providing a loving

The author with his daughter, whom he and his wife adopted from China in 1996.

home for the child. In fact, some couples who choose to adopt are fertile and have become, or are capable of becoming, pregnant and having biological children to enjoy a diverse family.

Your Sexual Philosophy

Conception, Pregnancy, and Birth

Conception, pregnancy, and childbirth should hold an important place in your sexual philosophy. If you still doubt this after reading this chapter, just ask people who have been pregnant, given birth, chosen to have an abortion, or dealt with impaired fertility. They will tell you how much these experiences changed their lives. And most of them will tell you they wish they could have been better prepared for them in one way or another. Anyone who is trying to develop a rational, knowledgeable, educated map of their future—one that is harmonious with his or her desires, goals, and personal values—must prepare answers in advance to address questions raised by this chapter, such as:

- Do you want children, and if so, when?
- How can you best ensure that you will be able to parent healthy children?
- What would you do about an unwanted pregnancy?
- Where would you prefer your child to be born—a hospital, birthing center, or home?
- Would you prefer a physician or midwife to deliver your baby?
- How do you feel about an elective C-section birth?
- How will you deal with the possibility of postpartum depression?
- What actions do you think you would take if you or your partner were found to be infertile?

You may not be at a stage in your life that requires immediate answers to all these questions. But that's true of most of life's choices relating to sexuality. What's important is to give these questions some careful consideration now, to think about them *before* you are under the real-life pressure of decision making. The information in this chapter relates to major life events and decisions that are much more difficult to handle when you are actually faced with them and your rational thinking skills may be clouded by your emotions. Even if your answers to the questions change as you live and learn and share your thoughts with a partner, you will be equipped with more effective tools to address all these issues, to make choices that work for you, and to plan ahead—provided that you make them part of your ongoing sexual philosophy.

Have You Considered?

1. Looking back at "Self-Discovery: Are You Destined to Be a Parent?" near the beginning of this chapter, what do you feel are the *five most important* questions people should ask themselves to determine whether they are ready to have children? Explain why you picked those five.
2. What would you say are three reasons that people should have as clear an understanding as possible of the events that lead up to the fertilization of an ovum by a sperm? Explain your answer.
3. Do you see any way to defuse the emotionally charged and sometimes violent debate

over abortion? If so, explain your ideas. If not, explain why you feel there may be no resolution to the debate.

4. Imagine that you are expecting a child for the first time. What choices do you think you would make regarding where you would have the baby; whom you would want to deliver your baby; what pain medication you would prefer, if any; and your feelings about a C-section versus a vaginal birth. Explain your answers.
5. Explain briefly how you think you would react if, now or in the future, you discovered that you were infertile. What do you think you would do about having children? Why?
6. Discuss your attitudes and opinions about gay and lesbian couples having children through assisted fertility techniques or through adoption.

Summary

Historical Perspectives

The Pain of Giving Birth

- A century ago, pain control for birth was usually achieved using general anesthesia. Today's pain control techniques allow a woman to be awake for, participate in, and assist in the birthing process.

Deciding Whether or Not to Have a Child

- Many factors must be considered in the decision to have a child. For some couples, having children is a top priority; others make a conscious choice to be child-free.
- Parents may now select to have a boy or a girl using sex-preselection technology. But the use of such technology is controversial due to the ethics of people's reasons for making such choices.

Conception

- In puberty, the female's ovaries begin to produce mature eggs, or ova. Once ovulation is established, most women will release one ovum per month for about 40 years. The ovum will enter the fallopian tube, where it is available for fertilization for only 12 to 24 hours. If not fertilized, the ovum is shed with the uterine lining during menstruation.
- Sperm cells generated in a man's testicles are stored in the epididymis until they are mature and ready to be ejaculated. The average ejaculate contains about 400 million sperm cells, but only one is necessary for conception.
- Hormonal changes in the woman's body signal whether or not conception has occurred. The fertilized ovum, or zygote, enters the uterus and attempts to implant in the rich lining that has built up on the walls of the uterus. If implantation is successful, pregnancy begins, and the zygote becomes an embryo.

Pregnancy

- Pregnancy is typically divided into three trimesters. During the first trimester (the first three months of pregnancy), growth of the embryo continues until, at eight weeks, the embryo becomes a fetus. At 12 weeks, the fetus is about 3 inches long.
- During the second trimester, the fetus begins more complex movements, and by the sixth month, it is approximately a foot long and weighs about 2 pounds. The fetus, if born prematurely at this point, has a chance of survival with modern intensive medical care.
- During the third trimester, the fetus grows quickly, develops, and prepares for birth.

Potential Problems in Pregnancy

- Potential problems in pregnancy and birth include ectopic pregnancy, spontaneous abortion, preterm birth, and fetal abnormalities. These problems are the exception, not the rule, but any of them may cause serious health difficulties and may threaten the life of the fetus or the mother.
- Teratogens are substances that may harm a developing fetus. Hundreds of teratogens exist, including radiation, alcohol and other drugs, and various toxic environmental substances.

Abortion

- Abortion procedures may be either surgical or medical (using medications to induce abortion).

- Abortion is one of the most socially, religiously, and politically contentious issues in the world today. The two polarized sides of the abortion controversy are commonly referred to as "pro-life" and "pro-choice." Most people fall into one camp or the other, but a few see some middle ground.
- The goal of reducing the number of abortions performed each year is shared by those on both sides of the controversy. However, the methods toward reaching that goal vary greatly. As birth control becomes more widely accepted and is used correctly, the number of abortions usually declines.

Birth

- The process of labor and birth is divided into three stages. In stage one, contractions begin, the cervix dilates, and the top of the baby's head becomes visible. During stage two, the muscles of the uterus and abdomen work to push the fetus out through the cervix and vagina. In stage three, the placenta, or afterbirth, is expelled from the uterus.
- Parents must make many choices about pregnancy and birth: whether to use a physician or midwife, where to have the birth, what pain medication, if any, they want to use or have available, and whether to have a cesarean section or a vaginal delivery.

Postpartum Issues

- Serious postpartum (after childbirth) depression occurs in a small percentage of women. This depression should be treated because it can have major and even life-threatening consequences on the mother and infant.
- Breastfeeding is probably the most important postpartum activity for the health and wellbeing of the infant and mother. Although ovulation may be suspended during breastfeeding, this is not a reliable form of contraception and barrier methods should be used when intercourse resumes.

Impaired Fertility

- Approximately 10% of couples who wish to become pregnant find that they cannot. Infertility is due to many factors and may be a female or a male problem or a combination of both. In 15% of cases, the cause of infertility is unknown.
- Many solutions to infertility are available to couples. These include simply continuing to try to conceive over a longer time period or using one or more assisted reproductive technology methods (such as in vitro fertilization).
- Another solution to impaired fertility is to adopt a child, either domestically or abroad. With so many children in need of loving homes, this solution is very gratifying to many prospective parents.

Your Sexual Philosophy

Conception, Pregnancy, and Birth

- A complete and accurate understanding of conception, pregnancy, and birth is crucial to everyone's sexual philosophy of life, whether or not they have children or plan to have them. Reproduction is, after all, one of our most basic functions as sexual beings.

Chapter 10
Gender
Expectations, Roles, and Behaviors

Learning Objectives

After you read and study this chapter you will be able to:

10.1 Explain how the characteristics of sex and gender are separate, distinct concepts.

10.2 Summarize and discuss the development of a person's biological sex including variations and disorders of physical sexual development.

10.3 Review the development of a person's gender identity from both the biological and environmental perspectives.

10.4 Summarize the topic of transgender identity including variations in gender expression, the issue of sex reassignment surgery, and the challenges faced by transgender children.

10.5 Analyze how gender identity and sexual orientation are two distinct characteristics and how they may interact.

10.6 Discuss the meaning of gender stereotypes and review how they develop, the manifestations of them in society, and whether gender stereotypes are based in fact or imagined differences.

10.7 Summarize the nature (biological) and nurture (environmental) influences on the origins of gender.

10.8 Examine how the concept of androgyny may reconcile the extremes of masculine and feminine gender including the two-dimensional model, how androgyny may be measured, and positive versus negative androgyny.

Since You Asked

1. I saw pictures of hermaphrodites in a magazine. Do they really exist? How does this happen? (see page 371)
2. Are male and female behaviors determined by genetics, or are they learned from society? (see page 373)
3. I have a friend who has a seven-year-old son who always wants to play with girls and dress up in girls' clothes. He does not like to play with other boys. She wonders, is this just a stage or something else? (see page 383)
4. I saw a commercial for a talk show with "male lesbians." Is that for real? (see page 385)
5. Why is it that men are always supposed to be the aggressors, the ones to take the lead in relationships? (see page 386)
6. Everyone always says that men are more aggressive than women, but I'm a woman, and I think I'm usually more aggressive than most men. Is everyone wrong, or am I just weird? (see page 390)
7. Why is it that men seem to want sex more than women? (see page 392)
8. Why do guys feel that they must "play the field," even after they've made a commitment to someone? (see page 392)
9. Why are women who have sex a lot seen in negative ways, while men who do the same are seen in a positive light? (see page 392)
10. I've tried talking to my boyfriend about how I feel and what I want, but he never seems to understand, even though he says he does. How can I make him understand how I really feel? (see page 393)
11. Can someone be both male and female, or masculine and feminine, at the same time? I feel as if I don't really fit into either masculine or feminine. (see page 397)

Here are some personal questions most people don't hear very often: Are you male or female? Are you a man or a woman? Are you masculine or feminine? These are three seemingly similar questions, and for most of you, your answers probably came to you quickly and easily. Yet the range of *possible* answers may surprise you. The first question is the one with the most certain answer: It asks only for a purely biological answer based on a person's chromosomes, hormones, and sexual anatomical structures. Most of you also have little trouble answering the second question: Virtually all people are quite sure about their **gender identity**, that is, which sex they perceive themselves to be, and they've known the answer since they were about four years old ("I'm a boy"; "I'm a girl"; "I'm always going to be this"). Odds are good that, as adults, none of you had to stop and think about whether you perceive yourself to be a man or a woman.

gender identity
The sex (male or female) that a person identifies himself or herself to be.

However, the third question might not be so easy to answer. Different people possess varying amounts of "maleness" and "femaleness," or masculinity and femininity. These characteristics are more focused on what we commonly refer to as sex roles or **gender roles**. Every culture has a set of generalized expectations about what are considered masculine or feminine behaviors, emotions, and attitudes. If you think about people you know, you can think of some you would place on the extremely feminine side of the scale (these are more likely to be women, but not always), others who fit on the extremely masculine side (these are more likely to be men, but not always), and some who seem to fall somewhere in between the two (these may be men or women). The categories are not intended to be judgmental—good or bad—they simply define one important variation among people. **Gender** refers to this masculinity-femininity dimension of our basic nature as humans. It is one of the most powerful forces of human personality, and it forms the basis of our discussion in this chapter.

gender roles
A set of behaviors, attitudes, and emotions that are generally socially expected for men and women in a given culture.

gender
The masculinity-femininity dimension of our basic nature as humans.

Focus on Your Feelings

The title of this chapter provides a clue to how emotions might become involved with the concept of gender. The world, as most people know it, is divided into two, and only two, genders into which everyone is expected to fit: male and female. Yet in reality, not all people can be placed neatly into one of those two categories. When a person defies gender expectations, he or she tends to make others uncomfortable. Try a simple experiment. The next time you meet someone or encounter a person you know, try to relate to that person as if he or she were a member of his or her opposite sex. Most likely, you will not be able to do it. You will find it impossible to switch a person's gender in your mind! This illustrates just how basic the notion of gender is to our view of the social world around us. When someone fails to adhere to our usual ideas about gender roles and behaviors, we are at a loss to know how to respond. And whenever our expectations about people are violated, we react emotionally: we might feel confused, uncomfortable, or even annoyed.

This chapter helps you recognize that gender is not a simple matter of dividing all people into two neat groups, male and female. Many shades, gradations, and combinations of gender exist in humans. And they are all *normal*. Suddenly, through the process of learning about the complexities of gender, we are able to see a whole new world of human sexual diversity. The concepts of male or female, man or woman, and masculine or feminine take on new dimensions of meaning and understanding. As if looking through a wide-angle lens, we are able to realize and, ideally, be more comfortable with the idea that people cannot always be neatly and conveniently stereotyped as male or female. Rather, each individual must be presumed to occupy a unique place on the gender continuum.

Gender and your gender identity are more about who you are as a person than simply about what sex you are. Few human characteristics define us more than our gender. Think about it for a moment. Imagine how you would feel if you were to wake up tomorrow morning with no idea whether you were a man or a woman, male or female. How do you think you would feel? Confused? At the least. Anxious and insecure? No doubt. You might even think to yourself, "I don't know who I am!" This is how important gender is to your identity as an individual; it is a fundamental part of who you are as a person.

Gender is equally important in helping us understand others. Most of us tend to interact differently with individuals depending on whether we perceive them to be male or female. You can probably think of individuals you may know or have met who do not fit into the culture's traditional gender molds. For example, many celebrities—David Bowie, Grace Jones, RuPaul, Peaches, Chris Colfer, to name a few—are often referred to as "gender-benders," in that their dress, actions, and attitudes challenge our social and cultural preconceptions of what it means to be male or female, man or woman, masculine or feminine.

Gender is a complex component of our study of human sexuality. In this chapter, we will discuss how a person's *sex* and *gender* differ (they are *not* the same), how your *sexual identity* as a man or a woman is biologically and environmentally influenced, the important distinction between gender identity and sexual orientation, gender stereotypes, scientific evidence of true gender differences, how gender has been shown to influence communication in intimate relationships, and the concept of *androgyny* (the idea that people may possess both masculine and feminine gender identities).

Historically, the world received one of its first lessons about the complexities of gender identity from a shy, slightly built World War II army veteran named George Jorgensen, who, at the age of twenty-six, became Christine Jorgensen.

Historical Perspectives

The Story of Christine Jorgensen

Here is the opening paragraph of an obituary that appeared in *Newsday* on Friday, May 5, 1989:

> It was meant to be a private affair, a quiet series of operations that would change the 26-year-old Bronx photographer into a woman and, in the process, exorcize the personal demons that had haunted him since childhood. But even before she left

George Jorgensen before sex reassignment surgery.

Christine Jorgensen after sex reassignment surgery.

the Copenhagen hospital in February, 1953—transformed from George Jorgensen, Jr., the 98-pound ex-GI, into Christine Jorgensen, "the convertible blonde"—word had leaked out. Overnight, it became the most shocking, celebrated surgery of the century. And even if the furor eventually waned, the curiosity lingered, following Jorgensen to her death Wednesday at San Clemente General Hospital after a 2-year battle with bladder and lung cancer. She was 62. (Ingrassia, 1989)

George Jorgensen was born a biological male. Throughout his early adult life, he never really had any doubt about his biological sex, but he never thought of himself as a man. Psychologically, he was a woman. The demons mentioned in the obituary referred to the emotional pain he felt in being forced into a gender role that was contrary to his personal gender identity. Imagine for a moment if, to avoid being socially ostracized and humiliated, you were required to begin to live your life as a member of the other sex, hiding your true gender from everyone. How do you think that would feel? You would likely feel that every minute of every day you were living was a lie. This is how George Jorgensen felt until he undertook, literally, to change his sex.

Prior to his travels to Denmark seeking surgery that would effect the transformation from George to Christine, Jorgensen was taking the female hormone estradiol, which caused his breasts to enlarge, his skin to soften, his beard to stop growing, and his appearance to become softer and more feminine. Although his was not the first surgical sex-change operation, very little was known about such surgery, and no surgeon in the United States was willing to perform one at the time, in the early 1950s. Moreover, the surgery was significantly more primitive at that time than today's gender reassignment surgical techniques (to be discussed later in this chapter). Basically, the only procedures Jorgensen received were a bilateral orchiectomy (removal of both testicles), removal of the scrotum, and penectomy (removal of the penis). No reconstructive surgery was done on Jorgensen at the time to create female genitals. Several years later, Christine returned for more surgery to construct a vagina, but that surgery was only moderately successful (Bullough, 2001).

Jorgensen's sex-change operation was the first to receive worldwide media attention, and she did not shy away from the publicity. The *New York Daily News* broke the story with the headline: "Ex-GI Becomes Blond Bombshell." Christine wanted publicity, chose to capitalize on her experience, and became one of the best-known names of the twentieth century. She was a performer; she had a singing and dancing road show that included monologues about herself, and at times earned as much as $5,000 per week performing throughout the United States. She also became a spokesperson for transgender, gay, and lesbian causes. Most researchers credit her with pioneering the development of more effective treatments and procedures for thousands of transgender people uncomfortable with their biological sex (we will return to this topic later in this chapter).

Most of us take our gender identity for granted. However, Christine Jorgensen's story points out just how powerful a force gender is in our lives. She demonstrated (as many others have) that our gender is such a fundamental part of who we are as humans that we will go to extremes to express it.

The Distinctions Between Sex and Gender

10.1 Explain how the characteristics of sex and gender are separate, distinct concepts.

In popular usage, the term *gender* is often used interchangeably with *sex*. However, in the world of sexuality research and education, the terms reflect separate aspects of human existence. Whereas *sex* is biologically determined, your *gender* is something

Some pop culture celebrities, such as *American Idol* star Adam Lambert or TV's *Glee* star Chris Colfer, are often considered "androgynous," in that their dress, actions, and attitudes challenge our social and cultural preconceptions of what it means to look and be male or female.

learned, developed, or constructed based on your self-concept and social and cultural experiences in childhood and throughout life: your *gender identity*. Those experiences are typically shaped by our biological sex and perhaps other biological or genetic factors, but biology does not solely determine gender. Sex and gender identity are two *separate* expressions of who we are as sexual beings. As demonstrated by Christine Jorgensen's story, some people's gender identity is the opposite of their biological sex: a biological male may perceive his gender as female, or a biological female may perceive her gender as male. These individuals are referred to as *transgender*.

As mentioned earlier, gender is more a personality characteristic than a biological attribute. To say that someone's gender is male or female is similar to saying that a person is, say, shy or outgoing; optimistic or pessimistic. Human personality probably develops through a combination of genetic, biological, and experiential influences. So it is with gender as well. Also, most personality characteristics are stable over time and across situations, as is our gender identity. It is precisely because gender affects our lives so strongly that many people resist the idea that gender and sex are separate (see Diamond, 2006). For most people, their biological sex is the *same* as their gender identity, and to see themselves, or anyone else, in any other way is unimaginable. But the two are *not* always the same.

The Development of Biological Sex

10.2 Summarize and discuss the development of a person's biological sex including variations and disorders of physical sexual development.

As we discuss in Chapter 9, "Conception, Pregnancy, and Birth," a person's biological sex is determined at the moment of conception based on the combination of *chromosomes* that result from fertilization. Normally the egg, or *ovum*, from the woman's ovary contains a single X sex chromosome, and the sperm cell that fertilizes the ovum carries either an X or a Y sex chromosome. Two X chromosomes combine to produce a female, but an XY combination produces a male. In the presence of the Y sex chromosome, male hormones called *androgens* are secreted, causing the fetus to develop testicles, a penis, and male internal sexual anatomy. With two X chromosomes, androgens are not

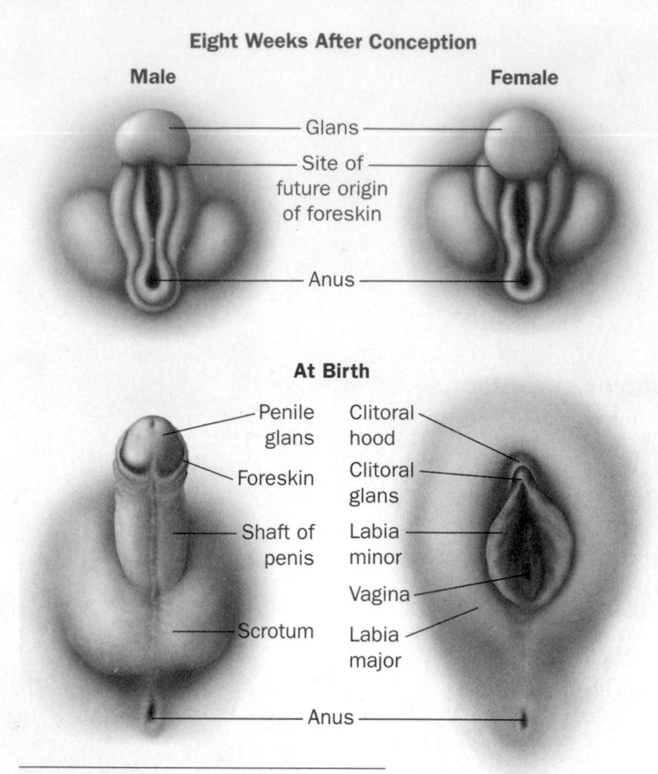

Figure 10.1 The genitals of normal male and female fetuses develop from the same anatomical tissues and are indistinguishable early in pregnancy, as shown at eight weeks. At about twelve weeks of pregnancy, the fetus's genitals begin to differentiate, and by birth, they are fully developed as male or female.

produced, and the fetus develops a vulva, ovaries, a uterus, and other internal female sexual anatomy. The genitals of male and female human fetuses are indistinguishable until about the twelfth week of pregnancy. At that point, they have developed to the point that the biological sex of the fetus is recognizable (see Figure 10.1).

The chromosomal development of sex is evolution's way of creating an approximate balance between the sexes to help ensure survival of the human species. The development of male or female genitals sets the stage, in most humans, for the development of gender identity in childhood. As noted earlier, however, biology is not the sole determinant of a child's gender, as you will see later in our discussion of socialization influences on gender identity development. Furthermore, not all fetuses contain only XX or XY chromosomes, and not all fetuses develop clearly male or female genitals.

Variations in Biological Sex

This may seem like an odd chapter heading to you. You may be thinking, a person is either male or female, so how are *variations* in biological sex possible? The fact is that biological sex is not as simple as it may at first appear. Not all fetuses have only XX or XY sex chromosomes. Two of the most common sex-chromosomal variations in humans are Klinefelter syndrome and Turner syndrome (O'Connor, 2008). In addition, some fetuses do not develop clear male or female genitals and are born with what are commonly referred to as "ambiguous genitalia." This does not imply that they have both male and female genitals but that, upon examination, the genitals are not readily identifiable as male or female and may have some features of both. These genital variations are probably due to hormonal imbalances in the uterus during fetal development.

Klinefelter Syndrome

Klinefelter syndrome
A male genetic condition characterized by a rounded body type, lack of facial hair, breast enlargement in puberty, and smaller-than-normal testicles.

Turner syndrome
A female genetic condition characterized by short stature, slow or no sexual development at puberty, heart abnormalities, and lack of ovarian function.

Approximately one out of every 500 male babies is born with an additional X sex chromosome (Wattendorf & Muenke, 2005). These males, instead of the typical XY pairing, have an XXY chromosome configuration and are referred to in the medical community as *XXY males*. This unusual genetic makeup may result in a condition called **Klinefelter syndrome**, named after the physician who discovered it in 1942. Not all XXY males will develop the characteristics of Klinefelter syndrome, some may not be diagnosed until puberty, and some may never even know they have the extra X chromosome. However, when the syndrome is activated, common physical signs and symptoms include a rounded body type, lack of facial hair, breast enlargement in puberty (often temporary), smaller-than-normal testicles, osteoporosis, and a tendency to be taller and heavier than average, as shown in Figure 10.2 ("Klinefelter Syndrome," 2010; Nahata, 2013).

Most, but not all, XXY males fail to produce enough sperm in adulthood to be fertile. They also appear to have a somewhat higher risk of autoimmune diseases such as diabetes and lupus. XXY males who develop breast tissue have a risk of breast cancer equal to that of women, which is twenty to fifty times greater than the risk to normal XY men (Amory et al., 2000; "Klinefelter Syndrome," 2010).

An early developmental concern for XXY males is that they often display delayed development of language and may have learning difficulties, especially in reading and writing. These children are not mentally retarded and eventually learn to speak and converse normally. Furthermore, their learning difficulties are treatable with proper guidance and attention.

Virtually all the symptoms of Klinefelter syndrome are treatable to various degrees, and males with the disorder are usually able to live normal, relatively healthy lives. Learning problems can be minimized with therapy; surgery, assisted fertility techniques, and adoption can help Klinefelter syndrome men have families; hormone therapy can enhance masculine development in puberty and throughout life; and breast enlargement often reverses itself naturally or, if desired, may be corrected surgically (Mehta, Clearman, & Paduch, 2014; "Studies Examine," 2008).

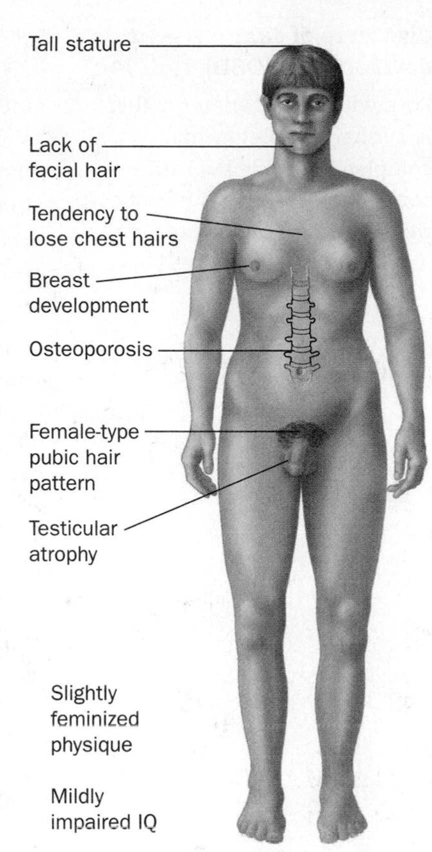

Figure 10.2 Characteristics of Klinefelter Syndrome

Turner Syndrome

In female infants, instead of an extra chromosome, **Turner syndrome**, named for the physician who discovered it in 1938, is caused by a lack of or damage to one of the pair of X chromosomes. This condition is far less common than Klinefelter syndrome, affecting only one in 2,000 to 2,500 female births. Nearly all cases of Turner syndrome (99%) result in miscarriage of the afflicted fetus in the first or second trimester of pregnancy (Ranke & Saenger, 2001; Wolf, vanDyke, & Powell, 2010).

For the fetuses who survive, the symptoms and effects of Turner syndrome tend to be more physically and psychologically serious than those of Klinefelter syndrome, although a wide range of symptomology exists, from mild to severe. The most common conditions associated with Turner syndrome (see Figure 10.3) are short stature (average height of 4 feet, 7 inches in adulthood), slow or no sexual development at puberty, puffy hands and feet, kidney malformations, hearing problems, extra folds of skin at the sides of the neck, heart abnormalities, lack of ovarian function (hormone and ovum production), and soft upturned fingernails (Collin, 2006; Morgan, 2007).

Because of hormonal abnormalities, virtually all Turner syndrome individuals are infertile. As they age, women with the syndrome are at a significantly increased risk of bone thinning (osteoporosis). Underdevelopment of the kidneys or the lack of one kidney is also quite common. Most worrisome are the 20% of Turner syndrome individuals who have heart valve and vessel malformations. These conditions may be life-threatening and require careful monitoring and treatment throughout life. As with Klinefelter syndrome, Turner syndrome patients have twice the incidence of diabetes of the general population. Cognitively, Turner syndrome individuals are of normal intelligence but may have learning difficulties, especially associated with math and spatial relationships (Morgan, 2007). Although the dangers of Turner syndrome may be serious, with awareness and proper treatment, most of the effects can be controlled and mitigated successfully.

Since You Asked

1. I saw pictures of hermaphrodites in a magazine. Do they really exist? How does this happen?

Figure 10.3 Characteristics of Turner Syndrome

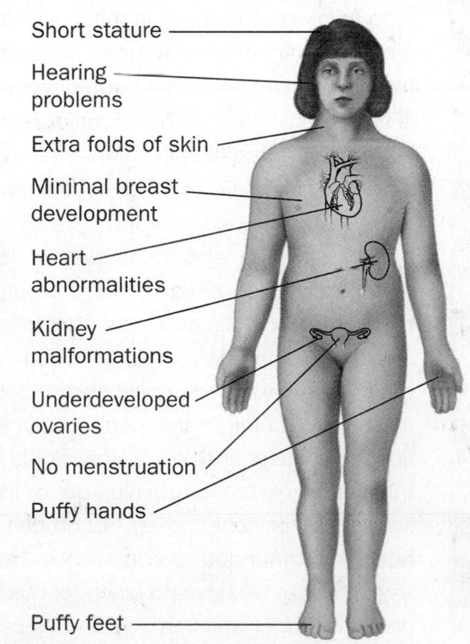

Disorders of Sex Development ("Intersex")

Not only can we identify various "degrees" of masculinity and femininity, but even biological sex is not always neatly divided into male and female. Most sexuality and biomedical researchers today agree that a small percentage of people (just how small is a topic of debate, but most studies show the percentage to be between 1% and 2% depending on exactly how it is defined) are born with a sexual anatomy that is neither completely male nor completely female but is rather ambiguous or some combination, with features of both (Greenberg, 2012). People with these characteristics are formally diagnosed with **disorders of sex development (DSD)**, which is replaced by the term ***intersex*** (Ahmed & Rodie, 2010; Fausto-Sterling, 2000;).

Most of you are probably aware of the stereotype of the *hermaphrodite*, usually portrayed as a person with a functional penis, a vagina, and female breasts. These individuals are often part of the mythology of pornographic literature, but some males do choose to alter their breasts with hormones or implants and leave their penises intact. It is also possible for an adult man to have both a penis and female-like breasts through naturally occurring hormone imbalances, such as may occur in Klinefelter syndrome. However, these are not examples of DSD individuals.

disorders of sex development (DSD)
Born with sexual anatomy that is neither completely male nor completely female but rather a combination with features of both that cannot be categorized as male or female.

Babies born with DSD typically have one of three conditions (see Gurney, 2007; Sax, 2002): (1) They are genetically male (XY chromosomes) but have external genitals that are completely female (vagina and clitoris); (2) they are genetically female (XX chromosomes) but have external genitals that are completely male (penis and testicles); or (3) they are genetically female but have external genitals that are ambiguous, somewhere in between male and female. Often in the first two conditions, if no genetic testing has been done, doctors and parents may be unaware of the condition until the child reaches puberty and begins to develop the secondary sexual characteristics of the sex opposite from how the child was raised. In the third case, however, when the baby's genitals are ambiguous, the usual response by parents and doctors has been to alter the child's external sexual anatomy surgically to resemble either a "normal" male or female. The selection of genital sex in these cases has typically been based on a combination of genetic sex and the child's anatomy so that the minimal amount of surgical reconstruction is required. The child is then raised as the sex that matches the surgically reconstructed genitals.

In recent years, the practice of routine surgical alteration of DSD babies has become increasingly controversial. Critics of the procedure, including many DSD adults, claim that altering DSD babies against their will and without their consent is ethically and morally wrong and is done simply because society is unable to accept the notion that a person might not be either male or female but somewhere in between. In addition, new research has demonstrated that surgical reconstruction of ambiguous genitals cannot predictably determine a person's later gender identity (Bomalaski, 2005; Reiner & Reiner, 2012). These critics propose that one solution may be to leave all babies intact as they come into the world and until they reach adulthood, when they can decide their sex for themselves (which may be male, female, or a combination). For more discussion, see "Sexuality, Ethics, and the Law: The Pitfalls of Surgically 'Changing' DSD Babies."

As is evident from the many variations discussed here, biological sex is far more complex than most people realize. This, of course, plays a role in the development of gender identity, a separate but interrelated process we turn to now.

Sexuality, Ethics, and the Law

The Pitfalls of Surgically "Changing" DSD Babies

As we discuss in this chapter, DSD does not usually mean having both male and female genitals; rather, it refers to people who are born with genitals that are somewhat ambiguous, not clearly either male or female. Although this characteristic is more common than most people realize, the existence of DSD is not widely known. Why? A major reason is that medical specialists have traditionally intervened in such cases soon after birth and surgically altered these infants to have the external appearance of one sex or the other. Such surgery is performed mainly because society is unable to accept the concept of a person who does not fall into one of two categories of biological sex: male or female. We do not even have a vocabulary to accommodate an in-between, ambiguous sexual concept. Which restroom would such a person use? Which locker room? Whom might this person marry? What sexual orientation would he or she have? How would he or she dress? Would this person try out for the high bar or the uneven parallel bars? Ice hockey or field hockey? Such cultural and social worries, however unfounded, go on and on. Basically, what society is saying is that we have no place for this person (see the section on transgender children later in this chapter). So DSD infants are surgically "fixed" without any regard, activists assert, for issues of consent or for what the person may want later in life.

The story of David Reimer in the book *As Nature Made Him* brought the issue of the treatment of DSD babies into popular culture.

The consent for this surgery is given by the parents, who usually have little knowledge (and are provided with very little medical information from health professionals) about the medical or psychological consequences (see Bomalaski, 2005). This, activists contend, amounts to a blatant violation of their rights. The intersex community and scientific researchers alike are beginning to speak with one voice about what may be regarded as radical surgery without informed consent: a basic violation of human rights and medical ethics (Gurney, 2007).

The issue of altering the sex of babies received international attention in the early 2000s with the publication of the book *As Nature Made Him*, about a boy named Bruce who, because of a substandard circumcision that destroyed his penis, was surgically altered and raised as a girl, Brenda, without his knowledge (Colapinto, 2001; Hass, 2004). However, Brenda never felt or behaved like a girl and experienced many emotional, psychological, and behavioral problems. As a teenager, Brenda was given hormone injections to promote breast development and other typical female characteristics. However, he continued to feel confused, depressed, and eventually became suicidal. Finally, he was told of the accident and surgery in infancy. He immediately recognized the source of his turmoil, chose surgery to change sex once again, and became David Reimer. Although David was not born a DSD baby per se, his story illustrates clearly that gender is far more than anatomy and physiology. David was never able to overcome the tumult that had accumulated over his lifetime. After many unsuccessful attempts to commit suicide, David ended his life on May 4, 2004, at the age of thirty-eight.

The Development of Gender Identity

10.3 Explain the development of a person's gender identity from both the biological and environmental perspectives.

Your genetic, biological sex was determined by your DNA at conception, but your gender and gender identity—your maleness or femaleness, masculinity or femininity—developed during the months and years after conception, guided in part by biological, inborn factors and influenced to some degree by your environment. To study the various influences on gender development, we turn to one of the oldest debates in psychology: nature versus nurture. Today, widespread agreement exists among behavioral and biological scientists that most human traits and characteristics are influenced by an interaction between *nature*, meaning biological factors, and *nurture*, referring to environmental, experiential forces. This appears to be the case for gender development as well, although recent research is leaning toward the biological (nature) side of the discussion.

Since You Asked

2. Are male and female behaviors determined by genetics, or are they learned from society?

Biology and Gender Development

As mentioned earlier in this chapter, the presence or absence of specific male and female hormones produced during pregnancy, depending on the chromosomal sex of the fetus (XX or XY), triggers the development of either male or female genitals and internal sex organs (for more details on this process, see Chapter 9, "Conception, Pregnancy, and Birth"). Some evidence suggests that hormones may also play a role in the development of a person's sexual orientation (see Chapter 11, "Sexual Orientation"). But do sex hormones also influence the characteristics in humans that are usually associated with gender? The answer appears to be that they do, but hormonal effects are difficult to study systematically in humans.

Obviously, we cannot ethically subject human fetuses, children, or adults to various amounts and types of sex hormones simply to study what effects they might have on the person's gender profile. Consequently, researchers have relied primarily on observations of people with hormonal disorders, studies of very young children whose behavior has not yet had much opportunity to be shaped by environmental forces, or animal research. Studies employing these methods provide a convincing, if not airtight, case for the influence of biology and genetics on gender identity.

In addition to the hormone disorders discussed in the preceding section, a condition known as *congenital adrenal hyperplasia*, or CAH, causes the adrenal glands of affected individuals to produce large amounts of male hormones beginning before birth and continuing throughout their lives (see Gurney, 2007; Finkielstain et al., 2012). Boys may have some health issues due to CAH but typically develop relatively normally. In girls, however, CAH usually produces ambiguous external genitalia, and the imbalance of male hormones must be controlled with hormone treatments (Donohoue, Poth, & Speiser, 2010).

DSD infants are born with genitals that are not clearly male or female.

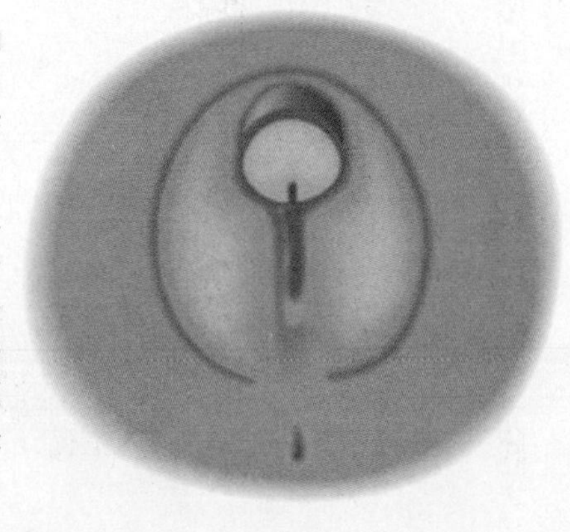

Researchers studying the *gender* effects of CAH on children have found that boys with the condition appear to develop the usual patterns of sex-typed behaviors, same as their peers. CAH girls, however, have been found to engage in many sex-atypical behaviors. These girls tend to prefer toys that are typically associated with boys (trucks, fire trucks, toy guns, etc.), seek out boys as playmates rather than girls, engage in more rough-and-tumble play activities than non-CAH girls, and reject traditional female childhood pretend roles of wife and mother. Findings such as these have been reported rather consistently in studies over several decades (Alexander, 2003; Boyse, & Sands, 2011; Berenbaum & Snyder, 1995; Ehrhardt & Baker, 1974). These studies lend further support to the idea that hormones play a significant role in the development of gender-typed behavior. Other evidence of this comes from systematic observations of early childhood development of normal children.

A well-established pattern of behavior in most children is a preference for friends, playmates, and activity-preferences of their own sex (Bem, 2008; Maccoby, 1988; Maccoby & Jacklin, 1987). This preference is seen as early as age three or four and extends throughout childhood until puberty, when sexual interest in the other sex awakens (if the child is heterosexual). But does this observation imply that the preference for same-sex playmates is biologically based? Not necessarily. You could easily argue that by age three or four, children have already been exposed to many potentially powerful environmental behavioral influences from parents, siblings, peers, and the media, which could have molded their play and peer tendencies completely separately from hormonal or genetic effects. However, several other factors lend strength to the hormonal argument for this gender-based behavior.

The degree of the preference for same-sex friends is very strong. Anyone who has ever observed young children over time has seen that girls and boys seem automatically to form totally separate groups in play settings and rarely allow a child of the opposite sex to participate in the group's activities. You can see one obvious example of this gender segregation at children's birthday parties: The guests, with very few exceptions, are all girls or all boys, and the mere suggestion that the child invite even one guest of the opposite sex is soundly rejected. Moreover, by age four, children are spending three times as many hours with same-sex peers as with other-sex peers, and by age six, they are spending as much as eleven hours with same-sex friends for every hour spent with opposite-sex peers (Lake, 2013; Maccoby & Jacklin, 1987; Pasterski et al., 2011).

Also, the propensity for same-sex playmates is seen in nearly every culture around the world. Because we know that culture exerts extremely powerful influences on human behavior, whenever we encounter a behavior that appears to be universal, existing in all or nearly all cultures, we can assume that it may be of *biological* origin rather than learned through interactions with the social environment.

These and other related studies demonstrate how strong a role biology plays in the development of gender identity (i.e., Martin et al., 2012). Some researchers are suggesting that gender identity may be preprogrammed, at least in part, into each person's brain prior to birth (Hembree et al., 2009). Clearly, we cannot simply ignore the influences of the environment in gender behaviors. From the moment you are born, and sometimes even *before*, your society and culture influence how you develop your male or female behaviors through expectations, modeling, and societal norms for gender roles.

Socialization and Gender Identity

Society influences and molds the behavior of its members through the process known as socialization. This process affects your development in many areas of your life, but perhaps none more significantly than your gender identity and gender behavior. From the moment of birth, society perceives you and places different expectations on you based, to a large extent, on your sex. These expectations create what are referred to as gender roles: the distinctive behaviors society expects and encourages each person to engage in

"Which cupcakes are for boys?"

depending on his or her sex. These perceptions and expectations often begin even before birth, when parents discover the sex of the fetus during pregnancy through genetic testing (such as amniocentesis) or ultrasonic imaging. They may then paint their baby's room the "right" color, purchase gender-appropriate clothes and toys, and begin to talk and fantasize about the unique joys and challenges of raising either a boy or a girl.

Gender differences are so socially ingrained in most people that they have trouble violating those expectations even if they are made acutely aware of them. For example, imagine you have been invited to visit some friends, a couple, whom you haven't seen in several years. You've kept in touch, so you know they have a child who has recently turned three years old. You want to bring a present for the child, so you make a trip to the toy store at your local mall. Take a moment to make a mental list of the gifts you might consider if (a) the child is a girl and if (b) the child is a boy. If you are like most people, your mental lists are not the same. Although some items may appear on both lists, many will appear on only one list, depending on the sex of the child; for example, a toy truck or a football for a boy versus a baby doll or a jump rope for a girl. Almost no one would place a baby doll on the boy's list or a toy truck on the girl's list. Gender differences are so deeply embedded in our perceptions of the world that we often follow them automatically, without conscious thought.

One of the reasons gender is so solidly rooted in society's perceptions of the social world is that, rightly or wrongly, gender is one of the most prominent human differences we use to try to understand others. Knowing a person's gender allows you to predict, or *think* you can predict, a great deal about that person's behavior, which in turn helps to make you feel more comfortable in social settings. Thus differences in gender are constantly reinforced through powerful social influences. Probably the most important of these influences include parents, peers, teachers, and the media. We will consider briefly each of these societal influences on gender.

PARENTS Parents' assumptions about the gender of their children are reflected in how the parents treat and interact with their sons and daughters and how their children behave

Many parents take steps to ensure that their babies are not mistaken for the "wrong" gender, as evidenced by this baby with a bow taped to her head.

in response (Epstein & Ward, 2011). Studies have shown that parents describe their infants in different ways depending on the sex of the baby. Parents tend to describe their newborn girls as soft, fine-featured, petite, delicate, and beautiful, and their boys as strong, big, and determined. They also describe infant girls as little, beautiful, pretty, cute, and resembling their mothers, but describe their infant sons primarily as *big* (Fausto-Sterling, Coll, & Lamarre, 2013; Rubin, Provenzano, & Luria, 1974). Keep in mind that these variations in perceptions of infants occur independently of any real, objective differences among the babies in any relevant measures such as weight, length, or activity level.

The number and strength of gender-based attributions made by parents appear to be decreasing as people have become more educated about gender issues and as fathers, especially stay-at-home dads, play a larger role in the birthing and parenting process (Karraker, Vogel, & Lake, 1995; Tucker, 2005). Nevertheless, parental expectations of their infants based on the baby's gender persist (see Kane, 2012). It is not difficult to find examples of these attitudes surrounding infants. One story tells of a new mother who was very concerned that people be able to tell at first glance that her baby was a girl. However, the baby had no hair yet, and even though the mother dressed her in feminine baby clothes, some people would mistake her for a boy. To avoid this problem, the mother took to taping a pink bow to the child's bald head. This, it turned out, was the ideal gender "marker"; no one mistook the child's sex again. Even the cards sent out to announce births are typically gender-specific, and infant gift websites often have separate "boy" and "girl" buttons to zero in on "appropriate" gifts.

In part, children are directed into gender-appropriate activities and attitudes throughout childhood by the choices parents make for toys, room décor, and clothing the child wears. Perhaps even more important, children are rewarded by the subtle or not-so-subtle reactions of parents and others to the children's behaviors. Most parents are uncomfortable if their child engages in activities and behaviors that are gender-inappropriate. For example, if a young boy enjoys playing with baby dolls, the parents may direct the child away from the dolls and into play that is seen as more gender-appropriate (Cherney & London, 2006; Wood, Desmarias, & Gugula, 2002). Parents and other important individuals in a child's life reward behaviors that conform to gender expectations and either withhold rewards for or punish behaviors that appear to violate those expectations. For example, one study found that if a child asks for a birthday gift that is gender-appropriate, he or she is far more likely to receive the desired item than if the request is for a gender-inappropriate item (Robinson & Morris, 1986).

Mothers tend to interact with greater emotional warmth and responsiveness with girls but encourage greater independence in boys. Fathers typically spend more time and engage in more physical activity with their sons than with their daughters (Karraker et al., 1995). Mothers tend to be more tuned in emotionally to daughters compared to fathers, and mothers are more sensitive to daughters than to sons (Schoppe-Sullivan et al., 2006) These parental influences on children's development continues throughout childhood. Among school-age children, many parents maintain a distorted perception of their children's academic skills based on gender. For example, one study demonstrated that parents rated their daughters as more competent in English but rated their sons as better in math and sports. In reality, the sons and daughters in the study performed equally well in all of these areas, but the parents based their ratings on established stereotypes (Eccles, Jacobs, & Harold, 1990; see also, Tiedemann, 2000). This line of research has suggested that parental perceptions are communicated to the children in ways that influence the children's self-perceptions and, consequently, their choices of activities throughout childhood and perhaps beyond. As children grow up, these childhood gender-based experiences lead males and females to pursue different educational and professional paths, further reinforcing cultural gender stereotypes (these stereotypes will be discussed in detail later in this chapter).

Boys and girls, without any encouragement from adults, display a strong preference for same-sex playmates beginning in early childhood.

PEERS Gender messages do not come from parents exclusively. Children typically spend a great deal of time with other children. These interactions may place equally strong or perhaps even stronger expectations on attitudes and behavior than those of parents. As noted earlier, children normally segregate themselves into same-sex groups in early childhood. Once this happens, very little social interaction occurs between the groups of boys and girls. You can imagine how strong peer pressure becomes on each child within his or her group to behave in ways that are appropriate for the sex of the group. Boys and girls who do not follow these behavioral norms are often ostracized by their same-sex group and may find it difficult to gain acceptance from opposite-sex peers as well (Leaper & Friedman, 2007; Legewie & DiPrete, 2012; Riley & Jones, 2007).

Within each same-sex group, very different types of behavior are rewarded through praise, imitation, and various verbal and nonverbal indications of approval. For instance, the tactics of persuasion for boys and for girls become significantly different as children interact with their same-sex peer groups (Underwood, 2007). Boys learn from their peers "controlling tactics"; that is, they use commands, threats, and physical strength to gain compliance from other children. Girls, by contrast, develop "obliging strategies" involving quieter and more refined methods for obtaining what they want, such as polite requests and other forms of subtle verbal persuasion (Riley & Jones, 2007; Strough, Swensen, & Cheng, 2001).

Another difference typically observed in male and female peer groups relates to the quality of their same-sex friendships. Boys tend to have larger groups of friends, but the quality of their relationships is typically more distant and less emotionally involved. Girls typically have fewer friends, but their friendships are warmer, and they place a higher value on trust and emotional closeness. These qualitative differences in peer relationships appear to become deeply entrenched and can be seen in adult friendships as well. Moreover, they may relate to gender differences in adult romantic and sexual desires and attitudes, which we will discuss shortly.

TEACHERS Social scientists have long recognized that children's teachers often reinforce and strengthen the gender-based attitudes that exist in the culture at large. Classroom participation from elementary school to college classes appears to be influenced by student gender (Crombie et al., 2003; Jobe, 2002). Because gender is such a powerful influence on society's fundamental view of who people are, teachers often reflect those views in their teaching styles and classroom management strategies without being aware that they are doing so. In general, teachers tend to give boys more time and attention in the classroom. Often, this is because boys display a higher level of activity, which, if not controlled, is disruptive to the classroom environment. Teachers may attempt to deal with boys' difficult behavior by involving them more in class activities and discussions and allowing them to break the rules more often and more blatantly than girls before administering discipline (Garrahy, 2001; Huang et al., 1998; Leaper & Friedman, 2007).

One influential study found that teachers also tend to interact differently with boys and girls during the teaching process itself. When students have difficulties in problem solving, teachers tend to guide girls quickly toward the answer or simply give them the answer, whereas they encourage boys to keep working until they can reach a solution on their own. This has the effect of undermining girls' confidence in their academic abilities (Huang et al., 1998). Furthermore, teachers often come into the classroom with the same gender-based learning biases held by parents and other nonteachers (Gunderson et al., 2012). For example, in math and science, teachers tend to overestimate boys' abilities, maintain higher expectations for boys, and have more positive attitudes about male students overall (Hinnant, O'Brien, & Ghazarian, 2009; Jobe, 2002). If you ask teachers, most will say with great confidence that they are "gender-blind" when it comes to treating their students equally. However, when their teaching practices are examined carefully, subtle yet important biases are often found throughout the culture of their classrooms (Garrahy, 2001; Geist & King, 2008).

THE MEDIA In most cultures throughout the world, children are bombarded by media images and messages from movies, storybooks, music, advertising, and especially television and video and computer games (Williams et al., 2009). You have probably heard the following TV-watching statistics before, but they deserve repeating. In the United States, preschool children watch TV an average of 30 hours per week, and many spend more time watching TV than any other activity except sleeping. This amount of viewing exposes children to approximately 400 advertisements per week. At age 16, children have watched more hours of television than they have spent attending school (Aulette, 1994; Leaper & Friedman, 2007; Witt, 2000). Consequently, when we discuss the many influences on the development of gender roles and identity, television must feature prominently. In her review of research on the influence of television on gender development, Witt (2000, p. 322) summarizes the process as follows:

> Children's ideas about how the world works come from their experiences and from the attitudes and behaviors they see around them. The young child who believes that only women are nurses and only men are doctors may have developed this understanding because the first doctor he or she saw was a man, who was assisted by a female nurse. This "man as doctor, woman as nurse" idea may have been reinforced further by parents, books, conversations with friends, and television. If the child frequently meets such gender biases... this knowledge will be incorporated into their future [sex-role] perceptions...
>
> Children who witness female characters on television programs who are passive, indecisive, and subordinate to men, and who see this reinforced by their environment, will likely believe that this is the appropriate way for females to behave. Female children are less likely to develop autonomy, initiative, and industriousness if they rarely see those traits modeled [by females]. Similarly, because male characters on television programs are more likely to be shown in leadership roles and exhibiting assertive, decisive behavior, children learn this is the appropriate way for males to behave.

Indeed, gender messages are sent out to all viewers of TV programs. Although the television networks and cable outlets have become more aware of gender bias in recent decades, bias persists. If you watch TV with this in mind, you will see this bias often on virtually all channels. Although exceptions exist, here is a list of the central findings of research on the content of most television programming, including commercials, as it relates to gender (Halim et al., 2011; Pike & Jennings, 2005):

- Men are usually more dominant than women in male–female interactions.
- Men are often portrayed as rational, ambitious, smart, competitive, powerful, stable, violent, and tolerant; women are portrayed as sensitive, romantic, attractive, happy, warm, sociable, peaceful, fair, submissive, and timid.
- Television programming emphasizes male characters' strength, performance, and skill; for women, it focuses on attractiveness and desirability.
- Marriage and family are not as important to men as to women in television programs. One study of TV programming found that for nearly half the men, it wasn't possible to tell if they were married, a fact that was true for only 11% of the women.

- Television ads for boy-oriented products focus on action, competition, destruction, and control; television ads for girl-oriented products focus on limited activity, feelings, and nurturing.
- Approximately 65% of the characters in television programs are male (even most of the Muppets have male names and voices).
- Men are twice as likely as women to come up with solutions to problems.
- Women are depicted as sex objects more frequently than men.
- Men are shown to be clumsy and inept in dealing with infants and children.
- Saturday morning children's programs typically feature males in dominant roles, with females in supporting or peripheral roles.

Do these gender messages on TV actually affect the gender-role development in children? Evidence suggests that the answer is yes (Gerding, 2014; Pike & Jennings, 2005). For one thing, studies have indicated that children who grow up without television tend to be less stereotyped in their attitudes about gender. In addition, children who watch programs that violate traditional gender roles—for example, those that show women in traditional male roles such as lawyers or police officers or men in more traditional female roles such as stay-at-home caregivers or schoolteachers—tend to be less traditional in their gender roles in the culture (Witt, 2000).

Although the strength of various internal and external influences on gender development is clear, this does not automatically imply that all boys and girls will develop a gender identity that corresponds to their biological sex or that conforms to society's expectations for "girl-ness" or "boy-ness." A small percentage of people perceive that their biological sex and their gender identity are in conflict. These individuals are referred to as transgender.

Transgender Identity

10.4 Summarize the topic of transgender identity including variations in gender expression, the issue of sex reassignment surgery, and the challenges faced by transgender children.

The story of Christine Jorgensen at the start of this chapter illustrates how some people experience extreme discomfort with their biological sex and feel that their physical self is at odds with their gender identity (which sex they *perceive* themselves to be). A **transgender** individual may be a biological male who perceives herself as partially or fully female and is typically uncomfortable, to varying degrees, with her male sexual body and society's expectations that she "behave like a male" (referred to as "male-to-female," or MTF, transgender). A transgender male (female-to-male, or FTM) perceives that his female body and societal pressures to "act like a female" are at odds with his personal gender self-perceptions.* In clinical practice, these attitudes and feelings of emotional distress over one's gender, in both children and adults, are referred to as **gender dysphoria** (DSM-5, 2013). However, many transgender individuals do *not* experience distress with their gender identity and would not present any reason to be "diagnosed" with gender dysphoria.

transgender
Individuals whose gender identity varies from their biological sex.

gender dysphoria
Refers to stress or discomfort stemming from the self-knowledge that one's biological sex does not conform to, or is the opposite of, his or her personal gender identity.

The number of transgender people is difficult to estimate accurately because so many keep their true gender identity hidden due to fear of the ridicule or violence that is often directed at transgender individuals. The transgender community estimates the prevalence of transgender individuals at about 1 in 3,000 for MTF and 1 in 10,000 for FTM (Conway, 2003b). Official medical estimates are about one-tenth of these numbers. Awareness of transgender issues has become far more mainstream in

---

* People are referred to as the sex they perceive themselves to be regardless of their "biological sex" (he, she, his, hers, etc.).

Western cultures, as exemplified by feature films such as *Flawless*, *Boys Don't Cry*, and *Transamerica*, in addition to the introduction of transgender characters on popular TV programs such as *Glee*, *Orange Is the New Black*, and *South Park*.

Some transgender individuals choose to suppress and hide their true gender identity and conform to society's expected gender-appropriate behavior as best they can. However, the emotional toll of "living a lie" often proves too difficult for many transgender individuals, and they may eventually decide to become more open about their true gender identity (Conway, 2003a). The process of becoming openly transgender may be smooth and easy for some, but terrifying and even dangerous (from the threats of prejudice, discrimination, and violence) to others (Bocktine et al., 2013; Kane-DeMaios & Bullough, 2006). To help combat these attitudes, some municipalities and states are adding *gender identity* or *transgender* to the categories that are protected by antidiscrimination laws, but this is still the exception rather than the rule. As of 2010, only 13 states, Washington, DC, and 109 cities or counties in the United States had enacted such laws, although they are under consideration in several other states and cities ("Non-discrimination Laws," 2010).

Others who are confused, anxious, or depressed about their gender identity (sometimes diagnosed with gender dysphoria) may require counseling or therapy to help them understand and resolve their gender-related discomfort (Reitman et al., 2013). This does not imply, however, that the goal of therapy is to "cure" a person's transgender identity. In fact, such a therapeutic approach with transgender individuals virtually always fails because a person's identity appears to be as much a part of who they are as their race, height, or eye color. Therapy for people experiencing gender anxiety typically revolves around helping clients understand and accept themselves for who they are and to emerge from the therapeutic process as contented and healthy individuals (Lev, 2006). That said, however, *most* transgender individuals are no more confused about their gender identity than anyone else; they are quite clear about their "femaleness" or "maleness." They are "different" only in that, unlike the majority of people, their gender identity does not happen to conform to their biological sex.

transsexual

A transgender person who has transitioned or is transitioning from his or her biological sex to his or her self-identified gender through actions, dress, hormone therapy, or surgery.

Some transgender people engage in wearing the clothes of the opposite sex, typically because this allows them to express their gender identity outwardly and achieve gender comfort for themselves. For some, wearing a few items of opposite-sex attire from time to time might satisfy their gender comfort needs; others may choose to dress fully as a member of the opposite sex most or all of the time and may, at least sometimes, seek to be perceived by society as a member of that sex. This form of "cross-dressing" is not considered a *fetish*, as it is when a man or a woman wears opposite-sex clothes for the goal of sexual gratification. Rather, this is the mode of dress that makes the transgender person feel most like him- or herself (fetishes, such as "transvestism," are discussed in Chapter 14, "Paraphilic Disorders: Atypical Sexual Behaviors"). For the transgender individual, choices in dress are about gender comfort, not sexual thrills.

Chaz Bono (son of the 1960s pop duo Sonny and Cher) has been willing to discuss publically his process of transitioning from his birth sex (female) to his true, male, gender identity.

The term **transsexual** is sometimes used to describe a specific subgroup of transgender individuals who, in varying degrees, feel trapped in the wrong-sex body and desire to transition from their inborn biological sex to the sex that conforms to their gender identity. Transsexual individuals may choose to remain at various stages in their sexual transition process.

Some transsexual individuals may choose to alter their physical sex in part biologically, through hormone treatments. Opposite-sex hormones, administered via injection or skin patch, produce significant changes in the bodies of both men and women. When male-to-female transsexuals receive the female hormones estrogen and progesterone, their bodies begin to change and become more feminine. Their breasts enlarge, fat is deposited on the hips, and the growth of facial hair decreases or stops altogether. When female-to-male individuals receive the male hormone testosterone, their bodies masculinize: the voice deepens, body hair growth increases in the usual male pattern, overall musculature enlarges, and the clitoris usually grows larger. In male-to-female individuals, hormone treatments do not produce a change in voice pitch, but most people will work to change voice patterns and inflections to sound more female.

Sex Reassignment Surgery

Another strategy some transsexuals turn to for relief from their gender conflicts is **sex reassignment surgery**, more commonly known as a sex-change operation. The thought of undergoing extensive surgery on one's genitals and other sexual anatomical structures in the search of gender identity relief is difficult for most people to imagine. However, the emotional pain many transsexuals feel in being "trapped" in the wrong-sex body often outweighs any hesitation they may feel, and the surgery becomes a deeply desired goal in their lives. As one illustration of this commitment and of the flexibility of the gender concept in humans, the feature on the next page, "Evaluating Sexual Research: Can a *Man* Give Birth?" (on the following page) discusses how a transgender man, during his sex-reassignment process, was able to become pregnant and deliver a healthy baby girl.

sex reassignment surgery
Surgical procedures used to transform physically an individual from one sex to the other, commonly known as a sex-change operation.

Over many decades, surgical procedures have been developed to allow transgender individuals to alter their sexual anatomy to resemble that of the opposite sex. The surgery is lengthy and expensive (in the United States, the cost is anywhere from $7,000 to $25,000 for MTF, depending on the procedures selected, and may cost over $50,000 for FTM) and is, for all intents and purposes, irreversible. Hospitals in foreign countries such as Thailand are offering sex reassignment surgeries for approximately half the cost or less in modern hospital facilities staffed by surgeons trained in major medical schools in the West. This is part of a new trend referred to as "medical tourism" (Horowitz & Rosensweig, 2007; Vyas, 2004).

Transgender individuals and medical professionals alike must be sure that surgery is the best course of action before it takes place. Psychological counseling, hormone therapy, and a presurgical gender transition period are typically required prior to undertaking sexual reassignment surgery. The World Professional Association for Transgender Health has proposed the following criteria for sex reassignment surgery (Coleman et al., 2012):

- Minimum legal age (age eighteen in the United States)
- Twelve months of continuous sex hormone therapy
- Twelve months of successful, continuous, full-time, real-life experience, living as the opposite sex
- Regular, responsible participation in counseling
- Clear knowledge about cost, length of hospitalization, complications, and rehabilitation relating to the surgery
- Knowledge of the availability of competent surgeons

Modern medical techniques, combined with continuous hormone therapy, are now capable of altering physical sex with remarkably accurate visual results. Often, sexual nerve pathways are rerouted to preserve the ability to respond sexually and have orgasms in the vast majority of postoperative individuals (De Cuypere et al., 2006; Lawrence, 2005). Theoretically, you could meet, get to know, date, have sex with, and even marry a postoperative transsexual (especially male-to-female) and be unaware of

Evaluating Sexual Research

Can a *Man* Give Birth?

In 2008, the headlines featured the story of the *man* who became pregnant and gave birth to a healthy baby girl. For many, this sounded confusing, or even akin to something out of a science fiction movie. But it was neither. Thomas Beatie, now in his early forties, is a transsexual male from Oregon who had been receiving treatments for ten years to change his biological, female body to become more male. In his twenties, he had legally changed his birth name of Tracy Langondino. He had been taking male hormones and his breasts had been surgically altered to flatten his chest to be more like that of a man. However, he had not yet undergone any surgeries to alter his reproductive organs, which were, therefore, still female. He still had ovaries, a uterus, a cervix, and a vagina. Temporarily discontinuing his use of hormones and employing artificial insemination techniques, he was able to conceive, carry a fetus to term, and deliver a 9-pound, 5-ounce baby girl vaginally. His wife, Nancy, the mother of two grown daughters from a previous marriage, was no longer able to become pregnant due to a prior hysterectomy. However, through a process of "induced lactation" combining hormones and physical stimulation, Nancy was able to breast-feed the new baby, named Susan Juliette. In the next two years, Thomas and Nancy had two more children, a son and another daughter.

In early 2008, Thomas Beatie, a transsexual man in the hormonal and breast surgery phases of transitioning to a male physical body, still had ovaries, a uterus, a cervix, and a vagina. Through artificial insemination, he was able to conceive, carry the fetus to term, and deliver a healthy 9-pound, 5-ounce baby girl. In June of 2009, the couple (shown here with wife, Nancy, and first child, Susan Juliette), announced that Thomas had given birth to a second child, a healthy baby boy. In July 2011, Thomas gave birth to the couple's third child, their second healthy baby boy.

This method of creating a family for transgender or transsexual individuals is the exception, not the rule (adoption and surrogacy are more common), but it is a striking example of the physical and emotional realities that accompany the complexities of human sex and gender.

the person's transgender status, unless he or she were to inform you of it. As you can see in Figure 10.4, the genitals of postoperative transsexuals appear quite natural.

Male-to-female sexual reassignment surgery may involve some or all of the following procedures:

- *Penectomy* (removal of the penis)
- *Uroplasty* (rerouting of the urethra)
- *Orchiectomy* (removal of the testicles)
- *Vaginoplasty* (the use of penile skin to construct labia and a vagina)
- *Breast implants* (if the patient feels that enlargement through hormone therapy has been inadequate)
- *Chondrolaryngoplasty* (optional procedure to reduce the size of the Adam's apple)
- *Phonosurgery* (optional procedure to raise voice pitch)

Female-to-male sexual reassignment surgery may include some or all of these procedures:

- *Mastectomy* (removal of the breasts and, optionally, reduction in nipple size)
- *Hysterectomy* (removal of the uterus, fallopian tubes, and ovaries)
- *Metadioplasty* (creation of an erectile phallus from the clitoris)
- *Phalloplasty* (formation of a penis from tissue taken from other areas of the body and transplanted using microsurgical techniques in the genital area; requires a penile implant for erection)

- *Uroplasty* (rerouting of the urethra)
- *Scrotoplasty* (reshaping and stretching of the labia to resemble a scrotum and the insertion of silicone prosthetic testicles)

Obviously sex reassignment surgery is *not* minor surgery. A complete sex change requires counseling; hormone treatments; and, typically, numerous operations occurring in stages over a period of months or years. However, research has shown that most transsexuals who undergo this surgery are satisfied with the outcome, are better adjusted in life, and are more sexually satisfied (De Cuypere et al., 2005; Murad et al., 2010).

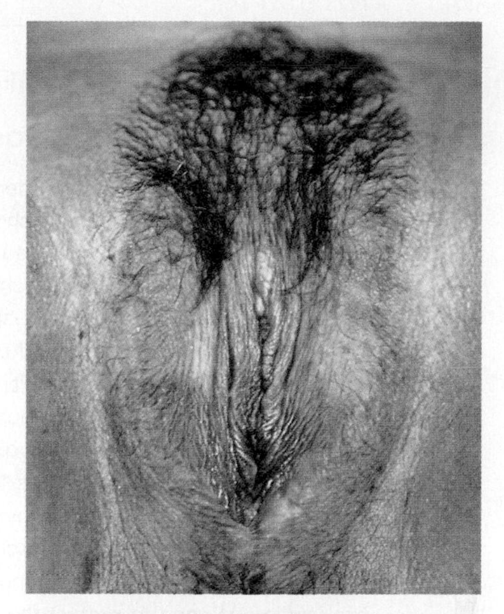

Male-to-Female Sex Reassignment

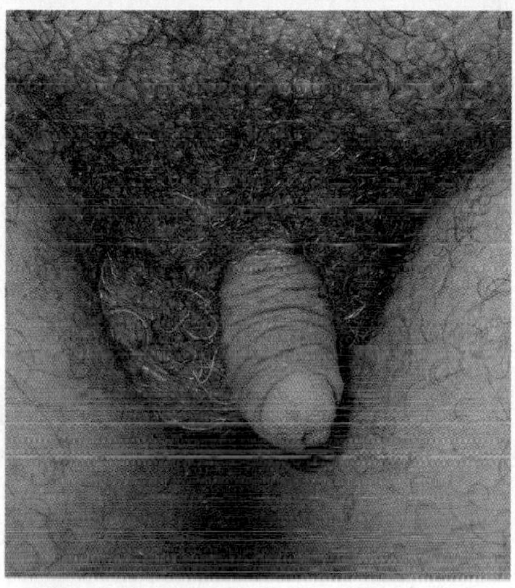

Female-to-Male Sex Reassignment

Figure 10.4 Outcomes of Sex Reassignment Surgery

The genitals after gender reassignment surgery are usually quite natural in both appearance and function.

Childhood Transgender Identity

Due primarily to social pressures, most transgender individuals suppress or hide their transgender identity until their teens or even later in life. However, over the past ten years or so, increased attention has been paid to transgender (or *trans*) identity in children and adolescents (e.g., see Luecke, 2011; Rosin, 2008). The newest findings have created a great deal of controversy, and these issues have been disturbing for some, and confusing for trans children and their parents. Transgender identity in children is often characterized by significantly more adamant and insistent behaviors than the boy who occasionally plays with a baby doll or the girl who will agree to join her brother in a game of computer football.

Some trans children are as convinced as any child can be of which sex they are, regardless of the fact that their gender identity is the opposite sex from their physical bodies. They do not like or want their given name if it implies the opposite sex from their self-identity, they shun sex-conforming toys and activities, will become frustrated and angry when called "son" or "daughter" by their parents, and hate the appearance of their genitals. From the time some of these trans children are able to speak, they will vehemently deny the designation of *son* or *daughter* and insist they are the opposite-sex child ("Mom! I'm *not* your son, I am you daughter!"). Trans girl children tend to be drawn to gender-typical girl activities, prefer to play with girls, dress like girls, and do whatever they can to be girls. The same is true in reverse for trans boys.

One story tells of a five-year-old transgender girl who, after a bath, tucked her penis between her legs and happily declared, "Look mom, I'm a girl!" When this same child was saying her prayers with her mother one evening, the mother said to her, "God made you a boy for a special reason." To this, she replied, "God made a mistake" (Rosin, 2008).

The number of trans children has been on the increase, but it is not clear whether this is primarily due to greater recognition and acceptance of these children's sexual identities. Doctors and therapists are working with more and more such children and their parents in an attempt to determine how the child should best be treated. When parents attempt to "force" gender-traditional toys and activities on these children, the children typically become angry, sad, and withdrawn. Their gender identity appears to be fixed and unyielding, and attempts by parents, doctors, and therapists to "change" them almost always fail.

In light of the ingrained nature of gender identity (in everyone), many parents are choosing to "allow" their transgender child to *be* his or her self-identified sex. For the young child this, of course, does not imply medical treatments of any kind, but the parents allow the child to become his or her desired sex. They make the decision to use a different name (usually of the child's choosing), allow the child to dress in the clothing of the "opposite" sex, change the pronoun ("he," "she") used to refer to the child's gender identity, and *accept* (often with great effort) that their daughter is actually their son or their son is actually their daughter. For many, this includes a shift to the child's "opposite" sex at school, with friends, with relatives, and with the trans child's world in general. Parents often remark how difficult this decision was, but the contentment

Since You Asked

3. I have a friend who has a seven-year-old son who always wants to play with girls and dress up in girls' clothes. He does not like to play with other boys. She wonders, is this just a stage or something else?

Sexuality, Ethics and the Law

Transgender Kids in School

The politics surrounding children's gender are evolving. In 2014, a new California law required public schools to accommodate K–12 students according to the sex each perceives him- or herself to be (Verdin, 2013). California schools are now required to allow students to participate in sex-specific programs and activities (such as various sports), and use bathroom and PE facilities which are consistent with their self-identified gender, regardless of their birth sex. Massachusetts, Connecticut, Washington, and Colorado have school policies that protect transgender students, but California is the first state to pass a statewide law. How do students "prove" what sex they are? They don't have to. The law requires school administrators to rely on the students' self-professed gender identity.

The law is designed to protect the rights of transgender students to an equal education, just as other laws have provided that guarantee based on race and religion (for example, allowing students not to participate in national or religious events that they are not a part of, or excuse them from school for events relating to their personal and family beliefs).

California's new law recognizes and reinforces what most psychologists and child development researchers already know: gender identity is fixed by age five and does not necessarily align with a child's biological sex.

The social fallout from this law is not known yet as the change is new in California's schools. The overall feeling is that parents are likely to be more concerned with this than the children who, at least in the early grades, adjust to such changes relatively easily. How the law will progress and the extent to which other states will follow suit should be clearer in the next several years.

In California, school children must be allowed to use whichever bathroom or PE facility they choose based on their self-identified gender.

Transgender children are not playing "dress up," but are deeply confident that they *are* the opposite sex from their birth sex. These are photos of Livvy (born male), with her mom, and Wren (born female).

and happiness they see in their child after their acceptance of him or her is more than worth it. The socio-political world is slowly coming to grips with transgender issues in children as well. "Sexuality, Ethics and the Law: Transgender Kids in School" discusses a potentially far-reaching transgender law recently passed in California.

But what happens when the child approaches puberty? As a biological boy or girl, the child will begin to develop secondary sex characteristics (breasts, penile enlargement, body and facial hair in either a male or female pattern, etc.). Furthermore, many trans children experience major negative psychological events as puberty approaches. One trans girl (born male) expressed her feelings this way as she approached puberty: "I feel like I'm getting empty. I'm not sad or worried any more, I just feel like I'm going to die. I'm going into a dark hole and it's getting narrower and narrower, and my life is going to end. I'm in the wrong place" (quoted in Luecke, 2011, p. 124).

At this point in the child's life (assuming the child has taken on the life of his or her transgender identity), doctors who specialize in gender issues may begin a procedure involving medications called "puberty blockers," which stop trans girls from growing facial hair, developing an Adam's apple, or experiencing a lowering of the voice (see "Primary Care Protocol," 2011). This procedure began in 2004. For trans boys, the medications increase height and prevent breast development, ovulation, and menstruation (Hembree et al., 2009; Rosin, 2008). These medications halt the development of physical features that would be much more difficult to alter surgically after they develop. So, with these medications, teen trans boys look like teenage boys and teen trans girls look like teenage girls.

Then, at about age sixteen, these medications are discontinued and cross-gender hormones are administered. The transgender teen then develops into a physical (as well as psychological) transgender adult (Hembree, 2009).

Gender and Sexual Orientation

10.5 Analyze how gender identity and sexual orientation are two distinct characteristics and how they may interact.

Just as a person's gender identity does not necessarily match up with his or her sexual anatomy, gender may not always be predictive of his or her **sexual orientation**, that is, which gender a person is primarily attracted to romantically, emotionally, and sexually (see Chapter 11, "Sexual Orientation," for a detailed discussion of sexual orientation). A person whose gender identity is male may be heterosexual, homosexual, or bisexual, just as a person whose gender identity is female may also be heterosexual, homosexual, or bisexual. Of course the vast majority of people who self-identify as male are attracted to those who self-identify as female, and vice versa, but this is not universally the case. Gender and sexual orientation tend to function as separate expressions of human sexuality. What does this separation of gender identity and sexual orientation mean in terms of the complexities of human sexual identity? It means that a person may live life as any one of many potential combinations of biological sex, gender, and sexual orientation (see Bieschke et al., 2006).

sexual orientation
Term specifying the sex of those to whom a person is primarily romantically, emotionally, and sexually attracted.

To take an example from the majority of cases, a biological male might have a gender identity of male and be heterosexual—sexually and romantically attracted to women. Similarly, a biological female might have a gender identity of female and be heterosexual—sexually and romantically attracted to men. In a small percentage of people, however, a biological male may have a gender identity of female and also be attracted to women. This person will probably consider herself to be lesbian (yes, a "male" lesbian—although this designation is a topic of some debate among some researchers and lesbian groups). Conversely, a transgender biological female who gender identifies as male and is attracted to women may feel heterosexual. Confusing? Maybe, but these complex combinations, although exceptions, show that most people's assumptions about human sexual identity are overly simplistic. Table 10.1 lists various combinations of sex, gender, and sexual orientation characteristics that may be found among the rich diversity of human sexual beings.

Since You Asked

4. I saw a commercial for a talk show with "male lesbians." Is that for real?

Table 10.1 The Complexities of Human Sexual Identity

Although the majority of people will identify with the characteristics in rows 1 and 2, the table illustrates that the characteristics of biological sex, gender identity, and sexual orientation are separate human characteristics and may exist in people in numerous combinations.

	Biological Sex	Gender Identity	Possible Primary Sexual or Romantic Attraction	Sexual Orientation
1[a]	♂	♂	♀	Heterosexual
2[a]	♀	♀	♂	Heterosexual
3	♂	♀[b]	♀	Lesbian[c]
4	♂	♀[b]	♂	Heterosexual
5	♂	♂	♂	Gay
6	♂	♂	♂ and ♀	Bisexual
7	♂	♀[b]	♂ and ♀	Bisexual
8	♀	♂[b]	♂	Gay
9	♀	♂[b]	♀	Heterosexual
10	♀	♀	♀	Lesbian
11	♀	♀	♂ and ♀	Bisexual
12	♀	♂[b]	♂ and ♀	Bisexual

[a] The majority of individuals fall into one of these first two rows.
[b] Transgender individuals.
[c] The validity of this designation is debated in some gay, lesbian, and transgender communities.

Gender Roles and Stereotypes

10.6 Discuss the meaning of gender stereotypes and review how they develop, the manifestations of them in society, and whether gender stereotypes are based in fact or imagined differences.

Since You Asked

5. Why is it that men are always supposed to be the aggressors, the ones to take the lead in relationships?

gender stereotypes

An assumption, usually negative, made about a person's appearance, behavior, or personality, based solely on his or her gender without regard for the person's individuality as a person.

Our discussion of the factors that influence the development of gender roles and identity leads us directly into an examination of what these forces may produce: **gender stereotypes**. A stereotype is the (false) assumption that all people who belong to a certain group share certain characteristics (i.e., "they're all the same"), regardless of the uniqueness of each individual group member. Therefore, a gender stereotype is an assumption about a person based solely on his or her gender, without regard for his or her individuality as a person.

We grow up learning from others and from society that certain behaviors and attitudes are "male" and others are "female." As these differences become psychologically attached to one gender or the other, we begin to incorporate them into our fundamental belief systems about men and women. The differences transform from "how a person behaves" into "who a person is." Once this happens, we come to *expect* men and women to behave differently, and we interact with them based on those expectations. When our expectations of individuals are based on their membership in a particular group, whether the group is a certain gender, race, religion, sexual orientation, profession, or any other characteristic, this is referred to as a stereotype.

That stereotypes exist is not open to debate; they do. To get an idea of the number of gender stereotypes and how widely they are believed, take a look at the list in "Self-Discovery: Traditional Gender Stereotypes." These stereotypes are common in Western cultures, and you can see signs of them nearly everywhere you look. They form the basis for bestselling books (the hugely popular "Venus and Mars" series was based largely on gender stereotypes) and hundreds of jokes (e.g., about men refusing to ask directions, or blonde females engaging in "dim-witted" behaviors). Stereotypes in and of themselves are not necessarily harmful. Rather, it is how they affect our attitudes and behavior that determines the consequences of stereotypes. Some stereotypes may be quite innocuous, such as your belief that your home-team players are more talented than members of other teams in a certain sport or that all cars made by a specific manufacturer are safe and dependable. When social scientists study stereotypes, they are primarily concerned with how stereotypes develop and to what extent they are based on true, rather than imagined, differences.

How Do Gender Stereotypes Develop?

Earlier, we discussed how society places many expectations on its members for different behavioral roles based on sex. These expectations are communicated to all of us very early in life. By the time children are three years old, they already have a sense of themselves and others as male or female. By age four or five, most have acquired the concept of *gender stability*; that is, they believe that when they grow up, they will be either a man or a woman, a daddy or a mommy, and most dream of traditional gender-based occupations (such as fireman for boys and ballerina for girls). By this age, they have also developed very strong preferences for gender-appropriate, sex-typed behaviors (e.g., most boys resist playing house, and most girls reject playing ninja warrior). Within another year or two, children understand *gender constancy*—that a person's gender stays the same even if that person violates expected, traditional sex-role behaviors (such as a man dressing up in women's clothes for a skit or a woman who is a plumber).

Gender stereotypes develop parallel with gender identity. By the time children are five years old, they have a remarkably clear sense of the difference between masculine and feminine sex roles in the culture, and they use people's gender as the main criterion to predict the behaviors of others. Even today, when society has developed an

Awareness of gender expectations, roles, and stereotypes is obvious in the Halloween costumes chosen by young boys and girls.

increased awareness of sexual inequalities, most preschoolers are confident that girls cannot be, say, firefighters and boys cannot be schoolteachers. If you watch young children pretending to go somewhere in a car, the boy nearly always drives. Interestingly, these stereotypes persist even when the child's personal experience provides examples of exceptions to the gender expectations, such as when the child's father is a schoolteacher, when the mother does all the driving, or when the child's books take a nonstereotypical approach to gender roles (Diekman & Murnen, 2004; Witt, 1997).

In the United States, you can easily see these gender beliefs in children by observing the costumes they choose for Halloween and other dress-up play. Although the costume might hide the child's true identity, 90% of children's costumes are conspicuously gender-appropriate, and only 10% are gender-neutral (Halim et al., 2013; Nelson, 2000). Girls dress up as beauty queens, princesses, brides, animals (butterfly, cat), and food items (lollipop, ice-cream cone). Boys are more likely to wear costumes representing police officers, warriors, villains, monsters, or symbols of death (Dracula, executioner, grim reaper). The gender identities and stereotypes in young children are so strongly formed that a child might well forgo all the candy Halloween promises rather than set out trick-or-treating in a clearly opposite-gender costume.

Self-Discovery

Traditional Gender Stereotypes

Below, in column 1, is a list of words often used to describe common human characteristics. The check marks in columns 2 and 3 indicate whether the trait has been traditionally seen as a masculine or feminine characteristic in most Western cultures. Columns 4 and 5 reflect society's views about whether each trait is usually perceived as desirable or undesirable for a person to possess. As you study the list, an interesting pattern emerges, doesn't it? Finally, the last two columns provide an opportunity for you to examine your own attitudes and see if you agree or disagree with society's traditional view of these male-female stereotypes.

Characteristic	Traditionally Seen as Male	Traditionally Seen as Female	Considered Desirable?	Considered Undesirable?	Do You Agree?	Do You Disagree?
Independent	✓		✓			
Assertive	✓		✓			
Strong	✓		✓			
Decisive	✓		✓			
Self-confident	✓		✓			
Submissive		✓		✓		
Passive		✓		✓		
Emotional		✓		✓		
Talkative		✓		✓		
Fearful		✓		✓		

As you can see, the pattern that emerges demonstrates how stereotypes typically seen as feminine have traditionally been viewed as undesirable. This suggests how stereotypes can lead (and have led) to prejudice and discrimination based on gender.

If you found yourself disagreeing with many of these, that is because you are *not* typical. You are more highly educated than the average person, and research has shown that as education level increases, belief in gender stereotypes declines.

As children move into the school years, gender stereotypes are further strengthened by systematic expectations about which subjects and activities are feminine and which are masculine. As early as second grade, children perceive math, sports, and various mechanical skills as masculine, whereas art, reading, and music are seen as feminine. This is not to say that boys refuse to produce art or play music or that girls reject math and sports. However, if you ask most children whether subjects such as math are "girl activities" or "boy activities," they are quite clear about the expected differences.

As children enter middle and high school, their gender stereotypes have already begun to mirror those held by adults. They see certain courses, extracurricular activities, recreational choices, and jobs as appropriate for one sex or the other. This stereotyped thinking then guides their social, educational, and professional choices throughout the teen years and into adulthood.

Gender-based stereotyped beliefs are so ingrained by adolescence that they tend to be an integral part of teens' view of the world. However, older children also tend to be more flexible in the gender-violating behaviors they will accept. They become increasingly willing, as most adults are, to judge people on other criteria in addition to their gender. They become aware that gender roles are social norms and that "breaking the rules" is sometimes acceptable (or even cool). However, this does not imply that as adults we become "gender blind." Various gender-role expectations and the resulting stereotypes remain strong throughout life. For example, many adults are still surprised—and some are uncomfortable or even disapproving—upon encountering, say, a female airline captain or a male nurse. This explains why many professions continue to be dominated by one sex or the other, as can be seen in Table 10.2.

The "Truth" About Gender Stereotypes

Accepting the reality that gender stereotypes exist brings us to another important question: Are they true? It is one thing for individuals or cultures to believe that certain characteristics and behaviors are appropriate for one gender or the other, but to what

Table 10.2 Occupations by Gender—United States

Gender stereotypes continue to influence people's choices of occupations.

Occupation	Number of Men	Number of Women
Airline pilot	79,000	3,000
Truck driver	2,307,000	79,000
Aerospace engineer	111,000	10,000
Dentist	21,000	20,000
Architect	85,000	36,000
Lawyer	435,000	234,000
Physician	416,000	189,000
College and university teacher	518,000	383,000
Psychologist	33,000	59,000
Occupational therapist	9,000	57,000
Waiter, waitress	287,000	538,000
Librarian	32,000	125,000
Legal assistant	46,000	261,000
Elementary school teacher	461,000	1,947,000
Registered nurse	207,000	1,970,000
Preschool and kindergarten teacher	16,000	524,000
Dental hygienist	2,000	50,000
Bank teller	35,000	297,000

SOURCE: Bureau of Labor Statistics (2010).

extent are gender stereotypes based on real differences between the sexes? Look again at "Self Discovery: Traditional Gender Stereotypes" on page 387. You may disagree with some or even most of the items listed as gender stereotypes. For example, your personal opinion may be that women are just as capable as men of being independent, or that men are just as sensitive as women. You may believe that some of the stereotypes should be reversed—that women are actually stronger than men in many ways or that men can be more talkative than women. The reason for these discrepancies is that two levels of gender stereotypes exist: cultural and personal. **Cultural gender stereotypes** are the beliefs about gender roles held by a majority of people in a given cultural setting and communicated through parenting, schooling, mass media, literature, and advertising, as discussed earlier. In addition, people hold **personal gender stereotypes**, which are beliefs about gender that are unique to each individual and may or may not agree with cultural stereotypes. Personal gender stereotypes are shaped by individual life experiences, such as those that caused you to disagree with some of the stereotypes in the Self-Discovery feature.

cultural gender stereotypes
Beliefs about gender roles held by a majority of people in a given cultural setting.

personal gender stereotypes
Beliefs about gender that are unique to each individual and may or may not agree with cultural stereotypes.

Who is right? The culture? You? Both? Neither? Social scientists have been working for many decades to tease apart scientific fact from cultural and personal beliefs about gender. Most people believe that every stereotype must contain a grain of truth or it would never have come into existence. Overall, research has shown this to be true for some gender stereotypes, but the differences between the sexes are usually smaller than most people expect. To illustrate, let's look at what the research tells us about two of the most common gender stereotypes: aggression and intuition. Then we will look a bit more closely at one stereotype that is very relevant to many of our discussions in this book: gender differences in sexual drive.

AGGRESSION One of the strongest and most widely accepted stereotypes is that male humans are more aggressive than female humans (Perry & Pauletti, 2011). A great deal of evidence seems to support that this stereotype represents a real gender difference (Coyne, Nelson, & Underwood, 2011; Knight et al., 2002). Men engage in significantly more aggressive behaviors in nearly all cultures worldwide. Among very young children, boys typically display higher levels of physical and verbal aggression than girls during play activities. As children enter school, this gender difference in aggression continues. As shown in Figure 10.5, based on data from one large, diverse urban school district, boys demonstrate significantly higher levels of aggressive behaviors than girls throughout the school years of first through seventh grade. In addition, Figure 10.6 demonstrates a clear difference in the types of aggression engaged in by boys and girls by the time they reach seventh grade. In adulthood, the greater level of physical aggression by

Figure 10.5 Boys' and Girls' Aggressive Behavior in Grades 1 Through 7

Boys tend to display consistently higher overall levels of aggression throughout childhood.

SOURCE: "Two Types of Aggression by Gender Among 7th Graders," based on data from the Institute for Teaching and Research on Women, 2002, Refer to www.towson.edu/womensstudies.

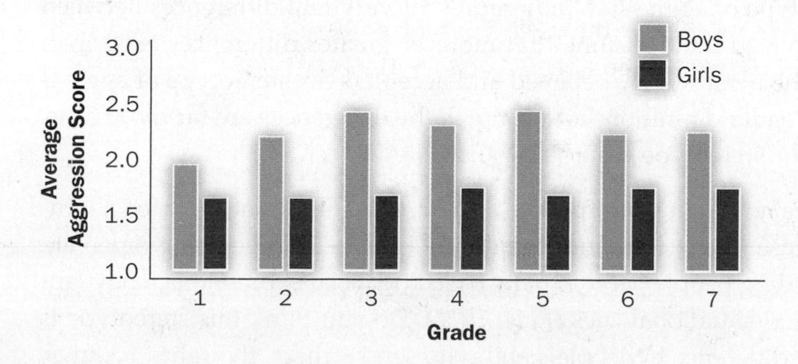

Figure 10.6 Aggression by Gender Among Seventh Graders

Boys have been shown to display greater levels of aggression across various behaviors.

SOURCE: Graph, "Boys' and Girls' Aggressive Behavior, Grades 1–7," from "Gender and Aggression: The Baltimore Prevention Study," Institute for Teaching and Research on Women, 2002. Refer to www.towson.edu/womensstudies.

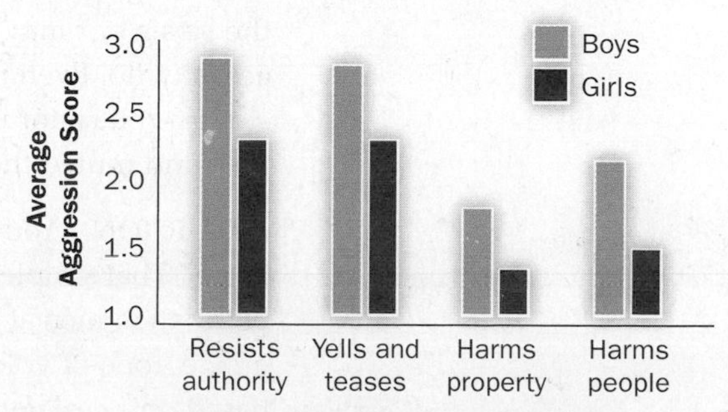

Females may be far more aggressive than males when the aggression involves relational aggression, such as social alienation.

men compared to women is exemplified by the fact that men commit the vast majority (80% to 95%) of all violent crimes, including robbery, aggravated assaults, and sexual assaults.

We must be cautious, however, and avoid jumping to conclusions (or stereotypes!) in how we interpret the research on aggression and gender. Although most researchers agree that men are, overall, more aggressive than women, a closer examination of some of the findings reveals that men may not always be more aggressive. In fact, when the *definition* of aggression is expanded and the *context* in which the aggression occurs is considered, gender differences diminish and may even reverse themselves (see Campbell & Muncer, 2007).

Most studies of aggression focus on verbal and physical aggression (often called *direct aggression*). If the definition of aggression is broadened to include *indirect aggression* (aggressive acts that are designed to hurt but without direct confrontation), the typical gender differences may be wrong (Wilbert, 2008).

One type of aggression that researchers have focused on more recently is a type of *relational aggression* called **social alienation** (Rudolph et al., 2014; Underwood, 2003). This is a more passive form of aggression that includes behaviors such as ("How Girls Bully," 2014):

Since You Asked

6. Everyone always says that men are more aggressive than women, but I'm a woman, and I think I'm usually more aggressive than most men. Is everyone wrong, or am I just weird?

social alienation

A passive form of aggression that includes behaviors such as malicious gossip, spreading negative rumors, and shunning.

- Playing jokes or tricks designed to embarrass and humiliate
- Deliberate exclusion of other kids for no real reason
- Whispering in front of other kids with the intent of making them feel left out
- Name calling, rumor spreading, and other malicious verbal interactions
- Being friends one week and then turning against a peer the next week with no incident or reason for the alienation
- Encouraging other kids to ignore or pick on a specific child
- Inciting others to act out violently or aggressively

When studies of children of all ages include these forms of aggression, girls have been found to engage in the same amount or even significantly more relational aggression than boys (Crick & Grotpeter, 1995; Marsee et al., 2014; Wilbert, 2008). Furthermore, this same gender difference in social alienation forms of aggression has been found in adult populations, including college students (Miller-Ott, & Kelly, 2014; Werner & Crick, 1999). Figure 10.7 graphically illustrates this profound difference in aggression styles by gender. It should be noted, however, that in the United States, researchers have recently focused on a new phenomenon: a significant increase in physical aggression and violence among girls, especially girl-on-girl violence (see Hernandez, 2007). This increase has been highlighted by the rise in arrests of girls for violent crime, the rise in girl-gang activity, popular books (such as *Odd Girl Out*) and feature films (such as *Mean Girls*) about violent girls, and the growth of Internet sites relating to girl-on-girl violence (Garbarino, 2006; Smolowe, 2008).

The gender stereotype of aggression may represent very real differences between the sexes but may also lead us to assume that more or greater differences exist than actually do. Even for the most widely believed and accepted sex stereotype of aggression, if you factor in broader definitions and context, the differences are far more complex and muted than the stereotype would have us believe.

INTUITION You have no doubt heard people make comments about "women's intuition." That simple phrase refers to another commonly held stereotype that women are better than men at "reading people" nonverbally based on facial expressions, body language, tone of voice, and situational cues (Fine, 2012). Do you think this stereotype is based on a real gender difference in people's ability to sense others' thoughts, feelings,

Figure 10.7 Gender Differences in Aggression in Fifth- and Eighth-Grade Children

Boys are far more aggressive than girls in overt forms of physical and verbal aggression, but for relational aggression, behaviors of social ostracism, and the spreading of malicious rumors, gender differences are reversed.

SOURCE: Table, "Gender Differences in Types of Aggression" (p. 1147), from "United States and Indonesian Children's and Adolescents' Reports of Relational Aggression by Disliked Peers," by D. French, E. Jansen, & S. Pidada, *Child Development, 73,* (2002), pp. 1143–1150. Copyright © 2002 Society for Research in Child Development. Reprinted by permission of the publisher.

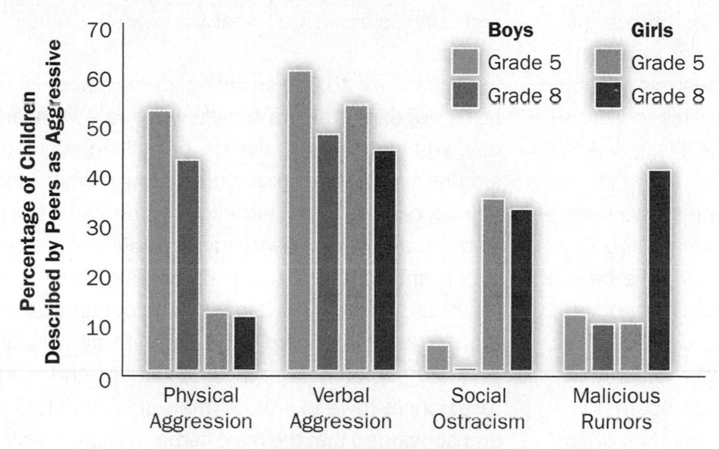

and desires and to decode nonverbal signals? Research has indicated that the answer is yes. This is not to say that men are never able to intuit the feelings of others, but women, on the whole, do indeed appear to be better at it (e.g., Naghavi & Redzuan, 2011).

Researchers have conducted many studies in which women and men are asked to interpret the emotions in films or videos of people in various emotional states and situations without any verbal cues. In audio studies, men and women listen to people's voices in recordings that have been filtered so that the words are removed and only the tone of voice remains. In the vast majority of these studies, women perform better than men (Hall, 1998; Plant, Kling, & Smith, 2004).

One important aspect of the overall skill of intuition is the ability to read facial expressions, referred to as *facial expression processing*. Over the past several decades, findings of studies looking at gender and facial expression processing have been mixed, with some research finding females performing better than males but other research showing no gender difference. However, a study from 2000 used a research technique called *meta-analysis* to attempt to resolve these contradictions in the literature. A meta-analysis gathers together the data from many previous studies, combines them, and reanalyzes them to see if any systematic differences emerge. "Evaluating Sexual Research: The Meta-Analysis" (following page) discusses this type of research in greater detail. When meta-analytic techniques were applied to gender effects on intuition, researchers found a clear female advantage in the ability to read nonverbal communication and facial expressions (Hall, 1978; McClure, 2000). Interestingly, we can see this advantage in facial expression processing from infancy all the way through adolescence. The fact that very young girls show greater skill in decoding nonverbal communication lends support to the notion that this gender difference may have a genetic, survival basis. We will discuss further the possible origins of gender differences later in this chapter.

SEXUAL ATTITUDES AND DESIRE Differences between males and females in various aspects of sexuality are included in discussions throughout this book. Clearly, many stereotypes exist about the sexual

"It's a guy thing."

Evaluating Sexual Research

The Meta-Analysis

In the field of human sexuality as well as other areas of research, a technique called *meta-analysis* has often been used to organize research and resolve disputes in the scientific literature. With the aid of computers, meta-analysis takes the results of many individual studies that may appear inconsistent and integrates them into a larger statistical analysis so that the evidence forms a more meaningful whole. This research strategy was first proposed in 1976 by Dr. Gene Glass of Arizona State University.

For example, to determine the validity of the stereotypes discussed in this chapter, such as gender effects on aggression or intuition, a multitude of individual studies have been conducted, yielding many conflicting findings. Subsequent meta-analyses combining hundreds of studies on male and female aggression, however, have consistently found that male humans are more physically and verbally aggressive than female humans. As for intuition, one meta-analysis of 75 gender and intuition studies found that in 68% of the studies, women were better at decoding nonverbal messages; in only 13% were men better; and 19% found no gender difference (Hall, 1978). If you were to read a random sampling of those 75 studies, you could end up quite confused or even draw incorrect conclusions about what the overall research has actually shown.

With over 40,000 scientific journals and new research articles appearing at the unbelievable rate of nearly 3,000 per day, you can see that the job of making sense out of the literature can be overwhelming. Through meta-analysis, many studies can be combined into one, and a larger, more organized picture of the research as a whole may be revealed.

It should be noted that, although meta-analytic studies often provide new insights into human behavior, they are not the "final word" and, as with all scientific research, they are sometimes open to argument and question. Some researchers have examined meta-analytic studies on gender and concluded that the male-female differences in many areas of sexuality may actually be smaller than commonly believed (e.g., Peterson & Hyde, 2011).

differences between men and women. Some of the differences people assume are based in fact. Specifically, research has tended to support the following gender differences in sexual attitudes and desire. Most of these probably won't surprise you; they may already be part of your gender expectations and personal stereotypes (Laumann et al., 1994; Peplau, 2003).

- Women focus more on the relationship aspects of sexual activities; men have a more physical or recreational orientation toward sexuality.
- Men are more tolerant than women of casual sexual encounters.
- A greater number of men than women masturbate, and men masturbate more frequently.
- Men are more likely than women to be accepting of premarital sex and tend to feel less guilty having it.
- Men are less disapproving than women of extramarital sexual behaviors.
- Men think about sex more often than women on a daily basis.
- Men are more likely than women to engage in intercourse without an emotional attachment.
- Men are more likely than women to assume that others are interested in sex.
- When describing sexual experiences, women are more likely to romanticize them, although men are more likely to describe them only in sexual terms.
- Women tend to see the goals of sex as building intimacy and expressing affection; men cite sexual variety and physical gratification as the goals of sexual activity.
- Men have stronger sexual drive and desire than women.

As with other gender stereotypes, however, we must consider these sexual differences with caution. For example, consider the last difference on the list. One of the most enduring stereotypes about human sexuality is that male sexual interest and desire is stronger than female sexual interest and desire. Virtually all the research on this difference tends to support this stereotype (Sine, 2014). However, if such a universal difference does exist, how large a difference is it? And how can we account for the

Since You Asked

7. Why is it that men seem to want sex more than women?
8. Why do guys feel that they must "play the field," even after they've made a commitment to someone?
9. Why are women who have sex a lot seen in negative ways, while men who do the same are seen in a positive light?

many heterosexual couples who report a higher sex drive for the woman than the man? The answer to these questions lies in how the differences in male and female sexual desire are statistically determined.

When researchers find gender differences—for example, that men's sexual interests and desires are greater than women's—they are basing their conclusion on *averages*. An average reveals little about how much the two groups may overlap. Although the *average* level of sexual desire among men is higher than that of women, the variation in sexual desire in both men and women is quite large and overlaps considerably. Men and women are not polarized or opposite in their sex drive. In other words, some women's sex drives are higher than some men's, and some men's sex drives are lower than some women's. Figure 10.8 graphically illustrates this point. You can see that the curve for male sexual desire is skewed ("slanted") toward the higher range of sexual interest (the blue area) and the female curve is skewed toward the lower range (the red area). However, the largest area under the two curves is the *overlapping* area (shown in purple), representing higher sex drive for women and lower for men. Two people in an intimate relationship might come from the red and blue ranges, respectively, but it is more likely they both fall somewhere in the purple range of sexual desire.

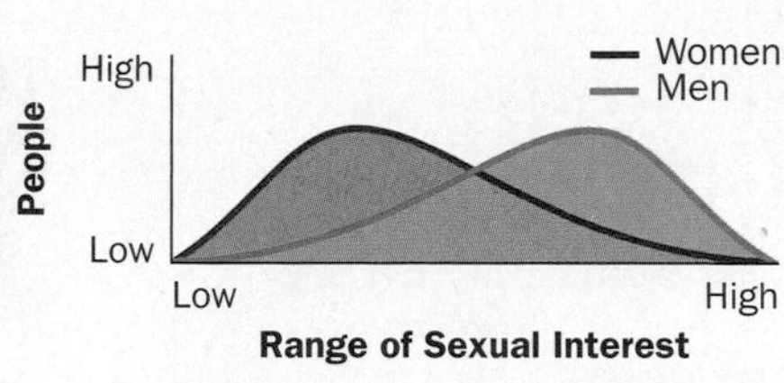

Figure 10.8 Male and Female Levels of Sexual Desire

Although levels of sexual desire are somewhat higher for men than for women overall, a great deal of overlap (purple area) exists between the sexes.

SOURCE: Baldwin & Baldwin (1997), p. 182.

The Overlapping Curve Model of Gender Differences

This **overlapping-curve model** of gender differences in sexual interest (as seen in Figure 10.8) can be applied to virtually all true gender differences in behavior. For example, although males are more physically aggressive than females overall, some women are more physically aggressive than some men, and some men display less aggression than some women. The male and female "curves of aggression" overlap, as do the gender curves for intuition and other sexual differences. The degree of difference between males and females determines the extent of the overlap. So the physical aggression curves probably overlap less than the sexual desire curves, meaning that we see a greater gender difference for physical aggression than for sexual desire. It is safe to say that most, if not all, gender differences have at least some overlap. Moreover, some research has reviewed numerous studies on gender and found that, for most human characteristics, gender differences are relatively small (Hyde, 2005; Peterson & Hyde, 2011). Such findings argue for more gender similarity than for major gender differences, overall.

Overlapping curve model
a graphic representation of how tow groups who differ on a particular attribute often share the characteristic to varying degrees and are not mutually exclusive

Gender and Intimate Communication

One area relating to sexuality in which gender plays a crucial role is communication. (This is discussed in more detail in Chapter 4, "Intimate Relationships.") As you may know from your personal experience, masculine and feminine patterns of communication differ in significant ways, and these differences can have far-reaching effects on intimate relationships.

A common stumbling block in heterosexual love relationships relates to fundamental gender differences in communication *styles*. An awareness of these differences can go a long way in helping a couple reach deeper and more effective levels of communication. The problem is not that men and women are purposely trying to communicate poorly with each other. On the contrary, both want the same thing: to be understood, to understand, and to arrive at mutually satisfying communication outcomes. However, a great deal of research has pointed to the unavoidable conclusion that men and women, especially in intimate relationships, are just not talking the same language.

Since You Asked

10. I've tried talking to my boyfriend about how I feel and what I want, but he never seems to understand, even though he says he does. How can I make him understand how I really feel?

You may be able to recognize the essence of this problem from one very common example that often occurs when male–female partners are attempting to discuss personal problems. Women tend to express personal problems to those they are close to because they want support and understanding for whatever it is they are going through. However, when men confide personal difficulties, they are typically looking for solutions. When a woman expresses her issue to her male partner, typically his first

Table 10.3 Gender and Communication Goals*

Men's Goals	Women's Goals
Give help; solve problems	Offer emotional support; empathize
Establish status	Seek harmony
Demonstrate authority	Strive for cooperation
Provide or receive information	Establish interaction
Avoid disagreement	Avoid being "cut off"
Gather information, formulate rules	Draw from others' personal experience
Avoid asking questions	Ask questions, explore, show interest
Provide information	Provide encouragement
Avoid private conversations	Seek private conversations
Avoid talking about feelings	Seek talking about feelings
Seek debate	Seek agreement
Seek respect	Seek acceptance
Challenge others' expertise	Accept others' expertise
Keep active and busy	Maintain companionship and interaction
Avoid personal discussions with friends	Discuss personal issues with friends
Obtain big picture	Obtain details
Display knowledge and expertise	Conceal knowledge and expertise
Seek opportunities to speak in public	Avoid public speaking
Find personal independence	Find sense of shared community

*These are general findings, and exceptions exist for both men and women.

SOURCE: Adapted from Tannen (1991).

reaction is to try to "fix it"—give advice and problem-solve—when she really wants support and understanding and may not want him to "fix it" at all. Conversely, when a man confides his problem to his female partner, she is likely to empathize—ask questions and offer support, understanding, and sympathy—which he may not want at all; he is looking for suggestions about how to "fix it." Both partners feel they are offering the other the help they are seeking, when in reality, they are offering what they themselves would want in the same situation. Often, this communication breakdown leads to frustration, distancing, and anger at the other's insensitivity, even when both have good intentions and are truly trying to be sensitive to each other's needs.

Barriers to communication such as this do not imply that men and women are consciously working to avoid mutual understanding, but rather that they usually have different basic and deeply ingrained goals for their interactions based on their gender. Even after learning about this "advice–support" communication disparity, men and women continue to have great difficulty learning and remembering to incorporate the knowledge into their intimate communications. This is only one example of many gender differences in communication styles that often interfere with mutual understanding. Table 10.3 lists some other gender differences in communication goals and strategies.

Origins of Gender Revisited

10.7 Summarize the nature (biological) and nurture (environmental) influences on the origins of gender.

Earlier in this chapter, we discussed the influences of biology and environment (socialization) on gender development. How do these forces affect the development of the gender differences that form the basis of gender stereotypes? To answer that question, we must return to the nature–nurture debate.

The Nature "Side"

The nature argument contends that gender differences are rooted in our biology and are passed down to us through evolution and our genetic heritage. Embedded in this position is the idea that these differences exist in humans as part of nature's grand design that has enabled humans to survive and evolve as a species. From this perspective, the reason that men are more aggressive, women are more intuitive, and men are more interested in sex is that these characteristics are survival strategies or *adaptive mechanisms.* Greater aggression in men stems from the evolutionary role of males to be protectors of females and infants, to fight for food and status, and to defend territory. Women are more intuitive because females, over the course of evolution, were placed in the role of caring for infants whose needs must be recognized without the benefit of language in order to survive and thrive. And the male's greater sexual drive relates to "reproductive strategies" and the biological need for the male to impregnate multiple partners to ensure continuation of his genes and the species in general. On the other hand, the female's best strategy for reproducing is to select the best, strongest, healthiest male and keep him around for protection and support. Although these mechanisms are probably no longer needed for our survival as a species, supporters of the genetic position contend that they continue to function and cause gender differences such as those we have discussed here. Interestingly, however, a few cultures worldwide draw the universality of these contentions into question (see "Sexuality and Culture: The Hijras: The Third Sex of India and Pakistan" (on the following page).

Many researchers contend that gender differences are learned through sociocultural expectations.

The Nurture "Side"

The nurture side of the debate downplays biological and genetic influences and focuses on sociocultural and other environmental factors as responsible for producing gender differences. This argument maintains that strong social and cultural factors, such as we discussed earlier, begin to mold male and female behavior nearly from the moment of birth and continue to exert pressure on the behavior of men and women throughout their lives. From this point of view, males are more aggressive than females because society *expects* and *allows* boys and men to display greater aggression but expects girls to behave in less (overtly) aggressive ways and does not allow them to display as much aggression. The female's greater intuitive skills may be explained by society's expectation that girls and women should be focused on the emotional side of social interactions and therefore learn to be skilled at reading nonverbal expressions of various emotional states. The nurture position might explain differences in sexual interest as stemming from the different messages boys and girls receive from the culture about sex. In most cultures, boys are "given permission" to be sexual, while girls receive the message that behaving in certain sexual ways is inappropriate and unacceptable. It follows, then, that boys and girls will grow up with different levels of sexual interest, desire, and behaviors.

Sexuality and Culture

The Hijras: The Third Sex of India and Pakistan

The Hijras, found primarily in India and Pakistan, are a religious sect of biological males who dress as and assume the role of women. Although they do not consider themselves male or female, but rather a "third sex" that is "neither man nor woman," the Hijras are generally referred to with feminine pronouns and treated by the culture as women. Their religious practices focus on the worship of the mother goddess Bahuchara Mata (Slijper, 1997). Some who consider themselves true Hijras undergo the ritualistic surgical removal of the penis and testicles, but research has shown that most have intact male genitals (Rehan, 2011). During their 600-year existence, the Hijras have been discriminated against, ridiculed, and marginalized as a low-class subculture (Kalra, 2012; Reddy, 2005). However, Hijras are believed by many to have the power to bestow fertility on others, so they often are hired to perform songs, dances, and blessings at weddings and the births of male infants.

Recently, Hijras have begun to organize and demand greater personal and civil rights. Hijras are running for and winning local, state, and national elections and are poised to become increasingly powerful politically. Moreover, they are not denying their cultural practices to do so. Instead, they are using their nonsexual identities for political advantage, as demonstrated by a Hijra campaign slogan: "You don't need genitals for politics; you need brains and integrity" (Reddy, 2005).

Hijras, shown here in Pune, India, define themselves as neither male nor female, man nor woman, but rather as asexual or a "third sex."

The central question becomes: Which side of the nature–nurture debate is right (or "more right") about gender? As you think about these two approaches, you may find yourself leaning more toward one or the other; most people do. However, you are probably also thinking that the truth may lie somewhere in between, that both genes and environment may be functioning together in some way. This is probably the case, and yet the debate over the relative influence of nature and nurture remains quite divided. In one study of gender differences in sexual interest, the authors explain the level of the controversy as follows:

> This debate often becomes so polarized that no compromises seem possible. Many people see women and men as opposite sexes and tend to explain the oppositeness in terms of biological differences in the form and function of the two sexes. Other people believe that the two sexes are basically similar, except for minor differences resulting from cultural factors.... When biological and cultural extremes are pitted against each other, discussions about female–male differences often turn into debates over "nature vs. nurture," and the advocates from each side attempt to explain all—or at least all the important—male–female differences in terms of *either* biological or cultural variables. Although behavioral scientists have been advised for decades to avoid the either-or arguments that pit nature against nurture and focus instead on developing models that unify nature and nurture, this goal is often not attained in analyzing sex and gender differences. (Baldwin & Baldwin, 1997, pp. 183–184)

Perhaps one way of turning down the volume of this controversy over gender differences is to return to the idea that genders overlap rather than directly oppose each other. Research has shown quite clearly that some people do appear to exhibit both masculine and feminine traits. These individuals are often referred to as **androgynous**.

androgynous
A person who embodies both masculine and feminine traits and behaviors.

Androgyny

10.8 Examine how the concept of androgyny may reconcile the extremes of masculine and feminine gender including the two-dimensional model, how androgyny may be measured, and positive versus negative androgyny.

Early theories of gender assumed a mutually exclusive view: that people have a gender-role identity that is either primarily masculine or primarily feminine and that masculinity and femininity are at opposite ends of a one-dimensional gender scale. If you were to complete a test measuring your gender identity based on this view, your score would place you somewhere along a single scale, either more toward the masculine or more toward the feminine end of the scale.

During the 1970s, psychologists proposed a groundbreaking **two-dimensional model of gender**. This approach allowed for the possibility that gender is not an either-or proposition but that people may manifest elements of both genders (Bem, 1974, 1994; Constantinople, 1973). This two-dimensional view of gender measures people on two *separate* scales, one for masculinity and one for femininity. Instead of being either masculine or feminine, a person can rate high on *both* masculinity and femininity. This may not sound very surprising to you now, but at the time, it was revolutionary. One of the leading figures in gender research, Sandra Bem, described people who perceive themselves as having both strong masculine and feminine traits as *androgynous* (from *andro*, meaning "male" or "masculine," and *gyn*, meaning "female" or "feminine"). Here is how she framed the issue in her now-famous 1974 article:

> Both in psychology and in society at large, masculinity and femininity have long been conceptualized as bipolar ends of a single continuum; accordingly, a person has had to be either masculine or feminine, but not both. This sex-role dichotomy has served to obscure two very plausible hypotheses: first that many individuals might be "androgynous"; that is, they might be both masculine and feminine, both assertive and yielding, both instrumental and expressive, depending on the situational appropriateness of these various behaviors; and conversely, that strongly typed individuals might be seriously limited in the range of behaviors available to them as they move from situation to situation. (Bem, 1974, p. 155)

two-dimensional model of gender
An approach to defining gender suggesting that gender is not an either-or proposition but that people may manifest elements of both genders simultaneously.

Since You Asked

11. Can someone be both male and female, or masculine and feminine, at the same time? I feel as if I don't really fit into either masculine or feminine.

Measuring Androgyny

Bem was not simply theorizing a new way of looking at gender, but was also suggesting that some *advantages* might exist for people who are less strongly sex-typed and more able to behave in either masculine or feminine ways depending on the situation. In her article, Bem developed a new instrument for measuring gender that incorporated her two-dimensional approach. The Bem Sex-Role Inventory contains a list of 60 characteristics that are masculine (e.g., acts as leader, ambitious, assertive, dominant, independent, self-reliant, willing to take risks), feminine (e.g., affectionate, childlike, sympathetic, understanding, yielding, shy), or gender-neutral (e.g., adaptable, conscientious, friendly, reliable, truthful) on which people can rate themselves on a 7-point scale (Bem, 1974). By examining the differences among the feminine, masculine, and gender-neutral scores, a person can determine his or her degree of masculine, feminine, or androgynous gender identity.

The concept of androgyny suggests that some people possess a balance of traditional masculine and feminine qualities.

Research on Androgyny

A great deal of research was generated by the new conceptualization of gender as two-dimensional, allowing for the existence of androgyny in addition to the traditional divisions of masculine and feminine. Prior to the 1970s, the prevailing belief was that people would be most well-adjusted in life if their "gender matched their sex." That is, boys and men should display masculine attitudes and behaviors, and girls and women should display feminine attitudes and behaviors. However, the recognition of androgyny shifted this focus.

Studies began to show that people who are more androgynous appear to be happier and better adjusted than those who are strongly sex-typed. For example, research has shown that androgynous children and adults tend to have higher levels of self-esteem and are more adaptable in diverse settings (Taylor & Hall, 1982). Other research has suggested that androgynous individuals have greater success in heterosexual intimate relationships, probably because of their greater ability to understand and accept the other person's differences and needs (Coleman & Ganong, 1985). More recent research has revealed that people with the most positive traits of androgyny tend to be psychologically healthier and happier overall (Woodhill & Samuels, 2004) and that androgynous individuals are more successful leaders in mixed-gender settings (Kark et al., 2102).

The basic theory of androgyny as developed by Bem and others has undergone various changes and refinements over the years. One finding has been that what are seen as masculine, feminine, and androgynous characteristics vary across cultures. Therefore, Bem's original methods of measuring androgyny may not apply to other, diverse cultures such as India, Taiwan, or Turkey, and may need to be adapted or reconceptualized to retain validity when used for research in nonwestern cultures (Peng, 2006; Turkum, 2005; Yim & Mahalingam, 2006).

Researchers has suggested that the psychological advantages experienced by people who score high in androgyny may be due more to the presence of masculine traits than to a balance between male and female characteristics (Whitley, 1983). If you think about it, this makes sense. Looking back at the Self-Discovery feature about gender stereotypes, you will notice that many traits associated with femininity are regarded by most Western societies as undesirable. We can therefore assume that people who possess more masculine than feminine characteristics will probably, in most circumstances (rightly or wrongly), receive more favorable treatment by others, which in turn creates greater levels of self-confidence and self-esteem in the individual.

Positive and Negative Androgynous Traits

Of course, not all masculine qualities are positive, nor are all feminine qualities negative. Positive and negative traits exist for both genders. The suggestion of positive and negative gender traits has led researchers to propose a further refinement of the androgyny concept to include *four* dimensions: *desirable femininity, undesirable femininity, desirable masculinity*, and *undesirable masculinity* (Ricciardelli & Williams, 1995; Woodhill & Samuels, 2003, 2004). Qualities such as firm, confident, and strong are seen as desirable masculine traits, whereas bossy, noisy, and sarcastic are undesirable masculine traits. On the feminine side, patient, sensitive, and responsible are desirable traits, and nervous, timid, and weak are undesirable traits. Depending on how someone's set of personality traits lines up, a person could be seen as positive masculine, negative masculine, positive feminine, negative feminine, positive androgynous, or negative androgynous. Table 10.4 illustrates these differences and how they could combine to produce positive and negative androgyny.

When gender characteristics are more carefully defined to consider both positive and negative traits, the advantages for *positive androgynous* individuals become even more pronounced (Woodhill & Samuels, 2004). People who combine the best of male and female gender qualities are more likely to be well-rounded, happier, more popular, better liked, more flexible and adaptable, and more content with themselves than those who are able to draw on only one set of gender traits or who combine negative aspects of both genders. Just imagine someone (male or female) who is patient,

Table 10.4 Sets of Gender Traits Creating Positive and Negative Androgyny: Some Examples

Positive Feminine Traits	Positive Masculine Traits	Negative Feminine Traits	Negative Masculine Traits
Patient	Firm	Worried	Bossy
Appreciative	Confident	Timid	Showing off
Loves children	Competitive	Self-critical	Noisy
Responsible	Strong	Nervous	Aggressive
Loyal	Outspoken	Bashful	Sarcastic
These two columns combine to create: Positive androgyny		**These two columns combine to create:** Negative androgyny	

NOTE: Some individuals may possess various combinations of positive and negative masculine or feminine traits, giving them various degrees of positive or negative androgyny.

SOURCE: Adapted from Ricciardelli & Williams (1995), pp. 644–645.

sensitive, responsible, firm, confident, and strong (positive androgyny) compared to a person who is nervous, timid, weak, bossy, noisy, and sarcastic (negative androgyny) and you'll easily get the idea behind this theory.

Your Sexual Philosophy

Gender: Expectations, Roles, and Behaviors

If you are like most people, you have probably not given very much thought to your gender or, for that matter, anyone else's. Most likely, you have simply made assumptions about gender all your life: People are male or female, masculine or feminine, man or woman. But the concept of gender is not nearly as simple as it seems, and the influence of gender on our understanding of human sexuality is powerful and far reaching. Gender plays a role in virtually every part of life, and an awareness of the complexities of masculinity and femininity is crucial to each individual's journey through life as a sexual being.

Consequently, gender also has a very important place in everyone's sexual philosophy. Although you may not have given a great deal of thought to your own or other people's gender in the past, you probably will now, after reading this chapter. One of the reasons an awareness of gender issues is important relates to the range of behaviors you perceive as available to you as an individual. As we discussed in this chapter, people who are most sex-typed—meaning those who display behaviors that are at the extremes of masculine or feminine characteristics—may be less flexible in their behavior in a given situation. In other words, very masculine or very feminine individuals may have fewer options available to them for dealing with the world around them.

For example, people who are mainly feminine might have difficulty being assertive or taking a leadership role but might be the ideal choice for caring for a child or an elderly person. On the other hand, highly masculine people might feel utterly lost if asked to care for a young child or help a friend with emotional difficulties but might be ready, willing, and able to assume the role of leader when necessary. Both of these skills are culturally desirable, and that is why it is important to understand that male and female, masculine and feminine, are not opposite ends of a single scale but rather two separate scales on which each person may be either high or low. The possibility exists for someone to embody both masculine and feminine sides of personality—to be androgynous. The flexibility of androgyny may allow for a fuller, more varied set of options for dealing effectively with many of life's complex situations.

Can people learn to be androgynous if they are not already? Or is it something people are born with or develop throughout life? The answer is yes to both questions. Many people seem to grow up feeling easy and comfortable displaying behaviors and characteristics commonly associated with both masculinity and femininity. Others enter adulthood with a very rigid, one-sided masculine or feminine self-concept. However, many of those who seem unable to accept androgynous behaviors and attitudes in themselves can learn to do so if they wish. And the learning begins with becoming educated about cultural expectations for gender-appropriate behavior and recognizing that violating those expectations, under the right conditions, can be seen as a strength and is often very rewarding. When the "macho man" discovers his capacity for tenderness, caring, and affection, or when the ultrafeminine woman realizes her ability to be strong, independent, and assertive, they both have grown and have become more effective people in life and in love.

Incorporating awareness and acceptance of gender diversity in yourself and others is a basic and compelling component of everyone's sexual philosophy. After all, the whole point of pursuing a clear personal sexual philosophy is to know who you are, know what you want and don't want for yourself in life, and plan ahead to be the most effective and satisfied person possible. Breaking free of some of society's limiting gender expectations is one of the many challenges along that path.

Have You Considered?

1. Think for a few minutes about a typical week in your life. List eight or ten activities that you would be likely to engage in during that week. Now imagine that your gender identity is suddenly reversed and you are a member of the opposite sex. Go back over your list to see how many of your usual activities might change and in what ways. What does this exercise tell you about gender?
2. Think of one gender-related behavior or activity that is likely to vary greatly from culture to culture. Now think of one that you would expect to be fairly consistent across cultures. Why do you think the first behavior varies while the other remains constant? What might this tell you about the two behaviors?
3. Picture groups of kindergarten boys and girls playing on the playground at recess. List at least three clear differences you are likely to see in the groups' activities, interactions, and behaviors. Based on what you've read in this chapter, what might explain these differences?
4. Imagine you are looking for a new romantic relationship and you meet someone with whom you really hit it off. You enjoy the same activities, you share similar attitudes and beliefs, and you find each other physically and sexually attractive. As the relationship progresses and begins to grow more intimate, your new partner confides in you that he or she is transgender and has undergone sex reassignment surgery. How do you think you might react to this news? What do you think might happen from that point on in the relationship?
5. Research finds that, overall, men display greater levels of sexual drive, desire, and interest than women. Why might this difference exist from both the nature and the nurture perspectives?
6. Think of three people you know who are very masculine, very feminine, or androgynous (no names, please!). For each person, list three aspects of their personality or behavior patterns that reflect their gender identity. Which of the three individuals do you feel is the best adjusted? Explain your answer.

Summary

Historical Perspectives

The Story of Christine Jorgensen

- George Jorgensen in 1953 became Christine Jorgensen as a result of the first widely publicized sex reassignment surgery. Through her willingness to go public with her transsexualism and sex reassignment process, Christine Jorgensen is today considered, by many, a pioneer in the social awareness of transgender issues.

The Distinctions Between Sex and Gender

- Gender and sex are two separate human dimensions.
- Biological sex is related to hormonal levels and genes.
- Gender and gender identity tend to be a more socially constructed characteristic (although they may have a biological foundation) relating to a person's degree of self-perceived masculinity or femininity.

The Development of Biological Sex

- A person's biological sex is determined at the moment of conception based on the combination of *chromosomes* that result from fertilization. Normally the egg, or *ovum*, from the woman's ovary contains a single X sex chromosome, and the sperm cell that fertilizes the ovum carries either an X or a Y sex chromosome. Two X chromosomes combine to produce a female, but an XY combination produces a male.
- In a small percentage of births, the biological sex of the infant is not always obvious. Some infants are born with ambiguous genitalia, a condition referred to as disorders of sex development (DSD). Abnormal levels of hormones may also produce sex- and gender-based disorders such as Klinefelter syndrome in males and Turner syndrome in females.
- In recent years, the practice of routine surgical alteration of DSD babies has become increasingly controversial. Critics of the procedure, including many DSD adults, claim that altering DSD babies against their will and without their consent is ethically and morally wrong.

The Development of Gender Identity

- Development of gender identity is believed by many experts to be influenced primarily by biological and environmental factors. These influences include genes and hormonal levels in the uterus during fetal development.
- Others hold that socialization primarily determines gender through society's overt and hidden influence on gender-appropriate behaviors, which are heavily influenced by parents, peers, teachers, television, and other media.

Transgender Identity

- Some individuals, referred to as transgender, feel that their biological sex does not accurately represent their gender identity, and some, often called transsexual individuals, take steps—including dress, hormone therapy, and surgery—to align their appearance with their gender identity. Current medical procedures for altering a person's physical sex are quite effective and produce natural-appearing and functional sexual structures.

Gender and Sexual Orientation

- Gender identity is not always predictive of a person's sexual orientation. A person who perceives his gender as male may self-identify as heterosexual, gay, or bisexual, just as a person whose gender identity is female may self-identify as heterosexual, lesbian, or bisexual. Most people who self-identify as male are attracted to those who self-identify as female, and vice versa, but this is not universally the case. Gender and sexual orientation may function as separate expressions of sexual identity.

Gender Roles and Stereotypes

- Gender stereotypes exert a powerful influence on a culture's perceptions of men and women. Stereotyped expectations of masculine and feminine behaviors begin in early childhood. Many gender stereotypes contain a grain of truth, but differences are smaller and more overlapping than typically believed.
- The stereotype that males are more aggressive than females is generally true for physical and verbal aggression. However, females are found to be more aggressive than males when the definition of aggression is broadened to include relational aggression such as social rejection and malicious gossip. Overall, girls and women are found to be better than boys and men at intuition, the reading of nonverbal communication signals. Males are found to have higher overall levels of sexual desire and interest, but the ranges of male and female sexual desire overlap significantly.
- Communication patterns and goals are different for men and women, often interfering with effective communication between the genders.

Origins of Gender Revisited

- The origins of sexual orientation tend to revolve around the larger nature–nurture controversy.
- The nature argument contends that gender differences are rooted in our biology and are passed down to us through our genetic heritage. Embedded in this position is the idea that these differences exist in humans as part of nature's grand design that has enabled humans to survive and evolve as a species.
- The nurture side of the debate downplays genetic influences and focuses on social and cultural factors as primarily responsible for producing gender differences. This position maintains that strong environmental factors begin to mold male and female behavior from the moment of birth and throughout life.
- Strong proponents of the nature and nurture sides of human development continue to polarize the debate over the exact origins of gender differences.

Androgyny

- In the 1970s, the concept of gender was redefined to include androgyny, the notion that both masculine and feminine characteristics may exist in the same person at the same time. Research suggests that androgynous individuals are generally happier and better adjusted. However, an androgynous person may possess masculine and feminine traits that are either socially positive or negative that combine to determine his or her overall gender personality.

Your Sexual Philosophy

Gender: Expectations, Roles, and Behaviors

- Gender is a major component of human sexual identity. Gender roles and behavioral expectations play a part in virtually every aspect of life. Awareness and understanding of the complexities of masculinity and femininity are crucial to each individual's journey through life as a sexual being.

Chapter 11
Sexual Orientation

Learning Objectives

After you read and study this chapter you will be able to:

11.1 Summarize the American Psychological Association's position on sexual orientation and review how sexual orientation (including bisexuality) has been defined for research purposes.

11.2 Review the central issues relating to nonheterosexual orientations including, the notion of choice of sexual orientation; same-sex marriage; sexual orientation and having children; and the characteristics of same-sex intimate relationships.

11.3 Analyze and evaluate various approaches to the origins of a person's sexual orientation, including biological views (nature), environmental theories (nurture), and how these opposing ideas might be reconciled.

11.4 Summarize the process timing, process, and difficulties of "coming out," the time when gay, lesbian, and bisexual individuals choose to acknowledge to themselves and to others their true sexual orientation which they had previously hidden due to societal pressures.

11.5 Discuss why and how the HIV/AIDS epidemic has affected the gay community over the past 30 years, including the myths and stigmas endured by gay individuals relating to the disease.

11.6 Examine societal prejudice and discrimination based on sexual orientation including the legal aspects, hate crimes, and violence targeted at gay and lesbian individuals.

Since You Asked

1. I have gay friends who often refer to something called "Stonewall," but I don't know what they are talking about and I don't want to ask and seem stupid or offend them. (see page 406)
2. Are gay and lesbian people ever attracted to members of the opposite sex? (see page 407)
3. Is bisexuality only a stage toward becoming gay or is it a sexual orientation in itself? (see page 408)
4. It seems to me that male sexual anatomy is designed to "fit" with female sexual anatomy, and vice versa. So I don't get why someone would be gay if their anatomies don't fit together so well. (see page 410)
5. How do people know or decide to be homosexual or heterosexual or bisexual? (see page 410)
6. I have no problem with homosexuals. I say live and let live. But I don't see why they are so concerned with getting married. I mean, if they can have their relationship, why do they have to get married? (see page 410)
7. Do gay and lesbian couples really love each other in the same way as straight couples? (see page 413)
8. How do gay and lesbian relationships differ from heterosexual ones, if they do? (see page 413)
9. Why are some people gay? Is it something they're born with or just bad experiences with their parents or the opposite sex? (see page 415)
10. How old are most people when they come to terms with or understand that they are homosexual? Do they usually tell people right away? Are they usually comfortable with the realization? (see page 423)
11. Do homosexual people feel comfortable holding hands and showing other public displays of affection with their partner? How often are they ridiculed or threatened? (see page 428)

The following is an excerpt from a second-year college student's paper written for a human sexuality class and reprinted here with his permission.

I have known I was gay for as long as I can remember. I hear stories of people saying that they didn't know until high school or college, but I really find that hard to believe. A person's sexuality is a very important part of his or her life, and I always think to myself, "How can you not know?" It has always been with me and I have always been aware of it.

Early in my life, I lived a lie. I behaved in ways that everyone would like and accept, and nobody, not even my parents, would suspect I was gay. I have to tell you I was miserable. Can you imagine going through life having to hide your true self? It was absolute torture. I was so upset because we only have one chance at life and I was given a major "handicap." I kept asking, "Did I do something to deserve this?" And I would have given anything to be like everyone else. I just knew that for the rest of my life, I would be forced to live this lie.

Never in my wildest dreams did I imagine that I would be able to talk freely and openly about my sexuality. My friends would have rejected me instantly if I had told them I was gay. When I entered college, everything changed. In college, I made new and wonderful friends who taught me it is all right to be who I am and I shouldn't worry about what other people think about me. But the real event that changed everything occurred over the summer. My best friend introduced me to a guy who was also gay. We hit it off right away, and one thing led to another and we started dating. This posed a lot of problems. I felt guilty about hiding it from my parents and everyone else, but I was afraid someone would find out. Soon, as my boyfriend and I became closer and closer, I realized that my happiness was more important than what other people thought. I was proud of who I was, and I didn't want to change.

I finally decided to tell my brother and my mom. Afraid of what might happen and hoping they would not reject me, it took a while for me to spit it out; but after I did, their response could not have been better. They were very accepting of the fact that I am gay and were sorry that I did not feel comfortable enough to tell them sooner. My boyfriend and I are still together, and we are doing great!

I must say, my life has been a roller coaster of emotional highs and lows. I would not change a thing, however, because I love the person I am today. I have realized that it doesn't matter what other people think about me as long as I am happy. I believe I am a stronger person because of what I have had to go through. Now, at 20 years old, I feel like my life is just beginning, and I know that it is going to be filled with happiness.

Sexual orientation refers to the sex of the individuals to whom a person is romantically, emotionally, and sexually attracted. The terms **Heterosexual** or **straight** refer to individuals who are primarily attracted in all those ways to members of the opposite sex. **Homosexual** applies to those whose primary attraction is to members of their own sex. **Gay** is generally preferred to refer to homosexual men or women, and **lesbian** refers to homosexual women. **Bisexual** individuals are those who are attracted to members of both sexes. The excerpt just presented recounts one student's journey as he discovered and accepted his sexual orientation.

Most of our attention in this chapter will be focused on nonheterosexual orientations (gay, lesbian, and bisexual). Why? Because most (but not all) people are not only heterosexual themselves but they also tend to be **heterocentric**. That is, they usually take for granted and assume people are heterosexual, and have difficulty understanding and accepting nonheterosexual orientations. Statistics vary considerably as to the percentage of people who have are gay, lesbian, or bisexual. Estimates range from as low as 2% to about 10%, depending on how sexual orientation is defined and where the statistics are gathered (see Gates, 2006; Leff, 2011). Overall, the figure of 10% is commonly used as a "convenience statistic" because, using self-report measures, it is approximately

sexual orientation
Term specifying the sex of those to whom a person is primarily romantically, emotionally, and sexually attracted.

heterosexual
A person who is attracted romantically and sexually primarily to persons of the opposite sex.

straight
Heterosexual.

homosexual
A person who is attracted emotionally, romantically, and sexually primarily to persons of his or her own sex.

gay
Homosexual; applied to both men and women.

lesbian
A female with a homosexual orientation.

bisexual
A person who is attracted emotionally, romantically, and sexually to members of both sexes.

heterocentric
The mistaken assumption of a "universal" heterosexual orientation.

Focus on Your Feelings

Like many of the topics throughout this book, issues of sexual orientation often provoke various intense emotions. Gay and lesbian individuals have very strong feelings about being able to live their lives freely without being marginalized or limited by society's attitudes. For some in the heterosexual community, the strongest emotions are triggered by the prejudice and discrimination that still exist in our society today. Although negative emotions relating to sexual orientation have begun to subside a bit, they are still far too common. These feelings may include confusion, sadness, fear, anger, outrage, hostility, and even violent intentions. As a culture, we are working through negative stereotypes, prejudices, and discriminatory behaviors targeted at gay and lesbian individuals. But events all around us make it clear that we still have a long way to go.

Reading about and studying sexual orientation in this book and in class is an opportunity for each of you to reveal (to yourself) and examine (within yourself) your feelings about this important topic. Some questions you might want to think about are these: Do you feel comfortable with your emotional reactions? Do your reactions truly reflect the attitudes you want to hold toward people of all sexual orientations? If you are unsure of or unhappy with your feelings as you read this chapter, can you identify where your attitudes came from and how they developed in you? How will you go about altering the feelings you do not like? And finally, after you have read the chapter, have your feelings about these topics changed? If so, how and why did they change? What do you think triggered those changes?

Remember, emotions are based on knowledge and experience, and it is a fact that many people simply have very little experience with or knowledge of people of various sexual orientations. Taking time to study and understand these topics, especially those that evoke the strongest emotions, can help us grow, change, and learn to respond to the world in productive and constructive ways. This is, as much as anything, what all education is about.

correct in most areas of the United States, and percentages are rarely higher than that, except in specific regions or cities such as San Francisco (15.4%), Seattle (12.9%), and Minneapolis-St. Paul (12.5%). A recent study found that percentages of adults who are gay, lesbian, or bisexual among a sampling of ten states found the highest totals ranged from 4.7% (Minnesota) to 8.1% (Washington, DC).

To understand nonheterosexual orientations better, we will examine what it means to be gay, lesbian, or bisexual in a heterocentric culture, including the current sociopolitical storm over the changing of laws in the United States pertaining to gay rights and same-sex marriage. We will analyze various theories about the origins of a person's sexual orientation. We will consider the challenges a person may face when acknowledging that he or she is not heterosexual, a process called *coming out*, including such topics as the stigma of HIV as it relates to people's sexual orientation (despite the fact that HIV is not a "gay disease"). We will conclude with a look at the disturbing nature of prejudice, discrimination, and violence against nonheterosexuals in our society and the emergence and growth of the gay rights movement.

Nonheterosexual orientations have been identified in writings and other media throughout the world for millennia. But in the next section, we will look back only a few decades to an event considered to be a turning point in the history of gay life in the United States: *The Stonewall Riot*.

Historical Perspectives

The Beginnings of the Gay Rights Movement: The Stonewall Riot

Since You Asked

1. I have gay friends who often refer to something called "Stonewall," but I don't know what they are talking about and I don't want to ask and seem stupid or offend them.

In 1969, homosexuality was against the law in the United States and most other countries. The prohibitions went so far as to include a ban on serving alcohol to homosexuals. At that time, the Stonewall Inn in New York City was one of the few night spots where gay people could meet each other, flirt, hang out together, order drinks, dance, and just be uninhibited or romantic. Once a month or so, the police would raid the bar, rough up and ridicule the patrons, make numerous arrests, and cart large numbers of patrons away to jail in police vans.

June 27, 1969, was just another evening at the Stonewall Inn. No one knows why, but when the police once again raided Stonewall in the early morning hours of June 28, the patrons fought back. They were fed up with being harassed, beaten, and arrested for simply being who they were. They began yelling at the police to get out and leave them alone. As the police punched, dragged, and otherwise forced the patrons out into the street toward the paddy wagons, the yelling turned into fist fighting, and then beer bottles and garbage cans were hurled at the police. Crowds began to gather outside the bar, and more people, seeing the patrons being victimized, joined the fight against the police. Astonished that "these kinds of people" would actually fight back, the police drew their guns and called in SWAT teams in full riot gear. Remarkably, by the time the dust cleared, no one had been killed or seriously injured, but the course of gay history had been changed forever. As word of the Stonewall riot spread, gays and lesbians across the country were emboldened to resist and even fight the prejudice and discrimination they had endured for so long. Four months after the riot, both *Time* and *Newsweek* ran cover stories on homosexuality in America, gay "be-ins" began to be held in New York's Central Park, and the gay rights movement had been born (Cusac, 1999).

Four months after the Stonewall riot, *Time* magazine published its first issue focusing on the rights of homosexuals. Below shows a plaque on the wall of the inn commemorating the event and the birth of the gay rights movement

Today, the Stonewall Inn still stands, although now as a men's clothing store. Most historians trace the true beginning of the now powerful gay rights movement to the Stonewall riot of 1969. Each year, parades and other celebrations are held around the end of June in nearly every large city in the United States commemorating the Stonewall Inn riot and the antidiscrimination campaign it started. Clearly, prejudice and discrimination based on nonheterosexual orientation still exist, but now the *discriminators* are the ones more likely to be breaking the law. We will return to today's ongoing issues about gay rights later in this chapter.

Straight, Gay, Lesbian, Bisexual: Distinct Groups?

11.1 Summarize the American Psychological Associations position on sexual orientation and review how sexual orientation (including bisexuality) has been defined for research purposes.

Regardless of your personal sexual orientation, you probably feel you have a pretty good idea of what it means if someone is referred to as "straight," "gay," "lesbian," or "bisexual." In fact, most of you (but not *all* of you) are quite sure that your own sexual orientation falls into one, and *only* one, of three categories: *heterosexual, homosexual,* or *bisexual*. However, these labels are not as clearly differentiated as you may think. A person's sexual orientation, just like his or her sexuality in general, is complex, multifaceted, and not always neat, clear-cut, or well defined.

The American Psychological Association (2014) describes sexual orientation in this way:

> Sexual orientation refers to an enduring pattern of emotional, romantic, and/or sexual attractions to men, women, or both sexes. Sexual orientation also refers to a person's sense of identity based on those attractions, related behaviors, and membership in a community of others who share those attractions. Research over several decades has demonstrated that sexual orientation ranges along a continuum, from exclusive attraction to the other sex to exclusive attraction to the same sex [to be discussed next]. However, sexual orientation is usually discussed in terms of three categories: heterosexual (having emotional, romantic, or sexual attractions to members of the other sex), gay/lesbian (having emotional, romantic, or sexual attractions to members of one's own sex), and bisexual (having emotional, romantic, or sexual attractions to both men and women). ... Sexual orientation is distinct from other components of sex and gender, including biological sex (the anatomical, physiological, and genetic characteristics associated with being male or female), gender identity (the psychological sense of being male or female), and social gender role (the cultural norms that define feminine and masculine behavior).

Sexual Orientation as Defined by Alfred Kinsey

Part of the description above includes the idea that sexual orientation is more complex than simply three categories: heterosexual, gay/lesbian, and bisexual. Research has shown that people's sexual orientation is often more nuanced and flexible when it is studied and analyzed carefully.

Perhaps just as important from a historical perspective as the Stonewall riot discussed in the chapter's opening pages, were Alfred Kinsey's early attempts to define sexual orientation in his famous reports on sexual behavior, published in the middle of the twentieth century. As noted in Chapter 1, Kinsey conducted large surveys of Americans in which participants were asked about virtually every aspect of their sexual practices. His reports on these interviews, *Sexual Behavior in the Human Male* (Kinsey, Pomeroy, & Martin, 1948) and *Sexual Behavior in the Human Female* (Kinsey et al., 1953), were intended for medical and other professionals, but as you might guess, they became runaway national bestsellers. People were eager to read about the sexual behaviors of others and compare themselves to the statistical findings

Since You Asked

2. Are gay and lesbian people ever attracted to members of the opposite sex?

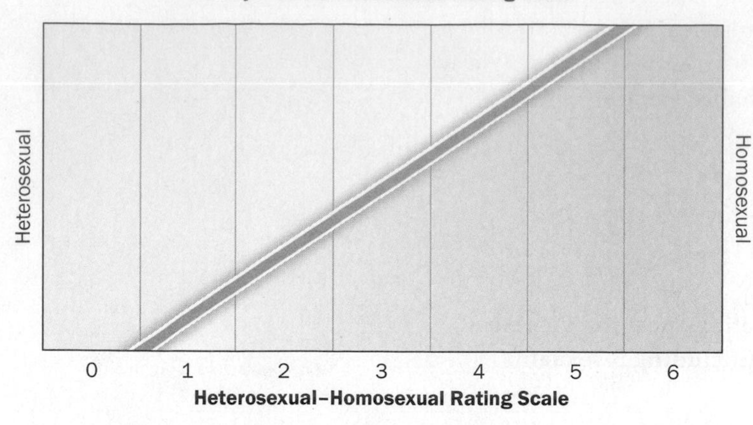

0 Exclusively heterosexual
1 Predominantly heterosexual, only incidentally homosexual
2 Predominantly heterosexual, but more than incidentally homosexual
3 Equally heterosexual and homosexual
4 Predominantly homosexual, but more than incidentally heterosexual
5 Predominantly homosexual, only incidentally heterosexual
6 Exclusively homosexual

Figure 11.1 Kinsey's Sexual Orientation Rating Scale

According to Kinsey's sexual orientation scale, most people cannot be classified as exclusively heterosexual or homosexual; rather, they fall somewhere in between, along a continuum.

SOURCE: "Kinsey's Sexual Orientation Rating Scale," from *Sexual Behavior in the Human Male*, by A. Kinsey, W. Pomeroy, & C. Martin. Copyright 1948. Reprinted by permission of The Kinsey Institute for Research in Sex, Gender, and Reproduction, Inc.

revealed in the Kinsey Reports. Kinsey's studies were conducted at a time in American history when homosexuality was hidden and rarely discussed, much less written about. Yet, there were the Kinsey researchers asking respondents the extent to which they had ever engaged in same-sex behaviors.

Based on responses from participants in the surveys, Kinsey asserted that very few people could be classified as totally heterosexual or totally homosexual. He maintained that most people fell somewhere in between or, along a *continuum*. Kinsey found that most people who identified themselves as heterosexual had at least some same-sex experiences (including fantasies, dreams, thoughts, emotions, and behaviors) and, conversely, that many of those who perceived themselves as homosexual had some experiences involving the opposite sex. Therefore, Kinsey developed a sexual orientation rating scale that places each person on a scale of 0 to 6, where 0 indicates "exclusively heterosexual" and 6 reflects "exclusively homosexual" (see Figure 11.1). According to Kinsey, most people would fall somewhere above 0 but below 6 (rather than exactly on 0 or 6).

Using this scale as a guideline, some people, based on their sexual feelings and experiences (rather than their overall self-identification), might be categorized as primarily homosexual (5 or 6 on the scale) or primarily heterosexual (0 or 1). Someone who falls in the middle categories (2, 3, or 4) might be categorized as bisexual. To allow you to develop a clearer idea of this type of analysis, "Self-Discovery: A Sexual Orientation Worksheet" is an opportunity for you to see where on Kinsey's scale you might fall.

Bisexuality

It may seem as if *gay*, *lesbian*, and *bisexual* all fit neatly into a single category of "non-heterosexual orientations." However, in some ways, bisexuality is a unique sexual orientation, different from being gay, lesbian, or heterosexual. Kinsey's efforts to place sexual orientation on a continuum notwithstanding, most people continue to see people as either straight or gay. It appears, however, that bisexuality is a clearly defined sexual orientation in which a person may be emotionally, psychologically, and physically attracted to members of either sex.

Historically, many people and even some researchers believed that individuals who self-identified as bisexual were in reality trying to hide the fact that they were actually gay or lesbian or were simply on their way to coming out as gay or lesbian. Consequently, they have been given little attention in the scientific research and have often been brushed aside or even outright rejected by both straight and gay groups (Baumgardner, 2008; Morgenstern, 2004). Many people mistakenly assume that when bisexual individuals are in a relationship with someone of the other sex, they must be straight, but when they are involved with someone of their own sex, they are gay or lesbian. Self-proclaimed bisexual individuals are subject to prejudice and discrimination by the heterosexual majority. But they encounter it in the gay world, too, with derisive phrases such as "Pick a lane" or "Choose a team." As one bisexual individual put it:

> We feel even more hurt when gays reject us than when straights do, since we feel that gay people know better. Many bisexuals have fought on the front lines of the gay rights movement, yet it seems like we have only token representation in the community; the word "bisexual" has been added to the masthead, but we don't feel truly included or accepted. (Morgenstern, 2004, p. 47)

Since You Asked

3. Is bisexuality only a stage toward becoming gay or is it a sexual orientation in itself?

Self-Discovery

A Sexual Orientation Worksheet

This worksheet is designed to help you explore your own sexual orientation as it relates to Kinsey's scale and to gain some insight into the idea that sexual orientation is rarely "all or nothing" in one direction or the other. Many of you may want to use a blank sheet of paper to complete this exercise. It is for your own self-discovery and needn't be shared with anyone (but you must respond honestly!). There are no right or wrong answers, and the scoring categories are rough approximations. You'll be able to understand the complexities of sexual orientation simply by looking at the distribution of your answers.

Check the box that applies best to you. Respond to the categories in the table. In accordance with the scoring key that follows the scale, a lower overall score indicate a heterosexual emphasis, higher scores relate to homosexual leanings, and scores in the middle may imply a bisexual orientation.

Behavioral Indicator	Other sex exclusively = 1	Other sex mostly = 2	Other sex some what more = 3	Both sexes about equally = 4	Same sex some what more = 5	Same sex mostly = 6	Same sex exclusively = 7
1. Sexual attraction throughout teen and adult years							
2. Sexual fantasies or dreams throughout teen and adult years							
3. Intimate or sexual experiences throughout teen and adult years							
4. Feelings of love and/or passion throughout teen and adult years							
5. When you imagine yourself in an intimate relationship							
6. When you imagine having a lifelong partner							

Scoring Interpretation

Possible scores range from 6 to 42 and are interpreted as follows to apply to Kinsey's scale on the previous page:

6–10 = Exclusively heterosexual

11–15 = Predominantly heterosexual, only incidentally homosexual

16–20 = Predominantly heterosexual, but more than incidentally homosexual

21–26 = Equally heterosexual and homosexual

27–31 = Predominantly homosexual, but more than incidentally heterosexual

32–36 = Predominantly homosexual, only incidentally heterosexual

27–42 = Exclusively homosexual

SOURCE: Adapted from: "Sexual Orientation Scale," from p. 38 in "Sexual Orientation: A Multi-Variable Dynamic Process," by F. Klein, B. Sepekoff, & T. Wolf, in *Two Lives: Bisexuality in Men and Women*, edited by F. Klein & T. Wolf. Copyright © 1985. Reprinted by permission of The Haworth Press.

Bisexual people are not more highly sexed or promiscuous than any other group of people (or in less delicate language, they are not out to "sleep with just anyone"). Their sexual orientation is not about being indiscriminate about their sexual or relationship partners. Some bisexuals might define themselves as more attracted to women than to men, or vice versa, while others feel an equal romantic and sexual attraction to both sexes. Regardless, most bisexual individuals will tell you that they feel open to relationships with a potential partner regardless of his or her biological sex, rather than wanting to have loving relationships with both sexes simultaneously. Bisexual individuals enter love relationships in the same manner as those of all sexual orientations, through dating, commitment, and, potentially marriage; the partner they select may be of the person's same or opposite sex. Once in a relationship, the couple is not threatened by the fact of the bisexual orientation of one or both partners to any greater degree than occurs in gay or straight relationships.

Nonheterosexual (Gay, Lesbian, Bisexual) Orientations

11.2 Review the central issues relating to nonheterosexual orientations including, the notion of choice of sexual orientation; same-sex marriage; sexual orientation and having children; and the characteristics of same-sex intimate relationships.

When many heterosexual individuals think about gay people, they tend to limit that thinking to one behavior: sex. But, if you really think about it, is your sexual orientation (whether you are straight or gay) determined entirely by whom you have sex with and how you have it? Of course not. For one thing, as discussed in Chapter 6, "Sexual Behaviors: Experiencing Sexual Pleasure," couples of any sexual orientation can choose to engage in nearly all sexual activities, so sexual behaviors cannot be divided into "gay sex" and "straight sex."

Since You Asked

4. It seems to me that male sexual anatomy is designed to "fit" with female sexual anatomy, and vice versa. So I don't get why someone would be gay if their anatomies don't fit together so well.

But more importantly, consider this: Imagine that, from this moment on, you decide to be celibate; never to have sex of any kind, for the rest of your life. (I know for many of you that's not easy to contemplate, but, just for a moment, try.) Would your decision to be celibate change your sexual orientation? No. If someone were to engage in zero sexual behavior, would the person cease to have a sexual orientation at all? No. The answers to these questions are obvious, but one of the most common mistakes many heterosexuals make about nonheterosexuals is assuming that what defines people as gay or lesbian or bisexual is merely their sexual behavior. However, straight people are not straight merely because they engage in sexual behaviors with members of the opposite sex. In the same way, gay and lesbian people are not gay and lesbian only because they engage in sexual behaviors with members of their own sex.

Sexual orientation is about you as a total person, including the sex of the people you want to date, fall in love with, have a romantic relationship with, perhaps spend the rest of your life with, and, all things being equal (which, of course, they are not in most places in the world), the sex of the person you would choose to marry.

Individuals of all sexual orientations experience similar attractions, emotions, and attitudes with respect to love, romance, and relationships (see LeVay, 2010).

Is Sexual Orientation a Choice?

We will shortly be discussing various theories about the causes of sexual orientation, which will shed some light on the question of whether sexual orientation is a conscious choice people make. But an important point should be made here: heterosexual individuals are rarely, if ever, asked if their sexual orientation was a choice. This question appears to be reserved for gay or lesbian people, and it is nearly always asked by heterosexuals—and the virtually universal answer is "no" (American Psychological Association, 2011). If you define yourself as heterosexual, as most of you do, try answering these questions: *When did you decide to become a heterosexual? Why did you choose your straight sexual orientation? At what age did you first realize you were heterosexual?* If you are like most heterosexual people, not only are the questions difficult to answer, but they seem nonsensical. Your answers were probably along the lines of: *I never really decided; I didn't choose to be straight, it's just who I am!* and *I don't recall any particular age; I just always sort of knew.*

Since You Asked

5. How do people know or decide to be homosexual or heterosexual or bisexual?

Since You Asked

6. I have no problem with homosexuals. I say live and let live. But I don't see why they are so concerned with getting married. I mean, if they can have their relationship, why do they have to get married?

This difference in perception is probably because nonheterosexual orientations are in the minority and majorities always tend to be curious about characteristics of minorities in general. So, a fascination exists about when, where, why, and how gay,

Table 11.1 A Heterosexual Questionnaire

Turnabout is fair play! Here is a list of questions that are frequently asked of gay and lesbian individuals, but the orientation in each question has been altered to heterosexual. It won't take long for you to realize how odd these questions must seem, unless you are gay, lesbian, or bisexual; then you already know.

1. When did you first decide you were a heterosexual
2. What do you think caused you to be heterosexual?
3. Is it possible your heterosexuality is just a phase you may grow out of?
4. If you've never slept with a person of the same sex, how do you know you wouldn't prefer that?
5. Why do you insist on being so obvious with public displays of affection? Can't you just be what you are and keep it quiet?
6. Whom have you informed about your heterosexual tendencies? How did they react?
7. Do heterosexuals hate or distrust others of their own sex? Is that what makes them heterosexual?
8. Why are heterosexuals so promiscuous?
9. There seem to be many unhappy heterosexuals. Techniques have been developed to help you change your sexual orientation if you want to. Have you considered trying conversion therapy?
10. Why do heterosexuals place so much emphasis on sex?
11. How can you enjoy a fully satisfying sexual experience with a person of the opposite sex when the physical, biological, and psychological differences between you are so great? How can a man possibly understand what pleases a woman sexually, or vice versa?

SOURCE: Developed by M. Rochlin, *The Language of Sex: The Heterosexual Questionnaire*, Changing Men (Spring 1982), Waterloo, ON: University of Waterloo.

straight, and bisexual orientations came to be. If you are straight, Table 11.1 will help you understand how difficult it is for gay and lesbian individuals to answer the kinds of questions often posed to them.

Same-Sex Marriage

Few fundamental social issues in the U.S. and many countries around the world have undergone such a radical change over a mere decade as same-sex marriage. Many gay and lesbian couples (just as heterosexual couples) who are deeply committed to each other have a strong desire to confirm that commitment personally, publicly, and legally through marriage. Beyond that personal and emotional desire, however, is the very real problem faced by nonheterosexual couples' inability to obtain the same legal rights and privileges that heterosexual couples take for granted. These marital rights number in the hundreds and include such rights as inheritance if a partner dies, participation in medical decisions, insurance coverage from a partner's employment, child custody rights, family leave benefits, domestic violence protection, and community property rights in the case of divorce, to name a few.

civil union
(also called domestic partnership) A legal contract between two members of the same sex that imparts all or most of the legal benefits of marriage but is not socially or religiously equated with heterosexual marriage.

In the U.S., the legalities of same-sex marriage are changing so rapidly you need to follow the news almost daily to keep current (see Figure 11.2 for a color-coded map as of November, 2014). It took 22 years for the first 10 states to recognize same-sex marriage; the next 7 states took the same action in less than 2 years. As of November of 2014, 32 states and the District of Colombia had legalized the issuance of marriage licenses for same-sex couples (shown in blue on the map).

Several states still have **civil union** laws that allow same-sex couples some or all of the rights and privileges that come with legal marriage (these are in green on the map). However, as has been proclaimed in many civil rights movements historically, "separate but equal" is not seen by most gay and lesbian individuals as truly equal to heterosexual marriage. States with remaining civil union laws are the most likely to legalize same-sex marriage in the near future. The legal status of same-sex marriage in the United States remains a work in progress that is changing virtually month-to-month.

Marriage in some parts of the United States (Massachusetts and California as of 2008) is no longer limited to "one man and one woman."

Figure 11.2 Same-Sex Marriage Laws in the U.S. as of November 2014

SOURCE: : http://gregstoll.dyndns.org/marriagemap/ (Used by permission.)

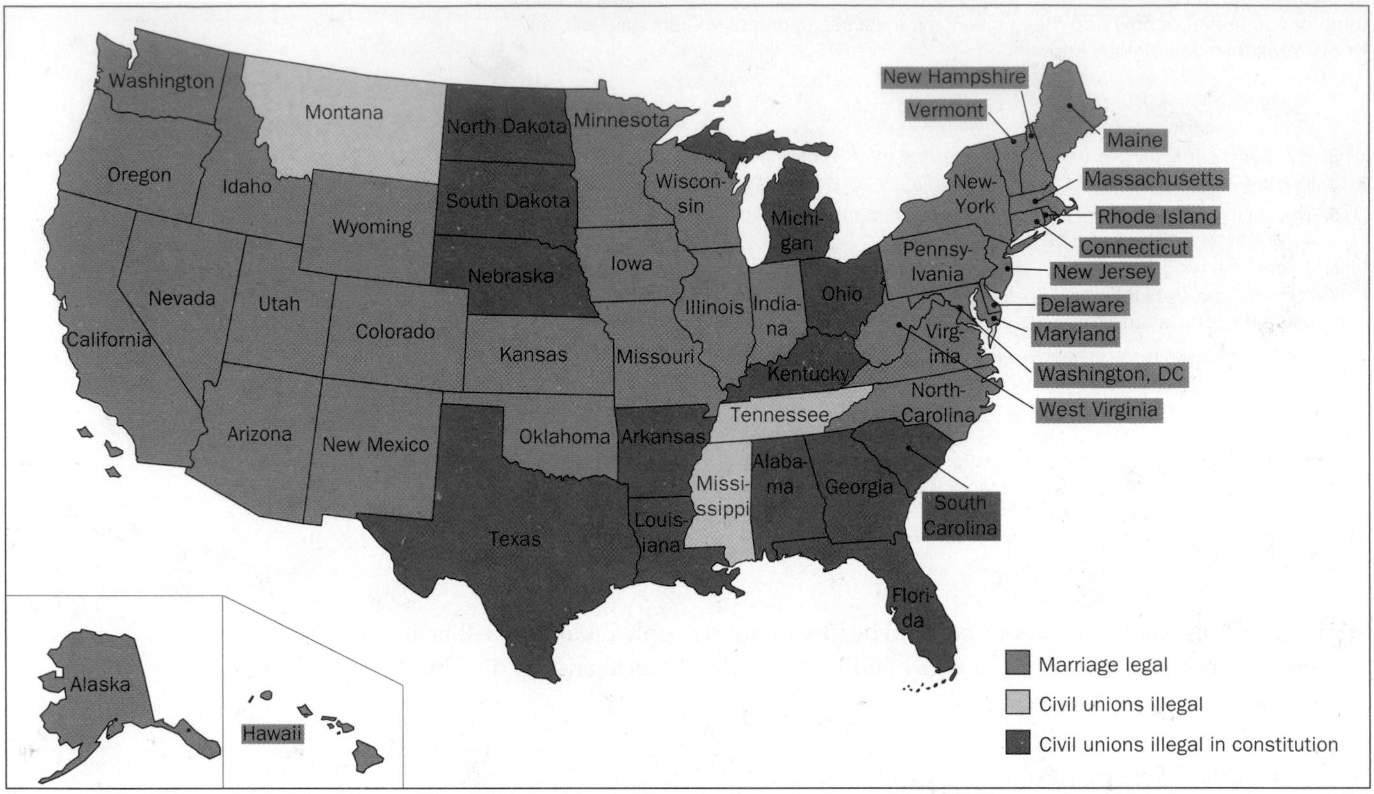

Legalization of same-sex marriage is moving swiftly in around the world as well. Below is a list, as of 2014, of the countries that have legalized same-sex marriage and the year the law was passed (Pew Research, 2014):

- The Netherlands: 2001
- Belgium: 2003
- Spain: 2005
- Canada: 2005
- South Africa: 2006
- Norway: 2009
- Sweden: 2009
- Portugal: 2010
- Iceland: 2010
- Argentina: 2010
- Denmark: 2012
- England and Wales: 2013
- France: 2013
- Brazil: 2013
- New Zealand: 2013
- Uruguay: 2013
- Scotland: 2014

Sexual Orientation and Having Children

One of the many judgmental and often prejudiced statements gay and lesbian couples hear goes something like this: "You can't have children, so what's the point of your relationship?" Of course, one of the many weaknesses in this particular antigay sentiment is that it fails to consider all the heterosexual couples who either cannot conceive due to infertility issues or choose to be child-free. One generally does not hear those same people saying, "What's the point of your relationship?" to them. The reality is that many gay and lesbian couples do have children and typically make excellent parents. Moreover, research has demonstrated that children of gay or lesbian couples show no differences in cognitive development, psychological adjustment, gender identity, or sexual partner preference (Crowel et al., 2008; Tasker, 2010; Tasker & Bigner, 2013).

You may be wondering how gay and lesbian couples have children. Sometimes individuals who come out or accept their sexual orientation as gay or lesbian later in life may have been in a heterosexual marriage and may have one or more children from that marriage. They do not cease to be parents when they come to awareness with their true sexual orientation. Beyond this, however, many gay and lesbian couples feel a deep desire to have and raise children of their own. Such couples cannot conceive children on their own, of course, but the various assisted reproduction technologies (ART) discussed in Chapter 9, "Conception, Pregnancy, and Birth," are now making it possible for them to conceive children, with assistance from others, just as would be the case for some straight couples.

This is not really as new an idea as you may think. One member of a gay or lesbian couple might have a friend help them achieve pregnancy through intercourse. However, because of the personal and interpersonal significance of asking and agreeing to such a suggestion, many gay and lesbian couples today prefer to rely on artificial insemination techniques to realize their desire for children. This may involve "do-it-yourself" artificial techniques involving collecting sperm from a donor and transferring it to the vagina of the designated mother-to-be using a needleless syringe or placing the man's semen into a diaphragm or cervical cap and inserting it into the vagina (Mallon, 2008; Mamo, 2007). However, many gay and lesbian couples are uncomfortable with such casual methods, do not have ready and willing friends or acquaintances, or feel that defining (or denying) the third party's parental rights is ethically too complicated. In these cases, they often turn to the various ART techniques carried out in a medical setting using donated sperm or eggs (see Chapter 9, "Conception, Pregnancy, and Birth," for more information on ART methods).

In many states, adoption for gay and lesbian individuals or couples is also an option ("Adoption Laws," 2008). Because the laws affecting nonheterosexual relationships vary greatly from state to state, gay and lesbian couples who decide to have children through any means should be sure to obtain the services of an attorney to ensure that both parents' rights are protected as well as the rights of the child. When gay and lesbian couples choose to conceive or adopt, many resources are available in most states and large cities to help them avoid legal problems that might arise in the future (see Brodzinsky & Pertman, 2011).

Interpersonal Qualities of Nonheterosexual Relationships

Many myths and misconceptions exist about the ways in which gay and lesbian relationships differ from heterosexual ones in the quality of the interpersonal interactions. For example, many people hold the following beliefs—all of them FALSE:

- FALSE: Most gays and lesbians are unable to form close, enduring romantic relationships and prefer to be promiscuous.
- FALSE: Gay and lesbian relationships are more likely than heterosexual relationships to be unhappy and dysfunctional.
- FALSE: In same-sex relationships, one partner chooses or is assigned the role of "husband" and the other takes on the role of "wife."
- FALSE: Most gay and lesbian couples are isolated from society and do not have meaningful social support networks.

All of these notions have been shown to be no truer of gay and lesbian couples than of straight couples. In 2001, the results of a 12-year study of gay and

Since You Asked

7. Do gay and lesbian couples really love each other in the same way as straight couples?
8. How do gay and lesbian relationships differ from heterosexual ones, if they do?

Table 11.2 Differences in Emotional Qualities Between Gay and Lesbian Versus Heterosexual Couples

Quality	Description
Gay and lesbian couples are more upbeat in the face of conflict.	Compared to heterosexual couples, gay and lesbian couples inject more affection and humor into their disagreements and conflicts, and same-sex partners are more positive in how they receive messages about conflict. Gay and lesbian couples are also more likely to remain positive about the relationship after a disagreement.
Gay and lesbian couples use fewer controlling or hostile emotional tactics.	Gay and lesbian partners display less belligerence, domineering behavior, and fear with each other than straight couples. They use fewer controlling tactics and recognize better the importance of fairness and a balance of power than heterosexual couples.
In a fight, gay and lesbian couples take things less personally.	In heterosexual couples, a partner may be more easily hurt by a negative comment and less likely to feel uplifted with a positive comment. The opposite seems to be the case in gay and lesbian couples. Gay and lesbian partners' positive comments have more impact on feeling good, and their negative comments are less likely to produce personal hurt feelings.
Gay and lesbian couples tend to show lower levels of physiological arousal during conflict.	Again, this is generally the opposite for heterosexual couples, for whom the physiological arousal (including elevated heart rate, sweaty palms, and jitteriness) produced by conflict signifies ongoing aggravation. The ongoing aroused state makes it difficult for the partners to calm down and move past the conflict. Gay and lesbian couples' lower level of psychological arousal allows them to soothe one another during or soon after the conflict.

SOURCE: Adapted from Gottman & Levinson (2001); also see Cloud (2008).

lesbian couples provided the best evidence to date of the similarities and differences between gay and straight romantic relationships (Gottman et al., 2003). This study, along with others (see Cloud, 2008; Frost, 2011; Gottman & Levenson, 2001), found that overall relationship satisfaction and quality are about the same regardless of the sex makeup of the couple. The main differences relate to the couples' emotional interactions, especially their handling of conflict, and to difficulties related to the amount of social support received (or withdrawn) from family members (e.g., Mohr et al., 2013).

For example, gay and lesbian couples tend to be generally more upbeat about their relationships than their heterosexual counterparts. They use more humor and withdraw affection less when conflicts arise, and they continue to view their relationships in more positive terms following discord (Gottman & Levinson, 2001). Gay and lesbian couples tend to be less hostile and controlling toward one another overall; they establish an equal balance of relationship power more readily. Another difference is that gay and lesbian couples appear to be able to calm down after conflict, soothe each other, and move past a problem more quickly compared to straight couples. Table 11.2 summarizes some of the differences between gay and straight romantic relationships.

Theories of the Origins of Sexual Orientation

11.3 Analyze and evaluate various approaches to the origins of a person's sexual orientation, including biological views (nature), environmental theories (nurture), and how these opposing ideas might be reconciled.

One of the hottest debates in the field of human sexuality has been about the *causes* of sexual orientation. This debate, in all its complexity, leads us back to the old, familiar nature-versus-nurture controversy (see Langstrom et al., 2010; Baily et al., 2014; Swaab & Garcia-Falgueras, 2009). Overall, this debate has not been particularly focused on the causes of heterosexuality or on sexual orientation in general, but rather on determining the causes of homosexuality. As you will see, none of the theories proposed thus far is convincing from a scientific perspective, so we will review the recent findings of relevant research. Finally, we will close this section with a brief look at the possible motivations behind this rather intense "search for a cause."

Since You Asked

9. Why are some people gay? Is it something they're born with or just bad experiences with their parents or the opposite sex?

Biological Influences on Sexual Orientation

Arguments that sexual orientation results from biological factors usually fall into one of three categories: brain structure and function, hormones, and genes. All have been implicated in the roots of sexual orientation, yet none has managed to lay claim to a central role. One issue appears clear from the vast majority of scientific research: your sexual orientation is not a choice and people's sexual orientation cannot be altered through so-called "restorative" or "reparative" therapy as claimed by some anti-gay groups, usually associated with specific religions (see the American Psychological Association, 2008; Haldeman, 2013).

BRAIN STRUCTURE AND FUNCTION In 1991, Simon LeVay, now an emeritus neurobiologist at the Salk Institute in La Jolla, California, made international headlines when he published in the journal *Science* that he had found a measurable, consistent difference in the brains of a small group of heterosexual and homosexual men. It was a minute difference in the *third interstitial nucleus*, a very small part within the brain structure called the *hypothalamus*, which is known to play a central role in emotions and sexual urges (LeVay, 1991; Pfaff, Frolich, & Morgan, 2002). The third interstitial nucleus is normally larger in men than in women. What LeVay found, in essence, was that this structure was larger in his sample of straight men than in gay men. The size of this section of the hypothalamus in his gay male sample was about the same as is typically found in women. Although this was not the first finding that the brains of men and women are different, it was the first to find a potential gay-straight variation.

LeVay's research did not *prove* that sexual orientation is *caused* by a brain difference, but it did suggest a biological difference that is, in his view, related in some way to a person's sexual orientation (Barinaga, 1991). The findings were seen by many gay individuals, gay activists, and their supporters as proof that sexual orientation, whether gay or straight, is a natural biological variation in humans and not a lifestyle choice, in the same way that one is male or female, nearsighted or farsighted, tall or short. Indeed, some gay rights supporters saw that if LeVay's findings could be interpreted as proof that sexual orientation is a biological, inborn characteristic, laws forbidding discrimination based on such differences might become easier to legislate by modeling legislation after similar statutes relating to race and gender.

The most common criticism of LeVay's research is related to the discussion earlier in this chapter about sexual orientation as a continuum rather than a series of divisions into discrete categories of gay and straight. In his brain research, LeVay seemed to be dividing his subjects along a strict gay–straight dichotomy (Fausto-Sterling & Balaban, 1993). But, as we discussed earlier, sexual orientation is difficult to divide so clearly into "camps." The reasoning behind this criticism goes like this: If people cannot be divided into purely heterosexual or homosexual groupings, then what do variations in brain structure really mean? Would bisexual individuals, for example, have interstitial nuclei that are of a size in between gay and straight brains? These questions have never been fully answered, and LeVay's findings on the gay male brain continue to this day. LeVay's early research and later writings have opened the scientific door to the possibility that sexual orientation might be rooted in biological or genetic, rather than environmental, sources.

Since LeVay's findings, other researchers have revealed various additional biological differences relating to sexual orientation (see Hines, 2010). For example, several studies have demonstrated that right- and left-handedness may be linked to sexual orientation (Bode, 2000; Lalumiere, Blanchard, & Zucker, 2000). Findings have shown that lesbians are 91% more likely to be left-handed than heterosexual women, and gay men are 34% more likely than straight men to be left-handed, although the *overall* percentages of left-handed people regardless of sexual orientation is only about 10% (Blanchard et al., 2006). Researchers have also found a connection between sexual orientation and how the brain processes sound (Jensen, 1998; McFadden, 2011). All people's ears make faint noises in response to clicking sounds, called "click-evoked otoacoustic emissions." However, lesbians' ears make significantly softer sounds than the ears of heterosexual women. In addition, research has demonstrated that visual-spatial abilities may also be related to measures of sexual orientation in men (Gulia & Mallick, 2010; Littrell, 2008).

As the psychobiological sciences have become technologically more sophisticated in recent years, new brain differences relating to sexual orientation are becoming increasingly open to exploration. When comparing MRI (magnetic resonance imaging) and PET (positron emission tomography) scans of the brains of homosexual and heterosexual individuals, researchers have found clear differences that are not likely to be due to environmental influences and, therefore, are probably present at birth. For example, in one study, the degree of symmetry between brain hemispheres and the neural connections among right and left brain structures are not the same in gay and straight people (Coghlan, 2008; Savic & Lindstrom, 2008). Overall, the findings indicate that the brains of gay men are similar to those of straight women, and gay women's brains are similar to straight men. In other words, the researchers found "sex-atypical cerebral asymmetry and functional connections in homosexual subjects" (Savic & Lindstrom, 2008, p. 9403).

These findings, taken together, continue to strengthen the position that sexual orientation has genetic and other biological foundations. This is an active area of research linking the fields of human sexuality and psychobiology (see Byne, 2014).

Hormones

Hormonal changes or imbalances in adulthood may change sexual behavior, sexual desire, or even male and female sexual anatomy (see Chapter 10, "Gender: Roles, Expectations, Roles, and Behaviors," for more detail on this topic), but they do not appear to alter a person's established sexual orientation. Hormonal explanations for gay, lesbian, and bisexual orientations focus on the exposure of the *fetus while in the uterus* to male or female hormones in combinations or levels that vary from those of heterosexuals (Friedman, 2007; Hines, 2011).

Separating hormonal influences from other biological factors is extremely difficult. Any physiological differences found in different sexual orientations, such as the brain differences discussed in the previous section, may be preprogrammed by genes (to be discussed next); influenced by behavior, environment, or illness; or caused by hormones during fetal development (Heino et al., 1995). In fact, the researchers who have found the spatial ability and auditory differences in gay versus straight orientations acknowledged the possibility of prenatal hormone exposure as a potential cause (Bao & Swaab, 2011; Cohen, 2002; Lalumiere et al., 2000).

One example of the hormonal hypothesis for the development of nonheterosexual orientations involves the fingers. You are probably not aware of this, but the length of the ring finger is more similar in length to the index finger in women than in men. In other words, the difference in length between the index finger and ring finger is greater in men than in women. This difference is measurable at age two and is stable into adulthood. You can now amaze your friends by bringing up this bit of trivia at your next party. Seriously, beyond this male-female difference, a similar difference exists between lesbian and straight women. That is, the index-to-ring-finger ratio in lesbian women is more similar to that of men. Research has also revealed that the ratio is even smaller (meaning the difference in finger lengths is less) in lesbians who self-identify as "butch" (more masculine) than in those who self-identify as "femme" (more feminine) (McFadden, 2011). Add to this the fact that the index-to-ring-finger ratio is a marker for fetal exposure to male hormones, and you have evidence for a potential hormonal cause of sexual orientation—not conclusive proof but an interesting, suggestive correlation.

Yet another study investigated women whose mothers were given the drug *diethylstilbestrol* (DES) while pregnant (Hines, 2011; Meyer-Bahlburg, Ehrhardt, & Gruen, 1995). DES is a form of synthetic estrogen that was thought to prevent miscarriage and premature labor and was widely prescribed to pregnant women between 1938 and 1971. Not only did the drug not work as advertised, but it caused an increased risk for cancer and infertility in the *daughters* of women who took the drug. The exposed daughters have been followed and studied extensively for years. One consistent finding has been that women exposed to DES in the womb are significantly more likely as adults to be bisexual or gay than nonexposed controls. This lends correlational evidence that hormonal exposure in the womb may indeed play a role in the development of sexual orientations.

Researchers have also discovered a new possible biological basis of sexual orientation that has become known as the "older brother effect" (Bogaert, 2006; LeVay, 2010; Swartz et al., 2009). A statistically significant association has been found demonstrating that as the number of older biological brothers a man has increases, so do his chances of being gay (regardless of the brothers' sexual orientations). This does not appear to be true of boys with older step- or adopted brothers. Each additional biological older brother increases the odds of a gay sexual orientation by as much as one-third. Various scientific studies across cultures have confirmed this finding. No such relationship has been found for lesbians regardless of how many older sisters or brothers they may have. What might be the reason for the "big brother effect"? Obviously, the male fetus in the womb hasn't the slightest idea how many brothers he has. But the child's mother's body does. Although the *correlation* between male birth order and sexual orientation is real, the reasons for it are more difficult to ascertain. The theory most often suggested by researchers relates to the immune system in women (Bogaert & Skorska, 2011). A few fetal cells can sometimes pass from the womb into the mother's bloodstream during pregnancy or the birthing process. Some researchers speculate that these cells interact with the mother's immune system and begin to develop antibodies against male fetal cells. The more male babies the mother has had, the stronger the immune response *against* them. Her body then, at least in some cases, may react by changing something (probably the hormonal balance) in the uterine environment for later-born males that in turn leads to a tendency toward a gay sexual orientation.

Genetics

The genetic theory proposes that a person's sexual orientation is *preprogrammed* at conception when the genes from the mother's egg and the father's sperm merge to create the blueprint for a new person. Historically, researchers had difficulty separating genetic from environmental influences on human characteristics because most people grow up in the same environment as their genetic donors, that is, their parents. Therefore, any behavior or personality similarities between children and parents might be either inherited or learned. However, over the past 30 years or so, research methods have become increasingly sophisticated, allowing researchers to tease apart the forces of environment and genes; consequently, researchers have been placing an increased focus on genes as a potential source of human behaviors and personality traits, including sexual orientation.

Research with identical twins has suggested clear genetic influence on sexual orientation.

Most influential genetic studies involve pairs of identical twins, some of whom were adopted at birth, or very soon after, into different families and therefore different environments. If two individuals with the same genetic makeup (i.e., identical or *monozygotic* twins) share certain personality characteristics far more often than fraternal (*dizygotic*) twins or nontwin brothers and sisters, this argues that genetic influences most probably account for those shared characteristics.

This avenue of research has been applied to sexual orientation, and the results have been quite consistent. Numerous studies have shown that when one member of a pair of identical twins is gay or lesbian, the chances are far higher (30% to 50%) that the other twin will share that same sexual orientation than would be expected among fraternal twins or nontwin siblings (LeVay, 2010; Selekman, 2007). In other words, as the degree of genetic relatedness increases, the similarity of siblings' sexual orientation also increases, regardless of their environment. Figure 11.3 summarizes representative findings from two of these twin studies. In 2008, researchers conducted the largest twin study up to that time, involving 7,600 pairs of Swedish twins (Långström et al., 2008). Findings from this study indicated a smaller influence of genetic factors (18% to 39%), but a very small influence of environmental factors shared by twins (0–17%). Each twin's individual experiences (such as uterine and childbirth conditions) appeared to influence choice of partner (for all sexual orientation) most strongly (approximately 60%).

Today, widespread scientific support exists that genes play an important role in sexual orientation. But has science uncovered a true "gay gene?" No. However, this has not prevented the idea of a "gay gene" from entering medical literature as an assumed cause (however erroneous) of sexual orientation (O'Riordan, 2012). Although researchers have demonstrated that sexual orientation is *influenced* by a person's genetic makeup, finding the actual gene or complex of genes responsible for that influence is far more difficult, if it exists at all. Furthermore, only about 40% to 50% of the variation in sexual orientation can be explained by genetic influences, so even if we accept the gay gene hypothesis, genes are not the whole story, and other influences cannot be ignored (see LeVay, 2010). Moreover, many scientists question the need to prove that a "gay gene" exists at all (we will return to this issue of the "necessity of causal conclusions" toward the end of this section).

Experiential and Environmental Influences on Sexual Orientation

Researchers appear to be closer to identifying meaningful potential biological causes of sexual orientation than to uncovering consistent evidence of specific environmental or experiential influences. Most of the environmental "causes" for nonheterosexual orientations that were popular during the second half of the twentieth century have been resoundingly rejected by more advanced research techniques. How many of these *myths* about family and environmental origins of homosexuality have you heard?

- Myth (false): Homosexual men are more likely to have had domineering or overly protective mothers and weak, distant fathers (and hence no masculine role model).
- Myth (false): Homosexual women were more likely to have had cold, unloving, rejecting mothers, and absent fathers.
- Myth (false): Homosexuals grew up hating or fearing members of the opposite sex, which caused them to turn to members of their own sex for love and intimacy.

Figure 11.3 Twin Studies and Sexual Orientation

As genetic relatedness increases, the similarity of same-sex siblings' sexual orientation also increases.

SOURCE: Alanko et al., 2010; Bailey & Benishay, 1993; Bailey & Pillard, 1991.

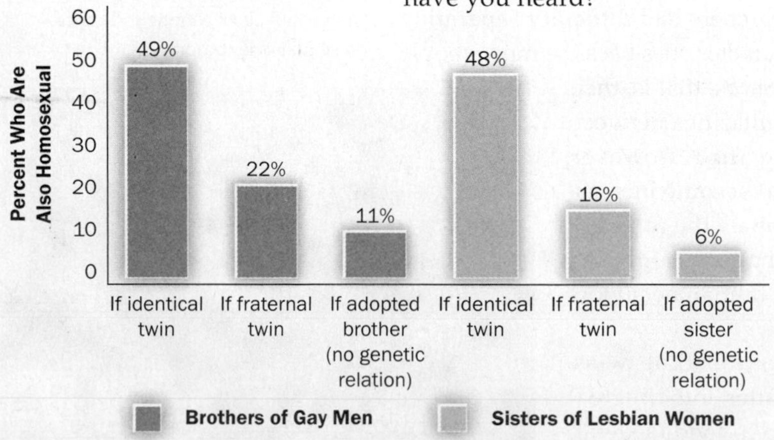

- Myth (false): Most homosexuals were molested during childhood by a same-sex adult, which turned them toward homosexuality as they became adults themselves.
- Myth (false): Homosexuals are more likely to have engaged in sex play as children with peers of their own sex.
- Myth (false): Homosexuals are more likely to have a gay mother or father.

All of these assumptions have been shown in numerous studies to be false. The methodology used to study the psychosocial origins of sexual orientation scientifically was fairly straightforward: Ask a large random sample of homosexual and heterosexual people about their experiences with all of these issues and compare the answers of the two groups. When researchers have done this, they find no significant differences among the groups for any of the influences on the list (Bell & Weinberg, 1978; Bell, Weinberg, & Hammersmith, 1981; Golombok & Tasker, 1996). In other words, nonheterosexual individuals are just as likely or unlikely as straight people to have experienced any of the listed events and circumstances during their lifetime.

Another childhood environmental issue that often arises relating to sexual orientation is abuse. The question is: Can sexual abuse in childhood lead to same-sex orientation in adulthood? The research evidence on this is inconclusive. Some studies have found that gay and lesbian adults are more likely to report childhood sexual abuse than heterosexual individuals (although it is a small percentage in both groups). The difficulty is determining if the abuse caused the sexual orientation or if the childhood behaviors stemming from an underlying gay or lesbian identity lead to abusive behaviors from parents or other adults (Roberts, Gloymore, & Koenen, 2013). Although we may never be able to untangle the direction of this link, the association between childhood abuse and sexual orientation is real and must be considered seriously in all efforts to reduce child abuse (Walker, Hernandez, & Davey, 2012) (see Chapter 13, "Sexual Aggression and Violence, Rape, Child Sexual Abuse, and Sexual Harassment" for more on this topic).

exotic becomes erotic (EBE) theory
Psychologist Daryl Bem's explanation for the interaction of biology and environment in determining a person's sexual orientation.

The question we must ask is, "Has scientific evidence clearly supported *any* purely environmental explanations for sexual orientation?" The research appears to show only one consistent finding: gay and lesbian adults appear to have engaged in greater *gender non conforming* behavior as children (LeVay, 2010).

The research that attempts to refute the biological evidence for sexual orientation does not claim any *exclusively* environmental causal model in its place. Rather, most of the criticism takes the position that the biological findings are not *yet* compelling enough to conclude that biology is solely responsible for sexual orientation and that any "complete" explanation of such a complex human quality will likely include an interaction between nature and nurture (Bem, 2000; Lippa, 2005; Veniegas & Conley, 2000). One such recent interactionist theory will be discussed next.

Children are naturally drawn to same-sex playmates and gender-conforming activities.

A Biology–Environment Interactive Approach

Most people have difficulty imagining that whether a person is straight, gay, lesbian, or bisexual could be a result of purely biological or environmental factors. Yet few have meshed these two sides of the discussion into a cohesive, interactionist theory to explain the development of sexual orientation as a combination of nature and nurture. One attempt to do so has received a great deal of attention over the past several years. Daryl Bem, a psychologist at Cornell University, has proposed what he has termed the **exotic becomes erotic (EBE) theory** of sexual orientation (Bem, 1996; 2000; 2008).

Bem suggests that people are not born with a genetic predisposition for a certain sexual orientation per se but rather for varying childhood temperaments such as aggression, activity level, or shyness. Indeed, these *temperamental variables* have received a great deal of scientific support outside the sexual orientation arena (e.g., Kagan, 2008; Zentner & Bates, 2008). According to Bem, these temperamental variations prime the child to prefer some early childhood activities over others. Usually, Bem points out, preferences in play behaviors track with the child's sex: Boys prefer male-typical activities such as rough-and-tumble play or competitive sports, whereas girls gravitate toward female-typical, quiet, cooperative play such as jacks or hopscotch. This is referred to as **gender-conforming behavior**.

gender-conforming behavior
Behavior that is consistent with traditional cultural expectations for a child's sex.

Moreover, Bem contends, children will seek out other children who share the same play preferences; that is, children who like to play jacks or hopscotch will seek out girls to play with, while children who prefer rough, competitive sports will prefer to play with boys. Sometimes, however, a child will engage in **gender-nonconforming behavior**, meaning that the child prefers play activities that are typical of opposite-sex children and will seek out those children as friends and playmates (see also Alanko et al., 2010).

gender-nonconforming behavior
Behavior that is inconsistent with traditional cultural expectations for a child's sex and considered more appropriate for children of the other sex.

Bem's theory proposes that nonconforming children will then feel increasingly different from their "outgroup" of peers. In other words, children who primarily have playmates of their own sex (the vast majority of children) will feel different from opposite-sex children, while children whose playmates are of the other sex (a small minority) will feel different from same-sex peers. Each will see the other group as strange and *exotic.* These feelings of difference create increased levels of emotional (not sexual) arousal toward children of the outgroup sex that might be expressed in statements such as "I hate girls" or "Boys are *so* weird!" This arousal may occur below conscious awareness, or sometimes it might be very obvious. "A particularly clear example," Bem suggests, "is the 'sissy' boy who is taunted by male peers for his gender-nonconformity and, as a result, is likely to experience the strong physiological arousal of fear and anger in their presence" (2000b, p. 533).

You may be thinking that this theory is heading in the wrong direction, that it will lead to rejection of the outgroup sex rather than attraction to it. However, Bem maintains that later, in puberty, as sexual feelings begin to surface, this perceived difference and "exoticness" in the gender outgroup and the accompanying emotional arousal transform into *erotic* feelings toward that group (see also Jenkins, 2010). Hence the theory's name, *exotic becomes erotic.* Figure 11.4 illustrates Bem's theory step by step.

After more than 15 years of research, Bem maintains that his EBE conceptualization is a major cause of sexual orientation in adults (Bem, 2008). Plus, his theory serves to link the biological (whether genetic or hormonal) and environmental influences that lead to a person's sexual orientation in adulthood. He also argues that even though a child's environment plays a major role in the formation of sexual orientation, this does not imply that sexual orientation is a choice or that it is "changeable" in adulthood any more than if it were found to be exclusively biological (Bem, 1997). Consistent childhood experiences may be just as deeply ingrained in a person as genetic heritage.

As you may imagine, the EBE theory has not been without controversy. It has been criticized on the grounds that it does not accurately represent the experiences of girls and women (as evidenced by his examples of playing jacks and hopscotch as female gender-conforming behavior), that it is not supported by the very scientific evidence Bem himself cites, and that it is too limited in its application of psychoanalytic theory because it omits the influence of early boyhood trauma (Nicolosi & Byrd, 2002; Peplau et al., 1998). Also, the theory does leave open the possibility (although Bem refutes this) of altering a child's future sexual orientation by promoting gender-appropriate behavior in childhood, and many top researchers in the area of sexual orientation doubt that this would be possible (LeVay, 2010). The validity of the EBE theory may never be completely settled, but as one of the few theories today attempting to integrate biological and experiential influences on sexual orientation, it is providing material for a worthwhile debate among social scientists on this complex issue.

Figure 11.4 Bem's Exotic Becomes Erotic Theory of Sexual Orientation

Bem's theory considers the interaction of biological and environmental factors in the development of sexual orientation.

SOURCE: "Exotic Becomes Erotic: Interpreting the Biological Correlates of Sexual Orientation," by D. Bem, in *Archives of Sexual Behavior, 29*, (2000), pp. 531–548. Copyright © 2000. Reprinted by permission of Springer Science and Business Media.

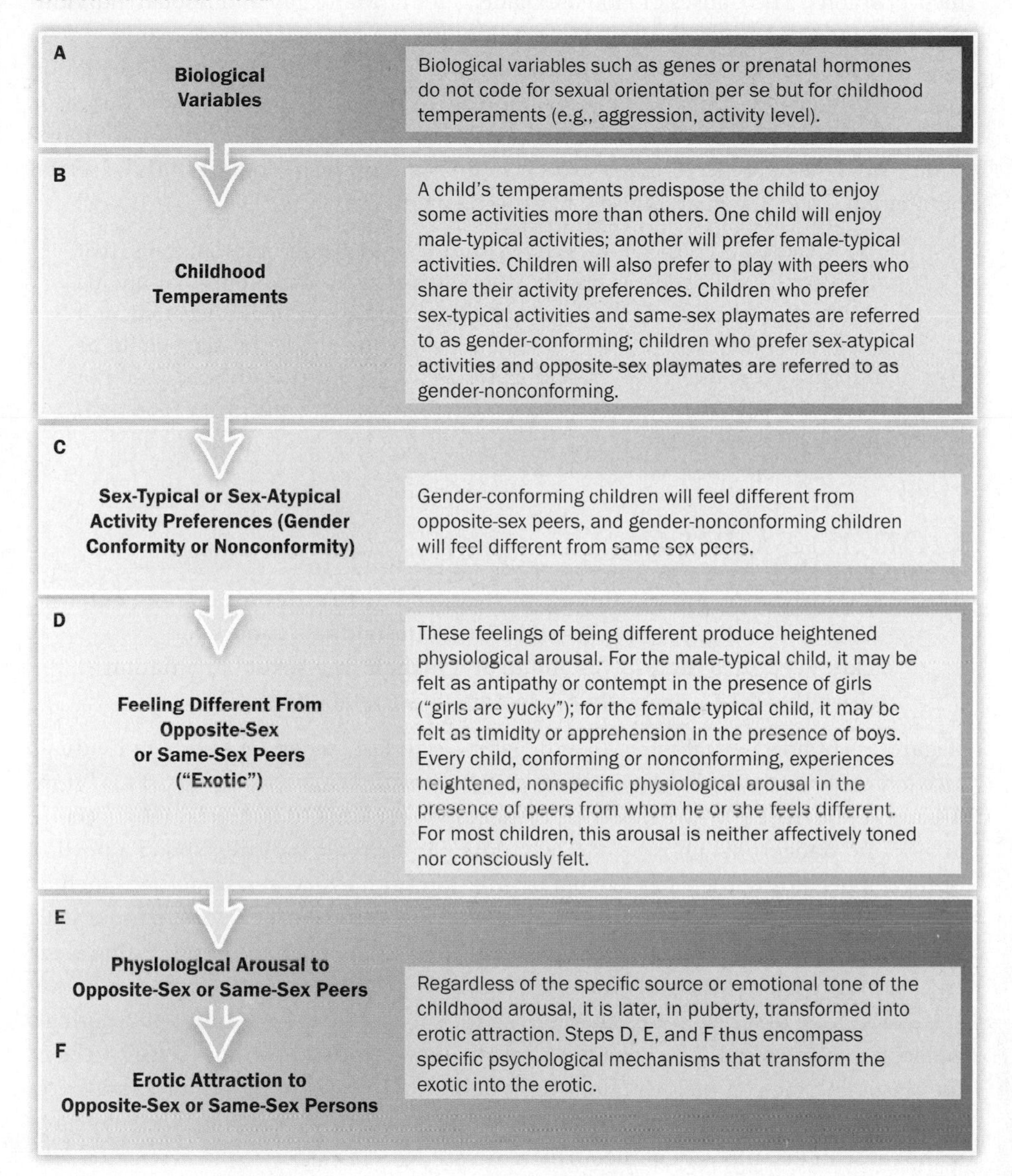

Do the Origins of Sexual Orientation Matter?

In the midst of the research and controversy over the question of the origins of sexual orientation, many experts are questioning the importance of finding the answer. Some researchers have suggested that the underpinnings of the search for the cause of sexual orientation lie in widespread prejudiced attitudes and hostile actions targeted at gays and lesbians in most Western cultures. It has been suggested that these negative attitudes and acts are based on four basic, unenlightened, and erroneous assumptions about homosexuality (Marmor, 1998): it is immoral and sinful, it is unnatural, it is a chosen behavior and therefore can be "unchosen," and it is potentially "contagious."

Some analysts assert that the basis for all of these assumptions can be challenged by demonstrating that sexual orientation is determined, by and large, naturally, biologically, and genetically, just like sex, hair color, and race. If you think about it, clear proof that sexual orientation is inborn could pose a dilemma for people who base their antigay attitudes on religious teachings, in that they also claim that humans are made in God's image. However, others contend that those who espouse such prejudice will find a way to distort the research, no matter what causes are finally established, to strengthen their position ("The Causes of Homosexuality," 2001). Many gay and lesbian individuals themselves downplay the importance of finding the cause of homosexuality.

Some researchers believe that placing people into categories of sexual orientation such as straight, gay, lesbian, or bisexual does all humans a disservice in that it implies a sameness or a lack of diversity among those in each group and in all the groups combined. One of the leading researchers in this field of the biological study of sexual orientation, and a gay man himself, has commented:

> The kaleidoscopic blend of gender-variant and gender typical traits that characterize gay people is exactly what enables us to make our own unique contributions to society. It's the reason we should be valued, celebrated, and welcomed into society rather than being merely tolerated. The aim should be to foster acceptance of gay people as we are, in all our rich diversity, and not to seek acceptance by shoe-horning ourselves into conformity with the straight majority. (LeVay, p. 295)

Originally, the network sitcom *Ellen* was not a show about a lesbian bookstore owner. It was a show about a straight woman who owned a bookstore and was looking for a relationship with a man. When the show's star, comedian and actress Ellen DeGeneres (shown here in 2008 with spouse Portia de Rossi), came out in real life, she made the decision that her character must do the same. When the Ellen "coming-out episode" aired on primetime TV (to a great deal of press and controversy), the impact on the attitudes and emotions of people of all sexual orientations was significant. It helped inspire tens of thousands of people to accept themselves for who they were, come out of the closet, and stop hiding their true sexuality (Ryan & Boxer, 1998).

Coming Out

11.4 Summarize the process timing, process, and difficulties of "coming out," the time when gay, lesbian, and bisexual individuals choose to acknowledge to themselves and to others their true sexual orientation which they had previously hidden due to societal pressures.

Regardless of how sexual orientation develops, gay, lesbian, and bisexual individuals face a life event that straight people never even imagine: *coming out*—short for "coming out of the closet"—which reflects the fact that in a society that is largely rejecting at best and dangerously hostile at worst, nonheterosexuals typically spend a portion of their lives hiding their true sexual orientation from everyone, sometimes including themselves. They keep their sexual identities "in the closet." Most people with nonheterosexual orientations eventually find that living such a lie and hiding their true sexuality is too stressful and emotionally unbearable (as was expressed in the student's essay that began this chapter), so they make the often painful and difficult decision to come out. If you are gay or lesbian, you already know how it feels to be in the closet and the complex emotions usually linked to the coming-out process. If you are heterosexual and have close friends or relatives who have come out, you may have some sense of the intensity of their experiences.

Heterosexuals never feel any need to hide their sexuality, of course, because society as a whole assumes that everyone is straight. They are able simply to be who they are, at least as it relates to their sexual orientation, and they are not in any sort of "closet" to begin with. Furthermore, heterosexuals often ask why gay and lesbian individuals need to come out at all, especially if it is so difficult and risks exposing them to the negative fallout of society's antigay attitudes. However, if you talk to nonheterosexuals who have come out, virtually all of them will tell you that the pain and frustration of "living a lie" about their true sexual identity was far more difficult, more painful, and more stressful than dealing with whatever adverse consequences may have accompanied their decision to come out (Kwon, 2013). And when gay people choose to live a life that is honest and open about their sexual orientation, they are happier, physically healthier, psychologically better adjusted to life in general, and better able to develop close and mutually satisfying friendships and romantic relationships with others (Horne et al., 2014; Scasta, 1998).

When Do People Come Out?

Coming out is not typically a single moment, a single decision, or a single act in someone's life. Usually, coming out as a gay, lesbian, or bisexual person is a process that takes place gradually, over time, in a step-by-step fashion (to be discussed in the next section). Most people begin to realize, or at least suspect, that they may have a non-heterosexual orientation in adolescence, around the time when puberty arrives and everyone begins to have erotic feelings toward others. This is not to say that sexual orientation begins in adolescence—that debate is ongoing, as we noted earlier in this chapter—or that most people will come out to themselves or others during adolescence. However, most self-aware gay and lesbian individuals are quite clear that they always knew they were "different" from the other kids in some fundamental ways for as long as they can remember.

Since You Asked

10. How old are most people when they come to terms with or understand that they are homosexual? Do they usually tell people right away? Are they usually comfortable with the realization?

In the famous "coming-out" episode of the 1990s sitcom *Ellen* (starring Ellen DeGeneres), one of Ellen's friends, a gay man named Peter, was helping Ellen gather the courage to tell her straight friends that she had finally realized she was gay. Peter tells her, "Ellen, it's *never* easy. I remember when I decided to tell my parents, I sat them down and said, 'Mom, Dad, I'm gay, and if you can't accept that it's too bad because it's who I am.' And the next year, when I entered kindergarten, they were behind me 100%!" This line produced a great deal of laughter from the audience, even though it is an exaggeration—but perhaps not *too* much of one. Young children are not aware of complex sexuality issues and probably are not able to analyze and recognize their sexual orientation, but looking back, most gay and lesbian adults will tell you they remember feeling "different." Once a person begins to acknowledge his or her gay or lesbian orientation consciously (typically in adolescence), that feeling of *difference* he or she felt throughout childhood begins to make a great deal more sense (refer back to our discussion of Bem's EBE theory in the previous section).

Still, no exact timeline exists for beginning the process of coming out. It might begin as early as the onset of puberty or as late as old age (Altman, 2000). Research has shown that the age of coming out is decreasing, probably due to increasing awareness of gay youth in the population and greater social support for doing so. The average coming out age today is 16, whereas in the 1980s, it was between 19 and 23 (AVERT, 2014). This does not mean it is an easy process for most high school students to decide to be openly gay. Although tolerance for sexual diversity is increasing in the U.S., these student are frequently subject to bullying, rejection by peers, and even violence. Schools that put in place education programs about sexual orientation and create policies that forbid such bullying provide a safer and less fearful place for teens to come out.

On the other hand, many people have lived heterosexual lives, dated members of the opposite sex, fallen in love with them, married them, and had families, only to realize later in life what it was that had been "missing" from their lives all those years; what had been making them feel unhappy, unfulfilled, and somehow "off track." They had been forcing their lives into a mold expected by society but one that was not true to themselves. They are among the many who will tell you that the intense relief and inner peace they felt when, finally, they came out far outweighed the fear and potential social difficulties of staying hidden (Floyd & Bakeman, 2006; Gardner, de Vries, B., & Mockus, 2014).

The Coming-Out Process

Various theories have been proposed to explain the process gay and lesbian individuals go through as they come out of the closet to themselves and to others. This typically occurs during adolescence through early adulthood (Floyd & Bakeman, 2006), but may happen at any age during a person's life (and some individuals choose never to reveal their true sexual identity). Also, the various stages may occur at significantly differing ages. Nearly all of the theories share certain common characteristics in that they assume a gradual, developmental process that involves various stages of thoughts, realizations, and behaviors that lead to a person's recognition of his or her nonheterosexual orientation and the decision to live openly as a gay person.

Two of the more widely accepted models of the coming-out process—one proposed by Vivienne Cass and another by Richard Troiden (Cass, 1984; Troiden, 1989)—reflect this concept of a gradual, step-like progression (see Floyd & Stein, 2002). Both models share certain basic stages and challenges that many gay individuals must confront on the path to coming out, such as "identity confusion" (feeling unsure about one's sexual orientation), "identity assumption" (acknowledging being gay to oneself), "identity acceptance" (becoming comfortable with one's sexuality), and "identity synthesis and commitment" (becoming openly gay and incorporating a gay sexual orientation into one's routine life with pride and without shame). Both models end with the person's acceptance and integration of his or her sexual orientation into an overall self-identity and healthy life adjustment.

The Dangers, Pitfalls, and Joys of Coming Out

Research mentioned earlier has shown that nonheterosexuals who come out live happier, better-adjusted lives. However, this does not imply that coming out is easy. On the contrary, many gay and lesbian individuals approach each step in the coming-out process with apprehension and even fear. It is no secret that being gay in a heterocentric society (one in which heterosexuality is the "norm") carries with it the potential for very real emotional, psychological, and physical harm. Some of the potentially negative consequences faced by individuals grappling with coming out include harassment and ridicule from peers, fellow students, or coworkers; rejection by friends, parents and other family members, and even one's church; eviction from and denial of housing; loss of current job, denial of access to military service, and other forms of prejudice and discrimination; and intimidation or physical violence that may result in destruction of property, serious injury, or even death (Davison, 2001, 2005; Gardner, de Vries, & Mockus, 2014; Murdoch & Price, 2001).

In the final stage of coming out, gay and lesbian individuals typically feel a high level of happiness, self-acceptance, and satisfaction with life.

Although few gay or lesbian individuals will experience all of these adversities, it's safe to say that most will encounter some of them as they move through the coming-out process. The reality or merely the expectation of these negative outcomes of choosing to live an openly gay life often takes a serious emotional toll (Bostwick et al., 2014). For example, teens who are struggling with the realization that they may be gay and the prospect of coming out to themselves and others have a significantly higher rate of psychological and adjustment problems, including depression, drug abuse, eating disorders, homelessness, and rates of STIs (Russell & Keel, 2002; Russell et al., 2011). The psychological effect of greatest concern is a significantly increased risk of suicide among gay and lesbian teens as they realize their true sexuality and face the prospect of how their sexuality will "play" in an intolerant world (this is discussed in greater detail next).

That said, if you ask any openly gay or lesbian person about the time in their lives when they made the choice to "come out," they will, almost to a person, tell you it was a relief, a weight had been lifted; they could finally be themselves; they felt "liberated"; and many other similar descriptions. These emotions virtually always feel stronger than the fear and problems they may face or are yet to face (e.g., see Vaughan & Waehler, 2009).

HOMOSEXUALITY AND SUICIDE Numerous studies have pointed to the unfortunate fact that gay and lesbian teens consider, attempt, and complete suicide in significantly greater proportions than straight teens (Almeida, 2009; Espelage et al., 2008; Harris, 2013). If you think about it, this is probably not so difficult to understand. Regardless of your own sexual orientation, imagine for a moment that you are subjected to ridicule, rejection, verbal abuse, and the threat of physical violence on a daily basis simply because of something about yourself over which you have no control, such as your height or the size of your feet. You can see how life under those circumstances might at times seem out of control and not worth living. This is often how young gay and lesbian individuals feel. As this risk has begun to be more widely recognized, various websites have been developed to help intervene in these difficult times for teens and young adults and to keep everyone up to date on changes in sexual orientation protection laws. Two of the most invaluable are the Gay, Lesbian, and Straight Education Network (www.glsen.org) and, focusing specifically on preventing gay, lesbian, bisexual, and transsexual teen suicide are *The Trevor Project* (www.thetrevorproject.org) and "It Gets Better" (www.itgetsbetter.org). If you or anyone you know is depressed and perhaps thinking of suicide, these resources are a tremendous help.

The path to developing a personal identity and a comfortable sense of self during adolescence is difficult enough for straight teens. But gay and lesbian adolescents must also face a society that is often hostile and rejecting of the central teen issue they are facing: their sexuality. Just at a time in our lives when we all want to be seen as individuals and begin to express ourselves as young adults, gay and lesbian youth represent an "invisible minority" in a culture that makes the assumption of heterosexuality. Most feel they must hide, at least temporarily, their true nature or face ostracism, isolation, loneliness, abuse, stigma, and oppression. These perceptions may understandably lead to feelings of anxiety, hopelessness, and depression, often accompanied by self-destructive behaviors such as alcohol and other drug abuse. Add to this the very real experiences of victimization and rejection by friends, family, and even teachers, and you have a strong recipe for hopelessness leading to suicidal thoughts, behaviors, and attempts.

When rates of attempted suicide among gay and straight teens are compared, the differences are striking. These alarming differences are summarized in Figure 11.5.

Figure 11.5 Suicidality: Heterosexual Versus Gay, Lesbian, Bisexual, and Unsure Adolescents, Grades 9-12 [During One-Year Period Prior to Survey]

SOURCE: CDC, "Sexual Identity" (2011).

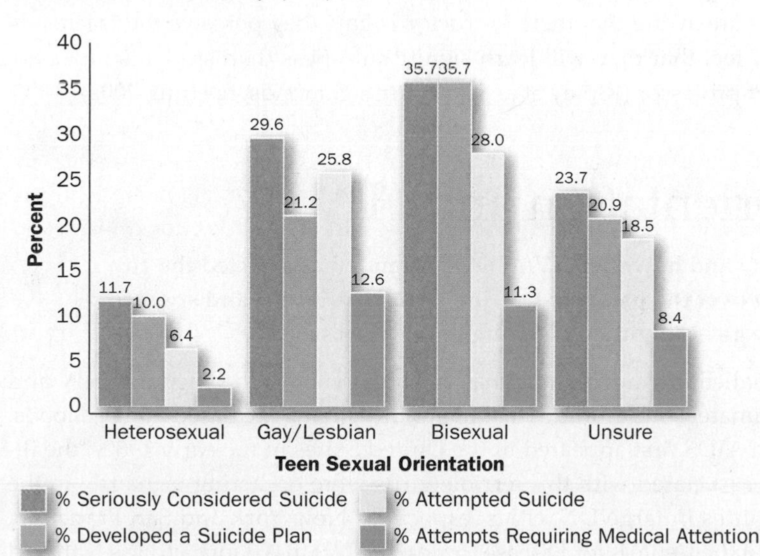

Teens who endure the stresses and strains of working through their nonheterosexual orientation during adolescence sometimes find that the process of coming out and being themselves becomes easier upon entering college. Many high schools and nearly all colleges and universities today are working to educate all students about nonheterosexual issues and to provide supportive and safe environments for gay, lesbian, and bisexual students. However, this is not to imply that it has become easy to be openly gay in school. Coming out and being openly gay on campus remain difficult challenges.

Coming Out in College

Many gay and lesbian students' experience of college life is different in some important and fundamental ways from that of their heterosexual counterparts. Most colleges and universities in the United States (including many of those affiliated with religions not known for their acceptance of nonheterosexual orientations) make genuine efforts to reduce homophobic and antigay attitudes and discriminatory behaviors within the educational community. You can get an idea of these programs simply by accessing the relevant pages devoted to gay, lesbian, and bisexual issues, support, and services on the universities' or colleges' websites (e.g., California State University at Long Beach's Lesbian, Gay, Bisexual, Transgender Student Resource Center; Gay, Lesbian, Bisexual, and Transgender Student Services at The Ohio State University; and PRIDE at the University of San Diego).

These efforts by colleges and universities to provide support and to encourage acceptance for students of all sexual orientations, though commendable and often helpful, may not always be adequate to solve the innumerable problems that nonheterosexual college students face. Research demonstrates that nonheterosexual students continue, in many cases, to experience hostile campus environments (Gortmaker & Brown, 2006; Howard & Stevens, 2000).

Specifically, nonheterosexual students tend to feel less accepted and respected by heterosexual students and by the campus environment in general. Lesbians have reported experiencing these feelings more strongly than gay men. Also, lesbian, gay, and bisexual students rate themselves as less confident than their heterosexual peers and perceive that they are treated less fairly in various academic settings, especially in dealings with administrators. Nonheterosexual students often feel the need to hide their orientation from other students, administrators, and campus health care providers (Gortmaker & Brown, 2006).

An antigay bias on campus appears to extend beyond attitudes directed toward gay, lesbian, and bisexual students to the teachers themselves. Some research has found that when students are aware that their instructor is gay, they perceive the teacher as less believable and feel that they will learn significantly less than students in a class taught by a straight professor (Ripley et al., 2012; Russ, Simonds, & Hunt, 2002).

Sexual Orientation and HIV

11.5 Discuss why and how the HIV/AIDS epidemic has affected the gay community over the past 30 years, including the myths and stigmas endured by gay individuals relating to the disease.

One of the most challenging hurdles that has faced gay men is the stigma of HIV and AIDS. And, unfortunately, the attitudes leading to this stigma are based on falsehoods. It is true that when AIDS first appeared in the United States in the early 1980s, the illnesses and deaths associated with this terrible virus were occurring primarily in the gay male communities in large U.S. cities, especially New York and San Francisco. When researchers at the Centers for Disease Control (CDC) in Atlanta and the National Institutes of Health (NIH) first began working to identify and isolate the cause of the growing epidemic, the illness was given the name *gay-related immune deficiency* (GRID). That designation lasted less than a year before heterosexual men and women began to become infected, and the official name of the disease was quickly changed to *acquired immune deficiency syndrome* (AIDS). Today, although the majority of existing cases are still among males, everyone knows that HIV and AIDS are clearly not limited to any one sex or sexual orientation. In 2009, nearly 24,000 new cases of HIV infection in the United States were among men who have sex with men, and just under 13,000 new cases were transmitted through heterosexual contact (CDC, "Basic Statistics," 2011). Anyone who still believes that HIV is only spread through male-to-male sexual contact needs to face the reality that more than *90%* of HIV infections *worldwide* are transmitted by heterosexual intercourse ("HIV Infection in Women," 2008).

Although HIV is not a "gay disease," gay people, regardless of whether or not they are HIV-positive, are often discriminated against based on irrational and misinformed beliefs about the illness. This form of discrimination has become known as **AIDS stigma**. The belief that gay individuals are to blame for the AIDS epidemic and are an ongoing threat via spreading the disease increases the overall level of prejudice and discrimination that already exists in society. "In Touch with Your Sexual Health: AIDS Myths and Stigma"

AIDS stigma
Prejudice and discrimination against nonheterosexual individuals based on the erroneous belief that gay individuals are solely to blame for the AIDS epidemic and are the primary threat for the continuing spread of the disease.

In Touch With Your Sexual Health

AIDS Myths and Stigma

AIDS stigma refers to a specific set of *false* prejudicial and discriminatory beliefs and practices that stem from ignorance about AIDS. These misguided notions are directed against people who are perceived to have AIDS or HIV and anyone who is associated with groups who are *believed* to be more likely to have or spread HIV, such as the gay male community. Below are the main prejudicial myths and the harmful, discriminatory effects of AIDS stigma, based on a survey of 7,500 adults in the United States.

Prejudicial Myths (Falsehoods) of AIDS Stigma (percent who subscribe to the belief)

- **Myth 1 (false):** People who are infected with HIV "get what they deserve" (18.7%).
- **Myth 2 (false):** All same-sex male sexual behavior transmits HIV, with or without the use of condoms (19%).
- **Myth 3 (false):** All same-sex male sexual behavior transmits HIV, even between two *uninfected* men without the use of condoms (47%).
- **Myth 4 (false):** HIV can be spread by a sterilized drinking glass used by someone with AIDS (27.1%).
- **Myth 5 (false):** You can become infected with HIV/AIDS by touching a sweater that has been worn by an infected person after it has been dry-cleaned and repackaged like new (27.5%).
- **Myth 6 (false):** You can become infected with HIV by being coughed or sneezed on by an infected person or by a gay man, whether he is infected or not (41.1%).
- **Myth 7 (false):** People with HIV/AIDS are personally responsible for their own illness (55.1%).

Negative Effects of AIDS Stigma

- **Harmful effect 1:** Increased discrimination toward gay individuals regardless of HIV status
- **Harmful effect 2:** Fear, delay, and avoidance of HIV testing
- **Harmful effect 3:** Avoidance and ostracism of people with HIV
- **Harmful effect 4:** Hiding infection from potential sexual partners
- **Harmful effect 5:** Less support for AIDS prevention programs
- **Harmful effect 6:** Support for "sodomy laws" (discussed later in this chapter)
- **Harmful effect 7:** Interpretation of gay and lesbian orientations as pathological (sick)
- **Harmful effect 8:** Reduced seeking of health care by gays and lesbians due to fear of disclosure and rejection
- **Harmful effect 9:** Isolation and lack of support for infected individuals regardless of sexual orientation
- **Harmful effect 10:** Calls for public disclosure of individuals with HIV
- **Harmful effect 11:** Calls for quarantining HIV-positive people

SOURCES: Carr & Grambling (2004); Herek & Capitanio (1999); Herek, Widaman, & Capitanio (2005); Lopes (2001).

summarizes some of the erroneous beliefs many people hold and the effects of AIDS stigma.

What can be done to ease the many difficult barriers typically faced by gay, lesbian, and bisexual individuals struggling with coming out? Perhaps the place to begin to answer that question is to educate as many people as possible about the prejudice, discrimination, and intolerance targeted at nonheterosexuals that continue to permeate society and then find ways of reducing or eliminating them. This is the focus of the next section.

Prejudice, Discrimination, and the Gay Rights Movement

11.6 Examine societal prejudice and discrimination based on sexual orientation including the legal aspects, hate crimes, and violence targeted at gay and lesbian individuals.

As you learned at the beginning of this chapter, the gay rights movement, which officially began with the Stonewall riot of 1969, already has multiple decades of history behind it. However, the gay rights movement is far more than a single event. It is a complex web of efforts by gay, lesbian, bisexual, and straight people using various social and political strategies to eliminate all forms of prejudice and discrimination based on sexual orientation. The fight for gay rights is being waged on several fronts, including working for formal antidiscrimination laws, fighting to overturn laws forbidding specific sexual acts that are aimed at nonheterosexuals, preventing violent attacks on gay and lesbian individuals, attempting to interact with and educate people about nonheterosexual orientations, and working to establish equal rights for nonheterosexuals with regard to marriage (as discussed earlier).

Since You Asked

11. Do homosexual people feel comfortable holding hands and showing other public displays of affection with their partner? How often are they ridiculed or threatened?

Laws Against Discrimination Based on Sexual Orientation

The campaign to establish antidiscrimination laws regarding sexual orientation is just beginning. Fewer than 10% of the world's countries have passed national legislation forbidding discrimination based on sexual orientation. Of those that have, the United States is conspicuously absent (although some states have such laws). Many political, social, and religious factors may account for the lack of federal U.S. laws designed to protect people from discrimination based on sexual orientation, but one that stands out is the debate over whether granting specific antidiscriminatory rights to nonheterosexual groups bestows on them "special rights."

The brutal abduction, beating, and killing of Wyoming college student Matthew Shepard in 1998 brought the issue of violence directed at those with nonheterosexual orientations to a higher level of public awareness in the United States and around the world. However, his was not an isolated case; numerous murders motivated by a victim's sexual orientation occur every year.

Proponents claim that people are routinely discriminated against in numerous ways, in various settings, solely because of their sexual orientation and therefore should be *specifically* protected by law. Legally naming nonheterosexuals as a "protected class" of people in the same vein as members of a particular sex, religion, and race is a hot political potato, and emotions on the subject run high. The official policy statement of the American Civil Liberties Union (2002) says, in part.

Although the other side of this debate has no specific policy statement, one opposing argument claims that the Fourteenth Amendment to the U.S. Constitution forbids *all* discrimination against *any* group, including, by default (but not stated), people of any sexual orientation. Therefore, the argument maintains, additional protections are unnecessary (see Badash, 2009). However, many lawmakers in both the federal and state governments do not consider nonheterosexual individuals one of the groups that are covered by this constitutional right. Therefore, supporters of gay rights point to many cases of discrimination that go unaddressed and argue that more specific language is necessary in state and local constitutions to guarantee nonheterosexuals the *same* rights afforded other protected groups. Where, within this polarized debate, does the "truth" lie? Of course, truth tends to be relative to one's convictions, but some statistics might help clarify the issue.

You are probably aware that discrimination in employment, housing, education, credit practices, and so on, is illegal according to various federal and state statutes. But the question that then arises is: Discrimination is illegal against *whom*? In general, state and federal antidiscrimination laws specify certain characteristics of groups that are protected under these laws, such as race, religion, sex (male or female), country of origin, age, or disability. These are referred to as **protected classes**. Approximately half of the states have passed laws adding sexual orientation to their list of protected classes that receive protections under the states' antidiscrimination laws. However, even in those states with legal protections based on sexual orientation, many have extended that protection to very specific settings, such as state employment or education, but not across the board, as is the case for other protected classes. Figure 11.6 identifies the states that have enacted laws prohibiting employment discrimination based on sexual orientation (some of these states also ban discrimination based on gender identity as discussed in Chapter 10, "Gender: Expectations, Roles, and Behaviors").

protected classes
Specific groups of people legally protected from discrimination under federal and state antidiscrimination laws, identified by race, religion, sex, age, sexual orientation, or other characteristics.

What does this mean for nonheterosexuals living in states that do not provide meaningful legal protections against discrimination based on sexual orientation? Depending on the state, it means that they can be denied or evicted from housing, denied a job, fired from a job, denied an equal education, denied credit, and subjected to many other discriminatory practices solely because they are gay, lesbian, or bisexual, and they have no legal recourse. You may be thinking to yourself, "Yeah, but these discriminatory acts never really happen, do they?" Yes, they happen frequently, but much more frequently in those states that do not have antigay discrimination laws (Sears, Mallory, & Hunter, 2009).

Figure 11.6 States with Laws Prohibiting Employment Discrimination on the Basis of Sexual Orientation, 2013

SOURCE: National Gay and Lesbian Taskforce (2013): State Non-Discrimination Laws in the U.S., www.thetaskforce.org/.

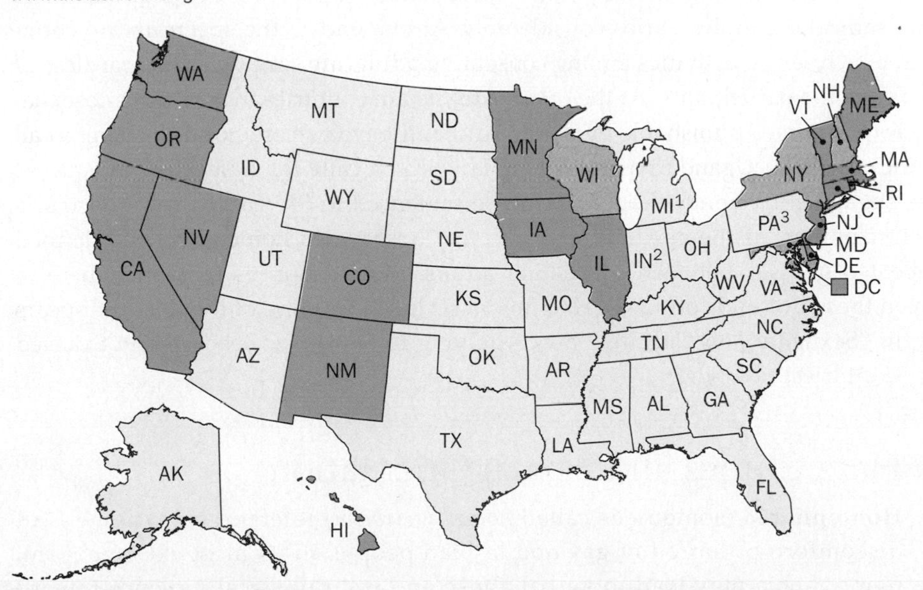

States banning discrimination based on sexual orientation and gender identity/expression (17 states and the District of Columbia)

Laws banning discrimination based on sexual orientation, but not gender identity (4 states)

Laws Prohibiting Gay and Lesbian Sexual Behaviors

sodomy laws
Laws prohibiting specific sexual activities between adults, even in private and with their consent.

In the past, all states enacted **sodomy laws**, which prohibited people from engaging in certain sexual acts that were deemed strange, deviant, or immoral. These laws referred to nonreproductive, noncommercial, consensual sexual acts between adults in private and included oral and anal sex at a minimum and in some cases virtually any sexual behavior other than sexual intercourse between a married man and woman in the "missionary position." During most of the twentieth century, these laws were rarely enforced, and when they were, they were usually applied only against gay male and sometimes lesbian individuals.

In 2003, a U.S. Supreme Court decision in effect negated all sodomy laws in the remaining 14 states. The case, known as *Lawrence v. Texas*, stemmed from a 1998 event in which the Houston police were called out on a domestic disturbance complaint (which later turned out to be a false report) by a neighbor of the defendants. When the police arrived, they did not find a disturbance but did discover two men, Geddes Lawrence and Tyron Garner, engaged in, according to the police report, "deviant sexual conduct, namely, anal sex," and arrested them for violating Texas's sodomy laws (Gibbs, 2003). The case was eventually heard by the Supreme Court, which in a 6 to 3 verdict struck down all of Texas's remaining sodomy laws (which had applied only to nonheterosexual couples, primarily gay men), basing their decision on the defendants' constitutionally guaranteed right to privacy. Writing for the majority, Justice Anthony Kennedy stated: "The petitioners are entitled to respect for their private lives. The state cannot demean their existence or control their destiny by making their private sexual conduct a crime" (Gibbs, 2003; Jurand, 2004).

This precedent-setting ruling had the effect of striking down *all* remaining sodomy laws *anywhere* in the United States for everyone, regardless of their sexual orientation. Following the ruling, a few states have attempted to continue to enforce their sodomy laws but have been unsuccessful in doing so. Some analysts suggest that the *Lawrence v. Texas* ruling could be overturned by a constitutional amendment or altered by an increasingly conservative Supreme Court. However, virtually all experts agree that both of these events are highly unlikely and sodomy laws in the United States are history.

As with antidiscrimination laws, the legal status worldwide of sexual behaviors among same-sex couples varies considerably. At one end of the spectrum are countries in which sexual activities among consenting adults are simply legal regardless of the sexes of the participants. At the other extreme are countries in which homosexual activities are crimes punishable by death. And anti-gay laws are not decreasing in all countries. In 2014, Uganda passed a new law which calls for first-time "offenders" (those engaging in "homosexual acts") to be sentenced to 14 years in jail. It also sets life imprisonment as the maximum penalty for "aggravated homosexuality," defined as repeated gay sex between consenting adults. Most countries fall somewhere in between these two ends of the legal/criminal scale. A sampling of these laws appears in "Sexuality and Culture: Laws Applying to Same-Sex Behaviors in Selected Countries."

homophobia
Extreme fear, discomfort, or hatred of nonheterosexual individuals.

Geddes Lawrence and Tyron Garner were the defendants in the case of *Lawrence v. Texas* that overturned the last of the sodomy laws in the United States.

Hate Crimes and Sexual Orientation

Homophobia (sometimes called *homonegativism*) refers to an extreme fear, discomfort, or hatred of gay and lesbian people. In its most extreme form, homophobia may lead to verbal abuse and even physical violence toward nonheterosexual individuals. In October 1998, Matthew Shepard, a gay college student at the University of Wyoming, was kidnapped by two men he had met

Sexuality and Culture

Laws Applying to Same-Sex Behaviors in Selected Countries

Country	Lesbian	Gay Male	Maximum Penalty
Africa			
Egypt	x	x	Various
Kenya	✓	x	14 years
Morocco	x	x	3 years, fine
Nigeria	✓	x	Death
South Africa	✓	✓	
Sudan	✓	x	Death
Uganda	x	x	14 years to life in prison
Americas			
Argentina	✓	✓	
Barbados	x	x	Unknown
Bermuda	✓	✓	
Brazil	✓	✓	
Canada	✓	✓	
Cuba	✓	✓	
Guatemala	✓	✓	
Jamaica	✓	x	10 years, hard labor
Mexico	✓	✓	
Nicaragua	x	x	3 years
Puerto Rico	✓	✓	
United States	✓	✓	
Asia-Pacific			
Afghanistan	x	✓	Death
Australia			
China	✓	✓	
India	x	x	Life in prison
Japan	✓	✓	
Nepal	x	x	Life in prison
Pakistan	x	x	Death
Philippines	✓	✓	
Singapore	x	x	Life in prison
South Korea	✓	✓	
North Korea	x	x	Unknown
Taiwan	✓	✓	

Country	Lesbian	Gay Male	Maximum Penalty
Europe			
Belgium	✓	✓	
Czech Republic	✓	✓	
Denmark	✓	✓	
France	✓	✓	
Germany	✓	✓	
Greece	✓	✓	
Italy	✓	✓	
Netherlands	✓	✓	
Poland	✓	✓	
Russia	✓	✓	
Spain	✓	✓	
Sweden	✓	✓	
Switzerland	✓	✓	
Turkey	✓	✓	
United Kingdom	✓	✓	
Vatican City	✓	✓	
Middle East			
Iran	x	x	Death
Iraq	✓	✓	Legal but "taboo"
Israel	✓	✓	
Kuwait	x	x	7 years
Lebanon	x	x	1 year
Palestine	x	x	10 years
Saudi Arabia	x	x	Death

Key: ✓ = legal; x = illegal

SOURCE: Adapted from Sodomy Laws (2007).

in a bar, driven to a deserted field, brutally beaten, tied to a fence in freezing temperatures, and left to die. Four months later, Billy Jack Gaither, a 39-year-old gay man who worked at a clothing company in Alabama, was brutally beaten to death. His throat was cut and his body beaten with an ax handle before he was thrown on top of a pile of automobile tires and set on fire. These horrible, ruthless murders focused renewed national attention on violence against gays and lesbians. However, many murders motivated by antigay attitudes occur every year in the United States and around the world. Between 1997 and 2007, nearly 200 *known* antigay killings occurred

Figure 11.7 Hate Crimes Against Gay, Lesbian, and Bisexual Individuals, 1991–2009

SOURCE: FBI (2011).

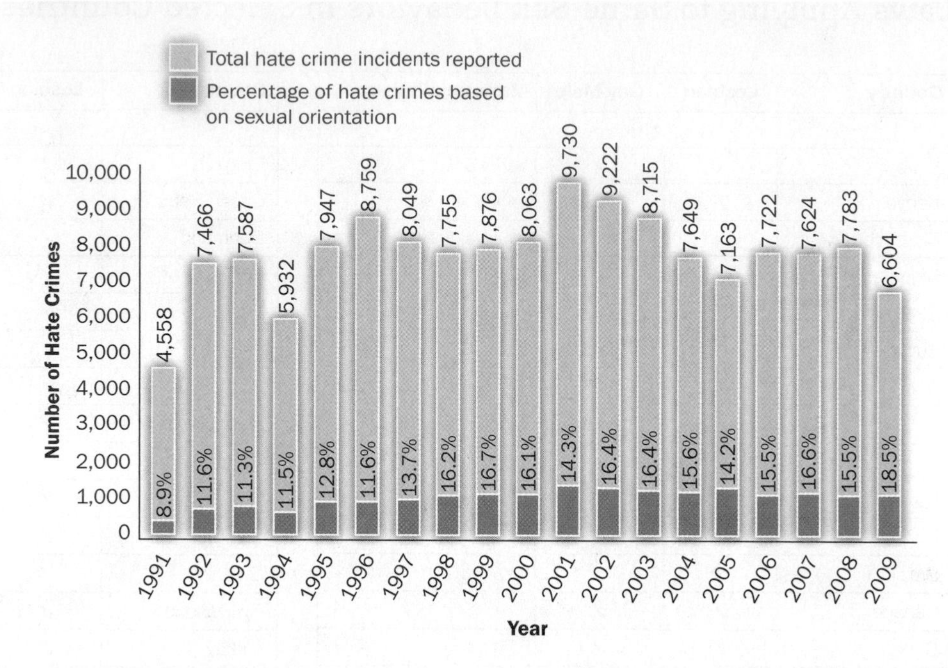

in the United States. The incidence of known murders based on sexual orientation or gender identify rose to an all-time high of 30 in 2011 and remained high at 24 in 2012 (NCAVP, 2013).

hate crimes
Violent crimes motivated by prejudice and discrimination, targeting specific groups of individuals.

hate crimes laws
Laws prescribing more stringent penalties for crimes motivated by bias or prejudice.

gay bashing
Criminal acts or violence motivated by homophobia and committed against nonheterosexual individuals.

In the U.S., violent crimes that target specific groups of individuals are called **hate crimes**. These violent crimes are motivated by strong feelings of fear and hate toward members of a certain "protected class" of people, such as members of a specific religion, racial group, or sexual minority (Cheng & Ickes, 2013). Typically, the perpetrators of these crimes are emotionally weak individuals who feel threatened by the mere existence of the group at which their violence is targeted. **Hate crimes laws** have allowed for more stringent penalties for violent crimes that can be shown to have been motivated by bias or prejudice toward a protected class of people, such as a racial or religious class. As you can see in Figure 11.7, extreme and violent forms of prejudice and discrimination targeting people of nonheterosexual orientations, often referred to as **gay bashing**, did not by any means begin or end with those high-profile stories from over ten years ago.

More than half of all hate crimes in 2012 were motivated by racial bias. Crimes against gay and lesbian individuals composed the second most common category of hate crimes total, 19% of such crimes in 2012 (FBI, 2013). Table 11.3 provides a breakdown

Table 11.3 FBI Hate Crime Statistics for 2012*

Category	Incidents	Percentage of All Hate Crimes
Race	3,467	48.5%
Sexual orientation	1,376	19.2%
Religion	1,340	18.7%
Ethnicity/national origin	866	12.1%
Disability (physical or mental)	102	.1.4%
Multiple bias incidents	13	0.09%
Total	7,164	100.0%

*All categories increased from 2009.

SOURCE: Adapted from FBI (2013), TABLE 1.

Figure 11.8 Hate Crimes Laws in the United States, 2009

SOURCE: National Gay and Lesbian Taskforce (2009): Hate Crimes Laws in the United States.

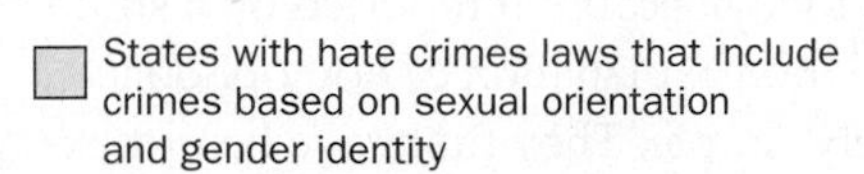

States with hate crimes laws that include crimes based on sexual orientation and gender identity

States with hate crimes laws that include crimes based on sexual orientation

States with hate crimes laws that do not include crimes based on sexual orientation or gender identity

States that do not have hate crimes laws that include crimes based on any characteristics

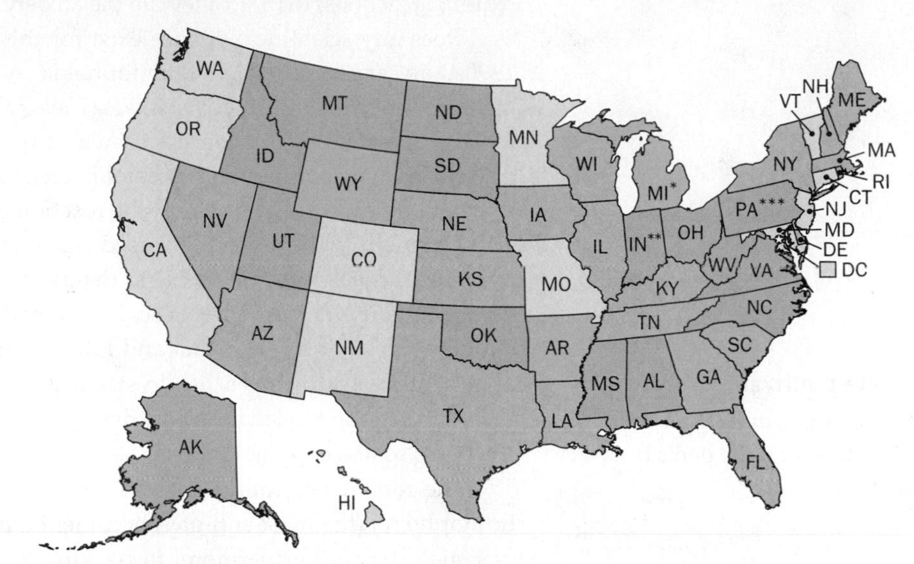

* Michigan's hate crimes penalty laws do not include sexual orientation, but hate crimes data collection laws do.

** Indiana has no hate crimes penalty laws, but does include sexual orientation in hate crimes data collection.

*** In 2008, Pennsylvania's highest court overturned the 2002 amendments to the hate crimes law that added sexual orientation, gender identity, ancestry, gender, and mental and physical disability, based on the procedural way the legislation was passed by the legislature, not the content of the law.

of the various categories of hate crimes in 2012. Thirty-two states have enacted hate crimes laws covering crimes motivated by antigay prejudice. Figure 11.8 (following page) shows the distribution of hate crimes laws in the United States.

The Psychology of Violent Crime Against Nonheterosexual Groups

Murders and beatings of gays, lesbians, and bisexuals tend to be among the most gruesomely violent of all violent crimes. Typically, the victims are tortured, brutally beaten, or stabbed numerous times. The degree of hate evidenced by these crimes is difficult for most people to comprehend. But anyone who is openly gay or lesbian will tell you that nonheterosexual individuals are routinely aware and vigilant of the possibility that they could be physically attacked at any moment. What is behind such extreme reactions to homosexuality that causes some people to resort to physical beatings and even murder to express their fear and hatred?

Most psychologists will agree that it is one thing to object to homosexuality on moral, philosophical, or religious grounds, but that these over-the-top reactions of violence and murder are vastly out of proportion to any real threat the victim poses. These overreactions suggest that deeper psychological mechanisms may be at work in people who commit violent acts against nonheterosexuals (often referred to as "gay-bashers"). One of the motivations for irrational antigay attitudes and behaviors that has been proposed is a **psychological defense mechanism** called a **reaction formation**, in which a person engages in extreme and exaggerated behaviors in the *opposite* direction of the person's unacceptable internal urges. How does this translate into violence toward gays and lesbians?

The theory says that when some men with strong anti-gay attitudes are forced to confront, consciously or unconsciously, their *own* homosexual urges—attractions they feel to other men—it creates an intense internal "fear of being gay" themselves. Those urges and the resultant fear drives them to beat or kill a homosexual person to "prove that

psychological defense mechanism

Originally suggested by Freud, a psychological distortion of reality serving to defend against personally unacceptable thoughts or urges.

reaction formation

A type of defense mechanism in which a person engages in exaggerated behaviors in the opposite direction of internal urges felt to be unacceptable or intolerable.

they could not possibly be gay." In other words, gay bashers may be in deep denial that they themselves are closeted homosexuals and will resort to extreme behaviors (i.e., form extreme reactions) to try to alleviate the anxiety their urges are creating in them.

Does any scientific evidence exist for this theory? The answer is yes. In the mid-1990s, an article titled "Is Homophobia Associated with Homosexual Arousal?" appeared in the *Journal of Abnormal Psychology* (Adams, Wright, & Lohr, 1996; Drescher, 2004; Gerstenfeld, 2013). The article created quite a stir in human sexuality and psychology circles and continues to be heatedly discussed today because it reported on a study that, still today, seems to support the reaction formation explanation of homophobia.

The method the researchers used was relatively simple. They first gave heterosexual male subjects a written scale to determine their level of negative attitudes toward homosexuality, a "homophobia scale." They then asked each subject to watch sexually explicit videos of heterosexual and homosexual activities and measured their degree of sexual arousal using a penile strain gauge called a **penile plethysmograph** (see Chapter 1 for an explanation and drawing of this device) that records blood flow into the penis to detect sexual arousal.

penile plethysmograph
An electronic device that records blood flow into the penis to detect sexual arousal.

The researchers found that only the men who scored highest on the measure of homophobia (the most anti-gay) became the most sexually aroused in response to the gay male video. Furthermore, those same homophobic men self-rated their level of sexual arousal to the gay male videos significantly lower than did the nonhomophobic participants. This implies that men with stronger antigay attitudes were experiencing homosexual feelings in themselves but denying them. It follows, then, that gay bashers, who clearly hold the strongest antigay feelings, may be resisting the strongest gay tendencies of all within themselves.

Your Sexual Philosophy

Sexual Orientation

We live in a heterocentric society in which most straight people place the assumption and expectation of heterosexuality onto most, if not all, others. If you are gay, lesbian, or bisexual, you already know this, perhaps all too well. But if your self-defined sexual orientation is heterosexual, you have probably not had to deal personally with many of the topics discussed here. However, nearly every straight person has friends, acquaintances, or family members who are gay, lesbian, or bisexual, whether they know it or not. A clear and accurate understanding of the issues they face is crucial to your personal journey through our sexually diverse world.

Issues relating to sexual orientation play important roles in everyone's basic sexual philosophy. An awareness of these issues contributes to the development of understanding and tolerance of sexual diversity. The material presented in this chapter relates to the development of your ability to feel comfortable with yourself and with those who may have sexual orientations different from yours.

If you are heterosexual, you may believe some of the misconceptions and stereotypes about gay, lesbian, and bisexual individuals examined in this chapter. If so, now that you have more accurate information about sexual orientation, your challenge is to begin to discard those erroneous beliefs and start becoming more tolerant of the richness of human sexual diversity. Tolerance of all kinds of human differences is usually a matter of education and learning to see others as individuals rather than solely as members of a stereotyped group. People with sexual orientations other than your own do not pose any serious threat to you, your sexual orientation, your health, or your life as a sexual being. As you acknowledge and accept these facts, which some of you have already, you will be able to plan ahead to interact with others, all others, from a position of tolerance, equality, and empathy, based on who they are as fellow humans and not as people with a certain sexual orientation.

If you are not heterosexual, your challenges are different but equally important. You have to be emotionally prepared for the unfortunate fact that others may view you through a lens of prejudice and discrimination and treat you accordingly. It is hoped, of course, that most of the people you care about in your life will not take these views, but for most of you, it is bound to happen, if it hasn't already, and you must be ready for it. How will you react to the homophobia that exists in others? Here is your chance to think about it now, analyze your position, develop strategies, plan ahead, and clarify your sexual philosophy about your sexual orientation and others' reactions. That will help you avoid being taken by surprise by prejudiced attitudes or discriminatory or violent acts. This is not to imply that you should not work to fight injustice and bigotry and to be comfortable and proud of who you are. Everyone should do that. You will also have the opportunity to educate others about sexual orientation, and education is the very best weapon against the ignorance on which bigotry is based.

Have You Considered?

1. Imagine for a moment that you are the parent of a son or daughter (your choice) who informs you that he or she is gay. How do you think you would react? What would be the three most important things you would want to tell your child after hearing this news?
2. Now turn the question on yourself. Imagine that you wake up tomorrow morning and your sexual orientation has changed (from straight to gay, from gay to straight, or from either to bisexual, etc.). Discuss how you might feel about such a change and how such a transformation would alter your life and your view of the world around you.
3. What do you think are the three best strategies for reducing prejudice and discrimination directed at nonheterosexual individuals? Explain your ideas.
4. Do you think it is important for researchers to get to the bottom of what causes homosexuality? Discuss at least three reasons for your answer.
5. Suppose you were placed in charge of gay, lesbian, and bisexual student services at your college or university. Discuss what actions you would take and what strategies you would propose for improving the climate on your campus for nonheterosexual students.
6. Do you think marriage should be legalized for nonheterosexual couples? Discuss at least three reasons for your answer.
7. What do you think are some strategies for reducing violence against gays and lesbians? Explain why your idea might be effective.

Summary

Historical Perspectives

The Stonewall Riot and the Beginning of the Gay Rights Movement

- The 1969 Stonewall riot sparked the beginning of the gay rights movement. Patrons of the Stonewall Inn, one of the country's few gay bars at that time, rose up and fought violently against police harassment. This event is marked by many as the beginning of the gay rights movement in the United States.

Straight, Gay, Lesbian, Bisexual: A Closer Look

- Sexual orientation is not categorical; straight and gay exist on a continuum. Most people have thoughts, feelings, fantasies, or experiences involving both sexes.
- Alfred Kinsey developed a rating scale for measuring degrees of sexual orientation rather than placing it into distinct categories.

Nonheterosexual Orientations: Issues and Attitudes

- Sexual orientation is about more than simply sexual behavior. Sexual orientation relates to which sex you are primarily attracted to, want to date, are likely to fall in love with, want to establish long-term relationships with, and perhaps marry.
- Historically, marriage has been legally defined in most states as a union between one man and one woman. In recent years, however, nearly 20 states have legalized gay marriage and this seemingly ingrained tradition is changing at a rate rarely seen for complex social issues.
- The quality of gay and lesbian relationships is similar to that of heterosexuals, but research shows that same-sex couples tend to experience less hostility and are generally more upbeat than straight couples overall.

Theories of the Origins of Sexual Orientation

- The scientific consensus is that sexual orientation is not a choice. Most anecdotal and scientific research indicates that sexual orientation is a basic and probably inborn characteristic.
- Researchers have found differences between heterosexual and homosexual brain structure and functioning. These biological findings, combined with studies of twins, have provided evidence that sexual orientation may be genetic in origin. Research also indicates that differences in the balance of hormones in the uterine environment during pregnancy may relate to a person's sexual orientation.
- Research has not found that gay men are more likely than straight men to have domineering mothers or absent fathers. The family backgrounds of gay individuals and of heterosexuals differ very little, if at all.
- The "exotic becomes erotic" theory attempts to integrate nature and nurture approaches to sexual orientation. It suggests that inborn tendencies, combined with early nonsexual play experiences with peers, may determine sexual orientation in puberty and adulthood.
- Some people argue that the origins of sexual orientation are unimportant and that nonjudgmental acceptance of nonheterosexual individuals is more important than finding an exact cause.

Coming Out

- Heterosexuals never have to "come out of the closet" because they are never "in the closet." Most cultures assume heterosexuality as the default orientation.
- Coming out is a gradual, step-by-step process, often occurring over years.
- Coming out is a risky and difficult, yet emotionally satisfying, process for most nonheterosexual individuals, who find "living a lie" more difficult than facing the possible prejudice, discrimination, and stigma of coming out.
- Suicide rates among gay youth are many times higher than among their heterosexual peers. Societal and peer pressures cause increased rates of serious depression among gay and lesbian teens.
- Gay, lesbian, and bisexual students experience college life more negatively than straight students. Bigoted peer attitudes and antigay campus climates may have a chilling effect on nonheterosexual students' college life.

Sexual Orientation and HIV

- AIDS stigma negatively affects gays and heterosexuals. This is one type of prejudice and discrimination based on gay and lesbian sexual orientation that often poses problems for those who have come out.

Prejudice, Discrimination, and the Gay Rights Movement

- The United States has failed to pass a law protecting citizens from discrimination based on sexual orientation. No federal law identifies sexual orientation as a characteristic of a protected class of individuals in cases of discrimination.
- Twenty-four states omit sexual orientation from antidiscrimination laws. Conversely, approximately half of all states have specifically added sexual orientation as a protected class in at least some of their discrimination laws.
- Sodomy laws have existed in all states in the past. However, the 2003 U.S. Supreme Court ruling in *Lawrence v. Texas* made all such laws unconstitutional.
- Violence against nonheterosexuals is the second most common hate crime in the United States. Only violence based on race is more common.
- Gay bashing may be a defense mechanism against the bashers' own homosexual desires. Research indicates that men who deeply fear their own gay tendencies may be the most likely to commit violent acts against gay men to "prove" that they are straight.

Your Sexual Philosophy

Sexual Orientation

- Developing an understanding of and sensitivity to the complex issues relating to sexual orientation is one of the keys to developing tolerance for sexual differences and appreciating human sexual diversity.

Chapter 12

Sexual Development Throughout Life

Learning Objectives

After you read and study this chapter you will be able to:

12.1 Explain how humans are "sexual beings" from birth through childhood and summarize the behaviors and potential difficulties that accompany childhood sexual development.

12.2 Review the complex issues associated with sexual development during adolescence including sex education, puberty, teen intimate relationships, teen dating, sexual behavior in adolescence, teen pregnancy, and the issue of STIs among adolescents.

12.3 Summarize the main issues relating to sexuality in college including sexual activities, STIs, and the connection between alcohol and sexual problems among college students.

12.4 Explain how intimate relationships in adulthood differ from adolescence and discuss the developmental milestones in adult sexuality including, cohabitation and marriage; sexual behavior possible sexual problems during adulthood, and the changes associated with menopause.

12.5 Address sexual development into later life including older adults' sexual behaviors, age-related changes in sexual responding, and intimate relationships among the elderly.

Since You Asked

1. Is it unhealthy to masturbate at an early age? (see page 442)
2. I recently caught my four-year-old daughter playing "doctor" with the neighbor boy (both naked). Is this a sign of a problem, or should I not be concerned? (see page 443)
3. If a child between the ages of five and seven acts out sexual acts, specifically lying down on top of another child and imitating movements of intercourse, does this indicate a problem, or is it part of normal childhood curiosity? (see page 445)
4. What prevents parents from talking to their kids about sex? My parents never had "the talk" with me, and I had to find out from friends, or even worse, boyfriends. Now, two abortions later, I'm finally taking this class. (see page 446)
5. I started puberty very late (around age 15). All my friends were already getting their periods and breasts and everything. It was very embarrassing. Why do some people start puberty so late? (see page 450)
6. In high school, a lot of my friends were having oral sex but not intercourse. Is it true that oral sex doesn't spread sexual diseases? (see page 456)
7. No one I know here has been tested for STDs. I have, but they say they don't want to or need to. How can I convince them to get tested? (see page 462)
8. My roommate went to a party and drank too much. She had sex with two guys (that she remembers), but she says she didn't really want to. Was this rape? (see page 464)
9. I've heard that when people get married, their sex life goes right down the tubes. Is this true? (see page 467)
10. At what age do most elderly people stop being interested in sex? (see page 471)

If you ask most people what they think of when you say "sexual development," they will probably respond with something relating to puberty, the time in our lives when we mature sexually and develop sexual characteristics such as breasts, pubic hair, and penis and testicular growth. However, this is only a part of the larger picture of sexual development. We are developing sexually throughout our lifespan. As has been noted often in this book, we are by nature sexual beings.

Sex, at its most fundamental level, is required for the human species to survive and not become extinct. Beyond the biological inevitability of certain sexual acts, we develop in many ways as sexual beings while surrounded by powerful social and cultural influences. These environmental forces combine with our genetic predispositions to mold our sexual identities, attitudes, and preferences. All of these factors combine to make us what are probably the most sexually complex creatures on earth.

As you read this chapter, you will notice that we touch on many of the topics from other chapters in this book. Virtually all the topics of human sexuality intersect with our lives at one or many points as we develop as sexual beings. This is easy to see if you simply turn to the table of contents and think for a moment about how each chapter topic has played a role in your sexual development or the sexual life of people you have known or learned about: sexual anatomy, sexual responding, love and intimacy, birth control, sexual activities, sexual problems, STIs, pregnancy, gender expectations, sexual orientation, sexual aggression and violence, atypical sexual behaviors, prostitution, and pornography. All of these topics influence, or are influenced by, human sexual development.

When social scientists study human development, whether focusing on sexuality or on other aspects of life, their goal is to examine the changes that unfold in all

Focus on Your Feelings

Children experience strong feelings relating to normal sexual exploration, either through masturbation or sex play with other children. They know it feels good, they want to experiment, and it's normal. But societal expectations, usually communicated by parents, often cause them to feel guilt and shame. Then, as children enter puberty, hormones surge and create new feelings of sexual attraction and desire. Most of you can probably remember the intensity of those early teen years, filled with highly charged emotions that are a normal part of adolescent sexual development but that sometimes may have felt too powerful and perhaps made you feel confused.

As you entered college, you were probably faced with new feelings (usually good ones!) that went along with sexually-charged university student cultures. Moreover, you were more often regarded as an adult (sometimes "suddenly") and expected to make your own decisions about life and about sex. College students must learn how to deal as an adult with the feelings that are generated by sexual relationships. This is an integral and a valuable part of the educational process for college students (although it's not in the catalog!), and it can be more challenging than any academic major.

The feelings associated with sexuality in adulthood can, at times, seem overwhelming. In early adulthood, most people will take one of the most important steps of their life: choosing a mate. The mix of emotions that comes with making such a commitment are all normal reactions to facing such a major life change. Then, moving through adulthood, most will face one or more emotionally charged issues such as sexual problems, marital discord, infidelity, separation, or divorce. Working through the emotions we feel during these experiences is part of human growth, learning, and developing and, although difficult, can make us stronger and more content in the long run.

Sexual responses and abilities change as we grow older. An awareness that certain changes are a normal part of aging can help us cope rationally with them and take whatever steps are necessary to resolve problems that arise. The process of aging, sexually speaking, is not something people should worry about, especially now that research is demonstrating that seniors often maintain a healthy sexual life well into their eighties and beyond!

of us, under normal circumstances, in a somewhat predictable way. In other words, they are not usually trying to focus on one particular individual and his or her specific development, but rather they are looking for universals—changes and milestones that happen to most of us at approximately the same age or stage in life. For example, most children acquire the concept of gender (that is, they know whether they are girls or boys and become aware of gender-role expectations) between the ages of three and five; children enter puberty with all of its accompanying physical, emotional, and social changes between ages ten and fifteen; most people form a loving, lifelong relationship (or try to) in their early twenties; and we all experience sexual changes as we age. These common and expected developmental events occur in fairly predictable patterns for all humans and form the basis of the topics in this chapter. Of course, many variations to the usual motifs can occur, but they tend to be exceptions.

We will proceed chronologically in our discussion of human sexual development, and that means we must begin not in infancy, but *before* birth, with sexual development and, believe it or not, sexuality in the womb. By the time this chapter is done, we will have traveled from the womb all the way to sexuality in old age, with a number of stops in between. First, we will take a brief look at some of the theories suggested by the person who first put the "sex" in sexual development: Sigmund Freud.

Historical Perspectives

Freud's Psychosexual Stages of Development: Oedipus and Electra?

In the late 1800s, Sigmund Freud was revolutionizing how the world thought about human nature. He conceptualized human personality as consisting of three dynamic entities: the *Id*, an unconscious force in all of us that drives us to fulfill our survival needs; the *Ego*, our conscious "self" that interacts consciously with the world around

Sigmund Freud believed that nearly all human development is sexually motivated.

us; and the *Superego*, Freud's interpretation of what most of us today recognize as our conscience, the psychological "judge" about what is right or wrong, what is moral.

Freud also saw the development of our personality—*who we are*—as occurring in childhood through a series of **psychosexual stages** during the first 12 years or so of life, each stage focusing on a different sexual body part. He called these stages the *oral stage* (birth to 18 months), the *anal stage* (18 months to 3 years), the *phallic stage* (3 to 6 years), the *latency stage* (6 years to puberty), and the *genital stage* (adolescence).

One major milestone in children's development, which Freud maintained happens during the phallic stage (around three or four years old), is the child's realization of his or her sex. This is a process called **gender identification**, which typically occurs between the ages of three and five and is discussed in detail in Chapter 10, "Gender: Expectations, Roles, and Behaviors." Freud also saw this as an important event in the developmental process. He believed that for a young boy to self-identify as male and learn to behave according to cultural expectations for men is a very difficult task. After all, the boy has been under the nurturing care of his mother since birth, so what possible motivation could he have, Freud asked, to leave her side and identify with his father? (If this sounds sexist to you, remember that Freud was living in late-1800s Victorian Europe, where such roles were very strictly delineated.) Freud contended that the boy's motivation must be very compelling, as evidenced in his theory of the **Oedipus complex**, named for the hero of the Greek tragedy *Oedipus Rex*, in which Oedipus unknowingly kills his father and marries his mother.

psychosexual stages
Freud's theory that the development of human personality occurs in a series of stages during childhood.

gender identification
A developmental stage in children between the ages of three and five during which they begin to understand which sex they are.

Oedipus complex
A Freudian notion explaining how a young boy comes to identify with his father.

In Freud's theory, all boys at about three or four years of age develop unconscious sexual desires for their mothers. They yearn, literally, though unconsciously, to have sexual intercourse with their mothers. However, every boy soon realizes (through unconscious processes at work) that he cannot have his mother in this way because someone else already does, namely, his father. He begins to see himself as competing with his father for the mother's affection, even though he knows he has no chance of winning because his father is so much bigger and more powerful. At the same time, the boy becomes terrified that the father might find out about the boy's feelings for the mother and punish him by cutting off the boy's penis (what Freud referred to as "castration anxiety"). The only way he can figure out to avoid this terrible fate is to identify with the father—to become "just like dad." By identifying with the father and imitating and emulating him, the boy comes to understand his gender identity as male, develops appropriate male social behaviors, and suppresses his desires for his mother until he can be as big and strong as his father, be able to do away with his father, and then possess his mother. Of course, by the time he becomes "as big as his father," those urges are fully repressed, and he seeks out a wife of his own (who, according to some interpretations of Freud, is as close a likeness to his mother as possible).

Gender identification is the process through which children discover their gender and begin to behave in socially prescribed, gender-appropriate ways.

You may be asking yourself, "What about girls?" Freud later developed a theory of female gender identification through what became termed the **Electra complex** (in the Greek tragedy *Electra*, the title character convinces her brother to kill her mother so that she can possess her father). In Freud's theory, a girl at about three years of age becomes aware that her father has a penis, something she and her mother are lacking. This feels unjust to her and makes her feel castrated (Freud called this "penis envy"). Her reaction, according to Freud, is to reject the mother and identify with the father until she realizes that she cannot possess his penis because it already "belongs" to her mother. So she returns to identify with her mother so that when she grows up, she will be able to find a penis of her own. And in some interpretations of Freud's notion of penis envy, when she finds the right penis, she marries it.

Electra complex
A Freudian notion explaining how a young girl comes to identify with her mother.

You can see why most modern-day psychologists reject the Oedipus and Electra explanations for gender identification. The theories are overly complex and not subject to scientific study that can prove or disprove them. Moreover, simpler and more testable theories exist that seek to explain gender identification in less fanciful ways. And Freud never studied children but developed his ideas through recollections of his adult patients in analysis, who were probably not typical of all human beings.

Freud's theories are fascinating to read and think about, but they do not carry a great deal of scientific weight. As we begin our discussion of *current* thinking about sexual development, we will see how modern psychological and sexuality research sheds more direct and believable light on this complex topic.

Sexuality in Infancy and Childhood

12.1 Explain how humans are "sexual beings" from birth through childhood and summarize the behaviors and potential difficulties that accompany childhood sexual development.

We do not think of babies and children as sexual people. However, we do come into this world with certain sexual characteristics already primed. Of course, very early sexuality is very unlike adolescents or adults; but it is an important and normal part of human development. Nevertheless, most adults are uncomfortable with the idea of infants and children being sexual beings.

Many parents and other adults become concerned when infants, toddlers, or young children engage in sexual behaviors such as masturbation or explorations of each other's sexual bodies, as in "playing doctor." Child development specialists, however, are quick to reassure them that such behaviors are usually normal signs of childhood curiosity and a healthy enjoyment of activities that feel good (Flanagan, 2011). Moreover, there is no evidence that early childhood sexual experimentation is predictive of any particular sexual attitudes or behaviors later in life. In fact, sexual behaviors in children have been shown to *decrease* as age increases. In other words, school-age children overall tend to engage in a decreasing number of sexual behaviors than two- or three-year-olds (Carpenter & DeLamater, 2012; Friedrich et al., 1998).

Part of the problem for parents may be the lack of opportunities to read about and discuss childhood sexuality. After all, it's not the most common topic of parental discourse at the playground! Therefore, a common misperception in most Western societies persists that humans are not "supposed" to be sexual until they enter puberty. This is simply not true. Although the interpersonal, emotional, and psychological components of sexuality begin to play a larger role during adolescence, many of the *physical* responses were always there. Studies examining childhood sexual experiences have found similar and very common sexual behaviors among most children between the ages of 2 and 12 (DeLamiter & Friedrich, 2002; Friedrich et al., 1998; Kellogg, 2010). However, lack of awareness of these norms among the general population means that parents sometimes perceive their children's sexual behaviors as abnormal, when they are not. Consequently, they often avoid talking to other parents or professionals about them out of fear of embarrassment or denial. However, if they did discuss them, they would find great comfort in the fact that most parents have experienced and dealt with similar issues and situations relating to childhood sexuality. Let's discuss some of these issues.

Childhood Masturbation

As noted earlier, the human body is designed to be sexual; it is normal for stimulation of sexual body parts to feel good. This is part of the evolution process to ensure our survival as a species. It should not be surprising that children begin to touch their genitals as soon as their motor development allows them to do so and to respond to those touches in sexual ways. Both male and female infants are capable of sexual arousal in response to stimulation soon after birth. Male infants have erections and female infants vaginally lubricate within the first 24 hours after birth (Carpenter & DeLamater, 2012). Five-month-old male infants have been found to have between five and forty erections per day.

Since You Asked

1. Is it unhealthy to masturbate at an early age?

In most Western societies, parents and other caregivers have become increasingly accepting of children's masturbation behaviors, but many still express surprise, shock, anger, and concern when they see it in their own children (Flanagan, 2011). Typically, a concerned parent will mention the behavior to the child's pediatrician or the family doctor and will be reassured that the behavior is normal and harmless. However, children need to learn that cultural expectations dictate that such behavior is personal and private, and should not be done in public. "Self-Discovery: Masturbation in Young Children" offers advice to parents or caregivers of young children who are masturbating.

The first signs of masturbation often appear in children between the ages of two and three (Reese, 2007; Viglianco-VanPelt & Boyse, 2009). By ages three to five, children understand that genital self-stimulation is pleasurable, and by this age, they are capable of experiencing orgasm through masturbation (Viglianco-VanPelt & Boyse, 2008). Based on adult memories of childhood sexual experiences, approximately 43% of men and 34% of women report exploring their genitals between the ages of six and ten, and 6% of men and 7% of women recall masturbating to orgasm during this age range (Larsson & Svedlin, 2002). Research relying on mothers' observations has shown that among children between two and ten years of age, 30% of boys and 21% of girls engage in hand-to-genital masturbation (Carpenter & DeLamater, 2012; Friedrich et al., 1998).

Perhaps the most important point frequently made in the research about childhood sexuality is that virtually all child development specialists, psychologists, physicians, and other professionals consider masturbation a natural, harmless, and extremely common childhood behavior. Furthermore, childhood masturbation itself has not been linked to any sexual problems in later life. On the contrary, some evidence suggests that children who are punished or made to feel shameful for their natural sexual feelings may be more uncomfortable with their bodies and less able to enjoy their sexuality as they get older (Zamosky, 2011).

Note that childhood masturbation *can*, in some cases, be a sign of a problem (see Viglianco-VanPelt & Boyse, 2009). Here is a list of physical or psychological concerns that may be related to a child's (otherwise normal) masturbatory behaviors:

- If the child seems to have an overly early understanding of the two-sidedness of the sex act (could be an indication of abuse).
- If the activity becomes compulsive and interferes with other normal activities or the child cannot be distracted easily from the masturbation.
- If the child simulates intercourse with another child (also implies too much knowledge that could be associated with abuse).
- If any penetration with another child is involved.

Self-Discovery

Understanding Masturbation in Young Children

Occasional masturbation is a normal behavior of many infants and preschoolers. Up to one-third of children in this age group discover masturbation while exploring their bodies, just as they explore all the parts of their bodies eventually. They find it feels good to touch their genitals and sometimes continue to do so. Genital or urinary infections do not cause masturbation; they cause pain or itching, inciting the child to scratch the area, but this is different from masturbation.

By age five or six, most children have learned that genital touching is not to be done in public places, and they masturbate only in private. Masturbation becomes increasingly common in puberty in response to the surges in hormones and sexual drive that occur at that time.

How should a parent deal with masturbation in their young children? Here are some suggestions:

1. Once your child has discovered and enjoys masturbation, it is not realistic to eliminate it entirely. A reasonable goal is to control where it occurs. Perhaps limit it to the bathroom or bedroom. Tell your child that it is something that should be done only in private. Don't ignore it completely; if you do, your child may think it's acceptable anytime and anyplace, which may result in criticism by adults and chiding by other children.
2. Ignore masturbation at naptime and bedtime. Keep in mind that this is often a self-comforting activity.
3. When masturbation occurs outside the child's bedroom, try distracting the child with a different activity. If this fails, remind the child that you know it feels good, but it is not allowed in front of other people.
4. Discuss your views on the behavior with others who may care for the child so that everyone is on the same page. Consistency among caregivers is a key element to success for all child behavior management.
5. Call the child's physician if you suspect that the masturbatory behavior may have been learned from someone else, if your child tries to masturbate others, or if your child continues to masturbate in front of others.

Masturbation does not cause physical injury to the body, later promiscuity, or sexual deviance later in life. Masturbation in children is normal. Masturbation is not a problem or considered excessive unless it is deliberately done in public places after the age of five or six. Masturbation generally leads to negative emotional consequences only if adults overreact to it and make it seem dirty or forbidden.

SOURCE: Adapted from Steele (2005).

- If the activity is intrusive or painful for the child.
- If the activity increases much above the original level, perhaps indicating the child is stressed about something and is trying to comfort him- or herself.
- If there is mouth to genital contact between your child and another child.
- If you feel your child is particularly unhappy or sad.
- If the behavior seems to be accompanied by trauma to the area from scratching or rubbing (your child may be infected with pinworms or have a bladder infection).

Childhood Sex Play

Although it is very common, masturbation is only one of many sexual behaviors normally seen in children (Flanagan, 2011; Kellogg, 2010). Not only are children curious about their own bodies, but beginning at about three years of age, they become very interested in other people's bodies as well, an interest that is evident until about age seven. This curiosity probably stems from children's developing cognitive awareness of their own gender and the realization that boys and girls are of different genders. They learn quickly that the most obvious difference between boys and girls is their genitals (Kellogg, 2010; McKee et al., 2010). Their natural curiosity leads them to want to see other children naked and engage in sexual games such as playing "doctor." Children also typically display a keen interest in adults' sexual bodies and will often make efforts to see their parents and other adults naked (Kellogg, 2009).

As with masturbation, when parents, teachers, or other caregivers discover that children are engaging in mutual sex play, they often become upset, worried, or angry (Flanagan, 2011; Kellogg, 2010). Their concerns typically revolve around the effect such

Since You Asked

2. I recently caught my four-year-old daughter playing "doctor" with the neighbor boy (both naked). Is this a sign of a problem, or should I not be concerned?

play will have on the children's emotional, psychological, and sexual development later in life. However, no evidence exists that early, normal sex play produces negative consequences for later sexual adjustment (Okami, Olmstead, & Abramson, 1997).

What should parents and caregivers do if they discover that their child is engaging in sex play with other children? Most childhood development experts recommend approaching the situation gently, avoiding any extreme emotional responses. Here are some general guidelines to help you handle this situation:

1. **Try not to over-react.** By understanding in advance that it is normal for children to explore their bodies and experiment with sexual play, you can avoid viewing the situation from your adult point of view and see it from theirs. This will allow you to react without shock and anger that might scare and confuse the children. This knowledge will give you the opportunity to take a moment to think about it. Then you can address the situation and work to teach the children what is acceptable play and what is not.
2. **Redirect the play.** Explain that playing without clothes is not OK. Say "Let's get you dressed," and steer them to a different, fun activity. Later, when the time is right, discuss the issue with the child and explain in simple terms about appropriate play and acceptable nudity, including which areas of the body are private (often referred to as "bathing suit areas") in terms of cultural expectations. If certain children continue with the naked play, increased monitoring of them may be in order.
3. **Become educated.** Consider reading one of the many good books that are available about early childhood sexual development. Sometimes sex play indicates that your child may have reached a level of maturity to learn more about sexuality. Help your child feel comfortable to talk to you and ask questions. Discuss issues and answer questions openly and in a matter-of-fact way. Children become increasingly curious about issues to which they cannot find explanations. This is also an opportunity for you to educate your child about good and bad touches from others and appropriate private versus public behaviors.
4. **Be alert to warning signs.** A preoccupation with sexual matters or sex play that seems too precocious in a child may be warning sign of other problems such as abuse. In addition, if the sex play involves a young child and a child who is significantly older, this may be problematic. Many sources of help and advice are out there to help you including your family physician, your child's pediatrician, and family and child therapists.

Although some exceptions exist (which we will discuss next), the vast majority of the sexual activities of children are an integral part of normal, healthy early development and should not be cause for concern. Moreover, many of these sexual behaviors decline or change on their own as children discover society's expectations of activities that are public versus private (Friedrich et al., 2001). Table 12.1 indicates the frequency with which children of varying ages engage in common sexual behaviors within view of their parents.

Now that we have discussed how most childhood sexual behaviors are part of normal and healthy child development, it is important to touch on behaviors that may not be normal or may signal unhealthy situations in a child's life. Again, these are exceptions, but an awareness of them on the part of all adults, whether parents or not, is important for protecting and helping children who may be experiencing difficulties.

A curiosity about other children's sexual bodies is a normal part of early childhood development.

When Childhood Sexual Behavior Is *Not* "Normal"

As parents and other caregivers become more relaxed and comfortable with healthy childhood sexuality, they must also be alert to certain forms of sexual acting out that may be a red flag that a problem exists relating to

Table 12.1 Frequency of Children's Observable (Not Hidden) Sexual Behaviors

The percentages in the table represent a sample of observations made by over 800 mothers of children between two and twelve years of age using a survey device called the Child Sexual Behavior Inventory. All of the children were screened to ensure the absence of any sexual abuse. Keep in mind that these are *parental* observations, so private behaviors are not reflected in the data. You can see that as they grow, children learn which behaviors are socially acceptable and which should be private.

	Boys Exhibiting Behavior (%)			Girls Exhibiting Behavior (%)		
Behavior	**2–5 Years**	**6–9 Years**	**10–12 Years**	**2–5 Years**	**6–9 Years**	**10–12 Years**
Touches own breasts	42.4	14.3	1.2	43.7	15.9	1.1
Touches genitals at home	60.2	39.8	8.7	43.8	20.7	11.6
Touches genitals in public	26.5	13.8	1.2	15.1	6.5	2.2
Attempts to see others undressed	26.8	20.2	6.3	26.9	20.5	5.3
Masturbates with hand	16.7	12.8	3.7	15.8	5.3	7.4
Shows genitals to adults	15.4	6.4	2.5	13.8	5.4	2.2
Shows genitals to children	9.3	4.8	0.0	6.4	2.4	1.1
Exhibits great interest in opposite sex	17.5	13.8	24.1	15.2	13.9	28.7
Touches other children's genitals	4.6	8.0	1.2	8.8	1.2	1.1
Masturbates with toy or object	3.5	2.7	1.2	6.0	2.9	4.3
Attempts to look at pictures of nude people	5.4	10.1	11.4	3.9	10.2	3.2
Touches adults' breasts or genitals	7.8	1.6	0.0	4.2	1.2	0.0
Tries to have intercourse	0.4	0.0	0.0	1.1	0.0	0.0

SOURCES: Friedrich (1998); see also, Carpenter & DeLamater (2012); Kellogg (2010).

sexual abuse (see Chapter 13, "Sexual Aggression and Violence," for more on this issue). But how are caregivers to know the difference? Unfortunately, the answer to that question is not as clear as we might wish (Carpenter & DeLamater, 2012; Flanagan, 2010). The best way parents and other adults in children's lives can care for and protect them is to become as educated as possible about sexual development and as skilled as they can in communicating with children. These two abilities combine to allow us to observe children intelligently and analytically for signs of problems and to talk and listen to them when they attempt to communicate with us in their own childlike ways.

Since You Asked

3. If a child between the ages of five and seven acts out sexual acts, specifically lying down on top of another child and imitating movements of intercourse, does this indicate a problem, or is it part of normal childhood curiosity?

Usually, normal childhood sexual behaviors and problematic sexual behaviors can be distinguished by the pervasiveness and intensity of the activities and by linking them to other troubling aspects of the child's behavior. For example, an excessive preoccupation with genital self-stimulation or the touching of other children, doing so in public, or refusing to stop or alter the behavior when told to do so may indicate that a child is feeling emotionally deprived or is imitating behaviors others may have inflicted on him or her. In such cases, "normalcy" boils down to a matter of degree and the context of the behavior. Also, certain sexual behaviors that signal a child knows "too much" about sexual acts should receive extra attention and concern. These include acts of oral sex; imitations of adult sexual acts; sexual contact with animals; sexual behaviors aimed at adults that cause discomfort for the adult involved; and sexual actions that produce fear, anxiety, or shame on the part of the child (Kellogg, 2009; "Sexual Development," 2009).

Another way to distinguish between healthy sexuality in children and sexual acting out that may be a sign of a problem is to consider the sexual behavior in the larger context of other problems the child may be manifesting. For example, when a young child appears to be engaging in troubling sexual behaviors, such as sexual behaviors that seem aggressive or sex play with children who are more than four or five years apart, caregivers should think about how the child seems to be adjusting in other areas of his or her life. Consider whether the child is showing an unusual interest (or

Table 12.2 Examples of Sexual Behavior Problems in Children

Behavior Type	Examples
Solitary	Behaviors that cause emotional distress, anxiety, or physical pain
	Repeated penetration of vagina or anus with an object or digit
	Behaviors that are persistent, and child becomes angry if distracted
	Behaviors associated with conduct disorders or aggression
	A variety of sexual behaviors displayed frequently or on a daily basis
Involving other persons	Sexual behaviors involving children four or more years apart in age
	One child coercing another into participating
	Explicit imitation of sexual intercourse
	Oral-genital contact
	Asking an adult to perform a specific sexual act

SOURCE: Kellogg, N. (2010). Sexual behaviors in children: evaluation and management. *American family physician, 82*(10), 1233–1238.

avoidance) of all things of a sexual nature; experiencing uncharacteristic sleep problems; withdrawing from friends or family; resisting going to school; portraying sexual molestation themes in drawings, games, or fantasies; engaging in excessive masturbation or unusual levels of aggression; displaying overly adult-like sexual behaviors or vocabulary; or expressing suicidal thoughts, ideas, statements, or behaviors (Putnam, 2003; "Sexual Development," 2009; "Sexual Abuse," 2002). Table 12.2 lists specific behaviors that may be warning signs to parents and care-givers that a sexual development problem may exist.

Sexual behaviors themselves form only one component of many factors that may combine to indicate a problem. The point here is that parents and caregivers need to make an effort to know the child and to view his or her sexual behaviors in the larger context of overall adjustment to better gauge if they might be a sign of a problem. However, the bottom line is that if an emotional problem or sexual abuse is suspected, professional help should be sought to both uncover and resolve the problem and to ensure that the child is safe and healthy.

How Children Learn about Sex

Virtually everyone, including most older children, agrees that parents should be the main source of sexual information for their children. However, this attitude assumes that parents are both willing and able to provide adequate and accurate information so that children will have the tools they need, when they need them, to deal with a complex sexual world. A vast majority of parents take on this as best they can. One 2011 study showed that 82% of parents say they are talking to their children about sex, but only 60% have discussed birth control and 57% are less than comfortable discussing becoming sexually active (Seicus, 2011). Other research evidencee suggests that many parents feel unprepared, factually or emotionally or both, to take on the role of sex educator for their children and choose to hand over the task to various other sources, including schools, the media, and peers (Robinson, 2012; Wilson et al., 2010).

Since You Asked

4. What prevents parents from talking to their kids about sex? My parents never had "the talk" with me, and I had to find out from friends, or even worse, boyfriends. Now, two abortions later, I'm finally taking this class.

Although understandable, parents' avoidance of responsibility for sex education may be contributing to an overall increase in young people's beliefs in sexual myths, poor sexual decision making, and reduced sexual knowledge among children and teens (Kellogg, 2010). Only half of all teenagers report having had a meaningful talk about sex with their parents during the preceding year, and in one study, while most parents (72%) claimed to have had a "good talk" with their teens about sex, only 45% of those teens felt that such a talk had occurred (M. Edwards, 2000). It appears that parents *think* they are providing the information their children need, but the children disagree.

Table 12.3 Teens' Sources of Sex Information

This table lists the percentage of 157 teenage boys and girls who reported receiving information from various sources on each of the sexual topics listed.

	Sources of Information for Teenage Boys and Girls				
Sexual Topic	School (%)	Peers (%)	Media (%)	Parents (%)	Professionals (Doctors, Nurses, etc.) (%)
Reproductive system	89	26	33	43	2
Father's part in conception	78	25	25	34	1
Menstruation	82	32	22	43	1
Nocturnal emissions	71	30	17	14	2
Masturbation	48	37	22	10	<1
Dating	57	48	26	44	<1
"Petting"	39	49	15	11	0
Sexual intercourse	82	46	35	34	<1
Birth control	66	30	29	29	1
Birth control use	27	15	8	16	1
Pregnancy risks	82	26	35	36	3
STIs	89	22	40	30	3
Love and/or marriage	48	29	33	43	2
Morality of premarital sex	42	27	19	32	3

SOURCE: Table, "Children's Sources of Sex Information," in "Does Source of Sex Information Predict Adolescents' Sexual Knowledge, Attitudes, & Behaviors?" by C. L. Somers and J. H. Gleason, *Education*, 121, (2001), pp. 674–681. Copyright © 2001. Reprinted by permission of Project Innovation, Mobile, AL, and C. L. Somers.

In another study, researchers asked boys and girls between the ages of 14 and 18 where they received their information about sex (Somers & Gleason, 2001). The participants were asked where they had received information about various sexual topics ranging from sexual anatomy (the reproductive system) to masturbation to sexually transmitted diseases. Their information sources were school, peers, media, parents, and professionals (doctors, nurses, counselors, etc.), in that order. For virtually every topic, school was the most common source of their current knowledge (see Table 12.3).

Although they might avoid the task, parents do *recognize* the value of talking to their children about sex. Research shows that parents are aware of the importance of talking to their children about sex, but many fail to do so (Wilson et al., 2010). One national survey found the vast majority of parents (82% to 96%) agreed that it is very important for parents to talk with their children about such issues as reproduction, becoming sexually active, pregnancy and STI prevention, and saying no to sex. And the same percentage said they feel comfortable talking to their kids about sex. However, the survey found that only about half of the parents had actually had such discussions with their children by age ten, and only 60% to 70% had done so by age twelve (Lake Snell, Perry et al., 2002). Why the gap between what parents feel they *should* do and what they actually do? Part of the reason may lie in the fact that many parents feel unsure that they possess adequate and correct knowledge about sex and about exactly what information to impart to their children (Elliot, 2010). This is especially true for parents of younger children. Therefore, as noted in Chapter 1 of this text, if parents (and prospective parents, such as most of you) first work to educate themselves as thoroughly as possible about all aspects of sexuality, they will be better prepared to discuss sex with their children at all stages of development. "In Touch with Your [Children's] Sexual Health: Guidelines for Talking to Children About Sex" offers guidelines for parents and caregivers for communicating effectively with children about sex.

In Touch with Your Children's Sexual Health

Guidelines for Talking to Children about Sex

The Sexuality Information and Education Council of the United States (SIECUS) offers the following guidelines to help parents and caregivers communicate with their children effectively about sexuality.

1. **You are the primary sexuality educator of your children.** They want to talk with you about sexuality and to hear your values.
2. **Find "teachable moments."** Make use of TV shows even if you believe they send the wrong message. Say, "I think that program sent the wrong message. Let me tell you what I believe."
3. **Remember that it is okay to feel uncomfortable.** It is often hard to talk about sexual matters. Relax and tell your children you are going to talk to them because you love them and want to help them.
4. **Don't wait until your children ask questions.** Many never ask. You need to decide what is important for them to know and then tell them before a crisis occurs.
5. **Be "ask-able."** Reward a question with "I'm glad you came to me." It will teach your children to come to you when they have other questions.
6. **Become aware of the "question behind the question."** The unspoken question, "Am I normal?" is often hiding behind questions about sexual development, sexual thoughts, and sexual feelings. Reassure your children as often as possible.
7. **Listen, listen, listen.** Ask them why they want to know and what they already know. That may help you prepare your answer.
8. **Remember that facts are not enough.** Share your feelings, values, and beliefs. Tell your children why you feel the way you do.
9. **Talk about the joys of sexuality.** Tell your children that loving relationships are the best part of life and that intimacy is a wonderful part of adult life.
10. **Remember that you are telling your children that you care about their happiness and well-being.** You are also sharing your values. This is one of the real joys of parenthood.
11. **Know what is taught about sexuality in your schools, faith communities, and youth groups.** Other groups can help. It is often helpful when professionals lead talks.

SOURCE: "Eleven Guidelines for Talking to Children about Sex," from *Families Are Talking. Special Supplement to SIECUS Report 1*, 2000.

Sexuality in Adolescence

12.2 Review the complex issues associated with sexual development during adolescence including sex education, puberty, teen intimate relationships, teen dating, sexual behavior in adolescence, teen pregnancy, and the issue of STIs among adolescents.

adolescence
The period of life between the ages of approximately 10 and 19 years, often a tumultuous, self-identity seeking stage of life.

As childhood draws to a close, an entirely new world (or perhaps it is better described as a "new galaxy") appears in sexual development: the galaxy known as adolescence. This phase in a child's life begins when the many physical and hormonal changes of puberty combine with the psychological, emotional, social, and cultural expectations that accompany the stormy, turbulent, and often wondrous transition from child to adult. **Adolescence** is usually defined as the period of life from 10 to 19 and is often divided into stages: about 10 to 13 years old is considered early adolescence ("tweens"), 14 to 15 years is considered middle adolescence, and 16 to 19 years is considered late adolescence (Windle et al., 2008). Adolescence is a self-identity seeking time of life. It is a time of psychological, emotional, and sexual changes as teens awaken sexually and experience the beginnings of romantic and sexual relationships and the complex feelings that accompany the experience of love.

If anything is more challenging for a child than the changes of adolescence itself, it is the culture's response to them. Western cultures, and especially U.S. society, often appear bewildered about how to nurture and support teens as they maneuver through those often-awkward years. Nevertheless, on their way through adolescence, teenagers must undertake to avoid many common sexual pitfalls, such as accidental pregnancy, sexually transmitted infections, and sexual coercion, to name just a few of the most common and most serious.

Adolescents probably receive a greater number of mixed and confusing messages about sex than any other age group. Some of the confusion no doubt stems from two opposing approaches to a long-standing debate about what to "teach" teens about sex (Levine, 2011). On the one side are people who suggest that adolescents should be encouraged to practice abstinence and, therefore, they should have little need for education about sexual behavior, contraception, or STIs. Often, these teens are encouraged to take what is known as a "virginity pledge." Consequently, the argument is, "If they're not going to do it, why teach them about it?" The effectiveness of this approach is clearly *not* supported by research (Hubler, 2004; Sparks, 2005).

"Abstinence-only" education was tried in the early 2000s and it failed to alter risky teen sexual behavior. School systems that tried it saw the rate of teen pregnancy, unprotected sexual intercourse, and STIs remain unchanged or increase. In fact, 50% of teens who take a "virginity pledge" do not keep it, and when they break the pledge, they are far less likely to use contraception compared to "non-pledgers" (Rosenbaum, 2009). In one large study to evaluate the effectiveness of abstinence programs, researchers evaluated over 2,000 students in four different government-approved and funded abstinence education programs (see Trenholm et al., 2007). The study found no difference in sexual activity, sexual risk-taking, use of condoms, or knowledge of STI prevention. Two examples of this study's findings are shown in Figure 12.1.

This makes sense when you consider that during the teen years, young adults are becoming analytical, thinking beings, and moving toward independence. Without meaningful, relevant information, they are unable to make educated, intelligent decisions about their sexual choices. In a way, this would be similar to finding your way to a desirable, yet unknown, destination in the dark, without a map. Think of all the chances for becoming lost!

The research is quite clear on this issue: Providing teens with comprehensive education and information about sexual behavior is a *solution* to helping adolescents navigate through the minefield of becoming sexual adults. When teens are well-educated about sex, their sexual choices are smarter and healthier and the negative outcomes associated with teen sexuality (especially unwanted pregnancy and STIs) decrease. A great many studies have supported these conclusions (i.e., Carter, 2012; Chin et al., 2012; Malone, 2011; Stanger-Hall & Hall, 2011).

Summarizing the complexities of adolescent sexuality in one brief section is a challenge. We will touch on some of the more important issues of this developmental

Figure 12.1 Ineffectiveness of Abstinence-Only Education Programs

Two examples of the lack of effectiveness of four different abstinence-only education programs instituted in selected public schools.

SOURCE: From "Impacts of Abstinence Education on Teen Sexual Activity, Risk of Pregnancy and Risk of Sexually Transmitted Diseases," by C. Trenholm et al., *Journal of Policy Analysis and Management*, Spring 2008, Volume 27, Issue 2, pp. 255–276. Copyright © 2008 by the Association for Public Policy Analysis and Management. Reprinted by permission of APPAM.

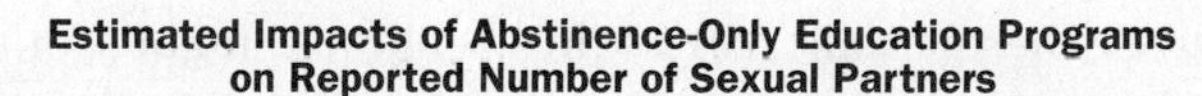

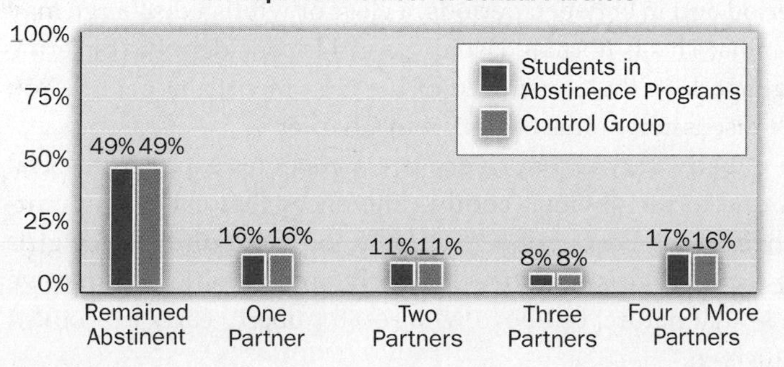

Estimated Impacts of Abstinence-Only Education Programs on Unprotected Sex, Last 12 Months

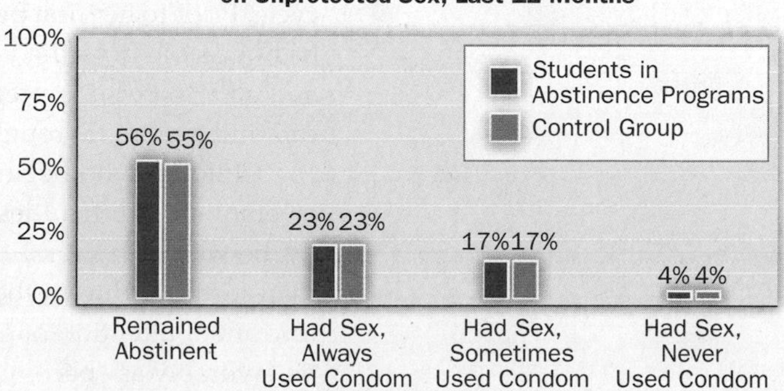

stage as we review the physical changes of puberty, the surge of sexual interest in teens' lives, the many sexual *firsts* that occur during this time, the use (or nonuse) of contraception, unwanted pregnancy, and teens' significant risks for contracting and transmitting STIs.

Puberty

puberty
A natural change in a child's body during early to middle childhood, during which secondary sexual characteristics develop and boys begin to produce sperm and girls begin menstruation.

pituitary gland
A gland in the brain that, at the onset of puberty, releases hormones necessary for the physical changes of puberty.

gonadotropin
A hormone released by the pituitary gland that signals the testes to release testosterone and the ovaries to release estrogen.

Puberty is different from adolescence in that puberty refers more to physical and biological changes, whereas adolescence is focused on social relationships. A nine-year-old who is showing signs of puberty (discussed shortly) would not yet be considered an adolescent.

You are already personally familiar with the physical changes that occur during puberty; you have experienced (or are experiencing them) yourself. Outward signs of puberty normally unfold between the ages of 11 and 16 for boys and 8 and 16 for girls. However, the exact age when puberty normally begins varies greatly and can be anytime within these ranges. Rarely, puberty may begin earlier (*precocious puberty*) or later (*delayed puberty*), but these are exceptions and may sometimes indicate a medical, hormonal condition requiring treatment. The onset of puberty is stimulated when the **pituitary gland** in the brain releases a hormone called **gonadotropin**, which in turn signals the testicles to release testosterone and the ovaries to release estrogen. Although these are the primary hormones influencing sexual development in boys and girls, small amounts of female hormones are present in boys and small amounts of male hormones are secreted in girls. The secretion of testosterone and estrogen are the primary causes of the internal and external physical changes associated with puberty.

Since You Asked

5. I started puberty very late (around age 15). All my friends were already getting their periods and breasts and everything. It was very embarrassing. Why do some people start puberty so late?

Internal biological changes in both sexes begin slightly before any of the familiar external, more obvious signs of puberty appear. In boys, as the testicles mature, they begin to manufacture and release testosterone. The testicles enlarge during puberty, and internal structures, such as the prostate gland and seminal vesicles, mature. The combination of these changes prepares the boy's body to manufacture sperm cells, produce semen, and ejaculate. In other words, during puberty boys become fertile and able to create a pregnancy through unprotected intercourse. However, it is common for a boy's early ejaculations, whether through masturbation or wet dreams (which typically begin at this time), to contain few sperm cells. However, this is no guarantee that he is not fertile during this time.

menarche
The beginning of menstruation during puberty; a girl's first period.

Internally, a girl's ovaries mature during puberty, and increases in estrogen secretion cause the building up of the uterine lining, leading to **menarche** (pronounced MEN-ar-ky), her first menstrual period. The age of menarche varies considerably among girls and may occur normally any time between 8 and 16 years of age; the average age is between 11 and 14 ("Menarche," 2013). Within a year after a girl's first period, hormonal changes also cause the ovaries to begin to release mature eggs, or ova. At first, the release of an ovum and the accompanying menstrual period may be very irregular both in frequency and amount of fluid. For example, a girl may have her first period and then no period for two or three months, then another period and skip a month, and so on, for the first year or so. Soon, however, barring illness or unhealthful levels of dieting or exercise, her periods will begin to follow a regular monthly cycle. Prior to her first period and in between periods, a clear or whitish discharge may be noticeable from the vagina. This is normal. Usually, a girl is considered to be fertile and able to become pregnant with the occurrence of her first menstrual period. You can find a more thorough discussion of menstruation in Chapter 2.

Children who begin puberty very early or very late sometimes experience social or emotional difficulties due to the obvious bodily differences that exist, temporarily, between them and their peers. Early puberty tends to be more difficult for girls than for boys due to the fact that girls who develop early are more likely to receive unwanted attention of a sexual nature, but boys who reach puberty early are looked on favorably as "becoming men."

Barring any physiological abnormalities, nearly all boys and girls will complete the physical changes of puberty by their 17th birthday. External physical changes, referred to as the development of **secondary sexual characteristics**, vary in timing and in sequence among boys and girls. Many teens who are entering, are in the midst of, or have recently completed this transition wonder what is considered "normal" development during puberty. To help answer this question, "In Touch with Your Sexual Health: The Physical Changes of Puberty" (see following page) outlines the physical process of the observable bodily changes associated with puberty.

secondary sexual characteristics
Physical changes not biologically related to reproduction that occur during puberty.

Romantic Relationships in Adolescence

The biological changes of puberty lead to a flood of many new emotions and attitudes (Blakemore, Burnett, & Dahl, 2010; Thomas, 2011). Typically, the most noticeable and powerful of these attitudes is the newly discovered interest in romantic relationships and sex. As puberty and adolescence occur, nature and nurture team up to facilitate adolescents' romantic and sexual feelings. The nature side relates to the hormonal and other biological changes just discussed. In addition to producing the physical developments of puberty, hormones also appear to affect cognitive factors that create sexual feelings and sexual attractions to others.

Biology is only part of the story, however. The social and cultural effects on adolescents—the nurture side—must not be underestimated. Adolescence is perceived by most societies throughout the world, and by adolescents themselves, as one of the most important stages of life: the passage from childhood to adulthood. Adolescence is a time of individuation, when young people are moving out of the orbit around their parents and becoming individuals in their own right, with their own personal beliefs, values, goals, and responsibilities. It is during the teen years when a sense of self develops, when people attempt to discover their identities and roles in life, when they begin to answer the question, "Who am I?"

Adolescence is also the stage in life when most people become sexually active. Part of the larger questions of overall personal identity is the more specific question of, "Who am I as a sexual person in relationships with others?" Most of the research on adolescent sexuality has focused on heterosexual intercourse and its accompanying consequences of accidental teen pregnancy and sexually transmitted infections. We will discuss these important issues shortly. However, teenagers are not, as many people believe, preoccupied with sex exclusively. Researchers are only now discovering what adolescents have known all along: that teen *romantic* relationships are not trivial instances of "puppy love" but constitute a very important and meaningful part of their lives, and that a great deal of teen sexual activity occurs in long-term, loving relationships (Blakemore, Burnett, & Dahl, 2010; Kaestle & Halpern, 2007).

Romantic partners become the most important relationships in adolescents' lives.

The potentially *negative* consequences of teen sexual activity do not exist independent of the interpersonal relationships in which the sexual behavior occurs. Dating and romantic relationships may be the most important interpersonal connections in the social lives of heterosexual adolescents, but until recently, research about this side of teenagers' sexual development has been virtually ignored (Furman, 2002; Kaestle & Halpern, 2007). Over half of all adolescents report having been involved in a romantic relationship during the previous 18 months, and by age 16, teens are socializing more frequently with romantic partners than with parents, siblings, or nonromantic friends (Furman, 2002; Patrick, Maggs,

In Touch with Your Sexual Health

The Physical Changes of Puberty

Puberty in Girls

Average Age*	Physical Features	Description of Changes
8–10	Growth, breasts, pubic hair	Height spurt begins. Breasts are prepubertal; no glandular tissue. Usually no pubic hair.
11–12	Breasts	The areola (pigmented area around the nipple) enlarges and becomes darker. It raises to become a mound with a small amount of breast tissue underneath. This is called a bud. Girls vary a great deal in the size, shape, and coloring of breast buds.
11–13	Pubic hair, growth	Maximum growth rate is reached. Body fat continues to increase normally.
11–13	Growth	Peak height velocity (maximum growth rate) is reached. Body fat continues to increase normally.
12–14	Breasts	Breast tissue grows to varying degrees past the edge of the areola.
12–14	Pubic hair	Hair is close to adult pubic hair in curliness and coarseness. Area of pubis covered is smaller than in adults, and there are no hairs on the middle surfaces of the thighs. Menarche occurs in 50% of girls.
13–14	Breasts	Continued development of breast tissue; areola and nipple protrude to varying degrees.
13–14	Growth	End of growth spurt. Normal body fat reaches adult proportions. After menstruation begins, girls gain at most 3 to 4 more inches in height, usually less.
14–15	Pubic hair, body fat	Adult levels. It is normal for some long pigmented hairs to grow on the inner thighs. Natural body fat stabilizes.
15–16	Breasts	Usually a girl's breasts have reached their adult size by age 16.

Puberty in Boys

Average Age*	Physical Features	Description of Changes
11–14	Growth of body hair and pubic hair	Hair begins to grow on various parts of the body. (Hair can continue to spread to other parts of the body until about age 20.) Sparse growth of slightly pigmented pubic hair at base of penis around age 12.4 years. Adult-type hair spreads to the inside of thighs but not up the abdomen yet (15.3 years). Hair begins to grow on the face, underarm area, pubic area, abdomen, chest, arms, legs, and buttocks. The amount and distribution of hair can vary considerably and may be genetic.
11–15	Voice changes	As a result of increased testosterone, vocal cords become longer and thicker, and the voice deepens. Although these changes are occurring, it is not unusual for the voice to change pitch abruptly or "crack" at times. Voice change begins around 13.5 years and is completed in about a year.
12–16	External genital development	Growth of penis and scrotum often starts about age 13 and continues until adult size is reached 2–3 years later. Thinning and reddening of scrotum begins around age 12 years.
11–20	Oil glands	Oil glands in the skin become more active. This can cause acne. Many people will have problems with acne into adulthood.
10–17	Growth, muscularity	The growth spurt in boys usually occurs about two years later than in girls. About age 12.5 years, the boy's body takes on a more muscular and angular shape when testosterone causes muscle mass to increase. The greatest effect can usually be seen in the upper chest and shoulder muscles. Testosterone also causes bones to lengthen, giving young men a heavier bone structure and longer arms and legs.
12–18	Penile erections	Males have spontaneous erections throughout their lives (even as infants). During puberty, boys tend to get erections more frequently. Erection can occur with or without any physical or sexual stimulation. Though sometimes embarrassing, it is a normal, spontaneous event.

*Average ages can vary widely; one or two years earlier or later than most of these listed ages are usually considered within the normal range.

SOURCES: Adapted from Coleman & Coleman (2002); "Sexual Development" (2008).

& Abar, 2007). These romantic relationships appear to be based on more emotional involvement than mere sexual interest and are often characterized by caring, friendship, and companionship.

Understanding the romantic aspect of heterosexual teenage relationships is vitally important because it is within these relationships that most adolescent sexual activity occurs. In fact, one of the most accurate predictors of first sexual intercourse is involvement in a romantic relationship during the previous 18 months

(Furman, 2002; Patrick, Maggs, & Abar, 2007). A national survey conducted in the mid-1990s found that for 35% of males and 86% of females, the primary reason for having intercourse for the first time was feelings of affection for their partner (Michael et al., 1994).

What all this means is that when we talk about teen sexual behavior, we must also acknowledge the context of the romantic relationship in which the sexual activity is most likely to occur.

That teens tend to form romantic relationships may come as a positive and hopeful sign that sexual behavior is more than a result of purely hormone-driven physical desires. Romantic experiences appear to facilitate adolescents' personal identity development and their understanding of intimacy. However, these same relationships also form the backdrop for early sexual activities, unwanted pregnancy, and the spread of STIs. We will discuss each of these shortly.

An unfortunate and sad consequence of some adolescent dating relationships is abuse and violence. Abusive and violent relationships in adolescence mirror those of adults (as we discuss in Chapter 4, "Intimate Relationships"), but can sometimes be even more devastating because teens are new to intimacy and often very emotionally fragile. "In Touch with Your Sexual Health: Teen Dating and Violence", on the following page, explores this important problem in detail.

Gay, Lesbian, and Bisexual Students

Studies have shown that as many as 75% of gay, lesbian, and bisexual students are being harassed and bullied in middle and high school and are being made the target of prejudicial slurs such as "faggot" or "dyke" (see "Anti-Gay Discrimination," 2011). Virtually all gay and lesbian students have heard the phrase "That is SO gay!" used to describe anything a straight student perceives as bad or unpleasant. Nearly one in five gay or lesbian students reports being physically assaulted in school because of his or her sexual orientation. These and other discriminatory, victimizing events have caused gay and lesbian students to perform worse in school overall and to skip going to school more often than their heterosexual peers. One of the leading researchers in the field of adolescent romantic relationships suggests that nonheterosexual teens may follow a somewhat different dating and sexual development path during early adolescence:

> Most have same-sex sexual experiences, but relatively few have same-sex romantic relationships because of both the limited opportunities to do so and the social disapproval such relationships may generate from families or heterosexual peers. Many sexual minority youths date other-sex peers; such experiences can help them clarify their sexual orientation or disguise it from others. (Furman, 2002, p. 178)

In many parts of the United States, families of teens and school districts are making an effort to create more open, accepting, and safe environments for teens who are gay, lesbian, transgender, or questioning their gender identity or sexual orientation. Some of these efforts include ("5 Steps," 2011; Fisher, 2012):

- establishing clear policies against antigay bullying
- providing support systems for gay students such as organizations or clubs that include both gay and straight students (called gay-straight alliances)
- identifying faculty and staff who are aware of and understand gay and lesbian issues (referred to as "allies") and who students know are safe to approach for help with problems related to their sexual orientation
- including respectful nonheterosexual issues and topics within the normal school curriculum so gay and lesbian students do not feel as marginalized
- advocating school-wide for local and state antidiscrimination laws.

In Touch with Your Sexual Health

Teen Dating and Violence

The following information about teen dating violence is based on a true story.

> Brenda is fifteen and has never had a boyfriend before. She recently started dating Frank. She thinks he is so cute. Her friends all tell her how lucky she is because she has a boyfriend. At first, Brenda thought it was sweet that Frank began calling her all the time. He always wants to know whom she is with, where she is, and when she'll be home. He has told her that she was meant to be with him and him only, forever.
>
> Recently, Frank has started belittling her in front of his friends, insulting her, and telling her she is fat. He doesn't want her to spend time with certain friends of hers—he thinks they are a bad influence. He threatens to break up with her if she won't do what he says and warns that no one else will ever want her. Brenda wants to make Frank happy. In fact, she'll do anything to keep her boyfriend. She thinks this is what being in a relationship is all about.

Unfortunately, many teens have faced or are dealing with situations similar to this. Relationship violence often starts as emotional or verbal abuse but can quickly escalate into physical or sexual violence ("Sexual Health," 2013). One in ten high school students reported having experienced dating violence. Eight percent of students (females and males) have been physically forced to have sexual intercourse; females (12%) are more likely than males (5%) to report the attack. Young women experience the highest rates of rape and sexual assault. More than 1 in 5 (22%) college women have been victims of physical abuse, sexual abuse, or threats of physical violence.

The following can help you better recognize if you or someone you know may be involved in a violent relationship

Relationship abuse and violence are not about having a disagreement or getting angry over something. What *is* relationship violence?

- A *pattern* of behavior used by someone to maintain control over his or her partner.
- Verbal, physical, emotional, or even sexual abuse.
- One partner being afraid of and intimidated by the other.

How often does it happen?

- Relationship violence is the number one cause of injury to women between the ages of fifteen and forty-four.
- Seventy percent of severe injuries and deaths occur when the victim is trying to leave or has already left the abusive relationship.
- Thirty-eight percent of date rape victims are young women between the ages of fourteen and seventeen.
- Seventy percent of pregnant teenagers are abused by their partners.

Who is involved?

- Relationship violence occurs between two people who are currently or formerly involved in a dating relationship.

- The abuse can begin at a very young age, as young as eleven or twelve years old.
- Friends of the couple are usually aware of the abuse but are unsure how to stop it.

Where can it happen?

- Relationship violence can occur at school—in the hall, in the classroom, in the parking lot, on the bus—at after-school activities, at a student's workplace, at a school dance, or at a student's home.
- Does one partner tell the other that no one else would ever go out with him or her?
- Is one partner being cut off from friends and family by the other partner?
- Does one partner feel that saying no to sexual activities will result in trouble or danger?
- Does one partner feel pushed or forced into sexual activity?
- Does one partner say that the other *caused* the abuse?
- Does one partner shove, grab, hit, pinch, hold down, or kick the other?
- Is one partner *really* nice sometimes and *really* mean at other times (almost like two different people or personalities)?
- Does one partner make frequent promises to change or say that he or she will never hurt the other again? Does one partner say that the other is "making too big a deal" out of the abuse?

Answering "yes" to any of the above questions probably indicates an abusive relationship.

What can you do to help someone you think may be in an abusive relationship?

- Remember, anyone can be a victim.
- If you or someone you know might be affected by relationship violence, look for resources that can provide help. School counselors, parents, teachers, women's shelters, and clergy are all potential sources of assistance.

- Tell someone you trust and who you feel can intervene, such as a local domestic violence agency. Try to get help as soon as possible before the violence increases.
- By reaching out, you may literally save your own or someone else's life.
- In teenage dating relationships, the abuse is often public, with peers witnessing the abuse; however, the abuse can also occur in private.

What are signs of an abusive teen dating relationship?

- Is one partner afraid of the other or scared to break up?
- Does one partner call the other names, make the other feel stupid, or tell the other that he or she cannot do anything right?
- Is one partner excessively jealous?
- Does one partner tell the other where he or she can and cannot go or whom he or she can and cannot be with or talk to?

SOURCES: Adapted from Washington State Medical Association (2003), "Teen Dating Violence" (2014).

Although more and more middle and high schools are working to implement policies such as these, antigay discrimination in schools continues across the United States, and antidiscrimination policies will take time to have their full intended effects.

The Beginnings of Sexual Activity

Most teens become sexually active prior to graduating from high school. The average age of first intercourse in the United States varies by study, but appears to be between ages 15 and 17 for both boys and girls (Mahoney, 2007; Wells & Twenge, 2005). About 30% will experience their first heterosexual intercourse in grade 9, and slightly over 60% will have had intercourse by grade 12. The overall percentages of high school students who have engaged in intercourse tend to run somewhat higher for African Americans than for Caucasians and Hispanics. Table 12.4 (on the following page) summarizes some of these statistics on rates of adolescent intercourse.

As mentioned earlier, the research literature about adolescent sexual behavior has been preoccupied with heterosexual intercourse. That this has been the focus is understandable, in that coitus is the behavior primarily responsible for accidental pregnancy and the transmission of many sexual infections. Moreover, most people, and this is especially true for teens, equate the word *sex* with intercourse. As has been pointed out frequently throughout this book, however, intercourse is only one form of sexual

Table 12.4 Percentage of High School Students Who Reported Ever Having Had Sexual Intercourse, by Grade and Race

Ever Had Sexual Intercourse United States High School Students by Sex And Race 2011

		Total (all grades)	9th grade	10th grade	11th grade	12th grade
Total (males and females)	**Total**	47.4	32.9	43.8	53.2	63.1
	Black	60.0	48.2	58.4	63.6	73.9
	Hispanic	48.6	36.8	46.5	56.0	60.0
	White	44.3	27.3	38.4	50.5	62.5
Female	**Total**	45.6	27.8	43.0	51.0	63.6
	Black	53.6	35.9	53.3	58.0	70.9
	Hispanic	43.9	28.7	42.4	52.1	59.0
	White	44.5	24.2	40.0	51.0	64.8
Male	**Total**	49.2	37.8	44.5	54.5	62.6
	Black	66.9	60.3	64.0	69.3	77.3
	Hispanic	53.0	44.9	50.3	59.7	61.0
	White	44.0	30.2	37.1	49.9	60.4

SOURCE: Adapted from: CDC, "Adolescent and School Health." (2014). Youth Risk Behavior Surveillance. Centers for Disease Control and Prevention. Retrieved from ccd.cdc.gov/youthonline.

behavior. A significant number of heterosexual adolescents engage in various sexual activities either in addition to, or instead of, penis–vagina intercourse. For example, the CDC reports that statistics from 2007 to 2010 show that among girls 15 to 19 years, 41% received and 43% gave oral sex to a boy (teen oral sex will be discussed in greater detail next). For boys during that same time frame, 47% received and 31% gave oral sex to a girl (Copen, Chandra, & Martinez, 2013). Thirteen percent of teens hand engaged in heterosexual intercourse.

Teen Oral Sex

The CDC reports that, overall, about half of males and females ages 15 to 19 have engaged in oral sex (Copen, Chandra, & Martinez, 2013). These statistics are worrisome, especially because 40% of teens claim that oral sex "doesn't count" as "having sex." If you ask most adolescents, "Are you having sex?" they may answer "no," being completely honest, even though oral sex may be a regular occurrence in their lives. Furthermore, when parents and schools have attempted to teach abstinence to teens, many adolescents take the message to heart and *substitute* oral sex for intercourse, thinking this constitutes abstinence as it has been explained to them.

Since You Asked

6. In high school, a lot of my friends were having oral sex but not intercourse. Is it true that oral sex doesn't spread sexual diseases?

The high incidence of oral sex among young teens may be, in part, related to an increased awareness among teens of the risks of STIs, especially HIV and AIDS, and unwanted teen pregnancy. Many teens have the *mistaken* belief that oral sex cannot spread STIs. Young girls may be performing oral sex on their boyfriends as a "substitute" for sexual intercourse, and they are not using condoms. Many girls feel pressured to be sexually active in order to be accepted and not rejected by boys. However, they may also feel they are too young and not yet ready for intercourse. Some researchers have suggested that teens see oral sex as a way of remaining a "technical virgin" while satisfying the demands of boyfriends.

More important, however, is the fact that many teens equate oral sex with safe sex. Of course, oral sex itself carries no risk of pregnancy, but many teens believe they are safe from STI infection if they refrain from vaginal or anal intercourse and limit their insertive sexual activities to oral sex. As we discuss in detail in Chapter 8, "Sexually Transmitted Infections/Diseases," this belief is seriously misguided. Even though HIV appears to have a low rate of transmission through oral-genital contact, virtually all other STIs, including gonorrhea, syphilis, herpes, hepatitis B, and human papilloma virus (genital warts), transmit readily through oral-genital contact. Genital warts have been diagnosed in the mouth and throat, and many cases of oral herpes are the cause of outbreaks on the genitals.

There does not seem to be an "epidemic" in oral sex among teens, but this does not mean that teens are not engaging in oral sexual behaviors; they are, and they are doing so in the numbers reported earlier in this section. However, a large study of over 2,000 male and female teens found that the vast majority of teens do not experience oral sex until *after* their first intercourse and that teens typically become sexually active in various ways at about the same time. For about a quarter of teens, oral sex may precede intercourse, but for the other 75%, oral sex occurs at the same time or within six months of first intercourse (Lindberg, Jones, & Santelli, 2008). Nevertheless, public health officials are calling for renewed efforts to educate teens about the dangers of nonintercourse sexual activities, especially oral sex. These efforts are necessary, not because sexual practices have changed drastically, but because widespread casual attitudes among adolescents about oral sex represent an

Nearly half of all teens believe that oral sex doesn't "count" as real sex.

important health danger. For example, interviews with junior high and high school students in California produced the following comments (Goldston & Wong, 2003):

> "It's a way to take the relationship to a different level without going all the way. It's a step in between" (boy).
>
> "I could have done it. I'll probably do it before I go to high school. I know girls who will do it" (boy).
>
> "In seventh grade we didn't even talk about oral sex, but I think things have evolved... Kids are introduced to things at a younger age" (girl).
>
> "I had oral sex when I was fifteen with a male friend. I was wanting to fit in, so you make the choice" (girl).
>
> "It's not like having sex. It's not as big a commitment" (boy).
>
> "It's not like being all committed. It's just basically friends with benefits" (girl).

Pregnancy in Adolescence

Teenage pregnancy is a long-standing problem in the United States and in many parts of the world. The rate of births among adolescents increased steadily through the 1980s, but the trend then reversed during the 1990s and early 2000s. As you can see in Figure 12.2, the overall teen pregnancy and birthrate (for ages 15 to 19) in the United States decreased significantly between 1991 and 2008–2011 ("Sexual Health," 2013). This decline is generally attributed to a combination of the development of more effective hormonal contraceptive methods and an increased use of condoms due to fears of STI transmission. The decline does *not* appear to be due to higher rates of abortion because, as the graph shows, abortion rates for teens also fell during the same years.

Beyond condom use, various new developments in contraceptive technology have helped reduce the rate of adolescent pregnancy. For example, in addition to the daily hormonal birth control pill, which carries the risk of inconsistent use, adolescents may choose an injectable contraceptive (Depo-Provera) that provides continuous protection for 90 days, or the hormonal implant contraceptive Implanon, which provides highly effective protection from pregnancy for up to three years. Teens are taking increased advantage of **emergency contraceptive pills (ECPs)**, also known as the "morning-after pill." This method of preventing pregnancy, though not a substitute for before-intercourse birth control, is more than 80% effective in preventing

emergency contraceptive pills (ECPs)
Hormonal contraceptives that help prevent pregnancy after an unprotected act of intercourse; also known as the *morning-after pill*.

Figure 12.2 United States: Teen birthrate 1991–2011; Pregnancy rate 1991–2008; Abortion rate: 2001–2008 for ages 15 to 19.

Teen pregnancies, births, and abortions have steadily declined to their lowest rates since 1991.

SOURCE: Kaiser Family Foundation. (2013). *Sexual Health of Adolescents and Young Adults in the United States*. http://kaiserfamilyfoundation.files.wordpress.com/2013/04/3040-06.pdf.

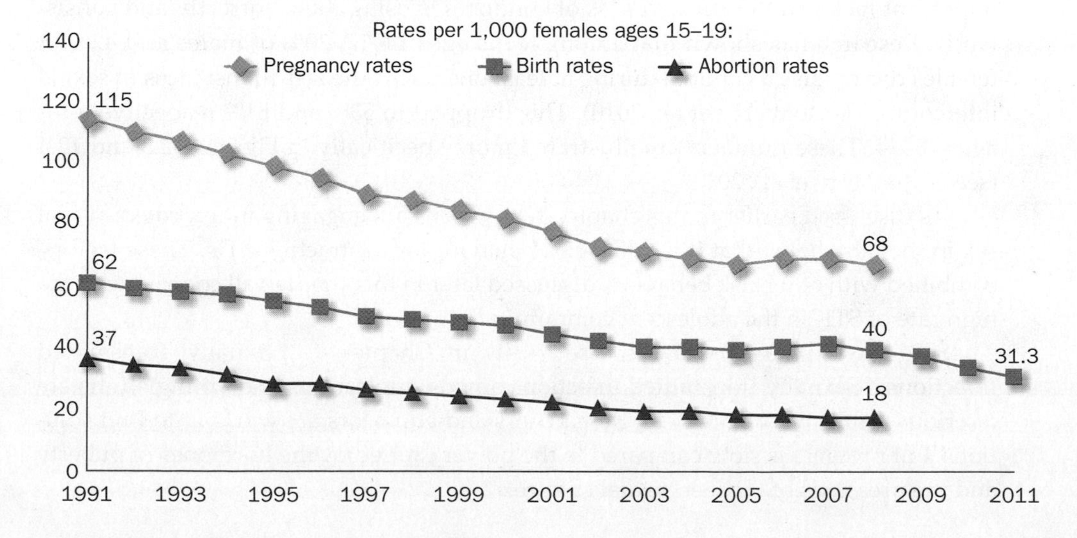

a pregnancy if taken correctly within 72 hours of intercourse. All of these methods of birth control are discussed in detail in Chapter 5, "Contraception: Planning and Preventing."

Nevertheless, over 320,000 teenage girls become pregnant each year. And the majority of teen pregnancies and births do not stem from the intention to become pregnant, nor do they lead to happy, healthy outcomes. Instead, they often begin with poor choices or emotional turmoil and end in adoption, abortion, or lead to major life difficulties. For example, 50% of teen girls who become pregnant do not receive a high school diploma by age 22, whereas 90% of girls who do not give birth in their teens receive a diploma by that age (CDC, "About Teen Pregnancy," 2011).

As mentioned earlier in this chapter, although many people believe that teaching teens about sexuality and contraception leads young people toward sexual experimentation, the opposite has been shown to be the case. In reality, **comprehensive sex education** for adolescents leads to better decision making, delay of first sexual intercourse, more conscientious contraception use, and decreases in unhealthy and risky sexual behaviors. On the other hand, **abstinence-only education** (which reached a peak in the U.S. school systems around 2007) appears to have failed to achieve any sexual health benefits and in some cases has increased teens' tendencies toward unhealthy and uninformed sexual decisions (e.g., see Boonstra, 2009; Hogben, Chessen, & Aral, 2010).

comprehensive sex education
Sexual education programs for adolescents that provide information about abstinence and about prevention of STIs and unwanted pregnancy.

abstinence-only education
A program of sexual education for adolescents that avoids teaching about contraception and STIs and stresses instead the importance of waiting until marriage to engage in sexual activities.

Furthermore, a clear link exists between teen depression and teen pregnancy; that is, teens who are depressed are more likely to experience an unintentional pregnancy. Some researchers believe that depressed teens (both boys and girls) are looking to sexual activity for acceptance and connectedness in order to feel less depressed. Others argue that the mediating factor between depression and pregnancy may be drug abuse. This position suggests that persons who use recreational drugs, especially alcohol, are much more likely to be depressed. The combination of drug use and depression then leads to poor sexual decision making (Koleva & Stuart, 2013; Mahoney, 2005).

Adolescents and STIs

The intersection of sexual activities and inconsistent condom use among adolescents leads directly to the problem of sexually transmitted infections among adolescents. Although condom use among teens has increased, the fact remains that, on average, 1 out of 5 adolescents between ages 14 and 17 did not use a condom during their most recent intercourse (Fortenberry et al., 2010). Moreover, these usage rates drop significantly in the 18–24 age group (discussed later under "Sexuality in College"). As discussed in Chapter 5, "Contraception: Planning and Preventing," the most important factor in the effectiveness of condoms is using them correctly and consistently. Research has shown that among teens ages 14–17, 20% of males and 42% of females did not use a condom during at least one out of the last 10 instances of sexual intercourse (Fortenberry et al., 2010). This dropped to 53% and 61% respectively for ages 18–24. These numbers are illustrated more specifically in Figures 12.3 and 12.4 (see Fortenberry et al., 2010).

As discussed earlier in this chapter, many teens are engaging in unprotected oral sex in the false belief that it is a low-risk behavior for contracting STIs. These factors, combined with other risk behaviors discussed later in this chapter, all contribute to the high rate of STIs in the adolescent community.

STIs are explained and discussed fully in Chapter 8, "Sexually Transmitted Infections." Sexually transmitted infections among adolescents and young adults are a serious public health problem. Researchers and educators are working to find solutions, but progress is slow compared to the power exerted by the hormones of puberty and the pressures of the teen sexual culture.

Figure 12.3 Condom Use among Adolescents During Most Recent Vaginal Intercourse

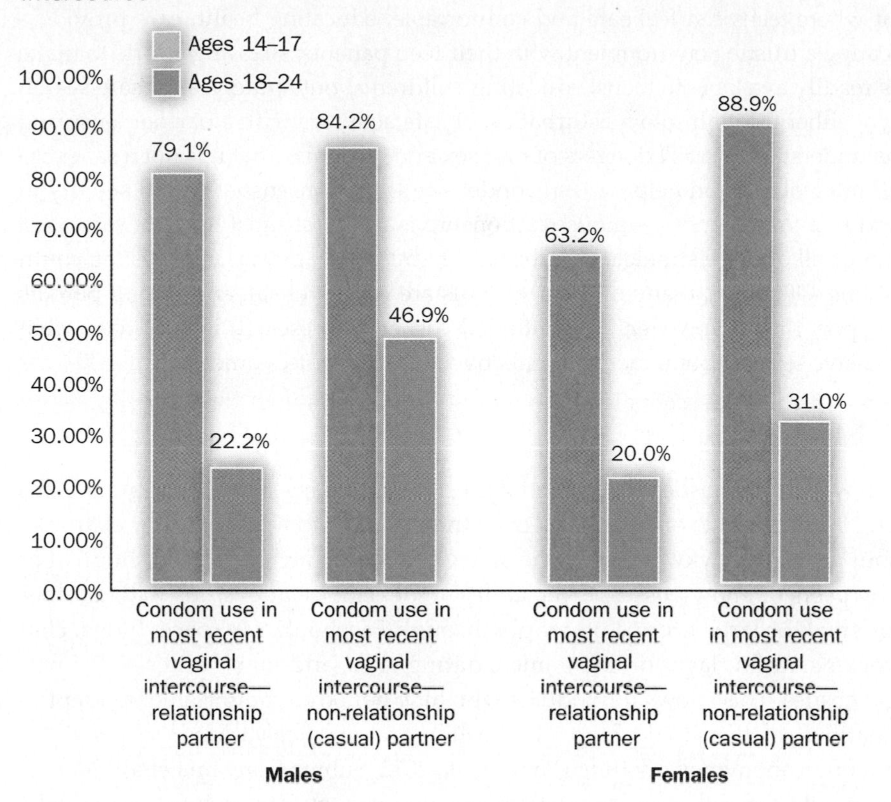

Numerous strategies for increasing STI awareness among teens have been suggested (e.g., "Campaign for Our Children," 2008). Some efforts include establishing sex and STI education programs in schools and communities for young people, while at the same time creating opportunities for educating parents about STIs and how to talk with

Figure 12.4 Consistent Condom Use among Adolescents and Young Adults During Past Ten Vaginal Intercourse Events

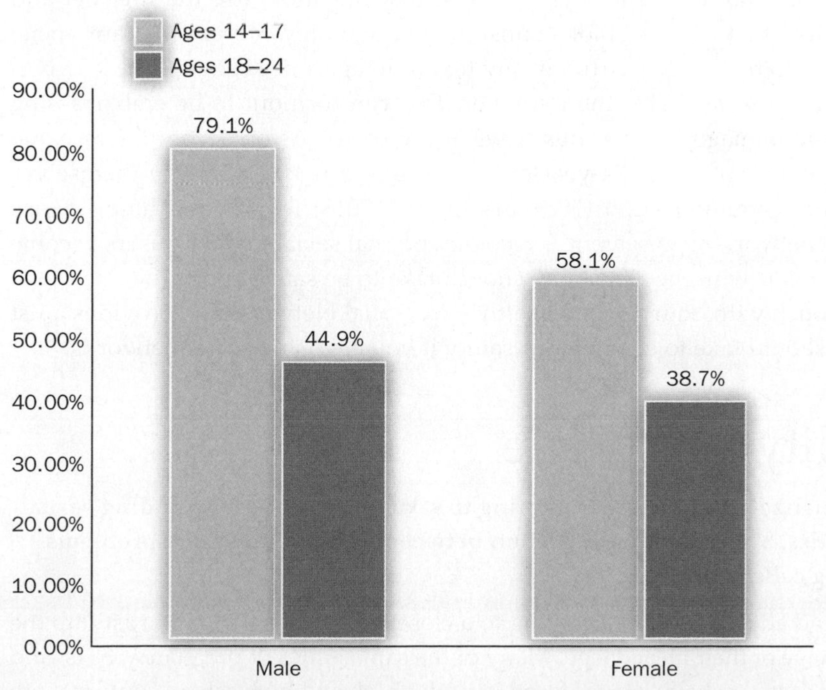

their children about them. Other efforts focus on reducing the spread of STIs among teens and young adults by establishing private, confidential avenues for STI testing and treatment where teens can feel safe and comfortable; educating health care providers about creating a trusting environment with their teen patients; stressing efforts to make condoms readily available to teens; educating children about strategies for safe sexual expression, either through solo masturbation or safe activities with a partner; ensuring that teens understand the STI dangers of oral sex and recognize that it is not a safe substitute for intercourse; and helping teens understand that consensual sexual activity in the context of a loving, monogamous relationship is the most satisfying *and* safest sex of all. Any or all of these strategies can be started by parent groups, schools, or youth organizations. Of course, many of the proposals are controversial, and not all parents would support them. However, as mentioned earlier, the research is quite clear that comprehensive sex education increases effective and safe choices among teens.

Social Networks

The digital world has drastically changed dating, relationships, and sexuality for teens and young adults (and everyone else!) over the past five to ten years. For example, think about the shy or awkward teen who, a few years ago, may have been shunned or rejected by peers but now possesses the ability to develop relationships with another person or small or large groups of people through Facebook, MySpace, blogs, chat rooms, forums, multiplayer online games, dating sites, and many other social networking websites. This allows individuals who may not be as comfortable or adept in social situations as others to develop relationships and practice various social skills in a safe, and often anonymous, setting (Papp et al., 2012; Subrahmanyama et al., 2008).

Many would argue that online relationships do not allow for "normal" social relationships to form, and some will hide behind them, thereby inhibiting their opportunities to develop face-to-face social skills that are expected in life. However, the research does not appear to confirm these fears, and in fact recent research has found numerous social benefits to such social networks (Baker & Oswald, 2010; Valkenburg & Peter, 2009).

The negatives surrounding social networks for teens and young adults are mostly related to the dangers inherent in the media themselves. As most people know, when you are socializing with someone online, you can't be absolutely sure if that person is who he or she says. Many sexual predators use the Internet and social media to find, track, and plot against their sexual prey. Moreover, many social networkers fail to realize that virtually any text or image, once conveyed into cyberspace, is likely to be available there forever. This can turn out to be embarrassing at the least and damaging in various ways at the worst. As one example, research has found that 13% of 14- to 24-year-olds have sent a naked photo of themselves through digital communication ("Sexual Health," 2013).To prevent these unwise uses of social networking, researchers recommend that social network users become intimately familiar with the steps they should take to be safe if they are to be social online. "In Touch with Your Sexual Health: Safe Social Networking" provides a list of steps you should take to ensure your safety if you are using social networks.

Sexuality in College

12.3 Summarize the main issues relating to sexuality in college including sexual activities, STIs, and the connection between alcohol and sexual problems among college students.

As puberty and adolescence are drawing to a close, most teens will be thrust into the next sexual stage of their life: college. Many of the same issues of pregnancy, STIs, and choices about sexual behavior teens faced in junior high and high school continue into

In Touch with Your Sexual Health

Safe Social Networking

Social networking sites such as those discussed above allow people to build and maintain online networks of friends. They often open up great new opportunities to share, communicate, and meet new people. However, social networking websites provide widely differing levels of protections, and it is important that you understand the safety and privacy protection of any service you use.

The following is a list of precautions you should take if you are involved or planning to be involved in social networking:

1. Maintain anonymity to protect your identity. Don't include your full name, phone number, where you work, or detailed location information in your profile or during early communications with online acquaintances. Stop communicating with anyone who presses you for this type of information.
2. Don't use your real name on your site (or anyone else's real name, either). Create a nickname or screen name that doesn't attract the wrong kind of attention or allow someone to find you.
3. Don't give information that puts you on the map. Don't mention such details as your address, school, where you work, even your town name (especially if it's a small town).
4. Don't reveal any information that gives away your age, such as your birth date or year of graduation.
5. Look at who's in a picture. If it shows friends or family members, you may be putting them at risk, too.
6. Use the e-mail system provided by the site rather than your own e-mail address to maintain your privacy.
7. Be smart about choosing profile pictures. Make sure that your images do not contain identifying information such as nearby landmarks or a T-shirt with your school or company logo.
8. Be realistic. Look for danger signs such as a display of anger, an attempt to control you, disrespectful comments, or any physically threatening or otherwise unwelcome behavior.
9. If a person becomes abusive, report it and block that person from contacting you again using the site settings.
10. If you decide to meet an online acquaintance in person, choose a safe environment. Keep first meetings short, and agree to meet in a public place during a busy time of day. Make sure somebody knows where you're going.
11. If an online acquaintance asks you for a loan or any financial information, it is virtually always a scam and you should report it.
12. Use caution when signing up on a social networking service. It is often very difficult to remove information from sites if you later regret the amount of information you have shared. It's best to be conservative in the information you share during the sign-up process; you can always add more later.
13. If your site or blog is not set to be private, anyone can visit and comment on what you're saying or posting. Often, you have to change the setting to make your blog private.
14. Periodically review who has access to your site and make changes if necessary. Friends change over time, and once trusted people may become less trusted.
15. Check out what your friends write about you in their personal sites or blogs. They may be giving out your address or real name, indicate the school you both go to, or perhaps have a photo of you on their site with a caption indicating who you are. Any of these actions may enable someone to find you. Check the comments friends leave on your blog to make sure they don't give away personal details.
16. Posting information about others is not okay—in comments, photos, or anywhere—unless they agree to share that information. When asking permission to share, make it clear who can see your site.
17. Before changing your settings to be more public, it is your obligation to seek permission from anyone you may expose. If they are not comfortable with additional exposure, remove any content about them from your site.

SOURCE: Adapted from "Socializing Online." (2008). Washington State Office of the Attorney General. www.atg.wa.gov/InternetSafety/SocializingOnline.aspx.

young adulthood and college, but the challenges become greater because college is typically a highly sexual environment. Young people who were considered children in the eyes of the culture a year earlier are suddenly, upon entering college, considered adults. They are increasingly independent of their parents' influence and must often make sexual decisions on their own. And in college, a pervasive influence is added to the sexual mix: *alcohol.* Although many high school students have experimented with alcohol and other drugs, on most college and university campuses, alcohol is an integral part of the social culture, and it is associated with sexual activities in a variety of very important ways, as we will discuss shortly.

Table 12.5 Sexual Activities in College-Age Adults During Previous Year (Males and Females Ages 18–24)*

Activity	Men (18–19)	Men (20–24)	Women (18–19)	Women (20–24)
Masturbation with partner	42%	44%	36%	36%
Received oral sex from female	54%	63%	4%	9%
Received oral sex from male	6%	6%	58%	70%
Gave oral sex to female	51%	55%	2%	9%
Gave oral sex to male	4%	7%	59%	74%
Vaginal intercourse	53%	63%	62%	80%
Insertive anal intercourse	6%	11%	—	—
Receptive anal intercourse	4%	5%	18%	23%

SOURCE: Data excerpted from Herbenick et al. (2010).

*For additional age ranges please refer to the *National Survey of Sexual Health and Behavior* at www.nationalsexstudy.indiana.edu.

Sexual Activity in College

As discussed in relation to adolescents, college sexual activity is clearly not limited to intercourse. Most college students, male and female, have also engaged in solo masturbation, sexual touching, oral sex, mutual masturbation, and a variety of other sexual activities (Herbenick et al., 2010). Table 12.5 summarizes the percentages of college-age adults who have engaged in various sexual activities over the previous year according to a 2010 national survey (see Herbenick et al., 2010).

Since You Asked

7. No one I know here has been tested for STDs. I have, but they say they don't want to or need to. How can I convince them to get tested?

With so many college students engaging in such a wide variety of sexual practices, you might suspect that sexually transmitted infections would be a major problem in this age group and on college and university campuses. Unfortunately, your suspicions are correct.

Sexually Transmitted Infections among College Students

Most high school students who go on to college have heard the message about contraception. A majority of college students have used some form of contraception, most commonly oral contraceptives (the pill). The pill and other hormonal contraceptives provide excellent protection against pregnancy, but they offer no protection from sexually transmitted infections (STIs/STDs). Although not perfect, the consistent use of male or female condoms provides the most reliable method for sexually active college students to reduce their risk of STIs. Recent research indicates that the majority of college-age adults do not use condoms when they engage in intercourse (refer back to Figures 12.3 and 12.4).

Young adults students make up about a quarter of all sexually active people in the U.S., but the 15–24 age group accounts for over half of the new STI infections annually (CDC, "STDs," 2013). Add to these statistics the fact that 25% of college students have had six or more sexual partners, and you can easily see why they have one of the highest rates for STIs of all groups in the United States. Figure 12.5 shows these alarming proportions for chlamydia, gonorrhea, and HIV infections. In addition to these STIs, other common STIs among college students are HPV (genital warts) herpes, and trichomoniasis.

Two of the most important reasons STIs spread so easily among college populations are: (1) the infections are often asymptomatic, so awareness of even having the infection is low (see Chapter 8, "Sexually Transmitted Infections/Diseases"); and (2) many students resist being tested for STIs, even though the tests themselves are not particularly painful or difficult. This is unfortunate because the majority of tests for STIs are negative, and the relief people feel makes their initial hesitation about being tested seem exaggerated. One study found that only 7% of sexually active university

Figure 12.5 Percentage of New STI/STD Cases by Age: Chlamydia, Gonorrhea, and HIV

SOURCE: Kaiser Family Foundation. (2013). Sexual Health of Adolescents and Young Adults in the United States. http://kaiserfamilyfoundation.files.wordpress.com/2013/04/3040-06.pdf.

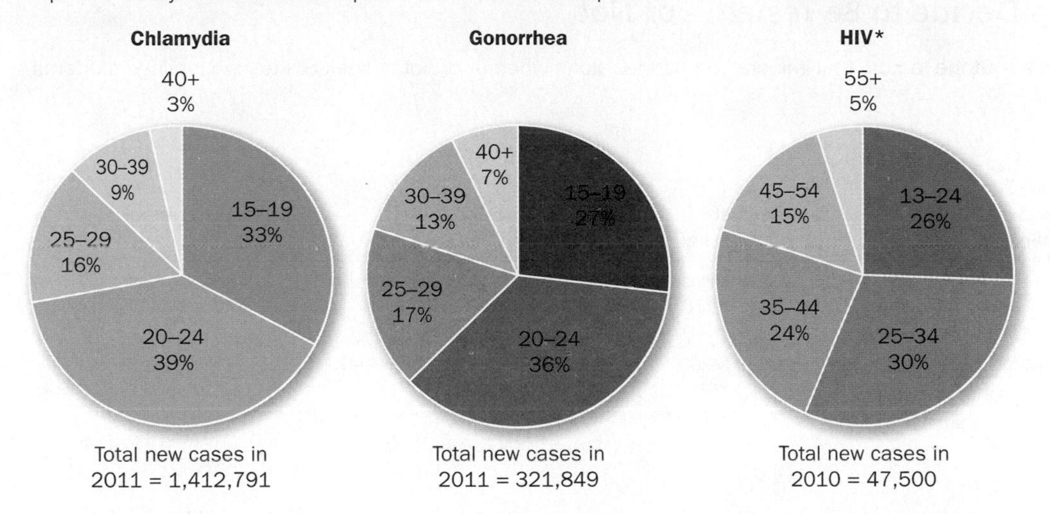

students who were tested for STIs received a positive result (Oswalt, Cameron, & Koob, 2005). On the other hand, those who do test positive are then able to obtain treatment to cure or control the symptoms and dangers of the infection sooner, and to reduce further spread. On the surface, it is difficult for anyone to defend a decision not to be tested, especially if that person is sexually active. Nevertheless, the number of students who seek STI testing is far lower, statistically, than the number who are at risk and *should* seek testing. Why does this discrepancy exist? Most people may feel some embarrassment about being tested, but many other issues about STI testing are in play.

A study examining the reasons behind college students' attitudes about seeking STI testing found that a number of factors influence their decisions. On the positive side are factors such as the relief from a negative test, the knowledge that treatments are available if a test is positive, and the desire to avoid infecting partners. However, a multitude of negative factors often override these positives. These include fear of a positive test, the social stigma associated with having an STI or simply being tested for one, denial, underestimation of vulnerability, and characteristics of the clinic and the provider (Barth et al., 2002; Mevissen, 2011). "In Touch with Your Sexual Health: STIs: Why Students Decide to Be Tested...or Not", on the following page, lists the various factors that influence college students' decisions about STI testing. When this list has been shown to other students in human sexuality classes, their response was nearly unanimous that the reasons for avoiding testing were too weak to justify such a decision. What do you think? Today, some colleges and universities are turning to social media to help inform students about STIs and encourage them to be tested. Studies have shown that these strategies are paying off and increasing awareness of STIs and rates of testing (Lim et al., 2008; Turner et. al., 2011).

Taking effective steps to avoid contracting STIs in the first place is significantly more effective than worrying about whether or not you may be infected. Most colleges and universities provide students with various resources on campus and on their websites to help inform and educate them about STIs and the behaviors that transmit them. Make a point to examine your own college's efforts to provide this information to students, and if you feel it is insufficient, consider taking action to improve it.

The Alcohol-Sex Link in College Students

Another very important reason for the high rate of STIs among college students is that sex and drugs, mainly alcohol, are intimately linked on college and university campuses. Alcohol is routinely a part of college social interactions, and sex is clearly one common form of social interaction. If a couple are drinking small to moderate

In Touch with Your Sexual Health

STIs: Why Students Decide to Be Tested...or Not

When asked during interviews about the factors that influenced their decisions whether or not to seek STI testing, college students cited the following factors.

Negative Perceptions or Personal Characteristics	Beneficial Personal Perceptions	Personal Vulnerability	Social Factors	Health System Factors
What others would think (88%)	Low severity of STIs vs. HIV (66%)	Characteristics of sexual partner(s) (93%)	Lack of knowledge about STIs (76%)	Gender of health provider (41%)
Embarrassment (61%)	Better to know for sure (44%)	Existing symptoms (88%)	Stigma of STIs (56%)	Health provider's knowledge (27%)
Fear of positive test result (56%)	Relief if test negative (42%)	Type of sexual encounter (80%)	Privacy about sex and STIs (49%)	Comfort with physician (20%)
Negative emotions (29%)	Health benefits (39%)	Past sexual history (73%)	Low media coverage of STI dangers (15%)	Reputation of test site (78%)
Fear of procedures (24%)	Concern for partner (10%)	Nonsexual exposure (10%)		Cost (60%)
Negative effect on future life (12%)				Confidentiality (60%)
Denial (29%)				Convenience (68%)
Would rather not know (22%)				Availability of testing services (37%)
				Not sure where to be tested (20%)

SOURCES: Barth et al. (2002); Shoveller (2010); Wolfers, de Zwart, Kok 2012.

amounts of alcohol in the course of an evening and, as that evening progresses, they mutually agree that they want to make love, that is their choice as consenting adults. The problem comes when larger amounts of alcohol (or some other drug) twist the scenario so that they make irresponsible sexual choices, such as unsafe, unprotected sexual activities. Alcohol may also render one member of the couple unable to consent freely (or legally) to sexual advances. Without that consent, if sexual penetration occurs, it is rape (incapacitating and raping women using alcohol and other drugs is discussed in detail in Chapter 13, "Sexual Aggression and Violence").

Alcohol use has been shown to contribute significantly to college students' decisions to engage in sexual intercourse and to participate in various indiscriminate sexual behaviors such as casual sex and sex with multiple partners (Certain, 2009). Also, studies have found that the influence of alcohol reduces the likelihood of safer sexual behaviors such as condom use or other contraceptive practices (Davis et al., 2010; Sakar, 2009). When intoxicated, some students tend to become single-minded, focusing on the pleasure and excitement of the sexual behaviors of the moment, and they lose sight of the long-term risks involved in those activities.

Since You Asked

8. My roommate went to a party and drank too much. She had sex with two guys (that she remembers), but she says she didn't really want to. Was this rape?

Sexuality in Adulthood

12.4 Explain how intimate relationships in adulthood differ from adolescence and discuss the developmental milestones in adult sexuality including, cohabitation and marriage; sexual behavior possible sexual problems during adulthood, and the changes associated with menopause.

As people enter their twenties, whether or not they are attending college, they begin the next stage of development, *adulthood.* Adult sexuality is characterized by less sexual experimentation and typically focuses more on lasting relationships, having children, and establishing a satisfying sexual life.

Adult Intimate Relationships

Adults perceive their romantic relationships in very different ways than do adolescents. You are probably thinking that adults look back fondly on the simplicity and carefree days of their adolescence, but just the opposite is true. Adults typically see their relationships as characterized by stability, mutual support, and mutual trust. Moreover, they find them more enjoyable and burdened with fewer problems than during adolescence (Shulman & Kipnis, 2001). Of course, adults have many more choices about their intimate relationships than do adolescents. Adults may freely choose to stay single, get married, live together, and start a family (or not), in almost any order and combination. One major decision many couples will make during early adulthood is whether to get married or to live together without being formally and legally married. The arrangement in which a romantically involved couple live together without marriage is called **cohabitation**.

cohabitation
Living together as if married without legally marrying.

COHABITATION A clear trend in Western societies is an increase in the number of heterosexual couples choosing to cohabit either before or instead of marrying. In 2010, approximately 11% of women and 12% of men were cohabiting, up from 9% in 2002 (CDC, "Key Statistics," 2013). Moreover, about half of all adults in the U.S. have cohabited with a member of the opposite sex at some point in their lives. The reasons researchers suggest for this striking increase probably won't surprise you. One is the desire of young people to wait longer before getting married in order to pursue personal or professional goals (see the next section on marriage patterns). Among those who wish to wait for marriage, living together becomes an increasingly attractive option as their relationship with a significant other grows. Another frequently cited reason for the trend toward more cohabitation relates to the examples of married life young people see around them in society today (Thorton, Axinn, & Xie, 2010). As one prominent researcher in the field noted, "High levels of marital disruption can increase the likelihood that people will cohabit as they learn either through observation or experience that marriage may not be permanent" (Smock, 2000, p. 5). In other words, when you see couples divorcing all around you, or when you have experienced a divorce yourself, cohabiting might begin to look more attractive than marriage. And what happens to couples who cohabit? There are three potential outcomes: Continue to cohabit over the long term, marry, or break up. Figure 12.6 summarizes the statistics on these events.

In terms of same-sex couples, of the approximately 7 million cohabiting couples in the United States, nearly 900,000 are same-sex couples for whom legal marriage was not an option, until very recently (in some states). Over 500,000 of those couples identified themselves as husband and wife or as spouses (Gates, 2009). As you know, the prohibitions on same-sex marriage are changing rapidly, so the statistics will likely be changing significantly within the next few years (see Chapter 11, "Sexual Orientation," for a discussion of same-sex marriage).

Most people say they consider cohabitation an acceptable alternative to marriage, at least for the short term. Many employers provide the same partner benefits for cohabiting couples as for those who are married. In 2009, 35% of employers in California offered domestic partnership benefits to same-sex couples, but this is likely to be increasing rapidly in the wake of legalization of same-sex marriage in that state and many others.

A question researchers have examined relates to whether cohabitation increases a couple's chances for a lifelong successful marriage. The conventional wisdom contends that living together is a way for couples to get to know each other in a "real-life," marriage-like setting and discover how compatible they are (or are not) prior to becoming legally married. This "trial run" should, many believe, pave the way for a better, happier marriage with less chance for discord and divorce.

Figure 12.6 Outcome of First Premarital Cohabitation: After One and Three years, Ages 15 to 44.

Outcome	By 1 year	By 3 years
Continue to cohabit	67.0%	32.2%
Marry	19.4%	40.3%
Break up	13.6%	27.4%

SOURCE: CDC, "Key Statistics," 2013. www.cdc.gov/nchs/nsfg/key_statistics/c.htm#chabitation.

Does the scientific research support this common view? In a word, the answer is no. Most research finds that people who cohabit have marriages that are less happy, less harmonious, and more likely to end in divorce. How can we explain this striking discrepancy between common sense and scientific "fact"?

The most likely reason for these findings is not related to cohabitation at all, but it is due to the characteristics of the individuals who choose to cohabit prior to marrying (Thorton, Axinn, & Xie, 2010; Woods & Emery, 2002). It is not much of a stretch to see that people who choose to cohabit may have different views about marriage from those who marry prior to living together. The fact that when cohabiters do marry they are less happy and more divorce-prone is not due to the fact that they previously cohabited but that they were not prime marriage candidates to begin with. They may, on average, be less supportive of the institution of marriage or have poorer relationship skills (Thorton, Axinn, & Xie, 2010; Woods & Emery, 2002). However, some evidence supports the notion that negative attitudes toward marriage may develop during the time a couple cohabits prior to marriage. This would indicate that the cohabitation experience itself may negatively influence marriage outcomes.

MARRIAGE Although cohabitation has risen dramatically and is therefore a topic of recent research interest and focus, marriage remains the most common lifestyle choice of adults. About 90% of all heterosexual adults in the United States will marry at some point in their lives, but couples are waiting longer, on average, to marry. In 1980, the median age for marriage was 24.7 years for men and 22.0 years for women. In 2010, those ages had risen dramatically to 28.2. and 26.1, respectively ("Infoplease," 2011). This trend reflects several possible causes, including a greater interest by women in establishing a career before committing to marriage, a greater social acceptance of staying single longer, the poor success rate of marriage based on rising divorce statistics, and greater social and economic support for cohabitation, as noted earlier.

At least 90% of all adults will marry at some point in their lives.

As for same-sex marriage, the statistics are more difficult to pin down. This is because, although many states have legalized same-sex marriage, others have civil-union or domestic-partnership laws, which may or may not be considered "marriage" from a legal or statistical perspective, but is by the partners. The recent fast-moving changes in same-sex marriage laws will lead to many statistical changes over the next few years. What we know now, officially, is that in 2010 there were 132,000 same-sex married couples in the U.S. and 515,000 same-sex cohabiting partners (U.S. Census Bureau, 2011). Please refer to Chapter 11, "Sexual Orientation," for more specific information about same-sex marriage.

Heterosexual married couples have always been the most common focus of research about sex. Most of the studies about sexual behavior and attitudes deal primarily with heterosexual married people. The reason for this is simply that, for now at least, heterosexual married couples are the single largest group of people who are dealing with the largest number of sexual behaviors and issues. We turn now to a discussion of marital sexual issues.

Sexual Behavior in Marriage

Among heterosexual married couples, approximately 20% of men between ages 18 and 40 report having sexual intercourse with their partner between 3 and 16 times per month. About 12% in this age group report more than four instances of intercourse per week, 20% report two to three times per week, and roughly 25% report having intercourse a few times per month or weekly (Reese et al., 2010). Married women in the same age group report the same frequency range as men. Among women, 2% report having intercourse four or more times per week, about 8% report a frequency of two to three times per week, and approximately 21% report intercourse occurs several time per month or weekly (Herbenick et al., 2010).

You must keep in mind, however, that "sex" within a marriage context is not limited to intercourse. Many married couples engage in a wide range of sexual activities either along with intercourse or, at times, instead of it. Table 12.6 summarizes the range of sexual expression among married couples from 18 to 70+ years of age.

Since You Asked

9. I've heard that when people get married, their sex life goes right down the tubes. Is this true?

Table 12.6 Sexual Activities Reported among Married Heterosexual Men and Women (activities occurred at least once in previous 90 days) Ages 10 to 70+

Sexual Activity	Age Range	Men%	Women%
Intercourse	18–24	96	79
	25–29	96	93
	30–39	92	90
	40–49	88	83
	50–59	76	68
	60–69	56	54
	70+	33	32
Masturbation (alone)	18–24	42	47
	25–29	71	51
	30–39	69	47
	40–49	61	44
	50–59	58	31
	60–69	48	26
	70+	27	12
Masturbation (with partner)	18–24	33	25
	25–29	34	40
	30–39	37	30
	40–49	30	24
	50–59	19	21
	60–69	17	9
	70+	8	7
Oral sex (received from partner)	18–24	58	59
	25–29	71	62
	30–39	70	59
	40–49	59	46
	50–59	46	36
	60–69	27	21
	70+	12	8
Oral sex (gave to partner)	18–24	33	71
	25–29	71	70
	30–39	64	64
	40–49	54	45
	50–59	43	40
	60–69	24	20
	70+	13	7
Anal intercourse	18–24	13	22
	25–29	18	13
	30–39	10	12
	40–49	11	10
	50–59	1	3
	60–69	5	1
	70+	0	1

SOURCES: Data from Herbenick et al. (2010); Reece (2010). National Survey of Sexual Health and Behavior. *The Journal of Sexual Medicine, 7*(S-5), 277–304.

The next important question concerns whether these couples who are satisfied and happy with the sexual aspects of their marriages are also happily married. Not surprisingly, the answer is yes. Married couples who report greater sexual satisfaction with their partner also report higher levels of happiness with the marriage (Brezsnyak & Whisman, 2004; Hendrick & Hendrick, 2002). The problem arises in attempting to determine which is causing which. Obviously, couples who get along well, enjoy each other's company, are more harmonious, and experience less discord are likely to feel more sexual attraction within the marriage. On the other hand, a satisfying sexual life may contribute to the couple's ability to get along, feel greater harmony, and find each other more sexually attractive.

One study of 5,000 married adults from forty-nine U.S. states examined various factors that might play a role in married couples' sexual satisfaction. The factors studied were couples' overall satisfaction with the marriage, their satisfaction with the nonsexual aspects of the marriage (i.e., shared goals, mutual respect, and shared recreational preferences), their frequency of orgasm during sex, their frequency of sexual activity, their degree of sexual experimentation, and their religious beliefs (Young et al., 1998). The factor that accounted for the greatest amount of the couples' sexual satisfaction was their rating of their overall satisfaction with the marriage in general, followed by their ratings of the nonsexual aspects of their relationship. Other factors also played significant yet less powerful roles, with the exception of couples' religious beliefs, which were not related to sexual satisfaction. Figure 12.7 summarizes the study's findings for the factors measured.

If sexual satisfaction and marital happiness are closely linked, is the opposite true? That is, are sexually inactive marriages *less* happy? Again, the answer appears to be yes, overall. In one study, marriages in which no sexual activity had occurred over the previous month (16% of those surveyed) were associated with less marital satisfaction, greater likelihood of separation, and a lack of shared activities (Donnelly, 1993). Another study found greater marital satisfaction among couples when their desired level of frequency was matched by their actual level (Santtila et al., 2008).

On the other hand, when one or both members of a couple felt that their level of sexual activity fell short of their desired level, marital satisfaction was lower. How many married couples feel that their sexual relationship is both physically and emotionally satisfying? One large survey of married couples conducted in the mid-1990s found that about 75% of men and 29% of women reported having an orgasm every time they make love with their spouse. When men were asked how physically and emotionally satisfied they were in their marriage, about 52% said they were extremely physically satisfied and 49% said they were extremely emotionally satisfied. For women, the numbers were 41% and 42%, respectively (Laumann et al., 1994).

Figure 12.7 Factors Predicting Sexual Satisfaction in Marriage

SOURCE: Figure, "Factors Predicting Sexual Satisfaction in Marriage," by M. Young et al., in *The Canadian Journal of Human Sexuality*, *7*, (1998), pp. 115–127. Copyright © 1998. Reprinted by permission of The Sex Information and Education Council of Canada.

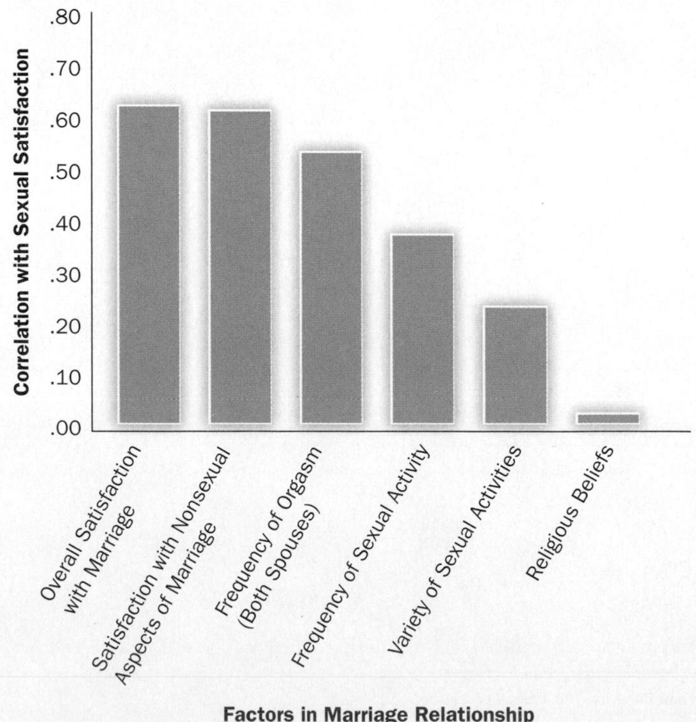

Sexual Problems in Adulthood

As people move from adolescence into adulthood and, in general, settle into more stable and meaningful intimate relationships, most expect to enjoy satisfying, relatively carefree sexual lives. This expectation stems from a common, yet mistaken, belief that sexual behavior and responses are naturally occurring events and that everything should "work" just as nature intended. Unfortunately, the majority of couples run into sexual problems of one sort or another at some point in their relationship. These sexual difficulties are typically the result of changing conditions in the couple's life, such as children, busy work schedules, stress, and fatigue. The most common sexual problems are a loss of sexual desire in one or both partners, problems achieving or maintaining an erection, lack of sexual arousal, lack of orgasm, delayed orgasm, rapid ejaculation, and painful sex. All of these problems are far more common than most people think, and some increase during the normal aging process. We discuss specific sexual problems and solutions in detail in Chapter 7, "Sexual Problems and Solutions."

The important point to make here is that couples need to know that whatever sexual problems they might experience are common, usually reversible, and do not signal the end of a satisfying sexual life. Approximately 43% of women and 31% of men report experiencing at least one sexual problem during the previous 12 months (Laumann, Paik, & Rosen, 1999). Figure 12.8 summarizes the frequency of occurrence of the most common sexual problems. Couples should understand that these and other sexual problems often resolve on their own when conditions in the couple's life become more accommodating to sexual intimacy. Moreover, those that do not resolve are nearly always successfully treated with counseling, medications, or a combination of the two (various treatments are also discussed in Chapter 7, "Sexual Problems and Solutions").

Finally, couples should realize that certain changes in sexual responding occur normally as part of the aging process. When we examine data on sexual problems by age groups, some clear trends emerge. For example, the percentage of men in their twenties and thirties who have difficulty maintaining erections is under 10%, whereas this problem is reported by about 20% of men in their fifties. For women, difficulty becoming physically aroused as indicated by vaginal lubrication is reported by approximately 18% of women in their twenties and thirties, and this increases to about 24% for women in their fifties (Laumann et al., 1994). Also, women experience naturally occurring hormonal changes, typically during the years between their mid-forties and mid-fifties, known as **menopause**. Menopause may lead to difficulties for some

Menopause

The normal, gradual change in a woman's life, typically occurring between age 45 and 55, when the ovaries produce decreasing amounts of female hormones and menstrual periods cease.

Figure 12.8 Common Sexual Problems in Adult Intimate Relationships

SOURCE: Figure, "Sexual Problems are Common," p. 369, in *The Social Organization of Sexuality*, by E. Laumann, J. Gagnon, R. Michael, & S. Michaels. Copyright © 1994. Reprinted by permission of The University of Chicago Press.

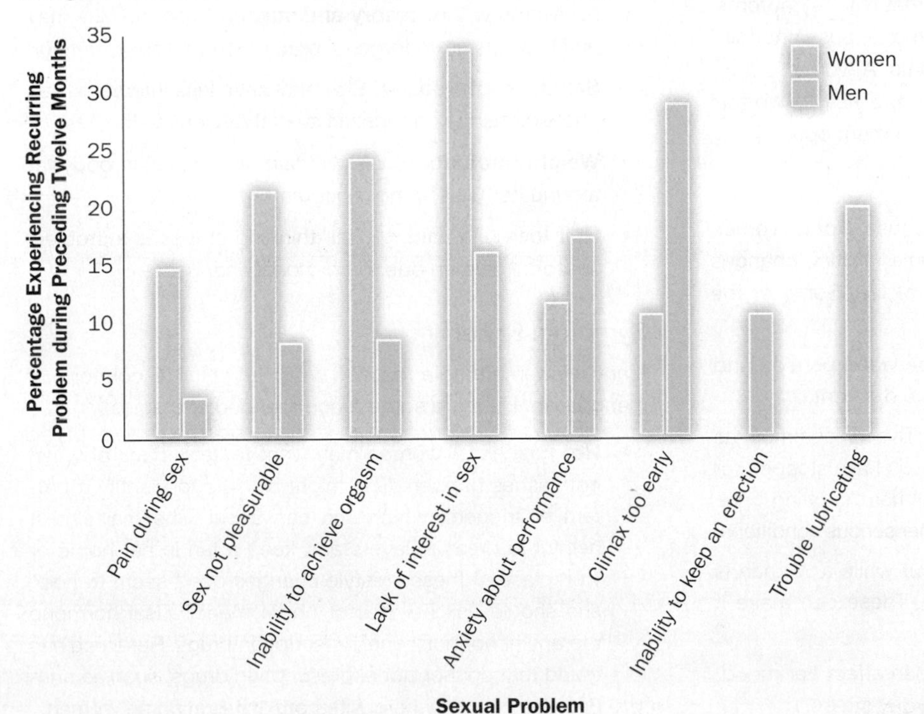

women relating to sexual feelings, sexual desire, sexual functioning, overall mood, and other health issues. Menopause, however, is far more of a gradual biological transition than most people realize, and virtually all sexual, physical, and emotional effects of menopause can be managed and treated effectively. "In Touch with Your Sexual Health: Menopause" summarizes this developmental process in women's lives.

Although the *frequency* of sexual activity for most people tends to decline with age, the desire for and enjoyment of sexual activity does not disappear by any means, even though many (especially younger) people think it does. Our ability to enjoy life as sexual beings continues into late adulthood and throughout the aging process. This chapter is called "Sexual Development *Throughout* Life" because our capacity for sexual responding, enjoyment, and pleasure is truly lifelong.

In Touch with Your Sexual Health

Menopause

The National Health Information Center, a division of the U.S. Department of Health and Human Services, through its website (http://www.womenshealth.gov/menopause), has developed guidelines to help women understand and deal with the developmental changes known as menopause.

What Is Menopause?

Menopause is a normal change in a woman's life when her period stops. It is often called the "change of life." During *perimenopause* (the time leading up to menopause), a woman's body gradually makes less of the hormones estrogen and progesterone. This typically begins between the ages of forty-five and fifty-five. A woman has reached menopause when she has not had a period for twelve months in a row and there are no other causes for this change. As she nears menopause, she may have symptoms relating to these biological changes. Many women wonder if these changes are normal, and many are confused about how to treat their symptoms. Women are reassured by learning about menopause and talking with their doctors about their symptoms. A woman's doctor can explain more about menopause and help a woman make well-informed choices about any treatment options.

Symptoms of Menopause

Every woman's period will stop at menopause. Some women have no other symptoms. But many women notice changes in body, mind, and mood at this stage of life. Some of the changes she might notice include:

- **Changes in her period.** The time between periods and the flow from month to month may be different.
- **Abnormal bleeding or "spotting."** This is common as she nears menopause. But if her periods have stopped for twelve months in a row, and she still has "spotting," she should talk to her doctor to rule out other serious conditions.
- **Night sweats.** Hot flashes that occur while a woman is sleeping and cause her to perspire. These can make it hard to get a good night's sleep.
- **Sleeping problems.** Lack of sleep can affect her mood, health, and ability to cope with everyday stress.
- **Vaginal changes.** The vagina may become dry and thin, and sex and vaginal exams may become uncomfortable or painful. She also might be prone to more frequent vaginal infections.
- **Thinning of the bones.** This may lead to loss of height and more fractures (osteoporosis).
- **Emotional changes.** These may include mood swings, sadness, tearfulness, and irritability. Although menopause does not cause depression, women are at a higher risk of depression in the years leading up to menopause. Some researchers think that the decrease in estrogen levels plays a role in the onset of depression in some women. Also, lack of sleep can strain a woman's emotional health.
- **Urinary problems.** She may have leaking, burning, or pain when urinating, or leaking when sneezing, coughing, or laughing.
- **Problems with memory and staying focused.** She may notice she is more forgetful or has trouble concentrating.
- **Sex drive decreases.** She may have less interest in sex and experience changes in sexual response.
- **Weight fluctuation.** Weight gain or increase in body fat around her waist is not uncommon.
- **Hair loss or thinning.** Hair thinning or loss is a problem for some women due to hormonal changes.

Symptom Relief

Many women are able to cope with the minor discomforts of menopause. Here are some recommended strategies:

- **Hot flashes.** A woman may want to keep track of when hot flashes happen. She might be able to identify a pattern or triggers, which she can avoid. She may find it helpful to dress in layers and keep a fan in her home or workplace. If these lifestyle changes don't seem to help, she should ask her doctor about menopausal hormone therapy or nonhormonal prescription drugs. Research has found that nonhormonal prescription drugs, such as antidepressants, may help with hot flashes in some women.

- **Vaginal dryness.** A woman can try over-the-counter, water-based vaginal lubricants. Prescription estrogen replacement creams and tablets also can help restore moisture and tissue health. If she is experiencing spotting or bleeding while using estrogen creams, she should discuss this with her doctor.
- **Problems sleeping.** Insomnia is a common problem for many people, not just women transitioning through menopause. The same strategies apply to anyone with sleep problems: get at least thirty minutes of physical activity on most days of the week; avoid physical activity close to bedtime; avoid alcohol, caffeine, large meals, and working right before bedtime; drink something warm, such as herbal tea or warm milk, before bedtime; keep the bedroom cool and dark; avoid napping during the day; and try to go to bed and get up at the same times every day.
- **Memory problems.** Some women complain about problems with memory and concentration. No evidence exists that menopause causes memory problems. If forgetfulness or other mental problems occur, she should talk to her doctor.
- **Mood swings.** A menopausal woman should discuss severe mood swings with her doctor. Some medications have been shown to help. For instance, menopausal hormone therapy might help if mood swings are related to disrupted sleep caused by night sweats. Also, her doctor can look out for signs of depression, which may warrant medical or psychological treatment.

Two additional, serious health problems can begin during menopause. These are often without symptoms, but a woman's doctor can perform simple tests for them and prescribe treatments.

- **Osteoporosis.** Day in and day out, a woman's body is breaking down old bone and replacing it with new healthy bone. Estrogen helps moderate bone loss. Decreasing estrogen production around the time of menopause causes many women to begin to lose more bone than is replaced. In time, bones can become weak and break easily. This condition is called osteoporosis.
- **Heart disease.** Heart disease becomes more likely in women after menopause. Changes in estrogen levels may be part of the cause, as is the aging process in general. As a woman ages, she may develop other health problems, such as high blood pressure or weight gain, which may or may not be related to menopause but put her at greater risk for heart disease.

SOURCE: The Office on Women's Health: U. S. Department of Health and Human Services (2008). *Understanding Menopause.*

Sexuality and Aging

12.5 Address sexual development into later life including older adults' sexual behaviors, age-related changes in sexual responding, and intimate relationships among the elderly.

As people live longer and healthier lives, an increasing proportion of the United States' population is over 65. In 2006, approximately 12% of the U.S. population was over 65, an increase of 15% from 1990 (U.S. Census Bureau, 2006). Since 1900, the number of U.S. residents over 65 has risen from 3 million to 37 million! As Americans are living longer, we are becoming increasingly aware that older people are still sexy! This is despite the traditional belief that as people age, their interest in love, intimacy, and passion fades, but that's wrong. One sign of this is the findings cited in Table 12.6 on page 467. Beyond the statistics, a great deal of evidence exists demonstrating that remaining romantically and sexually active into old age is far more common than most people would guess and that continuing to be interested in sex, romance, and dating provides a wide range of emotional and physical health benefits.

Since You Asked

10. At what age do most elderly people stop being interested in sex?

Intimacy and sexual interactions are important to most people *throughout their lives*. When you try to imagine your parents making love together, do you think, "Yuck!"? How about your grandparents? "Double yuck?!" Well, you might as well get used to it. Many older people continue to be sexual and enjoy sexual activities well into their fifties, sixties, seventies, eighties, and nineties. In fact, some senior citizens may be having more sex than some busy, two-career parents in their twenties! Very often, older couples find that being alone with their spouse after the kids have left the nest and no longer needing to be concerned about birth control allow them to rekindle a sexual life together that may have diminished over the years.

A major study in 2007 using interviews and questionnaires involving over 3,000 men and women between the ages of 57 and 85 asked numerous questions about

Table 12.7 Sexual Behaviors among Older Adults in the United States

Sexual Activity	Percentage among Men Ages 57–64	Percentage among Women Ages 57–64	Percentage among Men Ages 65–74	Percentage among Women Ages 65–74	Percentage among Men Ages 75–85	Percentage among Women Ages 75–85
Any sexual activity with a partner in previous 12 months	83.7	61.6	67.0	39.5	38.5	16.7
Activities among Those Reporting Sexual Activity During Last 12 Months:						
Sexual Activity	**Percentage among Men Ages 57–64**	**Percentage among Women Ages 57–64**	**Percentage among Men Ages 65–74**	**Percentage among Women Ages 65–74**	**Percentage among Men Ages 75–85**	**Percentage among Women Ages 75–85**
Sexual activity 2–3 times per month or more	67.5	62.6	65.4	65.4	54.2	54.1
Vaginal intercourse	91.1	86.8	78.5	85.4	83.5	74.4
Giving or receiving oral sex	62.1	52.7	47.9	46.5	28.3	35.0
Masturbation (solo)	63.4	31.6	53.0	21.9	27.9	16.4

SOURCE: Lindau et al., 2007, adapted from Table 2, pp. 766–768.

their sexual lives, activities, and feelings (Lindau et al., 2007). As it turned out, these seniors were, for the most part, relatively comfortable discussing their personal lives with the interviewers. In general, the findings confirmed that, as we age, we continue to feel sexy and engage in sexual activities. A sampling of the study's results are summarized in Table 12.7.

Women and men have the capacity for sexual desire and sexual activity throughout their lives and are emotionally and physically able to express their sexuality well beyond midlife. Women and men who have remained sexually active throughout life seem to be more sexually responsive in old age than those who have not. The key to maintaining sexual function in later years is to continue a pattern of regular sexual activity. This is one instance in which the "use it or lose it" rule applies.

Humans are sexual beings from birth to death, and most senior individuals continue to have sexual feelings; desires; and, in many cases, active sexual lives. There are few, if any, physical reasons that anyone should ever have to stop being sexual. The most common reason that older individuals stop having sex is that they buy into cultural expectations that it is inappropriate for the aged to "behave that way." Consequently, their sex lives diminish and they avoid discussing sexual issues with health care professionals (Laumann et al., 2009).

Apart from depriving older adults of the enjoyment and intimacy sexual interactions bring, these myths and expectations can also prevent older adults from receiving adequate sexual health care. For example, health care providers often neglect to deal with issues related to sexual problems when they are treating older clients, and their older patients are hesitant to bring these issues to the doctor's attention due to embarrassment, age differences between doctor and patient, and negative social attitudes about sex and aging (Lindau et al., 2007). Similarly, when dealing with older clients, doctors often do not consider the possible effects of chronic medical conditions and medications on sexual activities and response. Consequently, they may not anticipate that some of these older patients may become frustrated with the sexual side effects of prescribed medications and may discontinue treatment without notifying their doctor. This, of course, can have serious and even fatal consequences.

Intimacy and sexual interactions are important to us throughout our lives.

Normal Age-Related Changes in Sexual Responding

Although sexual activity can continue well into one's nineties and beyond, the aging process does have some predictable, and quite normal, effects on sexual responding. In general, the response cycle, as described by Masters and Johnson (see Chapter 3, "The Physiology of Sexual Response"), slows down. The stages of response take longer, the intensity of sensation may be reduced, and the genitals may become somewhat less sensitive. Although sexual excitement and orgasm may be slightly subdued, they are no less pleasurable. Indeed, for many people, the later years can offer a rich sex life without the worry of pregnancy and the inconvenience of contraception. However, it is important to remember that the risk of acquiring HIV and other STIs does not disappear with age.

In older women, menopause results in drops in estrogen and progesterone, causing physiological changes that affect sexual function. These hormone-related changes include thinning of the vaginal lining, reduced elasticity of the vagina, and decreased lubrication, sometimes resulting in discomfort or pain during intercourse. Urinary incontinence may also occur (because of reduced estrogen), as well as loss of libido (because of reduced testosterone). All of these changes are reversible and treatable if they are interfering with enjoyment of sexual activities. In men, the time to achieve erection increases, and the period of time between orgasms lengthens. Most men, as they age, require more direct stimulation of the penis for erection, whereas at younger ages, mere visual or fantasy images were sufficient. Normal physiological changes for men and women associated with aging may be divided into categories of sexual activity shown in Table 12.8 (following page).

Aging and Dating

Western cultures have always seen romance, passion, and sex as reserved for the young. However, the reality is that the group we commonly refer to as "senior citizens" are often as interested and active in dating and intimacy as people decades younger. Cultural attitudes about aging, romance, and sex are just plain wrong. Single adults over sixty report that dating and physical intimacy are important and meaningful parts of their lives (DeLamatera, Hyde, & Fonga, 2008).

You might think that when the elderly date, the feelings and behaviors associated with meeting someone new, going out, and becoming involved would be significantly different from dating among people in their teens and twenties. However, the emotions older people experience surrounding dating are remarkably similar. As the authors of one study of dating among those over sixty-five wrote (Bulcroft & O'Connor-Roden, 1986):

> One of our major findings was the similarity between how older and younger daters feel when they fall in love—what we've come to call the "sweaty palm syndrome." This includes all the physiological and psychological somersaults, such as a heightened sense of reality, perspiring hands, a feeling of awkwardness, inability to concentrate, anxiety when away from the loved one, and heart palpitations. A 65-year-old man told us, "Love is when you look across the room at someone and your heart goes pitty-pat."
>
> A widow, aged 72, said, "You know you're in love when the one you love is away and you feel empty." Or as a 68-year-old divorcee said, "When you fall in love at my age there's initially a kind of 'oh, gee!' feeling... and it's just a little scary." (p. 66).

Dating for many older individuals is not merely platonic companionship. Sex, including intercourse, may play an important role. Sometimes, even when a married couple moves into a nursing home, their sexuality may not be recognized by the staff. For example, a married couple moved into a nursing home room with two hospital beds. One spouse had to have a leg elevated, but it was on the same side as the partner's bed,

Table 12.8 Normal Aging-Related Changes in Sexual Response

Age-related changes that may interfere with a satisfying and pleasurable sexual life are normal and *can be treated effectively*. For many people, their later years can offer a rich sex life without the worry of pregnancy and the inconvenience of contraception. However, it is important to remember that the risk of contracting STIs does not disappear with age.

Normal Aging Sexual Changes	For Older Men	For Older Women
Desire	– A decrease in sexual desire may be experienced which may be due to changes in hormone levels (which are treatable) or cultural expectations about sex and aging. – Certain chronic illnesses or other physical conditions may require adjustments in the sexual behaviors and desires shared by older couples, but these rarely require giving up sex.	– A decrease in sexual desire, which may be due to changes in hormone levels (which are treatable) or cultural expectations about sex and aging. – Certain chronic illnesses or other physical conditions may require adjustments in the sexual behaviors and desires shared by older couples, but these rarely require giving up sex.
Sexual Arousal	– Erections are slower to develop with increasing age. A younger man may be able to experience an erection in a matter of seconds, yet some older men may need several minutes of direct stimulation to become erect. – Erections may be less firm.	– There is less blood engorgement of the genitals than in younger women. – Vaginal lubrication may take longer to occur (due to hormone changes during and following menopause), and the amount of lubrication may be reduced. Use of artificial lubricants may increase ease of intercourse. – Nipple erection is slower. – Elasticity of vagina tissues may decrease with age.
	– Older men may have less overall muscle tension during sexual arousal than younger men. – Complete erection may not be achieved until near end of the plateau phase. – The testes do not elevate as far up toward the body. – The older man is able to stay in the plateau phase longer. This may enhance his and his partner's sexual pleasure—a positive changes that occurs with aging.	– Enlargement of the inner vagina (tenting) is slightly less than in younger women. – The uterus elevates slightly less.
Orgasm	– Older men may experience a somewhat lowered level of intensity during orgasm, but not less pleasure. – Orgasm sometimes produces fewer contractions. – The seminal fluid may be thinner in consistency and somewhat reduced in volume.	– Women may experience fewer muscle contractions during orgasm, but no less pleasure. – Sexual flush may be less pronounced.
Resolution	– This phase progresses faster in older men (body returns to unaroused state more quickly). – The testes lower away from the body more rapidly. – Nipple erection lasts longer than in younger men. – The refractory period tends to lengthen.	– This phase occurs more rapidly (body returns to unaroused state more quickly). – Older women have less vasocongestion than younger women, so less time is needed for the body to return to the unaroused state. – Some older women may experience vaginal discomfort or pain if lubrication is insufficient.

SOURCES: Mayo Clinic (2011); Meston (1997); "Sex Info" (2003).

which made it hard for them to hold hands. Some staff members didn't see the importance of allowing the couple intimacy and said the problem couldn't be fixed (Jankowiak & Cornelison, 2008). But many nursing home residents see things quite differently. One 71-year-old widower told researchers, "You can talk about candlelight dinners and sitting in front of the fireplace, but I still think the most romantic thing I've ever done is to go to bed with her" (Kris & O'Connor-Roden, 1986, p. 66). For many in their sixties, seventies, and eighties, a dating partner is their only source of physical affection and touching. Just as is true throughout life, sexual intimacy is a way for people to continue to feel loved, attractive, and needed as they enter old age. As one 77-year-old woman said, "Sex isn't as important when you're older, but in a way you need it more" (p. 67).

The main roadblock older daters face is society's discomfort with romance and sexuality among the elderly. Many senior citizens who have a dating partner are not interested in marrying again. They are happy with the mix of independence and romance that they have established. However, being sexual outside marriage often goes against the values they believed throughout their lives and evokes embarrassment or guilt. In an ironic role reversal, some elderly daters hide their dating intimacy from their grown children. But that does not mean that love and intimacy are any less important to them. For many older people, dating is a happy and central part of their lives. In the words of one 65-year-old man, "I'm very happy with life right now. I'd be lost without my dating partner. I really would." Or as a 64-year-old woman summed it up, "I suppose that hope does spring eternal in the human breast as far as love is concerned. People are always looking for the ultimate, perfect relationship. No matter how old they are, they are looking for this thing called love" (Kris & O'Connor-Roden, 1986, p. 69).

Unfortunately, not all elderly people live independent lives that allow them to meet others and form dating relationships. Many who reach a point in life where they need extra care move to assisted living facilities or nursing homes. It may surprise you to learn that the human need, desire, and perhaps instinct for an intimate connection with another is not extinguished in these circumstances, either. Sexual relations among nursing home residents is a reality that requires sensitive understanding from management and nursing staff. "In Touch with Your Sexual Health: Sex and Nursing Homes" elaborates on this important issue.

In Touch with Your Sexual Health

Sex and Nursing Homes

Over 1.5 million elderly U.S. citizens reside in nursing homes. Talk to any nursing home staff member and you will hear stories about the sexual exploits of their clients. This is a controversial topic, due in part to society's expectation that "old people aren't supposed to be sexual." Families of residents are often shocked to discover that their aging relative has been spending several nights each week in another resident's room and bed. However, with few exceptions, these senior citizens are consenting adults with the right to express themselves sexually if they choose (Jankowiak & Cornelison, 2008).

Nursing homes are required by law to provide for the personal privacy needs of their cognitively competent residents, just as any other consenting adult would expect in U.S. culture. In a nursing home care setting, these needs may include:

> Training staff to understand what behaviors fall under their job responsibilities and which do not; taking steps to ensure the respect and understanding of the diversity of moral and religious beliefs among residents; dealing with family members who may be uncomfortable or opposed to sexual activities in the home; and providing residents with appropriate information about sexual activities and potential sexual problems (Bonifazi, 2000).

Senior citizens grew up during a time of changing and liberalizing sexual standard in the U.S. It's not unusual in today's nursing homes, for residents to be more open about and interested in sexual issues and activities and to seek out others with whom to be sexual or discuss sexual topics. One story out of many may exemplify this.

In a nursing home in Chicago, "Jason" met "Estelle" at the facility's video bowling game night. There were both in their late 70s. They began chatting and flirting and spending meals and most of their free time together. They fell in love. The staff enjoyed watching their blooming romance and

Jason and Estelle did not hide the fact that they had become physically intimate.

They began sharing a room most nights. Estelle was sure that her romantic and sexual days were over and was filled with joy at how Jason had reawakened those feelings in her. She became truly happy for the first time since her family had decided she should move into the assisted-living facility.

Jason spoke of how he had fallen for Estelle harder than ever before in his life, more than either of his two wives. He talked about how they were soul mates and destined to find each other and be together. He asked the staff for a dancing night once a month, even though neither was very steady on their feet and had to be careful. Many of the other residents joined in as music for 40 years ago played in the dining hall.

After a year, they were married in the facility; the wedding was attended by all the staff and many family members on both the bride's and groom's sides. The home provided them with a new double bed and they began living a happy married life in the home.

Your Sexual Philosophy

Sexual Development Throughout Life

We have come full circle. We have taken a look at our sexual development from before birth to the last stages of life. As mentioned at the beginning of this chapter, we are intended by nature to be sexual, to respond sexually, and to desire intimacy with others. For better or for worse, these human capacities are present in all of us throughout our entire lifespan.

This chapter relates directly to the importance of developing a sexual philosophy as early in life as possible. Everyone experiences the various phases of sexual development discussed in this chapter, and each sexual passage presents new and different challenges for healthy sexual adjustment. Even children will benefit from some guidance in developing their own version of a sexual philosophy that includes an understanding of appropriate and inappropriate sexual behavior in public, respect for their own and others' bodies, awareness of the difference between good and bad touches by an adult, and knowledge of how to communicate to trusted adults any uncomfortable sexual encounters they may experience.

Adolescence is perhaps the stage in most people's lives when focusing on their sexual philosophy becomes most important. There is probably no other time in life during which sexual issues exert more power over us. You know this is true from the way people talk about "raging hormones" causing teens to lose all control over their sexual desires and actions. However, adolescents' hormones do not, in reality, have the power to cause any specific sexual behaviors. They merely awaken natural sexual feelings and motivations. What a person does with those feelings still requires conscious choices. Teens who have developed at least the beginnings of a personal sexual philosophy are much better equipped to make safe and healthy choices when faced with sexual situations because they have had the opportunity to think about them before they are in the heat, and the peer pressure, of the moment.

The value of a well-developed sexual philosophy does not end with adolescence. Major life decisions will be made during adulthood involving relationships, living together, marriage, having children, sexual satisfaction, and sexual problems. This does not imply that you should have all the answers to these life events before you are faced with them. No one can predict the future, and everyone's sexual philosophy evolves and changes through learning and experience. However, just think how much more easily and effectively you will be able to approach sexual issues in adulthood if you have taken some time to think about them; to consider your own personal needs, wishes, and desires; and to become educated about them.

Finally, continuing to expand and develop your sexual philosophy is important as a lifelong process. Sexuality does not end with old age, but it does change in various ways. Preparing for those changes and knowing what is a normal part of aging will give you the best chance of staying sexually active throughout your later years, which will in turn make the aging process a healthier and happier experience.

Have You Considered?

1. Imagine you are babysitting for two five-year-old children. You notice they have become "too quiet" in the other room, and you go in to find the kids naked and "playing doctor." Describe how you think you would react to this. What actions you would take, if any?

2. Discuss when childhood masturbation is not a cause for concern and when it might indicate a problem.
3. Imagine that you are the parent of a precocious 13-year-old girl, and you are having a serious parent-daughter talk with her about sexual issues. She tells you that she has never had intercourse, but upon further discussion, she admits that she is having oral sex with boys. What do you think your response to this would be, and what information would you want to give her?
4. What do you think are the three most important sex education issues for young teens to understand as fully as possible? Explain your answer.
5. If you were a sexuality educator hired by a large urban school district to develop programs to reduce unwanted pregnancy and the spread of STIs among teens, discuss at least three strategies you would propose to accomplish these goals.
6. Discuss the various barriers elderly people face in maintaining sexually intimate and active lives. How would you suggest some of these barriers might be removed?

Summary

Historical Perspectives

Freud's Psychosexual Stages of Development: Oedipus and Electra?

- Freud contended that nearly all human development is sexually based. Although many of his ideas have not held up well to modern scientific research methods, Freud's five psychosexual stages of human development were widely accepted in the early 1900s.

Sexuality in Infancy and Childhood

- Childhood masturbation and sexual exploration are normal activities in childhood. When parents or other caretakers find children masturbating or "playing doctor," they usually need not be concerned. However, extreme sexual acting out in childhood may indicate emotional problems or sexual abuse issues.
- Most children do not receive adequate information about sex from their parents. Studies show that many children rely on school, peers, and the media for sex education.

Sexuality in Adolescence

- Puberty may begin at any time between about 8 and 15 years of age.
- Many teens report that an intimate partner is the most important relationship in their lives.
- Gay, lesbian, and bisexual teens face unique problems with prejudice, discrimination, and violence in middle and high school.
- Most schools are working on programs to help nonheterosexual teens feel safe and accepted.
- Most teens will experience their first intercourse by grade 12, but teen sex is not limited to intercourse. Oral sex among teenagers is becoming increasingly common, based on their false belief that oral sex is safe sex.
- Although the teen pregnancy rate is declining, the rates of sexually transmitted infections remain high among adolescents.

Sexuality in College

- Nearly all college students are sexually active. The rates of most sexual experiences among college students are similar for males and females.
- College students appear to be protecting themselves fairly well from unwanted pregnancy but not taking adequate precautions against STIs. Less that half of college students use condoms, and the majority of college students resist being tested for STIs.
- Sexual assault, acquaintance rape, and date rape on college campuses are often associated with the use of alcohol or other drugs.

Sexuality in Adulthood

- Adults report that intimate relationships become more enjoyable as they grow older, and they feel less conflicted about them than during adolescence.

- Research has shown a striking increase in the number of couples choosing to cohabit before or instead of marrying.
- Cohabitation is associated with less happy subsequent marriages and higher divorce rates. Over 90% of all adults in the United States will marry, although they are waiting longer to marry. The median age for marriage in the United States is now around 27 for men and 25 for women. Same sex couples, for whom cohabitation was the only option until recently, now comprise a significantly higher proportion of marriages in the U.S.
- Sex in marriage is not limited to intercourse among heterosexual couples; oral and anal sex are also common. A couple's sexual satisfaction and the happiness of their relationship are closely linked.

Sexuality and Aging

- Although sexual responses change as we age, few physical reasons exist for the elderly to forgo satisfying sexual interactions. Sexual intimacy among the elderly is an important part of their happiness, and research has shown that many seniors remain sexually active well into their eighties.

Your Sexual Philosophy

Sexual Development Throughout Life

- Sexuality is a lifelong part of human nature, which changes as we age.
- Understanding and preparing for sexual changes—knowing what to expect and what is normal—allows us to stay sexually healthy throughout life, which in turn makes our lives healthier and happier.

Chapter 13

Sexual Aggression and Violence

Rape, Child Sexual Abuse, and Sexual Harassment

Learning Objectives

After you read and study this chapter you will be able to:

13.1 Review the issue of rape including rape statistics, how it is defined, the perpetrators of rape, the issue of male rape, rape myths, the problem of rape on college campuses, and what can be done to reduce the incidence of rape.

13.2 Discuss the problem of child sexual abuse (CSA) including a summary of the perpetrators of this crime; the prevalence and incidence of CSA in the U.S. and globally; the psychological, emotional, and physical effects on victims; and the strategies to reduce and/or prevent CSA.

13.3 Explain what is meant by sexual harassment and review the settings in which harassment commonly occurs, the behaviors often involved, what a person should do as a victim, and the effects of sexual harassment on victims.

Since You Asked

1. My friend's husband wants to have children, but she is not ready. She told me he is forcing her to have sex and not using protection. I told her this is rape, but she doesn't think so because they are married. Is that true? (see page 482)
2. Why do guys feel that a woman's body is "theirs," but the guy's body is never the woman's? (see page 482)
3. I read a story about a man being raped by a woman. Is this possible? (see page 485)
4. If women don't want to be sexually attacked, why do they dress like they do? (see page 487)
5. I've heard that some women really want to be raped because that way they can have sex without taking any responsibility for it. (see page 488)
6. Someone told me that all men, if they got the chance and knew they would never be caught, would commit rape. Is that true? (see page 489)
7. My roommate went to a party last weekend and got totally wasted (drunk). Two of the guys at the party told her that two other guys had sex with her while she was passed out, but she has no memory of this. Could this really have happened? Why would guys want to have sex with someone who is passed out? (see page 490)
8. Is there any way to tell if someone has slipped a roofie into your drink (I mean, before you drink it)? (see page 491)
9. My sister was raped over three years ago, but she's still very upset and angry about it. What can she do to get over it? Will she ever get over it? (see page 499)
10. Can being sexually abused as a child prevent a person from having good relationships in adulthood? Do the effects of the abuse really last so long? (see page 506)
11. I heard that a professor was fired for sexual harassment. Could that be true? What did he do? (see page 516)

Marci had liked him. When she met Kirk two weeks ago after a party at a fraternity house, he was attentive, caring, and eager to know all about her. He was willing to talk about himself and his feelings. They had gone out twice during the next week, and on their second date, they had spent almost an hour kissing in his car in front of her campus apartment. Just kissing, that's all; she was pleased that he didn't seem particularly aggressive or interested just in sex like so many of the other guys she had dated. The next weekend, she invited Kirk over to her room for pizza and a movie. After dinner, they began kissing on the bed, and his hands began to wander. Marci stopped him and told him that she wasn't ready to be sexual with him. He said he was sorry, but as the kissing continued, he began to force his hands under her clothing. The more she protested, the more forceful he became. She told him no and pleaded with him to stop, but he was bigger and stronger, and with all his weight on top of her, she just felt powerless. She tried to fight but couldn't. For some reason, she didn't call out for help. He got her panties off and his pants down and pushed himself into her. When he was done, he told her how great it was and actually wanted to kiss her. When she told him to get out, he acted hurt, as if nothing wrong had happened.

* * *

It is a beautiful campus, framed by rolling hills and pine-covered mountains. The main entrance into the library, theater, student union, and classroom buildings is a red brick path through blue spruce and pine. On either side of the walk are benches and tables where students gather to eat, study, or just hang out, especially on warm days. Over the past couple of years, this walkway has become, at times, an obstacle course of sexual harassment. On hot days, groups of male students congregate on the benches to leer, gesture, and make inappropriate sexual remarks to female students as they walk by on their way to and from the campus. A few of the women have complained to trusted faculty and staff about this sexually harassing behavior. Two women students even cited this harassment as a reason for transferring to other colleges, explaining that their discomfort was so great that they did not want to come to the campus at all. Several women have been followed to their cars by one or more of the gawking men and asked out on dates. When the women refused, they were subjected to scorn and ridicule on subsequent trips along the walkway. The men involved have been warned by the administration, fraternity sponsors, and coaches, but the behavior has continued. Last week, the local police came onto campus and arrested 12 of the male students and charged them with sexual harassment. They have all been suspended from the college and are awaiting trial.

* * *

These stories, adapted from the author's files of actual events on college campuses, portray the dark side of human sexuality. Most people find it nearly impossible to comprehend that the human capacity for love and physical intimacy can, in some cases, become so corrupted as to produce its opposite side in the forms of rape, child sexual abuse, and sexual harassment. The reality is that these acts happen all too often. They are acts of violence and control and usually have little to do with intimacy or sexual expression. Yet the fear and trauma of these acts create painful obstacles to a survivor's ability to experience the joy of healthy sexual intimacy with a loving partner. That is why this chapter is so crucial to this book. We will examine each act—rape, child sexual abuse, and sexual harassment—in turn, but first we will discuss briefly a period in our recent history relating to sexual aggression that might just surprise you.

Sexual violence: The "dark side" of human sexuality.

Historical Perspectives

Marital Rape

If a man forces his wife to have sex, is it rape? If you are like most people, you are probably nodding an emphatic *yes*. You'll probably be surprised to learn that, until relatively recently, most states' rape laws did *not* prohibit forcible sexual acts by a husband upon his wife. Many state legal statutes, in fact, provided specific exceptions that protected husbands from

Focus on Your Feelings

Undoubtedly, most of the feelings you will experience while reading this chapter will not be particularly pleasant ones. High on the list will probably be anger, fear, sadness, and outrage. Considering the terrible acts discussed in this chapter, all of these emotional reactions are normal and understandable. Such strong emotions may serve a positive function in that they can help you identify with the seriousness of the problems of rape, child sexual abuse, and sexual harassment, even if you have not been victimized by any of them yourself. Furthermore, these strong negative emotions often incite people to take action aimed at reducing or eliminating their source, and everyone would agree that the world would be a better place without sexual aggression. From this perspective, what you do with your feelings can be recast in a reasonably positive light.

Strong emotional reactions alone are insufficient to bring about needed social change. For that we must channel our emotional energy into rational, organized actions. Movements such as "take back the night" marches against rape, "men against rape" organizations, Megan's law (requiring notifying communities of residential convicted sexual offenders), and the passage of antiharassment laws with real teeth are examples of how people's feelings about these terrible acts have been redirected into concrete social action for change. If some of you who read this chapter, upon feeling these emotions, choose to become involved in similar movements, society will continue to change for the better.

arrest and prosecution for forcing their wives to engage in any sexual act. It was not until 1996 that rape in the context of marriage became illegal in all 50 states.

Where does this twisted rape "double standard" come from? It stems from English common law, in which a wife was seen as the property of her husband. Such "ownership" included the husband's sexual rights over his wife; that is, as his legal wife, she was obligated to submit to sex on his terms, whenever and however he desired it. Within this framework, if she refused and he forced her, the blame for the assault was placed on *her,* not him. Moreover, if a woman was raped by someone other than her husband, it was considered a property crime, and the victim was not the woman, but the woman's husband! This notion of "wife as property" has been implied, if not actually stated, in rape laws in the United States. It probably sounds like ancient history to you, but these laws only began to change in the 1970s.

The current legal status in the fifty states regarding marital rape (also referred to as "wife rape" or "spousal rape") still varies, although all states now require that a wife give consent for sexual activity. Furthermore, all states provide for criminal penalties for rape within marriage, but to this day, only a minority view marital rape as indistinguishable from other forms of rape. Thirty states continue to make exceptions for husbands who rape their wives. These exceptions may include one or more of the following: the requirement that physical force was used, shorter jail or prison penalties for spousal rape, the requirement that the wife must be physically or mentally incapacitated when the rape occurred, the stipulation that the rape must have occurred during divorce proceedings or a marital separation, or a shorter statute of limitations (see Figure 13.1). In other words, some states still view the rape of a woman by her husband as somehow "less serious" than a rape committed by an acquaintance or a stranger.

Since You Asked

1. My friend's husband wants to have children, but she is not ready. She told me he is forcing her to have sex and not using protection. I told her this is rape, but she doesn't think so because they are married. Is that true?
2. Why do guys feel that a woman's body is "theirs," but the guy's body is never the woman's?

Figure 13.1 Marital Rape Laws in the United States

Although marital rape is a crime, similar or identical to any rape, some states still have laws that differentiate marital rape in one or more ways from other rapes. Refer to www.criminaldefenselawyer.com/marital-rape-laws.html, for legal analysis, state-by-state.

SOURCES: Marital Rape Laws by State (2011); Greenson (2008); National Clearinghouse on Marital and Date Rape (2005).

NOTES: In all states, federal statutes may apply in certain circumstances. All exemptions for spousal rape in the military courts were repealed in 2005.

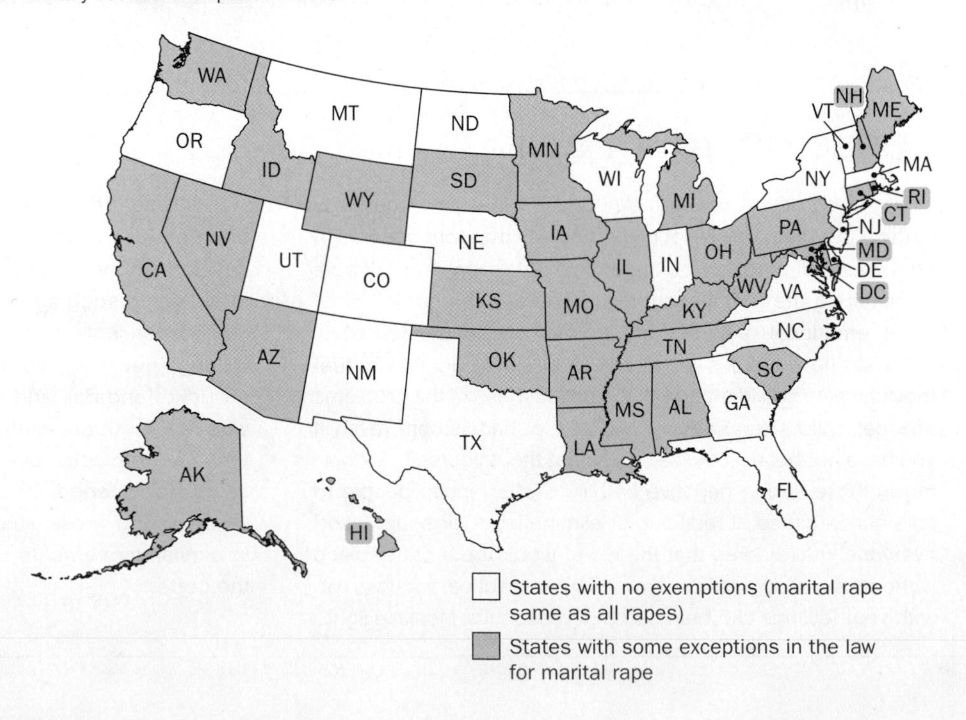

In reality, nothing could be further from the truth. Rapes occurring in the context of an ongoing intimate relationship can be equally or even more traumatic than rape committed outside a relationship. "Intimate rape" is less likely to be reported because the victims themselves may not interpret the act as rape or may be unaware that the act is against the law. Women often blame themselves for these attacks and assume that they are the only ones experiencing actually being raped by a man they "love." But this is not a rare event. Estimates of marital rape are typically under 15% of married women, and this number is probably low due to underreporting of the crime. Moreover, the psychological pain, physical injury, and loss of self-worth that accompany all forms of rape are at least as intense in marital rapes. Injuries associated with wife rape more often include broken bones, black eyes, knife wounds, miscarriages, infertility, depression, and various sexual dysfunctions. Often sexual assault and rape occur as part of a larger pattern of violence in the relationship. Many abusers use coercive sexual tactics along with other forms of physical and verbal abuse to maintain a position of power and control over their partners (Clinton-Sherrod & Walters, 2011; Dugan & Hock, 2009).

In spite of the significant strides made in the area of women's rights and sexual equality, as you are reading this chapter, many groups continue to work to convince state legislatures and governments around the world that marital rape should carry the same criminal severity and punishment consequences as any other kind of rape.

Today in the U.S., sexual aggression and violence refers to *any* sexual activity for which consent is not given by one of the parties involved. By definition, only adults who are in control of their faculties can legally give consent. In this chapter, we will focus on the current status of three all-too-common areas of sexual aggression and violence: rape, child sexual abuse, and sexual harassment.

Rape

13.1 Review the issue of rape including rape statistics, how it is defined, the perpetrators of rape, the issue of male rape, rape myths, the problem of rape on college campuses, and what can be done to reduce the incidence of rape.

Until the 1980s, most people thought of rape as a crime committed by a deranged stranger lying in wait for his female victim, attacking her, overpowering her or threatening her with a knife or gun, and forcing intercourse and other sexual acts on her. Of course, such rapes do occur, but they are much less common than rapes in which the victim has a personal relationship or connection with the rapist. The crime of rape is more common than most people believe. In 2009, the average number of rapes was over 240 per day in the United States alone (see Table 13.1 on following page). As is also evident in the table, the number of rapes has changed very little over the years. Rape results in about 32,000 pregnancies each year (CDC, "Sexual violence," 2012). Nearly 1 in 5 women and 1 in 70 men reported experiencing rape at least once.

Many researchers argue that, although there may be some sexual aspects to rape, it is not primarily a crime of sex, but one of violence; power; control over and humiliation of the victim; and extreme anger on the part of the rapist. The degree to which aggression/power/anger/sex are motivations is an ongoing debate in the study of sexually-related violence (see Bryden & Grier, 2011, for a review). One thing is sure, however: to the rape victim, it is never about sex; it is sheer terror.

Defining Rape

Rape encompasses certain unwanted sexual acts that fall under a larger category of *sexual assault*. The legal definition of rape is left to individual states and countries, and researchers have suggested a multitude of definitions over the years (see Sable et al.,

Table 13.1 FBI Reported Rape Statistics, United States, 2000–2012

Year	Number of Rapes
2000	90,178
2001	90,863
2002	95,235
2003	93,433
2004	94,635
2005	94,347
2006	94,782
2007	91,874
2008	90,479
2009	88,097
2010	85,593
2011	84,175
2012	84,376

SOURCE: FBI Uniform Crime Statistics. www.fbi.gov/about-us/cjis/ucr/crime-in-the-u.s.

2006; Young & Maguire, 2003). Therefore, an exact wording of the specific behaviors that define rape varies greatly.

In addition, the definition of rape has changed over time to reflect law enforcement's and society's broadening understanding of this crime.

Until recently, this was the FBI's official definition, established in 1927 (FBI, 2012): "The carnal knowledge of a female forcibly and against her will." The shortcomings of this definition are obvious and many have wondered why a change did not occur sooner. In 2011, the FBI's definition changed to: "The penetration, no matter how slight, of the vagina or anus with any body part or object, or oral penetration by a sex organ of another person, without the consent of the victim." Some would argue that this may be lacking, too, but this change in the definition of rape has come about partly through the recognition that forced sex may involve various acts, does not require a penis, and that the victims of rape are sometimes men (to be discussed shortly).[1] At the time of these changes, then attorney general Eric Holder stated:

> These long overdue updates to the definition of rape will help ensure justice for those whose lives have been devastated by sexual violence and reflect the Department of Justice's commitment to standing with rape victims. This new, more inclusive definition will provide us with a more accurate understanding of the scope and volume of these crimes. (FBI, 2012)

rape
Nonconsensual sexual penetration of the body using physical force, the threat of bodily harm or incapacitation with drugs or alcohol.

For our purposes, and to create a simple, yet comprehensive definition of rape, we can effectively summarize the act of **rape** as *nonconsensual sexual penetration of the body*. The word "nonconsensual" encompasses the use of physical force; the threat of bodily harm, or any other fear tactic; or incapacitation with alcohol or other drugs.

The key to identifying any sexual assault of any kind is *consent*. In the U.S., sexual acts are not illegal, not assaults, and not rape, if two *adults* consent to the behaviors. Conversely, if consent is not given by one of the parties, it is a crime. This may seem straightforward; however, in most cases, sexual assaults occur in private and the issue

statutory rape
In general, as defined by state laws, any sexual penetration by an adult of a minor child regardless of the minor's consent.

[1] Another type of rape or sexual assault (that may or may not involve force) is sexual activity engaged in between an adult and a minor. This is referred to as **statutory rape** because a minor does not have the legal right under the states' statutes to consent to sexual interactions with an adult. So the sexual activity is considered rape or sexual assault regardless of the minor's agreement or intentions (see child sexual abuse later in this chapter).

of consent often becomes one person's word against another's. The evidence collected can prove that a sexual act has taken place, but it usually cannot determine if consent was given.

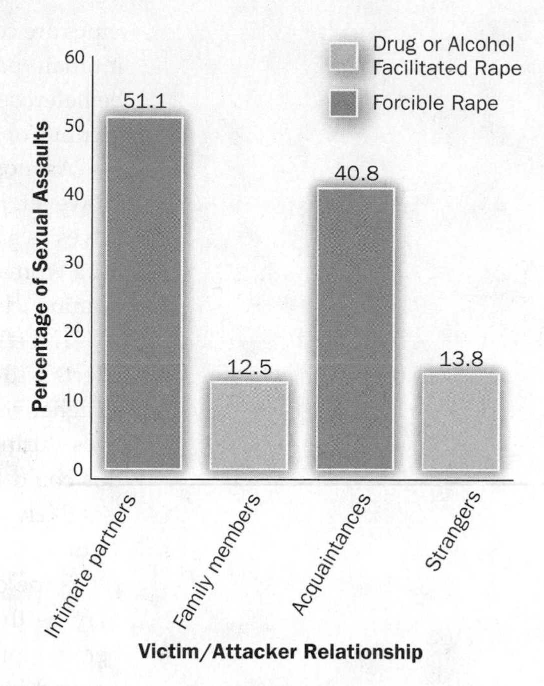

Figure 13.2 Relationships Between Victims of Sexual Assault and Their Attackers*

SOURCES: CDC, "Sexual Violence" (2012); Nett, Kirkpatrick, & Rude (2007).

*Rapes facilitated by drugs or alcohol will be discussed later in this chapter.

Categories of Rape

Rape can be divided into several categories of assault depending on the connection between the victim and the perpetrator (i.e., acquaintance, friend, partner, etc.) and how the attack is carried out: through physical force or through the use of alcohol or other incapacitating drugs (Kilpatrick et al., 2007). Figure 13.2 illustrates the percentages of rapes that fall into these various categories. You can see from the graph that rapes committed by assailants who are strangers to their victims comprise only a small percentage of the total number of rapes. However, this in no way implies that the trauma of being raped is any more or less based on who the attacker was or how the rape was committed.

Rape, regardless of who perpetrates the attack, is one of the most devastating traumas anyone can experience. Some women have stated that they would rather be murdered than raped. Rape is literally a brutal invasion of a person's most intimate territory: her or his body. Psychologists, sexuality educators, and counselors are in general agreement that the act of rape usually has little to do with sex but arises out of a misplaced, warped desire to overpower, aggress against, humiliate, control, and victimize another person. In other words, most rapists are not sexually deprived; they are violent criminals. This can be understood especially strongly when we hear about rape being used as a tactic by solders in countries at war in the Middle East and Africa.

The vast majority of rapes are committed by men against women. There are, of course, rapes of men by men and even rare cases of women raping men. But it is women who carry with them psychologically—some of them on a daily or even hourly basis the fearsome burden of the *possibility* of being raped (Pryor & Hughes, 2013). In a rape-awareness workshop held in 2000, university men and women were asked separately to list the steps they take on a daily basis to minimize the chance of their being sexually assaulted or raped (see Ottens, 2001). "In Touch with Your Sexual Health: Daily Precautions Taken to Prevent Being Sexually Assaulted or Raped" (on the following page) shows how the two sexes' lists looked at the end of the exercise. This eye-opening exercise demonstrates that women must maintain nearly constant vigilance against the threat of rape, although men rarely give it a thought. Most of the men who participated in this exercise reported that they were unaware of the steps women routinely take to avoid sexual assault (O' Brien, 2001).

Can a Man Be Raped?

The answer to that question is an unequivocal *yes.* Male rape is becoming an increasingly recognized crime, and over the past three decades, most states have revised their rape laws to be sex-neutral; in other words, rape may be perpetrated *against* either sex *by* either sex. Statistics on male rape are unreliable because men may feel a greater sense of shame and stigma than women do, and so even more men than women may choose not to report being raped. Keeping that in mind, recent statistics show that approximately 3% of men report being the victim of rape or attempted rape at some point during their lifetime, compared to approximately 20% of women (Tjaden & Thoennes, 2006; Stemple, 2009). Although the percentage for male rape is probably artificially low, it represents over 4 million men.

Since You Asked

3. I read a story about a man being raped by a woman. Is this possible?

The majority of male rapes are committed by other men and usually involve forced anal or oral penetration (or both). As is true for female rapes, only 15% of male rapes are committed by strangers; the remainder by acquaintances, family members or intimate partners (CDC, "Sexual violence," 2012). The perpetrators and victims may be heterosexual or gay, and the attacks may happen in various settings, including the victim's or assailant's home, in a car, or out of doors.

As most people are aware, male rape in prisons is a common and serious problem. However, it should be noted that most men in prison who rape do not perceive themselves as gay, but self-identify as heterosexual and perceive the victim as a "substitute" for a woman. The act of rape itself, in prisons or elsewhere, is an act of aggression and dominance (see Human Rights Watch, 2006).

The effects of rape on male victims tend to parallel those of female victims (these effects will be discussed later in this chapter). Men who are raped by other men are at a higher risk for greater physical injury and, as for female victims, often fear for their lives during the attack. They frequently feel guilt, shame, and confusion about "how this could happen to a man," that sometimes lasts for years after the attack. Men are less likely to report the crime or to seek physical or psychological help than are female victims.

Rape of men by women is the exception, not the rule, but it does happen. You may be thinking, "No way! No man would be able to have an erection if he was being violently or otherwise coerced, right?" Well, not exactly. Men have been coerced intononconsensual sex by women through many means including psychological pressure (such as threatening to accuse *him* of rape if he doesn't comply), physical force (holding down, tying up, etc.), threats with or use of weapons, the use of alcohol or rape drugs, and acting as an accomplice in the rape of a man by another man (Krahe, Waizenhofer, & Moller, 2003). Moreover, some research suggests that extreme anxiety or fear may, in some instances, lead to, rather than prevent an erection (e.g., Turchik & Edwards, 2012; Bancroft, 1989). The point is that a woman *can* rape a man, and all rape needs to be recognized and taken seriously and the perpetrators held accountable. The relatively small number of women who rape men (however serious this crime is) must not divert attention from the larger problem of men raping women.

In Touch with Your Sexual Health

Daily Precautions Taken to Prevent Being Sexually Assaulted or Raped

Studies have demonstrated (e.g., Ottens & Hotelling, 2001; Pryor & Hughes, 2013) that woman, but not men, take precautionary measures to reduce their risk being sexually assaulted or raped. When men and women are asked to list the precautions they frequently take to reduce their chance of sexual attack, the difference is, to say the least, striking. As you can see below, women carry the burden of potential sexual assault with them daily, but men don't really think about it.

Precautions Listed By Women		Precautions Listed By Men
• Always lock all doors in home.	• Carry keys as weapon in fisted hand.	
• Check backseat of car before entering.	• Alternate jogging routes to defeat rapist's ability to plan attack.	
• Avoid eye contact with unknown men.	• Cross to the other side of the street when a male is approaching.	
• Check around and under car before entering	• Lock all car doors when driving or sitting in car.	
• Dress conservatively to avoid appearing sexy.	• Carry a weapon (gun, knife, pepper spray)	
• Park in well-lit areas at night.	• Leave lights on at home, inside and outside.	
• Monitor drinks while out or at a party to avoid drug spiking.	• Notify friends of plans when going out.	
• Avoid wearing high heels when walking alone to enable running away.	• Go out with friends rather than alone	

SOURCE: Pryor, D. & Hughes, M. (2013). Fear of rape among college women: a social psychological analysis. *Violence and victims*, *28*(3), 443-465.

Rape Myths—Blaming the Victim

Sexy clothes are not an invitation for rape.

You have seen that nearly all areas of human sexuality have their share of myths and misunderstandings, and you may have discovered some of your own along the way. Unfortunately, rape is no different. Rape myths may be among the worst because most of them place the blame on the victim in a false attempt to absolve the perpetrator (see Edwards et al., 2011; "Myths and Facts," 2008). As a result, these ignorant beliefs support an overall acceptance of rape among some men that leads to a staggering number of victims of this violent crime. Let's go through some of the more common and dangerous rape myths and set the record straight. Keep in mind that not all rapists share all these beliefs, but all rapists believe at least *some* of them.

MYTH *Women encourage rape by their dress and actions.* "If she didn't want sex, why did she dress so hot?" "She kept touching my arm and resting her hand on my shoulder. We started kissing and she was totally into it. The signals were clear that she wanted it." "When she agreed to come over to my apartment for more drinks, I just assumed she wanted to have sex." These statements are typical of the thinking of some men who rape. The mistaken assumption they make is that certain clothing and behaviors on the part of a woman actually led to her being forced into intercourse by him. This myth allows some perpetrators to feel justified in forcing sex because he thought "she's asking for it" by her choice of clothing or other actions (e.g., Edwards et al., 2011).

Since You Asked

4. If women don't want to be sexually attacked, why do they dress like they do?

TRUTH Can you think of one reason anyone would *want* to be raped? Of course not. That's like asking if someone is asking to be mugged for wearing a Rolex watch, beaten up for wearing a street gang's color in the wrong neighborhood, or burglarized for not having bars on windows at home. A woman may dress and behave in certain ways because she is interested in appearing sexually attractive or wants to meet someone for consensual sex. But no one wants to be raped. All people (in the United States, at least) have the right to dress and behave as they choose without giving others permission to commit a violent crime against them. Conversely, women will often avoid certain modes of dress in potentially vulnerable situations, but this is not because they think they will "cause" a rape. Rather, they know that some men believe in the "victim as cause" myth and are taking steps to protect themselves. The only person who can *cause* a rape is, by definition, the rapist.

MYTH *Men who rape simply lose control over their sexual urges.* This misconception relates closely to the preceding one. If men are seen as compelled to rape by the strength of their sexual drive, then it seemingly becomes the woman's responsibility to avoid any behavior that might provoke him. Once provoked, this myth contends, he becomes so crazed by sexual desire that he is unable to control himself.

TRUTH There are two main problems with this way of thinking. First, as mentioned earlier, rape is not primarily a sexual act. It is the rapist's attempt to exert power and control over a victim through force, coercion, and violence. Most rapists are not sex-deprived, and many have an ongoing relationship with a consensual sexual partner. Furthermore, when people commit other crimes involving violence, we do not attempt to excuse their acts because, as men, they are unable to control their violent urges. Second, if men really were helpless puppets of their sexual urges, rape would be even more universal than it is.

MYTH *Men who rape are mentally ill.* This falsehood takes us back to the stereotype of the rapist as a deranged stranger hiding in the bushes or coming through a bedroom window in the middle of the night.

TRUTH Although some rapists may have psychopathic tendencies or other psychological disorders, mental illness among rapists is no more common that among the population in general (Eher et al., 2011). The vast majority of rapes, 80% to 90% in fact, are committed by someone the victim knows, is dating, or is involved with in a romantic relationship. Believing that rapists are mentally ill leads to a dangerous perception that potential rapists can be identified by their deviant behavior. This may lull potential victims into a false sense of safety with acquaintances, dates, or partners who are clearly not mentally ill, but with whom the threat of rape is, statistically, the greatest.

Since You Asked

5. I've heard that some women really want to be raped because that way they can have sex without taking any responsibility for it.

MYTH *Women secretly want to be raped.* This belief stems from fictional novels, movies, and magazine stories as well as from the fact that some women's sexual fantasies involve rape motifs. The reasoning is that if women fantasize about being raped, they must, "deep down," really want to be forced to have sex.

TRUTH Although it is true that some women have reported fantasies involving rape and that such fantasies are sometimes exploited in various media, this in no way implies that women want to be raped in real life (see Chapter 6, "Sexual Behaviors, Experiencing Sexual Pleasure" for a more detailed discussion of sexual fantasies). First of all, many people entertain fantasies, sexual or otherwise, that they would never want to happen in reality. Furthermore, in a fantasy of forced sex, the fantasizer knows that it is not real and that she is not at any real risk of bodily harm. In fantasy, the fantasizer is in control, whereas in a real rape, all control is taken away by the rapist. Most women who fantasize about being raped actually maintain control in the fantasy by luring or seducing the pretend "attacker" and playing the role of victim, knowing she is ultimately in control of the entire episode (Bivona et al., 2012; Strassberg & Lockerd, 1998).

MYTH *Any woman can resist if she really wants to.* This false belief assumes that a woman who truly wants to avoid being raped can do so by fighting, running, or repositioning her body ("she could have fought him off if she wanted to!"). This ties in directly with the misguided belief that women who are raped secretly want to give in and that clearly her saying no, repeatedly demanding that the attacker stop, and physically struggling and fighting against the attacker are only token resistance.

TRUTH This myth is full of flaws. First, men, on the whole, are physically stronger and heavier than women and are capable of pinning a woman down, forcibly removing her clothes, and penetrating her. Beyond this, however, rapists use threats of violence against her or her loved ones, actual violence such as hitting and punching, and weapons to create a situation in which fighting back is obviously futile or not even a choice. In addition, some rapists commonly prey on women who are unable to resist due to extreme alcohol or other drug intoxication or a disability.

Threats of physical harm or death may intimidate a sexual assault victim to submit rather than fight back.

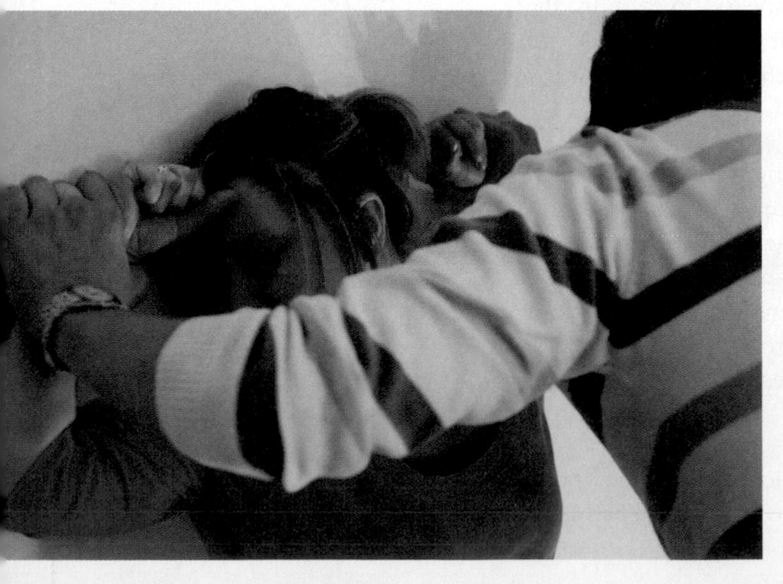

MYTH *Women falsely accuse men of rape.* The thinking here is that a woman may regret or feel guilty about having intercourse with a man, be angry with him, or be seeking revenge against him for some past offense, so she accuses him of raping her when no rape has taken place.

TRUTH The false reporting of rape is no more common than the false reporting of any violent crime; it is very rare. When it does happen, the woman nearly always recants before the crime is prosecuted. The exact opposite of this myth appears to be true. Many survivors of rape choose *not* to report the crime to authorities because of shame, guilt, fear of retaliation, or confusion over whether what happened to them was truly rape.

MYTH *Most rapes are committed by strangers.*

TRUTH As discussed earlier, this is simply not true. If you take another look at Figure 13.2, which identifies the relationships of women to their rapists, you can see that only a small fraction of rapes are committed by strangers. The vast majority of the assailants are known to the victim—they are friends, classmates, acquaintances, boyfriends, husbands (or ex-husbands), or other family members. Rapes committed by strangers are terrible and traumatic (as are all rapes), but they are the *exception*, not the rule. The perception that rape is primarily committed by strangers can interfere with the important awareness of the much larger problem of rapes committed by men victims already know.

MYTH *All men are capable of rape.* Again the subtext of this myth is that it is in men's nature to force sex and therefore it is up to the woman to protect herself from it.

TRUTH The vast majority of men have not and would never engage in any sexually assaultive behaviors, including rape. The number of victims of sexual assault is significantly higher than the number of perpetrators. We can infer from this that rapists have multiple victims. One study of university men found that 6% admitted to rape or attempted rape, and of those, 63% admitted to committing an average of six rapes each (Lisak & Miller, 2002). A disturbingly high proportion of college and university men engage in behaviors that comprise some sort of sexual assault, including rape (e.g., Parkhill & Abbey, 2008). We will discuss this high rate of rape on university campuses in the next section.

Since You Asked

6. Someone told me that all men, if they got the chance and knew they would never be caught, would commit rape. Is that true?

MYTH *Some women become aroused and have an orgasm during a rape, so they must enjoy it.*

TRUTH Part of sexual arousal and orgasm in women and men is made up of reflexes over which we do not have any intentional control. These responses may occur during a sexual attack, just as a man may experience an erection and ejaculate while being raped, as discussed earlier. These reflexes in no way imply than any victim "enjoyed" or consented to being raped (Levin & van Berlo, 2004).

Rape on Campus

Two facts about rape are clear: it is predominantly committed by a rapist who is acquainted, to some degree, with the victim, and the victims of rape are primarily young. Although women of any age can be and are raped, 40% of female victims are between the ages of 12 and 24 (see Figure 13.3).

Contributing to these statistics is the unfortunate reality that rape and sexual assaults are a widespread and serious problem on college campuses. During their college years, one in five college women experiences at least one completed sexual assault (CDC, "Sexual violence," 2012; Krebs et al., 2009). Moreover, stranger rape is even more uncommon on campus than in the general population; only 3% of college rapes are committed by strangers. It follows, then, that nearly all rapes on college and university campuses occur between people who know each other. Why? What is turning college life into an environment of rape? Research has focused on several factors that contribute to the problem of rape on college campuses.

Figure 13.3 Female Victims of Rape by Age Group

SOURCE: Data from the Bureau of Justice Statistics, National Crime Victims Survey (2009), http://bjs.ojp.usdoj.gov/content/pub/pdf/cv09.pdf.

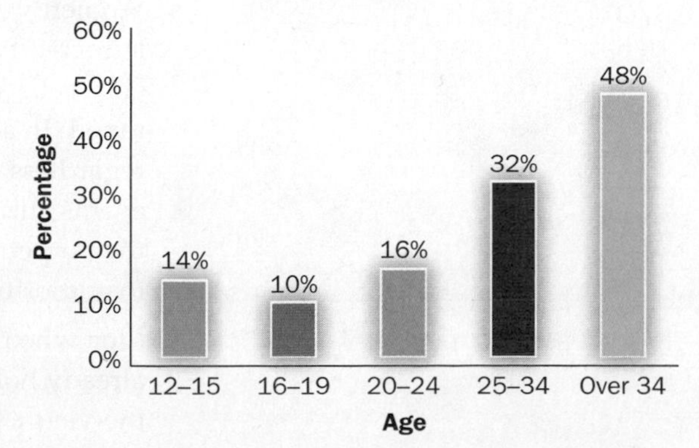

The Alcohol–Rape Connection

Alcohol should never be seen as *causing* rape. Clearly, many men and women drink alcohol as a routine part of their social interactions and

Since You Asked

7. My roommate went to a party last weekend and got totally wasted (drunk). Two of the guys at the party told her that two other guys had sex with her while she was passed out, but she has no memory of this. Could this really have happened? Why would guys want to have sex with someone who is passed out?

are never involved in sexual assaults. Conversely, rapes happen when neither the perpetrator nor the victim has had any alcohol at all.

However, alcohol plays a frequent and important role in many rapes, especially acquaintance and date rapes, and in college rapes in particular (as do other drugs, which will be discussed shortly). Studies have shown that alcohol use by the victim or perpetrator is involved in the majority of all college rapes (Testa & Livingston, 2009; Krebs et al., 2009). A great deal of research suggests that alcohol facilitates rape on college and university campuses through a combination of the following factors (Abbey, 2002; Abbey et al., 2001; Carr & Van Deusen, 2004; Krebs et al., 2009; McCauley, 2010; Mohler-Kuo et al., 2004):

- Drinking alcohol goes hand in hand with college life. The percentage of college students who drink in excess is significantly higher than in the general population.
- Most alcohol use occurs in social settings, primarily parties at off-campus apartments or fraternity houses, where men and women go with the express purpose of meeting others, having fun, and getting drunk.
- Alcohol, as a drug, inhibits brain centers that are responsible for judgment, problem solving, impulse control, and recognizing the future consequences of behavior. When drinking, men are more likely to interpret a woman's friendliness as sexual interest and become less able to choose nonaggressive routes to sexual satisfaction.
- Alcohol is increasingly considered a **date rape drug**, contributing to what is known as **incapacitated rape** (discussed specifically in the next section).

date rape drugs
Powerful sedatives that render a rapist's potential victim unconscious or otherwise unable to resist; also known as *club drugs*.

incapacitated rape
A rape that occurs when the victim has been rendered unable to take defensive action, usually accomplished through the use of alcohol or date rape drugs.

- Alcohol is known to increase the tendency toward all forms of interpersonal aggression, especially in men, who tend to be more aggressive than women in general.
- Alcohol increases a potential victim's vulnerability, mentally and physically. It interferes with a woman's ability to read a date's intentions and may cause her to assume he understands that she is not interested in sex when, also due to alcohol, he understands just the opposite. Furthermore, alcohol interferes with physical coordination and may limit a woman's ability to effectively resist a sexual attack. College judicial files are full of cases of rape in which the victim was too drunk to walk, had passed out, or was throwing up during the rape.
- Men use alcohol to exploit the vulnerability of potential victims. A common shared goal of men at fraternity parties is to get women drunk in order to loosen up their sexual inhibitions. Some male predators prey on women by keeping tabs on those who are drinking the most and becoming the most intoxicated and target them for sexual assault.
- Men believe that alcohol enhances their sexual prowess. Physiologically, of course, this is not true, but the *perception* of greater sexual abilities through alcohol use increases the likelihood of rape.
- Women who have been drinking at the time they were raped are often seen as less credible when reporting the assault. This is because many people mistakenly believe that if a woman was drinking, she shares the responsibility for being raped. In addition, alcohol in larger quantities often produces memory blackouts regardless of whether there was a loss of consciousness. A spotty memory of the events surrounding and during the rape makes investigation and prosecution of the perpetrator difficult. Knowing this, the victim is less likely to report the rape to authorities.
- Men who rape often use alcohol as an excuse to justify their behavior. A man who already holds stereotypic beliefs about rape and who believes in various "blame-the-victim" rape myths is more likely to mistakenly think that when a woman

is being friendly, she is, to his way of thinking, coming on to him. This perception may be enhanced by alcohol and provide him with an excuse to rape.

- Many sexually aggressive men believe that women who drink moderate to heavy amounts of alcohol are "fair game" for sexual coercion and that by drinking they bring the assault on themselves.

Alcohol use and abuse contributes significantly to sexual aggression on college campuses.

Date Rape Drugs

Alcohol is not the only drug that plays a role in incapacitated rape on college campuses, although it is the most common one. One of the most disturbing trends in sexual assault over the past 20 years has been the increased use of so-called "date rape drugs," also sometimes referred to as "club drugs." These are powerful sedatives that render a potential victim unconscious or unable to move. They are referred to as date rape drugs because a sexual predator will sometimes slip a dose into a date's drink, rendering her incapacitated and unable to defend herself against a sexual assault (see Womenshealth.gov, 2012). When combined with alcohol, these drugs can create an interaction that can prove fatal. These drugs dissolve quickly, are tasteless and odorless, and typically cannot be detected by the victim. You should be aware that the name "date rape drug" is somewhat deceiving in that sometimes the rapist is not the victim's date, and may not even be known to the victim. Therefore, these are also referred to as "predator drugs" and the crime as, "drug-facilitated sexual assault." Some of the most common rape drugs are reviewed next (see womenshealth.gov, 2012).

Since You Asked

8. Is there any way to tell if someone has slipped a roofie into your drink (I mean, before you drink it)?

Roofies. Rohypnol pills, commonly known as "roofies," "circles," or "R-2," contain the tranquilizer flunitrazepam, which suppresses the wakefulness and pain receptors in the brain. In Europe, Rohypnol is prescribed as a sleeping pill or surgical preanesthetic; in the United States, it is illegal but is the most common date rape drug in current use. In pill form, Rohypnol is white or off-white; when dissolved in liquid, it is impossible to detect by odor, taste, or color. It may also be obtained in a clear liquid form. In an effort to make the use of these drugs more discernible, the manufacturer now makes Rohypnol tablets that contain a blue dye and dissolve more slowly.

GHB. Gamma hydroxybutyrate (GHB), sometimes called "G" or "liquid ecstasy," is another drug that can be used to incapacitate a potential rape victim. This drug was legal in the United States, primarily as a bodybuilding food supplement (Marsa, 2002), until the early 1990s, when the FDA banned its sale after studies revealed that it could be deadly. GHB comes in the form of an odorless, colorless liquid that tastes salty or a white powder or capsule. The strength of the drug varies according to who is manufacturing it and how it is formulated, but overall the effects of GHB begin within ten minutes to one hour after ingestion. Low doses may have various effects, including sleepiness, increased sex drive, memory loss, hallucinations, headache, and loss of muscle reflexes. Larger amounts may lead to nausea, vomiting, difficulty breathing, seizures, unconsciousness, coma, and even death (especially when combined with alcohol).

Special K. Ketamine, or ketamine hydrochloride, sometimes called "Special K" or "K" on the streets, is yet another drug used for rape. It was developed in the 1970s as a surgical anesthetic for humans and animals; it is now illegal to use in the United States except under a doctor's supervision. Taken orally, ketamine begins working in 10 to 20 minutes and can last up to 48 hours. Ketamine causes increases in heart rate, blood pressure, and oxygen consumption (Long, Nelson, & Hoffman, 2002). It has increased in popularity as a club drug because it may also cause hallucinations, memory loss,

In Touch with Your Sexual Health

Have I Been Drugged and Raped?

How can I tell if I might have been drugged and raped?

It is often hard to tell. Most victims don't remember being drugged or assaulted. The victim might not be aware of the attack until eight or twelve hours after it occurred. These drugs also leave the body very quickly. Once a victim gets help, there might be no proof that drugs were involved in the attack. But there are some signs that you might have been drugged:

- You feel drunk and haven't drunk any alcohol—or, you feel like the effects of drinking alcohol are stronger than usual.
- You wake up feeling very hung over and disoriented or having no memory of a period of time.
- You remember having a drink, but cannot recall anything after that.
- You find that your clothes are torn or not on right.
- You feel like you had sex, but you cannot remember it.

If you think you may have been drug-raped, seek medical care right away. Call 911 or have a trusted friend take you to a hospital emergency room. Don't clean up the scene, urinate, douche, bathe, brush your teeth, wash your hands, change clothes, or eat or drink before you go. These things may give evidence of the rape. The hospital will collect physical evidence. Call the police from the hospital. Tell the police exactly what you remember. Be honest about all your activities. Remember, nothing you did—including drinking alcohol or doing drugs—ever justifies rape.

SOURCE: Womenshealth.gov, 2012. www.womenshealth.gov/publications/our-publications/fact-sheet/date-rape-drugs.html#c

dreaminess, numbness, paralysis, and out-of-body experiences. Recreational users of the drug call these out-of-body feelings "entering a K-hole."

In addition to sedating properties, all of these drugs produce an amnesia effect, so victims of a rape that occurred while under the drug's influence typically have no memory of the event. These are, obviously, dangerous drugs. They can lead to death, especially when used in combination with alcohol or other recreational drugs. When people ingest one of these drugs, they rarely know the exact dose they are receiving or what other dangerous chemicals may have been added, such as methamphetamines, LSD, or even drain cleaners.

However, because the drugs are metabolized quickly by the body and are difficult to detect in the bloodstream the following day, combined with the fact that the victims have very vague memories or no memory at all of the attack, prosecution of drug rapists has been difficult. If you or someone you know might have been victimized by a drug-rapist, it is important that immediate steps be taken. "In Touch with Your Sexual Health: Have I been Drugged and Raped?" discusses appropriate step to take.

Federal and state laws on the books provide severe penalties for the crime of drug-facilitated rape, and the drugs themselves are illegal substances in the United States. Laws are also being enacted on the state and federal levels to define alcohol as a rape drug; this could significantly increase prosecutions and penalties. "Sexuality, Ethics, and the Law: Date Rape Drugs" offers a closer look at these drugs and how women can reduce their risk of victimization.

UNDERREPORTING OF RAPE Rape is one of the most underreported crimes in the United States, and this is especially true on college campuses. Estimates of rapes that are not reported by victims range from 60% to 90% of the total rapes and on some college and university campuses, the number may even be higher. One study in the 1990s of 140 women at Skidmore College in New York found that of the women who had experienced rape at the college, not a single one reported it to the authorities (Finkelson & Oswalt, 1995).

Nonreporting is most common for acquaintance and date rapes. The reasons behind this underreporting are many and complex. Many victims assume, wrongly, that the assault "wasn't really rape, things just got out of hand," or that somehow she led him on or gave him "mixed signals." With alcohol involved in so many college

Sexuality, Ethics, and the Law

Date Rape Drugs

Jenna, a college sophomore, was having a great time at the off-campus party. Trevor, an attractive guy, was paying a lot of attention to her. She liked him and thought he might offer to drive her home.

"How about another beer?" Trevor asked.

"Sure," replied Jenna.

"Be right back with it," Trevor said.

As she sipped the drink Trevor gave her, Jenna chatted happily and was surprised when she started feeling dizzy and drunk. This was only her second beer, and she usually didn't feel this way so quickly.

"Are you OK?" Trevor asked. "You look a bit shaky."

Jenna held her head in her hands. "No... I don't feel well. I think I'd better go home."

"Why don't I give you a ride?" Trevor suggested.

"OK, thanks," Jenna said.

By the time they reached Jenna's house, she could barely stand up. Trevor carried her inside and put her on her bed. Then she passed out and woke up eight hours later feeling nauseated and weak.

Jenna was surprised to find her clothes on the floor. She also ached all over, as if someone had hit her repeatedly. She tried desperately to remember what had happened, but her last memory was of Trevor driving her home from the party.

"Oh no, he must have raped me!" she whispered. "But how can I prove it? I don't remember anything."

Jenna had indeed been raped. Trevor had slipped a tablet of the drug Rohypnol into her drink at the party. The tablet had dissolved quickly, leaving no taste, odor, or color. Knowing what the drug would do to Jenna, Trevor had offered to drive her home with the intent of raping her.

Unfortunately, Jenna's experience is happening all-too-frequently to other women since Rohypnol and the other so-called date rape drugs (GHB and ketamine) were introduced into the United States in the early 1990s.

In 1996, the U.S. federal government passed the Drug-Induced Rape Prevention and Punishment Act. This law makes it a felony to give an unsuspecting person a drug with the intent of committing violence, including rape, against him or her. The law also imposes penalties of large fines and up to twenty years in prison for importing or distributing more than one gram of these drugs. However, even with this law in effect, the use of date rape drugs is growing. These drugs are out there, and it's up to you to make sure you aren't a victim. Here are some suggestions to help stay safe from predator drugs, especially in college party settings:

- Avoid parties where you know everyone will be drinking a lot of alcohol. It's more likely you would encounter date rape drugs on such occasions.
- Always keep your eye on your drink at any party or on a date. Don't put the drink down and leave it unattended, even to go to the restroom or to greet a friend on the other side of the room.
- Don't accept a drink of any sort from someone else, particularly if you don't know the person well.
- Never drink anything from an open container (beer bottle, soda can) unless you opened it yourself.
- Avoid drinking from a punch bowl or other communal container; someone (or a group of people) may have drugged it.
- Go to parties with close friends and watch out for each other. If you leave the party, tell a friend where you're going and with whom.
- If you think someone drugged you, call 911 and get someone to take you to the hospital to be tested for drugs—and for rape treatment, evidence collecting, and counseling.

SOURCES: Excerpted from Abramovitz (2001), pp. 18–21; Womenshealth.gov, 2012. www.womenshealth.gov/publications/our-publications/fact-sheet/date-rape-drugs.html#c

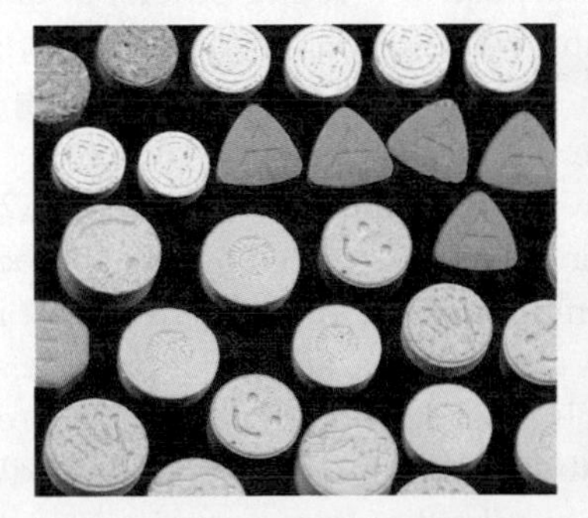

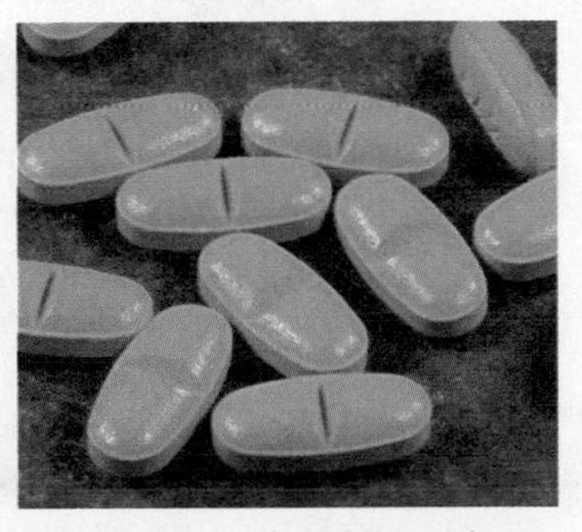

Date rape drugs come in many shapes and sizes, as well as in liquid form.

rapes, victims sometimes mistakenly blame themselves because they had been drinking. Victims know that their attacker will deny the rape (many acquaintance rapists even feel they have committed no sexual assault at all). Therefore, many victims see their case as a "she said, he said" situation and may fear becoming involved in a long, public judicial or legal court proceeding. They may also fear being harassed or threatened on campus by their assailant, or they may wish to avoid being judged negatively by friends, peers, or future potential dating partners. Many victims choose not to report rape out of a perhaps justifiable fear of being victimized again.

Whatever the reason, underreporting is a serious obstacle in the effort to reduce the incidence of rape on campus for several reasons. First, without accurate statistics about the prevalence of rape on campus, school officials will not likely give the problem the attention and resources it needs. Second, underreporting may lull potential victims into a false sense of security and cause them to fail to take the precautions necessary to protect themselves from sexual assault. Finally, and perhaps most important, men who rape rarely do so only once. For most of them, sexual violence represents a pattern of behavior over time and across situations. Obviously, as we have stated, not all men rape. But when a rape goes unreported, the rapist goes unpunished for his behavior. Thus, the rapist is not only free to engage in the same actions again but may even feel encouraged to do so, now that his belief that he has done nothing wrong has been tacitly supported.

The Culture of Sexual Coercion in Fraternities and Athletic Groups

Not all fraternities or male athletic groups, and certainly not all fraternity members or athletes individually, "support" rape by any means. Indeed, many fraternities and college sports programs have begun to take steps to reduce the incidence of rape both within their groups and throughout the larger college community. However, historically, fraternities, frat houses, and athletic groups have been notorious for tolerating or even encouraging rape (Bannon, Brosi, & Foubert, 2013; Humphrey & Kahn, 2000). Although the sexually aggressive culture of these groups is beginning to change, it persists at many colleges and universities.

Researchers have suggested that some fraternities and university athletic groups encourage a "culture of rape" by socially rewarding coercive sexual behavior directly or indirectly through membership, acceptance, camaraderie, and a sense of belonging (Brown, Sumner, & Nocera, 2002; Foubert & McEwan, 1998). Many studies have also found that men in fraternities and sometimes even women in sororities tend to accept the rape myths (discussed earlier) that blame the victim and, therefore, reduce the responsibility that should be placed on the male attackers leading to less condemnation of the crime of rape (Bannon, Brosi, & Foubert, 2013). Furthermore, the use of alcohol as a date rape "tool" may often be observed in the fraternity party setting. You can imagine how a fraternity party, where young adults are gathered for the purpose of meeting others, in an unsupervised location full of bedrooms and beds and where alcohol is flowing freely, would easily create an environment for rape.

Various programs for reducing rape within the Greek fraternity system have been suggested and evaluated (Anderson & Whiston, 2005; Foubert, Godin, & Tatum, 2010). For example, Binder (2001) pointed out that, to be effective, any rape prevention approach must be comprehensive, must address risk reduction strategies as well as alcohol and substance abuse issues, and must provide sexual assault awareness training. In addition, schools must develop clear policies, procedures, and consequences for sexual assault and distribute this information repeatedly and actively (i.e., not merely state it in the student handbook) to everyone in the college community. Colleges and universities must *anticipate* problem situations for sexual violence. Addressing the problem of rape on campus in proactive ways, such as requiring workshops on sexual aggression for high-risk groups—including first-year women, new fraternity members, and athletes—may help head off potential problems.

A national program on many campuses across the U.S. is designed to reduce the incidence of sexual assault focuses on all-male, high-risk groups, primarily fraternity members, and college athletes. *Mentors in Violence Prevention* (MVP) concentrates on raising awareness of sexual violence, altering beliefs about women and sex, providing opportunities for open dialogue about sexual violence issues, motivating participants to become leaders in reducing sexual assaults, and inspiring immediate, on-the-scene *bystander intervention* in potential rapes and sexual assaults (Banyard, Moynihan, & Crossman, 2009; Katz, 2011;). The following exercise occurs early in the Mentors in Violence Prevention program. It is a visualization in which the team, fraternity members, or other all-male group is given the following instructions:

> Take a few deep breaths.... Close your eyes and imagine the woman closest to you—your daughter, sister, mother, or girlfriend.... She may be at a party, in a room, or walking down the street when she is approached and sexually assaulted by a man.... Now imagine that there is a third person at the scene, a bystander who sees what's happening to the woman you love and is in a perfect position to help, but chooses to do nothing.... The bystander either watches or walks away. (O'Brien, 2001, p. 147)

The group is then asked: How did it feel to imagine the woman you care about being assaulted by that man? Ask yourself the same question. The responses most commonly given by participants in this workshop as to how they felt are "helpless," "vengeful," "furious," and "confused." When asked to describe the bystander, typical characterizations are "gutless," "just as guilty as the attacker," "a punk," "a coward," and "scared." Later, members of the group are asked to place themselves in various scenarios in which they become the bystander to a sexual assault. You can see how their view of such a situation is often changed by the first visualization.

Another program targeting high-risk groups, called *Male Athletes Against Violence* (MAAV), has been developed by Professor Sandra Caron at the University of Maine. MAAV is becoming a model for similar programs across the country in helping to raise awareness among male athletes and reduce sexual violence among this group of young men. The pledge signed by MAAV members is strong and very clear:

I pledge:

- To educate myself on issues surrounding violence while developing personal beliefs against the use of violence
- To be a positive role model for my community.
- To look honestly at my actions in regard to violence and make changes if necessary.
- To educate others on all issues surrounding violence.
- To take initiative! I will be prepared and have the courage to correct others regarding violence.
- To support women's groups. I may not be able to prevent all violence but I will help those who have been victimized.
- To be aware of pre-existing beliefs, stereotypes or rumors that may alter my judgments.
- To challenge the social norms. If no one ever challenges the norm, nothing will change.
- To learn to identify problematic situations to help victims (MAAV, 2014).

Sexual Violence in the Military

Over the past decade, a serious, widespread problem within the ranks of the U.S. military has come to light: Rape by male solders of female soldiers. This problem has grown since women have been able to serve in the armed forces in greater numbers and in a greater variety of roles. However, the extent of the problem and the efforts to stop it have come to the forefront in the military, the U.S. Congress, and the awareness of the public relatively recently (e.g., Baltrushes & Karnik, 2013; Carlson, 2013).

This problem in the military in general, and the individual reports of sexual violence from survivors have placed the armed forces under a critical and very unbecoming microscope, and they have been on the defensive. There appear to be several obstacles to addressing this growing scandal: (1) The military has always stressed a culture of "protecting our own," (2) criminal acts, including rape, are reported to, investigated, and disposed of by the soldiers' chain of command, that is, the vertical command structure under which the soldiers serve, (3) victims are unwilling to come forward due to fear of blame, retribution, or damage to their military careers, and (4) most of the ranking officers charged with working toward solving this problem have themselves been found to have engaged in sexual aggression (another clue as to just how widespread the problem). "Sexuality and Culture: Rape in the Military": discusses this problem in greater detail.

Can We Predict Who Will Rape?

In taking precautions to protect themselves from sexual assault, many women want to know if certain warning signs might alert them that an acquaintance or date is a potential rapist. The answer appears to be only maybe. Men who rape share some characteristics, attitudes, and behavior patterns, but "profiling" has not been shown to be an effective rape prevention strategy. Nevertheless, as part of enhancing a woman's overall awareness of the risk of rape, a general working knowledge of some of the possible warning signs of a "rapist personality" might be useful in raising a red flag of caution (Beauregard, 2010).

What follows are admittedly generalizations, but a woman should be extra careful if a man she's seeing seems to be a very impulsive or aggressive person overall (Bridon & Greier, 2011; Chapleau, Oswald, & Russell, 2008; Harney & Muehlenhard, 1991; Lisak, 2008; Prentky & Knight, 1991). For example, if he:

- Displays a need to dominate and control events around him;
- Is emotionally abusive, discounts her opinions, insults or belittles her;
- Is excessively and irrationally jealous; expresses any of the rape myths discussed earlier;
- Exhibits any physical violence toward her or others even if it's minor pushing, grabbing, or "pretend" fighting;
- Believes in very traditional male and female roles;
- Has a very short anger fuse;
- Intimidates her by crowding her, blocking her way, or touching her in intimate ways without her consent;
- Is fascinated with various weapons;
- Abuses alcohol or other drugs;
- Bullies others;
- Is cruel to animals or children

It goes without saying that a great many men have some of these characteristics but would never dream of sexually assaulting anyone. However, rapists seldom possess

Sexuality and Culture

Rape in the U.S. Military

The following excerpt is from a 2013 article titled, "The Military's culture of Sexual Violence," in the online news service, *Bloomberg View*.

> Last year [2012], the Pentagon received 3,374 reports of sexual assault, according to its Sexual Assault Prevention and Response Office. (The actual number of assaults is probably closer to 26,000, the office says.) Of those 3,374, almost 1,000 were deemed baseless or outside the military's jurisdiction and several hundred were dropped by commanders as unfounded or for other reasons...
>
> It's true, as the military is fond of saying, that the great majority of military officers are law-abiding. But when a fellow service member is accused, the law-abiding tend to side with the accused. Reporting a rape is never easy, but it's much harder when the perpetrator is of higher rank than the victim (50 percent of the time) and when the perpetrator is in the victim's chain of command (23 percent of the time). Join the military, where you may be more prone to sexual attack and you don't even get the protections, however flawed, you would get at your local police precinct, because the brass close ranks. (Carlson, 2013, p.1).

This quote sums up well the problem of sexual aggression within the US military. You can see for the often-cited percentages that only 13% of rapes in the military were reported. Although, as discussed above, rapes are often underreported, this percentage is far lower than in the civilian population. What should a female soldier do when she has been raped by someone of a higher rank, suspecting that any report she files may simply be lost in the chain-of-command procedures that are embedded in the "protect our own" military mentality? Moreover, because rape cases are often the perpetrator's word against the victim's, a lower ranking female knows that her accusation may carry very little weight against her higher ranking attacker's claims that sex was consensual (see Baltrushes & Karnik, 2013).

You can imagine how terrible it must be for a military woman to have to salute her attacker the next day, weeks, and months after the event. And if a jury in a military court marshal does convict a rapist, his commanding office may, under military law, overturn the conviction without explanation. Many of these women return from service bearing not only the trauma of serving in violent conflicts, but the lasting stress that accompanies sexual assault.

Various paths are being explored to reduce this problem. Structures and procedures within the military are being altered, the Secretary of Defense and the President of the United States are speaking out and sending the message that this must stop, and Congress is considering at least two major bills to increase the ability of victims to come forward without fear of negative consequences and maximize accountability and successful prosecution and punishment of perpetrators. One congressional effort focuses on greater oversight of the military's internal handling of sexual violence. The other seeks to remove military sexual assault cases from the military's justice system (take it out of the chain of command) and assign cases to civilian legal jurisdiction. As of 2015, these approaches were being debated in Congress (Lerman, 2013).

Victims of Rape in the Military have little recourse in a system that protects perpetrators and punishes whistleblowers.

only a few of these traits; they will usually exhibit many of them. A working knowledge of this list will help women avoid risky situations with potentially dangerous men.

What to Do If You Are Raped

Ideally, neither you nor anyone you know or care about will ever have to face the trauma of rape. However, the reality is that nearly a quarter of all college students, primarily women, will be sexually assaulted at some point by the end of their college years. Rape is a traumatic and terrifying event at the very least. Victims frequently have difficulty thinking clearly and are unsure what steps to take in the immediate aftermath of rape. What survivors *should* do is often quite different from what they probably *feel like* doing at the time. The first reaction of many victims is often denial and withdrawal. They want to be alone, shower or bathe, wash away any traces of the terrible event. The emotional jumble of confusion, pain, humiliation, embarrassment, anger, and powerlessness that typically accompanies rape often causes victims to wish they could just disappear and avoid seeing or talking to anyone. Although all of these reactions are understandable and normal, rape survivors should try to follow certain specific postrape guidelines (see "In Touch with Your Sexual Health: Guidelines for Rape Survivors"). Why? Because these guidelines can help survivors deal with and heal from the trauma of rape, and they may assist authorities in catching and prosecuting rapists and perhaps preventing them from victimizing others.

THE AFTERMATH OF RAPE Rape has profound effects on victims. Survivors of rape may experience shock over what happened to them; continuing fear of their attacker; terror over the idea of another attack; anxiety or depression over their loss of personal control over their own body; rage at the perpetrator or themselves that this could have

In Touch with Your Sexual Health

Guidelines for Rape Survivors

Every survivor has the personal prerogative of handling the trauma in whatever way he or she chooses. Many of the steps suggested here for survivors in the immediate aftermath of rape are difficult and often seem to be exactly the opposite of what the victim would be emotionally inclined to do. However, following these guidelines can both help a survivor cope more effectively with the trauma and assist in the arrest and prosecution of the rapist.

1. **REPORT.** Reporting the assault as soon as you are safe allows the authorities to take an immediate description and collect evidence. Most police departments today are trained in proper and sensitive handling of survivors of rape and will not pressure you into taking any actions against the rapist that you are hesitant to take.
2. **PRESERVE EVIDENCE.** Bathing, showering, douching (cleansing of the inside of the vagina with water or other liquids), washing hands, brushing teeth or hair, or changing bed linens or clothes may destroy evidence that will be important in charging, prosecuting, and punishing your attacker. Even if you feel you do not want to press charges or pursue the matter further, preserve evidence anyway, just in case you change your mind in the future (as many victims do).
3. **OBTAIN MEDICAL CARE.** Here are four very important reasons why a victim of rape should seek medical care as soon after the attack as possible:
   - A medical evaluation for physical or emotional injuries that may need professional attention.
   - The collection of various specimens and physical evidence that may be crucial to any charges that may be brought against the rapist.
   - Testing for STIs. Many STIs can be prevented (even without evidence of exposure) with quick treatment. In particular, anti-HIV medications have been shown to reduce the potential for transmission dramatically if received within 72 hours of exposure through rape (Roland et al., 2005).
   - Obtaining emergency contraception if the survivor is at all concerned about the possibility of pregnancy resulting from the attack.
4. **SEEK EMOTIONAL SUPPORT.** Remember: rape is *never* the victim's fault. Seek out support from your network of trusted friends, family, or religious advisers. If you feel uncomfortable talking about this experience to people close to you, call a local rape hotline in your area (hotlines are listed in the phone book) or consider seeing a college or other professional counselor. If you need more information on where to find support and counseling, you may contact the National Sexual Assault Hotline at (800) 656-HOPE (4673), or visit the Rape, Abuse, and Incest National Network website at www.rainn.org. The most difficult course of action is to try to cope with the trauma of rape alone.

Again, if you are a victim of rape, you will probably be tempted to ignore all of these steps. Your hesitancy is normal and understandable. But if you follow these guidelines, you'll be much more likely to enhance your healing, speed your recovery, and help prevent a rapist from preying on other victims.

happened; and embarrassment, guilt, and shame stemming from (their false) perception that the rape was somehow their fault (Barglow, 2014; Pryor & Hughes, 2013). All of these feelings are normal reactions to being raped and they can stay with a victim for weeks, months, years, and even a lifetime. When a survivor becomes consumed by these feelings for a prolonged period of time, professional counseling may be indicated, not to make it somehow OK, but to allow the survivor to move past the trauma in their own life.

Since You Asked

9. My sister was raped over three years ago, but she's still very upset and angry about it. What can she do to get over it? Will she ever get over it?

During the 1970s, researchers described a sequence of emotional, psychological, and physical reactions that survivors of rape typically experience following the attack, called the **rape trauma syndrome** (e.g., Burgess & Holmstrom, 1974). Although this is no longer a formal clinical diagnosis, the syndrome illuminated a framework for understanding some of the psychological and physiological phases victims must endure following rape (see Brodsky, 2010).

rape trauma syndrome

A two-stage set of symptoms that follow the trauma of being raped, consisting of physical, emotional, and behavioral stress reactions.

posttraumatic stress disorder (PTSD)

A pervasive psychological and emotional reaction to a traumatic event experienced or witnessed (such as war, natural disasters, rape, etc.), including symptoms of anxiety, flashbacks, hypervigilance, irritability, nightmares, and depression.

Rape trauma syndrome parallels the symptomology and progression of the current diagnosis of **posttraumatic stress disorder (PTSD)**, a very well-documented response in the aftermath of many types of major psychological, emotional, and/or physical traumas, including battlefield experiences, natural disasters, acts of terrorism, and victimization by violent crime, especially rape. PTSD is characterized by traumatic flashbacks to the event, avoidance of any situation that may remind the survivor of the event, denial that the event happened, depression and social isolation, self-blame for the event, hypervigilance (a heightened state of arousal and watchfulness, an increased startle response, and exaggerated emotional reactions to events that may provoke memories of the event), and difficulty in resuming normal functioning (DSM 5). Rape is the most frequent cause of PTSD in women (Cloitre, 2004; Zinzow et al., 2010).

The symptoms accompanying the trauma of rape may be divided into several phases ("Butcher, Hooley, & Mineka, 2013; "Phases of Traumatic Stress," 2014): the *anticipatory phase*, when the horrible realization sets in that a rape may happen; the *impact phase*, experiencing the terror of the event itself; the *recoil phase*, surviving the trauma in the hours and days immediately after; and the longer-lasting *reconstitution phase*, when the stress of the rape is assimilated, told to others, and perhaps counseling and legal steps are taken. The reconstitution phase may take many months or, in some cases, years.

Rape trauma involves physical, emotional, and behavioral stress reactions that result from facing the life-threatening, violent, and traumatic aspects of the act. In the anticipatory and impact phases, victims experience intense and overt emotions of fear, anxiety, and anger. Following the attack, during the recoil phase, survivors usually feel a deeper, less outwardly obvious shock and numbness, which may in some cases hide their true emotions. This phase also may include physical responses to the traumatic event, such as sleep disturbances, appetite and digestion problems, and various other pains and symptoms relating to the attack. As the recoil phase progresses, survivors begin to experience more complex emotions that may alternate unpredictably between fear and anger and humiliation, degradation, shame, and guilt.

As more time passes, survivors enter the reconstitution phase, which often includes changes in lifestyle—moving away, changing one's phone number, hiding out—anything that helps victims feel safer. They may get a dog; install alarms on their residence; stay with friends or family members; and even obtain weapons such as pepper spray, knives, or guns for personal protection.

During this phase, survivors typically experience frightening nightmares that may reenact the rape itself or portray other situations in which the victims feel helpless and at the mercy of others. Fortunately, as survivors heal emotionally and physically, the nightmares usually decrease, and as they become psychologically stronger, the images become less frightening over time. Survivors should know that it is OK to talk about these feelings and nightmares with friends and family or a counselor. Talking them through often reduces their power and renders them less frightening.

Finally, in the months or years that follow, survivors may develop phobias; irrationally strong fears of situations such as being in crowds, being alone, or beginning new, potentially intimate relationships. If the anxiety does not fade, and the stress and anxiety continue to control their lives, they need to seek continued professional help (Butcher, Hooley, & Mineka, 2013).

Preventing Rape

No one ever deserves to be raped for any reason, and, as has been stressed throughout this chapter, the violence is *never* the victim's fault. However, every one of us is a potential victim, regardless of behavior, lifestyle, or personal characteristics. Some danger signs to watch for that may offer clues to potential rapists were listed earlier in this chapter. Beyond that awareness, everyone should incorporate as many strategies into their daily lives as possible to reduce the chances of being sexually assaulted. "Self-Discovery: Staying Safe from Rape" presents actions you can take to minimize your risk in various situations. The goal here is not to frighten or to suggest that you are somehow to blame but simply to enhance your safety. You should also be aware that even if a woman follows all of the suggestions on the list, a rape cannot be prevented in every case.

Self-Discovery

Staying Safe from Rape

No one "asks" to be a victim of any crime. To suggest that a woman is somehow responsible for being sexually victimized is as ludicrous as claiming that you are responsible for a robber taking your wallet at gunpoint. Nevertheless, we all take certain measures in various situations to protect ourselves from being victimized by crime and injury. Below are a number of precautions suggested by researchers and women themselves that can help a person reduce the risk of becoming a victim of a sexual assault. **Of course,** not **taking these precautions** never **implies the victim is at fault if an assault does occur**, but incorporating some of these suggestions into life routines may help to increase a person's safety from sexual attack.

Safety While Dating	Safety at Home	Safety While Walking	Safety While Driving
Be aware of your decreased ability to judge and react under the influence of alcohol or drugs; *stay sober.*	If a door or window is found forced open while you were away, do not enter. Silently leave and use a neighbor's phone or a cell phone to call the police. Wait until they arrive before entering.	Get to know the areas where you are likely to be walking. Know which stores and restaurants are open late in the evening. Watch for homes with lights on. If an attempt is made to attack you, run to these places.	Check your car's backseat before entering. When parking, select a place that will be well lighted if you return after dark. After dark, always try to walk to your car with others. If no one is available, call campus security for an escort.
Go to parties and other social events with friends, and agree to keep an eye on each other. Never leave your friend alone at a party, especially if she has been drinking.	List only your last name and initials in the phone directory and on your mailbox.	*Be aware and alert.* Always look around to see if you are being followed. If someone suspicious is behind or ahead of you, cross the street. Walk toward other people or lights. Don't be afraid to run.	When possible, travel on well-lighted streets.
Never leave a gathering with a man you do not know well. If you do leave with someone, be sure to tell a friend you are leaving and with whom.	Be sure to lock your doors when you are at home, even during the day.	Make eye contact with strangers. Don't look away or down.	Keep windows closed and doors locked at all times.
Make sure someone knows where you are, and check in with that person at a prearranged time.	Never open the door to your home immediately after a knock; ask who it is. Use a deadbolt and door viewer for identification.	Walk near the curb, and avoid passing close to shrubbery, dark doorways, and other places of possible concealment.	Lock your doors immediately upon entering your car. This is a common moment of attack.
Avoid isolated areas. Be cautious of areas such as empty houses, abandoned buildings, "parking spots," and so on.	Leave a light on at night. Set a timer so a light will be on if you will be returning home after dark.	Avoid shortcuts through isolated areas.	Keep your car in gear while stopped at traffic lights and stop signs.

(continued)

Safety While Dating	Safety at Home	Safety While Walking	Safety While Driving
Trust your intuition. If someone scares you or creeps you out in any way, steer clear.	If a stranger asks to use your phone, do not let the person enter your home. Offer to make the call on his or her behalf.	Have your keys ready in your hand so that your house or car can be opened immediately. Lock the door *immediately* upon entering.	If you believe you are being followed, pull over where people are present and let the car pass. If the other car stays behind you, drive to a safe, populated location.
If you have concerns, consider double-dating the first few times you go out with someone you do not know well.	Do not leave your name or any unnecessary details about yourself or whereabouts on your voice mail. Consider using a message in a male's voice and use "we" instead of "I."	When arriving home by taxi or in someone else's car, ask the driver to wait until you are safely inside.	If you are followed into your driveway at night, stay in your car with the doors locked until you can identify the occupants of the other car. If you are unsure, call the police.
Never be afraid or hesitant to communicate *"no"* firmly and forcefully. When you say *no*, mean it, and don't change your mind. Set sexual limits, and communicate them clearly to your date.	Secure all doors and windows against forced entry as best you can. Keep emergency phone numbers posted for easy access. Call 911 if you fear someone is trying to break in.	If someone in a car pulls alongside and orders you to get in the car, don't go, even if the assailant has a weapon. Yell, scream, and run away in a zigzag pattern toward a barrier such as a tree or another car.	Never leave your keys in the ignition. Even if you park for only a short time, take them with you and be sure your car is locked.
Find out where you are going and what time you will be home, and then tell someone.	Ask local law enforcement to make regular drive-bys of your home for increased security.	Carry your cell phone at all times. If the cost of a cell phone is a problem, low-cost phones for emergency-only use are available.	Park where you can get to your car without a potential assailant following or seeing you.
If you meet someone online through a dating service or social network, use great care in agreeing to any in-person meeting.	Never give any online person, no matter how well you seem to have gotten to know each other, any information about your true identity or location.		Meet in a public place at a crowded hour and if anything appears suspicious about the person or circumstances, leave.

SOURCES: Houston Area Women's Center (2008); Powell (1996).

Clearly, the issues of sexual assault and rape are seen far differently by most men and women. The precautions each sex takes (men appear to take none) and the attitudes and emotions of each sex surrounding these issues are as different as night and day. In fact, on this point, men and women have what could be described as separate *cultures*. Women have taken various actions to raise awareness of rape and to reduce its incidence (self-defense training, "take back the night" marches, alcohol and other drug education, etc.). The necessity of taking so many safety precautions against sexual assault may seem exasperating and terribly unjust. The frequency of these crimes, however, turns the cliché "better safe than sorry" into an unfortunate truism.

Men's Role in Preventing Rape

It is important to add that it is not only women who should work to reduce sexual violence. Men can play an important role as well. In fact, strong, caring men are often in the best position to help reduce the occurrence of rape. Men who rape are typically weak and insecure (in addition to violent) and believe in the myth that sexual aggression is a way to express their masculinity to themselves and to their peers. Furthermore, as discussed earlier, some all-male settings, such as fraternities and men's athletic teams, tend to promote a culture of rape acceptance. Here are some suggested guidelines that men might use to reduce or prevent sexual assaults and rape (see Powell, 1996, p. 150).

1. **Get to know your date as a person** (what a concept!). There is always time for sex; it doesn't have to be tonight. Besides, sex is better and safer when two people know each other first.
2. **Do *not* make comments that treat women like sex objects**; don't brag about sexual activities, and do not accept other men doing so. Call them on it, ignore them, or say, "I really do not want to hear that crap!"

3. **Do not try to get women drunk for sexual manipulation, and don't allow others to do so either.** Watch out for women's welfare at parties or other occasions where alcohol is present.
4. **If your date is drunk, do not have sex with her.** She is legally unable to consent to sex (or any other agreement) while intoxicated, and intercourse under those conditions may be considered rape, no matter what she says or does.
5. **Remember that *no* means *no*,** and if you're not sure, ask her what she does and does not want to do.
6. **Never *assume* that she wants the same amount of sexual intimacy as you do**. To you, kissing and touching may communicate a desire to move on to intercourse, but it may not mean that at all to her. Ask, "Are you comfortable with how things are going?" "How far do you want this to go?" It may sound corny, but most women love feeling honored and respected. In other words, don't assume that silence means "yes." Ask.
7. **Communicate with words what you want.** Are you unwilling to go out with her if there is no possibility of sexual intercourse? If not, you need to let her know that. If you are willing to wait for sex in the relationship, she needs to know that, too.
8. **If she doesn't want to have intercourse, that doesn't mean she doesn't want you**. She is making choices about her sexual behavior, which may be very different from her romantic feelings about you.
9. **You *can* stop**! Things with your date may be hot and heavy; your desire for sex may be at a fever pitch. But if there is any doubt about her full consent, just stop. It might be frustrating, but it won't kill you. Consider this: imagine that your mother were to burst in on you when your desire was at its peak. You could stop then, right?
10. **Understand and be sensitive to the size and strength advantage you probably have over most women.** Some women may feel intimidated and frightened by your physical presence without your even being aware of it. Some might *feel* forced, even if you have no intention of forcing.
11. **Spread this information to other men.** Not only is rape violent and wrong, but it also carries major criminal penalties. Determining and proving the identity of a rapist today is as simple as a DNA test. Most rapists today and in the future will be caught, tried, and convicted. The days of "getting away with it" are over.
12. **Support organizations of men that adopt antirape policies**, and work to promote antirape policies in groups of men to which you belong.
13. **If you know or suspect that a rape is being committed or may be committed, do everything in your power to intervene and STOP it.**

If more men understand how they can help in toppling the systems and attitudes that support rape, some of the motivation for committing rapes will likely disappear.

Child Sexual Abuse

13.2 Discuss the problem of child sexual abuse (CSA) including a summary of the perpetrators of this crime; the prevalence and incidence of CSA in the U.S. and globally; the psychological, emotional, and physical effects on victims; and the strategies to reduce and/or prevent CSA.

Although child sexual abuse is not just about rape, consider these sobering statistics: 42% of female rape victims were first raped when they were under 18; 30% were first raped between the ages of 11 and 17; and 12% of female victims and 28% of male victims were first raped then they were under age 11! (CDC, "Sexual

Violence," 2012). Tragically, rape is only one of many coercive sexual behaviors perpetrated by adults against children, both boys and girls. The sexual crimes committed on children by adults include inappropriate kissing, fondling of the child's genitals, forcing the child to fondle the genitals of the adult, oral sex on the child, forcing the child to perform oral sex on the adult, vaginal penetration with the child, and anal sex with the child.

Child sexual abuse (CSA) typically falls into three main categories, all of which involve the sexual victimization of children. **Pedophilia** is a psychological disorder in which a person experiences uncontrollable sexual compulsions involving children (see Chapter 14, "Paraphilic Disorders: Atypical Sexual Behaviors"). **Child molestation** refers to sexual acts with a child by an adult or a much older child, regardless of a clear diagnosis of pedophilia. The third type of child sexual abuse, **incest**, involves a perpetrator who is a relative, such as parent, aunt, uncle, grandparent, brother, or sister.

child sexual abuse (CSA)
The sexual victimization of a child by an adult or a significantly older child.

pedophilia
Uncontrollable sexual compulsions involving children.

child molestation
Any sexual act performed with a child by an adult or a much older child.

incest
Molestation of a child by a blood relative such as parent, aunt, uncle, grandparent, brother, or sister.

Sexual abusers of children typically do not use violence or physical force to achieve their goals; rather, they rely on their position of greater power and threats of punishment if the child refuses or threatens to expose the adult's acts. Like rape, child sexual abuse is not about sex so much as an expression of power and control over someone of weaker status physically, emotionally, and socially. And also, as is true of rape, the vast majority of abusers are male, and they are almost always members of the victim's family or nonrelatives whom the child knows well. Table 13.2 details the breakdown of the adult abusers' relationship to their child victims. You can see that only 7% of total reported cases involved a perpetrator who is a stranger to the child, similar to the percentage of rapes by a stranger discussed earlier in this chapter.

As you might imagine, child victims of sexual coercion typically experience terrible fear, trauma, guilt, and the sad loss of innocence that accompanies such a betrayal by adults that they know, trust, and love. Let's take a closer look at this horrendous crime.

Abusers' Characteristics

Do sexual abusers of children share any predictive characteristics other than the high likelihood that they know or are related to their victims? The answer is yes, but no single set of characteristics appears to be consistent enough to predict *for sure* who will prey sexually on children. You can probably remember news reports in which

Table 13.2 Abusers' Relationship to Their Victims, Reported Cases, United States

Abusers' Relationship to Child Victim	Percentage of Victims	Total Number
Parent	26.2	17,235
Other relative	29.1	19,113
Foster parent	0.3	207
Day care provider	1.8	1,167
Unmarried parental partner	6.1	3,979
Legal guardian	0.1	73
Child residential facility staff	0.3	168
Other professionals	0.5	312
Friends or neighbors	4.4	2,860
Stranger	6.9	4,547
Other	24.4	16,000

SOURCE: USHHS (2008), data from Table 3-19, p. 63.

someone known as a "pillar of the community," a "wonderful parent," someone "trusted with children," or someone in another respected position was discovered to be a child molester. When these stories appear, they invariably shake our trust in our ability to identify adults who might victimize our children.

With that caution, here is what we do know about common characteristics among perpetrators of child sexual assault ("Characteristics," 2009; Hanson & Morton-Bourgon, 2005; Murray, 2000):

- Most are shy and relatively immature males who probably were sexually abused themselves as children.
- They usually report that they are sexually aroused by children of both sexes, and they fear sexual relationships with adults.
- Most suffer from a number of psychological problems, especially anxiety disorders and depression.
- Typically, they know their victims and are usually family members or friends of the victim's family.
- Child sexual abusers live in relative isolation or with a parent and may abuse alcohol or other drugs.
- They are often unemployed or work in menial, low-paying jobs.
- It is not unusual for them to also engage in other atypical sexual behaviors, such as exhibitionism or sadomasochism.
- Some are devoutly religious and moralistic, as evidenced, in part, by the widely publicized child sex-abuse scandals in the Catholic Church (child abuse by Catholic priests is discussed in detail in Chapter 14, "Paraphilic Disorders: Atypical Sexual Behaviors").
- The vast majority have engaged in a *pattern* of child sexual abuse with more than one victim; often with many victims, sometimes 100 or more (e.g., Bourke & Hernandex, 2009).

If you apply these general descriptors to construct an overall profile of a typical child molester, you may be better prepared to suspect if someone in your child's adult-contact circle is likely to victimize your child. But because predicting child sexual abuse is far from an exact science, suspecting innocent adults of such behavior is unjustifiable and harmful. The best advice is to be alert and attentive to the warning signs and never allow children to be alone with any adult you are suspicious of or do not know reasonably well. Also, be sure to monitor children's Internet use and social networking and educate your child about the safe use of these digital channels to avoid molestation (see McCarthy 2010). Awareness of the profile is worthwhile for parents, child care workers, and other adults responsible for the safety of children.

How Common is Child Sexual Abuse?

In a human sexuality class, during a discussion of childhood sexual abuse, a female student stated that she felt very lucky she had not suffered incest or molestation in her lifetime. Several other women in the class agreed that they also felt fortunate to have escaped the trauma of sexual abuse. At that moment, the discussion abruptly ceased while the implications of what these women had said dawned on most of the class (a real discussion stopper!). Think about it: they felt lucky *not* to have been the victim of a crime that no child should ever have to face. Their feeling of being lucky grew out of the fact that they all knew at least one other woman who had been molested as a child (one student knew four). This was a nonscientific, yet powerful, demonstration of just how common child sexual abuse is.

Sexuality and Culture

Global Prevalence of Childhood Sexual Abuse (CSA) by Continent

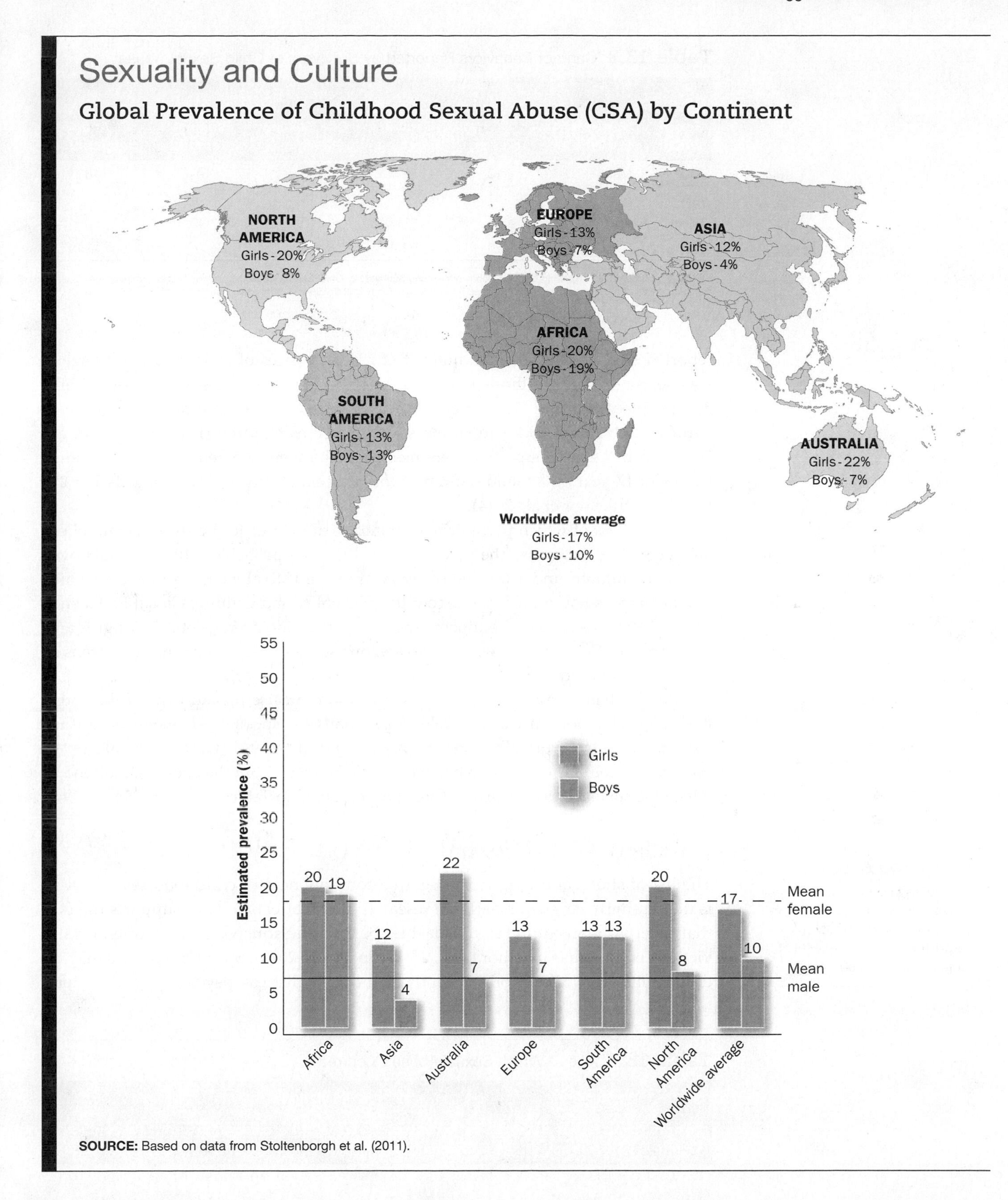

SOURCE: Based on data from Stoltenborgh et al. (2011).

Statistics for child molestation do exist, but as is true for adult rape, researchers agree that the truth about child sexual abuse is distorted due to underreporting and variations in definitions. With that in mind, here are some estimates for the prevalence of child sexual abuse. The U.S Department of Health and Human Services (HSS)

Table 13.3 Contact Behaviors Reported by Survivors of Child Sexual Abuse

Sex of Victim	Sex of Perpetrator	Percentage Reporting: Kissing	Genital Touching	Oral Sexual Activities	Vaginal Intercourse	Anal Intercourse
Male	Male	10	82	30	–	18
Female	Male	31	90	10	14	1
Male	Female	64	82	10	42	0
Female	Female	25	92	0	–	–

SOURCE: "Percentage of Specific Contact Behaviors Among Survivors of Child Sexual Abuse," from *The Social Organization of Sexuality*, by E. Laumann, J. Gagnon, R. Michael, & S. Michaels. Copyright © 1994. Reprinted by permission of The University of Chicago Press.

reported that in 2012, the total number of reported cases of child abuse and neglect was nearly 680,000. Of those, 63,000, or 9.3% were instances of sexual abuse ("Child Maltreatment," 2013). This percentage was borne out in a large national survey of adults which found 10.4% reporting sexual abuse as children (Pérez-Fuentes et al., 2013). Within that group, 25% were men and 75% were women. However, another survey of 17-year-olds found self-reports of past sexual abuse of 11% for girls, but 2% for boys (Finkelhor et al., 2014).

Regardless of which percentage or numbers are closer to the truth, the numbers are staggering. If we use the percentage of 10% as a probable estimate, that's over 7 million children under the age of eighteen in the U.S. alone. And of course, child sexual abuse is not limited to this country. "Sexuality and Culture: Global Prevalence of Child Sexual Abuse by Continent" shows a worldwide average of 20% for girls and 7% for males. Even in Asia, where rates are lowest, these percentages represent tens of millions of children.

Rather than overall numbers of victims, more precise breakdowns of the abuse itself might be more instructive. Table 13.3 details the specific behaviors comprising the sexual contact from a national survey published in 1994 (Laumann et al., 1994). From that same survey, we also have some information about the ages in childhood at which the sexual abuse occurred. These are contained in Table 13.4.

Effects of Child Sexual Abuse on Victims

Since You Asked

10. Can being sexually abused as a child prevent a person from having good relationships in adulthood? Do the effects of the abuse really last so long?

The list of short- and long-term negative consequences of childhood sexual abuse is depressingly lengthy. Nearly all research on the effects of CSA supports the fact that reverberations of the trauma last far beyond the abusive events—often for the victim's entire life (Finkelhor et al., 2013; Pipe et al., 2013). Specific reactions may be divided into several categories, including emotional distress, psychological disorders,

Table 13.4 Age at Which Sexual Abuse Occurred

Sex of Victim	Sex of Abuser	Victim's Age: Percentage 6 Years and Under	Percentage 7 to 10 Years	Percentage 11 to 13 Years
Female	Male	33	40	27
Male	Male	30	46	24
Male	Female	31	26	43

NOTE: There were too few reports of abuse of females by females for meaningful statistical analysis.

SOURCE: Table, "Age at Which Sexual Abuse Occurred," from *The Social Organization of Sexuality, by E. Laumann*, J. Gagnon, R. Michael, & S. Michaels. Copyright © 1994. Reprinted by permission of The University of Chicago Press.

relationship problems, and physical or medical complaints. Although no survivor of CSA is likely to exhibit all of the effects discussed here, most will experience at least one; many will suffer from a combination of symptoms.

EMOTIONAL DISTRESS Most of the emotional disturbances manifested by CSA survivors stem from the perception, however false, that the abuse was somehow the victim's fault. Many victimized children are haunted into adulthood with the notion that they somehow allowed the abuse to happen; that they didn't report the abuser and that, due to their silence, no one rescued them; or that perhaps in some way they had enjoyed it. Many survivors see themselves as bad, immoral, or sexually perverted—"damaged goods," not worthy of being loved or treated well by an intimate partner. In many cases, this attitude may lead them into abusive relationships in adulthood.

The emotional distress experienced by survivors is usually increased when they keep the abuse a closely guarded secret over time, often many years (Finkelhor et al., 2014; Pipe et al., 2007; Smith et al., 2000). One study found that the average time span between the end of the abuse and the survivor's disclosure of it was 14 years (Roesler, 2000). Failure to disclose the abuse does not imply, of course, that it was in any way the child's fault. On the contrary, the child's motivation to keep the secret may be an integral part of the abuser's tactics. Abusers will convince children that telling anyone will cause great trouble, and the child (or the child and the abuser, if the abuser is a beloved family member) will be punished. Or the abuser will threaten the child with physical harm or harm to other family members or the family pets if the child tells anyone.

The burden of feeling at fault combined with minimal support over many years almost always leads to emotional problems for the victim, which may include various combinations of the following (McGuigan & Middlemiss, 2005; Ray, 2001; Roesler, 2000):

Anger—toward oneself, the abuser, or some other party who should have stopped the abuse but did not.

Poor self-esteem—due to the feeling that something must be wrong or bad in their character to have allowed the abuse to occur and to continue.

Self-blame—because the abuser was a beloved family member who "could do no wrong," and to blame the perpetrator would destroy the victim's ideal image of the abuser.

Shame—stemming from self-blame.

Guilt—overengaging in forbidden, immoral, sinful acts, the failure to tell anyone, or the perception of causing the abuse.

Isolation and loneliness—stemming from the belief that the victim is unworthy of love, closeness, and intimacy with another person.

PSYCHOLOGICAL DISORDERS In addition to emotional suffering, the trauma of childhood sexual abuse leaves long-lasting psychological scars years and decades after the abuse has ended. Childhood sexual abuse, like rape, robs victims, both males and females, of their personal power and sense of control over their lives and their bodies. When victims feel powerless and unable to control their own destinies, they suffer from a wide range of psychological disorders. The prevalence of psychological diagnoses is strikingly higher in those with a history of CSA. You can clearly see this difference in Figure 13.4.

The trauma of child sexual abuse typically continues far into the child's future life.

Figure 13.4 Psychological Disorder Diagnoses: CSA versus Non-CSA Adults

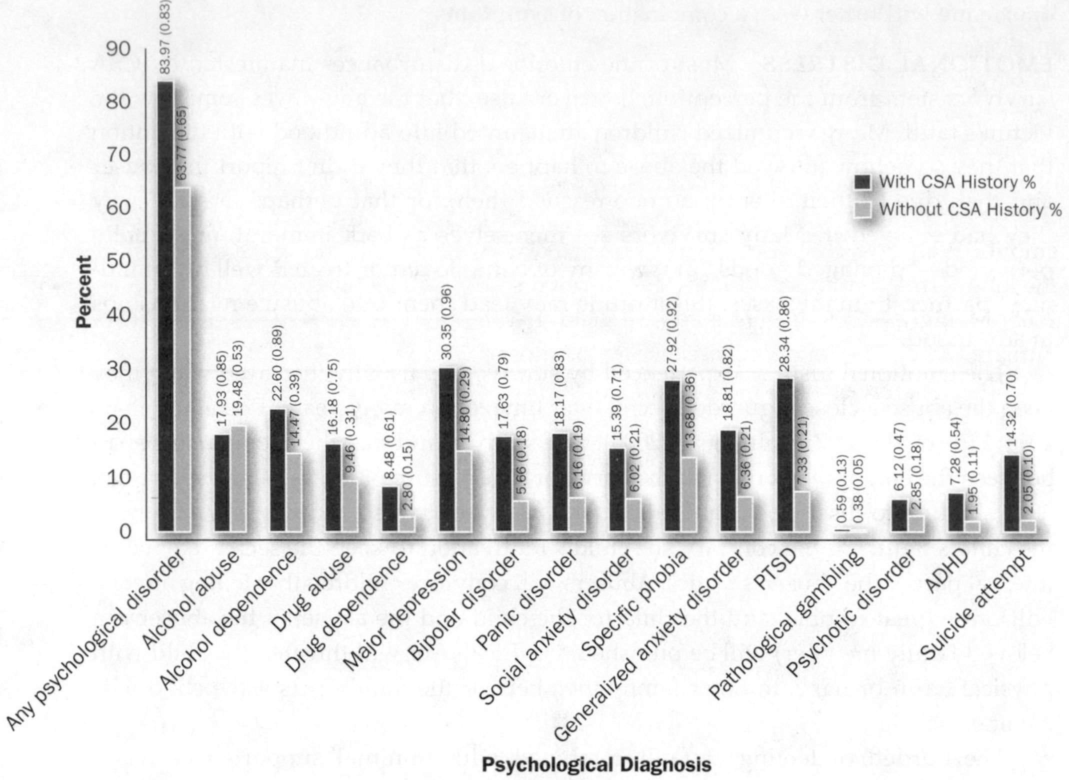

ADULT INTIMATE RELATIONSHIP PROBLEMS Not surprisingly, survivors of CSA often experience difficulty with close relationships in adulthood. Intimate relationship problems are the most commonly encountered aftereffects of CSA (physical and medical complaints run a very close second). Interpersonal problems experienced by former victims include engaging in unsafe sexual behaviors, various sexual disorders, and sexual violence victimization. Numerous relationship issues are more common among CSA survivors than among the general population (Cantón-Cortés & Cantón, 2010; Dube, 2005; Dugan & Hock, 2006; Loeb et al., 2011; Laumann et al., 1994; Maniglio, 2009; Ray, 2001; Roesler, 2000; Walsh & Dilillo, 2011). These include:

Revictimization—a significantly greater chance of becoming a victim of sexual violence such as rape and partner abuse as adults.

Emotional distance—from intimate partners and often friends and family.

Less trust—in others in general and in intimate partners in particular.

Feelings of danger in intimate relationships—an inability to feel safe and secure in an intimate relationship for fear of betrayal or harm.

Lack of enjoyment in sex—which, for many survivors, reactivates the emotional trauma of the abuse, resulting in low or absent sexual desire. Moreover, the body may react in ways that inhibit sexual responding altogether.

Anxiety over sexual performance—due to memories and trauma of abuse.

Specific sexual problems—such as those discussed in Chapter 7, "Sexual Problems and Solutions": erectile disorder, inability to achieve orgasm, inhibited sexual arousal, vaginismus, and dyspareunia.

Forcing sex on a partner—including sexual assaults, rapes, and child sexual abuse.

Increased promiscuity—due at least in part to such factors as low self-esteem, lack of trust in intimate relationships, emotional distance, and lack of feelings of a secure attachment.

MEDICAL CONSEQUENCES Of course, if the abusive behaviors are particularly aggressive and violent, physical injury of the genitals, anus, rectum, mouth, and throat, as well as trauma to other parts of the body, may result, sometimes requiring medical treatment immediately or at some point later in life. Just as victims will work to keep the abuse a secret, they will suffer physical injuries in silence for fear of being "found out."

Beyond these acute medical injuries, recent research has demonstrated a variety of additional long-term medical consequences of CSA. This does not necessarily imply that the abuse caused these problems directly; some of them may be secondary to the emotional and psychological issues associated with the abuse. Nevertheless, most of the long-term after effects among CSA survivors are medical (Berkowitz, 1998; Dube, 2005; Maniglio, 2009; Kendall-Tackett, 2000; Laumann et al., 1994; Trickett, Noll, & Putnam, 2011). The following are among the most common:

Alcoholism—possibly stemming from depression, anxiety, or low self-esteem.

Obesity—perhaps secondary to low self-worth, self-blame, avoidance of romantic or sexual entanglements, or self-destructive behavior.

Tobacco use—likely due to such issues as low self-worth, self-blame, avoidance of romantic or sexual entanglements, stress, or self-destructive behavior.

STIs—probably related to the combination of the greater number of sexual partners and the self-destructive practice of unsafe sexual behaviors.

Gastrointestinal problems—such as irritable bowel syndrome and others commonly precipitated by chronic anxiety.

Gynecological disorders—possibly related to the combination of a greater number of sexual partners and the avoidance of gynecological exams due to reactivation of past sexual abuse experiences.

Chronic pain—such as headache, stomach pain, and pelvic pain, for which CSA survivors are treated by physicians and hospitalized in exceptionally high numbers.

Insomnia—often linked to stress, anxiety, and depression.

Eating disorders—perhaps arising from issues of powerlessness and lack of control, poor self-esteem and body image, self-destructive behavior, or depression.

Asthma—which may be triggered by psychological factors.

One additional and very important health-related consequence of child sexual abuse is *avoidance of routine medical care*, especially gynecological exams or other exams such as breast exams, mammograms, dental exams, prostate exams and colonoscopies, and tests that require any type of confinement, restriction, or other situations of powerlessness, such as MRIs and CT scans. Paradoxically, some survivors of CSA tend to overuse other medical services. This relates to a phenomenon known as *somatization*, in which traumatic psychological issues are expressed through physical complaints that produce real symptoms. These health differences between survivors and nonvictims are not subtle. One study of women members of an HMO found that among those who reported a history of sexual abuse, 83% suffered from depression, 35% were obese, 64% reported frequent gastrointestinal problems, 45% suffered from chronic headaches, and 13% had been diagnosed with asthma (Felitti, 1991; Maniglio, 2009). All of these rates were approximately twice those for nonabused women.

Reducing Child Sexual Abuse

The consensus regarding sexual abusers of children is that the abusive behaviors are deeply ingrained and compulsive. That is, perpetrators find it very difficult to stop victimizing children, even if they sincerely want to. Talk to an average citizen and you will likely hear the opinion that repeat child molesters should be executed. Our justifiably emotional reaction to this crime against the most innocent members of society is understandable. Some states have considered adding repeat child molestation to their list of capital offenses. However, in 2008, the U.S. Supreme Court, in a 5–4 (and publicly controversial) split decision, found that a Louisiana law allowing execution for the most serious child sexual offense, child rape, was cruel and unusual punishment and ruled it unconstitutional (Oliphant, 2008).

Typically, when a child molester is arrested and convicted, the "treatment" is jail or prison time. Very few penal systems offer therapy to offenders, and after their release, offending may recur. Estimates of the number of child molesters who reoffend vary greatly due to methods of data gathering and definitions of offenses. Some studies look at rearrests of child molesters for any criminal offense, others consider all sexual reoffending, and still others attempt to look only at rearrests for sexual crimes involving children. In addition, studies examine differing time frames for rearrest. For example, one study found that within a period of 10 years following release from prison, 30% of child molesters are rearrested, and after 20 years, approximately 50% have become repeat offenders (Prentky et al., 1997). By contrast, a 2007 U.S. Department of Justice report found that 3.3% of child sexual molesters released from prison were rearrested for another sexual crime against children in the following three years (DOJ, 2007). Wherever the statistical "truth" lies, most people would agree that even one repeat child molester is too many.

Success in the treatment of child molesters has been notoriously poor. How can society prevent child molesters from victimizing more children? In the United States, three approaches are usually taken, either individually or in combination: (1) alter the abuser's behavior; (2) prevent abusers from gaining access to potential victims; and (3) help children understand how to recognize, avoid, resist, and report attempted molestation by an adult.

SURGICAL AND CHEMICAL CASTRATION Because child molesters tend to deny their crimes and because they are often ordered, against their will, into treatment by the courts, therapy and counseling have shown poor treatment success overall. Consequently, states have begun to get tough. In 1996, California passed a law *requiring* surgical or chemical castration for repeat child molesters in order for them to be granted parole from prison. The law took effect in January 1997, and since that time, Florida, Georgia, Iowa, Louisiana, Montana, Oregon, Texas, and Wisconsin have passed or are considering similar statutes.

These laws provide that any convicted, repeat child molester must agree to surgical or chemical castration as a condition of parole, suspended sentence, or probation. *Surgical castration* involves, as you probably know, the medical removal of the testicles, which are the glands that produce testosterone in men (see Chapter 2, "Sexual Anatomy," for more detailed information). In **chemical castration**, a man receives regular doses of female hormones that effectively block the production of testosterone, producing a similar effect to surgical castration (usually, the hormone is Depo-Provera, which is also a popular method of contraception for women). Unlike surgical castration, however, a man's normal production of testosterone returns when the hormones given in chemical castration are discontinued. In most states with castration laws, the perpetrator has the option of surgical or chemical castration, or, if the felon chooses, he may serve out his entire sentence, forgoing parole, rather than submit to any castration requirement.

chemical castration
Blocking the production of testosterone in repeat sex offenders through the regular injection of female hormones.

Some studies have demonstrated that the hormone injections are helpful, especially in conjunction with counseling, in reducing the risk that child sexual abusers and rapists will offend again (e.g., Hall & Hall, 2007; Kafka, 2007). However, the research is incomplete and mixed (Rice & Harris, 2011). For example, we must consider the possibility that men who agree to chemical castration may be more motivated to change their predatory behaviors. If this deterrent is effective, it is probably because blocking testosterone reduces the male offenders' drives toward sex and aggression, which in turn helps them to control their violent urges toward children. Newer drugs, such as triptorelin, which also produce testosterone-blocking effects but have fewer negative side effects compared to Depo-Provera, have also been shown to be effective for some repeat offenders (Schober et al., 2005).

As you might imagine, these forms of treatment for child molesters are extremely controversial, and many legal rights groups, lawyers, and ethicists have denounced any form of castration as cruel and unusual punishment, no matter how horrific society views the crime (see Daley, 2008). Nevertheless, the castration laws enacted thus far have not undergone any serious legal challenges (Stinneford, 2006). On the other side of the debate, victims and victims' families argue that these criminals must be stopped, and the welfare of their potential victims far outweighs any concern for the molesters' rights. Moreover, sexual offenders themselves have often acknowledged that testosterone-inhibiting treatment helps them control their antisocial sexual urges.

ISOLATING CHILD MOLESTERS FROM CONTACT WITH CHILDREN Child molesters are prevented from contact with children in two ways: imprisonment and notification laws. Repeat child molesters are often sentenced to lengthy prison terms. This, of course, removes them from society at large so that they are unable to victimize more children. However, in general, no attempt is made during incarceration to rehabilitate molesters, largely because of the widespread belief that such attempts are usually ineffectual. Some molesters themselves have admitted that if and when they are released, they will almost surely molest again.

The crime of child sexual abuse, if it does not involve extreme physical violence, is usually punishable with prison terms of five years or less. A disturbing question, therefore, is what should society do when these criminals have served their time and are released? Currently, in the United States, one answer is *Megan's Law.* This law was named after Megan Kanka, a seven-year-old girl who in the early 1990s was raped and killed by a convicted child molester on parole who had moved into her neighborhood without anyone's knowledge. Since then, states have enacted laws that require local communities to be notified when sex offenders are living in their residential area. The idea of this is not necessarily to punish the perpetrator but to allow residents to be aware of and alert to the offenders' presence and protect their children accordingly. For more on Megan's Law in your town or city, contact your state's office of the Attorney General, which should include a link to Megan's Law or your local police department, which in turn will allow you to access the list of offenders residing in your area (see Chapter 14, "Paraphilic Disorders: Atypical Sexual Behaviors," for more about Megan's Law).

Another law that has been passed in most state legislatures is called "Jessica's Law," named after nine-year-old Jessica Lunsford who, in 2005, was kidnapped, sexually assaulted, and murdered by a convicted sex offender in Florida. Approximately half of all states have passed similar, but not identical, versions of Jessica's Law. This legislation goes a step further than Megan's Law and requires (depending on the specific state statutes) continual, lifelong tracking of the location of felony child sex offenders with GPS devices that they must wear at all times (these are locked around an ankle and cannot be removed without triggering an alarm); forbids offenders to live near settings where children may be present (schools, parks, playgrounds, etc.); and/or establishes longer minimum prison sentences for first-time offenders.

For example, in California the law prohibits registered sex offenders from living within 2,000 feet of schools, parks, and other "safe zones" established by local governments and mandates a *minimum* 25-year prison sentence for all felony sex offenders. These laws are controversial because some contend that the established "safe zones" may include nearly all of the residential areas in a town or city, making it impossible

In all situations, children should be made aware of specific "safe adults"—those the parents or care-givers trust—to whom the children can go to for help and protection if they ever feel threatened or uncomfortable due to actions of others.

for sex offenders to find housing. This may not seem like a bad idea to many people, but as a result, many sex offenders become homeless transients, which may actually increase their chances of reoffending. Moreover, keeping convicted molesters away from children seemingly makes sense, but, as we have already discussed, most abusers do not prey on strangers but rather on children they already know or to whom they are related. Although the law keeps molesters away from places where children congregate, it does not prevent them from seeking out new victims (Butts, 2014). The overall success of Jessica's Law in reducing child sex offenses is, as yet, unclear (Butts, 2014; Payne, DeMichelle, & Button, 2008; Peirce, 2008).

TEACHING CHILDREN TO PROTECT THEMSELVES Many believe that child molestation may be effectively prevented in many cases through increased and improved education of children and their parents. Studies have suggested that when children participate in school- or home-based child sexual abuse prevention programs, they are less likely to be victimized (Kenny & Wurtele, 2010; Putnam, 2003). One comparative study found that among college women, 8% of those who had participated in such a program had experienced sexual abuse, compared to 14% who were not involved in prevention education (Gibson & Leitenberg, 2000).

Prevention programs typically include "good touch, bad touch" awareness and helping children understand that they have the right to say no if anyone attempts to touch them inappropriately. These programs also impress upon children the importance of telling a trusted adult if anyone touches them, asks them to do something, or behaves in a way that makes them uncomfortable, and they insist that the child should tell even if the offending adult is a relative or warns the child not to tell anyone (Kenny & Wurtele, 2010).

Children also need to know that it is OK to take action to get away or to find help if a situation feels dangerous or uncomfortable: yell, scream, run away, seek help from a trusted "safe" adult. This information should be at an appropriate age level for the child, but parents should start them at a very young age, probably three or four years of age at the oldest. Then information should be updated and expanded on a yearly basis as the child and the potential dangers grow. Of course, many parents find such discussions difficult and fear scaring their child away from healthy relationships and touching. Many resources exist in books and on the Internet to assist parents and educators in this awareness education process. Two of the best websites on CSA are from the U.S. Department of Health and Human Services Child Welfare Information Gateway at www.childwelfare.gov and the Massachusetts Citizens for Children at http://enoughabuse.org.

Sexual Harassment

13.3 Explain what is meant by sexual harassment and review the settings in which harassment commonly occurs, the behaviors often involved, what a person should do as a victim, and the effects of sexual harassment on victims.

The third and final category of sexual victimization in this chapter is sexual harassment (relationship abuse and domestic violence are discussed in Chapter 4, "Intimate Relationships"). Sexual harassment has become a common problem in work and educational environments throughout North America and elsewhere in the world. Some people might argue that compared to other forms of sexual aggression, sexual harassment is less serious and less damaging to victims. Although this may be true in some cases, many individuals who have been the victim of sexual harassment in school or on the job would disagree. The trauma and life disruption caused by sexual harassment may affect an individual in ways comparable to those discussed in relation to rape and child sexual abuse.

Defining Sexual Harassment

The crime of **sexual harassment** as defined by the United States government's Equal Employment Opportunity Commission (EEOC) may be summarized as follows:

> Unwelcome advances, requests for sexual favors, and other verbal and physical conduct of a sexual nature constitute sexual harassment when (1) submission to such conduct is made explicitly or implicitly a term or condition of an individual's academic access, instruction, or employment; (2) submission to or rejection of such conduct by an individual is used as a basis for academic or employment decisions affecting such individuals; or (3) such conduct has the purpose or effect of substantially interfering with an individual's work or academic performance or creating an intimidating, hostile, or offensive work or academic environment. ("Commission on Human Rights," 2011; "What Is Sexual Harassment?" 2011)

sexual harassment
A pattern of unwelcome sexual advances, requests for sexual favors, or other verbal and physical conduct that is coercive or creates a hostile work or educational environment.

If you stop for a moment to analyze this legal definition, you will notice that, although it covers many behaviors and actions, we can boil it down to two categories of sexual harassment. The first is the requirement of sexual favors in exchange for some beneficial event (a job promotion, a raise, a good job evaluation, etc.). This form of sexual harassment is commonly described as illustrating **quid pro quo**, a Latin phrase meaning "this for that," or something required in exchange for a benefit or reward. The second category is the creation of a **hostile environment** due to unwanted, overt, or covert sexually related activities. These definitions apply equally to educational environments in which an instructor or administrator may demand a quid pro quo arrangement with a student for academic decisions (admissions, grades, letters of recommendation, etc.), or in which an offensive or hostile educational environment has been created through words or actions. Notice that the definition is not gender-specific and applies to a person of either sex harassing a person of either sex.

quid pro quo
Something given in exchange for a benefit or reward; with reference to sexual harassment, a situation in which a person in a position of power over another requires sexual favors in exchange for some beneficial outcome for the victim.

hostile environment
A distressing work or educational environment resulting from overt or covert sexually related activities or intrusions.

The difficulty here, as with most definitions, lies in interpreting the components of the definitions for a specific situation. For example, when do friendly or seemingly *consensual* interactions cross the line into sexual harassment territory? In an attempt to clarify this sticky problem, various refinements of the definition have been offered (Stoddard et al., 2000; "What Is Sexual Harassment?" 2011). Conduct of a sexual nature is more likely to be considered harassing if:

- It is unwelcome, unsolicited, and offensive to the victim(s).
- It is repeated or becomes a pattern of behavior, particularly after a warning from the receiver that it was unwelcome or offensive.
- The behavior involves any supervisor–subordinate relationship where the harasser is in a position of power over some aspect of the receiver's educational or professional career.
- The conduct is extremely and flagrantly verbally hostile, physically abusive, disruptive, continuous, pervasive, or provoking.
- Preferential treatment of some individuals in the workplace or classroom, based on their sexual actions, has a negative impact on others in the same environment.
- The behaviors involve physical touching, groping, massaging, or other intimate behaviors that the recipient (of the same or opposite sex) finds uncomfortable or offensive.
- A "reasonable person" (in the legal sense) would likely be affected negatively by similar conduct in a similar situation ("Preventing Sexual Harassment," 2011; "What Is Sexual Harassment?" 2011).

If you consider these various guidelines, you can see that a charge of sexual harassment is rarely made for the occasional inappropriate sexual joke, an innocent flirtation, or a single friendly touch. Most sexual harassment charges are in response to very clear-cut and usually blatant offenses. By way of example, here are two case summaries of sexual harassment, both based on actual events (from author's files). As you will see, sexual harassment is a serious crime and the U.S. court system takes cases of sexual harassment very seriously. The first demonstrates a quid pro quo harassment situation, and the second demonstrates sexual harassment based on the creation of a hostile classroom environment.

The Bookstore

Henry, a gay man, was hired at a bookstore in the town where he attended the state university, majoring in history. He knew and loved books and quickly became popular with the shop's customers for the advice and assistance he could offer them. Hester Grant [not her real name], the store's owner and manager, seemed pleased with him and his work until she discovered his sexual orientation. Henry had never tried to hide that he was gay, but he had never seen any need to mention it to his boss. One day, Henry introduced Hester to his partner when he stopped by the store to go to lunch with Henry. From that day on, Hester's attitude toward Henry underwent a clear change. She became less flexible about his working hours, stopped seeking his advice about books to stock that would be in the high-demand area, and began spending less time in the store when Henry was there. Paradoxically, she also became flirtatious toward him. She would brush her hand across his back when she walked past and make remarks such as "Heeeey, Henry…" in a tauntingly inviting tone of voice or compliment him on his "sexy" clothes. Henry would try to make light of these events with comments such as "Watch it, Hester!" or "Hester, give me a break, would you?" and even, "Careful, Hester. That's sexual harassment, you know." But her teasing escalated. She began to suggest that the only reason he liked men was that he had never been with a really talented woman—like her. "Come on, Henry. Come to my apartment tonight; I'll show you just what you're missing." Henry finally quit his job, sued Hester for sexual harassment, and was awarded $700,000 in compensatory and punitive damages.

The Classroom

Professor Richard Flemming (not his real name) taught contemporary literature part time at a small private college in New England. One of his in-class activities was to have students read specific passages of novels and plays for class discussion and interpretation. He would call on individual members of the class and ask them to begin reading on a certain page and continue until he stopped them. He would then lecture or lead a discussion of the passage. Usually, his selection of students to read appeared to be random, except when a particular passage contained sexually explicit material. In these cases, he would invariably call on one of several attractive women students in the class. As the woman was reading the sexual passage aloud, he would smile and wink and raise his eyebrows at the men in the class, especially the athletes, who would smile, chuckle, and exchange glances with one another in response. As this pattern became established over several weeks, some of the women became so uncomfortable with the hostile environment in the classroom that they either stopped attending or dropped the course. Near the end of the semester, two of the female students who had stayed in the class reported Flemming's behavior to the sexual harassment mediators on campus. A review board was convened so that the identity of the students would be concealed because they feared reprisals from the instructor. The instructor was ordered to stop the behavior in question and undergo eight weeks of sexual harassment training before being allowed to teach another class at the college. He left the college at the end of that semester and never returned to his part-time position.

Little doubt or gray area may be found in these accounts of sexual harassment. Typically, most sexual harassment cases involve a similar degree of clarity.

Sexual Harassment Settings

As awareness of sexual harassment as a form of victimization has grown over the past 20 years, the variety of settings in which it is commonly found has increased as well. Harassment used to be seen as primarily occurring between two people of unequal social power, such as supervisor–subordinate or professor–student. However, today it is generally accepted that sexual harassment may also occur between peers of equal or nearly equal power, regardless of the setting. Although the fact of one person sexually harassing another may occur virtually anywhere, several specific venues should be discussed. These are K–12 schools, colleges and universities, and the workplace.

SEXUAL HARASSMENT IN SCHOOLS Recently, awareness has been growing that adults are not the only ones to harass or to suffer from sexual harassment (considered by many to be one common form of bullying). Today, sexual harassment situations have been identified and supported by state and federal court findings in elementary, middle, and high schools. Sometimes the harassment takes the form of teachers or administrators discriminating against certain students because of their sex, but more often sexual harassment in middle and high school (and sometimes elementary school as well) is about kids sexually harassing other kids. Believe it or not, approximately 80% of adolescent boys and girls report experiencing sexual harassment on almost a daily basis from their peers (Shute, Owens, & Slee, 2008). Even among children as young as sixth-graders, 29% reported being sexually harassed in the previous 90 days and 11% experienced harassment at least weekly (Ashbaughma & Cornella, 2008).

Other common forms of peer harassment in schools include sexual graffiti about specific individuals, sexual verbal taunting, rumor spreading about sexual activities, sexual teasing, intimidation, unwanted touching, sexual assault, and rape. These forms of harassment create significant negative impacts on students' emotional, psychological, and even physical well-being, and on their academic performance (Gruber & Fineran, 2008).

Sexual harassment in schools gained national attention in 1999 when the U.S. Supreme Court ruled that a school district that fails to prevent, or is "deliberately indifferent" to, student-on-student sexual harassment may be held financially responsible for the emotional and psychological damage caused by the harassing events (Taylor, 1999). The original suit was brought by the mother of LaShondra Davis against a school board in Georgia for the suffering her daughter endured over a five-month period during the fifth grade, caused by a fifth-grade boy. The boy allegedly touched the girl's breasts, rubbed against her suggestively, and repeatedly told her he wanted to have sex with her (although those were not the exact words he used). Yet even when notified of this behavior, the school took no action.

The Supreme Court decision placed all K–12 school districts in the United States on notice that they must take affirmative steps to protect their students, regardless of age, from possible peer sexual harassment. Peer sexual harassment is relatively uncommon in the elementary grades but appears to increase dramatically through the middle and high school years (Timmerman, 2004). Schools are legally mandated to enact antiharassment and antibullying measures for preventing, identifying, intervening in, and punishing harassing conduct (American Association of University Women, 2002; "Sexual Harassment at School," 2011). However, even with such measures in place, sexual harassment in schools continues and many researchers contend that schools, especially at the secondary level, need to do more to stop the behavior and address the causes and consequences (see Agger & Day, 2011).

SEXUAL HARASSMENT AT COLLEGES AND UNIVERSITIES Sexual harassment in higher education (colleges and universities) may take either form, quid pro quo or the creation of a hostile environment. It may be perpetrated by individuals in positions of

Since You Asked

11. I heard that a professor was fired for sexual harassment. Is that true? What did he do?

power (usually faculty members), on those with less power (the students), or, more commonly, it takes the form of peer harassment and the creation of a hostile environment (Bursik & Gefter, 2011). The problem is widespread. One major national survey conducted in 2005 found that over 60% of both men and women students reported at least one incident of being sexually harassed at their college or university (Hill & Silva, 2005).

Reports of sexual relations between faculty (usually male) and students (usually female) are innumerable. Sometimes these liaisons grow out of an honest attraction between two people who just happen to be an instructor and a student. Unfortunately, however, some college professors prey on their vulnerable students. Today, most colleges and universities have specific policies that limit or outright forbid faculty from engaging in such relationships. Some researchers, instructors, and college students themselves believe that it is not the university's business. They argue that if two consenting adults wish to become involved voluntarily in a relationship of any sort, it is their choice and indeed their right to do so (e.g., Paglia, 1998).

Figure 13.5 College Students' Reports of Specific Harassing Behaviors Experienced (percent)

SOURCE: Adapted from Hill & Silva (2005). Drawing the Line: Sexual Harassment on Campus (p. 15).

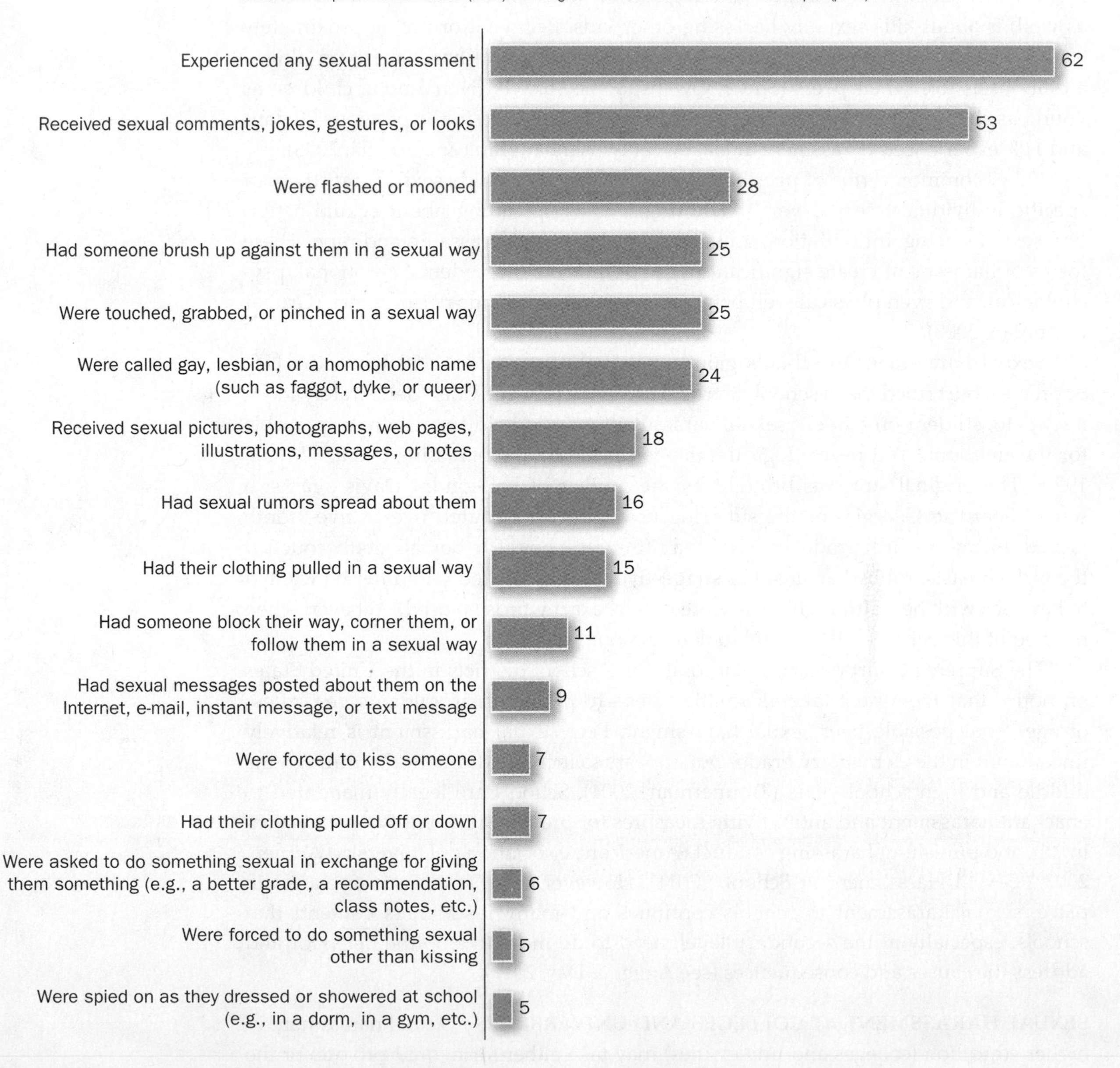

Base = All qualified respondents (n = 2,036); 1,096 female and 940 male college students ages 18 to 24.

However, the arguments against such relationships focus on the question of whether such relationships are *truly* voluntary; does the student feel completely free to consent or refuse? If the student is currently in the faculty member's class; may be in the future; or may need to rely on the faculty member at some point for advice, academic counseling, letters of recommendation, and so on, the possibility always exists that the student will feel pressured, either subtly or explicitly, to submit to an instructor's advances. This is a clear example of sexual harassment based on different degrees of social power. If someone has the power to control your future, how free are you to say no to that person's request of you? Therefore, rather than try to judge who is being harassed and who is choosing freely, colleges and universities tend to avoid that labyrinth and simply make policies forbidding such relationships.

Far more common is peer sexual harassment on college campuses (sometimes referred to as *bullying*) (Angelone et al., 2005; Stein, 2007). Both college men and women may be victims of peer sexual harassment, and it may be male-on-male or female-on-female, but the majority of cases reported involve a male harasser and a female victim. Typical behaviors associated with student-to-student harassment include such behaviors as continually asking for dates even when the answer has been a clear no; an unwelcome pattern of waiting outside classrooms, dining halls, and residence halls to "walk her home" or simply to "say hi"; or repeatedly stopping by her room, classroom, or apartment without consent. In addition, the harasser may make unwelcome sexual remarks about the woman's appearance or body and may suggestively rub against her and even touch her breasts or buttocks. Other harassing behaviors may include unwanted kissing; intimidating her by leering, making suggestive gestures, or blocking her freedom of movement; telephoning at all hours of the day or night; spreading unflattering sexual rumors about her ("bad-mouthing"); stalking her; sending numerous unwanted e-mails or voice mails; verbally abusing her following rejection; and rape or attempted rape. Figure 13.5 lists percentages for various types of harassing behavior.

Most colleges and universities have instituted policies that specifically prohibit sexual harassment between students, and a student perpetrator may be suspended, expelled, or arrested. The problem here is that most sexual harassment on campuses is hidden and insidious, so proving it is often difficult. However, if potential victims understand what constitutes sexually harassing behaviors, they will be in a much better position to prevent or stop such behaviors.

Moreover, potential harassers need to be better informed about exactly what comments and behaviors are likely to be considered sexual harassment. Often men will make comments such as "I just can't say or do anything around here without someone accusing me of sexual harassment!" But this is usually far from the truth. Such statements simply underscore the lack of understanding many men still have about sexual harassment. "Self-Discovery: Evaluating Sexually Harassing Behaviors" (on the following page) offers guidelines for judging whether your behavior toward another person might be harassment.

In some settings, victims of sexually harassing actions may be unsure that what they are experiencing would be defined as sexual harassment. If, after reading through the material in this chapter, you feel that you have been, or are being, sexually harassed, you probably are. To those who understand the issue, sexual harassment is rarely subtle or vague. Everyone has the right to work, go to school, or engage in any other activity in an environment that is safe from any form of unwanted sexual behaviors. In addition, sexual harassment is a crime punishable by civil or criminal penalties (these vary from state to state and country to country), which, if proven, will be levied against the individual, the organization in which the offenses occurred, or both. "Sexuality, Ethics, and the Law: Steps to Take if You Are the Victim of Sexual Harassment", on the following page, offers guidelines and effective actions a person can take in response to sexual harassment.

Self-Discovery

Evaluating Sexually Harassing Behaviors

Sometimes a person may unintentionally engage in a sexually harassing behavior. Ask yourself the following questions to help you determine if your or another's comments or actions might be interpreted as unwanted, inappropriate, or sexually harassing.

1. Would I want my comments and/or behaviors to appear in the newspaper or on TV so that my family and friends would know about them?
2. Is this something I would say or do if my mother, father, girlfriend, boyfriend, sister, brother, wife, or husband were present?
3. Would I be comfortable if someone else said or did this to my mother, father, girlfriend, boyfriend, sister, brother, wife, or husband?
4. Is this something I would say or do in front of the other person's boyfriend, girlfriend, wife, or husband?
5. Is there a commonly accepted difference in social status or power between me and the other person? Am I the person's boss, teacher, professor, counselor, priest?
6. Am I physically bigger or stronger than the other person?

If you are unsure about a certain action or comment, applying these questions before doing or saying it will probably help you make the right choice. If you answered yes to any of these questions, it should raise a red flag of warning. Remember, avoiding harassing behavior is about respect for others' feelings and comfort level, and when you substitute people you care deeply about for the target person, the appropriateness of your actions quickly becomes clearer.

SOURCE: Excerpt from Sexual Harassment and Teens: A Program for Positive Change, by Susan Strauss. Copyright © 1992. Published by Free Spirit Publishing. Reprinted by permission of Susan Strauss, http://www.straussconsult.com.

Cyberharassment

Cyberharassment
sexual harassment using various forms of electronic an ddigital media such as Facebook, Twitter, and Instagram.

As you can see in Figure 13.5, 9% of college students reported being sexually harassed through "web pages," email, and text messages. Keep in mind that the data in the table are from 2005. You can imagine that with today's lightning-fast advances in the power of social media, this form of victimization, known as **Cyberharassment**, has taken on much larger proportions. Although social networking has numerous positive used, it can is also being used to sexually harass and bully. Part of the reason for this is the sheer reach of these sites to reach huge numbers of people and, consequently, create a greater impact than any other form of harassment can even approximate. And with that greater impact comes the potential for far more emotional

Sexuality, Ethics, and the Law

Steps to take if you are the Victim of Sexual Harassment*

1. Say "NO!" in very clear terms. This is not always possible due to fear of reprisals or social power differences, but if you let the harasser or harassers know that the behavior is not OK with you and is unwelcome, it should, by law, stop immediately.
2. If the harassment is occurring within an organization such as an employer or college or university, determine what the official policy is regarding sexual harassment. Most organizations are required to develop written policies that adhere to the law (for an excellent example, see Stanford University's policy at http://harass.stanford.edu). The policy should include the procedure that victims of harassment should take. (Note that it is also illegal for an organization to engage in any form of punishment or retribution toward anyone who reports sexual harassment.)
3. Talk to others whom you trust about your experiences. This is important because you may find personal support in them and, in some cases, you may discover that one or more of them has experienced the harassment as well.

4. Keep a record of every incident of sexual harassment, including dates, times, locations, names, and any witnesses, no matter how small some of the events may seem.
5. Save any notes, gifts, text message records, voice mails, and the like, from your harasser.
6. Report the harassment to the appropriate official in your workplace or college or university. This official should be identified in the organization's sexual harassment policy. You may be allowed to make such a report anonymously, and the designated official is required to take action on your complaint. In some cases, you are not required to meet face to face with or make your name available to your harasser.
7. If you prefer, you may bypass your specific organization entirely and report your experience directly, in person or in writing, to the nearest office of the U.S. Equal Employment Opportunity Commission (EEOC) or the Office for Civil Rights (OCR) at the U.S. Department of Education (see www.eeoc.gov/contact or).
8. Consult a lawyer. A lawyer will be an effective advocate for you and can advise you of your legal rights and possible courses of action. You may be able to find a lawyer for a small or no fee through the American Bar Association (www.abanet.org) or legal-aid services (listed in the phonebook) and law schools in your area.

*NOTE: Laws and practices may vary by state, but typically, for you to file a complaint, it is not required that the harassment be targeted directly at you. If you feel sexually harassed by someone's behavior toward others, you are considered a victim as well and may proceed with these steps.

damage to victims. Add to this the ability for people to harass anonymously by creating names, handles, and sites using false names, and extreme harassment can occur without fear of punishment or retribution for the perpetrator. You have undoubtedly been aware of the many news stories about Internet or social media harassment stories resulting in the destruction of teen's and young adult's reputations, self-esteem, and even leading to suicides. Just two examples will demonstrate this growing threat.

In 2010, a Rutgers University students committed suicide by jumping off the George Washington Bridge after two fellow students secretly recorded him in his dorm room having a sexual encounter with a man and posted the video online (Parker, 2013). In 2013, an event was reported about a 17-year-old girl, who suffers from seizures due to a brain injury she at birth, has overcome her disorder, is active in sports and is a cheerleader. Rather than appreciating her courageousness, fellow students began to bully and harass her through social media in the cruelest of ways. One of the posted texts from a fellow student read, "[She] should just have one of her f***ing seizures and die because people [here] don't want her. That's the reason she has seizures, because that's karma for [being born] a freaky slut" (Fiorella, 2013).

As these types of disturbing cases continue to increase, middle schools, high schools, and universities are taking steps to add or strengthen policies and consequences related to cyberharassment. In addition, most states have anti-bullying laws in place and more and more states have added or have proposed specific language addressing cyberbullying (Hinduja & Patchin, 2014). Social networking sites have become increasingly sensitive to these uses of their services and are removing anonymous posting and making tracking of bullying posts easier. Moreover, fighting fire with fire, new technologies are being used to by law enforcement to track and arrest the perpetrators even in an anonymous environment.

Cyberharassment and cyberbullying are a new, insidious form of intimidation that often allows cowardly harassers to hide their identity.

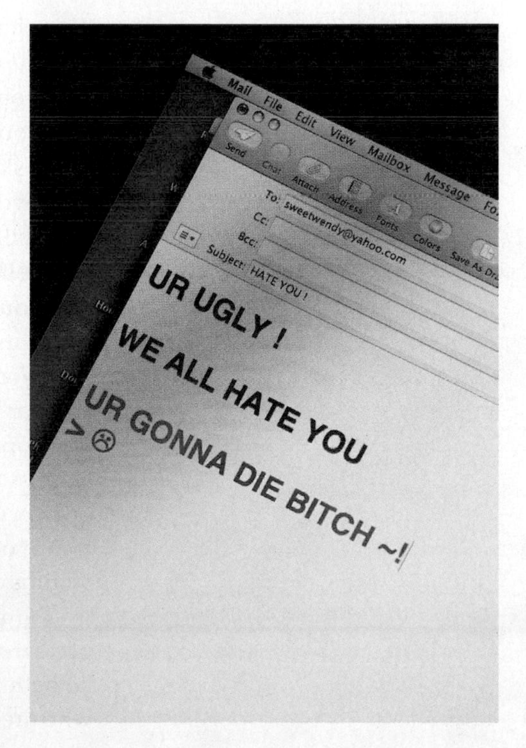

Sexual Harassment at Work and in the Military

The workplace and military were the most visible places where sexual harassment awareness originally began (refer back to the earlier discussion on rape in the military). Typically, the harassment usually took the form of a superior (a boss, supervisor, or ranking officer) using his or her position of power to intimidate and coerce subordinates into providing sexual favors as a condition of or in return for raises, promotions, better working conditions or schedules, or keeping a desirable position. Although this quid pro quo form of power-based sexual harassment continues to be a problem in these settings, the circumstances that may be defined as creating a hostile, intimidating, or offensive work environment have received greater attention. Most now are aware that anyone, at *any* level, in any organization may engage in behaviors that create a hostile environment, including military officers, executives, managers, supervisors, and peers.

As you might imagine, laws prohibiting sexual harassment on the job or in colleges and universities, together with the penalties accompanying them, have begun to sink into the national consciousness. The courts' advice to business, the military, and schools is this: "Have a clearly stated policy against sexual harassment, widely publicize it, and rigorously implement it, following up on all complaints.... In taking these steps, you're sending an unmistakable message:... Sexual harassment will not be tolerated" (McGarvey, 1998, pp. 86–87). Many U.S. businesses, corporations, the armed forces, and higher education institutions now enact strict antiharassment policies and engage in extensive education programs for executives, managers, workers, and students to prevent harassment or, if it happens, to intervene and stop it as soon as possible (T. S. Nelson, 2003).

Effects of Sexual Harassment on the Victim

Sexual harassment has many effects similar to those of rape and child sexual abuse discussed earlier in this chapter, but because it differs from other forms of sexual victimization in many ways, its overall effects are unique. Victims of sexual harassment, regardless of the setting in which it occurs, experience a wide range of negative emotional and psychological repercussions, including a greater likelihood of depressive and anxiety disorders, higher rates of alcohol and drug abuse, insomnia, poor self-esteem, anger, poorer work or academic performance, greater absenteeism at work or in the classroom, and an increased probability of resigning from a job or dropping out of school (Collinsworth, Ftizgerald, & Drasgow, 2009; D. Smith, 2001; Willness, Steel, & Lee, 2007).

Your Sexual Philosophy

Sexual Aggression and Violence: Rape, Child Sexual Abuse, and Sexual Harassment

This chapter has been about a dark side of human sexuality, and now you know just how dark it can be. However, there is a brighter side to all this; it's called *prevention*. It's probably quite clear to you how the material in this chapter can be incorporated into your sexual philosophy. All you need to do is ask yourself a few basic questions: What steps will you take to help ensure that you or anyone you care about will never (or never again) be the victim of sexual aggression? How can you be totally sure that you will never (or never again) sexually victimize anyone else? How can you help yourself or others who are victims of sexual aggression or have been in the past? What will you do, personally, to contribute to reducing rape, child sexual abuse, and sexual harassment in our culture and around the world?

Working to develop a clear sexual philosophy about the disturbing topics in this chapter will help you predict when a sexually dangerous situation exists or could develop. You will be able to make the best possible choices in situations where sexual aggression might unfold, and you will be more likely to take effective steps to prevent or intervene in it, whether you or someone else is the potential target. Those who are aware of and vigilant about sexual aggression are significantly less likely to become victims or to victimize others.

You will also be more attuned to the possibility that child sexual abuse may be occurring within a family or other setting where children might be vulnerable. You'll be better able to notice suspicious signs of abuse and take action to protect children who are or might be sexually abused.

Finally, you will be able to spot sexual harassment wherever it might occur: in the classroom, on the campus, in the workplace, or in schools. Incorporating this knowledge into your sexual philosophy will not only reduce the chances that you will be sexually harassed but also help you determine what to do about it if you are. Furthermore, you will be far less likely to sexually harass others, you will be in a position to help those who are being harassed, and you will be able to contribute to the creation of policies in your school or workplace that reduce and eliminate sexual harassment.

Although all chapters in this book contain information that is crucial to your sexual philosophy, it's just possible the material in this chapter about *preventing* violence and abuse may be the most critical.

Have You Considered?

1. Imagine you are at a party and you notice a woman you do not know drinking beer after beer and a man whom she has met at the party encouraging her to drink more. He is watching her, touching her, and cheering her on. A little later, you see them off in a corner where he is rubbing against her and trying to kiss her. She is laughing but attempting, without much success, to push him away. Discuss how you think you would react to this situation. What are at least two courses of actions you might take?
2. If you were asked to give a workshop to first-year college *men* about dating violence and date rape, what three topics would you want to cover in the fullest detail? Why did you pick those three?
3. If you were asked to give a workshop to first-year college *women* about dating violence and date rape, what three topics would you want to cover in the fullest detail? Why did you pick those three?
4. How do you think society should deal with repeat sexual child molesters? Explain your answer.
5. Discuss what actions you might take if you suspected that someone in your neighborhood could be a pedophile.
6. Create a realistic scenario of student-on-student sexual harassment in a college setting in which the harasser is female and the victim is male. Include in your scenario suggestions on how the problem might be resolved.

Summary

Historical Perspectives

Marital Rape

- Forced sex in marriage was not considered rape in the past. Not until 1993 did the last of the fifty states include husband-wife rape in laws against rape. Prior to 1973, husbands who forced sexual acts on their wives were exempt from rape laws, as wives were considered their husband's "property."

Rape

- Rape is defined by the U.S. Department of Justice as "The penetration, no matter how slight, of the vagina or anus with any body part or object, or oral penetration by a sex organ of another person, without the consent of the victim." Other nonconsensual sexual acts are often referred to as sexual assault.
- Most rapists are known to their assailants. By far the most common types of rapes (approximately 70%) occur between a perpetrator and victim who know each other through dating, friendship, or an intimate relationship.
- Rapists and victims may be of either sex, but most rapes are committed by men against women. Rapes of men by men and rapes of men by women, though

rare, do occur. Studies have found that approximately 22% of women and 2% of men have been victims of a forced sexual act.

- Rape myths typically blame the victim and remove responsibility for the crime from the perpetrator. Two common examples are: "If she didn't want it, how come she dressed like that?" and "Women secretly want to be raped."
- The vast majority of rapes on college campuses occur in an atmosphere of heavy use of alcohol by the victim and usually by the perpetrator as well.
- A diabolical trend on the dating scene is rapists who slip date rape drugs into women's drinks, rendering them unconscious and unable to resist or even remember the sexual assault that follows.
- Some women who are victims of rape, in an attempt to put the trauma behind them, resist reporting the crime and having tests to collect evidence to arrest and convict their attackers.
- Unfortunately, the underreporting of rape is widespread and leaves rapists free to prey on more victims.
- A disproportionately large number of acquaintance rapes occur at fraternity house parties.
- Studies reveal that some college fraternities and athletic groups turn a blind eye to their "culture of rape," in essence encouraging rape through subtle or overt social rewards to their members for sexual assaults.
- Rape in the military is a widespread and serious problem because it is a male-dominated culture, and fear of retribution prevents most reporting of rape and creates an atmosphere that rape will not lead to punishment.
- Many emotional and psychological difficulties stem from rape. Effects on the victim are similar to post-traumatic stress disorder and may include shock, depression, shame, anger, guilt, embarrassment, and difficulties in forming and maintaining relationships with others.
- In a positive trend, more men are joining rape prevention efforts. Men may have a great deal of power to discourage behaviors among their peers and discourage environments that contribute to the incidence of rape.

Child Sexual Abuse

- The vast majority of child molesters are relatives of the child or people the child knows.
- The effects of child sexual abuse tend to be long-lasting and include emotional distress, psychological disorders, future relationship problems, and numerous medical conditions.
- Several states now have laws requiring repeat sexual offenders to undergo surgical or chemical castration as a condition of receiving parole.

Sexual Harassment

- The requiring of sexual favors by a person in a position of power in exchange for a positive event or outcome for the subordinate is known as quid pro quo sexual harassment. When a person is made to feel uncomfortable in a work, military, or school setting due to a pattern of unwanted sexual behaviors, advances, or language, this form of sexual harassment is said to create a hostile environment. Both types are illegal, and perpetrators are subject to criminal prosecution and civil remedies.
- The characteristics of bullying and sexual harassment have been shown to be similar, and both have been found in primary schools as well as in higher grade levels.
- Student-on-student sexual harassment is common on college campuses. When one student persists in unwanted advances toward another student, this often creates a hostile environment for the victim and is considered illegal under sexual harassment laws.
- Cyberharassment has become a serious problem leading to major psychological problems and suicide of victims of harassment through social media.
- The effects of sexual harassment are often similar to those of rape.

Your Sexual Philosophy

Sexual Aggression and Violence: Rape, Child Sexual Abuse, and Sexual Harassment

- Rape, child sexual abuse, and sexual harassment represent the dark side of human sexuality.
- Incorporating an understanding of factors related to sexual aggression and violence into your sexual philosophy may at first seem unnecessary. But such knowledge is power—in this case, the power to help stop the violence.

Chapter 14

Paraphilic Disorders

Atypical Sexual Behaviors

Learning Objectives

After you read and study this chapter you will be able to:

14.1 Review the basic and clinical (APA) criteria for defining and diagnosing paraphilias and paraphilic disorders, summarize the well-known paraphilias, and discuss the related issues of sexual addiction and how society judges the severity of a paraphilia.

14.2 Explain why paraphilias are seen far more commonly in men compared to women.

14.3 Summarize the various theories that have been proposed to explain the development of paraphilias and how the differing theories can be combined and reconciled.

14.4 Analyze the differences between victimizing and non-victimizing paraphilic disorders, discuss each of the disorders that fall into each category, and summarize the sexual abuse scandal in the Catholic Church.

14.5 Evaluate the treatment of paraphilic disorders including reasons for seeking treatment, the types of treatments available, and their effectiveness.

Since You Asked

1. Sometimes my girlfriend and I like to do some things that are a little strange (like tying each other up and stuff). Is this really weird and perverted? (see page 526)
2. I get so angry reading about all the child abusers out there! Why don't we just lock them all up for life? (see page 530)
3. Last year, there was this guy who kept looking into girls' windows in the dorms (Peeping Tom). What is that all about? (see page 535)
4. I saw this movie (not porno) where a guy got off by rubbing up against women at concerts when everyone was crowding and pushing up to the stage. What is up with that? (see page 538)
5. I was sexually molested as a child. I never told anyone (until now). With all these stories about priests molesting kids, I think about it a lot and wonder why it happened. Do you think I should get some counseling to work through it? (see page 540)
6. I met a woman at a party. She told me that "on the side," meaning not her main job, I guess, she's a dominatrix. Does that mean what I think it means? (see page 542)
7. I saw a movie from the 1990s on cable called *Rising Sun*. In the movie, a man strangles a woman during sex. It seemed that she wanted him to do it, but he did it too hard and killed her. Why would anyone want to be choked during sex? (see page 544)
8. My mom has this friend who shops for shoes constantly. It's like if she's not eating or sleeping, she's shoe shopping. Is this what is meant by a shoe fetish? (see page 545)
9. My boyfriend likes to put on a woman's long blond wig (he's got a dark beard!) just to make people laugh. Is he being a transvestite? (see page 547)
10. Everyone in my neighborhood has received a notice that a convicted sex offender has been released from prison and will be moving near us. Of course, this news is scary. Can sex offenders be cured (I think he's an exhibitionist)? What are the chances he will do those acts again? (see page 550)

All individuals are unique in the sexual behaviors and specific types of stimulation that arouse them. Some might be drawn to a certain body type, a special fragrance, a particular sexual activity that they find especially exciting, or any number of personal sexual preferences. These differences are what we might call "normal" human differences in sexual tastes, and they are quite varied. However, if we are to become aware of and understand the *full* range of human sexuality, we must extend our discussion beyond the boundaries of normal sexual behavior. We must include in our study the sexual activities that are practiced by a relatively small percentage of people, activities that most people would consider odd, deviant, or even pathological. In this chapter, we explore a group of behaviors called **paraphilias**: compulsive, out-of-the-ordinary sexual practices that may fascinate, repel, or anger you. The prefix *para*, means "faulty" or "abnormal," and *philia* means "attraction" or "love." Thus *paraphilia* literally means "abnormal attraction."

Paraphilias
compulsive sexual activities that are practiced by a small percentage of people and that most members of a given culture would consider abnormal, deviant, or pathological.

We will explore in detail the current criteria that define paraphilias and explain when a paraphilia becomes a paraphilic disorder, including the origins and the role of gender in these behaviors; we will consider each of the paraphilic disorders individually within the context of two categories, those that involve a nonconsenting victim and those that are nonvictimizing; we will examine available treatments to help people overcome paraphilic disorders; and we will close with a look at how an awareness of paraphilias can be incorporated into your personal sexual philosophy. Before all that, however, let's take a brief look at reports from over a hundred years ago describing behaviors that are known today as paraphilias.

Focus on Your Feelings

Discussions about atypical sexual behaviors known as paraphilias usually stir up strong emotional reactions in most of us. These feelings may include embarrassment that anyone would ever want to try such an activity, shock at how "abnormal" the behavior seems, anger at the people who engage in the activity, repulsion over how disgusting the behavior sounds, and perhaps even interest—because the *idea* of some of the activities, no matter how strange, may sound exciting to some people. Any of these and many other reactions to the material in this chapter are common and perfectly normal. Remember, thoughts and fantasies are not the same as actual behaviors.

As you read this chapter, you may experience these or other strong feelings. If you do, think about which paraphilias evoke your reactions and why you feel the way you do. Understanding your reactions to unusual or "abnormal" sexual behaviors is extremely valuable if you ever find yourself confronting that particular behavior in your life. You can learn a lot about yourself by analyzing your reactions and thinking about whether they are appropriate and effective responses. This self-knowledge is also essential to understanding how you will incorporate learning about paraphilias into your sexual philosophy, which is discussed at the end of this chapter.

Historical Perspectives

An Early Account of Paraphilias

One of the first detailed accounts of paraphilic behaviors was in a medical text written in 1886 called *Psychopathia Sexualis*, by psychiatrist Richard von Krafft-Ebing. His 600-plus-page book consisted of 237 case studies of the kinds of sexual activities we will discuss in this chapter. *Psychopathia Sexualis* was a groundbreaking scientific study of sexual pathology that influenced many medical and psychiatric writers and researchers of his day, including Sigmund Freud. Most of the terms for various unusual sexual activities you will see in this chapter were developed by Krafft-Ebing. The book was republished in its entirety in 1998 (Rosen-Molina, 2001). Here are just two examples of the cases he describes (Krafft-Ebing & Klaf, 1998). The first involves a paraphilia called *zoophilia*:

> In a provincial town a man was caught having intercourse with a hen. He was thirty years old, and of high social position. The chickens had been dying one after another, and the man causing it had been "wanted" for a long time. To the question of the judge, as to the reason for such an act, the accused said that his genitals were so small that coitus with women was impossible. Medical examination showed that actually the genitals were extremely small. The man was mentally quite sound. There were no statements concerning any abnormalities at the time of puberty. (case 229)

Another case from *Psychopathia Sexualis* is about a type of behavior we now call a *fetish*:

> Z began to masturbate at the age of 12. From that time he could not see a woman's handkerchief without having orgasm and ejaculation. He was irresistibly compelled to possess himself of it. At that time he was a choir boy and used the handkerchiefs to masturbate within the bell tower close to the choir. But he chose only such handkerchiefs as had black and white borders or violet stripes running through them. At age 15, he had coitus. Later on he married. As a rule, he was potent only when he wound such a handkerchief around his penis. Often he preferred coitus between the thighs of a woman where he had placed a handkerchief. Whenever he espied a handkerchief, he did not rest until he was in possession of it. He always had a number of them in his pockets and around his penis. (case 110)

From these examples, you can see that our knowledge of very atypical behaviors, referred to as *paraphilias,* is not at all new. However, exactly what behaviors qualify as paraphilias has changed markedly throughout history. Although many of Krafft-Ebing's sexual pathologies are still seen as paraphilias, others that he strongly condemned, including masturbation, oral sex, anal sex, and nonheterosexual activities, are seen today by nearly all sexuality educators, psychologists, and medical professionals as *normal* variations in sexual behaviors (Szasz, 2000).

Defining Paraphilias

14.1 Review the basic and clinical (APA) criteria for defining and diagnosing paraphilias and paraphilic disorders, summarize the well-known paraphilias, and discuss the related issues of sexual addiction and how society judges the severity of a paraphilia.

Since You Asked

1. Sometimes my girlfriend and I like to do some things that are a little strange (like tying each other up and stuff). Is this really weird and perverted?

Paraphilias are far more complex than merely "atypical sexual acts" or the word's literary definition of "abnormal attraction" implies. However, in general it is fairly easy, using a few simple criteria, to analyze a particular behavior to see if it is likely to be a paraphilia. Beyond this, society applies standards to unusual sexual behaviors to determine how pathological (disordered), or deviant, they are. In addition, mental health professionals have their own set of form al criteria for diagnosing paraphilias.

Basic Criteria for a Paraphilia

For the purposes of our discussion here, a sexual activity may be considered a paraphilia if it meets *all three* of the following criteria (these are shown graphically in Figure 14.1):

1. The behavior is engaged in for the purpose of sexual arousal or gratification.
2. The behavior tends to be compulsive and recurrent.
3. A clear majority of people in a given cultural setting would consider the behavior to be strange, deviant, **pathological**, or abnormal.

pathological
any behavior seen as caused by sickness or disease.

The first criterion is obvious: at least one of the primary reasons a person is engaging in the behavior is *because* it is a sexual turn-on. If a man dresses up as a woman

Figure 14.1 Criteria for Determining a Paraphilia

Behavioral Criterion	MEETS CRITERION?	MEETS CRITERION?
The behavior is engaged in for the purpose of sexual arousal or gratification.	NO: NOT A PARAPHILIA	YES
The behavior tends to be compulsive and recurrent.	NO: NOT A PARAPHILIA	YES
The behavior is considered by society to be deviant, pathological, or abnormal.	NO: NOT A PARAPHILIA	YES
	NOT A PARAPHILIA	**PROBABLY A PARAPHILIA**

for a part in a play, his cross-dressing is probably unrelated to sexual arousal or gratification and is unlikely to be defined as a paraphilia (transvestism). Many of you are aware of the ritual in New Orleans at Mardi Gras where women expose their breasts so that revelers on parade floats will throw strings of shiny beads to them. This would not be an example of a paraphilia (exhibitionism) because the women, presumably, are not engaging in the behavior for the purpose of sexual arousal.

The second criterion of paraphilias is that they tend to constitute **compulsive behavior**. Any behavior, sexual or not, may for some people become compulsive. In essence, a compulsive behavior is one that controls the person instead of the other way around. Usually, it is a behavior that the person would like to stop doing but feels powerless to control. The destructive effects of compulsions may be seen in many behaviors, including compulsive drinking, compulsive gambling, and compulsive drug use.

compulsive behavior
any behavior, sexual or otherwise, that a person is unable to control regardless of repeated attempts to do so.

This compulsive component of paraphilias is manifested in the inability to stop the undesirable behavior; even if the person is distressed by it and wants to stop it, he or she is unable to control it. Some people with paraphilias find themselves tormented after each occurrence of the act, vowing and promising themselves that they will never do that again. But then, as time passes, the obsessive thoughts creep back, the desire and need escalate, and they are irresistibly drawn back to the behavior once again. This compulsive aspect of paraphilias typically causes distress and dismay at being unable to resist the impulse. However, if a person engages in an unusual sexual behavior once or twice and never has the desire to do so again or simply chooses not to, the behavior would not be characterized as a paraphilia.

The third criterion of paraphilias is what sets them apart from other sexual behaviors (even those that might meet the first two criteria). Paraphilias are sexual behaviors that would strike most people in a particular cultural setting as bizarre or abnormal. This judgment is, of course, usually very subjective. You may be thinking, "What's deviant to one person is mainstream to another," or "Who's to say what's abnormal anyway?" This is true of many human behaviors, but defining paraphilias is not as tricky as it seems. It is important to take into account the culture in which the behavior occurs because societies are often surprisingly diverse in their social and sexual customs. That said, if you ask 100 people about virtually any of the paraphilias discussed in this chapter, the vast majority will agree that they are, indeed, deviant or abnormal. Here we are not discussing a formal, diagnostic meaning of abnormal but rather a cultural one (more on this in the next section).

With these criteria, you have the tools necessary to analyze any sexual behavior to decide if it meets the informal criteria for a paraphilia. Questions from students (usually in the anonymous "sex questions box") often ask if a particular behavior they have heard about or engaged in is a paraphilia. When the class applies the three criteria discussed here, usually (but not always) the behavior in question does not make the cut. The "Self-Discovery: Do *You* Have a Paraphilia?" (on the following page) applies these criteria to paraphilias and will help you obtain a clearer idea of how to judge for yourself whether a certain sexual behavior might be defined as a paraphilia (Cormier, 1993).

Clinical Criteria for Paraphilic Disorders

If a person's behavior meets all three of the criteria for a paraphilia, does that imply that the person may be diagnosed with a psychological disorder? Although many or even most members of a society might feel that certain paraphilias are abnormal or "sick," formally diagnosing a psychological disorder is a more rigorous process.

The *Diagnostic and Statistical Manual of Mental Disorders* of the American Psychiatric Association, referred to as the DSM-5, distinguishes between a paraphilia and a **paraphilic disorder**. Engaging in atypical sexual behaviors, even as described above, may or may not meet the formal criteria for a paraphilic disorder. Here are the main distinctions put forth in the DSM-5. First, a person may have a paraphilia

paraphilic disorder
a paraphilia involving distress to the diagnosed person and/or lack of consent of the victim

Self-Discovery

Do You Have a Paraphilic Disorder?

First of all, you probably *don't* have a paraphilic disorder. These are not at all common. However, many people worry that a sexual activity they enjoy, either in reality or in fantasy, might be "abnormal." Here is a self-test to help you get an idea if the sexual behavior you are concerned about might be diagnosable as a paraphilic Disorder. You can substitute any sexual behavior into the questions to get an idea if it might "qualify."

Paraphilic disorder Self-Test

Question	Yes	No
1. Is the sexual behavior (or urge) extremely sexual arousing for you?	______	______
2. Have you acted on the sexual urges or thoughts and engaged in the behavior itself?	______	______
3. Have you felt guilt, shame, anxiety, depression, remorseful, or significantly distressed over your urges, thoughts, or acts relating to this behavior?	______	______
4. Have you been fearful and anxious that you will be "found out" by others that you are engaging in the behavior?	______	______
5. Do you worry that the behavior is seen by society as very deviant, abnormal, forbidden, or psychologically sick?	______	______
6. Have these sexual urges or acts interfered with your ability to form and maintain *mutually* satisfying intimate or sexual relationships with a partner?	______	______
7. Have you engaged in the behavior with non-consenting partners or underage victims?	______	______
8. Have you ever been arrested or in trouble with the authorities for the behavior?		

Scoring: If you answered *yes* to question 3; and *yes* to any one of questions 4, 5, 6, 7, or 8 you may have a paraphilic disorder or a tendency toward developing one.

that is *not* considered to be a psychological disorder when it is freely engaged in without personal distress or victimization (meaning, either there is no involvement by another person, or the other person is an adult and participates with full consent).

The behaviors, however, become a *paraphilic disorder* when: (1) the person feels personal distress about the fantasies, urges, or behaviors (not merely distress resulting from society's disapproval) and this distress continues over time (at least 6 months); or (2) the person displays a sexual desire or behavior that involves another person's psychological distress, injury, or death, or a desire for sexual behaviors involving unwilling persons or persons unable to give legal consent (i.e., incapacitated or underage) (APA, "Paraphilic Disorders," 2013). If someone acts on a paraphilia with a nonconsenting person (including anyone underage), the behavior then rises to the diagnosis of a paraphilic disorder, and also a crime. Either or both of these two criteria must be met for an official diagnosis of a paraphilic disorder to be made (Addis, 2014).

The DSM has for decades identified and listed specific criteria for eight specific paraphilias: *voyeurism, exhibitionism, fetishism, frotteurism, pedophilia, sexual masochism, sexual sadism,* and *transvestic fetishism*. The DSM-5 has slightly updated these names to reflect the emphasis on separating a paraphilia from a paraphilic disorder. They are: *voyeuristic disorder, exhibitionistic disorder, fetishistic disorder, frotteuristic disorder, pedophilic disorder, sexual masochism disorder, sexual sadism disorder,* and *transvestic disorder* (DSM-5, 2013). These are listed in Table 14.1, along with some other paraphilic disorders that are not, as yet, specifically listed in the DSM.

sexual addiction
A term often used to refer to strong compulsive sexual behavior—more correctly called "hypersexual disorder."

hypersexual disorder
Usually defined as nonparaphilic sexual compulsivity; an obsessive preoccupation with and compulsive need for sexual activity.

Is Sexual Addiction a Paraphilic Disorder?

In a word: No. Although the compulsive nature of paraphilias may remind some people of what is commonly known as **sexual addiction**, this is not defined as a paraphilia. The DSM-5 calls this form of sexual compulsivity **hypersexual disorder** and

Table 14.1 Some Well-Documented Paraphilias

Name of Disorder	Target of Sexual Interest	Type of Activity	Included in DSM-5
Voyeuristic	Without consent, watching others undress or engage in sexual behavior	Victimizing	YES
Exhibitionistic	Displaying genitals or sexual acts to others without consent	Victimizing	YES
Pedophilic	Children	Victimizing	YES
Fetishistic	Nonliving objects or nonsexual body parts	Victimless	YES
Frotteuristic	Rubbing genitals against another person without consent	Victimizing	YES
Sexual Masochism	Receiving pain	Nonvictimizing	YES
Sadism	Inflicting pain on another	Victimizing	YES
Transvestic	Dressing in clothes of opposite sex	Nonvictimizing	YES
Autoerotic asphyxiation	Self-strangulation during masturbation	Self-victimizing	NO
Formicophilia	Insects or other small crawling creatures	Victimless (unless insects seen as victims)	NO
Telephone scatalogia	Obscene telephone calling	Victimizing	NO
Zoophilia	Animals	Animal victims	NO
Gerontophilia	Elderly people	Usually nonvictimizing (role play)	NO
Infantilism	Being treated as an infant	Nonvictimizing (role play)	NO
Klismaphilia	Enemas	Victimless	NO
Necrophilia	Human corpses	Victimizing	NO
Stigmatophilia	Piercing or scarring; tattooing	Self-victimizing	NO
Urophilia	Urine	Usually nonvictimizing	NO
Coprophilia	Feces; excrement	Usually nonvictimizing	NO

it is in the appendix, meaning that it is not a clear diagnosis (yet), but is worthy of further research and analysis. An obsessive preoccupation and compulsive need for sexual activity (meaning that a person cannot stop the thoughts or behaviors even if they are harmful or if he or she wishes to) may present serious problems for a person and may lead to psychological intervention. However, these factors are not adequate to define sexual addiction as a paraphilia.

So-called "sex addicts" attempt to satisfy their constant and overpowering sexual cravings by taking excessive time planning and engaging in sexual behavior. Often, these overpowering urges occur in response to feelings of anxiety, depression, or other stressful events. As mentioned, these cravings are obsessive—out of the person's control—and he or she has been unable to stop them despite repeated attempts. These cravings may become so strong that they can harm the person or others (DSM-5, 2013). However, these behaviors may not always meet the third paraphilic criterion of being seen as abnormal or deviant (such as consensual sex with one or more partners or masturbation). However, many who are diagnosed (or diagnosable) with hypersexual disorder may be drawn into sexual behaviors that are clearly paraphilic disorders, including exhibitionistic disorder, voyeuristic disorder, pedophilic disorder, or other socially unacceptable or illegal behaviors such as extramarital affairs, sexual harassment, or rape.

Society's Criteria for Judging Paraphilic Disorders

Alongside scientific or formal diagnostic criteria for paraphilias are the judgments made for human behaviors by the society in which they occur. As a member of society, you are constantly making judgments about behavior—your own and that of others. But have you ever really thought about exactly what criteria you are applying when you make such judgments? When a culture or society looks at any human behavior, it does so through a lens of broadly shared norms and standards. The majority of members

Since You Asked

2. I get so angry reading about all the child abusers out there! Why don't we just lock them all up for life?

in a given society apply those criteria in making a judgment about whether or not the behavior, in the eyes of the broader social context, is normal, moral, and acceptable (Bhugra, Popelyuk, & McMullen, 2010; Blachère & Cour, 2013; Goodman, 2001). And it is safe to say that few, if any, categories of behavior exist onto which all societies place greater social judgment than *sex*.

Within the category of sex, the paraphilias probably receive among the most intense negative social judgments of all human behaviors. All societies invoke numerous, often unspoken, guidelines for judging "abnormal" sexual practices. However, for the purposes of our discussion here, a paraphilia will be more likely to be judged negatively by a *society* when it is seen as meeting one or more of the following four criteria (refer as necessary to Table 14.1 for the specifics of the paraphilias mentioned here):

1. **The behavior is harmful or destructive to the person engaging in it.** When a paraphilia is seen by society as harmful to the person doing it, either physically or emotionally, that behavior will be judged in significantly more negative ways. Examples of paraphilias that fit into this category are a person who participates in extreme sexual masochism to the point of serious injury, scarring, or broken bones; a man who dresses in women's clothes in public and is ridiculed, scorned, rejected by his family, and fired from his job; or a teenager involved in autoerotic asphyxiation behaviors (strangulation during sexual activity).
2. **The behavior is illegal.** Laws are one way a society makes overt statements about acceptable and unacceptable behavior. It makes sense that if the paraphilia involves behavior that is against the formal, written laws of the state, people will tend to judge it more harshly than if it is outside the mainstream, but not illegal. Illegal paraphilias generally involve victimization, as in the case of voyeurism, exhibitionism, pedophilia, and frotteurism.
3. **The behavior interferes with the person's ability to form and maintain loving, intimate, and sexual relationships with others.** In other words, if a person engages in paraphilic behavior *instead of* forming healthy intimate relationships with others, the sexual activity will be judged more negatively. On the other hand, if a person engages in certain *victimless* paraphilias and is able to develop and maintain a close loving relationship, he or she is typically judged less harshly by society for his or her atypical sexual behavior.
4. **The behavior involves another person without that person's consent.** This standard may be the most important in determining how most societies will judge the "wrongness" of a paraphilia. As you will see when we discuss specific paraphilias, some are typically **consensual**, meaning entered into voluntarily by a partner, or **victimless**, meaning not harming or even involving a partner at all. An example of a typically consensual paraphilia is sexual masochism disorder; by contrast, fetishistic disorder and transvestic disorder are victimless paraphilias. Nonconsensual paraphilias, by definition, involve victimization or coercion of another person; examples include voyeuristic, exhibitionistic, frotteuristic, and pedophilic disorder. Nonconsensual paraphilic behaviors are likely to be illegal, whereas the legal status of consensual and victimless paraphilias varies from state to state and country to country.

consensual

A behavior entered into voluntarily by all involved parties.

victimless

An activity that harms no one.

Men and Women and Paraphilic Disorders

14.2 Explain why paraphilias are seen far more commonly in men compared to women.

You may notice as you read this chapter that whenever the sex of a person with a paraphilia has been mentioned, it is virtually always a male. The reason for this is that *nearly* all those who engage in paraphilias are, indeed, men (Dawson, Bannerman, & Lalumière, 2014; Hall & Hall, 2007; Katz, 2012). The one possible exception to this rule is sexual masochism (sexual gratification from receiving pain), which is found in women more often than other paraphilias but is still more common in men (American Psychiatric Association, 1994). Why do we see such a gender gap for these behaviors?

Many sexuality theorists throughout history, including Krafft-Ebing (discussed in "Historical Perspectives"), have applied the overly simplistic explanation that men have a stronger sexual drive than women do, and this "overdeveloped" sexual appetite leads them uncontrollably into unusual and compulsive sexual behaviors. However, this explanation fails to address the question of why it is that if men have a high sex drive, only a small minority become involved in paraphilias, and most men are content with more mainstream sexual behaviors.

The relative prevalence of paraphilias among men as compared to women may be understood better from a larger cultural perspective. In most Western cultures, males are given greater "permission" to be sexual in their thinking, language, and behaviors. Most societal attitudes consider overt sexuality in men not only normal but also desirable, something to be encouraged and rewarded. You can easily see this in the language used to talk about sexuality in men and women. A man who is highly sexually oriented is typically referred to in positive terms such as "stud" or "hottie," whereas a woman who talks or behaves in similar ways is usually labeled in negative terms such as "slut" or "whore." It may be that as males are given more latitude in Western cultures to be sexual, some of them will learn to express that freedom in paraphilic ways.

This is not to imply that women never engage in paraphilias (e.g., Dawson, Bannerman, & Lalumière, 2014). As noted, a small but significant percentage of women practice sexual masochism. And one study found a few examples (twelve in all) of women pedophiles, exhibitionists, and sexual sadists (Fedoroff, Fishell, & Fedoroff, 1999). However, by all accounts, these cases represent the exception, not the rule. In addition, some research has indicated that in cultures where sexual activity is regarded as serving primarily a reproductive function (as opposed to an attraction-based function, as in the United States), paraphilias among men and women are significantly less common (Bhugra, 2000).

It is important to understand that all research into paraphilias is extremely challenging, to say the least. Take, for example, trying to determine the relative frequency of paraphilic behavior in men versus women or among various diverse cultures. How can a researcher gather data on such personal and socially forbidden behaviors? The answer is that doing so is not impossible, but we must always think analytically about such findings, as explained in "Evaluating Sexual Research: How Biased Are Sexual Research Data?"

Evaluating Sexual Research

How Biased Are Sexual Research Data?

When researchers are trying to obtain data about sexual behavior differences between males and females or among diverse cultures, the only ethical way to do so is to ask people what they do and how often they do it (we cannot, of course, ethically peep into windows or set up hidden camcorders!). But when you ask people to tell you about their intensely personal, perhaps embarrassing, or socially questionable sexual activities, you have to assume that many, if not most, are likely to fudge their answers in ways that make them look as socially acceptable as possible. We discussed this problem, called the *social desirability bias*, in Chapter 6, as it relates to the content of people's sexual fantasies. You can imagine that if people tend to be less than truthful about their sexual fantasies, then they are much more likely to distort the truth when asked about their paraphilias! It is important to keep in mind that self-report data always have the potential to be biased, especially when interpreting research findings showing, for example, that a certain percentage of people participate in a certain paraphilia, that few women appear to engage in paraphilic behavior compared to men, or that paraphilias are practiced more or less frequently in various cultures. This does not imply that all self-report sexual research is worthless; rather, it must always be examined with a critical eye.

In many countries, however, we have a source of data other than self-reports for paraphilic disorders that we do not have for many other sexual behaviors: criminal justice records. Because many paraphilic disorders victimize others and are therefore illegal, some perpetrators are caught and arrested. When we examine those arrest records, we find that almost all the perpetrators are males. That would suggest that dramatic gender differences do indeed exist for nonconsensual paraphilic disorders. Still, we must ask ourselves whether those crime statistics might reflect a cultural bias in the criminal justice system that focuses on male offenders and casts a blind eye toward female perpetrators of nonconsensual paraphilic disorders. For example, consider the following: Would a man looking into a window where a woman is undressing be viewed in the same way as a woman looking into a window where a man is undressing? Or might the man be perceived in the first case as a voyeur and in the second case as an exhibitionist, and the woman as the victim in both cases? It's something to think about.

The Development of Paraphilias

14.3 Summarize the various theories that have been proposed to explain the development of paraphilias and how the differing theories can be combined and reconciled.

Upon first hearing or reading about paraphilias, your first question might be, "How could someone acquire such an unusual sexual compulsion?" Explanations exist, but the research, or lack of it, points to one inescapable conclusion: Nobody knows for sure what causes paraphilias. About all we can do here is consider some of the theories proposed to explain the development of paraphilias and let you draw your own conclusions.

First, it is probably safe to start with the assumption that no one is born with any particular sexual eccentricities but that unusual sexual attractions develop in one way or another during a person's life. Genetics may play a role, in that some people may inherit a predisposition, a slight push from nature, toward compulsive behaviors in general, which in some may manifest in sexual acts. However, at this point in genetic research, it is doubtful that anyone is, say, a "genetic exhibitionist."

That said, at least three theories—psychodynamic, behavioral, and biological—have been suggested to account for the development of a paraphilia (Barbaree, Marshall, & McCormick, 1998; Brannon, 2002; Friedrich & Gerber, 1994; Price et al., 2001; Wise, 1985).

Psychodynamic Theories of Paraphilias

Psychodynamic theories of human nature rest on the theories of Sigmund Freud, who contended that all psychological problems in adulthood, sexual or otherwise, stem from traumatic events that occurred in early childhood. The fundamental assumption here is that major developmental traumas or conflicts you may have encountered during your early formative years, from birth through about age ten, have been repressed into your unconscious in order to protect you from their disturbing nature. Because you repressed them and pushed them into your unconscious, you are even now totally unaware that those conflicts exist. However, in adulthood, the theory maintains, those repressed traumas may exert powerful forces on your behavior and cause you to engage in socially unacceptable behaviors as your unconscious tries to resolve the tensions they create.

In simplistic terms, for example, the psychodynamic approach might attempt to explain how a man who engages in voyeurism may have been severely punished or abused when, as a young child, he unknowingly walked into his parents' bedroom when they were naked or engaging in sexual activity. The theory claims that, although the voyeur is unaware of the unconscious conflict caused by that childhood trauma, his compulsive desire to observe people undressing or engaging in sexual interactions may stem directly from it. Of course, any psychoanalyst would be quick to point out that the connections between childhood traumas and dysfunctional adult behavior are rarely so simplistic (e.g., Kaplan, 1997; Schott, 1995; Seth, 1997). Here is a summary of an article about a pantyhose fetish that appeared in a psychoanalytic professional journal called *Gender and Psychoanalysis* that will help you appreciate the flavor of the psychoanalytic approach to explaining paraphilias:

> In this paper...it is suggested that the wearing of pantyhose by males serves a range of functions, including, but not limited to, repairing psychic structure, an expression and defense against underlying aggression, enabling the development of symbol formation, allaying annihilation and separation anxiety, acting as a type of transitional object in terms of serving as a "second skin" component, soothing primitive anxieties related to the damaged body ego, and serving an erotic function involving desire and excitement. The function of pantyhose as a kind of "magic skin" or "second skin" suggests a relationship between perverse symptom formation and the *development* of a "skin ego." Implications for the symbolizing function of this "magic or second skin" as a bridge to the mother figure is discussed in terms of theoretical and treatment issues. The functional use of pantyhose is viewed as a creative solution to a seemingly irresolvable interpersonal-intrapsychic dilemma. (Lothstein, 1997, p. 103)

A strict psychoanalytic approach to studying and explaining any human behavior has faded from favor as modern scientific psychology has grown and developed over the past century. Most modern-day psychologists find little validity in most of Freud's conceptualizations of human nature. Nevertheless, when sexuality researchers and educators discuss the possible causes of paraphilic compulsions, many allow that they may be rooted, at least to some extent, in certain childhood experiences (e.g., Berner, 2012).

Behavioral Theories of Paraphilias

The behavioral approach to explaining the origins of paraphilias rests on the various components of classical and operant conditioning. If you have studied these learning theories in other psychology classes, you'll recall that in *classical conditioning*, a response to a particular stimulus is learned (or conditioned) when that stimulus is pairedin the brain with another event that naturally produces a particular response. An example of this as it relates to sexual behavior would be a person who finds a particular perfume fragrance to be an intense turn-on. This seemingly involuntary (or reflexive) physical response, from a behaviorism perspective, is not due to anything inherent in the perfume itself (no matter what the manufacturer would like you to think), but rather it came about through repeated pairings of the perfume with sexually arousing events, such as sexual touching, oral sex, or intercourse. And probably no one needs to tell you that these can be *very* powerful sexual stimuli. Behaviorists contend that our personal, individual, sexual experiences determine each person's unique set of sexual likes and dislikes, preferences, and characteristic sexual inclinations. This, they propose, is why no two people are sexually aroused by exactly the same patterns of situations and events. Behaviorists believe that this same process explains why some people engage in paraphilias. Somehow, somewhere during their sexual life, their experiences caused them to associate strong, highly pleasurable sexual arousal with some sort of atypical sexual stimulus.

Behavior theories propose that paraphilias begin with the pairing of a particular event with sexual arousal.

According to other behaviorism theories, the sexual learning process is further strengthened through *operant conditioning*, in which a *voluntary* behavior in a certain sexual setting is learned because it is followed by a rewarding consequence, referred to as *reinforcement*. This principle is, in essence, quite simple: Any behavior that occurs in a particular setting and is followed by a rewarding event (reinforcement) will be more likely to reoccur in the future in a similar setting (the reward is associated with the setting and produces the behavior). It's probably obvious to you how this form of conditioning plays a role in the development of paraphilias. Few events in life are more rewarding (reinforcing) than an orgasm!

Combining these two basic principles of behaviorism provides a pretty clear picture of how they might account for the tendency of some people toward paraphilias. First, a person learns to associate sexual arousal with objects or unusual activities, and then those interactions are reinforced strongly with sexual pleasure and orgasm. The result? A paraphilia (or a paraphilic disorder). For example, imagine a teenage boy sitting at his desk in his bedroom who happens to see a woman who forgot to close her blinds undressing in a window across the street. He becomes sexually aroused as he watches, and he masturbates to orgasm. After this first experience, he begins to watch and wait for the woman to undress again, and he masturbates again. He may then begin to fantasize about spying on the woman when he masturbates at other times as well. Over the next few months or years, his desire to watch unsuspecting women undress may become increasingly strong, and he may begin to seek out opportunities to satisfy his urges. He may, for instance, prowl neighborhoods at night looking in windows for victims and, upon finding them, may masturbate on the spot as he watches, or he may use the experience for later masturbatory fantasies. He has now acquired the paraphilia we call voyeurism (to be discussed in more detail later in this chapter).

Biological Theories of Paraphilias

Biological theories of paraphilias derive from the assumption that something physiological has malfunctioned and is leading to the person's abnormal and compulsive behaviors. These theories focus primarily on imbalances in two biochemical systems: hormones and neurotransmitters (Assumpção et al., 2014; Rosler & Witztum, 2000). The hormone hypothesis suggests that compulsive sexual behavior may be related to an overproduction of male hormones, mainly testosterone (Bradford, 2001), but other hormones, such as epinephrine (adrenaline) and norepinephrine, have been implicated as well. The neurotransmitter serotonin, which is known to be involved in various psychological disorders, especially depression, anxiety, and obsessive-compulsive disorder (OCD), has also been connected to paraphilias (Garcia et al., 2013).

Most studies on the relationship among hormones, neurotransmitters, and paraphilias have provided evidence of a link by administering medications designed to change the balance of these chemicals in men with uncontrollable paraphilic urges (Assumpção et al., 2014). By blocking the action of, say, testosterone or enhancing the effects of serotonin with medications, some individuals with paraphilias find that their compulsions are reduced to more controllable levels, and they are better able to resist acting on their urges. This finding fits with discoveries that these same medications have also been successful in treating nonsexual obsessive-compulsive disorders. Later in the chapter, we will discuss the use of these kinds of drugs in treating paraphilias.

As an example of this theory, one study measured levels of adrenaline in men with pedophilic disorder and compared them to those of normal men (Maes et al., 2001). The researchers found significantly higher levels of the hormone in the pedophilia group in the three hours following the administration of an adrenaline-enhancing agent (see Figure 14.2). This suggests a biological foundation for pedophilia

Figure 14.2 Adrenaline Levels in Pedophiles Versus Normal Controls

Convicted pedophiles produced more adrenaline than a normal control group, which may provide evidence for a biological foundation in pedophilic disorder.

SOURCE: Adapted from data in Maes et al. (2001).

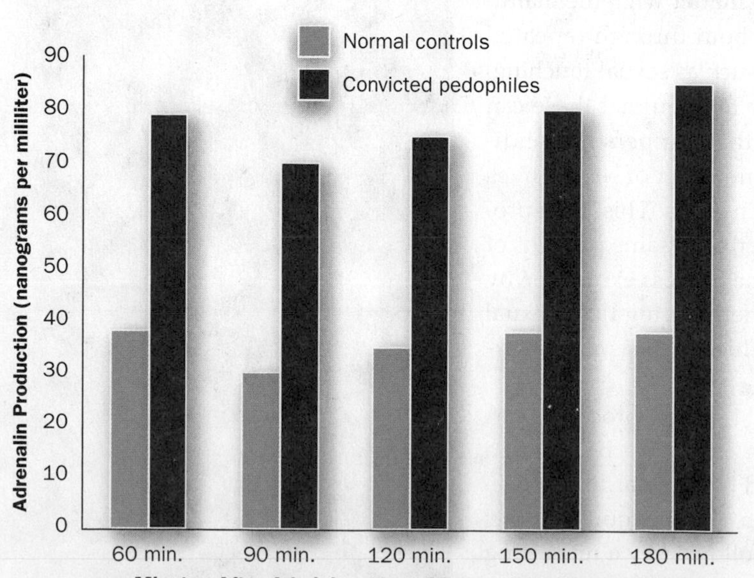

associated with the sympathetic nervous system, which triggers the release of adrenaline in highly emotional situations and may be overly active in pedophiles.

Reconciling the Three Theoretical Approaches

Which theory is the correct one, you ask? As with most competing opinions about anything, it depends on whom you ask. However, if you think about it, any two or even all three approaches to explaining paraphilias may be combined into a more global, cohesive explanation. For example, few psychologists would dispute that certain experiences in childhood cause us to pay more or less attention to and react in more positive or negative ways to specific sexual events, activities, and situations in our adult life. These tendencies may in turn lead us to be attracted to, enjoy, and engage in a personally unique set of sexual behaviors (psychodynamic theory). As we do so, we learn to associate those activities with sexual arousal and are rewarded for engaging in them with sexual pleasure and orgasm. Clearly, the range of sexual associations is so great that for a minority of people, the connections formed may be seen as bizarre within the culture (behavioral theory). Add to this set of events one or more biological or genetic factors that create in some people obsessive-compulsive tendencies (biological theory). If those obsessive-compulsive propensities become focused on certain atypical sexual inclinations, the ones that have been conditioned through association and reinforcement, what do you have? Correct! A paraphilia.

Many behavioral scientists and sex researchers would resist such a blending of views. However, little scientific research supports or even suggests a single, clear, definitive answer to the causes of paraphilic disorders. Finding ways of combining proposed theories simply allows us to broaden, rather than restrict, our thinking on this subject.

Specific Paraphilic Disorders

14.4 Analyze the differences between victimizing and non-victimizing paraphilic disorders, discuss each of the disorders that fall into each category, and summarize the sexual abuse scandal in the Catholic Church.

Now that you have an overall picture of the important definitions, characteristics, and features of paraphilic disorders in general, let's take a closer look at some of the more common ones. We will limit this detailed discussion, with one important exception, to the paraphilic disorders that are defined formally in the DSM-5, which is used to help mental health professionals identify psychological disorders. Each discussion opens with a brief case study that is fairly typical of the paraphilic disorder, followed by a description of the behavior. Again, keep in mind that some paraphilic disorders are clearly coercive and victimizing, whereas others are relatively victim-free and engaged in either alone or by mutual consent. We will divide our discussion of the more common paraphilic disorders into these two categories.

victimizing paraphilic disorders
Sexual activities involving an unsuspecting, nonconsenting, or unwilling person as the target of the atypical, compulsive behavior.

Victimizing Paraphilic Disorders

Victimizing paraphilic disorders involve an unsuspecting, nonconsenting, or unwilling victim who is the target of the compulsive, sometimes violent behavior. Indeed, the presence of a nonconsenting victim is often a motivating factor, part of the "thrill" component of the pathology experienced by the person engaging in the paraphilic act. It is primarily due to the coercive nature of these behaviors that most states in the United States and most countries worldwide have enacted laws against these paraphilic disorders.

Since You Asked

3. Last year, there was this guy who kept looking into girls' windows in the dorms (Peeping Tom). What is that all about?

VOYEURISTIC DISORDER *J.W. is a twenty-two-year-old college student. For several years, he has been having persistent fantasies of secretly watching various female classmates in their dorm rooms or apartments undressing or engaging in sexual activities. He finds these*

The voyeur achieves sexual arousal through watching unsuspecting others undressing or engaging in sexual activities.

fantasies especially exciting because of the fact that the women are completely unaware that he is watching them. Over the past couple of semesters, J.W. has been taking walks through campus housing areas at night looking for open blinds or cracks in curtains through which to sneak a peek. Lately, he has found several "watching posts" with views into windows and surrounded with trees and bushes that protect him from being seen. He has taken to going to these places two or three nights a week to sit, watch, and masturbate. After each foray, he feels guilty and swears never to do it again, but within a few days, he is drawn back to the behavior once more.

Voyeuristic disorder
Secretly watching others undress or engage in sexual activities without their knowledge or consent for the purpose of achieving sexual arousal.

Voyeuristic disorder refers to a recurring urge to watch others undress or engage in sexual activities without their knowledge or consent. You have undoubtedly heard of this rather common paraphilia; you may know someone who was watched, or perhaps you discovered that someone had been watching you. The common term for this person is a "Peeping Tom." Remember, this peeping is victimizing; it is done without the victim's consent. If someone is sexually aroused watching his partner undress prior to having sex, going to clubs with sex shows, or watching sexually explicit videos, these acts would *not* qualify as voyeurism. The true voyeur's sexual experience is defined and *enhanced* by the nonconsensual nature of the act.

For the victim, voyeurism can be extremely upsetting. Discovering that you have been secretly watched during your most private moments, sometimes over long periods of time, can be very frightening and may increase feelings of vulnerability in other areas of life. Some researchers contend that voyeurism is one of many forms of sexual assault and exploitation and may carry with it some of the effects of more physical forms of sexual assault such as rape (Allies Program, 2000; Wisconsin Coalition, 2000). Therefore, if you discover that a voyeur is active in your residential area, you should report it to law enforcement immediately.

Although the practice of voyeurism is centuries old, recent developments in technology and the Internet have sharply refocused attention on this paraphilic disorder. New and relatively cheap, tiny spy cameras and "web cams" are allowing voyeuristic predators alarming ease of access to their victims' private worlds. Websites exist that claim to contain images that are the product of nonconsensual voyeurism, secretly capturing people in intimate acts with these new technological tools. An increasing number of reports are appearing of apartment or hotel managers, store employees, women's dates and acquaintances, or other voyeurs setting up hidden video cameras in hotel rooms, apartment bedrooms, bathrooms, locker rooms, and dressing rooms to spy on unsuspecting victims, either for the voyeur's own pleasure or for distribution for profit over the Internet. Throughout the U.S. and the world, laws are being passed to make this high-tech form of voyeurism a crime to the same or greater extent as peeping through windows.

Exhibitionistic disorder
Achieving sexual arousal and gratification by displaying one's genitals to others without the victims' consent.

The exhibitionist is sexually aroused by exposing his genitals to unsuspecting victims.

EXHIBITIONISTIC DISORDER *S.N. is a forty-one-year-old male. He is married with one child and has a successful accounting business. Over the past several years, he has been exposing himself while masturbating as he tries to make eye contact with unsuspecting women in relatively public places such as the beach, the park, or the mall. This behavior has become so compulsive that he can now become sexually aroused only by exposing himself or by fantasizing about his most recent episode of exhibitionistic disorder. S.N. feels totally unable to control this impulse and has been arrested twice for indecent exposure and lewd and lascivious acts. He is separated from his wife and is currently forbidden by the court to visit his daughter except under strict supervision by a court-appointed social worker.*

Exhibitionistic disorder is the flip side of voyeurism in that the focus for sexual arousal and gratification is displaying one's

genitals to others without their consent. A common term for someone who engages in this paraphilic disorder is a "flasher." The stereotypical image of an exhibitionist is a "dirty old man" in a knee-length raincoat, throwing the coat open like a cape to reveal his genitals. In reality, most exhibitionists are men between the ages of sixteen and forty who experience high levels of sexual gratification from exposing their genitals to women they do not know, in various settings, and who usually masturbate at the same time or later, while fantasizing about the event. Some exhibitionists may also engage in voyeurism or obscene telephoning (Kolarsky, 2006). As with voyeurism, part of the arousal of this paraphilia is the shock value of the behavior on the victim. Most exhibitionists do not engage in other predatory or violent acts such as rape and usually desire no physical contact at all with their victims, but they are sexual predators nonetheless.

New technology and the Internet are playing an increasing role in exhibitionistic disorder, just as they are in voyeurism. Some people choose to place cameras in their homes and allow others to watch them online in a sort of consensual "exhibitionism-voyeurism" way of satisfying their curiosity about how others live and love. Also, sexually explicit websites are beginning to appear that allow anyone to engage in sexual behaviors and voluntarily be observed by others. Whether or not this is technically an exhibitionistic disorder is open for debate because those watching and being watched are presumably doing so voluntarily; so, arguably, no true victim of the acts exists.

In most Western cultures, people take their clothes off in front of others for a wide variety of reasons, ranging from exotic dancing to acting in film or theater to undressing in provocative ways for their partners. However, as you now understand, these people are not engaging in the paraphilic disorder of exhibitionism (or voyeurism) because the undresser and the observer are consenting to the act and, in some cases, the undresser is doing so for reasons other than his or her own sexual arousal and gratification.

If you are ever confronted by an exhibitionist, the experience can be frightening. However, it is very rare for exhibitionists to escalate the behavior into any actions that may be more violent or dangerous. Usually, your best "defense" against the exhibitionist is to ignore him, turn, and walk away. Of course, this paraphilic disorder is a crime, and you should report the event to campus security or local law enforcement officials as soon as possible.

FROTTEURISTIC DISORDER *E.S. is obsessed with rush hour in Manhattan. The crowded subways are the highlight of his day. At peak traffic time, E.S. enters the subway, spots a woman he is attracted to, and follows her onto the train. He finds a way to get close enough in the crush of people to rub his penis against her while attempting to maintain a casual, matter-of-fact outward demeanor. It arouses him even more to know that the act is his secret; no one knows what he is doing except his trapped victim. He fantasizes that he and his victim share a special sexual bond, and he often has an orgasm through this experience. He also uses these interactions as a basis for fantasies while masturbating.*

As you can tell from the case of E.S., **frotteuristic disorder** is the recurring compulsion to rub one's genitals against a nonconsenting person for sexual arousal and, typically, orgasm. The perpetrator usually fantasizes that he has a close, caring, loving relationship with his victim, but he will make every effort to run away and escape detection following the act. Virtually all frotteurists are men. Often when a woman learns of this paraphilic disorder, she can remember a time or two in her life when a man had crowded her too closely in a situation tightly packed with people. It made her very uncomfortable at the time, and now that creepy, victimized feeling makes a lot more sense! If a

Frotteuristic disorder
Rubbing one's genitals against a nonconsenting person for sexual arousal and, typically, orgasm.

Frotteurism involves rubbing one's genitals against a nonconsenting victim.

woman suspects she is a victim of frotteurism, she should move away from the situation as soon as possible and warn others in the setting about the perpetrator. Frotteurism generally falls under sexual assault laws, and, although sometimes difficult to prove, the perpetrator may be criminally charged for these acts.

Typically, men who become involved in this paraphilic disorder are most active between the ages of fifteen and twenty-five and tend to decrease and stop the behavior altogether as they age (Miller, 2000). Frotteurists often believe that due to a lack of any response, their victims are unaware of or enjoying the act, but in truth they are usually too shocked, afraid, or embarrassed to take action, or the act happens so fast that they do not have a chance to react at all. However, this, as with the other victimizing paraphilic disorders discussed here, is an illegal act and is considered a form of sexual assault.

Since You Asked

4. I saw this movie (not porno) where a guy got off by rubbing up against women at concerts when everyone was crowding and pushing up to the stage. What is up with that?

SEXUAL SADISM *A.R. is a thirty-five-year-old electronics salesman. He is unable to become sexually aroused unless he is causing intense pain, suffering, and humiliation to another person. These acts are not in any way playful, and his victims' lives have at times been in danger. Last year, he was finally caught and arrested on charges of kidnapping, assault, sexual assault, and rape. His latest victim was found in A.R.'s basement, caged like an animal, with evidence of sexual and other physical abuse. A.R. has stated that for all of his adult life, he had persistent fantasies about abducting, binding, cutting, sexually assaulting, and humiliating women. He gave in to his most extreme urges and is now facing life in prison for his crimes.*

sexual sadism disorder
Inflicting pain, injury, or humiliation on another person for the sexual gratification of the person performing the action.

sadomasochism
Sexual activities that combine sadism and masochism.

BDSM
Sexual activities that combine bondage, discipline, and sadomasochism.

Sexual sadism disorder, the recurring, compulsive urge to inflict pain and humiliation for sexual gratification, is named for the Marquis de Sade, an eighteenth-century French author and aristocrat who spent much of his adult life imprisoned for engaging in and writing about his particularly violent form of sexual expression.

The clinical definition of sexual sadism disorder clearly involves innocent victims and is often one of the most violent paraphilic disorders, as you can see from the case of A.R. Sadism takes many forms, including dominating, restraining, tying up, blindfolding, beating, cutting, whipping, strangling, mutilating, and even killing another person, usually during sexual activities. Part of the attraction and excitement for the sadist is the terror experienced by the victim. Sexual sadism disorder is seen in the pattern of behaviors of some serial rapists.

Less severe forms of sexually sadistic activities also exist; usually they are combined with masochistic behavior (which is discussed with the nonvictimizing paraphilic disorders). This combination is referred to as **sadomasochism** or **BDSM** (for bondage, discipline, and sadomasochism). Some people are sexually aroused by bondage—tying up or being tied up by a partner (with consent); discipline—dominating and humiliating or being dominated or humiliated by a sexual partner (with consent); or light sadism or masochism—giving or receiving minor pain such as spankings.

However, these acts, though appearing to be paraphilic in nature, are done as sexual play, usually with strict rules and agreements to ensure that the activities never cross any undesired pain or comfort boundaries for either partner (see "In Touch with Your Sexual Health: BDSM Explained").

"Every man wants to be a tyrant when he fornicates." This quote has been attributed to the Marquis de Sade (1740–1814), from whose name we derive the word *sadism*.

PEDOPHILIC DISORDER *O.L. is twenty-four, and his only sexual interest is in young children. He has never been sexually attracted to anyone beyond the age of puberty. O.L. is a teacher at a day care center and loves his work. The center has no knowledge of his paraphilic disorder. He enjoys the company of children and has never been able to interact very well, sexually or otherwise, with people his own age. He has used his job as a way to be near children and to receive the only kind of love he feels safe in accepting: the love of children. More than once, this closeness has crossed the line into sexual expression. He has encouraged the children in his care to play touching games with one another while he watches, and he sometimes involves himself in the touching games as well.*

O.L. suffers tremendous guilt about his actions and is terrified of being found out, fired from his job, arrested, and ostracized. Yet he has little control over his impulses. O.L. himself was sexually abused at an early age, and he has recently sought treatment to attempt to overcome his sexual compulsions toward children. He desperately desires to be able to develop a healthy expression of his sexuality with an appropriately aged partner. So far, however, the therapy has been unsuccessful.

Since You Asked

5. I was sexually molested as a child. I never told anyone (until now). With all these stories about priests molesting kids, I think about it a lot and wonder why it happened. Do you think I should get some counseling to work through it?

In Touch with Your Sexual Health

BDSM for "Fun"

Columbia University in New York City has a rather unusual support group that is probably not found on many college campuses. Here are just a few excerpts from its website, http://conversiovirium.com/aboutcv:

> Conversio Virium (CV) is the Columbia University student BDSM education and discussion group. Conversio Virium (CV) is a Columbia University student organization for discussion, education, and peer support concerning BDSM issues. Membership is open only to Columbia students, faculty, staff, alumni, and other affiliates, although CV's meetings are open to any interested person of any race, ethnicity, or sexual orientation. Conversio Virium is dedicated to discussion, education, and peer support concerning BDSM issues. All meetings are strictly confidential.
>
> Conversio Virium does not promote, support, or engage in violence of any sort. What CV does promote and support is safe, sane, and consensual (SSC) BDSM play. The only things that attendees engage in at CV's meetings are discussion, listening, and learning.
>
> CV is an education-only group. The presence of BDSM play would disrupt our meetings and events and turn attention away from the goals of the group in favor of personal interactions. In addition, some of those present might feel personally uncomfortable with such play, and therefore hesitate to attend, or participate in, CV meetings. Accordingly, BDSM play will not be a part of, nor will be permitted at, any CV meetings or events.

Mild, consensual sadomasochism has become a major worldwide industry.

BDSM Explained

BDSM: bondage and discipline (BD), domination and submission (DS), and sadomasochistic (SM) is considered a form of enjoyable and arousing sexual interaction between responsible and consenting adults. BDSM is not considered a psychological disorder. Paraphilias of sexual sadism disorder and sexual masochism disorder are defined by official criteria that differentiate them from BDSM. BDSM, when practiced responsibly, may create intense sexual experiences but not physical or emotional damage. Here are some of the terms used by those engaging in this sexual activity.

bondage: use of physical restraints to achieve various degrees of immobilization.

discipline: practice that produces discomfort for the purpose of sexual or emotional arousal.

domination: imposing one's will on another for the enjoyment of both partners; high level of psychological play.

submission: yielding to another's will for the enjoyment of both partners; high level of psychological play.

sadomasochism/SM: intense physical sensations enjoyed by the masochist and given by the sadist; this term makes many people uncomfortable because of its association with historical figures of disputable moral character.

scene: the period of time, defined either beforehand or during the course of the interaction, where BDSM roles and activities take place; as a verb, *scene* refers to engaging in BDSM activity. "The scene" refers to people who feel they are part of a BDSM community.

negotiation: the process of expressing interest in a BDSM scene, exchanging information about preferences and limits, and deciding whether or not to play and for what duration of time.

limit: activities, words, or scenarios that the person does not wish to experience for either physical or emotional reasons.

safeword: a code word, often a word not used in everyday or sexual contexts, indicating that the interaction needs to stop or that an activity needs to be changed.

top: one who takes control of the activities of a scene; can refer to both physical and psychological play.

bottom: one who gives over a degree of power to another; can refer to both physical and psychological play.

Facts about BDSM

- *Mutual consent* is what distinguishes BDSM from abuse and assault, just as consent distinguishes sex from rape.

- *Context* is what determines whether or not pain is experienced as pleasurable, though the context depends on the individual. An example of "good" pain may be getting scratched during sex; an example of "bad" pain may be stubbing your toe.
- Not all BDSM play is between heterosexual couples.
- People who are submissive with their partner in a BDSM scene may not be necessarily submissive in other aspects of their lives.
- Ligature marks around wrists or ankles cause safety questions to be raised. Warn patients about erotic asphyxiation. Choking play or hanging play is very dangerous but common.
- Accidents can happen in BDSM, just as in any other physical activity, but this isn't abuse.
- Partners who know each other very well may sometimes negotiate a scene without a safeword. This is still not abuse but a matter of profound trust.

Pedophilic disorder
Uncontrollable sexual compulsions involving children.

This paraphilic disorder is probably the most reviled victimizing "sexual" behavior of all. **Pedophilic disorder** refers to the pattern of exploitation of prepubescent children for an adult's sexual purposes. Most people cannot begin to fathom what would cause someone to victimize children in this way. However, as everyone is acutely aware, it is an all too common occurrence. Because pedophilia is an abusive and sexually aggressive act perpetrated against children, the specific characteristics of child sexual molesters and the effects of this crime on their victims are detailed in Chapter 13, "Sexual Aggression and Violence: Rape, Child Sexual Abuse, and Sexual Harassment." Here we focus on child sexual abuse as a paraphilic disorder called pedophilic disorder, a diagnosable psychological condition.

pedophile
An adult, usually male, whose sexual focus is on children.

No single profile of a typical **pedophile** exists, although there are some commonalities (de Silva, 1999). Most pedophiles are men, but numerous individual cases of women engaging in an extensive pattern of child molestation have been reported. The majority of pedophiles are heterosexual (regardless of the sex of their victims), and many are married with children of their own. Pedophiles commonly experience marital difficulties, and alcohol addiction is frequently found among this group. One of the most notable characteristics shared by the majority of pedophiles is that they were themselves sexually abused as children (Lee et al., 2002; de Silva, 1999). Pedophiles tend to be depressed, see themselves as lonely, exhibit low self-esteem and feelings of inferiority, and are emotionally immature (Hall & Hall, 2007). The number of pedophiles has been estimated at 4% of the population (Cloud, 2002). In the United States, that percentage represents well over a million people.

The aftereffects victims experience vary greatly, depending on the nature and duration of the sexual abuse, the age of the child, and other psychological variables. However, research has found that those who were sexually abused as children are at greater risk for depression, suicide, alcohol addiction, posttraumatic stress disorder, anxiety disorders, rape, subsequent sexual victimization as teenagers and as adults, and divorce (for more detail on the effects of pedophilia on victims, see the discussion of *child sexual abuse* in Chapter 13, "Rape, Child Sexual Abuse and Sexual Harassment").

In 2011, a total of 61,500 children were estimated to have been sexually abused in the United States (CDC, "Child Maltreatment," 2008). This is a staggeringly large number. However, on a positive note, over the past 20 years child sexual abuse has decreased 62% in the U.S. (Finkelhor, Jones, & Shattuck, 2013). Figure 14.3 shows a twenty-year trend (1990–2011) in cases of child sexual abuse. The reasons for this consistent and significant decline are not well understood. Some researchers contend the decrease may be due to a combination of increased criminal punishments for abusers; greater parental awareness and vigilance in relation to the problem; new treatments for offenders; or how data are reported (Finkelhor, Jones, & Shattuck, 2013). Another possible explanation is that Megan's Law (requiring that the public be provided with photographs, locations, and descriptive information on sex offenders) has been enacted in all fifty states (beginning with California in 1996) and may have served as a deterrent to sexual offenders in general and child sexual abusers specifically (Barnoski, 2005; Hall & Hall, 2007). See Chapter 13, "Sexual Aggression

Figure 14.3 Child Sexual Abuse Rates in the U.S., 1990–2011

SOURCE: Data from Finkelhor, Jones, & Shattuck (2013).

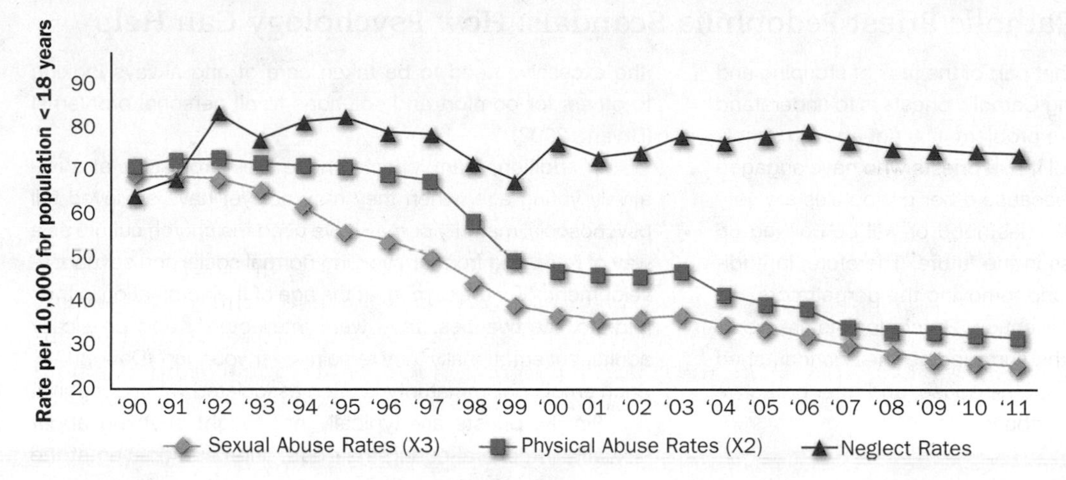

and Violence: Rape, Child Sexual Abuse, and Sexual Harassment" for more information on Megan's Law.

PEDOPHILIC DISORDER IN THE CATHOLIC CHURCH Although the decline in cases of pedophilic disorder strikes an optimistic note, we should remind ourselves that during those same years and continuing to the present day, the dismaying flood of reports of pedophilia among Catholic priests has brought the horrors of this paraphilic disorder to a new level of public awareness (see Terry & Akerman, 2008). The loss of trust in individuals long thought to be the most trustworthy has shaken many people's faith in human nature and in the Catholic Church as well. Around the world, accusations and convictions of Catholic priests who engaged in child sexual abuse have become front-page news, and the reports continue (i.e., Heilprin & Winfield, 2014; Medina & Goodstein, 2013). Estimates place the number of Catholic priests who have been sexually involved with minors over the past fifty years at between 3,000 and 5,000 in the U.S., which represents approximately 4% to 5% of the total number of Catholic priests. The number of boys (now men) who were abused by these priests is estimated at approximately 11,000 (Goodstein, 2004; Plante, 2004) and could reach more than 20,000 victims worldwide. The Catholic Church has paid out hundreds of millions of dollars to settle lawsuits brought by the victims of the abuse, causing several Catholic dioceses to declare bankruptcy. In April 2008, Pope Benedict XVI, on his first visit to the United States, acknowledged that the scandals of sexual abuse of children by priests has created "deep shame" within the church and caused "enormous suffering" to victims and their communities, and he deemed it "gravely immoral behavior." He met and prayed in private with several victims of the abuse (now adults) and pledged to keep pedophiles out of the priesthood (Simpson, 2008).

Still, however, the church and its leadership continue to face high hurdles in finding effective ways of preventing such abuse in the future and regaining the trust and confidence of a disillusioned Catholic world. New cases of child molestation by Catholic priests continue to come to light worldwide as ever more accusers come forward.

Many believe that psychology will play a key role in developing solutions to this complex and painful problem. One suggestion is that psychologists must view the Catholic Church and the priesthood as cultures unto themselves and adopt a multicultural perspective, just as one might do in helping any cultural subgroup overcome an internal problem that threatens its internal social structure. "Sexuality and Culture: Overcoming the Catholic Priest Pedophilia Scandals: The Role of Psychologists" (on the following page) dispels some of the common myths about the scandals and discusses specific strategies that psychology might offer for developing solutions to this Catholic "culture-wide" crisis.

Sexuality and Culture

Overcoming the Catholic Priest Pedophilia Scandals: How Psychology Can Help

Most psychologists agree that part of the task of stopping and preventing pedophilia among Catholic priests is to understand the origins and causes of the problem. It is not enough simply to uncover, report, and expel those priests who have engaged in sexual abuse of minors because other pedophiles are very likely either presently in the priesthood or will be coming up through the training process in the future. Therefore, in addition to stopping the abuse and removing the perpetrators, we must focus on long-term prevention. Psychologists are working to dispel the many myths surrounding the scandal, shed light on the possible causes of the abuse, and offer proactive strategies for avoiding future abuse.

Two Persistent Myths

Two common beliefs about this problem are not grounded in fact and have taken the focus away from real solutions. The first is that the pedophilic behavior of priests has been due to the church's culture of celibacy (the renunciation of marriage and the vow of chastity). You can see that this argument is misguided in that although celibacy may cause a buildup of sexual tension for some people (whether in the priesthood or not), most of those individuals will not make children the focus of their desire or their sexual acting out.

The second myth is that the church scandals are related to a culture of homosexuality in the church, in that men with homosexual tendencies are drawn to the all-male culture of the priesthood, which in itself is not typically based in fact for most priests, and the crimes of child molestation have been attributed to more situational factors by many (Terry & Akerman, 2008). Moreover, the notion that the sexual orientation of priests is connected to the church's pedophilia scandals is founded in the belief that most child molesters are gay males, a belief that is simply untrue. The vast majority of adults who sexually abuse children (including priests) are heterosexual, regardless of the sex of their victims. In fact, a study in the 1990s found that in 82% of child molestation cases, the molester was the heterosexual partner of a close relative of the child, and the chance of a child molester's being a homosexual male was between 0% and 3.1% (Jenny, Roesler, & Poyer, 1994).

Commonalities

We do see several common characteristics in many or most of the priests who have abused children. Usually, as with most pedophiles, abusing priests were typically abused themselves as children. Psychologically, various characteristics and disorders are found among pedophile priests in significantly larger proportions than in the general population, including poor impulse control; underdeveloped social skills; substance abuse; depressive and bipolar disorders; and various personality disorders, especially narcissistic disorder (grandiose sense of self-importance, constant need for external adoration, and marked lack of empathy for others) and dependent disorder (the excessive need to be taken care of and always looking to others for comfort and solutions to all personal problems) (Bryant, 2002).

In addition, many priests chose their profession at a relatively young age, when they may not yet have achieved full psychosocial maturity or may have used the church culture as a way of retreating from or avoiding normal social and sexual development. "For these men, at the age of their ordination in their mid- to late-twenties, they were intellectually and physically adults, but emotionally they remained far younger" (Daw, 2002). Such emotional immaturity is often associated with pedophilia.

Finally, priests are typically not taught anything about sexuality in general during their education and training at the seminary. Seminaries make the often erroneous assumption that they have already dealt with those issues prior to their decision to become priests (Bill Mochon, in Daw, 2002). Priests who have not dealt with sexuality issues are left to their own devices to figure it all out, and some, obviously, end up making very inappropriate choices.

Possible Solutions and Prevention Strategies

Psychology has developed various treatments for paraphilias in general in the form of psychotherapy and medication. These will be discussed in greater detail later in this chapter. Beyond this, psychologists working alongside criminal justice experts and those who are part of the culture of the church may assist in the prevention and treatment of sexual abuse among priests in a number of ways (Daw, 2002; Terry & Akerman, 2008):

- Develop strategies to help practicing priests and those in training to develop healthy interpersonal relationships that are nonsexual. In some cases at least, psychological intimacy with others may reduce or eliminate the need for priests with pedophilic inclinations to turn to sexually abusing children.
- Employ more effective screening strategies for high-risk individuals entering the priesthood.
- Increase surveillance of priests when they are interacting with children.
- Provide valid psychological assessment and evaluation instruments that may be used prior to admission to seminary or ordination to screen out prospective priests who may have compulsive tendencies toward pedophilia.
- Offer effective consulting, counseling, and therapy services for priests, bishops, and parishes. These services may involve one-on-one counseling with at-risk clergy, priest wellness or therapy groups, or community education with parishioners about keeping children safe or identifying warning signs of potential sexual misconduct.
- Reduce one-on-one time for priests and children.

- Provide training workshops for priests on maintaining appropriate boundaries with parishioners; raising awareness of warning signs of potential problems; working with challenging clients; and dealing with stress, anxiety, or conflict.
- Establish specific codes and rules of conduct that set rules for unacceptable behavior of priests with minors.
- Engage in substantive scientific research to enhance understanding, prediction, and treatment of sexual abuse within the ranks of the clergy.

SOURCE: Based on Daw (2002); Terry & Akerman (2008).

Nonvictimizing Paraphilic Disorders

If you look back at the characteristics of paraphilic disorders discussed earlier in this chapter, you will see that the presence of a nonconsenting victim is not a necessary criterion in the definition. Many paraphilic disorders (refer to Table 14.1) are nonvictimizing in that they are solitary activities or involve adults who have given their consent to participate. Therefore, these paraphilias are often considered victimless, unless you support the notion, as many people do, that the person engaging in the paraphilic act is victimizing himself or herself due to the degrading or even potentially dangerous nature of the behavior. However, this use of the term *victim* is not how it is usually applied to sexual acts. It is because these acts do not *compel* others to participate against their will that most authorities refer to them as victimless and nonvictimizing.

Since You Asked

6. I met a woman at a party. She told me that "on the side," meaning not her main job, I guess, she's a dominatrix. Does that mean what I think it means?

SEXUAL MASOCHISM *T.W. is male, forty-seven, a banker, stockbroker, real estate tycoon, and millionaire. At least once or twice a month, T.W. feels an overwhelming desire for a sexual partner who will make him feel worthless, powerless, and insignificant. He often pays a dominatrix, a woman who knows exactly how to engage in behaviors that humiliate, demean, and hurt him. His favorite and most sexually exciting acts are being tied up, blindfolded, spanked with a wooden paddle, and having hot wax dripped onto his penis and scrotum. T.W. reports that after a session with his "mistress," he often feels relieved and "in balance," ready to face the world as a power broker. His desire for such a masochistic session tends to increase as the stress in his life rises.*

sexual masochism disorder
Sexual arousal and gratification that is associated with acts or fantasies of being hurt, humiliated, or otherwise made to suffer.

bondage
Being bound, tied up, or otherwise restrained during sexual activity; typically, a consensual activity.

Sexual masochism disorder refers to sexual arousal and gratification that is associated with acts or fantasies of being hurt, humiliated, or otherwise made to suffer. The masochist has a recurring pattern of engaging in such acts voluntarily and often seeks them out. As in the case of T.W., a wide range of behaviors are enticing and sexually arousing to masochists, including being tied up or otherwise restrained (called **bondage**); piercing with pins or needles; affixing clothespins or clamps to nipples or genitals; hanging weights from genitals; being spanked, whipped, beaten, strangled, slapped, electrically shocked, or verbally degraded; branding the skin; and applying ice to the bare skin (Santtila et al., 2002).

Sadomasochists are sexually aroused by games involving pain, bondage, and humiliation.

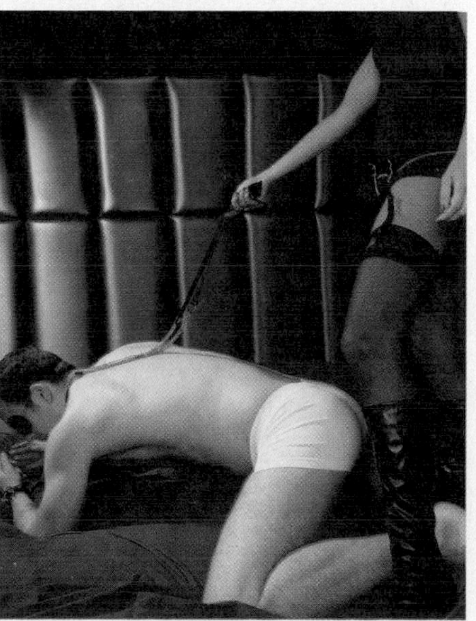

As mentioned in our discussion of sexual sadism, the sadomasochism subculture typically engages in very ritualized behavior in which the partners make agreements about the acts they desire, assign specific roles to be enacted, and carry out "scripted" sequences of behaviors, from very simple acts such as making love while blindfolded or in handcuffs, to complicated and complex scenarios that may include role-playing, elaborate costumes, and a wide variety of painful or humiliating acts (Baumeister, 1995; Santtila et al., 2002).

Usually the humiliation and pain involved do not draw blood, break bones, or leave permanent scars. Because these acts are carried out by adults with the consent of all involved, they are typically not illegal. However, such "consenting adult laws" have been tested from time to time, especially when the sadomasochistic acts are more extreme. For example, in England in 1990, several members of a sadomasochistic sex club were arrested for behaviors that the prosecutors claimed violated the United Kingdom's Offenses Against the Person Act of 1861, which states, "Whoever shall unlawfully and maliciously wound or inflict any grievous bodily harm on any other person, either with or without any weapon or instrument, shall be liable to imprisonment."

The specific acts for which the men were arrested were videotaped for distribution to other members of the club and consisted of maltreatment of sexual body parts, such as dripping hot wax on the nipples and penis, safety pins piercing the scrotum, sandpaper used on sensitive body parts, fish hooks through nipples and penile skin, and ritualistic beatings either with bare hands or implements, including spiked belts and cat-o'-nine tails. There was heat branding and infliction of injuries causing bleeding. All of these acts were consented to by all men in attendance and no activity caused infection or any permanent physical damage (Green, 2001).

The men involved were convicted and sentenced to four years in prison. However, the convictions were later reversed and the case was dismissed by the appeals court, which compared the consensual injurious behavior with the sport of boxing. If you think about it, boxing is consensual pain and beating (sometimes) with each opponent intent on inflicting as much injury on his opponent as possible, including cutting and rendering the other unconscious. And each boxer *freely chooses* this. These consensual events could make a masochism club appear mild in comparison (Green, 2001).

This case provides an excellent example of the confusion many people feel about consensual sadomasochism. To many, it *feels* somehow wrong that people should be allowed to inflict pain and injury on each other in the name of sexual gratification. However, many others acknowledge that if these are truly consenting adults, it is probably no one's business but their own if they choose to participate.

Since You Asked

7. I saw a movie from the 1990s on cable called *Rising Sun*. In the movie, a man strangles a woman during sex. It seemed that she wanted him to do it, but he did it too hard and killed her. Why would anyone want to be choked during sex?

AUTOEROTIC ASPHYXIATION (HYPOXYPHILIA) *R.G. is nineteen years old and is on his college wrestling team. He has recurrent fantasies of self-strangulation. He has lost consciousness several times during a stranglehold on the wrestling mat and has found the sensation of falling, weightlessness, and darkness to be extremely compelling. Several years ago, he began to experiment with ways of limiting oxygen to his brain during sexual fantasy and masturbation. He is very aroused by the risk he takes in courting possible death, and he loves the light-headed feeling that he gets when he cinches his belt around his neck, attaches it to his headboard, and strangles himself almost to the point of unconsciousness while he masturbates to orgasm.*

This is the one exception mentioned earlier of a paraphilia that does not have its own classification in the DSM-5. It is included separately here because it can be a part of some people's compulsive masochistic acts, it is more common than most people think, and it is dangerous. To say that asphyxophilia is a "victimless" paraphilia is controversial because, although it is a solo act, it can be fatal. It is the most personally dangerous of all the paraphilic disorders.

Although some paraphilic disorders may be *psychologically* damaging to the person performing them and, in the case of victimizing paraphilic disorders, to a potential victim, most do not carry with them a high risk of death. This paraphilic disorder, however, is often fatal. **Autoerotic asphyxiation** involves depriving the brain of oxygen, usually through some form of strangulation or hanging, during masturbation (Janssen et al., 2005; A. P. Jenkins, 2000).

autoerotic asphyxiation
A form of sexual masochism involving depriving the brain of oxygen, usually through some form of strangulation or hanging; also referred to as *asphyxophilia*.

If this behavior is engaged in by a couple, where one chokes the other, with consent, during sexual activity, it is usually called *sexual hypoxia* or *hypoxyphilia*; but this is far less common than the autoerotic form. This behavior is engaged in mostly by boys and men between the ages of fifteen and thirty; this is the group in which it is most often fatal as well. Autoerotic asphyxia has also been found in a small number of women (Martz, 2003; Sheleg & Ehrlich, 2006).

The motivation for this behavior appears to be the belief that oxygen deprivation enhances the arousal and orgasmic sensations during masturbation. However, loss of oxygen to the brain only causes lightheadedness and dizziness and is not in reality connected to enhanced sexual arousal. In other words, *do not try this at home!* It is extremely dangerous and potentially deadly. It is estimated that nearly one-third of all deaths by strangulation among teenagers are due to this activity. The total number of cases is estimated to be well over 1,000 a year in the United States, and this figure is without doubt low due to underreporting. Often such deaths are reported as simple suicides to avoid the potential embarrassment and stigma that may accompany the family's already intense grief (Gosink & Jumbelic, 2000; A. P. Jenkins, 2000). However, autoerotic asphyxiation should not be confused with suicide because death linked to this behavior is nearly always accidental.

As with all paraphilic disorders, people who engage in autoerotic asphyxiation do so as a pattern, repeatedly, over time. The usual method is for the person to arrange some sort of hanging or other strangulation device that will cut off the supply of oxygen just up to, but not including, complete loss of consciousness. The danger, as you no doubt have guessed, is that sometimes these devices are too effective and cause unintentional unconsciousness and death.

Typically, victims of autoerotic asphyxiation devise some sort of apparatus that will cut off oxygen but (hopefully) release prior to full unconsciousness using a rope (hanging), suffocation with a plastic bag, or inhalation of oxygen-inhibiting chemicals.

Obviously, this is a potentially tragic sexual compulsion. Usually, no one other than the victim is aware of the behavior because it is nearly always performed in secret and in solitary, private locations. Sadly, the first sign the family has that this behavior is occurring is the discovery of the body. Consequently, intervening in the behavior of those at risk is extremely difficult.

Warning signs that someone might be engaging in this potentially deadly behavior include: a combination of the presence of pieces of short rope, padded ropes, or neckties with unusual knots; unusual marks around the neck; bloodshot eyes; marks on bedposts, headboards, or closet rods from ropes or belts; plastic bags that may be used for suffocation or possession of oxygen-depriving drugs; locks on bedroom doors; signs elsewhere on the body suggesting participation in bondage or other masochistic activities; and recent visits to websites on "scarfing" or "asphyx." Your awareness of these warning signs will help you to recognize if someone you know may be engaging in this dangerous behavior and to take appropriate steps to help the person stop through education and counseling.

FETISHISTIC DISORDER *B.R. is a thirty-eight-year-old naval officer. He loves women's high-heeled shoes. He is very sexually aroused by seeing women wear them, by touching them, and by collecting them. He routinely asks his sexual partners to wear them when they make love. In fact, he would just as soon have sex with the shoes as with any of his partners. Lately, he has taken to stealing his preferred type of shoes from his girlfriends in order to masturbate with them.*

The first time B.R. can remember becoming sexually aroused was at a young age watching his aunt remove her high heels, before taking off the rest of her clothes in front of him. As he matured, high heels were always a significant part of B.R.'s sexual fantasies and experiences. In recent years, he has become dependent on high heels for sexual arousal and is unable to achieve orgasm with a partner unless high heels are a component of the act.

Since You Asked

8. My mom has this friend who shops for shoes constantly. It's like if she's not eating or sleeping, she's shoe shopping. Is this what is meant by a shoe fetish?

Everyone has a unique set of sexual preferences, desires, and turn-ons. What is a sexy or exciting behavior, an arousing setting, or a provocative item of clothing or part of the body to one person might inspire boredom or even avoidance in someone else. Such differences are part of the normal diversity of human sexuality. However, when a sexual preference intensifies to the point that a person obsesses almost exclusively on a nonhuman object or sometimes on a body part that most members of a culture do not find sexy, such as feet or toes, this is called a **fetish** (de Silva, 1999; Seligman & Hardenberg, 2000).

Fetishistic disorder

A sexual preference for a nonhuman object or a body part that most members of a culture do not consider sexual.

Sometimes the line between a "normal" sexual preference and a fetish may seem blurred. However, if you apply the guidelines for defining a paraphilia discussed at the beginning of this chapter, the distinction will probably become clearer. For example,

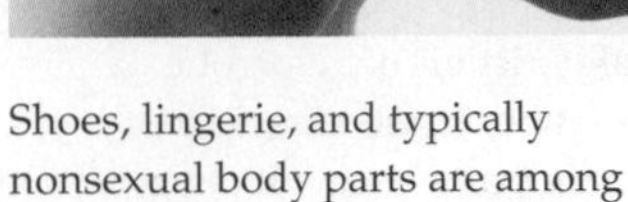

Shoes, lingerie, and typically nonsexual body parts are among the most common fetish objects.

take the case of B.R. The progression of his fetish may have gone something like this: At first, he found women's high-heeled shoes to be a very attractive part of female attire. He enjoyed it very much whenever a woman would wear extra-high heels on a date, and he found that he was sexually aroused more easily and more intensely when she did. Soon he began asking his sexual partners to leave their shoes on when they made love and he found this very exciting. Although this preference may seem a little odd to some, it is doubtful that his degree of preference for high heels would be considered a fetish. However, over time, his fascination for high-heeled shoes increased until he reached the point that he was unable to become sexually aroused unless high-heeled shoes were worn by his partner or were at least present on the bed. At that point, the fetish object, in this case a certain type of shoe, had become a *necessity* for sex and had moved more clearly into the category of a fetish. He then advanced to the most extreme level of a fetish when his obsession for high-heeled shoes became so strong and primary that he actually began preferring sexual interaction with his partner's shoes to making love with her.

Most sexual fetishes are probably not particularly harmful, and a fetish, by definition, is nonvictimizing. However, in its most extreme form, a fetish may become a psychological or social problem. If B.R.'s partners are comfortable with his insistence on high-heeled shoes during sex and he is able to form close, loving relationships in spite of his fetish, his obsession with high heels would probably not be considered a psychological disorder. However, if his fetishism is causing discord in his relationships, eliciting anger and resentment from his partners, preventing him from forming healthy relationships, provoking extreme feelings of guilt and self-hate, or creating other difficulties, some form of therapy might be appropriate.

The development of fetishes probably stems from experiences in childhood or early adulthood in which some body part, article of clothing, or object was paired with sexual arousal and orgasm (in keeping with the behavioral theories regarding paraphilias discussed earlier in this chapter). After one or more pairings of the object and sexual responding, the object itself becomes a trigger for sexual feelings.

The most common fetish objects are shoes and boots, women's underwear and lingerie, rubber or latex objects, and feet and toes (Freund, Seto, & Kuban, 1996; Seligman & Hardenberg, 2000). If you think about it, you can see why these objects would be more likely to become associated with sexual arousal and orgasm in someone's sexual development. Theoretically, however, *anything* can become the object of a sexual fetish given the relevant sexual experiences. For example, one study reported

Cross-dressing for show business as does the famous cross-dresser RuPaul, is not considered a transvestic disorder.

on a very unusual fetish focusing on, believe it or not, a particular small car: the Austin Metro Mini. The man was described as very shy with very little social interaction, sexual knowledge, or sexual experience and whose main sexual outlet was masturbating in, on, or behind this particular make of car (de Silva & Pernet, 1992).

TRANSVESTIC DISORDER *J.T. is a factory foreman, a deacon in his church, and a respected member of the community. He is quiet, seemingly well adjusted, and happily married. Once a month, J.T. tells his wife he's going fishing for the weekend. Instead, he books a hotel room in the next town and spends the weekend lounging in women's lingerie. He loves the feel of the delicate lace, the smooth, soft silk, and the sheer texture of stockings. He sips champagne, watches television, and masturbates. When he returns, he feels refreshed and happy, but his happiness is tinged with guilt and anxiety. He is constantly concerned that his wife will find out about his cross-dressing, assume he must be gay or crazy, and leave him. He doesn't like sneaking around, and he wishes he could share this side of himself with his wife, but he's terribly afraid she'll be disgusted with his behavior and file for divorce.*

You are probably aware that the term **cross-dressing** refers to wearing clothes traditionally associated with the opposite sex. But cross-dressing itself is not a paraphilia; it is not the same as *transvestism*. Remember, paraphilias are behaviors that have a *compulsive* component and are engaged in for sexual arousal and gratification. People cross-dress for many nonsexual reasons, such as a Halloween costume, a disguise, a costume for a movie, or for laughs on stage.

Someone with a **transvestic disorder**, however, is a person who obtains *sexual gratification* by wearing female clothing. The psychological term for this paraphilic disorder is **transvestic disorder**. The women's clothing is, in essence, a fetish object, but in this case, it is the act of actually wearing it, not just touching or fondling it, that produces the sexual thrill. One of the most pervasive myths about human sexuality is that transvestites are gay men. This is, for the most part, simply wrong. Studies indicate that most transvestites (70% to 90%) are heterosexual men, not gay or bisexual (Bullough & Bullough, 1997; Docter & Prince, 1997). It is true that some gay men cross-dress, but this is typically for fun, for political statement, or for show, not for sexual arousal. Approximately 80% of transvestites either are or have been married and come from a wide variety of religions, professions, and family backgrounds. Most say that cross-dressing allows them to express different parts of their personalities and excites them sexually.

Transvestites are typically not interested in changing their behavior, but most hide their cross-dressing from wives, friends, and family and are afraid of being caught or found out. The fears transvestites cite include being rejected by the people close to them and by society (47%); being labeled a "sissy," "queer," or "faggot" (23%); being seen as mentally ill (22%); and sinning against God (5%) (Bullough & Bullough, 1997).

Another common misconception is that men who dress as women for sexual purposes do so because they want to be women. Again, this is almost always false. Ninety-five percent of transvestites perceive themselves, without hesitation, as male.

It should be stressed here that transvestism is *not* to be confused with *transgenderism*. **Transgender** individuals are *not* engaging in a paraphilia. These individuals typically self-identify with a gender different from their biological sex. That is, **transsexual** men and women often feel that they have been born in the wrong body, that the gender they feel themselves to be is not the same as the sex of the body with which they were born. They are not confused or in doubt about the gender, even though the bodies they were born with did not physically reflect their sexual identity. Again, transgenderism is *not* a paraphilia. If a transgender person dresses in the clothing of the opposite sex, it is not primarily for a sexual thrill but rather because he or she feels that the clothes reflect his or her gender identity. Many transgender individuals see "sex" as not categorical but more fluid, and that a person may see him- or herself as falling anywhere between male and female. The issue of transgenderism is discussed in greater detail in Chapter 10, "Gender: Expectations, Roles, and Behaviors."

Since You Asked

9. My boyfriend likes to put on a woman's long blond wig (he's got a dark beard!) just to make people laugh. Is he being a transvestite?

cross-dressing
Dressing in clothes traditionally associated with the opposite sex.

transvestite
A man who obtains sexual satisfaction by wearing female clothing.

transvestic disorder
The paraphilic disorder characterized by dressing in the clothing of the other sex for sexual arousal

transgender
Individuals who do not self-identify as the gender that conforms to their biological sex; this is not a paraphilia.

transsexual
A transgender person who has transitioned or is transitioning from his or her biological sex to his or her self-identified gender through actions, dress, hormone therapy, or surgery.

Most transvestites feel that cross dressing allows them to express an important part of their sexual personality.

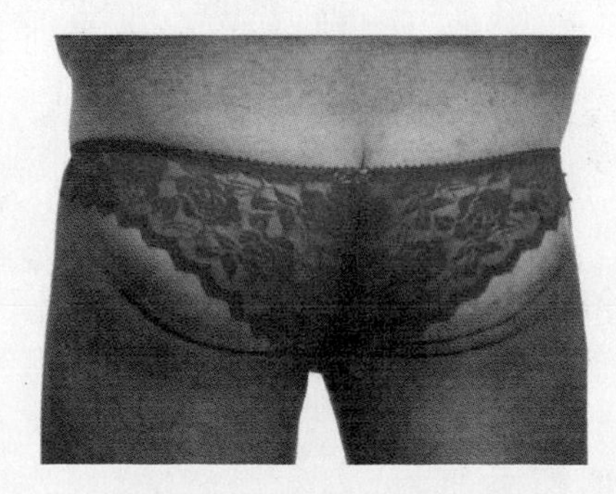

Treatment of Paraphilic Disorders

14.5 Evaluate the treatment of paraphilic disorders including reasons for seeking treatment, the types of treatments available, and their effectiveness.

Not everyone who has a paraphilia wants or even needs to alter the paraphilic behavior. However, paraphilic disorders, by definition, create significant problems for the perpetrator, victims, friends and family members, and society at large. For this reason, sexuality educators and mental health professionals have developed many strategies for intervening and helping clients control their dysfunctional and destructive paraphilic behaviors.

Who Seeks Treatment

People who seek treatment for a paraphilic disorder may do so for a variety of reasons and come from a wide range of life situations. Their reason for obtaining help usually falls into one or more of the following four categories (de Silva, 1999):

- **Court-ordered treatment.** Individuals who have engaged in illegal acts involving paraphilic disorders are often ordered by the court to undergo treatment as part of their sentence or probation. These individuals may or may not be motivated to make the changes the judge is demanding, so the potential success of treatment is highly variable.
- **Self-distress.** These individuals are embarrassed by behaviors they cannot seem to control; are worried that they will be caught or arrested; feel that they are unworthy, sick, or unnatural; or are afraid that they may be risking their life and career if they continue the behavior. These clients, by definition, are motivated to change, and although success rates vary (to be discussed shortly), they have a higher likelihood of successful treatment.
- **Relationship distress.** Many seek help for a paraphilia because it is creating discord or other problems in their primary relationship. The client is distressed because of the emotional pain the behavior is causing the partner and the discord it is causing in the relationship. In general, these clients come from long-term, stable relationships and are very motivated to change.
- **Distress due to sexual dysfunction.** When a paraphilia becomes well established, it often begins to interfere with sexual functioning. For example, as we discussed under fetishes earlier, some fetishists are unable to become sexually aroused or cannot achieve an orgasm without touching or interacting with the fetish object. Another example might be a transvestite who loses interest in making love with his partner because he cannot, in those situations, wear women's clothing without revealing his paraphilia. These individuals may seek treatment for the sexual dysfunction, which of necessity will involve treating the paraphilic disorder that underlies it.

Types of Treatment

Various treatment approaches may be used in an effort to modify unwanted paraphilic behaviors. In the past, the goal of such therapy was to extinguish the behavior completely. This strict and inflexible goal, combined with the limited techniques available in the past, usually led to disturbingly poor success rates. Today, although success rates are far from perfect for most paraphilic disorders, the effectiveness of newer therapies is improving. Obviously, for some of the most dangerous and coercive paraphilic disorders discussed in this chapter, rapid and successful treatment is of utmost importance.

Under current international standards, the three goals in the treatment of paraphilic disorders are: (1) to control paraphiliac fantasies and behaviors in order to decrease the risk of recidivism; (2) to control sexual urges; (3) to decrease the level of distress of the paraphilic individual (Thibaut et al., 2014). Therapeutic strategies for paraphilic disorders include a variety of approaches. The method employed depends on what paraphilic disorder is being treated, the severity or danger of the behavior, and the characteristics of the client (Assumpção et al., 2014; Garcia & Thibaut., 2010). Treatment approaches include psychotherapy, orgasmic reconditioning, and drug therapies.

PSYCHOTHERAPY Of course standard, well-known forms of talk therapy may be used in treating paraphilic disorders. The most successful approach appears to be **cognitive-behavioral therapy**, which combines techniques designed to modify unwanted behaviors with altering the client's thinking patterns that lead to the unwanted behaviors (Krueger & Kaplan 2001; Thibaut, 2010; Kafka, 2007). The behavior component of this strategy has sometimes included **aversion therapy**, in which behaviors and images associated with the paraphilia are paired with an unpleasant stimulus such as an electric shock or a noxious odor. Recent research, however, has shown this approach to be of limited effectiveness (Beech & Harkins, 2012). Because of ethical considerations, therapists today more often employ a technique called **covert sensitization**, in which, instead of shocks or bad odors, the client repeatedly fantasizes the unwanted behavior and then adds an extremely aversive event to the fantasy. For example, a pedophile might engage in various fantasies of sexual behavior with a child and imagine that, just before the event, a squad of police officers bursts in, guns drawn, and takes him to jail. This technique has been shown in some cases to be effective with certain paraphilic disorders, including exhibitionistic disorder and sexual sadism disorder (Brannon, 2002).

cognitive-behavioral therapy
A therapeutic approach designed to gradually eliminate specific thoughts and associated behaviors that may be contributing to sexual problems.

aversion therapy
Psychotherapy in which unwanted sexual behaviors and images are reduced by associating them with an unpleasant stimulus such as electric shocks or noxious odors.

covert sensitization
A type of aversion therapy in which, instead of actual shocks or bad odors, the client repeatedly fantasizes the unwanted behavior and adds an extremely aversive event to the fantasy.

Other forms of psychotherapy that have found some success alone or in combination with other techniques include *group therapy* (which helps break down denial and avoidance by surrounding each client with others who engage in the same behaviors), *social-skills training* to help the client learn to develop healthy relationships with others, and 12-step programs similar to *Alcoholics Anonymous* based on the assumption that paraphilic disorders have at their core a compulsiveness similar to that of alcohol addiction (Brannon, 2002).

ORGASMIC RECONDITIONING A specific form of psychotherapy for paraphilic disorders is called **orgasmic reconditioning**. Sexual arousal and orgasm are very powerful reinforcements for whatever behavior becomes associated with them. If a person who desires to modify his paraphilic behavior could be conditioned to become aroused and have orgasms in settings that bear no resemblance to the paraphilia, his maladaptive associations and behaviors might be altered. Here's how orgasmic reconditioning typically works: The client is told to masturbate using objects, fantasies, or images related to his paraphilia, but as orgasm approaches, he is instructed to switch completely to objects, fantasies, and images of more mainstream, socially acceptable sexual behaviors. Eventually, if successful, his sexual desires and urges will become redirected, or reconditioned, toward the new, more acceptable activities (see Kahn et al., 2011).

orgasmic reconditioning
A type of therapy for paraphilias in which a person is conditioned to become aroused and have orgasms in socially acceptable settings that bear no resemblance to the paraphilia.

DRUG THERAPY As noted in our discussion of biological theories of paraphilias, scientists now suspect that biochemical factors are involved in the development and maintenance of paraphilic disorders. If an imbalance in brain and glandular chemicals is leading to unwanted, inappropriate sexual behavior, it makes sense that it might be treated with medications that alter, in some way, the production or effects of those chemicals. Today, drug-based therapies, especially in combination with psychotherapeutic interventions, are showing the most promise of all the treatments for paraphilic disorders (Garcia et al., 2013).

Historically, drug treatment for paraphilias consisted of medications that would reduce the action of testosterone. The rationale was that if male hormones are responsible for sex drive, less of it should decrease sex drive and paraphilic behavior along with it. Hormone-blocking drugs such as Depo-Provera and Lupron have been used with some success (Assumpção et al, 2014; Kafka, 2007; Malin & Saleh, 2007). Although such drugs do decrease paraphilic behavior in most men, they also inhibit sex drive and sexual functioning altogether, and they have the potential for other side effects (such as breast growth and erectile difficulties), which cause many men to discontinue treatment.

Selective serotonin reuptake inhibitors (SSRIs), originally developed as antidepressants, have shown promise in treating young adults and more resistant and serious cases of paraphilic disorders (Assumpção et al., 2014; Bradford, 2000). SSRIs with trade names you may recognize, such as Prozac, Zoloft, Paxil, and Lexapro, were originally developed and approved for the treatment of depression. However, the effective use of these drugs has been expanding to include such psychological conditions as obsessive-compulsive disorder, attention deficit disorder, and various anxiety disorders. It is logical to assume that paraphilic behaviors may be linked to some or all of these other disorders, and, therefore, these drugs have been used in paraphilic disorder treatment regimens. Several studies on SSRI treatments have shown positive outcomes, including reduced paraphilic fantasies, impulses, and behaviors (Assumpção et al., 2014; Birken, Hill, & Berner, 2003; Williams, 2008).

Since You Asked

10. Everyone in my neighborhood has received a notice that a convicted sex offender has been released from prison and will be moving near us. Of course, this news is scary. Can sex offenders be cured (I think he's an exhibitionist)? What are the chances he will do those acts again?

Effectiveness of Treatment for Paraphilic Disorders

Success rates of the various therapies for stopping or limiting paraphilic compulsions are difficult to determine with a high degree of accuracy. The primary source of statistics is the criminal justice system, where sex offenders who have been arrested are usually sent to jail, placed in treatment, or both. If, after serving their sentence, they reoffend and are arrested again, those treatment failures can be tracked. Beyond that, we have reports from therapists about the extent to which their clients say that the treatment has been successful.

Overall studies examining the effectiveness of treatments for paraphilic disorders report two main findings: Treatment methods are marginally effective at best, and the research into treatment methods is inadequate (Balon, 2013; Beech & Harkins, 2012). A critical analysis of the available literature indicates that, in a broad sense, *some form* of treatment is better than no treatment in reducing unwanted behaviors. For example, one study found that, overall, 11% of *treated* sex offenders reoffended, compared to 18% of those who received no treatment. Overall, the success rate for most paraphilic disorders is not optimistic. High rates of reoffending are common following most treatment regimens. The most successful treatments appear to be found among individuals who are treated early in their history of paraphilic behaviors and who are the most self-motivated to stop the deviant behavior and involve themselves in normal adult sexual relationships (Seligman & Hardenberg, 2000).

As biomedical, psychological, and sexual research continues to refine treatment methods and strategies, the outlook remains hopeful for increasingly effective therapies for paraphilic disorders (Garcia & Thibaut, 2011; Rosler & Witztum, 2000). However, we must keep in mind that sexual arousal and response (orgasm) are powerful motivators that have the power to lead people either to heights of great pleasure or to the depths of antisocial and dangerous sexual actions. Power that strong is never easy to overcome.

Your Sexual Philosophy

Paraphilic Disorders: Atypical Sexual Behaviors

As mentioned at the beginning of this chapter, our study of human sexuality would not be complete without discussing the world of nonmainstream sexual practices, and by now you may understand them better than you ever wanted to! But paraphilias are a very real component in the landscape of human sexuality. You already know about the rich diversity of human nature; now you also have some insight into the farthest reaches of the complexity and variation of human sexuality.

How on earth, you may be thinking, can these very strange and sometimes victimizing sexual behaviors possibly find a place in my sexual philosophy? If you consider them carefully, you'll see that a knowledge of paraphilias is an important piece in the complex puzzle that allows you to know who you are, to be clear about what you want and don't want in your life as a sexual person, and to plan ahead so that you can make the choices that are right for you.

For example, you might at some point in your life find yourself in a situation involving a person who may want to engage in one of the behaviors discussed in this chapter. You will need to figure out, maybe without a lot of time for reflection, how you feel about that. Is the behavior OK with you? Are you willing to be a participant? Do you need to remove yourself from the situation quickly? Do you need to take immediate actions (such as intervening or calling the police) to protect yourself or others? You should have some ready answers to these questions instead of trying to figure out how you feel and what to do in the highly pressurized moment when they may arise.

Moreover, what if you find yourself attracted to sexual acts that would be considered paraphilias? How would you handle that? A clearer understanding of what these behaviors are and which ones are dangerous, victimizing, or just "different" is crucial in your decision to act on your impulses, work to reject them from your life, or seek professional help to stop or avoid them.

Finally, the material in this chapter *must* be part of your sexual philosophy so that you are as well educated and equipped as you can be to protect others, such as friends, relatives, and especially children, from sexual victimization at the hands of individuals who seem to be unable to control their coercive sexual compulsions.

Have You Considered?

1. Which of the four criteria listed in the section titled "Society's Criteria for Judging Paraphilias" do *you* feel are the most important in determining whether a certain sexual behavior is abnormal? Explain your answer.
2. Of the various theories of the origins of paraphilic disorders, which one makes the most sense to you? Why, in your opinion, does that one seem to explain paraphilic disorders better than the others?
3. Explain why, in your opinion, paraphilic disorders are much more common in men than in women.
4. How do you think the Catholic Church should handle the sex scandals involving pedophile priests in the future? How can the church try to ensure that these abuses of power and position do not happen again?
5. In the chapter's discussion of masochism, you read about the arrests made in England of members of a masochism club. The appeals court, in overturning their arrests, compared sexual masochism to the sport of boxing. Does this comparison make sense to you? Why or why not?
6. Imagine that you are sexually and romantically involved with someone who has a paraphilia. Would you be willing to try incorporation therapy as a treatment strategy? If so, explain which paraphilias you think it would be suited for and why. If not, why not?

Summary

Historical Perspectives

An Early Account of Paraphilias

- One of the earliest written accounts of the unusual behaviors known as paraphilias dates back to the late 1800s, when Richard von Krafft-Ebing published his 600-page book, *Psychopathia Sexualis*. His work detailed 237 case studies of atypical sexual activities, including reports of a man who had intercourse with a hen and a teenage boy who had what we would now call a handkerchief fetish.
- Some of the behaviors Krafft-Ebing condemned as pathological are today considered common and

normal by most health professionals and people in general; these include masturbation, oral sex, and nonheterosexual orientations.

Defining Paraphilias

- Paraphilia, which literally means "abnormal love," refers to acts that most people would consider deviant or abnormal that are performed for the purpose of sexual gratification and tend to be compulsive, meaning the person is helpless to stop the behavior.
- Societies tend to gauge the degree of abnormality of a paraphilia using various criteria, including the legality of the act; whether or not the act victimizes others; how harmful the act is to the person performing it and to others; and whether the person engaging in the act is able to maintain normal, intimate relationships with others.
- Currently, in professional practice, a sexual behavior may be diagnosed as a paraphilic disorder if the person has experienced recurring, intense, sexually arousing fantasies, sexual urges, or behaviors related to the paraphilia for a period of at least six months *and* the person has acted on those urges *and* the urges have caused marked distress or interpersonal difficulties.
- When a paraphilia victimizes others, it may be diagnosed as a paraphilic disorder regardless of whether or not the person engaging in the behavior feels distressed by it.

Gender and Paraphilias

- Nearly all paraphilics are men. The one possible exception is sexual masochism (sexual gratification from receiving pain), which is found in both men and women but is nevertheless far more common in men.
- The greater prevalence of paraphilias in men may be due to the fact that in most Western cultures, males are given greater "permission" to be sexual in their thinking, language, and behaviors and therefore have a greater opportunity to experiment with unusual sexual activities.
- Determining exact numbers of people who engage in any specific paraphilia is difficult due to the necessity of relying on self-report data, which is easily subject to bias when dealing with such socially condemned behaviors.

Origins of Paraphilias

- Although the exact cause of paraphilias is not known, various theories have been proposed to account for why some people engage in these sexual anomalies. These include psychodynamic theories that contend that paraphilias stem from early childhood conflict; behavioral theories that propose a learning theory approach to the development of paraphilic behavior; and biological hypotheses that place the source of paraphilic behavior within human biology, focusing on brain chemical or hormonal imbalances.

Specific Paraphilic disorders

- Paraphilic disorders may be divided into two definitive categories: victimizing and nonvictimizing. Victimizing paraphilic disorders involve the nonconsensual inclusion of another person. These include voyeurism, exhibitionistic disorder, frotteurism, sexual sadism disorder, and pedophilic disorder and are punishable by law.

Nonvictimizing Paraphilic disorders

- Nonvictimizing paraphilic disorders are either individual (solo) activities or involve other *consenting* adults. These include masochism, fetishism, transvestism, and autoerotic asphyxiation (self-strangulation during masturbation).

Treatment of Paraphilic disorders

- Paraphilic behaviors are among the most difficult behavior problems to treat or cure among adults. This is probably because the sexual arousal and orgasm that typically accompany the behaviors are among the most rewarding events a person can experience, and the behavior therefore becomes deeply ingrained.
- Types of therapy include incorporation therapy, cognitive-behavior therapy, aversion therapy, covert sensitization, orgasmic reconditioning, and medications. Success rates for treatment are generally poor, with high rates of repeat arrests of paraphilic sexual offenders.

Your Sexual Philosophy

Paraphilic Disorders: Atypical Sexual Behaviors

- The information in this chapter must be part of your sexual philosophy so that you are as well educated and equipped as possible to deal with atypical sexual behaviors, should you be confronted with them, and so that you are in the best position possible to protect others, such as friends, relatives, and especially children, from sexual victimization at the hands of individuals unable to control their coercive sexual compulsions.

Chapter 15

The Sexual Marketplace

Prostitution and Pornography

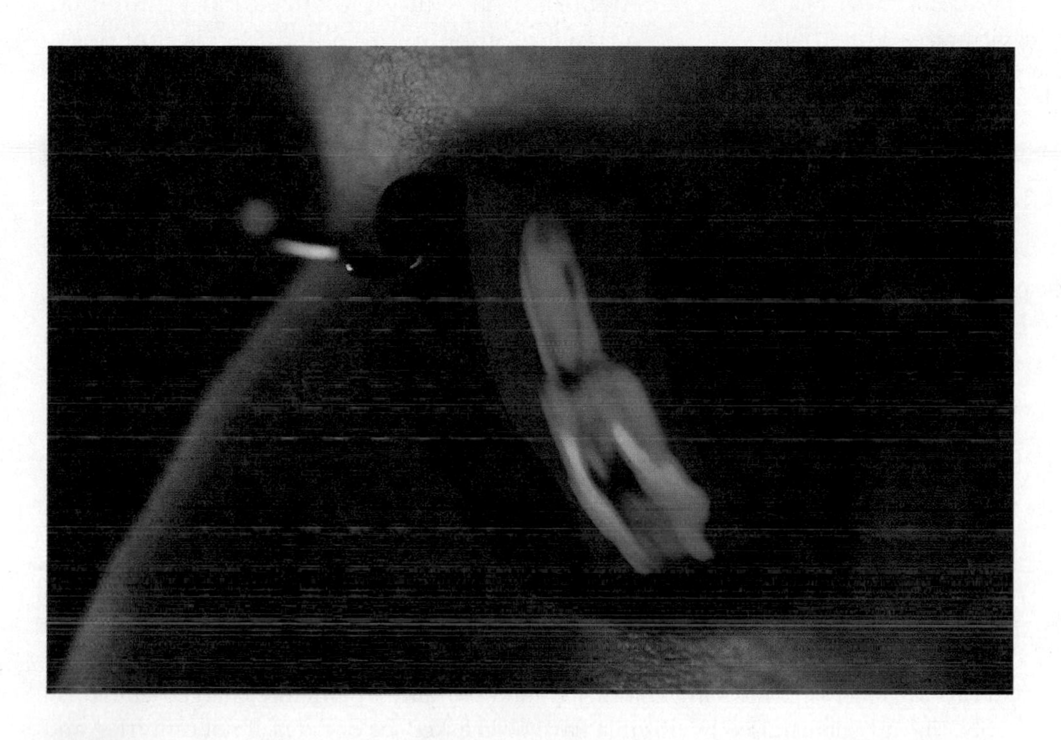

Learning Objectives

After you read and study this chapter you will be able to:

15.1 Review and analyze the complex cultural, social, and individual aspects of sex trade workers and their clients.

15.2 Summarize the various aspect of pornography including definitions; distinctions between pornography and erotica; the research on the effects of pornography; and discuss the global problem of child pornography.

Since You Asked

1. Why is prostitution called the world's oldest profession? Is it really? (see page 555)
2. My roommate is sleeping with a guy so that he will do her math homework for her. Isn't that the same as being a prostitute? (see page 557)
3. Is prostitution legal in all of Nevada or just in Las Vegas? How does it work there? (see page 558)
4. I cannot understand why any woman would ever become a prostitute. Is there something psychologically wrong with these women? (see page 562)
5. Is it true that AIDS and other diseases are spread mostly by prostitutes? (see page 566)
6. What are the main reasons men go to prostitutes? Is it usually because they can't get sex any other way? (see page 569)
7. I like to look at magazines like *Playboy* and *Hustler*, but my girlfriend says they are pornographic trash. I think pornography refers to much worse kinds of magazines and videos. Who is right? (see page 574)
8. Everybody's always saying how bad pornography is because it is violent toward women. But I've seen a lot of XXX videos that don't have any violence or anything bad against women in them at all. So is that not pornography? (see page 576)
9. Can looking at pornography cause guys to want to do weird or sick things sexually? (see page 581)
10. I read that child pornography is everywhere on the Internet. I think this is just so sick. What is being done to stop it? (see page 587)

Selling sex is big business in the United States and throughout the world. A staggering amount of money changes hands both legally and illegally in the buying and selling of sexual services and sexually explicit materials. Both prostitution and pornography (often referred to today as the commercial sex trades) are controversial topics because they weave together sexuality, law, censorship, and morality in a highly charged emotional mix. The questions raised by this mix have been asked for decades, if not centuries, and continue to be asked today. Should states enact laws that determine what people can and cannot do with their own bodies? Is prostitution a "victimless" crime? What are the factors that cause someone to become a *sex trade worker*? Why does someone choose to pay for sexual activities? Would decriminalizing prostitution help reduce the spread of HIV and other STIs? How do sexually explicit materials fit into the freedom-of-speech protections in the First Amendment to the U.S. Constitution? Is there a difference between *erotica* and *pornography*? Should sexually explicit films, videos, magazines, and books be censored or banned? If so, which ones? Where should societies draw the line between what is legal and illegal? Does pornography discriminate against certain groups? Does it promote rape and other sexual violence? What can be done about child pornography? Do sexually explicit materials ever serve a socially positive purpose?

These and countless other questions are the focus of some of the most heated controversies in the field of human sexuality. This is why these two topics alone make up the entirety of one chapter in this text. Chances are that these debates may never be resolved fully because history has shown us that attitudes toward pornography and prostitution are continuously reshaped by ever-changing cultural influences. For example, as you will read in this chapter, at a time in history when the debate over the legalization of prostitution was relatively quiet, the feminist movement in the 1960s and the frightening pandemic of HIV/AIDS in the 1980s and 1990s reenergized and redirected that dialogue. When at last some fairly reasonable and effective systems had

Focus on Your Feelings

Few topics in this book will generate as complex a mixture of emotions as those in this chapter: prostitution and pornography. Why? Because they create a crossroads where sex, morality, law, victimization, and children all meet. That's a powerful mix of emotion-producing themes.

Virtually no one is indifferent when it comes to prostitution, or what has become known in research and social services as the **commercial sex trades**. You may feel anger as you read of the uncaring people and circumstances that sometimes conspire to force a girl or woman into prostitution and keep her there. That anger may turn to distress when it becomes clear that drug addiction and coercion by pimps, and even sometimes by boyfriends and husbands, leaves some prostitutes little choice about exchanging their bodies for money or drugs, or that a vast international sex-slave-trafficking market, focusing on children, is thriving despite efforts to stamp it out. And what about your reactions to the men who choose to buy sex from prostitutes? You may feel moral outrage that those men must abuse others to satisfy their own selfish sexual urges. You may be surprised to learn that the characteristics of clients of prostitutes match quite closely the characteristics of men in general. A fierce debate has been ongoing worldwide for centuries about the legal status of prostitution. Whatever side you may find yourself on, one fact is sure: the controversy is likely to be emotionally supercharged.

Emotional reactions to pornography tend to run equally high. When you read about the difficulties in defining pornography, you may feel confused that people can't seem to agree on just what this thing everyone calls "pornography" actually means. Moreover, the difficulty in defining pornography leads directly to the debate over what sexually explicit materials, if any, should be censored, that is, legally banned. Opinions run the gamut, from the argument that all explicit sexual images should be illegal to the belief that nothing at all should be censored for adults. Whatever you believe, you can easily find someone to debate it with you.

This chapter covers what may be the most emotionally wrenching topic of all : child pornography. It is rampant on the Internet, yet virtually everyone not involved in that depraved business reacts to such materials with revulsion. So, get ready. Of all the "Focus on Your Feelings" boxes in this text, this one may be the most important in preparing you for some personal feelings you probably won't be able to escape.

been developed in the United States for dealing with the complex legalities of sexually explicit materials, along came the Internet to draw them into question once again. Therefore, we can probably assume that these issues and controversies are here to stay in one incarnation or another. That fact may be the best possible reason for all of us to continue to pursue the most complete understanding of them that we can have.

With that in mind, we should perhaps start at the beginning. Prostitution has often been referred to as the world's oldest profession. That claim is probably arguable (wouldn't hunting and gathering have come first?), but it is a cliché that has become inescapably linked with prostitution. In the first half of this chapter, we will explore the current state of the **commercial sex trades**, analyzing what prostitution is and the profiles of prostitutes and their clients. We will also consider the risks of STIs and other dangers in the sex trades and some of the ongoing legal controversies that surround prostitution. Next, we will turn to the topic of pornography, its definition, and what we know about its possible effects, especially in the cases of child pornography and cyberporn. As always, we will close with a consideration of how and why the topics of this chapter are important to your personal sexual philosophy.

commercial sex trades
Selling or trading sexual goods or activities for money or other items of value.

Historical Perspectives

The "Oldest Profession's" Oldest Brothel?

How old is the profession of prostitution? What evidence do we have to say that prostitution is the "oldest profession," as many claim? A recent archaeological find might help answer these questions. In 1996, archaeologists were excavating the ruins of an ancient Roman forum in Salonika, in the ancient area of northern Greece known as Macedonia. As the digging took the scientists underneath the forum, they happened upon a remarkable discovery: a bathhouse and brothel from more than 2,000 years ago. The artifacts, rooms, and decor they found left very little doubt about the purpose of the structure.

Since You Asked

1. Why is prostitution called the world's oldest profession? Is it really?

Archeological discoveries from ancient Greece, suggest that prostitution may indeed be one of the world's "oldest professions."

Experts dug out and pieced together hundreds of rare clay artifacts, glass objects, and erotic paraphernalia from the first century BC. On them are hundreds of depictions of sexual intercourse in myriad forms and combinations, including human orgies with goats. The treasure trove also included a clay dildo, two red-and-green-colored clay masks, and a glass cup trimmed with a relief depicting *Fortuna*, a fertility deity and capricious dispenser of good and bad luck. Lighting up the scientists' eyes—in shock, they say—were at least 700 oil lamps decorated with explicit sexual scenes, all pieced together from millions of clay shards found on the establishment's ground floor. The designs and the abundance of lamps provided additional proof that the complex operated heavily during evening and during the nighttime hours (Bird, 1999).

Many other similar finds have demonstrated that prostitution is indeed a very old, if not the "oldest," profession. From the archaeological evidence, social scientists are able to piece together vivid scenarios that go beyond the visible, concrete discoveries to the actual behaviors of the people who visited these facilities devoted to sexual and sensual pleasure. Brothels such as the one discovered in Salonika were apparently not disreputable, seedy, back-street sex parlors. On the contrary, they were high-class social clubs that were visited as a normal part of a day in the life of Roman men in ancient times, in much the same way as, say, a health club might be frequented today (although by both sexes). Sexual activities were only one of a variety of pleasures available to patrons. There were lavish spreads of food, over 25 "hot tubs" and ice baths (with food and drink available while bathing), and the prostitutes, who were refined members of society and treated as equals by their clients, would perform sexually explicit plays, recite poetry, and otherwise entertain the patrons.

Many researchers believe that both boys and girls; men and women, were available to patrons for sexual favors. The Salonika brothel site was opened for public viewing in 1999.

Prostitution

15.1 Review and analyze the complex cultural, social, and individual aspects of sex trade workers and their clients.

Hooker, hustler, cocotte, cruise, whore, harlot, bawd, tart, cyprian, fancy woman, working girl, streetwalker, strumpet, trollop, sporting lady, woman of the street, chuspanel, chippy—the list of slang and street terms for female and male prostitutes seems never-ending and unrelentingly derogatory. As we have discussed elsewhere in this book, language reflects the attitudes of a society. Clearly, most cultures throughout the world condemn the profession of prostitution, as evidenced by language, laws, and punishments, which range from no criminal penalties (Nevada) to misdemeanors (San Francisco and New York), flogging (Iran), and even public beheadings (pre-2003 Iraq). In this section, we will look at the commercial sex trade from various perspectives, including how it is defined, the many and varied approaches sex trade workers take in practicing their profession, the characteristics of the workers and their customers, and the important issues of drug abuse and STIs in the sex trades. We will also look at the laws surrounding prostitution and the debate over its legalization.

Artifacts from ancient bath houses in Greece and Italy leave little doubt as to the sexual nature of these establishments.

Defining Prostitution

As discussed earlier, most researchers and social welfare agencies are using the term *sex worker* instead of prostitute. In this discussion, at times we will use the word *prostitute* or *prostitution*. Why? Because that is the term that continues to be used by most people worldwide. If you look up "sex worker" on the Internet, you will find approximately 2 million sites using that term. Entering the word *prostitute* returns 33 million sites.

The definition of *prostitution* depends on the context in which it is used. For example, various legal jurisdictions may use differing definitions, psychologists and sociologists may have specific definitions for research purposes, and even individuals rarely agree on a definition if you try to pin them down on specifics. Most people would say that prostitution means "sex for money." However, this straightforward interpretation is probably oversimplified and may be misleading. What exactly is meant by "sex"? Sexual intercourse? Manual stimulation? Oral sex? Kissing? Stripping? Posing for sexual photographs? All of the above?

Since You Asked

2. My roommate is sleeping with a guy so that he will do her math homework for her. Isn't that the same as being a prostitute?

Moreover, payment of money itself for sexual acts is not technically necessary to be considered prostitution. Sexual services exchanged for virtually anything—money, valued objects, personal property, drugs, shelter, or another service of some sort—probably qualify as prostitution. In fact, if you think about it, the only exchange that would likely never be defined as prostitution is sex for sex.

Taking all these factors into account, a general, working definition of **prostitution**, for our purposes, is "providing or receiving sexual acts by specific agreement between a prostitute, the client, and sometimes the prostitute's employer ('pimp') in exchange for money or some other form of remuneration." Prostitutes and their clients may be male or female and of the same or opposite sex.

prostitution
Providing or receiving sexual acts, between a prostitute and a client, in exchange for money or some other form of remuneration. Also referred to as the "sex trades."

This definition describes the overall practice of prostitution and says nothing about the legal definition or criminal penalties related to providing or partaking of such practices. You will also notice that the definition does not specify how much money or what the other forms of remuneration might be. It does, however, apply the definition of prostitution to both the prostitute and the client and clarifies that those involved in prostitution are not limited to any particular combination of sexes of the participants. This is an important point because historically, prostitution has been thought of and often legally defined as "acts furnished by a female to a male."

streetwalker
A prostitute who sells sexual services on the street, typically to customers who are driving by, soliciting sex.

Types of Prostitution

Individuals who work in the sex trades are often divided into various categories depending on where they work, how they procure their clients, and what activities they engage in. Without being overly specific, three main categories cover most of those who are engaged in the profession of prostitution: streetwalkers, call girls, and brothel workers. These typically apply to female prostitutes; we will discuss male prostitution later in this chapter.

STREETWALKERS As the name implies, **streetwalkers** sell their services on the street, typically to customers driving up in cars. The women in this category typically make the least money overall and per customer, and for most, a large percentage of their earnings go to their pimp. Estimating the earning of sex trade workers is virtually impossible due to the illegal, hidden nature of the activity as well as wide variations in income depending on geographical location and the nature of the individual sex worker's conditions. Streetwalker sex workers, however, generally earn low or very low incomes, and many barely earn enough to afford the drugs they need and basic life necessities.

Street-walker prostitutes sell sexual services on the street, typically to customers who are driving by, seeking sexual services.

Streetwalkers are in the greatest danger of all sex workers for arrest, violence, and contracting STIs. They tend to be less educated and are nearly always involved with illegal drug abuse (Brawn & Roe-Sepowitz, 2008; Cobbina, & Oselin, 2011). The sexual services provided by streetwalkers are typically less varied and of shorter duration than those provided by other prostitutes. The most common services are manual masturbation, oral sex, or quick vaginal intercourse.

Call girls maintain a list of regular clients and are often very highly paid for their services.

call girl

A prostitute who is contacted in private by clients when her services are desired and who generally charges more and provides a wider range of services than streetwalkers, such as serving as an escort or offering overnight stays.

Since You Asked

3. Is prostitution legal in all of Nevada or just in Las Vegas? How does it work there?

brothel worker

A prostitute who works in a house of prostitution—such as a brothel, bordello, or massage parlor—and receives paying customers.

House prostitute

A brothel worker.

gigolo

A male prostitute who is paid or otherwise compensated for providing sexual services to women.

male escort

A man hired by women as a companion; not necessarily for sexual purposes.

CALL GIRLS The **call girl**, as the name implies, is usually contacted by a client when her services are desired. Typically, call girls maintain a list of clients whom they see on a regular basis, ranging from once or twice a week to once a year or less. Some call girls work for a "madam," who introduces them to clients and schedules dates for a percentage of the fee. Call girls will typically see far fewer clients than streetwalkers, often no more than one or two per day, but will charge more per client. High-priced call girls can earn from several hundred dollars for a date of an hour or two, to several thousand or more for an entire night with one client. Many call girls earn hundreds of thousands of dollars per year. For example, here is a (disguised but accurate) sample of information compiled from some call girls websites:

> My charges are based on hourly rates. We can spend as little as one hour together or all day or all night. These prices are firm, so please do not try to negotiate with me. Pricing information is available for longer periods of time or for travel, so please give me a call or e-mail me.
>
> *Rates:* $300 first + $200 every hour thereafter; $400 dinner + 1 hr; $550 dinner + 2 hrs; $1,500 overnight
>
> *Prepaid Rates:* Prepaid rates can save you money by purchasing your dates with me in advance. When you purchase my hours, you can divide them up however you like. For example, if you purchase 12 hours, you could have two 2-hour appointments and eight 1-hour appointments. These prepaid hours are good for up to one year after purchase. 3 hrs—$750 (up to 3 one-hour shows); 6 hrs—$1,350 (up to 6 one-hour shows); 9 hrs—$1,800 (up to 9 one-hour shows); 12 hrs—$2,200 (up to 12 one-hour shows); 18 hrs—$3,150 (up to 18 one-hour shows)
>
> *Monthly Rates:* Monthly rates and discounts are available for regular clients. Please call me for more information.

Once a call girl has a consistent list of regular clients, she typically no longer needs to "hustle" to find customers. Call girls are much less likely to be drug abusers and are at significantly less risk of violence, especially once they get to know their regular customers. Some call girls are listed in the phone book or place ads in magazines under the heading "Escort Services."

BROTHEL WORKERS In all states, prostitution in any form is a crime. One important exception to this is the state of Nevada, where prostitution is not a crime in many counties but is under state control. Nevada is one of the few areas in the world where prostitution is not outlawed or prohibited in some formal way. The special case of Nevada is discussed in detail in "Sexuality and Culture: Prostitution in Nevada."

Brothel workers or **house prostitutes** are women who work in "houses" that receive paying customers. These houses of prostitution are referred to by many names and euphemisms, including brothels, bordellos, and massage parlors. In Nevada, where brothels are legal and state-controlled, prostitutes typically work under the safest conditions of all sex trade workers. These brothels are closely monitored for any signs of danger to the workers, and clients are screened before the outside gate to the entrance is opened.

House prostitutes, who share fees with the house manager or madam, typically earn more money than streetwalkers but considerably less, on average, than call girls. Owing to the requirements of the state, in the case of Nevada, or of the brothel itself, house prostitutes are far more likely than streetwalkers to receive regular medical exams and use condoms consistently. Therefore, they are less likely to contract STIs, even though they may see as many as or more clients than the typical streetwalker.

Sexuality and Culture

Prostitution in Nevada

Prostitution is illegal in the United States and most (but not all) other countries. It is true that laws forbidding prostitution exist in all 50 states. However, in parts of one state, Nevada, prostitution is not a crime and is practiced in state-licensed brothels. Prostitution of any type—including streetwalking, escort services, and massage parlors—is illegal in all other states. Virtually all prostitutes in Nevada brothels are women, and all customers are men. The reason for this is not that women would not go to male prostitutes (although that point is arguable), but that Nevada state law requires brothel workers to have a cervical cell exam (a Pap test) every three months. And because men obviously do not have cervixes, this would be impossible. However, as of early 2010, one county voted to allow one brothel (the "Shady Lady") to add male prostitutes (Powers, 2010). The intended customer for these men is women, not other men. The men will be required to undergo tests for STIs on a regular basis, just as the women brothel workers do now. As an experiment, the Shady Lady's offering of male prostitutes was a failure. The demand simply was not there. Nevertheless, in 2012, another Nevada brothel, the "Kit Kat Guest Ranch," hired four males to service women; whether it catches on this time remains to be seen.

Nevada currently has nineteen licensed brothels. This is significantly fewer than in the mid-1980s, when there were over thirty. The economic downturn and the availability of sex on the Internet have cut deeply into Nevada's brothel business (Vekshin, 2013). Interestingly, none of them is where one might expect to find them—Las Vegas or Reno; in fact, these famous gambling spots are located in parts of Nevada where prostitution is *not* legal. Because counties are allowed to determine whether to make brothels legal, some have opted to restrict them. Although prostitutes may look for business in the state's two major gambling towns, official, state-sanctioned brothels are not found there.

Nevada brothels are regulated by the state in terms of location. They may not be in highly populated areas, on a main traffic thoroughfare, or anywhere near a school or church. These regulations imply that Nevada brothels are hidden from Nevada society in general, but in reality, they are often quite obvious and even brazenly conspicuous.

One of the most common concerns raised about legalized prostitution is the potential for spreading STIs—HIV, HPV (genital warts), gonorrhea, syphilis, and so on. In Nevada, all customers are required to use condoms for all insertive sexual activities, including vaginal, oral, and anal sex. Furthermore, Nevada has very strict laws requiring all brothel workers be tested for gonorrhea and chlamydia once a week, and every month for HPV, HIV, and syphilis. Brothel owners may be held liable if customers become infected with HIV after a prostitute has tested positive for the virus. Any positive STI result prohibits sex work until the infection is treated and cured. If any Nevada sex worker tests positive for HIV, he or she is no longer allowed to work legally as a brothel prostitute.

Of course, none of this means that everyone in Nevada, or certainly anywhere else, approves of legal prostitution in Nevada. Every few years, some Nevada lawmakers introduce new bills designed to outlaw all forms of prostitution in the state. Furthermore, many women's groups argue that prostitution degrades and harms all women sex workers regardless of its legal status. However, legal prostitution is deeply ingrained in Nevada's social, cultural, and economic heritage, dating back to the silver mining boom of the 1800s (although brothels were not officially licensed until 1971); it is unlikely that it will ever be outlawed. As discussed later in this chapter, the debate over decriminalizing or legalizing prostitution in other states is a much more hotly debated issue than any attempt to ban brothels in Nevada.

In Nevada, not only is prostitution legal, but it has become a big business. Although brothels in Nevada are required by law to be inconspicuous, often the opposite is true. One brothel, the "Shady Lady," in Nye County, Nevada, may require a name change now that it has been approved to add male prostitutes to service female customers.

Although brothel prostitutes typically appear to enter this form of work voluntarily, and it is legal in specific Nevada counties, many believe prostitution is a high-risk occupation and that the women are subjected to discrimination and coercion based on the idea that they are selling *themselves;* have little choice in what behaviors they must

Brothel workers or house prostitutes work in an environment where they are usually protected from violence and STIs, but this environment, although voluntary and legal, is nevertheless seen by many as degrading toward women.

engage in and with whom; and, in reality, are vulnerable, powerless, and often feell degraded—whether it is legal or not (Sullivan, 2007).

MALE PROSTITUTES The vast majority of prostitutes are women who service men. This is primarily due to long-standing cultural norms and expectations that (rightly or wrongly) give men more "permission" to engage in a greater range of sexual activities and assume that women have no need to pay for sex owing to the number of "willing" male sexual partners in the general population. If we examine the differences between male and female prostitutes, this appears to be true.

Many of the characteristics and activities involved in male prostitution are unique and require some separate discussion. First of all, unlike female prostitutes, male prostitutes sell their services primarily to other males. Male prostitutes who service women are rare and are usually referred to as **gigolos or male escorts**. These men tend to function in a similar fashion to call girls as discussed earlier, but usually with more emphasis on the escort role and less on mere sexual acts.

The vast majority of male prostitutes employ tactics that are very similar to those of female streetwalkers. These men are most commonly referred to as **hustlers**. Most will hang out on street corners or in bars waiting for customers to approach them. The sexual interactions typically take place in a car, a public restroom, an outdoor location hidden from public view, a motel room, or less often, in the prostitute's or client's residence. The most common behaviors are oral sex, masturbation, and anal intercourse. The actual number of male prostitutes is difficult to gauge, but they account for more than a third of all the arrests made for prostitution in the United States.

hustler
A male prostitute who services male clients and employs professional tactics similar to those of female streetwalkers.

Although male prostitutes are selling sex to other males, only 18% self-identify as gay. The rest self-identify as either heterosexual (46%) or bisexual (36%). However, their sexual relations with women (if any) tend to occur outside their professional activities (Altman & Aggleman, 1999; Weitzer, 2005).

Male hustling is clearly linked to various drug abuse and health issues (as is true of female prostitutes, to be discussed next). Studies have found significantly higher than average rates of all STIs, including HIV (28%), syphilis (37%), herpes (67%), and hepatitis B (38%), among active male sex workers (McGrath-Lone, 2013; Mindel & Estcourt, 2001). Moreover, in related findings, rates for drug abuse are very high among this group of men, with between 40% and 53% reporting a history of injection drug use, 76% cocaine abuse, and 61% crack abuse (Reitmeijer et al., 1998; Timpson et al., 2007).

Male prostitutes, referred to as hustlers, account for more than a third of all prostitution arrests in the United States.

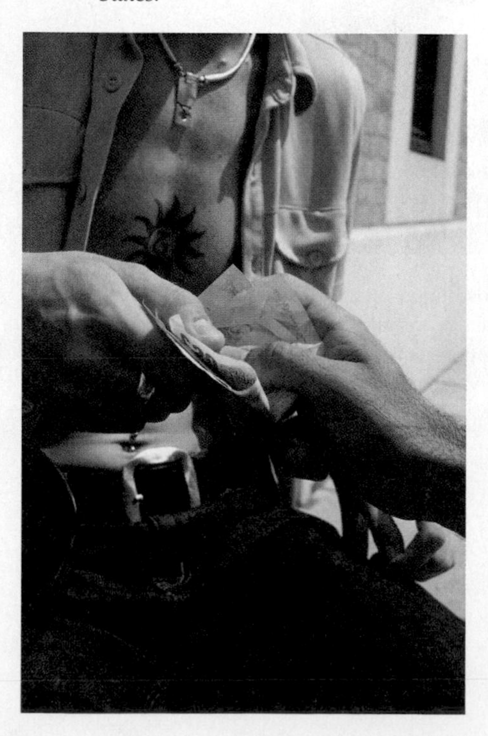

Who Becomes a Prostitute?

In general, the personal and family lives of sex trade workers prior to becoming prostitutes are different in some significant ways from those of nonprostitutes. However, these differences are not the whole story because *most* men and women who have experienced the same life situations as sex workers do *not* become prostitutes. Life experiences do not by any means fully explain people's motivations toward prostitution. Also, these life differences do not imply that prostitutes enter the trade of their own free will; many do not, as we will discuss shortly. With those points in mind, the three most salient factors associated with working in the commercial sex trades are poverty, early sexual experiences and abuse, and drug addiction. These factors do not function separately but tend to be linked in important ways.

Throughout the world, commercial sex trade workers tend to come from the lowest economic levels of society (Sanders, O'Neill, & Pitcher, 2009). This makes a certain amount of sense in that choosing a profession generally fraught with great social stigma and the risks of disease, violence, and drug addiction would appear to imply a certain financial desperation. Some women feel that prostitution is their only option for financial and even physical survival. When we factor in drug abuse and addiction, the links between poverty and prostitution become even more compelling.

Figure 15.1 Age of Drug Use and Prostitution Onset among Sex Trade Workers

As shown in the data for one study, the age at which young women begin intravenous drug use corresponds closely to entry into prostitution.

SOURCE: "Age of Drug Use and Prostitution Onset among Sex Trade Workers," from "Pathways to Prostitution: The Chronology of Sexual and Drug Abuse Milestones," by J. Potterat et al., in *Journal of Sex Research*, (1998), pp. 333–340. Copyright © 1998. Reprinted by permission of The Society for the Scientific Study of Sex.

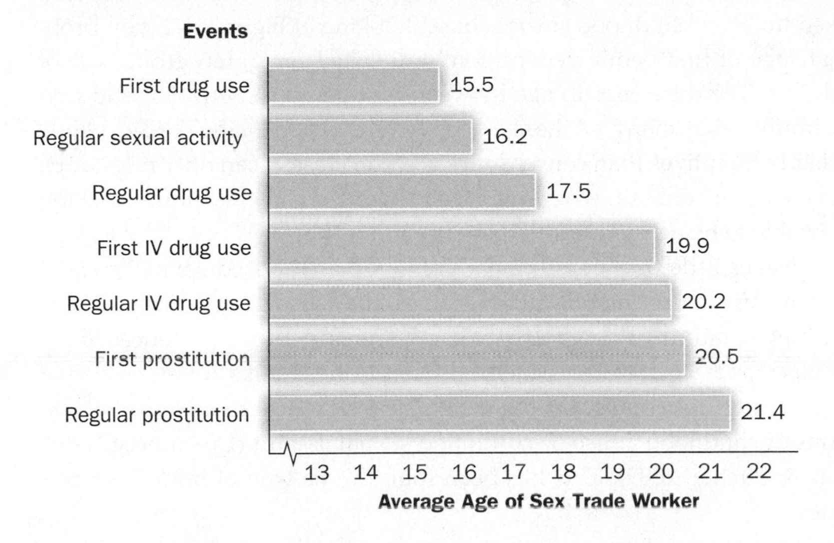

One question often asked in research on prostitution concerns the chronological sequence of drug use and prostitution. In other words, do poor young women and men generally become involved with drugs and then turn to prostitution to support their addiction? Or are girls living in poverty first introduced to sex work, which provides access to drugs and the drug culture; become addicted; and then remain in the profession to support the addiction?

One study that examined this issue produced some interesting findings (Potterat et al., 1998; see also Weber et al., 2004). The research revealed that the vast majority of prostitutes began using drugs prior to their first experience with prostitution. This finding is graphically illustrated in Figure 15.1. You will notice that the age of first drug use among prostitutes was approximately 15, but the first prostitution experience was not until age 20.5. Examining the data more closely, however, you can see that these women began their sex work within months, on average, after first using intravenous drugs and were regularly engaged in prostitution within 14 months after beginning regular use of IV drugs. We should note that the age of first drug use did not differ significantly among a matched group of nonprostitutes. This implies that women who later became prostitutes did not start using drugs sooner in life, but once they became addicted, they may have entered prostitution as a means of obtaining or buying the drugs.

One stereotype of prostitutes depicts a teenager who becomes sexually active early and whose early sexual experience plays a role in the eventual entry into the sex trades. Thus precocious sexual activity is portrayed as a gateway to prostitution, either through the young person's initial enjoyment of sex or perhaps through desensitization of the youth to concerns about having sexual relations with many partners. However, do we have evidence that prostitutes do indeed experience sexual activity earlier than nonprostitutes? Some research does support this idea. In the same study by Potterat and colleagues (1998) cited earlier, the sexual histories of a sample of prostitutes was compared to those of a matched group of women seeking care at a free health clinic. These results are summarized in Figure 15.2. You can see that the age of first sexual intercourse, age of first consensual sex, and age of beginning regular sexual

Figure 15.2 Age of Onset of Sexual Activity: Prostitutes versus Comparison Group

Age of first sexual intercourse, first consensual sex, and the beginning of regular sexual activity was significantly earlier for girls who entered prostitution.

SOURCE: "Age of Onset of Sexual Activity," from "Pathways to Prostitution: The Chronology of Sexual and Drug Abuse Milestones," by J. Potterat et al., *Journal of Sex Research*, (1998), pp. 333–340. Copyright © 1998. Reprinted by permission of The Society for the Scientific Study of Sex.

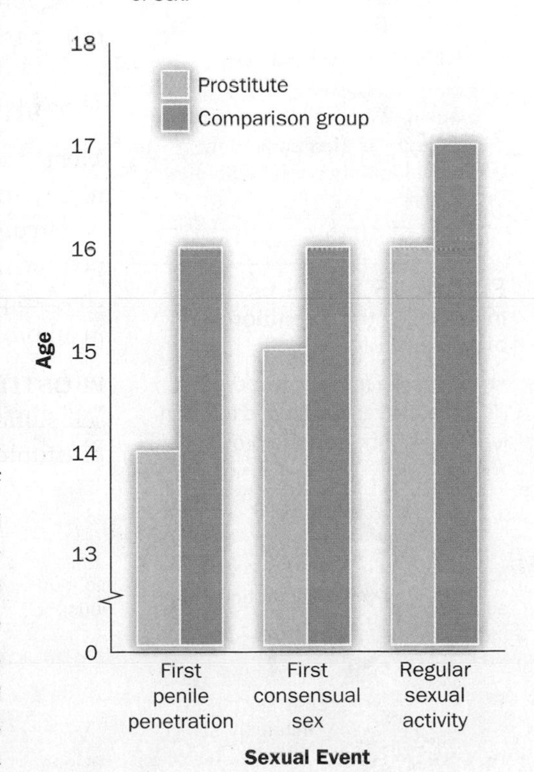

activity all differed significantly for the two groups. The prostitutes began all these activities 1.3 to 2.5 years earlier in life than the nonprostitutes (see also Dodsworth, 2012).

These are, of course, correlational findings, so we cannot assume that age of first sexual activity *caused* the women in the study to enter the sex trade. It may well be that these women were sexually precocious for other reasons in their lives that may have contributed to their eventual entry into prostitution. One possible cause that cannot be overlooked involves childhood sexual abuse. Looking at Figure 15.2, you probably noticed that "age of first penile penetration" was listed separately from "age of first consensual sex." The difference in age between first penile penetration and first consensual sex implies that many of these women were experiencing forced penile penetration earlier in their lives than consensual sex. "Forced sex" can only mean rape or childhood sexual abuse, or both, which may lead some girls to leave home, making them more vulnerable to being lured into prostitution.

The research leaves little question that children who are abused sexually are at greater risk for various developmental difficulties as they grow into adults (this is discussed in greater detail in Chapter 13, "Sexual Aggression and Violence: Rape, Child Sexual Abuse, and Sexual Harassment"). One of these risks involves sexual adjustment later in life. Prostitutes, on the whole, are significantly more likely to have been victims of childhood abuse or childhood sexual assault (Denenberg, 1997; Sanders, O'Neill, & Pitcher, 2009). This has been found to be true of both male and female prostitutes.

What conclusions can we draw from these factors in the lives of prostitutes? It is confusing, but probably the best interpretation is that no single variable can be isolated as a "cause" of a person's becoming a prostitute. All of the factors discussed here, along with others, probably play a role in a complex set of circumstances that determine how sex trade workers develop throughout their lives, how they view themselves, and how their experiences come together to move them along the path to prostitution. Childhood abuse, poverty, drug addiction, and sexual experiences in adolescence may form an interactive model, as shown in Figure 15.3, that explains some, but probably not all, of the forces at work in the development of a commercial sex trade worker. More importantly, the ability to choose prostitution as opposed to being forced into the sex trades is, for many, the most powerful factor in understanding a person's entry into prostitution.

Since You Asked

4. I cannot understand why any woman would ever become a prostitute. Is there something psychologically wrong with these women?

Prostitution: Usually *Not* a Profession of Choice

Most streetwalker prostitutes are not in the sex trades by choice. They are, for the most part, forced into prostitution through the coercion and intimidation of others or by circumstances in their lives (such as poverty) that they perceive as leaving them no other choice. An ongoing debate in this area relates to whether or not *anyone* truly chooses prostitution or whether some elements of coercion or entrapment are present in all prostitutes' lives.

PROSTITUTION BY CHOICE? Would all prostitutes, given the opportunity to be self-sufficient in another career, choose to give up sex work? The answer is no. Some prostitutes say they are satisfied with the profession they have chosen.

Figure 15.3 Factors Influencing the Evolution of a Prostitute

An interactive model may best illustrate the various factors that can lead a young person into prostitution.

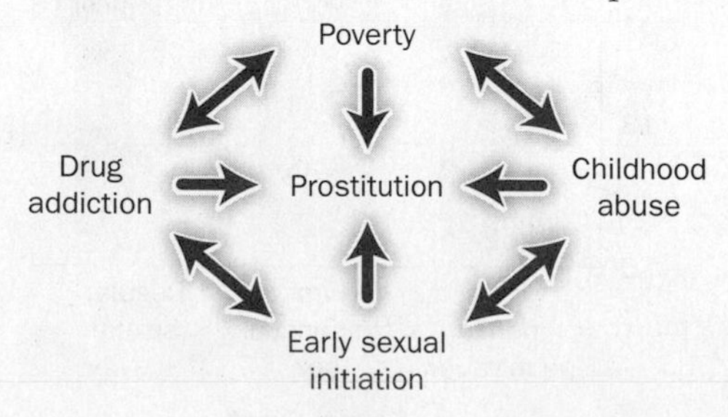

In the brothels in Nevada, for example (as discussed earlier), many prostitutes are able to choose their work hours and yearly schedules. Some will work four or five months out of the year and make enough money to live without working the rest of the time. Some of those women claim that they would be unable to have the life they want in any other profession and would not want to stop what they are doing. In other cases, especially among some highly paid call girls who are earning six-figure incomes, the women feel they could not make as much money in any other profession.

Some writers and researchers contend that prostitution is *never* truly a choice, that it is invariably sexist and degrading, and that some degree of coercion exists in every act of prostitution. Nevertheless, some sex trade workers say they enjoy their work. Most deny that they receive sexual pleasure from their interactions with clients, but some are content with the knowledge that they are helping and giving pleasure to others, as well as making a good living (Brewis & Linstead, 2000; Owens & Lopes, 2003). Prostitutes with such positive attitudes typically work in brothels or fall into the category of call girls or escorts. To get a feel for how someone might choose to enter into prostitution and to analyze if it is truly voluntary, see "Sexuality and Culture: A Personal Look at the Life of a Prostitute."

Sexuality and Culture

A Personal Look at the Life of a Sex Worker

Dear Dr. Raj Persaud, Please help me with this dilemma. I am 25, unemployed, a single mother with three children all under 8, and I desperately want to be able to buy them some proper clothes and take them on a short holiday. I could only afford to do this last year after dabbling in prostitution, and I am thinking of returning to it now. Could you give me any advice about whether I should?

Delia

Dr. R: *How did you get involved in prostitution the first time around?*

Delia: I saw an ad in a local newspaper for sauna staff, and the owner interviewed me in a pub—he asked whether I knew what kind of work it is and I answered I could take a good guess. I was desperate for the money, I felt so low and terrible at not being able to provide basics for my children. The girls at the sauna were really helpful—they ran through important safety tips, like you must always be on top as it is more difficult then for the man to turn aggressive or be in control than when he is on top of you, and never let him have sex from behind as then he could reach forward and strangle you. I found having sex surprisingly easy, and I really enjoyed the power—you are the one in control. I did three six-hour stints and earned over $300. But I stopped because it began to be difficult to explain to my parents, who were babysitting, where I had been. Also, I had split up with my boyfriend back then, and now we are together again and he is bound to be suspicious if I went out for so long again.

Dr. R: *Given these difficulties, how are you thinking of getting involved in prostitution again?*

Delia: This time I would advertise in phone booths and run it from home—seeing the men in local hotels. I would buy a mobile phone so they wouldn't call me on the home phone. If I did it that way, I would only be out of the house for around an hour, so my parents would think I was just visiting a friend. The problem with the sauna is you had to be there for too long; I could never organize babysitting for that period.

Dr. R: *Sounds as though you have it all pretty well thought out. What is stopping you from going ahead?*

Delia: I am worried that seeing men alone in hotel rooms might be less safe than seeing them at the sauna. I wondered, as you are a psychiatrist and see people from all walks of life, what your opinion is.

Dr. R: *It certainly doesn't sound very safe seeing men alone in a hotel room, particularly if it is a secret from everyone you know and no one else knows where you are. Have you discussed it with any prostitutes? What was their advice?*

Delia: No, the only prostitutes I know are the ones at the sauna, and I haven't been back there for a year. I suppose I could ask them. I did tell my two best friends about the escapade last year—they were really shocked and said if I was thinking of doing it again, I should discuss it with them first.

Dr. R: *And have you?*

Delia: No, I know they would disapprove and try to talk me out of it.

Dr. R: *Are you not worried about your parents or anyone else finding out? Could you not earn the money some other way?*

Delia: I would have to be very careful. Obviously, if the neighbors found out, I would have to move. The way my benefits are at the moment, there is no job I could get, even if there was work—which there isn't—that would leave me better off than I am at the moment. I am taking a course in secretarial skills, but even after that, I don't think I will earn as much as being a prostitute, particularly when you take the expense of child care into account.

Dr. R: *So it sounds like the only two issues that might stop you are your concerns about personal safety and other people finding out.*

Delia: But I don't think if I am careful anyone will find out.

Dr. R: *Well, that depends. But you still seem somehow very drawn to being a prostitute.*

Delia: Yes, I am. When I was there in the sauna, I was happy for the first time in a very long time. It was an escape for 12 hours—I had no contact with anyone from my neighborhood or home, no parents or anyone who knew me. It was an escape from reality of my life, where for a short period I had no responsibility or pressure.

Dr. R: *I get the feeling you really contacted me so that I might give you permission to do it. You really do seem to have made up your mind already. I am worried that you are naive about the downward spiral prostitutes get caught in.*

Dr. Persaud went on to comment on the potential pitfalls of prostitution, including the potential loss of self-respect and the dangers of drugs, violent pimps, and STIs. In addition, the stress of always being vigilant to keep her prostitution secret might affect Delia to the point that she would not be as warm and loving with her children; and her attitude toward men might change to the point that she would find it difficult to form a loving, long-term relationship with a man.

Dr. Raj Persaud's column, "The Session," appears monthly in *Cosmopolitan.*

SOURCE: Reprinted by permission of Dr. Raj Persaud, www.rajpersaud.com.

FORCED PROSTITUTION AND SEX TRAFFICKING The majority of prostitutes, especially streetwalkers and hustlers, are forced into a lifestyle they hate and from which they can see no escape. Many sex workers are pressured to engage in prostitution for any one or a combination of reasons, including: (1) threats of bodily harm if they do not prostitute themselves to earn money for their husbands or boyfriends; (2) coercion into streetwalking by pimps who "rescued" them from homelessness and who now maintain total financial control over them; (3) drug dependence or addiction, especially to crack cocaine (often initiated by a pimp), that can only be supported by exchanging sex for money or for the drugs themselves; (4) a life of extreme poverty, often supporting children, and lacking the education or training to find other forms of work; and (5) being lured, sold, or kidnapped into the sex trades by international sex worker traffickers (Kara, 2008). These girls, boys, women, and men begin selling their bodies out of desperation or fear, but the lifestyle takes on a weight of its own, causing them to spiral down to greater depths of poverty and addiction until escape feels literally impossible (Dodsworth, 2012; McClelland, 2001).

Prostitutes are at a significantly increased risk of rape and other sexual violence (Sullivan, 2007). Some people believe the myth that if someone is in the business of selling sex, that makes sex with them *always* consensual, and that a prostitute by definition cannot be raped. However, as discussed in Chapter 13, "Sexual Aggression and Violence: Rape, Child Sexual Abuse, and Sexual Harassment," rape and sexual assault are *forced, nonconsensual* sexual attacks, and this violence is just as victimizing and traumatic (as well as illegal) for sex workers as it is for anyone. One study of sex workers in Washington, DC, reported that 50% of prostitutes reported having been raped, compared to 9% in the general population (Valera et al., 2000). Adding to this victimization is the fact that sex workers are less likely than others to report rapes to the police because they often feel powerless with law enforcement and fear that authorities will not take them seriously or that they themselves may be arrested for prostitution (Sullivan, 2007).

In what has become a worldwide tragedy, millions of young girls from impoverished countries throughout the world have been coerced, forced, and sold into sexual slavery in the United States and other Western countries.

The illegal international trafficking of sex workers has become a major worldwide tragedy involving, by some estimates, as many as 4 million victims and an annual monetary exchange of as much as $15 billion (Marcus et al., 2014; Smits, 2004). Children, mostly girls, as young as nine or ten years of age are lured, bought from their parents, or kidnapped from their homes in many of the less developed countries and parts of the world, including Thailand, eastern Europe, China, and western Africa. These victims are often promised a good job, a better life, and other "dreams come true" in a Western country, but instead are literally sold into sexual slavery and imprisonment (Buffington & Guy, 2013; Kara, 2008; Power, Radcliffe, & MacGregor, 2003).

International laws protecting sex trafficking victims have not kept pace with the scope of the problem. In the United States, the number of victims remains shockingly high, but prosecutions and punishments for traffickers, as well as help for victims, are increasing. In 2000, the U.S. Congress passed the Victims of Trafficking and Violence Prevention Act (VTVPA), which established new felony laws designed to combat sex trafficking and authorized a wide range of social services and protections for victims of trafficking. The law also provided for temporary U.S. visas, allowing victims to remain in the United States for their protection and to help law enforcement officials capture and prosecute traffickers. Nearly 15 years later, the available evidence is that, overall, the VTVPA has not significantly stemmed the growing tide of human sex trafficking (i.e., George, 2012). Moreover, for many countries, human trafficking, as horrendous as it is, remains a relatively low priority on their list of difficult social problems, and the problem continues to increase internationally.

PROSTITUTION AND DRUG ABUSE Drugs and the commercial sex trades often go hand in hand. Drug abuse plays a dominant role in the lives of most prostitutes. For many, it is the gateway into the trade and the primary force keeping them there. At first, prostitution may serve as a way for poor women and men to finance a drug habit. However, as they become trapped in a life of prostitution, drug use may become a coping mechanism to deal with the many stresses and anxieties encountered as a result of their occupation (Burnette et al., 2008; Potterat et al., 1998). Some prostitutes report using drugs to enhance their perceptions of personal self-confidence and to decrease their feelings of guilt and lack of sexual intimacy in their lives (Young, Boyd, & Hubbell, 2000).

In addition to alcohol and marijuana, prostitutes frequently abuse cocaine, methamphetamines, and heroin (Shannon et al., 2011). However, one particular highly addictive drug exposes those involved in the commercial sex trades to the greatest risk: **crack**. And it's nearly universal. One study estimates that as many as 95% of sex trade workers in the U.S. are addicted to crack cocaine (Patton, Snyder, & Glassman, 2013). Crack produces an intense but relatively short-lived high and for is known for its rapidly addictive qualities. Prostitutes who are addicted to crack therefore need more frequent fixes and are more willing to engage in a wide range of unsafe sexual practices in exchange for the drug or to earn the money to buy it. Crack use thus increases these sex workers' risk and vulnerability to the most negative consequences of prostitution. One large study by the National Institute on Drug Abuse found that among crack smokers and injectable drug users, over the previous 30 days, 80% had engaged in unprotected intercourse, 24% had exchanged sex for drugs, and 23% had had sex with an injectable drug user (Booth, Kwiatkowski, & Chitwood, 2000; Burnette et al., 2008).

crack

A powerful, processed form of cocaine that is smoked or sometimes injected, known for its intense but relatively short high and its highly addictive qualities.

When a sex worker decides to leave the trade, he or she is often unable to support an expensive drug habit and is drawn inescapably back to prostitution.

Prostitution and STIs

The commercial sex trade is often singled out as playing a major role in the spread of sexually transmitted infections, especially HIV/AIDS. Although accurate statistics are difficult to obtain because of the illegal nature of their profession, prostitutes *overall* have significantly higher rates of sexually transmitted diseases than the population in general. However, the reasons behind this are more complicated than you may think. You are aware, of course, that one of the highest risks of contracting STIs is having multiple sexual partners, and prostitutes, obviously, have more sexual partners than virtually any other group of sexually active people. That factor alone places them, as a group, at greater risk. Moreover, the other two highest-risk behaviors for the transmission of STIs are unprotected sexual activities (sex without a condom) and the sharing of drug paraphernalia, especially needles. In the world of the commercial sex trades, these risks tend to occur together.

Many sex workers become addicted to drugs and see prostitution as the only way they can support their habit.

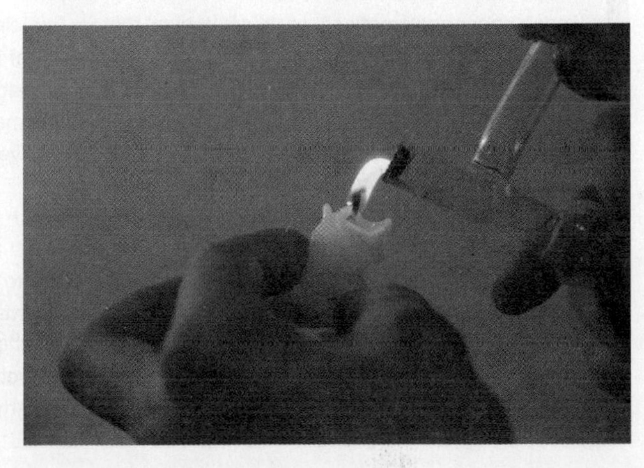

UNPROTECTED SEX, DRUGS, AND STIs To the extent that a prostitute chooses to engage in unprotected oral, vaginal, or anal intercourse and abuses drugs, his or her risk of infection escalates accordingly. These unprotected behaviors are far more common among streetwalker sex workers than among brothel workers or call girls. However, in addition to their greater risk of infection, prostitutes also may pose an extra risk of STI spread. If a prostitute with an STI engages in unprotected intercourse with a client, the infection may be transmitted to the customer, who may in turn transmit it to other sexual partners; and those sexual partners may in turn spread the infection to additional partners.

One group of sex workers who are especially likely to spread STIs to the general population are male transvestite prostitutes who impersonate women and engage in sexual interactions with male customers. "In Touch with Your Sexual Health: Male Transvestite Prostitutes and Sexually Transmitted Infections" describes the possibility for this exponential spreading of STIs to the general population. Moreover, many prostitutes are at increased risk of transmitting STIs among themselves if they share needles and other drug supplies, thereby further exacerbating the spread of these infections.

Since You Asked

5. Is it true that AIDS and other diseases are spread mostly by prostitutes?

Streetwalkers and hustlers are among the groups with the highest rates of STIs worldwide. As we discussed earlier, these are the groups in the sex trades that are most likely to be drug abusers and struggling financially. Their often desperate need to make

In Touch with Your Sexual Health

Male Transvestite and Transgender Prostitutes and Sexually Transmitted Infections

One group of male sex workers is particularly susceptible to sexually transmitted infections: male transvestite prostitutes (MTPs). As the name implies, these are men who dress up as women and sell sexual services to other men. Although both male and female prostitutes, especially streetwalkers, are at a significantly increased risk of STIs, male transvestite hustlers face the greatest danger. Research has shown that over 60% of MTPs are HIV positive, over 80% test positive for syphilis, and over 75% have contracted hepatitis B. Rates for these infections in nontransvestite male prostitutes are 27%, 22%, and 58%, respectively. What accounts for such an alarming difference?

Several factors have been suggested to account for this difference. First, clients may be looking for and demanding to receive anal sex more frequently from MTPs than the clients of other male prostitutes. In an effort to obtain clients and to charge as much as possible, MTPs may be engaging more often in unprotected anal intercourse, which is known to be the easiest route of transmission for HIV and other STIs. Second, MTPs may experience emotional or psychological factors in their personal or professional lives that put them at greater risk of intravenous drug use, which also carries a high risk of bloodborne disease transmission. And third, the increased social stigma and prejudice experienced by MTPs may cause them to have less contact with public health care agencies and providers.

You may be wondering what this has to do with your sexual health. To the extent that MTPs are infected with STIs, they are also potentially transmitting those infections to their customers. Moreover, many of their customers self-identify as heterosexual and have intimate heterosexual relationships; if these customers keep their activities with MTPs secret and continue to have unprotected sex with their female partners, they are putting their female partners at high risk for STIs. This is how the widespread STI infection among MTPs and other sex trade workers provides a vector for the spread of STIs throughout the general population. The diagram at the left illustrates the pervasiveness of the risk.

The risk is especially high for transvestite male prostitutes to spread STIs throughout the general, nonprostitute population.

SOURCES: Inciardi & Surratt (1997); Schepel (2011); Wellings (2012); Wiessing et al. (1999).

money to support their drug habit motivates them to agree to engage in high-risk sexual behaviors. The majority of crack-addicted prostitutes routinely agree to sexual acts without the use of a condom, usually for a higher price, and are especially likely to do so if business has been slow and they are increasingly desperate for drug money.

The widespread use of crack among street-based sex workers also plays a role in higher rates of STI transmission. The reason for this appears to be that the frequent smoking of crack damages the lining of the mouth and gums. Street prostitutes are more likely to engage in unprotected fellatio than vaginal or anal intercourse because of their belief that it carries a lower risk of disease transmission and because it is the most common unporotected activity requested by clients of sex workers (Adriaenssens & Hendrickx, 2012; Monto, 2001). However, the tissue damage in the mouth due to crack use may increase the risk for STI transmission more than is commonly believed (Wallace, Weiner, & Bloch, 1996; Weitzer, 2009).

BROTHEL WORKERS AND STIs Prostitutes who work in brothels, especially where they are legal, as in Nevada, or in countries where sex work has been decriminalized, appear to be at much less risk of STIs because they are less likely to be drug abusers and are much more likely to use condoms consistently (Rodgers, 2010; Wohlfeiler, & Kerndt, 2013). One study (Albert, Warner, & Hatcher, 1998) found that clients in Nevada brothels who were reluctant to use condoms (about 3% of all customers) ended up with one of three outcomes: They changed their mind and used them (72%), they chose a nonpenetrative sexual activity (12%), or they left "unfulfilled" (16%). If you do the math, that's 100%. With these kinds of statistics, we can assume that Nevada brothel workers' insistence on condom use is keeping both them and their clients better protected from STIs. Interestingly, however, the same study found that the brothel workers were far less likely to use condoms with their sexual partners *outside the brothel*. It has been speculated that not using a condom is one way that the women and their partners signal to each other the difference between a "real-life" personal relationship and a paid, professional one.

On the other hand, women who work in noncontrolled houses, such as massage parlors, "saunas," or strip clubs, may be at an even greater risk of STIs than streetwalkers (Ibbison, 2002). In these venues, customers are more likely to request unprotected vaginal or anal intercourse because the sexual encounters are less hurried than on the streets, and the prostitutes may agree more readily because they feel that repeat customers are more familiar and therefore present less risk. If there are no laws or house rules compelling them to use condoms and if prostitutes are able to charge significantly more money for these risky activities, they may also be more likely to comply. Consequently, these sex workers may be placing themselves at higher risk of some STIs than streetwalkers, even without the added risk factor of drug abuse.

You can see that the link between the commercial sex trades and STIs is extremely complex. The simple fact of working as prostitutes does not automatically imply that they will contract STIs, nor does it necessarily mean that they will pass them on to their customers. The transmission of STIs in the sex trades depends on the very same behaviors that spread the infections in all populations. The relatively recent recognition of this fact is assisting public health efforts to reduce the incidence of STIs among prostitutes and nonprostitutes worldwide, without necessarily singling out one particular group as any more "at fault" than another.

PROMOTING SEXUAL HEALTH AMONG SEX WORKERS During the 1980s and 1990s, as the HIV epidemic was expanding at an alarming rate, commercial sex workers (primarily streetwalkers and workers in unregulated brothels) were often the

In brothels where prostitution is legal, condom use is usually strictly enforced. The text under the "Open" sign on this brothel says: "Customers must use a latex condom during all sexual activities."

main focus of efforts to slow the spread of the deadly virus. Ibbison (2002) observed that the problem with those early approaches was that they generally took a "blame the prostitute" stance in which sex workers were seen as pools of HIV and were responsible for spreading it to the general population via their male customers. As HIV and other STI prevention strategies developed, they were originally not motivated by a concern for the sex workers themselves, but targeted at limiting the spread to the men and their partners. Today, public health and medical prevention strategies take a nonjudgmental and inclusive approach rather than one of stigmatizing and blame, leading to greater effectiveness in reducing the overall spread of HIV.

Recently the World Health Organization released some general guidelines for the prevention of the spread of STI among and by sex workers. They include (WHO, 2012):

- Enhanced strategies that encourage and promote the use of condoms
- Periodic, routine screening for STIs that often present no symptoms
- Greater availability of voluntary counseling and testing for STIs
- Greater access to antiretroviral therapy to help treat and reduce transmission of HIV
- Increased availability for STI for HPV, hepatitis B, and others as they are developed

When you read over these guidelines, you can see how they might apply to virtually any group of people whose behavior might place them at risk of STIs. These illustrate how prevention today is nonjudgmental and avoids stigmatizing and stereotyping.

The Clients of Prostitutes

We have given quite a lot of attention to people who work in the sex trades, but what about the men who buy their services? As you probably know, the clients of prostitutes are usually called johns or tricks. In most places where selling sex is illegal, buying it is against the law, too. Typically, soliciting the services of a prostitute is a misdemeanor and carries a possible sentence of thirty days in jail to years in prison. However, it is extremely rare that a male customer of a prostitute spends even one night in jail. Historically, prostitutes have been the ones who have been targeted by the criminal justice system with much more energy than their customers. The assumption behind these police tactics is that if the supply of prostitutes is reduced, the overall amount of illegal prostitution will decrease. However, this approach has proved to be nearly universally ineffective in areas where it has been attempted and now, with the added Internet connection, enforcement becomes even more challenging. If the demand for sex services remains constant, prostitutes will find a way to fill it. A crackdown on prostitution in one area of a city typically reduces sex work activity temporarily in that location as it moves to another neighborhood. After a while, the sex trade picks up once again in the previous location as police turn their attention elsewhere. Therefore, more attention is now being paid to punishing— and in some cases, educating—johns as a way of reducing prostitution through the demand side rather than the supply side (see "Sexuality, Ethics, and the Law: John School").

CHARACTERISTICS OF CLIENTS OF PROSTITUTES A study of more than 1,200 men arrested while attempting to buy the services of prostitutes (who, to the men's distress, turned out to be police decoys) in three major U.S. cities (San Francisco; Portland, Oregon; and Las Vegas) sheds a great deal of light on the johns' characteristics (Monto, 2001). The myth that prostitutes' customers are depraved, deviant, drug-addicted men who are social outcasts with no other sexual outlets was soundly rejected. In fact, most johns are employed, middle- to upper-class, educated, heterosexual, married males. In other words, these are "normal," average, mainstream men.

Sexuality, Ethics, and the Law:

"John School"

Some cities have established *diversion programs* for men who are arrested for soliciting prostitution. The idea behind this tactic is to require these men to attend what amounts to **john school**. "John school" is a generic term that is used to describe a wide range of programs that involve an education or treatment component. A useful working definition for john school is: "An education or treatment program for men arrested for soliciting illegal commercial sex. To that basic definition, one could add that in order for an education program to be considered a john school, it must cover a range of topics to discourage men from "buying sex." (Shively et al., 2012). Similar to the concept of traffic school, men arrested for solicitation may opt for the program in lieu of criminal sentencing. The school may be a one-day session to as many at ten meetings and costs, on average, about $400. As you may imagine, the focus of the intervention is to educate johns about the legal and health consequences of the behavior and the impact of prostitution on the girls and women who are enslaved and trafficked in the sex trades. This approach to curbing rates of sex trade activity is based on a list of assumptions about the clients of prostitutes. These include (Shively et al., 2012, p. 2):

1. The belief that the risk of arrest and legal sanction are low.
2. Denial or ignorance of the risk of contracting STDs or HIV through purchased sex.
3. Ignorance of the risk of being robbed or assaulted by prostitutes or pimps.
4. Denial or ignorance of the negative impact prostitution has on the neighborhoods in which it occurs.
5. Ignorance of the links between street prostitution and larger, organized systems of sex trafficking.
6. Denial or ignorance of what motivates them to solicit prostituted women or girls (e.g., addictions, compulsions, unmet social or sexual needs).
7. Denial or ignorance of the negative impact of prostitution on "providers."
8. Denial or ignorance of the fact that money is the only reason prostituted persons have sex with them.
9. The mistaken belief that the women they hire care about them, and that they are in some kind of relationship with them.
10. Denial or ignorance of the anger, revulsion, or indifference that many prostituted women have while they are having sex with johns.
11. Ignorance about how to have the healthy relationships that could replace their reliance upon commercial sex.

Do these programs work to reduce men's solicitation tendencies? The answer appears to be, "We don't really know, yet." Only a few studies have been undertaken to demonstrate the effectiveness of john schools. When a man is arrested for solicitation, the arrest and accompanying fine and embarrassment serve as powerful deterrents. It is as yet unclear whether the requirement of attending john school significantly adds to reducing recidivism rates. One study found that the rate of re-arrest was very small and not significantly different for men who attended the diversion program and those who did not (Monto & Garcia, 2001). Further research with larger samples of men will be necessary to determine with confidence if john schools offer a viable way to reduce sex trade activity.

Table 15.1 summarizes the demographic and behavioral findings for the men in the study. Looking them over, it is possible that if you were to randomly select 1,200 men from the same cities, their profiles would look fairly similar to the sample of johns, except that your random sample would probably have a higher percentage of gay and married men.

Twenty percent of these men claimed that this was the first time they had ever attempted to hire a prostitute, 59% reported having sex with a prostitute at least once over the past year, and 10% admitted to buying sex from prostitutes at least once a month during the past 12 months (Monto, 2001).

Since You Asked

6. What are the main reasons men go to prostitutes? Is it usually because they can't get sex any other way?

john school

a court mandated diversion program for men who have been arrested for soliciting prostitution.

What Do Prostitutes' Clients Want?

Many people believe that men seek out prostitutes for experiences that are strange, vastly different from the mainstream, or unlikely to be shared with their partner or wife. However, this does not appear to be the case. As you can see in Table 15.1 (on the following page), by far the most commonly purchased activity is fellatio. Fifty percent of customers of had engaged in oral sex with prostitutes, compared to vaginal intercourse (14%), both oral and vaginal sex (10%), and manual masturbation (6%) (Monto, 2009).

Table 15.1 Characteristics of Clients of Prostitutes in the United States

Characteristic	Percentage
Ethnicity	
White	58
Latino	20
Asian	13
Black	5
Other or combination	4
Level of education	
Less than high school	11
High school	18
Some college	36
Bachelor's degree	24
Graduate degree	11
Marital status	
Married	42
Never married	35
Divorced	15
Separated	6
widowed	2
Employment status	
Working full-time	81
Working part-time	7
Student	2
Other	11
Used a condom with prostitute	
Never	4
Seldom	3
Sometimes	8
Often	11
Always	74
Age at first encounter	
9–17	18
18–21	33
22–25	21
26–35	20
36–45	6
46+	2
Most common sexual activity with prostitute	
Oral sex (fellatio)	50
Vaginal intercourse	14
Oral and vaginal sex	10
Manual sex	6
Other	3
Two or more	17

SOURCE: Based on data from Monto (2009).

Another common belief is that clients of prostitutes are dissatisfied with their sexual life with their partner or prefer the sex that they have with prostitutes over partner sex. This also appears to be a myth. Only 18% of the men studied said they strongly or somewhat preferred sex with a prostitute over sex in a conventional relationship. Moreover, nearly 60% of the men surveyed disagreed that the sexual activities they paid for were different from the activities engaged in with their regular partners. Only 8% of the men in the study reported that their personal sexual interests were very different from those of their wife or partner. Other than specific sexual activities, what other psychological or emotional motives lead men to seek out prostitutes? One reason often cited is that paying for sex allows men to experience a brief, uncomplicated sexual encounter and to avoid any emotional or relationship responsibilities that typically accompany sex with a regular partner. In other words, sex with a prostitute is "easier," sexually *and* psychologically. Other common reasons offered by the johns themselves include loneliness, sexual problems at home, curiosity about sex with various women, desire for sexual behaviors refused by the primary sexual partner, lack of sex overall, and satisfaction of an overly strong sex drive (Lowman, Atchinson, & Fraser, 1997; Pitts et al., 2004; Weitzer, 2009).

Whatever the true reasons behind people becoming sex workers and men paying them for sexual services, the fact remains that the "contract" that is agreed to and carried out between the sex worker and the client remains a major law enforcement issue in the United States and many other countries. Many people believe that enforcing laws against adults charging and paying for consensual sex is a futile exercise and a waste of tax revenues and police time and effort. This has led some people to question criminal approaches to reducing the prevalence of prostitution.

Prostitution and the Law

On occasion, in the early history of the United States (and in some other countries), prostitution was legalized in some situations. These instances usually occurred in wartime, such as during the Civil War in the United States or World War II in Japan, when the "welfare" of an all-male military was of great concern to governments. To help soldiers avoid contracting sexually transmitted infections (then called *venereal diseases*), which, of course, would reduce their effectiveness on the battlefield, women were paid to service these men sexually and were checked regularly for any signs of infection. Today, the evidence is clear that these practices virtually always involved filthy living conditions, rape, and other inhumane treatment of the women who were forced into this degrading service; they were *not* volunteering their "services to their country."

In Amsterdam, in the Netherlands, where prostitution is legal, fewer of the criminal, health, and human trafficking problems exist than in countries (including the United States) where prostitution is illegal.

Today, as mentioned earlier in this chapter, prostitution is illegal and carries criminal penalties in most countries and in most states in the United States. The exact laws vary considerably from state to state and country to country in terms of seriousness of the crime, punishments for the sex workers and their customers, and the exact definition of prostitution. Nevertheless, prostitutes and their customers can be, and are, arrested for this activity. Exceptions exist to this illegal status, as exemplified by the brothel business in Nevada (discussed earlier) and legalized prostitution in countries such as Germany and the Netherlands. One of the many heated controversies in the field of human sexuality and law enforcement is whether prostitution should remain a criminal offense or be *legalized* or *decriminalized*. You may be wondering what the difference is between legalization and decriminalization. To those in the sex trades, the difference is an important one.

In simple terms, the result of **legalization of prostitution** would be similar to the current system of brothels in Nevada, where prostitution is limited and controlled by the state and violation of state limits and controls, such as health, safety, and financial statutes, may be criminal offenses. **Decriminalization of prostitution** would remove all criminal laws relating to prostitution and equate the sex trade with any other commerce that a person chooses to pursue, as long as it adheres to the same laws governing all legal business practices. People arguing for legalizing or decriminalizing sex work contend that it is not the morality or immorality but the *illegality* of prostitution that causes the drug abuse, violence, and STI spread related to street-level sex work.

legalization of prostitution
Regulation of prostitution by state laws, with statutes defining where, when, and how prostitution may take place.

decriminalization of prostitution
Repeal of all laws against consensual adult sexual activity in both commercial and noncommercial contexts.

Those in favor of the decriminalization or legalization of prostitution (e.g., C. Lee, 2005) contend that doing so would create the following positive outcomes:

- Reduce or eliminate the rampant victimization of prostitutes by pimps, johns, and the other criminal elements now associated with prostitution.
- Help control STIs (among prostitutes and the general public) by requiring sex workers to obtain regular checkups and health certificates.
- Remove many prostitutes from street work and allow them to work in a safe, controlled environment.
- Redirect more of our limited law enforcement resources into fighting other, more serious crimes.
- Eliminate the profit motive in human trafficking and much of the motivation for pimps to addict young women (and men) to drugs in order to force them into prostitution.
- Eradicate a significant source of income for criminal networks.
- Allow governments to receive tax benefits from the regulation of sex trade businesses.

Although some of these benefits have been partially realized in Nevada and in those countries where prostitution is legal, the other side of the debate contends that legal prostitution creates the following negative outcomes (Farley, 2004; Farley & Kelly, 2000; Hodge, 2008; Otchet, 1998):

- Has increased trafficking and victimization of children because illegal sex trade activities can be concealed behind the legal ones.
- Is predominantly violent, not truly consensual, and invariably victimizes, exploits, and injures girls and women in countless ways.
- Forces women to prostitute themselves as a last resort for economic survival.
- Leads to physical illness (such as multiple STIs and drug addiction) as well as psychological harm (especially posttraumatic stress disorder).
- Is, in reality, a form of "paid rape."
- Is fundamentally immoral and removing it from our criminal laws would be a move in the wrong direction for society and could lead to a rampant sex industry in the United States, victimizing even more girls and women.

The debate over the legal status of prostitution in the United States and elsewhere in the world is often extremely contentious. The chance of calming these waters is unlikely in the near future. In the meantime, owing in large part to the level of emotion generated by the debate, politicians completely avoid the issue whenever they can. One exception is in Texas, which has enforced some of the harshest prostitution laws in the country. Legislators there are considering measures that would reduce criminal penalties for prostitution from a felony to a misdemeanor and provide for counseling and treatment for offenders instead of jail time. The reasoning behind this is that fines and incarceration do little to curb prostitution and only serve to further victimize sex workers who are already victims of the industry (Forsyth, 2013.)

To sum up, studying prostitution, or the commercial sex trades, is a challenging task. We have discussed the difficulties of studying various sexual issues elsewhere in this book; just imagine the added burden of research into areas of sexual behavior that are illegal. It is no small challenge. Whether or not it is the world's oldest profession is far less important than the fact that it may be one of the world's *least understood* professions. It is safe to say that people will continue attempting to study it, debate it, pass laws about it, work in it, and patronize it, regardless of whether or not we ever fully understand it.

Prostitution is connected to pornography, our next topic, in a way that surprises many people. The root of the word *pornography* means "writing about prostitutes." It derives from the Greek words *porne*, meaning "prostitute" or "harlot," and *graphein*, meaning "to write." Of course, as you will see, our current use of the word means substantially more than this formal etymological definition.

Pornography

15.2 Summarize the various aspect of pornography including definitions; distinctions between pornography and erotica; the research on the effects of pornography; and discuss the global problem of child pornography.

pornography
In legal terms, any sexually explicit work deemed obscene according to legal criteria and therefore exempt from freedom of speech protections.

As a lead-in to class discussions on the topic of **pornography**, I ask students in my human sexuality classes to write down, anonymously, on 3-by-5 cards, the answer to the following question: "What is the most pornographic thing you've ever seen?" The cards are then collected and read aloud. The purpose of this exercise is threefold. First, because no further information is given with the question, it reveals how most people, at least most of the students in the classes, define pornography. Second, the range of answers given provides an excellent example of the variety of what is commonly perceived of as pornography. And third, it reminds the class of the diversity of personal experience with pornography that exists among their peers.

If students were each asked to write their own definition of pornography, the number of different definitions would probably equal the number of students in the class. Why? Because each of you, if asked to think critically about it, would likely draw the line differently between acceptable sexually explicit materials and those that you would call pornographic. Is *Playboy* magazine pornographic? How about R-rated movies (such as 2014's *Fifty Shades of Gray*, or *About Last Night*)? Many of you would disagree, perhaps forcefully, about this, primarily because you would disagree about the definition of the term *pornography*. Our first task, then, is to determine what we mean when we talk about this vague abstraction we call pornography.

Searching for a Definition

One approach to a definition for pornography might be to turn to the dictionary. General dictionaries are not particularly useful for scientific definitions; that's not their primary purpose. However, for an idea of how a word is *commonly* used, they can, at times, be helpful. The problem here is that if you were to look up *pornography*, you would find yourself in one of those "dictionary scavenger hunts," looking up more and more words trying to understand the words used in each of the previous definitions. For example, your dictionary scavenger hunt might go something like this (feel free to try this yourself):

Pornography: 1. *Written or graphic material intended to excite lascivious feeling.*

Lascivious: 1. *Lewd*. 2. *Lecherous*.

Lewd: 1. *Licentious*. 2. *Lustful*. 3. *Obscene*.

Lecherous: *Given to inordinate sexual indulgence* (Inordinate: *Exceeding reasonable limits;* Indulgence: *Going unpunished for sinful acts*).

Lustful: 1. *Excited by lust*. 2. *Lecherous* (again).

Lust: 1. *Intense or unbridled sexual desire* (Unbridled: *Freed from all restraint*). 2. *Lasciviousness*.

Licentious: 1. *Lacking moral discipline or sexual restraint*. 2. *Lewd* (again).

Obscene: 1. *Disgusting to the senses*. 2. *Containing language regarded as taboo in polite usage*. 3. *Repulsive by reason of a crass disregard of moral or ethical principles*. 4. *Offensive to accepted standards of decency*.

You see the problem here? Every underlined word or phrase in the definition above is "loaded" or subjective, meaning each person would have a unique definition of it, as they would for "pornography." Clearly, pornography is not an easy word for anyone to define. However, our dictionary perseverance finally paid off somewhat in the last part of the exercise, as the word *obscene* comes closest to capturing the most formal (and legal) definition of pornography developed thus far. It is called the "moral standard" and was not devised by sexuality researchers or psychologists but by judges sitting on the U.S. Supreme Court. This definition does not apply to all sexually explicit materials but is designed as a legal "test" for whether a *specific* sexually explicit work can be considered obscene and, consequently, censored.

THE MORAL STANDARD Throughout the history of the United States, a tension has existed between the First Amendment to the Constitution and citizens' rights to sell, distribute, look at, or otherwise participate in sexually explicit materials. The exact wording of the first amendment regarding our discussion is, "Congress shall make no law...abridging the freedom of speech, or of the press." The drafters of the Constitution saw this right as the most basic and most fundamental, along with freedom of religion, assembly, and the right to petition the government for change.

Virtually all legal scholars agree that, from the perspective of basic constitutional rights in the United States, nothing is more inviolate than the constitutional right to freedom of speech. However, several well-established exceptions exist to this basic right, including the defamation of another's character; speech that causes panic, such as shouting "Fire!" in a crowded building; treasonous speech, such as advocating the overthrow of the government; and most pertinent to our discussion, **obscenity**, which has been applied to certain sexually explicit works.

obscenity
Sexually explicit works that meet specific legal criteria that render them exempt from First Amendment protections and may be declared illegal.

Tension is bound to be created when something of a sexual nature, which some people may find offensive and immoral, is produced for public consumption. Those who produce the work and those who want access to it will invoke their First Amendment protections, as those who want to censor it and make it illegal will claim that the First Amendment does not protect, and was never intended to protect, that particular form of speech because, in their interpretation, it is *obscene*. The problem lies in determining who should or can decide what works are to be labeled obscene and allowed to be censored. Historically, where do you suppose these disagreements over sexually explicit materials have gone for legal resolution? That's right: to the courts.

Since You Asked

7. I like to look at magazines like *Playboy* and *Hustler*, but my girlfriend says they are pornographic trash. I think pornography refers to much worse kinds of magazines and videos. Who is right?

Typically, as these cases went through the lower courts, the losing side would appeal the decision. Many of those appealed cases were subsequently appealed again until they reached the U.S. Supreme Court, whose job is to interpret the legal meaning of the Constitution—in this case, the limits of the First Amendment. As hundreds of obscenity cases relating to sexually explicit materials were heard and decided by the highest court over many years, the difficulty of this task became increasingly clear. How could one panel of nine justices, representing an entire nation, determine whether a particular sexually explicit work should be denied free-speech protection in one town or city when citizens in another town or city might have an entirely different set of standards? The answer is, it couldn't.

It was not that they didn't try. Here is how three past justices' definitions have been described (Woodward & Armstrong, 2005):

> [Justice Byron] White: *No erect penises, no intercourse, no oral or anal sodomy*. For White, no erections and no insertions equaled no obscenity... [Justice William J.] Brennan, like White, had his own private definition of pornography: *No erections*. He was willing to accept penetration as long as the pictures passed what his clerks referred to as the "limp dick" standard. Oral sex was tolerable if there was no erection... [Justice Potter] Stewart conceded the subjective nature of any definition: "I shall not today attempt to further define the kind of materials I understand to be [hard core pornography]...*but I know it when I see it*..." He had seen it during World War II when he served as a Navy lieutenant. In Casablanca, as watch officer for his ship, he had seen his men bring back locally produced pornography. From this experience, he believed he knew the difference between true "hard core" materials and other, less extreme materials that came to the Court. He called it his "Casablanca Test." (pp. 232–234, emphases added)

Even among themselves, the Supreme Court justices could not agree on what was obscene and should be censored. Then, in 1973, in a landmark case known as *Miller v. California*, the Court took a new approach. The case involved a mass mailing of sexually explicit brochures that were described in the Court's decision as follows:

> The brochures advertise four books entitled "Intercourse," "Man-Woman," "Sex Orgies Illustrated," and "An Illustrated History of Pornography," and a film entitled "Marital Intercourse." While the brochures contain some descriptive printed material, primarily they consist of pictures and drawings very explicitly depicting men and women in groups of two or more engaging in a variety of sexual activities, with genitals often prominently displayed. (Burger, 1973, p. 18)

In their opinion, the justices abandoned their long-standing attempts to separate obscene from nonobscene sexually oriented materials. Instead, they developed a procedure for determining if a work is obscene that could be applied across the nation, but was, by its very nature, *not* intended to yield consistent results. That may sound strange, but it has been the law of the land for over a third of a century. The Court provided three "basic criteria," all of which must be met before a work may be deemed to be obscene and legally banned:

1. Whether the work depicts or describes, in a patently offensive way, sexual conduct specifically defined by the applicable state law.
2. Whether the work, taken as whole, lacks serious literary, artistic, political, or scientific value.
3. Whether an average person, applying contemporary community standards, would find that the work, taken as a whole, appeals to a **prurient interest** in sex.

The Supreme Court, led by Chief Justice Warren Burger, drafted the 1973 definition of obscenity, which continues to guide legal action on sexually explicit materials today.

prurient interest
An excessive focus on exclusively sexual matters.

Remember, in applying this standard, it is necessary that *all three* prongs of what has become known in legal circles as the *Miller* test must be present for a ruling of obscenity. In essence, these standards attempt to provide guidance for judging the morality of a sexually explicit work.

APPLYING THE MORAL STANDARD As you read through the three-pronged test of obscenity, you may have been thinking that it fixes nothing; it is still full of *loaded*, subjective wording that is open to individual interpretation: "offensive," "serious value," "community standards," "prurient interest." How, exactly, did this opinion by the Supreme Court fix the problem? The court attempted to do so by developing these standards not for the Supreme Court itself, but to guide *local* governments and jurisdictions in resolving disputes before they ever leave the community in which the complaint originated. As Chief Justice Burger (1973, p. 32) noted in his opinion:

> It is neither realistic nor constitutionally sound to read the First Amendment as requiring that the people of Maine or Mississippi accept public depiction of conduct found tolerable in Las Vegas or New York City.... People in different States vary in their tastes and attitudes, and this diversity is not to be strangled by the absolutism of imposed uniformity.

In essence, what the Court was saying was, "Please stop bringing these cases to us! We believe that no single, national standard of morality, as it relates to sexually explicit materials, can be determined or defended. You must decide for yourself, *in your own county or city*, whether sexual works are obscene or not."

This legal standard for defining *obscene* sexual materials has not been without its critics over the past 40 years. Many of these critics' words echo Justice William O. Douglas (1973, p. 44) in his dissenting opinion in the *Miller* case: "The idea that the First Amendment permits punishment for ideas that are 'offensive' to the particular judge or jury sitting in judgment is astounding." This is the same criticism discussed earlier: that the language of the moral standard is too vague and subjective to allow any court, at any level, to find a work obscene while protecting the people's First Amendment rights.

Another criticism frequently raised about this moral standard relates not to whether the criteria are proper for defining obscenity but rather to the idea that the entire foundation on which the standard is based is misplaced. These critics maintain that the legal right to censor a sexually explicit work rests not on its morality (which is unique to each individual) but on the extent to which the work *discriminates* against a particular group of people—in this case, women.

Catherine MacKinnon contends that pornography is a weapon that discriminates against women and justifies a culture of rape.

THE DISCRIMINATION STANDARD A suggested *alternative* approach for dealing with the legality of sexually explicit materials is to view it from the perspective of *discrimination* rather than morality. That is, under this view, a sexually explicit work could be legally banned if it is proved to discriminate against and victimize a particular group of people. Most of what is currently referred to as "pornography" is intended for male audiences, and most of it portrays women merely as sexual objects or "sexual servants" whose sole purpose is to provide sexual outlets for men, no matter what the women themselves may want. Some pornographic material depicts humiliation and violence toward women in sexual contexts, glorifies rape, and may even show women being killed at the hands of their male oppressors. The discrimination standard for obscenity maintains that if these works victimize women in the making of the materials, in the viewing of them by men, or in the potential subsequent victimization of women by men who view them, they are obscene and should be censored.

Catherine MacKinnon, at the University of Michigan, has advocated the following basic ideas of this approach (MacKinnon, 1996; 2000; Richards, 2003):

1. Pornography, by its vary nature, is discrimination against women. The distinction between sexually explicit materials and the actions they depict is meaningless.
2. Pornography reflects and constitutes male domination and dehumanization of women.
3. Pornography objectifies women by portraying women as nothing more than tools of male pleasure.
4. Pornography represents men's power over women in society and embodies the dehumanization of women in an already male-dominated world.
5. Pornography promotes discrimination against women, which in turn advocates and justifies rape.

This approach to defining obscenity has not yet made its way into the legal statutes governing pornography in the United States. However, it does provide a provocative perspective for examining this highly charged issue. Moreover, this perspective raises the question of whether extremely sexually explicit materials, such as those often referred to as pornography, can ever be seen as *not* discriminating against women. One argument in support of this possibility concerns the distinction between pornography and **erotica**.

erotica
Sexually explicit works expressing physical desire, passion, and attraction among people who freely and equally choose to engage in sexual activities together.

Pornography or Erotica? Although people will never agree about issues relating to pornography and obscenity, most adults will allow that some sexually explicit materials, including those designed with the clear intent to be sexually arousing, are more acceptable than others and may even have desirable or beneficial effects. In Chapter 7, "Sexual Problems and Solutions," we discussed how sexually oriented materials can be useful in the treatment of certain sexual disorders in couples and individuals. However, if some people feel that pornography is obscene or that it discriminates against or victimizes women, how can we reconcile that belief with sexual materials that have positive effects?

One way is by drawing a distinction between *pornography* and *erotica.* Some people may argue that this is simply a matter of semantics, and "one person's erotica is another's pornography." However, most researchers in this area see a clear difference between the two. One researcher summarized the various distinctions that have been suggested as follows (Levinson, 2005, p. 229):

1. The erotic and the pornographic are both concerned with sexual stimulation or arousal.
2. While the term "erotic" is neutral or even approving, the term "pornographic" is pejorative or disapproving.

Since You Asked

8. Everybody's always saying how bad pornography is because it is violent toward women. But I've seen a lot of XXX videos that don't have any violence or anything bad against women in them at all. So is that not pornography?

3. While "erotic art" is a familiar, if somewhat problematic, notion, "pornographic art" seems an almost oxymoronic one.
4. Whereas pornography has a paramount aim, namely, the sexual satisfaction of the viewer, erotic art, even if it also aims at sexual satisfaction on some level, includes other aims of significance.
5. Whereas we *appreciate* (or relish) erotic art, we *consume* (or use) pornography. In other words, our interactions with erotic art and pornography are fundamentally different in character, as reflected in the verbs most appropriate to the respective engagements.

In others words, erotica is about *sexual sharing*, whereas pornography is about *sexual using*.

In the early 1980s, Gloria Steinem, who has been a leading figure in the women's movement for the past 40 years, was among the first to suggest that pornography and erotica are fundamentally different.

A cogent statement on the distinction between erotica and pornography was published in 1980 in an essay by Gloria Steinem, one of the most influential and articulate founders of the women's movement. Her essay, "Erotica and Pornography: A Clear and Present Difference," describes the differences between these two forms of sexual content as follows:

> Sex as communication can send messages as different as life and death; even the origins of "erotica" and "pornography" reflect that fact. After all, "erotica" is rooted in "eros" or passionate love, and thus in the idea of positive choice, free will, the yearning for a particular person. "Pornography" begins with a root "porno," meaning "prostitution" or "female captives," thus letting us know that the subject is not mutual love, or love at all, but domination and violence against women. The message of pornography is one of violence, male dominance and conquest, of sex being used as a tool or weapon to create an inequality between men and women.... Erotica celebrates the equality of persons sharing the positive aura of sexuality and mutual pleasure. Erotica stresses freedom of choice. (Steinem, 1980, p. 50)

In the final analysis, everyone must decide individually where to draw the lines between obscenity, pornography, and erotica, and some people may feel that such lines are unnecessary because they feel that virtually all sexually explicit materials are morally equal, whether good, bad, or in between. However, if consenting adults are to be allowed the freedom to view, read, or otherwise use sexual materials, especially if the materials' effects on people and society and culture are to be studied, such a distinction may be inescapable.

The Effects of Pornography

Beyond the issues involved in defining pornography, human sexuality researchers and other social scientists are interested in the effects of sexually explicit materials on the behaviors, attitudes, and feelings of the individuals who view them. Over the many decades of research in this area, several questions have been of primary concern: What is the relationship between sexually explicit materials and sexual arousal for men and women? Can sexually explicit materials enhance the human sexual experience? Does pornography contribute to prejudice and discrimination against women? Does pornography lead to greater violence against women? Does pornography create attitudes conducive to rape? Let's examine what the research into pornography has found in attempting to answer these questions.

PORNOGRAPHY AND SEXUAL AROUSAL Does reading or viewing sexually explicit materials turn you on? Most of you probably answered with something along the lines of, "It depends on the materials." In spite of all of the inherent flaws, problems, and controversies that we will discuss in the next sections, the fact remains that pornography and erotica exist primarily for the purpose of sexually arousing people. If no one found these materials sexually exciting, there would be little reason for their existence.

Most researchers today draw a distinction between *pornography*, in which women are used and victimized by men, and *erotica*, such as shown here, which portrays mutual desire and equality in sexual activities, even though the sexual content may be equally explicit.

Exactly what types of sexually explicit images or words arouse each individual is as varied as the real-life sexual activities people find exciting. You might be aroused by reading a certain passage in a sexually explicit book, seeing a sexy R-rated movie, or watching a specific pornographic video that another person might find repugnant. Interestingly, however, research has demonstrated that even when people are offended by sexual images, some sexual arousal may occur (i.e., Davis & McCormick, 1997; Laan et al., 1994). This is not surprising, in that our sexual responses, as discussed in Chapter 3, are, in part, reflexive and not necessarily under our conscious control at all times. Also, the fact that something arouses a person sexually does not necessarily imply that the person likes the material or wants to imitate the behaviors involved (more on this in a moment).

Of course, people are more sexually aroused by materials they find acceptable, pleasing to look at or read, and not offensive to their personal moral values (e.g., Scotta & Cortez, 2011). For example, people who tend to be religiously and politically conservative and have a strong authoritarian belief system feel more guilt and react more negatively to sexually explicit materials in general (Nelson, Padilla-Walker, & Carroll, 2010). Also, studies have found that men who score lower in intelligence and who exhibit greater antisocial and aggressive tendencies prefer more violent forms of pornography (Bogaert, 2001).

Interestingly, although men and women differ in some ways in their arousal reactions to sexually explicit materials, they are also similar in many ways (Scotta & Cortez, 2011). Not surprisingly, some studies have found that upon viewing pornographic videos depicting themes degrading women, men reported more arousal and rated the videos as more acceptable and less degrading compared to the ratings of women viewers (e.g., Gardos & Mosher, 1999). However, acceptance and enjoyment ratings decreased significantly as the level of degrading content increased. Another study compared pornographic videos made by men for men (explicit sex scenes only) with videos designed by women for women (explicit sex with more plot and partner interaction). The men in the study rated all the videos more positively than did the women, but as you might expect, the women rated the videos designed for female audiences significantly more positively than those made by men (Mosher & MacIan, 1994). In a similar vein, a study compared women's responses to sexually explicit videos designed for women or for couples versus standard videos aimed at male audiences (Pearson & Pollack, 1997). The women viewers rated their level of sexual arousal higher in response to the couples video than to the traditional male-oriented video.

On the other hand, some studies have failed to find this male–female difference. One study showed women erotic films made by women and erotic films made by men and measured sexual arousal using the vaginal plethysmographic device explained in Chapter 1 (Laan et al., 1994). Although sexual arousal was substantial, the level of physiological excitement did not differ between the two types of films. However, the women *reported* feeling more sexual arousal with the woman-made film and more feelings of shame, guilt, and aversion for the male-centered film. The contradictory findings among these studies probably reflect the fact that sexual feelings and arousal are complex reactions that involve both physical and psychological processes. Our bodies may react to sexual stimuli that emotionally and cognitively we may not interpret as particularly arousing.

Table 15.2 Sexual Arousal in Response to Explicit Videos with and Without Romantic Themes

	Romantic Content in Video*	
Sexual Arousal	Low	High
Males	3.22	4.07
Females	3.11	4.20

*Larger numbers indicate greater sexual arousal.

SOURCE: Based on data from Quackenbush, Strassberg, & Turner (1995).

Before we condemn all men as emotionless, indiscriminate pornography lovers, some studies have uncovered a more romantic side in men's pornography preferences, too. (see Table 15.2). For example, one study showed college men and women highly explicit sexual videos with either a romantic theme or a nonromantic theme (Quackenbush, Strassberg, & Turner, 1995). Findings revealed that both men and women rated the more highly explicit video that contained the high romantic story as significantly more arousing than the

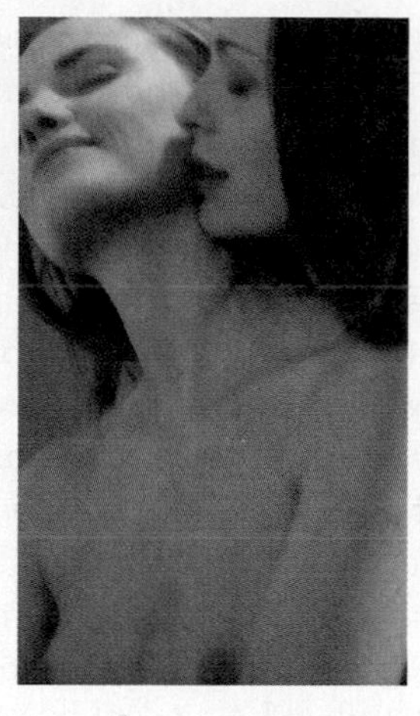

Research has shown that women tend to be sexually aroused by explicit sexual scenes of all sexual orientations, but men are primarily aroused by depictions of their personal sexual orientation.

highly explicit video with the less romantic story. This was opposite to the researchers' predictions.

What about the effects of viewing pornography on relationships? Most of you are aware that the use of pornography can cause various problems in an intimate relationship. When one partner (usually the man in heterosexual couples) views pornography alone on a regular basis, a wall is sometimes created between the partners, causing the other to feel angry and betrayed if and when the behavior is discovered (Manning, 2006). With the easy access to pornography via the Internet today, this is becoming a common cause of relationship conflicts and even breakups.

However, some studies have found that when sexually explicit materials are incorporated into a relationship, the effect may be positive. Research suggests that when couples view pornography together, rather than alone, relationship satisfaction and intimacy may be enhanced and couples report feeling closer (Maddox, Rhoades, & Markman, 2011; Manning, 2006). This assumes, however, that the materials viewed are enjoyable for both partners.

Although both women and men are sexually aroused by various sexually explicit materials, some forms of pornography may have harmful and even dangerous effects. Most of the concerns over pornography relate to negative, degrading, demeaning, and in some cases, violent portrayals of women that exists in the majority of sexually explicit materials on the market.

PORNOGRAPHY AND DISCRIMINATION AGAINST WOMEN As you are undoubtedly aware, many people believe that the availability of pornography plays a significant role in discrimination against and victimization of women in society. This is one of the central arguments in the frequent calls throughout the United States to censor pornography. Others disagree, contending that adults in a free society should be able to view whatever sexually explicit materials they choose, and the mere act of viewing pornography is generally unrelated to real-life violence toward and victimization of women (see Stans, 2007). This side of the argument often points out that many men who victimize women are not pornography users, and, conversely, most pornography users do not victimize women.

However, public opinion on either side of the debate is, by definition, not science, and it is often based on purely emotional arguments or myths that may bear little resemblance to the "truth." In the 1970s and 1980s, two U.S. presidential commissions were established to study the meaning and effects of pornography. The first, formed by President Lyndon Johnson and called "The President's Commission on Obscenity and Pornography," published its findings in 1970 ("The Report," 1970). Fifteen years later, President Ronald Reagan ordered his attorney general, Edwin Meese, to form what became "The Attorney General's Commission on Pornography" (often referred to as "The Meese Report"), which released its findings (all 2,000 pages of them) in 1986. Ultimately, these two highly publicized reports have only served to add fuel to the public debate about the effects of pornography because, wouldn't you know, they reached completely opposite conclusions.

The Johnson Commission (which included trained social science researchers) reported that the available evidence was too weak to support a clear link between pornography and discrimination against women, and recommended educating children about healthy sexual attitudes and behaviors and restricting their access to pornography, but removing all such restrictions for adults. The Meese Commission, on the other hand (which was comprised primarily of politicians, most of whom already held antipornography positions) concluded that pornography is dangerous on many levels, increases discrimination and violence against women, and helps to fund organized crime. Many social scientists and others criticized the report as unscientific, biased, and distorted, or said it ignored research findings that found minimal negative effects of pornography (see Wilcox, 1987).

This controversy over the effects of pornography may never lead to a definitive conclusion. However, if we examine the scientific evidence, we will at least have the tools to develop an informed, knowledgeable opinion for ourselves. That evidence will be summarized next.

One researcher has suggested that the basic themes of most pornography are as follows:

1. All women at all times want sex from all men.
2. Women enjoy all the sexual acts that men perform or demand.
3. Any woman who does not at first realize this can be easily turned with a little force, though force is rarely necessary because most of the women in pornography are the imagined "nymphomaniacs" about whom many men fantasize (Jensen, 2004).

Although some romantic and intimate themes may be found, most video pornography contains material that is objectifying, dehumanizing, or violent toward women and devoid of intimacy (Jensen, 2004; Monk-Turner & Purcell, 1999). In large part, pornography demeans and/or victimizes women and portrays them as objects whose only reason for existence is to be used by men, purely for the men's sexual pleasure (Davis & McCormick, 1997; Vega & Malamuth, 2007). The majority of pornographic materials depict men as having complete power over women; contain scenes that perpetuate rape myths (such as women enjoy being raped and women are responsible for being raped); or include scenes of overt sexual aggression, physical violence, or rape (Bridges et al., 2010; Davis & McCormick, 1997).

Depicting and encouraging violence against a particular group of people—in this case, women—is an indisputable example of discrimination. Now we must examine the specific forms this discrimination takes.

PORNOGRAPHY, VIOLENCE, AND RAPE To the extent that pornography contains depictions of victimization of and violence toward women, many people are convinced that pornography, including sexually explicit materials that are becoming more mainstream and available through DVDs and Internet-based providers, also play a role in causing real-world rape and other forms of aggression against women (see Scully, 2013). Many studies have attempted to demonstrate such a connection (e.g., MacKinnon, 2005; Malamuth, Hald, & Koss, 2012; Malamuth & Huppin, 2005; Vega & Malamuth, 2007). All of these investigations employed either correlational or experimental laboratory methods, both of which contain important limitations in what we may conclude from the findings. (These research limitations are discussed in greater detail in "Evaluating Sexual Research: The Lab Experiment–Correlation Trade-Off.")

correlational research
a research method that reveals how two observations (variables) are related to each other (how they co-relate), but does not determine if one variable is causing the other.

The main problem in this area of social-scientific research is the leap from research findings to real-world applications. As yet, no solid scientific research exists that

Evaluating Sexual Research

The Lab Experiment–Correlation Trade-Off

Research on the connection between pornography and violence against women provides an ideal example of the perplexing methodological difficulties often encountered in studying weighty social issues. Fundamentally, researchers only have two methods of study at their disposal: correlational research and experiments. Both are legitimate forms of scientific inquiry, yet either may lead us to erroneous conclusions.

As we have discussed elsewhere in this text, **correlational research** draws on existing data about the subjects being studied and attempts to find meaningful relationships among those data. To examine pornography and violence against women using correlational techniques, we might, for example, look for a link between convicted sex offenders and their exposure to pornography compared to a similar group of nonoffenders, or perhaps we could measure men's attitudes toward rape and their experiences with pornography to see if a predictable association exists. However, we must approach the findings that come from such studies with caution. Why? Because correlation does not imply causation. That is, the fact that sexually explicit materials may be *associated* with negative attitudes or violent behavior does not allow us to assume that the pornography *caused* the attitudes or behavior. Perhaps the causal link is in the entirely opposite direction: Men with more callous attitudes toward women may be drawn to pornography. Or maybe men who have more aggressive personalities are more likely to be violent toward women *and* read violent pornography. No matter how logical it may seem to make a causal connection, this is true for virtually any correlational finding. Results based on correlations *may* reflect a causal relationship between the two variables, but we cannot simply assume causation from a correlational finding.

If we wish to determine with some degree of confidence that one variable does indeed cause another, we would need to conduct an **experiment**. In an experiment, the researchers control what happens to the participants and the setting in which it happens. This is completely different from a correlational study, in which no control whatsoever is possible because the data are already "attached to" the participants. When we conduct an experiment, the data are gathered under controlled conditions, so we may draw cause-and-effect conclusions. An example would be to see if the effect of watching pornographic videos causes men to develop hostile attitudes or violent attitudes toward women. In such an experiment, male participants would be divided randomly into two or more groups and shown different pornographic videos over a specified amount of time. One group would be shown, say, violent pornography; another, nonviolent pornography; and perhaps another, nonpornographic videos. All other potential influences would be held constant for all the groups (such as sex of the researcher, video-watching environment, and video production quality). Then the attitudes of men toward women from all three groups would be measured. If the group exposed to the violent videos displayed significantly more negative attitudes or behaviors toward women, we could then conclude with reasonable certainty that the pornography was indeed the *cause* of the negative attitudes.

Although we may have now revealed a cause-and-effect relationship between pornography and negative reactions toward women, we cannot automatically assume that our findings can be applied to the real world outside the controlled laboratory setting. The experimental setting was contrived by the experimenters and may not have accurately resembled how pornography is used in the actual lives of men who are exposed to it. We are left with a research conundrum: If we examine data from the real world, we cannot conclude cause and effect, but if we exert scientific control to reveal causal relationships, we give up some accuracy in applying our findings to the real world. What's a scientist to do?

Unfortunately, there is no ideal answer to that question. Obviously, we cannot go out and do a "real-world experiment" on the connection between pornography and violence (we would have to control people's real-life exposure to pornography and "allow" violence; we cannot do that). All we can do is try to control for as many extraneous variables in our correlational research as possible and try to make our experimental laboratory conditions as realistic as possible. And always, when we engage in or read about any research, we must be skeptical, think critically, and keep our minds open to other interpretations of the findings.

proves a cause-and-effect relationship between viewing pornography and real-world violence, sexual assault, or rape of women. However, this is not to say such a causal relationship does *not* exist; as you can imagine, definitive research to demonstrate this connection would be difficult, if not impossible, to conduct for both methodological and ethical reasons (we cannot do a study to see if men who view pornography will actually commit rape).

Most of the research in this area has drawn a distinction between pornography that portrays violence or degradation toward women, and pornography that is equally explicit but without an overt degrading or violent component. Table 15.3 offers a wider view by summarizing a number of studies carried out since the early 1980s on the effects of pornography on violence and rape.

Since You Asked

9. Can looking at pornography cause guys to want to do weird or sick things sexually?

Experiment

a scientific research method that examines the causal link between two variables by controlling for all other extraneous variables

Table 15.3 Pornography and Aggression Toward Women: What the Research Has Shown (Representative Sample of Research)

Study	Method	Main Findings
	Correlational Studies	
Garcia (1986)	Gathered data from 115 undergraduate male college students.	Subjects having greater exposure to violent sexual materials were found to express more negative attitudes toward women in the area of sexual behavior.
Smith & Hand (1987)	Presented a pornographic movie on a college campus; studied partner aggression reported by 230 college women during the week prior to and after the movie.	No differences in reported aggression were found before versus after the showing or between males who viewed the movie and those who did not.
Demare, Briere, & Lips (1988)	Surveyed 222 male undergraduates on pornography use, attitudes toward women, and self-reported likelihood of rape or use of sexual violence.	Use of sexually violent pornography combined with acceptance of interpersonal violence against women predicted greater self-reported likelihood of rape and sexual violence.
Padgett, Brislin-Slutz, & Neal (1989)	Assessed attitudes toward women of college students and patrons of X-rated theaters.	Hours of viewing pornography did not predict attitudes toward women for either subject group.
Vega & Malamuth (2007)	At the University of California, Los Angeles, 102 male college students were given wide-ranging sexual attitudes, general hostility, and aggression scales, along with pornography consumption and sexual aggression self-report surveys.	Pornography consumption contributed significantly to sexual aggression scores; this effect is most pronounced in men scoring high on other factors of aggression and hostility and negative sexual attitudes toward women.
	Experimental Studies	
Donnerstein (1980)	One hundred twenty college males saw an aggressive pornographic film, a "slasher" film (nonsexual violence against woman), or a consensual sexually explicit film; were then angered by a female confederate; and aggression (level of "fake" shocks delivered to that female) was measured.	Only aggressive pornography increased aggression toward the target female.
Donnerstein & Berkowitz (1981)	College males watched film of (1) a violent rape in which the victim was portrayed as "enjoying" the rape; (2) a rape in which the victim was not "enjoying" the rape; (3) sexually explicit, consensual sex; or (4) a nonsexual nature; some subjects were insulted by a female confederate prior to seeing the films.	Group 1 subjects displayed aggression to the female confederate whether insulted or not; group 2 displayed increased aggression to the insulting female; and no aggression was found for the other groups.
Zillman & Bryant (1982)	Eighty men and 80 women were exposed to either 36 or 18 nonviolent pornographic films, nonsexual control films, or no films, and then read a rape case, after which recommended punishment for rapists and support for the women's rights movement were measured.	Subjects in the massive-exposure group (36 films) recommended half the prison sentence (5 years versus 10 years) for rapists and expressed half as much support for the women's movement compared to the low-exposure (18 films) group. (*Note:* This is the only study to find such results for *nonviolent* pornography.)
Fisher & Grenier (1994)	Seventy-nine undergraduate males were exposed to violent pornography, nonviolent pornography, or neutral videos; then attitudes toward women were measured.	No antiwoman aggression, fantasies, or attitudes were found.
Mullin & Linz (1995)	Subjects were exposed to a sequence of sexually violent films over several days, and then were compared with a no-film group for sensitivity to victims of domestic violence.	Three days following exposure to the final film, subjects were found to be less sensitive than the control group to domestic violence victims and rated their injuries as less severe. However, five days after the final film exposure, subjects' sensitivity to the domestic violence victims rebounded to control group levels.
Milburn, Mather, & Conrad (2000)	Male and female college students (137 total) were exposed to either video clips sexually objectifying women or to sexually neutral clips and were then shown a news magazine article about a stranger or acquaintance rape.	The male participants who viewed the film clips that portrayed sexual degradation and objectification of women were found, on subsequent testing, to minimize the suffering of the victim and were more likely to express rape myths such as: the woman being raped had enjoyed the experience and had wanted it to happen.
Martin Hald, Malamuth, & Yuen (2010)	A meta-analysis (a method of combining many diverse studies conducted over time) of correlational studies on the link between men's viewing of pornography and their attitudes relating to violence toward women.	Results revealed a positive association between pornography use and attitudes supporting violence against women. These attitudes were found to be especially strong when linked with sexually violent pornography compared to nonviolent pornography.

CORRELATIONAL STUDIES Correlational studies have generally come up with very mixed and confusing results when attempting to link pornography and violence toward women. For example, one study compared sex offenders with a matched control group of non-sex-offending men from the same community in Canada (Langevin et al., 1988). The results indicated that the control group of men reported *greater* exposure to sexually explicit materials than the sex offenders, and the type of pornography both groups had been exposed to was about the same. A study comparing adolescent sexual offenders with matched nonoffenders found that the sexual offenders were more likely to have read pornographic *magazines*, but the nonoffenders were more likely to have viewed pornographic *videos* (Zgourides, Monto, & Harris, 1997). Another correlational study of more than 380 college men found that those who already harbored antiwoman attitudes and who had also viewed violent pornography (compared to those with similar attitudes, but who had not viewed pornography) scored higher on self-reported measures of likelihood to rape or use sexual force and were more likely to say they had used force or coercion to obtain sex in the past (Demare, Lips, & Briere, 1993). However, this same study found no association between nonviolent pornography and any measures of violence toward women, even among subjects with antiwoman attitudes (for a review of these studies, see Martin Hald, Malamuth, & Yuen, 2010).

Other studies have focused on links between pornography laws and other sex offenses. One study looked at four states (Maine, North Carolina, Pennsylvania, and Washington) where pornography laws had at one time been strictly enforced but were later suspended (Winick & Evans, 1996). The rates of other sex crimes, including rape and prostitution, were compared for the periods of enforcement and nonenforcement of pornography statutes. The results indicated that lifting the enforcement of antipornography laws in all four states had no significant effect on the rates of other sex crimes.

One study has suggested that access to pornography via the Internet has actually served to *reduce* the incidence of rape (see Kendall, 2007). Kendall argues that easy access to pornography online may serve as a "substitute" sexual outlet for men who might have rape tendencies, thereby reducing their motivation to act on their rape impulses. This suggestion is based on comparing rates of rape in various states as Internet use became more easily accessible and frequent. Overall, the study found that a 10% increase in Internet use was accompanied by a 7.3% decrease in incidence of rape (Kendall, 2007). Kendall further supports this claim with the fact that Internet use did not reduce any other crime rate, except prostitution. In another correlational study, male undergraduates viewed either violent and degrading pornography or nonsexual materials on the Internet (Isaacs & Fisher, 2008). All participants then completed questionnaires on rape-myth acceptance and attraction to sexual aggression. Results indicated no significant differences between the men in the two groups, indicating that the pornography caused no increased acceptance of rape myths or greater attraction to sexual aggression.

In light of these findings, we cannot automatically assume that laws against pornography will reduce sex-related crime. In fact, these studies suggest that greater access to sexually explicit material may decrease sex crimes, and that men sexually assault and rape for reasons and motivations that are different and more complex than the effects of viewing pornography.

EXPERIMENTAL RESEARCH On the laboratory, experimental side of the research on the effects of pornography, studies have looked at how viewing various types of sexually explicit materials affects men's attitudes and behaviors toward women. Over the decades of research, the findings have become reasonably clear: Nonviolent sexually explicit materials, no matter how explicit they are, do not appear to produce negative effects on men's attitudes or behaviors toward women. However, when the pornography depicts violence toward or degradation of women, negative changes in men's attitude, behaviors, and intentions toward women may occur.

Variations on a widely used investigation strategy have been employed by researchers in laboratory settings. This method exposes participants, usually college students, to differing types and amounts of sexually explicit materials and then measures the subjects on some response of interest, such as callous attitudes toward women, belief in rape myths, feelings of aggression toward women, appropriate punishment for a rapist, or willingness to punish a woman who had insulted them (for a review of these studies, see Linz & Donnerstein, 1990; and C. Hunter, 2000). Two of the leading researchers in this field summarized the findings of these various studies as follows:

> Overall, this line of research found that men who viewed violent pornography displayed greater levels of aggressive behavior toward women. How did they know this? They asked the men to deliver electric shocks to a female "victim" in a laboratory setting and measured the number of volts they chose to deliver. Of course, no one was being shocked, but the male par-ticipants believed they were real. The men's attitudes also changed as measured by their ten-dency to view a rape victim presented to them later as less injured, less traumatized, and more responsible for her assault than did men who had not viewed the violent pornography. In addition, the violent pornography group was more likely to agree with the idea that women secretly enjoy sexual assault. (Linz & Donnerstein, 1992, pp. B3–4)

As noted earlier, these findings tend to disappear in nearly all studies when subjects are exposed to nonviolent pornography, even when the exposure is extreme. For example, one study exposed participants to five straight days of nonviolent pornography and then measured their attitudes toward women and rape. The results indicated no difference in subjects' attitudes before and after the "pornography-fest" and no difference between subjects who had viewed the pornography and a group of controls who viewed non-sexually-explicit materials (Padgett, Brislin-Slutz, & Neal, 1989). Again, it is important to remember that regardless of whether laboratory research reveals a causal relationship between sexually explicit materials and negative attitudes, we still cannot assume that those findings signify such a relationship in the real world, outside the controlled setting where the research took place (see "Evaluating Sexual Research: The Lab Experiment–Correlation Trade-Off").

To the extent that pornography causes, or fails to cause, negative responses toward women may depend on other characteristics of the man viewing it and on the images he chooses to view. When college men were given a choice among various sexually explicit videos to watch, 51% declined to watch any at all, 15% selected nonviolent pornographic materials, and only 4% chose videos containing sexual violence (Fisher & Barak, 2001). In addition, a leading researcher in the field, in a review of the relevant literature, concluded that if a person has "built-in" aggressive sexual tendencies due to past personal experiences, viewing pornography may tap into those predispositions and reactivate and strengthen his preexisting violent and coercive thoughts and behaviors (Malamuth, Addison, & Koss, 2000).

Therefore, a laboratory setting, where participants are *assigned* different types of images for viewing and then measured on attitude or behavioral tests, may produce deceptive results because many of those participants may have no interest in those materials when they have the freedom to choose. If these studies offer additional evidence to the link between pornography and real-life violence toward women, the reason may lie in the fact that men who are already predisposed to aggressive attitudes or violence toward women are drawn to that type of pornography, not the other way around. This we simply cannot know.

THE VIOLENCE-NOT-SEX HYPOTHESIS As study after study continued to find a link between antiwoman attitudes and violent, but not nonviolent, pornography, some researchers began to wonder if the *violence* might be playing a larger role in the effects of pornography than the sexual content. Some research evidence exists to support this idea (e.g., Donnerstein, Linz, & Penrod, 1987; Malamuth, Addison, & Koss, 2000; Wilson et al., 2002). Furthermore, as with sexual aggression, discussed in Chapter 13 ("Sexual Aggression and Violence: Rape, Child Sexual Abuse, and Sexual Harassment"), alcohol use may enhance the possibility of male violence against women in the context of pornography (see Davis et al., 2006).

As we have discussed previously, many studies have demonstrated a link between violent pornography and callousness or violent tendencies toward women, but some studies have found the same link for violent materials *without* the sexually explicit content. For example, in one study, three groups of subjects were each shown a different version of the same film (Donnerstein & Linz, 1986). The versions were as follows:

1. One group saw a film containing a sexual aggression scene in which a woman is violently tied up, threatened with a gun, and raped.
2. A second group saw the exact same film and the same scene, but only the violence was shown, and the sexually explicit rape sequence was removed.
3. A third group saw the same film with the same scene, but only the sexually explicit rape scene was shown, and the accompanying violence was deleted.

After all groups had viewed the films, the researchers measured the attitudes of the men toward rape. The members of group 2, who had seen the violence without the sexual content, were found to have the most insensitive attitudes about rape and the largest percentage of subjects who admitted some likelihood of raping (if they would not be caught) or using sexual force. Subjects in group 3 (sexual content only) were least callous or likely to rape, and those in group 1 (sexual and violent content) fell in between the other two groups.

The distinction between violent materials and violent pornography is becoming increasingly important as both violence and sex become more prevalent and more easily accessible on TV, DVDs, the Internet, and in video and computer games. Clearly violence, whether or not in a sexual context, has the ability to influence men's attitudes toward women and may translate into greater violence toward, victimization of, and discrimination against women. However, as scientific research has made fairly clear, a free society must take great care in the limitations its institutions place on rights of free speech if clear, rational evidence fails to demonstrate harm convincingly.

One type of sexually explicit material, however, instead of inflaming controversy and debate over its harmful effects, brings forth solidarity in the fight against it: child pornography.

cyberporn
Sexually explicit materials accessible online, through the Internet.

Internet Pornography

You probably already know just how rampant Internet pornography (often referred to as **cyberporn**) has become. Thirty percent of all Internet data is pornography and pornographic Websites are estimated to have more monthly visitors than Netflix, Amazon, and Twitter combined. Moreover, consider the following selected estimates of pornography online (Ropelato, 2008). These are the most recent statistics and have likely increased considerably since they were reported:

Pornographic websites	12.6 million (12% of total)
Pornographic pages	420 million
Daily pornographic search engine requests	68 million (25% of total)
Daily pornographic e-mails	2.5 billion (8% of total)
Websites offering illegal child pornography	100,000+
Monthly pornographic Internet downloads	1.5 billion (35% of total)
Pornography revenues (2006)	$27.4 billion (China)
Pornography revenues (2006)	$25.7 billion (South Korea)
Pornography revenues (2006)	$20 billion (Japan)
Pornography revenues (2006)	$13.3 billion (U.S.)

There are over 12 million pornography sites on the Internet and 1.5 billion pornographic downloads each month.

The definitions and obscenity laws rear their heads once again when XXX-rated adult websites are discussed. The problem of defining Internet-based pornography is that the notion of "contemporary community standards," mentioned earlier in our discussion of defining pornography, loses all of its meaning in cyberspace.

The U.S. Congress and governing bodies around the world are attempting to write laws to control sexually explicit materials on the Internet. In the United States, such laws have invariably been struck down by the courts on the grounds that they violate the protections of the First Amendment and do not meet the three-pronged test for obscenity developed in *Miller v. California*, discussed earlier in this chapter. The main problem is that no single "community standard" exists that can be used as a basis for judging whether or not sexually explicit material is offensive and should be deemed obscene. This is obviously a problem when nearly everything on the World Wide Web is available to anyone with an Internet connection, anywhere in the world. As one judge put it: "I believe that 'indecent' and 'patently offensive' are inherently vague, particularly in light of the government's inability to identify the relevant [Internet] community by whose standards the material will be judged" (Sloviter, 1996, p. 854).

The debate on whether or how to control sexually explicit materials on the Internet is likely to proceed unabated for decades (we discuss this more in the next section on child pornography). One thing is clear, however: People are using the Internet (a lot) to view sexually explicit materials. One study found that about half of all college students had used the Internet for sexual purposes of one kind or another (Boies, 2002). Tables 15.4 and 15.5 summarize how and why college students access sexual information online.

Clearly, use of the Internet for sexually explicit purposes is very common and will probably continue to grow as the number and variety of sites increase and more people begin to access them for their own personal reasons.

In many ways, the most important discussion in this chapter may be about the use of children in pornography. When children are victimized for the sexual purposes of adults, whether involving prostitution, pornography, or sexual abuse, people stop worrying about defining it or arguing about whether or not it should be stopped. Instead, the discussion turns to the horrors of such victimization of innocent children and how it can be stopped immediately. Child pornography is a huge business, estimated at over $3 billion per year in the United States. We will discuss this terrible problem next.

Table 15.4 Online Sexual Activities Engaged in by College Students

	Percentage of Students		
Online Sexual Activity (Previous 12 Months)	**Men**	**Women**	**Total**
Relationship-focused			
Sought new contacts*	48.2	38.6	41.8
Used online dating	18.0	9.7	12.5
Information-focused			
Sought sexual information or advice*	67.6	45.1	52.5
Entertainment-focused			
Viewed sexually explicit materials (SEM)*	72.0	24.1	40.1
E-mailed SEM to others*	52.2	35.8	41.2
Received SEM from others	87.8	86.1	86.9
Used sex chat rooms*	13.4	7.0	9.3
Masturbated while online*	71.6	22.1	38.5
Found online sex partners	9.9	7.3	8.2

*Items for which sex differences were statistically significant.

SOURCE: "Sexually-Related Online Activities of College Students," from "University Students' Uses of and Reactions to Online Sexual Information and Entertainment: Links to Online and Offline Sexual Behavior," by S. Boies, from *The Canadian Journal of Human Sexuality, 11*, (2002), pp. 77–90. Copyright © 2002. Reprinted by permission of The Sex Information and Education Council of Canada.

Table 15.5 College Students' Reactions to Sexual Materials Viewed Online

	Percentage of Students	
Reaction to Sexually Explicit Materials Viewed Online	**Males**	**Females**
Sexually excited*	91.9	70.1
Satisfied curiosity*	71.1	57.8
Learned new sexual techniques*	71.0	54.5
Images were disturbing	54.0	61.0
Fulfilled sexual fantasies*	65.2	42.2
Improved personal sexual relationships offline*	50.8	36.4
Bored	36.5	43.2
Satisfied sexual need*	25.9	16.6

*Reactions for which sex differences were statistically significant.

SOURCE: "College Students' Reactions to Online Sexual Activities," from "University Students' Uses of and Reactions to Online Sexual Information and Entertainment: Links to Online and Offline Sexual Behavior," by S. Boies, from *The Canadian Journal of Human Sexuality, 11,* (2002), pp. 77–90. Copyright © 2002. Reprinted by permission of The Sex Information and Education Council of Canada.

Child Pornography

During the 1980s, child pornography was a dying menace. Child pornographers were isolated, hunted individuals, and access to pornography involving children was relegated to the deepest, darkest recesses of society. In short, child pornography had been all but eradicated in most of the civilized world. But then along came the Internet, and the world of child pornography changed. Today, law enforcement around the world must fight a continuous and uphill battle to intervene and arrest individuals who produce and distribute child pornography on the web. In addition, parents, schools, teachers, and public libraries are constantly wrestling with the problem of preventing access to child pornography (and all other pornography) on Internet-connected computers that have become an integral part of our lives. Anyone who wishes to can download hundreds of thousands of pornographic images of children, often engaged in unimaginable activities with other children or with adults. It is estimated that more than a million pornographic images of children on the Internet at any one moment and a single child pornography site may receive more than 400,000 hits per month. Between 50,000 and 100,000 pedophiles are involved in organized child pornography rings globally with a third of these originating in the U.S. (Wortley & Smallbone, 2006).

Since You Asked

10. I read that child pornography is everywhere on the Internet. I think this is just so sick. What is being done to stop it?

LEGAL ASPECTS OF CHILD PORNOGRAPHY Any child pornography activity, whether it involves producing it, looking at it, selling it, or distributing it, is illegal throughout the world. In the United States, the Supreme Court has developed guidelines for child pornography that make it easier to find these materials obscene and ban them (Silver, 2001). Ten years after creating the moral standard for pornography in *Miller v. California* (discussed earlier), the Court found that:

> Recognizing and classifying child pornography as a category of material outside the protection of the First Amendment is incompatible with our earlier decisions...A trier of fact need not find that the material appeals to the prurient interest of the average person; it is not required that sexual conduct portrayed be done so in a patently offensive manner; and the material at issue need not be considered as a whole. (*New York v. Ferber,* 1982)

With over 100,000 child pornography websites currently online internationally, going after child pornography, child pornographers, and child pornography users is a major global endeavor (Luders, 2007). International

The availability of child pornography has exploded since the advent of the Internet in the 1980s, fueling an insidious worldwide industry. To circumvent child pornography laws, a common strategy in the pornography business is to use models of legal age posing as "minors," as this photo depicts.

Sexuality, Ethics, and the Law

The Child Pornography Fighters—One Example of an International Success

During the first decade of 2000, an insidious and despicable child pornography ring, based in the U.S., grew until it reached across five continents and at least 14 countries. The site was called "Dreamboard." It was one of the most flagrant and vile online child pornography in the history of the Web (see Frieden, 2011). Then Homeland Security Secretary, Janet Napolitano, reported that, at its height, the Website contained as many as *123 terabytes of child pornography*, equivalent to 16,000 DVDs!

As difficult as it is to believe, membership in the site required a person to upload sexually explicit photos of children under age 12 and the more images each person uploaded the greater the access to images was granted. Videos of the members themselves molesting children were given highest value and "prestige" on the site. The site went so far as to post unspeakably abominable videos of adults engaging in violent sex acts and other abuse with very young children. If a member failed to add more images every month or so, they could be expelled from the site. When members uploaded large amounts of child pornography, they actually acquired an elite "honorary" status of VIP, Super VIP, or "SuperVIPdot."

In 2011, top law enforcement from numerous countries, following a several month investigative effort busted the global ring. Four of the perpetrators pleaded guilty, and all received sentences of more than 20 years in prison. Many more convictions have followed are the cases are still ongoing. To underscore the scope of this ring, those arrested in the U.S. were from Illinois, Alabama, Florida, and Kentucky. Perpetrators captured outside the U.S. were from Canada, Denmark, Ecuador, France, Germany, Hungary, Kenya, the Netherlands, the Philippines, Qatar, Serbia, Sweden, and Switzerland.

Then Attorney General Eric Holder was quoted as saying, "It's hard for me to imagine that there will ever be a penalty that could appropriately deal with this kind of conduct. Twenty to 30 years that the people have gotten... is barely sufficient to handle what they have done in damaging the lives of these young people."

police agencies, FBI, Interpol, and other law enforcement organizations are becoming increasingly skilled at finding, tracing, and arresting adults who are engaged in child pornography practices online (see "Sexuality, Ethics, and the Law: The Child Pornography Fighters"). Moreover, criminal justice systems are attempting to keep pace by increasing penalties for these illegal activities. For example, in England no distinction is drawn between downloading one pornographic child image or 10,000; the legal penalties are the same: a maximum of five years in jail. Similarly strict laws are being enacted in the United States. In 2007, the U.S. Congress passed a new anti-child-pornography law called SAFE (Securing Adolescents from Exploitation Online) that tripled the penalties for Internet service providers, Wi-Fi broadcasters, and all public electronic communications providers who fail to report child pornography distributed through their channels. The new fines are $150,000 per image *per day* for the first offense. For repeat offenders, the fine shoots up to $300,000 per image, per day (H.R. 3791: SAFE Act, 2007). In 2003, the U.S. Congress passed a bill nationalizing what has become known as the "Amber Alert" law, which, in addition to establishing a system for enlisting the public's help in recovering kidnapped children, requires that those arrested for violating child pornography laws be placed on public sex offender registries in all states and strengthens criminal penalties for all child predatory crimes. The various digital methods child predators employ are shown in Table 15.6.

Specific child pornography laws and penalties vary from state to state, but the federal government's laws pertaining to sexual exploitation of children contains clear language relating to child pornography, as follows:

> **Sexual Offenses: Federal Law 18 U.S.C. 2251: Sexual Exploitation of Children:**
>
> It is forbidden for any person to employ, use, persuade, induce, entice, or coerce any minor or to... engage in, any sexually explicit conduct if such person knows or has reason to know that such visual depiction was produced using materials that have been mailed, shipped, or transported in interstate or foreign commerce by any means, including by computer, or if such visual depiction has actually been transported in interstate or foreign commerce or mailed.

It is forbidden for any parent, legal guardian, or person having custody or control of a minor to knowingly permit such minor to engage…in sexually explicit conduct for the purpose of producing any visual depiction of such conduct if the parent, legal guardian, or person knows or has reason to know that such visual depiction will be transported in interstate or foreign commerce or mailed, if that visual depiction was produced using materials that have been mailed, shipped, or transported in interstate or foreign commerce by any means, including by computer, or if such visual depiction has actually been transported in interstate or foreign commerce or mailed.

It is forbidden for any person to knowingly make, print, or publish, or cause to be made, printed, or published, any notice or advertisement seeking or offering

- to receive, exchange, buy, produce, display, distribute, or reproduce, any visual depiction, if the production of such visual depiction involves the use of a minor engaging in sexually explicit conduct and such visual depiction is of such conduct, or
- to participate in any act of sexually explicit conduct by or with any minor for the purpose of producing a visual depiction of such conduct, if such person knows or has reason to know that such notice or advertisement is or will be transported in interstate or foreign commerce by any means including by computer or mail. ("Sex Laws," 2008)

Table 15.6 Strategies Incorporating the Internet Used by Child Sexual Predators

Electronic Method	How Method Is Used by Cyber-Predators
Web pages and websites	Specific child pornography websites may be created, or child pornography images may be embedded in general pornography sites. However, there is debate about how much child pornography is available on the web. Some argue that it is relatively easy to find images. Others argue that, because of the vigilance of ISPs and police in tracking down and closing child pornography websites, it is unlikely that a normal web search using key words such as child porn would reveal much genuine child pornography. Instead, the searcher is likely to find legal pornographic sites with adults purporting to be minors, "sting" operations, or vigilante sites. One strategy of distributors is to post temporary sites that are then advertised on pedophile bulletin boards. To prolong their existence, these sites may be given innocuous names (e.g., volleyball) or other codes (e.g., ch*ldp*rn) to pass screening software. The websites may be immediately flooded with hits before they are closed down. Often, the websites contain Zip archives, the password for which is then later posted on a bulletin board.
Web cam	Images of abuse may be broadcast in real time. In one documented case of a live broadcast, viewers could make online requests for particular sexual activities to be carried out on the victim.
E-mail	E-mail attachments are sometimes used by professional distributors of child pornography, but more frequently, they are used to share images among users, or they are sent to a potential victim as part of the grooming/seduction process. This method is considered risky by seasoned users because of the danger in unwittingly sending e-mails to undercover police posing as pedophiles or as potential victims.
E-groups	Specific child pornography e-groups exist to permit members to receive and share pornographic images and exchange information about new sites. Some of these groups appear on reputable servers and are swiftly shut down when they are detected. However, they may use code names or camouflage child pornography images among legal adult pornography to prolong their existence.
Newsgroups	Specific child pornography newsgroups provide members with a forum in which to discuss their sexual interests in children and to post child pornography. This is one of the major methods of distributing child pornography. Some child pornography newsgroups are well known to both users and authorities (for example, the abpep-t or alternative binaries pictures erotica pre-teen group). Most commercial servers block access to such sites. Some servers do provide access to them, but a user runs the risk of having his or her identity captured either by the credit card payments required for access, or the record kept by the server of his or her IP address. However, a computer-savvy user can access these groups by using techniques that hide his or her identity by concealing his or her true IP address.
Bulletin board systems (BBS)	Bulletin boards may be used legally to host discussions that provide advice to seekers of child pornography, including the URLs of child pornography websites and ratings of those sites. These bulletin boards may be monitored by system administrators to exclude bogus or irrelevant postings, such as from vigilantes.
Chat rooms	Chat rooms may be used to exchange child pornography and locate potential victims. Chat rooms may be password-protected. Open chat rooms are avoided by seasoned child pornographers because they are often infiltrated by undercover police.
Peer-to-peer (P2P)	P2P networks facilitate file sharing among child pornography users. These networks permit closed groups to trade images.

SOURCE: Center for Problem Oriented Policing (2011). www.popcenter.org/problems/child_pornography.

Table 15.7 Federal Law Governing Child Pornography and Enticement Crimes (selected sections)

Section	Prohibits	Mandatory Minimum Penalty	Maximum Penalty
18 U.S.C. § 2251(a)	Employing, using, or enticing a minor to engage in sexually explicit conduct for the purpose of producing a visual depiction of that conduct	15 years (first offense); 25 years (second offense); 35 years (third offense)	30 years (first offense); 50 years (second offense); life (third offense)
18 U.S.C. § 2251(b)	Parent or guardian permitting a minor to engage in sexually explicit conduct for the purpose of producing a visual depiction of that conduct	Same as above	Same as above
18 U.S.C. § 2252(a)(2)	Receiving or distributing a visual depiction of a minor engaging in sexually explicit conduct	5 years (first offense); 15 years (second offense)	20 years (first offense); 40 years (second offense)
18 U.S.C. § 2252(a)(4)	Possessing a visual depiction of a minor engaging in sexually explicit conduct	None (first offense); 10 years (second offense)	10 years (first offense); 20 years (second offense)
18 U.S.C. § 2252A(a)(1)	Transporting child pornography	5 years (first offense); 15 years (second offense)	20 years (first offense); 40 years (second offense)
18 U.S.C. § 2252A(a)(5)	Possessing child pornography	None (first offense); 10 years (second offense)	10 years (first offense); 20 years (second offense)

SOURCE: U.S. Department of Justice (2006), Project Safe Childhood, pp. 39–43.

Perhaps more importantly, these laws provide a wide range of criminal punishments for any violation of this law, depending on the severity of the crime. Table 15.7 provides a sampling of the various sections in the law and the associated criminal penalties.

Despite some significant arrests of major child pornographers, these laws and penalties appear to be having little effect on the amount of child pornography available on the Internet. This is because as fast as law enforcement can work to track down the perpetrators, the methods used to hide the operations throughout the world become increasingly sophisticated.

TECHNOLOGICAL SOLUTIONS TO CHILD PORNOGRAPHY In addition to criminal justice strategies to attempt to limit child pornography online or limit access or exposure to it, Internet filtering software has been developed, some of which can screen out sexually explicit materials (McIntyre, 2013; Wortley & Smallbone, 2012). The problem, however, is that, although the software may block pornographic sites, it also restricts the user's access to some nonpornographic sites that are part of the all people's routine use of the Internet for legitimate and necessary purposes. This has become a much-debated trade-off: protection from child victimization versus freedom of access to the vast digital world of information (Ramirez, 2014). For example, filters that use keywords to limit access to pornographic sites key on words, phrases, or graphics that the software interprets as forbidden; however, these filters may block legitimate and important information such as some fine arts sites, legitimate sexuality education sites, and, perhaps most unfortunately, health-related sites. One study tested seven of the most popular pornography-filtering programs and found that at the most restrictive settings, 700 out of 3,000 Internet health sites were blocked, and one of every three sexual health sites was blocked even at the least restrictive settings (C. R. Richardson et al., 2002). These were not minor web-based sources of health information; they included sites from the Food and Drug Administration, the National Library of Medicine, and the Centers for Disease Control and Prevention.

This exchange of filtering unwanted sites with losing access to legitimate Internet resources may be acceptable in some settings, such as families with children who are on the Internet routinely. However, this approach is rarely acceptable for adults who rely on the World Wide Web to gather health-related and other information for their own purposes, and it is totally unacceptable for professionals who use the Internet for research or to augment information for their clients and patients.

New technologies are being created to filter out child pornography without focusing only on text. One strategy uses visual recognition techniques—programs that actually "see" images and identify suspect sites (i.e., Ulges & Stahl, 2011). Although this strategy remains a challenge in terms of accuracy, error rates as low as 11–24% have been achieved thus far.

One possible solution to this dilemma would change the Internet addresses of all pornographic websites in a way that makes it easy to block them out. Since 2000, the Internet Corporation for Assigned Names and Numbers (ICANN), which directs the naming of Internet sites through approval of allowed top-level domain names (the letters following the "dot"), considered adopting a new domain suffix, ".xxx" (Kruger, 2013). The idea behind this is that if all websites containing adult, sexually explicit materials for "entertainment purposes" were ensconced in the xxx domain, they would all be essentially "in one place." This would accomplish two goals: they would be easier to find, and they would also be easier to filter out. Finally, ICANN, in a split decision, approved the .xxx domain name in 2011. This seems like a good idea; however, use of the .xxx domain is voluntary, not required, and only about 10% of all Internet pornography is found there. Moreover, the illegality of and penalties for child pornography make it highly unlikely that anyone involved in that loathsome business would register (or be allowed to register) an xxx domain name.

GOOGLE'S ANTI-CHILD-PORNOGRAPHY PLAN Perhaps the world's leading providers of web content have the best chance of actually reducing the availability of child pornography on the Internet. And they would likely have the technological expertise to accomplish this seemingly impossible task. Google, the world's most frequently used search engine, announced in 2013 that it will begin isolating child pornography on the Internet into a single database in an effort to eradicate child sexual abuse entirely from the web. The company plans to spend at least $5 million to fight child pornography online.

"Behind these images are real, vulnerable kids who are sexually victimized and victimized further through the distribution of their images," Google executive Jacquelline Fuller said. "It is critical that we take action as a community—as concerned parents, guardians, teachers and companies—to help combat this problem. . . . We can do a lot to ensure it's not available online—and that when people try to share this disgusting content, they are caught and prosecuted" (Kerr, 2013).

The technology Google uses is called "hashing," which allows Google's technology to find offensive images, tag them, and then wipe all copies away without anyone needing to view them again (Kleinman, 2013). Moreover, Google is sharing this technology and its results worldwide and in 150 languages with other search providers, private companies, law enforcement, and charities, so that a broad-based "seek-and-destroy" method can be implemented. Google has also created a multimillion dollar fund to assist companies in developing child protection technology to fight online child exploitation.

Google is developing technological tools for finding and destroying images of child exploitation across the World Wide Web.

FIGHTING CHILD PORNOGRAPHY THROUGH *GREATER* ACCESSIBILITY? One intriguing and courageous suggestion has been made by Philip Jenkins, professor of history and religious studies at Pennsylvania State University. Jenkins, a strong advocate of total elimination of child pornography on the Internet, suggests that the legal crackdown on child pornography, as ineffective as it is, prevents serious researchers from studying and uncovering how this illicit trade functions and how best to attack it. Although Jenkins estimates that only about one-tenth of 1% of Internet child pornography distributors and users are ever caught, arrested, and prosecuted, no serious social scientist or legal scholar would ever want to take the risk of accessing the material for legitimate research purposes and then try to use his or her research goals as a defense if caught in a law enforcement action. As Jenkins states:

> Once I had found the child-porn culture on the Internet, the next question was how to study it, given a legal environment in which virtually any contact with the material can lead to a federal prison sentence. "Children," for legal purposes, means anyone below the age of 18, and "pornography" includes depictions that would be only mildly indecent if adult subjects were involved. Moreover, one "possesses" an electronic image merely by downloading it, by clicking on a Web link. (P. Jenkins, 2002, p. B16)

Jenkins is suggesting that we, as a society, think about making some sort of exception to the current laws to allow for legitimate access and study of the Internet child pornography subculture, with the goal and intent of destroying it. The reason, Jenkins asserts, is that most people do not have the slightest notion of just how horrible and widespread the problem is:

> The absence of previous studies—journalistic or academic—explains why most writers have failed even to notice the existence of this burgeoning subculture. We search the literature in vain for references even to core child-porn newsgroups. If the public does not even know that the newsgroups and bulletin boards exist, then the police face no pressure to eradicate them, with all the complex international cooperation that such an effort would demand.... The main reason people don't fight child pornography is that most of them have never seen it. Actual exposure to this material would galvanize public opinion—and, incidentally, would make clear the huge difference in potential harm between child porn and even the hardest of hard-core adult images.... Opening this avenue would raise the possibility of better exposing the trade, creating public awareness about its key institutions, and pressuring politicians to act against major suppliers and trafficking institutions, rather than just hapless individual consumers. (P. Jenkins, 2002, p. B16)

Jenkins's ideas may make sense to some people but may be difficult to put into practice. Who would determine what constitutes "legitimate research"? How would a researcher go about obtaining such an exception from legal prosecution? How could we prevent abuse of such a system? These are all legitimate questions, but if, as a culture, we are to develop an effective defense against the victimization of children at the hands of Internet pornographers, bold new approaches will be necessary.

Your Sexual Philosophy

The Sexual Marketplace: Prostitution and Pornography

We have taken a fairly detailed look at two important ways human sexuality functions as a business: prostitution, in which sexual interactions between individuals are bought and sold; and pornography, in which various depictions of those interactions are bought and sold. Some would claim that a world of difference exists between these two sexual marketplaces, but others would argue that they are in reality quite similar in terms of who sells, who buys, who profits, and who is victimized. Whether prostitution and pornography are distinct industries or represent two aspects of the same continuum of commercial sex, one point is clear: these are among the most controversial of sexually related issues worldwide.

You may be wondering why you would need to add the topics of prostitution and pornography to your developing sexual philosophy. After all, you are probably not planning to become a professional pornographer or a prostitute, right? Probably not. However, you should be aware that both of these issues *will* come up in your life, if they haven't already, in one way or another. This is just one more reason why it's easier and less stressful to have thought about such matters before they have a chance to cause a problem for you or your partner. As stated earlier, your sexual philosophy is about knowing who you are, what you want and don't want, and planning accordingly.

As the debate about prostitution laws continues, it is possible that the lawmakers in your state, city, or town may at some point consider reducing legal control over commercial sex work. You may think it could never happen in *your* town, but that's what people said of lotteries, casinos, and other forms of legalized gambling only a couple of decades ago. So you never know. As an involved and concerned citizen, you will need to understand the issues relating to prostitution to participate intelligently in whatever controversy may arise. Furthermore, as discussed in this chapter, prostitution is here to stay. You are bound to see it, read about it, see ads for it (escort services, massage parlors, etc.), and maybe even be solicited by a prostitute. How will you react to these events if they happen to you or your partner? Chances are, after reading this chapter and thinking about the issues relating to prostitution and sex workers, you will be able to respond rationally and thoughtfully and avoid *overreacting* and creating more of a problem than might be necessary. In addition, now that you understand better the factors so often involved in the path to prostitution, you may one day want to become involved in helping young girls avoid the trap of prostitution or helping women already entangled find their way out.

Pornography is even more likely to be an issue in your life someday, and for some of you, it may be already. A common problem in relationships is that one partner (usually the man, in heterosexual couples) becomes interested in viewing sexually explicit materials, but the other partner either is not interested or is offended and put off by the idea. This may raise difficult issues that the couple must address with knowledge and understanding to prevent them from threatening the relationship. Even more likely is the possibility that you will encounter sexually explicit materials yourself and you will need to analyze your feelings about them. Compared to several decades ago, access to sexually explicit materials has changed dramatically. Before VCRs and DVDs, the only way most people ever saw a pornographic film was to slip into an "adult theater" and hope no one they knew spotted them (or even worse, ran into them in the theater!). Today, however, it is a simple matter to walk into the local video store for a huge selection of X-rated videos, check some out, and watch them in the privacy of your own home. Furthermore, the Internet now makes readily accessible every kind of sexually explicit material imaginable—and you don't even have to leave the house to see it. In the information age, it has become more necessary than ever to understand what pornography is, what you want to view, and what you are unwilling to view. By understanding this now and planning ahead for situations in your life involving sexually explicit materials, you will be better equipped to handle these situations effectively, with confidence, control, and self-assurance, when they arise.

Have You Considered?

1. Discuss three strategies that might be effective in helping women who are involved in street prostitution find their way out of the business and into a healthier, happier life.
2. Do you believe that prostitution should be legalized or decriminalized? Explain your answer.
3. Discuss three reasons why you think male prostitutes for female customers are so rare.
4. Imagine that you are a social worker hired by a medium-size city to develop a program to prevent young girls from entering prostitution. List at least three strategies you might employ in your new job, and explain why you think they would be effective.
5. Do you think some sexually explicit materials (other than child pornography) should be legally censored? If so, what materials and why? If not, why not?
6. Discuss your opinion about using the current "moral standard" in judging whether sexually explicit materials are obscene. Do you think it is effective? Why or why not?

Summary

Historical Perspectives

The Oldest Brothel?

- Archaeological findings reveal that prostitution is one of the oldest professions. While digging under the ruins of an ancient Roman forum in Greek Macedonia, scientists discovered a bathhouse and brothel dating back to more than 2,000 years ago.

Prostitution

- Prostitution today, also termed the *commercial sex trades*, is defined as sexual acts exchanged for anything of value to the sex workers or their "managers" or pimps, including money, valued objects, personal property, drugs, shelter, or other services.
- Sex workers and their clients may be male or female and of the same or opposite sexes.
- Prostitution is illegal in most countries of the world and all U.S. states, with the exception of certain counties in Nevada, where prostitution is legal and controlled.
- Various types of prostitutes exist, the most common of which are referred to as streetwalkers, brothel workers, and call girls.
- Male prostitutes' customers are typically other men. When males are hired by women, the men tend to function more as escorts, and their interaction may or may not include sex.
- Some women enter the profession of prostitution by choice, but most are forced into it through international human trafficking, drug abuse and addiction, or coercion by pimps. Studies show that most girls enter prostitution to pay for their addiction to crack or injectable drugs. Prostitutes' age at first intercourse, whether voluntary or not, is significantly younger than that of nonprostitutes.
- Rates of STIs are high among sex trade workers, owing to the same risk factors faced by the general population: multiple sex partners, unprotected sex, and drug abuse. Rates of STIs among male prostitutes and transgender female prostitutes are particularly high.
- The profile of clients of prostitutes ("johns" or "tricks") does not differ significantly from males in general. The most common activity requested by clients is oral sex.
- The debate over the legalization or decriminalization of prostitution in the United States is often a heated one, but states show few signs of changing the legal status of prostitution.

Pornography

- Overall, college students have more experience with viewing pornography than is commonly believed.
- The current legal definition of pornography, called the "moral standard," was developed by the U.S. Supreme Court in 1973 and requires that local, contemporary community standards be applied to determine if a sexually explicit work should be deemed obscene and be exempt from First Amendment protections of freedom of expression. The sale, ownership, or viewing of material found to be obscene may be made illegal.

- A competing definition of pornography contends that sexually explicit materials that discriminate against a particular group of people, such as women, may be defined as obscene. This definition, however, is not used in current legal practice.
- Most researchers draw a distinction between pornography and erotica. The term *erotica* is often applied to sexually explicit works expressing physical desire, passion, and attraction among people who freely choose to engage in sexual activities together. Erotica typically goes beyond "raw sex" and includes psychological and emotional factors as well. Erotica is seen as *sexual sharing*, whereas pornography is interpreted as *sexual using*.
- Research has demonstrated that women tend to be sexually aroused by a wider variety of sexually explicit images than are men.
- The link between pornography and discrimination and aggression toward women has yet to be demonstrated clearly in scientific research. Studies show little effect of nonviolent pornography on men's attitudes toward women, but violent pornography may increase men's callousness toward women. Violent and degrading pornography has also been shown to increase men's self-reported likelihood of rape, but this has not been demonstrated to translate into actual rape behavior. Overall, the violence may be more important than the sexual content in the effects of violent pornography on men's attitudes toward women.
- Child pornography is a huge business, estimated at over $3 billion a year in the United States alone. Child pornography was decreasing in the 1980s as law enforcement was becoming increasingly effective in arresting and stamping out makers and users. However, today child pornography is growing at an alarming rate on the Internet, and worldwide attempts at controlling it have been minimally effective.
- The Internet has made all forms of sexually explicit materials easily available to anyone, anywhere. An estimated 420 million pages of pornography exist on the Internet (often called cyberporn), and 35% of all Internet downloads each month are from pornographic sites. That equals over 1.5 billion downloads.
- Much of the Internet use for sexually explicit materials is by college students. Students use the Internet to view pornography for a variety of reasons, ranging from seeking sexual excitement to satisfying their curiosity about pornography to learning new sexual techniques.

Your Sexual Philosophy

The Sexual Marketplace: Prostitution And Pornography

- If you have not already, you will encounter in your life, in one way or another, the main topics of this chapter: prostitution and pornography. Having thought about them and analyzed your feelings and attitudes will help prevent them from catching you by surprise and causing discomfort for you or a problem for you and your partner. Remember, as we have discussed throughout this book, your sexual philosophy is about knowing who you are, what you want and don't want, and planning ahead.

Glossary

abortion Termination of a pregnancy before week 37; in common usage, assumed to be the result of an intentional act as opposed to a miscarriage. (p. 337)

abstinence-only approach The decision to avoid teaching adolescent students about sexual activity, STIs, contraception, etc., based on the theory that such education is unnecessary if students are taught to abstain from sexual behavior. (pp. 7; 458)

acceptable level of risk The level of risk one is willing to accept when making behavioral choices about one's health and well-being. (p. 305)

acquired immune deficiency syndrome (AIDS) A gradual failure of the immune system, leading to serious, opportunistic infections, which, in turn, may lead to death. (p. 282)

adolescence The period of life between the ages of approximately 10 and 19 years, often a tumultuous, self-identity seeking stage of life. (p. 448)

agape love A style of love focused on giving the partner whatever he or she may want or need without the expectation of receiving anything in return. (p. 117)

AIDS stigma Prejudice and discrimination against nonheterosexual individuals based on the erroneous belief that gay individuals are solely to blame for the AIDS epidemic and are the primary threat for the continuing spread of the disease. (p. 427)

alcohol myopia theory The belief that under the influence of alcohol, people are more likely to focus on immediate, "feel-good" behaviors (such as sexual arousal) and ignore future negative consequences. (p. 268)

amenorrhea Cessation of a woman's period. (p. 162)

anal intercourse A sexual position in which the penis is inserted through the partner's anus into the rectum. (p. 208)

androgynous A person who embodies both masculine and feminine traits and behaviors. (p. 396)

anilingus Oral stimulation of the anus. (p. 208)

anonymous testing Tests administered without collecting any personal information about clients, who are identified only by an assigned code number. (p. 290)

antibiotic-resistant strain A strain of bacteria that has mutated and is no longer treatable with standard antibiotic therapy. (p. 303)

antiherpetics Medications developed to treat (reduce or prevent but not cure) outbreaks of the herpes virus. (p. 275)

antiretroviral therapy (ART) A combination of several medications prescribed for people who are HIV-positive to delay the onset of AIDS. (p. 291)

anus The end of the digestive tract and outlet for bodily excretions. It is also a sexually stimulating area for some people. (p. 43)

APGAR score A test that analyzes infant health at birth on the basis of skin color, pulse, reflexes, movement, and breathing. (p. 346)

aphrodisiac Mythical substances that are thought to enhance sexual arousal and desire. (p. 77)

areola The darker skin encircling each nipple; actually part of the skin of the nipple. (p. 56)

assisted reproductive technology (ART) Various treatments to help infertile women or couples to become pregnant and have a child. (p. 359)

asymptomatic Having no noticeable symptoms despite the presence of an infectious agent. (p. 264)

asymptomatic shedding Release of infectious virus particles when no symptoms of infection are present. (p. 275)

autoerotic asphyxiation A form of sexual masochism involving depriving the brain of oxygen, usually through some form of strangulation or hanging; also referred to as *asphyxophilia*. (p. 544)

aversion therapy Psychotherapy in which unwanted sexual behaviours and images are reduced by associating them with an unpleasant stimulus such as electric shocks or noxious odours. (p. 549)

barrier method Any contraceptive method that protects against pregnancy by preventing live sperm from entering the woman's reproductive tract and fertilizing the egg. (p. 170)

BDSM Sexual activities that combine bondage, discipline, and sadomasochism. (p. 538)

birth defect A physical abnormality or metabolic dysfunction that is present at birth and may result in physical or mental deficits. (p. 333)

birthing center A hospital-like facility with basic medical care equipment, focusing on a natural, family-centered approach to the birth process in a home-like setting. (p. 347)

bisexual A person who is attracted romantically and sexually to members of both sexes. (p. 405)

blastocyst The developing zygote, with cells surrounding a fluid-filled core, upon entering the uterus and before implanting in the uterine wall. (p. 326)

bondage Being bound, tied up, or otherwise restrained during sexual activity; typically, a consensual activity. (p. 543)

breech birth Delivery of a fetus emerging with buttocks or legs first rather than head first. (p. 345)

brothel worker A prostitute who works in a house of prostitution—such as a brothel, bordello, or massage parlor—and receives paying customers. (p. 558)

call girl A prostitute who is contacted in private by clients when her services are desired and who generally charges more and provides a wider range of services than streetwalkers, such as serving as an escort or offering overnight stays. (p. 558)

celibacy Choosing to engage in no sexual activities whatsoever. (pp. 152; 310)

celibate choosing to forego all sexual activities. (p. 6)

cervical cap A device similar to the diaphragm that fits more snugly over the cervix. (p. 171)

cervix The lower end of the uterus that connects it to the vagina. (p. 61)

cesarean section Removal of a fetus from the mother's uterus surgically, through an incision in her abdomen; also called a *C-section birth*. (p. 349)

chancre A sore that typically appears at the site of infection with syphilis. (p. 300)

chemical castration Blocking the production of testosterone in repeat sex offenders through the regular injection of female hormones. (p. 510)

child molestation Any sexual act performed with a child by an adult or a much older child. (p. 503)

child sexual abuse (CSA) The sexual victimization of a child by an adult or a significantly older child. (p. 503)

chlamydia A sexually transmitted bacterium, often causing a thick, cloudy discharge from the vagina or penis; may be asymptomatic, especially in women. (p. 296)

circumcision Removal of the foreskin of the penis. (p. 37)

cirrhosis of the liver A potentially serious liver disease that may lead to liver cancer. (p. 281)

civil union (also called domestic partnership) A legal contract between two members of the same sex that imparts all or most of the legal benefits of marriage but is not socially or religiously equated with heterosexual marriage. (p. 411)

clitoral glans The outer end or tip of the clitoris. (p. 49)

clitoral hood Tissue that partially or fully covers the clitoral glans. (p. 49)

clitoris An erectile sexual structure consisting of the clitoral glans and two shafts (*crura*) that is primarily responsible for triggering orgasm in most women. (p. 49)

cognitive-behavioral therapy A therapeutic approach designed to gradually eliminate specific thoughts and associated behaviors that may be contributing to sexual problems. (pp. 236; 549)

cohabitation Living together as if married without legally marrying. (p. 465)

coitus Penis–vagina intercourse. (p. 209)

combination pill An oral contraceptive containing a combination of estrogen and progestin. (p. 161)

commercial sex trades Selling or trading sexual goods or activities for money or other items of value. (p. 555)

companionate love Love based on true intimacy and commitment but lacking passion; the partners are companions more than lovers. (p. 111)

complaining Expressing an unmet need for something a person desires but is not receiving from a partner. (p. 119)

comprehensive sex education Sexual education programs for adolescents that provide information about abstinence and about prevention of STIs and unwanted pregnancy. (p. 458)

compulsive behavior Any behavior, sexual or otherwise, that a person is unable to control regardless of repeated attempts to do so. (p. 527)

conception The moment a single sperm cell penetrates the wall of an ovum and the sperm's and ovum's DNA fuse together. (p. 144)

confidential testing Testing in which recipients' names are kept on file in the lab's records, but with the assurance of full confidentiality. (p. 290)

consensual A behavior entered into voluntarily by all parties. (p. 530)

consummate love Love that encompasses intimacy, passion, and commitment simultaneously. (p. 111)

contempt Disrespect, disgust, or hate expressed when the positive feelings partners once had for each other have dissipated. (p. 120)

contraception The process of preventing sperm cells from fertilizing an ovum. (p. 144)

contraceptive patch A stick-on patch that delivers a precise dose of two hormones into a woman's body through the skin, preventing ovulation. (p. 167)

contraceptive ring A colorless, flexible, transparent silicone ring about 2 inches in diameter that is inserted into the vagina and releases a continuous, low dose of estrogen- and progestin-like hormones into the blood-stream. (p. 167)

control group The participants in an experiment who receive no treatment and are allowed to behave as usual, for the purposes of comparison to an experimental group; also known as the *comparison group*. (p. 26)

corona The raised edge at the base of the penile glans. (p. 37)

corpora cavernosa Two parallel chambers that run the length of the penis and become engorged with blood during erection. (p. 37)

corpus spongiosum A middle chamber running the length of the penis into the glans that engorges with blood during erection. (p. 37)

correlational research A research method that reveals how two observations (variables) are related to each other (how they co-relate), but does not determine if one variable is causing the other. (p. 24; 580)

covert sensitization A type of aversion therapy in which, instead of actual shocks or bad odors, the client repeatedly fantasizes the unwanted behavior and adds an extremely aversive event to the fantasy. (p. 549)

Cowper's glands Small glands near the penile urethra that produce a slippery, mucus-like substance during male sexual arousal (also referred to as the *bulbourethralglands*). (p. 47)

crack A powerful, processed form of cocaine that is smoked or sometimes injected, known for its intense but relatively short high and its highly addictive qualities. (p. 565)

criticism Verbal fault-finding, such as commenting on a character flaw in the partner. (p. 119)

cross-dressing Dressing in clothes traditionally associated with the opposite sex. (p. 547)

cultural gender stereotypes Beliefs about gender roles held by a majority of people in a given cultural setting. (p. 389)

cunnilingus Oral sex performed on a female. (p. 205)

cyberharassment A type of bullying or harassment (including sexual harassment) that is perpetrated through digital means, including social networking, texting, and email. (p. 518)

cyberporn Sexually explicit materials accessible online, through the Internet. (p. 585)

cycle of abuse The repetitive pattern of stages that define most abusive and violent relationships, cycling through the honeymoon stage, the tension-building phase, and the explosion of violence, followed by a return to the honeymoon stage and the beginning of a new cycle. (p. 134)

date rape drugs Powerful sedatives that render a rapist's potential victim unconscious or otherwise unable to resist; also known as *club drugs*. (p. 490)

debriefing Explanations of the purpose and potential contributions of the findings given to participants at the end of a study. (p. 29)

decriminalization of prostitution Repeal of all laws against consensual adult sexual activity in both commercial and noncommercial contexts. (p. 571)

delayed ejaculation Unwanted delay in reaching orgasm or absence of orgasm regardless of amount of stimulation. (p. 246)

dependent variable The result of an experiment, evaluated to determine if the independent variable actually caused a change in the experimental group of participants. (p. 26)

Depo-Provera A hormonal contraceptive in the form of an injection that provides 90 days of protection from conception. (p. 166)

diaphragm A flexible ring of latex or silicone inserted into the vagina that impedes conception by preventing sperm from getting past the cervix. (p. 170)

dilation and evacuation (D&E) A method of abortion commonly used when a pregnancy has progressed beyond the first trimester, involving scraping of the uterine walls and suctioning out of the contents. (p. 340)

directed masturbation A sex therapy strategy in which the therapist advises the client on how to use masturbation activities to help overcome a sexual problem. (p. 233)

disorders of sex development (DSD) Born with sexual anatomy that is neither completely male nor completely female but rather a combination with features of both that cannot be categorized as male or female. (p. 372)

Dual Control Model of Sexual Response A theory that sexual arousal is controlled by a combination of excitatory and inhibitory processes. (p. 92)

ectopic pregnancy A pregnancy complication in which a fertilized ovum attaches and begins to grow outside the uterus, most commonly in the fallopian tube, which is called a *tubal pregnancy*. (pp. 63; 330)

ejaculation Expulsion of semen through the penis. (p. 43)

ejaculatory duct A continuation of the tube that carries semen into the urethra for ejaculation. (p. 45)

ejaculatory inevitability In males, the sensation produced during the emission phase of ejaculation that expulsion of semen is imminent, reflexive, and cannot be stopped; often referred to as the "point of no return." (p. 82)

Electra complex A Freudian notion explaining how a young girl comes to identify with her mother. (p. 441)

embryo A blastocyst that has implanted in the uterine wall. (p. 326)

embryonic period The initial eight weeks of pregnancy following fertilization. (p. 327)

emergency contraceptions (EC) Hormonal contraceptive that helps prevent pregnancy after an unprotected act of intercourse; also known as the "morning-after" pill. (p. 169)

emergency contraceptive pills (ECPs) Hormonal contraceptives that help prevent pregnancy after an unprotected act of intercourse; also known as the *morning-after pill*. (p. 457)

emission In males, the buildup of sperm and semen in the urethral bulb just prior to being expelled through the urethra. (p. 82)

empty love Love based on commitment but lacking intimacy or real passion. (p. 111)

endometriosis A potentially painful and dangerous medical condition caused by endometrial cells migrating outside the uterus into the abdominal cavity. (p. 62)

endometrium The tissue lining the uterus that thickens in anticipation of pregnancy and is sloughed off and expelled during menstruation. (p. 62)

epidemic A sudden occurrence of a disease characterized by far more cases than would be expected in a community or region; less widespread than a pandemic. (p. 264)

epididymis A crescent-shaped structure on each testicle where sperm cells are stored as they mature. (p. 43)

epididymitis A painful swelling and inflammation of the epididymis, the structure at the back of each testicle that stores maturing sperm; often caused by one or more untreated STIs. (p. 297)

episiotomy Surgical cutting of the perineum during childbirth, a procedure that was believed to allow for easier passage of the infant and less tearing of the vaginal opening. Found to be ineffective, it is rarely performed today. (p. 56)

EPOR model Masters and Johnson's approach to explaining the process of human sexual response, encompassing four arbitrarily divided phases: excitement, plateau, orgasm, and resolution. (p. 76)

erectile disorder (ED) Recurring or persistent difficulty in achieving or maintaining an erection. (p. 237)

erection Rigidity of the penis or clitoris resulting from an inflow of blood during sexual arousal. (p. 37)

eros love An erotic, passionate style of love often characterized by shortlived relationships. (p. 115)

erotica Sexually explicit works expressing physical desire, passion, and attraction among people who freely choose to engage in sexual activities together. (p. 576)

erotic stimulus pathway theory A model of human sexual response based on the psychological and cognitive stages of seduction, sensations, surrender, and reflection. (p. 91)

erotic touch Intimate or sexual touching between partners, usually with the hands, for the purpose of sexual arousal and sharing sexual or sensual pleasure. (p. 201)

erotophilia An attitude toward sexuality in which individuals are comfortable with sexual issues, seek out sexual information, enjoy sexual behavior, and respond with positive reactions to sexual topics. (p. 188)

erotophobia An attitude toward sexuality in which individuals are generally uncomfortable with sexual topics, respond negatively and uncomfortably to sexual issues, and tend to avoid sexual information and activities. (p. 188)

estrogen The female hormone responsible for regulating ovulation, endometrial development, and the development of female sexual characteristics. (p. 63)

excitement phase The first phase in the EPOR model, in which the first physical changes of sexual arousal occur. (p. 77)

exhibitionistic disorder Achieving sexual arousal and gratification by displaying one's genitals to others without the victims' consent. (p. 536)

exotic becomes erotic (EBE) theory Psychologist Daryl Bem's explanation for the interaction of biology and environment in determining a person's sexual orientation. (p. 419)

experiment A research method that examines the link between two or more variables while controlling for all others to determine cause and effect relationships among variables. (p. 581)

experimental group The participants in an experiment who are subjected to a variable of research interest. (p. 26)

experimental method A type of scientific research in which variables of interest are changed while all other unrelated variables are held constant to determine cause-and-effect relationships among variables. (p. 25)

expulsion In males, the contraction of pelvic muscles that force semen through the urethra and out of the body through the penis. (p. 83)

fallopian tubes The tubes that carry the female ovum from the ovaries to the uterus and in which fertilization occurs. (p. 62)

fatuous love Love based on passion and commitment but lacking intimacy; a foolish or pointless love. (p. 111)

fellatio Oral sex performed on a male. (p. 205)

female-superior position A position for heterosexual intercourse in which the woman is sitting on or crouching over the male. (p. 210)

female condom A tube or pouch of thin polyurethane with a flexible ring at each end. One end is sealed and the other is open. The condom is inserted into the vagina to protect against pregnancy and the transmission of STIs. (p. 158)

female genital mutilation (FGM) Removing part or most of the vulva to prevent sexual stimulation or pleasure; a cultural practice in many countries, especially in Africa. (p. 53)

female orgasmic disorder A sexual problem in which a woman rarely or never reaches orgasm or orgasms are delayed; also known as *inhibited female orgasm* or *anorgasmia*. (p. 250)

female sexual interest/arousal disorder A woman's frequent or persistent inability to attain or maintain sexual arousal. (p. 241)

fertility awareness A method of contraception based on ovulation prediction and the viability of sperm; intercourse is timed to avoid fertile days, or a barrier method is used during those days. (p. 173)

fetal alcohol spectrum disorders A variety of disorders that may occur in a person whose mother drank alcohol during pregnancy. (p. 335)

fetishistic disorder A sexual preference for a nonhuman object or a body part that most members of a culture do not consider sexual. (p. 545)

fetus An embryo after eight weeks of pregnancy. (p. 327)

field of eligibles All the individuals who meet a person's criteria as a potential romantic partner. (p. 102)

flirting Subtle behaviors designed to signal sexual or romantic interest in another person. (p. 107)

follicle-stimulating hormone (FSH) A hormone that stimulates the development of a mature ovum. (p. 64)

follicular phase The early period during a woman's monthly fertility cycle when the pituitary gland secretes *follicle-stimulating hormone* (FSN) to enhance ovum development. (p. 322)

foreskin A layer of skin covering the glans of the penis. (p. 37)

frenulum The band of tissue connecting the underside of the penile glans with the shaft of the penis. (p. 37)

frotteuristic disorder Rubbing one's genitals against a nonconsenting person for sexual arousal and, typically, orgasm. (p. 537)

G-spot In some women, an area of tissue on the anterior (upper) wall of the vagina that, when stimulated, may cause a woman to experience enhanced sexual arousal and more intense orgasms. (pp. 58; 85)

gay Homosexual; often applied to both men and women. (p. 405)

gay bashing Criminal acts or violence motivated by homophobia and committed against nonheterosexual individuals. (p. 432)

gender-conforming behavior Behavior that is consistent with traditional cultural expectations for a child's sex. (p. 420)

gender-nonconforming behavior Behavior that is inconsistent with traditional cultural expectations for a child's sex and considered more appropriate for children of the other sex. (p. 420)

gender The masculinity-femininity dimension of our basic nature as humans. (p. 366)

gender dysphoria Refers to stress or discomfort stemming from the self-knowledge that one's biological sex does not conform to, or is the opposite of, his or her personal gender identity. (p. 379)

gender identification A developmental stage in children between the ages of three and five during which they begin to understand which sex they are. (p. 440)

gender identity The sex (male or female) that a person identifies himself or herself to be. (pp. 4; 366)

gender identity disorder A strong cross-gender identification characterized by the desire to be the other sex, combined with persistent discomfort about one's biological sex or culturally prescribed gender role. (p. 400)

gender roles A set of behaviors, attitudes, and emotions that are generally socially expected for men and women in a given culture. (p. 366)

gender stereotypes An assumption, usually negative, made about a person's appearance, behavior, or personality, based solely on his or her gender without regard for the person's individuality as a person. (p. 386)

genital herpes An STI caused by the herpes simplex virus (type 2) and characterized by painful sores and blisters, usually in the genital or anal area. (p. 273)

Genito-Pelvic Pain/Penetration Disorder (GPPPD) Intercourse-related pain that interferes physically or emotionally with sexual activity and/or enjoyment. (p. 253)

gigolo A male prostitute who is paid or otherwise compensated for providing sexual services to women. (p. 558)

gonadotropin A hormone released by the pituitary gland that signals the testes to release testosterone and the ovaries to release estrogen. (p. 450)

gonads Organs that produce cells (ova or sperm) for reproduction. (p. 42)

gonorrhea A sexually transmitted bacterium typically producing pain upon urination and a thick, cloudy discharge from the penis or vagina; often asymptomatic, especially in women. (p. 297)

hate crimes Violent crimes motivated by prejudice and discrimination, targeting specific groups of individuals. (p. 432)

hate crimes laws Laws prescribing more stringent penalties for crimes motivated by bias or prejudice. (p. 432)

hepatitis B virus (HBV) A virus that may be sexually transmitted and may lead to inflammation and impaired functioning of the liver. (p. 279)

heterocentric The assumption of a "universal" heterosexual orientation. (p. 405)

heterosexual A person who is attracted romantically and sexually primarily to persons of the opposite sex. (p. 405)

homophobia Extreme fear, discomfort, or hatred of nonheterosexual individuals. (p. 430)

homosexual A person who is attracted romantically and sexually primarily to persons of one's own sex. (p. 405)

hostile environment A distressing work or educational environment resulting from overt or covert sexually related activities or intrusions. (p. 513)

house prostitute A brothel worker. (p. 558)

human immunodeficiency virus (HIV) The virus that causes AIDS. (p. 282)

human papilloma virus (HPV) A sexually transmitted virus that is typically characterized by warts in the genital or anal area and may lead to some forms of cancer; also known as *genital warts*. (pp. 61; 276)

human sexuality An area of research and study focusing on all aspects of humans as sexual beings. (p. 2)

hustler A male prostitute who services male clients and employs professional tactics similar to those of female streetwalkers. (p. 560)

hymen A ring of tissue surrounding, partially covering, or fully screening the vaginal opening. (p. 53)

hymenorrhaphy A medical procedure, common in some cultures, to reconstruct or repair the hymen to allow a woman to appear "virginal"; also known as *hymenoplasty*. (p. 54)

hypersexual disorder Usually defined as nonparaphilic sexual compulsivity; an obsessive preoccupation with and compulsive need for sexual activity. (p. 528)

hypoactive sexual desire (HSD) A persistently low level or lack of sexual fantasies or desire for sexual activity; also known as *inhibited sexual desire*. (p. 91)

incapacitated rape A rape that occurs when the victim has been rendered unable to take defensive action, usually accomplished through the use of alcohol or date rape drugs. (p. 490)

incest Molestation of a child by a blood relative such as parent, aunt, uncle, grandparent, brother, or sister. (p. 503)

incidence The number of new cases of a disease in a given population over a specific time period. (p. 283)

incubation period The time between infection and the appearance of physical symptoms of illness. (p. 281)

independent variable The variable of interest in an experiment that is allowed to change between or among groups while all other variables are held constant. (p. 26)

infatuation Love based on passion but lacking intimacy and commitment; usually very sexually charged but shallow and devoid of much meaning. (p. 110)

infertility A failure to conceive for 12 consecutive months despite persistent attempts. (p. 355)

informed consent Agreeing to participate in an experiment only after having been provided with complete and accurate information about what to expect in the study. (p. 29)

intrauterine device (IUD) A small plastic device in the shape of a T that is inserted by a doctor into the uterus through the cervix, via the vagina. It then remains in place one to ten years, during which time pregnancy is effectively prevented. (p. 176)

jaundice A symptom of hepatitis characterized by a deep yellowing of the skin and eyes. (p. 281)

john school A court-mandated diversion program for men who have been arrested for soliciting prostitution. (p. 569)

Kaplan's three-stage model An alternative to Masters and Johnson's EPOR model of human sexual response developed by Helen Singer Kaplan that features the three stages of desire, excitement, and orgasm. (p. 90)

Klinefelter syndrome A male genetic condition characterized by a rounded body type, lack of facial hair, breast enlargement in puberty, and smaller-than-normal testicles. (p. 370)

labia majora Folds of skin and fatty tissue that extend from the mons down both sides of the vulva, past the vaginal opening to the perineum. (p. 48)

labia minora The smooth, hairless, inner lips of the vulva. (p. 49)

laparoscopy A surgical procedure in which a tube with a tiny camera and light is inserted through a small incision in the abdomen. (pp. 178; 359)

legalization of prostitution Regulation of prostitution by state laws, with statutes defining where, when, and how prostitution may take place. (p. 571)

lesbian A female with a homosexual orientation. (p. 405)

ludus love A style of love that focuses on the excitement of forming a relationship more than the relationship itself and typically moves rapidly from one relationship to another. (p. 115)

luteal phase The later period of a woman's monthly fertility cycle when the lining of the uterus thickens in preparation for receiving a fertilized ovum if conception has occurred. (p. 323)

luteinizing hormone (LH) A hormone that acts in concert with follicle-stimulating hormone to stimulate ovulation and the release of estrogen and progesterone. (p. 64)

male condom A thin sheath of latex (rubber), polyurethane (plastic), or animal tissue that is placed over an erect penis prior to intercourse. (p. 153)

male escort A man hired by women as a companion, not necessarily for sexual purposes. (p. 558)

male hypoactive sexual desire (MHSDD) A persistently low level or lack of sexual fantasies or desire for sexual activity. (p. 235)

mammogram Low-dose X-ray of the breast to detect tumors. (p. 59)

mania love A possessive, dependent, and often controlling style of love. (p. 116)

masturbation Any sexual activity performed on oneself by oneself, typically focusing on manipulation of the genitals to orgasm. (p. 195)

matching hypothesis The theory that people tend to seek romantic and sexual partners who possess a level of physical attractiveness similar to their own. (p. 105)

medical abortion A method of abortion using drugs rather than surgery to terminate a pregnancy. (p. 340)

menarche The beginning of menstruation during puberty; a girl's first period. (pp. 64; 322; 450)

menopause The normal, gradual change in a woman's life, typically occurring between age 45 and 55, when the ovaries produce a decreasing amount of female hormones and menstrual periods cease. (pp. 68; 469)

menstrual cycle The hormone-controlled reproductive cycle in the human female. (p. 64)

mere exposure effect The psychological principle that humans appear to have a natural and usually unconscious tendency to grow fonder of a "novel stimulus" the more often they are exposed to it. (p. 106)

midwife A person (usually a woman) who has been trained in most aspects of pregnancy, labor, and delivery, but who is not a physician or registered nurse. (p. 347)

midwifery The practice of trained midwives assisting women through normal pregnancy and childbirth. (p. 317)

minipill An oral contraceptive containing progestin only. (p. 161)

miscarriage The loss (without any purposeful intervention) of an embryo or fetus during the first 20 weeks of pregnancy; also called *spontaneous abortion*. (p. 330)

mons veneris A slightly raised layer of fatty tissue on the top of a woman's pubic bone, usually covered with hair on an adult. (p. 48)

morals A person's individual, unique attitudes about what constitutes right and wrong. (p. 2)

multiple orgasms More than one orgasm at relatively short intervals as sexual stimulation continues without a resolution phase or refractory period in between orgasms. (p. 82)

mutual masturbation Partners' touching of each other's genitals, often to orgasm and enjoyed as a sexually intimate and satisfying activity. (p. 204)

mycoplasma genitalium (M. genitalium) A sexually transmitted bacterium that is responsible, along with chlamydia, for the largest percentage of nongonococcal urethritis. (p. 299)

neonate A newborn infant. (p. 346)

new view of women's sexual problems A model of female sexual response incorporating a larger variety of factors than previous models, including physical, cognitive, social, and relationships issues. (p. 92)

nocturnal penile tumescence Erection of the penis while a man is asleep. (p. 240)

nongonococcal urethritis (NGU) A sexually transmitted bacterial infection of the urethra characterized by urethral inflammation and discharge, but not caused by the gonorrhea bacterium. (p. 299)

normal jealousy Jealousy based on a real threat to the relationship, as when one partner discovers that the other has been sexually unfaithful. (p. 128)

nurse-midwife A registered nurse who has completed an accredited midwifery program and has been certified by the state to deliver babies. (p. 347)

OB/GYN Short for obstetrician-gynecologist, a physician specializing in pregnancy and childbirth. (p. 347)

obscenity Sexually explicit works that meet specific legal criteria that render them exempt from First Amendment protections and may be declared illegal. (p. 574)

observational research Gathering behavioral data through direct or indirect observation using scientific techniques. (p. 22)

oedipus complex A Freudian notion explaining how a young boy comes to identify with his father. (p. 440)

oocyte An immature reproductive egg, or ovum. (p. 322)

opportunistic infections Diseases that establish themselves in the human body only when the immune system is weakened and incapable of fighting them off. (p. 282)

oral contraceptives Tablets containing female hormones that are ingested every day. They constitute the most popular reversible contraception method used by women in the United States; also known as birth control pills. (p. 160)

orgasm The peak of sexual arousal. (pp. 46; 82)

orgasmic phase The third stage in the EPOR model, during which sexual excitement and pleasure reach a climax. (p. 82)

orgasmic reconditioning A type of therapy for paraphilias in which a person is conditioned to become aroused and have orgasms in socially acceptable settings that bear no resemblance to the paraphilia. (p. 549)

os The very narrow passageway through the cervix, from the vagina to the uterus. (p. 61)

outercourse A form of abstinence in which a couple chooses to engage only in sexual behaviors that are unlikely to result in pregnancy or infection and to avoid all others, such as vaginal or anal intercourse and oral sex. (p. 152)

ovarian cyst A fluid-filled sac on the surface of the ovary, formed during normal ovulation; sometimes cysts may swell and cause pain and abnormal bleeding. (p. 63)

ovaries The female organs that produce sex hormones such as estrogen and progesterone and where follicle cells are stored and mature into ova. (p. 63)

overlapping-curve model A description of differences between males and females where the averages vary, but the distribution of the characteristic includes both males and females. (p. 393)

ovulation The release of an egg, or ovum, from the ovary into the fallopian tube. (p. 64)

ovum The female reproductive cell stored in the ovaries; usually, one ovum is released approximately every 28 days between menarche and menopause. The plural is *ova*. (p. 62)

pandemic A sudden outbreak of a disease (an epidemic) of major proportions that affects a large region, a continent, or spreads worldwide. (p. 263)

pap test A routine test in which cells from the cervix are examined microscopically to look for potentially cancerous abnormalities. (p. 61)

paraphilias Compulsive sexual activities that are practiced by a small percentage of people and that most members of a given culture would consider abnormal, deviant, or pathological. (p. 524)

pathological Any behavior seen as caused by sickness or disease. (p. 526)

pathological jealousy Jealousy felt within one partner despite the fact that no threat to the relationship actually exists. (p. 128)

pedophile An adult, usually male, whose sexual focus is on children. (p. 540)

pedophilia Uncontrollable sexual compulsions involving children. (p. 503)

pedophilic disorder Uncontrollable sexual compulsions involving children. (p. 540)

pelvic inflammatory disease (PID) A painful condition in women marked by inflammation of the uterus, fallopian tubes, and ovaries; typically caused by one or more untreated STIs. (p. 296)

penile glans The end or tip of the penis, its most sexually sensitive part. (p. 36)

penile plethysmograph An electronic device that records blood flow into the penis to detect sexual arousal. (p. 434)

penile shaft The area of the penis between the glans and the abdomen. (p. 37)

penis The primary male anatomical sexual structure. (p. 36)

perimenopausal changes The physical and psychological changes many women experience during the decade leading up to menopause. (p. 68)

perineum The area of skin in the female between the vulva and the anus, and in the male between the scrotum and the anus. (p. 54)

personal gender stereotypes Beliefs about gender that are unique to each individual and may or may not agree with cultural stereotypes. (p. 389)

personal sexual philosophy A person's unique foundation of knowledge, attitudes, and actions relating to what the person wants and who he or she is as a sexual being. (p. 2)

pituitary gland A gland in the brain that, at the onset of puberty, releases hormones necessary for the physical changes of puberty. (p. 450)

placenta An organ that develops on the uterine wall during pregnancy and joins the developing embryo to the mother's biological systems, transferring nourishment, oxygen, and waste products between the fetus and the mother. (p. 326)

planned home birth Delivery of an infant in a private home setting, usually with necessary equipment and personnel provided by a professional service. (p. 348)

plateau phase The second phase in the EPOR model, during which sexual arousal levels off (reaches a plateau) and remains at an elevated level of excitement. (p. 77)

pornography In legal terms, any sexually explicit work deemed obscene according to legal criteria and therefore exempt from freedom of speech protections. (p. 572)

postpartum Literally "following birth"; typically refers to the months or first year following the birth of a child. (p. 351)

postpartum depression (PPD) A psychological depressive disorder that begins within four weeks after childbirth. (p. 352)

postpartum psychosis A severe postpartum psychological disorder that may include delusions, hallucinations, and extreme mental disorganization. (p. 352)

posttraumatic stress disorder (PTSD) A pervasive psychological and emotional reaction to a traumatic event experienced or witnessed (such as war, natural disasters, rape, etc.), including symptoms of anxiety, flashbacks, hypervigilance, irritability, nightmares, and depression. (p. 499)

pragma love A love style in which partners are selected in a businesslike way on the basis of rational, practical criteria. (p. 116)

pre-ejaculate The fluid produced by the Cowper's glands. (p. 47)

pregnancy The period of growth of the embryo and fetus in the uterus. (pp. 144; 326)

premature ejaculation (PE) A man's tendency to have an orgasm suddenly with little penile stimulation, typically just before, upon, or shortly after penetration of the penis into the vagina; also referred to as *rapid* or *early ejaculation*. (p. 247)

premenstrual dysphoric disorder (PMDD) A significantly more intense and debilitating form of PMS. (p. 66)

premenstrual syndrome (PMS) A set of symptoms that may occur during the days just before and during the start of a woman's period, which includes irritability, depressed mood, and feelings of physical bloating or cramping. (p. 66)

preterm birth Birth of an infant less than 37 weeks after conception. (p. 331)

prevalence The total cumulative number of cases of a disease in a given population. (p. 283)

pro-choice The belief that a woman has the moral and legal right to choose freely to abort her pregnancy. (p. 338)

pro-life The belief that voluntary abortion is akin to murder and that it should be illegal. (p. 338)

prodromal symptoms Warning signs, such as itching, burning, or pain, that an outbreak of an infection such as herpes may be impending. (p. 274)

progesterone The female hormone responsible for the release of ova and implantation of the fertilized egg in the uterine wall. (p. 63)

prostate gland A gland in males surrounding the urethra that produces the largest proportion of seminal fluid (ejaculate). (p. 46)

prostatitis An uncomfortable or painful inflammation of the prostate gland, usually caused by bacteria. (p. 46)

prostitution Providing or receiving sexual acts, between a prostitute and a client, in exchange for money or some other form of remuneration. Also referred to as the "sex trades." (p. 557)

protected classes Specific groups of people protected under federal and state antidiscrimination laws, identified by race, religion, sex, age, or other characteristics. (p. 429)

proximity effect The theory that the closer you are to another person in geographical distance, the greater the probability that you will grow to like or even love the person. (p. 106)

prurient interest An excessive focus on exclusively sexual matters. (p. 575)

psychological defense mechanism Originally suggested by Freud, a psychological distortion of reality serving to defend against personally unacceptable thoughts or urges. (p. 433)

psychosexual stages Freud's theory that the development of human personality occurs in a series of stages during childhood. (p. 440)

puberty A natural change in a child's body during early to middle childhood, during which secondary sexual characteristics develop and boys begin to produce sperm and girls begin menstruation. (p. 450)

pubic lice Small, bug-like parasites, usually sexually transmitted, that infest the genital area, causing extreme itching; often referred to as "crabs" because of their resemblance to a sea crab. (p. 304)

quickening The first movement of the fetus that is felt by the mother. (p. 328)

quid pro quo Something given in exchange for a benefit or reward; with reference to sexual harassment, a situation in which a person in a position of power over another requires sexual favors in exchange for some beneficial outcome for the victim. (p. 513)

random sampling A method of selecting a sample of participants in such a way that each member of the population has an equal chance of being selected. (p. 20)

rape Nonconsensual sexual penetration of the body using physical force or the threat of bodily harm. (p. 484)

rape trauma syndrome A two-stage set of symptoms that follow the trauma of being raped, consisting of physical, emotional, and behavioral stress reactions. (p. 499)

reaction formation A type of defense mechanism in which a person engages in exaggerated behaviors in the *opposite* direction of internal urges felt to be unacceptable or intolerable. (p. 433)

rear-entry position A position for heterosexual intercourse in which the penis is inserted into the vagina while the man is behind the woman. (p. 211)

reciprocity of attraction The idea that someone you like or love likes or loves you back—reciprocates your feelings—with approximately the same degree of intensity. (p. 109)

refractory period A period of time following orgasm when a person is physically unable to become aroused to additional orgasms. (p. 88)

reliability the extent to which a measurement is consistent over repeated administrations. (p. 27)

resolution phase The fourth and last stage in the EPOR model, during which sexual structures return to their unaroused state; also referred to as *detumescence*. (p. 87)

respondents Individuals selected to respond to a researcher's request for information. (p. 18)

retrovirus A type of virus, such as HIV, that survives and multiplies by invading and destroying the DNA of normal body cells and then replicating its own DNA into the host cell's chromosomes. (p. 283)

romantic love Love based on intimacy and passion but lacking commitment. (p. 111)

sadomasochism Sexual activities that combine sadism and masochism. (p. 538)

safe-sex fatigue A loss of tolerance for the necessity of practicing safer sex behaviors. (p. 285)

sample A subset of the target population selected by researchers to represent the entire population under study. (p. 20)

scrotum The sac of thin skin and muscle containing the testicles in the male. (p. 42)

secondary sexual characteristics Physical changes not biologically related to reproduction that occur during puberty. (p. 451)

selective abstinence Choosing to engage in or avoid certain sexual behaviors on the basis of their risks of STIs or pregnancy. (pp. 152; 310)

self-selection bias The effect of allowing members of a target population under study to volunteer to participate in the study; it may compromise the randomness and validity of the research. (p. 21)

semen The fluid produced primarily by the prostate gland and seminal vesicles that is ejaculated with the sperm cells by men during orgasm. (p. 45)

seminal vesicle A structure that produces fluid that becomes part of the semen that is expelled during ejaculation. (p. 45)

seminiferous tubules Tightly wound microscopic tubes that comprise the testicles in the male, where sperm cells are generated. (pp. 43; 324)

sensate focus A sex therapy technique that requires a couple to redirect emphasis away from intercourse and focus on their capacity for mutual *sensuality*. (p. 230)

sex flush A darkening or reddening of the skin of the chest area that occurs in some people during sexual arousal. (p. 77)

sex reassignment surgery Surgical procedures used to transform physically an individual from one sex to the other; commonly known as a sex-change operation. (p. 381)

sexual addiction A term often used to refer to strong compulsive sexual behavior—more correctly called "hypersexual disorder." (p. 528)

sexual harassment A pattern of unwelcome sexual advances, requests for sexual favors, or other verbal and physical conduct that is coercive or creates a hostile work or educational environment. (p. 513)

sexual health A general concept referring to physical, emotional, psychological, and interpersonal well-being. (p. 13)

sexually transmitted diseases (STDs) A group of viral, bacterial, and other infections that are spread primarily by sexual behaviors (another name for sexually transmitted infections). (p. 261)

sexually transmitted infections (STIs) A group of viral, bacterial, and other infections that are spread primarily by sexual behaviors; sometimes also called sexually transmitted diseases (STDs). (p. 261)

sexual masochism disorder Sexual arousal and gratification that are associated with acts or fantasies of being hurt, humiliated, or otherwise made to suffer. (p. 543)

sexual orientation Term specifying the sex of those to whom a person is primarily romantically, emotionally, and sexually attracted. (pp. 4; 385; 405)

sexual sadism disorder Inflicting pain, injury, or humiliation on another person for the sexual gratification of the person performing the action. (p. 538)

sexual self-disclosure Revealing private sexual thoughts and feelings to another person. (p. 124)

Skene's glands In the female, a pair of glands on either side of the urethra that in some women may produce a fluid that is expelled during orgasm; also known as the *paraurethral glands*. (p. 86)

social alienation A passive form of aggression that includes behaviors such as malicious gossip, spreading negative rumors, and shunning. (p. 390)

sodomy laws Laws prohibiting specific sexual activities between adults, even in private and with their consent. (p. 430)

spectatoring Mentally observing and judging oneself during sexual activities with a partner; may cause sexual problems. (p. 238)

spermatic cords Supporting each testicle and encasing the vas deferens, nerves, and muscles. (p. 42)

spermicide Any substance containing a chemical (most commonly nonoxinol-9) that kills sperm cells, thereby preventing them from fertilizing an egg. (p. 172)

SSRIs Selective serotonin reuptake inhibitors; drugs administered to treat depression that may cause various sexual side effects, especially inhibited or delayed arousal or orgasm. (p. 226)

stage one labor The first stage of the birth process, involving the beginning of contractions of the uterus. (p. 344)

stage three labor The final stage of the birth process, when the placenta is expelled from the uterus with the umbilical cord attached. (p. 346)

stage two labor The stage of the birth process in which contractions occur closer together than in stage one, involve the muscles of the abdomen as well as the uterus, and continue until the infant has been expelled from the mother's body. (p. 345)

standard days method A fertility awareness technique for tracking fertile and infertile days during a woman's menstrual cycle. (p. 173)

start-stop method A technique used in the treatment of premature ejaculation involving intermittent increases and decreases in arousal (p. 249)

statutory rape In general, as defined by state laws, any sexual penetration by an adult of a minor child regardless of the minor's consent. (p. 484)

sterilization Any surgical alteration that prevents the emission of sperm or eggs; also referred to as voluntary surgical contraception. (p. 178)

stonewalling Relying on a passive form of power and aggression by being unresponsive (erecting a metaphorical "stone wall") when disagreements and disputes erupt. (p. 120)

storge love A love style characterized by caring and friendship. (p. 116)

straight Heterosexual. (p. 405)

streetwalker A prostitute who sells sexual services on the street, typically to customers who are driving by, soliciting sexual acts. (p. 557)

styles of love Lee's theory that people follow individual psychological motifs or styles in relating to a love partner. (p. 115)

survey The scientific collection of data from a group of individuals about their beliefs, attitudes, or behaviors. (p. 18)

symptothermal method A fertility awareness method based on monitoring a woman's cervical secretions and internal body temperature upon awakening in the morning. (p. 175)

syphilis A sexually transmitted bacterium characterized by a sore, or chancre, at the point of infection; untreated, it may progress to more serious stages and even death. (p. 300)

target population The entire group of people to which a researcher is attempting to apply a study sample's findings. (p. 19)

tenting A widening of the inner two-thirds of the vagina during sexual arousal. (p. 80)

teratogen Any agent that has the potential to cause a fetal abnormality. (p. 333)

testicles Oval structures approximately 1.0 to 1.5 inches in length made up of microscopic tubes in which sperm cells and testosterone are produced in the male. (p. 42)

testosterone The male sex hormone responsible for male sexual characteristics and the production of sperm cells. (p. 42)

transgender Individuals who do not self-identify as the gender that conforms to their biological sex; this is not considered a paraphilia. (pp. 379; 547)

transition phase The end of stage one labor as the fetus begins moving through the cervix and down the birth canal. (p. 345)

transsexual A transgender person who has transitioned or is transitioning from his or her biological sex to his or her self-identified gender through actions, dress, hormone therapy, or surgery. (pp. 380; 547)

transvestic disorder The paraphilic disorder characterized by dressing in the clothing of the other sex for sexual arousal. (p. 547)

transvestite A man who obtains sexual satisfaction by wearing female clothing. (p. 547)

treatment The action performed on or by a group in an experiment. (p. 25)

triangular theory of love Sternberg's theory that three fundamental components of love—intimacy, passion, and commitment—in various combinations, define the qualities of a relationship. (p. 110)

trichomoniasis A common sexually transmitted protozoan parasite causing symptoms in women, including genital irritation, painful urination, and a foul-smelling vaginal discharge; infected men are typically asymptomatic, yet contagious. (p. 303)

trimester One of three periods of about three months each that make up the phases of a full-term pregnancy. (p. 327)

tubal ligation A permanent method of contraception involving tying, cutting, clipping, or otherwise blocking the fallopian tubes to prevent passage of an ovum. (p. 178)

Turner syndrome A female genetic condition characterized by short stature, slow or no sexual development at puberty, heart abnormalities, and lack of ovarian function. (p. 370)

twoday method A fertility awareness technique that relies on careful observation of secretions from the cervix to predict ovulation. (p. 174)

two-dimensional model of gender An approach to defining gender suggesting that gender is not an either-or proposition but that people may manifest elements of both genders simultaneously. (p. 397)

umbilical cord A structure approximately 22 inches in length, consisting of one large vein and two arteries that transport nutrients, oxygen, and fetal waste products back and forth between the fetus and the placenta. (p. 326)

urethra The tube extending from the bladder to the urethral opening, which carries urine out of the body in both women and men, as well as semen in men. (p. 37)

urethral bulb The prostatic section of the urethra that expands with collected semen just prior to expulsion, creating the sensation of ejaculatory inevitability. (p. 46)

urethral opening An opening in the midsection of the vulva, between the clitoral glans and the vagina, that allows urine to pass from the body. (p. 53)

urethritis A painful inflammation of the urethra; often caused by one or more untreated STIs. (p. 297)

urinary tract infection (UTI) An infection of the urethra, bladder, or other urinary structure, usually caused by bacteria. (p. 53)

uterus A very flexible organ with strong muscle fibers where a fertilized egg implants and an embryo and fetus grow, from a few days after fertilization until birth. (p. 62)

vacuum aspiration A method of abortion in which a small tube is inserted through the cervix to extract the contents of the uterus, including the endometrium lining and embedded embryo. (p. 338)

vagina A flexible, muscular canal or tube, normally about 3 to 4 inches in length, that extends into the woman's body at an angle toward the small of the back, from the vulva to the cervix. (p. 58)

validity The extent to which a measurement accurately reflects the concept being measured. (p. 27)

vas deferens A tube extending from the testicle (epididymis) into the male's body for the transport of mature sperm cells during ejaculation. (p. 43)

vasectomy Cutting and tying off or sealing each vas deferens so that sperm produced by the testicles can no longer mix with semen in the ejaculate. (p. 178)

vasocongestion The swelling of erectile tissues due to increased blood flow during sexual arousal. (p. 77)

victimizing paraphilic disorder Sexual activities involving an unsuspecting, nonconsenting, or unwilling person as the target of the atypical, compulsive behavior. (p. 535)

victimless Harming no one, with the possible exception of the person performing the action. (p. 530)

viral shedding The release of virus particles that can potentially spread the infection to others. (p. 275)

voyeuristic disorder Secretly watching others undress or engage in sexual activities without their knowledge or consent for the purpose of achieving sexual arousal. (p. 536)

vulva The female external genitals. (p. 48)

withdrawal method Removing the penis from the vagina just prior to ejaculation—a usually unreliable method of contraception; also called *coitus interruptus* and "pulling out." (p. 159)

zygote A fertilized ovum (or egg) moving down the fallopian tube. (p. 325)

References

5 steps to safer schools. (2011). Southern Poverty Law Center: Teaching Tolerance. Retrieved from **www.tolerance.org/activity/5-steps-safer-schools**

Abbey, A. (2002). Alcohol-related sexual assault: A common problem among college students. *Journal of Studies on Alcohol, 63*, S118–S128.

Abbey, A., Zawacki, T., Buck, P., Clinton, A., & McAuslan, P. (2001). Alcohol and sexual assault. *Alcohol Research and Health, 25*, 43–51.

Abdo, C. (2013). Treatment of premature ejaculation with cognitive behavioral therapy. In Jannini, E., McMahon, A., Waldinger, C., & Marcel, D. (Eds.). *Premature Ejaculation: From Etiology to Diagnosis and Treatment*. New York, NY: Springer.

Abramovitz, M. (2001, March). The knockout punch of date rape drugs. *Current Health*, pp. 18–21.

Abusharaf, R. (1998). Unmasking tradition: A Sudanese anthropologist confronts female "circumcision" and its terrible tenacity. *Sciences, 38*, 22–28.

Acevedo, B., & Aron, A. (2009). Does a long-term relationship kill romantic love? *Review of General Psychology, 13*, 59–65.

ACOG. (2007). *Birth control*. American College of Obstetricians and Gynecologists.

ACOG. (2013). Cesarean delivery on maternal request: Committee opinion No. 559. American College of Obstetricians and Gynecologists. *Obstetrics and Gynecololgy, 121*, 904–907.

Adam, T., & Bathija, H. (2010). Estimating the obstetric costs of female genital mutilation in six African countries. *Bulletin of the World Health Organization, 88*, 281–288.

Adams, H., Wright, L., & Lohr, B. (1996). Is homophobia associated with homosexual arousal? *Journal of Abnormal Psychology, 105*, 440–446.

Addis, J. (2014, January 21). A mental health primer on paraphilias in DSM-5. *The National Psychologist*. Retrieved from **www.nationalpsychologist.com/2014/01/a-mental-health-primer-on-paraphilias-in-dsm-5/102417.html**

Adoption laws. (2008). Adoption laws: State by state. *The Human Rights Campaign*. Retrieved from **www.hrc.org/issues/parenting/adoptions/8464.htm**

Adriaenssens, S., & Hendrickx, J. (2012). Sex, price and preferences: accounting for unsafe sexual practices in prostitution markets. *Sociology of health & illness, 34*(5), 665–680.

Agger, C., & Day, K. (2011). Accessible information and prevention strategies related to student sexual harassment: A review of students harassing students. *The High School Journal, 94*(2), 77–78.

Ahmadi, A. (2013). Ethical issues in hymenoplasty: Views from Tehran's physicians. *Journal of Medical Ethics*. Retrieved from **www.jme.bmj.com/content/early/2013/06/12/medethics-2013-101367.full.pdf+htmldoi:10.1136/medethics-2013-101367**

Ahmed, S., & Rodie, M. (2010). Investigation and initial management of ambiguous genitalia. *Best Practice & Research Clinical Endocrinology & Metabolism, 24*(2), 197–218.

Ahrold, T. K., Farmer, M., Trapnell, P. D., & Meston, C. M. (2011). The relationship among sexual attitudes, sexual fantasy, and religiosity. *Archives of sexual behavior, 40*(3), 619–630.

"AIDS Signs." (2011). AIDS signs and symptoms. UCSF Medical Center.

Alanko, K., Santtila, P., Harlaar, N., Witting, K., Varjonen, M., Jern, P., . . . & Sandnabba, N. (2010). Common genetic effects of gender atypical behavior in childhood and sexual orientation in adulthood: A study of Finnish twins. *Archives of Sexual Behavior, 39*(1), 81–92.

Albaugh, J., & Kellogg-Spadt, S. (2002). Sensate focus and its role in treating sexual dysfunction (Intimacy Issues). *Urologic Nursing, 22*, 402–403.

Albert, A. E., Warner, D. L., & Hatcher, R. A. (1998). Facilitating condom use with clients during commercial sex in Nevada's legal brothels. *American Journal of Public Health, 88*, 643–646.

Aldhous, P. (2008). Depressed dad spells trouble for kid's vocabulary. *New Scientist, 198*, 12.

Alessi, E. J., & Martin, J. I. (2010). Conducting an internet-based survey: Benefits, pitfalls and lessons learned. *Source Social Work Research*, 34, 2, 122–128.

Alexander, G. (2003). An evolutionary perspective of sex-typed toy preferences: Pink, blue, and the brain. *Archives of Sexual Behavior, 32*, 7–14.

Allies Program. (2000). *Sexual harassment and sexual assault*.

Almeida, J., Renee, M., Johnson, H., Corliss, B., & Azrael, D. (2009). Emotional distress among LGBT youth: The influence of perceived discrimination based on sexual orientation. *Journal of Youth and Adolescence, 38*, 1001–1014. doi:10.1007/ s10964-009-9397-9

Alteri, C. J., Hagan, E. C., Sivick, K. E., Smith, S. N., & Mobley, H. L. (2009, September). Mucosal immunization with iron receptor antigens protects against urinary tract infection. *PLoS Pathogens, 5*(9), e1000586. doi: 10.1371/journal.ppat.1000586

Althof, S. (2006). Psychological approaches to the treatment of rapid ejaculation. *Journal of Men's Health & Gender, 3*, 180–186.

Altman, C. (2000, February-March). Gay and lesbian seniors: Unique challenges of coming out in later life. *SEICUS Report*, pp. 14–17.

Altman, D., & Aggleman, P. (1999). *Men who sell sex: International perspectives on male prostitution and HIV/AIDS*. Philadelphia: Temple University Press.

Altman, L. K. (2008, October 7). Discoverers of AIDS and cancer viruses win Nobel. *New York Times*. Retrieved from **www.nytimes.com/2008/10/07/health/07nobel.html**

American Association of University Women. (2002). A free resource for preventing sexual harassment in schools: The American Association of University Women (AAUW) Educational Foundation offers a new guide online. *Curriculum Review, 42*, 2–3.

American Cancer Society. (2010). *Cervical cancer: Prevention and early detection*. Retrieved from **www.cancer.org/Search/index?QueryText=pap+smear**

American Cancer Society. (2013a). Breast cancer survival rates by stage. Retrieved from **www.cancer.org/cancer/%20breastcancer/detailedguide/breast-cancer-survival-by-stage**

American Cancer Society. (2013b). Can breast cancer be found early? Retrieved from **www.cancer.org/cancer/breastcancer/detailedguide/breast-cancer-detection**

American Cancer Society. (2013c). What is ovarian cancer? Retrieved from **www.cancer.org/cancer/ovariancancer/detailedguide/ovarian-cancer-key-statistics**

American Cancer Society. (2014). What are the key statistics about testicular cancer? Retrieved from **www.cancer.org/cancer/testicularcancer/detailedguide/testicular-cancer-key-statistics**

American Pregnancy Association. (2014). Miscarriage. Retrieved from **www.americanpregnancy.org/pregnancycomplications/miscarriage.html**

American Psychiatric Association. (1994). *Diagnostic and statistical manual of mental disorders* (4th ed.) (DSM-IV). Washington, DC: Author.

American Psychological Association. (2008). Report on the APA Task Force on Mental Health and Abortion. Retrieved from **www.apa.org/releases/abortion-report.pdf**

American Psychological Association. (2011). Sexual orientation and homosexuality: Is sexual orientation a choice? Retrieved from **www.apa.org/helpcenter/sexual-orientation.aspx**

American Psychological Association. (2014). What is sexual orientation? Retrieved from **www.apa.org/topics/lgbt/orientation.aspx?item=2**

American Society for Aesthetic Plastic Surgery. (2013). Cosmetic Surgery National Data Bank. Retrieved from **www.surgery.org/sites/default/files/ASAPS-2012-Stats.pdf**

Amory, J., Anawalt, B., Paulson, A., & Bremmer, W. (2000). Klinefelter's syndrome. *Lancet, 356*, 333–335.

Anagrius, C., Lore, B., & Jensen. J. (2005). Mycoplasma genitalium: Prevalence, clinical significance, and transmission. *Sexually Transmitted Infections, 81*(6), 458–462. doi:10.1136/sti.2004.012062

Anderson, D. J. (2003). The impact on subsequent violence of returning to an abusive partner. *Journal of Comparative Family Studies, 34*, 93–125.

Anderson, L., & Whiston, S. (2005). Sexual assault education programs: A meta-analytic examination of their effectiveness. *Psychology of Women Quarterly, 29*, 374–388.

Anderson, M., Foster, C., McGuigan, M. R., Seebach, E., & Porcari, J. P. (2004). Training vs. body image: Does training improve subjective appearance ratings? *Journal of Strength and Conditioning Research, 18*, 255–259.

Angelone, D., Hirschman, R., Suniga, S., Armey, M., & Armelie, A. (2005). The influence of peer interactions on sexually oriented joke telling. *Sex Roles: A Journal of Research, 52*, 187–199.

Annon, J. S. (1974). *The behavioral treatment of sexual problems.* Honolulu: Kapiolani Health Services.

Anti-gay discrimination in schools. (2011). Southern Poverty Law Center: Teaching Tolerance.

APA, "Paraphilic Disorders." (2013). Paraphilic disorders. The American Psychiatric Association. Retrieved from **www.dsm5.org/Documents/Paraphilic%20Disorders%20Fact%20Sheet.pdf**

APA. (2008). Answers to your questions: For a better understanding of sexual orientation and homosexuality. Washington, DC: American Psychological Association. Retrieved from **www.apa.org/topics/sorientation.pdf**

APA. (2013). American Psychiatric Association. *Diagnostic and statistical manual of mental disorders* (5th ed.). Arlington, VA: American Psychiatric Publishing.

Approved Treatments for Hepatitis B. (2010). *Hepatitis B Foundation.* Retrieved from **www.hepb.org/patients/hepatitis_b_treatment.htm**

Arévalo, M., Jennings, V., Nikula, M., & Sinai, I. (2004). Efficacy of the new TwoDay Method of family planning. *Fertility and Sterility, 82*, 885–892.

ARHP. (2010). *Diaphragm.* The Association of Reproductive Health Professionals. Retrieved from **www.arhp.org/MethodMatch/details.asp?productId=9**

Ashbaughma, L., & Cornella, D. (2008). Sexual harassment and bullying behaviors in sixth-graders. *Journal of School Violence, 7*, 21–38. doi:10.1300/J202v07n02_03

ASRM. (2014). Is in vitro fertilization expensive? American Society for Reproductive Medicine. Retrieved from **www.reproductivefacts.org/detail.aspx?id=30232014**

Assumpção, A., Garcia, F., Garcia, H., Bradford, J., & Thibaut, F. (2014). Pharmacologic Treatment of paraphilias. *Psychiatric Clinics of North America, 37*, 173–181.

Astbury-Ward, J. (2008). Emotional and psychological impact of abortion: A critique of the literature. *Journal of Family Planning and Reproductive Health Care, 34*(3), 181–184.

AUA guideline. *American Urological Association Education and Research.* Retrieved from **www.guideline.gov/content.aspx?id=37281**

Aulette, J. (1994). *Changing families.* Belmont, CA: Wadsworth.

AVERT. (2014). Coming out. Averting HIV and AIDS. Retrieved from **www.avert.org/coming-out.htm**

Baber, K., & Murray, C. (2001). A postmodern feminist approach to teaching human sexuality. *Family Relations, 50*, 23–33.

Badash, D. (2009). Does the U.S. Constitution already make gay marriage legal? *Legal Issues: Marriage.*

Bagarozzi, D. (1990). Marital power discrepancies and symptom development in spouses: An empirical investigation. *American Journal of Family Therapy, 18*, 51–65.

Baggaley, R., White, R., & Boily, M. (2010). HIV transmission risk through anal intercourse: Systematic review, meta-analysis and implications for HIV prevention. *International Journal of Epidemiology, 39*, 1048–1063. doi:10.1093/ije/dyq057

Bailey, D. H., Ellingson, J., & Bailey, J. (2014). Genetic Confounds in the Study of Sexual Orientation: Comment on Roberts, Glymour, and Koenen (2014). *Archives of sexual behavior*, 1–3.

Bainbridge, D. (2003). *Making babies: The science of pregnancy.* Boston: Harvard University Press.

Baker, L., & Oswald, D. (2010). Shyness and online social networking services. *Journal of Social and Personal Relationships, 27*(7), 873–889.

Bakken, I., Skjeldestad, F., Lydersen, S., & Nordbo, S. (2007). Births and ectopic pregnancies in a large cohort of women tested for *Chlamydia trachomatis. Sexually Transmitted Diseases, 34*, 739–743.

Baldwin, J., & Baldwin, J. (1997). Gender differences in sexual interest. *Archives of Sexual Behavior, 26*, 181–210.

Baldwin, J., & Baldwin, J. (2000). Heterosexual anal intercourse: An understudied high-risk behavior. *Archives of Sexual Behavior, 29*, 357–373.

Balon, R. (2013). Controversies in the diagnosis and treatment of paraphilias. *Journal of sex & marital therapy, 39*(1), 7–20.

Balon, R., & Seagraves, R. (2009). *Clinical manual of sexual disorders.* Washington, DC: American Psychiatric Publishing.

Baltrushes, N., & Karnik, N. (2013). Victims of military sexual trauma—you see them, too. *Journal of Family Practice, 62*(3), 120–125.

Bancroft, J., Graham, C., Janssen, E., & Sanders, S. (2009). The dual control model: Current status and future directions. *Journal of Sex Research, 46*(2–3), 121–42. doi:10.1080/00224490902747222

Bancroft, J. (1989). *Human sexuality and its problems.* Edinburgh: Churchill Livingstone, pp. 128–130.

Bancroft, J. (2002). Biological factors in human sexuality. *Journal of Sex Research, 39*, 15–21.

Bancroft, J., Loftus, J., & Long, S. (2003). Distress about sex: A national survey of women in heterosexual relationships. *Archives of Sexual Behavior, 32*, 193–208.

Banister, J. (2004). Shortage of girls in China today. *Journal of Population Research, 21*, 19–45.

Banyard, V., Moynihan, M., & Crossman, M. (2009). Reducing sexual violence on campus: The role of student leaders as empowered bystanders. *Journal of College Student Development, 50*(4), 446–457.

Bao, A., & Swaab, D. (2011). Sexual differentiation of the human brain: relation to gender identity, sexual orientation and neuropsychiatric disorders. *Frontiers in neuroendocrinology, 32*(2), 214–226.

Barbaree, H., Marshall, W., & McCormick, J. (1998). The development of deviant sexual behaviour among adolescents and its implications for prevention and treatment. *Irish Journal of Psychology, 19*, 1–31.

Barelds, D., & Barelds-Dijkstra, P. (2010). Humor in intimate relationships: Ties among sense of humor, similarity in humor and relationship quality. *International Journal of Humor Research, 23*, 447–465. doi: 10.1515/humr.2010.021

Barglow, P. (2014). Numbing after rape, and depth of therapy. *American Journal of Psychotherapy, 68*(1), 117–139.

Barinaga, M. (1991). Is homosexuality biological? *Science, 253*, 956–957.

Barnoski, R. (2005). Sex offender sentencing in Washington State: Has community notification reduced recidivism? Washington State Institute for Public Policy, December, 2005. Retrieved from **www.wsipp.wa.gov/rptfiles/05-12-1202.pdf**

Bartels, R. M., & Gannon, T. A. (2011). Understanding the sexual fantasies of sex offenders and their correlates. *Aggression and Violent Behavior, 16*(6), 551–561.

Barth, K., Cook, R., Downs, J., Switzer, G., & Fischhoff, B. (2002). Social stigma and negative consequences: Factors that influence college students' decisions to seek testing for sexually transmitted infections. *Journal of American College Health, 50*(4), 153–159.

Bartlik, B., & Goldberg, J. (2000). Female sexual arousal disorder. *Principles and practice of sex therapy, 3*, 85–117.

Bartz, D., & Greenberg, A. (2008). Sterilization in the United States. *Review of Obstetrics and Gynecology, 1*(1), 23 -32.

Basdekis-Jozsa, R., Turner, D., & Briken, P. (2013). Pharmacological treatment of sexual offenders and its legal and ethical aspects. In K. Harrison & B. Rainey (Eds). *The Wiley Blackwell Handbook of Legal and Ethical Aspects of Sex Offender Treatment and Management* (pp. 302–320). New York, NY: Wiley-Blackwell.

Basson, R. (2001). Female sexual response: The role of drugs in the management of sexual dysfunction. *Obstetrics and Gynecology, 9*, 350–353.

Battaglia, C., Nappi, R. E., Mancini, F., Alvisi, S., Del Forno, S., Battaglia, B., & Venturoli, S. (2010). 3-D volumetric and vascular analysis of the urethrovaginal space in young women with or without vaginal orgasm. *Journal of Sexual Medicine, 7*(1), 1445–1453.

Bauer, W., Shams, H., & Bauer, R. (2013). Vaccination against urinary tract infections caused by E. coli. *Advances in Bioscience and Biotechnology, 4*, 487–492.

Baumeister, R. F. (1995, November–December). An inside look at S&M. *Psychology Today*, pp. 47–52.

Baumgardner, J. (2008, February 26). Objects of suspicion: Love a man, love a woman, either way they distrust you. What is it about bisexual women that lesbians hate so much? *The Advocate, 1002*, 24–27.

Beauregard, E. (2009). Rape and sexual assault in investigative psychology: The contribution of sex offenders' research to offender profiling. *Journal of Investigative Psychology and Offender Profiling, 7*, 1–13.

Beckman, L., Harvey, S., Thorburn, S., Maher, J., & Burns, K. (2006). Women's acceptance of the diaphragm: The role of relationship factors. *Journal of Sex Research, 43*, 297–306.

Bedaiwy, M., Barakat, E., & Falcone, T. (2011). Robotic tubal anastomosis: Technical aspects. *JSLS, 15*(1), 10–15. doi: 10.4293/108680810X12924466009041

Beech, A., & Harkins, L. (2012). DSM-IV paraphilia: Descriptions, demographics and treatment interventions. *Aggressive and Violent Behavior, 17*, 527–39. doi:10.1016/j.avb.2012.07.008

Bélanger, C., Laughrea, K., & Lafontaine, M. F. (2001). The impact of anger on sexual satisfaction in marriage. *Canadian Journal of Human Sexuality, 10*, 91–99.

Belden, H. (2006). Novel contraceptive inserted as a single rod. *Drug Topics, 150*, 16.

Bell, A., & Weinberg, M. (1978). *Homosexualities: A study of diversity among men and women*. New York: Simon & Schuster.

Bell, A., Weinberg, M., & Hammersmith, S. (1981). *Sexual preference: Its development in men and women*. Bloomington: Indiana University Press.

Belzer, E., Whipple, B., & Moger, W. (1984). On female ejaculation. *Journal of Sex Research, 20*, 403–406.

Bem, D. J. (1996). Exotic becomes erotic: A developmental theory of sexual orientation. *Psychological Review, 103*, 320–335.

Bem, D. J. (1997, August). *Exotic becomes erotic: Explaining the enigma of sexual orientation*. Address at the annual meeting of the American Psychological Association, Chicago.

Bem, D. J. (2000). Exotic becomes erotic: Interpreting the biological correlates of sexual orientation. *Archives of Sexual Behavior, 29*, 531–548.

Bem, D. J. (2008). Is there a causal link between childhood gender nonconformity and adult homosexuality? *Journal of Gay & Lesbian Mental Health, 12*, 61–79.

Bem, S. (1974). The measurement of psychological androgyny. *Journal of Consulting and Clinical Psychology, 42*, 155–162.

Bem, S. (1994). *The lenses of gender*. New Haven, CT: Yale University Press.

Berenbaum, S. A., & Snyder, E. (1995). Early hormonal influences on childhood sex-typed activity and playmate preferences: Implications for the development of sexual orientation. *Developmental Psychology, 31*, 31–42.

Berer, M. (2002). Making abortions safe: A matter of good public health policy and practice. *Reproductive Health Matters, 10(5), 31–44.*

Berer, M. (2005). Medical abortion: Issues of choice and acceptability. *Reproductive Health Matters, 13*, 25–34.

Berga, S. L. (2007, October). PMDD—Diagnostic challenges. *Medscape Ob/Gyn, Women's Health.*

Bergen, R. (1999). Marital rape. Minnesota Center Against Violence and Abuse.

Berkowitz, C. (1998). Medical consequences of child sexual abuse. *Child Abuse and Neglect, 22*, 541–550.

Berle, J., & Spigset, O. (2011). Antidepressant use during breastfeeding. *Current Women's Health Reviews, 7*(1), 28–34.

Berner, W. (2012). Pleasure Seeking and the Aspect of Longing for an Object in Perversion. A Neuropsychoanalytical Perspective. *American Journal of Psychotherapy, 66*(2), 129–150.

Bernstein, D., Bellamy, A., Hook, E., Levin, M., Wald, A., Ewell, M., Wolff, P., ... Belshe, R. (2013). Epidemiology, clinical presentation, and antibody response to primary infection with herpes simplex virus type 1 and type 2 in young women. *Clinical Infectious Diseases, 56*(3), 344–351. doi:10.1093/cid/cis891

Besharat, M. (2003). Relation of attachment style with marital conflict. *Psychological Reports, 92*, 1135–1140.

Bhathena, R., & Guillebaud, J. (2008). Intrauterine contraception: An update. *Journal of Obstetrics & Gynaecology, 28*(3), 262–265. doi:10.1080/01443610802042266

Bhugra, D. (2000). Disturbances in objects of sexual desire: Cross-cultural issues. *Sexual and Relationship Therapy, 15*, 67–78.

Bhugra, D., Popelyuk, D., & McMullen, I. (2010). Paraphilias across cultures: Contexts and controversies. *Journal of sex research, 47*(2–3), 242–256.

Bianchi, D., Parker, R., Wentworth, J., Madankumar, R., Saffer, C., Das, A., & Sehnert, A. (2014). DNA sequencing versus standard prenatal aneuploidy screening. *New England Journal of Medicine, 370*, 799–808.

Bieschke, K., Perez, R., & DeBord, K., (2006). *Handbook of counseling and psychotherapy with lesbian, gay, bisexual, and transgender clients, 2nd edition.* Washington, DC: American Psychological Association.

Bird, M. (1999, January 11). Glories that were Rome. *Time International*, p. 52.

Birken, R., Hill, A., & Berner, W. (2003). Pharmacotherapy of paraphilias with long-acting agonists of luteinizing hormone-releasing hormone. *Journal of Clinical Psychiatry, 64*, 890–897.

Blachère, P., & Cour, F. (2013). Deviant sexual behaviors, paraphilias, perversions. *Progres en Urologie: Journal de l'Association Francaise d'Urologie et de la Societe Francaise d'Urologie, 23*(9), 793–803.

Blakemore, S., Burnett, S., Dahl, R. (2010). The role of puberty in the developing adolescent brain. *Human Brain Mapping: Special Issue: Challenges and Methods in Developmental Neuroimaging, 31*, 926–933. doi:10.1002/hbm.21052

Blanc, A. (2001). The effect of power in sexual relationships on sexual and reproductive health: An examination of the evidence. *Studies in Family Planning, 32*, 189–213.

Blanchard, R., Cantor, J., Bogaert, A., Breedlove, S., & Ellis, L. (2006). Interaction of fraternal birth order and handedness in the development of male homosexuality. *Hormones and Behavior, 49*, 405–414.

Blanchflower, D. & Oswald, A. (2004). Money, sex and happiness. An empirical study. *Scandinavian Journal of Economics, 106* (3), 393–415.

Bloomberg News. Retrieved from **www.bloomberg.com/news/2013-12-24/merck-sues-actavis-unit-over-patent-for-generic-nuvaring.html**

Bock, R. (1993, August). *Understanding Klinefelter syndrome: A guide for XXY males and their families.* NIH Publication No. 93-3202.

Bockting, W., Miner, M., Swinburne Romine, R., Hamilton, A., & Coleman, E. (2013). Stigma, mental health, and resilience in an online sample of the US transgender population. *American Journal of Public Health, 103*(5), 943–951.

Bode, N. (2000, November-December). Sexuality at hand. *Psychology Today*, p. 11.

Boeringer, S. (1999). Associations of rape-supportive attitudes with fraternal and athletic participation. *Violence against Women, 5*, 81–90.

Bogaert, A. (2001). Personality, individual differences, and preferences for sexual media. *Archives of Sexual Behavior, 30*, 29–53.

Bogaert, A. (2006). Biological versus nonbiological older brothers and men's sexual orientation. *Proceedings of the National Academy of Sciences, 103*(28), 10771–10774.

Bogaert, A., & Skorska, M. (2011). Sexual orientation, fraternal birth order, and the maternal immune hypothesis: A review. *Frontiers in Neuroendocrinology, 32*(2), 247–254.

Bogart, L., Cecil, H., Wagstaff, D., Pinkerton, S., & Abramson, P. (2000). Is it sex? *Journal of Sex Research, 37*, 108–116.

Boies, S. (2002). University students' uses of and reactions to online sexual information and entertainment: Links to online and offline sexual behavior. *Canadian Journal of Human Sexuality, 11*, 77–90.

Bomalaski, D. (2005). A practical approach to intersex. *Urologic Nursing, 25*(1), 11–18, 23–24.

Bonifazi, W. (2000). Somebody to love. *Contemporary Long-Term Care, 23*, 22–28.

Boonstra, H. (2006). Meeting the sexual and reproductive health needs of people living with HIV. *Guttmacher Policy Review, 9*, 17–22.

Boonstra, H. (2009). Advocates call for a new approach after the era of "abstinence-only" sex education. *Guttmacher Policy Review, 12*(1), 6–11.

Boonstra, H. (2010). Key questions for consideration as a new federal teen pregnancy prevention initiative is implemented. *Guttmacher Policy Review, 13*, 1–7.

Booth, R., Kwiatkowski, C., & Chitwood, D. (2000). Sex-related HIV risk behaviors. *Drug and Alcohol Dependence, 58*, 219–226.

Bornstein, R. (1989). Exposure and affect: Overview and meta-analysis of research. *Psychological Bulletin, 106*, 613–628.

Bostwick, W., Boyd, C., Hughes, T., West, B., & McCabe, S. (2014). Discrimination and mental health among lesbian, gay, and bisexual adults in the United States. *American Journal of Orthopsychiatry, 84*(1), 35.

"Botox Vaginismus Treatment." (2011). The medical center for female sexuality. Retrieved from **http://www.centerforfemalesexuality.com/vaginismus-botox.html**

Bouchard, T. (1999). Genes, environment, and personality. In S. J. Ceci & W. M. Williams (Eds.), *The nature-nurture debate: The essential readings* (pp. 97–103). Malden, MA: Blackwell.

Bourke, M., & Hernandez, A. (2009). The "Butner Study" redux: A report of the incidence of hands-on child victimization by child pornography offenders. *Journal of Family Violence, 24*, 183–191. doi:10.1007/s10896-008-9219-y

Boyse, K., & Sands, T. (2011). Congenital Adrenal Hyperplasia (CAH) effects on girls and boys. University of Michigan Health System. Retrieved from **www.med.umich.edu/yourchild/topics/caheffects.htm**

Bradford, J. (2000). The treatment of sexual deviation using a pharmacological approach. *Journal of Sex Research, 37*, 248–257.

Bradford, J. M. W. (2001). The neurobiology, neuropharmacology, and pharmacology treatment of the paraphilias and compulsive sexual behavior. *Canadian Journal of Psychiatry, 46*, 26–34.

Brannon, Y. N., Levenson, J. S., Fortney, T., & Baker, J. N. (2007). Attitudes about community notification: A comparison of sexual offenders and the non-offending public. *Sexual abuse: a journal of research and treatment, 19*(4), 369–379.

Brassil, D., & Keller, M. (2002). Female sexual dysfunction: Definitions, causes, and treatment. *Urologic Nursing, 22*, 237–245.

Brawn, K., & Roe-Sepowitz, D. (2008). Female juvenile prostitutes: Exploring the relationship to substance use. *Children and Youth Services Review, 30*(12), 1395–1402.

Breech position and breech birth. Retrieved from **www.babies.sutterhealth.org/health/healthinfo/index.cfm?A=C&hwid=hw179937#hwTop**

Brewis, J., & Linstead, S. (2000). "The worst thing in the screwing": Context and career in sex work. *Gender, Work, and Organization, 7*, 168–180.

Brezsnyak, M., & Whisman, M. A. (2004). Sexual desire and relationship functioning: The effects of marital satisfaction and power. *Journal of Sex and Marital Therapy, 30*, 199–217.

Bridges, A., Wosnitzer, R., Scharrer, E., Sun, C., & Liberman, R. (2010). Aggression and sexual behavior in best-selling pornography videos: A content analysis update. *Violence Against Women, 16*, 1065–1085. doi:10.1177/1077801210382866

Broder, J. (2012). Pap tests less frequent under new guidelines, *Medscape Medical News.* Retrieved from **www.medscape.com/viewarticle/773282**

Brodsky, A. (2010). A decade of feminist influence on psychotherapy. *Psychology of Women Quarterly, 34*, 331–344.

Brodzinsky, D., & Pertman, A. (Eds.). (2011). *Adoption by lesbians and gay men: A new dimension in family diversity.* New York: Oxford University Press.

Brown, B. (2001). Sexual intercourse and orgasm during late pregnancy may have a protective effect against preterm delivery. *Family Planning Perspectives, 33*, 185–187.

Brown, T., Sumner, K., & Nocera, R. (2002). Understanding sexual aggression against women an examination of the role of men's athletic participation and related variables. *Journal of Interpersonal Violence, 17*(9), 937–952.

Bruckner, H., & Bearman, P. (2005, April). After the promise: The STD consequences of adolescent virginity pledges. *Journal of Adolescent Health*, pp. 271–278.

Bryant, C. (2002). Psychological treatment of priest sex offenders. *America, 186*, 14–18.

Bryden, D., & Grier, M. (2011). The search for rapists' "real" motives. *The Journal of Criminal Law & Criminology, 101*, 177–278.

Buffington, R., & Guy, D. (2013). Sex trafficking. In R. Buffington, E. Luibhéid, &, D. Guy (Eds.). *A Global history of sexuality: The modern era* (pp. 151–194). Hoboken, NJ: Wiley-Blackwell.

Bulcroft, K., & O'Connor-Roden, B. (1986, June). Never too late: Single people over 65 who are dating and sexually active belie the notion that passion and romance are only for the young. *Psychology Today*, pp. 66–69.

Bullough, B., & Bullough, V. (1997). Are transvestites necessarily heterosexual? *Archives of Sexual Behavior, 26*, 1–11.

Bullough, V. (2001). Christine Jorgensen: A personal autobiography. *Journal of Sex Research, 38*, 379–380.

Bumgarner, A. (2007). A right to choose? Sex selection in the international context. *Duke Journal of Gender Law & Policy, 14*, 1289–1309.

Bunker, C. (2010). Skin conditions of the male genitalia. *Medicine, 38*, 294–299.

Burger, W. (1973). Miller v. California. Supreme Court of the United States, 413 U.S. 15 (1972).

Burgess, A., & Holmstrom, L. (1974). Rape trauma syndrome. *American Journal of Psychiatry, 131*, 981–986.

Burnette, M., Lucas, E., Ilgen, M., Frayne, S., Mayo, J., & Weitlauf, J. (2008). Prevalence and health correlates of prostitution among patients entering treatment for substance use disorders. *Archives of General Psychiatry, 65*, 337–344.

Bursik, K., & Gefter, J. (2011). Still stable after all these years: Perceptions of sexual harassment in academic contexts. *The Journal of Social Psychology, 151*(3), 331–349.

Burstein, G., Snyder, M., Conley, D., Boekeloo, B., Quinn, T., & Zenilman, J. (2001). Adolescent chlamydia testing practices and diagnosed infections in a large managed care organization. *Sexually Transmitted Diseases, 28*, 477–483.

Butcher, J. (2003, June 4). A psychosexual approach to managing dispareunia. *Practitioner*, pp. 484–487.

Butcher, J., Hooley, J., & Mineka, S. (2013). *Abnormal psychology*. Boston, MA: Pearson.

Butts, S. (2014). Sex offender residency restrictions serve no purpose. Golden Gate University School of Law Digital Commons: The Legal Scholarship Repository. Retrieved from **www.ggulawreview.org/2013/10/29/sex-offender-residency-restrictions-serve-no-purpose**

Butzer, B., & Campbell, L. (2008). Adult attachment, sexual satisfaction, and relationship satisfaction: A study of married couples. *Personal Relationships, 15*, 141–154.

Buunk, B., & Hupka, R. B. (1987). Cross-cultural differences in the elicitation of sexual jealousy. *Journal of Sex Research, 23*, 12–22.

Byers, E. S., & Grenier, G. (2003). Premature or rapid ejaculation: Heterosexual couples' perceptions of men's ejaculatory behavior. *Archives of Sexual Behaivor, 32*, 261–270.

Byers, E. S., Purdon, C., & Clark, D. A. (1998). Sexual intrusive thoughts of college students. *Journal of Sex Research, 35*, 359–369.

Byers, E., & Demmons, S. (1999). Sexual satisfaction and sexual self-disclosure within dating relationships. *Journal of Sex Research, 36*(2), 180–189.

Byne, W. (2014). Science and belief: Psychobiological research on sexual orientation. In J. De Cecco & D. Parker, (Eds.) *Sex, cells, and same-sex desire: the biology of sexual preference* (pp. 303–344). New York: Routledge.

Cairns, G. (2011). Partners study expands our knowledge of HIV transmission risk infectiousness and viral load. *NAM Publications*. Retrieved from **www.aidsmap.com/page/1685273/#item1685275**

Calvert, W., & Bucholz, K. (2008). Adolescent risky behaviors and alcohol use. *Western Journal of Nursing Research, 30*, 147–148.

Campaign for Our Children. (2008). Teen guide. Campaign for Our Children, Inc. Retrieved from **www.cfoc.org/index.php/teen-guide**

Campbell, A., & Muncer, S. (2007). An intent to harm or injure? Gender and the expression of anger. *Aggressive Behavior, 33*, 1–12.

Campbell-Yesufu, O., & Gandhi, R. (2011). Update on human immunodeficiency virus (HIV)-2 Infection. *Clinical Infectious Diseases, 52*(6), 780–787.

Cannold, L. (2003). Do we need a normative account of the decision to parent? *International Journal of Applied Philosophy, 17*, 277–290.

Cantón-Cortés, D., & Cantón, J. (2010). Coping with child sexual abuse among college students and post-traumatic stress disorder: The role of continuity of abuse and relationship with the perpetrator. *Child Abuse & Neglect, 34*(7), 496–506.

Carey, M. (1998). Cognitive-behavioral treatment of sexual dysfunctions. In V. Caballo (Ed.), *International handbook of cognitive and behavioural treatments of psychological disorders*. Oxford: Pergammon/Elsevier.

Carlson, M. (2013, May 21). The military's culture of sexual violence. *Bloomberg View*. Bloomberg.com. Retrieved from **www.bloombergview.com/articles/2013-05-21/the-military-s-culture-of-sexual-violence**

Carpenter, D., Janssen, E., Graham, C., Vorst, H., & Wicherts, J. (2008). Women's scores on the Sexual Inhibition/Sexual Excitation Scales (SIS/SES): Gender similarities and differences (Report). *Journal of Sex Research, 45*, 36–48.

Carpenter, L., & DeLamater, J. (2012). *Sex for Life: From Virginity to Viagra, How Sexuality Changes Throughout Our Lives*. New York: NYU Press.

Carr, J. L., & Van Deusen, K. M. (2004). Risk factors for male sexual aggression on college campuses. *Journal of Family Violence, 19*, 279–289.

Carter, D. (2012). Comprehensive sex education for teens is more effective than abstinence. *The American Journal of Nursing, 112*(3), 15. doi: 10.1097/01.NAJ.0000412622.87884.a3

Cartwright, R., Elvy, S., & Cardozo, L. (2007). Do women with female ejaculation have detrusor overactivity? *Journal of Sexual Medicine, 4*, 1655–1658.

Carusi, D. A. (2008). Can safety and efficacy go hand in hand? Contraception for medically complex patients. *OBG Management, 20*, 42–48.

Casey, P., & Pruthi, S. (2008). The latest contraceptive options: What you must know. *Journal of Family Practice, 57(12), 797–805*.

Cass, V. (1984). Homosexual identity formation: Testing a theoretical model. *Journal of Homosexuality, 20*, 143–167.

Castleman, M. (2009). *Great sex*. Emmaus, PA: Rodale Books.

Catalano, C. (2013). Intimate partner violence: attributes of victimization, 1993–2011: U.S.

Catalano, S., Smith, E., Snyder, H., & Rand, M. (2009). *Female victims of violence*. U.S. Department of Justice; Office of Justice Programs; Bureau of Justice Statistics.

Catania, J. A. (1999). A framework for conceptualizing reporting bias and its antecedents in interviews assessing human sexuality. *Journal of Sex Research, 36*, 25–38.

Cato, S., & Leitenberg, H. (1990). Guilt reactions to sexual fantasies during intercourse. *Archives of Sexual Behavior, 19*, 49–63.

CDC, "Condoms and STDs." (2013). Condoms and STDs: Fact sheet for public health personnel Centers for Disease Control and Prevention.

CDC, "Diagnoses of HIV Infection." (2013) Diagnoses of HIV infection in the United States and dependent areas. Retrieved from **www.cdc.gov/hiv/library/reports/surveillance/2011/surveillance_Report_vol_23.html**

CDC, "Discontinuation." (2008). Discontinuation of contraceptive methods. Centers for Disease Control and Prevention. Retrieved from **www.cdc.gov/nchs/about/major/nsfg/abclist_d.htm**.

CDC, "Genital HPV Infection." (2014). Human papillomavirus (HPV) fact sheet. Centers for Disease Control and Prevention. Retrieved from **www.cdc.gov/std/HPV/STDFact-HPV.htm**

CDC, "Gonorrhea." (2014). 2012 sexually transmitted diseases surveillance: Gonorrhea. Centers for Disease Control and Prevention. Retrieved from **http://www.cdc.gov/std/stats12/gonorrhea.htm**

CDC, "Hepatitis C." (2014). Hepatitis C information for the public. Centers for Disease Control and Prevention. Retrieved from **www.cdc.gov/hepatitis/C/cFAQ.htm#cFAQ31**

CDC, "HPV and Men." (2012). HPV and men fact sheet. Centers for Disease Control and Prevention.

CDC, "Parasites, Treatment." (2013). Parasites - Lice - Pubic "Crab" Lice Treatment. Centers for Disease Control and Prevention. Retrieved from **www.cdc.gov/parasites/lice/pubic/treatment.html**

CDC, "Reported STDs." (2014). CDC FACT SHEET Reported STDs in the United States, 2012 – National data for chlamydia, gonorrhea, and syphilis. Centers for Disease Control and Prevention. Retrieved from **www.cdc.gov/nchhstp/newsroom/docs/STD-Trends-508.pdf**

CDC, "Sexual Identity." (2011). Sexual identity, sex of sexual contacts, and health-risk behaviors among students in grades 9–12—Youth risk behavior surveillance, selected sites, United States, 2001–2009. Retrieved from **www.cdc.gov/mmwr/pdf/ss/ss60e0606.pdf**

CDC, "Sexually Transmitted Diseases." (2010, December 17). Sexually transmitted diseases treatment guidelines: Recommendations and reports. *Morbidity and Mortality Weekly Report, 59*. Retrieved from **www.cdc.gov/std/treatment/2010/STD-Treatment-2010-RR5912.pdf**

CDC, "Adolescent and School Health." (2014). Youth Risk Behavior Surveillance. Centers for Disease Control and Prevention. Retrieved from ccd.cdc.gov / youthonline.

CDC, "Chlamydia." (2014). Chlamydia - CDC Fact Sheet. Centers for Disease Control and Prevention. Retrieved from **www.cdc.gov/std/chlamydia/STDFact-Chlamydia.htm**

CDC, "Key Statistics." (2013). Key Statistics from the National Survey of Family Growth. Centers for Disease Control and Prevention. Retrieved from **www.cdc.gov/nchs/nsfg/key_statistics/c.htm**

CDC, "Parasites." (2013). Parasites - lice - pubic "crab" lice. Centers for Disease Control and Prevention. Retrieved from **www.cdc.gov/parasites/lice/pubic/index.html**

CDC, "STDs." (2013). 2012 Sexually Transmitted Diseases Surveillance. Centers for Disease Control and Prevention. Retrieved from **www.cdc.gov/std/stats12/default.htm**

CDC, "Syphilis Fact Sheet." (2014). Syphilis - CDC Fact Sheet. Centers for Disease Control and Prevention **www.cdc.gov/std/syphilis/STDFact-Syphilis.htm**

CDC, "Trichomoniasis." (2012). Trichomoniasis - CDC fact sheet. Centers for Disease Control and Prevention. Retrieved from **www.cdc.gov/std/trichomonas/STDFact-Trichomoniasis.htm**

Center for Disease Control and Prevention. (2012). Sexual Violence. Retrieved from **www.cdc.gov/ViolencePrevention/pdf/sv-datasheet-a.pdf**

Centers for Disease Control and Prevention. (2012). *National ART Success Rates* http: / / nccd.cdc.gov / DRH_ART / Apps / NationalSummaryReport.aspx

Centers for Disease Control and Prevention. (2006). STD surveillance report (Figures 7 & 18).

Centers for Disease Control and Prevention. (2008). Child maltreatment: Facts at a glance. Retrieved from **www.cdc.gov/ncipc/dvp/CM_Data_Sheet.pdf**

Centers for Disease Control and Prevention. (2014). Fetal alcohol spectrum disorders (FASDs). Retrieved from **http://www.cdc.gov/ncbddd/fasd/research-tracking.html**

Centers for Disease Control and Prevention. (2010). Pelvic inflammatory disease (PID)—CDC fact sheet. Retrieved from **www.cdc.gov/std/PID/STDFact-PID.htm**

Centers for Disease Control and Prevention. (2011). *Assisted reproductive technology success rates: National summary and fertility clinic reports*. American Society for Reproductive Medicine, Society for Assisted Reproductive Technology, 2008. U.S. Department of Health and Human Services, 2010.

Centers for Disease Control and Prevention. (2012). HIV / AIDS statistics and surveillance: Basic statistics. Retrieved from **www.cdc.gov/hiv/topics/surveillance/basic.htm#hivest**

Centers for Disease Control and Prevention. (2013). Assisted. *Reproductive technology national summary report*. Retrieved from **www.cdc.gov/art/ART2011/PDFs/ART_2011_National_Summary_Report.pdf**

Centers for Disease Control and Prevention. (2014). Tobacco Use. *Tobacco use and pregnancy*. Retrieved from **www.cdc.gov/reproductivehealth/TobaccoUsePregnancy**

Certain, H. E., Harahan, B. J., Saewyc, E. M., & Fleming, M. F. (2009). Condom use in heavy drinking college students: The importance of always using condoms. *Journal of American College Health, 58*(3), 187–194. Retrieved from EBSCOhost.

Chachkin C. J. (2007). What potent blood: Noninvasive prenatal genetic diagnosis and the transformation of modern prenatal care. *American Journal of Law & Medicine, 33*(1), 9–53.

Chalker, R. (2002). *The clitoral truth: The secret world at your fingertips.* New York: Seven Stories Press.

Chan, C., Harting, M., & Rosen, T. (2009). Systemic and barrier contraceptives for the dermatologist: A review. *International Journal of Dermatology, 48*(8), 795–814. doi:10.1111 / j.1365-4632.2009.04148.x

Chandra, A., Mosher, W., Copen, C., Sionean, C. (2011). Sexual behavior, sexual attraction, and sexual identity in the United States: Data from the 2006–2008 National Survey of Family Growth. *National health statistics reports; no. 36*. National Center for Health Statistics. **Retrieved from www.cdc.gov/nchs/data/nhsr/nhsr036.pdf**

Chapleau, K., Oswald, D., & Russell, B. (2008). Male rape myths: The role of gender, violence, and sexism. *Journal of Interpersonal Violence, 23*, 600–615.

Characteristics. (2009). Characteristics and behavioral indicators of adults who molest children. *Minnesota Department of Corrections.*

Chen, L., Murad, M., Paras, M., Colbenson, K., Sattler, A., Goranson, E., . . . & Zirakzadeh, A. Sexual abuse and lifetime diagnosis of psychiatric disorders: systematic review and meta-analysis. (2010, July). *Mayo Clinic Proceedings, 85*, 618–629.

Cheng, W., Ickes, W., & Kenworthy, J. (2013). The phenomenon of hate crimes in the United States. *Journal of Applied Social Psychology, 43*(4), 761–794.

Cherney, I., & London, K. (2006). Gender-linked differences in the toys, television shows, computer games, and outdoor activities of 5- to 13-year-old children. *Sex Roles: A Journal of Research, 54*, 717–726.

Chesson, H., Ekwueme, D., Saraiya, M., & Markowitz, L. (2008). Cost-effectiveness of human papillomavirus vaccination in the United States. *Emerging Infectious Diseases, 14,* 244–251.

Chesson, H., Harrison, P., & Stall, R. (2003). Changes in alcohol consumption and in sexually transmitted disease incidence rates in the United States, 1983–1998. *Journal of Studies on Alcohol, 64,* 623–630.

Chida, Y., & Mao, X. (2009). Does psychosocial stress predict symptomatic herpes simplex virus recurrence? A meta-analytic investigation on prospective studies. *Brain, Behavior, and Immunity, 23,* 917–925.

Child Sexual Abuse Committee of the National Child Traumatic Stress Network National Center on Sexual Behavior of Youth. (2009, April). Sexual development and behavior in children.

Child Trends. (2010). *Preterm births.*

Chillot, R. (2002). Overcome your greatest sexual fear. *Prevention, 54,* 144–149.

Chin, H., Sipe, T., Elder, R., Mercer, S., Chattopadhyay, S., Jacob, V.,... & Santelli, J. (2012). The effectiveness of group-based comprehensive risk-reduction and abstinence education interventions to prevent or reduce the risk of adolescent pregnancy, human immunodeficiency virus, and sexually transmitted infections: two systematic reviews for the Guide to Community Preventive Services. *American Journal of Preventive Medicine, 42*(3), 272–294.

Chu, M., & Lobo, R. (2004). Formulations and use of androgens in women. *Journal of Family Practice, 53,* S3–S5.

Chudnovsky, A., & Niederberger, C. (2007). Gonadotropin therapy for infertile men with hypogonadotropic hypogonadism. *Journal of Andrology, 28*(5): 644–6.

Chughtai, B., Sawas, A., O'Malley, R., Naik, R., Khan, S., & Pentyala, S. (2005). A neglected gland: A review of Cowper's gland. *International Journal of Andrology, 28*(2), 74–77. doi: 10.1111/j.1365-2605.2005.00499.x

Clark, L., Jackson, M., & Allen-Taylor, L. (2002). Adolescent knowledge about sexually transmitted diseases. *Sexually Transmitted Diseases, 29,* 436–443.

Clayton, A., Warnock, J., & Kornstein, S. (2004). Bupropion SR for SSRI-induced sexual dysfunction. *Psychopharmacological Update, 15,* 5–6.

Clements, M., Stanley, S., & Markman, H. (2004). Before they said "I do": Discriminating among marital outcomes over 13 years. *Journal of Marriage and the Family, 66,* 613–626.

Clinton-Sherrod, A., & Walters, J. (2011). Marital rape and sexual violation by intimate partners. In T. Bryant-Davis (Ed.). *Surviving sexual violence: a guide to recovery and empowerment.* (pp. 48–58). New York: Rowman & Littlefield.

Cloitre, M. (2004). Trauma and PTSD. *CNS Spectrums, 9,* 4–5.

Cloud, J. (2002, April 29). Pedophilia. *Time,* pp. 42–46.

Cloud, J. (2008, January 28). Are gay relationships different? (Annual Mind & Body Special Issue) (The Science of Romance; Gay Marriage). *Time, 171*(4), 78.

Cobbina, J., & Oselin, S. (2011). It's not only for the money: An analysis of adolescent versus adult entry into street prostitution. *Sociological inquiry, 81*(3), 310–332.

Coghlan, A. (2008, June 21). Gay or straight, it's decided at birth. *New Scientist, 198,* 10.

Cohen, J. (2010). Immunology painful failure of promising genital herpes vaccine. *Science, 330*(15), 304. doi:10.1126/science.330.6002.304

Cohen, K. M. (2002). Relationships among childhood sex-atypical behavior, spatial ability, handedness, and sexual orientation in men. *Archives of Sexual Behavior, 31,* 129–143.

Colapinto, J. (2001). *As nature made him: The boy who was raised as a girl.* New York: Perennial Press.

Coleman, E., Bockting, W., Botzer, M., Cohen-Kettenis, P., DeCuypere, G., & Zucker, K. (2012). Standards of care for the health of transsexual, transgender, and gender-nonconforming people. World Professional Association for Transgender Health (wpath). Retrieved from **www.wpath.org/uploaded_files/140/files/Standards%20of%20Care,%20V7%20Full%20Book.pdf**

Coleman, M., & Ganong, L. (1985). Love and sex role stereotypes. Do macho men and feminine women make better lovers? *Journal of Personality and Social Psychology, 49,* 170–176.

Collin, J. (2006). An introduction to Turner syndrome. *Pediatric Nursing, 18,* 38–43.

Collinsworth, L., Fitzgerald, L., & Drasgow, F. (2009). In harm's way: Factors related to psychological distress following sexual harassment. *Psychology of Women Quarterly, 33,* 475–490.

Colombo, D. F. (2002). Predicting spontaneous preterm birth: Fetal fibronectin and ultrasonography help rule out labor, not rule it in. *British Medical Journal, 325,* 289–290.

Comer, R. (1998). *Abnormal psychology* (3d ed.). New York: Freeman.

Commission on human rights. (2011). Connecticut commission on human rights and opportunities sexual harassment posting and training regulations. Retrieved from **www.ct.gov/chro/cwp/view.asp?a=2527&q=333112**

Conley, T. D., & Collins, B. E. (2005). Differences between condom users and condom nonusers in their multidimensional condom attitudes. *Journal of Applied Social Psychology, 35,* 603–620.

Consedine, N., Krivoshekova, Y., & Harris, C. (2007). Bodily embarrassment and judgment concern as separable factors in the measurement of medical embarrassment: Psychometric development and links to treatment-seeking outcomes. *British Journal of Health Psychology,* 12(3), 439–462. doi:10.1348/135910706X118747

Constantinople, A. (1973). Masculinity-femininity: An exception to a famous dictum? *Psychological Bulletin, 80,* 389–407.

Conway, L. (2003a). Basic TG/TS/IS information: Gender basics and transgenderism.

Conway, L. (2003b). How many of us are there? Retrieved August 15, 2003, from **www.gendercentre.org.au/44article4.htm**

Copen, C., Chandra, A., & Martinez, G. (2013). Prevalence and timing of oral sex with opposite-sex partners among females and males aged 15–24 years: United States, 2007–2010. *National Health Statistics Reports, 56.* National Center for Health Statistics. Retrieved from **www.cdc.gov/nchs/data/nhsr/nhsr056.pdf**

Corinna, H. (2010). What's the typical use-effectiveness rate of abstinence? *RH Reality Check.*

Cornog, M. (2004). The decloseting of masturbation? *Journal of Sex Research, 41,* 310–312.

Corona, G., Jannini, E., Lotti, F., Boddi, V., De Vita, G., Maggi, M. (2011). Premature and delayed ejaculation: Two ends of a single continuum influenced by hormonal milieu. *International Journal of Andrology, 34,* 41–48. doi: 10.1111/j.1365-2605.2010.01059.x

Corty, E., & Guardiani, J. (2008). Canadian and American sex therapists' perceptions of normal and abnormal ejaculatory latencies: How long should intercourse last? *The Journal of Sexual Medicine, 5*(5), 1251–1256.

Cowan, F., Copas, A., Johnson, A., Ashley, R.M, Corey, L., & Mindel, A. (2002). Herpes simplex virus type 1 infection: A sexually transmitted infection of adolescence? *Sexually Transmitted Infections 7,* 346–348. doi:10.1136/sti.78.5.346

Coyne, S., Nelson, D., & Underwood, M. (2011). Aggression in children. In P. Smith and C. H. Hart (Eds.). *The Wiley-Blackwell handbook of childhood social development, second edition* (491–509). Oxford, UK: Wiley-Blackwell. doi:10.1002/9781444390933.ch26

Crayford, T., Campbell, S., Bourne, T., Rawson, H., & Collins, W. (2000). Benign ovarian cysts and ovarian cancer: A cohort study with implications for screening. *The Lancet, 355,* 1060–1063.

Crick, N., & Grotpeter, J. (1995). Relational aggression, gender, and social-psychological adjustment. *Child Development, 66*, 710–722.

Crombie, G., Pyke, S., Silverthorn, N., Jones, A., & Piccinin, S. (2003). Students' perceptions of their classroom participation and instructor as a function of gender and context. *Journal of Higher Education, 74*, 51–76.

Crowl, A., Ahn, S., & Baker, J. (2008). A meta-analysis of developmental outcomes for children of same-sex and heterosexual parents. *Journal of GLBT Family Studies, 4*(3), 385-407.

Crowley, T., Goldmeier, D., & Hiller, J. (2009). Diagnosing and managing vaginismus. *Clinical Review BMJ, 338*, b2284.

C-section. (2010). March of Dimes Foundation. Retrieved from **www.marchofdimes.com/pregnancy/faq_csectionreasons.html**

Cunningham, J., Yonkers, K., O'Brien, S., & Eriksson, E. (2009). Update on research and treatment of premenstrual dysphoric disorder. *Harvard Review of Psychiatry, 17*(2), 120–137. doi:10.1080/10673220902891836

Cusac, A. (1999). The promise of Stonewall. *Progressive, 6*, 10–13.

Cvencek, D., Meltzoff, A., & Greenwald, A. (2011). Math–gender stereotypes in elementary school children. *Child Development* (in press). doi:10.1111/j.1467-8624.2010.01529.x

D'Emilio, J., & Freedman, E. (1988). *Intimate matters: A history of sexuality in America*. New York: Harper & Row.

Dailard, C. (2003, December). Understanding "abstinence": Implications for individuals, programs and policies. *The Guttmacher Report on Public Policy, 6*(5). Retrieved from **www.guttmacher.org/pubs/tgr/06/5/gr060504.html**

Daley, A. (2009). Exercise and premenstrual symptomatology: A comprehensive review. *Journal of Women's Health, 18*(6), 895–899. doi:10.1089/jwh.2008.1098

Daley, M. (2008). Flawed Solution to the Sex Offenders Situation in the United States: The Legality of Chemical Castration for Sex Offenders, A. *Indiana Health Law Review, 5*, 87–122.

Dalton, J. (2008). *Conservative management of foreskin conditions*. New York: Springer Publishing.

Darling, C. A., Davidson, J. K., & Jennings, D. A. (1991). The female sexual response revisited: Understanding the multiorgasmic experience in women. *Archives of Sexual Behavior, 20*, 527–540.

Das, A. (2007). Masturbation in the United States. *Journal of Sex & Marital Therapy, 33*, 301–317.

Davidson, J., & Moore, N. (1994). Masturbation and premarital sexual intercourse among college women: Making choices for sexual fulfillment. *Journal of Sex and Marital Therapy, 20*, 179–199.

Davidson, K., & Hoffman, L. (1986). Sexual fantasies and sexual satisfaction: An empirical analysis of erotic thought. *Journal of Sex Research, 22*, 184–205.

Davis, C., & McCormick, N. (1997). What sexual scientists know about pornography. *Sexual Science, 3*.

Davis, K., Kirkpatrick, A., Levy, M., & O'Hearn, R. (2013). Stalking the elusive love style: Attachment styles, love styles and relationship development. In R. Erber & R. Gilmour (Eds.), *Theoretical frameworks for personal relationships* (digital edition). New York: Psychology Press.

Davis, K., Norris, H., George, W., Martell, J., & Heiman, J. (2006). Men's likelihood of sexual aggression: The influence of alcohol, sexual arousal, and violent pornography. *Aggressive Behavior, 32*, 581–589.

Davis, K., Norris, J., Hessler, D. M., Zawacki, T., Morrison, D. M., & George, W. H. (2010). College women's sexual decision making: Cognitive mediation of alcohol expectancy effects. *Journal of American College Health, 58*(5), 481–489.

Davison, G. C. (2001). Conceptual and ethical issues in therapy for the psychological problems of gay men, lesbians, and bisexuals. *Journal of Clinical Psychology, 57*, 695–703.

Davison, G. C. (2005). Issues and nonissues in the gay-affirmative treatment of patients who are gay, lesbian, or bisexual. *Clinical Psychology: Science and Practice, 12*, 25–28.

Daw, J. (2002). Can psychology help a church in crisis? *Monitor on Psychology, 33* (electronic edition). Retrieved from **www.apa.org/monitor/jun02/church.html**

Dawson, S., Bannerman, B., & Lalumière, M. (2014). Paraphilic Interests: An Examination of Sex Differences in a Nonclinical Sample. *Sexual Abuse: A Journal of Research and Treatment*. doi:1079063214525645

De Bro, S., Campbell, S., & Peplau, L. (1994). Influencing a partner to use a condom. *Psychology of Women Quarterly, 18*, 165–182.

De Cuypere, G., Tsjoen, G., Beerten, R., (2006, March–April). Sexual and physical health after sex reassignment surgery. *Pain Digest, 16*, 106–107.

De Cuypere, G., Tsjoen, G., Beerten, R., Selvaggi, G., De Sutter, P., Hoebeke, P., Monstrey, S., Vansteenwegen, A., & Rubens, R. (2005). Sexual and physical health after sex reassignment surgery. *Archives of Sexual Behavior, 34*, 679–690.

de Silva, P., & Pernet, A. (1992). Pollution in "Metroland": An unusual paraphilia in a shy young man. *Sexual and Marital Therapy, 7*, 301–306.

de Silva, W. (1999). Sexual variations. *British Medical Journal, 318*, 654–656.

Decarlo, P., Alexander, P., & Hus, H. (1996). What are sex workers' HIV prevention needs?

Decuyper, M., De Bolle, M., & Fruyt, F. (2012). Personality similarity, perceptual accuracy, and relationship satisfaction in dating and married couples. *Personal Relationships, 19*, 128–145. doi:10.1111/j.1475-6811.2010.01344.x

Del Mar, C. (2010). Urinary tract infections in healthy women: A revolution in management? *BMC Family Practice, 11*(42). doi:10.1186/1471-2296-11-42

DeLamater, J., & Friedrich, W. N. (2002, February). Human sexual development. *Journal of Sex Research, 39*(1), 10–14.

DeLamatera, J., Hyde, J. S., & Fonga, M. (2008). Sexual satisfaction in the seventh decade of life. *Journal of Sex & Marital Therapy, 34*(5), 439–454.

Dell'Amore, C. (2010). Scientists discover how men produce 1500 sperm in a second.

Delvin, D., & Webber, C. (2005). Sex after giving birth. Retrieved from **www.netdoctor.co.uk/sex_relationships/facts/sexdelivery.htm**

Demare, D., Lips, H., & Briere, J. (1993). Sexually violent pornography, antiwomen attitudes, and sexual aggression: A structural equation model. *Journal of Research in Personality, 27*, 285–300.

Dempsey, A., Gebremariam, A., Koutsky, L., & Manhart, L. (2008). Using risk factors to predict human papillomavirus infection: Implications for targeted vaccination strategies in young adult women. *Vaccine, 26*, 1111–1117.

Denenberg, R. (1997). Childhood sexual abuse as an HIV risk factor in women. *Treatment Issues, 11*, 7–8.

DeNoon, D. (2008). HIV death gap closing: near-normal life span with early HIV detection, treatment. *WebMD Health News*.

DeNoon, D. (2010). Gardasil approved for anal cancer prevention. WebMD.

Department of Justice Office of Justice Programs; Bureau of Justice Statistics; Special Report. BJS, NCJ 243300. Retrieved from **www.bjs.gov/content/pub/pdf/ipvav9311.pdf**

Deptula, D. P., Henry, D. B., & Schoeny, M. E. (2010). How can parents make a difference? Longitudinal associations with adolescent sexual behavior. *Journal of Family Psychology, 24*(6), 731.

Desiderato, L., & Crawford, H. (1995). Risky sexual behavior in college students: Relationships between number of sexual partners, disclosure of previous risky behavior, and alcohol use. *Journal of Youth and Adolescence, 24*, 55–68.

Deu, N., & Edelmann, R. J. (1997). The role of criminal fantasy in predatory and opportunist sex offending. *Journal of Interpersonal Violence, 12*, 18–29.

Diamond, M. (2006). Biased-Interaction theory of psychosexual development: "How does one know if one is male or female?" *Sex Roles: A Journal of Research, 55*, 589–601.

DiCenso, A., Guyatt, G., Willan, A., & Griffith, L. (2002). Interventions to reduce unintended pregnancies among adolescents: Systematic review of randomized controlled trials. *British Medical Journal, 324*, 1426–1430.

Dickens, B. (2002). Can sex selection be ethically tolerated? *Journal of Medical Ethics, 28*, 335–336.

Didymus, John Thomas. (2012, May 25). Father who went for kidney stones discovers he is a woman. Retrieved from **www.digitaljournal.com/article/325471#ixzz2fYnKizcq**

Diekman, A., & Murnen, S. (2004). Learning to be little women and little men: The inequitable gender identity in nonsexist children's literature. *Sex Roles, 50*, 373–385.

Dixon, B. K. (2006). N-9 spermicides safe for those at low risk of STIs. *Family Practice News, 36*, 40.

Docter, R., & Prince, V. (1997). Transvestism: A survey of 1032 cross-dressers. *Archives of Sexual Behavior, 26*(6), 589–605.

Dodson, B. (2002, December). Getting to know me: A primer on masturbation. Ms., pp. 27–29. Reprinted from *Esquire*, August 1974, pp. 106–109.

Dodsworth, J. (2012). Pathways through sex work: childhood experiences and adult identities. *British Journal of Social Work, 42*(3), 519–536.

Does Facebook bring out the green-eyed monster of jealousy? *CyberPsychology & Behavior, 12*(4), 441–444. doi:10.1089/cpb.2008.0263

Donnelly, D. (1993). Sexually inactive marriages. *Journal of Sex Research, 30*, 171–179.

Donnerstein, E. I., & Linz, D. G. (1986, December). The question of pornography. *Psychology Today*, pp. 56–59.

Donnerstein, E., Linz, D., & Penrod, S. (1987). *The question of pornography: Research findings and policy implications.* New York: Free Press.

Donohoue, P., Poth, M., & Speiser, P. (2010). Congenital adrenal hyperplasia. *Journal of Clinical Endocrinology & Metabolism, 95*(2).

Doucleff, M. (2013). Gates Foundation says it's time for a snazzier condom. *National Public Radio.*

Douglas, K. (1997). Results from the 1995 national college health risk behavior survey. *Journal of American College Health, 46*, 55–66.

Down syndrome screening expanded. (2007). *Clinician Reviews, 17*, 18.

Dragon, W., & Duck, S. (2005). *Understanding research in personal relationships: A text with readings.* Thousand Oaks, CA: Sage Publications.

Drescher, J. (2004, October 1). The closet: Psychological issues of being in and coming out. *Psychiatric Times (Electronic Version)*, 21–32. Retrieved from **www.psychiatrictimes.com/display/article/10168/47271?pageNumber=2**

Driedger, S. (1996). Mystical passion: Tantric sex. *Macleans*, 38, 44–46.

DSM-5. (2010). The future of psychiatric diagnosis. *American Psychiatric Association.* Retrieved from **www.dsm5.org/Pages/Default.aspx**

DSM-5. (2013). American Psychiatric Association *Diagnostic and statistical manual of mental disorders* (5th ed.). Arlington, VA: American Psychiatric Publishing.

DSM-5. (2013). *Diagnostic and statistical manual of mental disorders* (5th ed.). American Psychiatric Association. Arlington, VA: American Psychiatric Publishing.

Dube, S., Anda, R., Whitfield, C., Brown, D., Felitti, V., Dong, M., & Giles, W. (2005). Long-term consequences of childhood sexual abuse by gender of victim. *American Journal of Preventive Medicine, 2*, 430–438.

Dugan, M., K., & Hock, R. R. (2006). *It's my life now: Starting over after an abusive relationship or domestic violence* (2d ed.). New York: Routledge

Dugan, J. (2001). Kissin' cousins. *Colonial America.*

Dugan, M. K., & Hock, R. R. (2006). *It's my life now: Starting over after an abusive relationship or domestic violence* (2d ed.). New York: Routledge.

Dugdale & Zieve (2010). Urinary tract infections – Adults.

Dunaway, A. (2008). When the patient asks. *Journal of the American Academy of Physicians Assistants, 21*, 63.

Dunn, M. S., Bartee, R. T., & Perko, M. A. (2003). Self-reported alcohol use and sexual behaviors of adolescents. *Psychological Reports, 92*, 339–348.

Dunne, E., Klein, N., Naleway, A., Baxter, R., Weinmann, S., Riedlinger, K.,... & Unger, E. (2013). Prevalence of HPV types in cervical specimens from an integrated healthcare delivery system: baseline assessment to measure HPV vaccine impact. *Cancer Causes & Control, 24*(2), 403–407.

Dunne, E., Unger, E., & Sternberg, M. (2007). Prevalence of HPV Infection among females in the United States. *JAMA, 297*(8), 813–819. doi:10.1001/jama.297.8.813.

Durex Corp. (2002b). What is the average penis size? Retrieved from **www.durex.com**

Dutton, J. (2009). *How we do it: How the science of sex can make you a better lover.* New York: Crown Archetype.

Dye, C., & Upchurch, D. M. (2006). Moderating effects of gender on alcohol use: Implications for condom use at first intercourse. *Journal of School Health, 76*, 111–116.

East, L., Jackson, D., Peters, K., & O'Brien, L. (2010). Disrupted sense of self: Young women and sexually transmitted infections. *Journal of Clinical Nursing, 19*, 1995–2003.

Eccles, J., Jacobs, J., & Harold, R. (1990). Gender role stereotypes, expectancy effects, and parents' socialization of gender differences. *Journal of Social Issues, 46*(2), 183–201

Ectopic pregnancy. (2009). Mayo Foundation for Medical Education and Research. Retrieved from **www.mayoclinic.com/print/ectopic-pregnancy/DS00622/DSECTION=all&METHOD=print**

Edell, D. (2001). As big as that? Retrieved from **www.healthcentral.com/drdean/deanfulltexttopics.cfm?ID=50890**

Edwards, J., & Moore, A. (1999). Implanon: A review of clinical studies. *British Journal of Family Planning, 4*, 3–16.

Edwards, K., Turchik, J., Dardis, C., Reynolds, N., & Gidycz, C. (2011). Rape myths: History, individual and institutional-level presence, and implications for change. *Sex Roles* (published online). doi:10.1007/s11199-011-9943-2

Edwards, M. (2000). Adolescents would prefer parents as primary sexuality educators. *Families are talking*, special supplement to *Seicus Report, 1*.

Effects on jealousy and relationship happiness. *Journal of Computer-Mediated Communication, 16*(4), 511–527. doi: 10.1111/j.1083-6101.2011.01552.x

Eher, R., Miner, M., & Pfafflin, F. (2011). *International perspectives on the assessment and treatment of sexual offenders: Theory, practice and research.* New York: Wiley.

Ehrhardt, A., & Baker, S. (1974). Fetal androgens, central nervous system differentiation, and behavior sex differences. In R. Friedman, R. Richart, & R. Vandewiele (Eds.), *Sex differences in behavior* (pp. 33–51). New York: Wiley.

Eich, T. (2010). A tiny membrane defending "us" against "them": Arabic Internet debate about hymenorraphy in Sunni Islamic law. *Culture, Health & Sexuality: An International Journal for Research, Intervention and Care, 12*(7), 755–769. doi:10.1080/13691051003746179

Eisenberg, E., & Brumbaugh, K. (2012). Infertility. Office on Women's Health in the Office of the Assistant Secretary for Health, U.S. Department of Health and Human Services. Womenshealth.gov. Retrieved from **http://www.womenshealth.gov/publications/our-publications/fact-sheet/infertility.html?from=AtoZ**

Eisenman, R. (2001). Penis size: Survey of female perceptions of sexual satisfaction. *BMC Women's Health, 1*. doi:10.1186/1472-6874-1-1.

Eliason, M. J. (1997). The prevalence and nature of biphobia in heterosexual undergraduate students. *Archives of Sexual Behavior, 26*, 317–326.

Elliot, W. (2004). Valacyclovir reduces genital herpes transmission. *Infections Disease Alert, 23*, S1.

Elliot, S. (2010). Talking to teens about sex: Mothers negotiate resistance, discomfort, and ambivalence. *Sexuality Research and Social Policy, 7*(4), 310–322.

Ellis, M. (2013). Planned home births carry significant risks, study shows. *Medical News Today*. Retrieved from **www.medicalnewstoday.com/articles/266244.php**

Elmes, D. G., Kantowitz, B. H., & Roediger, I. H. L. (2011). Research methods in psychology. Retrieved from **www.CengageBrain.com**

Emefiele, M., Aziken, A., Orhue, A., Okpere, E., & Wilson, O. (2007). Sexual intercourse and preterm delivery: Any correlation? *Journal of Turkish-German Gynecological Association, 8*(2), 177–183.

"Engender Health," 2003a Sexual response and aging.

Epstein, M., & Ward, L. (2011). Exploring parent-adolescent communication about gender: Results from adolescent and emerging adult samples. *Sex roles, 65*(1–2), 108–118.

Epstein, R., Klinkenberg, W., Wiley, D., & McKinley, L. (2001). Ensuring sample equivalence across Internet and paper-and-pencil assessments. *Computers in Human Behavior, 17*, 339–346.

Espelage, D., Aragon, S., Birkett, M., & Koenig, B. (2008). Homophobic teasing, psychological outcomes, and sexual orientation among high school students: What influence do parents and schools have? (Report). *School Psychology Review, 37*, 202–216.

Evans, D., Young, K., Bulman, M., Shenton, A., Wallace, A., & Lalloo, F. (2008). Probability of BRCA1/2 mutation varies with ovarian histology: Results from screening 442 ovarian cancer families. *Clinical Genetics*, *73*(4), 338–345. doi:10.1111/j.1399-0004.2008.00974.

Evans, W. (2003, September 6). Taboo topics: Where should line be drawn on cartoons? *Sacramento Bee*. Retrieved from **www.sacbee.com**

Facts on induced abortion. (2011). In brief: Fact sheet. *Facts on induced abortion in the United States*. Guttmacher Institute. Retrieved from **www.guttmacher.org/pubs/fb_induced_abortion.html**

Fairchild, A. L., & Bayer, R. (1999). Uses and abuses of Tuskegee. *Science, 284*, 919–921.

Farley, M. (2004, October 1). Prostitution is sexual violence (Clinical reflections). *Psychiatric Times, 21* (electronic version). Retrieved from **www.psychiatrictimes.com/display/article/10168/48311**

Farley, M., & Kelly, V. (2000). Prostitution: A critical review of the medical and social sciences literature. *Women and Criminal Justice, 11*, 29–64.

Farmer, M., & Meston, C. (2006). Predictors of condom use self-efficacy in an ethnically diverse university sample. *Archives of Sexual Behavior, 35*, 313–326.

Farrington, J. (2002). Sexual myth or fact? Do you know what's true? *Current Health, 28*, SS1–SS3.

Faulkner, S. (2003). Good girl or flirt girl? Latinas' definitions of sex and sexual relationships. *Hispanic Journal of Behavioral Sciences, 25*, 174–205.

Fausto-Sterling, A. (2000). *Sexing the body: Gender politics and the construction of sexuality*. New York: Basic Books.

Fausto-Sterling, A., & Balaban, E. (1993). Genetics and male sexual orientation, *Science, 261*, 1257.

Fausto-Sterling, A., Coll, C., & Lamarre, M. (2012). Sexing the baby: Part 2 applying dynamic systems theory to the emergences of sex-related differences in infants and toddlers. *Social science & medicine, 74*(11), 1693–1702.

FBI. (2012). Attorney general Eric Holder announces revisions to the uniform crime report's definition of rape. U.S. Department of Justice. Office of Public Affairs. Retrieved from **www.fbi.gov/news/pressrel/press-releases/attorney-general-eric-holder-announces-revisions-to-the-uniform-crime-reports-definition-of-rape**

FDA. (2013). U.S. Food and Drug Administration Antiretroviral drugs used in the treatment of HIV infection. Retrieved from **www.fda.gov/ForConsumers/ByAudience/ForPatientAdvocates/HIVandAIDSActivities/ucm118915.htm**

Federal Bureau of Investigation. (2013). Uniform crime reports. *Hate crime statistics, 2012*. Retrieved from **www.fbi.gov/about-us/cjis/ucr/hate-crime/2012/topic-pages/victims/victims_final**

Fedoroff, J. P., Fishell, A., & Fedoroff, B. (1999). A case series of women evaluated for paraphilic sexual disorders. *Canadian Journal of Human Sexuality, 8*, 127–140.

Feingold, A. (1988). Matching for attractiveness in romantic partners: A meta-analysis and theoretical critique. *Psychological Bulletin, 104*, 226–235.

Feldhaus-Dahir, M. (2009). The causes and prevalence of hypoactive sexual desire disorder. *Urologic Nursing, 29*, 259–263.

Feldman, M., Goldstein, I., Hatzichristou, D., Krane, R., & McKinlay, J. (1994). Impotence and its medical and psychosocial correlates: Results of the Massachusetts Male Aging Study. *Journal of Urology, 151*, 54–61.

Feldman, R., Eidelman, A., Sirota, L., & Weller, A. (2002). Comparison of skin-to-skin (kangaroo) and traditional care: Parenting outcomes and preterm infant development. *Pediatrics, 110*, 16–26.

Feldman, R., Rosenthal, Z., & Eidelman, A. I. (2014). Maternal-preterm skin-to-skin contact enhances child physiologic organization and cognitive control across the first 10 years of life. *Biological psychiatry, 75*(1), 56–64.

Felitti, V. J. (1991). Long-term medical consequences of incest, rape, and molestation. *Southern Medical Journal, 84*, 328–331.

Ferentz, K. S. (2007, January). A guide to switching antidepressant therapy (Clinical report). *Patient Care for the Nurse Practitioner*, 16–21.

Fergusson, D., Horwood, L., & Ridder, E. (2005). Abortion and subsequent mental health. *Journal of Child Psychology and Psychiatry, 47*, 16–24.

Fetal development: How your baby grows. (2005). Retrieved from **www.babycenter.com**

Fine, C. (2012). Explaining, or sustaining, the status quo? The potentially self-fulfilling effects of "hardwired" accounts of sex differences. *Neuroethics, 5*(3), 285–294.

Finer, L. (2010). Unintended pregnancy among US adolescents: accounting for sexual activity. *Journal of Adolescent Health, 47*(3), 312–314.

Finger, W., Lund, M., & Slagle, M. (1997). Medications that may contribute to sexual disorders. *Journal of Family Practice, 44*, 33–43.

Finkelhor, D., & Jones, L. (2004, January). Explanations for the decline in child sexual abuse. *Juvenile Justice Bulletin*, pp. 1–12.

Finkelhor, D., & Yllo, K. (1985). *License to rape: Sexual abuse of wives*. New York: Holt, Rinehart and Winston.

Finkelhor, D., Jones, L., & Shattuck, A. (2013). Updated trends in child maltreatment, 2011. *Durham, NH: Crimes Against Children*

Research Center. Retrieved from **www.unh.edu/ccrc/pdf/CV203_Updated%20trends%202011_FINAL_1-9-13.pdf**

Finkelhor, D., Shattuck, A., Turner, H., & Hamby, S. (2014). The lifetime prevalence of child sexual abuse and sexual assault assessed in late adolescence. *Journal of Adolescent Health, 54*, 16–27.

Finkelhor, D., Turner, H. A., Shattuck, A., & Hamby, S. (2013). Violence, crime, and abuse exposure in a national sample of children and youth: An update. *JAMA Pediatrics, 167*(7), 614–621.

Finkelson, L., & Oswalt, R. (1995). College date rape: Incidence and reporting. *Psychological Reports, 77*, 526.

Finkielstain, G., Kim, M., Sinaii, N., Nishitani, M., Van Ryzin, C., Hill, S., & Merke, D. (2012). Clinical characteristics of a cohort of 244 patients with congenital adrenal hyperplasia. *The Journal of Clinical Endocrinology & Metabolism, 97*(12), 4429–4438.

Fiorella, S. (2013, October 18). Cyber-bullying, social media, and parental responsibility. *The Huffington Post*. Retrieved from **www.huffingtonpost.com/sam-fiorella/cyber-bullying-social-media-and-parental-responsibility_b_4112802.html**

Fisher, E. (2012). *Responsive school practices to support lesbian, gay, bisexual, transgender, and questioning students and families*. New York, NY: Routledge.

Fisher, T. D., Moore, Z. T., & Pittenger, M. (2012). Sex on the brain?: An examination of frequency of sexual cognitions as a function of gender, erotophilia, and social desirability. *The Journal of Sex Research, 49*, 69–77.

Fisher, W. A. (1988). The sexual opinion survey. In C. Davis, W. Yarber, and S. Davis (Eds.), *Sexually related measures: A compendium* (pp. 34–38). Lake Mills, IA: Graphic Publishing.

Fisher, W. A., Byrne, D., White, L. A., & Kelley, K. (1988). Erotophobia-erotophilia as a dimension of personality. *Journal of Sex Research, 25*, 123–151.

Fisher, W., & Barak, A. (2001). Internet pornography: A social psychological perspective on Internet sexuality. *Journal of Sex Research, 38*, 312–333.

Fissell, M. (1999a). The paradox of twilight sleep. *Women's Health in Primary Care, 2*, 972.

Fissell, M. (1999b). Removing the curse of Eve. *Women's Health in Primary Care, 2*, 908.

Fitzgerald, K. (2013). Postpartum depression affects one in seven new moms. *Medical News Today*. Retrieved from **www.medicalnewstoday.com/articles/257704.php**

Flanagan, P. (2010). Making molehills into mountains: Adult responses to child sexuality and behaviour. *Explorations: An E-Journal of Narrative Practice, 1*, 57–69.

Flanagan, P. (2011). Making sense of children's sexuality: Understanding sexual development and activity in education contexts. *Waikato Journal of Education, 16*(3), 69–79.

Flannery, D., & Ellingson, L. (2003). Sexual risk behaviours among first year college students, 2000–2002. *Californian Journal of Health Promotion, 1*(3), 93–104.

Flannigan, J. (2008). HIV and AIDS: Transmission, testing and treatment. *Nursing Standard, 22*, 48–56.

Floyd, F., & Bakeman, R. (2006). Coming-out across the life course: Implications of age and historical context. *Archives of Sexual Behavior, 35*(3), 287–296.

Floyd, F., & Stein, T. (2002). Sexual orientation identity formation among gay, lesbian, and bisexual youths: Multiple patterns of milestone experiences. *Journal of Research on Adolescence, 12*, 167–191.

Foldes, P., & Buisson, O. (2009). The clitoral complex: A dynamic sonographic study. *The Journal of Sexual Medicine, 6*, 1223–1231.

Forsyth, J. (2013, August 17). State on verge of making major changes in prostitution laws. *WOAI, Clear Channel Media*. Retrieved from **www.woai.com/articles/woai-local-news-119078/state-on-verge-of-making-major-11199383#ixzz34vNfbS1Z**

Fortenberry, J., Schick, V., Herbenick, D., Sanders, S., Dodge, B., & Reece, M. (2010). Sexual behaviors and condom use at last vaginal intercourse: A national sample of adolescents ages 14 to 17 years: Findings from the National Survey of Sexual Health and Behavior (NSSHB). *The Journal of Sexual Medicine Volume, 7*(S5), 305–314.

Foubert, J. D., & McEwan, M. K. (1998). An all-male rape-prevention peer education program: Decreasing fraternity men's behavioral intent to rape. *Journal of College Student Development, 36*, 548–556.

Foubert, J., Godin, E., & Tatum, J. (2010). In their own words: Sophomore college men describe attitude and behavior changes resulting from a rape prevention program two years after their participation. *Journal of Interpersonal Violence, 25*, 2237–2257.

Fowler, F. (2008). *Survey research methods (applied social research methods)* (4th ed.). Thousand Oaks, CA: Sage Publications, Inc.

Fox, K. (1997). Mirror, mirror: A summary of the findings on body image. Retrieved from **www.sirc.org/publik/mirror.html**

Fox, M., & Hayes, J. (2007). Cervical preparation for second-trimester surgical abortion prior to 20 weeks of gestation. *Society of Family Planning and Contraception, 76*, 486–495.

Franiuk, R. (2007). Discussing and defining sexual assault: A classroom activity. *College Teaching, 55*(3), 104–107.

Frank, J., Mistretta, P., & Will, J. (2008). Diagnosis and treatment of female sexual dysfunction. *American Family Physician, 77*, 635–642.

Franzini, L., & Sideman, L. (1994). Personality characteristics of condom users. *Journal of Sex Education and Therapy, 20*, 110–118.

Frederick, D., Peplau, L., & Lever, J. (2006). The swimsuit issue: Correlates of body image in a sample of 52,677 heterosexual adults. *Body Image, 3*, 413–419.

Freund, K., Seto, M. C., & Kuban, M. (1996). Two types of fetishism. *Behaviour Research and Therapy, 34*, 687–694.

Frieden, T. (2011, August 3). 72 charged in online global child porn ring. *CNN Justice*. Retrieved from **www.cnn.com/2011/CRIME/08/03/us.child.porn.ring/index.html**

Friedman, R. (2007). Sexual orientation: Neuroendocrine and psychodynamic influences. *Psychiatric Times, 24*, 47–52.

Friedrich, W., Fisher, J., Dittner, C., (2001). Child Sexual Behavior Inventory: normative, psychiatric, and sexual abuse comparisons. *Child Maltreatment, 6*(1), 37–49.

Friedrich, W., & Gerber, P. (1994). Autoerotic asphyxia: The development of a paraphilia. *Journal of the American Academy of Child and Adolescent Psychiatry, 33*, 970–974.

Friedrich, W., Fisher, J., Broughton, D., Houston, M., & Shafran, C. (1998). Normative sexual behavior in children: A contemporary sample [Electronic version]. *Pediatrics, 101*, 9–16.

Friis-Moller, N., et al. (2003). Combination antiretroviral therapy and the risk of myocardial infarction. *New England Journal of Medicine, 349*, 1993–2003.

Frost, D. M. (2011). Similarities and differences in the pursuit of intimacy among sexual minority and heterosexual individuals: A personal projects analysis. *Journal of Social Issues, 67*(2), 282–301.

Furman, W. (2002). The emerging field of adolescent romantic relationships. *Current Directions in Psychological Science, 11*, 177–181.

Gabelnick, N., Schwartz, J., & Darroch, J. (2008). Contraceptive research and development. In A. Nelson, W. Cates Jr., F. Stewart, & D. Kowal (Eds.), *Contraceptive Technology* (19th rev. ed, pp. 433–450). New York: Ardent Media.

Gadpayle, A., Kumar, N., Duggal, A., Rewari, B., & Ravi, B. (2012). Survival trend and prognostic outcome of AIDS patients according to age, sex, stages, and mode of transmission—A retrospective study at ART centre of a tertiary care hospital. *Journal of the Association of Physicians of India, 13*(4), 291–298.

Gagnon, J., & Simon, W. (2011). *Sexual conduct: The social sources of human sexuality*. Piscataway, NJ: Transaction Publishers.

Gallagher, K. (2006). Vaginal delivery in breech position and breech birth. Treatment overview.

Garbarino, J. (2007). *See Jane hit: Why girls are growing more violent and what we can do about it*. New York: Penguin Press.

Garcia, F. D., & Thibaut, F. (2011). Current concepts in the pharmacotherapy of paraphilias. *Drugs, 71*(6), 771–790.

Garcia, F., & Thibaut, F. (2010). Sexual addictions. *The American journal of drug and alcohol abuse, 36*(5), 254–260.

Garcia, F., & Thibaut, F. (2011). Current concepts in the pharmacotherapy of paraphilias. *Drugs, 71*(6), 771–790.

Garcia, F., Delavenne, H., Assumpção, A., & Thibaut, F. (2013). Pharmacologic treatment of sex offenders with paraphilic disorder. *Current psychiatry reports, 15*(5), 1–6.

Garcia, S., Khersonsky, D., & Stacey, S. (1997). Self-perceptions of physical attractiveness. *Perceptual and Motor Skills, 84*, 242–248.

Gardner, A., de Vries, B., & Mockus, D. (2014). Aging out in the desert: Disclosure, acceptance, and service use among midlife and older Lesbians and Gay men. *Journal of homosexuality, 61*(1), 129–144.

Gardos, P., & Mosher, D. (1999). Gender differences in reactions to viewing pornographic vignettes: Essential or interpretive? *Journal of Psychology and Human Sexuality, 11*, 65–83.

Garenne, M. (2008). Long-term population effect of male circumcision in generalised HIV epidemics in sub-Saharan Africa African. *Journal of AIDS Research 2008, 7*(1), 1–8.

Garnett, G., Wilson, D., Law, M., Grulich, A., Cooper, D., & Kaldor, J. (2008). Relation between HIV viral load and infectiousness: A model-based analysis. *Lancet, 372*(9635), 270–271, 314–320.

Garrahy, D. A. (2001). Three third-grade teachers' gender-related beliefs and behavior. *Elementary School Journal, 102*, 81–94.

Gates, G. (2006). Same-sex couples and the gay, lesbian, bisexual population: New estimates from the American Community Survey. Williams Institute: UCLA School of Law.

Gates, G. (2009). Same-sex spouses and unmarried partners in the American Community Survey, 2008. *The Williams Institute*. Retrieved from **http://williamsinstitute.law.ucla.edu/wp-content/uploads/Gates-ACS2008FullReport-Sept-2009.pdf**.

Gates, G., & Sonenstein, F. (2000). Heterosexual genital sexual activity among adolescent males: 1988 and 1995. *Family Planning Perspectives, 32*, 295–297, 304.

Gazeteer for Scotland. (1995). Sir James Young Simpson. Retrieved from **www.geo.ed.ac.uk/scotgaz/people/famousfirst60.html**

Geiger, A., & Foxman, B. (1996). Risk factors of vulvovaginal candidiasis: A case-control study among university students. *Epidemiology, 7*, 182–187.

Geist, E. A., & King, M. (2008). Different, not better: Gender differences in mathematics learning and achievement (Report). *Journal of Instructional Psychology, 35*, 43–52.

George, S. (2012). The strong arm of the law is weak: how the trafficking victims protection act fails to assist effectively victims of the sex trade. *Creighton Law Review, 45*(3), 563–580.

George, W., Davis, K., Norris, J., Heiman, J., Stoner, S., Schacht, L., & Kajumulo, K. (2009). Indirect effects of acute alcohol intoxication on sexual risk-taking: The roles of subjective and physiological sexual arousal. *Archives of Sexual Behavior, 38*(4), 498–513. doi:10.1007/s10508-008-9346-9

Gerding, A. (2014). Gender roles in tween television programming: A content analysis of two genres. *Sex Roles, 70*(1/2), 43–46. doi:10.1007/s11199-013-0330-z

Gerstenfeld, P. B. (2013). *Hate crimes: Causes, controls, and controversies*. New York: Sage.

Gibbs, N. (2003, July 7). A yea for gays: The Supreme Court scraps sodomy laws, setting off a hot debate. *Time*, pp. 38–39.

Gibson, L., & Leitenberg, H. (2000). Child sexual abuse prevention programs: Do they decrease the occurrence of child sexual abuse? *Child Abuse and Neglect, 24*, 1115–1125.

Giles, G., Severi, G., English, D., McCredie, M., Borland, R., Boyle, P., & Hopper, J. (2003). Sexual factors and prostate cancer. *British Journal of Urology International, 92*, 211–216.

Gillespie, R. (2003). Childfree and feminine: Understanding the gender identity of voluntarily childless women. *Gender and Society, 17*, 122–136.

Giraldi, A., Rellini, A. H., Pfaus, J., & Laan, E. (2013). Female sexual arousal disorders. *The journal of sexual medicine, 10*(1), 58–73.

Glasier, A. (2007). Pregnancy in women over 45: Should this be encouraged? *Menopause International, 13*, 6–7.

Glenn, D. (2004, April 30). A dangerous surplus of sons? *Chronicle of Higher Education*, pp. A14–A15.

Glynn, S. (2012). New test can predict preterm delivery. *Medical News Today*. Retrieved from **www.medicalnewstoday.com/articles/250330.php**

Glynn, S. (2013). Men more depressed than women if childless. *Medical News Today*. Retrieved from **www.medicalnewstoday.com/articles/258531.php**

Goldbaum, G. M., Yu, T., & Wood, R. W. (1996). Changes at a human immunodeficiency virus testing clinic in the prevalence of unsafe sexual behavior among men who have sex with men. *Sexual Transmission of Disease, 23*, 109–114.

Goldmeier, D., Garvey, L., & Barton, S. (2008). Does chronic stress lead to increased rates of recurrences of genital herpes? A review of the psychoneuroimmunological evidence. *International Journal of STDs, 19*, 359–362. doi:10.1258/ijsa.2007.00730419

Goldstein, I., Meston, C., Davis, S., & Traish, A. (2006). *Women's sexual function and dysfunction: Study, diagnosis, and treatment*. New York: Taylor & Francis.

Goldston, L., & Wong, N. (2003, June 13). Teens talk frankly about sex, themselves. *San Jose Mercury News*, p. B3.

Golombok, S., & Tasker, F. (1996). Do parents influence the sexual orientation of their children? Findings from a longitudinal study of lesbian families. *Developmental Psychology, 32*, 3–11.

Gomez, A. (2004). Gender and the HIV/AIDS pandemic. *Women's Health Journal, 2–3*, 104–112.

Gonorrhea diagnosis. (2011). Department of Health and Human Services. National Institutes of Health. Retrieved from **www.niaid.nih.gov/topics/gonorrhea/understanding/Pages/diagnosis.aspx**

Goodboy, A., & Booth-Butterfield, M. (2009). Love styles and desire for closeness in romantic relationships. *Psychological Reports, 105*, 191–197.

Goodman, R. E. (2001). Beyond the enforcement principle: Sodomy laws, social norms, and social panoptics. *California Law Review, 89*, 643–740.

Goodson, P., Suther, S., Pruitt, B., & Wilson, K. (2003). Defining abstinence. *Journal of School Health, 73*, 91–96.

Goodstein, L. (2004, February 27). Two studies cite child sex abuse by 4% of priests. *New York Times*, p. A1.

Gortmaker, V., & Brown, R. (2006). Out of the college closet: Differences in perceptions and experiences among out and closeted lesbian and gay students. *College Student Journal 40*, 606–619.

Gosink, P. D., & Jumbelic, M. I. (2000). Autoerotic asphyxiation in a female. *American Journal of Forensic Medicine and Pathology, 21*, 114–118.

Gottlieb, A. (2008). Intimate partner violence: A guide for clinicians. *Sexuality, Reproduction & Menopause, 6*, 10–14.

Gottlieb, G. (2013). Changing HIV epidemics: what HIV-2 can teach us about ending HIV-1. *AIDS, 27*(1), 135–137. doi:10.1097/QAD.0b013e32835a11a4

Gottman, J. M. (1998). Psychology and the study of marital processes. *Annual Review of Psychology, 49*, 169–197.

Gottman, J. M., & Levenson, R. W. (2001). 12-year study of gay and lesbian couples. Gottman Institute.

Gottman, J. M., & Levenson, R. W. (2002). A two-factor model for predicting when a couple will divorce: Exploratory analyses using 14-year longitudinal data. *Family Process, 41*, 83–95.

Gottman, J. M., et al. (2003). Correlates of gay and lesbian couples' relationship satisfaction and relationship dissolution. *Journal of Homosexuality, 45*, 23–40.

Gottman, J., & Carrere, S. (2000, September). Welcome to the Love Lab. *Psychology Today, 33*, 42–48.

Gottman, J., & Gottman, J. (2008). Method couple therapy. In A. Gurman (Ed.), *Clinical Handbook of Couple Therapy*, 4th edition (pp. 138–164). New York: The Guilford Press.

Gottman, J., & Silver, N. (2000). *The seven principles for making marriage work*. New York: Three Rivers Press.

Graber, B. (1993). Medical aspects of sexual arousal disorders. In W. O'Donohue & J. Geer (Eds.), *Handbook of sexual dysfunctions* (pp. 103–156).

Grady, W. R., Klepinger, D. H., Billy, J. O. F., & Tanfer, K. (1993). Condom characteristics: The perceptions and preferences of men in the United States. *Family Planning Perspectives, 25*(2), 67–73.

Graham, C. (2003). A new view of women's sexual problems. *Journal of Sex and Marital Therapy, 29*, 325–327.

Graham, R. (2001, November 21). *College students shunning condoms*. Retrieved from **www.intelihealth.com**

Grand Challenges Explorations Grants. (2013). *Grand Challenges in Global Health*. Retrieved from **www.grandchallenges.org/explorations/Pages/grantsawarded.aspx?Topic=Contraception&Round=all&Phase=all**

Green, R. (2001). (Serious) sadomasochism: A protected right of privacy? *Archives of Sexual Behavior, 30*, 543–549.

Greenberg, J. A. (2012). *Intersexuality and the law: why sex matters*. New York: NYU Press.

Greenberg, J., Bruess, C., & Haffner, D. (2002). *Exploring the dimensions of sexuality*. Sudbury, MA: Jones & Bartlett.

Greenlee, R. T., Kessel, B., Williams, C. R., Riley, T. L., Ragard, L. R., Hartge, P., & Reding, D. J. (2010). Prevalence, incidence, and natural history of simple ovarian cysts among women > 55 years old in a large cancer screening trial. *American Journal of Obstetrics and Gynecology, 202*(4), 373e1–373e9.

Grenier, G., & Byers, E. S. (1995). Rapid ejaculation: A review of the conceptual, etiological, and treatment issues. *Archives of Sexual Behavior, 24*, 447–472.

Grenier, G., & Byers, E. S. (1997). The relationship among ejaculatory control, ejaculatory latency, and attempts to prolong sexual intercourse. *Archives of Sexual Behavior, 26*, 27–47.

Griebel, C. P., Halvorsen, J., & Goleman, T. B. (2005, October 1). Management of spontaneous abortion. *American Family Physician, 72*(7), 1243–1250.

Griebling, T. (2005). Urologic diseases in America project: Trends in resource use for urinary tract infections in women. *Journal of Urology, 173*(4), 1281–1287.

Griensven, F., de Lind van Wijngaarden, J. W., Baral, S., & Grulich, A. (2009, July). The global epidemic of HIV infection among men who have sex with men. *Current Opinion in HIV & AIDS, 4*(4), 300–307. doi:10.1097/COH.0b013e32832c3bb3

Griffin, J., Umstattd, M., & Usdan, S. (2010). Alcohol use and high-risk sexual behavior among collegiate women: A review of research on Alcohol Myopia Theory. *Journal of American College Health, 58*(6), 523–532.

Griffing, S., Ragin, D. F., Morrison, S. M., Sage, R. E., Madry, L., & Primm, B. J. (2005). Reasons for returning to abusive relationships: Effects of prior victimization. *Journal of Family Violence, 20*(5), 341–348.

Gruber, J., & Fineran, S. (2008). Comparing the impact of bullying and sexual harassment victimization on the mental and physical health of adolescents. *Sex Roles, 59*(1–2), 1–13.

Guillette, E. (2002). Unambiguous results: When Mexico's Yaqui Indians split into two different agricultural camps in the 1950s, their children became an unusually perfect test group for the effects of pesticide exposure. *Alternatives Journal, 28*, 24–26.

Gulia, K., & Mallick, H. (2010). Homosexuality: A dilemma in discourse. *Indian Journal of Physiological Pharmacology, 54*(1), 5–20.

Gulledge, A., Gulledge, M., & Stahmann, R. (2003). Romantic physical affection types and relationship satisfaction. *American Journal of Family Therapy, 31*, 233–241.

Gunderson, E., Ramirez, G., Levine, S., & Beilock, S. (2012). The role of parents and teachers in the development of gender-related math attitudes. *Sex Roles, 66*(3/4), 153–166. doi:10.1007/s11199-011-9996-2

Gurney, K. (2007). Sex and the surgeon's knife: The family court's dilemma—informed consent and the specter of iatrogenic harm to children with intersex characteristics. *American Journal of Law & Medicine, 33*, 625–661.

Guttmacher Institute. (2013). *Facts on Induced Abortion in the United States*. Retrieved from **www.guttmacher.org/pubs/fb_induced_abortion.html**

H.R. 3791: SAFE Act of 2007. (2007). *GovTrack.us (database of federal legislation)*. Retrieved from **www.govtrack.us/congress/bill.xpd?bill=h110-3791**

Ha, T., van den Berg, J., Engels, R., & Lichtwarck-Aschoff, A. (2012). Effects of attractiveness and status in dating desire in homosexual and heterosexual men and women. *Archives of Sexual Behavior, 41*(3), 673–682.

Haas, K. (2004). Who will make room for the intersexed? *American Journal of Law & Medicine, 30*, 41–77.

Haddad, L., & Nour, N. (2009). Unsafe abortion: unnecessary maternal mortality. *Reviews in obstetrics and gynecology, 2*(2), 122.

Haldeman, D. (2013). The pseudo-science of sexual orientation conversion therapy: Clinical and social implications. Angles: *The Policy Journal of the Institute for Lesbian and Gay Strategic Studies*, 4(1), 1–4.

Halim, M., Ruble, D., & Amodio, D. (2011). From pink frilly dresses to "one of the boys": A social-cognitive analysis of gender identity development and gender bias. *Social and Personality Psychology Compass, 5*(11), 933–949.

Hall, J. (1978). Gender effects in decoding nonverbal cues. *Psychological Bulletin, 85*(4), 845–857. doi:10.1037/0033-2909.85.4.845

Hall, J. A. (1998). How big are nonverbal sex differences? The case of smiling and sensitivity to nonverbal cues. In D. J. Canary & K. Dindia (Eds.), *Sex differences and similarities in communication: Critical essays and empirical investigations of sex and gender in interaction* (pp. 157–177). Mahwah, NJ: Erlbaum.

Hall, R., & Hall, R. (2007). A profile of pedophilia: Definition, characteristics of offenders, recidivism, treatment outcomes, and forensic issues. *Mayo Clinic Proceedings, 82*, 457–481.

Hall-Flavin, D. (2011) Mayo Clinic Diseases and Conditions. Retrieved from **www.mayoclinic.org/diseases-conditions/depression/expert-answers/antidepressants/FAQ-20058104**

Halpern, C. J. T., Udry, J. R., Suchindran, C., & Campbell, B. (2000). Adolescent males' willingness to report masturbation. *Journal of Sex Research, 3*, 327–332.

Hamilton, B., Martin, J., & Ventura, S. (2010). Births: Preliminary data for 2009. National Center for Health Statistics. *National Vital Statistics Reports Web Release, 59*. Retrieved from **www.cdc.gov/nchs/data/nvsr/nvsr59/nvsr59_03.pdf**

Hampton, T. (2008). Abstinence-only programs under fire. *Journal of the American Medical Association, 299*, 2013–2015. doi:10.1001/jama.299.17.2013

Haning, R., O'Keefe, S., Beard, K., Randall, E., Kommor, M., & Stroebel, S. (2008). Empathic sexual responses in heterosexual women and men. *Sexual & Relationship Therapy, 23*(4), 325–344. doi:10.1080/14681990802326743

Hans, J., Gillen, M., & Akande, K. (2010). Sex redefined: The reclassification of oral-genital contact. *Perspectives on Sexual & Reproductive Health, 42*(2), 74–78. doi: 10.1363/4207410

Hanson, R., & Morton-Bourgon, K. (2005). The characteristics of persistent sexual offenders: a meta-analysis of recidivism studies. *Journal of consulting and clinical psychology, 73*(6), 1154.

Hargitai, B., Marton, T., & Cox, P. M. (2004). Examination of the human placenta. *Journal of Clinical Pathology, 57*, 785–792. doi:10.1136/ jcp.2003.014217

Haritaworn, J., Lin, C., & Klesse, C. (2006). *Sexualities, 9*, 515–529. doi: 10.1177/1363460706069963

Hariton, E. (1973, March). The sexual fantasies of women. *Psychology Today*, pp. 39–44.

Harmon, K. (2010, May 18). Fact or fiction: Fathers can get postpartum depression. *Scientific American*. Retrieved from **www.scientificamerican.com/article.cfm?id=fathers-postpartum-depression**

Harney, P. A., & Muehlenhard, C. L. (1991). Factors that increase the likelihood of victimization. In A. Parrot & L. Bechhofer (Eds.), *Acquaintance rape: The hidden crime* (pp. 159–175). New York: Wiley.

Harper, B., & Tiggemann, M. (2008). The effect of thin ideal media images on women's self-objectification, mood, and body image. *Sex Roles, 58*(9–10), 649–657.

Harris, C. (2003). A review of sex differences in sexual jealousy, including self-report data, psychophysiological responses, interpersonal violence, and morbid jealousy. *Personality & Social Psychology Review, 7*(2), 102–128.

Harris, K. (2013). Sexuality and suicidality: Matched-pairs analyses reveal unique characteristics in non-heterosexual suicidal behaviors. *Archives of sexual behavior, 42*(5), 729–737.

Hartmann, K., Palmieri, R., Gartlehner, G., Thorp, J., Lohr, K., & Viswanathan, M. (2005). Outcomes of routine episiotomy. *Journal of the American Medical Association, 293*, 2141–2148.

Hatcher, R. A. (1998). Depo-Provera, Norplant, and progestin-only pills (minipills). In R. A. Hatcher, J. Trussell, F. Stewart, W. Cates Jr., G. K. Stewart, F. Guest, & D. Kowal (Eds.), *Contraceptive Technology* (17th rev. ed., pp. 467–509). New York: Ardent Media.

Hatcher, R. A. et al. (1994). *Contraceptive Technology* (16th rev. ed.). New York: Irvington.

Hatcher, R. A., & Nelson, A. L. (2004). Combined hormonal contraceptive methods. In R. A. Hatcher, J. Trussell, F. Stewart, W. Cates Jr., A. L. Nelson, F. Guest, & D. Kowal (Eds.), *Contraceptive Technology* (18th rev. ed., pp. 391–460). New York: Ardent Media.

Hatcher, R. A., Trussell, J., Nelson, A., Cates, W., Stewart, F., Kowal, D. (2008). *Contraceptive Technology* (19th rev. ed.). New York: Ardent Media.

Hatcher, R., Trussell, J., Nelson, A., Cates, W., & Kowal, D. (Eds). (2011). *Contraceptive Technology* (20th rev. ed.). New York: Ardent Media.

Hatzimouratidisa, K., Amarb, E., Eardleyc, I., Giulianod, F., Hatzichristoua, D.,... Vardif, Y. (2010). Guidelines on male sexual dysfunction: Erectile dysfunction and premature ejaculation. *European Urology, 57*, 804–814.

Healthwise. (2004). Choices for an abortion. Retrieved from **www.my.webmd.com/hw/womens_conditions**

Heeney, J., Dalgleish, A., & Weiss, R. (2006, July 28). Origins of HIV and the evolution of resistance to AIDS. *Science, 313*(5786), 462–466.

Hegazy, A., & Al-Rukban, M. (2012). Hymen: facts and conceptions. *The Health, 3*, 109–115.

Heilprin, J. & Winfield, N. (2014, May 23). UN panel slams Vatican on priest sex-abuse scandal. *The Seattle Times*. Retrieved from **www.seattletimes.com/html/nationworld/2023676376_apxunited-nationsvaticanabuse.html?syndication=rss**

Heim, L. J. (2001, April 15). Evaluation and differential diagnosis of dyspareunia. *American Family Physician, 63*, 1535–1544.

Heiman, J. R. (2002). Sexual dysfunction: Overview of prevalence, etiological factors, and treatments. *Journal of sex research, 39*(1), 73–78.

Heino, F. L., et al. (1995). Prenatal estrogens and the development of homosexual orientation. *Developmental Psychology, 31*, 12–21.

Hembree, W., Cohen-Kettenis, P., Delemarre-van de Waal, V., Gooren, L., Meyer, W., Spack, N., Tangpricha, V., & Montori, V. (2009). Treatment of transsexual persons: An endocrine society clinical practice guideline. *The Journal of Clinical Endocrinology & Metabolism, 94*, 3132–3154.

Hendrick, S. (2004). Close relationship research: A resource for couple and family therapists. *Journal of Marital and Family Therapy, 3*, 13–32.

Hendrick, S. S., & Hendrick, C. (2002). Linking romantic love with sex: Development of the perceptions of love and sex scale. *Journal of Social and Personal Relationships, 19*, 361–378.

Henshaw, S. (1998). Unintended pregnancy in the United States. *Family Planning Perspectives, 30*, 24–29, 46.

Hepatitis B Vaccine. (2009). *Hepatitis B Foundation*. Retrieved from **www.hepb.org/hepb/vaccine_information.htm**

Herbenick, D., Reece, M., Schick, V., & Sanders, S. A. (2014). Erect penile length and circumference dimensions of 1,661 sexually active men in the United States. *The journal of sexual medicine, 11*(1), 93–101.

Herbenick, D., Reece, M., Schick, V., Sanders, S. A., Dodge, B., & Fortenberry, J. D. (2010a). Sexual behaviors, relationships, and perceived health status among adult women in the United States: Results from a national probability sample. 291 Findings from the National Survey of Sexual Health and Behavior (NSSHB). *The Journal of Sexual Medicine, 7*(supplement 5 impact factor: 4.884), 277–290.

Herbenick, D., Reece, M., Schick, V., Sanders, S., Dodge, B., & Fortenberry, J. (2010b). Sexual behavior in the United States: Results from a national probability sample of men and women ages 14–94. *The Journal of Sexual Medicine, 7*(5), 255–265.

Herbenick, D., Reece, M., Schick, V., Sanders, S., Dodge, B., & Fortenberry, D. (2010a). Sexual behaviors, relationships, and perceived health among adult men in the United States: Results from a national probability sample. *Journal of Sexual Medicine, 7*(5), 277–290.

Herbert, B. (2007, September 11). Fantasies, well meant. *New York Times* (electronic version).

Hernandez, J. (2007, Summer-Fall). Disruptive girlhoods: Books on aggression in girls. *Feminist Collections: A Quarterly of Women's Studies Resources, 28*, 23–28.

Hesketh, T., Lu, L., & Xing, Z. (2011). The consequences of son preference and sex-selective abortion in China and other Asian countries. *Canadian Medical Association Journal, 183*(12), 1374–1377.

Heubeck. E. (2007). Sharing your sex fantasies with your partner: Sizzler or fizzler? *Health & Sex. WebMD*. Retrieved from **www.webmd.com/sex-relationships/features/sharing-your-sex-fantasies**

Heyman, R., Hunt, A., Malik, J., & Smith Slep, A. (2009). Desired change in couples: Gender differences and effects on communication. *Journal of Family Psychology, 23*(4), 474–484. doi: 10.1037/a0015980

Hicks, C. (2010). Menstrual manipulation: Options for suppressing the cycle. *Cleveland Clinic Journal of Medicine, 77*(7), 445–453.

Hill, C. (2002). Gender, relationship stage, and sexual behavior: The importance of partner emotional investment within specific situations. *Journal of Sex Research, 39*, 228–240.

Hill, C., & Silva, E. (2005). Drawing the line: sexual harassment on campus. American Association of University Women Educational Foundation. Retrieved from **www.aauw.org/files/2013/02/drawing-the-line-sexual-harassment-on-campus.pdf**

Hinduja, S., & Patchin, J. (2014). A brief review of state cyberbullying laws and policies. Cyberbullying Research Center—State Cyberbullying Laws. Retrieved from **www.cyberbullying.us/Bullying_and_Cyberbullying_Laws.pdf**

Hines, M. (2010). Sex-related variation in human behavior and the brain. *Trends in cognitive sciences, 14*(10), 448–456.

Hines, M. (2011). Prenatal endocrine influences on sexual orientation and on sexually differentiated childhood behavior. *Frontiers in neuroendocrinology, 32*(2), 170–182.

Hinnant, J., O'Brien, M., & Ghazarian, S. (2009). The longitudinal relations of teacher expectations to achievement in the early school years. *Journal of Educational Psychology, 101*(3), 662–670. doi:10.1037/a0014306

Hipp, L., Kane Low, L., & van Anders, S. (2012). Exploring Women's Postpartum Sexuality: Social, Psychological, Relational, and Birth-Related Contextual Factors. *The Journal of Sexual Medicine, 9*, 2330–2341.

Hite, S. (1976). *The Hite report*. New York: Macmillan.

Hite, S. (1981). *The Hite report on male sexuality*. New York: Random House.

Hock, R. 2009. *Forty studies that changed psychology: Explorations into the history of psychological research*. Upper Saddle River, NJ: Pearson.

Hodge, D. (2008). Sexual trafficking in the United States: A domestic problem with transnational dimensions. *Social Work, 53*, 143–152.

Hoff, T., Greene, L., & Davis, J. (2003). *National survey of adolescents and young adults: Sexual health knowledge, attitudes and experiences*. Menlo Park, CA: Kaiser Family Foundation.

Hogben, M., Chesson, H., & Aral, S. O. (2010). Sexuality education policies and sexually transmitted disease rates in the United States of America. *International Journal of STD & AIDS, 21*, 293–297. doi: 10.1258/ijsa.2010.009589

Hollander, D. (2007). Behavioral risk factors HIV imperil millions. *Perspectives on Sexual and Reproductive Health, 39*(1), 4–5.

Hollander, E., & Rosen, J. (2000). Impulsivity. *Journal of Psychopharmacology, 14*, S39–S44.

Hollingsworth, L. (2005). Ethical considerations in prenatal sex selection. *Health and Social Work, 30*, 126–135.

Holman, T., & Jarvis, M. (2003). Hostile, volatile, and validating couple-conflict types: An investigation of Gottman's couple-conflict types. *Personal Relationships, 10*, 267–282.

Holmes, L., Escalante, C., & Garrison, O. (2008). Testicular cancer incidence trends in the United States (1975–2004): Plateau or shifting racial paradigm? *Public Health, 122*(9), 862–872.

Holmes, H. (2010). Human teratogens: Update 2010 Birth Defects Research Part A. *Clinical and Molecular Teratology, 91*, 1–7.

Holvey, N. (2014). Supporting women in the second stage of labour. *British Journal of Midwifery, 22*(3), 182–186.

Hong, J. (1984). Survival of the fastest: On the origins of premature ejaculation. *Journal of Sex Research, 20*, 109–112.

Horne, S., Puckett, J., Apter, R., & Levitt, H. (2014). Positive psychology and LGBTQ populations. *Cross-Cultural Advancements in Positive Psychology, 7*, 189–202.

Horowitz, M., & Rosensweig, J. (2007). Medical tourism: Health care in the global economy. *Physician Executive, 33*, 24–29.

Horowitz, R., Aierstuck, S., Williams, E., & Melby, B. (2010). Herpes simplex virus infection in a university health population: Clinical manifestations, epidemiology, and implications. *Journal of American College Health, 59*(2), 69–74.

Hosenfeld, C., Workowski, K., Berman, C., Zaidi, A., Dyson, J., & Bauer, H. (2009). Repeat infection with chlamydia and gonorrhea among females: A systematic review of the literature. *Sexually Transmitted Diseases, 36*, 478–489.

"How Girls Bully." (2014). How girls bully. Nottingham Bully Help. Retrieved from **www.nottinghambullyhelp.com/2010/09/how-girls-bully/**

Howard, K., & Stevens, A. (2000). *Out and about campus: Personal accounts of lesbian, gay, bisexual, and transgendered college students*. Los Angeles: Alyson Publications.

Hsu, B., Kling, A., Kessler, C., Knapke, K., Diefenbach, P., & Elias, J. (1994). Gender differences in sexual fantasy and behavior in a college population: A ten-year replication. *Journal of Sex & Marital Therapy, 20*(2), 103–118.

Huang, A., Ring, A., Toich, S., & Torres, T. (1998, March 1). Gender inequalities in education. Gender relations in educational applications of technology.

Huang, P. (2009). Problems of the foreskin and glans penis. *Anesthesiology Clinics of North America, 10*, 56–59.

Hubler, D. (2004, December 1). New report finds abstinence-only programs mislead students. *Education Daily*, p. 3.

Hudson, T. (2003). Research and medical update. *Townsend Letter for Doctors and Patients, 245*, 138–139.

Hughes, S., & Kruger, D. (2011). Sex differences in post-coital behaviors in long- and short-term mating: An evolutionary perspective. *The Journal of Sex Research, 48*, 496–505. doi:10.1080/00224499.2010.501915

Human Rights Watch. (2006). No escape: Male rape in U.S. prisons. Retrieved from **www.hrw.org/reports/2001/prison/report.html#_1_2**

Humphrey, S., & Kahn, A. (2000). Fraternities, athletic teams, and rape: Importance of identification with a risky group. *Journal of Interpersonal Violence, 15*, 1313–1322.

Humphries, A., & Cioe, J. (2009). Reconsidering the refractory period: An exploratory study of women's post-orgasmic experiences. *The Canadian Journal of Human Sexuality*. Retrieved from **www.highbeam.com/doc/1G1-210595139.html**

Hunter, C. (2000). The dangers of pornography? A review of the effects literature. Retrieved from **www.asc.upenn.edu/usr/chunter/porn_effects.html**

Hussein, J. (2007). Celebrating progress toward safer pregnancy. *Reproductive Health Matters, 15*, 216–218.

Hyde, J. S. (2005). The gender similarities hypothesis. *American Psychologist, 60*, 581–592.

Ibbison, M. (2002). Out of the sauna: Sexual health promotion with "off-street" sex workers. *Journal of Epidemiology and Community Health, 56*, 903–904.

Immanuel, F., & Phill, M. (2011). Sex therapy: A cognitive behavioural approach. Retrieved from **www.nursingplanet.com/pn/sex_therapy.html**

Inciardi, J., & Surratt, H. (1997). Male transvestite sex workers and HIV in Rio de Janeiro, Brazil. *Journal of Drug Issues, 27*, 135–146.

Incidence of abortion. (2008). *Facts on induced abortion in the United States*. Guttmacher Institute. Retrieved from **www.guttmacher.org/pubs/fb_induced_abortion.html**

Infertility. (2010). Infertility: Causes. Mayo Foundation for Medical Education and Research. Retrieved from **www.mayoclinic.com/health/infertility/DS00310/DSECTION=causes**

Infoplease. (2011). Median age at first marriage, 1890–2010. Infoplease, Part of Family Education Network, Pearson Education. Retrieved from **www.infoplease.com/ipa/A0005061.html**

Ingrassia, M. (1989, May 5). In 1952 she was a scandal. *Newsday*, p. 2.

IPPF (International Planned Parenthood Federation). (2010). *9,000 free condom vending machines set up in Shanghai.*

Irish, C., & Savage, R. (2008). Ultrasound for diagnosis of ectopic pregnancy. *Practical Summaries in Acute Care, 3*(10), 73–80.

Isaacs, C., & Fisher, W. (2008). A computer-based educational intervention to address potential negative effects of Internet pornography (Author abstract) (Report). *Communication Studies, 59*, 1–17.

Jacobson, T. Z., Barlow, D. H., Garry, R., & Koninckx, P. (2013). Laparoscopic surgery for pelvic pain associated with endometriosis. *Cochrane Database of Systematic Reviews*, (4), 1–12.

Jaeger, F., Caflisch, M., & Hohlfeld, P. (2009). Female genital mutilation and its prevention: A challenge for paediatricians. *European Journal of Pediatrics, 168*(1), 27–33. doi:10.1007/s00431-008-0702-5

James, C. W., & Szabo, S. Infections in the 21st century. (2006, February). *Journal of Pharmacy Practice, 19*(1), 3–4. doi:10.1177/0897190005283224

Jamieson, D. (2007). Poststerilization regret: Findings from India and the United States. *Indian Journal of Medical Sciences, 61*, 359–360.

Jancin, B. (2000, October 1). STD chemoprophylaxis reduces preterm birth. *Family Practice News*, pp. 24–25.

Jankowiak, M., & Cornelison, L. (2008). Research helps nursing homes deal with residents' sexual expression. College of Human Ecology, Kansas State University.

Jannini, E., Gravina, G., Brandetti, P., Martini, P., Carosa, E., Di Stasi, S., Morano, S., & Lenzi, A. (2008). In vivo measurement of the human G-spot. *Sexologies, 17*, S52–S53.

Jannini, E., Whipple, B., Kingsberg, S., Buisson, O., Foldès, P., & Vardi, Y. (2010). Who's afraid of the G-spot? *Journal of Sexual Medicine, 7*, 25–34.

Janssen, E. (2007). *The psychophysiology of sex.* Bloomington, IN: Indiana University Press.

Janssen, E., Vorst, H., Finn, P., & Bancroft, J. (2002). The Sexual Inhibition (SIS) and Sexual Excitation (SES) Scales: I. Measuring sexual inhibition and excitation proneness in men. *Journal of Sex Research, 39*, 114–126.

Janssen, P., Saxell, L., Page, L., Klein, C., Liston, M., & Lee, S. (2009). Outcomes of planned home birth with registered midwife versus planned hospital birth with midwife or physician. *CMAJ: Canadian Medical Association Journal, 181*(6/7), 377–383. doi:10.1503/cmaj.081869

Janssen, W., Koops, E., Anders, S., Kuhn, S., & Puschel, K. (2005). Forensic aspects of 40 accidental autoerotic deaths in Northern Germany. *Forensic Science International, 1*, S61-S64.

Janus, S., & Janus, C. (1993). *The Janus report on sexual behavior*. New York: Wiley.

Jeng, C., Wang, L., Chou, C., Shen, J., & Tzeng, C. (2006). Management and outcome of primary vaginismus. *Journal of Sex & Marital Therapy, 32*, 379–387.

Jenkins, G. (2006, September 8). Issues relating to infertility. *Practice Nurse*, 14–17.

Jenkins, P. (2002, March 1). Bringing the loathsome to light. *Chronicle of Higher Education*, pp. B16–B17.

Jenkins, W. (2010). Can Anyone Tell Me Why I'm Gay? What Research Suggests Regarding The Origins of Sexual Orientation. *North American Journal of Psychology, 12*(2), 279–295.

Jennings, Ashley. (2012, May 25). Man admitted to hospital for kidney stone, discovers he's a woman. Retrieved from **www.abcnews.go.com/blogs/health/2012/05/25/man-admitted-to-hospital-for-kidney-stone-discovers-hes-a-woman**

Jennings, C. (2013). Future Treatments for Erectile Dysfunction. Retrieved from **www.webmd.com/erectile-dysfunction/guide/future-treatments-ed**

Jennings, C. (2013). Prostatitus. WebMD.

Jennings, V., & Landy, H. (2006). Explaining ovulation awareness-based family planning methods: These easily taught birth control methods may be just what some patients are looking for. Nonhormonal approaches like the Standard Days and the TwoDay Methods hinge on identifying a woman's fertile window: the days during her cycle when pregnancy is likely. *Contemporary OB/GYN, 51*, 48–53.

Jenny, C., Roesler, T. A., & Poyer, K. (1994). Are children at risk of sexual abuse by homosexuals? *Pediatrics, 94*, 41–43.

Jensen, M. (1998). Heterosexual women have noisy ears. *Science News, 153*, 151.

Jensen, R. (2004). Pornography and sexual violence. *Minnesota Center against Violence and Abuse.* Retrieved from **www.vadv.org/secPublications/pandsv.pdf**

Jick, S., Hagberg, K., Hernandez, R., & Kaye, J. (2010). Postmarketing study of ORTHO EVRA and levonorgestrel oral contraceptives containing hormonal contraceptives with 30 mcg of ethinyl estradiol in relation to nonfatal venous thromboembolism. *Contraception, 81*(1), 16–21.

Joanning, H., & Keoughan, P. (2005). Enhancing marital sexuality. *Family Journal, 13*(3), 351–355. doi:10.1177/1066480705276194

Jobe, D. (2002). Helping girls succeed. *Educational Leadership, 60*, 64–66.

Johns Hopkins, (2014). Mouth infections. Health System Health Library. Retrieved from **www.hopkinsmedicine.org/healthlibrary/conditions/adult/oral_health/mouth_infections_85,P00888**

Johnson, K. (2004, December 15). Testosterone patch may relieve sexual dysfunction in surgically menopausal women. *Family Practice News*, p. 84.

Johnson, K. C., & Daviss, B.-A. (2005, June 18). Outcomes of planned home births with certified professional midwives: Large prospective study in North America. *British Medical Journal, 330*(7505), 1416–1420.

Johnson, L. (2002). Sexuality after childbirth. *Baby Parenting.*

Johnson, N. (2010, September). For-profit hospitals performing more C-sections. *California Watch.*

Jones, J., & Barlow, D. (1990). Self-reported frequency of sexual urges, fantasies, and masturbatory fantasies in heterosexual males and females. *Archives of Sexual Behavior, 19*, 269–279.

Jones, K. (2011). Peer-Led sex education program for teens in China. MedIndia. Retrieved from **www.medindia.net/news/Peer-Led-Sex-Education-Program-for-Teens-in-China-85684-1.htm**

Jose-Miller, A., Boyden, J., & Frey, K. (2007). Infertility. *American Family Physician, 75*(6), 849–856.

Jurand, S. H. (2004). Sexual privacy is not a right in Eleventh Circuit, despite Lawrence. *Trial, 40*, 87.

Jütte, R. (2008). *Contraception: A history*. Malden, MA: Polity Books.

Kaestle, C., & Halpern, C. (2007). What's love got to do with it? Sexual behaviors of opposite-sex couples through emerging adulthood. *Perspectives on Sexual & Reproductive Health, 39*(3), 134–140. doi:10.1363/3913407

Kafka, M. (2007, December 1). Sexual impulsivity disorders. *Psychiatric Times, 24*(14), 15.

Kagan, J. (2008). The biological contributions to temperaments and emotions. *European Journal of Developmental Science, 2*, 38–51.

Kahr, B. (2008). *Who's been sleeping in your head? The secret world of sexual fantasies.* New York: Basic Books.

Kalb, C. (2003, February 3). Farewell to "Aunt Flo": A new version of the birth-control pill would limit menstruation to four times a year. Are women ready? *Newsweek*, 48.

Kalb, C., Nadeau, B., & Schafer, S. (2004, February 2). Brave new babies: Parents now have the power to choose the sex of their children. *Newsweek International*, pp. 38–40.

Kalick, S. (1988). Physical attractiveness as a status cue. *Journal of Experimental Social Psychology, 24*, 469–489.

Kalra, G. (2012). Hijras: the unique transgender culture of India. *International Journal of Culture and Mental Health, 5*(2), 121–126.

Kane, E. (2012). *The gender trap: parents and the pitfalls of raising boys and girls*. New York: NYU Press.

Kane-DeMaios, A., & Bullough, V. (2006). *Crossing sexual boundaries: Transgender journeys, uncharted paths*. Amherst, NY: Prometheus Books.

Kaplan, A. (2004). Experts publish sexual dysfunction guidelines. *Psychiatric Times, 21*, n.p.

Kaplan, C. (2007). Placental examination. *LabMedicine, 38*, 624–628. doi:10.1309/g4lqxrfbdr9q3ce1

Kaplan, H. S. (1974). *The new sex therapy: Active treatment of sexual dysfunctions*. New York: Brunner/Mazel.

Kaplan, H. S. (1979). *Disorders of sexual desire*. New York: Simon & Schuster.

Kaplan, L. J. (1997). Clinical manifestations of the perverse strategy. *Journal of Psychoanalysis and Psychotherapy, 14*, 79–89.

Kaplan, M., & Krueger, R. (2012). Cognitive-Behavioral Treatment of the Paraphilias. *Israel Journal of Psychiatry & Related Sciences, 49*(4), 291–296.

Kapp, N., Whyte, P., Tang, J., Jackson, E., & Brahmi, D. (2013). A review of evidence for safe abortion care. *Contraception, 88*(3), 350–363. doi:10.1016/j.contraception.2012.10.027

Kara, S. (2008). *Sex trafficking: inside the business of modern slavery*. New York: Columbia University Press.

Karabinus, D. (2009). Volume flow cytometric sorting of human sperm. *MicroSort clinical trial update, 71*, 74–79. doi: 10.1016/j.theriogenology.2008.09.013

Kardas-Nelson, M. (2012). Criminalising condom possession by sex workers is a global trend. *NAM-aidsmap*. Retrieved from **www.aidsmap.com/Criminalising-condom-possession-by-sex-workers-is-a-global-trend/page/2448677/</**

Kark, R., Waismel-Manor, R., & Shamir, B. (2012). Does valuing androgyny and femininity lead to a female advantage? The relationship between gender-role, transformational leadership and identification. *The Leadership Quarterly, 23*(3), 620–640.

Karraker, K., Vogel, D., & Lake, M. (1995). Parents' gender-stereotyped perceptions of newborns: The eye of the beholder revisited. *Sex Roles, 33*, 687–701.

Katz, E. (2002). *Biographical sketch: Mary Louise Higgins (Sanger)*.

Katz, J. (2011). Mentors in violence prevention (MPV). *Gender Violence Prevention Education & Training*. Retrieved from **www.jacksonkatz.com/mvp.html**

Katz, M. (2012). Paraphilias. *Sexual Conditions Health Center*. Retrieved from **www.webmd.com/sexual-conditions/guide/paraphilias-overview?page=3**

Kaunitz, A., Grimes, D., & Stier, E. (2007). Intrauterine contraception and noncontraceptive benefits. *OBG Management, 19*, SS4–SS11.

Kayhan, S., Baig, S., Mehmi, H., & Basra, A. (2010). How does the media influence our thoughts on body image? *York University: The Bibliography Project*.

Kelderhouse, J., & Smith, T. (2013). A review of treatment and management modalities for premenstrual dysphoric disorder. *Nursing for Women's Health, 17*, 294–305.

Keller, M., von Sadovsky, V., Pankratz, B., & Hermsen, J. (2000). Self-disclosure of HPV infection to sexual partners. *Western Journal of Nursing Research, 22*, 285–302.

Kellogg, N. (2010). Sexual behaviors in children: evaluation and management. *American family physician, 82*(10), 1233–1238.

Kellogg, N. D. (2009). *Clinical report: The evaluation of sexual behaviors in children*. Committee on Child Abuse and Neglect. *Pediatrics, 124*, 992–998. doi: 10.1542/peds.2009-1692

Keltner, N., McAfee, K., & Taylor, C. (2002). Mechanisms and treatments of SSRI-induced sexual dysfunction. *Perspectives in Psychiatric Care, 38*, 111–116.

Kendall, T. (2007). Pornography, rape, and the Internet. Clemson University. John E. Walker Department of Economics. Retrieved September 15, 2008, from **www.toddkendall.net/internetcrime.pdf**

Kendall-Tackett, K. (2000). Physiological correlates of childhood abuse: Chronic hyperarousal in PTSD, depression, and irritable bowel syndrome. *Child Abuse and Neglect, 24*, 799–811.

Kenny, M. C., & Wurtele, S. K. (2010). Children's abilities to recognize a "good" person as a potential perpetrator of childhood sexual abuse. *Child Abuse and Neglect, 34*, 490–495.

Kerckhoff, A. (1962). Value consensus and need complementarity in mate selection. *American Sociological Review, 27*, 295–303.

Kerr, D. (2013, June 16). Google plans to wipe child porn from the Web. *CNET*. Retrieved from **www.cnet.com/news/google-plans-to-wipe-child-porn-from-the-web**

Kestleman, P., & Trussell, J. (1991). Efficacy of the simultaneous use of condoms and spermicide. *Family Planning Perspectives, 23*, 226–227.

Khan, M., Mukhtar, M., Bajwa, M., Alwi, M., Gul, M., & Niaz, S. (2011). Role of orgasmic conditioning and reconditioning in psychiatry. *European Psychiatry, 26*, 1546.

Kilchevsky, A., Vardi, Y., Lowenstein, L., & Gruenwald, I. (2012). Is the female G-Spot truly a distinct anatomic entity? *The Journal of Sexual Medicine, 9*(3), 719–726.

Killick, S., Leary, C., Trussell, J., & Guthrie, K. (2012). Sperm content of pre-ejaculatory fluid. *Human Fertility, 14*, 48–52. doi:10.3109/14647273.2010.520798

Killick, S., Leary, C., Trussell, J., & Guthrie, K. (2011). Sperm content of pre-ejaculatory fluid. *Human Fertility, 14*(1), 48–52. doi:10.3109/14647273.2010.520798

Kilpatrick, D., Resnick, H., Ruggiero, K., Conoscenti, L., & McCauley, M. (2007). Drug-facilitated, incapacitated, and forcible rape: A national study. National Crime Victims Research & Treatment Center: National Criminal Justice Reference Service, U.S. Department of Justice. Retrieved August 24, 2008, from **www.ncjrs.gov/pdffiles1/nij/grants/219181.pdf**

King, M., & Woolett, E. (1997). Sexually assaulted males: 115 men consulting a counseling service. *Archives of Sexual Behavior, 26*, 579–588.

Kinnaman, D. (2007). Well, they're gonna have sex anyway: Twisted logic, screwy decisions, and mixed messages. *District Administration, 43*(12), 88–89.

Kinsey, A. C., Pomeroy, W. B., & Martin, C. E. (1948). *Sexual behavior in the human male*. Philadelphia: Saunders.

Kinsey, A., Pomeroy, W., Martin, C., & Gebhard, P. (1953). *Sexual behavior in the human female*. Philadelphia: Saunders.

Klein, A. (2012, November 13). How ".XXX," porn's domain name, made hundreds of millions. *The Daily Beast*. Retrieved from **www.thedailybeast.com/articles/2012/11/13/how-xxx-porn-s-domain-name-made-hundreds-of-millions.html**

Klein, N., Naleway, A., & Baxter, R. (2013) Prevalence of HPV types in cervical specimens from an integrated healthcare delivery system:

baseline assessment to measure HPV vaccine impact. *Cancer Causes & Control, 24*, 403–407.

Kleinman, A. (2013, June 17). Google vs. child porn: The search giant will try to eradicate all of the Internet's child porn. *The Huffington Post. Retrieved from* **www.huffingtonpost.com/2013/06/17/google-vs-child-porn_n_3453456.html**

Kleinplatz, P. (2007). Coming out of the sex therapy closet: Using experiential psychotherapy with sexual problems and concerns. *American Journal of Psychotherapy, 61*(3), 333–348.

Kleinplatz, P., Ménard, A., Paquet, M., Paradis, N., Campbell, M., Zuccarino, D., Mehak, L. (2009). The components of optimal sexuality: A portrait of "great sex." *Canadian Journal of Human Sexuality, 18*, 1–13.

Klinefelter Syndrome. (2010). Mayo Foundation for Medical Education and Research. Retrieved from **www.mayoclinic.com/health/klinefeltersyndrome/DS01057/METHOD=print**

Kling, J. (2010, November 22). Novel spermicide equivalent to Nonoxynol-9. *Medscape Medical News*. Retrieved from **www.medscape.com/viewarticle/732964**

Klitsch, M. (1994). Proportion of high school students receiving AIDS instruction increases, while risky behavior declines. *Family Planning Perspectives, 26*, 144–145.

Klotter, J. (2002, January). Pregnancy and older women. *Townsend Letter for Doctors and Patients*, pp. 17–18.

Knight, G. P., Guthrie, I. K., Page, M. C., & Fabes, R. A. (2002). Emotional arousal and gender differences in aggression: A meta-analysis. *Aggressive Behavior, 28*, 366–394.

Knox, D., Zusman, M. E., Mabon, L., & Shriver, L. (1999). Jealousy in college student relationships. *College Student Journal, 33*, 328–329.

Knox, D., Zusman, M., & McNeely, A. (2008). University student beliefs about sex: Men vs. women. *College Student Journal, 42*, 181–185.

Kolarsky, A. (2006). Beyond sex offenses. *Archives of Sexual Behavior, 35*, 251–253.

Koleva, H., & Stuart, S. (2013). Risk factors for depressive symptoms in adolescent pregnancy in a late-teen subsample. *Archives of women's mental health*, 1–4.

Komisaruk, B. R., & Whipple, B. (1995). The suppression of pain by genital stimulation in females. *Annual Review of Sex Research, 6*, 151–186.

Komisaruk, B., Beyer-Flores, C., & Whipple, B. (2006). *The science of orgasm*. Baltimore, MD: Johns Hopkins University Press.

Komisaruk, B., Whipple, B., Nasserzadeh, S., & Beyer-Flores, C. (2009). *The orgasm answer guide*. Baltimore, MD: Johns Hopkins University Press.

Kowal, D. (2008). Abstinence and the range of sexual expression. In A. Nelson, W. Cates Jr., F. Stewart, & D. Kowal (Eds.), *Contraceptive Technology*, 19th Revised Edition (pp. 81–85). New York: Ardent Media.

Krafft-Ebing, R., & Klaf, F. (1998) *Psychopathia sexualis, with especial reference to the antipathic sexual instinct: A medico-forensic study*. New York: Arcade. (Originally published 1886)

Kraft, S. (2011). Study: Best if mothers administer their own epidural levels. *Medical News Today*. Retrieved from **www.medicalnewstoday.com/articles/216322.php**

Krahe, B., Waizenhofer, E., & Moller, I. (2003). Women's sexual aggression against men: Prevalence and predictors. *Sex Roles, 49*, 219–232.

Krebs, C., Lindquist, C., Warner, T., Fisher, B., & Martin, S. (2009). College women's experiences with physically forced, alcohol-or other drug-enabled, and drug-facilitated sexual assault before and since entering college. *Journal of American College Health, 57*(6), 639–649.

Krueger, R. B., & Kaplan, M. S. (2001). Depotleuprolide acetate for treatment of paraphilias: A report of twelve cases. *Archives of Sexual Behavior, 30*, 409–422.

Krugger, L. (2014). Internet governance and the domain name system: issues for congress. Amazon.com: CreateSpace Independent Publishing Platform.

Kuehn, B. (2006). Study downgrades amniocentesis risk. *Journal of the American Medical Association, 296*, 2663–2664.

Kun Suk Kim & Jongwon Kim. (2012). Disorders of Sex Development. *Korean Journal of Urology, 53*(1), 1–8. doi:10.4111/kju.2012.53.1.1

Kunkel, A., & Burleson, B. (2003). Rational implications of communication skill evaluations and lovestyles. *Southern Communication Journal, 68*, 181–197.

Kusseling, F., Wenger, N., & Shapiro, M. (1995). Inconsistent contraceptive use among female college students: Implications for intervention. *College Health, 43*, 191–195.

Kwon, P. (2013). Resilience in Lesbian, Gay, and Bisexual Individuals. *Personality and Social Psychology Review, 17*(4), 371–383.

Laan, E., Everaerd, W., van Bellen, G., & Gerritt, H. (1994). Women's sexual and emotional responses to male- and female-produced erotica. *Archives of Sexual Behavior*, 23, 153–169.

Laan, E., Rellini, A. H., & Barnes, T. (2013). Standard operating procedures for female orgasmic disorder: Consensus of the International Society for Sexual Medicine. *The journal of sexual medicine, 10*(1), 74–82.

Lacy, R., Reifman, A., Pearson, J., Harris, S., & Fitzpatrick, J. (2004). Sexual-moral attitudes, love styles, and mate selection. *Journal of Sex Research, 41*, 121–128.

LaHood, A., & Bryant, C. (2007). Outpatient care of the premature infant (Clinical report). *American Family Physician, 76*, 1159–1166.

Lake Snell Perry & Associates. (2002). Lower income parents on teaching and talking with children about sexual issues: Results from a national survey. *Sexuality Education and Information Council of the U.S. (SEICUS)*.

Lake, Jaboa. (2013). Same-sex vs. mixed-sex playgroup participation in young children. *UC Merced Undergraduate Research Journal, 4*(2). Retrieved from **www.escholarship.org/uc/item/9j50h9cq**

Lalumiere, M., Blanchard, R., & Zucker, K. J. (2000). Sexual orientation and handedness in men and women: A meta-analysis. *Psychological Bulletin, 126*, 575–592.

Lampiao, F. (2014). Coitus Interruptus: Are there spermatozoa in the pre-ejaculate? *International Journal of Medicine and Biomedical Research, 3*(1), 1–4.

Langevin, R., Lang, R., Wright, P., Handy, L., Frenzel, R., & Black, E. (1988). Pornography and sexual offenses. *Annals of Sexual Research, 1*, 335–362.

Långström, L., Rahman, Q., Carlström, E., & Lichtenstein, P. (2008, June). Genetic and environmental effects on same-sex sexual behavior: A population study of twins in Sweden. *Archives of Sexual Behavior* (electronic edition). Retrieved from **www.springerlink.com/content/2263646523551487**

Långström, N., Rahman, Q., Carlström, E., & Lichtenstein, P. (2010). Genetic and environmental effects on same-sex sexual behavior: A population study of twins in Sweden. *Archives of Sexual Behavior, 39*(1), 75–80.

Lankveld, J. (2009). Self-help therapies for sexual dysfunction. *Journal of Sex Research, 46*(2/3), 143–155. doi: 10.1080/00224490902747776

Laqueur, T. W. (2004). *Solitary sex: A cultural history of masturbation*. Cambridge, MA: MIT Press.

Larsson, I., & Svedlin, C. (2002). Sexual experiences in childhood: Young adults' recollections. *Archives of Sexual Behavior, 31*, 263–273.

Laumann, E. O., Glasser, D. B., Neves, R. C. S., & Moreira, E. D. (2009, February 26). A population-based survey of sexual activity, sexual problems and associated help-seeking behavior patterns in mature adults in the United States of America. *International Journal of Impotence Research, 21*, 171–178. doi:10.1038/ijir.2009.7.

Laumann, E., Gagnon, J., Michael, R., & Michaels, S. (1994). *The social organization of sexuality: Sexual practices in the United States.* Chicago: University of Chicago Press.

Laumann, E., Paik, A., & Rosen, R. (1999). Sexual dysfunction in the United States: Prevalence and predictors. *Journal of the American Medical Association, 281*, 537–544.

Lawrence, A. (2005). Sexuality before and after male-to-female sex reassignment surgery. *Archives of Sexual Behavior, 34*, 147–166.

Leaper, C., & Friedman, C. (2007). The socialization of gender. In J. Grusec & P. Hastings (Eds.), *Handbook of socialization: Theory and research* (pp. 561–587). New York: Guilford Publications.

Lee, A. (2001). The mere exposure effect: An uncertainty reduction explanation revisited. *Personality & Social Psychology Bulletin, 27*, 1255–1266.

Lee, C. (2005). Common myths about prostitution.

Lee, J. A. (1973). *The colors of love: An exploration of the ways of loving.* Don Mills, Ontario: New Press.

Lee, J. A. (1977). A typology of styles of loving. *Personality and Social Psychology Bulletin, 3*, 173 182.

Lee, J. A. (1988). Love-styles. In R. J. Sternberg & M. L. Barnes (Eds.), *The Psychology of Love* (pp. 38–67). New Haven, CT: Yale University Press.

Lee, J., Jackson, H., Pattison, P., & Ward, T. (2002). Developmental risk factors for sexual offending. *Child Abuse and Neglect, 26*, 73–92.

Lee, L., Loewenstein, G., Ariely, D., Hong, J., & Young, J. (2008). If I'm not hot, are you hot or not? Physical-attractiveness evaluations and dating preferences as a function of one's own attractiveness. *Psychological Science, 19*, 669–677. doi:10.1111/j.1467-9280.2008.02141.x

Leff, L. (2011, April 8). Study: 4 million gay adults in U.S. *The Press Democrat*, B4.

Legewie, J., & DiPrete, T. (2012). School context and the gender gap in educational achievement. *American Sociological Review, 77*(3), 463–485.

Lehr, A., & Geher, G. (2006). Differential effects of reciprocity and attitude similarity across long versus short-term mating contexts. *Journal of Social Psychology, 146*(4), 423–439.

Leiblum, S., & Rosen, R. (Eds.). (2000). *Principles and practice of sex therapy.* New York: Guilford Press.

Leichliter, J., Chandra, A., Liddon, N., Fenton, K., & Aral, S. (2007). Prevalence and correlates of heterosexual anal and oral sex in adolescents and adults in the United States (Survey). *Journal of Infectious Diseases, 196*(12), 1852–1859.

Leitenberg, H., & Henning, K. (1995). Sexual fantasy. *Psychological Bulletin, 117*, 469–496.

Lemey, P., Pybus, O., Wang, B., Saksena, N., Salemi, M., and Vandamme, A. (2003). Tracing the origin and history of the HIV-2 epidemic. *Proceedings of the National Academy of Sciences, 100*, 6588–6592.

Lemieux & Hale, 2002 Cross-sectional analysis of intimacy, passion, and commitment: Testing the assumptions of the triangular theory of love. *Psychological Reports, 90*, 1009–1115.

Lerman, D. (2013, May 7). Obama Vows Action After Air Force Officer Assault Arrest. Retrieved from **www.bloomberg.com/news/2013-05-07/head-of-air-force-sex-assault-prevention-arrested.html**

Lesnewski, R., & Prine, L. (2006). Initiating hormonal contraception. *American Family Physician, 74*, 105–112.

Letourneau, E., & O'Donohue, W. (1993). Sexual desire disorders. In W. O'Donohue & J. Geer (Eds.), *Handbook of sexual dysfunctions* (pp. 53–81). Needham Heights, MA: Allyn & Bacon.

Lev, A. (2006). Disordering gender identity and gender identity disorder in the DSM-IV-TR. *Journal of Psychology & Human Sexuality, 17*, 35–69.

LeVay, S. (1991). A difference in hypothalamic structure between heterosexual and homosexual men. *Science, 253*, 1034–1037.

LeVay, S. (2010). *Gay, straight, and the reason why: The science of sexual orientation.* Cary, NC: Oxford University Press, USA.

Levin, R. (2005). Sexual arousal—its physiological roles in human reproduction. *Annual Review of Sex Research, 16*, 154–189.

Levin, R. (2007). Sexual activity, health and wellbeing—the beneficial roles of coitus and masturbation. *Sexual & Relationship Therapy, 22*(1), 135–148. doi: 10.1080/146819906011491

Levin, R. J. (2006). The breast/nipple/areola complex and human sexuality. *Sexual & Relationship Therapy, 21*(2), 237–249. doi: 10.1080/14681990600674674

Levin, R., & van Berlo, W. (2004, April). Sexual arousal and orgasm in subjects who experience forced or non-consensual sexual stimulation—A review. *Journal of Clinical Forensic Medicine, 11*, 82–88.

Levin, R., & Meston, C. (2006). Nipple/breast stimulation and sexual arousal in young men and women. *Journal of Sexual Medicine, 3*, 450–454.

Levin, R., & Riley, I. (2007). The physiology of human sexual function. *Psychiatry, 6*, 90–94.

Levine, J., & Elders, J. M. (2002). *Harmful to minors: The perils of protecting children from sex.*

Levine, S. (2011). Facilitating parent-child communication about sexuality. *Pediatrics in Review, 32*,129–130. doi:10.1542/pir.32-3-129.

Levinson, J. (2005). Erotic art and pornographic pictures. *Philosophy and Literature, 29*(1), 228–240.

Li, Y., Cottrelli, D., Wagner, D., & Ban, M. (2004). Needs and preferences regarding sex education among Chinese college students: A preliminary study. *International Family Planning Perspectives, 30*, 1–11.

Lilienfeld, S., Lynn, S., Ruscio, J., & Beyerstein, B. (2010). Mythbusting in introductory psychology courses: The whys and the hows. In S. Meyers & J. Stowell (Eds.), *Essays from excellence in teaching* (Vol. 9, pp. 55–61). Retrieved from **http://teachpsych.org/resources/e-books/eit2009/index.php**

Lim, M., Hocking, J., Hellard, M., & Aitken, C. (2008). SMS STI: a review of the uses of mobile phone text messaging in sexual health. *International journal of STD & AIDS, 19*(5), 287–290.

Lin, K., & Kirchner, J. (2004). Hepatitis B. *American Family Physician, 69*, 75–82.

Lindau, S. T., Schumm, L. P., Laumann, E. O., Levinson, W., O'Muircheartaigh, C. A., & Waite, L. J. (2007). A study of sexuality and health among older adults in the United States. *New England Journal of Medicine, 357*, 22–34.

Lindberg, L., Jones, R., & Santelli, J. (2008). Noncoital sexual activities among adolescents. *Journal of Adolescent Health, 42*, 44–45.

Lindberg, L., Jones, R., Santelli, J. (2008). Noncoital sexual activities among adolescents. *Journal of Adolescent Health, 43*(3), 231–238.

Linfei, Wu. (2012). 60% Chinese Teenagers Open-minded Towards Premarital Sex. *Women's Foreign Language Publications of China.* Retrieved from **www.womenofchina.cn/html/womenofchina/report/142586-1.htm**

Lingappa, J., Lambdin, B., & Bukusi, B. (2011). Infected partner's plasma HIV-1 RNA level and the HIV-1 set point of their heterosexual seroconverting partners. Eighteenth conference on retroviruses and opportunistic infections, Boston, *Abstract 134*.

Linz, D. G., & Donnerstein, E. I. (1990, April–May). Sexual violence in the media. *World Health*, 26–27.

Lippa, R. (2005). *Gender, nature, and nurture.* New York: Routledge.

Lisak, D., & Miller, P. (2002). Repeat rape and multiple offending among undetected rapists. *Violence and Victims, 17*(1), 73–84.

Lisak, D. (2008). Understanding the predatory nature of sexual violence. *Retrieved from Harvard Kennedy School Ash Center for Democratic Governance and Innovation website* **www.innovations.harvard.edu/showdoc.html**

Littrell, J. (2008). Incorporating information from neuroscience and endocrinology regarding sexual orientation into social work education. *Journal of Human Behavior in the Social Environment, 18*, 101–128. doi:10.1080/10911350802285854

Liu, K., & Case, A. (2011). Advanced reproductive age and fertility. *Journal of Obstetrics and Gynaecology Canada, 33*(11), 1165–1175.

Llibre, J. M., Falco, V., Tural, C., Negredo, E., Pineda, J., Mu´noz, J., & Clotet, B. (2009). The changing face of HIV/AIDS in treated patients. *Current HIV Research, 7*(4), 365–377.

Lloyd, E. (2006). *The case of the female orgasm: Bias in the science of evolution.* Cambridge, MA: Harvard University Press.

Lockwood, C. (2002). Predicting premature delivery—no easy task. *New England Journal of Medicine, 346*, 282–285.

Loeb, T., Gaines, T., Wyatt, G., Zhang, M., & Liu, H. (2011). Associations between child sexual abuse and negative sexual experiences and revictimization among women: Does measuring severity matter? *Child abuse & neglect, 35*(11), 946–955.

Lohse, N., Eg Hansen, A., Pedersen, G., Kronborg, G., Gerstoft, J., Sorensen, H., Vaeth, M., & Obel, N. (2007). Survival of persons with and without HIV infection in Denmark, 1995–2005. *Fifth Annals of Internal Medicine, 146*, 87–95.

Long, H., Laack, N., & Gostout, B. (2007). Prevention, diagnosis, and treatment of cervical cancer. *Mayo Clinic Proceedings, 82*, 1566–1574.

Long, H., Nelson, L., & Hoffman, R. (2002). Ketamine medication error resulting in death. *Journal of Toxicology: Clinical Toxicology, 40*, 614.

Lorius, C. (2008). Sexologies. *Abstracts of the 9th Congress of the European Federation of Sexology, 1*, S69–S70.

Lothian, J. (2011). Lamaze breathing: What every pregnant woman needs to know. *Journal of Perinatal Education, 20*(2), 118–120. doi: 10.1891/1058-1243.20.2.118

Lothstein, L. M. (1997). Pantyhose fetishism and self cohesion: A paraphilic solution? *Gender and Psychoanalysis, 2*, 103–121.

Lott, D. A., & Veronsky, F. (1999, January-February). The new flirting game. *Psychology Today*, pp. 42–45.

Lowman, J., Atchinson, C., & Fraser, L. (1997). Sexuality in the 1990s: Survey results: The Internet client sample.

Lubman, S. (2000, March 17). Infanticide detailed in China. *San Jose Mercury News*, p. 1A.

Lucchetti, A. (1999). Deception in disclosing one's sexual history: Safe-sex avoidance or ignorance? *Communication Quarterly, 47*, 300–314.

Luders, W. (2007, July 1). Child pornography websites: Techniques used to evade law enforcement. *The FBI Law Enforcement Bulletin.*

Ludwig, M. (2008). Diagnosis and therapy of acute prostatitis, epididymitis and orchitis. *Andrologia, 40*, 76–80. doi:10.1111/j.1439-0272.2007.00823.x

Ludeke, M. (2009), November. Transgender youth. *Principal Leadership*, 12–16.

Luecke, J. (2011). Working with transgender children and their classmates in pre-adolescence: Just be supportive. *Journal of LGBT Youth, 8*, 116–156.

Lunde, I., Larson, G. K., Fog, E., & Garde, K. (1991). Sexual desire, orgasm, and sexual fantasies: A study of 625 Danish women born in 1910, 1936, and 1958. *Journal of Sex Education and Therapy, 17*, 111–116.

Luo, S., & Zhang, G. (2009). What leads to romantic attraction: Similarity, reciprocity, security, or beauty? Evidence from a speed-dating study. *Journal of Personality, 77*, 933–964.

Luzzi, G. (2003). Male genital pain disorders. *Sexual and Relationship Therapy, 18*, 225–235.

Macapagal, K., & Janssen, E. (2011). The valence of sex: Automatic affective associations in erotophilia and erotophobia. *Personality and Individual Differences, 51*, 699–703.

Maccoby, E. E. (1988). Gender as a social category. *Developmental Psychology, 24*, 755–765.

Maccoby, E. E., & Jacklin, C. N. (1987). Gender segregation in childhood. In E. H. Reese (Ed.), *Advances in child development and behavior* (vol. 20, pp. 239–288). New York: Academic Press.

MacDorman, M., Mathews, T., & Declercq, E. (2014). Trends in out-of-hospital births in the United States, 1990–2012. Center for Disease Control and Prevention. Retrieved from **www.cdc.gov/nchs/data/databriefs/db144.htm**

Mackey, R., Diemer, M., & O'Brien, B. (2000). Psychological intimacy in the lasting relationships of heterosexual and same-gender couples. *Sex Roles, 43*, 201–227.

MacKinnon, C. (1996). *Just words.* Cambridge, MA: Harvard University Press.

MacKinnon, C. (2000). Not a moral issue. In D. Cornell (Ed.), *Feminism and pornography* (pp. 169–197). Oxford, UK: Oxford University Press.

MacKinnon, C. (2005, May 20). X underrated. *Times Higher Education Supplement, 1692*, 18–19.

Macnair-Semands, R., Cody, W., & Simono, R. (1997). Sexual behaviour change associated with a college HIV course. *AIDS Care, 9*(6), 727–738. doi:10.1080/09540129750124759

MacNeil, S., & Byers, E. (2009). Role of sexual self-disclosure in the sexual satisfaction of long-term heterosexual couples. *Journal of Sex Research, 46*(1), 3–14.

Maddox, A., Rhoades, G., & Markman, H. (2011). Viewing sexually-explicit materials alone or together: Associations with relationship quality. *Archives of sexual behavior, 40*(2), 441–448.

Maes, M., De Vos, N., Van Hunsel, F., Van West, D., Westenberg, H., Cosyns, P., & Neels, H. (2001). Pedophilia is accompanied by increased plasma concentrations of catecholamines, in particular, epinephrine. *Psychiatry Research, 103*, 43–49.

Mah, K., & Binik, Y. (2002). Do all orgasms feel alike? Evaluating a two-dimensional model of the orgasm experience across gender and sexual context. *Journal of Sex Research, 39*, 104–113.

Mahoney, D. (2005). Teen sex, drugs may be catalyst for depression. *Family Practice News, 35*, 41.

Mahoney, D. (2007). Pap test guidelines may miss teens at high risk. (Women's Health). *Family Practice News, 37*(12), 1–2.

Majeroni, B., & Ukkadam, S. (2007). Screening and treatment for sexually transmitted infections in pregnancy. *American Family Physician, 76*, 265–270.

Malacad, B., & Hess, G. (2010). Oral sex: Behaviours and feelings of Canadian young women and implications for sex education. *The European Journal of Contraception and Reproductive Health Care, 15*(3), 177–185.

Malamuth, N., & Huppin, M. (2005). Pornography and teenagers: The importance of individual differences. *Adolescent Medicine, 16*, 315–326.

Malamuth, N., Addison, T., & Koss, M. (2000). Pornography and sexual aggression: Are there reliable effects and can we understand them? *Annual Review of Sex Research, 11*, 26–91.

Malamuth, N., Hald, G., & Koss, M. (2012). Pornography, individual differences in risk and men's acceptance of violence against women in a representative sample. *Sex Roles, 66*(7–8), 427–439.

"Male Infertility." (2014). Male infertility. Mayo Clinic. Retrieved from **www.mayoclinic.org/diseases-conditions/male-infertility/basics/causes/CON-20033113**

Male Athletes Against Violence. (2008). University of Maine, Orono. Retrieved from **www.umaine.edu/maav**

Malin, H., & Saleh, F. (2007, April 15). Paraphilias: Clinical and forensic considerations. *Psychiatric Times, 24*, 32–35.

Malone, P. (2011). Comprehensive sex education vs. abstinence-only-until-marriage programs. *Human Rights, 38*(2), 5–22.

Mamo, L. (2007). Negotiating conception: Lesbians' hybrid technological practices. *Science, Technology, and Human Values 32*, 369–393. doi:10.1177/0162243906298355

Mangan, D. (2003). New oral contraceptive means fewer menstrual periods. *RN, 66*, 95.

Manhart, L., Holmes, K., Hughes, J., Houston, L., & Totten, P. (2007). *Mycoplasma genitalium* among young adults in the United States: An emerging sexually transmitted infection. *American Journal of Public Health, 97*, 1118–1125.

Maniglio, R. (2009). The impact of child sexual abuse on health: A systematic review of reviews. *Clinical psychology review, 29*(7), 647–657.

Manning, J. (2006). The impact of Internet pornography on marriage and the family: A review of the research. *Sexual Addiction & Compulsivity, 13*(2–3), 131–165.

Mao, A., & Anastasi, J. (2010). Diagnosis and management of endometriosis: The role of the advanced practice nurse in primary care. *Journal of the American Academy of Nurse Practitioners, 22*(2), 109–116. doi: 10.1111/j.1745-7599.2009.00475.x

March of Dimes update: Taking action against prematurity. (2003). *Contemporary OB/GYN, 48*, 92–97.

Marchione, M. (2014, March 5). Doctors hope for cure in a 2nd baby with HIV. Associated Press. Retrieved from **http://bigstory.ap.org/article/doctors-hope-cure-2nd-baby-born-hiv**

Marcus, A., Horning, A., Curtis, R., Sanson, J., & Thompson, E. (2014). Conflict and Agency among Sex Workers and Pimps: A Closer Look at Domestic Minor Sex Trafficking. *The ANNALS of the American Academy of Political and Social Science, 653*(1), 225–246.

Marelich, W., Lundquist, J., Painter, K., & Mechanic, M. (2008). Sexual deception as a social-exchange process: Development of a behavior-based sexual deception scale. *Journal of Sex Research, 45*, 27–35.

Marjoribanks, J., Brown, J., O'Brien, P., Wyatt, K. (2013). Selective serotonin reuptake inhibitors for premenstrual syndrome. *Cochrane Database of Systematic Reviews, Issue 6*. Art. No.: CD001396. doi: 10.1002/14651858.CD001396.pub3.

Markey, P., & Markey, C. (2007). Romantic ideals, romantic obtainment, and relationship experiences: The complementarity of interpersonal traits among romantic partners. *Journal of Social and Personal Relationships, 24*, 517–533. doi: 10.1177/0265407507079241

Markman, H., Stanley, S., & Blumberg, S. (2010). *Fighting for your marriage*. Hoboken, NJ: Josey-Bass.

Marmor, J. (1998). Homosexuality: Is etiology really important? *Journal of Gay and Lesbian Psychotherapy, 2*, 19–28.

Marsa, L. (2002). Don't let this drug trap you (GHB). *Health, 16*, 110–113.

Marsee, M., Frick, P., Barry, C., Kimonis, E., & Aucoin, K. (2014). Profiles of the forms and functions of self-reported aggression in three adolescent samples. *Development and psychopathology*.

Marshall, S., & Gilbert, W. (2010a). Neonatalperinatal medicine. Healthwise, Sutter Health.

Marston, C., & Cleland, J. (2003). Relationships between contraception and abortion: A review of the evidence. *International Family Planning Perspectives, 29*. Retrieved from **www.guttmacher.org/pubs/journals/2900603.html**

Martin Hald, G., Malamuth, N., & Yuen, C. (2010). Pornography and attitudes supporting violence against women: Revisiting the relationship in nonexperimental studies. *Aggressive Behavior, 36*, 14–20.

Martin, C., DiDonato, M., Clary, L., Fabes, R., Kreiger, T., Palermo, F., & Hanish, L. (2012). Preschool children with gender normative and gender non-normative peer preferences: Psychosocial and environmental correlates. *Archives of Sexual Behavior, 41*(4), 831–847.

Martin, J., Hamilton, B., Sutton, P., & Ventura, S. (2009). Births: Final data for 2006. National Center for Health Statistics. *National Vital Statistics Reports, 57*. Retrieved from **www.cdc.gov/nchs/data/nvsr/nvsr57/nvsr57_07.pdf**

Martin, W. E. (2001, September-October). A wink and a smile: How men and women respond to flirting. *Psychology Today*, pp. 26–27.

Martz, D. (2003). Behavioral treatment for a female engaging in autoerotic asphyxiation. *Clinical Case Studies, 2*, 236–242.

Marx, V., & Lawton, G. (2008). Cut! *New Scientist, 199*(2665), 40–43.

Masters, W. H., & Johnson, V. E. (1966). *Human sexual response*. Boston: Little, Brown.

Masters, W. H., & Johnson, V. E. (1970). *Human sexual inadequacy*. Boston: Little, Brown.

Masters, W. H., & Johnson, V. E. (1974). *The pleasure bond*. Boston: Little, Brown.

Masters, W. H., & Johnson, V. E. (1979). *Homosexuality in perspective*. Boston: Little, Brown.

Masters, W. H., Johnson, V. E., & Kolodny, R. C. (1977). *Ethical issues in sex therapy and research*. Boston: Little, Brown.

Masters, W. H., Johnson, V. E., & Kolodny, R. C. (1994). *Heterosexuality*. New York: HarperCollins.

Masters, W. H., Johnson, V. E., & Kolodny, R. C. (1995). *Human sexuality* (5th ed.). New York: Minneapolis, MN: University of Minnesota Press.

Mayer, K., & Venkatesh, K. (2010). Antiretroviral therapy as HIV prevention: Status and prospects. *American Journal of Public Health, 100*(10), 1867–1876. doi:10.2105/AJPH.2009.184796

Mayo Clinic. (2011). Sexual health and aging: Keep the passion alive. *Sexual health*. Mayo Foundation for Medical Education and Research. Retrieved from **www.mayoclinic.org/healthy-living/sexual-health/in-depth/sexual-health/art-20046698**

McCarthy, B. W. (1992). Erectile dysfunction and inhibited sexual desire: Cognitive-behavioral strategies. *Journal of Sex Education and Therapy, 18*, 22–34.

McCarthy, B. W. (1995). Bridges to sexual desire. *Journal of Sex Education and Therapy, 21*, 132–141.

McCauley, J. (2010). Binge drinking and rape: A prospective examination of college women with a history of previous sexual victimization. *Journal of Interpersonal Violence, 25*, 1655–1668.

McClelland, S. (2001, June 25). A way out. *Maclean's*, pp. 44–46.

McClure, E. (2000). A meta-analytic review of sex differences in facial expression processing and their development in infants, children, and adolescents. *Psychological Bulletin, 126*, 424–453.

McDowell, M., Brody, D., & Hughes J. (2007). Has age at menarche changed? Results from the National Health and Nutrition Examination Survey (NHANES) 1999–2004. *Journal of Adolescent Health. 40*(3), 227–231.

McFadden, D. (2011). Sexual orientation and the auditory system. *Frontiers in neuroendocrinology, 32*(2), 201–213.

McGarvey, R. (1998). Hands off! How do the latest Supreme Court decisions on sexual harassment affect you? *Entrepreneur, 26*, 85–87.

McGowin, C., & Anderson-Smits, C. (2011). Mycoplasma genitalium: An emerging cause of sexually transmitted disease in women. *PLoS Pathogens, 7*(5). Retrieved from **www.plospathogens.org/article/info%3Adoi%2F10.1371%2Fjournal.ppat.1001324**. doi: 10.1371/journal.ppat.1001324

McGrath-Lone, L., Marsh, K., Hughes, G., & Ward, H. (2013). The sexual health of male sex workers in England: Analysis of cross-sectional data from genitourinary clinics. *Sexually Transmitted Infections, 90*, **38–40**. doi:10.1136/sextrans-2013-051320, 2013

McGreal, S. (October 16, 2012). The pseudoscience of race differences in penis size. Retrieved from **www.psychologytoday.com/blog/unique-everybody-else/201210/the-pseudoscience-race-differences-in-penis-size**

McGuigan, W., & Middlemiss, W. (2005). Sexual abuse in childhood and interpersonal violence in adulthood: A cumulative impact on depressive symptoms in women. *Journal of Interpersonal Violence, 20*, 1271–1287.

McHugh, M. (2006). What do women want? A new view of women's sexual problems. *Sex Roles: A Journal of Research, 54*, 361–369.

McIntyre, T. (2013). Child abuse images and cleanfeeds: Assessing internet blocking systems. In I. Brown (Ed.). *Research handbook on governance of the Internet* (pp. 277–308). Northampton, MA: Edward Elgar Publishers.

McKain, L. (2002). How long should we wait before having oral sex after childbirth? Retrieved from **www.babycenter.com/expert/baby/postpartumsex**

McKay, A. (2005). Sexuality and substance use: The impact of tobacco, alcohol, and selected recreational drugs on sexual function. *Canadian Journal of Human Sexuality, 14*(1/2), 41–56.

McKay, A. (2006). Chlamydia screening programs: A review of the literature. Part 1: Issues in the promotion of chlamydia testing of youth by primary care physicians. *Canadian Journal of Human Sexuality, 15*, 1–10.

McKee, A., Albury, K., Dunne, M., Grieshaber, S., Hartley, J., Lumby, C., & Mathews, B. (2010). Healthy sexual development: a multidisciplinary framework for research. *International Journal of Sexual Health, 22*(1), 14–19.

Mechcatie, E. (2006). Data are mixed on thrombosis from Ortho Evra patch vs. pills. *Family Practice News, 36*, 5.

Medical abortion. (2010). Medical abortion. Mayo Foundation for Medical Education and Research. Retrieved from **www.mayoclinic.com/health/medical-abortion/MY00819**

Medical Student Clinical Journal. Retrieved from **medworld.stanford.edu/features/review/**

Medina, J., & Goodstein, L. (2013, February 1). Diocese papers in Los Angeles detail decades of abuse. *The New York Times*. Retrieved from **www.nytimes.com/2013/02/02/us/church-documents-released-after-years-of-resistance.html?pagewanted=all&_r=2&**

Meeks, J., Sheinfeld, J., & Eggener, S. (2012). Environmental toxicology of testicular cancer. *Urological Oncology, 30*(2), 212–215. doi: 10.1016/j.urolonc.2011.09.009

Mehta, A., Clearman, T., & Paduch, D. (2014). Safety and Efficacy of Testosterone Replacement Therapy in Adolescents with Klinefelter Syndrome. *The Journal of Urology, 191*(5), 1527–1531.

Meltzer-Brody, S., Payne, J., & Rubinow, D. (2008). Postpartum depression: What to tell patients who breast-feed. *Current Psychiatry 7*, 87–95.

Menarche. (2013). Menarche—Topic overview. WebMD. Retrieved from **www.webmd.com/parenting/tc/menarche-topic-overview**

Merke, D. (2009). Facts about CAH: (Congenital Adrenal Hyperplasia). National Institute of Child Health and Human Development. NIH Clinical Center. Retrieved from **www.cc.nih.gov/ccc/patient_education/pepubs/cah.pdf 2009**

Merrill, D., & Zieve, D. (2010). Postpartum depression. U.S. Department of Health and Human Services. National Institutes of Health. MedLine Plus. Retrieved from **www.nlm.nih.gov/medlineplus/ency/article/007215.htm**

Mertz, G. (2008). Asymptomatic shedding of herpes simplex virus 1 and 2: implications for prevention of transmission. *Journal of Infectious Diseases, 198*, 1098–1100. doi:10.1086/591914

Meston, C., Levin, R., Sipski, M., Hull, E., & Heiman, J. (2004). Women's orgasm. *Annual Review of Sex Research, 15*, 173–257.

Mevissen, F., Ruiter, R., Meertens, R., Zimbile, F., & Schaalma, H. (2011). Justify your love: Testing an online STI-risk communication intervention designed to promote condom use and STI-testing. *Psychology and Health, 26*(2), 205–221.

Meyer-Bahlburg, H., Ehrhardt, A., & Gruen, R. (1995). Prenatal estrogens and the development of homosexual orientation. *Developmental Psychology, 31*, 12–21.

Michaels, S. (2013). Sexual Behavior and Practices: Data and Measurement. In *International Handbook on the Demography of Sexuality* (pp. 11–20). Springer Netherlands.

Midwives. (2011). American Pregnancy Association. Retrieved from **www.americanpregnancy.org/labornbirth/midwives.html**

Miller, J. (2000). *What are paraphilias?* Retrieved from **www.athealth.com/Consumer/disorders/Paraphilias.html**

Miller-Ott, A., & Kelly, L. (2014). Communication of female relational aggression in the college environment. *Qualitative Research Reports in Communication, 14*(1), 19–27.

Millet, G. (2007). Circumcision status and HIV infection among Black and Latino men who have sex with men in 3 U.S. cities. *Acquired Immune Deficiency Syndrome, 46*(5), 643–650.

Mindel, A., & Estcourt, C. (2001). Sexual health education for male sex workers. *The Lancet, 357*, 1148.

"Miscarriage." (2014). Miscarriage. American Pregnancy Association. Retrieved from **www.americanpregnancy.org/pregnancycomplications/miscarriage.html**

Moalem, S., & Reidenberg, J. (2009). Does female ejaculation serve an antimicrobial purpose? *Medical Hypotheses, 73*, 1069–1071.

Mofenson, L., Taylor, A., Rogers, M., Campsmith, M., Ruffo, N., Clark, J., Lampe, M., Nakashima, A., & Sansom, S. (2006). Reduction in perinatal transmission of HIV infection—United States, 1985–2005. *Morbidity and Mortality Weekly Report, 55*, 592–597.

Mohler-Kuo, M., Dowdall, G. W., Koss, M. P., & Wechsler, H. (2004). Correlates of rape while intoxicated in a national sample of college women. *Journal of Studies on Alcohol, 65*, 37–45.

Mohr, J. J., Selterman, D., & Fassinger, R. E. (2013). Romantic attachment and relationship functioning in same-sex couples. *Journal of counseling psychology, 60*(1), 72.

Mollen, D. (2006). Voluntarily childfree women: Experiences and counseling considerations. *Journal of Mental Health Counseling, 28*, 269–282.

Moniz, M., & Beigi, R. (2012). Prevention of sexually transmitted diseases. In R. Beigi (Ed.) *Sexually transmitted diseases*. Hoboken, NJ: Wiley-Blackwell.

Monk-Turner, E., & Purcell, H. C. (1999). Sexual violence in pornography: How prevalent is it. *Gender Issues, 17*(2): 58–67.

Montesi, J., Conner, B., Gordon, E., Fauber, R., Kim, K., & Heimberg, R. (2013). On the relationship among social anxiety, intimacy, sexual communication, and sexual satisfaction in young couples. *Archives of Sexual Behavior, 42*, 81–91.

Monto, M. (2009). Customer motives and misconceptions. In Ronald Weitzer (Ed.), *Sex for sale: Prostitution, pornography, and the sex industry, 2nd ed.* (pp. 233–254). New York: Routledge.

Monto, M., & Garcia, S. (2001). Recidivism among the customers of female street prostitutes: Do intervention programs help? *Western Criminology Review, 3*.

Moore, D. (2002). *Postpartum depression*. Retrieved from **www.drdonnica.com/display.asp?article=154**

Moore, T., Eisler, R., & Franchina, J. (2000). Causal attributions and affective responses to provocative female partner behavior by abusive and nonabusive males. *Journal of Family Violence, 15*, 69–80.

Morgan, T. (2007). Turner syndrome: Diagnosis and management. (Disease/Disorder overview). *American Family Physician, 76*, 405–410.

Morgenstern, J. (2004, May-June). Myths of bisexuality. *Off Our Backs*, pp. 46–48.

Moses, S. (2009). Male circumcision: A new approach to reducing HIV transmission. *Canadian Medical Association 181*, E134–E135. doi:10.1503/cmaj.090809

Mosher, W., & Jones, J. (2010). Use of contraception in the United States: 1982–2008. National Center for Health Statistics. *Vital Health Statistics, 23.* Retrieved from **www.cdc.gov/NCHS/data/series/sr_23/sr23_029.pdf**

Mosher, W., Chandra, A., & Jones, J. (2005). *Sexual behavior and selected health measures: Men and women 15–44 years of age, United States, 2002.* National Center for Health Statistics. Retrieved from **www.cdc.gov/nchs/data/ad/ad362.pdf**

Mosher, D. L., & MacIan, P. (1994). College men and women respond to X-rated videos intended for male or female audiences: Gender and sexual scripts. *Journal of Sex Research, 31*, 99–113.

Moszynski, P. (2007). Unhygienic circumcisions may increase risk of HIV in Africa. *British Medical Journal, 334*, 498.

Mudur, G. (2002). India plans new legislation to prevent sex selection. *British Medical Journal, 324*, 385.

Muise, A., Christofides, E., & Desmarais, S. (2009). More information than you ever wanted: Does Facebook bring out the green-eyed monster of jealousy? *CyberPsychology & Behavior, 12*(4), 441–444.

Murad, M., Elamin, M., Garcia, M., Mullan, R., Murad, A., Erwin, P., & Montori, V. (2010). Hormonal therapy and sex reassignment: a systematic review and meta-analysis of quality of life and psychosocial outcomes. *Clinical Endocrinology, 72*(2), 214–231.

Murdoch, J., & Price, D. (2001, July 9). A sorry history of anti-gay bias. *National Law Journal, 23*, A24.

Murdock, M. (2004). Female fertility drug can be used for men with low sperm count.

Murphy, F., & Merrell, J. (2009). Negotiating the transition: Caring for women through the experience of early miscarriage. *Journal of Clinical Nursing, 18*, 1583–1591.

Murray, J. B. (2000). Psychological profile of pedophiles and child molesters. *Journal of Psychology, 134*, 211–224.

Murray, S. L., Holmes, J. G., Bellavia, G., Griffin, D. W., & Dolderman, D. (2002). Kindred spirits? The benefits of egocentrism in close relationships. *Journal of Personality and Social Psychology, 82*, 563–581.

Murray, S. L., Holmes, J. G., Griffin, D. W., Bellavia, G., & Rose, P. (2001). The mismeasure of love: How self-doubt contaminates relationship beliefs. *Personality and Social Psychology Bulletin, 27*, 423–436.

Murthy, A. (2010). Obesity and contraception: Emerging issues. *Seminar on Reproductive Medicine, 28*(2), 156–163.

Myths and Facts about Rape. (2008). Dartmouth College, Center for Women and Gender. Retrieved from **www.dartmouth.edu/~cwg/archives/whiteribbon/myths.html**

Naghavi, F., & Redzuan, M. (2011). The relationship between gender and emotional intelligence. *World Applied Sciences Journal, 15*(4), 555–561.

Nagler, R. (2009). Saliva as a tool for oral cancer diagnosis and prognosis. *Oral Oncology, 45*, 1006–1010.

Nahata, L., Rosoklija, I., Richard, N. Y., & Cohen, L. E. (2013). Klinefelter Syndrome: Are We Missing Opportunities for Early Detection? *Clinical pediatrics, 52*(10), 936–941.

National Cancer Institute. (2013). Breast Cancer. Retrieved from **http://www.cancer.gov/cancertopics/types/breast**

National Cancer Institute. (2012). Pap and HPV testing. *U.S. Department of Health and Human Services. National Institutes of Health.* Retrieved from **www.cancer.gov/cancertopics/factsheet/detection/Pap-HPV-testing**

National Clearinghouse on Marital and Date Rape. (2005). State law chart. Retrieved from **www.members.aol.com/ncmdr/state_law_chart.html**

National Coalition of Anti-Violence Programs. (2013). Lesbian, gay, bisexual, transgender, queer and HIV-affected hate violence in 2012. Retrieved from **www.avp.org/storage/documents/ncavp_2012_hvreport_final.pdf**

National Gay and Lesbian Taskforce (2013): State non-discrimination laws in the U.S. Retrieved from **www.thetaskforce.org/downloads/reports/issue_maps/non_discrimination_6_13_color.pdf**

National Institute of Allergies and Infectious Diseases. (2008). HIV Infection in Women. Department of Health and Human Services. National Institutes of Health.

National Institute of Child Health and Human Development (NICHD). (2010). Research on miscarriage and stillbirth: Miscarriage. *Nebraska Symposium on Motivation, 54*, 113–139.

Neff, K., Kirkpatrick, K. L., & Rude, S. (2007). Self-compassion and adaptive psychological functioning. *Journal of Research in Personality, 41*, 139–154.

Neighbors, C., Walker, D., Mbilinyi, L., Lyungai, F., O'Rourke, A., Edleson, J., Zegree, J., & Roffman, R. (2010). Normative misperceptions of abuse among perpetrators of intimate partner violence. *Violence Against Women, 16*, 370–386.

Nelson, A. (2000). The pink dragon is female: Halloween costumes and gender markers. *Psychology of Women Quarterly, 24*, 137–144.

Nelson, E., Heath, A., & Madden, P. (2002). Association between self-reported childhood sexual abuse and adverse psychosocial outcomes: Results from a twin study. *Archives of General Psychiatry, 59*, 139–146.

Nelson, L., Padilla-Walker, L., & Carroll, J. (2010). "I believe it is wrong but I still do it": A comparison of religious young men who do versus do not use pornography. *Psychology of Religion and Spirituality, 2*, 136–147. doi:10.1037/a0019127

Nelson, T. S. (2003). *For love of country: Confronting rape and sexual harassment in the U.S. military.* New York: Haworth Maltreatment and Trauma Press.

Nettleman, M. (2014). HIV testing. MedicineNet. Retrieved from **www.medicinenet.com/hiv_testing/page3.htm 2014**

New View Campaign. (2008). Challenging the medicalization of sex.

New York v. *Ferber*, 458 U.S. 747 (1982).

Newbury, R., Hayterb, M., Wyliec, K., & Riddelld, J. (2012). Sexual fantasy as a clinical intervention. *Sexual and Relationship Therapy, 27*, 358–371. doi:10.1080/14681994.2012.733816

Newton, D., & McCabe, M. (2008). Effects of sexually transmitted infection status, relationship status, and disclosure status on sexual self-concept. *The Journal of Sex Research, 45*, 187–192. doi:10.1080/00224490802012909

NIAID HIV vaccine research: A year in review, looking ahead. Department of Health and Human Services. National Institutes of Health. Retrieved from **www.niaid.nih.gov/topics/HIVAIDS/Research/vaccines/Pages/vacResearchYearInReview.aspx**

Nichols, C. (2013). Active Surveillance is the preferred approach to clinical stage I testicular cancer. *Journal of Clinical Oncology, 31*(28), 3490–3493.

Nicollette, J. (1996). The female condom: Where method and user effectiveness meet. *Stanford Medical Student Clinical Journal.* Retrieved from **http://medworld.stanford.edu/features/review/nicolette.html#ref**

Nicolosi, J., & Byrd, D. (2002). A critique of Bem's "exotic becomes erotic" theory of sexual orientation development. *Psychological Reports, 90*, 931–945.

Nicolson, P., & Burr, J. (2003). What is "normal" about women's (hetero) sexual desire and orgasm? A report of an in-depth interview study. *Journal of Social Science and Medicine, 57*, 1735–1745.

Nihira, M. (2009). Pregnancy and conception. WebMD. Retrieved from **www.webmd.com/baby/guide/understanding-conception**

Nihira, M. (2010). Pregnancy and sexually transmitted diseases. WebMD. Retrieved from **www.webmd.com/baby/pregnancy-sexually-transmitted-diseases**

Nilson, R., & de Melo, R. (2010). Estrogen-free oral hormonal contraception: Benefits of the progestin-only pill. *Women's Health, 6*(5), 721–735. doi:10.2217/whe.10.36

NOCIRC. (2010). Answers to your questions about premature (forcible) retraction of your young son's foreskin. Retrieved from **www.nocirc.org/publish/pamphlet6.html**

Non-discrimination laws. (2010). Non-discrimination laws that include gender identity and expression. *Transgender Law and Policy*. Retrieved from **www.transgenderlaw.org/ndlaws/index.htm#public**

Nordenberg, T. (1996). Looking for a libido lift? The facts about aphrodisiacs. *FDA Consumer, 30*(1).

Nordqvist, C. (2011). What is amniocentesis? *Medical News Today*. Retrieved from **www.medicalnewstoday.com/articles/215965.php**

Nordqvist, J. (2013a). What is a miscarriage? What causes a miscarriage? *Medical News Today*. Retrieved from **www.medicalnewstoday.com/articles/262941.php**

Nordqvist, J. (2013b). C-Section rates steady for first time in decades. *Medical News Today*. Retrieved from **www.medicalnewstoday.com/articles/262605.php**

Norris, J. (1994). Alcohol and female sexuality: A look at expectancies and risks. *Alcohol and Research World, 18*, 197–201.

Nurnberg, G., Hensley, P., Heiman, J., Croft, H., Debattista, C., & Paine, S. (2008). Sildenafil treatment of women with antidepressant-associated sexual dysfunction: A randomized controlled trial. *JAMA, 300*(4), 395–404. doi:10.1001/jama.300.4.395

Nurnberg, H. G. (2007, August 1). Options for management of serotonin reuptake inhibitor-induced sexual dysfunction. *Psychiatric Times, 24*, 60–64.

Nurnberg, H., Hensley, P., Heiman, J., Croft, H., Debattista, C., & Paine, S. (2008). Sildenafil treatment of women with antidepressant-associated sexual dysfunction: A randomized controlled trial. *JAMA, 300*(4), 395–404. doi:10.1001/jama.300.4.395

O'Connell, H., Sanjeevan, K., & Hutson, J. (2005). Anatomy of the clitoris. *The Journal of Urology, 174*, 1189–1195. doi: 10.1097/01.ju.0000173639.38898.cd

O'Connell, R. (2001). *Implanon*.

O'Connor, C. (2008). Chromosomal abnormalities: Aneuploidies. *Nature Education, 1*(1). Retrieved from **www.nature.com/scitable/topicpage/Chromosomal-Abnormalities-Aneuploidies-290**

O'Connor, G. (2007, May). "I have an STD. Now what?" When these women were diagnosed, there was anger, fear, guilt, and finally, acceptance. With total frankness, they relate what they went through. *Cosmopolitan, 243*, 206–210.

O'Donohue, W., Letourneau, E., & Geer, J. (1993). Premature ejaculation. In W. O'Donohue & J. Geer (Eds.), *Handbook of sexual dysfunctions* (pp. 303–333). Needham Heights, MA: Allyn & Bacon.

O'Reilly, D. (2011). APGAR. U.S. Department of Health and Human Services, National Institutes of Health MedLine Plus. Retrieved from **www.nlm.nih.gov/medlineplus/ency/article/003402.htm**

O'Riordan, K. (2012). The life of the gay gene: from hypothetical genetic marker to social reality. *Journal of sex research, 49*(4), 362–368.

Oattes, M., & Offman, A. (2007). Global self-esteem and sexual self-esteem as predictors of sexual communication in intimate relationships. *Canadian Journal of Human Sexuality, 16*, 89–100.

Okami, P., Olmstead, R., & Abramson, P. (1997). Sexual experiences in early childhood: 18-year longitudinal data from the UCLA Family Lifestyles Project. *Journal of Sex Research, 34*, 339–347.

Oliphant, J. (2008). Death penalty for child rape banned. *Los Angeles Times* (electronic edition).

Ompad, D. C., Strathdee, S. A., Celentano, D. D., Latkin, C., & Poduska, J. M. (2006). Predictors of early initiation of vaginal and oral sex among urban young adults in Baltimore, Maryland. *Archives of Sexual Behavior, 35*, 53–65.

Oncale, R., & King, B. (2001). Comparison of men's and women's attempts to dissuade sexual partners from the couple using condoms. *Archives of Sexual Behavior, 30*, 379–397.

Oral sex, masturbating—but not men—linked to recurrent infections. (2004, January 5). *OBGYN and Reproduction Week*, p. 12.

Ostrzenski, A. (2012). G-Spot anatomy: A new discovery. *The Journal of Sexual Medicine, 9*, 1355–1359. doi: 10.1111/j.1743-6109.2012.02668.x

Oswalt, S., Cameron, K., & Koob, J. (2005). Sexual regret in college students. *Archives of Sexual Behavior, 34*, 663–669.

Otchet, A. (1998, December). Should prostitution be legal? *UNESCO Courier*, pp. 37–39.

Oths, K., & Robertson, T. (2007). Give me shelter: Temporal patterns of women fleeing domestic abuse. *Human Organization, 66*, 249–260.

Ottens, A. J. (2001). The scope of sexual violence on campus. In A. J. Ottens & K. Hotelling (Eds.), *Sexual violence on campus: Policies, programs, and perspectives* (pp. 1–29). New York: Springer.

Owen, P. R., & Laurel-Seller, E. (2000). Weight and shape ideals: Thin is dangerously in. *Journal of Applied Psychology, 30*, 979–990.

Owens, T., & Lopes, A. (2003). Recommendations for political policy on prostitution and the sex industry. International Union of Sex Workers.

Paddock, C. (2009). Air traffic pollution linked to increased risk of preeclampsia and preterm birth in southern California. Retrieved from **www.medicalnewstoday.com/articles/155436.php**

Padgett, V. R., Brislin-Slutz, J. A., & Neal, J. A. (1989). Pornography, erotica, and attitudes toward women: The effects of repeated exposure. *Journal of Sex Research, 26*, 479–491.

Paglia, C. (1998, March 23). A call for lustiness: Just say no to the sex polices. *Time*, p. 54.

Pallone, S., & Bergus, G. (2009). Fertility awareness-based methods: Another option for family planning. *The Journal of the American Board of Family Medicine, 22*(2), 147–157. doi: 10.3122/jabfm.2009.02.080038

Palmer, N. (2004). "Let's talk about sex, baby": Community-based HIV prevention work and the problem of sex. *Archives of Sexual Behavior, 33*, 271–275.

Palmquist, S. (2004, July-August). Handsome ambitions. *Psychology Today*, p. 33.

Panel on Antiretroviral Guidelines for Adults and Adolescents. (2011, January 10). *Guidelines for the use of antiretroviral agents in HIV-1-infected adults and adolescents*. Department of Health and Human Services. Pages 1–166. Retrieved from **www.aidsinfo.nih.gov/ContentFiles/AdultandAdolescentGL.pdf**

Papp, L., Danielewicz, J., & Cayemberg, C. (2012). "Are we Facebook official?" Implications of dating partners' Facebook use and profiles for intimate relationship satisfaction. *Cyberpsychology, behavior, and social networking, 15*(2), 85–90. doi:10.1089/cyber.2011.0291

Parker, I. (2013, February 6). The story of a suicide: Two college roommates, a webcam, and a tragedy. *The New Yorker*. Retrieved from **www.newyorker.com/reporting/2012/02/06/120206fa_fact_parker**

Parkhill, M., & Abbey, A. (2008). Does alcohol contribute to the confluence model of sexual assault perpetration? *Journal of Social & Clinical Psychology, 27*(6), 529–554.

Parrot, & L. Bechhofer (Eds.), *Acquaintance rape: The hidden crime* (pp. 159–175). New York: Wiley.

Pasterski, V., Geffner, M., Brain, C., Hindmarsh, P., Brook, C., & Hines, M. (2011). Prenatal hormones and childhood sex segregation: Playmate and play style preferences in girls with congenital adrenal hyperplasia. *Hormones and behavior, 59*(4), 549–555.

Pastor, Z. (2013). Female ejaculation orgasm vs. coital incontinence: a systematic review. *The Journal Of Sexual Medicine, 10*(7), 1682–1691.

Patrick, M., Maggs, J., & Abar, C. (2007). Reasons to have sex, personal goals, and sexual behavior during the transition to college (Report). *Journal of Sex Research, 44*, 240–249.

Patton, R., Snyder, A., & Glassman, M. (2013). Rethinking substance abuse treatment with sex workers: How does the capability approach inform practice? *Journal of substance abuse treatment, 45*(2), 196–205.

Patton, W., & Mannison, M. (1994). Investigating attitudes towards sexuality: Two methodologies. *Journal of Sex Education and Therapy, 20*, 185–197.

Payne, B., DeMichele, M., & Button, D. (2008, January–February). Understanding the electronic monitoring of sex offenders. *Corrections Compendium, 33*, 1–4.

Pazol, K., Creanga, A., Burley, K., Hayes, B., & Jamieson, D. (2013). Abortion surveillance-United States, 2010. *Morbidity and mortality weekly report. Surveillance summaries, 62*, 1–44.

Pazol, K., Zane, S., Parker, W., Hall, L., Gamble, S., Hamdan, S., Berg, C., & Cook, D. (2011). Abortion surveillance. United States, 2007. Division of Reproductive Health, National Center for Chronic Disease Prevention and Health Promotion, CDC. *Surveillance Summaries 60* (ss01), 1–39.

Pearson, S. (2013, December 24). Merck sues Actavis unit over patent for generic NuvaRing. *Bloomberg View.*

Pearson, S., & Pollack, R. (1997). Female responses to sexually explicit films. *Journal of Psychology and Human Sexuality, 9*, 73–88.

Pedersen, C., McGrath, J., Mortensen, P., & Petersen, L. (2013). The importance of father's age to schizophrenia risk. *Molecular psychiatry, 19*, 530–531.

Peipert, J., Madden, T, Allsworth, J., & Secura, G. (2013. Preventing unintended pregnancies by providing no-cost contraception. *Obstetrics & Gynecology, 120*, 1291–1297. doi: 0.1097/aog.0b013e318273eb56

Peirce, N. (2008, February 11). Celebrated sex offender laws doing more harm than good. *Nation's Cities Weekly, 31*, 2.

Peng, T. (2006). Construct validation of the Bem Sex Role Inventory in Taiwan. *Sex Roles: A Journal of Research, 55*, 843–851.

Peplau, L. (2003). Human sexuality: How do men and women differ? *Current Directions in Psychological Science, 12*, 37–40.

Peplau, L., Garnets, L., Spalding, R., Conley, D., & Veniegas, C. (1998). A critique of Bem's "exotic becomes erotic" theory of sexual orientation. *Psychological Review, 105*, 387–394.

Pereira, V., Arias-Carrión, O., Machado, S., Nardi, A., & Silva, A. (2013). Sex therapy for female sexual dysfunction. *International archives of medicine, 6*(1), 37.

Pérez-Fuentes, G., Olfson, M., Villegas, L., Morcillo, C., Wang, S., & Blanco, C. (2013). Prevalence and correlates of child sexual abuse: a national study. *Comprehensive psychiatry, 54*(1), 16–27.

Perper, T. (1985). *Sex signals: The biology of love*. New York, NY: ISI Press.

Perper, T. (1986). *Sex signals: The biology of love*. Philadelphia: ISI Press.

Perper, T., & Fox, V. S. (1980, June). *Flirtation and pickup patterns in bars*. Paper presented at the Eastern Conference on Reproductive Behavior, Saratoga Springs, NY.

Perper, T., & Fox, V. S. (1981). *Flirtation behavior in public settings: Final report*. New York: Harry Frank Guggenheim Foundation.

Perper, T., & Weiss, D. (1987). Proceptive and rejective strategies of U.S. and Canadian college women. *Journal of Sex Research, 23*, 455–480.

Perry, D., & Pauletti, R. (2011). Gender and adolescent development. *Journal of Research on Adolescence, 21*(1), 61–74.

Peterson, J. & Hyde, J. (2011). Gender differences in sexual attitudes and behaviors: A review of meta-analytic results and large datasets. *Journal of Sex Research, 48*, 149–165.

Peterson, Z., & Muehlenhard, C. (2007). What is sex and why does it matter? A motivational approach to exploring individuals' definitions of sex. *Journal of Sex Research, 44*(3), 256–268.

Petrozza, J., & Styer, A. (2006, July 20). Assisted reproduction technology. *eMedicine*. Retrieved from **www.emedicine.com/med/topic3288.htm**

Pettijohn, T. F. II, & Dunlap, A. V. (2010). The effects of a human sexuality course on college students' sexual attitudes and perceived course outcomes. *Electronic Journal of Human Sexuality, 13*. Retrieved from **www.ejhs.org/volume13/sexclass.htm**

Pew Research. (2014). Gay Marriage around the World. Pew Research Center's Religion & Public Life Project. Retrieved from **www.pewforum.org/2013/12/19/gay-marriage-around-the-world-2013/#allow**

Pfaff, D., Frolich, J., & Morgan, M. (2002). Hormonal and genetic influences on arousal—sexual and otherwise. *Trends in Neurosciences, 25*, 45–50.

Phases of traumatic stress reactions in a disaster. PTSD: National Center for PTSD. Retreived from **www.ptsd.va.gov/professional/trauma/disaster-terrorism/phases-trauma-reactions.asp**

Pike, J., & Jennings, N. (2005). The effects of commercials on children's perceptions of gender appropriate toy use. *Sex Roles: A Journal of Research, 52*, 83–91.

Pilcher, C. D. (2003, February 10–14). *Acute HIV infection*. Paper presented at the Tenth Conference on Retroviruses and Opportunistic Infections, Boston. Retrieved from **www.natap.org/2003/Retro/day20.htm**

Pilkinton, C., Kern, W., & Indest, D. (1994). Is safer sex necessary with a "safe" partner? Condom use and romantic feelings. *Journal of Sex Research, 31*, 203–210.

Pines, A. M. (2005). *Falling in love*. New York, NY: Routledge.

Pipe, M., Lamb, M., Orbach, Y., & Cederborg, A. (2007). *Child sexual abuse disclosure, delay, and denial*. New York, NY: Psychology Press.

Pipe, M., Lamb, M., Orbach, Y., & Cederborg, A. (Eds.). (2013). *Child sexual abuse: Disclosure, delay, and denial*. New York, NY: Psychology Press.

Pitts, M., Smith, A., Grierson, J., O'Brien, M., & Misson, S. (2004). Who pays for sex and why? An analysis of social and motivational factors associated with male clients of sex workers. *Archives of Sexual Behavior, 33*, 353–358.

Plant, E., Ashby, K., Kling, C., & Smith, G. (2004). The influence of gender and social role on the interpretation of facial expressions. *Sex Roles: A Journal of Research, 51*, 187–196.

Plante, T. (2004). Another aftershock: What have we learned from the John Jay Report? *America, 190*, 10–11.

Poels, S., Bloemers, J., van Rooij, K., Goldstein, I., Gerritsen, J.,... Tuiten, J. (2013). Toward personalized sexual medicine (part 2): Testosterone combined with a PDE5 inhibitor increases sexual satisfaction in women with HSDD and FSAD, and a low sensitive system for sexual cues. *The Journal of Sexual Medicine, 10*, 810–823. doi: 10.1111/j.1743-6109.2012.02983.x

Polan, M. (2007, May). Androgens in women: To replace or not? Androgen therapy can improve sexual desire and response—here's how. *OBG Management, 19*, 72–78.

Policy Statement. (2012). Circumcision policy statement. *Pediatrics, 130*(3), 585–586. doi: 10.1542/peds.2012–1989

Pollack, M. (2009) Circumcision: If it isn't ethical, can it be spiritual? In C. Denniston, F. M. Hodges, & M. Milos (Eds). *Circumcision and Human Rights* (189–194). New York: Springer. doi: 10.1007/978-1-4020-9167-4_17

Polonsky, D. (2000). Premature ejaculation. In S. Leiblum & R. Rosen (Eds.), *Principles and practice of sex therapy* (pp. 305–332). New York: Guilford Press.

Popkin, J. (1994, December 19). A case of too much candor. *U.S. News and World Report*, p. 31.

Potterat, J., Rothenberg, R., Muth, S., Darrow, W., & Phillips-Plummer, L. (1998). Pathways to prostitution: The chronology of sexual and middle-aged and young adults. *International Journal of Aging and Human Development, 41*, 281–297.

Powell, E. (1996). *Sex on your terms*. Needham Heights, MA: Allyn & Bacon.

Power, C., Radcliffe, L., & MacGregor, K. (2003, November 17). Preying on children. *Newsweek International*, pp. 34–35.

Powers, A. (2010, January 6). Male prostitution is Nevada's newest legal profession. *Los Angeles Times*.

Prentky, R., & Knight, R. (1991). Identifying critical dimensions for discriminating among rapists. *Journal of Counseling and Clinical Psychology, 59*, 643–661.

Prentky, R., Lee, A., Knight, R., & Cerce, D. (1997). Recidivism rates among child molesters and rapists: A methodological analysis. *Law and Human Behavior, 21*, 635–659.

Preventing sexual harassment. (2011). *U.S. Department of Transportation*.

Previous *Chlamydia trachomatis* infections adversely affect male fertility. (2004, March 29). *OBGYN and Reproduction Week*, p. 35.

Price, J., & Stevens, S. (2009). Partners of veterans with PTSD: Research findings. *U.S. Department of Veterans Affairs*. Retrieved from **www.ptsd.va.gov/professional/pages/partners_of_vets_research_findings.asp**

Price, M., Kafka, M., Commons, M., Gutheil, T., & Simpson, W. (2002). Telephone scatologia: Comorbidity with other paraphilias and paraphilia-related disorders. *International Journal of Law and Psychiatry, 25*(1), 37–49.

Prinstein, M., Meade, C., & Cohen, G. (2003). Adolescent oral sex, peer popularity, and perceptions of best friends' sexual behavior. *Journal of Pediatric Psychology, 28*, 243–249.

"Primary Care Protocol." (2011). Primary care protocol for transgender patient care. Center of Excellence for Transgender Health, University of California, San Francisco, Department of Family and Community Medicine. Retrieved from **www.transhealth.ucsf.edu/trans?page=protocol-00-00**

Pryor, D. W., & Hughes, M. R. (2013). Fear of rape among college women: a social psychological analysis. *Violence and victims, 28*(3), 443–465.

Puente, S., & Cohen, D. (2003). Jealousy and the meaning (or nonmeaning) of violence. *Personality and Social Psychology Bulletin, 29*, 449–451.

Puppo, V., & Guenwald, I. (2012). Does the G-spot exist? A review of the current literature. *International Urogynecologocal Journal, 23*, 1665–1669. doi: 10.1007/s00192-012-1831-y

Puri, S., & Nachtigall, R. (2010). The ethics of sex selection: a comparison of the attitudes and experiences of primary care physicians and physician providers of clinical sex selection services. *Fertility and sterility, 93*(7), 2107–2114.

Putnam, F. (2003). Ten-year research update review: Child sexual abuse. *Journal of the American Academy of Child and Adolescent Psychiatry, 42*, 269–278.

Qiaoqin, M., Ono-Kihara, M., Cong, L., Xu, G., Xiaohong, P., Zamani, S.,...Kihara, M. (2009). Early initiation of sexual activity: A risk factor for sexually transmitted diseases, HIV infection, and unwanted pregnancy among university students in China. *BMC Public Health*, 9,111. doi: 10.1186/1471-2458-9-111

Quackenbush, D. M., Strassberg, D. S., & Turner, C. W. (1995). Gender effects of romantic themes in erotica. *Archives of Sexual Behavior, 24*(1), 21–35.

Quina, K., Harlow, L., Morokoff, P., Burkholder, G., &. Deiter, P. (2000). Sexual communication in relationships: When words speak louder than actions. *Sex Roles, 42*, 523–549.

Radio. Retrieved from **www.npr.org/blogs/health/2013/03/25/175258772/gates-foundation-says-its-time-for-a-snazzier-condom**

Rai, R., & Regan, L. (2006, August 12). Recurrent miscarriage. *The Lancet, 368*, 601–611.

RamaRao, S., Friedland, B., & Townsend, J. (2007). A question of ethics: Research and practice in reproductive health. *Studies in Family Planning, 38*(4), 229–241.

Ramirez, J. (2014). Propriety of Internet restrictions for sex offenders convicted of possession of child pornography: Should we protect their virtual liberty at the expense of the safety of our children? *Ave Maria Literature Review, 12*, 123–149.

Ramisetty-Mikler, S., Caetano, R., Goebert, D., & Nishimura, S. (2004). Ethnic variation in drinking, drug use, and sexual behavior among adolescents in Hawaii. *Journal of School Health, 74*, 16–22.

Ramos, M. (2013). *Masturbation and relationship satisfaction*. Denton, Texas. UNT Digital Library. Retrieved from **digital.library.unt.edu/ark:/67531/metadc271884**

Rattue, P. (2011). Sex of fetus can often be verified by testing DNA from mother's blood. Retrieved from **www.medicalnewstoday.com/articles/232528.php**

Ray, S. (2001). Male survivors' perspectives of incest/sexual abuse. *Perspectives in Psychiatric Care, 37*, 49–59.

Rea, L. M., & Parker, R. A. (2005). *Designing and conducting survey research: A comprehensive guide*. Hoboken, NJ: John Wiley & Sons.

Reddy, G. (2005). *With respect to sex: Negotiating Hijra identity in south India*. Chicago, IL: University Of Chicago Press.

Reece, M. (2010). Condom use rates in a national probability sample of males and females ages 14 to 94 in the United States. *Journal of Sexual Medicine, 7*(5), 266–276.

Regan, P., & Medina, R. (2001). Partner preferences among homosexual men and women: What is desirable in a sex partner is not necessarily desirable in a romantic partner. *Social Behavior & Personality: An International Journal, 29*(7), 625–633.

Rehan, N. (2011). Short report: Genital examination of Hijras. *JPMA-Journal of the Pakistan Medical Association, 61*(7), 695.

Reilly, T., Woodruff, S., Smith, L., Clapp, J., & Cade, J. (2010). Unsafe sex among HIV positive individuals: Cross-sectional and prospective predictors. *Journal of Community Health, 35*(2), 115–123.

Reiner, W., & Townsend Reiner, D. (2012). Thoughts on the nature of identity: how disorders of sex development inform clinical research about gender identity disorders. *Journal of Homosexuality Special Issue: The Treatment of Gender Dysphoric/Gender Variant (GD/GV) Children and Adolescents, 59*(3), 434–449. doi: 10.1080/00918369.2012.653312

Reinholtz, R., & Muehlenhard, C. (1995). Genital perceptions and sexual activity in a college population. *Journal of Sex Research, 32*, 155–165.

Reinisch, J. M. (1990). *The Kinsey Institute new report on sex: What you must know to be sexually literate*. New York: St. Martin's Press.

Reis, H., Mancini, M., Caprariello, P., Eatwick, P., & Finkel, E. (2011). Familiarity does indeed promote attraction in live interaction. *Journal of Personality and Social Psychology, 101*(3), 557–570. doi: 10.1037/a0022885

Reiss, I. (1995, November). *The new sex survey: Paradise lost or found?* Address at the annual meeting of the Society for the Scientific Study of Sexuality, San Francisco.

Reissing, E., Binik, Y., Khalifé, S., Cohen, D., & Amsel, R. (2004). Vaginal spasm, pain and behaviour: An empirical investigation of the diagnosis of vaginismus. *Archives of Sexual Behavior, 33*, 1–13.

Reitman, D., Austin, B., Belkind, U., Chaffee, T., Hoffman, N., Moore, E., & Ryan, C. (2013). Recommendations for promoting the health and well-being of lesbian, gay, bisexual, and transgender adolescents: A position paper of the Society for Adolescent Health and Medicine: Society for Adolescent Health and Medicine. *Journal of Adolescent Health, 52*(4), 506–510.

Reitmeijer, C., Wolitski, R., Fishbein, M, Corby, N., & Cohn, D. (1998). Sex hustling, injection drug and nongay identification among men who have sex with men: Associations with high-risk sexual behaviors and condom use. *Sexually Transmitted Diseases, 25*, 353–360.

Renaud, C. A., & Byers, S. E. (1999). Exploring the frequency, diversity, and content of university students' positive and negative sexual cognitions. *Canadian Journal of Human Sexuality, 8*, 17–31.

Renaud, C., & Byers, E. (2001). Positive and negative sexual cognitions: Subjective experience and relationships to sexual adjustment. *Journal of Sex Research, 38*, 252–262.

Renaud, C., & Byers, S. (2005). Relationship between sexual violence and positive and negative cognitions of sexual dominance. *Sex Roles: A Journal of Research, 53*, 253–260.

Resnick, S. (2002). Sexual pleasure: The next frontier in the study of sexuality. *SEICUS Report, 30*, 6–11.

Ribner, D. (2003). Modifying sensate focus for use with Haredi (ultra-Orthodox) Jewish couples. *Journal of Sex and Marital Therapy, 29*, 165–171.

Ricciardelli, L., & Williams, R. (1995). Desirable and undesirable traits in three behavioral dimensions. *Sex Roles, 33*, 637–655.

Rice, M., & Harris, G. (2011). Is androgen deprivation therapy effective in the treatment of sex offenders? *Psychology, Public Policy, and Law, 17*(2), 315.

Richards, M. (2003). Obscenity. Retrieved from **www4.gvsu.edu/richardm/PLS307/OBSCENITY.htm**

Richardson, C. R., Resnick, P. J., Hansen, D. L., Derry, H. A., & Rideout, V. J. (2002). Does pornography-blocking software block access to health information on the Internet? *Journal of the American Medical Association, 288*, 2887–2894.

Richters, J., de Visser, R., Rissel, C., & Smith, A. (2006). Sexual practices at last heterosexual encounter and occurrence of orgasm in a national survey. *Journal of Sex Research, 43*, 217–226.

Riela, S., Rodriguez, G., Aron, A., Xu, X., & Acevedo, B. P. (2010). Experiences of falling in love: Investigating culture, ethnicity, gender, and speed. *Journal of Social and Personal Relationships, 27*(4), 473–493.

Rietmeijer, C., Van Bemmelen, R., Judson, F., & Douglas, J. (2002). Incidence and repeat infection rates of chlamydia trachomatis among male and female patients in an STI clinic: Implications for screening and rescreening. *Sexually Transmitted Diseases, 29*(2), 65–72.

Riley, J., & Jones, R. (2007). When girls and boys play: What research tells us (Review of research). *Childhood Education, 84*, 38–43.

Ringheim, K. (1995). Ethical issues in social science research with special reference to sexual behavior research. *Social Science and Medicine, 40*, 1691–1697.

Ripley, M., Anderson, E., McCormack, M., & Rockett, B. (2012). Heteronormativity in the university classroom: Novelty attachment and content substitution among gay-friendly students. *Sociology of Education, 85*(2), 121–130.

Roberts, A., Glymour, M., & Koenen, K. (2013). Does maltreatment in childhood affect sexual orientation in adulthood? *Archives of sexual behavior, 42*(2), 161–171.

Roberts, D., & Oliva, E. (2006). Clinical significance of placental examination in perinatal medicine. *Journal of Maternal-Fetal and Neonatal Medicine, 19*, 255–264. doi: 10.1080/14767050600676349

Roberts, J., Wolfer, L., & Mele, M. (2008). Why victims of intimate partner violence withdraw protection orders. *Journal of Family Violence, 23*, 369–375. doi: 10.1007/s10896-008-9161-z

Robertson, M. (2007, January–February). You want to give birth where? At first critical of homebirth, a thoughtful expectant dad comes to understand why women choose to labor at home. *Mothering, 140*(5), 60.

Robinson, C. C., & Morris, J. T. (1986). The gender stereotyped nature of Christmas toys received by 36-, 48-, and 60-month-old children: A comparison between requested and nonrequested toys. *Sex Roles, 15*, 21–32.

Robinson, J., & Burke, A. (2013). Obesity and Hormonal Contraceptive Efficacy: Executive Summary. *Women's Health, 9*(5), 463–476.

Robinson, K. (2012). "Difficult citizenship": The precarious relationships between childhood, sexuality and access to knowledge. *Sexualities, 15*(3–4), 257–276.

Robinson, P. (1976). *The modernization of sex*. London, England: Paul Elek Publishers.

Rodgers, D., (2010). The viability of Nevada's legal brothels as models for regulation and harm reduction in prostitution. *Electronic Theses, Treatises and Dissertations.* Paper 1834.

Rodriguez-Diaz, C., Clatts, M., Jovet-Toledo, G., Vargas-Molina, R., Goldsamt, L., & García, H. (2012)More than foreskin: Circumcision status, history of HIV/STI, and sexual risk in a clinic-based sample of men in Puerto Rico. *Journal of Sexual Medicine, 9*, 2933–2937.

Roesler, T. A. (2000, October). Adult's reaction to child's disclosure of abuse will influence degree of permanent damage. *Brown University Child and Adolescent Behavior Letter*, pp. 1–2.

Roland, M., Neilands, T., Krone, M., Katz, M., Franses, K., Grant, R., Busch, M., Hecht, F., Shacklett, B., Kahn, J., Bamberger, J., Coates, T., Chesney, M., & Martin, J. (2005). Seroconversion following nonoccupational postexposure prophylaxis against HIV. *Clinical Infectious Diseases, 41*, 1507–1513.

Romito, K., Marshall, S., & Jones, K. (2010a). Rupture of the membranes. *Healthwise,* Sutter Health.

Romito, K., Marshall, S., & Jones, K. (2010b). Labor, delivery, and postpartum period: Active labor, first stage. *Healthwise,* Sutter Health. Retrieved from **www.babies.sutterhealth.org/health/healthinfo/index.cfm?section=healthinfo&page=article&sgml_id=tn9759&seq_id=6**

Romito, K., Marshall, S., & Jones, K. (2010c). Episiotomy or perineal tear. *Healthwise,* Sutter Health. Retrieved from **www.babies.sutterhealth.org/health/healthinfo/index.cfm?section=healthinfo&page=article&sgml_id=hw194808**

Ropelato, J. (2008). *Internet filter review. Internet pornography statistics.*

Rosen, R., & Barsky, J. (2006). Normal sexual response in women. *Obstetrics and Gynecology Clinics of North America, 33*(4), 515–526.

Rosen, R., & Laumann, E. (2003). The prevalence of sexual problems in women: How valid are comparisons across studies? Commentary on Bancroft, Loftus, and Long's (2003) "Distress about sex: A national survey of women in heterosexual relationships." *Archives of Sexual Behavior, 32*, 209–211.

Rosen, R., & Leiblum, S. (1995). Treatment of sexual disorders in the 1990s: An integrated approach. *Journal of Consulting and Clinical Psychology, 63*, 107–121.

Rosen, R., Taylor, J., Leiblum, S., & Bachman, G. (1993). Prevalence of sexual dysfunction in women: Results of a survey study of 329 women in an outpatient gynecological clinic. *Journal of Sex and Marital Therapy, 19*, 171–188.

Rosenbaum, J. (2009). Patient teenagers? A comparison of the sexual behavior of virginity pledgers and matched nonpledgers. *Pediatrics, 123*, 10–20.

Rosen-Molina, M. (2001, April 2). Study provides insight into odd sexual kinks. *Daily Bruin Online.*

Rosenthal, A., Melvin, C., & Barker, D. (2006). Treatment of tobacco use in preconception care. *Maternal and Child Health Journal, 10*, 147–148.

Rosin, H. (2008, November). A boy's life. *The Atlantic*. Retrieved from **www.theatlantic.com/magazine/toc/2008/11**

Rosler, A., & Witztum, E. (2000). Pharmacotherapy of paraphilias in the next millennium. *Behavioral Science and the Law, 18*, 43–56.

Ross, M., & Williams, M. (2001). Sexual behavior and illicit drug use. *Annual Review of Sex Research, 12*, 290–310.

Ross, M., & Williams, M. (2002). Effective targeted and community HIV/STI prevention programs. *Journal of Sex Research, 39*, 58–62.

Rothblum, E. (2009). Contemporary perspectives on lesbian, gay, and bisexual identities. *Sociological Review, 27*, 295–303.

Rowland, D., & Motofei, I. (2007). The aetiology of premature ejaculation and the mind-body problem: Implications for practice. *International Journal of Clinical Practice, 61*, 77–82.

Rowniak, S. (2009). Safe sex fatigue, treatment optimism, and serosorting: New challenges to HIV prevention among men who have sex with men. *Journal of the Association of Nurses in AIDS Care, 20*, 31–38.

Roy, R., Schumm, W., & Britt, S. (2014). Voluntary versus involuntary childlessness. In R. Roy & W. Schumm, *Transition to parenthood* (pp. 49–68). New York, NY: Springer.

Rubarth, L. (2012). Back to basics: the apgar score: simple yet complex. *Neonatal Network, 31*(3), 169–177.

Rubin, Z., Provenzano, F., & Luria, Z. (1974). The eye of the beholder: Parents' views on sex of newborns. *American Journal of Orthopsychiatry, 44*, 512–519.

Rudolph, K., Lansford, J., Agoston, A., Sugimura, N., Schwartz, D., Dodge, K., & Bates, J. (2014). Peer victimization and social alienation: predicting deviant peer affiliation in middle school. *Child Development, 85*(1), 124–139. doi:10.1111/cdev.12112

Russ, L., Simonds, C., & Hunt, S. (2002). Coming out in the classroom: An occupational hazard? The influence of sexual orientation on teacher credibility and perceived student learning. *Communication Education, 51*, 311–323.

Russell, C., & Keel, P. (2002). Homosexuality as a specific risk factor for eating disorders in men. *International Journal of Eating Disorders, 31*, 300–306.

Russell, D. E. H. (1990). *Rape in marriage*. Indianapolis: Indiana University Press.

Russell, S., Ryan, C., Toomey, R., Diaz, R., & Sanchez, J. (2011). Lesbian, gay, bisexual, and transgender adolescent school victimization: Implications for young adult health and adjustment. *Journal of School Health, 81*(5), 223–230.

Ryan, S., Franzetta, K., Manlove, J., & Holcombe, E. (2007). Adolescents' discussions about contraception or STDs with partners before first sex (sexually transmitted diseases). *Perspectives on Sexual and Reproductive Health, 39*, 149–157.

Rye, B., & Meaney, G. (2007). The pursuit of sexual pleasure. *Sexuality & Culture, 11*, 28–33.

Rye, B., Meaney, J., Yessis, J., & Mckay, A. (2012). Uses of the "Comfort with Sexual Matters for Young Adolescents" scale: A measure of erotophobia-erotophilia for youth. *Canadian Journal of Human Sexuality, 21*(2), 91–100.

Rye, B.J., Meaney, G.J., & Fisher, W.A. (2011). The *Sexual Opinion Survey*. In T.D. Fisher, C.M. Davis, W.L. Yarber, & S.L. Davis (Eds.), *Handbook of Sexuality-Related Measures* (3rd ed.) (Rev. ed.; pp. 231–236). New York, NY: Routledge.

Sable, M., Danis, F., Mauzy, D., & Gallagher, S. (2006). Barriers to reporting sexual assault for women and men: Perspectives of college students. *Journal of American College Health, 55*, 157–162.

Sadovsky, R. (2005). Selecting patients appropriate for treatment. *Journal of Family Practice, 54*, SS5–14.

Saey, T. (2008). Dad's hidden influence: A father's legacy to a child's health may start before conception and last generations. *Science News, 173*, 200–201.

Saigal, C., Wessells, H., Pace, J., Schonlau, M., Wilt, T. (2006). Predictors and prevalence of erectile dysfunction in a racially diverse population: Urologic diseases in America project. *Archives of Internal Medicine, 166*, 207–212.

Salerian, A., Deibler, W., Vittone, B., & Geyer, S. (2000). Sildenafil for psychotropic-induced sexual dysfunction in 31 women and 61 men. *Journal of Sex and Marital Therapy, 26*, 133–140.

Saletan, W. (2009). The beauty of artificial virginity. *Slate; Washington Post, Newsweek Interactive.*

Sananes, N., Langer, B., Gaudineau, A., Kutnahorsky, R., Aissi, G., Fritz, G., & Favre, R. (2014). Prediction of spontaneous preterm delivery in singleton pregnancies: Where are we and where are we going? A review of literature. *Journal of Obstetrics & Gynaecology, 40*, 1–5.

Sanders, T., O'Neill, M., & Pitcher, J. (2009). *Prostitution: Sex work, policy and politics.* Thousand Oaks, CA: Sage Publications.

Sanfilippo, J. S., Cox, J. T., & Wright, T. C. (2007, October). What you need to know about cervical cancer, genital warts, and HPV (human papillomavirus) (Disease/Disorder overview). *OBG Management, 19*(10), S1(2).

Santtila, P., Sandnabba, N., Alison, L., & Nordling, N. (2002). Investigating the underlying structure in sadomasochistically oriented behavior. *Archives of Sexual Behavior, 31*, 185–200.

Santtila, P., Wagera, I., Wittinga, K., Harlaara, N., Jerna, P., Johanssona, A., Varjonena, M., & Sandnabbaa, N. K. (2008). Discrepancies between sexual desire and sexual activity: Gender differences and associations with relationship satisfaction. *Journal of Sex & Marital Therapy, 34*(1), 31–44.

Saslow, D., Solomon, D., Lawson, H., Killackey, M., Kulasingam, S., & Cain, J. (2012). American Cancer Society, American Society for Colposcopy and Cervical Pathology, and American Society for Clinical Pathology Screening Guidelines for the Prevention and Early Detection of Cervical Cancer. *CA: A Cancer Journal for Clinicians, 62*(3), 147–172. doi: 10.3322/caac.21139

Savic, I., & Lindstrom, P. (2008). PET and MRI show differences in cerebral asymmetry and functional connectivity between homo- and heterosexual subjects. *Proceedings of the National Academy of Sciences of the United States 105*, 9403–9408.

Sax, L. (2002). How common is intersex? A response to Anne Fausto-Sterling. *Journal of Sex Research, 39*, 174–179.

Scasta, D. (1998). Issues in helping people come out. *Journal of Gay and Lesbian Psychotherapy, 2*, 87–97.

Schaffir, J. (2006). Sexual intercourse at term and onset of labor. *Obstetrics & Gynecology, 107*, 1310–1314.

Schepel, E. (2011). *A Comparative Study of Adult Transgender and Female Prostitution*. Doctoral dissertation, Arizona State University.

Schmerling, R. (2004). In Search of the elusive aphrodisiac: Sex, food and myth. *Aetna Intelihealth; Harvard Medical School*. Retrieved from **www.intelihealth.com**

Schneider, M. (2008). FDA requires warning labels for nonoxynol 9. *Family Practice News, 38*, 7.

Schnurr, P., Lunney, C., Forshay, E., Thurston, V., Chow, B., Resick, P., & Foa, E. (2009). Sexual function outcomes in women treated for posttraumatic stress disorder. *Journal of Women's Health, 18*(10), 1549–1557. doi:10.1089/jwh.2008.1165

Schober, J., Kuhn, P., Kovacs, P., Earle, J., Byrne, P., & Fries, R. (2005). Leuprolide acetate suppresses pedophilic urges and arousability. *Archives of Sexual Behavior, 34*, 691–705.

Schoppe-Sullivan, S., Diener, M., Mangelsdorf, S., Brown, G., McHale, J., & Frosch, C. (2006). Attachment and sensitivity in family context: The roles of parent and infant gender. *Infant and Child Development, 15*(4), 367–385.

Schott, R. (1995). The childhood and family dynamics of transvestites. *Archives of Sexual Behavior, 24*, 309–337.

Scotta, C., & Cortez, A. (2011). No longer his and hers, but ours: Examining sexual arousal in response to erotic stories designed for both sexes. *Journal of Sex & Marital Therapy, 37*, 165–175.

Scully, D. (2013). *Understanding sexual violence: A study of convicted rapists* (Vol. 3). New York, NY: Routledge.

Sears, B., Mallory, C., & Hunter, N. (2009). Initiatives to repeal or prevent laws prohibiting employment discrimination against LGBT people, 1974–present. Documenting discrimination on the basis of sexual orientation and gender identity in state employment. *The Williams Institute, UCLA School of Law*. Retrieved from **www.escholarship.org/uc/item/58j4w7k3**

Sedgh, G., Singh, S., Shah, I., Ahman, E., Henshaw, S., & Bankole, A. (2012). Induced abortion: incidence and trends worldwide from 1995 to 2008. *Lancet, 379*(9816), 625–632.

Segraves, R., & Althof, S. (1998). Psychotherapy and pharmacotherapy of sexual dysfunctions. In P. Nathan & J. Gorman (Eds.), *A guide to treatments that work* (pp. 447–465). New York: Oxford University Press.

Seidel, K. (2011). Healthy & safe pregnancy after 40: Special risks and some precautions to take. *Pregnancy Today*.

Selekman, J. (2007). Homosexuality in children and/or their parents. *Pediatric Nursing, 33*, 453–458.

Seligman, L., & Hardenberg, S. (2000). Assessment and treatment of paraphilias. *Journal of Counseling and Development, 78*, 107–113.

Senecal, D., & Morelli, J. (2007). Hepatitis C virus infection: A current review. The standard therapy for hepatitis C cures 40% to 50% of infections, but new treatments now being studied in clinical trials may improve these numbers. *Journal of the American Academy of Physicians Assistants, 20*, 21–25.

Sepilian, V., & Wood, E. (2011). Ectopic Preganancy. eMedicine. Retreived from **www.emedicine.medscape.com/article/258768**

Seth, W. (1997). The disavowal of desire: A relational view of sadomasochism. *Psychoanalysis and Psychotherapy, 14*, 107–123.

Sex Laws. (2008). *Sexual offenses. Federal law 18 U.S.C. 2251 1. Sexual exploitation of children*. Retrieved from **www.sexlaws.org/18usc 2251**

Sexual abuse. (2002). Retrieved from **www.counselingcorner.net/disorders/sexual-abuse.html**

Sexual harassment at school. (2011). *Equal Rights Advocates*.

"Sexual Health." (2013). Sexual health of adolescents and young adults in the United States. *Kaiser Family Foundation's publication (#3040-06)*. Retrieved from **kaiserfamilyfoundation.files.wordpress.com/2013/04/3040-06.pdf**

Sexuality Information and Education Council of the United States. (2011). New Poll on Parent-Child Communication Released for Let's Talk Month. Retrieved from **www.siecus.org/index.cfm?fuseaction=Feature.showFeature&featureid=2065&parentid=478**

Sexuality Information and Education Council of the United States. (2000). "Eleven Guidelines for Talking to Children about Sex," from *Families Are Talking: Special Supplement to SIECUS Report 1*, 2000.

Shaeer, O., & Shaeer, K. Z. (2013). The Global Online Sexuality Survey (GOSS): The United States of America in 2011 penile size and form among English speakers. *Human Andrology, 3*(2), 46–53.

Shafik, I., Shafik, A., Sibai, O., & Shafik, A. A. (2009). An electrophysiologic study of female ejaculation. *Journal of Sex & Marital Therapy, 35*, 337–346.

Shah, N., & Christopher, J. (2002). Can shoe size predict penile length? *BJU International, 90*(6), 586–587.

Shamloul, R. (2010). Natural aphrodisiacs. *The Journal of Sexual Medicine, 7*, 39–49. doi: 10.1111/j.1743-6109.2009.01521.x

Shannon, K., Strathdee, S., Shoveller, J., Zhang, R., Montaner, J., & Tyndall, M. (2011). Crystal methamphetamine use among female street-based sex workers: Moving beyond individual-focused interventions. *Drug and alcohol dependence, 113*(1), 76–81.

Sharlip, I., Belker, A., Honig, S., Labrecque, M., Marmar, J., & Sokal, D. (2012, May). Vasectomy: AUA guideline. *American Urological Association Education and Research.*

Shaughnessy, K., Byers, S., & Walsh, L. (2011). Heterosexual Students: Gender Similarities and Differences. *Archives of Sexual Behavior, 40*, 419–427.

Shaw, J. (2008). Diagnosis and treatment of testicular cancer. *American Family Physician, 77*(4), 469.

Shaw, M. (1998). Delight in the world: Tantric Buddhism as a path of bliss. *Parabola, 23*, 39–44.

Shelby, L. (2003, February-March). Youth first: An integrated sexuality education program for preadolescents. *SIECUS Report*, pp. 31–32.

Sheleg, S., & Ehrlich, E. (2006). *Autoerotic asphyxiation: Forensic, medical, and social aspects*. Tucson, AZ: Wheatmark.

Shively, M., Kliorys, K., Wheeler, K., & Hunt, D. (2012). A national overview of prostitution and sex trafficking demand reduction efforts. The National Institute of Justice. Office of Justice Programs, U.S. Department of Justice. Retrieved from **www.ncjrs.gov/pdffiles1/nij/grants/238796.pdf**

Shoveller, J., Knight, R., Johnson, J., Oliffe, J., & Goldenberg, S. (2010). "Not the swab!" Young men's experiences with STI testing. *Sociology of health & illness, 32*(1), 57–73.

Shulman, J., & Horne, S. (2006). Guilty or not? A path model of women's sexual force fantasies. *Journal of Sex Research, 43*, 368–377.

Shulman, S., & Kipnis, O. (2001). Adolescent romantic relationships: A look from the future. *Journal of Adolescence, 24*, 337–352.

Shute, R., Owens, L., & Slee, P. (2008). Everyday victimization of adolescent girls by boys: Sexual harassment, bullying or aggression? *Sex Roles, 58*(7–8), 477–489. doi:10.1007/s11199-007-9363-5

Silber, S. (2007). How to get pregnant. The Infertility Center of St. Louis. St. Luke's Hospital.

Silver, J. (2001). Movie day at the Supreme Court: "I know it when I see it." Retrieved from **www.coollawyer.com/webfront/pdf/Obscenity%20Article.pdf**

Simons, J., & Carey, M. (2001). Prevalence of sexual dysfunctions: Results from a decade of research. *Archives of Sexual Behavior, 30*, 177–219.

Simpson, B. (2008, April 18). Pope Benedict prays with victims of clergy sex abuse scandal. *International Business Times* (electronic edition).

Sinai, I., Lundgren, R., Arevalo, M., & Jennings, V. (2006). Fertility awareness-based methods of family planning: Predictors of correct use. *International Family Planning Perspectives, 32*, 94–100.

Sine, R. (2014). Sex drive: How do men and women compare? Experts say men score higher in libido, while women's sex drive is more "fluid." *Sexual Health Center WebMD*. Retrieved from **www.webmd.com/sex/features/sex-drive-how-do-men-women-compare**

Sit, D., Rothschild, A., & Wisner, K. (2006). A review of postpartum psychosis. *Women's Health, 15*(4), 352–68.

Sked, A. (1999, June 5). The rehabilitation of Onan. *Spectator*, p. 12.

Slattery, D. (2003). Africa rhino conservation. *PSA Journal, 69*(7), 24.

Slijper, F. (1997). Neither man nor women: The Hijras of India. *Archives of Sexual Behavior, 26*, 450–453.

Slob, A. K., van Berkel, A., & van der Werff ten Bosch, J. J. (2000). Premature ejaculation treated by local penile anaesthesia in an uncontrolled clinical replication study. *Journal of Sex Research, 37*, 244–251.

Sloviter, D. (1996). *Janet Reno* v. *American Civil Liberties Union*, 929 F. Supp. 824.

Smith, D. (2001). Harassment in the hallways. *Monitor on Psychology, 32*(8), 38–40.

Smith, D., Letourneau, E., Saunders, B., Kilpatrick, D., Resnick, H., & Best, C. (2000). Delay in disclosure of childhood rape: Results from a national survey. *Child Abuse and Neglect, 24*, 273–287.

Smith, G., Frankel, S., & Yarnell, J. (1997). Sex and death: Are they related? Findings from the Caerphilly cohort study. *British Medical Journal, 315*, 1641–1645.

Smith, R., Smith, J., & Bisits, A. (2010). The endocrine regulation of human labor, in preterm birth: prevention and management. In V. Berghella (Ed.), *Preterm* Sex trade workers. (2001). *Alberta Report, 28*, 41.

Smits, M. (2004, November 28). Sex slave's story told. *Sophia Echo*. Retrieved from **www.sofiaecho.com/article/sex-slaves-story-told/id_10416/catid_29**

Smock, P. J. (2000). Cohabitation in the United States: An appraisal of research themes, findings, and implications. *Annual review of Sociology, 26*, 1–20.

Smolowe, J. (2008, April 28). Mean girls. *People Weekly, 69*(16), 104.

Smyth, A. (2002). Sexual problems overview.

Somers, C., & Gleason, J. (2001). Does source of sex education predict adolescents' sexual knowledge, attitudes, and behaviors? *Education, 121*, 674–681.

South, S. (1991). Sociodemographic differentials in mate selection preferences. *Journal of Marriage and the Family*, 53, 928–940.

Sparks, S. (2005). Study casts new doubt on abstinence-only approach: Teens participating in Texas program found to be more sexually active, not less. *Education Daily*, pp. 4–5.

Spector, I., & Carey, M. (1990). Incidence and prevalence of the sexual dysfunctions: A critical review of the empirical literature. *Archives of Sexual Behavior, 19*, 389–408.

Sprecher, S. (2002). Sexual satisfaction in premarital relationships: Associations with satisfaction, love, commitment, and stability. *Journal of Sex Research, 39*, 190–197.

Stacey, D. (2012). Obesity and contraception effectiveness: The link between weight and birth control failure. *About.com*. Retrieved from **contraception.about.com/od/unplannedpregnancy/a/Obesity-And-Contraception-Effectiveness.htm**

"State Policies." (2014). *State facts about abortion: State policies in brief*. Guttmacher Institute. Retrieved from **www.guttmacher.org/statecenter/spibs/spib_OAL.pdf**

Stalburg, C. (2008). "Doctor, I want a C-section." How should you respond? Is she motivated by a fear of childbirth or a true wish for C-section? Here's how to identify candidates. *OBG Management, 20*, 58–64.

Stanger-Hall, K., & Hall, D. (2011) Abstinence-only education and teen pregnancy rates: Why we need comprehensive sex education in the U.S. *PLoS ONE, 6*(10). Retrieved from **www.plosone.org/article/info%3Adoi%2F10.1371%2Fjournal.pone.0024658**. doi: 10.1371/journal.pone.0024658

Stans, M. (2007). *The effect of pornography consumption on males' propensity to rape*. Master's Thesis, University of Auburn.

Stayton, W. (2002). A theology of sexual pleasure. *SIECUS Report, 30*(4), 27–29.

Stayton, W., Haffner, D., & McNiff, S. (2007). Sexuality and reproductive health of men and women. In B. Boyer & M. Paharia (Eds.), *Comprehensive handbook of clinical health psychology* (pp. 425–452). New York: Wiley.

Steen, R., & Dallabetta, G. (2003). Sexually transmitted infection control with sex workers: Regular screening and presumptive treatment augments efforts to reduce risk and vulnerability. *Reproductive Health Matters, 11*, 74–90.

Stein, N. (2007, Fall). Bullying, harassment and violence among students. *Radical Teacher, 80*, 30–35.

Steinem, G. (1980). Erotica and pornography: A clear and present difference. In L. Lederer (Ed.), *Take back the night: Women on pornography* (pp. 35–39). New York: Morrow.

Stemple, L. (2009). Male rape and human rights. *Hastings Law Journal, 60*, 622–625.

Stephenson, K., & Meston, C. (2013). The conditional importance of sex: Exploring the association between sexual well-being and life satisfaction. *Journal of Sexual Medicine, 10*(9), 2177–2189.

Sternberg, R. J. (1986). A triangular theory of love. *Psychological Review, 93*, 119–135.

Sternberg, R. J. (1988). *The triarchic mind: A new theory of human intelligence*. New York: Viking.

Sternberg, R. J. (1997). Construct validation of a triangular love scale. *European Journal of Social Psychology, 27*, 313–335.

Sternberg, R. J. (1998). *Cupid's arrow: The course of love*. New York: Cambridge University Press.

Stevenson, M., Crownover, B., & Mackler, L. (2009). When to suggest this OC alternative (Cover story). *Journal of Family Practice, 58*(4), 207–210.

Stinneford, J. (2006). Incapacitation through maiming: Chemical castration, the Eighth Amendment, and the denial of human dignity. *University of Saint Thomas Law Journal* (Spring), 559–599.

Stinson, R. (2009). The behavioral and cognitivebehavioral treatment of female sexual dysfunction: How far we have come and the path left to go. *Sexual & Relationship Therapy, 24*(3/4), 271–285. doi:10.1080/14681990903199494

Stobbe, M., & Cheng, M. (2011). Drug stops HIV among hetero couples, not just gays. *BioScience Technology*. Retrieved from **www.biosciencetechnology.com/News/FeedsAP/2011/07/drug-stopshiv-among-hetero-couples-not-just-gays**

Stockwell, F., & Moran, D. (2013). A relational frame theory account of the emergence of sexual fantasy. *Journal of Sex and Marital Therapy. [ePub]* doi:10.1080/0092623X.2012.736921

Stockwell, F., & Moran, D. (2014). A relational frame theory account of the emergence of sexual fantasy. *Journal of sex & Marital Therapy, 40*(2), 92–104.

Stoddard, J., Anderson, M., Berkowitz, C., Britton, C., Nordgren, R., Pan, R.,... & Tunnessen, W. (2000). Prevention of sexual harassment in the workplace and educational settings. *Pediatrics, 106*(6), 1498–1499.

Stolberg, M. (2003). A woman down to her bones: The anatomy of sexual difference in the sixteenth and early seventeenth centuries. *Isis, 94*, 274–300.

Stoltenborgh, M., van IJzendoorn, M., Euser, E., & Bakermans-Kranenburg, M. (2011). A global perspective on child sexual abuse: Meta-analysis of prevalence around the world. *Child Maltreatment, 16*(2), 79–101.

Stoltz, J. E. (2013). Postpartum Psychosis: What Happens When the Bough Breaks? *Journal of Obstetric, Gynecologic, & Neonatal Nursing, 42*(s1), S96–S96.

Stone, G. (2009, October 3). Sex and Sin. *Huffington Post*. Retrieved from **www.huffingtonpost.com/geoffrey-r-stone/sex-and-sin_b_308732.html**

Stoppler, M., & Lee, D. (2008). Ovarian cancer symptoms, early warning signs, and risk factors. Retrieved from **www.medicinenet.com/ovarian_cancer/article.htm**

Strassberg, D. S., & Lockerd, L. K. (1998). Force in women's sexual fantasies. *Archives of Sexual Behavior, 27*, 403–414.

Strauss, L., Herndon, J., Chang, J., Parker, W., Bowens, S., Zane, S., & Berg, C. (2004). Abortion Surveillance—United States, 2001. Division of Reproductive Health National Center for Chronic Disease Prevention and Health Promotion. Retrieved from **www.cdc.gov/mmwr/preview/mmwrhtml/ss5309a1.htm#tab19**

Striar, S., & Bartlik, B. (1999). Stimulation of the libido: The use of erotica in sex therapy. *Psychiatric Annals, 29*, 60–62.

Strong, B. & Cohen, T. (2013). *The marriage and family experience: Intimate relationships in a changing society*, 12th *ed*. New York: Cengage.

Strough, J., Swenson, L. M., & Cheng, S. (2001). Friendship, gender, and preadolescents' representations of peer collaboration. *Merrill-Palmer Quarterly, 47*, 475–499.

Studies examine infertility related to extra X chromosome. (2008, April). *AORN Journal, 87*, 779.

Study refutes benefit of "sperm-saving." (2003). Retrieved from **www.intelihealth.com/IH/ihtPrint/EMIHC276/333/22002/366291.html?d=dmtICNNews&hide=t&k=basePrint**

Subrahmanyama, K., Reichc, S. M., Waechterb, N., & Espinozab, G. (2008). Online and offline social networks: Use of social networking sites by emerging adults. *Journal of Applied Developmental Psychology, 29*(6), 420–433.

Sulak, P. (2007). The demise of 21/7 contraceptive regimens. *Family Practice News, 37*, 9.

Sulak, P., Smitln, V., & Coffee, A. (2008). Frequency and management of breakthrough bleeding with continuous use of the transvaginal contraceptive ring: A randomized controlled trial. *Obstetrics and Gynecology, 112*, 563–571.

Sullivan, B. (2007). Rape, prostitution and consent. *Australian and New Zealand Journal of Criminology, 40*, 127–143.

Sullivan, M. (2013). *Sexual minorities: Discrimination, challenges and development in America*. New York: Routledge.

Sunger, M. (1999). Cultural factors in sex therapy: The Turkish experience. *Sexual and Marital Therapy, 14*, 165–171.

Sutton, K. S., Boyer, S. C., Goldfinger, C., Ezer, P., & Pukall, C. F. (2012). To lube or not to lube: experiences and perceptions of lubricant use in women with and without dyspareunia. *The journal of sexual medicine, 9*(1), 240–250.

Swaab, D., & Garcia-Falgueras, A. (2009). "Sexual differentiation of the human brain in relation to gender identity and sexual orientation." *Functional Neurology, 24*(1), 17–28.

Swamy, G., Ostbye, T., & Skjaerven, R. (2008). Association of preterm birth with long-term survival, reproduction, and next-generation preterm birth (Clinical report). *Journal of the American Medical Association*, 299, 1429–1434.

Swartz, C. (2009). Assortative mating. In H. Reis & T. Sprecter (Eds.), *Encyclopedia of Human Relationships* (pp. 123–125). Thousand Oaks, CA: Sage Publications.

Sweeney, J., & Bradbard, M. (1988). Mothers' and fathers' changing perceptions of their male and female infants over the course of pregnancy. *Journal of Genetic Psychology, 149*, 393–404.

Szasz, T. (2000). Remembering Krafft-Ebing. *Ideas on Liberty, 50*, 31–32.

Tasker, F. (2010). Same-sex parenting and child development: Reviewing the contribution of parental gender. *Journal of Marriage and Family, 72*, 35–40.

Tasker, F., & Bigner, J. (2013). *Gay and lesbian parenting: New directions*. New York: Routledge.

Taylor, S. (1999). Harassment by kids: Are more lawsuits the answer? *National Journal, 231*, 1512–1513.

TCRC. (2012). Testicular Cancer Resource Center. Retrieved from **http://tcrc.acor.org/tcexam.html**

Teal, S. (2011). Postpartum contraception. In D. Shoupe (Ed.), *Contraception*. New York, NY: Wiley.

"Teen Dating Violence." (2014). Talking About Dating Violence. Washington State Medical Association. Retrieved from **www.atg.wa.gov/ProtectingYouth/TeenDatingViolence/BreakingTheSilence.aspx#.U_5V8U1OV9Aom**

Teich, M. (2007). A man's shelf life. *Psychology Today, 40*(5), 90–95.

Templeton, D. J., Millett, G. A., & Grulich, A.E. (2010, February). Male circumcision to reduce the risk of HIV and sexually transmitted infections among men who have sex with men. *Current Opinion in Infectious Diseases, 23*(1), 45–52. doi:10.1097/QCO.0b013e328334e54d

Terris, M. (2013). Urethritis medication. *Medscape*. Retrieved from **www.medscape.com/public/about**

Terry, K., & Ackerman, A. (2008). Child sexual abuse in the catholic church how situational crime prevention strategies can help create safe environments. *Criminal Justice and Behavior, 35*(5), 643–657.

Testa, M., & Livingston, J. A. (2009). Alcohol consumption and women's vulnerability to sexual victimization: Can reducing women's drinking prevent rape? *Substance Use & Misuse, 44*(9–10), 1349–1376.

Thabet, S. (2013). New findings and concepts about the G-spot in normal and absent vagina: Precautions possibly needed for preservation of the G-spot and sexuality during surgery. *Journal of Obstetrics and Gynaecology Research, 39*(8), 1339–1346.

The causes of homosexuality. (2001, September 11). *Advocate*, p. 6.

"The Report." (1970). president's commission on obscenity and pornography. Report of the commission on Obscenity and Pornography. Washington, D.C.: U. S. Government Printing Office.

Thibaut, F., Barra, F., Gordon, H., Cosyns, P., & Bradford, J. (2010). The World Federation of Societies of Biological Psychiatry (WFSBP) guidelines for the biological treatment of paraphilias. *World Journal of Biological Psychiatry, 11*(4), 604–655.

Thibaut, F., Drakulić, A. M., Ilanković, A., Damjanović, A., Ilanković, V., Filipović, B.,... & Lanfranco, R. (2014). Acute treatment of schizophrenia: introduction to the Word Federation of Societies of Biological Psychiatry guidelines. *Psychiatria Danubina, 26*(1), 2–11.

Thomas, I. (2011). *Why do I have periods? Menstruation and puberty (Inside My Body)*. Chicago, IL: Heinemann Raintree.

Thompson, M., & Thompson, N. (2010). BPH. *Bay Area Medical Information*. Retrieved from **www.bami.us/GU/BPH.html**

Thorton, A., Axinn, W., & Xie, Y. (2010). *Marriage and cohabitation*. Chicago, IL: University of Chicago Press.

Tiedemann, J. (2000). Parents' gender stereotypes and teachers' beliefs as predictors of children's concept of their mathematical ability in elementary school. *Journal of Educational Psychology, 92*(1), 144–151.

Tiefer, L. (2001). A new view of women's sexual problems: Why new? Why now? *Journal of Sex Research*, 38, 89–96.

Tiefer, L. (2010). Still resisting after all these years: An update on sexuomedicalization and on the New View Campaign to challenge the medicalization of women's sexuality. *Sexual & Relationship Therapy, 25*(2), 189–196. doi: 10.1080/14681991003649495

Timmerman, G. (2004). Adolescents' psychological health and experiences with unwanted sexual behavior at school. *Adolescence, 39*, 17–25.

Timpson, S., Ross, M., Williams, M., & Atkinson J. (2007). Characteristics, drug use, and sex partners of a sample of male sex workers. *American Journal of Drug and Alcohol Abuse, 33*(1), 63–69.

Tjaden, P., & Thoennes, N. (2006). Extent, nature, and consequences of rape victimization: Findings from the national violence against women survey.

Tobin, H. (2008). Confronting misinformation on abortion: Informed consent, deference, and fetal pain laws. *Columbia Journal of Gender and Law, 17*, 111–151.

Trenholm, C., Devaney, B., Fortson, K., Quay, L., Wheeler, J., & Clark, M. (2007, April). Impacts of four Title V, Section 510, Abstinence Education Programs: Final report. Mathematica Policy Research, Inc. Retrieved from **www.mathematica-mpr.com/publications/PDFs/impactabstinence.pdf**

Triandis, H. C. (1989). The self and social behavior in differing cultural contexts. *Psychological Review, 96*, 506–520.

Triandis, H. C., Bontempo, R., Villareal, M. J., Asai, M., & Lucca, N. (1988). Individualism and collectivism: Cross-cultural perspectives on self-in-group relationships. *Journal of Personality and Social Psychology, 54*, 323–338.

Trickett, P., Noll, J., & Putnam, F. (2011). The impact of sexual abuse on female development: Lessons from a multigenerational, longitudinal research study. *Development and Psychopathology, 23*(2), 453.

Trimble, L. (2009). Transformative conversations about sexualities pedagogy and the experience of sexual knowing. *Sex Education, 9*(1), 51–64. doi:10.1080/14681810802639954

Troiden, R. (1989). The formation of homosexual identities. *Journal of Homosexuality, 17*, 43–71.

Tronstein, E., Johnston, C., Huang, M., Selke, S., Magaret, A., Warren, T.... Wald, A. (2011). Genital shedding of herpes simplex virus among symptomatic and asymptomatic persons with HSV-2 infection. *JAMA. 305*(14), 1441–1449. doi:10.1001/jama.2011.420

Trudel, G., Marchand, A., Ravart, M., Aubin, S., Turgeon, L., & Fortier, P. (2001). The effect of a cognitive-behavioral group treatment on hypoactive sexual desire in women. *Sexual and Relationship Therapy, 16*, 145–164.

Trussell, J., & Wynn, L. (2008). Reducing unintended pregnancy in the United States. *Contraception, V77*, 1–5.

"Tubal Ligation Reversal." (2012). WebMD. Infertility & Reproduction Health Center. Retrieved from **www.webmd.com/infertility-and-reproduction/guide/tubal-ligation-reversal**

Tucker, M. (2008). Specific symptoms flag endometriosis diagnosis. *Family Practice News, 38*(1), 24–25.

Tucker, M. E. (2003, May 15). Trichomoniasis tricky to diagnose and treat: Better tests needed. *Family Practice News*, p. 30.

Tucker, P. (2005, September–October). Stay-at-home dads: At-home dads can benefit children and mothers. *The Futurist, 39*, 12–13.

Turchik, J. A., & Edwards, K. M. (2012). Myths about male rape: A literature review. *Psychology of Men & Masculinity, 13*(2), 211.

Turkum, A. (2005). Who seeks help? Examining the differences in attitude of Turkish university students toward seeking psychological help by gender, gender roles, and help-seeking experiences. *Journal of Men's Studies, 13*, 389–390.

Turley, K., & Rowland, D. (2013). Evolving ideas about the male refractory period. *BJU International, 112*, 442–452.

Turner, A., Comston, T., Davis, J., Nasrin, Z., & Vaughn, J. (2011). Social media and chlamydia testing by university students: a pilot study. *Sexually Transmitted Infections, 87*(Suppl. 1), A89–A90.

U.S. Census Bureau. (2006). *State and county Quick-Facts.*

U.S. Census Bureau. (2011). Census bureau releases estimates of same-sex married couples. Retrieved from **www.census.gov/newsroom/releases/archives/2010_census/cb11-cn181.html**

U.S. Department of Health and Human Services, Administration for Children and Families, Administration on Children, Youth and Families. Children's Bureau. (2012). Child Maltreatment. Retrieved from **www.acf.hhs.gov/programs/cb/research-data-technology/statistics-research/child-maltreatment**.

U.S. Department of Health and Human Services, Office on Women's Health. (2010). Menstruation and the menstrual cycle. Retrieved from **www.womenshealth.gov/faq/menstruation.cfm#e**

U.S. Department of Justice. (2007). Criminal offender statistics. *Bureau of Justice Statistics.*

Ulges, A., & Stahl, A. (2011, July). Automatic detection of child pornography using color visual words. In *Multimedia and Expo (ICME), 2011 IEEE International Conference on* (pp. 1–6). IEEE.

UN AIDS report on the global AIDS epidemic. (2010). *Joint United Nations Programme on HIV/AIDS (UNAIDS)*. Retrieved from **www.unaids.org/globalreport/documents/20101123_GlobalReport_full_en.pdf**

UNAIDS – Global Report. (2013). Global report: UNAIDS report on the global AIDS epidemic 2013. Global Joint United Nations Programme on HIV/AIDS (UNAIDS). Retrieved from **www.unaids.org/en/media/unaids/contentassets/documents/epidemiology/2013/gr2013/UNAIDS_Global_Report_2013_en.pdf**

Underhill, K., Montgomery, P., & Operario, D. (2007). Sexual abstinence only programmes to prevent HIV infection in high income countries: Systemic review (Clinical report). *British Medical Journal, 335*(7613), 248–252.

Underwood, M. (2003). *Social aggression among girls*. New York: Guilford Press.

Underwood, M., (2007, July). Introduction to the special issue on gender and children's friendships: Do girls' and boys' friendships constitute different peer cultures, and what are the trade-offs for development? *Merrill-Palmer Quarterly, 53*, 319–324.

United Nations Children's Fund. (2013). *Female Genital Mutilation/Cutting: A statistical overview and exploration of the dynamics of change*. New York: UNICEF.

Urinary Tract Infections (2010). Urinary Tract Infections in Adults. National Institute of Diabetes and Digestive and Kidney Diseases. Retrieved from **www.kidney.niddk.nih.gov/kudiseases/pubs/utiadult/index.aspx**

Utz, S., & Beukeboom, C. (2011). The role of social network sites in romantic relationships: Effects on jealousy and relationship happiness. *Journal of Computer-Mediated Communication, 16*(4), 511–527.

Vahabi, S., Haidari, M., Akbari Torkamani, S., & Gorbani Vaghei, A. (2010). New assessment of relationship between APGAR score and early neonatal mortality. *Minerva Pediatrics, 62*(3), 249–252.

Valera, R., Sawyer, R., & Schiraldi, G. (2000). Violence and posttraumatic stress disorder in a sample of inner-city street prostitutes. *American Journal of Health Studies, 16*, 149–155.

Valkenburg, P. M., & Peter, J. (2009, February). Social consequences of the Internet for adolescents: A decade of research. *Current Directions in Psychological Science, 18*(1), 1–5.

Van Lankveld, J., & Sykora, H. (2008). The sexual self-consciousness scale: Psychometric properties. *Archives of Sexual Behavior, 37*, 925–933. doi:10.1007/s10508-007-9253-5

Vance, E. B., & Wagner, N. N. (1976). Written descriptions of orgasm: A study of sex differences. *Archives of Sexual Behavior, 5*, 87–98.

Vares, T., Potts, A., Gavey, N., & Grace, V. (2003). Hard sell, soft sell: Men read Viagra ads. *Media International Australia, 108*, 101–114.

Vaughan, M., & Waehler, C. (2009). Coming out growth: Conceptualizing and measuring stress-related growth associated with coming out to others as a sexual minority. *Journal of Adult Development, 17*, 94–109.

Vega, V., & Malamuth, M. (2007). Predicting sexual aggression: The role of pornography in the context of general and specific risk factors. *Aggressive Behavior, 33*, 1–14.

Vekshin, A. (2013, August 28). Brothels in Nevada suffer as web disrupts oldest trade. *Bloomberg News*. Retrieved from **www.bloomberg.com/news/2013-08-28/brothels-in-nevada-shrivel-as-web-disrupts-oldest-trade.html**

Veniegas, R., & Conley, T. (2000). Biological research on women's sexual orientations: Evaluating the scientific evidence. *Journal of Social Issues, 56*(2), 267–282.

Venning, R., & Cavanah, C. (2003). *Sex toys 101: A playfully uninhibited guide*. New York: Fireside.

Verdin, T. (2013, August 12). California's transgender-student law: Kids can choose bathrooms, sports teams. *The Christian Science*

Monitor. Retrieved from **www.csmonitor.com/USA/Latest-News-Wires/2013/0812/California-s-transgender-student-law-Kids-can-choose-bathrooms-sports-teams**

Viglianco-VanPelt, M., & Boyse, K. (2008). Your child topics: Masturbation. University of Michigan Health System. Retrieved from **www.med.umich.edu/1libr/yourchild/masturb.htm**

Viglianco-VanPelt, M., & Boyse, K. (2009). Masturbation. University of Michigan Health System U-M Medical School. Retrieved from **www.med.umich.edu/yourchild/topics/masturb.htm**

Von Sadovszky, V., Keller, M., & McKinney, K. (2002). College students' perceptions and practices of sexual activities in sexual encounters. *Journal of Nursing Scholarship, 34*(2), 133–138.

Vorvick, L., & Storck, S. (2010). Ectopic pregnancy. MedlinePlus 2010. U.S. Department of Health and Human Services National Institutes of Health.

Vyas, K. (2004, September 12). Thais plan first live sex-change operation.

Waldinger, M. (2005). Lifelong premature ejaculation: Current debate on definition and treatment. *Journal of Men's Health & Gender, 2*, 333–338.

Waldinger, M. D., (2007, August 1). New insights in premature ejaculation (Disease/Disorder overview). *Psychiatric Times, 24*, 52–54.

Walker, M., Hernandez, A., & Davey, M. (2012). Childhood sexual abuse and adult sexual identity formation: Intersection of gender, race, and sexual orientation. *The American journal of family therapy, 40*(5), 385–398.

Wallace, J., Weiner, A., &, Bloch, D. (1996, July). *Fellatio is a significant risk activity for acquiring AIDS in New York City streetwalking sex workers*. Paper presented at the Eleventh International Conference on AIDS, Vancouver, BC.

Wallen, K., & Lloyd, E. (2008). Clitoral variability compared with penile variability supports nonadaptation of female orgasm. *Evolution & Development, 10*, 1–2.

Walling, A. D. (1999). Etiology and management of cystitis. *American Family Physician, 59*, 1314.

Walsh, K., & DiLillo, D. (2011). Child sexual abuse and adolescent sexual assault and revictimization. *The psychology of teen violence and victimization, 1*, 203–220.

Warner, D. L., & Hatcher, R. (1998). Male condoms. In R. A. Hatcher, J. Trussell, F. Stewart, W. Cates Jr., G. K. Stewart, F. Guest, & D. Kowal (Eds.), *Contraceptive technology* (17th rev. ed., pp. 325–355). New York: Ardent Media.

Warren, Harvey, M., & Henderson, J. (2010). Do depression and low self-esteem follow abortion among adolescents? Evidence from a national study. *Perspectives on Sexual and Reproductive Health, 42*, 230–235. doi: 10.1363/4223010

Waterman, A., Reid, J., Garfield, L., & Hoy, S. (2001). From curiosity to care: Heterosexual student interest in sexual diversity courses. *Teaching of Psychology, 28*, 21–26.

Wattendorf, D., & Muenke, M. (2005). Klinefelter syndrome. *American Family Physician, 72*, 2559– 2562.

Weber, A., Boivin, J., Blais, L., Haley, N., & Roy, L. (2004). Predictors of initiation into prostitution among female street youths. *Journal of Urban Health: Bulletin of the New York Academy of Medicine, 81*, 584–595.

Weber, B. (2013). Preterm birth rate drops to 15-year low. *Medical News Today*. Retrieved from **www.medicalnewstoday.com/articles/268339.php**

Weeks, G. R., & Gambescia, N. (2000). *Erectile dysfunction: Integrating couple therapy, sex therapy, and medical treatment*. New York: Norton.

Weeks, R., & James, J. (1999). *Secrets of the superyoung: The scientific reasons some people look ten years younger than they really are—and how you can, too*. New York: Berkley.

Weitzer, R. (2005). New directions in research on prostitution. *Crime, Law and Social Change, 43*(4–5), 211–235.

Wellings, K. (2012). *Sexual health: A public health perspective*. New York, NY: Open University Press.

Wells, B., & Twenge, M. (2005). Changes in young people's sexual behavior and attitudes, 1943–1999: A cross-temporal meta-analysis. *Review of General Psychology, 9*, 249–261.

Wells, D., & Fragouli, E. (2013). Preimplantation genetic diagnosis. In K. Coward & D. Wells, (Eds.) *Textbook of Clinical Embryology*, (pp. 346–349). Cambridge: **Cambridge University Press.**

Werner, N., & Crick, N. (1999). Relational aggression and social psychological adjustment in a college sample. *Journal of Abnormal Psychology, 108*, 615–623.

Wessells, L., & McAninch, J. (1996). Penile length in the flaccid and erect states: Guidelines for penile augmentation. *Journal of Urology, 156*, 995–997.

What is sexual harassment? (2011). Title IX Sexual Harassment Office, University of California, Santa Cruz. Retrieved from **www2.ucsc.edu/title9-sh/whatissh.htm**

Where HIV began. (2006, June 3). *New Scientist, 190*.

Whipple, B. (1999, November 4). *Beyond the G-spot: Where do we go from here?* Address delivered at the joint annual conference of the American Association of Sex Educators, Counselors, and Therapists and the Society for the Scientific Study of Sexuality, St. Louis, MO.

Whipple, B. (2000). Beyond the G-spot. *Scandinavian Journal of Sexology, 3*, 35–42.

Whipple, B. (2002). Review of Milan Zaviacic's book, *The human female prostate: From vestigial Skene paraurethral glands and ducts to woman's functional prostate. Archives of Sexual Behavior, 31*, 457–458.

Whitehead, L. (2007). Methodological and ethical issues in Internet-mediated research in the field of health: An integrated review of the literature. *Social Science & Medicine, 65*(4), 782–791.

Whiteman, H. (2014). Passive smoking increases risk of miscarriage, stillbirth and ectopic pregnancy. *Medical News Today*. Retrieved from **www.medicalnewstoday.com/articles/273167.php**

Whitley, B., (1983). Sex role orientation and self-esteem. A critical meta-analytic review. *Journal of Personality and Social Psychology, 44*, 765–778.

Wiederman, M. (1998). The state of theory in sex therapy. *Journal of Sex Research, 35*, 88–99.

Wiessing, L., van Roosmalen, S., Koedijk, P., Bieleman, B., & Houweling, H. (1999). Silicones, hormones and HIV in transgender street prostitutes. *AIDS, 13*, 2315–2316.

Wight, D., Henderson, M., Raab, G., Abraham, C., Buston, K., Scott, S., & Hart, G. (2000). Extent of regretted sexual intercourse among young teenagers in Scotland: A cross-sectional survey. *British Medical Journal, 320*, 1243–1244.

Wilbert, C. (2008). Boys, girls equal at social aggression: Study shows boys as likely as girls to use "indirect" aggression. WebMD Health News.

Wilcox, B. (1987). Pornography, social science, and politics: When research and ideology collide. *American Psychologist, 42*, 941–943.

Willaimson, S., & Nowak, R. (1998) The truth about women. *New Scientist, 159*, 34–35.

Williams, D., Martins, N., Consalvo, M., & Ivory, J. (2009). The virtual census: representations of gender, race and age in video games. *New Media & Society, 11*(5), 815–834. doi:10.1177/1461444809105354

Williams, M. (2008). Sexual compulsivity: Defining paraphilias and related disorders. *Psychoactive drug treatments*. Brain Physics Mental Health Resource. Retrieved from **www.brainphysics.com/paraphilias.php**

Williams, S. (2007). Mother knows all: Next generation of prenatal tests finds clues to baby's health in mother's blood. *Science News, 172*, 295–297.

Williams, S. P. (2001, August 12–15). *Living positively: A formative study of HIV+ seroconcordant heterosexual couples*. Presentation at the 2001 National HIV Prevention Conference, Atlanta.

Willness, C., Steel, P., & Lee, K. (2007). A metaanalysis of the antecedents and consequences of workplace sexual harassment. *Personnel Psychology, 60*, 127–162.

Wilson, B., Holm, J., Bishop, K., & Borowiak, D. (2002). Predicting responses to sexually aggressive stories: The role of consent, interest in sexual aggression, and overall sexual interest. *Journal of Sex Research, 39*, 275–283.

Wilson, E., Dalberth, B., Koo, H., & Gard, J. (2010). Parents' perspectives on talking to preteenage children about sex. *Perspectives on Sexual & Reproductive Health, 42*(1), 56–63. doi:10.1363/4205610

Wilson, J., & Young, S. (2014, March 6). Second baby possibly "cured" of HIV. CNN. Retrieved from **www.cnn.com/2014/03/06/health/hiv-baby-cured/index.html**

Wimpissinger, F., & Stifter, K. (2007). The female prostate revisited: Perineal ultrasound and biochemical studies of female ejaculate. *The Journal of Sexual Medicine, 4*, 1388–1393.

Wimpissinger, F., Stifter, K., Grin, W., & Stackl, W. (2007). The female prostate revisited: Perineal ultrasound and biochemical studies of female ejaculate. *Journal of Sexual Medicine, 4*, 1388–1393.

Wimpissinger, F., Tscherney, R., & Stackl, W. (2009). Magnetic resonance imaging of female prostate pathology. *The Journal of Sexual Medicine, 6*, 1704–1711.

Wincze, J. P., & Carey, M. P. (2012). *Sexual dysfunction: A guide for assessment and treatment*. New York, NY: Guilford Press.

Windle, M., Spear, L., Fuligni, A., Angold, A., Brown, J., Pine, D.,... Dahl, R. (2008). Transitions into underage and problem drinking: developmental processes and mechanisms between 10 and 15 years of age. *Pediatrics, 121*, S273–S289. doi:10.1542/peds.2007-2243C).

Winick, C., & Evans, J. (1996). The relationship between nonenforcement of state pornography laws and rates of sex crime arrests. *Archives of Sexual Behavior, 25*, 439–453.

Winks, C., & Semans, A. (2002). *The good vibrations guide to sex*. Pittsburgh: Cleis Press.

Wisconsin Coalition Against Sexual Assault. (2000). Campus sexual assault.

Wise, T. (1985). Fetishism: Etiology and treatment: A review from multiple perspectives. *Comprehensive Psychiatry, 26*, 249–257.

Withnal, A. (2014). Police call for condoms to be banned from saunas in Edinburgh. *The Independent.*

Witt, S. (1997). Parental influence on children's socialization to gender roles. *Adolescence, 126*, 253–259.

Witt, S. (2000). The influence of television on childrens' gender role socialization. *Childhood Education, 76*, 322–324.

Wlodarski, R., & Dunbar, R. (2013). Examining the possible functions of kissing in romantic relationships. *Archives of Sexual Behavior, 42*, 1415–1423.

Wohlfeiler, D., & Kerndt, P. (2013). Personal Risk and Public Impact: Balancing Individual Rights with STD and HIV Prevention. In S. Aral, K. Fenton, & J. Lipshutz (Eds.). *The New Public Health and STD/HIV Prevention* (pp. 111–125). New York, NY: Springer.

Wolfers, M., de Zwart, O., & Kok, G. (2011). Adolescents in the Netherlands underestimate risk for STI and deny the need for STI testing. *STI, 25*(5), 49.

Womenshealth.gov. (2012). Date rape drugs. Office on Women's Health in the Office of the Assistant Secretary for Health at the U.S. Department of Health and Human Services. Retrieved from **www.womenshealth.gov/publications/our-publications/fact-sheet/date-rape-drugs.pdf**

Woo, J., Brotto, T., Gorzalka, L., & Boris, B. (2011). The role of sex guilt in the relationship between culture and women's sexual desire. *Archives of Sexual Behavior, 40*, 385–394.

Wood, D., & Brumbaugh, C. (2009). Using revealed mate preferences to evaluate market force and differential preference explanations for mate selection. *Journal of Personality & Social Psychology, 96*(6), 1226–1244.

Wood, E., Desmarais, S., & Gugula, S. (2002). The impact of parenting experience on gender stereotyped toy play of children. *Sex Roles, 47*(1–2), 39–49.

Woodhill, B., & Samuels, C. (2003). Positive and negative androgyny and their relationship with psychological health and well-being. *Sex Roles, 48*(11–12), 555–565.

Woods, L., & Emery, R. (2002). The cohabitation effect on divorce: Causation or selection? *Journal of Divorce & Remarriage, 37*(3–4), 101–122.

Woodward, B., & Armstrong, S. (2005). *The brethren: Inside the Supreme Court*. New York: Simon and Schuster.

Woodworth, M., Freimuth, T., Hutton, E. L., Carpenter, T., Agar, A. D., & Logan, M. (2013). High-risk sexual offenders: An examination of sexual fantasy, sexual paraphilia, psychopathy, and offence characteristics. *International journal of law and psychiatry, 36*(2), 144–156.

World Health Organization. (2010). World Health Organization fact sheet: Female genital mutilation. Retrieved from **www.who.int/mediacentre/factsheets/fs241/en**

World Health Organization. (2012). Prevention and treatment of HIV and other sexually transmitted infections for sex workers in low-and middle-income countries: Recommendations for a public health approach.

Wortley, R., & Smallbone, S. (2006). The problem of internet child pornography: Child pornography on the Internet: Guide No. 41. *Center for Problem-Oriented Policing.* Retrieved from **www.popcenter.org/problems/child_pornography**

Wortley, R., & Smallbone, S. (2012). *Internet Child Pornography: Causes, Investigation, and Prevention*. Santa Barbara, CA: Praeger, ABC-CLIO. Retrieved from **www.unaids.org/en/media/unaids/contentassets/documents/epidemiology/2013/gr2013/20130923_FactSheet_Global_en.pdF**

Wylie, K., & Eardley, I. (2007). Penile size and the "small penis syndrome." *BJU International, 99* (6), 1449–1455.

Wylie, K., & Mimoun, S. (2009). Sexual response models in women. *Maturitas, 63*(2), 112–115.

Yim, J., & Mahalingam, R. (2006). Culture, masculinity, and psychological well-being in Punjab, India. *Sex Roles, 55*(9–10), 715–724.

Young, A., Boyd, C., & Hubbell, A. (2000). Prostitution, drug use, and coping with psychological distress. *Journal of Drug Issues, 30*(4), 789–800.

Young, M., Luquis, R., Deny, G., & Young, T. (1998). Correlates of sexual satisfaction in marriage. *Canadian Journal of Human Sexuality, 7*, 115–127.

Young, S., & Maguire, K. (2003). Talking about sexual violence. *Women and Language, 26*(2), 40–51.

Yunting, Liu. (2013, August 30). 90% of China's New HIV Infections through Sex. *Women's Foreign Language Publications of China*. Retrieved from **www.womenofchina.cn/html/node/155970-1.htm**

Zajonc, R. B. (1968). Attitudinal effects of mere exposure [Monograph]. *Journal of Personality and Social Psychology, 9*(Suppl.), 1–27.

Zajonc, R. B. (2001). Mere exposure: A gateway to the subliminal. *Current Directions in Psychological Science, 10*, 225–228.

Zamora, R., Winterowd, C., Koch, J., & Roring, S. (2013). The relationship between love styles and romantic attachment styles in gay men. *Journal of LGBT Issues in Counseling, 7*, 200–217. doi:10.1080/15538605.2013.812927

Zamosky, L. (2011, January 4). Caught your kid masturbating? Here's what to do and what to say—because how you react matters. *WebMD the Magazine*. Retrieved from **www.webmd.com/parenting/features/caught-your-kid-masturbating**

Zaviacic, M. (2002a). Female urethral expulsions evoked by local digital stimulation of the G-spot: Differences in the response patterns. *Journal of Sex Research, 24*, 311–318.

Zaviacic, M. (2002b). *The human female prostate: From vestigial Skene paraurethral glands and ducts to woman's functional prostate*. Bratislavia, Slovakia: Slovak Academic Press.

Zeliadt, S., Moinpour, C., Blough, D., Penson, D., Hall, I., Smith, J.... Ramsey, S. (2010). Preliminary treatment considerations among men with newly diagnosed prostate cancer. *American Journal of Managed Care, 16*(5), 121–130.

Zentner, M., & Bates, J. (2008). Child temperament: An integrative review of concepts, research programs, and measures. *European Journal of Developmental Science, 2*, 7–37.

Zgourides, G., Monto, M., & Harris, R. (1997). Correlates of adolescent male sexual offense: Prior adult contact, sexual attitudes, and use of sexually explicit materials. *International Journal of Offender Therapy and Comparative Criminology, 41*, 272–283.

Zhuhong. (2010, August 25). China to Carry Out Sexual Health Education Program. *Women's Foreign Language Publications of China*. Retrieved from **www.womenofchina.cn/html/womenofchina.cn/html/node/108338-1.htm**

Zieve, D., & Miller, S. (2010). Semen analysis. MedLine Plus. U.S. Department of Health and Human Services. National Institutes of Health. Retrieved from **www.nlm.nih.gov/medlineplus/ency/article/003627.htm**

Zimmerman, T., Haddock, S., Current, L., & Ziemba, S. (2003). Intimate partnership: Foundation to the successful balance of family and work. *American Journal of Family Therapy, 31*, 107–114.

Zini, A. (2010). Vasectomy update. *Canadian Urology Association Journal, 4*(5), 306–309.

Zinzow, H., Resnick, H., Amstadter, A., McCauley, J., Ruggiero, K., & Kilpatrick, D. (2010). Drug- and alcohol-facilitated, incapacitated, and forcible rape in relation to mental health among a national sample of women. *Journal of Interpersonal Violence, 25*(12), 2217–2236.

Zoler, M. (2010). U.S. newborn male circumcision rate dropped sharply. *Elsevier Global Medical News*.

Zukerman, Z., Weiss, D., Orvieto, R. (2003). Does preejaculatory penile secretion originating from Cowper's gland contain sperm? *Assisted Reproductive Genetics, 20*(4), 157–159.

Zukerman, Z., Weiss, D., & Orvieto, R. (2003). Does preejaculatory penile secretion originating from Cowper's gland contain sperm? *Journal of Assisted Reproductive Genetics, 20*, 157–159.

Zurbriggen, E., & Yost, M. (2004). Power, desire, and pleasure in sexual fantasies. *Journal of Sex Research, 41*, 288–300.

Credits

Photographs and Cartoons

Frontmatter p. xxv Roger R. Hock, Ph.D.

Chapter 1 Page 1 Anson 0618/Shutterstock; p. 11 Photos 12/Alamy; p. 12 Michael Reynolds/EPA/Newscom; p. 15 (top) Courtesy of Stevie Crecelius; p. 15 (bottom) Helen H. Richardson/Getty Images; p. 17 (left) Bettmann/Corbis; p. 17 (right) Fox Searchlight/Everett Collection; p. 20 Marka/Alamy; p. 21 Pearson Education; p. 23 Bettmann/Corbis; p. 24 Ocean/Corbis; p. 29 Gary Conner/PhotoEdit.

Chapter 2 Page 33 (left) Jim Dowdalls/Getty Images; p. 33 (right) Jim Dowdalls/Getty Images; p. 36 Library of Congress Prints and Photographs Division; p. 37 (bottom left) Custom Medical Stock Photo; p. 37 (bottom middle) John Henderson/Alamy; p. 37 (bottom right) Ansell Horn/Newscom; p. 38 (top) Blickwinkel/Alamy; p. 40 Prisma Archivo/Alamy; p. 42 (top) Publiphoto/Science Source; p. 43 Science Photo Library/Alamy; p. 45 John Henderson/Alamy; p. 46 Robin Lazarus/Custom Medical Stock Photo; p. 49 (top left) Daniel Sambraus/Science Source; p. 49 (top middle) University of Oregon Library; p. 49 (top right) Wolfgang Weinhaupl/Science Source; p. 51 Friedrich Stark/Alamy; p. 57 (top left) Jules Selmes and Debi Treloar/Dorling Kindersley; p. 57 (top right) Jules Selmes and Debi Treloar/Dorling Kindersley; p. 57 (top left) Jules Selmes and Debi Treloar/Dorling Kindersley; p. 57 Dr. P. Marazzi/Science Source; p. 57 (bottom) Ethan Miller/Thomson Reuters; p. 60 (top left) All Over photography/Alamy; p. 60 (top right) Science Photo Library/Custom Medical Stock Photo; p. 61 (bottom left) Science Photo Library/Custom Medical Stock Photo; p. 61 (bottom right) Dr. P. Marazzi/Science Source; p. 62 CNRI/Science Source.

Chapter 3 Page 71 Bikerider London/Shutterstock; p. 74 Doug Menuez/Photodisc/Getty Images; p. 76 (top left) Bettmann/Corbis; p. 76 (top right) Michael Desmond/Showtime/Everett Collection; p. 78 Werner Forman/Art Resource, NY; p. 90 Bernard Gotfryd/Premium Archive/Getty Images.

Chapter 4 Page 98 Image Source/Getty Images; p. 101(top) The Colonial Williamsburg Foundation; p. 101(bottom) Queerstock/Alamy; p. 103 YAY Media AS/Alamy; p. 104 (top) Pictorial Press Ltd/Alamy; p. 104 (bottom) Inga Marchuk/Shutterstock; p. 105 (bottom) Kablonk!/Golden Pixels/Alamy; p. 106 (top) Xi Xin Xing/Shutterstock.com; p. 108 (top) Moodboard/Corbis; p. 114 (top) Bridgeman-Giraudon/Art Resource, NY; p. 114 (middle) Michael Jung/Shutterstock; p. 114 (bottom) ABM/Corbis; p. 116 (top) Bikeriderlondon/Shutterstock; p. 116 (bottom) Bloom image/Getty Images; p. 118 (top) Supri Suharjoto/Shutterstock; p. 118 (bottom) Superstock; p. 119 Ana Blazic Pavlovic/Fotolia; p. 123 Catchlight Visual Services/Alamy; p. 127 (top) Merzzie/Shutterstock; p. 127 (bottom) Zdorov Kirill Vladimirovich/Shutterstock; p. 131 Photodisc/Getty Images; p. 139 Nyul/Fotolia.

Chapter 5 Page 142 Charles Thatcher/The Image Bank/Getty Images; p. 145 Bettmann/Corbis; p. 146 Bettmann/Corbis; p. 154 (top) Jonathan A. Meyers/Science Source; p. 154 (bottom) Pichi Chuang/Reuters/Landov; p. 155 Courtesy of Origamicondoms.com; p. 156 David J. Green/Alamy; p. 157 (top left, right) Richard Megna/Fundamental Photographs; p. 158 Scott Camazine/Science Source; p. 161 Ted Morrison/Getty Images; p. 166 (top Phanie/Science Source; p. 166 (bottom) Siu Biomed Comm Custom Medical Stock Photo/Newscom; p. 167 Tomasz Trojanowski/Shutterstock; p. 168 Foto Begsteiger/Vario Images/Alamy; p. 170 Kristoffer Tripplaar/Alamy; p. 171 (top) Jules Selmes and Debi Treloar/DK Images; p. 171 (bottom) Courtesy of FemCap; p. 175 Pearson Education; p. 179 Feature Photo Service/Newscom.

Chapter 6 Page 184 Piotr Marcinski/Alamy; p. 187 Piotr Marcinski/Alamy; p. 195 Mathew Sturtevant/Alamy; p. 197 Mary Evans Picture Library/Alamy; p. 200 Piotr Marcinski/Fotolia; p. 201 Ruth Jenkinson/Dorling Kindersley; p. 202 PhotoStock-Israel/Getty Images; p. 203 B2M Productions/Getty Images.

Chapter 7 Page 218 Kelvin Murray/Getty Images; p. 225 Image Source/Getty Images; p. 227 Wavebreak Media ltd/Alamy; p. 231 Radius Images/Alamy; p. 239 Robert Mankoff/Bios/The New Yorker/The Cartoon Bank; p. 240 Robert Mankoff/Bios/The New Yorker/The Cartoon Bank.

Chapter 8 Page 260 Science Photo Library/Custom Medical Stock Photo; p. 262 Paul J. Richards Agence France Presse/Newscom; p. 264 Barbara Smaller/The New Yorker/The Cartoon Bank; p. 268 Michael Stevens/AGE Fotostock; p. 274 (left) Biophoto Associates/Science Source; p. 274 (right) Custom Medical Stock Photo; p. 277 (left) Dr. P. Marazzi/Science Source; p. 277 (right) DR P. Marazzi/Science Photo Library; p. 281 Medical-on-Line/Alamy; p. 282 Eightfish/Alamy; p. 287 DCPhoto/Alamy; p. 296 (left) Dr. P. Marazzi/Science Source; p. 296 (right) Sebastian Kaulitzki/Alamy; p. 298 Centers for Disease Control and Prevention; p. 300 (top) Chris Bjornberg/Science source; p. 300 (bottom left) SPL/Science Source; p. 300 (bottom right) CNRI/Science Source; p. 301 Lester V. Bergman/Encyclopedia; p. 303 David M. Phillips/The Population Council/Science Source; p. 304 E. Gray/Science Source; p. 307 Noel Hendrickson/Photodisc/Getty Images; p. 309 John Nebraska Design; p. 311 BSIP SA/Alamy.

Chapter 9 Page 314 Stockbyte/Getty Images; p. 317 Interphoto/Alamy; p. 322 From Tulandi's Atlas of Laparoscopy and Hysteroscopy Technique, W.B. Saunders, London, 1999 (Reproduced with permission); p. 324 (bottom left) CNRI/Science Source; p. 324 (bottom right) Science Pictures Ltd/Science Source; p. 325 Pascal Goetgheluck/Science Source; p. 327 (top left) Mehau Kulyk/Science Source; p. 327 (top middle) Mopic/Alamy; p. 327 (top right) BSIP SA/Alamy; p. 327 (bottom) Petit Format/Science Source; p. 327 (bottom) Sebastian Kaulitzki/Shutterstock; p. 328 (top) Biophoto Associates/Science Source; p. 328 (middle) Chris Downie/Getty Images; p. 328 (bottom) Nucleus Medical Art, Inc./Alamy; p. 331 All Over Photography/Alamy; p. 335 Dr. Najeeb Layyous/Science Source; p. 344 (top, bottom,) Michele Davidson/Pearson; p. 348 Illini Hospital Operated by Genesis Health System; p. 349 Andrea Quillien/Alamy; p. 352 Golden Pixels LLC/Alamy; p. 354 (top) Select-Stock/Getty Images; p. 354 (bottom) B2M Productions/Getty Images; p. 360 (top) 67photo/Alamy; p. 360 (bottom) Mauro Fermariello/Science Source; p. 361 Courtesy of Roger Hock.

Chapter 10 Page 365 Rob Walls/Alamy; p. 368 (top, bottom) Bettmann/Corbis; p. 369 (top left) Helga Esteb/Shutterstock; p. 369 (top right) AF archive/Alamy; p. 372 Reuters/Corbis; p. 375 Barbara Smaller/The New Yorker Collection/The Cartoon Bank; p. 376 Artiga Photo/Masterfile; p. 377 (top) Horizon International Images Limited/Alamy; p. 377 (bottom) Maskot/Alamy; p. 380 Kevan Brooks/AdMedia/Newscom; p. 382 Joerg Carstensen/EPA/Newscom; p. 383 (top, bottom) Courtesy of Gary J. Alter, M.D.; p. 384 (bottom) Steve Meddle/Presselect/Alamy; p. 384 (bottom) AP Images/Jason Franson; p. 387 Philip Game/Alamy; p. 390 Big Cheese Photo/SuperStock; p. 391 Donald Reilly/The New Yorker Collection/The Cartoon Bank; p. 395 (top) Andia/Alamy; p. 395 (bottom) Blend Images/Alamy; p. 396 Philip Bigg/Alamy; p. 397 Jupiter Images/Getty Images.

Chapter 11 Page 403 Image Source/Images.com; p. 406 Tracey Whitefoot/Alamy; p. 411 ArrowStudio, LLC/Shutterstock; p. 417 Imagestate Media Partners Limited - Impact Photos/Alamy; p. 419 (top) Andreas

Gradin/Shutterstock; p. 419 (bottom) Kablonk! RM/Golden Pixels LLC/Alamy; p. 422 David Longendyke/Everett Collection/Alamy; p. 424 Maskot/Alamy; p. 428 Mike Stewart/Sygma/Corbis; p. 430 Erich Schlegel/Newscom.

Chapter 12 Page 437 Stockbyte/Getty Images; p. 440 (top) Pictorial Press Ltd/Alamy; p. 440 (bottom left) Tony Freeman/PhotoEdit; p. 440 (bottom right) WavebreakMediaMicro/Fotolia LLC; p. 444 Petro Feketa/Fotolia; p. 451 Mode Images/Alamy; p. 454 Kelly Boreson/Getty Images; p. 456 Elie Bernager/Photodisc/Getty Images; p. 466 (top) Nick Hanna/Alamy; p. 466 (bottom) Corbis; p. 472 Norbert Schaefer/Corbis; p. 475 Fotolia.

Chapter 13 Page 479 Tim Gainey/Alamy; p. 481 Ian Thraves/Alamy; p. 487 Jim Arbogast/Digital Vision/Getty Images; p. 488 Natalie Jezzard/Alamy; p. 491 Philip Baran/Alamy; p. 493 (left, right) David Hoffman Photo Library/Alamy; p. 497 MC Images/Alamy; p. 507 Susan Burrell/Alamy; p. 512 Jules Selmes/Pearson Education; p. 518 Andrew Drysdale/Rex Features/Presselect/Alamy; p. 519 Doug Steley/Alamy.

Chapter 14 Page 523 Image Source/Corbis; p. 533 Cultura Limited/SuperStock; p. 536 (top) Doram/Getty Images; p. 536 (bottom) Masaaki Toyoura/Getty Images; p. 537 Bob Zahn/The Cartoon Bank; p. 538 (top) Digital Vision/Getty Images; p. 538 (bottom) The Marquis de Sade (1740–1814) in Prison (engraving), French School, (19th century)/Private Collection/The Bridgeman Art Library; p. 539 DK Images; p. 543 PBNJ Productions/Blend Images/Getty Images; p. 545 Masson/Fotolia; p. 546 (left) AGE Fotostock/SuperStock; p. 546 (middle) AISPIX/Shutterstock; p. 546 (right) Lisa F. Young/Alamy; p. 546 (bottom) ZUMA Press, Inc./Alamy; p. 547 Shutterstock.

Chapter 15 Page 553 PSL Images/Alamy; p. 556 (top) Julian Money-Kyrle/Alamy; p. 556 (bottom) Everett Collection Historical/Alamy; p. 557 Vibrant Image Studio/Shutterstock; p. 558 Terry J Alcorn/Getty Images; p. 559 Ximena Griscti/Alamy; p. 560 (top) John Van Hasselt/Corbis; p. 560 (bottom) Bob Jones/Alamy; p. 564 Dan Vincent/Alamy; p. 565 Patrick Durand/Sygma/Corbis; p. 566 Seth Resnick/Passage/Corbis; p. 567 John Van Hasselt/Corbis; p. 571 Yadid Levy/Alamy; p. 575 John Rous/AP Images; p. 576 Time & Life Pictures/Getty Images; p. 577 Mario Anzuoni/Reuters; p. 578 Agencja Fotograficzna/Alamy; p. 579 (left) Mocker/123 RF; p. 579 (middle) Jean-Claude Marlaud/Getty Images; p. 579 (right) Uwe Krejci/Corbis; p. 585 Laurence Dutton/Getty Images; p. 587 Oleksiy Maksymenko/Alamy; p. 591 Phillip Bond/Alamy.

Figures and Tables

Chapter 1 p. 5 Maya, first-year student; p. 5 Rick, sophomore; p. 7 Fischer, G. (1986). College students' attitudes toward forcible date rape: Changes after taking a human sexuality course. Journal of Sex Education and Therapy, 12, 42–46; Flores, S., & Hartlaub, M. (1998). Reducing rapemyth acceptance in male college students: A metaanalysis of intervention studies. Journal of College Student Development, 39, 438–448.; Patton, W., & Mannison, M. (1994). Investigating attitudes towards sexuality: Two methodologies. Journal of Sex Education and Therapy, 20, 185–197; Table 1.2 Boonstra, H. (2009). Advocates call for a new approach after the era of "abstinence-only" sex education. Guttmacher Policy Review, 12(1). Retrieved from http://www.guttmacher.org/pubs/gpr/12/1/gpr120106.html; p. 13 Klein, M. (1993, May). Am I sexually normal? New Woman, (5), 49–52; Table 1.3 Based on data from King, B., Parisi, B., & O'Dwyer, K. (1993). College sexuality education promotes future discussions about sexuality between former students and their children. Journal of Sex Education and Therapy, 19, 285–293; p. 15 Ashley Jennings, Man admitted to Hospital for Kidney Stone, Discover He's a Woman, May 26, 2012; Table 1.6 Adapted from Clement, U. (1990). Surveys of heterosexual behavior. Annual Review of Sex Research, 1, 45–74; Morokoff, P. (1986). Volunteer bias in the psychophysiological study of female sexuality. Journal of Sex Research, 22, 35–51; Strassberg, D. S., & Lowe, K. (1995). Volunteer bias in sexuality research. Archives of Sexual Behavior, 24, 369–382.; Wiederman, M. (1993). Demographic and sexual characteristics of nonresponders to sexual experience items in a national survey. Journal of Sex Research, 30, 27–35; and Wol-chik, S., Spencer, S. L., & Lisi, L (1983). Volunteer bias in research employing vaginal measures of sexual arousal. Archives of Sexual Behavior, 12, 399–408; p.23 Masters, W. H., & Johnson, V. E. (1966, p. 4). Human sexual response. Boston: Little, Brown; p. 23: Masters, W. H., & Johnson, V. E. (1966, pp. 22–23). Human sexual response. Boston: Little, Brown.

Chapter 2 Page 36 Leonardo da Vinci, Leonardo on the Human Body (New York: Schuman, 1952); pp. 39–40 Policy Statement (2012) Circumcision policy statement. Pediatrics, 130(3), 585–586. doi: 10.1542/peds.2012-1989; Table 2.1 Eisenman, R. (2001). Penis size: Survey of female perceptions of sexual satisfaction. BMC Women's Health, 1. Doi: 10.1186/1472-6874-1-1; Masters, W. H., & Johnson, V. E. (1966). Human sexual response. Boston: Little, Brown; McGreal, S. (October 16, 2012). The pseudoscience of race differences in penis size. Retrieved from, www.psychologytoday.com/blog/unique-everybody-else/201210/the-pseudoscience-race-differences-in-penis-size; Shah, N. Christopher J. (2002). Can shoe size predict penile length? BJU International, 90 (6), 586–587; Wessells, H., LuePenile, T. (1996). Length in the flaccid and erect states: Guidelines for penile augmentation. The Journal of Urology 156, 995–997.; Wylie, K. & Eardley, I. (2007). Penile size and the 'small penis syndrome.' BJU International, 99 (6), 1449–1455; p. 50 Williamson & Nowak, 1998, The truth about women: Anatomical studies show there is far more to the clitoris than anyone guessed, New scientist. no. 2145, (1998): 34; p. 51 Cruellest cut outlawed (female circumcision abolished). (2007). New Scientist, 194, 7; p. 52 United Nations Children's Fund. http://www.unfpa.org/webdav/site/global/shared/documents/news/2012/FGM-C_highlights.pdf; Table 2.2 American Cancer Society, Breast Cancer Survival Rates by Stage, 2013; p. 60 American Cancer Society. (2013b). Can breast cancer be found early? Retrieved from www.cancer.org/cancer/breast-cancer/detailedguide/breast-cancer-detection.

Chapter 3 Page 74 Masters, W. H., & Johnson, V. E. (1966, p. 21). Human sexual response. Boston: Little, Brown; Figure 3.1 Adapted from Masters, W. H., & Johnson, V. E. (1966). Human sexual response. Boston: Little, Brown; Figure 3.9 Based on data in Mah, K., & Binik, Y. (2002), p. 110. Do all orgasms feel alike? Evaluating a two-dimensional model of the orgasm experience across gender and sexual context. Journal of Sex Research, 39, 104–113; p. 84 Whipple, B. (1999, November 4). Beyond the G-spot: Where do we go from here? Address delivered at the joint annual conference of the American Association of Sex Educators, Counselors, and Therapists and the Society for the Scientific Study of Sexuality, St. Louis, MO; p. 85 Kilchevsky et al., 2012, p. 719. Is the Female G-Spot Truly a Distinct Anatomic Entity?, The Journal of Sexual Medicine, Vol. 9 (3), pages 719–726; p. 86 Giles, G., Severi, G., English, D., McCredie, M., Borland, R., Boyle, P. & Hopper, J. (2003). Sexual factors and prostate cancer. British Journal of Urology International, 92, 211–216; "His and Hers" (2001); Komisaruk, B. R., & Whipple, B. (1995). The suppression of pain by genital stimulation in females. Annual Review of Sex Research, 6, 151–186.; Resnick, S. (2002). Sexual pleasure: The next frontier in the study of sexuality. SEICUS Report, 30, 6–11; Komisaruk, B., Beyer-Flores, C., & Whipple, B. (2006). The science of orgasm. Baltimore, MD: Johns Hopkins University Press; Levin, R. (2007). Sexual activity, health and well-being-the beneficial roles of coitus and masturbation. Sexual & Relationship Therapy, 22(1), 135–148. doi: 10.1080/14681990601149l; Smith, G., Frankel, S., & Yarnell, J. (1997). Sex and death: Are they related? Findings from the Caerphilly cohort study. British Medical Journal, 315, 1641–1645.; Weeks, R., & James, J. (1999). Secrets of the superyoung: The scientific reasons some people look ten years younger than they really are-and how you can, too. New York: Berkley.; Whipple, B. (2000). Beyond the G-spot. Scandinavian Journal of Sexology, 3, 35–42; Table 3.1 Adapted from Hock (2012); Table 3.2 Adapted from Levin, R. (2005). Sexual arousal-its physiological roles in human reproduction. Annual Review of Sex Research, 16, 154–189; Table 3.3 Based on Bancroft, J., Graham, C. A.,

Janssen, E. and Sanders, S. (2009). 'The Dual Control Model: Current Status and Future Directions', Journal of Sex Research, 46: 121–42; p. 93 Tiefer, L. (2001), p. 93. A new view of women's sexual problems: Why new? Why now? Journal of Sex Research, 38, 89–96; p. 93 "FSD Alert," 2008; p. 94 Tiefer, L. (2001), p. 94. A new view of women's sexual problems: Why new? Why now? Journal of Sex Research, 38, 89–96.

Chapter 4 Page 104 Forsterling, F., Preikschas, S., & Agthe, M. (2007). Ability, luck, and looks: An evolutionary look at achievement ascriptions and the sexual attribution bias. Journal of Personality and Social Psychology, 92, 775–788; Ha, T., van den Berg, Engels, R., & Lichtwarck-Aschoff, A. (2012). Effects of attractiveness and status in dating desire in homosexual and heterosexual men and women. Archives of Sexual Behavior, 41, 673–682.; Regan, P., & Medina, R. (2001). Partner preferences among homosexual men and women: What is desirable in a sex partner is not necessarily desirable in a romantic partner. Social Behavior & Personality: An International Journal, 29(7), 625–633; Wood, D., & Brumbaugh, C. (2009). Using revealed mate preferences to evaluate market force and differential preference explanations for mate selection. Journal of Personality & Social Psychology, 96(6), 1226–1244; p. 107 Murray, S. L., Holmes, J. G., Bellavia, G., Griffin, D. W., & Dolderman, D. (2002), p. 563. Kindred spirits? The benefits of egocentrism in close relationships. Journal of Personality and Social Psychology, 82, 563–581; p. 108 Martin, W. E. (2001, September–October). A wink and a smile: How men and women respond to flirting. Psychology Today, pp. 26–27; Perper, T. (1986). Sex signals: The biology of love. Philadelphia: ISI Press; Perper, T., & Fox, V. S. (1980, June). Flirtation and pickup patters in bars. Paper presented at the Eastern Conference on Reproductive Behavior, Saratoga Springs, NY; Figure 4.1 Figure from The Triangle of Love by Robert J. Sternberg. Copyright © 1998. Published by Basic Books. Reprinted by permission of Dr. Robert J. Sternberg; p. 112–113 Scale, "The Triangle of Your Love" from The Triangle of Love by Robert J. Sternberg. Copyright © 1998. Published by Basic Books. Reprinted by permission of Dr. Robert J. Sternberg.; Figure 4.2 Based on student analysis; p. 115 Adapted from Hendrick, C., & Hendrick, S. (1986). A theory and method of love. Journal of Personality and Social Psychology, 50, 392–40; pp 121–122 Adapted from Markman, H., Stanley, S., & Bloomberg, S. (2010, pp. 21–28). Fighting for your marriage. Hoboken, NJ: Josey-Bass; Table 4.2 Adapted from Notarius, C., & Markman, H.J. (1993). We can work it out: Making sense of marital conflict. New York: Putnam; p. 126 Adapted from Michigan State University Counseling Center (2003). Sex and relationships: Establishing guidelines for decisions about sexual expression. Retrieved from www.couns.msu.edu/self-help/sexguide.htm; p. 130, Utz, S., & Beukeboom, C. (2011). The role of social network sites in romantic relationships: Effects on jealousy and relationship happiness. Journal of Computer-Mediated Communication, 16(4), 511–527. doi: 10.1111/j.1083-6101.2011.01552.x; Figure 4.3 "Power & Control In Dating Relationships" and "Lesbian/Gay Power And Control Wheel" as published by Domestic Abuse Intervention Programs. Reprinted by permission of DAIP, 202 E. Superior Street, Duluth, MN 55802; p. 136 Adapted from publications of Project Sanctuary, Ukiah and Fort Bragg, California; p. 137 Dugan & Hock, 2009, Neu anfangen nach einer Misshandlungsbeziehung (Bern: Huber, 2009); Table 4.3 Adapted from The Project for Victims of Family Violence, Inc. Reprinted by permission of The Project for Victims of Family Violence, Inc.; p. 139 Adapted from Dugan, M. K., and Hock, R. R. (2006, p. 242). It's my life now: Starting over after an abusive relationship or domestic violence (2d ed.) New York: Routledge.

Chapter 5 Page 148 From Contraceptive Technology 18th edition, R. Hatcher, J. Trussel, F. Stewart, W. Cates, G. Stewart, F. Guest & D. Kowal (Eds.). Copyright © 2004 by CTC, Inc. Reprinted by permission of the publisher, Ardent Media, Inc.; Table 5.2 Based on Hatcher, R., Trussell, J., Nelson, A., Cates, W., & Kowal, D. (Eds). (2011). Figure 3-1, p. 52. Contraceptive Technology, Twentieth Revised Edition. New York: Ardent Media; Table 5.3 Chan, C., Harting, M., & Rosen, T. (2009). Systemic and barrier contraceptives for the dermatologist: A review. International Journal of Dermatology, 48(8), 795–814, doi: 10.1111/j.13654632.2009.04148.x; Corinna, H. (2010).what's the typical use-effectiveness rate of abstinence? RH Reality Check, Retrieved from http://rhrealitycheck.org/article/2010/03/05/whats-typical-useeffectiveness-rate-abstinence.; Hatcher, R., Trussell, J., Nelson, A., Cates, W., & Kowal, D. (Eds). (2011). Figure 3-1, p. 52. Contraceptive Technology, Twentieth Revised Edition. New York: Ardent Media; Pallone, S., & Burgess, G. (2009). Fertility awareness based methods: Another option for family planning. The Journal of the American Board of Family Medicine, 22(2), 147–157. doi: 10.3122/jabfm.2009.02.080038; p. 153 Adapted from Kowall, D. (2008). Abstinence and the range of sexual expression. In A. Nelson, W. Cates Jr., F. Stewart, & D. Kowal (Eds.), Contraceptive Technology, 19th Revised Edition. New York: Ardent Media; p. 155 Grand Challenges Explorations Grants. (2013). Grand Challenges in Global Health. Retrieved from www.grandchallenges.org/explorations/Pages/grantsawarded.aspx?Topic=Contraception&Round=all&Phase=all; p. 157 Adapted from Warner & Steiner, (2008). Male condoms. In Contraceptive Technology 18th Revised Edition, R. Hatcher, J. Trussel, F. Stewart, W. Cates, G. Stewart, F. Guest, & D. Kowal (Eds.), (pp. 297–316). Reprinted by permission of the publisher, Ardent Media, Inc; pp. 163–164 www.plannedparenthood.org (2010).; Hatcher, R., Trussell, J., Nelson, A., Cates, W., & Kowal, D. (Eds). (2011). Contraceptive Technology, Twentieth Revised Edition. New York: Ardent Media; p. 165 Adapted from Hatcher, R. A., Trussell, J., Nelson, A., Cates, W., Stewart, F., & Kowal, D. (2008). Contraceptive Technology, 19th rev. ed. New York: Ardent Media, pp. 254–255; Figure 5.8 Based on Institute for Reproductive Health (2008). Research to practice: The Two-Day Method. Retrieved from www.irh.org/RTP-TDM.htm; p. 181 Guttmacher Institute, Facts on Induced Abortion in the United States, December 2013. Retrieved from www.guttmacher.org/pubs/fb_induced_abortion.html.

Chapter 6 Page 187 Walling, W. H. (1904, pp. 38–39). Sexology, Philadelphia: Puritan; p. 187 Walling, W. H. (1904, pp. 42, 46). Sexology, Philadelphia: Puritan; Table 6.1 Data from Barak, A., Fisher, W. A., Belfry, S., & Lashambe, D. (1999). Sex, guys, and cyberspace: Effects of Internet pornography and individual differences on men's attitudes toward women. Journal of Psychology and Human Sexuality, 11, 63–91; Byers, E. S., Purdon, C., & Clark, D. A. (1998). Sexual intrusive thoughts of college students. Journal of Sex Research, 35, 359–369; Fisher, W. A., Byrne, D., White, L. A., & Kelley, K. (1988). Erotophobia-crotophilia as a dimension of personality. Journal of Sex Research, 25, 123–151; Forbes, G., Adams-Curtis, L., Hamm, N., & White, K. (2003). Perceptions of the woman who breastfeeds: The role of crotophobia, sexism, and attitudinal variables. Sex Roles, 49, 379–388; Garcia, L. (1999). The certainty of the sexual self-concept. Canadian Journal of Human Sexuality, 8, 263–270; Geer, J., & Robertson, G. (2005). Implicit attitudes in sexuality: Gender differences. Archives of Sexual Behavior, 34(6), 671–677. doi: 10.1007/s10508-005-7923-8; Humphreys, T., & Newby, J. (2007). Initiating new sexual behaviours in heterosexual relationships. Canadian Journal of Human Sexuality, 16, 77–88; p. 190 Janssen, E., Vorst, H., Finn, P., & Bancroft, J. (2002).p. 126. The Sexual Inhibition (SIS) and Sexual Excitation (SES) Scales: I. Measuring sexual inhibition and excitation proneness in men. Journal of Sex Research, 39, 114–126; Table 6.2 Data adapted from: D. Herbenick, et al. (2010). National Survey of Sexual Health and Behavior, Indiana University, 261–263; Table 6.4 Alfonso, V., Allison, D., & Dunn, G. (1992). Sexual fantasy and satisfaction: A multidimensional analysis of gender differences. Journal of Psychology and Human Sexuality, 5, 19–37; Byers, E. S., Purdon, C., & Clark, D. A. (1998). Sexual intrusive thoughts of college students. Journal of Sex Research, 35, 359–369; Carpenter, D., Janssen, E., Graham, C., Vorst, H., & Wicherts, J. (2008). Women's scores on the Sexual Inhibition/Sexual Excitation Scales (SIS/SES): Gender similarities and differences (Report). Journal of Sex Research, 45, 36–48; Cato, S., & Leitenberg, H. (1990). Guilt reactions to sexual fantasies during intercourse. Archives of Sexual Behavior, 19, 49–63; Davidson, K., & Hoffman, L. (1986). Sexual fantasies and sexual satisfaction: An empirical analysis of erotic thought. Journal of Sex Research, 22, 184–205; Leitenberg, H., & Henning, K. (1995). Sexual fantasy. Psychological Bulletin, 117, 469–496; Masters, W. H., Johnson, V. E., & Kolodny, R. C. (1995). Human sexuality (5th ed.). New York: HarperCollins; Reinisch, J. M. (1990). The Kinsey Institute new report on sex: What you must know to be sexually literate. New York: St. Martin's Press; Renaud, C. A., & Byers, S. E.

(1999). Exploring the frequency, diversity, and content of university students' positive and negative sexual cognitions. Canadian Journal of Human Sexuality, 8, 17–31; Striar, S., & Bartlik, B. (1999). Stimulation of the libido: The use of erotica in sex therapy. Psychiatric Annals, 29, 60–62; Wilson, G. (2010). Measurement of sex fantasy. Sexual & Relationship Therpay, 25(1), 57–67. doi: 10.1080/14681990903550134; Zurbriggen, E., & Yost, M. (2004). Power, desire, and pleasure in sexual fantasies. Journal of Sex Research, 41, 288–300; Table 6.5 Adapted from Hsu, B., Kling, A., Kessler, C., Knapke, K., Diefenbach, P., & Elias, J. E. (1994). Gender differences in sexual fantasy and behavior in a college population: A ten-year replication. Journal of Sex & Marital Therapy, 20(2), 103–11; p. 197 Popkin, J. (1994, December 19). A case of too much candor. U.S. News and World Report, p.31; Table 6.6 Davidson, J., & Moore, N. (1994). Masturbation and premarital sexual intercourse among college women: Making choices for sexual fulfillment. Journal of Sex and Marital Therapy, 20, 179–199; Kay, D. S. G. (1992). Masturbation and mental health: Uses and abuses. Sexual and Marital Therapy, 7, 97–107; Kelly, M., Strassberg, D., & Kircher, J. (1990). Attitudinal and experimental correlates of anorgasmia. Archives of Sexual Behavior, 19, 165–177; Levin, R. (2007). Sexual activity, health and well-being-the beneficial roles of coitus and masturbation. Sexual & Relationship Therapy, 22(1), 135–148. doi: 10.1080/146819906011491; Mahoney, S. (2006, February). How love keeps you healthy. Prevention, 58(2), 164–213; Masters, W. H., & Johnson, V. E. (1974). The pleasure bond. Boston: Little, Brown; "The Politics of Masturbation" (1994). Lancet, 344, 1714–1715; Tiefer, L. (1998). Masturbation: Beyond caution, complacency, and contradiction. Sexual and Marital Therapy, 13, 9–14; p. 206 Hite, S. (1976, p. 234). The Hite report. New York: Macmillan.; p. 206 Hite, S. (1981, p. 539). The Hite report on male sexuality. New York: Random House; Table 6.7 Adapted from: Reece, M., Herbenick, D., Schick, V., Sanders, S. A., Dodge, B., & Fortenberry, J. D. (2010), Table 5, p. 299. Sexual behaviors, relationships, and perceived health among adult men in the United States: Results from a national probability sample findings from the National Survey of Sexual Health and Behavior (NSSHB). The Journal of Sexual Medicine, 7 (supplement 5 impact factor: 4.884), 291–304; Herbenick, D., Reece, M., Schick, V., Sanders, S. A., Dodge, B., & Fortenberry, J. D. (2010), Table 5, p. 285. Sexual behaviors, relationships, and perceived health status among adult women in the United States: Results from a national probability sample. 291 findings from the National Survey of Sexual Health and Behavior (NSSHB). The Journal of Sexual Medicine, 7 (supplement 5 impact factor: 4.884), 277–290; Table 6.8 Adapted from the Durex Corp. (2005). Global Sex Survey, 2005. Retrieved from www.durex.com/us/gss2005results.asp.

Chapter 7 Table 7.1 American Psychiatric Association. (2013). Diagnostic and statistical manual of mental disorders (5th ed.). Arlington, VA: American Psychiatric Publishing; Table 7.2 Data from Balon, R., & Seagraves, R. (2009). Clinical manual of sexual disorders. Washington, DC: American Psychiatric Publishing; Hayes, R., Dennerstein, L., Bennett, C., & Fairley, C. (2008). Epidemiology: What is the "true" prevalence of female sexual dysfunctions and does the way we assess these conditions have an impact? [Original research]. The Journal of Sexual Medicine, 5(4), 777–787; Laumann, E., Gagnon, J., Michael, R., & Michaels, S. (1994). The social organization of sexuality: Sexual practices in the United States. Chicago: University of Chicago Press; Mischianu & Pemberton (2007); Montorsi, F. (2005). Prevalence of premature ejaculation: A global and regional perspective. Journal of Sexual Medicine, 2(S-2), 96–102; Saigal, C., Wessells, H., Pace, J., Schonlau, M., Wilt, T. (2006). Predictors and prevalence of erectile dysfunction in a racially diverse population: Urologic diseases in America Project. Archives of Internal Medicine, 166, 207–212; Townsend, M. C. (2006). Psychiatric mental health nursing: concepts of care in evidence-based practice. Philadelphia, PA: E.A. Davis Company; Reissing, Laliberte, & Davis (2005); Table7.4 A. (2005). Sexuality and substance use: The impact of tobacco, alcohol, and selected recreational drugs on sexual function. Canadian Journal of Human Sexuality, 14(1/2), 41–56.; Finger, W., Lund, M., & Slagle, M. (1997). Medications that may contribute to sexual disorders. Journal of Family Practice, 44, 33–43; p. 231 Significant Differences for Expectations for Romantic Behavior: University Students in Puer to Rico and the United States (Total Participants = 395); p. 233 Kaplan, H. S. (1974, p. 213). The new sex therapy: Active treatment of sexual dysfunctions. New York: Brunner/Mazel; p. 233 Kaplan, H. S. (1974). The new sex therapy: Active treatment of sexual dysfunctions. New York: Brunner/Mazel; p. 237 DSM-5. (2013). Diagnostic and statistical manual of mental disorders (5th ed.). American Psychiatric Association. Arlington, VA: American Psychiatric Publishing; Table 7.6 "Understanding Male Arousal and Erections," from "Erectile Dysfunction and Inhibited Sexual Desire: Cognitive-Behavioral Strategies," by B. McCarthy, in Journal of Sex Education and Therapy, 18, (1992), pp. 22–34. Copyright © 1992. Reprinted by permission of American Association of Sex Educators, Counselors & Therapists; Table 7.7 Kaplan, H. S. (1974). The new sex therapy: Active treatment of sexual dysfunctions. New York: Brunner/Mazel; p. 245 McCarthy, B. W. (1995, p. 140). Bridges to sexual desire. Journal of Sex Education and Therapy, 21, 132–141; p. 246 McCarthy, B. W. (1995, p. 133). Bridges to sexual desire. Journal of Sex Education and Therapy, 21, 132–141; Table 7.8 "Cognitive-Behavioral Approach to Hypoactive Sexual Desire," by B. McCarthy, in Journal of Sex Education & Therapy, 21, (1995), pp. 132–141. Copyright © 1995. Reprinted by permission of American Association of Sex Educators, Counselors & Therapists; p. 247 Polonsky, D. 2000, p. 310. Premature ejaculation. In S. Leiblum & R. Rosen (Eds.) Principles and practice of sex therapy (pp. 305–332). New York: Guilford Press; Table 7.9 Data from Grenier, G., & Byers, E. S. (1995). Rapid ejaculation: A review of the conceptual, etiological, and treatment issues. Archives of Sexual Behavior, 24, 447–472; Hong, J. (1984). Survival of the fastest: On the origins of premature ejaculation. Journal of Sex Research, 20, 109–112; Masters, W. H., & Johnson, V. E. (1970). Human sexual inadequacy. Boston: Little, Brown; O'Donohue, W., Letourneau, E., & Geer, J. (1993). Premature ejaculation. In W. O' Donohue & J. Geer (Eds.), Handbook of sexual dysfunctions (pp. 303–333). Needham Heights, MA: Allyn & Bacon; Polonsky, D. 2000, Premature ejaculation. In S. Leiblum & R. Rosen (Eds.) Principles and practice of sex therapy (pp. 305–332). New York: Guilford Press; Waldinger, M. D., (2007, August 1). New insights in premature ejaculation (Disease/Disorder overview). Psychiatric Times, 24, 52–54; Wincze, J. & Carey, M. (2012) Sexual dysfunction, second edition: A guide for assessment and treatment. NY: The Guilford Press; Table 7.10: Data from Kaschak, A., & Tiefer, L. (Eds.) (2002). A new view of women's sexual problems. Binghamtom, NY: Haworth Press.; McHugh, M. (2006). What do women want? A new view of women's sexual problems. Sex Roles: A Journal of Research, 54, 361–369; New View Campaign (2008). Challenging the medicalization of sex. Retrieved from www.fsdalert.org; Wood, J., Koch, P., & Mansfield, P. (2006). Women's sexual desire: A feminist critique. Journal of Sex Research, 43, 236–244; "FSD Alert," (2008); Teifer (2010).

Chapter 8 Figure 8.1 CDC – "Reported STDs," (2014). http://www.cdc.gov/nchhstp/newsroom/docs/STD-Trends-508.pdf; pp. 266–267 Troth, T., & Peterson, C. (2000). Factors predicting safe-sex talk and condom use in early sexual relationships. Health Communication, 12, 195–212; p. 277 Broder, J. (2012). Pap tests less frequent under new guidelines, Medscape Medical News. Retrieved from http://www.medscape.com/viewarticle/773282.; National Cancer Institute. (2012). Pap and HPV testing. U.S. Department of Health and Human Services. National Institutes of Health. Retrieved from www.cancer.gov/cancertopics/factsheet/detection/Pap-HPV-testing; Saslow, D., Solomon, D., Lawson, H., Killackey, M., Kulasingam, S., Cain, J...Stoler, M., Schiffman, M., Castle, P., Myers, H...Lawson, K...(2012). American Cancer Society, American Society for Colposcopy and Cervical Pathology, and American Society for Clinical Pathology screening guidelines for the prevention and early detection of cervical cancer. Journal Of Lower Genital Tract Disease, 16(3), 175–204; p. 280 Based on Vamos, C., McDermott, R., & Daley, E. (20 08). The HPV vaccine: Framing the arguments for and against mandatory vaccination of all middle school girls. Journal of School Health, 78, 302–309; Figure 8.5 UNAIDS–Global Report (2013). www.UNAIDS.org. www.unaids.org/en/media/unaids/contentassets/documents/epidemiology/2013/gr2013/UNAIDS_Global_Report_2013_en.pdf; p. 285 UNAIDS – Global Fact Sheet (2013).

www.UNAIDS.org; Table 8.4 CDC, "Diagnoses of HIV Infection," (2013); Figure 8.6 CDC, "Risk Chart." Adapted from www.cdc.gov/hiv/topics/treatment/PIC/pdf/chart.pdf; p. 291 CDC, "HIV Testing," 2010): Centers for Disease Control and Prevention. (2010e). HIV testing basics for consumers. Retrieved from www.cdc.gov/hiv/topics/testing/resources/qa/index.htm; Figure 8.7 NIH,(2012). DrugFacts: HIV/AIDS and Drug Abuse: Intertwined Epidemics. National Institutes of Health, National Institute on Drug Abuse. www.drugabuse.gov/publications/drugfacts/hivaids-drug-abuse-intertwined-epidemics; p. 293 Marchione, M. (2014, March 5). Doctors hope for cure in a 2nd baby with HIV. Associated Press. Retrieved from http://bigstory.ap.org/article/doctors-hope-cure-2nd-baby-born-hiv.; Wilson, J. & Young, S. (2014, March 6). Second baby possibly 'cured' of HIV. CNN. Retrieved from www.cnn.com/2014/03/06/health/hiv-baby-cured/index.html; p. 302 (CDC, "PID," 2010) Centers for Disease Control and Prevention. (2010). Pelvic inflammatory disease (PID)—CDC fact sheet. Retrieved from www.cdc.gov/std/PID/STDFact-PI.

Chapter 9 Page 319 Adapted from "Evaluate Your Parenting Readiness" (2006). www.babycenter.com/0_evaluate-your-parenting-readiness_7311.bc; Table 9.1 "Childless by Choice" (2001, November 1). American Demographics, pp. 44–50; Connidis, I., & McMullin, J. (1999). Permanent childlessness: Perceived advantages and disadvantages. Canadian Journal on Aging, 18, 447–465; Hollander, D. (2007). Women who are fecund but do not wish to have children outnumber the involuntarily childless. Perspectives on Sexual and Reproductive Health, 39, 120; Kopper, B., & Smith, S. (2001). Knowledge and attitudes toward infertility and childless couples. Journal of Applied Social Psychology, 31, 2275–2291; La Mastro, V. (2001). Childless by choice? Attributions and attitudes concerning family size. Social Behavior and Personality, 29, 231–243; Mcquillan, J., Greil, A., Shreffler, K., & Tichenor, V. (2008). The importance of motherhood among women in the contemporary United States. Gender & Society 22, 477–496. doi: 10.1177/0891243208319359; Mollen, D. (2006). Voluntarily childfree women: Experiences and counseling considerations. Journal of Mental Health Counseling, 28, 269–282; Morell, C. (2000). Saying no: Women's experience with reproductive refusal. Feminism & Psychology, 14, 1–13; Table 9.3 Adapted from Chang, 2010; Obstetrics-Gynecology-Infertility Group, 2008. Common discomforts of pregnancy: Safe simple remedies. Retrieved from www.ob-gyn-infertility.com/Common%20discomforts.html; p. 331,: Nordqvist, J. (2013a). What is a miscarriage? What causes a miscarriage? Medical News Today. Retrieved from www.medicalnewstoday.com/articles/262941.php; Figure 9.12 Based on data from Draper et al. (2003); Ross, M. (2007, May 31). Preterm labor. eMedicine. Retrieved from www.emedicine.com/med/topic3245.htm; Tyson, J., Parikh, N., Langer J., Green C., & Higgins, R. (2008). Intensive care for extreme prematurity: Moving beyond gestational age. New England Journal of Medicine, 358, 1672–1681; Table 9.4 Adapted from CDC, "Birth Defects" (2011); March of Dimes Foundation, National Perinatal Statistics (2008). Retrieved from www.marchofdimes.com/aboutus/680_2164.asp; March of Dimes foundation, (2010). Birth defects: What they are and how they happen. Retrieved from www.marchofdimes.com/Baby/birthdefects.html; p. 335 CDC, "FASDs," 2010. Centers for Disease Control and Prevention. (2010). Fetal alcohol spectrum disorders (FASDs). Retrieved from www.cdc.gov/ncbddd/fasd/facts.html; p. 337 "Facts on Induced Abortion," 2011, www.guttmacher.org/pubs/fb_induced_abortion.html; p. 339 Sedgh, G., Singh, S., Shah, I., Ahman, E., Henshaw, S., & Bankole, A. (2012). Induced abortion: incidence and trends worldwide from 1995 to 2008. Lancet, 379(9816), 625–632; Table 9.5 Berer, M. (2005). Medical abortion: Issues of choice and acceptability. Reproductive Health Matters, 13, 25–34; Kapp, N., Whyte, P., Tang, J., Jackson, E., & Brahmi, D. (2013). A review of evidence for safe abortion care. Contraception, 88(3), 350–363. doi: 10.1016/j.contraception.2012.10.027; Healthwise (2004). Choices for an abortion. Retrieved from http://my.webmd.com/hw/womens_conditions; Hollander, D. (2000). Most abortion patients view their experience favourably, but medical abortion gets a higher rating than surgical. Family Planning Perspectives, 32, 264; pp. 342–343 "State Policies." (2014). State facts about abortion: State policies in brief. Guttmacher Institute. Retrieved from www.guttmacher.org/statecenter/spibs/spib_OAL.pdf; p. 350 ACOG. (2013) Cesarean delivery on maternal request: Committee opinion No. 559. American College of Obstetricians and Gynecologists. Obstetrics and Gynecololgy, 121, 904–7; Table 9.7 Adapted from World Health Organization, World health statistics 2010, Part II, Table 4, pp. 92–96 (partial); p. 352–353 Based on D. Moore, 2002, Postpartum depression. Retrieved from http://www.drdonnica.com/display.asp?article=154; Figure 9.19 Adapted from Petrozza, J., & Styer, A. (2006, July 20). Assisted reproduction technology eMedicine. Retrieved from www.emedicine.com/med/topic3288.htm.; p. 360, Data from CDC, "ART," 2011; Eisenberg, E., & Brumbaugh, K. 2009. Infertility Office on Women's Health in the Office of the Assistant Secretary for Health, U.S. Department of Health and Human Services. Womenshealth.gov. Retrieved from http://womenshealth.gov/faq/infertility.cfm; Figure 9.24 CDC. Assisted Reproductive Technology Surveillance, United States, 2006. Figure 2, p. 15.

Chapter 10 Page 378 Data from Witt, S. (2000). The influence of television on childrens' gender role socialization. Childhood Education, 76, 322–324.p. 322; pp. 378–379 Data from Halim, M., Ruble, D., & Amodio, D. (2011). From pink frilly dresses to 'one of the boys': A social-cognitive analysis of gender identity development and gender bias. Social and Personality Psychology Compass, 5(11), 933–949.; Pike, J., & Jennings, N. (2005). The effects of commercials on children's perceptions of gender appropriate toy use. Sex Roles: A Journal of Research, 52, 83–91; p. 381 Coleman, E., Bockting, W., Botzer, M., Cohen-Kettenis, P., DeCuypere, G. … Zucker, K. (2012) World professional association for transgender health (wpath). Standards of care for the health of transsexual, transgender, and gender-nonconforming people. Retrieved from http://www.wpath.org/uploaded_files/140/files/Standards%20of%20Care,%20V7%20Full%20Book.pdf; p. 383 Rosin, H. (2008, November). A boy's life. The Atlantic. Retrieved from www.theatlantic.com/magazine/toc/2008/11; p. 384 Quoted in Luecke, J. (2011). Working with transgender children and their classmates in pre-adolescence: Just be supportive. Journal of LGBT Youth, 8, 116–156., p. 124; Table 10.2 Bureau of Labor Statistics. (2010). Median weekly earnings of full-time wage and salary workers by detailed occupation and sex. Retrieved from ftp://ftp.bls.gov/pub/special.requests/lf/aat39.txt.; Figure 10.5 "Two Types of Aggression by Gender Among 7th Graders," based on data from the Institute for Teaching and Research on Women, 2002, Refer to www.towson.edu/womensstudies; Figure 10.6 Graph, "Boys' and Girls' Aggressive Behavior, Grades 1–7," from "Gender and Aggression: The Baltimore Prevention Study," Institute for Teaching and Research on Women, 2002. Refer to www.towson.edu/womensstudies; p. 390 Hardcastle, M. (2014). how girls bully: The covert tactics used when girls bully About.com. Retrieved from http://teenadvice.about.com/od/violencebullying/a/girlbullies.htm; Figure 10.7 Table, "Gender Differences in Types of Aggression" (p. 1147), from "United States and Indonesian Children's and Adolescents' Reports of Relational Aggression by Disliked Peers," by D. French, E. Jansen, & S. Pidada, Child Development, 73, (2002), pp. 1143–1150. Copyright © 2002 Society for Research in Child Development. Reprinted by permission of the publisher; Figure 10.8 Baldwin, J., & Baldwin, J. (1997). Gender differences in sexual interest. Archives of Sexual Behavior, 26, 181–210, p. 182; Table 10.3 Adapted from Tannen, D. (1991). You just don't understand: Men and women in conversation. New York: Ballantine Books; p. 396 Reddy, G. (2005). With respect to sex: Negotiating Hijra identity in south India. Chicago, IL: University of Chicago Press; p. 396 Baldwin, J., & Baldwin, J. (1997). Gender differences in sexual interest. Archives of Sexual Behavior, 26, 181–210, pp. 183–184; p. 397 Bem, S. (1974, p. 155). The measurement of psychological androgyny. Journal of Consulting and Clinical Psychology, 42, 155–162; Table 10.4 Adapted from Ricciardelli, L., & Williams, R. (1995). Desirable and undesirable traits in three behavioral dimensions. Sex Roles, 33, 637–655, pp. 644–645.

Chapter 11 Figure 11.1 "Kinsey's Sexual Orientation Rating Scale," from Sexual Behavior in the Human Male, by A. Kinsey, W. Pomeroy, & C. Martin. Copyright 1948. Reprinted by permission of The Kinsey Institute for Research in Sex, Gender, and Reproduction, Inc.; p. 408 Morgenstern, J. (2004, May-June, p. 47). Myths of bisexuality. Off Our

Backs, pp. 46–48; p. 409 Adapted from: "Sexual Orientation Scale," fromp. 38 in "Sexual Orientation: A Multi-Variable Dynamic Process," by F. Klein, B. Sepekoff, & T. Wolf, in Two Lives: Bisexuality in Men and Women, edited by F. Klein & T. Wolf. Copyright © 1985. Reprinted by permission of The Haworth Press; p. 410 Boyd, M. (1997, June 24). Male bonding. Advocate, p. 11; Table 11.1 Developed by M. Rochlin, The Language of Sex: The Heterosexual Questionnaire, Changing Men (Spring 1982), Waterloo, ON: University of Waterloo; Figure 11.2 http://gregstoll.dyndns.org/marriagemap/ (Used by permission.); Table 11.2 http://gregstoll.dyndns.org/marriagemap/ (Used by permission.); Table 11.2 Adapted from Gottman, J. M., & Levenson, R. W. (2001). 12-year study of gay and lesbian couples. Gottman Institute. Retrieved from www.gottman.com/research/projects/gaylesbian; p. 416 Savic, I., & Lindstrom, P. (2008, p. 9403). PET and MRI show differences in cerebral asymmetry and functional connectivity between homo- and hetero-sexual subjects. Proceedings of the National Academy of Sciences of the United States 105, 9403–9408; Figure 11.3 Alanko, K., Santtila, P., Harlaar, N., Witting, K., Varjonen, M., Jern, P., Johansson, A., von der Pahlen, B., & Sandnabba, N. (2010). Common genetic effects of gender atypical behavior in childhood and sexual orientation in adulthood: A study of Finnish twins. Archives of Sexual Behavior, 39, 81–92; Bailey, J., & Benishay, D. (1993). Familial aggregation of female sexual orientation. American Journal of Psychiatry, 150, 272–277; Bailey, J., & Pillard, R. (1991). A genetic study of male sexual orientation. Archives of General Psychiatry, 48, 1089–1096; p. 420 Bem, D. J. (2000b, p. 533). Exotic becomes erotic: Interpreting the biological correlates of sexual orientation. Archives of Sexual Behavior, 29, 531–548; Figure 11.4 "Exotic Becomes Erotic: Interpreting the Biological Correlates of Sexual Orientation," by D. Bem, in Archives of Sexual Behavior, 29, (2000), pp. 531–548. © 2000. Reprinted by permission of Springer Science and Business Media, p. 422 LeVay, S. (2010, p. 295). Gay, straight, and the reason why: The science of sexual orientation. Cary, NC: Oxford University Press, USA.; Figure 11.5 CDC, "Sexual Identity" (2011); p. 427 Carr, R., & Grambling, L. (2004). Stigma: A health barrier for women with HIV/AIDS. Journal of the Association of Nurses in AIDS Care, 15, 30–39; Herek, G. M., & Capitanio, J. P., (1999). AIDS stigma and sexual prejudice. American Behavioral Scientist, 42, 1126–1143; Herek, G. M., Widaman, K. F., & Capitanio, J. P. (2005). When sex equals AIDS: Symbolic stigma and heterosexual adults inaccurate beliefs about sexual transmission of AIDS. Social Problems, 52, 15–37; Lopes, S. (2001). AIDS-related stigma. Retrieved from www.asc.upenn.edu; p. 430 Gibbs, N. (2003, July 7). A yea for gays: The Supreme Court scraps sodomy laws, setting off a hot debate. Time, pp. 38–39; Figure 11.7 Federal Bureau of Investigation. (2011). Uniform crime reports. U.S. Department of Justice. Retrieved from www.fbi.gov/about-us/cjis/ucr/ucr#ucr_nibrs; p. 431 Adapted from Sodomy Laws. (2007). Laws around the world. Retrieved from www.glapn.org/sodomylaws/world/world.htm; Table 11.3 Adapted from FBI. (2013). Uniform crime reports. Hate crime statistics, 2012. Retrieved from www.fbi.gov/about-us/cjis/ucr/hate-crime/2012/topic-pages/victims/victims_final, Table 1.

Chapter 12 Table 12.1 Carpenter, L., & DeLamater, J. (2012). Sex for Life: From Virginity to Viagra, how Sexuality Changes Throughout Our Lives. New York: NYU Press.; Friedrich, W., Fisher, J., Broughton, D., Houston, M., & Shafran, C. (1998). Normative sexual behavior in children: A contemporary sample [Electronic version]. Pediatrics, 101, 9–16; Kellogg, N. (2010). Sexual behaviors in children: evaluation and management. American family physician, 82(10), 1233–1238; Table 12.2 Kellogg, N. D. (2010). Sexual behaviors in children: evaluation and management. American family physician, 82(10), 1233–1238; Table 12.3 Table, "Children's Sources of Sex Information," in "Does Source of Sex Information Predict Adolescents' Sexual Knowledge, Attitudes, & Behaviors?" by C. L. Somers and J. H. Gleason, Education, 121, (2001), pp. 674–681. Copyright © 2001. Reprinted by permission of Project Innovation, Mobile, AL, and C. L. Somers; p. 448 "Eleven Guidelines for Talking to Children about Sex," from Families Are Talking: Special Supplement to SIECUS Report 1, 2000. Copyright © 2000 by SIECUS. Reprinted by permission of Sexuality Information and Education Council of the United States (SIECUS); Figure 12.1 From "Impacts of Abstinence Education on Teen Sexual Activity, Risk of Pregnancy and Risk of Sexually Transmitted Diseases," by C. Trenholm et al., Journal of Policy Analysis and Management, Spring 2008, Volume 27, Issue 2, pp. 255–276. Copyright © 2008 by the Association for Public Policy Analysis and Management. Reprinted by permission of APPAM; p. 452 Adapted from Coleman, L., & Coleman, J. (2002). The measurement of puberty: A review. Journal of Adolescence, 25, 535–550; "Sexual Development" (2008). Sexual development and reproduction. U.S. Department of Health & Human Services: 4Parents.gov. Retrieved from http://4parents.gov/sexdevt/index.html; pp. 454–455 Adapted from Washington State Medical Association (2003), "Teen Dating Violence" 2005. Retrieved from www.atg.wa.gov/violence/DVAC_brochure_10_2003.pdf; p. 453 Furman, W. (2002, p. 178). The emerging field of adolescent romantic relationships. Current Directions in Psychological Science, 11, 177–181; p. 453 Data from "5 Steps," 2011; Fisher, E. (2012). Responsive school practices to support lesbian, gay, bisexual, transgender, and questioning students and families NY: Routledge; Table 12.4 Adapted from: CDC, "Adolescent and School Health." (2014). Youth Risk Behavior Surveillance. Centers for Disease Control and Prevention. Retrieved from ccd.cdc.gov/youthonline; p. 457 Goldston, L., & Wong. N. (2003, June 13). Teens talk frankly about sex, themselves. San Jose Mercury News, p. B3; Figure 12.2 Kaiser Family Foundation. (2013). Sexual Health of Adolescents and Young Adults in the United States. http://kaiserfamilyfoundation.files.wordpress.com/2013/04/3040-06.pdf; Figure 12.3 Fortenberry, J., Schick, V., Herbenick, D., Sanders, S., Dodge, B. & Reece, M. (2010). Sexual behaviors and condom use at last vaginal intercourse: A national sample of adolescents ages 14 to 17 years: Findings from the National Survey of Sexual Health and Behavior (NSSHB). The Journal of Sexual Medicine Volume, 7 (S5), 305–314; Figure 12.4 Fortenberry, J., Schick, V., Herbenick, D., Sanders, S., Dodge, B. & Reece, M. (2010). Sexual behaviors and condom use at last vaginal intercourse: A national sample of adolescents ages 14 to 17 years: Findings from the National Survey of Sexual Health and Behavior (NSSHB). The Journal of Sexual Medicine Volume, 7 (S5), 305–314; p. 461 Adapted from "Socializing Online." (2008). Washington State Office of the Attorney General. www.atg.wa.gov/InternetSafety/SocializingOnline.aspx; Table 12.5 Data excerpted from Herbenick, D., Reece, M., Schick, V., Sanders, S. A., Dodge, B., & Fortenberry, J. D. (2010). Sexual behaviors, relationships, and perceived health status among adult women in the United States: Results from a national probability sample. 291 Findings from the National Survey of Sexual Health and Behavior (NSSHB). The Journal of Sexual Medicine, 7(supplement 5 impact factor: 4.884), 277–290; Figure 12.5 Kaiser Family Foundation. (2013). Sexual Health of Adolescents and Young Adults in the United States. http://kaiserfamilyfoundation.files.wordpress.com/2013/04/3040-06.pdf; p. 464 Data from Barth et al., 2002; Shoveller, J., Knight, R., Johnson, J., Oliffe, J., & Goldenberg, S. (2010). 'Not the swab!'Young men's experiences with STI testing. Sociology of health & illness, 32(1), 57–73.; Wolfers, M., de Zwart, O., & Kok, G. (2011). Adolescents in the Netherlands underestimate risk for STI and deny the need for STI testing. STI, 25(5), 49; p. 465 Smock, Pamela J. 2000. "Cohabitation in the United States: An Appraisal of Research Themes, Findings, and Implications." Annual Review of Sociology. Vol.26, pp.1–20.p. 5; Figure 12.6 CDC – "Key Statistics," 2013. www.cdc.gov/nchs/nsfg/key_statistics/c.htm#chabitation); Table 12.6 Data from Herbenick, D., Reece, M., Schick, V., Sanders, S. A., Dodge, B., & Fortenberry, J. D. (2010). Sexual behaviors, relationships, and perceived health status among adult women in the United States: Results from a national probability sample. 291 Findings from the National Survey of Sexual Health and Behavior (NSSHB). The Journal of Sexual Medicine, 7(supplement 5 impact factor: 4.884), 277–290; Reece et al. (2010). National Survey of Sexual Health and Behavior. The Journal of Sexual Medicine, 7(S-5), 277–304; Figure 12.7 Figure, "Factors Predicting Sexual Satisfaction in Marriage," by M. Young et al., in The Canadian Journal of Human Sexuality, 7, (1998), pp. 115–127. Copyright © 1998. Reprinted by permission of The Sex Information and Education Council of Canada; Figure 12.8 Figure, "Sexual Problems are Common,"p. 369, in The Social Organization of Sexuality, by E. Laumann, J. Gagnon, R. Michael, & S. Michaels. Copyright © 1994. Reprinted by permission of The University of Chicago Press; p. 470–471 The Office on Women's Health: U. S. Department of Health and Human Services (2008).

Understanding Menopause; Table 12.7 Lindau, S., Schumm, P., Laumann, E., Levinson, W., O'Muircheartaigh, C., & Waite, L. (2007). A study of sexuality and health among older adults in the United States (Clinical report). New England Journal of Medicine, 357, 762–774., adapted from Table 2, pp. 766–768; p. 473 Bulcroft, K., & O'Connor-Roden, B. (1986, June, p.66). Never too late: Single people over 65 who are dating and sexually active belie the notion that passion and romance are only for the young. Psychology Today, pp. 66–69; Table 12.8 Mayo Clinic (2011). Sexual health and aging: Keep the passion alive. Sexual health. Mayo Foundation for Medical Education and Research. Retrieved from www.mayoclinic.org/healthy-living/sexual-health/in-depth/sexual-health/art-20046698; Meston, C. (1997). Aging and sexuality. Western Journal of Medicine, 167, 285–290; "Sex Info" (2003); p. 474 Kris Bulcroft & Margaret O'Connor-Roden, "Never Too Late," Psychology Today June 1986, pp. 66–69; p. 475 Kris Bulcroft & Margaret O'Connor-Roden, "Never Too Late," Psychology Today June 1986, pp. 66–69. p.69.

Chapter 13 Figure 13.1 Marital Rape Laws by State (2011). Marital rape laws by state. Criminal defense lawyer (online). Retrieved from www.criminaldefenselawyer.com/resources/criminal-defense/sex-crimes/marital-rape-laws-state.htm; Greenson, T. (2008, March 23). An evolution of law: Spousal rape recently prosecutable. Times-Standard (electronic edition). Retrieved from www.times-standard.com/ci_8668359?source=most_viewed; National Clearinghouse on Marital and Date Rape (2005). State law chart. Retrieved from http://members.aol.com/ncmdr/state/law_chart.html; Table 13.1 FBI Uniform Crime Statistics. www.fbi.gov/about-us/cjis/ucr/crime-in-the-u.s; p. 484 FBI, 2011; p. 484 FBI. (2012). Attorney general Eric Holder announces revisions to the uniform crime report's definition of rape. U.S. Department of Justice. Office of Public Affairs. Retrieved from www.fbi.gov/news/pressrel/press-releases/attorney-general-eric-holder-announces-revisions-to-the-uniform-crime-reports-definition-of-rape;Figure13.2CDC–"Sexual Violence." (2012). Sexual violence. Center for Disease Control and Prevention. Retrieved from www.cdc.gov/ViolencePrevention/pdf/sv-datasheet-a.pdf; Neff, Kirkpatrick, & Rude (2007); Figure 13.3 Data from the Bureau of Justice Statistics, National Crime Victims Survey (2009), http://bjs.ojp.usdoj.gov/content/pub/pdf/cv09.pdf; p. 492 Womenshealth.gov, 2012. www.womenshealth.gov/publications/our-publications/fact-sheet/date-rape-drugs.html#c; p. 493: Excerpted from Abramovitz (2001), pp. 18–21; Womenshealth.gov, 2012. www.womenshealth.gov/publications/our-publications/fact-sheet/date-rape-drugs.html#c; p. 495 O'Brien, J. (2001). "The MVP Program: Focus on Student-Athletes," in A. Ottens & K. Hotelling (Eds.), Sexual Violence on Campus: Policies, programs, and perspectives (141–161). New York: Springer. p. 147; p. 497 Source: Data from Carlson, M. (2013, May 21). The military's culture of sexual violence. Bloomberg View. Retrieved from www.bloombergview.com/articles/2013-05-21/the-military-s-culture-of-sexual-violence; pp. 500–501 Houston Area Women's Center. (2008). Personal safety awareness. Retrieved from www.hawc.org Powell, E. (1996). Sex on your terms. Needham Heights, MA: Allyn & Bacon; pp. 501–502 Powell, E. (1996, p.150). Sex on your terms. Needham Heights, MA: Allyn & Bacon; Table 13.2 USHHS (2008), data from Table 3-19, p. 63; p. 505 Based on data from Stoltenborgh, M., van IJzendoorn, M., Euser, E., & Bakermans-Kranenburg, M. (2011). A global perspective on child sexual abuse: Meta-analysis of prevalence around the world. Child Maltreatment, 16(2), 79–101; Table 13.3 "Percentage of Specific Contact Behaviors Among Survivors of Child Sexual Abuse," from The Social Organization of Sexuality, by E. Laumann, J. Gagnon, R. Michael, & S. Michaels. Copyright © 1994. Reprinted by permission of The University of Chicago Press; Table 13.4 Table, "Age at Which Sexual Abuse Occurred," from The Social Organization of Sexuality, by E. Laumann, J. Gagnon, R. Michael, & S. Michaels. Copyright © 1994. Reprinted by permission of The University of Chicago Press."; Figure 13.4 Finkelhor, D., Shattuck, A., Turner, H., & Hamby, S. (2014). The lifetime prevalence of child sexual abuse and sexual assault assessed in late adolescence. Journal of Adolescent Health.54, 16–27 Chen, L., Murad, M., Paras, M., Colbenson, K., Sattler, A., Goranson, E.,... & Zirakzadeh, A. (2010, July). Sexual abuse and lifetime diagnosis of psychiatric disorders: systematic review and meta-analysis. Mayo Clinic Proceedings, 85, 618–629); p. 513 Commission on human rights. (2011). Connecticut commission on human rights and opportunities sexual harassment posting and training regulations. Retrieved from www.ct.gov/chro/cwp/view.asp?a=2527&q=333112 What is sexual harassment? (2011). Title IX Sexual Harassment Office, University of California, Santa Cruz. Retrieved from www2.ucsc.edu/title9-sh/whatissh.htm; Figure 13.5 Adapted from Hill & Silva (2005). Drawing the Line: Sexual Harassment on Campus (p. 15). www.aauw.org/files/2013/02/drawing-the-line-sexual-harassment-on-campus.pdf; p. 518 Excerpt from Sexual Harassment and Teens: A Program for Positive Change, by Susan Strauss. Copyright © 1992. Published by Free Spirit Publishing. Reprinted by permission of Susan Strauss, http://www.straussconsult.com; p. 519 Fiorella, S. (2013, October 18). Cyber-bullying, social media, and parental responsibility The Huffington Post. Retrieved from www.huffingtonpost.com/sam-fiorella/cyber-bullying-social-media-and-parental-responsibility_b_4112802.html; p. 520 McGarvey, R. (1998). Hands off! How do the latest Supreme Court decisions on sexual harassment affect you? Entrepreneur, 26, 85–87.

Chapter 14 Page 525 Krafft-Ebing,R; von, & Klaf, F. (1998). Psychipathia sexualis, with especial reference to the antipathic sexual instinct: A medico-forensic study. New York: Arcade. (Originally published 1886); p. 528 APA – "Paraphilic Disorders," 2013; p. 533 Lothstein, L. M. (1997). Pantyhose fetishism and self cohesion: A paraphilic solution? Gender and Psychoanalysis, 2, 103–121. P.103; Figure 14.2 Adapted from data in Maes, M; De Vos, N; Van Hunsel, F; Van West, D; Westenberg, H, Cosyns, P; & Neels, creased plasma concentrations of catecholamines, in particular, epinephrine. psychiatry Research, 103, 43–49; p. 538 Marquis de Sade (1971). "Philosophy in the Bedroom". Translated by Richard Seaver & Austryn Wainhouse, Grove Press; pp. 539–540 Adapted from www.columbia.edu/cu/cv.; Figure 14.3 Data from Finkelhor, Jones, & Shattuck (2013); p. 542 Based on Daw, J. (2002). Can psychology help a church in crisis? Monitor on psychology, 33.Retrieved from www.apa.org/monitor/june02/church.html. Terry, K, & Ackerman, A. (2008). Child sexual abuse in the Catholic Church: How situational"; p. 544 United Kingdom's Offenses Against the Person Act of 1861; p. 548 De Silva, W. (1999). Sexual variations. British Medical Journal, 318, 654–656; p. 548 Thibaut, F., Barra, F., Gordon, H., Cosyns, P., & Bradford, J. (2010). The World Federation of Societies of Biological Psychiatry (WFSBP) guidelines for the biological treatment of paraphilias. World Journal of Biological Psychiatry, 11(4), 604–655..

Chapter 15 Figure 15.1 "Age of Drug Use and Prostitution Onset Among Sex Trade Workers," from "Pathways to Prostitution: The Chronology of Sexual and Drug Abuse Milestones," by J. Potterat et al., in Journal of Sex Research, (1998), pp. 333–340. Copyright © 1998. Reprinted by permission of The Society for the Scientific Study of Sex; Figure 15.2 "Age of Onset of Sexual Activity," from "Pathways to Prostitution: The Chronology of Sexual and Drug Abuse Milestones," by J. Potterat et al., Journal of Sex Research, (1998), pp. 333–340. Copyright © 1998. Reprinted by permission of The Society for the Scientific Study of Sex; p. 563–564 Reprinted by permission of Dr. Raj Persaud, www.rajpersaud.co; p. 566 Data from Inciardi, J; & Surratt, H. (1997). Male transvestite sex workers and HIV in Rio de Janeiro, Brazil. Journal of Drug Issues, 27, 135–146. Schepel, E. (2011). A Comparative Study of Adult Transgender and Female Prostitution Doctoral dissertation, Arizona State University. Retrieved from http://hdl.handle.net/2286/1y7uaci726n Wellings, K. (2012). Sexual health: A public health perspective. NY: Open University Press Wiessing, L; van Roosmalen, S; Koedijk, P; Bieleman, B; & Houweling, H. (1999). Silicones, hormones and HIV in transgender street prostitutes. AIDS, 13, 2315–2316; p. 569 Shively, M., Kliorys, K., Wheeler, K., & Hunt, D. (2012). A national overview of prostitution and sex trafficking demand reduction efforts. The National Institute of Justice. Office of Justice Programs, U.S. Department of Justice. Retrieved from www.ncjrs.gov/pdffiles1/nij/grants/238796.pdf; p. 569 Shively, M., Kliorys, K., Wheeler, K., & Hunt, D. (2012). A national overview of prostitution and sex trafficking demand reduction efforts. The National Institute of Justice. Office of Justice Programs, U.S. Department of Justice. Retrieved from www.ncjrs.gov/pdffiles1/nij/grants/238796.pdf; Table 15.1 Based on data from Monto, M. (2009).

Customer motives and misconceptions. In Ronald Weitzer (Ed.), Sex for sale: Prostitution, pornography, and the sex industry, 2nd ed. (pp. 233–254). New York: Routledge; p. 573 First Amendment to the United States Constitution; p. 574 Woodward, B., & Armstrong, S. (2005). The brethren: Inside the Supreme Court. New York: Simon & Schuster. Originally published in 1979. (pp. 232–234, emphases added); p. 574 Chief Justice Burger, Miller v. California. 413 U.S. 5, 93 S.Ct. 2607, 37 L.Ed. 2d 419 (1973). P.18; p. 575 Chief Justice Burger, Miller v. California. 413 U.S. 5, 93 S.Ct. 2607, 37 L.Ed. 2d 419 (1973). P.32; p. 575 Douglas, W. O. (1973, p. 44). Miller v. California, 413 U.S. 15, 93 S.Ct. 2607, 37 L.Ed. 2d 419; p. 576 Source: MacKinnon, C. (1996). Just words. Cambridge, MA: Harvard University Press. MacKinnon, C. (2000). Not a moral issue. In D. Cornell (Ed.). Feminism and pornography (pp. 169–197). Oxford, UK: Oxford University Press. Richards, M. (2003). Obscenity. Retrieved from http://www4.gvsu.edu/richardm/PLS307/OBSCENITY.htm; pp. 576–577 Levinson, J. (2005). Erotic art and pornographic pictures. Philosophy and Literature, 29(1), 228–240. p. 229; p. 577 Steinem, G. (1980). Erotica and pornography: A clear and present difference. In L. Lederer (Ed.), Take back the night: Women on pornography (pp. 35–39). New York: Morrow. P.50; Table 15.2 Based on data from Quakenbush, D., Strassberg, D., & Turner, C. (1995). Romantic themes in x-rated videos: Gender effects of romantic themes in erotica. Archives of Sexual Behavior, 24, 21–35; p. 580 Based on data from Quakenbush, D., Strassberg, D., & Turner, C. (1995). Romantic themes in x-rated videos: Gender effects of romantic themes in erotica. Archives of Sexual Behavior, 24, 21–35; p. 584 Jensen, R. (2004). Pornography and sexual violence. Minnesota Center against Violence and Abuse. Retrieved from, www.vadv.org/secPublications/pandsv.pdf; p. 585 Data from Ropelato, J. (2008). Internet filter review. Internet pornography statistics. Retrieved from www.internet-filter-review.toptenreviews.com/internet-pornography-statistics.html; p. 586 Sloviter, D. (1996). Janet Reno v. American Civil Liberties Union, 929 F. Supp. 824.p. 854; Table 15.4 "Sexually-Related Online Activities of College Students," from "University Students' Uses of and Reactions to Online Sexual Information and Entertainment: Links to Online and Offline Sexual Behavior," by S. Boies, from The Canadian Journal of Human Sexuality, 11, (2002), pp. 77–90. Copyright © 2002. Reprinted by permission of The Sex Information and Education Council of Canada; Table 15.5 "College Students' Reactions to Online Sexual Activities," from "University Students' Uses of and Reactions to Online Sexual Information and Entertainment: Links to Online and Offline Sexual Behavior," by S. Boies, from The Canadian Journal of Human Sexuality, 11, (2002), pp. 77–90. Copyright © 2002. Reprinted by permission of The Sex Information and Education Council of Canada; p. 587 New York v. Ferber, 458 U.S.747 (1982); Table 15.6 New York v. Ferber, 458 U.S.747 (1982); pp. 588–589 Sex Laws (2008). Sexual Offenses: Federal Law 18 U.S.C. 2251: Sexual Exploitation Of Children. Retrieved from www.sexlaws.org/18usc_2251; Table 15.7 U.S. Department of Justice (2006), Project Safe Childhood, pp. 39–43; p. 588 Excerpted from Frieden, (2011). Seventy-two charged in online global child porn ring. CNN Justice. www.cnn.com/2011/CRIME/08/03/us.child.porn.ring/index.html; p. 591 Kerr, D. (2013, June 16). Google plans to wipe child porn from the Web. CNET. Retrieved from www.cnet.com/news/google-plans-to-wipe-child-porn-from-the-web. CBS News; p. 592 Jenkins, P. (2002, March 1). Bringing the loathsome to light. Chronicle of Higher Education, pp. B16–B17; p. 592 Jenkins, P. (2002, March 1). Bringing the loathsome to light. Chronicle of Higher Education, pp. B16–B17. B16.

Author Index

C

H

L

T

U

V

Subject Index

A

B

I

T